MICROWAVE ENGINEERING

Fundamentals, Design and Applications

Subal Kar
Professor and Former Head
Institute of Radio Physics and Electronics
University of Calcutta, Kolkata, India

Universities Press

MICROWAVE ENGINEERING
Fundamentals, Design and Applications

UNIVERSITIES PRESS (INDIA) PRIVATE LIMITED
Registered Office
3-6-747/1/A & 3-6-754/1, Himayatnagar, Hyderabad 500 029, Telangana, India
info@universitiespress.com: www.universitiespress.com

Distributed by
Orient Blackswan Private Limited

Registered Office
3-6-752 Himayatnagar, Hyderabad 500 029, Telangana, India

Other Offices
Bengaluru, Bhopal, Chennai, Guwahati, Hyderabad, Jaipur, Kolkata,
Lucknow, Mumbai, New Delhi, Noida, Patna, Vijayawada

First published 2016

ISBN: 978-81-7371-989-9

Typeset by
Bookcraft Publishing Services (I) Pvt Ltd, Chennai

Printed at
Rajiv Book Binding House Delhi

Published by
Universities Press (India) Private Limited
1/24, Asaf Ali Road, New Delhi 110 002

Dedicated to

my Late Father and Mother

Foreword

Rarely does one encounter an author who can combine his/her lifelong experience in pedagogy, research, collaborations and passion for innovation, into a labour of love as Prof. Subal Kar has done in his current book, ***Microwave Engineering—Fundamentals, Design and Applications***. As a professor in the Institute of Radio Physics and Electronics of the University of Calcutta, India, with long experience as a teacher, researcher, research administrator and scientific leader, Prof. Kar's teaching has always been enriched with the latest developments and innovations in research. What has been an enabling force in this desirable trait are Prof Kar's very active research profile and his local, national and international research collaborations. As a Fulbright scholar at Lawrence Berkeley National Laboratory of the University of California (USA) and Visiting Scientist to other world-class institutions such as Cockcroft Institute (UK) and John Adams Institute of Oxford University (UK) and various other laboratories in Japan and around the world, Prof. Kar has been continuously applying and expanding his knowledge of microwave principles in science and engineering to such innovative fields as the science and technology of charged particle accelerators and lately to the design of microwave components using the latest novel innovation of 'metamaterials'.

This book introduces the student to a systematic, methodical and rigorous treatment of microwave engineering principles and applications. The readers' conceptual journey would also find its natural culmination in the useful design examples using both the classical analytical techniques and the modern simulation techniques in a number of chapters. Starting from the very fundamentals of Maxwell's equations in vacuum with boundary conditions, microwave signal transmission/propagation via guided media and antennae has been discussed. Impedance matching and filter design which are extremely important for microwave engineers have also been delved into, with significant coverage from a practical point of view. Microwave active and passive devices covered in this book will benefit the students and practicing engineers alike. The most basic and important application of microwave engineering is in radar and the author has devoted one chapter exclusively to radar principles and systems which will also benefit students in a meaningful way. Inclusion of microwave integrated circuit and EMI–EMC as independent chapters also indicates that the book is very up-to-date.

The treatise does not stop there however—as the final two chapters of the book exemplify. The author has exposed the reader/student to the fascinating concept of Left-Handed Materials

(LHM) or metamaterials. The author proudly displays the first Indian-made plasmonic metamaterial to inspire the students. This subject is as current and relevant as it can be, being the subject of many articles and reports in *Science* and *Nature* with such path-breaking applications as electromagnetic cloaking and sub-wavelength imaging.

Finally the book introduces the reader to the use and utility of microwaves in daily life—from broadband mobile communication to the microwave oven and industrial applications in drying, vulcanisation including medical applications like microwave hyperthermia and ablation as well as radiofrequency monitoring and detection of life.

The book is particularly very suitable as a modern text for students in multi-year integrated undergraduate degree programs such as B Tech./B E in Electrical and Electronics Engineering with microwaves as a full paper. Its advanced chapters are also highly recommended for post-graduate curriculum and for students interested in microwave research. This book is not a stereotype microwave engineering book available in the market, but a scholarly book of microwave principles and associated devices and circuits including design and applications. In this respect, it will be equally valuable to scientists and engineers alike.

I wish the book a healthy and enthusiastic readership and future editions which will surely continue to excel in their perfection and impact, but more importantly in their relevance at the frontier of microwave principles and applications.

Prof. Swapan Chattopadhyay F Inst. P, FRSA
Distinguished Scientist, Fermilab, USA
Professor and Director of Accelerator Research, Northern Illinois University, USA
Visiting Professor, University of Oxford, UK
Visiting Scholar, University of California at Berkeley, USA
Formerly Sir John Cockcroft Chair of Physics and Inaugural Director,
Cockcroft Institute, UK

Foreword

I have great pleasure in writing the foreword for ***Microwave Engineering—Fundamentals, Design and Applications*** authored by Prof. Subal Kar. First, I would like to congratulate Prof. Kar for having taken keen interest in writing this book. Prof. Kar is a highly renowned academician in the field of Microwaves. I had the opportunity to interact with him very closely for the last 15 years when I was heading the SAMEER Kolkata Centre and later as Director, SAMEER. He was particularly involved with the scientists of SAMEER, Kolkata in a few important R&D projects as an Expert Member. He was the prime motivator and guide to the newly recruited scientists at the centre way back in 2006 and helped in developing a core design team which has later on successfully designed many microwave and millimetre-wave components, sub-assemblies and systems.

This book contains all the important chapters required to be studied to acquire knowledge in microwave and millimetre-wave technology. The chapters are well written in a very lucid manner with review questions at the end of each chapter including a number of solved problems and exercises in most of the chapters, to enable proper understanding of concepts. The book also contains many design tips with state-of-the-art simulation tools to enable designers involved in developing components for various microwave systems.

Prof. Kar's R&D experience while working with SAMEER as well as in his own university and abroad is reflected in the book in terms of the designs examples and the application oriented practical tips given in different chapters. His lecture notes while teaching in the university have become part of the book and this will really benefit the student community. Particularly, I would like to mention the chapter on ***metamaterials*** which contains invaluable insight gathered from research work carried out by Prof. Kar and his research team along with the scientists of SAMEER and this has resulted in certain experimental facilities created at SAMEER Kolkata centre and a few good publications.

I have great pleasure in recommending this book both for undergraduate and post-graduate students in Engineering and Science, as well as researchers, academics, scientists and practicing engineers. I am sure this book will benefit the readers and find wide acceptance among all. I wish this book great success.

Dr Ananta Lal Das
Ex-Director
Society for Applied Microwave Electronics Engineering and Research (SAMEER)
Department of Electronics and Information Technology
Ministry of Communications and Information Technology
Government of India
Mumbai

Preface

A number of text books on microwave engineering are available in the market—then where is the need for another book of that kind? Well, here is my modest answer—*it is not another book of that kind*! This book on ***Microwave Engineering*** emphasises the fundamental concepts of microwave engineering, numerous design examples and application view points in a markedly different manner when compared with all the existing books on the subject.

Chapter 2 of this book gives a quick review and an overview of the applied aspects of *electromagnetics* so that students can approach the subject with new vigour. **Chapter 3** deals with *high frequency behaviour of transmission lines*; this is basically an extension of power transmission lines (used at power line frequency of 50/60 Hz) with distributed circuit analysis. Transmission line analysis at higher RF/microwave frequency also follows this approach, however, with some special differences at the microwave frequency of operation. **Chapter 4** deals with *guiding structures: waveguides and cavity resonators* used as microwave guiding structures and microwave energy storage systems. The scattering or S-parameters defined in terms of incident, reflected and transmitted wave variables are used for microwave circuit/device/system analysis; the basic foundation of *microwave network and scattering matrix* is discussed in **Chapter 5**. Both waveguide-based and planar transmission line-based *microwave passive circuit components* including lumped circuit elements and microstrip based distributed circuit elements at microwave frequency are covered in **Chapter 6.**

Impedance matching in transmission lines and waveguides has been discussed with a number of design examples in **Chapter 7.** Filters find numerous applications in RF/microwave communication and radar systems. Thus, in **Chapter 8** *microwave filters* and their design using both the conventional analytical synthesis approach as well as electronic design automation (EDA) simulation tools available commercially are covered. The details of *microwave active devices* (both tube and solid-state type) together with oscillator, amplifier and power combiner design that use such devices have been discussed in **Chapter 9. Chapter 10** covers in a concise manner, the various types of *microwave antennae.* Microwave and radar are inseparable twins; **Chapter 11** is devoted to *radar principle and systems* which includes pulse and CW radar, surveillance and tracking radar, moving target indicator (MTI) and so on. **Chapter 12** deals with some basic *microwave measurement techniques* and covers two basic yet sophisticated microwave measuring instruments—network analyser and spectrum analyser.

Microwave integrated circuit (MIC) – both hybrid MIC and monolithic MIC – have been covered in **Chapter 13** highlighting design steps, material requirements and fabrication processes involved. *Electromagnetic interference* (EMI) *and electromagnetic compatibility* (EMC) which are still developing subjects are dealt with in **Chapter 14. Chapter 15** is devoted to some fundamental knowledge of *metamaterials/left-handed materials* (LHM) which are emerging research and development fields. **Chapter 16** gives a brief note on possible applications; however, all through the book, applications have been addressed as and when required.

The main purpose of writing this book was to address in a holistic way the fundamentals, design techniques and application view-points of microwave engineering so that this inclusive approach can benefit the present day microwave engineering teaching in a befitting way. I hope this inclusive approach covers the undergraduate syllabus of almost all Indian institutions where microwave engineering is taught and to some extent a part of the post-graduate curriculum too.

In order to make the book students-friendly, review questions are included which are not stereotypical but help in *self-assessment*. The chapter recapitulation serves to reinforce the fundamentals. These two elements may also be useful for practicing engineers who cannot spare the time to go through the chapter as a whole but can immediately pick-up the necessary concept/ formula needed for interpreting a result or to make an engineering design. Further, Electronic Design Automation (EDA) simulation tools and 3D electromagnetic field simulators have been used to solve a number of design problems which is the order of the day for microwave design of filters, matching networks, oscillators, amplifiers and so on. The inclusion of MATLAB in generating many characteristic curves and solving problems is essential for learners and has thus been included.

It is the honest duty of an author to acknowledge the help received from all those without which the endeavour could not have been successful to the level expected by the author. First, I would like to acknowledge from the core of my heart, the help received from my wife Rina and son Sauradeep who have been my constant inspiration through thick and thin, and for providing unfailing encouragement. I acknowledge the immense help I have received from some of the scientists of SAMEER Kolkata Centre—Mrs Paramita Banerjee, Mr Amitesh Kumar, Mr Sukhendu Bhanja, Mrs Souma Chatterjee, Ms Madhuja Ghosh and Mr Aritra Banerjee. I also acknowledge the help received from my M. Tech. students—Ms Amrita Chakarborty and Ms Sujata Ghosh. I must not fail to acknowledge the help received from Dr Ananta Lal Das, the Ex-Director of SAMEER, and Mr Arijit Majumder, Scientist-in-Charge of SAMEER Kolkata centre, for allowing me to use their simulation facilities and experimental arrangements for making the CAD solutions of design problems and to carry out some experiments included in this book. I would also like to thank the reviewers who made some constructive suggestions while reviewing the manuscript of the book. I must not fail to acknowledge the help I have derived from various textbooks and reference books while writing this textbook. Finally, I would like to thank Prof. Swapan Chattopadhyay, Distinguished Scientist, Fermilab, USA and Dr A L Das, Ex-Director SAMEER, India for writing forewords for my book and also Dr Gita S Dattatri and her team at Universities Press Pvt Ltd for all the editorial help and suggestions to polish the book into its final form.

Any suggestions from the users of this book to improve its quality will be gratefully accepted and highly appreciated; since, at the end of the day the whole purpose of the book is to serve the genuine needs of those for whom it has been written.

Subal Kar
Professor and Former Head
Institute of Radio Physics and Electronics
University of Calcutta, Kolkata, India

Contents

1 Introduction

Nature permits and nature prohibits
—You just have to understand the limits set by her

This chapter begins with an introduction to the microwave frequency range. This is usually subdivided into different bands for commercial and military use by microwave engineers. A discussion on the application potential of microwaves for scientific, commercial, domestic and military usage follows. In addition, the importance of microwaves compared to other radio frequency waves on the lower frequency side and infrared, optical waves on the higher frequency side of the electromagnetic spectrum are also mentioned. The proposition that microwave engineering is a special subject, as well as a brief history of the development of microwave engineering, are discussed.

1.1 Microwaves and Their Applications

Microwaves are electromagnetic waves/radio frequency (RF) waves spanning the frequency range from 300 MHz to 300 GHz (Fig. 1.1a shows the electromagnetic spectrum indicating specifically the RF and microwave spectrum, Fig. 1.1b shows the important applications). From an engineering point of view, however, the frequency range covering 3.0–30 GHz (wavelengths from 10 cm down to 1 cm) may be called the centimetric microwave frequency range or SHF (Super High Frequency) range while on its lower side is the metric microwave frequency range or UHF (Ultra High Frequency) range and on its higher side are the millimetric and sub-millimetric waves or the EHF (Extra High Frequency) range.

The ability of microwaves (radio waves) to propagate through clouds, smoke, dust and fog, especially under zero visibility conditions (night time) makes it a logical choice over the IR (infrared) and optical wavelengths. However, in the millimetre-wave range, selective atmospheric absorption due to oxygen and water vapour (Fig. 1.2) allows RF communication only through some narrow windows of frequency bands known as millimetre-wave window frequencies centred around 34 GHz, 94 GHz, 140 GHz and so on.

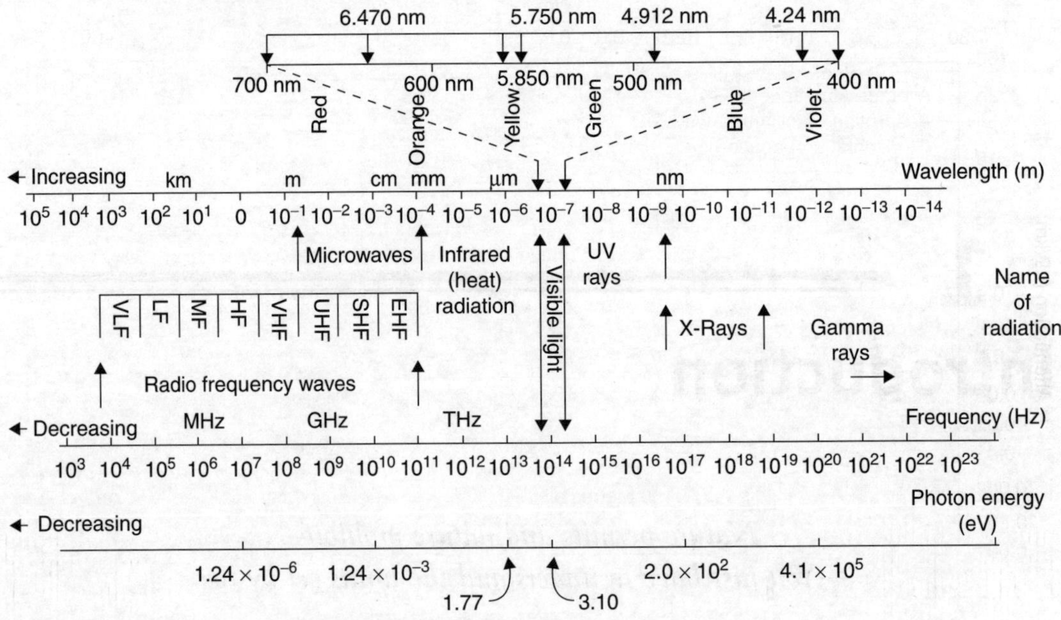

Fig. 1.1(a) Electromagnetic spectrum showing the microwave frequency range.

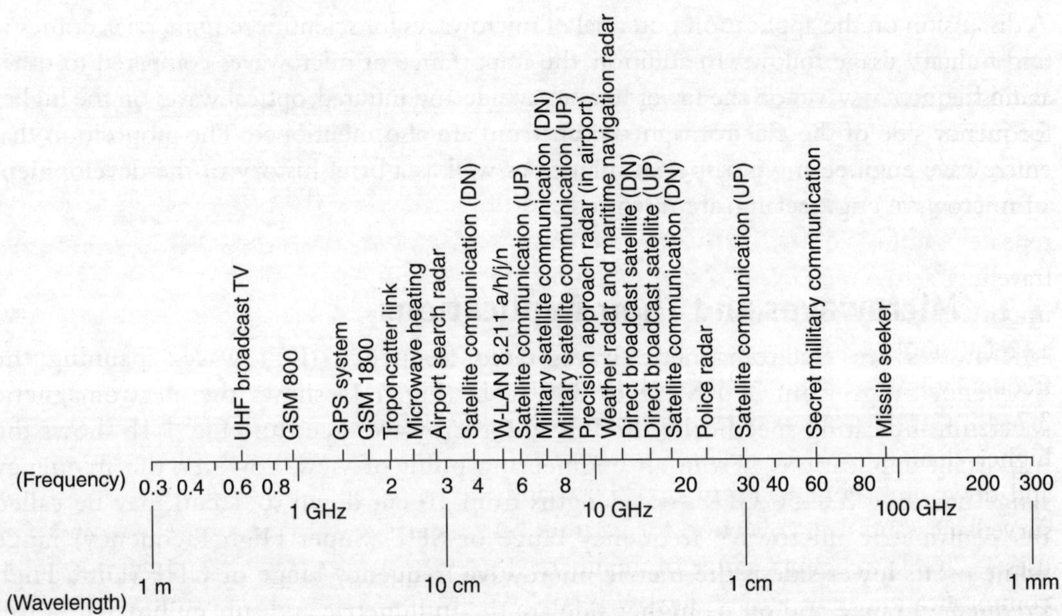

Fig. 1.1(b) Microwave frequency spectrum with applications.

It has become customary to designate the whole of the microwave frequency range in terms of various frequency bands. A detailed chart for commercial and military designations are given in Table 1.1.

Application of microwaves started since World War II with the need for high resolution RADAR (RAdio Detection And Ranging) capable of detecting and precisely locating enemy

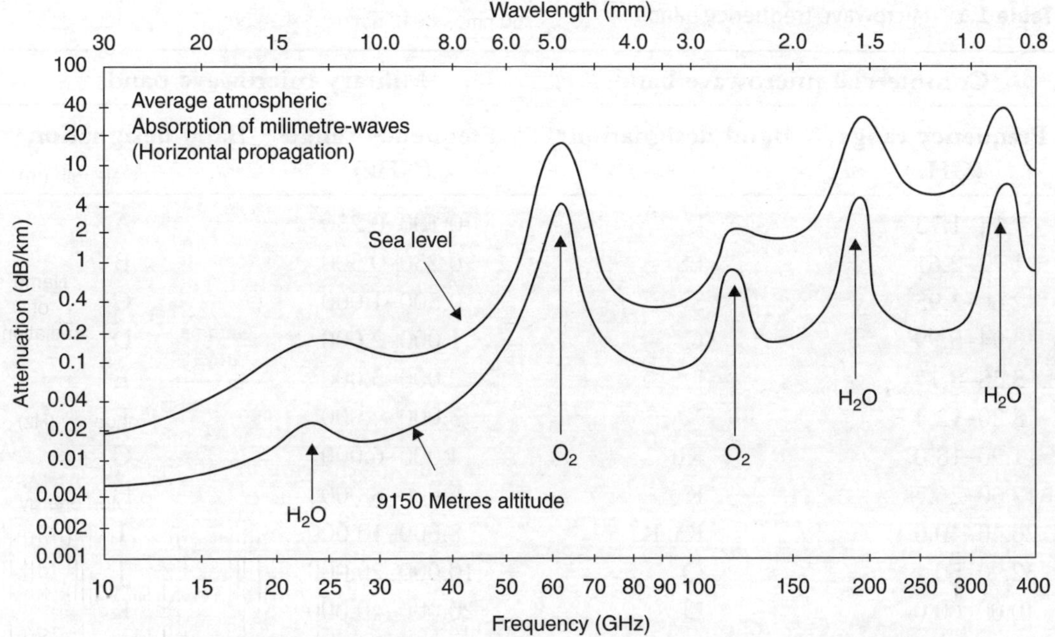

Fig. 1.2 Average atmospheric absorption of millimetre-waves (horizontal propagation). Millimetre-wave window frequencies are shown at 34 GHz, 94 GHz and 140 GHz.

targets, fighter aircrafts and war ships. Even today radar is used in many advanced forms such as missile guidance/tracking, airborne warning and control system (AWACS) applications for military needs. In civil applications, air traffic control radar, police radar for speed limit enforcement, collision avoidance radar and so on, operate at microwave frequency.

A common application of microwaves in these days is in communications links via repeater stations located in microwave towers on hilltops (a familiar sight to motorists travelling across the country) used by utility companies and television networks. Another important means of communication using microwave frequency is the use of satellites as relay stations for national and global communication. Existing old systems use the frequency in the range of 5.9–6.4 GHz (J-band) as the up-link while the frequency range 3.7–4.2 GHz (C-band) is used as the down-link frequency. However, new systems with higher channel capacity required for both civil and military applications prefer to use up and down link frequencies in the X, Ku and Ka bands. Such satellites are also used for surveillance and data collection for weather forecasting. Global Positioning System (GPS) uses 1–2 GHz frequency range of the microwave frequency spectrum. In addition to these communication applications, microwaves are also used for basic and applied research. Even in our day-to-day lives we use microwave ovens for cooking (usually operates at 2.45 GHz, in the ISM band). Sensitive microwave receivers are used in radio astronomy to detect and study the electromagnetic radiation emitted from the sun and a number of radio stars that emit such radiation in the microwave frequency band. Microwave radiometers (that create temperature images of objects) are used to map atmospheric temperature profiles, moisture conditions in soils and crops and for other remote sensing applications as well. Further, in modern day communication, cell phone networks that allow us mobile communication

Table 1.1 Microwave frequency bands

Commercial microwave bands		Military microwave bands	
Frequency range (GHz)	**Band designation***	**Frequency range (GHz)**	**Band designation****
1.14–1.73	L	0.100–0.250	A
1.72–2.61	LS	0.250–0.500	B
2.60–3.95	S	0.500–1.000	C
3.94–5.99	C	1.000–2.000	D
5.38–8.17	J	2.000–3.000	E
8.20–12.4	X	3.000–4.000	F
11.90–18.0	Ku	4.000–6.000	G
17.60–26.7	K	6.000–8.000	H
26.40–40.0	Ka, R	8.000–10.000	I
32.90–50.1	Q	10.000–20.000	J
40.00–60.0	U	20.000–40.000	K
50.00–75.0	V	40.000–60.000	L
60.00–90.0	E	60.000–80.000	M
90.00–140	W		
110.00–170	D		
170.00–260	Y		

* *Used by Hewlett Packard, Narda and other companies.*
** *New US military designation.*

and computing, operate at microwave frequency (GSM band: Tx: 880–915 MHz and Rx: 925–960 MHz). Use of microwave frequency extends the range of wireless internet service via WIMAX (Worldwide Interoperability with Microwave Access); conventional WiFi coverage is limited to 30 m while WIMAX can easily provide a 30 km coverage. Microwave radiation finds medical applications in hyperthermia or localised heating of tumours and microwave medical equipments are also used to monitor heartbeat, lung water detection and so on.

In industry, microwaves are used for processing of food, rubber, plastic and forest products. It is also used to dry inks, leather, textiles, tobacco and grains. Microwaves find applications in non-destructive testing of materials, process control—moisture and thickness monitoring and so on.

In recent times, artificially structured electromagnetic material known as metamaterial that can exhibit negative refractive index has opened up a new dimension in the research and practical developments using microwave frequency. Metamaterials have application potential for cloaking and stealth technology including the capability for improved design of microwave components such as filters, phase shifters and antennae.

1.2 Importance of Microwaves

A common question that needs to be addressed is: why are microwaves so important?

- Lower frequency bands like HF, VHF and others, have been in use since the early 20th century, which saw the beginning of radio communication, leaving no scope for accommodating further new communication bands desired by emerging new technologies. With the ever-increasing communication traffic load we need wider available transmission bandwidth to be able to transmit a large number of information channels (audio, video or data), may be using cost-effective multiplexing technique, which is possible only if the carrier frequency is high enough, that is, in the microwave range.
- Highly directive antenna beams can be obtained with realistic antenna dimensions at microwave frequencies because the gain and directivity of antenna is dependent on the relative size of the radiating aperture in terms of wavelengths. This has opened up new applications in point-to-point and covert communication including high resolution radar tracking. Also, the airborne and space borne applications have become practically possible with the advent of microwave technology.
- Microwave signals can penetrate through different layers of the ionosphere with negligible absorption and reflection and hence space and satellite communication with microwave signals has become possible with its inherent wider available transmission bandwidth compared to HF, VHF and so on. Also, at microwave frequencies, fading of radio wave signals is less compared to lower RF frequencies making microwave communication more reliable and predictable.
- Various molecular, atomic and nuclear resonances occur at microwave frequencies. This has created a new class of applications in basic sciences, remote sensing, medical diagnostics and treatment.

1.3 Why is 'Microwaves' a Special Subject?

This is usually considered by many as a special subject. The specialty or difficulty, if any, arises from the fact that we have to deal with very small wavelengths in this frequency range (of the order of centimeters and millimeters).

At such short wavelengths of the microwave signal, the active device and passive component dimensions become comparable with the wavelength and hence the phase of the oscillating voltage or current changes significantly over the physical extent of the device. The microwave components are thus considered as the '*distributed elements*', and hence the lumped circuit approximations of low frequency circuit theory are not applicable at microwave frequencies. In fact, an adequate description of a microwave circuit can be carried out in terms of the electric and magnetic fields associated with the device. Microwave engineering is thus logically based on applied electromagnetics, and one begins with Maxwell's electromagnetic field equations and their solutions. It is in the nature of these equations that mathematical complexities arise, since Maxwell's equations involve vector differential or integral operations on vector field quantities and these fields are functions of spatial coordinates.

At much lower frequencies than the microwaves, the wavelength is large enough and hence there is a very insignificant phase variation across the dimensions of a component. Such circuits are adequately described in terms of mesh currents and node voltages and are analysed with the help of *Z*, *Y* or *h* parameters. As the wavelength of the signal becomes comparable with the circuit dimensions, the propagation effect dominates and the current and voltage become derived quantities of the magnetic and electric fields. At microwave frequencies, the measurable quantities are the incident, reflected and transmitted electromagnetic waves and in terms of these wave variables, a new set of circuit parameters is defined, called *scattering parameters* (*S*-parameters). The microwave circuit analysis and design is done with the help of these wave parameters.

The physical properties of passive components such as inductors, capacitors and resistors that store magnetic energy, store electric energy and dissipate electric energy remain the same irrespective of the frequency of operation. However, the geometric forms with which they are realised at the microwave frequency are quite different from their lower frequency counterparts. For example, the use of a coil of wire as inductance at microwave frequency is not possible because at these high frequencies the inter-turn capacitance will be very significant making the coil into an undesirable resonant circuit. And the current circulating in this unshielded structure (with its dimension comparable to the wavelength of the signal) will unduly radiate electromagnetic signals. The LC tuned circuit at microwave frequency is reliably realised with a closed conducting enclosure, called the *resonator cavity*.

An essential requirement in a microwave circuit is to transport signals from one point to another without radiation loss. The wire used to transport electrical signals at the power-line frequencies (50/60 Hz) are unusable due to the radiation problem. This means that the electromagnetic energy at microwave frequency has to be transported from one point to another as a propagating wave in a shielded structure. A variety of structures have been developed that can guide electromagnetic waves from one point to another. *Twin-wire lines* and *coaxial lines* are used for lower microwave frequencies, while *waveguides* are used for frequencies above 3.0 GHz. Various forms of *planar transmission lines* and *dielectric integrated guides* are used for microwave and millimeter-wave integrated circuits.

Further, at microwave frequencies, the inductive reactance due to the connecting leads of active devices and the capacitive reactance due to the distributed or stray capacitance become significant. Also the transit-time effect limits the high frequency operation of conventional low frequency tubes and solid-state devices. Today's technology could not improve the transit-time limitation of conventional tubes and solid-state devices beyond a certain frequency limit. Using *disk-seal triode* and *lighthouse tube*, frequencies up to 10 GHz could be achieved and the *interdigitated transistor* and *Schottky-barrier gate* or *MESFET* can operate at maximum frequencies of 20 GHz and 40 GHz respectively. Ultimately, technologists decided to utilise transit-time for device operation. So a whole new class of microwave tube devices like *klystron*, *TWT*, *magnetron* and solid-state devices like *Gunn*, *IMPATT* and so on, have been developed using the transit-time effect to favour the operation of the active microwave devices.

It may now be appreciated that microwave is a special subject but by no means a difficult subject!

1.4 A Brief History of the Development of Microwave Engineering

The history of development of microwave engineering is enmeshed with the evolution of electromagnetic fields and waves. In 1873, the great Scottish physicist and mathematician James Clerk Maxwell hypothised four sets of mathematical equations expressing the fundamental laws for electromagnetic fields. Maxwell's equation unified the partial theories of electricity and magnetism and predicted that there could be wave-like disturbances in the combined electromagnetic field. These electromagnetic field expressions along with Newton's laws, the three laws of thermodynamics, the quantum theory, and the theories of relativity form the basis for our current understanding of the physical universe. Maxwell's electromagnetic field equations appear to be fundamental in that, they are not modified by relativity as Newton's equations of motion are when particle velocities approach the velocity of light. Further, Maxwell's equations are equally applicable for natural materials (also known as right-handed materials) and metamaterials (also known as left-handed materials).

Heinrich Rudolf Hertz, a German professor of physics and a gifted experimentalist, carried out a set of experiments during the period 1887–1891 that validated Maxwell's prediction of electromagnetic waves. Hertz used an oscillating spark gap near 66 cm wavelength to induce similar oscillations in a distant wire loop. It is interesting to observe that this is an instance of discovery occurring after a prediction has been made on theoretical grounds—a characteristic of many of the major discoveries throughout the history of science.

Following Hertz, a number of scientists like J C Bose (India), Oliver Lodge (UK), A S Popoff (USSR) carried out some significant experimental research on the characteristics and applications of electromagnetic waves in the microwave and millimetre-wave frequency spectrum. J C Bose was successful with his 5 mm wavelength modified Hertzian spark gap generator to trip a relay which threw a heavy iron ball and fired a pistol with a wireless signal. Bose is also credited for developing the first metal–semiconductor diode for millimetre-wave detection and a horn antenna for millimetre-wave signal reception.

However, the Italian born engineer Guglielmo Marconi, working in England (in his own words 'an amateur in radio') was the first to achieve transmission of intelligible signals by means of Hertzian waves (around 350 m wavelength). On December 12, 1901, Marconi received the radio signal (corresponding to the Morse code 'S') at Signal Hall in Newfoundland, which was sent by his team located at Cornwall, England, 1700 miles away on the other side of the Atlantic Ocean. History has accredited him with the invention of an early form of radio telegraphy (wireless).

The radio technology in the early part of 20th century utilised primarily the high frequency (HF) and very high frequency (VHF) range, and none in the microwave range due to the lack of reliable microwave sources. The urgent need for radar development during World War II triggered substantial interest in the development of microwave theory and technology. The refreshed vigour in microwave technology started with the development of the klystron by the Varian brothers just before World War II. In the Radiation Laboratory of Massachusetts Institute of Technology (MIT), US, intensive research was done both on the theory and experimental developments of waveguide components and antennae. These developments are well documented in the 28 volumes of Radiation Laboratory Series of books [1] written by the members of the MIT Radiation Laboratory and they are still considered to be the classic texts for microwave engineers.

Review Questions

1.1 What are microwaves?

1.2 What are the advantages of microwaves/radio waves over IR and optical waves?

1.3 What are millimetre-wave window frequencies? Mention a few of them.

1.4 Mention the frequency ranges for X-band and W-band.

1.5 Mention a few military and civilian applications where microwave is used.

1.6 Mention the up-link and down-link microwave frequencies (both in the existing and new systems) used for satellite communication.

1.7 At what frequency does a microwave oven work?

1.8 Mention two common applications of microwave frequency in modern day communication.

1.9 What are the applications of microwave radiometers?

1.10 To obtain larger coverage for wireless internet connectivity would you prefer WiFi or WIMAX?

1.11 Mention one medical application of microwave radiation.

1.12 Metamaterial is an emerging application of microwaves—justify.

1.13 Mention three salient points to indicate the importance of microwaves.

1.14 Microwave components are often referred to as distributed circuit elements—justify.

1.15 Microwave engineering is based on applied electromagnetics—justify.

1.16 Microwave circuit analysis has to be done in terms of wave parameters, the *S*-parameters—justify.

1.17 Coil of wire used as inductance at low frequency cannot be used at microwave frequency—why?

1.18 Why can't power-line wires be used to transport microwave frequency signals?

1.19 Name a few microwave transmission lines. What should be their special characteristics?

1.20 Mention the basic limitations which prohibit the use of conventional tubes and solid state devices at microwave frequency. Name a few microwave tube and solid-state devices.

1.21 Mention the most important outcome of Maxwell's equations.

1.22 Who validated Maxwell's prediction of electromagnetic (em) waves? What wavelength of em signal was used by him?

1.23 Name the first Indian scientist who could generate a 5 mm wavelength signal.

1.24 History has accredited him with the invention of an early form of radio telegraphy (wireless)—name the inventor.

1.25 They are still classic texts for microwave engineers—which texts are referred to?

2 Electromagnetics Revisited

Identity always comes with a difference
—when compared with others

2.1 Introduction

It was mentioned in Chapter 1 that microwaves are treated in terms of electric (*E*) and magnetic (*H*) fields instead of using current and voltage, which are normally used at lower frequencies. It was J C Maxwell who first established through his set of four electromagnetic equations that time-varying electric and magnetic fields form coupled electric and magnetic field pairs that give rise to propagating electromagnetic waves. Though readers are familiar with the fundamentals of electromagnetics, a quick recapitulation of some important topics of applied electromagnetics, particularly Maxwell's equations, propagation of electromagnetic waves in unbounded and interfacing media including power flow, polarisation and radiation of electromagnetic waves will be worthwhile as these are the foundations for microwave engineering.

2.2 Time-varying Fields and Maxwell's Equations

Maxwell's equations are the fundamental pillars of applied electromagnetics – and hence microwave engineering – based on Gauss's law, Faraday's law of electromagnetic induction and his (Maxwell's) own correction of Ampere's circuital law for time-varying fields [2]. Prior to Maxwell, electricity and magnetism were treated as different branches of physics in their own right. But Maxwell's work led to the fact that whenever the electric (*E*) or magnetic (*H*) field changes with time, they become intimately and inseparably related with each other and this coupled *E*–*H* field gives rise to the *electromagnetic field*. Faraday's law of electromagnetic induction established that a magnetic field changing with time gives rise to a curling (rotational) electric field. Thus, James Clerk Maxwell argued: If the converse can also be true—will it be possible to get a curling magnetic field from a time-varying electric field? Maxwell noticed an inconsistency in the current continuity equation when he tried to apply the Ampere's circuital law to find the magnetic field at a point outside a capacitor connected to a time-varying current. However, he found that the inconsistency could be removed by introducing an additional current component, known as the *displacement current* (originating from a time varying electric field), to modify Ampere's circuital law for a time-varying situation. The most important implication that emerges from Maxwell's equations is the existence of electromagnetic waves; which are the coupled time-varying electric and magnetic fields that propagate in space. The speed of these waves, according to these equations, turns out to be very close to the velocity of light, as obtained from optical measurements. This led to the remarkable conclusion that light is an electromagnetic wave. Maxwell's work thus unified the domain of electricity, magnetism and light in a path breaking way.

For static electric and magnetic fields the two divergence equations are:

$$\nabla \cdot \vec{D} = \rho \qquad \text{(Gauss' law)} \tag{2.1a}$$

$$\nabla \cdot \vec{B} = 0 \qquad \text{(Solenoidal static magnetic field, non-existent monopole)} \tag{2.1b}$$

The curl equations for the static electric and magnetic fields are:

$$\nabla \times \vec{E} = 0 \qquad \text{(Irrotational static electric field)} \tag{2.1c}$$

$$\nabla \times \vec{H} = \vec{J} \qquad \text{(Ampere's circuital law)} \tag{2.1d}$$

For the time-varying case, the divergence equations remain unaffected but the curl equations have to be modified to include the time dependence of the field. In the above equations $\vec{D}\ (= \varepsilon\vec{E})$ is the electric displacement, $\vec{B}\ (= \mu\vec{H})$ is the magnetic flux density, ρ is the electric charge density, $\vec{E}$ and $\vec{H}$ are respectively the electric and magnetic field intensities.

The experimental work of the English physicist and chemist Michel Faraday provided an empirical formula relating the electromotive force generated in a loop when it is moved in a fixed magnetic field. From reciprocity, this empirical formula can also be interpreted by saying that a time-varying magnetic field is capable of producing a rotational (that is, curling) electric field. Thus, for a time-varying situation, the electric field is no longer irrotational and according to the Faraday's law:

$$\nabla \times \vec{E} = -\frac{\partial \vec{B}}{\partial t} \tag{2.2}$$

With the background of the laws of Gauss, Ampere and Faraday, Maxwell started theoretical investigations on the time-varying properties of the electric and magnetic fields. He observed that Ampere's circuital law, which is the backbone of magnetostatic fields, shows inconsistency for time-varying currents. Maxwell included a new quantity called the *displacement current* ($\vec{J}_d = \partial\vec{D}/\partial t$) in Ampere's circuital law to make it consistent with the time-varying current continuity equation.

Displacement current derives its name from the displacement vector $\vec{D}$ and dimensionally $\partial\vec{D}/\partial t$ represents a current. Physically, this is the current that flows through a dielectric when subjected to a time-varying electric field (as in the case of current flow through a capacitor when an AC signal is applied across it). In fact, when a time-varying electric field is applied across a dielectric, the polarisation vector $\vec{P}$ inside the dielectric goes on changing its direction periodically in synchronisation with the variation of the time-varying electric field and this *disturbance* is propagated as displacement current via the dielectric from one end to the other (resembling the propagating disturbance on free electrons which is the flow of *conduction current* in a conductor). In either case, for an electromagnetic signal, the said disturbance travels through the dielectric/conductor with the velocity of light.

For time-varying fields, the set of four equations given by Eq. 2.1 are modified and we have the divergence and curl equations for time-varying field representing the *Maxwell's equations in the differential or point form* as:

$$\nabla \cdot \vec{D} = \rho \tag{2.3a}$$

$$\nabla \cdot \vec{B} = 0 \tag{2.3b}$$

$$\nabla \times \vec{E} = -\frac{\partial \vec{B}}{\partial t} \tag{2.3c}$$

$$\nabla \times \vec{H} = \vec{J} + \frac{\partial \vec{D}}{\partial t} \tag{2.3d}$$

It may be seen that even for the time-varying case, the magnetic field remains *solenoidal* (that is, the lines of magnetic flux close upon themselves having the appearance of a tube or pipe, meaning the Greek word *sol-een*), while both the electric and magnetic field become *rotational*, that is, they have non-zero curl and hence it becomes a *vortex field*. It may further be noted from Eqs 2.1c, 2.1d, 2.3c and 2.3d, that when the curl of the vector field happens to be zero, it is a *conservative* field; otherwise it is *non-conservative*.

These four equations together with the current continuity equation and Lorentz's force equation $[\vec{F} = q\,(\vec{E} + \vec{v} \times \vec{B})]$ form the foundation of electromagnetic theory. These equations can be used to explain and predict all macroscopic electromagnetic phenomena. The universal importance and impressive beauty of Maxwell's equations led Boltzmann to say '*Was it God who wrote these equations?*'.

A conceptual illustration of propagating electromagnetic field that comes out of an antenna as electromagnetic radiation may now be given on the basis of Maxwell's curl equations. The time-varying current flowing through the conductor of an antenna gives rise to a charge separation field $\vec{D}^{(1)} = \varepsilon\vec{E}^{(1)}$ which varies with time and thus gives rise to a displacement current: $\varepsilon\partial\vec{E}^{(1)}/\partial t$. This in addition to the current $\vec{J}^{(0)}$ gives rise to a rotating or curling magnetic field: $\nabla \times \vec{H}^{(1)} = \vec{J}^{(0)} + \varepsilon(\partial\vec{E}^{(1)}/\partial t)$. (According to Maxwell's magnetic curl equation, Eq. 2.3d, the direction of magnetic field with respect to current is given by the right-hand screw rule.) Then, according to Maxwell's electric curl equation (Eq. 2.3c), the electric field $\vec{E}^{(2)}$ winds or curls in left-hand screw direction around the time-varying magnetic field $\vec{H}^{(1)}$, when: $\nabla \times \vec{E}^{(2)} = -\mu(\partial\vec{H}^{(1)}/\partial t)$. This way, coupled $\vec{E}$ and $\vec{H}$ fields curling around each other results in the propagating electromagnetic field/wave in space (Fig. 2.1).

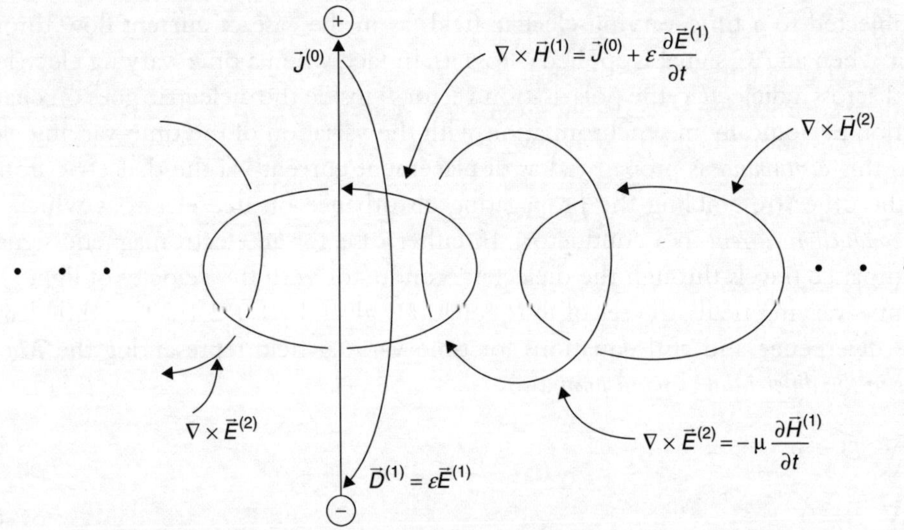

Fig. 2.1 Conceptual illustration of propagating electromagnetic wave in space.

Maxwell's equations in the differential form are valid at all points in space, that is, they permit us to study directly the interaction between the field vectors and their relationships with source densities at individual points. However, in explaining electromagnetic phenomena in a physical environment, we must deal with finite objects of specified shapes and boundaries. Thus, to deal with a practical physical situation, we need to convert Maxwell's equations in their differential forms into their integral form equivalents.

Taking the volume integral of both sides of the divergence Eqs 2.3a and 2.3b over a volume V with a closed surface S and using the divergence theorem while taking the surface integral of both sides of the curl Eqs 2.3c and 2.3d over an open surface S bounded with a contour C and applying Stokes' theorem, we get the *Maxwell's equations in the integral form* as:

$$\oint_S \vec{D} \cdot d\vec{s} = \int_V \rho_v \, dv \tag{2.4a}$$

$$\oint_S \vec{B} \cdot d\vec{s} = 0 \tag{2.4b}$$

$$\oint_C \vec{E} \cdot d\vec{l} = -\int_S \frac{\partial \vec{B}}{\partial t} \cdot d\vec{s} \tag{2.4c}$$

$$\oint_C \vec{H} \cdot d\vec{l} = \int_S \left(\vec{J} + \frac{\partial \vec{D}}{\partial t} \right) \cdot d\vec{s} \tag{2.4d}$$

The Maxwell's equations we have discussed so far hold good for electromagnetic quantities with arbitrary time dependence. The actual type of time functions for the field quantities depend on the source functions ρ_v and $\vec{J}$. In engineering applications, sinusoidal time functions are commonly used.

For time-harmonic (steady-state sinusoidal) fields, it is convenient to use a phasor notation. The time harmonic field $\tilde{\vec{E}}(r, t)$ may be expressed in terms of the corresponding phasor quantity $\vec{E}_s(r)$ as: $\tilde{\vec{E}}(r \cdot t) = \text{Re}\left[\vec{E}_s(r)\, e^{j\omega t}\right]$, In which the symbol (~) has been placed over the time-varying quantity to distinguish it from the phasor quantity. The phasor equation may be derived from the time-varying equation by replacing each time-varying quantity with a phasor quantity [that is, $\tilde{\vec{E}}(r, t)$ by $\vec{E}_s(r)$, which varies only with the space coordinates x, y, z] and each time derivative with a $j\omega$ factor.

Thus, for *time harmonic fields, Maxwell's equations* may be expressed *in phasor form* as:

$$\nabla \cdot \vec{D} = \rho \qquad \oint_S \vec{D} \cdot d\vec{s} = \int_V \rho_v \, dv \tag{2.5a}$$

$$\nabla \cdot \vec{B} = 0 \qquad \oint_S \vec{B} \cdot d\vec{s} = 0 \tag{2.5b}$$

$$\nabla \times \vec{E} = -j\omega\vec{B} \qquad \oint_C \vec{E} \cdot d\vec{l} = -\int_S j\omega\vec{B} \cdot d\vec{s} \tag{2.5c}$$

$$\nabla \times \vec{H} = \vec{J} + j\omega\vec{D} \qquad \oint_C \vec{H} \cdot d\vec{l} = \int_S (\vec{J} + j\omega\vec{D}) \cdot d\vec{s} \tag{2.5d}$$

The above equations contain the *equation of continuity*:

$$\nabla \cdot \vec{J} = -j\omega\rho \quad \text{or} \quad \oint_S \vec{J} \cdot d\vec{s} = -\int_V j\omega\rho_v \, dv \tag{2.5e}$$

The *constitutive relations* for homogeneous, isotropic medium retain their forms, being:

$$\vec{D} = \varepsilon\vec{E}, \quad \vec{B} = \mu\vec{H}, \quad \vec{J} = \sigma\vec{E} \tag{2.5f}$$

EXAMPLE 2.1

A source-free non-magnetic material ($\mu_r = 1$) having $\varepsilon_r = 10$, is excited by a time-dependent magnetic field $\tilde{\vec{B}} = \vec{a}_x\, 20 e^{j(10^4 t + 10^{-4} z)}$. Using Maxwell's equations find:

(i) the electric field intensity in the material, (ii) the electric flux density and the magnetic field intensity in the material.

Solution

(i) For source-free material, $\vec{J} = 0$, thus Maxwell's magnetic curl equation (Eq. 2.5d) is given by: $\nabla \times \vec{H} = j\omega\vec{E}$.

The LHS of this equation can be written as:

$$\nabla \times \vec{H} = \frac{1}{\mu} \nabla \times \vec{B} = \vec{a}_y \frac{1}{\mu} \frac{\partial \vec{B}_x}{\partial z} \quad \text{(since only the } y \text{ component of } \vec{B} \text{ has to be retained)}$$

$$= \vec{a}_y \frac{1}{\mu} \frac{\partial (10 e^{j(10^{-2} z)})}{\partial z} = \vec{a}_y\, j\, \frac{10 \times 10^{-2}\, e^{j(10^{-2} z)}}{\mu}$$

When this is equated with the RHS of the above curl equation, using $\mu = \mu_0$, $\omega = 10^{-2}$ and $\varepsilon = 10\varepsilon_0$, we have:

$$\vec{E} = \vec{a}_y \frac{10 \times 10^{-2}\, e^{j(10^{-2} z)}}{10 \times 10^2\, \mu_0\, \varepsilon_0} = \vec{a}_y\, 10^{12}\, e^{j(10^{-2} z)} \text{ (V/m)}$$

(ii) Electric flux density ($\vec{D}$) and the magnetic field intensity ($\vec{H}$) are respectively given by:

$$D = 10\, \varepsilon_0\, E = \vec{a}_y \frac{10^{-3}\, e^{j(10^{-2} z)}}{\mu_0} = \vec{a}_y\, 795\, e^{j(10^{-2} z)} \text{ (C/m)}$$

$$\vec{H} = \frac{\vec{B}}{\mu_0} = \vec{a}_x \frac{10\, e^{j(10^{-2} z)}}{\mu_0} = \vec{a}_y\, 7.95 \times 10^6\, e^{j(10^{-2} z)} \text{ (A/m)}$$

2.3 The Wave Equation and Plane Electromagnetic Waves in Unbounded Media

The waves of electric and magnetic fields are immediately possible to be derived from Maxwell's two curl equations leading to the so called *wave equation*. The solution of these wave equations gives the propagating electromagnetic waves. The general solution of the wave equation is of little practical value; particular solutions are much more important and contribute a great deal to our understanding of certain physical phenomena. One simple solution of wave equation corresponds to a *plane wave*; its importance stems from the fact that complicated fields can be constructed from plane wave solutions by the application of Fourier's theorem. Further a plane wave can be produced experimentally to a fair degree of accuracy in the shape of a parallel beam of an electromagnetic wave.

The general form of the so-called *wave equations* obeyed by the $\vec{E}$ and $\vec{H}$ fields in source-free unbounded media are given by:

$$\nabla^2 \tilde{\vec{E}} - \mu\sigma \frac{\partial \tilde{\vec{E}}}{\partial t} - \mu\varepsilon \frac{\partial^2 \tilde{\vec{E}}}{\partial t^2} = 0 \qquad (2.6a)$$

$$\nabla^2 \tilde{\vec{H}} - \mu\sigma \frac{\partial \tilde{\vec{H}}}{\partial t} - \mu\varepsilon \frac{\partial^2 \tilde{\vec{H}}}{\partial t^2} = 0 \qquad (2.6b)$$

These wave equations govern the electromagnetic fields in a linear, homogeneous and isotropic medium (conducting or non-conducting) in which the charge density is zero (that is, source free). The mathematical symmetry of Eqs 2.6a and 2.6b implies that the solutions of both will have the same mathematical form. Hence we can solve either and infer the solution to the other.

The wave equations for the time-harmonic fields may be immediately written [by replacing $\tilde{\vec{E}}\,(r, t)$ by $\vec{E}_s\,(r)$ and each time derivative in Eq. 2.6 by a $j\omega$ factor] as:

$$\nabla^2 \vec{E}_s + (\omega^2 \mu\varepsilon - j\omega\mu\sigma)\vec{E}_s = 0$$

Or $$\nabla^2 \vec{E}_s - \gamma^2 \vec{E}_s = 0 \tag{2.7}$$

where $\gamma^2 = j\omega\mu(\sigma + j\omega\varepsilon)$. Again $\gamma = \alpha + j\beta$, with:

$$\alpha = \omega\sqrt{\frac{\mu\varepsilon}{2}}\left\{\sqrt{1+\left(\frac{\sigma}{\omega\varepsilon}\right)^2} - 1\right\}^{1/2} \tag{2.7a}$$

$$\beta = \omega\sqrt{\frac{\mu\varepsilon}{2}}\left\{\sqrt{1+\left(\frac{\sigma}{\omega\varepsilon}\right)^2} + 1\right\}^{1/2} \tag{2.7b}$$

Equation 2.7 is known as the *Vector Helmholtz equation.* A similar equation also holds for the $\vec{H}$ field. It may be noted from Eqs 2.7a and 2.7b that both the attenuation constant (α) and the propagation constant (β) of a general lossy medium are non-linear functions of frequency (ω).

For free space which is an unbounded dielectric medium with zero conductivity (a loss-less medium) the wave equation represented by Eq. 2.7 simplifies to:

$$\nabla^2 \vec{E}_s + k^2 \vec{E}_s = 0 \tag{2.8}$$

Or $$\nabla^2 \vec{E}_s + \frac{\omega^2}{c^2}\vec{E}_s = 0$$

where $$k = \omega\sqrt{\mu\varepsilon} = \beta \tag{2.8a}$$

In Eq. 2.8a, k is called the *wave number* and β is known as the *propagation constant.* These are used interchangeably in literature and represent the same physical quantity; physicsts perfer to use k while engineers prefer to use β. It may be noted from Eq. 2.8a that for a loss-less medium β (that is, k) is a linear function of frequency.

The most easily treated solutions of the wave equation are known as the plane wave solutions. A *plane wave* is defined as a wave in which the phase of the electric field (and the magnetic field) at any instant of time are constants over infinite planes (xy-plane) orthogonal to the direction of propagation (z-direction). If the equiphase surface is also the equiamplitude surface, it is a uniform plane wave (non-uniform plane waves are encountered in waveguides).

It is known from electromagnetic theory that the phase velocity of an electromagnetic wave is given by:

$$v_p = \frac{\omega}{k} = \frac{1}{\sqrt{\mu\varepsilon}} \tag{2.9}$$

Which in loss-less free space (with $\mu = \mu_0$ and $\varepsilon = \varepsilon_0$) was found to be equal to the velocity of light. However, for waves propagating in a lossy medium or those that propagate in guided structures this does not hold true, which will be seen subsequently.

The propagation constant β bears a definite relation to the wavelength:

$$\beta = \omega\sqrt{\mu\varepsilon} = \frac{\omega}{v_p} = \frac{2\pi}{\lambda} = k \tag{2.10}$$

Thus, k is a measure of the number of wavelengths in a span of 2π, hence the name *wave number.*

The total electric and magnetic field strengths of an electromagnetic field bears a constant ratio given by:

$$\frac{E}{H} = \eta = \sqrt{\frac{j\omega\mu}{\sigma + j\omega\varepsilon}} \tag{2.11}$$

where η is known as the *wave impedance* or the *intrinsic* (*characteristic*) *impedance* of the medium. For a purely dielectric medium with zero conductivity ($\sigma = 0$), $\eta = \sqrt{\mu/\varepsilon}$ and thus for free space $\eta_0 = 120\ \omega = 377\ \omega$. It is interesting to note that for DC and power frequency cases where voltage (V) and current (I) are measurable quantities, impedance is defined as the ratio of V to I while for RF and microwave frequency where V is derived from E field and I is derived from H field we talk in terms of *wave impedance.*

The mutual orientation of the electric and magnetic field vectors and the direction of propagation of uniform plane waves are such that the $\vec{E}$ and the $\vec{H}$ fields are at right angles to each other and the direction of propagation (z-direction) is perpendicular (transverse) to the plane (wave front) containing the E and H field vectors. Thus, a uniform plane wave propagates through the free space as a *transverse electromagnetic* (TEM) wave (Fig. 2.2).

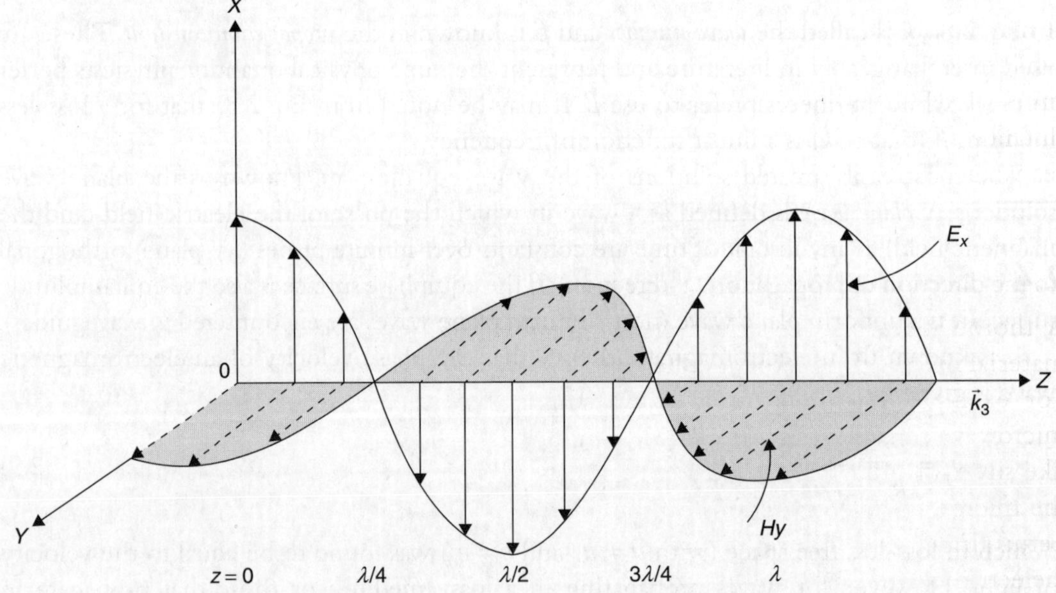

Fig. 2.2 Transverse electromagnetic (TEM) wave propagating in the +z direction in free space (full lines showing the *E*-field and dashed lines with shading is the *H*-field.

EXAMPLE 2.2

An electromagnetic wave of frequency 4 GHz propagates downward from a satellite and falls on sea water having relative permittivity of 72. Calculate the phase velocity, wavelength and wave number in air and in sea water.

Solution

The phase velocity of the electromagnetic wave in air and in water is given by:

$$v_{pa} = \frac{1}{\sqrt{\mu_0 \varepsilon_0}} = \frac{1}{\sqrt{4\pi \times 10^{-7} \times 8.854 \times 10^{-9}}} = 3 \times 10^8 \text{ (m/s)}$$

$$v_{psw} = \frac{v_{pa}}{\sqrt{\varepsilon_r}} = \frac{3 \times 10^8}{\sqrt{72}} = 3.535 \times 10^7 \text{ (m/s)}$$

$$\lambda_a = \frac{v_{pa}}{f} = \frac{3 \times 10^8}{4 \times 10^9} = 0.075 \text{ (m)}$$

$$\lambda_{sw} = \frac{v_{psw}}{f} = \frac{3.535 \times 10^7}{4 \times 10^9} = 8.83 \times 10^{-3} \text{ (m)}$$

$$k_a = \frac{2\pi}{\lambda_a} = \frac{2\pi}{0.075} = 83.775 \text{ (rad/m)}$$

$$k_{sw} = \frac{2\pi}{\lambda_{sw}} = \frac{2\pi}{8.83 \times 10^{-3}} = 711.5 \text{ (rad/m)}$$

It may be noted that the phase velocity in sea water is lowered by factor of $\sqrt{\varepsilon_r} = 8.485$, while the wavelength is shortened by a factor of 8.485 and the wave number is 8.485 times larger. Since the phase of the signal is represented by wave number multiplied by the distance z (in the direction of propagation), the phase of the wave changes much faster in sea water than in air.

2.4 Electromagnetic Wave Propagation in Material Media

A thorough knowledge of the characteristics of electromagnetic wave propagation in material media (conductor and dielectric) is of great importance to microwave engineers. For example, the inner walls of waveguides (used for guided wave propagation of microwave signal) should be perfect conductors and thus are usually coated with metals like silver or gold. Most food items and biological media are lossy dielectrics and thus the microwave oven functions on the principle of dielectric heating in lossy dielectrics. Let us first discuss in brief the wave propagation in conductors and then the principle of dielectric heating in detail in sub-section 2.4.1.

For good conductors, the loss tangent $\tan\delta_c$ $(= \sigma/\omega\varepsilon) >> 1$ and thus neglecting 1 in comparison with $\sigma/\omega\varepsilon$, we have from Eqs 2.7a and 2.7b: $\alpha = \beta = \sqrt{\omega\mu\sigma/2}$. Since for good conductors $\sigma \to \infty$, attenuation $\alpha \to \infty$ and thus the field can hardly penetrate in a good conductor. The lossy nature of a good conductor is commonly assessed in terms of *skin depth* δ which indicates the degree to which an electromagnetic wave can penetrate into a good conductor. The skin depth is defined as the depth into which the incident electromagnetic wave upon impinging into a good conductor is attenuated to $1/e$ (or ~36.8%) of its original intensity. The skin depth is thus given by:

$$\delta = \frac{1}{\alpha} = \frac{1}{\sqrt{\pi f \mu \sigma}} \tag{2.12a}$$

For good conductors, the skin depth is quite small compared to the wavelength of the electromagnetic wave in the medium (for copper at 300 MHz, $\delta = 0.00382$ mm $= 3.82\ \mu$m). Thus, in a typical transmission line (say a coaxial line) the electromagnetic wave propagates in the dielectric region between the conductors at the velocity of light in free space (or at the phase velocity appropriate for the dielectric medium, $v_{pD} = v_{p0}/\sqrt{\varepsilon_r}$). The conductor structure simply guides the electromagnetic wave (satisfying the boundary conditions for the electric and magnetic field for electromagnetic wave propagation) and the penetration of the wave into the conductor is only to the extent permitted by the imperfection of the conductor and frequency of operation as dictated by Eq. 2.12a.

The inverse frequency dependence of the skin depth ($\delta\ \alpha f^{-1/2}$) in good conductors leads to many important applications. At a typical microwave frequency of 10 GHz, the skin depth for good conductors like silver and gold are less than a micron (μm). Consequently, the difference of performance between a pure silver component and a silver plated brass component would be practically negligible. This is indeed the case, and the plating (either with silver or gold) technique is used to reduce the material cost of high quality waveguide components made of brass or copper.

The wave impedance or intrinsic impedance of a good conductor ($\sigma \to \infty$) can be written from Eq. 2.11 as:

$$\eta_c \cong \sqrt{\frac{j\omega\mu}{\sigma}} = \sqrt{\frac{\omega\mu}{\sigma}} \cdot \frac{(1+j)}{\sqrt{2}} \tag{2.12b}$$

[since $\sqrt{j} = (e^{j\pi/2})^{1/2} = e^{j\pi/4} = (1+j)/\sqrt{2}$].

Thus, in a good conductor the wave impedance is extremely small (since σ is very large) which means that an ideal conductor (a metal) acts as a perfect short.

It may be noted that for wave propagation in dielectric medium (for which the loss tangent $\tan\delta_c$ ($= \sigma/\omega\varepsilon$) << 1, with the approximation: $\sqrt{1 + \frac{\sigma^2}{\omega^2 \varepsilon^2}} \cong \left(1 - \frac{\sigma^2}{2\omega^2 \varepsilon^2}\right)$, we can write the modified form of Eq. 2.11 to represent the intrinsic impedance of a lossy dielectric as:

$$\eta_d \cong \sqrt{\frac{\mu}{\varepsilon}} \left(1 + j\frac{\sigma}{2\omega\varepsilon}\right) \tag{2.12c}$$

This indicates that the presence of finite conductivity in a dielectric adds a reactive component to its intrinsic impedance, making it a lossy dielectric. In the next section we will see that the presence of any non-zero conductivity and a high frequency dependent effective conductivity is the cause of dielectric heating in a lossy dielectric at high RF, especially at microwave frequency.

2.4.1 Principle of Dielectric Heating and Microwave Oven

The physical principle responsible for the heating process in lossy dielectric material can be understood as follows. When an external time-varying field is applied to a dielectric material, bound charges (that is, electron cloud and protons in the nucleus which are elastically bound to each other) are displaced, giving rise to polarisation density (or electric polarisation, or simply polarisation) $\vec{P}$. Since electrons and protons have finite mass, it is clear that the amount of polarisation must be a function of the frequency of the applied field. At sinusoidal steady state, the polarisation $\vec{P}$ varies at the same frequency as the applied field $\vec{E}$. At low frequencies, $\vec{P}$ is in phase with $\vec{E}$ and both quantities reach their maxima and minima at the same points in the radio frequency cycle. As the frequency increases, the inertia of the charged particles (being attached to their molecules) tends to prevent the particle displacements from keeping in phase with the field changes. Thus $\vec{P}$ falls out of phase with $\vec{E}$, and work must be done against the frictional damping force causing the applied field to lose power, and this power is dissipated in the medium as heat. This phenomena of $\vec{P}$ being out-of-phase with $\vec{E}$ in a dielectric medium when subjected to a high frequency electromagnetic field is known as *dielectric heating* and can be characterised by a complex electric susceptibility and hence a complex permittivity. The imaginary part of the *complex permittivity* ε_c also includes ohmic losses due to non-zero conductivity σ for lossy dielectric in addition to the power dissipation in the dielectric medium due to dielectric heating mentioned above due to the high frequency field; ε_c is given by:

$$\varepsilon_c = \varepsilon' - j\varepsilon'' \tag{2.13a}$$

Following the Debye equation for dielectric relaxation, it can be shown that:

$$\varepsilon' = \varepsilon_\infty + \frac{\Delta\varepsilon}{1+(\omega/\omega_0)^2} \quad \text{and} \quad \varepsilon'' = \frac{(\omega/\omega_0)\,\Delta\varepsilon}{1+(\omega/\omega_0)^2} \tag{2.13b}$$

where $\Delta\varepsilon = (\varepsilon_s - \varepsilon_\infty)$ with ε_s and ε_∞ representing respectively the static permittivity (permittivity at $\omega = 0$) and permittivity of the medium at frequencies far-off resonance, and ω_0 is the resonant frequency of the relaxation process. The frequency dependent variation of ε' and ε'' are shown in Fig. 2.3a and 2.3b respectively.

It may be observed that the real part of complex permittivity (which just stores the electric energy) starts off at low frequencies at the static value (ε_s) and starts falling with frequency attaining a value $\Delta\varepsilon/2$ [where $\Delta\varepsilon = (\varepsilon_s - \varepsilon_\infty)/2$] at resonance then drops to ε_∞ as the frequency increases further. While the imaginary part (responsible for absorption of energy as it relates to the loss part) peaks at the resonance (ω_0) of the relaxation process (causing maximum absorption of electromagnetic energy by the dielectric medium at this frequency) and falls off sharply to zero on either side of the resonant frequency. Out of the several absorption mechanisms (like orientational or dipolar, ionic and electronic) of electromagnetic signal in dielectric media, the dipolar relaxation process (which relates to

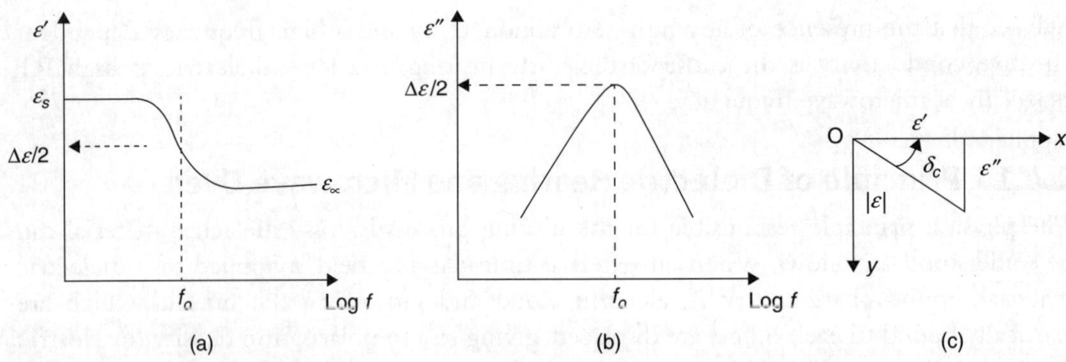

Fig. 2.3 (a) Plots of ε' against frequency (b) Plot of ε'' against frequency (c) Polar form of plot of complex permittivity.

the above discussion of dielectric heating) exhibits a resonant peak of ε'' around the UHF and microwave frequency region. In fact, for water molecules the highest absorption peak is at 2.45 GHz; thus, the ISM (Industrial-Scientific-Medical) band standardisation specified the international frequency standard for a microwave oven as 2.45 GHz.

In literature, loss tangent $\tan\delta_c$ is often used to represent the dissipation in a dielectric at higher RF (analogous to power factor in low frequency circuit theory). For a lossy dielectric, Eq. 2.13a may be expressed in polar form as: $\varepsilon' - j\varepsilon'' = |\varepsilon_c|\, e^{-j\delta_c}$; where $|\varepsilon| = \sqrt{(\varepsilon')^2 + (\varepsilon'')^2}$ and the arg $(\varepsilon) = -\delta_c$. The polar from of plot of complex permittivity in terms of its real and imaginary parts is shown in Fig. 2.13c and the tangent of the phase angle δ_c, loss of the complex permittivity, that is, the loss tangent is given by:

$$\tan\delta_c = \varepsilon''/\varepsilon' \tag{2.14}$$

For all practical purposes, air is considered to be a perfect dielectric having loss tangent equal to zero. Commonly used dielectrics like RT-Duroid 5880 ($\varepsilon_r = 2.2$) and FR4 ($\varepsilon_r = 4.35$) respectively, have loss tangents: 0.0004 @ 1 MHz; 0.0009 @ 10 GHz and 0.02 @ 100 MHz; 0.023 @ 2 GHz (the loss tangent is frequency dependent due to the frequency dependency of ε'').

Let us now proceed to derive the expression for the microwave power dissipated in a lossy dielectric due to dielectric heating of the material by incident microwave signals. For lossy dielectric materials (with $\vec{J} = \sigma\vec{E} \neq 0$) the Maxwell's Eq. 2.5d can be written as:

$$\nabla \times \vec{H} = \sigma\vec{E} + j\omega(\varepsilon' - j\varepsilon'')\,\vec{E} = (\omega\varepsilon'')\,\vec{E} + j\omega\varepsilon'\,\vec{E} \tag{2.15}$$

(where non-zero σ is included in ε'').

It thus appears that the imaginary part of ε_c (namely, ε'') leads to a volume current density term that is in phase with the electric field, as if the material had an effective conductivity $\sigma_{\text{eff}} = \omega\varepsilon''$. At low frequencies, $\omega\varepsilon''$ is small, because ω is small and ε'' itself is small, so the losses are largely negligible. However, at high frequencies close to resonance, $\omega\varepsilon''$ increases abruptly and produces the same macroscopic effect as if the dielectric had effective conductivity $\sigma_{\text{eff}} = \omega\varepsilon''$. When a steady current $\vec{J} = \sigma\vec{E}$ flows in a conducting material in response to constant applied field $\vec{E}$, the electric power per unit volume dissipated in the material is given by $\vec{E}\cdot\vec{J}=\sigma\vec{E}^2$. Similarly, when a dielectric is excited at frequencies high enough for $\omega\varepsilon''$ to be appreciable, an alternating current density $\omega\varepsilon''\vec{E}$ flows (in addition to the displacement current density given by $j\omega\varepsilon'\vec{E}$) in response to an applied alternating field $\vec{E}$, leading to an instantaneous power dissipation of $[(\omega\varepsilon''\vec{E})\cdot\vec{E}]=\omega\varepsilon''|\vec{E}|^2$. When the electric field is time harmonic, that is, $\tilde{\vec{E}}(z,t)=\vec{E}\cos(\omega t\pm\beta z)$, the time average power dissipated in the material per unit volume at any arbitrary point (say at $z=0$) is given by:

$$P_{av}=\frac{1}{T_P}\int_0^{T_P}\omega\varepsilon''\vec{E}^2(t)\,dt$$

$$=\frac{1}{T_P}\int_0^{T_P}\omega\varepsilon''\vec{E}^2\cos^2(\omega t)\,dt=\frac{1}{2}\omega\varepsilon''|\vec{E}|^2 \tag{2.16}$$

This power dissipated in the material is the basis of *microwave heating* of lossy dielectric materials.

A few points may be noted here in connection with microwave heating. As of today, the foods cooked with microwave heating are not known to cause any harm whatsoever. But microwave heating caused by the microwave signal of cell phone may be harmful to human tissues on long time exposure and with microwave signal strength above the specified value. Hence cell phone manufacturers must design it as per the internationally specified SAR (Specific Absorption Rate) value (which is a measure of the amount of radio frequency energy absorbed by the body when using a mobile phone). As per European standards, a safe SAR value is 2 W/kg averaged over 10 g of human tissue while as per US standard it is 1.6 W/kg averaged over 1 g of tissue (in India, the US standard is followed).

Microwave heating has many industrial and medical applications. In fact many industrial processes involve one or more stages of product drying, which is often very costly. Microwaves offer what is sometimes the most viable alternative in terms of energy efficiency. For example, microwave heating is used in paper and printing industries, leather and textile industries, drying of tobacco, in vulcanisation, polymerisation, in nuclear waste treatment and so on. Microwaves have important therapeutic applications in tissue heating and the most promising of these microwave heating applications is hyperthermia for the treatment of cancer. This technique seems to have no side effects. It continues to be effective because the body does not get accustomed to it, and it produces only a minimum of discomfort for the patient. The results in themselves are very positive and become remarkable when microwaves are combined with radiotherapy or chemotherapy.

EXAMPLE 2.3

A microwave oven cooks food with microwave power generated by a magnetron which produces a sinusoidal electric intensity of amplitude 250 V/m at a frequency of 2.45 GHz. Assuming the dielectric constant of the food item to be cooked is 40 with a loss tangent of 0.35 at the operating frequency of the magnetron, calculate the average power per cubic metre.

Solution

In this problem, it is given that $\tan\delta_c = 0.35$ and ε_r (dielectric constant or relative permittivity) of the food item to be cooked = 40.

Since, $\tan\delta_c = (\varepsilon''/\varepsilon') = [\varepsilon''/\varepsilon_0\,\varepsilon_r]$

$$\varepsilon'' = \tan\delta_c\,\varepsilon_0\,\varepsilon_r' = 0.35 \times (10^{-9}/35\pi) \times 40 = 0.1237 \times 10^{-9}$$

Now the average power dissipated per unit volume of the food item from Eq. 2.16 is,

$$P_{av} = \frac{1}{2}\,\omega\varepsilon''\,|\vec{E}|^2 = 59.5\ (\text{kW/m}^3)$$

EXAMPLE 2.4

A bakelite slab having relative permittivity of 5 is placed inside a microwave oven operating at 2.45 GHz and producing a sinusoidal electric field intensity with amplitude 400 V/m inside the bakelite dielectric. Assuming that the electric field intensity is uniform in the dielectric slab and that it is perpendicular to the surface of the slab, calculate the displacement current density in the bakelite slab. Will there be a displacement current in the air around the dielectric, if so, calculate this displacement current density too.

Solution

The displacement current density is defined as the time derivative of the electric flux density $\vec{D}$ $(= \varepsilon\vec{E})$, where $\vec{E}$ is the sinusoidal electric field intensity in the material. Thus displacement current density is given by:

$$\vec{J}_d = \frac{\partial\vec{D}}{\partial t} = \varepsilon\frac{\partial\vec{E}}{\partial t} = \varepsilon_0\varepsilon_r\,\frac{d}{dt}(E_0\,\sin\omega t) = \varepsilon_0\varepsilon_r\,E_0\,\omega\,\cos\omega t$$

$$= 8.854\times10^{-12}\times5\times400\times2\times\pi\times2.45\times10^9\times\cos(2\times\pi\times4.45\times10^9)\,t$$

$$= 272\,\cos(4.9\times10^9\,\pi)\,t\ (\text{A/m}^2)$$

It is known from the boundary conditions to be satisfied at the interface of two media (discussed later in this chapter) that the normal component of electric flux density is continuous across the interface, that is, $D_{n(\text{air})} = D_{n(\text{bakelite})}$. Thus, the electric flux density and hence the displacement current density in air around the bakelite dielectric will be the same (J_d) as that calculated above.

2.5 Electromagnetic Power Flow and Poynting Vector

When an electromagnetic wave propagates from one point in a system to another, there is a transfer of energy between the two points in the system. A direct relationship is expected to exist between the rates of this energy transfer (that is, electromagnetic power flow) and the amplitudes of the electric and magnetic field strengths of the electromagnetic wave. This is because the units of $\vec{E}$ are volts per metre and the units of $\vec{H}$ are amperes per metre, thus the product of their magnitudes is VA/m^2 or watts per square metre. The analytical relation between the field vectors and the power flow for electromagnetic wave can be derived from Maxwell's equation, which is given below:

$$\tilde{\vec{P}} = \tilde{\vec{E}} \times \tilde{\vec{H}} \tag{2.17}$$

where $\tilde{\vec{P}}$ represents the *instantaneous power density* vector associated with the electromagnetic field at a given point and is known as a Poynting vector [3]. The direction of the vector $\tilde{\vec{P}}$ indicates the direction of the instantaneous power flow at the point. It may be noted that $\tilde{\vec{P}}$ is normal to both $\tilde{\vec{E}}$ and $\tilde{\vec{H}}$ and is therefore along the direction of propagation; however, if either one of $\tilde{\vec{E}}$ or $\tilde{\vec{H}}$ is reversed in direction then $\tilde{\vec{P}}$ is reversed, but if both $\tilde{\vec{E}}$ and $\tilde{\vec{H}}$ reverse direction, $\tilde{\vec{P}}$ preserves its original sense.

The integration of a Poynting vector over any closed surface S enclosing a volume gives the net power flowing out of that surface:

$$\vec{P}_o = \oint_S \tilde{\vec{P}} \cdot d\vec{s} = \int_S (\tilde{\vec{E}} \times \tilde{\vec{\mathrm{H}}}) \cdot d\vec{s} \tag{2.18}$$

In most applications we are interested in the Poynting vector averaged over time rather than in its instantaneous value. Considering the expressions for the instantaneous electric and magnetic field of a time-harmonic uniform plane wave in the general case of a lossy medium respectively, $\tilde{\vec{E}}(z,t) = \vec{a}_x E_0 e^{-\alpha z}\cos(\omega t - \beta z)$ and $\tilde{\vec{H}}(z,t) = \vec{a}_y (E_0/\eta) e^{-\alpha z}\cos(\omega t - \beta z - \varphi_\eta)$, the *time average power* can be obtained from Eq. 2.17 by integrating over one period (T_p) of the sinusoidal variation as:

$$\vec{P}_{av}(z) = \frac{1}{T_p}\int_0^{T_p} \tilde{\vec{P}}(z,t)dt = \vec{a}_z \frac{E_0^2}{2|\eta|} e^{-2\alpha z}\cos\varphi_\eta \tag{2.19a}$$

However, such a time averages Poynting flux may not be always necessary to evaluate for sinusoidal signals since the *average Poynting vector* $\vec{P}_{av}$, can be directly obtained in terms of the phasor fields $\vec{E}$ and $\vec{H}$ as:

$$\vec{P}_{av} = \frac{1}{2}\mathrm{Re}\,(\vec{E} \times \vec{H}^*) \tag{2.19b}$$

This follows from the fact that we can write the *complex Poynting vector* $\vec{P} = \frac{1}{2}(\vec{E} \times \vec{H}^*) = \vec{P}_{av} + j\vec{P}_{\text{ractive}}$, where $\vec{P}_{av}$ is given by Eq. 2.19b above in analogy with the vector counterpart of the average power calculation in sinusoidal, steady state circuit analysis problem. That is, $P_{av} = 1/2[\mathrm{Re}(VI^*)]$, where V and I are the phasor voltage and current associated with a two terminal circuit element and P_{av} is the time-average power delivered to the element.

Certain important observations may be made with reference to the Poynting vector and the electromagnetic power flow. Firstly, though power flow through any closed surface is correctly given by Eq. 2.18, it does not necessarily follow that $\tilde{\vec{P}} = \tilde{\vec{E}} \times \tilde{\vec{H}}$ represents correctly the power flow at each point of that surface. In other words, the Poynting vector $\vec{E} \times \vec{H}$ is meaningful only if it is integrated over the surface of the volume but it cannot necessarily be applied to a localised portion of the surface. Secondly, any constant vector $\vec{X}$ which is the curl of another vector, say $\vec{Y}$, can be added to $\vec{E} \times \vec{H}$ without violating the surface integral of the Poynting vector given by Eq. 2.18, because $\nabla \cdot \nabla \times \vec{Y} \equiv 0$. In other words, an infinite number of vectors may be possible to represent electromagnetic energy flow in lieu of $\vec{E} \times \vec{H}$. Further, even though it is possible to write an expression in terms of a Poynting vector that gives the net flow of power through a closed surface, it is still not possible to state just where the energy resides. However, this philosophical issue is not much bothered with from an engineering point of view and we will continue with Poynting's hypothesis unless it is challenged by experimental evidence against it.

EXAMPLE 2.5

(i) Let a uniform axial current of density J flow through a long straight cylindrical wire of electrical conductivity σ and radius r. Find the magnitude and direction of the Poynting vector at the surface of the wire.

(ii) Let a plane electromagnetic wave with peak amplitude of the electric filed E_0 be incident on a conducting slab of conductivity σ ($>> \omega\varepsilon_0$). Find the time average Poynting vector within the slab. If the conductor is a perfect one ($\sigma \to \infty$), find the value of the average Poynting vector everywhere in space.

Solution

(i) Let the wire be oriented along the z-axis of the cylindrical coordinate system (r, φ, z) and let the current J flow along the $+z$ direction. Thus we have the electric field inside the wire as $\vec{E} = \frac{J}{\sigma}\vec{a}_z$. According to the continuity of the tangential component of $\vec{E}$ across the interface (Section 2.7.1) on the outside surface of the wire, the same electric field exists. Using Ampere's circuital law, we have the magnetic field intensity near the surface of the wire as $\vec{H} = \frac{I}{2\pi r}\vec{a}_\varphi = \frac{Jr}{2}\vec{a}_\varphi$, since $J = I/\pi r^2$. Thus, the Poynting vector at the surface of the wire is:

$$\vec{P} = \vec{E} \times \vec{H} = \frac{J}{\sigma}\vec{a}_z \times \frac{Jr}{2}\vec{a}_\varphi = -\frac{J^2 r}{2\sigma}\vec{a}_r$$

(ii) For simplicity let us assume that the direction of wave propagation is parallel to the normal to the surface of the slab, that is, along the +ve z-direction. Then using the result of Eq. 2.19a and the magnitude and phase of η_c given by Eq. 2.12b we get the time average Poynting vector as:

$$\vec{P}_{av}(z) = \frac{1}{2}\sqrt{\frac{\sigma}{\omega\mu_0}}\cos\left(\frac{\pi}{4}\right)E_0^2\, e^{-2\alpha z}\vec{a}_z = 0.353\sqrt{\frac{\sigma}{\omega\mu_0}}E_0^2\, e^{-2\alpha z}\vec{a}_z,$$

Now, since the conductor is a perfect one, that is, $\sigma \to \infty$, and $\alpha \to \infty$ when $\sqrt{\sigma}\, e^{-2\alpha z}$ $\to 0$. Thus the P_{av} inside the conducting slab will be zero causing the wave to be totally reflected at the surface of the slab and would interfere with the incident signal in the outside region of the slab resulting in a standing wave. Thus P_{av} will be zero everywhere.

2.6 Polarisation of Electromagnetic Wave

The phenomenon of restricting the oscillations of waves to just one direction in the transverse plane is known as *polarisation of waves*. Polarisation is a property specific to electromagnetic waves, and is not found in other types of wave propagation such as acoustic waves. This is because electromagnetic waves are transverse waves (that is, they have vibrations in all possible directions and thus can be restricted in a plane perpendicular to the direction of propagation) allowing it to be polarised; while acoustic waves are longitudinal waves (that is, they have vibrations of particles occurring only along the direction of propagation) and thus cannot be polarised. By convention, the polarisation of electromagnetic field refers to the wave electric field.

Polarisation is a very important property of electromagnetic waves and hence of radio wave (RF)/microwave communication and antenna systems. Amplitude Modulation (AM) radio broadcasts operating in relatively lower RF range use a long vertical metallic pole above the ground as transmitting antenna producing vertically polarised signals, with the wave electric field oriented perpendicular to the ground. Thus, for maximum reception of the AM radio signal, the receiver antenna should be parallel to the electric field. Most AM radio receivers (for the sake of using miniaturised antenna), however, use horizontally oriented 'ferrite-rod' antenna to detect the wave magnetic field (instead of using vertical wire antenna to intercept the wave electric field); this is possible since electric and magnetic fields are orthogonal to each other for free space TEM wave. FM broadcast stations normally utilise circular polarisation, so the orientation of an FM receiving antenna (which is usually a 'whip' antenna) need not be critical as long as the antenna lies in a plane orthogonal to the direction from which the signal arrives. In case of TV broadcast, horizontally polarised waves are used (as man-made noise caused by automobiles, rotating motors and so on is usually vertically polarised); this is the reason why typical roof top antenna (Yagi-Uda type) used for the reception of TV signals are horizontal. Further, telecommunication systems take advantage of dual-polarisation (vertical–horizontal) to achieve isolation between the transmitted and received signals. For example, the blockage effect due to the sub-reflector in Cassegrain feed type antenna is suppressed by designing the reflector surface in such a way that it rotates the plane of polarisation of the microwave signal from the primary feed by 90°. In satellite communication, the use of two orthogonal linear polarisations (alternating between vertical and horizontal polarisations) permits doubling the information carrying capacity in a given frequency band utilising the so called 'frequency-reuse' technique.

The polarisation of a uniform plane wave is conventionally understood to be the time-varying behaviour of the electric field intensity vector $\vec{E}$ at a given point in space. Let a uniform plane wave be travelling in the +ve z direction with the $\vec{E}$ and $\vec{H}$ field vectors lying in the xy plane (Fig. 2.2). If E_x is zero and E_y is present, the wave is said to be polarised in the y-direction while if E_x was present and E_y was zero, the wave would have been polarised in the x-direction. If both E_x and E_y are present, and are in phase (that is, if they reach their maximum values at the same instant of time) then the resultant electric field has a definite direction dependent on the relative magnitudes of E_x and E_y. The angle θ, which this direction makes with the x-axis, is given by: $\tan^{-1}(E_y/E_x)$ and this angle will be constant with time (Fig. 2.4a). In all the above cases in which the direction of the resultant vector is constant with time, the wave is said to be *linearly polarised*. If E_x and E_y are not in phase, that is, if they reach their maximum values at different instants of time, then the direction of the resultant electric vector will vary with time. When E_x and E_y are of unequal amplitude and maintain a constant phase difference of 90°, the $\vec{E}$ field describes an ellipse in space and the wave is said to be *elliptically polarised* (Fig. 2.4b). In the particular case, when E_x and E_y have equal magnitudes and a 90° phase difference, the locus of the resultant $\vec{E}$ field is a circle and the wave is *circularly polarised* (Fig. 2.4c).

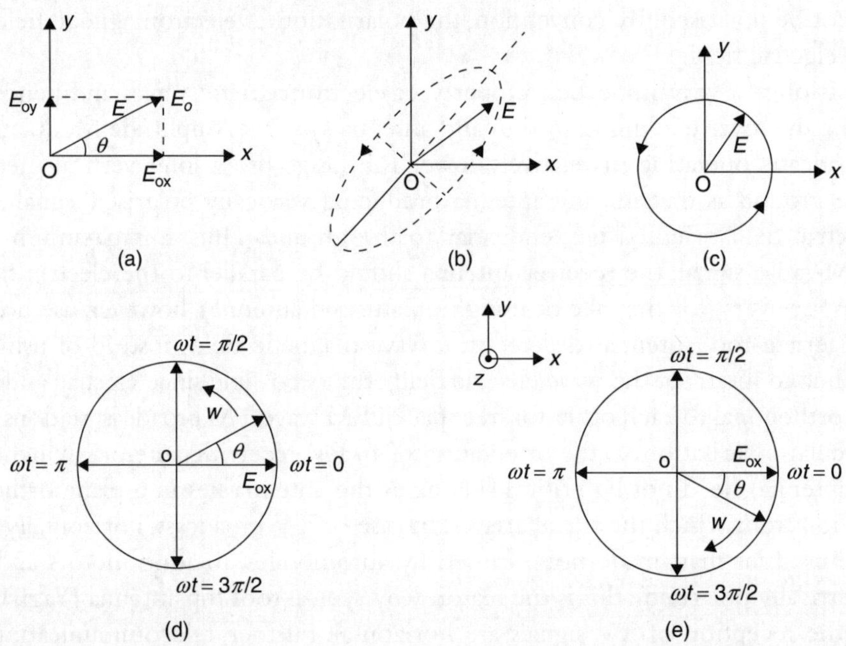

Fig. 2.4 Polarisation of electromagnetic wave (a) Linearly polarised, (b) Elliptically polarised, (c) Circularly polarised (d) Right-handed circularly polarised (RHCP), (e) Left-handed circularly polarised (LHCP).

In the case of circular polarisation, if the resultant $\vec{E}$ field rotates at a uniform rate with an angular velocity ω in a *counterclockwise* direction when viewed looking towards the –ve z direction, the wave is said to be *right-handed circularly polarised* (RHCP) (Fig. 2.4d). This is as per IEEE (Institute of Electrical and Electronic Engineers, USA) convention. According to this, when the fingers of the right hand follow the direction of rotation of $\vec{E}$ field, the thumb points to the direction of propagation of wave (in this case the +ve z direction).

However, most physicists specialising in optics, prefer to have the thumb point to the source (in the –ve z-direction), that is, from where the wave is coming, when the RHCP wave becomes *clockwise polarised.* This is termed as *positive polarisation.* Further, when $\vec{E}$ field rotates with an angular velocity ω in a clockwise direction, this is a *left-hand circularly polarised* (LHCP) wave (Fig. 2.4e). With the alternative convention as mentioned above, LHCP becomes *anti-clockwise polarised* and is termed as *negative polarisation.*

EXAMPLE 2.6

The electric field intensity of the electromagnetic wave received from a satellite transponder operating at a frequency of 4 GHz is $\vec{E} = (4\vec{a}_x - j5\vec{a}_y)\, e^{-j\beta z}$ V/m. Determine the nature and sense of polarisation of the wave.

Solution

The instantaneous expression for the electric field intensity is given by:

$$\tilde{\vec{E}} = \text{Re}\,\{\vec{E}\,(z)\, e^{j\omega t}\} = \text{Re}\,\{(4\vec{a}_x - j5\vec{a}_y)\, e^{-j\beta z}\, e^{j\omega t}\} = \text{Re}\,\{4\vec{a}_x\, e^{j(\omega t - \beta z)}\} - \text{Re}\,\{j5\vec{a}_y\, e^{j(\omega t - \beta z)}\}$$

$$= 4\vec{a}_x \cos(\omega t - \beta z) + 5\vec{a}_y \sin(\omega t - \beta z)$$

Thus the x- and y-component of the electric field intensity are respectively given by:

$$E_x(z, t) = 4\cos(\omega t - \beta z) \quad \text{and} \quad E_y(z, t) = 5\sin(\omega t - \beta z)$$

In the $z = 0$ plane, these components become:

$$\frac{1}{4} E_x(0, t) = \cos\omega t \quad \text{and} \quad \frac{1}{5} E_y(0, t) = \sin\omega t$$

Which when squared and added:

$$\frac{1}{16} E_x^2(0, t) + \frac{1}{25} E_y^2(0, t) = 1$$

This is the equation of an ellipse, thus the received field is *elliptically polarised* with the major axis along the y-direction and the minor axis along the x-direction (Fig. 2.4d).

To determine the sense of polarisation, we set $\omega t = 0$ (at $t = 0$), when we have: $E_x(0, 0) = 4$ and $E_y(0, 0) = 0$, and the tip of the $\vec{E}$ field is on the positive x-axis. But when $\omega t = \pi/2$ (at $t = \pi/2\omega$) we have: $E_x(0, \pi/2\omega) = 0$ and $E_y(0, \pi/2\omega) = 5$, when the tip of the $\vec{E}$ field has moved to a point on the positive y-axis; thus the rotation is anti-clockwise as shown in Fig. 2.4d. As per IEEE convention this represents a *right-handed* elliptically polarised wave.

2.7 Wave Propagation in Interfacing Media

In Section 2.4 we have discussed in brief the characteristics of electromagnetic wave propagation in unbounded material medium (that is, a particular medium, conductor or dielectric, not being interfaced with another medium). But in many practical situations we come across dielectric–conductor interface (waveguides, RF shields and so on), dielectric–dielectric interface (low-reflection microwave loads), multiple dielectric interfaces (radome, RAdar DOME) and others. In all such cases, wave propagation takes place in two or more media with different constitutive parameters $(\varepsilon, \mu, \sigma)$ that are being interfaced with each other. To study the problem of wave propagation in such an interfacing medium, at first, we have to know the *boundary conditions* which must be satisfied by the electric and magnetic fields in an interface between two media. This will be followed by analytical treatment of the phenomena of reflection and transmission of electromagnetic waves from the interface of two or more media both for normal and oblique incidence which are of great practical importance.

2.7.1 Electromagnetic Boundary Conditions

The electromagnetic boundary conditions are determined by applying the integral form of Maxwell's equation (the differential form cannot be used because they involve space derivatives and so are not applicable at points of discontinuities in the medium). The boundary conditions that the field vectors $\vec{E}$, $\vec{H}$, $\vec{D}$ and $\vec{B}$ must satisfy at the boundary between two media are summarised below:

(i) Applying Eq. 2.4c, which follows from Faraday's law, to the contour ABCD as shown in Fig. 2.5a(i) we have:

$$\oint_{ABCDA} \vec{E} \cdot d\vec{l} = E_{1t} \cdot \Delta l + E_{2t}(-\Delta l) = -\frac{\partial \vec{B}}{\partial t} \Delta l \cdot \Delta h$$

where AB = DC = Δl and BC = AD = Δh. Since we are evaluating at the boundary surface, $\Delta h \to 0$, thus from the above equation we get:

$$E_{1t} = E_{2t} \tag{2.20a}$$

This means that *the tangential component of the electric field intensity* $\vec{E}$ *is continuous across the boundary between two media.*

(ii) Similarly applying Eq. 2.4d, which follows from Ampere's circuital law duly modified by Maxwell, to the contour ABCD in Fig. 2.5b(i) we can show that:

$$H_{1t} = H_{2t} \tag{2.20b}$$

That is, *the tangential component of magnetic field intensity* $\vec{H}$ *is continuous at the boundary between two media* (provided either of the medium is not a perfect conductor. Otherwise surface current J_s will be present and Eq. 2.20b will need modification (which will be discussed shortly).

(iii) A consequence of applying Eq. 2.4b, which follows from Gauss's law for magnetic field, to the small pillbox-shaped surface S which intersects the boundary between the two media [Fig. 2.5b(ii)] we have:

$$\oint_S \vec{B} \cdot d\vec{s} = \vec{B}_1 \cdot \hat{n}_2 + \vec{B}_2 \cdot \hat{n}_1 = (\vec{B}_1 - \vec{B}_2) \cdot \hat{n}_2 = 0$$

Since $\hat{n}_2 = -\hat{n}_1$, where $\hat{n}_2$ and $\hat{n}_1$ are outward units normal from the medium 2 and 1 respectively [Fig. 2.5b(ii)], when we have:

$$B_{1n} = B_{2n} \tag{2.20c}$$

Thus, *the normal component of the magnetic flux density $\vec{B}$ is continuous across the boundary between two media.*

(iv) Similarly, it can be shown that on applying Eq. 2.4a, which follows from Gauss's law, to the pillbox-shaped surface S in Fig. 2.5a(ii), that:

$$D_{1n} = D_{2n} \tag{2.20d}$$

Hence, the *normal component of the electric flux density $\vec{D}$ is continuous at the boundary between two media* (provided either of the medium is not a perfect conductor otherwise surface charge density ρ_s will be present and Eq. 2.20d will need modification—which will be discussed shortly).

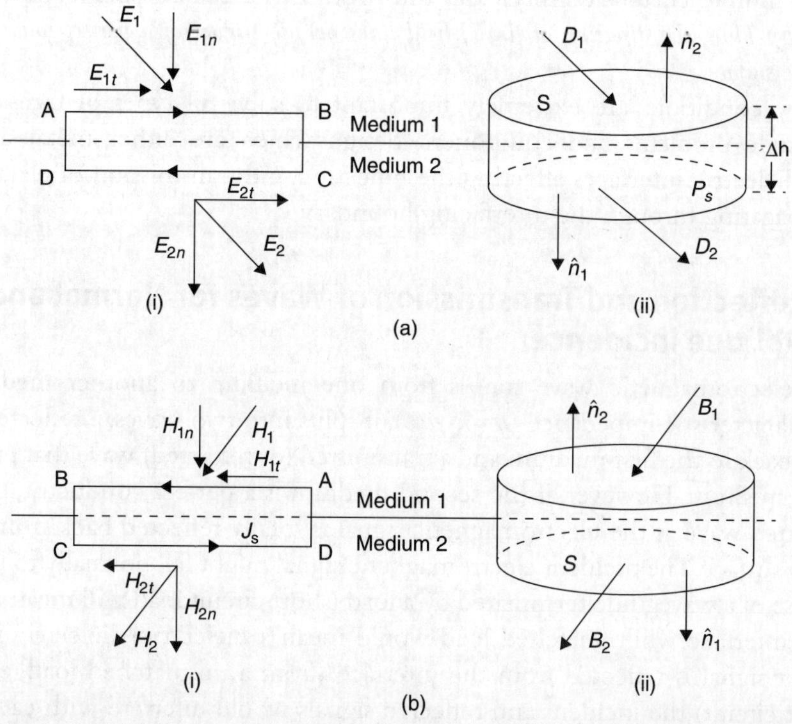

Fig. 2.5 Interface between two media to derive boundary conditions for (a): (i) Tangential component of electric field intensity (*E*), (ii) Normal component of electric flux density (*D*); (b) (i) Tangential component of magnetic field intensity (*H*), (ii) Normal component of magnetic flux density (*B*).

A few observations may be made from Eqs 20a–d. The continuity of the normal component of magnetic flux density is always maintained at the boundary between two media but the tangential magnetic field is continuous *only when there is no surface current*. While the tangential electric field at the boundary between two media is always continuous but the electric flux density is continuous *only when there is no surface charge density*.

But, if surface current is present then referring to Fig. 2.5a, Eq. 2.20b will be modified to:

$$H_{1t} = H_{2t} = J_s \tag{2.20f}$$

And if surface charge density ρ_s is present, then from Fig. 2.5b, Eq. 2.20d will be modified to:

$$D_{1n} = D_{2n} = \rho_s \tag{2.20e}$$

Now, surface current and surface charge is possible only in case one of the two media is a conductor (as high frequency current flows on the surface of a conductor and charge resides on the surface of a conductor). If we consider that medium 2 is a perfect conductor then $H_2 = B_2 = 0$ (since magnetic field inside a perfect conductor is zero) when we get from Eqs 2.20f and 2.20c that $H_{1t} = J_s$ and $B_{1n} = 0$. *This means that the magnetic field should be tangential to the surface of a perfect conductor.* Again with medium 2 as perfect conductor, $E_2 = D_2 = 0$ (as E and D is identically zero inside a perfect conductor, otherwise it would produce an infinite current density), and thus from Eqs 2.20a and 2.20e we have $E_{1t} = 0$ and $D_{1n} = \rho_s$. *Thus, the direction of the E-field lines on the surface of a perfect conductor must be normal to the surface.*

Boundary conditions are extremely important to solve all types of transmission line and waveguide problems and component design which are either metal–dielectric or dielectric–dielectric interfaces affecting the reflection and transmission of electromagnetic signals propagating through the interfacing boundary.

2.7.2 Reflection and Transmission of Waves for Normal and Oblique Incidence

When an electromagnetic wave travels from one medium to another medium having different characteristic impedance ($\eta = \sqrt{\mu/\varepsilon}$), it splits into two waves, a reflected wave that propagates back to the first medium and a transmitted (or refracted) wave that proceeds into the second medium. However, if the second medium is a perfect conductor, then there is no transmitted wave as the electromagnetic signal is totally reflected back from a perfectly conducting surface.The incident electromagnetic signal might fall normally to the interface as in the case of a waveguide terminated by a load (short circuit load will mean a dielectric–conductor interface while matched load would mean a dielectric–dielectric interface). If the incident signal is reflected from the interface (from a mismatched load, extreme case being short circuit) the incident and reflected signals would interfere with each other and would form a *standing wave*.The wave propagation in waveguides used for guiding microwave signals from one point to another involves multiple reflections of the electromagnetic waves back and forth from the interface of dielectric (normally air)-conductor (metal walls of the waveguide) with the electromagnetic signal falling obliquely every time on the dielectric–conductor interface. The phenomenon of total internal reflection takes place when an electromagnetic wave is incident at an oblique angle from a dense medium onto a less dense medium as in the case of a dielectric waveguide or optical fibres.

2.7.2.1 Normal incidence on plane boundaries

Let us consider a uniform plane wave in medium 1 (with constitutive parameters: $\sigma_1, \varepsilon_1, \mu_1$) that is incident normally (at right angles to the interface) with the propagation vector along the z-direction while $\vec{E}$ and $\vec{H}$ field vectors are along the x and y-directions respectively on medium 2 (with constitutive parameters: $\sigma_2, \varepsilon_2, \mu_2$,) as shown in Fig. 2.6.

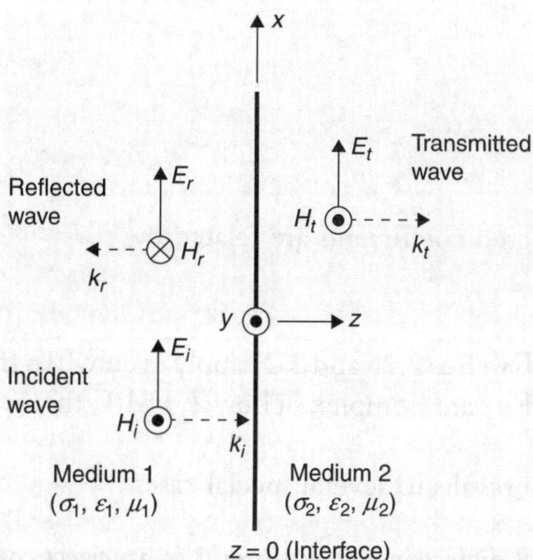

Fig. 2.6 Normal incidence of uniform plane wave to the interface between two media.

The incident electromagnetic field on the interface of two media will, in general, be partially transmitted and partially reflected. The incident, transmitted and reflected wave fields are given by:

Incident field: $$\vec{E}_i(z) = \vec{a}_x E_{i0}\, e^{-\gamma_1 z} \quad \text{and} \quad \vec{H}_i(z) = \vec{a}_y \frac{E_{i0}}{\eta_1} e^{-\gamma_1 z} \tag{2.21}$$

Transmitted: $$\vec{E}_t(z) = \vec{a}_x E_{t0}\, e^{-\gamma_2 z} \quad \text{and} \quad \vec{H}_t(z) = \vec{a}_y \frac{E_{t0}}{\eta_2} e^{-\gamma_2 z} \tag{2.22}$$

Reflected: $$\vec{E}_r(z) = \vec{a}_x E_{r0}\, e^{\gamma_1 z} \quad \text{and} \quad \vec{H}_r(z) = -\vec{a}_y \frac{E_{r0}}{\eta_1} e^{\gamma_1 z} \tag{2.23}$$

where η_1 and η_2 are the intrinsic impedances of medium 1 and medium 2 respectively, and γ_1 and γ_2 are respectively the propagation constants of the electromagnetic signals in the two media. The negative sign on the right hand side of the magnetic field of Eq. 2.23 is due to the fact that either the electric or magnetic field experienced a phase reversal as per the right-hand rule followed by the Poynting vector.

Applying the boundary conditions on the $\vec{E}$ and $\vec{H}$ fields at the interface (at $z = 0$) we have:

$$\vec{E}_i(0) + \vec{E}_r(0) = \vec{E}_t(0) \quad \text{or} \quad E_{i0} + E_{r0} = E_{t0} \tag{2.24}$$

$$\vec{H}_{i0} + \vec{H}_{r0} = H_{t0} \quad \text{or} \quad \frac{1}{\eta_1}(E_{i0} - E_{r0}) = \frac{E_{t0}}{\eta_2} \tag{2.25}$$

Solving the above two equations we have the *transmission coefficient* ($T = E_{t0}/E_{i0}$) and *reflection coefficient* $\Gamma = (E_{r0}/E_{i0})$ as:

$$T = \frac{2\eta_2}{\eta_2 + \eta_1} \tag{2.26}$$

$$\Gamma = \frac{\eta_2 - \eta_1}{\eta_2 + \eta_1} \tag{2.27}$$

The transmitted and reflected coefficients are related by:

$$T = 1 + \Gamma \tag{2.28}$$

The definitions for T and Γ in Eqs 2.26 and 2.27 apply even when the media are dissipative—that is, even when η_1 and η_2 are complex. Thus, T and Γ themselves are complex in the general case.

Let us now apply these results to several special cases.

(i) ***Medium 1 is a perfect dielectric and medium 2 is a perfect conductor*:** Since medium 2 is a perfect conductor, the electromagnetic field cannot exist inside medium 2 (that is, there will be no transmitted signal) and the signal will be totally reflected from medium 2 with the electric field experiencing a phase reversal. Thus, the total electric field inside medium 1 under this situation is:

$$\vec{E}_1(z) = \vec{E}_i(z) + \vec{E}_r = \vec{a}_x E_{i0}(e^{-j\beta_1 z} - e^{j\beta_1 z}) = -\vec{a}_x 2jE_{i0} \sin\beta_1 z \tag{2.29a}$$

where we have substituted $\gamma_1 = +j\beta_1$; since medium 1 is a perfect dielectric, $\sigma_1 = 0$ and therefore $\alpha_1 = 0$. While the instantaneous value of the total electric field intensity phasor in medium 1 is given by:

$$\vec{E}_1(z,t) = \text{Re}\{\vec{E}_1(z)e^{j\omega t}\} + \vec{E}_r = \vec{a}_x 2E_{i0} \sin\beta_1 z \sin\omega t \tag{2.29b}$$

The total magnetic field in medium 1 is:

$$\vec{H}_1(z) = \vec{H}_i(z) + \vec{H}_r = \vec{a}_y \frac{E_{i0}}{\eta_1}(e^{-j\beta_1 z} + e^{j\beta_1 z}) = \vec{a}_y 2\frac{E_{i0}}{\eta_1}\cos\beta_1 z \tag{2.30a}$$

The instantaneous value of the total magnetic field intensity phasor in medium 1 is:

$$\vec{H}_1(z,t) = \text{Re}\{\vec{H}_1(z)e^{j\omega t}\} + \vec{H}_r = \vec{a}_y 2\frac{E_{i0}}{\eta_1}\cos\beta_1 z \cos\omega t \tag{2.30b}$$

The space–time behaviour of the total field in medium 1 (represented by Eqs 2.29b and 2.30b) when compared with that of the incident wave (given by:

$\vec{E}_i(z,t) = \text{Re}\,[\vec{E}_i(z)\,e^{j\omega t}] = \vec{a}_x\,E_{i0}\cos(\omega t - \beta_1 z)$ indicates that the total field is a *standing wave* (unlike the incident wave which is a *travelling wave*).

The plot of a standing wave in Fig. 2.7 shows that the magnitude of the electric field varies sinusoidally with distance in medium 1 (in the region $z < 0$) starting from the reflecting surface at $z = 0$.

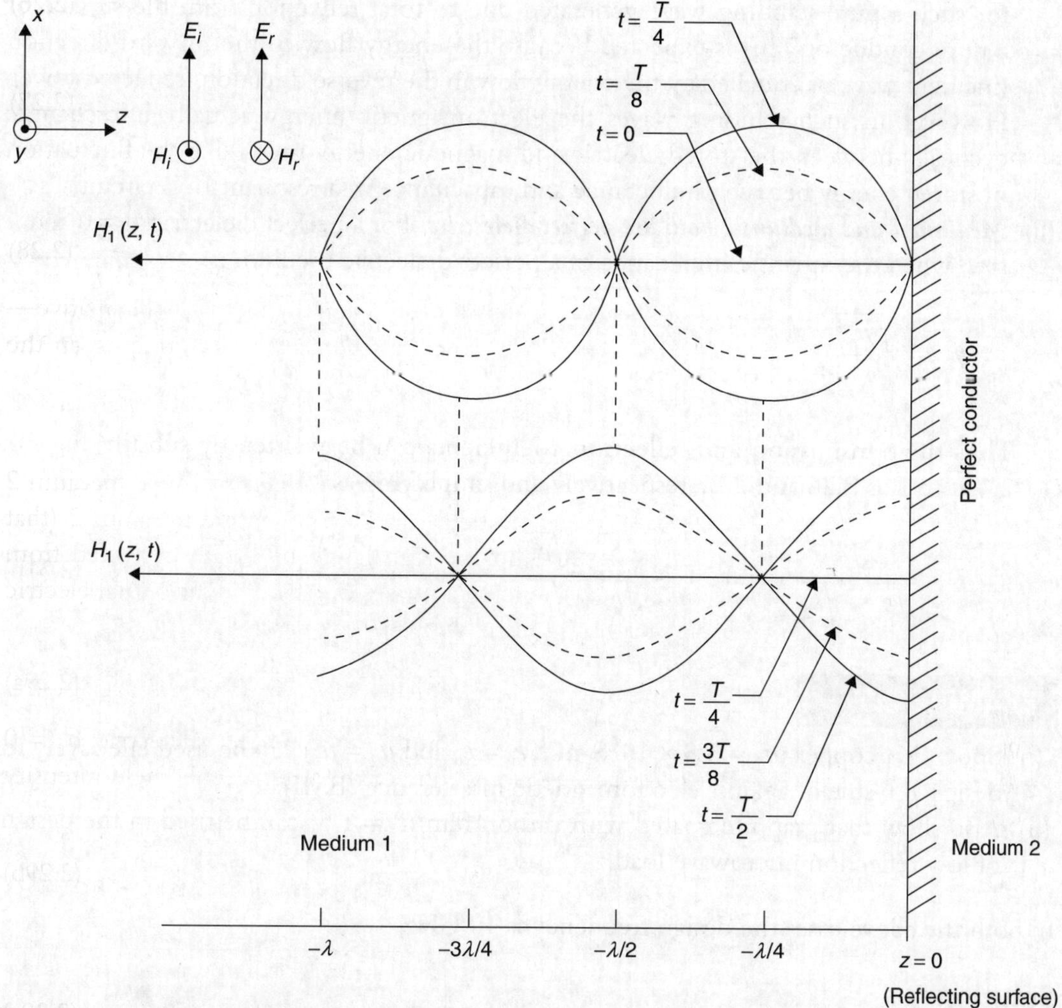

Fig. 2.7 Standing wave pattern for electric (*E*) and magnetic (*H*) field in the medium 1 when medium 2 is a perfect conductor (reflecting surface)

The magnitude of $\vec{E}_i(z,t)$ is zero at the reflecting surface and at the planes located at multiples of half wavelength from that surface (that is, at $z = n\lambda/2$, where $n = 0, 1, 2\ldots$). It has a maximum value of twice the electric field of the incident wave at the planes located at odd multiples of a quarter wavelength, that is, at $z = (n+1)\lambda/4$. The total magnetic field in medium 1 also represents a standing wave. In this case, however, the field has a maximum value at the surface of the conductor, at $z = 0$, and at locations where $z = n\lambda/2$, while field nulls occur at $z = (n + 1)\lambda/4$.

It may be noted that the electric and magnetic fields were in time phase in the incident plane wave. But the total electric and magnetic fields given by Eqs 2.29 and 2.30 are in time quadrature ($\vec{E}_1$ lags behind $\vec{H}_1$ by 90° because of the $-j$ factor). Thus, the average power calculated from the Poynting vector turns out to be zero for such a pure standing wave generated due to total reflection from the surface of a pure conductor. This is expected because the energy flow in the forward direction (incident wave) is equalled by the energy flow in the reverse direction (reflected wave). However, in such standing waves the electromagnetic energy actually interchanges cyclically between the stored electric and magnetic energy much like the fluctuation of stored energy between inductance and capacitance in a resonant L-C circuit.

(ii) ***Medium 1 and medium 2 both are perfect dielectrics***: For a perfect dielectric, $\sigma = 0$ and $\mu = \mu_0$ when the intrinsic impedance of a perfect dielectric medium is:

$$\eta_{1,2} = \sqrt{\frac{\mu_0}{\varepsilon_{1,2}}} \tag{2.31a}$$

Thus the transmission and reflection coefficients can be written by substituting Eq. 2.13a to Eqs 2.26 and 2.27 respectively and simplifying as:

$$T = \frac{2\sqrt{\varepsilon_1}}{\sqrt{\varepsilon_1} - \sqrt{\varepsilon_2}} \quad \text{and} \quad \Gamma = \frac{\sqrt{\varepsilon_1} - \sqrt{\varepsilon_2}}{\sqrt{\varepsilon_1} + \sqrt{\varepsilon_2}} \tag{2.31b}$$

EXAMPLE 2.7

(i) Show that copper ($\sigma_2 = 5.8 \times 10^7$ S-m^{-1}, $\varepsilon_2 = \varepsilon_0$ and $\mu_2 = \mu_0$) can be used effectively to design RF shields against electromagnetic interference (EMI).

(ii) Also show that graphite coated with carbon film ($\varepsilon_r = 1.6$) can be used in the design of low reflection microwave load.

In both the above cases the design frequency is 10 GHz.

Solution

(i) Here medium 1 is air, for which the intrinsic impedance $\eta_1 = 377\ \Omega$. Medium 2 is copper whose intrinsic impedance at 10 GHz is given by $\eta_2 = (\mu\omega/\sigma)^{1/2} e^{j45°} = 0.0370 \angle 45°\ \omega$. Thus, the copper sheet can be considered to be practically a zero impedance surface.

The magnitude of transmission and reflection coefficients can be calculated form Eqs 2.27 and 2.26 as:

$$|\Gamma| = \left|\frac{\eta_2 - \eta_1}{\eta_2 + \eta_1}\right| = 0.9999, \quad |T| = \left|\frac{2\eta_2}{\eta_2 - \eta_1}\right| = 0.000196$$

Thus, copper as RF shield will reflect almost all the signals with very little or practically no transmission through the copper enclosure that will be used as RF shield.

(ii) We can calculate the transmission and reflection coefficients for air–graphite coated with carbon film interface from Eqs 2.31b as:

$$T = \frac{2\sqrt{\varepsilon_1}}{\sqrt{\varepsilon_1}-\sqrt{\varepsilon_2}} = 0.8830, \quad \Gamma = \frac{\sqrt{\varepsilon_1}-\sqrt{\varepsilon_2}}{\sqrt{\varepsilon_1}+\sqrt{\varepsilon_2}} = -0.1169$$

Thus, the incident electromagnetic signal is found to be grossly transmitted through the dielectric medium and since its loss tangent at 10 GHz is also reasonably high ($\tan\delta_c = 0.036$, that is, a highly lossy dielectric), the use of graphite coated with carbon film will act as a good microwave load with low reflection.

2.7.2.2 Oblique incidence on plane boundaries

When an electromagnetic wave is incident at any arbitrary angle on the interface of two media having different constitutive properties, we refer to it as an oblique incidence (Fig. 2.8). The normal incidence is thus a special case of oblique incidence with an angle of incidence $\theta_i = 0$.

The reflection coefficient and transmission coefficient is independent of the polarisation of the incident wave in the case of normal incidence at the interface between two media as both the $\vec{E}$ and $\vec{H}$ fields of a normally incident plane wave are always tangential to the interface boundary regardless of the wave polarisation. This is not the case for oblique incidence at an angle $\theta_i \neq 0$ where the $\vec{E}$ field of the incident wave can make any angle with the interface boundary plane (x–y plane). Two special cases are of practical importance—the *perpendicular polarised* (or *s*-polarised, *senkrecht*, German for perpendicular) and *parallel polarised* (or *p*-polarised) waves. If the $\vec{E}$ field is normal to the plane of incidence (x–z plane), we refer to it as a *perpendicularly polarised wave* (Fig. 2.8a), in which the $\vec{E}$ field is in the y-direction. But if the $\vec{E}$ field lies in the plane of incidence, we call it a *parallel polarised wave* (Fig. 2.8b). Let us first consider the perpendicular polarisation case.

CASE I Perpendicular polarisation (s-polarised): For such polarisation of the incident plane wave having the $\vec{E}$ field in the y-direction (Fig. 2.8a) we can write the incident and reflected electric and magnetic field intensity phasors in medium 1 as:

Incident wave: $$\vec{E}^i_\perp(x, z) = \vec{a}_y\, E^i_{\perp 0}\, e^{\gamma_1 x_i} = \vec{a}_y\, E^i_{\perp 0}\, e^{\gamma_1 (x \sin\theta_i + z\cos\theta_i)} \tag{2.32a}$$

$$\vec{H}^i_\perp(x, z) = \vec{a}_{y_i} \frac{E^i_{\perp 0}}{\eta_1} e^{\gamma_1 x_i} = \frac{E^i_{\perp 0}}{\eta_1}(-\vec{a}_x \cos\theta_i + \vec{a}_z \sin\theta_i)\, e^{\gamma_1 (x \sin\theta_i + z\cos\theta_i)} \tag{2.32b}$$

[Since $x_i = x\sin\theta_i + z\cos\theta_i$ and $\vec{a}_{y_i} = -\vec{a}_x \cos\theta_i + \vec{a}_z \sin\theta_i$]

Reflected wave: $$\vec{E}^r_\perp(x, z) = \vec{a}_y\, E^r_{\perp 0}\, e^{\gamma_1 x_r} = \vec{a}_y\, E^r_{\perp 0}\, e^{\gamma_1 (x \sin\theta_r - z\cos\theta_r)} \tag{2.33a}$$

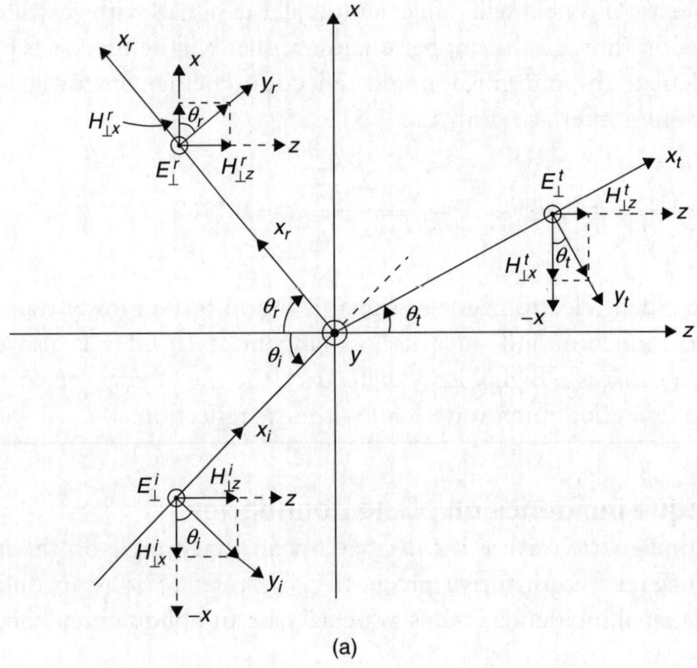

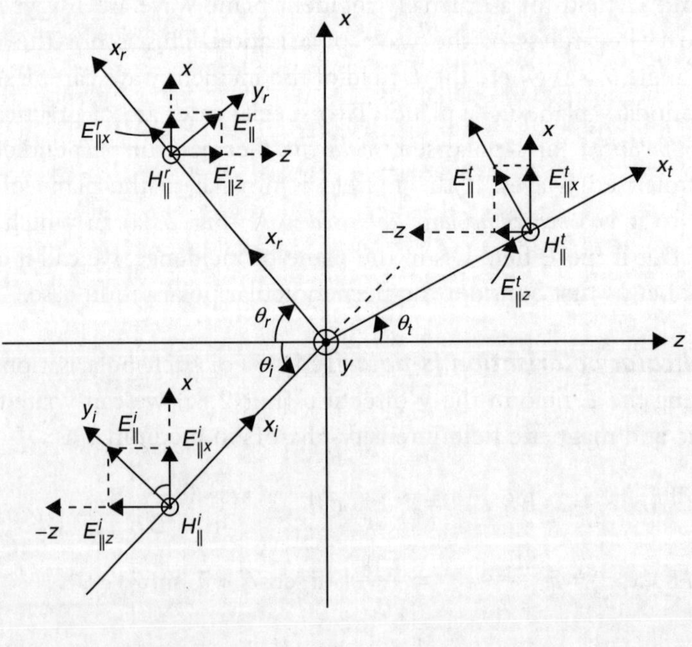

Fig. 2.8 Oblique incidence of an electromagnetic wave on the interface between two media (a) Perpendicular polarisation (that is, the incident signal is an *s*-polarised wave), (b) Parallel polarisation (that is, the incident signal is a *p*-polarised wave).

$$\vec{H}^r_\perp(x,z) = \vec{a}_{y_r}\frac{E^r_{\perp 0}}{\eta_1}e^{\gamma_1 x_r} = \frac{E^r_{\perp 0}}{\eta_1}(\vec{a}_x\cos\theta_r + \vec{a}_z\sin\theta_r)\,e^{\gamma_1(x\sin\theta_r - z\cos\theta_r)} \tag{2.33b}$$

(Since $x_r = x\sin\theta_r - z\cos\theta_r$ and $\vec{a}_{y_r} = \vec{a}_x\cos\theta_r + \vec{a}_z\sin\theta_r$)

The transmitted electric and magnetic field intensity phasors in medium 2 can be written as:

Transmitted wave: $$\vec{E}^t_\perp(x,z) = \vec{a}_y E^t_{\perp 0}e^{\gamma_2 x_t} = \vec{a}_y E^t_{\perp 0}e^{\gamma_2(x\sin\theta_t + z\cos\theta_t)} \tag{2.34a}$$

$$\vec{H}^t_\perp(x,z) = \vec{a}_{y_t}\frac{E^t_{\perp 0}}{\eta_2}e^{\gamma_2 x_t} = \frac{E^t_{\perp 0}}{\eta_2}(-\vec{a}_x\cos\theta_t + \vec{a}_z\sin\theta_t)\,e^{\gamma_2(x\sin\theta_t + z\cos\theta_t)} \tag{2.34b}$$

(Since $x_t = x\sin\theta_t + z\cos\theta_t$ and $\vec{a}_{y_t} = -\vec{a}_x\cos\theta_t + \vec{a}_z\sin\theta_t$)

The total electric field in medium 1 is the sum of the incident and reflected electric fields $\vec{E}^1_\perp = \vec{E}^i_\perp + \vec{E}^r_\perp$, and the same is true for the total magnetic filed in medium 1 $\vec{H}^1_\perp = \vec{H}^i_\perp + \vec{H}^r_\perp$. Boundary conditions state that the tangential components of $\vec{E}$ and $\vec{H}$ each must be continuous across the boundary plane (x–y plane) between the two media. Since the electric fields in medium 1 and 2 have only the y-component, the boundary condition for $\vec{E}$ demands that $(\vec{E}^i_{\perp y} + \vec{E}^r_{\perp y})_{z=0} = (\vec{E}^t_{\perp y})_{z=0}$. Using Eqs 2.32a, 2.33a and 2.34a, and setting $z = 0$ at the boundary, we have:

$$E^i_{\perp 0}e^{\gamma_1 x\sin\theta_i} + E^r_{\perp 0}e^{\gamma_1 x\sin\theta_r} = E^t_{\perp 0}e^{\gamma_2 x\sin\theta_t} \tag{2.35a}$$

Again, the magnetic fields in medium 1 and 2 have only the x-component and hence the boundary conditions for $\vec{H}$ demands $(\vec{H}^i_{\perp x} + \vec{H}^r_{\perp x})_{z=0} = (\vec{H}^t_{\perp x})_{z=0}$. On using Eqs 2.32b, 2.33b and 2.34b, we have at the interface boundary,

$$-\frac{E^i_{\perp 0}}{\eta_1}\cos\theta_i\,e^{\gamma_1 x\sin\theta_i} + \frac{E^r_{\perp 0}}{\eta_1}\cos\theta_r\,e^{\gamma_1 x\sin\theta_r} = -\frac{E^t_{\perp 0}}{\eta_2}\cos\theta_t\,e^{\gamma_2 x\sin\theta_t} \tag{2.35b}$$

Since the conditions represented by Eqs 2.35a and 2.35b must be satisfied for all values of x, the argument of all the three exponentials must be equal, leading to the phase matching condition:

$$\gamma_1 x\sin\theta_i = \gamma_1 x\sin\theta_r = \gamma_2 x\sin\theta_t \tag{2.36}$$

The first equality in Eq. 2.36 leads to Snell's law of reflection ($\theta_r = \theta_i$) and the second equality leads to Snell's law of refraction ($\sin\theta_t/\sin\theta_i$) = ($\gamma_1/\gamma_2$).

Applying the equality given by Eq. 2.36, we get from Eq. 2.35, the following:

$$E^i_{\perp 0} + E^r_{\perp 0} = E^t_{\perp 0} \tag{2.37a}$$

$$(-E^i_{\perp 0} + E^r_{\perp 0})\frac{\cos\theta_i}{\eta_1} = -E^t_{\perp 0}\frac{\cos\theta_t}{\eta_2} \tag{2.37b}$$

These two equations can be solved simultaneously to yield the following expressions for reflection and transmission coefficients (known as *Fresnel's equations*) in the case of perpendicular polarisation:

$$\Gamma_{\perp} = \frac{E_{\perp 0}^{r}}{E_{\perp 0}^{i}} = \frac{\eta_2 \cos\theta_i - \eta_1 \cos\theta_t}{\eta_2 \cos\theta_i + \eta_1 \cos\theta_t} \tag{2.38a}$$

$$T_{\perp} = \frac{E_{\perp 0}^{t}}{E_{\perp 0}^{i}} = \frac{2\eta_2 \cos\theta_i}{\eta_2 \cos\theta_i + \eta_1 \cos\theta_t} \tag{2.38b}$$

It may be observed from the expressions of $\Gamma_{\perp}$ and $T_{\perp}$ given by Eqs 2.38a and 2.38b that:

$$1 + \Gamma_{\perp} = T_{\perp} \tag{2.39}$$

This is similar to Eq. 2.28 for normal incidence.

Let us now consider some of the special cases concerning two media interfacing with each other.

(i) ***Dielectric–conductor interface:*** As medium 2 is a (perfect) conductor, the transmission coefficient is zero (as no electromagnetic field can exist inside a perfect conductor). Thus, from Eq. 2.39, $\Gamma_{\perp} = -1$, which implies that $E_{\perp 0}^{r} = -E_{\perp 0}^{i}$ (from Eq. 2.38a), and from Snell's law: $\theta_r = \theta_i$. Thus, the total electric field at any point in the dielectric medium (medium 1) is from Eqs 2.32a and 2.32b as:

$$\vec{E}_{1T}(x, z) = \vec{a}_y\, E_{\perp 0}^{i}\,(e^{-j\beta_1 z \cos\theta_i} - e^{+j\beta_1 z \cos\theta_i})\, e^{-j\beta_1 x \sin\theta_i}$$

$$= -\vec{a}_y\, j2E_{\perp 0}^{i} \sin(\beta_1 \cos\theta_i)\, e^{-i\beta_1 x \sin\theta_i} \tag{2.40a}$$

where β_1 is the propagation constant in medium 1 which is a loss-less ($\gamma = +j\beta$, since $\alpha = 0$) dielectric medium. The total magnetic field in medium 1 is obtained from Eqs 2.32b and 2.33b as:

$$\vec{H}_{1T}(x, z) = -\frac{2E_{\perp 0}^{i}}{\eta_1}\{\vec{a}_x \cos\theta_i \cos(\beta_1 z \cos\theta_i) + \vec{a}_z\, j \sin\theta_i \sin(\beta_1 z \cos\theta_i)\}\, e^{-j\beta_1 x \sin\theta_i} \tag{2.40b}$$

The average power density in the dielectric medium (medium 1) is then:

$$\vec{P}_{av} = \frac{1}{2}\,\text{Re}\left\{\vec{E}_{1T}(x, z) \times \vec{H}_{1T}^{*}(x, z)\right\}$$

$$= \frac{1}{2}\,\text{Re}\left\{-\vec{a}_z\, j\, \frac{2(E_{\perp 0}^{i})^2}{\eta_1} \cos\theta_i \sin(2\beta_1 z \cos\theta_i) + \vec{a}_x\, \frac{4(E_{\perp 0}^{i})^2}{\eta_1} \sin\theta_i \sin^2(\beta_1 z \cos\theta_i)\right\}$$

$$= \vec{a}_x\, \frac{2(E_{\perp 0}^{i})^2}{\eta_1} \sin\theta_i \sin^2(\beta_1 z \cos\theta_i) \tag{2.40c}$$

The above equation indicates that there is no power flow in the z-direction (which is expected because no average power is absorbed by a perfect conductor). However, there will be power flow in the x-direction, parallel to the interface, which is the basis for the design of the parallel plate waveguide, as may be seen below.

The time-harmonic E and H fields in medium 1 can be derived from Eqs 2.40a and 2.40b as:

$$\tilde{\tilde{E}}_{1y}(x, z, t) = -E_0 \sin(\beta_1 z \cos\theta_i)\sin(\omega t - \beta_1 x \sin\theta_i) \tag{2.41a}$$

$$\tilde{\tilde{H}}_{1z}(x, z, t) = -\frac{2E_{i0}}{\eta_1}\cos\theta_i \cos(\beta_1 z \cos\theta_i)\cos(\omega t - \beta_1 x \sin\theta_i) \tag{2.41b}$$

$$\tilde{\tilde{H}}_{1z}(x, z, t) = -\frac{2E_{i0}}{\eta_1}\sin\theta_i \sin(\beta_1 z \cos\theta_i)\sin(\omega t - \beta_1 x \sin\theta_i) \tag{2.41c}$$

As $\tilde{\tilde{E}}_{1y}$ and $\tilde{\tilde{H}}_{1z}$ are in phase both in space and time, there will be a propagating wave with phase velocity:

$$v_{p1x} = \frac{\omega}{\beta_{1x}} = \frac{\omega}{\beta_1 \sin\theta_i} = \frac{v_{p1}}{\sin\theta_i} \tag{2.42}$$

Since $\sin\theta_i \leq 1$, it is expected that $v_{p1x} \geq v_{p1}$. In other words, if the medium is free space, we can expect that the phase velocity of the electromagnetic wave in the x-direction will be greater than the velocity of light. This is possible as a point of constant phase on the wave always appears to move faster when viewed along a direction other than its direction of propagation. The electromagnetic energy, however, propagates with group velocity which is always less than the velocity of light. Thus, causality is not violated.

The wavelength of the propagating wave in the x-direction is given by:

$$\lambda_{1x} = \frac{2\pi}{\beta_{1x}} = \frac{\lambda_1}{\sin\theta_i} \tag{2.43}$$

where $\lambda_1 = 2\pi/\beta_1$, is the wavelength of the incident wave. It may be noted that the propagating wave in the x-direction is a *non-uniform plane wave*. It is non-uniform because the electric field amplitude of the wave varies as a function of position (z) over any plane of constant phase, however, the wave is still a plane wave since the phase fronts are planar (as: $x \sin\theta_i$ = constant, describing planes parallel to the yz plane).

It may further be noted from Eq. 2.41a, that the total electric field in medium 1 is identically zero at a discrete set of planes (parallel to xy plane) located at:

$$z = -\frac{p\lambda_1}{2\cos\theta_i} \quad (p = 1, 2, 3 \ldots) \tag{2.44}$$

It is thus possible to place a conducting sheet at any of these locations [say at $a = \lambda_1/(2\cos\theta_i)$] at which the electric field is nulled without affecting the field pattern to the left of the sheet (Fig. 2.9).

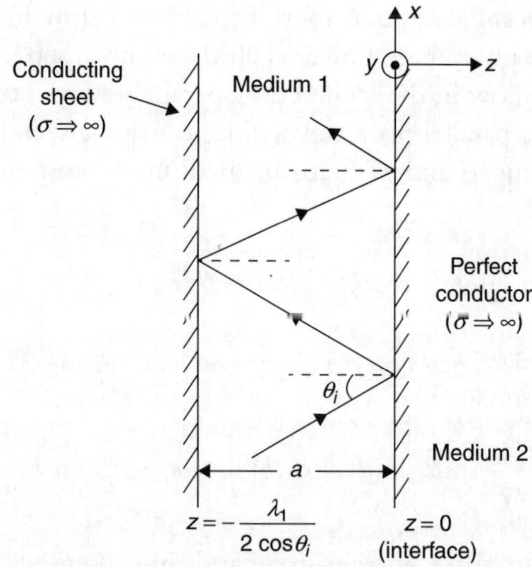

Fig. 2.9 A parallel-plate waveguide is formed when a conducting sheet is placed at a distance $z = -(\lambda_1/2\cos\theta_i)$ from the interface of the two media.

(ii) ***Dielectric–dielectric interface*:** When both the media are perfect dielectrics following Eq. 2.31a, we can write: $(\eta_1/\eta_2) = (\varepsilon_{r1}/\varepsilon_{r2})^{1/2}$ and Snell's law of refraction as $\sin\theta_t = (\varepsilon_{r1}/\varepsilon_{r2})^{1/2}\sin\theta_i$. With these substitutions in Eq. 2.38a and after simplification we get:

$$\Gamma_{\perp} = \frac{\cos\theta_i - \sqrt{\dfrac{\varepsilon_{r2}}{\varepsilon_{r1}} - \sin^2\theta_i}}{\cos\theta_i + \sqrt{\dfrac{\varepsilon_{r2}}{\varepsilon_{r1}} - \sin^2\theta_i}} \tag{2.45}$$

Practical applications of guiding of electromagnetic waves in dielectric waveguides or optical fibres demand the incidence of the wave from a dense medium 1 to a less dense medium 2, that is, a case when $\varepsilon_{r1} > \varepsilon_{r2}$. Under this situation, the reflection coefficient is real as long as $\sin\theta_i \leq (\varepsilon_{r2}/\varepsilon_{r1})^{1/2}$. Considering the equality sign and calling $\theta_i = \theta_{ic}$, the *critical angle of incidence*, we observe that $\Gamma_{\perp}$ becomes unity and $\theta_t = 90°$. Under this condition, the transmitted wave will propagate entirely parallel to the interface and the reflected power density will be equal to the incident power density, causing *total internal reflection* as depicted in Fig. 2.10.

It may be observed that at large angles of incidence when $\theta_i > \theta_{ic}$, we have total internal reflection, that is, we have the propagating signal confined to medium 1 (the denser medium) as in an optical fibre. However, it may be understood with detailed electromagnetic analysis that the nature of wave propagation in this case is such that it propagates along the x-direction, that is, along the interface and quickly decays exponentially along the z-direction. These types of electromagnetic waves are termed as *evanescent waves* and as they are tightly bound to the interfacing surface of the two media—they are also known as *surface waves*.

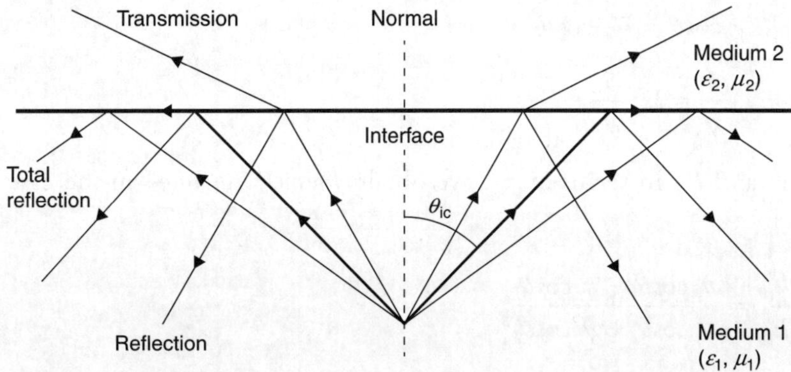

Fig. 2.10 Total internal reflection due to electromagnetic signal propagation from a relatively dense dielectric medium 1 to a less dense medium 2.

CASE II Parallel polarisation (p-polarised): When a uniform plane wave with parallel polarisation is incident obliquely on a plane boundary, the electric field lies in the plane of incidence while the associated magnetic fields are perpendicular to the plane of incidence (Fig. 2.8b). The incident and reflected electric and magnetic field intensity phasors in medium 1 are given by:

Incident wave: $$\vec{E}^i_{\|}(x, z) = \vec{a}_{y_i}\, E^i_{\|0}\, e^{\gamma_1 x_i} = (\vec{a}_x \cos\theta_i - \vec{a}_z \sin\theta_i)\, E^i_{\|0}\, e^{\gamma_1 (x \sin\theta_i + z \cos\theta_i)} \tag{2.46a}$$

$$\vec{H}^i_{\|}(x, z) = \vec{a}_y \frac{E^i_{\|0}}{\eta_1} e^{\gamma_1 x_i} = \vec{a}_y \frac{E^i_{\|0}}{\eta_1} e^{\gamma_1 (x \sin\theta_i + z \cos\theta_i)} \tag{2.46b}$$

(Since $x_i = x \sin\theta_i + z \cos\theta_i$ and $\vec{a}_{y_i} = \vec{a}_x \cos\theta_i - \vec{a}_z \sin\theta_i$)

Reflected wave: $$\vec{E}^r_{\|}(x, z) = \vec{a}_{y_r}\, E^i_{\|0}\, e^{\gamma_1 x_r} = (\vec{a}_x \cos\theta_r + \vec{a}_z \sin\theta_r)\, E^i_{\|0}\, e^{\gamma_1 (x \sin\theta_r - z \cos\theta_r)} \tag{2.47a}$$

$$\vec{H}^r_{\|}(x, z) = -\,\vec{a}_y \frac{E^i_{\|0}}{\eta_1} e^{\gamma_1 x_r} = -\,\vec{a}_y \frac{E^i_{\|0}}{\eta_1} e^{\gamma_1 (x \sin\theta_r - z \cos\theta_r)} \tag{2.47b}$$

(Since $x_r = x \sin\theta_r - z \cos\theta_r$ and $\vec{a}_{y_r} = \vec{a}_x \cos\theta_r + \vec{a}_z \sin\theta_r$)

The transmitted electric and magnetic field phasors in medium 2 are given by:

Transmitted wave: $$\vec{E}^t_{\|}(x, z) = = \vec{a}_{y_t}\, E^t_{\|0}\, e^{\gamma_2 x_t} = (-\,\vec{a}_x \cos\theta_t + \vec{a}_z \sin\theta_t)\, E^t_{\|0}\, e^{\gamma_2 (x \sin\theta_t + z \cos\theta_t)} \tag{2.48a}$$

$$\vec{H}^t_{\|}(x, z) = \vec{a}_y \frac{E^t_{\|0}}{\eta_2} e^{\gamma_2 x_t} = \vec{a}_y \frac{E^t_{\|0}}{\eta_2} e^{\gamma_2 (x \sin\theta_t + z \cos\theta_t)} \tag{2.48b}$$

(Since $x_t = x \sin\theta_t + z \cos\theta_t$ and $\vec{a}_{y_t} = \vec{a}_x \cos\theta_t - \vec{a}_z \sin\theta_t$)

Continuity requirements of $\vec{E}$ and $\vec{H}$ fields at $z = 0$ lead again to Snell's law as well as the following two equations:

$$(E^i_{\parallel 0} + E^r_{\parallel 0}) \cos\theta_i = E^t_{\parallel 0} \cos\theta_t \tag{2.49a}$$

$$(E^i_{\parallel 0} + E^r_{\parallel 0}) \frac{1}{\eta_1} = E^t_{\parallel 0} \frac{1}{\eta_2} \tag{2.49b}$$

Solving for $E^r_{\parallel 0}$ and $E^t_{\parallel 0}$ in terms of $E^i_{\parallel 0}$, we obtain *Fresnel's equations* in the case of parallel polarisation:

$$\Gamma_{\parallel} = \frac{E^t_{\parallel 0}}{E^i_{\parallel 0}} = \frac{\eta_2 \cos\theta_t - \eta_1 \cos\theta_i}{\eta_2 \cos\theta_t + \eta_1 \cos\theta_i} \tag{2.50a}$$

$$T_{\parallel} = \frac{E^t_{\parallel 0}}{E^i_{\parallel 0}} = \frac{2\eta_2 \cos\theta_i}{\eta_2 \cos\theta_t + \eta_1 \cos\theta_i} \tag{2.50b}$$

Thus, $1 + \Gamma_{\parallel} = T_{\parallel} (\cos\theta_t/\cos\theta_i)$, which is different from the case of perpendicular polarisation. However, when $\theta_i = \theta_r = \theta_t = 0$, we have the expression for the case of normal incidence.

From Eq. 2.50a, an expression for the incident angle θ_i, called the *Brewster angle* $\theta_{B\parallel}$, may readily be found which makes $\Gamma_{\parallel} = 0$, for no reflection. This requires $\eta_2 \cos\theta_t = \eta_2 \cos\theta_{B\parallel}$. Applying Snell's law of refraction $(\sin\theta_t/ \sin\theta_{B\parallel}) = (\varepsilon_{r1}\, \mu_{r1}/\varepsilon_{r2}\, \mu_{r2})^{1/2}$, it can be evaluated that:

$$\sin\theta_{B\parallel} = \left\{ \frac{1-(\mu_{r2}\, \varepsilon_{r1})/(\mu_{r1}\, \varepsilon_{r2})}{1-(\varepsilon_{r1}/\varepsilon_{r2})^2} \right\}^{1/2} \tag{2.51a}$$

For non-magnetic materials $(\mu_1 = \mu_2 = \mu_0)$, the Brewster angle may be written as:

$$\theta_{B\parallel} = \sin^{-1}\left(1+\frac{\varepsilon_{r1}}{\varepsilon_{r2}}\right)^{-1/2} = \tan^{-1}\left(\frac{\varepsilon_{r2}}{\varepsilon_{r1}}\right)^{1/2} = \tan^{-1}\left(\frac{n_2}{n_1}\right) \tag{2.51b}$$

where n_1 and n_2 are the refractive indices of the two media. Thus, for a parallel polarised incident signal, there is no reflected wave from the interface if the wave is incident at the Brewster angle.

A question that naturally arises is: does a perpendicularly polarised wave also have a Brewster angle for no reflection? This may be investigated as follows. Setting the numerator of Eq. 2.38a to be zero, we have: $\eta_2 \cos\theta_{B\perp} = \eta_1 \cos\theta_t$, which in conjunction with Snell's law of refraction yields:

$$\sin\theta_{B\perp} = \left\{ \frac{1-(\mu_1\, \varepsilon_{r2}/\mu_2\, \varepsilon_{r1})}{1-(\mu_1/\mu_2)^2} \right\}^{1/2} \tag{2.51c}$$

If we substitute $\mu_1 = \mu_2 = \mu_0$, then $\theta_{B\perp}$ does not exist.

This information related to the Brewster angle, regarding parallel and perpendicular polarisation, is of practical importance. When an unpolarised electromagnetic signal (which has both parallel and perpendicularly polarised components) is incident on an interface of two dielectric media at the Brewster angle, only the component with the perpendicular polarisation will be reflected. Also a circularly or elliptically polarised wave with the angle of incidence equal to the Brewster angle upon reflection from the interface will be converted into a linearly polarised wave. Thus, the Brewster angle is sometimes called the *polarising angle*. That is why, at the ends of laser tubes, the quartz windows are set at the Brewster angle to control the polarisation of the emitted light signal.

2.8 Radiation of Electromagnetic Waves and Antennae

An antenna (US) or aerial (UK) basically *acts as a transducer* that helps in converting the electrical signal into an electromagnetic signal at the transmitting end while reciprocally converting the electromagnetic signal into an electrical signal at the receiving end of a wireless system. Most of the antennae used for communication and radar are *directive antennae*, which *act as spatial filters* in the sense of concentrating electromagnetic power in certain directions at the expense of decreasing power radiated in other directions. Antennae transmit and receive certain frequencies better than other frequencies thus acting as *frequency filters*. Antennae that respond to a very small range of frequencies are known as narrow band antennae (for example, a resonant antenna such as a dipole antenna). Antennae that respond over wide range of frequencies are known as broad band antennae (for example, helical antenna). If bandwidth is more than or equal to 10% then it is considered a broad band antenna. Ultra Wide Band (UWB) antennae are those having a BW ≥ 25% (as per Defence Advanced Research Projects Agency, DARPA, US) or ≥ 20% (as per Federal Communications commission, FCC, US).

Electromagnetic radiation is produced whenever a charge undergoes acceleration due to a time-varying electric current. This is because when a time-varying current is applied to a conductor, free electrons are accelerated and travel in the intervening spaces between atoms with a back and forth motion in a synchronised manner with the time variation of the exciting signal. The time-dependent electron motion results in a time-varying magnetic field (according to Faraday's law) which eventually generates a time-varying electric field (according to Maxwell's magnetic curl equation, the modified Ampere's circuital law) and so forth. The overall effect of acceleration applied to the electrons is, therefore, to generate an outward-propagating electromagnetic wavefront that expands with time. Thus, any system of conductors in open space carrying a time-varying current radiates electromagnetic energy. This is precisely the physical principle behind the electromagnetic radiation produced by a (time-varying) current carrying conductor.

The first antenna was designed and used by Heinrich Hertz and operated at about 455 MHz. However, it was Guglielmo Marconi who first put antennae into practical use for long distance wireless communication in 1901. Nowadays, antennae are integral part of our everyday environment and can be of various shapes and sizes and are used in radio and television broadcasting and reception, radiowave and microwave communication systems, cellular telephones, global positioning systems, radar and satellite communication systems and so on.

2.8.1 Retarded Vector Potential for Antenna Analysis

It may be recalled that in the study of electrostatic fields and steady magnetic fields it was found possible, and often simpler, first to set up potentials (scalar electric potential U, in terms of charges and magnetic vector potential $\vec{A}$, in terms of currents) and then to obtain the electric or magnetic fields from these potentials. Similarly for the radiated electromagnetic field it appears to be simpler first to set up potentials that are related to charges or currents of the source of excitation, and then to obtain the $\vec{E}$ and $\vec{B}$ of the radiated field from these potentials. But for a radiated field (which is caused by time-varying current flowing through the antenna element) one cannot use magnetic vector potential used for static field (which will be applicable only if the antenna size is assumed to be extremely small so that propagation effects do not jeopardise the physical situation). The problem may be solved using the wave equation but that approach is very complicated even with simple antenna geometry. However, a logical approach that supports the physical situation yet allows us to obtain the correct solution without solving the wave equation directly, is the use of *retarded magnetic vector potential.* The idea may be elucidated as follows. Suppose the (time-varying) current carrying antenna element generates the magnetic vector potential in the space surrounding it, the solution must obey the wave equation, thus the waves generated at the source (the antenna element) propagate in all directions with a constant and finite phase velocity ($v_p = 1/\sqrt{\mu\varepsilon}$). An observer located at a distance r from the source would receive the signal after a time $t = r/v_p$. Thus, the wave that is received at the observer's location is a retarded wave and we have to modify the magnetic vector potential expression of the static field with the inclusion of this time delay and obtain the expression for *retarded vector potential* as:

$$\vec{A}(\vec{r}, t) = \frac{\mu}{4\pi} \int_V \frac{\vec{J}(r', t - r/v_p)}{r} dv' \tag{2.52}$$

Thus, the vector potential at the observation point at a time t due to the time-varying current flowing through the antenna element has been generated by a current at time $(t - r/v_p)$ at the source point (that is, at the antenna). In other words, the current itself gets retarded by a factor $(t - r/v_p)$ as this current for radiation to take place is a time-varying one.

2.8.2 Antenna Characteristic Parameters

An antenna acting as a radiating element exhibits different electromagnetic field characteristics at different distances from it, which is related to the wavelength of operation and the physical dimension (aperture) of the antenna as shown in Fig. 2.11.

Closest to the antenna extending up to $2D^2/\lambda$ (commonly known as *Raleigh distance*) is the *near-field region* (which is further sub-divided into *reactive or evanescent near-field region* and *radiating near field or Fresnel region*). It is found that the power density in the evanescent near-field region is purely reactive. In most situations we are interested in the radiation properties in the *far-field* or *Fraunhofer region*, that is, beyond the *Raleigh distance*, $2D^2/\lambda$ [6]. The Raleigh distance is determined by considering the transmitting antenna as a point source and allowing maximum phase deviation across the aperture of the antenna to be less than $\pi/8$ radians (Fig. 2.12).

From this figure we can find $R^2 + (D/2)^2 = (R + \lambda/16)^2$ (since $\pi/8$ radians is equivalent to $\lambda/16$ in terms of path difference), from which it is easy to see that $R \approx 2D^2/\lambda$; where

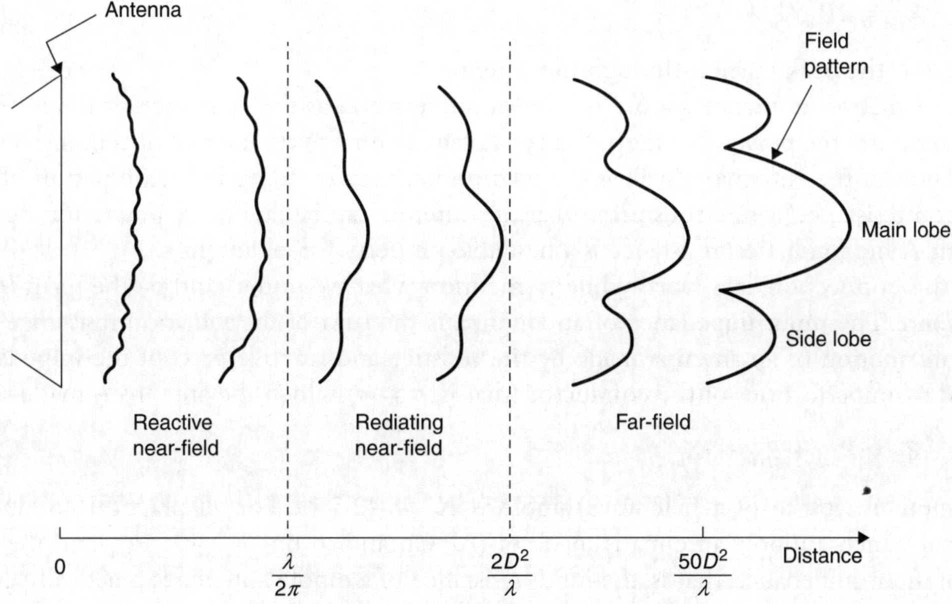

Fig. 2.11 Antenna radiation pattern in near-field and far-field regions (D = largest dimension of antenna and λ = operating wavelength).

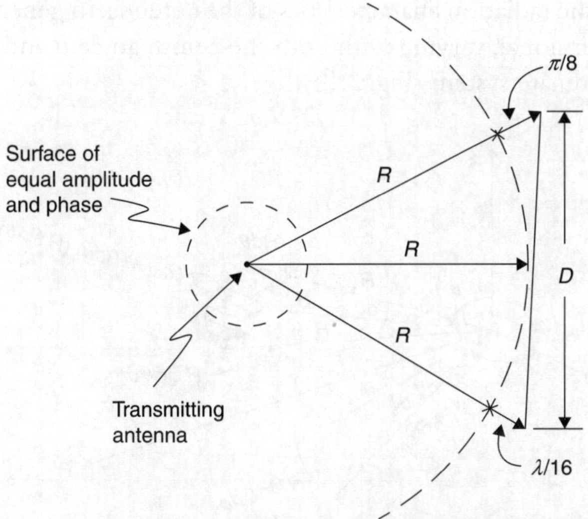

Fig. 2.12 Schematic for determining the Raleigh distance ($R \approx 2D^2/\lambda$).

R is the distance of the receiving antenna from transmitting antenna and D is the largest dimension of the receiving antenna.

For antennae, we quite often come across a term called ***radiation resistance*** which is understood to be an equivalent resistance that absorbs the total time-average power being radiated by the antenna. In other words, it is a fictitious or hypothetical resistance that would dissipate an amount of power equal to the radiated power.

$$R_{\text{rad}} = 2P_{\text{rad}}/I_0^2 \tag{2.53}$$

where I_0 is the peak current through the antenna.

The radiation resistance should not be misunderstood as the resistance of the antenna but it indicates the power that the antenna can radiate for a given current depending on the dimension of the antenna as well as the environment surrounding it. Maximisation of the radiation resistance would thus mean that the antenna can radiate more power for a given current. A high radiation resistance is a desirable property for an antenna.

In this connection, it is worthwhile to mention what we understand by the term ***input impedance***. The input impedance of an antenna is the sum of the radiation resistance and the contribution of an antenna made by the resistive and reactive parts of the impedance caused by imperfections of the conductor (that is, $\sigma \neq \infty$) which the antenna is made of:

$$Z_{\text{in}} = R_{\text{loss}} + R_{\text{rad}} + jX_{\text{in}} \tag{2.54}$$

The input reactance of a half-wave dipole is $X_{\text{in}} = 42.5\ \omega$. For all practical conductor material which a dipole antenna is made of (Al, Cu and so on) $R_{\text{loss}} << R_{\text{rad}}$, and the real part of input impedance, that is, the input resistance of a dipole antenna is practically equal to its radiation resistance.

The ***radiation pattern*** $R(\theta, \phi)$ that describes the relative far-zone field strength versus direction at a fixed distance from the antenna is an important parameter for all types of antennae as it shows the radiation characteristics of the antenna. In general, antenna radiation pattern is three-dimensional, varying with both the zenith angle θ and the azimuthal angle ϕ in a spherical coordinate system (Fig. 2.13).

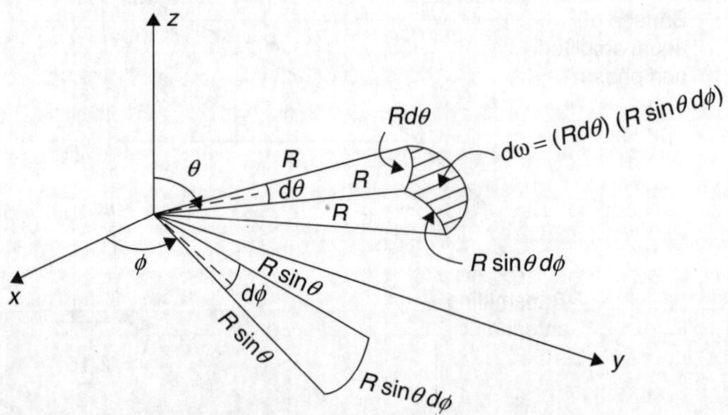

Fig. 2.13 Spherical coordinate system showing the solid angle ($d\omega$).

To avoid the complexities associated in interpreting the three-dimensional plots, it is customary to separately plot the magnitude of the normalised (with respect to the peak value) field strength or radiated power in two principal planes, namely, the E and H planes. Typical E field pattern in the E and H planes for a Hertzian dipole antenna is shown in Fig. 2.14.

A commonly used parameter to measure the overall ability of an antenna to direct radiated power in a given direction is the ***directive gain***, which can be defined in terms

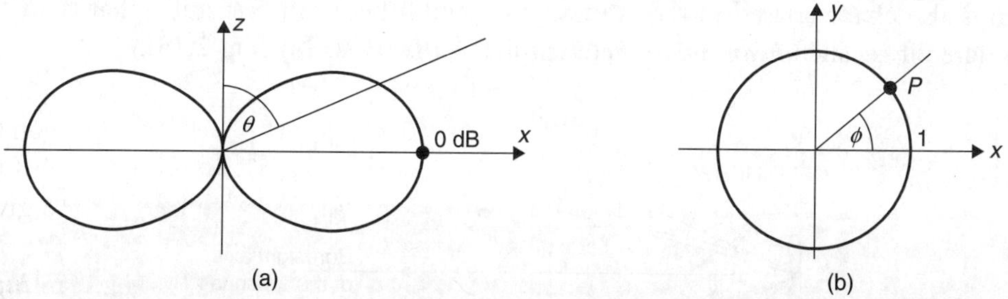

Fig. 2.14 Radiation pattern of a Hertzian dipole showing electric filed pattern (a) In *E*-plane (b) In *H*-plane.

of radiation intensity. The ***radiation intensity*** $I_r(\theta, \phi)$ in a given direction is defined as the power per unit solid angle in that direction. The directive gain, $G_D(\theta, \phi)$ of an antenna pattern is the ratio of the radiation intensity in the direction (θ, ϕ) to the average radiation intensity which spreads over the surface of the sphere of radius *r*.

$$G_D(\theta, \phi) = \frac{4\pi I_r(\theta, \phi)}{\oint I_r \, d\omega} \tag{2.55}$$

where I_r is the radiation intensity and $d\omega = \sin\theta d\theta d\phi$ is the differential solid angle (***Note***: solid angle is a measure of surface on a sphere, (Fig. 2.12), just as a plane angle is the measure of arc length on a circle). The maximum directive gain of an antenna is known as the ***directivity*** ($G_{D(\max)}$). Thus, directivity is a constant (as for example for omnidirectional or isotropic antenna it is 1 or 0 dB, for Hertzian dipole it is 1.5 or 1.76 dB, and for half-wave dipole it is 1.64 or 2.15 dB); while directive gain is a function of angle (θ, ϕ), that is, being specified for a particular direction. However, in common usage the terms are used interchangeably.

The directivity provides the theoretical gain possible with an antenna. But in practice, there may invariably be some losses in the antenna depending on its type and frequency of operation and the practical gain, called ***power gain*** *G*, which is always less than the directive gain G_D, that is, directivity $G_{D(\max)}$, and the two are related by:

$$G = \eta_a G_{D(\max)} \tag{2.56}$$

where η_a is known as the radiation efficiency, which is ideally 1 but for practical antennae it may be 0.5 to 0.7, typically 0.6.

While discussing the antenna radiation pattern, it may be noted that antennae are designed to direct most of their radiated power in a preferred direction called the *main lobe*, however, in the case of all practical antennae, the pattern also has several *side lobes* and *back lobes*. For most applications, these extra or spurious lobes are undesirable as they result in wasted energy for transmitting antennae and potential interference directions for receiving antennae. To characterise the width of the main lobe of the antenna in a given plane, a term called ***half-power beam width*** (HPBW) is used. It is the angular width in degrees, measured on the power radiation pattern of the main lobe, between points where the radiated power has fallen to half of its maximum value (that is, 3 dB down). Sometimes the radiation pattern is also characterised in terms of angular width between first nulls (that is, first side lobes) when it is called ***beam width between first nulls*** (BWFN) (Fig 2.15).

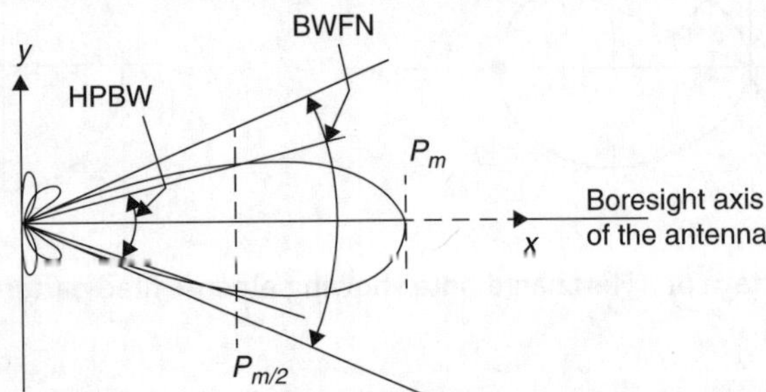

Fig. 2.15 *E*-plane radiation pattern of an antenna showing half-power beam width (HPBW) and beam width between first nulls (BWFN).

An antenna can be used either as a transmitting antenna or a receiving antenna. A receiving antenna receives (captures) the electromagnetic waves incident on it and extracts power. In discussing receiving antennas, it is thus convenient to define a quantity called ***effective aperture*** or ***effective area*** or ***receiving cross-section***.

The effective aperture A_e, is defined as the ratio of the average power delivered to the load (to which the receiving antenna is connected) to the power density of the incident (received) electromagnetic wave at the antenna and is given by:

$$A_e = \frac{\lambda^2}{4\pi} G_D(\theta, \phi) \tag{2.57}$$

The effective aperture of the receiving antenna is quite important as it defines the antenna in terms of how much power it can transfer to the load even if the antenna has 'no area', in a practical sense. It is interesting to note herewith that maximum directivity of both elemental (Hertzian) dipole (1.5) and half-wave dipole (1.642) are close to each other and hence their effective area. But, Hertzian dipoles radiate very little (for example, radiation resistance of a 0.01 λ Hertzian dipole is only 0.79ω) compared to half-wave dipole (R_{rad} = 73ω) as a transmitting antenna but such a Hertzian dipole would be an useful receiving antenna. In fact, this property is often used in car radios for AM/FM reception using a vertical 'whip' antenna of 0.5–1 m long instead of a half-wave antenna which needs to be approximately 100 m long (!).

For a transmission–reception system (in link design), ***Friss' transmission equation*** is important from a communication engineer's point of view. It relates the power received to the power transmitted between two antennae placed beyond each other's Fraunhofer zone (that is, separated by a distance $R > 2D^2/\lambda$), where D is the largest dimension of either antenna). The formula is given by [5]:

$$\frac{P_r}{P_t} = A_{er} A_{et} \left(\frac{1}{\lambda r}\right)^2 \tag{2.58}$$

[The above formula follows from the fact that the power density at the receiver in the direction of the main beam of the transmitting antenna of gain G_t and power output P_t is given by: $P_r = \left(\frac{P_t}{4\pi r^2}\right) \cdot G_t \cdot A_{er}$, where r is the distance between the transmitter and receiver and A_{er} is the effective aperture of the receiver antenna. Since the effective aperture of transmitting antenna is given by: $A_{et} = \frac{\lambda^2}{4\pi} G_t$; thus Eq. 2.58 follows].

This formula indicates that for a given transmitted power, the power received by the receiver is directly proportional to the effective apertures of both transmitting and receiving antennae and inversely proportional to the wavelength and distance between the them. It means that if we require a large power at the receiving antenna for a given transmitter power, we can either make one or both antennae with larger effective apertures, increase the frequency of transmission (decrease the wavelength) or transmit to shorter distances. Hence this is a useful link design equation for communication engineers that help to trade off between antenna size, frequency of operation and the distance between transmitter and receiver. This simple relation explains why we use very large antennae whenever there is a need for transmission to a large distance (for example, for deep-space exploration) or when we need to receive signals from relatively weak sources (ground stations for satellites).

Finally, ***antenna noise temperature*** is a useful parameter for communication/microwave engineers dealing with antennae. An antenna has an inherent thermal noise caused by its physical temperature T_p and also picks up noise power from the background due to background temperature T_b; the latter being dependent on earth's ground, sun, the microwave cosmic background (that is, the deep-space galactic background) and so on. For an antenna with radiation efficiency $\eta_a < 1$, the antenna noise temperature is given by:

$$T_a = \eta_a T_b + (1 - \eta_a) T_p \tag{2.59}$$

If the antenna is highly lossy, $\eta_a = 0$ and hence the gain G of the antenna becomes zero [though it may have finite directivity, since $G = \eta_a G_{D(\max)}$], it is not capable to pick up any noise from the background except being affected by only its physical temperature T_p (when it acts only as a resistor maintained at a temperature T_p); thus from Eq. 2.59, $T_a = T_p$. But if the antenna is an ideal one, that is, perfectly loss-less ($\eta_a = 1$) then from Eq. 2.59, $T_a = T_b$ (being loss-less, its representative resistance is zero and hence the thermal noise due to T_p will not be present). An isotropic antenna would have its noise temperature affected by an averaged background temperature due to all possible natural thermal environments around the antenna; however, in the case of a directional antenna, the antenna noise temperature will depend on the environment temperature only in which direction the antenna is 'looking'.

EXAMPLE 2.8

A half-wave dipole is used to transmit 150 kW power at 99.7 MHz from an FM transmitter. A whip antenna in the form of a Hertzian dipole, $\lambda/50$ m in length, is used to receive the signal in a car located 30 km away from the FM transmitting station. Find the maximum power received by the whip antenna.

Solution

The transmitted power (P_t) and received power (P_r) is related by Friss' formula (Eq. 2.48), when:

$$P_r = P_t A_{er} A_{et} \left(\frac{1}{\lambda^2 r^2} \right)$$

The maximum value of P_r is obtained by using maximum effective aperture for both transmitting and receiving antenna, which is obtained by using Eq. 2.46 when $G_D(\theta, \varphi)$ is respectively 1.642 and 1.5 for half-wave dipole and Hertzian dipole. Thus, $A_{er(\max)}$ and $A_{et(\max)}$ are given by:

$$A_{er\,(\max)} = 1.5 \frac{\lambda^2}{4\pi} = \frac{1.5 \times 3.009^2}{4 \times 3.1415} = 1.0807\,\text{m}^2$$

$$A_{et\,(\max)} = 1.642 \frac{\lambda^2}{4\pi} = \frac{1.642 \times 3.009^2}{4 \times 3.1415} = 1.18309\,\text{m}^2$$

Thus, the maximum received power at the whip antenna is:

$$P_r = 150 \times 10^3 \times 1.0807 \times 1.18309 \left(\frac{1}{3.009^2 \times 30{,}000^2} \right) = 2.224 \times 10^{-5}\ \text{W}$$

Recapitulation

Time-varying fields and Maxwell's equations

- For static electric and magnetic fields the two divergence equations are:

  $\nabla \cdot \vec{D} = \rho$ (*Gauss's law*)

  $\nabla \cdot \vec{B} = 0$ (*Solenoidal static magnetic field, non-existent monopole*)

  The curl equations for the static electric and magnetic fields are:

  $\nabla \times \vec{E} = 0$ (*Irrotational static electric field*)

  $\nabla \times \vec{H} = \vec{J}$ (*Ampere's circuital law*)

  For the time-varying case, the divergence equations remain unaffected but the curl equations are modified to include the time dependence of the field:

  $$\nabla \times \vec{E} = -\frac{\partial \vec{B}}{\partial t}; \qquad \nabla \times \vec{H} = \vec{J} + \frac{\partial \vec{D}}{\partial t}$$

  Thus, even for the time-varying case the magnetic field remains *solenoidal* while both the electric and magnetic field become *rotational*, that is, they have non-zero curl and hence become the *vortex field*.

  When the curl of the vector field happens to be zero, it is a *conservative* field; otherwise it is *non-conservative*.

- In explaining electromagnetic phenomena in a physical environment, that is, to deal with a practical physical situation, we need to convert *Maxwell's equations* in differential forms into their *integral form* equivalents which are given by:

$$\oint_S \vec{D} \cdot d\vec{s} = \int_V \rho_v \, dv; \qquad \oint_S \vec{B} \cdot d\vec{s} = 0$$

$$\oint_C \vec{E} \cdot d\vec{l} = -\int_S \frac{\partial \vec{B}}{\partial t} \cdot d\vec{s}; \qquad \oint_C \vec{H} \cdot d\vec{l} = \int_S \left(\vec{J} + \frac{\partial \vec{D}}{\partial t} \right) \cdot d\vec{s}$$

- In engineering applications, sinusoidal time functions occupy a unique position. Thus for time-harmonic (steady-state sinusoidal) fields it is convenient to use a phasor notation. *Maxwell's equations* for time-harmonic fields may be expressed in *phasor form* as:

$$\nabla \cdot \vec{D} = \rho \qquad \oint_S \vec{D} \cdot d\vec{s} = \int_V \rho_v \, dv$$

$$\nabla \cdot \vec{B} = 0 \qquad \oint_S \vec{B} \cdot d\vec{s} = 0$$

$$\nabla \times \vec{E} = -\, j\omega\vec{B} \qquad \oint_C \vec{E} \cdot d\vec{l} = -\int_S j\omega\vec{B} \cdot d\vec{s}$$

$$\nabla \times \vec{H} = \vec{J} + j\omega\vec{D} \qquad \oint_C \vec{H} \cdot d\vec{l} = \int_S (\vec{J} + j\omega\vec{D}) \cdot d\vec{s}$$

The wave equation and plane electromagnetic waves in unbounded media

- The general form of the so-called *wave equations* obeyed by the $\vec{E}$ and $\vec{H}$ fields in unbounded media are given by:

$$\nabla^2 \tilde{\vec{E}} - \mu\sigma \frac{\partial \tilde{\vec{E}}}{\partial t} - \mu\varepsilon \frac{\partial^2 \tilde{\vec{E}}}{\partial t^2} = 0; \qquad \nabla^2 \tilde{\vec{H}} - \mu\sigma \frac{\partial \tilde{\vec{H}}}{\partial t} - \mu\varepsilon \frac{\partial^2 \tilde{\vec{H}}}{\partial t^2} = 0$$

These wave equations govern the electromagnetic fields in a linear, homogeneous and isotropic medium (conducting or non-conducting) in which the charge density is zero (source free).

- The wave equations for the time-harmonic fields (known as the *Vector Helmholtz equation*) may be immediately written [by replacing $\tilde{\vec{E}}\,(r, t)$ by $\vec{E}_s\,(r)$ and each time derivative by a $j\omega$ factor] as:

$$\nabla^2 \vec{E}_s + (\omega^2 \mu\varepsilon - j\omega\mu\sigma)\, \vec{E}_s = 0 \quad \text{or} \quad \nabla^2 \vec{E}_s - \gamma^2 \vec{E}_s = 0$$

Where $\gamma^2 = j\omega\mu(\sigma + j\omega\varepsilon)$. Again $\gamma = \alpha + j\beta$, with:

$$\alpha = \omega\sqrt{\frac{\mu\varepsilon}{2}} \left\{ \sqrt{1 + \left(\frac{\sigma}{\omega\varepsilon}\right)^2} - 1 \right\}^{1/2} \quad \text{and} \quad \beta = \omega\sqrt{\frac{\mu\varepsilon}{2}} \left\{ \sqrt{1 + \left(\frac{\sigma}{\omega\varepsilon}\right)^2} + 1 \right\}^{1/2}$$

It may be noted that both the attenuation constant (α) and the propagation constant (β) of a general lossy medium are non-linear functions of frequency (ω).

- For free space which is an unbounded dielectric medium with zero conductivity (a loss-less medium) the wave equation simplifies to:

$$\nabla^2 \vec{E}_s + k^2 \vec{E}_s = 0 \quad \text{or} \quad \vec{E}_s + \frac{\omega^2}{c^2} n^2 \vec{E}_s = 0; \quad \text{Where } k = \omega\sqrt{\mu\varepsilon} = \beta$$

k is called the *wave number* and β is known as the *propagation constant* which are used interchangeably in literature representing the same physical quantity. It may be noted that for a loss-less medium β (k) is a linear function of frequency.

- It is known from electromagnetic theory that the phase velocity of electromagnetic wave is given by:

$$v_p = \frac{\omega}{k} = \frac{1}{\sqrt{\mu\varepsilon}}$$

which, in loss-less free space (with $\mu = \mu_0$ and $\varepsilon = \varepsilon_0$) was found to be equal to the velocity of light.

- The propagation constant β bears a definite relation to the wave length: $\beta = \omega\sqrt{\mu\varepsilon} = \frac{\omega}{v_p} = \frac{2\pi}{\lambda} = k$. Thus k is a measure of the number of wavelengths in a span of 2ω, hence the name *wave number.*
- The total electric and magnetic field strengths of an electromagnetic field bears a constant ratio given by:

$$\frac{E}{H} = \eta = \sqrt{\frac{j\omega\mu}{\sigma + j\omega\varepsilon}}$$

where η is known as the *wave impedance or the intrinsic* (*characteristic*) *impedance* of the medium. For a purely dielectric medium with zero conductivity ($\sigma = 0$), $\eta = \sqrt{\mu/\varepsilon}$ and thus for free space $\eta_0 = 120\ \pi = 377\omega$.

- A uniform plane wave propagates through the free space as a *transverse electromagnetic* (TEM) wave in which the $\vec{E}$ and the $\vec{H}$ fields are at right angles to each other and the direction of propagation is perpendicular (transverse) to the plane (wavefront) containing the E and H field vectors.

Electromagnetic wave propagation in material media

- The skin depth is defined as the depth into which the incident electromagnetic wave upon impinging into a good conductor is attenuated to $1/e$ (or ~36.8%) of its original intensity. The skin depth is thus given by: $\delta = \frac{1}{\alpha} = \frac{1}{\sqrt{\pi f \mu\sigma}}$.
- The wave impedance or intrinsic impedance of a good conductor ($\sigma = \infty$) is given by:

$$\eta_c \cong \sqrt{\frac{j\omega\mu}{\sigma}} = \sqrt{\frac{\omega\mu}{\sigma}} \cdot \frac{(1+j)}{\sqrt{2}}$$

While in dielectric medium, it is given by:

$$\eta_d \cong \sqrt{\frac{\mu}{\varepsilon}}\left(1 + j\frac{\sigma}{2\omega\varepsilon}\right)$$

- The phenomenon of the polarisation vector $\vec{P}$ falling out-of-phase with the externally applied $\vec{E}$ in a dielectric medium when subjected to a high frequency electromagnetic field causes $\vec{E}$ to lose power in doing work against frictional damping forces; resulting in *dielectric heating* which is characterised by a complex permittivity. The imaginary part of the complex permittivity ε_c also includes ohmic losses due to nonzero conductivity σ for a lossy dielectric in addition to the power dissipation in the dielectric medium due to dielectric heating mentioned above due to the high frequency field; and ε_c is given by $\varepsilon_c = \varepsilon' - j\varepsilon''$. The time average power dissipated in the lossy dielectric material per unit volume is given by $P_{av} = \frac{1}{2}\omega\varepsilon''\left|\vec{E}\right|^2$. This dissipated power in the material is the basis of *microwave heating* of lossy dielectric materials.

Electromagnetic power flow and Poynting vector

- The analytical relation between the field vectors and the power flow for electromagnetic wave can be derived from Maxwell's equation, which is given by $\tilde{\vec{P}} = \tilde{\vec{E}} \times \tilde{\vec{H}}$, where $\vec{P}$ represents the instantaneous power density vector associated with the electromagnetic field at a given point and is known as Poynting vector. The direction of the vector $\tilde{\vec{P}}$ indicates the direction of the instantaneous power flow at the point. The average power for time harmonic field is given by $\vec{P}_{av} = \frac{1}{2}\,\mathrm{Re}\,(\vec{E} \times \vec{H}^*)$.

Polarisation of electromagnetic waves

- The phenomenon of restricting the oscillations of waves to just one direction in the transverse plane is known as polarisation of electromagnetic waves. Polarisation is a property specific to electromagnetic waves because these are transverse waves. Polarisation is a very important issue related with radio wave and microwave communication and antenna systems. The polarisation of a uniform plane wave is conventionally understood to be the time-varying behaviour of the electric field intensity vector $\vec{E}$ at a given point in space. For propagation in the z-direction if E_x and E_y are in time phase the resultant vector is constant we have a linearly polarised wave. But if E_x and E_y are not in phase and maintain a constant phase difference of 90° having equal amplitude we have a circularly polarised wave and if they are of unequal amplitude we get an elliptically polarised wave. In the case of circular polarisation, if the resultant $\vec{E}$ field rotates at a uniform rate with an angular velocity ω in a *counterclockwise* direction when viewed looking towards the –ve z direction, the wave is said to be *right-handed circularly polarised* (RHCP) wave and ω with *clockwise* rotation we have *left-handed circularly polarised* (LHCP) wave; as per IEEE convention.

Wave propagation in interfacing media

- The boundary conditions that the field vectors $\vec{E}, \vec{H}, \vec{D}$ and $\vec{B}$ must satisfy at the interface between two media are:

$E_{1t} = E_{2t},\quad B_{1n} = B_{2n}$

$D_{1n} = D_{2n}$ (if there is no surface charge density)

$H_{1t} = H_{2t}$ (if there is no surface current)

But if one of the media is a perfect conductor when surface charge density (ρ_s) and surface current (J_s) will be present and under that situation:

$D_{1n} - D_{2n} = \rho_s,\quad H_{1t} - H_{2t} = J_s$

Under this condition (when medium 2 is a perfect conductor) with $E_2 = D_2 = 0$, $E_{1t} = 0$ and $D_{1n} = \rho_s$; and with $H_2 = B_2 = 0$, $H_{1t} = J_s$ and $B_{1n} = 0$. This indicates that the *E*-field lines are normal while the *H*-field lines are tangential to the surface of a perfect conductor medium.

- For normal incidence at plane boundaries between two media having different constitutive parameters ε, μ, σ the reflection coefficient (Γ) and transmission coefficient (T) are given by:

$$\Gamma = \frac{\eta_2 - \eta_1}{\eta_2 + \eta_1}, T = \frac{2\eta_2}{\eta_2 + \eta_1}, \text{ when } T = 1 + \Gamma; \text{ with } \eta = \sqrt{\mu/\varepsilon}$$

If medium 1 is a pure dielectric and medium 2 is a pure conductor then the electric field on the surface of the conductor will be zero while the magnetic field will be maximum on the surface of the conductor. A *standing wave* will be formed in medium 1 due to the interference of the totally reflected signal from medium 2 and the incident signal in medium 1. The average power carried by such a pure standing wave is zero and the electromagnetic energy actually interchanges cyclically between the stored electric and magnetic energy in the standing wave.

If both the media are pure dielectrics then $\Gamma = \dfrac{\sqrt{\varepsilon_1} - \sqrt{\varepsilon_2}}{\sqrt{\varepsilon_1} + \sqrt{\varepsilon_2}}$ and $T = \dfrac{2\sqrt{\varepsilon_1}}{\sqrt{\varepsilon_1} - \sqrt{\varepsilon_2}}$.

- For oblique incidence on plane boundaries we have two special cases of practical interest—the *perpendicular polarised wave* (when $\vec{E}$ field is normal to the plane of incidence) and *parallel polarised wave* (when $\vec{E}$ field lies in the plane of incidence).

In the case of a perpendicularly polarised wave:

$$\Gamma_\perp = \frac{\eta_2 \cos\theta_i - \eta_1 \cos\theta_t}{\eta_2 \cos\theta_i + \eta_1 \cos\theta_t} \quad \text{and} \quad T_\perp = \frac{2\eta_2 \cos\theta_i}{\eta_2 \cos\theta_i + \eta_1 \cos\theta_t}$$

where θ_i and θ_t are the angle of incidence and angle of transmission respectively of the electromagnetic wave from the interface. When $\theta_i = 0$, making $\theta_r = \theta_t = 0$, these expressions reduce to those for normal incidence. The dielectric–conductor interface with perpendicular polarisation wave forms the basis for understanding the principle of *parallel plate waveguide*; while the dielectric–dielectric interface with perpendicular polarisation wave forms the understanding of *dielectric waveguide* and of *optical fibre*.

For parallel polarisation waves:

$$\Gamma_{\parallel} = \frac{\eta_2 \cos\theta_t - \eta_1 \cos\theta_i}{\eta_2 \cos\theta_t + \eta_1 \cos\theta_i} \quad \text{and} \quad T_{\parallel} = \frac{2\eta_2 \cos\theta_i}{\eta_2 \cos\theta_t + \eta_1 \cos\theta_i}$$

which is different from the case of perpendicular polarisation; however, when $\theta_i = \theta_r = \theta_t = 0$, we have the expression for the case of normal incidence.

For a parallel polarised incident signal, there is no reflected wave from the interface if the wave is incident at the *Brewster angle*; but this not the case with a perpendicular polarised wave. This information helps to control the polarisation of emitted light signal from laser tubes.

Radiation of electromagnetic waves and antennae

- Electromagnetic radiation is produced whenever a charge undergoes acceleration due to a time-varying electric current. This is because when a time-varying current is applied to a conductor, free electrons are accelerated and travel in the intervening spaces between atoms with a back and forth motion in synchronisation with the time variation of the exciting signal. The time-dependent electron motion results in a time-varying magnetic field (according to Faraday's law) which eventually generates a time-varying electric field (according to Maxwell's magnetic curl equation, the modified Ampere's circuital law) and so forth. The overall effect of acceleration applied to the electrons is, therefore, to generate an outward-propagating electromagnetic wave front that expands with time. This is precisely the principle by which electromagnetic energy is radiated from the antenna/aerial (any system of conductors in open space carrying a time-varying current).
- Antenna analysis is done in terms of *retarded vector potential* given by:

$$\vec{A}(\vec{r}, t) = \frac{\mu}{4\pi} \int_V \frac{\vec{J}(r', t - r/v_p)}{r} dv'$$

This resembles the expression for magnetic vector potential of magnetostatics but here the current gets retarded by factor $(t - r/v_p)$ as the observer (located at the receiving point) receives the signal from the source point (location of the transmitting antenna which carries a time-varying current) after a time $t = r/v_p$.

- An antenna acting as a radiating element exhibits different electromagnetic field characteristics at different distances from it. In most situations, we are interested in the radiation properties in the far-field or *Fraunhofer region* that is beyond the *Raleigh distance* $2D^2/\lambda$, where D is physical dimension (aperture) of the antenna and λ is the operating wavelength.
- *Radiation resistance* is considered to be an equivalent resistance that absorbs the total time-average power being radiated by the antenna. In other words, it is a fictitious or hypothetical resistance that would dissipate an amount of power equal to the radiated power. $R_{rad} = 2P_{rad}/I_0^2$, where I_0 is the peak current through the antenna.

The *directive gain*, $G_D(\theta, \phi)$ of an antenna pattern is the ratio of the radiation intensity in the direction (θ, ϕ) to the average radiation intensity which spreads over the surface of the sphere of radius r, given by: $G_D(\theta, \phi) = 4\pi I_r(\theta, \phi)/\oint I_r \, d\omega$ where I_r is the radiation intensity and $d\omega = \sin\theta d\theta d\phi$ is the differential solid angle. The directive gain of an *isotropic* or *omni-directional* antenna (an antenna that radiates uniformly in all directions) is unity. The maximum directive gain of an antenna is known as the *directivity*. For omnidirectional or isotropic antenna, the directivity is 0 dB, for Hertzian dipole it is 1.76 dB, and for a half-wave dipole, the directivity is 2.15 dB. The directive gain provides the theoretical gain possible with an antenna. But in practice there may invariably be some losses in the antenna depending on its type and frequency of operation and the practical gain, called *power gain* G, is always less than the directive gain G_D, and the two are related by $G = \eta_a G_D$, where η_a is known as the antenna efficiency, which is ideally 1 but for practical antennae it may be 0.5 to 0.7, typically 0.6.

To characterise the width of the main lobe of an antenna in a given plane a term called *half-power beam width* (HPBW) is used. It is the angular width in degrees, measured on the power radiation pattern of the main lobe, between points where the radiated power has fallen to half of its maximum value (3 dB down). Sometimes, the radiation pattern is also characterised in terms of the angular width between first nulls (first side lobes) when it is called the *beam width between first nulls* (BWFN).

- In discussing receiving antennas, it is convenient to define a quantity called *effective aperture* (A_e) or *effective area* or *receiving cross section* given by:

$$A_e = \frac{\lambda^2}{4\pi} G_D(\theta, \phi)$$

Effective aperture of a (receiving) antenna is independent of the dimensions of the antenna, however, it is dependent on the directivity of the antenna. It may be noted that the Herzian dipole is not useful as a transmitting antenna because of its very small radiation resistance but it can be used as a receiving antenna of practical size (as whip antenna of length 1 m or so in a car) because its directivity is close to that of half-wave dipole.

Friss' transmission equation relates the power received to the power transmitted between two antennae placed in each other's Fraunhofer zone and is given by:

$$\frac{P_r}{P_t} = A_{er} A_{et} \left(\frac{1}{\lambda r}\right)^2$$

This is a useful link design equation for communication engineers that help to trade-off between antenna size, frequency of operation and the distance between transmitter and receiver.

- *Antenna noise temperature* is a useful parameter for communication/microwave engineers dealing with antennae. An antenna has an inherent thermal noise caused by its physical temperature T_p and also picks up noise power from the background due to background temperature T_b; the latter being dependent on the earth's ground, the sun, the microwave cosmic background (the deep-space galactic background) and so on. For an antenna with radiation efficiency $\eta_a < 1$, the antenna noise temperature is given by:

$$T_a = \eta_a T_b + (1 - \eta_a)\, T_p$$

An isotropic antenna would have its noise temperature affected by an averaged background temperature due to all possible natural thermal environments around the antenna; however, in case of a directional antenna, the antenna noise temperature will depend on the environment temperature only in which direction the antenna is 'looking'.

REVIEW QUESTIONS

2.1 What are electromagnetic fields? Who first theoretically hypothesised the existence of such field?

2.2 A static electric field is irrotational while for a time-varying situation the electric field is rotational—justify.

2.3 A magnetic field is solenoidal both for static and time-varying situations—justify.

2.4 Maxwell modified Ampere's circuital law of magnetostatics with the inclusion of displacement current for a time-varying electric field—justify.

2.5 For a time-varying case, the electromagnetic field is a vortex field—justify.

2.6 Explain the concept of propagating electromagnetic waves on the basis of Maxwell's curl equations.

2.7 What information about the field do we get from the differential form of Maxwell's equation? Why do we need the integral form?

2.8 For engineering applications we need the time-harmonic or the phasor form of Maxwell's equations—justify.

2.9 What are Vector Helmhotz equations?

2.10 The attenuation constant and propagation constant of an electromagnetic wave propagating in a general lossy medium is a non-linear function of frequency—justify.

2.11 What is wave number? What is its physical significance?

2.12 What are plane waves? Why are plane wave solutions sought for wave equations?

2.13 Phase velocity of an electromagnetic wave is given by the velocity of light—justify.

2.14 What is wave impedance? What is its value for free space?

2.15 A uniform plane wave propagating through free space is a TEM wave—justify.

2.16 What is skin depth? Explain how inverse frequency dependence of skin depth helps to reduce material costs of high performance waveguides with silver/gold plating.

2.17 The principle of operation of microwave oven is based on dielectric heating—explain.

2.18 Why is 2.45 GHz in ISM band used for microwave ovens?

2.19 Mention a few industrial applications of microwave heating.

2.20 What is Poynting vector? What are its limitations?

2.21 Write the expression for time-average power when the electromagnetic field is time harmonic.

2.22 What do you understand by polarisation of electromagnetic waves? Acoustic waves cannot be polarised—explain.

2.23 Why was the Yagi-Uda antenna with horizontal polarisation used in earlier days for reception of TV signals?

2.24 Dual polarisation is used in Cassegarin feed type antenna—explain.

2.25 What are linearly, circularly and elliptically polarised waves?

2.26 What are LHCP and RHCP waves?

2.27 State the boundary conditions (bc) to be satisfied by E_t, H_t, B_n and D_n at the interface between two dielectric media. If one medium is a conductor or a lossy dielectric how are the bc for H_t and D_n modified?

2.28 Show that the transmission and reflection coefficients are related by $T = 1 + \Gamma$.

2.29 In a dielectric–conductor interface, standing waves are formed in the dielectric medium—explain.

2.30 Discuss the nature of the total electric and magnetic fields of the standing wave mentioned in question RQ 2.28.

2.31 The average power carried by a pure standing wave is zero but the electromagnetic energy interchanges cyclically between the stored electric and magnetic energy—explain.

2.32 Copper, being a very good conductor, can be used effectively as an RF shield while graphite coated with carbon film, being a highly lossy dielectric, can act as a practical matched load at microwave frequency—justify.

2.33 What do you understand by perpendicularly polarised and parallel polarised waves in case of oblique incidence of electromagnetic wave in an interface?

2.34 Discuss the principle of parallel plate waveguide in terms of dielectric–conductor interface with the incident electromagnetic field being perpendicularly polarised.

2.35 The quartz windows at the end of laser tubes are set at the Brewster's angle—why?

2.36 A conductor carrying time-varying electric current gives rise to electromagnetic radiation—explain.

2.37 Retarded magnetic vector potential is used for antenna analysis—explain.

2.38 What do you understand by Fresnel region, Fraunhofer region and Raleigh distance in connection with antenna characterisation?

2.39 What do you understand by the radiation resistance of an antenna? What is the value of radiation resistance of a half-wave dipole antenna?

2.40 What do you understand by *E*-plane pattern and *H*-plane pattern of the radiated field from an antenna?

2.41 What do you understand by directive gain and power gain of an antenna?

2.42 What do you understand by HPBW and BWFN?

2.43 What do you understand by effective aperture of a receiving antenna? Explain how a whip antenna, even being a Hertzian dipole, serves as a practical receiving antenna.

2.44 Explain how Friss' transmission equation serves as a useful designer equation for communication link design.

2.45 What do you understand by antenna noise temperature?

Numerical Problems

2.1 Show that the formulae $\vec{E} = \nabla \times \nabla \times \vec{\Psi}$ and $\vec{B} = \frac{1}{c^2} \nabla \times \left(\frac{\partial \vec{\Psi}}{\partial t} \right)$ satisfy Maxwell's equations in free space provided that the vector function $\vec{\Psi}$ in rectangular coordinate is a solution of the wave equation $c^2 \nabla^2 \vec{\Psi} = \partial^2 \vec{\Psi}/\partial t^2$. Further verify that $A\vec{a}_z \sin px \sin(\omega t - \Phi z)$ is a possible form of $\vec{\Psi}$, if p, Φ and ω satisfy the condition $p^2 + \Phi^2 = \omega^2/c^2$; where A, p, Φ are real constants and $\vec{a}_z$ is a unit vector along the z-axis. Find the condition that such a solution should represent a field between two perfectly conducting planes at $x = 0$ and $x = a$, and prove that if the field is that of a wave propagated along the z-axis then $\omega > n\pi c/a$, where n is an integer. Also show that the time average energy flow is parallel to the z-axis and is given by $\omega \Phi a p^2 A^2/(4\mu_0)$ per unit width.

2.2 In free space, $\tilde{\vec{E}}(z, t) = 100 \cos(\omega t - \beta z)\, \vec{a}_x$ V/m. Using the Poynting vector find the average power crossing a circular area of radius 3 m in the plane z = constant.

[**Ans:** 375 W]

2.3 A voltage source V is connected to a pure resistance R via a length of a coaxial cable. Deriving the electric and magnetic fields in the dielectric between the two conductors of the coaxial cable, show that the total instantaneous power over the cross-section of the dielectric calculated with the Poynting vector is equal to the instantaneous power loss in the resistor obtainable from circuit theory result.

[**Ans:** P_{av} is zero inside and outside the slab when $\sigma \to \infty$]

2.4 A geopositioning satellite orbiting at an altitude of 1500 km transmits a total power of 60 kW isotropically at a down link frequency of 187.5 MHz. The electric field of the received wave on the earth at a point immediately below the satellite is given by: $\vec{E}(z) = C_1(-j\vec{a}_x + \vec{a}_y)\, e^{-j\beta z}$. Find the polarisation of the electromagnetic wave. Calculating the values of β and C_1 write the expression for the magnetic field.

[**Ans:** The wave is left-handed circularly polarised.
$\vec{H}(z) = 3.34\,(-\vec{a}_x - j\vec{a}_y)\, e^{-j\,3.92z}$]

2.5 The plane $z = 0$ marks the boundary between the free space and a dielectric medium having a dielectric constant of 40. The $\vec{E}$ field next to the interface in free space is $\vec{E} = 13\vec{a}_x + 40\vec{a}_y + 50\vec{a}_z$ V/m. Determine the $\vec{E}$ field on the other side of the interface.

[**Ans:** $\vec{E} = 13\vec{a}_x + 40\vec{a}_y + 1.25\vec{a}_z$]

2.6 Two magnetic materials one having relative permeability $\mu_{r1} = 4$ and the other has $\mu_{r2} = 2$, meet at an interface. The magnetic flux density in material 1 is given by $\vec{B} = (\vec{a}_x 0.1 + \vec{a}_y 0.2 + \vec{a}_z 0.1)$ weber/m^2. The tangential component of $\vec{H}$ in material 2 is known to be zero. Find (a) the surface current density that must exist on the interface for this condition to be satisfied and (b) the magnetic field intensity in the material 2.

[**Ans:** (a) The surface current has two components: one in x-direction ($= 10^6/16\pi$) and the other one in y-direction ($= 10^6/8\pi$) in A/m^2
(b) $B_2 = 0.1$ weber/m^2 in the z-direction]

2.7 A current sheet $\vec{J} = 5\vec{a}_y$ A/m lies on the dielectric interface $x = 0$ between two dielectric media with $\varepsilon_{r1} = 4, \mu_{r1} = 2$ in region 1 ($x < 0$) and $\varepsilon_{r2} = 4, \mu_{r2} = 4$ in region 2 ($x > 0$). If the magnetic field in region 1 at $x = 0$ is $\vec{H}_1 = (5\vec{a}_x + 50\vec{a}_y)$ A/m, find the magnetic field in region 2 at $x = 0^+$.

[**Ans:** $\vec{H}_2 = (2.5\vec{a}_x + 50\vec{a}_y - 50\vec{a}_z)$ A/m]

2.8 Determine the polarisation of the electromagnetic wave if the electric field intensity of the wave is given by $\vec{E} = (3\vec{a}_x + j4\vec{a}_y)\, e^{-0.2z}\, e^{-j0.5z}$ V/m.

[**Ans:** The wave is left-handed elliptically polarised]

2.9 Show that a charge moving in a circle with a constant speed v emits no radiation in the plane of the circle in directions inclined at an angle $\cos^{-1}(v/c)$ to the direction of motion.

2.10 A half wave dipole antenna radiates 10 kW at a frequency of 100 MHz. A short dipole antenna situated at a distance of 25 km is used as a receiving antenna. If both antennae are symmetrically placed in the xy plane and the medium is free space, determine the effective area of each antenna and power absorbed by the receiving antenna.

[**Ans:** $A_{et} = 1.175$, $A_{er} = 1.074$, $P_r = 2.243$ mW]

2.11 A satellite earth station operating at 6.0 GHz has transmitter power of 150 W and antenna gain of 40 dB. The receiving antenna of the transponder placed at a synchronous orbit (at 36,000 km from the earth's surface) has a gain of 30 dB. Calculate the received power at the transponder.

[**Ans:** $P_r = 1.834 \times 10^{-11}$ W]

3

High Frequency Behaviour of Transmission Lines

The way you think
—that determines what you become

3.1 Introduction

The transmission line is a generic term used for describing a structure that is used for the transport of electrical or electromagnetic signals from one point in the system to another. All of us are familiar with the two-wire overhead electrical power transmission lines used for the transport of electric energy from the generating station to households or industrial sites. However, the twin-wire line may also be used for voice signal transmission as in telephones or as a flat lead-in line from a roof top antenna to a television receiver. The *twin-wire line* (Fig. 3.1a) may be used satisfactorily to transport electromagnetic energy up to 300 MHz (at higher frequencies it becomes difficult to maintain the small distance between the two conductors for reasonable coupling that is required for guided wave RF signal transmission via such lines). Further, at higher frequencies, the radiation losses become far too much to be tolerated. However, the radiation loss in a two-conductor transmission line may be made negligible with a different configuration where a circular cylindrical conducting rod is placed coaxially with an outer conducting tube (Fig. 3.1b), which is known as a *coaxial transmission line*. The fields in the coaxial line remain confined to the annular region between the inner and the outer conductors (which is filled with a dielectric material to keep the conductors in position). Thus, the signal cannot radiate out and, pick-up of spurious signals cannot take place in such 'shielded' two-conductor coaxial transmission lines. The principal mode of propagation either in twin-wire line or in the coaxial line is transverse electromagnetic (*TEM*), which is also the mode for electromagnetic wave propagation in free space, with the electric and magnetic field distribution on the lines as indicated in Fig. 3.1c.

A fundamental difference between the signal transmission at power line frequency and that at radio frequency (RF) frequency is that, the wavelength of the signal at power line frequency (50/60 Hz) is very much larger than the size of the electrical networks. Thus, the circuit elements in an electrical network are considered to be '*discrete*' and are usually said to be described by '*lumped parameters*'. It is assumed that currents flowing in lumped-circuit elements do not vary spatially over the elements and that no standing waves exist. But at radio frequency, the component size is a considerable fraction of the wavelength, especially at microwave frequency it may be many wavelengths long. A radio frequency transmission line thus said to behave as a *distributed-parameter network*; with current, voltage and impedance varying all along the length of the line. A standing wave exists in the radio frequency transmission line, except when it is terminated with a matched load. Details of these issues related to high frequency (HF) behaviour of transmission lines will be discussed in this chapter.

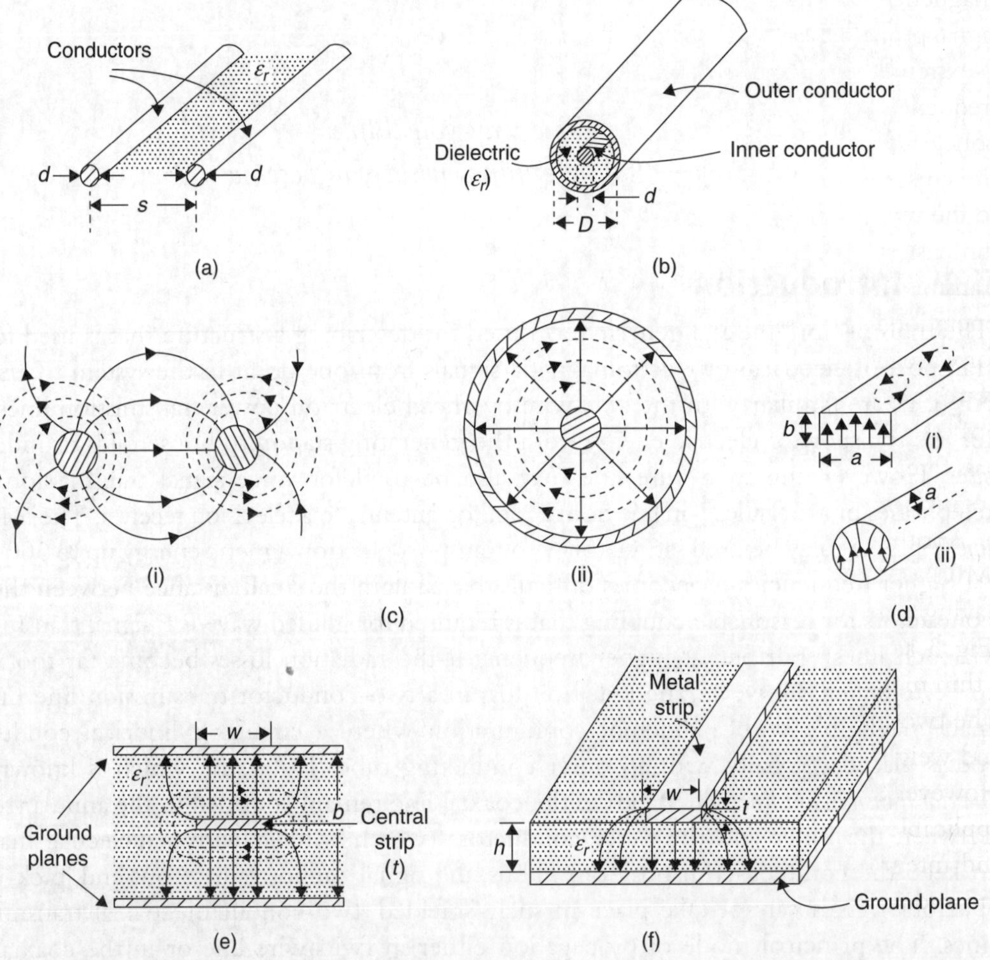

Fig. 3.1 Different types of high-frequency transmission lines (a) Twin-wire transmission line ($s >> d$), (b) Coaxial transmission line, (c) Field lines ($E \rightarrow$ full lines, $H \rightarrow$ dashed lines) for (i) Twin-wire line (ii) Coaxial line, (d) Waveguide (i) Rectangular (with TE_{10} mode) (ii) Circular (with TE_{11} mode), (e) Strip-line (cross-sectional view with field lines), (f) Microstrip line.

The coaxial line is used as a transmission structure for electromagnetic signals up to 3 GHz (with a specialised design, it can work up to 40 GHz, but becomes very expensive) above which the losses in the dielectric (that increases with frequency) and also in the conductor become appreciable and cannot be neglected. The telephone and TV cables and the input cables to microwave precision measuring instruments are generally the coaxial lines. The losses incurred in the coaxial line may be significantly reduced by dispensing with the centre conductor and hence the dielectric support, using merely a hollow metal tube of rectangular or circular cross-section (Fig. 3.1d). Such a system is known as a *waveguide*. A waveguide differs from the twin-wire or the coaxial transmission line (which, in principle, is also a wave-guiding structure for RF signal transmission) in its construction in that it is a single conductor system. The field analysis of the waveguide (given in detail in Chapter 4) shows that the boundary conditions to be satisfied by the fields inside a hollow metal waveguide is such that the TEM-mode cannot exist. Instead, either a transverse electric (TE) or transverse magnetic (TM) mode propagates down a waveguide. The attenuation of the electromagnetic signal propagating through an air-dielectric (which is normally the case) waveguide structure is extremely low. Although waveguides are the best transmission line structures at microwave frequencies (covering the range 3–300 GHz) so far as the efficiency of power transmission is concerned, they cannot be practically used at lower microwave frequencies. This is because the cross-sectional dimensions of such metal pipes or waveguides have to be directly related to the wavelength of operation (broad dimension approximately half-wavelength and narrow dimension approximately a quarter-wavelength in the case of a rectangular waveguide) for transmission of power to take place through them. It is therefore only when the frequency of operation is a few gigahertz (say 10 GHz) that the cross-sectional dimensions are small enough for waveguides to be convenient for practical use. So far as the propagation of electromagnetic waves along the waveguides is concerned, in contrast to the coaxial line, the waveguides possess a low frequency cut-off and exhibit a high-pass property or more specifically, a band pass filter (thus comparatively narrow band), whereas the operation of coaxial line in principle begins from DC and goes up to a few GHz (hence a broadband structure).

Another class of transmission lines that is important for Microwave Integrated Circuit (MIC) is *planar transmission lines*. Phenomenologically, when a coaxial line is thought of as having planar symmetry, we get the fundamental planar transmission line—the strip-line (Fig. 3.1e). A common type of practical planar transmission line is the microstrip, in which a thin metal strip is deposited on a dielectric substrate backed by a ground plane (Fig. 3.1f). The two main advantages obtained with planar transmission lines are, the reduction in size and weight and the ability to use planar integrated circuit technology for their fabrication. However, the power handling capability of such structures is comparatively low and is thus applicable for low and medium power applications but when demand for miniaturisation and integration is a practical need, such planar transmission lines are exclusively used. At the end of this chapter, a brief discussion on planar transmission lines has been included.

3.2 Analysis of HF Transmission Lines

Analysis of HF transmission lines can be done either by the solution of Maxwell's field equations or by the method of distributed circuit theory. As far as single conductor transmission lines (waveguides) are concerned, Maxwell's equations with appropriate boundary conditions are used to derive guided wave solutions (as will be discussed in Chapter 4). For two-wire transmission structures such as the twin-wire line and coaxial line, distributed circuit analysis is conventionally used. This is because in the cross-section of a TEM transmission line, the distribution of the electromagnetic field is the same as that for the static field. Thus, we can introduce the concepts of equivalent voltage, equivalent current and equivalent impedance. However, to ensure that this method is applicable, the geometrical length l of the transmission line should be comparable to the wavelength λ, as is usually the case with any high frequency/radio frequency (RF)/microwave transmission line. A disadvantage of this method is that it reveals little about the electromagnetic field pattern or other possible modes of propagation (which can be obtained with rigorous field solution). However, it does describe the propagation characteristics and the impedance properties of the line for the principal mode (TEM) of transmission and hence is of considerable engineering importance.

3.2.1 Transmission Line Equations

Our analysis is for uniform transmission lines, in which the dimensions and the electrical properties of the line are identical at all planes transverse to the direction of signal propagation. A two-conductor uniform transmission line at RF may be modelled as a lumped element circuit with a very large number of elements (Fig. 3.2).

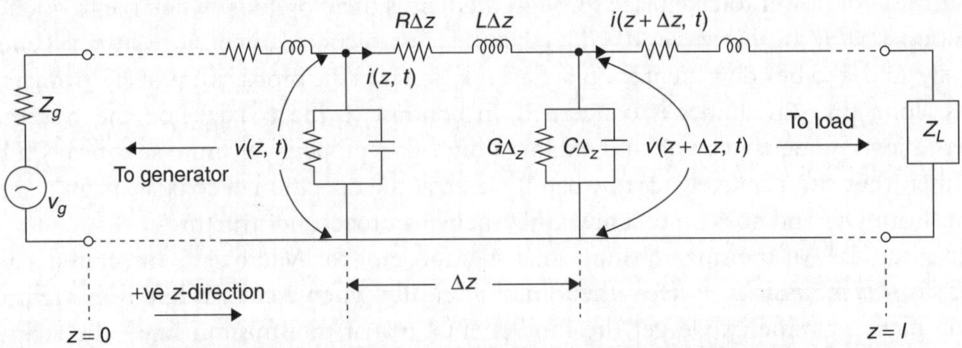

Fig. 3.2 Lumped element circuit representation of a transmission line at high frequency (RF).

The total line length l, can be divided into an infinite number of sections of length Δz ($<< \lambda$), and thus the line may be analysed by using ordinary AC circuit theory. Since $\Delta z \rightarrow 0$, the results are valid at all frequencies. The lumped circuit components L, C, R and G are the inductance, capacitance, resistance and conductance per unit length of the line which are the distributed parameters of the line. The transmission line parameters for twin-wire and coaxial line are given in Table 3.1 (the derivations of which are available in standard text books).

The physical relevance of these lumped components may be visualised with respect to a coaxial line in which L represents the series inductance of the centre conductor, C represents the centre conductor's capacitance with the outer (shield) conductor, R represents

Table 3.1 Transmission line parameters for twin-wire line and coaxial line

Twin-wire transmission line (Fig. 3.1a)	Coaxial transmission line (Fig. 3.1b)
$R = \dfrac{2}{\pi d \delta \sigma_c}$	$R = \dfrac{1}{\pi \delta \sigma_C}\left[\dfrac{1}{D} + \dfrac{1}{d}\right]$
$L = \dfrac{\mu}{\pi}\cosh^{-1}\dfrac{s}{d}$	$L = \dfrac{\mu}{2\pi}\ln\dfrac{D}{d}$
$G = \dfrac{\pi\sigma}{\cosh^{-1}(s/d)}$	$G = \dfrac{2\pi\sigma}{\ln(D/d)}$
$C = \dfrac{\pi\varepsilon}{\cosh^{-1}(s/d)}$	$C = \dfrac{2\pi\varepsilon}{\ln(D/d)}$

[where δ is the skin depth, σ_c is the conductivity of the conductor, σ, μ, ε are respectively the conductivity, permeability and permittivity of the dielectric between the conductors. All the above quantities are expressed in per unit length of the line.]

the resistance of the centre conductor and shield, and G represents the leakage resistance of the dielectric between the two conductors.

Referring to Fig. 3.2, let $v(z, t)$ and $i(z, t)$ represent the time-varying voltage and current at the input of the line section, while $v(z + \Delta z, t)$ and $i(z + \Delta z, t)$ represent the output values. The wave propagation is assumed to be in the +ve z direction.

Applying Kirchhoff's voltage and current law to the line section yields:

$$v(z,t) = R\Delta z i(z,t) + L\Delta z \frac{\partial i(z,t)}{\partial t} + v(z+\Delta z, t) \tag{3.1}$$

$$i(z,t) = G\Delta z v(z+\Delta z, t) + C\Delta z \frac{\partial v(z+\Delta z, t)}{\partial t} + i(z+\Delta z, t) \tag{3.2}$$

Rearranging, dividing by Δz and letting the limiting case $\Delta z \to 0$, we get:

$$-\frac{\partial v(z,t)}{\partial z} = Ri(z,t) + L\frac{\partial i(z,t)}{\partial t} \tag{3.3}$$

$$-\frac{\partial i(z,t)}{\partial z} = Gv(z,t) + C\frac{\partial v(z,t)}{\partial t} \tag{3.4}$$

Equations 3.3 and 3.4 are a pair of first-order partial differential equations in $v(z, t)$ and $i(z, t)$. They are the *generalised transmission line equations* in the time domain and are popularly known as *Telegrapher's equations*.

For harmonic time dependence, the use of phasors simplifies the transmission line equations to ordinary differential equations. For a cosine reference we may write:

$$v(z,t) = \text{Re}[V(z)e^{j\omega t}] \tag{3.5a}$$

$$i(z, t) = \text{Re}[I(z)e^{j\omega t}] \quad (3.5b)$$

where $V(z)$ and $I(z)$ are the functions of the space coordinate z only and both may be complex. Substitution of Eqs 3.5a and 3.5b in Eqs 3.3 and 3.4 yields the following ordinary differential equations for phasors $V(z)$ and $I(z)$:

$$-\frac{dV(z)}{dz} = (R + j\omega L)\, I(z) \quad (3.6a)$$

$$-\frac{dI(z)}{dz} = (G + j\omega C)\, V(z) \quad (3.6b)$$

Equations 3.6a and 3.6b are *time-harmonic transmission-line equations.*

3.2.2 Characteristic Parameters of HF Transmission Lines

The coupled time-harmonic transmission line Eqs 3.6a and 3.6b, can be combined to obtain the following equations which may be solved for $V(z)$ and $I(z)$.

$$\frac{d^2 V(z)}{dz^2} = \gamma^2 V(z) \quad (3.7a)$$

and $$\frac{d^2 I(z)}{dz^2} = \gamma^2 I(z) \quad (3.7b)$$

where $$\gamma = \alpha + j\beta = \sqrt{(R + j\omega L)(G + j\omega C)} \quad (3.8)$$

where γ is the *propagation constant* whose real and imaginary parts, α and β are respectively the *attenuation constant* and *phase constant* of the line.

The general solution for Eqs 3.7a and 3.7b are:

$$V(z) = V^+ e^{-\gamma z} + V^- e^{\gamma z} \quad (3.9a)$$

$$I(z) = I^+ e^{-\gamma z} + I^- e^{\gamma z} \quad (3.9b)$$

where the $e^{-\gamma z}$ term represents a wave that propagates in the $+z$ direction, and the $e^{\gamma z}$ term represents wave propagation in the $-z$ direction. $V^{\pm}$ and $I^{\pm}$ are the amplitude constants defined in terms of the boundary conditions, that is, dependent on the line terminations. Substituting Eq. 3.9a into Eq. 3.6a, we can get an expression for current on the line as:

$$I(z) = \frac{\gamma}{R + j\omega L}(V^+ e^{-\gamma z} - V^- e^{\gamma z}) \quad (3.9c)$$

We can equate the right hand sides of Eqs 3.9b and 3.9c setting the $e^{\gamma z}$ factor as zero (that is, with no reflected waves) and also set $z = 0$, and solve the resulting equation for an impedance term at the load plane (at $z = 0$) given by:

$$Z_0 = \frac{V^+}{I^+} = \frac{R + j\omega L}{\gamma} = \sqrt{\frac{R + j\omega L}{G + j\omega C}} \tag{3.10}$$

This impedance Z_0 is known as the *characteristic impedance* of the line which means a specific impedance with which if the transmission line is terminated at the load plane ($z = 0$) there will be no reflected wave ($e^{\gamma z} = 0$) in the line and the line is said to be *matched* at the load end. Ideally, if a line is infinitely long then no signal returns from the load end (resembling a straight railway track in an open field in the village side which appears to be a never-ending track) but practically a finite length transmission line terminated with characteristic impedance behaves like an *infinite line* or *matched line* with no reflected signal from the load end. In a more general way, Eq. 3.10a can be rewritten as

$$Z_0 = \frac{V^+}{I^+} = -\frac{V^-}{I^-} = \frac{R + j\omega L}{\gamma} = \sqrt{\frac{R + j\omega L}{G + j\omega C}} \tag{3.11}$$

The characteristic impedance is, in general, frequency dependant and thus at higher RF/microwave frequencies (when Ω is very large), $R << \omega L$ and $G << \omega C$, and we can assume the lines to be low-loss or practically loss-less. Under this situation, we can derive simplified expressions for the propagation constant and characteristic impedance. Referring to Eq. 3.8, we can use the binomial expansion when the propagation constant can be expressed as:

$$\gamma \cong \frac{1}{2}\left(R\sqrt{\frac{C}{L}} + G\sqrt{\frac{L}{C}}\right) + j\omega\sqrt{LC} \tag{3.12a}$$

$$= j\omega\sqrt{LC} \text{ (when } R = G = 0\text{, for loss-less line)}$$

Therefore the attenuation and phase constants are, respectively, given by:

$$\alpha \cong \frac{1}{2}\left(R\sqrt{\frac{C}{L}} + G\sqrt{\frac{L}{C}}\right) \tag{3.12b}$$

$$= 0 \text{ (for loss-less line)}$$

and $\beta \cong \omega\sqrt{LC}$ (3.12c)

whence the phase velocity is:

$$v_p = \frac{\omega}{\beta} = \frac{1}{\sqrt{LC}} = \frac{1}{\sqrt{\mu_0\,\varepsilon_0}} = c \text{ (velocity of light)} \tag{3.12d}$$

Similarly, the characteristic impedance (Eq. 3.10b) is found to be:

$$Z_0 = \sqrt{\frac{L}{C}} \tag{3.12e}$$

It may be observed that the two-conductor transmission line at higher RF/microwave frequency possesses characteristic impedance that has a constant real value (resistive). As the phase constant is a linear function of frequency (directly proportional), the phase velocity of electromagnetic waves propagating down the two-conductor transmission line will be constant (independent of frequency) and is found to be equal to the phase velocity of electromagnetic waves in free space, that is, the velocity of light *c*. Thus a two-conductor RF/microwave transmission line will be *non-dispersive*. This is the reason why, at microwave frequency, the coaxial line (and not the waveguide, even though flexible ones are available at higher cost) is safely usable to connect the measuring instruments to the main waveguide bench systems. Further, a two-conductor transmission line at higher RF/microwave frequency will also behave as a *distortionless line* (that is, a practical communication signal with different frequency components will reach the destination point at the same phase velocity as the phase velocity is independent of frequency and therefore, the signal at the destination will be undistorted). But, it may be noted that in the case of a lossy line at lower radio wave frequency, the phase constant β is a non-linear function of frequency (ω), Eq. 2.7b, when the Heaviside condition $\frac{R}{L} = \frac{G}{C}$ has to be used to realise distortionless performance of the line. In fact, intercontinental telephone lines are designed on the basis of the said condition.

The characteristic impedance of twin-wire line and coaxial line are given in Table 3.2.

Table 3.2 Characteristic impedance of twin-wire line and coaxial line

Twin-wire transmission line (Fig. 3.1a)	**Coaxial transmission line (Fig. 3.1b)**
$Z_0 = \frac{276}{\sqrt{\varepsilon_r}} \log_{10}\left(\frac{2s}{d}\right)\Omega,$	$Z_0 = \frac{138}{\sqrt{\varepsilon_r}} \log_{10}\left(\frac{D}{d}\right)\Omega,$
[typically (250–300) Ω]	[typically 50 Ω/75 Ω]

($\varepsilon_r \rightarrow$ dielectric constant of the dielectric)

3.2.3 Terminated HF Transmission Lines

In all practical applications, a transmission line is terminated with suitable load impedance. Depending on the load end termination, the incident signal from the source might either be reflected back to the source or be absorbed at the load itself. It is thus important to study the properties of a transmission line terminated in an arbitrary load Z_L, which will serve to illustrate how the forward and backward propagating waves can be combined to satisfy the boundary conditions at the termination. A transmission line terminated in a load impedance Z_L is shown in Fig. 3.3, the line has a characteristic impedance Z_0 and it is assumed to be loss-less (at higher RF/microwave frequency) having a propagation constant $\gamma = j\beta$.

To discuss the characteristics of terminated transmission lines (which is a finite transmission line with a generator at one end and the load at the other), we first need to choose suitable reference points with which to relate all our calculations. We could choose

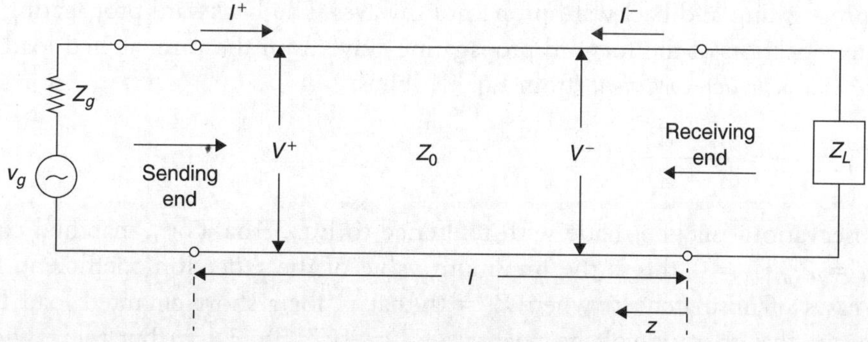

Fig. 3.3 Transmission line terminated in a load impedance ($Z_L \neq Z_0$).

either the generator or the load for this purpose; however, it is customary to use the load as the reference point. Thus +ve z-direction is towards the generator. However, note that the positive direction of power flow, in principle, remains from the generator to the load, only the forward propagating wave ($V^+e^{j\beta z}$) is now in the −ve z-direction and the backward propagating wave ($V^-e^{-j\beta z}$) is in the +ve z-direction. Following this convention for the sign of z, we have from Eq. 3.9, the voltage and current travelling waves on the line as:

$$V(z) = V^+ e^{j\beta z} + V^- e^{-j\beta z} \tag{3.13a}$$

$$I(z) = I^+ e^{j\beta z} + I^- e^{-j\beta z} \tag{3.13b}$$

While the current expressed in terms of voltage is given by:

$$I(z) = \frac{V^+}{Z_0} e^{j\beta z} - \frac{V^-}{Z_0} e^{-j\beta z} \tag{3.13c}$$

(Eq. 3.10 for the expression for Z_0).

The ratio of voltage to current at the receiving end is the load impedance:

$$Z_L = \frac{V_L}{I_L} \tag{3.14a}$$

where V_L and I_L are the load voltage and load current given by Eqs 3.13; setting $z = 0$. Thus,

$$Z_L = \frac{V(0)}{I(0)} = \frac{V^+ + V^-}{I^+ + I^-} = \frac{V^+ + V^-}{\left(\frac{V^+}{Z_0}\right) - \left(\frac{V^-}{Z_0}\right)} = Z_0 \frac{V^+ + V^-}{V^+ - V^-} \tag{3.14b}$$

It may be noted from Eq. 3.14b that if only forward propagating waves exist ($V^- = 0$), the load impedance must be equal to the characteristic impedance of the line. This condition defines a *match* between the load and the line. Under the matched condition ($Z_L = Z_0$), there are no backward propagating waves. On the other hand, if $Z_L \neq Z_0$, there will be both forward propagating and backward propagating waves. The backward propagating wave is due to the reflection of the forward propagating wave from the mismatched load and we define the *load reflection coefficient* (from Eq. 3.14b) as:

$$\Gamma_L = \frac{V^-}{V^+} = \frac{Z_L - Z_0}{Z_L + Z_0} \tag{3.15a}$$

A few observations may be made with reference to Eq. 3.15a. For a matched condition when $Z_L = Z_0$, $\Gamma_L = 0$, this is the minimum value of the reflection coefficient. For two extreme cases of mismatch: (i) when $Z_L = 0$, that is, for a short-circuited load $\Gamma_L = -1$; which means that there is voltage reversal ($V^+ = -V^-$, Eq. 3.15a) but there will be no current reversal ($I^+ = I^-$, Eq. 3.10b) as expected for a short-circuit, while (ii) when $Z_L = \infty$, that is, for an open-circuited load $\Gamma_L = +1$, meaning that there will be no voltage reversal ($V^+ = V^-$) but there will be current reversal ($I^+ = -I^-$) which is also expected for an open-circuit. Physically, this would mean that from a short-circuited load, the voltage wave will get reflected back to the generator side while for open-circuited load the voltage wave will continue past the load against the generator direction and will give rise to a radiated signal caused by the fringing out of the electric field lines. However, in both cases, the magnitude of the maximum value of Γ_L is 1; thus $0 \le |\Gamma_L| \le 1$.

It is important to remember that Γ_L given by Eq. 3.15a is the reflection coefficient at the load, and is thus denoted with a subscript *L*. At any other location on the line, the reflection coefficient is, in general, different and so subsequently we would define a generalised reflection coefficient. The reflection coefficient is, in general, a complex quantity since it is the ratio of the complex amplitudes V^- and V^+ and we prefer to write:

$$\Gamma_L = |\Gamma_L|\, e^{j\theta_\Gamma} \tag{3.15b}$$

where $|\Gamma_L|$ is the magnitude and θ_Γ is the phase of the reflection coefficient. In terms of the definition of Γ_L given by Eq. 3.15a, the voltage equation in Eq. 3.13a can be written as:

$$V(z) = V^+ \left(e^{j\beta z} + \Gamma_L e^{-j\beta z}\right) \tag{3.16a}$$

We can define a term called *transmission coefficient* as $T = \frac{V_{tr}}{V^+}$, when Eq. 3.16a will be:

$$V(z) = V^+ \, T e^{j\beta z} \tag{3.16b}$$

where the transmitted wave V_{tr} from the first line (with characteristic impedance Z_0) delivers the signal to a second line (with characteristic impedance Z_L) which is matched at its load end and thus does not give rise to any reflected signal. In other words, there is a through transmission of signal from the first line to the second line without the transmitted signal being reflected back towards the first line (which precisely represents the physical situation for a transmitted signal from the first line to the second line). From Eqs 3.16a, 3.16b and 3.15a we get:

$$T = 1 + \Gamma = \frac{2Z_L}{Z_L + Z_0} \tag{3.16c}$$

The transmission coefficient between two points in a circuit is often expressed in decibels when it is called the *insertion loss*, given by:

$$L_{\text{insertion}} = -20 \log |T| \text{ (in dB)} \tag{3.16d}$$

From Eq. 3.16d it is understood that 0 dB insertion loss (with $T = 1$) means there is perfect transmission with no reflection (when $Z_L = Z_0$, Eq. 3.16c) while with less and less transmission, that is, with poor impedance matching, results in large reflection thereby increasing insertion loss.

For a mismatched line ($Z_L \neq Z_0$) we need to know the line impedance at any location z on the line which is very important from an engineering point of view as it will help us to carry out our calculations to achieve matching in transmission line systems. The line impedance at any point z may be written by using Eqs 3.13a, 3.13b and 3.15a as:

$$Z(z) = \frac{V(z)}{I(z)} = Z_0 \frac{e^{j\beta z} + \Gamma_L e^{-j\beta z}}{e^{j\beta z} - \Gamma_L e^{-j\beta z}} \tag{3.17a}$$

It may also be expressed as:

$$Z(z) = Z_0 \frac{Z_L + jZ_0 \tan \beta z}{Z_0 + jZ_L \tan \beta z} \tag{3.17b}$$

[At lower RF, this equation is given by $Z(z) = Z_0 \dfrac{Z_L + jZ_0 \tanh \gamma z}{Z_0 + jZ_L \tanh \gamma z}$. When we set $\gamma = j\beta$, for loss-less line (as $\alpha = 0$) at microwave frequency, we get Eq. 3.17b.]

This is a very important equation for terminated transmission line characterisation and is known as the *transmission line impedance equation*, which represents the input impedance of the line looking towards the load. It may be noted that as for the voltage and current, the impedance looking into the line also varies with the length of the transmission line. When we set $z = 0$, we have $Z(0) = Z_L$, that is, the impedance at the receiving terminal (the load itself), and when $z = l$, we get the input impedance of the line as seen from the sending end. In other words, a load impedance Z_L gets transformed into this input impedance when it is viewed through the transmission line of length l with a characteristic impedance Z_0.

The reflection coefficient at point z on the line, that is, the generalised reflection coefficient is given by:

$$\Gamma(z) = \frac{V^-(z)}{V^+(z)} = \frac{V^+ \Gamma_L e^{-j\beta z}}{V^+ e^{j\beta z}} = \Gamma_L e^{-2j\beta z} = |\Gamma_L| e^{j\theta_\Gamma - 2j\beta z} \tag{3.18}$$

It may be observed that $\Gamma(z)$ becomes Γ_L given by Eq. 3.15b when we set $z = 0$, the load end condition in Eq. 3.18.

Analogous to the voltage reflection coefficient Γ_V, a current reflection coefficient Γ_I may also be defined as $\Gamma_I = (I^-/I^+)$, which is the negative of Γ_v (this follows from Eq. 3.10b). However, it is customary to do all transmission line analysis in terms of voltage reflection coefficient; hence Γ would implicitly mean the voltage reflection coefficient (Γ_V). The value of the magnitude of Γ as observed earlier, is given by: $0 \leq |\Gamma| \leq 1$.

For a matched line ($\Gamma = 0$), the magnitude of the voltage on the line from Eq. 3.13a is $V_z = |V^+|$, which has a constant value and is thus a matched line, is said to be a *flat line* or infinite line as it transmits all power as a travelling wave and does not reflect and hence acts as a *non-resonant line*. However, when the line is mismatched, there will be both the incident wave and the reflected wave and the two together will interfere on the line giving rise to the so called *standing wave* (that is, a wave that 'stands' locally and does not travel). Under this situation, it is seen that the magnitude of the voltage (current) on the line will not be constant, but will vary along the length of the line (this is a specialised characteristic of RF transmission lines unlike the power lines). Thus, a line terminated with $Z_L \neq Z_0$ reflects some or all power, forming a standing wave and thus acts as a *resonant line* behaving as an LC circuit at RF. The analytical characterisation of the standing wave on a microwave transmission line can be done by using the expression for the travelling wave voltage on the line given by Eq. 3.13a. We may write this equation as:

$$V(z) = V^+ e^{j\beta z} (1 + \Gamma_L e^{-j2\beta z}) \tag{3.19a}$$

Taking the magnitude on both sides and manipulating (that is, putting $|e^{j\beta z}| = 1$, and $\Gamma_L = |\Gamma_L| e^{j\theta_\Gamma}$) we have:

$$|V_z| = |V^+| |1 + |\Gamma_L| e^{-j(\theta_\Gamma - 2\beta z)}| \tag{3.19b}$$

$$= |V^+| \{(1 + |\Gamma_L| \cos \Psi)^2 + (|\Gamma_L| \sin \Psi)\}^{1/2}$$

Therefore, $$|V_z| = |V^+| (1 + |\Gamma_L|^2 + 2|\Gamma_L| \cos \Psi)^{1/2} \tag{3.20}$$

where $\Psi = (\theta_L - 2\beta z)$.

We can now find the positions and values of the maximum and minimum of the standing wave voltage/current pattern as follows.

Maxima occur when:

$$\cos\psi = +1, \text{ that is, with } (\theta_\Gamma - 2\beta z) = \pm 2m\pi, \quad \text{where} \quad m = 0, 1, 2, 3, \ldots \tag{3.21a}$$

Giving: $$|V(z)|_{max} = |V^+| (1 + |\Gamma_L|) \tag{3.21b}$$

Minima occur when:

$$\cos\psi = -1, \text{ that is, with } (\theta_\Gamma - 2\beta z) = \pm(2m + 1)\pi, \text{ where } m = 0, 1, 2, 3, \ldots \tag{3.22a}$$

Giving: $$|V(z)|_{min} = |V^+| (1 - |\Gamma_L|) \tag{3.22b}$$

The amplitude of the current maxima and minima are also given by similar expressions as for the voltage. However, since the current reflection coefficient is the negative of the voltage reflection coefficient, the current wave subtracts (destructively interferes) where the voltage wave adds (constructively interferes) and vice versa. Further, it follows from Eqs 3.21a and 3.22a that the difference between any two successive maxima or minima (voltage or current) is one half wavelength (π radians as is expected from physical intuition).

The ratio of the maximum of the standing wave pattern to the minimum is defined as the *standing-wave ratio*, designated by ρ which may also be expressed in terms of the reflection coefficient:

$$\rho = \frac{|V(z)|_{\max}}{|V(z)|_{\min}} = \frac{1+|\Gamma_L|}{1-|\Gamma_L|} \tag{3.23}$$

The standing wave ratio for a *pure travelling wave* or one with a matched termination (when $\Gamma = 0$) is 1 while that for a *pure standing wave* or for one with a perfect mismatch (when $\Gamma = 1$) it is ∞. In other words, the bounds of the standing wave ratio is given by $1 \le \text{SWR} \le \infty$. Because of the wide range of the value of ρ, it is customary to express it on a logarithmic scale, $20 \log_{10}(\rho)$ (in dB). Note that a high value of ρ on a line is undesirable as it results in large power losses.

Next, we can determine the reflection coefficient and impedance at a voltage maximum/minimum of the standing wave pattern at any location on the line.

Using Eq. 3.17, we can write Eq. 3.19a as:

$$|V(z)| = |V^+|\,|1+\Gamma(z)| \tag{3.24}$$

If we compare this with Eqs 3.21b and 3.22b, we get:

$$\Gamma(z) = |\Gamma_L|, \quad \text{at a voltage maximum}$$

$$\Gamma(z) = -|\Gamma_L|, \quad \text{at a voltage minimum} \tag{3.25}$$

Thus, the impedances at voltage maximum and minimum can be obtained from 3.14b as:

$$Z_{\text{in}} = Z_0 \frac{1+|\Gamma_L|}{1-|\Gamma_L|} = Z_0\,\rho, \quad \text{at a voltage maximum} \tag{3.26a}$$

and $$Z_{\text{in}} = Z_0 \frac{1-|\Gamma_L|}{1+|\Gamma_L|} = Z_0/\rho, \quad \text{at a voltage minimum} \tag{3.26b}$$

The standing wave pattern for a generalised load is shown in Fig. 3.4. It may be noted that, in general, $V_{\min}$ is non-zero and $V_{\max}$ is less than $2\,|V^+|$.

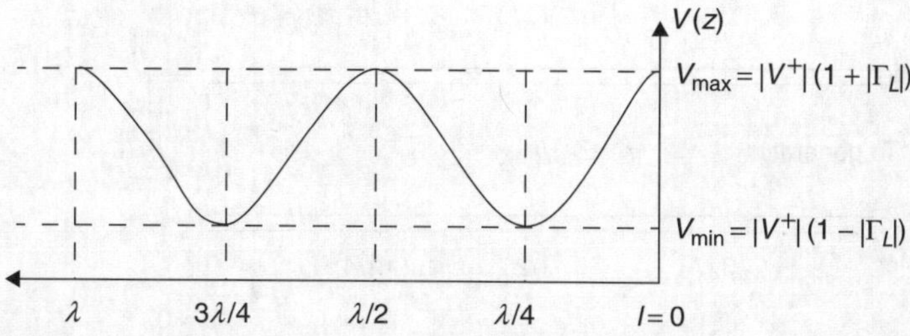

Fig. 3.4 Standing wave pattern for a generalised load.

However, for extreme cases when $Z_L = 0$ or ∞ or the termination is a reactive load, that is, $Z_L = \pm jX$, we find that $\Gamma_L = 1$, and when $V_{min} = 0$, and $V_{max} = 2|V^+|$. The standing wave pattern for different load terminations is given in Fig. 3.5.

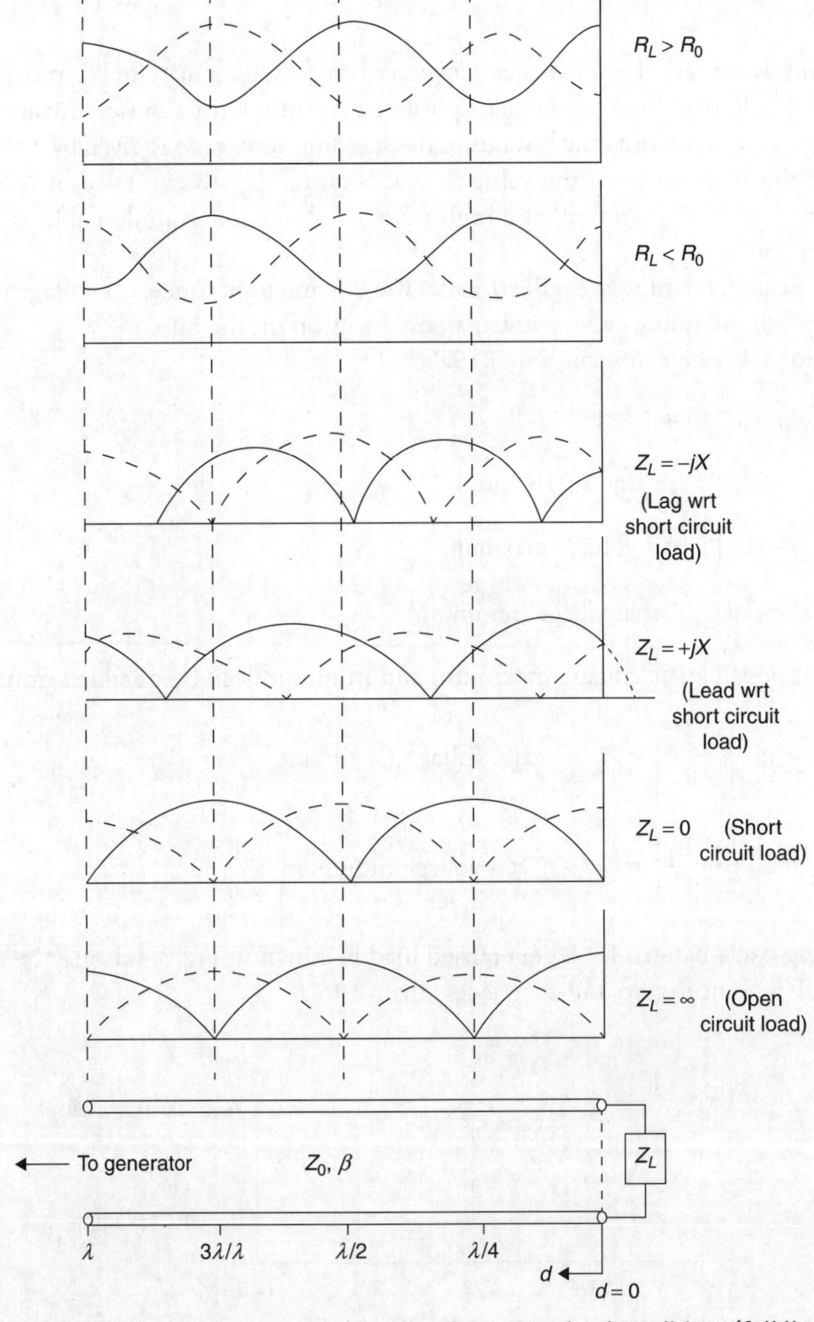

Fig. 3.5 Standing wave pattern on transmission line for various load conditions (full lines for voltage and dashed lines for current).

When we terminate the load end with a pure resistive load we have $Z_L = R_L$ and $Z_0 = R_0$ and using Eqs 3.15a, 3.23 and 3.25, we can evaluate the nature of the standing wave and the value of the standing wave ratio. When $R_L > R_0$, $|\Gamma_L|$ is positive (from Eq. 3.15a) and thus a voltage maximum (or a current minimum) occurs at the load end (Eq. 3.25) and we have the SWR $= R_L/R_0$ (from Eqs 3.23 and 3.25a). But when $R_L < R_0$, $|\Gamma_L|$ is negative and then a voltage minimum (or current maximum) occurs at the load plane and the SWR is R_0/R_L. For pure reactive termination $\Gamma_L = (jx - 1)/(jx + 1)$, where $x = X/Z_0$ is the normalised load reactance. For example, if $x = +1$, $\Gamma_L = 1\angle 90°$, while for $x = -1$, $\Gamma_L = 1\angle -90°$. Thus, for a pure reactive load $|\Gamma_L| = 1$ and is independent of x. Thus, the SWR is infinite. The physical interpretation is that in the steady state, a pure reactance cannot absorb power and hence all the incident power must be returned or reflected back. This same argument applies to the short and open-circuited load conditions. Consequently, $|\Gamma_L| < 1$, only when Z_L has a resistive component.

It may be observed from Eq. 3.15a that when $Z_L = Z_0$, the reflection coefficient $|\Gamma_L| = 0$. In this case, all the power in the incident wave is transmitted to the load and none of it is reflected back to the generator, the load is then said to be *matched* to the transmission line. In other words, a matched generator (whose impedance $Z_g = Z_0$) is said to be delivering power to an infinite transmission line (from whose load end no signal can ever return back). Again, when $Z_L = 0$ (short-circuit), $Z_L = \infty$ (open-circuit) or even when $Z_L = \pm jX$, the magnitude of the reflection coefficient is 1. In general, $|\Gamma_L| \leq 1$. When $Z_L \neq Z_0$, the load is said to be *mismatched* to the line and a reflected wave is produced. The power delivered to the load may be obtained by calculating the average Poynting vector as:

$$P_{av} = \frac{1}{2}\,\mathrm{Re}\,\{V(0)\,I(0)^*\} \quad \text{(since the load plane is at } z = 0)$$

$$= \frac{1}{2}\,\mathrm{Re}\left\{(V^+ + V^-)\left(\frac{V^+}{Z_0} - \frac{V^-}{Z_0}\right)^*\right\} \quad \text{(Eqs 3.13a, 3.13c)}$$

$$= \frac{|V^+|^2}{2Z_0}\,\mathrm{Re}\,\{(1+\Gamma_L)(1-\Gamma_L)^*\}$$

$$P_{av} = \frac{|V^+|^2}{2Z_0}(1 - |\Gamma_L|^2) \tag{3.27}$$

It is physically obvious because the average power transmitted to the load (P_{tr}) is equal to the incident power P_{in} $(= |V^+|^2/2Z_0)$ minus the reflected power P_{refl} $(= |V^+|^2|\Gamma_L|^2/2\,Z_0)$. It can be shown that $T^2 = (Z_L/Z_0)(1 - |\Gamma_L|^2)$. If $\Gamma = 0$, maximum power is delivered to the load, while no power is delivered for $\Gamma = 1$.

When the load is mismatched, then, not all of the available power from the matched generator is transmitted (delivered) to the load, a portion is reflected back. An estimate of this loss is given by *return loss* which is defined as:

$$L_{return} = -10\log_{10}\left(\frac{P_{ref}}{P_{in}} = |\Gamma_L|^2\right) = -20\log_{10}|\Gamma_L| = 20\log_{10}\left(\frac{1}{|\Gamma_L|}\right) \tag{3.28}$$

From this definition, it is clear that the return loss is nothing but the number of decibels by which the reflected power is below the level of the incident power at the load end. A perfectly matched load ($\Gamma_L = 0$) exhibits a return loss of ∞ dB (no reflected power), whereas a perfect mismatch or total reflection ($\Gamma_L = 1$) has a return loss of 0 dB (all incident power is reflected). Thus, for a practically good match condition we need a very high value for the return loss (may be 30–50 dB).

In addition to the term return loss which takes care of the power being reflected from the mismatched load, another term of practical importance for the transmission line having $Z_L \neq Z_0$ is *reflection loss* or the *mismatch loss* which gives an estimate of the power being absorbed/dissipated in the load (due to the reflection caused by the mismatch of the load to the characteristic impedance of the line). The reflection loss is given by:

$$L_{\text{reflection}} = -10 \log_{10} \left\{ \frac{P_{tr}}{P_{\text{in}}} = \frac{(P_{\text{in}} - P_{\text{ref}})}{P_{\text{in}}} = (1 - |\Gamma_L|^2) \right\} = 10 \log_{10} \frac{1}{(1 - |\Gamma_L|^2)} \quad (3.29)$$

EXAMPLE 3.1

A transmission line of characteristic impedance 50 Ω is terminated by a load impedance of (100 – *j*50) Ω and is fed by a matched generator. The measured voltage amplitude at the load terminal is 100 volts. Find: (i) The reflection coefficient and VSWR on the line. (ii) Location of all voltage maxima and minima of the standing wave pattern, starting from the load plane, and the impedance at these voltage maxima and minima. (iii) The magnitude of the maximum and minimum voltages on the line, power transmitted to the load and also the return and reflection losses.

Solution

(i) The reflection coefficient at the load is:

$$\Gamma_L = \frac{Z_L - Z_0}{Z_L + Z_0} = \frac{50 - j50}{150 - j50} = \frac{1 - j1}{3 - j1} = \frac{2 - j1}{5} = \frac{1}{\sqrt{5}} e^{-0.1475\pi}$$

The standing wave ratio is then:

$$\rho = \frac{1 + |\Gamma_L|}{1 - |\Gamma_L|} = \frac{1 + \frac{1}{\sqrt{5}}}{1 - \frac{1}{\sqrt{5}}} = \frac{\sqrt{5} + 1}{\sqrt{5} - 1} = 2.618$$

(ii) Following Eqs 3.21a and 3.22a, the positions of the first voltage maximum and minimum from the load plane and the successive ones are given by:

$$z_{\max 1} = \frac{\lambda}{4\pi}(\theta_\Gamma + 2\pi) \quad \text{and} \quad z_{\min 1} = \frac{\lambda}{4\pi}(\theta_\Gamma + \pi), \text{ with } \theta_\Gamma = -0.1475\pi$$

Thus $z_{\max 1} = 0.463\lambda$, with successive maxima at 0.963λ, 1.463λ, 1.963λ … while $z_{\min 1} = 0.213\lambda$, with successive minima at 0.713λ, 1.213λ, 1.713λ…

The impedance at voltage maxima and minima are give from Eqs 3.26a and 3.26b as:

$$Z_{max} = Z_0\rho = 50 \times 2.618 = 130.9\ \Omega \quad \text{and} \quad Z_{min} = \frac{Z_0}{\rho} = \frac{50}{2.618} = 19.09\ \Omega$$

(iii) To calculate the maximum and minimum voltages on the line from Eqs 3.21b and 3.22b, we first need to calculate the forward wave amplitude in terms of total voltage amplitude at the load.

From Eq. 3.18, putting $z = 0$, we have $V_L = V^+(1 + \Gamma_L)$, when:

$$V^+ = \frac{V_L}{1+\Gamma_L} = \frac{100}{\{1+(2-j1)/5\}} = 10\,(7 + j1)$$

The magnitude of forward voltage is then:

$$|V^+| = |10\,(7 + j1)| = 10\sqrt{50} = 70.71\text{ V}$$

Thus, the maximum and minimum voltages on the line are:

$$V_{max} = |V^+|\left(1+\frac{1}{\sqrt{5}}\right) = 70.71(1 + 0.4472) = 102.33\text{ V}$$

$$V_{min} = |V^+|\left(1-\frac{1}{\sqrt{5}}\right) = 70.71(1 - 0.4472) = 39.09\text{ V}$$

The power transmitted to the load can be calculated from Eq. 3.27 as:

$$P_L = \frac{|V^+|}{2Z_0}(1-|\Gamma_L|^2) = \frac{(70.71)^2}{2\times 50}\left\{1-\left(\frac{1}{\sqrt{5}}\right)^2\right\} = 50 \times 0.8 = 40\text{ watts}$$

This is about 80% of the power in the incident wave:

$P_{in} = \dfrac{|V^+|^2}{2Z_0}$ (= 49.99 W) and thus 20% is reflected from the load.

The return loss is given from Eq. 3.28 as:

$$L_{return} = 20\log_{10}\left(\frac{1}{|\Gamma_L|}\right) = 20\log_{10}(\sqrt{5}) = 6.98\text{ dB}$$

while the reflection loss from Eqn. 3.29 is given as:

$$L_{reflection} = 10\log_{10}\frac{1}{(1-|\Gamma_L|^2)} = 10\log_{10}\frac{1}{(1-0.2)} = 0.96\text{ dB}$$

3.2.4 Transmission Line as HF Circuit Elements

At higher RF and microwave frequencies, conventional low frequency lumped circuit elements cannot be used. For example, a coil inductor would start to behave as a resonator with inter-turn capacitance becoming very significant at such high frequency. Though microwave lumped circuit elements are available up to 20 GHz, and with limited values even up to 60 GHz (as will be discussed in chapter 6), for most applications, transmission line type distributed RF circuit elements are very common at higher RF and microwave frequencies. Sections of transmission lines with proper termination at the load end can be designed to give inductive or capacitive impedance and are also used to match an arbitrary load to the internal impedance of a generator for maximum power transfer. The required length of such lines as circuit elements becomes practical in the ultra high frequency (UHF) range. At frequencies much lower than 300 MHz, the required lines tend to be too long, whereas at frequencies higher than 3 GHz the physical dimensions become inconveniently small, and it would be advantageous to use waveguide components.

The input impedance of a loss-less line (for which $Z_0 = R_0$) of length l follows from Eq. 3.17b as:

$$Z(z) = R_0 \frac{Z_L + jR_0 \tan\beta z}{R_0 + jZ_L \tan\beta z} \tag{3.30}$$

Let us now consider some important cases.

Case 1: *Open-circuit ($Z_L \to \infty$) termination*: For open-circuit termination, it follows from Eq. 3.30 that:

$$Z_{\text{inoc}} = jX_{\text{inoc}} = -\frac{jR_0}{\tan\beta l} = -jR_0 \cot\beta l \tag{3.31a}$$

Equation 3.31a shows that the input impedance of an open-circuited loss-less line is purely reactive. This impedance obtained with the line section can be either capacitive or inductive because the function cot βl can be either positive or negative, depending on the value of βd $(= 2\pi d/\lambda)$. Figure 3.6a shows the plot of $X_{\text{inoc}} = -jR_0 \cot\beta t$ vs l. It may be observed that X_{inoc} can assume all values from $-\infty$ to $+\infty$.

When the length of an open-circuited line is very short in comparison with a wavelength, $\beta l << 1$, we can obtain a very simple formula for its capacitive reactance by noting that tan $\beta l \approx \beta l$, and from Eq. 3.31a we get:

$$Z_{\text{inoc}} = jX_{\text{inoc}} \cong -\frac{jR_0}{\beta l} = -j\frac{\sqrt{L/C}}{(\omega\sqrt{LC})\, l} = -j\frac{1}{\omega Cl} \tag{3.31b}$$

which is the impedance of a capacitance of Cl farads.

Case 2: *Short circuit ($Z_L \to 0$) termination*: For short-circuit termination it follows from Eq. 3.30 that:

$$Z_{\text{insc}} = jX_{\text{insc}} = j\, R_0 \tan\beta l \tag{3.32a}$$

Since tan βl can range from $-\infty$ to $+\infty$, the input impedance of a short-circuited loss-less line can also be either purely inductive or capacitive, depending on the value of βl. Figure 3.6b shows the plot of X_{insc} vs l.

Comparing Figs 3.6a and 3.6b, we observe that in the range where X_{inoc} is capacitive X_{insc} is inductive, and vice versa. The input reactance of open-circuited and short-circuited loss-less transmission lines are the same if their lengths differ by an odd multiple of $\lambda/4$.

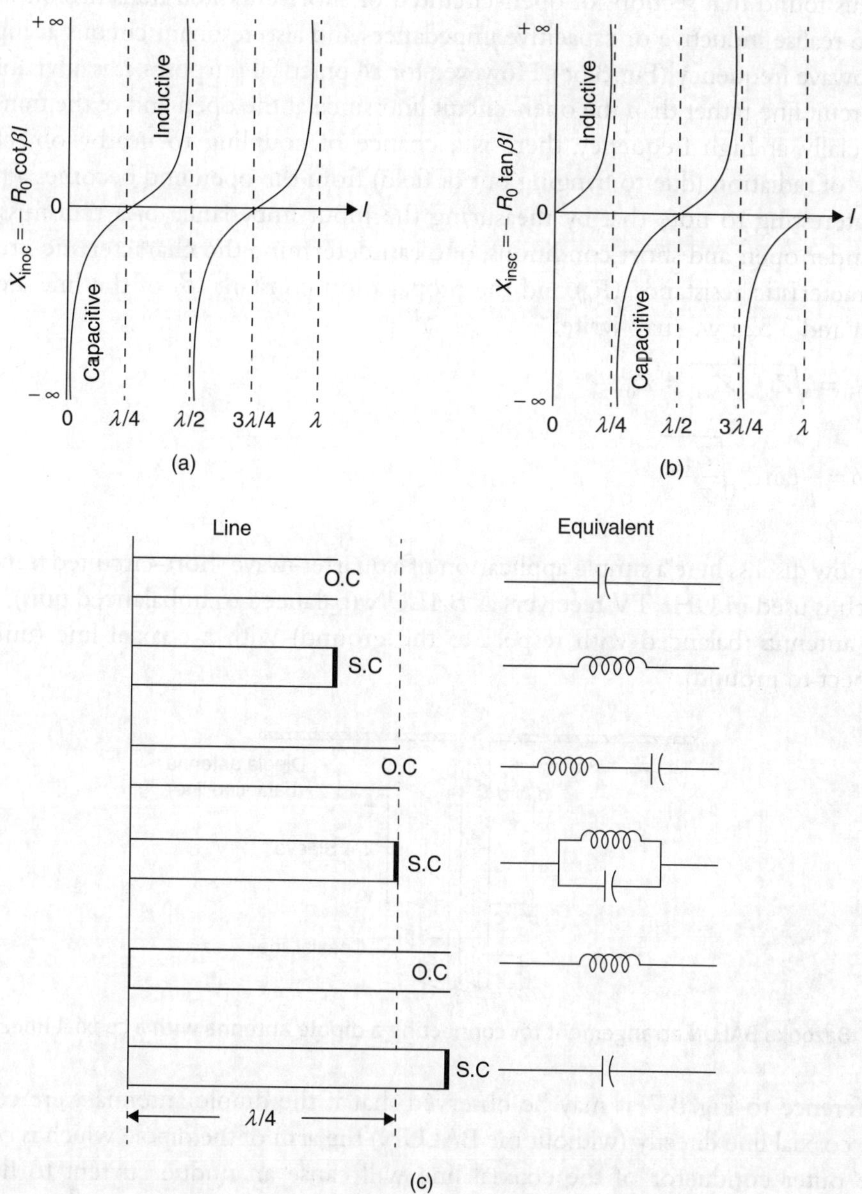

Fig. 3.6 (a) Input reactance of open-circuited transmission line vs length of the line (b) Input reactance of short-circuited transmission line vs length of the line (c) Transmission line sections and their LC equivalents.

When the length of a short-circuited line is very short in comparison with a wavelength, $\beta l << 1$, Eq. 3.32a becomes approximately:

$$Z_{\text{insc}} = jX_{\text{insc}} \cong jR_0\, \beta l = j\sqrt{\frac{L}{C}}\,\omega\sqrt{LC}\,l = j\omega Ll \tag{3.32b}$$

This is the impedance of an inductance of Ll henries.

It is thus found that sections of open-circuited or short-circuited transmission lines may be used to realise inductive or capacitive impedances and also resonant circuits at higher RF and microwave frequency (Fig. 3.6c). However, for all practical purposes, it is advisable to use a short-circuit line rather than the open-circuit line; since at the open end of the transmission line, especially at high frequency, there is a chance of coupling to nearby objects and a possibility of radiation (due to fringing out of field) from the open end becomes a problem.

It is interesting to note that by measuring the input impedance of a transmission line section under open and short conditions, one can determine the characteristic impedance (Z_0)/characteristic resistance (R_0) and the propagation constant (β) of the line. Following Eqs 3.31a and 3.32a we may write:

$$R_0 = \sqrt{Z_{\text{inoc}}\, Z_{\text{insc}}} = Z_0 \tag{3.33a}$$

and $$\beta = \frac{1}{l}\tan^{-1}\sqrt{\frac{Z_{\text{insc}}}{Z_{\text{inoc}}}} \tag{3.33b}$$

We may now discuss here a simple application of a quarter-wave short-circuited transmission line which is used in UHF TV receivers as *BALUN* (balanced to unbalanced unit) to match a dipole antenna (balanced with respect to the ground) with a coaxial line (unbalanced with respect to ground).

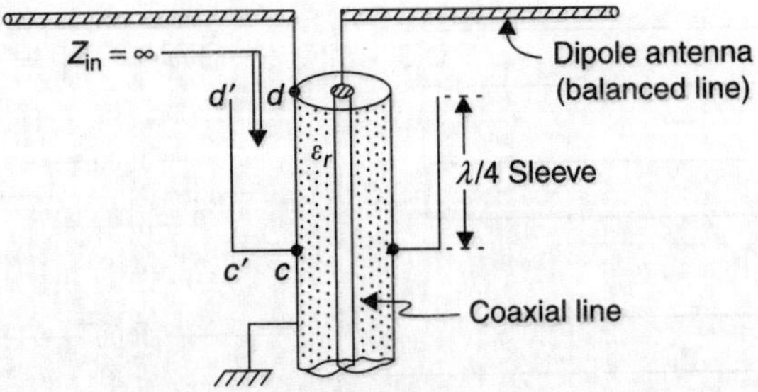

Fig. 3.7 **Bazooka BALUN arrangement for connecting a dipole antenna with a coaxial line.**

With reference to Fig. 3.7, it may be observed that if the dipole antenna were connected with the coaxial line directly (without the BALUN) the arm of the dipole which is connected with the outer conductor of the coaxial line will cause an undue current to flow from the antenna through the coaxial line to the ground (due to the potential difference existing between the two). But with the BALUN [which is just a short-circuited quarter-wave ($\lambda/4$) transmission line sleeve on the outer conductor of the coaxial line] connected between the dipole antenna and the coaxial line, this problem gets solved. The quarter-wave sleeve transmission line is seen to be short-circuited across cc', thus the impedance seen across dd' ($\lambda/4$ away from cc' line) looking down into the transmission line formed of the sleeve and the outer conductor of the coaxial line is now infinite (and hence no longer has zero impedance to ground at c as was the case without the sleeve). Thus, one of the arms of the dipole antenna can now safely be connected to it without the fear of being short-circuited to the ground, the other arm of the dipole antenna being connected to the inner conductor of the coaxial line. As the presence of BALUN prevents undue current flow from the dipole antenna to the ground via the coaxial outer conductor, it is known as choke type balun; the structure also resembles a bazooka, the weapon used to fire rockets against military tanks, and is thus also known as *bazooka balun*. The bazooka arrangement (sleeve arrangement around the outer conductor of the coaxial cable) allows a standing wave formation within the sleeve, thus allowing a minimal amount of undue radiation that would otherwise have taken place if one arm of the dipole antenna were connected directly at quarter-wavelength distance to the outer conductor of coaxial line to ground.

Case 3: *Quarter-wave ($l = \lambda/4$) line section*: When the length of the line is an odd multiple of $\lambda/4$, $l = (2n - 1)\ \lambda/4$, ($n = 1, 2, 3, \ldots$), we have:

$$\beta l = \frac{2\pi}{\lambda}(2n-1)\frac{\lambda}{4} = (2n-1)\frac{\pi}{2}, \text{ with } \tan\beta l = \tan\{(2n-1)\ \pi/2\} \rightarrow \pm\infty$$

and thus Eq. 3.30 becomes:

$$Z_{\text{in}} = \frac{R_0^2}{Z_L} \tag{3.34}$$

Hence a quarter-wave loss-less line transforms the load impedance to the impedance at its input terminals when the square of its characteristic resistance is multiplied by the inverse of the load impedance. In other words, it acts as an impedance inverter and is often referred to as a *quarter-wave impedance transformer.* Thus, an open-circuited quarter-wave line would appear as a short-circuit while a short-circuited quarter-wave line would appear as an open-circuit at its input terminals.

Case 4: *Half-wave ($l = \lambda/2$) line section*: When the length of a line is an integral multiple of $\lambda/2$, $l = n\ \lambda/2$, ($n = 1, 2, 3, \ldots$), we have:

$$\beta l = \frac{2\pi}{\lambda}\left(\frac{n\lambda}{2}\right) = n\pi, \text{ with } \tan\beta l = 0,$$

and Eq. 3.30 reduces to:

$$Z_{\text{in}} = Z_L \tag{3.35}$$

This states that a half-wave loss-less line transfers the load impedance to the input terminals without change, thus acting as an *one to one impedance transformer.*

Thus, when we desire to use a length of transmission line to interpose between two ports of a microwave network, the length of the line should be a multiple of half-wavelength. The short and open-circuited transmission lines of quarter or half wave lengths are also very much in use as 'stubs' for impedance matching.

EXAMPLE 3.2

The short-circuit and open-circuit impedances measured at the input terminals of a loss-less transmission line of length 10 cm, are $+j154\ \Omega$ and $-j16.3\ \Omega$, respectively. Find (i) Z_0 and β of the line. (ii) Determine the input impedance of a short-circuited line that is twice the given length at the same operating frequency. (iii) If the short-circuited line has to exhibit an open-circuit at the input terminals then what should be its length?

Solution

(i) We have from Eq. 3.33a, $R_0 = \sqrt{Z_{inoc}\,Z_{insc}} = \sqrt{-j16.3\,(j154)} = 50\ \Omega$
and from Eq. 3.33b,

$$\beta = \frac{1}{l}\tan^{-1}\sqrt{\frac{Z_{insc}}{Z_{inoc}}} = \frac{1}{10}\tan^{-1}\sqrt{\frac{j\,154}{-j\,16.3}} = \frac{j}{10}\tan^{-1}(3.07) = j0.1255\ \text{rad/cm}$$

(ii) A short-circuited line twice as long as the previous one will have:

$$\beta l = j0.1255 \times 20 = 2.51\ \text{rad}$$

Thus, the input impedance of this short-circuited line from Eq. 3.32a is:

$$Z_{insc} = jR_0 \tan \beta l = j50\tan(143.8°) = -j36.59\ \Omega$$

It may be noted that the Z_{insc} for 20 cm line is now a capacitive reactance, where as that for the 10 cm line it was an inductive reactance. We may conclude from Fig. 3.6b that 10 cm $< \lambda/4 <$ 20 cm.

(iii) In order for a short-circuit line to exhibit open-circuit at the input terminals, it should be an odd multiple of a quarter-wavelength length:

$$l = \frac{(2n+1)\,\lambda}{4}, \text{ (where } n = 0, 1, 2, 3\ldots) \text{ with } \lambda = \frac{2\pi}{\beta} = \frac{2\pi}{0.1255} = 50\ \text{cm}$$

EXAMPLE 3.3

Show that the microwave circuit shown in Fig. 3.8a is capable of preventing the even harmonic (that is, $2f$) signal from the generator to reach the antenna while the circuit shown in Fig. 3.8b is capable of preventing the odd harmonic (that is, $3f$) signal from the generator to reach the antenna, where f is the frequency of the fundamental signal.

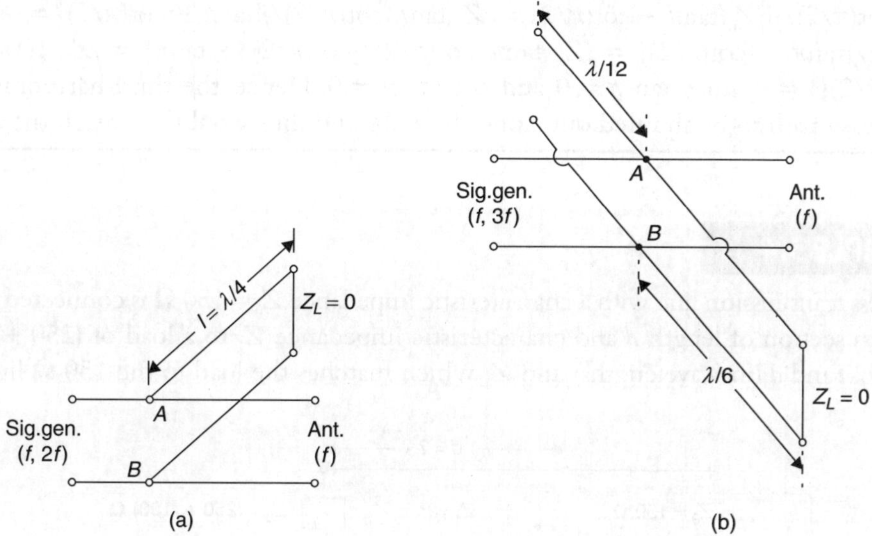

Fig. 3.8 (a) Even harmonic suppression with short-circuited stub (b) Odd harmonic suppression with a combination of open and short-circuited stubs.

Solution

From Fig. 3.8a, at a fundamental frequency f, the quarter-wave short-circuited stub presents infinite impedance looking into the stub connected across AB terminal of the transmission line. Thus, the fundamental frequency cannot be shunted along the stub and as it finds comparatively low impedance (the characteristic impedance) path, looking along the line it will pass through the line unhindered and reach the antenna. While, for the second harmonic frequency ($2f$), the same stub will behave as a half-wave short-circuited line and will present a very low impedance (ideally zero impedance) at its input terminals (AB). Thus, the $2f$ signal will be shunted out along the stub, thereby preventing it from reaching the antenna.

Next, from Fig. 3.8b, the two shunt stubs (one $\lambda/6$ short-circuited and another $\lambda/12$ open-circuited) will have their input impedances $Z_{\text{in(s.c)}}$ and $Z_{\text{in(o.c)}}$ appearing in parallel across the AB terminals and their equivalent impedance Z_{eq} would determine the passage/prevention of f and $3f$ to the antenna. For $\lambda/6$ short-circuited line at the fundamental frequency f, $Z_{\text{in(s.c)}} = jZ_0 \tan\beta l = j\sqrt{3}\, Z_0$; while for $\lambda/12$ open-circuited line at the fundamental frequency f, $Z_{\text{in(o.c)}} = -jZ_0 \cot\beta l = -j\sqrt{3}\, Z_0$. Thus, Z_{eq} across the AB terminals at the fundamental frequency is: $(j\sqrt{3}\, Z_0)(-j\sqrt{3}\, Z_0)/(j\sqrt{3}\, Z_0 - j\sqrt{3}\, Z_0) = \infty$. Hence, the fundamental cannot get a passage through the stubs as equivalent impedance looking into the stubs for fundamental frequency is infinite and it will thus reach the antenna as that would be a lower impedance (characteristic impedance) path for the fundamental signal. But at the third harmonic frequency $3f$, the short-circuited $\lambda/6$ line will behave as a $\lambda/2$ line while the open-circuited $\lambda/12$ line will behave as a $\lambda/4$ line. Now for a $\lambda/2$ short-circuited line (at third harmonic frequency $3f$, where f is the fundamental frequency), $Z_{\text{in(s.c)}} = jZ_0 \tan\beta l = jZ_0 \tan\pi$, while for a $\lambda/4$ open-circuited line (at third harmonic frequency $3f$), $Z_{\text{in(o.c)}} = -jZ_0 \cot\beta l = -jZ_0 \cot(\pi/2)$. Thus, Z_{eq} across the AB terminals at the third harmonic is: $(jZ_0 \tan\pi)$

$(-jZ_0\cot(\pi/2)/)[jZ_0(\tan\pi - \cot(\pi/2)) = Z_0\tan\pi\cot(\pi/2)/j[\tan\pi - \cot(\pi/2)] = -jZ_0\tan\pi\cot(\pi/2)/[\tan\pi - \cot(\pi/2)] = jZ_0\tan\pi\cot(\pi/2)/[\cot(\pi/2) - \tan\pi] = jZ_0/\{(1/\tan\pi) - [1/\cot(\pi/2)]\} = 0$, since $\tan\pi = 0$ and $\cot(\pi/2) = 0$. Hence, the third harmonic would find an easy path to be shunted out along the stubs and thus would not reach the antenna.

EXAMPLE 3.4

A loss-less transmission line with a characteristic impedance $Z_0 = 150\ \Omega$ is connected through a loss-less section of length d and characteristic impedance Z_1 to a load of $(250 + j100)\ \Omega$ (Fig. 3.9). Find d (in wavelengths) and Z_1 which matches the load to the $150\ \Omega$ line.

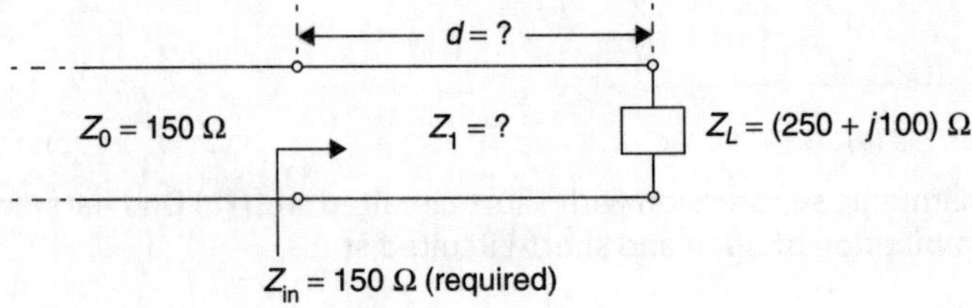

Fig. 3.9 Matching of complex load to a transmission line of given characteristic impedance (Z_0) with a transmission line of length d and characteristic impedance Z_1.

Solution

For the specified problem we can use Eq. 3.17b and can write:

$$150 = Z_0 \frac{Z_L + jZ_0\tan\beta d}{Z_0 + jZ_L\tan\beta d}$$

Putting $Z_L = R + jX$ and simplifying we get:

$$150\,Z_1 - 150\tan\beta d - Z_1 R = 0 \quad \text{and} \quad 150R\tan\beta d - Z_1 X - Z_1^2\tan\beta d = 0$$

Substituting the values of $R = 250\ \Omega$ and $X = 100\ \Omega$, it can be evaluated that $Z_1 = 229\ \Omega$ and $d = 0.342\lambda > \lambda/4$. The extra length of $(0.342 - 0.25)\lambda = 0.092\lambda$ of transmission line is required to take care for the matching of the imaginary part of the load. It will be seen in Chapter 7 that for matching a complex load terminating a transmission line, the principle of stub matching is popularly used.

3.3 Smith Chart and its Applications

Transmission line problems may, in general, be solved by evaluating analytical equations which often involve tedious manipulation of complex numbers. To avoid this, a number of well structured graphical techniques were developed to solve transmission line problems of which the best known and most widely used one is the *Smith chart*, developed by

Philips H Smith of Bell Telephone Laboratories in 1939. Essentially, the chart is the complex impedance plane mapped on to the complex reflection coefficient plane. Thus, the Smith chart is a graphical plot of normalised resistance and reactance functions in the complex reflection coefficient plane, which provides an intuitive understanding of how individual variables in the system affect the end result. Instead of having separate Smith charts for transmission lines with different characteristic impedances, it is preferable to have just one (where all impedances are normalised with respect to the characteristic impedance Z_0) that can be used for any line. In modern microwave engineering, this impedance chart has become indispensable in the solution of transmission line problems and has become an integral part of most of the current computer-aided design software and test equipment. As a matter of fact, most vector network analysers (useful for microwave circuit characterisation), discussed in Section 12.9, have the Smith chart display as one of the primary screen options. The Smith chart has many applications in solving microwave circuit problems including the solution of stub matching problems (Chapter 7), amplifier design (Chapter 9) and impedance measurement at microwave frequency (Chapter 12).

The development of the Smith chart begins with the fundamental equation for reflection coefficient:

$$\Gamma = \frac{Z_L - Z_0}{Z_L + Z_0} \tag{3.36}$$

This may be solved for Z_L, which when normalised with respect to the characteristic impedance Z_0 yields:

$$z_L = \frac{Z_L}{Z_0} = \frac{1+\Gamma}{1-\Gamma} \tag{3.37}$$

Putting $z_L = r + jx$ and $\Gamma = u + jv$, we get from Eq. 3.37:

$$r + jx = \frac{1 + u + jv}{1 - u - jv} \tag{3.38}$$

Rationalising the right hand side of Eq. 3.38 and equating the real and imaginary parts on both sides we have:

$$r = \frac{(1-u)^2 - v^2}{(1-u)^2 + v^2} \tag{3.39a}$$

$$x = \frac{2v}{(1-u)^2 + v^2} \tag{3.39b}$$

These equations can be rearranged so that by completing squares we obtain:

$$\left(u - \frac{r}{1+r}\right)^2 + v^2 = \frac{1}{(1+r)^2} \tag{3.40a}$$

$$(u-1)^2+\left(v-\frac{1}{x}\right)^2=\frac{1}{x^2} \tag{3.40b}$$

These two equations generate r and x circles on the complex $(u - v)$ plane. The r-circles have radii equal to $1/(1 + r)$ and the centre at $[u = r/(1 + r), v = 0]$ while the x-circles have radii equal to $1/|x|$ and the centre at $[u = 1, v = 1/x]$. Since x is the normalised reactance, x can have either +ve or −ve values. Figures 3.10a and 3.11a respectively show tables for several values of r and x while Figs 3.10b and 3.11b show corresponding r and x circles in the complex $(u - v)$ plane, that is, the Γ-plane.

r	Centre at		Radius
	u	v	
0	0	0	1
1/2	1/3	0	2/3
1	1/2	0	1/2
2	2/3	0	1/3
∞	1	0	0

(a)

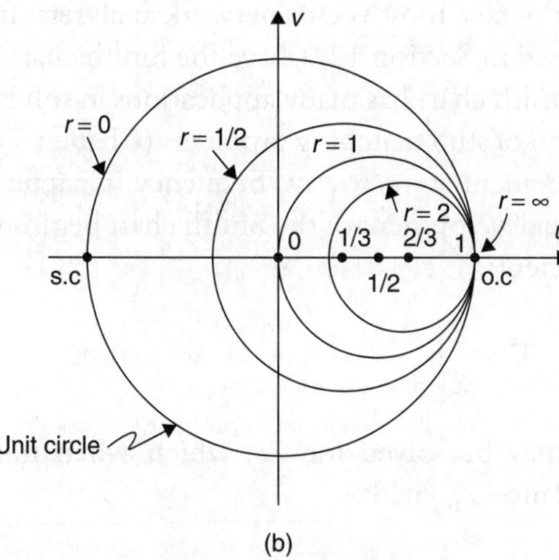

(b)

Fig. 3.10 Mapping of constant r-circles in the complex Γ-plane (a) Centre and radius values for different values of r (b) r-circles in complex $(u - v)$, that is, Γ-plane.

x	Centre at		Radius
	u	v	
0	1	∞	∞
0.5	1	2	2
1	1	1	1
2	1	1/2	1/2
∞	1	0	0

(a)

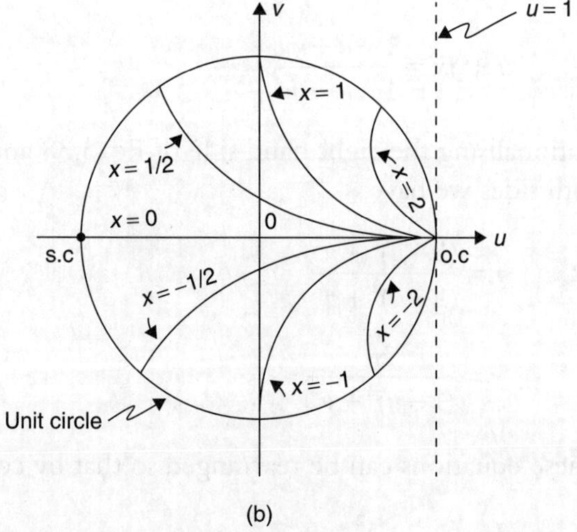

(b)

Fig. 3.11 Mapping of constant x-circles in the complex Γ-plane (a) Centre and radius values for different values of x (b) x-circles in complex $(u - v)$, that is, Γ-plane.

Let us now make a few observations.

1. The centre of all r-circles lie on the u-axis, the $r = 0$ circle (having unity radius and centred at the origin) is the largest. The r-circles become progressively smaller as r increases from 0 (short-circuit) towards ∞, ending at the ($u = 1$ and $v = 0$) point for open-circuit (o.c). All r-circles pass through this point marked as o.c.
2. The centres of all x-circles lie on the $u = 1$, line; those for $x > 0$ (inductive reactance) lie above the u-axis, and those for $x < 0$ (capacitive reactance) lie below the u-axis, the $x = 0$ circle is along the u-axis. The x-circles become progressively smaller as $|x|$ increases from 0 towards ∞, ending at the ($u = 1$ and $v = 0$) point for open-circuit, and all x-circles pass through this o.c point.
3. $r = 0$ means pure reactive load (when $z_L = \pm jx$), that is, complete reflection ($\Gamma = 1$). Thus the Smith chart is bounded by the $r = 0$ circle.
4. The point $r = 1, x = 0$ (when $Z_L = Z_0$, that is, perfectly matched condition) corresponds to $u = 0$ and $v = 0$ (the origin of the Γ plane, $|\Gamma| = 0$).

Essentially, a Smith chart is thus a chart of r and x-circles in the complex Γ-plane for $|\Gamma| \le 1$ as shown in Fig. 3.12. The r and x-circles are orthogonal to each other everywhere.

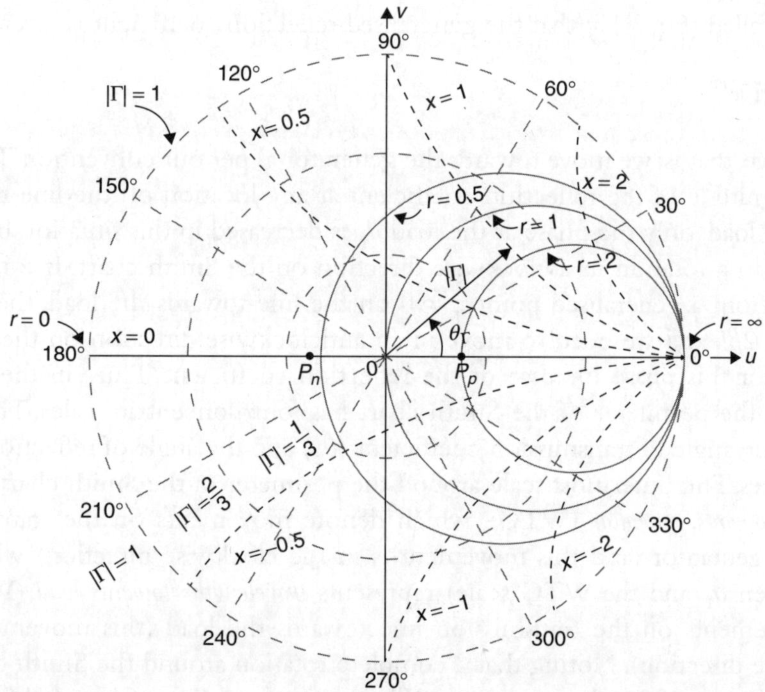

Fig. 3.12 **Smith chart in polar coordinates with r and x-circles shown together in the Γ-plane.**

The intersection of an r-circle and an x-circle defines a point P that represents a normalised load impedance $z_L = r \pm jx$, which when multiplied by the characteristic impedance of the line gives the actual load impedance. For all practical purposes, a Smith chart is marked with polar coordinates such that every point in the Γ-plane is specified by a magnitude $|\Gamma|$ and a phase angle θ_Γ. If we draw a circle centred at the origin through the load point P (which is a constant Γ-circle), the fractional distance (compared to unity radius to the edge of the chart) from the centre to the point is equal to the magnitude of reflection coefficient ($|\Gamma|$) and the angle that the line to the point makes with the real axis is the phase of the reflection coefficient (θ_Γ).

Each $|\Gamma|$-circle intersects the real axis at two points, the P_p on the positive real axis and P_n on the negative real axis (Fig. 3.12). Since $x = 0$ along the real axis, P_p and P_n both represent situations with a purely resistive load, $Z_L = R_L$. Thus, we observe that, $R_L > R_0$ at P_p where $r > 1$ and $R_L < R_0$ at P_n where $r < 1$. We know that for a purely resistive load, when $R_L > R_0$, the VSWR (ρ) = R_L/R_0, which also happens to be equal to r. This implies that the value of the r-circle passing through the point P_p is numerically equal to the standing-wave ratio ρ. Similarly, the value of the r-circle passing through the point P_n on the negative real axis is numerically equal to $1/\rho$.

So far, we have discussed the construction of the Smith chart based on defining the reflection coefficient at the load ($z = 0$). However, it may be required (actually it is required in many transmission line problems) to move either towards the generator or towards the load from a generalised point $z \neq 0$, say at $z = l$ on the transmission line. In such situations it may be recalled (Eq. 3.17) that the generalised reflection coefficient is given by:

$$\Gamma_l = |\Gamma| e^{j(\theta - 2\beta l)}$$

where z is +ve, that is, we move towards the generator as per our convention. This indicates that the magnitude of the reflection coefficient at any location on the line remains same as that at the load, only the phase at the load θ_Γ is decreased in this situation by $2\beta l$, which corresponds to a rotation in a clockwise direction on the Smith chart. If z is –ve, that is, if we move from a generalised point $z \neq 0$ on the line towards the load, then θ_Γ will be increased by $2\beta l$, and we need to move in an anticlockwise direction on the Smith chart to account for this phase increase of the reflection coefficient. Thus, in the commercial Smith chart, the perimeter of the Smith chart has four concentric scales. The two inner scales label the angle of transmission coefficient (θ_T) and the angle of reflection coefficient (θ_Γ) in degrees. The outermost scale around the perimeter of the Smith chart is called the *wavelengths towards generator* (WTG), which denote movements on the transmission line towards the generator (and this movement is in the clockwise direction) while a fourth scale (between θ_Γ and the WTG scale) represents *wavelengths towards load* (WTL), which denote movements on the transmission line towards the load (this movement is in the anticlockwise direction). Noting that a complete rotation around the Smith chart is equal to a phase change of 2π, then equating $2\beta l$ with 2π for a line of length l for which the round-trip phase change is 2π we can find that $l/\lambda = 0.5$. Thus, the WTG and WTL scales are expressed in terms of the length of transmission line (l) normalised with respect to wavelength (λ) and hence the scales start from 0 and ends at 0.5 for a complete phase rotation of 2π (Fig. 3.13).

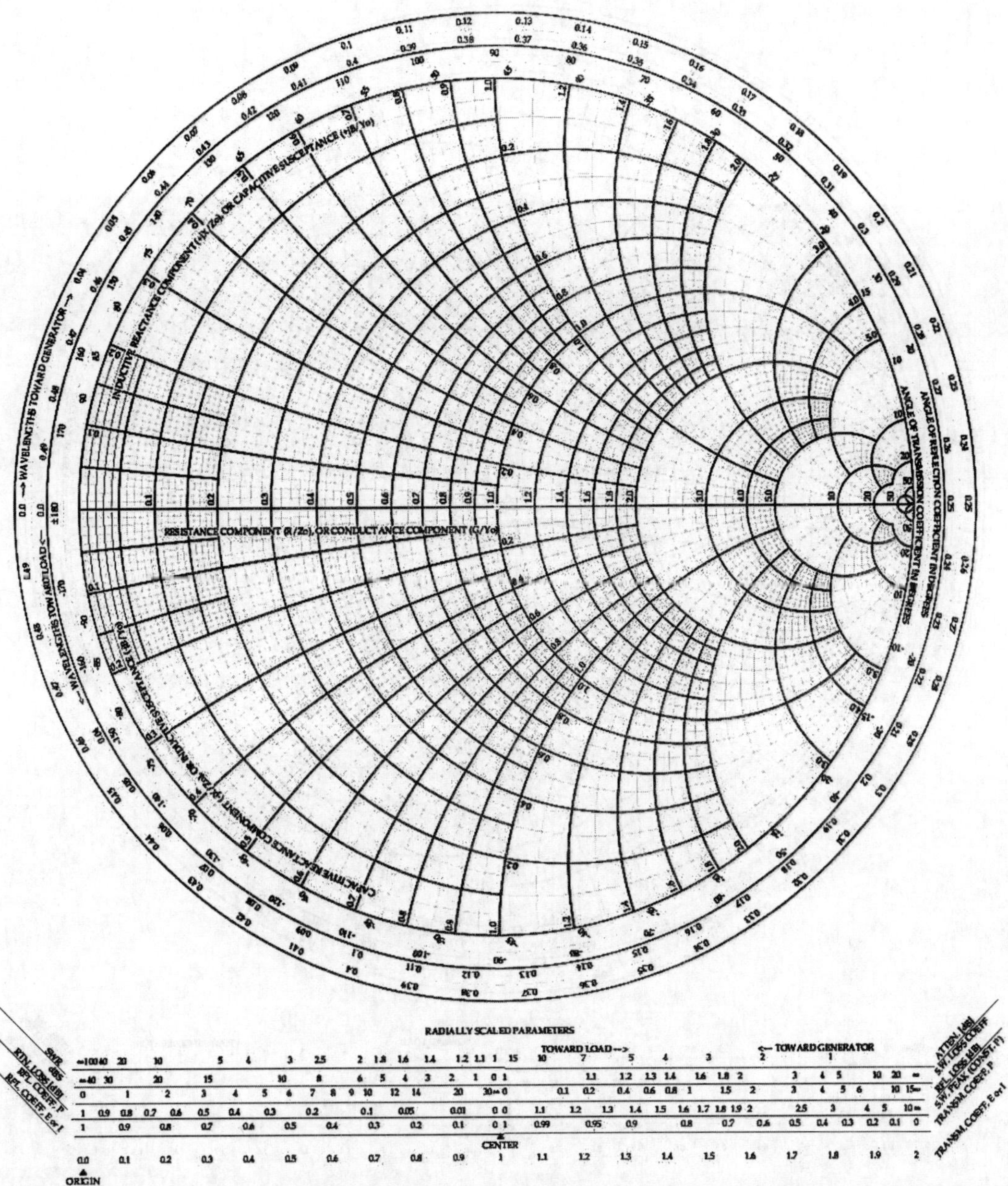

Fig. 3.13 A typical Smith chart (impedance chart) showing wavelength towards generator (WTG), wavelength towards load (WTL) and radially scaled parameters.

In solving certain type of transmission line problems (for example in stub matching), it is often more convenient to work with admittances rather than with impedances. The Smith chart can be used for normalised admittance ($y_L = g \pm jb$) in the same way that it is used for normalised impedance ($z_L = r \pm jx$), and can also be used to convert between

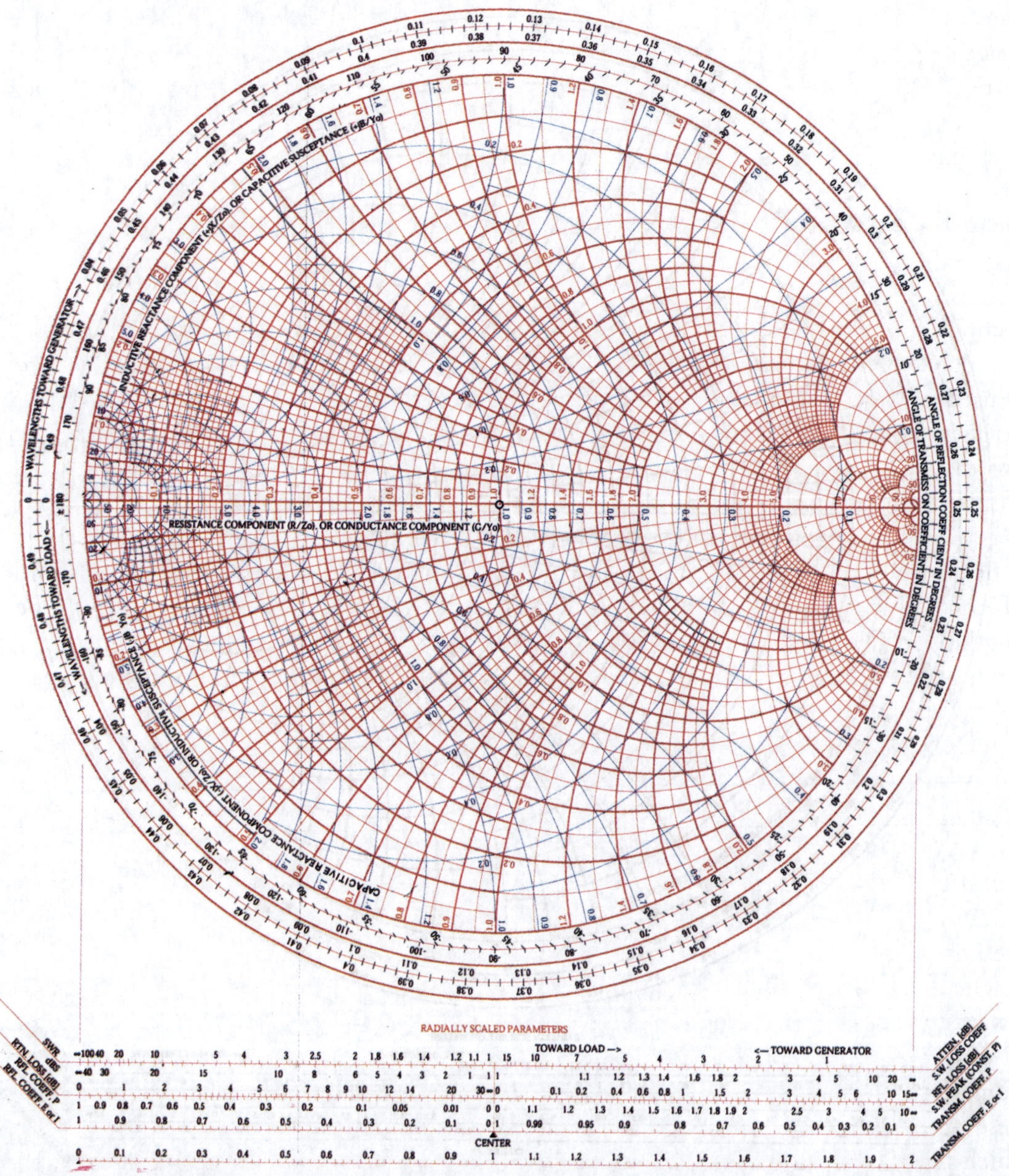

Fig. 3.14 Smith and Jones chart (that is, impedance-admittance Smith chart in dual grid: *Z-Y* format).

impedance and admittance. It may be observed that, in the normalised form, the input impedance of a load z_L connected to a $\lambda/4$ line is from Eq. 3.34 is $z_{\text{in}} = 1/z_L$, which has the effect of converting a normalised impedance to normalised admittance. Since a complete revolution around the Smith chart corresponds to a length of $\lambda/2$, a $\lambda/4$

transformation is equivalent to rotating the chart by 180°, this effectively means we can image a given impedance (or admittance) point across the centre of the chart to obtain the corresponding admittance (or impedance) point. Thus, the Smith chart can be used either as an impedance or admittance chart. As an impedance chart, the Smith chart consists of r_L and x_L circles, representing the normalised resistance and reactance of a normalised load impedance z_L. When used as an admittance chart, the r_L circles become g_L circles and x_L circles become the b_L circles, where g_L and b_L are the normalised conductance and susceptance of the normalised load admittance y_L respectively. However, in the solution of some typical transmission line problem (impedance matching with reactive elements, Section 7.5), it may be necessary at different stages of solving to use the impedance and admittance charts simultaneously; in such cases an *impedance–admittance Smith chart* needs to be used where the regular impedance chart is superimposed with another Smith chart that has been rotated 180° (which represents the admittance chart) and two scales have normally two different colours, the standard impedance chart is in red and the admittance chart is in blue (Fig. 3.14). This dual grid format (Z–Y) chart is also referred to as the *Smith and Jones chart*.

So far, we have discussed the use of the Smith chart for loss-less transmission lines in which the magnitude of the reflection coefficient term does not change with the line length and that we can find input impedance of the line from the load impedance, and vice versa, simply by moving along the $|\Gamma|$ circle by an angle equal to $2\beta l$. But on a lossy line, the reflection coefficient at any point on the line will be given by $\Gamma_l = |\Gamma|\, e^{-2\alpha l + j(\alpha - 2\beta l)}$, in which the loss term $2\alpha l$ is now non-negligible. Hence to find z_{in} from z_L, we cannot simply move along the $|\Gamma|$ circle; auxiliary calculations are necessary to account for the $e^{-2\alpha l}$ factor. The implication of this loss factor is that as we now move from the load towards the generator, the magnitude of the reflection coefficient decreases continuously and hence we now need to move from the load point to the generator point in a spiralling path directed towards the centre. Many practical charts have convenient scales (such as the transmission coefficient scale in the radially scaled parameters of the Smith chart shown in Fig. 3.13) attached to them, so that the amount of spiralling is readily obtained by moving either towards the generator or towards the load depending on the situation.

When working on the design of oscillators/amplifiers using negative resistance device, the magnitude of the reflection coefficient happens to be more than unity, $|\Gamma| > 1$. Under these circumstances, we cannot use the usual form of the Smith chart discussed so far which is used for networks where $\mathrm{Re}\,(Z) \geq 0$ which corresponds to mapping of only the positive half of the impedance plane ($z = r \pm jx$) into a circle in the reflection coefficient plane with radius $|\Gamma| \leq 1$ (as shown in Fig. 3.13), in which r takes the value from 0 (s.c) to ∞ (o.c). But to cater for negative resistance values, we need a more general form/open form of the Smith chart known as the *compressed Smith chart* which takes care of both positive and negative values of the real part of impedance ($z = \pm r \pm jx$) (Fig. 3.15). The compressed Smith chart is obtained when the entire impedance plane is mapped on a one-to-one basis onto the reflection coefficient plane. This mapping of impedance with both positive and negative real parts in the Γ plane helps to include the region of $|\Gamma| > 1$. In the open form, the compressed chart is of an infinite extent, however, for all practical purposes it is a standard practice to bind the chart by imposing a limit on the maximum value of reflection coefficient that will exist in the design. A value of 3.16 is usually selected for the reflection coefficient as this allows a significant portion of the negative resistance contours that is generally encountered in practical design problems.

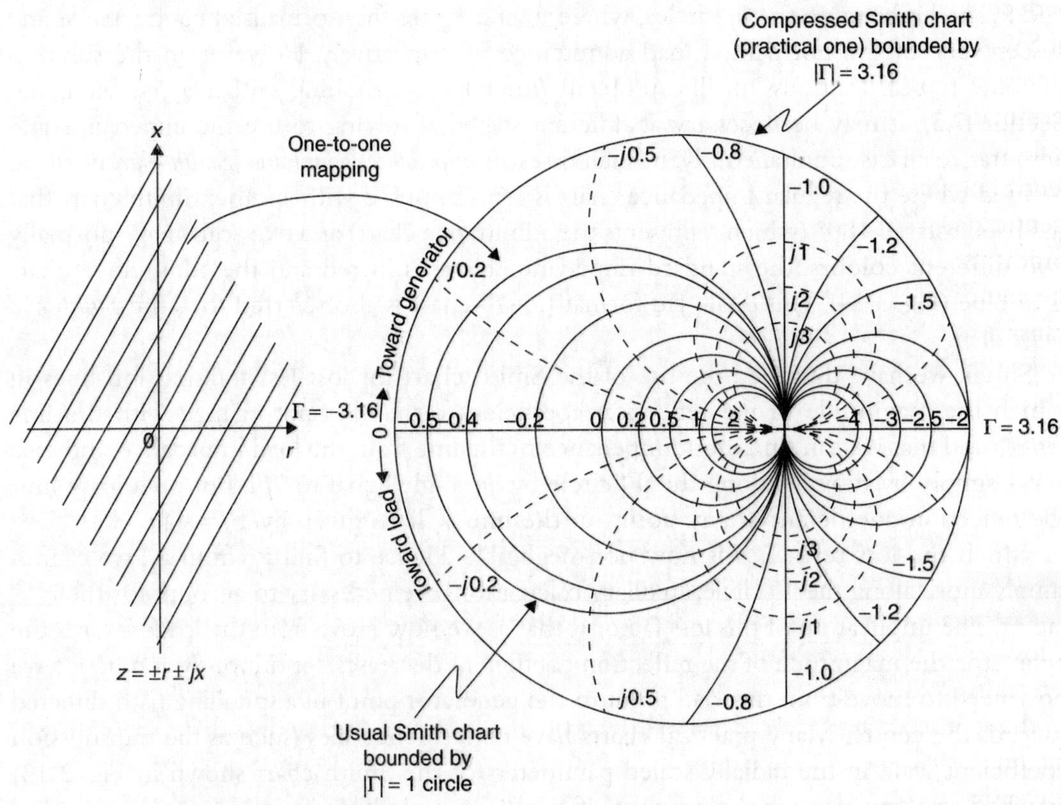

Fig. 3.15 Compressed Smith chart.

Finally, let us look at the 'radially scaled parameters' at the bottom of the Smith chart. The scales marked RFL. COEFF, E or I on the left half of the scale is for $|\Gamma| = |V^-/V^+|$ or $|I^-/I^+|$, the magnitude of reflection coefficient and the adjacent scale on top of it marked RFL. COEFF, P gives the power reflection coefficient that is $|\Gamma|^2 = |V^-/V^+|^2$. Both the scales start from the centre of the Smith chart (where the magnitude of the reflection coefficient is zero, perfectly matched condition) and ends at the periphery of the chart [where the magnitude of the reflection coefficient is maximum (equal to 1), complete reflection]. On top of this scale on the left is RTN. LOSS (dB) and along the same line on the right of the centre of the Smith chart is RFL. LOSS (dB) (Eqs 3.28 and 3.29 respectively). The 'return loss' scale starts from 0 (corresponding to $|\Gamma| = 1$) at the periphery of the chart and approaches infinity at the centre of the chart (where $|\Gamma| = 0$, perfectly matched); which implies that with better load matching, the reflection from the load will be less and hence the return loss will be higher. The 'reflection loss' scale starts with its zero at the centre of the chart (meaning that for perfectly matched condition, with $|\Gamma| = 0$, total power is absorbed in the load) and approaches infinity as $|\Gamma|$ approaches unity at the periphery of the chart (meaning that for complete reflection $|\Gamma| = 1$, there is no absorbed signal by the load). The top left scale of radially scaled parameters give the 'standing wave' expressed either in ratio or in dB (with unity at the centre for ratio scale and zero for dB scale). For the perfectly matched condition, the minimum value of SWR is 1; higher and higher values of SWR are approached as one moves towards the periphery of the chart when mismatch increases, leading to infinity for a perfect mismatch as for open, short and purely reactive terminations. The other major subheading on the right side of the radially scaled parameter scale is for 'transmission coefficient' (TRANSM. COEFF, P) which is seen to be 1 at the centre and 0 at the periphery of the Smith chart. This is expected since zero reflection (centre of Smith chart) leads to full transmission while complete reflection (at the rim of Smith chart) leads to zero transmission. If a transmission line is lossy and we move from the source to the load, then following the TOWARD LOAD direction, we find that the transmission coefficient decreases (as expected) while a move from the load to source, that is, in the TOWARD GENERATOR direction, the transmission coefficient increases. This is because for a lossy line, the closer we are to the generator plane, the less is the transmission loss. The closer we are to the load plane, the larger is the transmission loss since it includes the lossyness of the line. The use of the Smith chart along with its radially scaled parameters will be clear when we solve the following examples.

EXAMPLE 3.5

A loss-less transmission line of length 0.624λ having a characteristic impedance of 50 Ω is terminated in an impedance $(150 + j50)\,\Omega$. Find (i) the voltage reflection coefficient, (ii) the standing wave ratio, (iii) the input impedance, (iv) the location of all possible voltage maxima and minima from the load end, and (v) the power reflected from the load and the power absorbed in the load if the input power to the line is 200 mW.

Solution

(i) We find normalised load impedance $z_l = Z_l/R_0 = 3.0 + j1.5$ and locate it on the Smith chart (that is, point P_1 in Fig. 3.16). Then with the centre as the origin of the Smith chart and radius $OP_1 = |\Gamma| = 0.56$, a circle is drawn. When the straight line OP_1 is extended, it cuts the wavelength towards generator (WTG) scale at P_1'. We read 0.227 on the WTG scale. Thus, $(l/\lambda) = (0.250 - 0.227) = 0.023$ has been travelled from the load plane towards the generator which corresponds to a phase angle of 16.56° [since $(l/\lambda) = 0.250$ around the Smith chart corresponds to a phase change of 180°].

$$\Gamma = |\Gamma| e^{j\theta} = 0.56 \angle 16.56°$$

(ii) The $|\Gamma| = 0.56$ circle intersects the negative real axis $OP_{s.c}$ at $r = \rho = 3.6$.

Thus VSWR = 3.6.

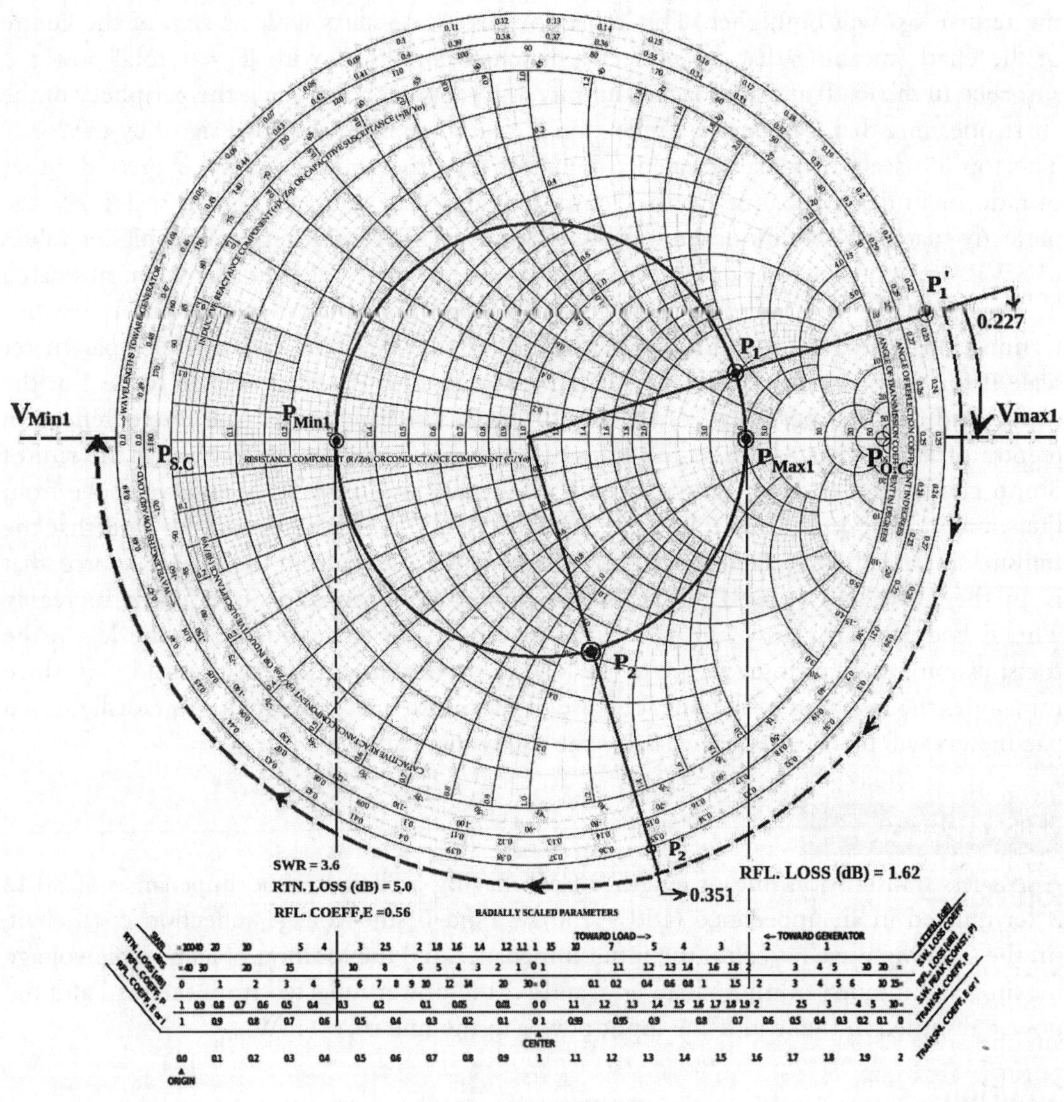

Fig. 3.16 Smith chart showing the solution of Example 3.5.

(iii) To find the input impedance we proceed as follows. We move from P_1' at 0.227 by a total of $l = 0.624$ in the WTG scale, first to 0.500 (same as 0.000) and then a further 0.351 [(0.500 – 0.227) + 0.351 = 0.624] to P_2'. Joining O and P_2' by a straight line we find that the line intersects the $|\Gamma| = 0.56$ circle at P_2. We read $r = 0.67$ and $x = 1.36$ at P_2. Thus $Z_i = Z_0 z_i = 50(0.67 + j1.36) = (33.5 + j68)\ \Omega$.

(iv) In going from P_1 to P_2, the $|\Gamma| = 0.56$ circle intersects the positive real axis $OP_{o.c}$ at $P_{max\,1}$, which is the location of the first voltage maximum located at $(0.25 - 0.227)\lambda$ or 0.023λ from the load end. The first voltage minimum is at $P_{min\,1}$ at $(0.023 + 0.25)\lambda = 0.273\lambda$, while the second voltage maximum is at $(0.273 + 0.25)\lambda = 0.523\lambda$ from the load end. So these are the possible voltage maxima and minima of the standing wave on the line of length 0.64λ.

(v) With respect to $|\Gamma| = 0.56$ circle we read Return loss = 5.0 dB and Reflection loss = 1.62 dB from the 'radially scaled parameters'.

Now as per definition of return loss: $L_{\text{return}} = -10\log_{10}\left(\dfrac{P_{\text{ref}}}{P_{\text{in}}}\right)$

Thus, $P_{\text{ref}} = P_{\text{in}}\,(10^{-L_{\text{return}}/10}) = 200\,(10^{-0.50}) = 63$ mW

while reflection loss is $L_{\text{reflection}} = -10\log_{10}\left(\dfrac{P_{tr}}{P_{\text{in}}}\right)$

Thus, $P_{tr}\,(= P_{\text{absorbed}}) = P_{\text{in}}\,(10^{-L_{\text{reflection}}/10}) = 200\,(10^{-0.162}) = 137$ mW

Thus, out of 200 mW of input power, 63 mW is reflected and 137 mW is absorbed in the load.

EXAMPLE 3.6

A transmission line is of length $l = 0.80\lambda$ and has a line loss of 3.0 dB. The normalised load impedance is $z_l = 0.15 + j0.5$. Locate the normalised input impedance on the Smith chart. Given $Z_0 = 50\ \Omega$, what is the input impedance? Find the percentage reduction in VSWR while going from the load to the source.

Solution

First locate the normalised load impedance $z_l = 0.15 - j0.5$ on the Smith chart (Fig. 3.17). Next we draw a circle with centre at the origin O of the Smith chart and radius as OP_L.

Next, let us drop a vertical line from point P_L on the 'TRANSM. COEFF, P' scale, which cuts the scale at 0.4. We now need to move to the 0.8 mark on this scale (TOWARDS GENERATOR, as we need to find input impedance, z_i) to take care of 3 dB line loss, when we reach the point marked P. A perpendicular is dropped back from this point P on transmission coefficient scale to the $x = 0$ line (u-axis of the complex u–v plane) on the Smith chart. Let it cut the $x = 0$ line at P_S. A circle is drawn with OP_S as radius.

The normalised input impedance lies on this circle. It is only necessary to move (starting from the load point, indicated by z_l) by 0.80λ on the chart following the direction of the WTG scale to locate z_i. To do this, we first need to move from z_l in the clockwise direction by 0.075, then 0.500 and finally a further 0.225 [as 0.075 + 0.500 + 0.225 = 0.80].

The VSWR locus from the load to the generator is a spiral path (which is expected as the line is lossy) and we reach the point P_S (by swinging clockwise from P_L with progressively reduced radius caused by continuously decreasing the VSWR value); z_i is located on this constant VWR circle. We read at z_i, $r = 2.4$ and $x = 0.8$. Thus,

$$Z_i = 50(2.4 + j0.8) = (120 + j40)\ \Omega$$

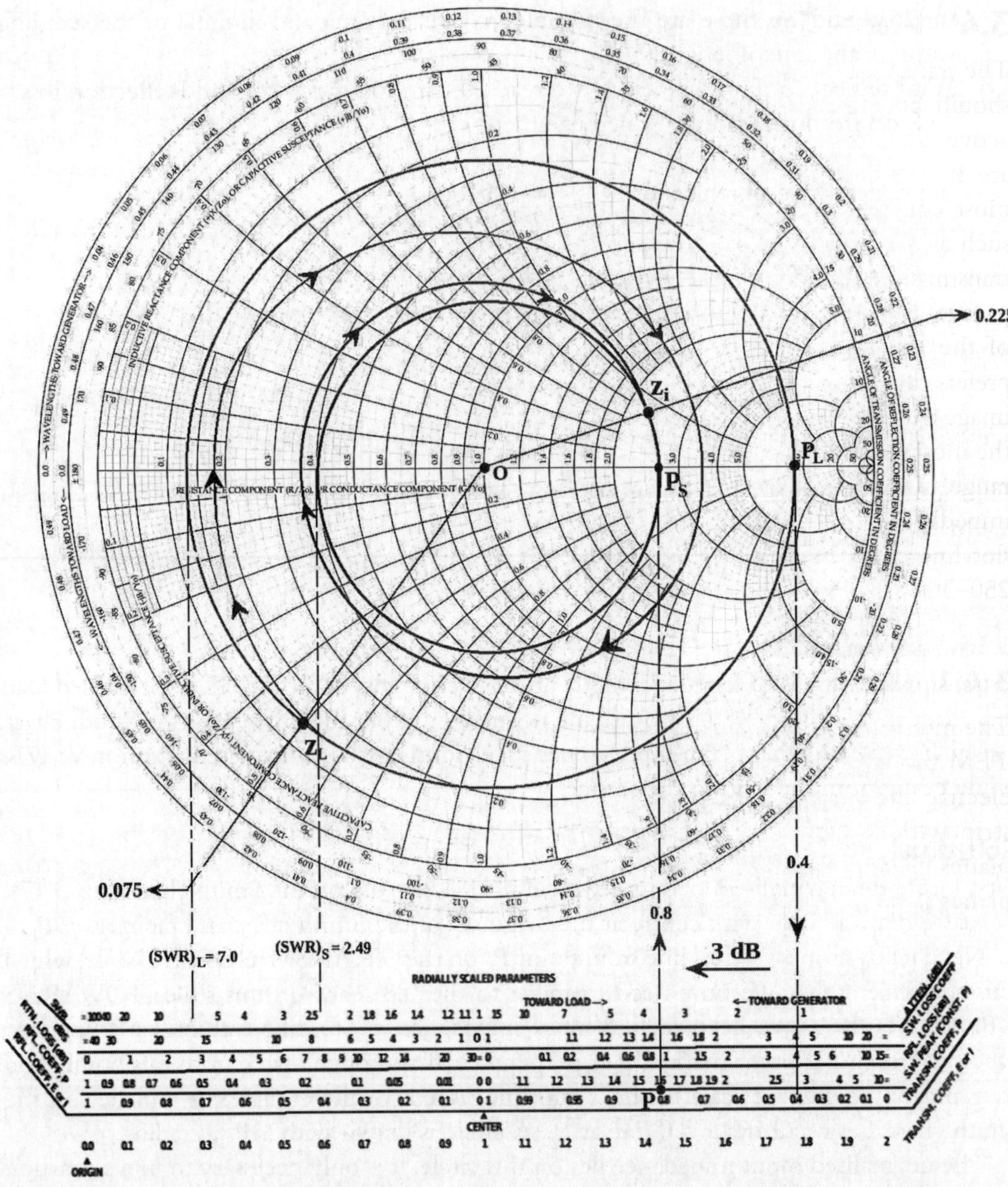

Fig. 3.17 Smith chart showing the solution of Example 3.6.

If we read the VSWR (ρ) at the load (P_L) and at the source (P_S) the values are $\rho_L = 7.0$ and $\rho_S = 2.49$ respectively. Thus, the percentage reduction in VSWR from load to source is:

$$\left(\frac{2.49}{7}\right) \times 100\% = 33.57\%$$

3.4 Planar Transmission Lines

The transmission line structures that are suitable for Microwave Integrated Circuits (MIC) should be of '*planar configuration*' as this will allow integration of both the passive and active devices via planar technology used for integrated circuit design. Various structures are in use as far as distributed planar transmission lines for MICs are concerned, the most common ones being the microstrip line. Variants of microstrip are also possible, such as, slot-line, coplanar-line, fin-line and image guide. The actual choice of the planar transmission line structure depends upon several factors like the type of circuit (active or passive) to be mounted and its operating frequency including the characteristic impedance of the line demanded by circuit applications. For example, series mounting of device prefers the use of microstrip line whereas shunt mounting is preferred in the slot-line. The image guide is suitable for millimetre wavelength applications. Though the microstrip is the most commonly used MIC transmission line structure for frequencies in the microwave range, in some applications such as for filter or matching network design, the characteristic impedance realisable with a microstrip (20–120 Ω) is inadequate and one has to resort to slot-line or fin-line, the latter being capable of having characteristic impedance as high as 250–300 Ω.

3.4.1 Microstrip Line

The microstrip is actually developed from a strip-line. The strip-line is a three conductor TEM mode planar transmission line. Its cross-sectional configuration and sketches for electric and magnetic field are shown in Fig. 3.18. It consists of a central conducting strip with its width W much greater than its thickness t, ($W >> t$) and two ground planes closing the structure from the top and bottom. The space between the ground planes is filled with a dielectric medium of dielectric constant ε_r and the central metal

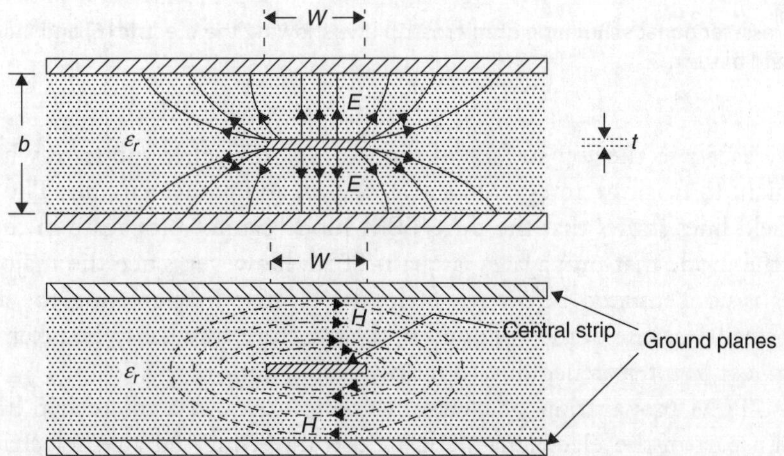

Fig. 3.18 Cross-sectional schematic of strip-line showing the electric (*E*) and magnetic (*H*) field pattern.

strip is embedded in this dielectric substrate. The strip-line is not a useful structure for MICs (however, it can be used for the design of planar circuit components useful in discrete circuit design applications) because active and passive components cannot be mounted on this transmission line structure (closed at both the top and bottom by ground planes).

A practical variant of the strip-line is the microstrip line. The first use of the microstrip line configuration was reported by engineers at the Federal Telecommunication Laboratories sometime after 1949.

A cross-sectional view of a microstrip line is shown in Fig. 3.19, together with the sketches for electric and magnetic fields. This line consists of a dielectric substrate (of thickness h) with a ground plane on one side and a conductor strip on the other face (note that the top ground plane of the strip-line is not present here).

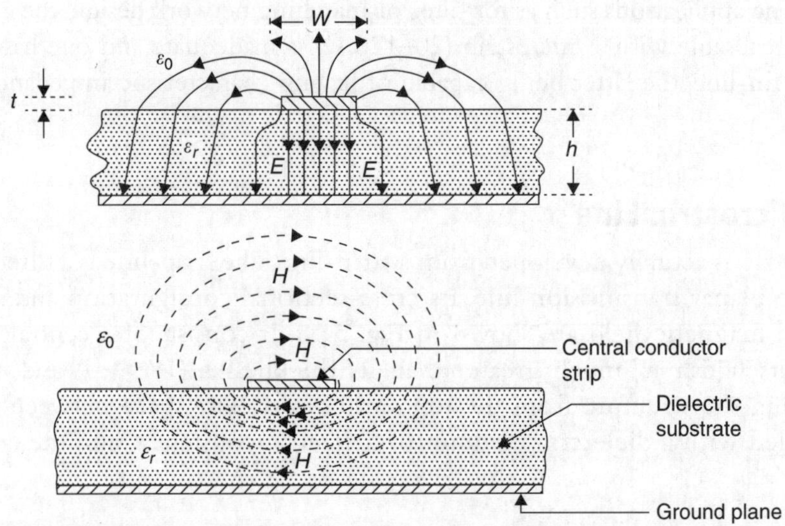

Fig. 3.19 Cross-sectional schematic of microstrip line showing the electric (E) and magnetic (H) field pattern.

There is easy access to the top surface of the exposed dielectric substrate which makes it very convenient to fabricate integrated active devices and passive circuits. The distribution of electric field lines shows that the pure TEM mode cannot propagate in the microstrip line, rather, the mode that propagates is quasi-TEM. However, since the major portion of E-field lines is concentrated below the strip, the electric flux crossing the air-dielectric boundary is small. Because of the openness of the structure, care has to be taken to minimise the radiation loss or interference due to nearby conductors.

The quasi-TEM propagation of electromagnetic waves in a microstrip line is due to the fact that a part of the electromagnetic field is traversing the air–dielectric composite in an inhomogeneous medium. The propagation velocity of the electromagnetic wave in the two dielectric media is different and thus an effective dielectric constant has to be considered. However, it is advisable to use a substrate with a high dielectric constant [alumina ($\varepsilon_r = 8 - 10$), sapphire ($\varepsilon_r = 9.4$)] so that the electromagnetic field will be confined more effectively in the substrate dielectric and hence near the strip. Further, substrates with high dielectric constants are advantageous since they reduce the phase velocity (and thus guide wavelength) and consequently the proportional circuit dimensions would be reduced too. The quasi-TEM propagation means, a hybrid mode propagation being characterised by a surface-wave mode propagating in the inhomogeneous air–dielectric media. To understand this, let us consider a parallel-plate waveguide partially filled with a dielectric [7] (Fig. 3.20a).

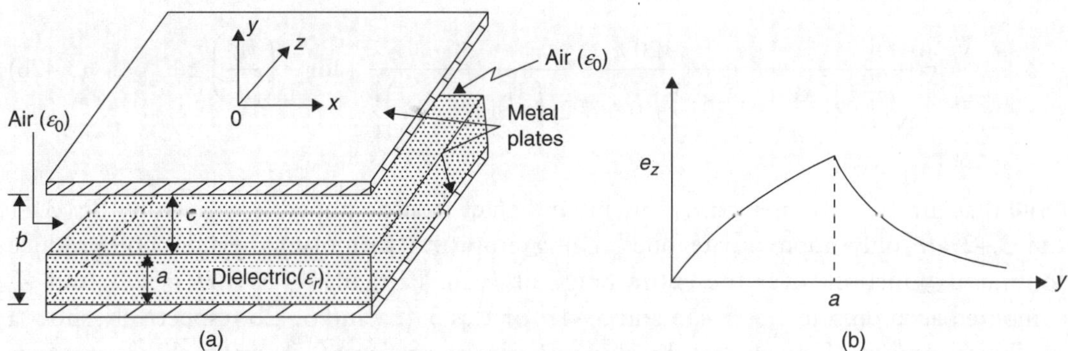

Fig. 3.20 (a) Parallel-plate waveguide partially filled with a dielectric (b) Axial electric field for surface-wave mode.

The analysis shows that the electromagnetic wave decays exponentially away from the air–dielectric interface and does not depend on the b dimension as long as $k_0 c = k_0(b - a)$ is large; where k_0 is the cut-off wave number for the dominant mode in air. This field is guided by the dielectric sheet in contact with the ground plane (resembling the structure of the microstrip) even if the upper plate is removed to infinity. This type of mode propagating in the air–dielectric interface supported by the dielectric is the *surface-wave mode*, because its field is confined close to the surface of the dielectric guiding structure. The axial electric field (e_z) for the surface-wave mode is shown in Fig. 3.20b.

Without going into the derivation of the formulae, details of which are given in suitable texts [8], we provide here the important ones useful for understanding the characteristics of the microstrip line and those that are time and again required for the design of microstrip-based circuits and components.

In the case of a microstrip line, an effective relative permittivity ($\varepsilon_{\text{reff}}$) is used, which is needed because the effective dielectric medium in such structures is ideally inhomogeneous (having air and other dielectric regions of the substrate) but for all practical purposes is considered to be homogeneous possessing an effective relative permittivity (dielectric constant) given by:

$$\varepsilon_{\text{reff}} = \frac{\varepsilon_r + 1}{2} + \frac{\varepsilon_r - 1}{2}\left\{\left(1 + 12\frac{h}{W}\right)^{-1/2} + 0.04\left(1 - \frac{W}{h}\right)^2\right\} \quad \text{for} \quad \left(\frac{W}{h}\right) \le 1 \qquad (3.41\text{a})$$

$$= \frac{\varepsilon_r + 1}{2} + \frac{\varepsilon_r - 1}{2}\left(1 + 12\frac{h}{W}\right)^{-1/2} \quad \text{for} \quad \left(\frac{W}{h}\right) \ge 1 \qquad (3.41\text{b})$$

The characteristic impedance is given by:

$$Z_0 = \frac{60}{\sqrt{\varepsilon_{\text{reff}}}} \ln\left(\frac{8h}{W} + \frac{W}{4h}\right) \quad \text{for} \quad \left(\frac{W}{h}\right) \le 1 \qquad (3.42\text{a})$$

$$= \frac{120\,\pi}{\sqrt{\varepsilon_{\text{reff}}}\left\{\left(\frac{W}{h} + 1.393\right) + 0.667 \ln\left(\frac{W}{h} + 1.444\right)\right\}} \quad \text{for} \quad \left(\frac{W}{h}\right) \ge 1 \qquad (3.42\text{b})$$

Note that the effective dielectric constant and characteristic impedance given by Eqs 3.41 and 3.42 are only approximate ones (being empirical formulae) they do not produce continuous functions over the entire range of W/h. Particularly, at $W/h = 1$, ε_r and Z_0 computed according to Eqs 3.41a and 3.41b, or Eqs 3.42a and 3.42b respectively, show a small discontinuity. However, as this discontinuity error is just 0.5% or so; these equations yield quite good results for all practical purposes.

From Eqs 3.41 and 3.42, it is seen that both, the effective dielectric constant and characteristic impedance, vary with the geometry (W/h) of the microstrip line. Computed plots for Z_0 vs W/h and $\varepsilon_{\text{reff}}$ vs W/h for different parametric values of dielectric constant (ε_r) of the substrate material have been shown in Figs 3.21a and 3.21b respectively. The parameter range of W/h and ε_r are chosen such that they span the domain typically encountered in practically relevant circuit values.

The wavelength of the propagating wave in the microstrip line will also be different from the free space wavelength (λ) due to the presence of the air–dielectric composite having an effective dielectric constant $\varepsilon_{\text{reff}}$, and thus the effective wavelength of propagating wave in a microstrip line is given by:

$$\lambda_{\text{eff}} = \frac{\lambda}{\sqrt{\varepsilon_r}}\left\{\frac{\varepsilon_r}{1 + 0.6\,(\varepsilon_r - 1)\,(W/h)^{0.0297}}\right\} \quad \text{for} \quad \left(\frac{W}{h}\right) < 0.6$$

$$= \frac{\lambda}{\sqrt{\varepsilon_r}}\left\{\frac{\varepsilon_r}{1 + 0.63\,(\varepsilon_r - 1)\,(W/h)^{0.1255}}\right\} \quad \text{for} \quad \left(\frac{W}{h}\right) > 0.6 \qquad (3.43)$$

Further, for design purposes, we need to have a relation that allows us to compute W/h ratios based on the given characteristic impedance Z_0 and dielectric constant ε_r of the substrate materials which are given below:

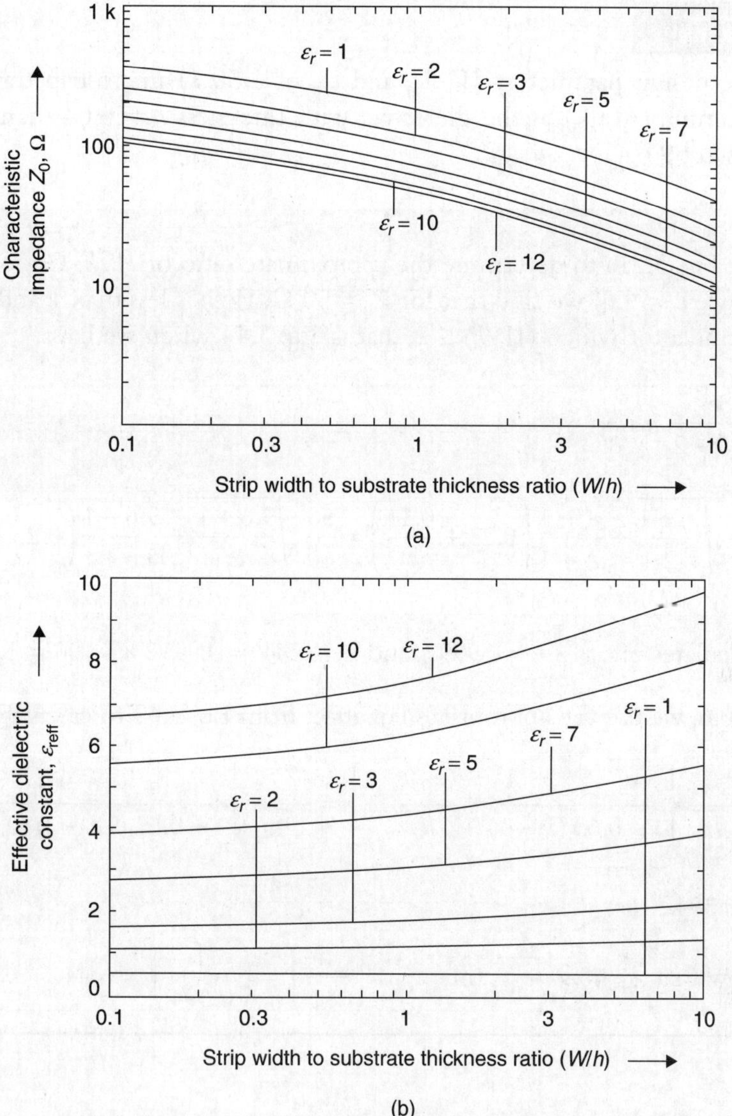

Fig. 3.21 Computed plots for (a) Z_0 vs W/h and (b) ε_{reff} vs W/h for different parametric values of dielectric constant (ε_r) of the substrate material.

$$\frac{W}{h} = \frac{8e^{A}}{e^{2A} - 2} \quad \text{for} \quad \left(\frac{W}{h}\right) \le 2$$

$$= \frac{2}{\pi}\left[B - 1 - \ln(2B - 1) + \frac{\varepsilon_r - 1}{2\varepsilon_r}\left\{\ln(B - 1) + 0.39 - \frac{0.61}{\varepsilon_r}\right\}\right] \quad \text{for} \quad \left(\frac{W}{h}\right) \ge 2 \quad (3.44)$$

where $A = \frac{Z_0}{60}\sqrt{\frac{\varepsilon_r + 1}{2}} + \frac{\varepsilon_r - 1}{\varepsilon_r + 1}\left(0.23 + \frac{0.11}{\varepsilon_r}\right)$ and $B = \frac{377\,\pi}{2Z_0\sqrt{\varepsilon_r}}$

EXAMPLE 3.7

Determine the design parameters W, λ_{eff} and ε_{eff} of a 50 Ω microstrip transmission line which uses alumina (Al_2O_3) as the dielectric with thickness h = 0.04 inch and having a dielectric constant 9.6.

Solution

At first we use Fig. 3.21a to determine the approximate ratio of W/h. Choosing the curve close to 10 (since $\varepsilon_r = 9.6$) we find that for $Z_0 = 50\ \Omega$, W/h is less than 1 and thus we need to take the formula for which $(W/h) \le 2$, that is, Eq. 3.44 when we have:

$$\frac{W}{h} = \frac{8e^{A}}{e^{2A} - 2}, \text{ with}$$

$$A = \frac{Z_0}{60}\sqrt{\frac{\varepsilon_r + 1}{2}} + \frac{\varepsilon_r - 1}{\varepsilon_r + 1}\left(0.23 + \frac{0.11}{\varepsilon_r}\right) = \frac{50}{60}\sqrt{\frac{9.6 + 1}{2}} + \frac{9.6 - 1}{9.6 + 1}\left(0.23 + \frac{0.11}{9.6}\right) = 2.11$$

Thus W/h calculates out as $\dfrac{W}{h} = 0.9992$, and hence $W = 0.9992 \times 1.106 = 1.1051$ mm.

As $W/h > 0.6$, we use the appropriate equation from Eq. 3.43 to calculate λ_{eff}:

$$\lambda_{\text{eff}} = \frac{\lambda}{\sqrt{\varepsilon_r}}\left\{\frac{\varepsilon_r}{1 + 0.63\,(\varepsilon_r - 1)\,(W/h)^{0.1255}}\right\} = \frac{\lambda}{\sqrt{9.6}}\left\{\frac{9.6}{1 + 0.63\,(9.6 - 1)\,(0.9992)^{0.1255}}\right\}^{1/2}$$

or, $\lambda_{\text{eff}} = 0.394\lambda$

Again we know that $\lambda_{\text{eff}} = \lambda/\sqrt{\varepsilon_{\text{eff}}}$, thus $\varepsilon_{\text{eff}} = \left(\dfrac{\lambda}{\lambda_{\text{eff}}}\right)^2 = \left(\dfrac{1}{0.394}\right)^2 = 6.44$

3.4.1.1 Even and odd mode in coupled transmission lines

In this context, it is worthwhile discussing the concept of *even- and odd-mode* excitation in coupled transmission lines [9] which we come across in the analysis of a number of microwave components such as coupled-line coupler, short-slot hybrid, Wilkinson power divider/combiner and so on (Chapter 6) including coupled line filters (Chapter 8). This situation arises in any three-conductor transmission structure such as, coupled strip-line (either edge-coupled or broadside-coupled) or coupled microstrip line (Fig. 3.22a), in which two conductors of the transmission structure are close enough to each other and the electromagnetic field between them has a chance to get significantly coupled.

For *even-mode excitation* (Fig. 3.22b), the instantaneous voltages at the two strip conductors of the coupled transmission line are at the same potential (say $+V$) so the

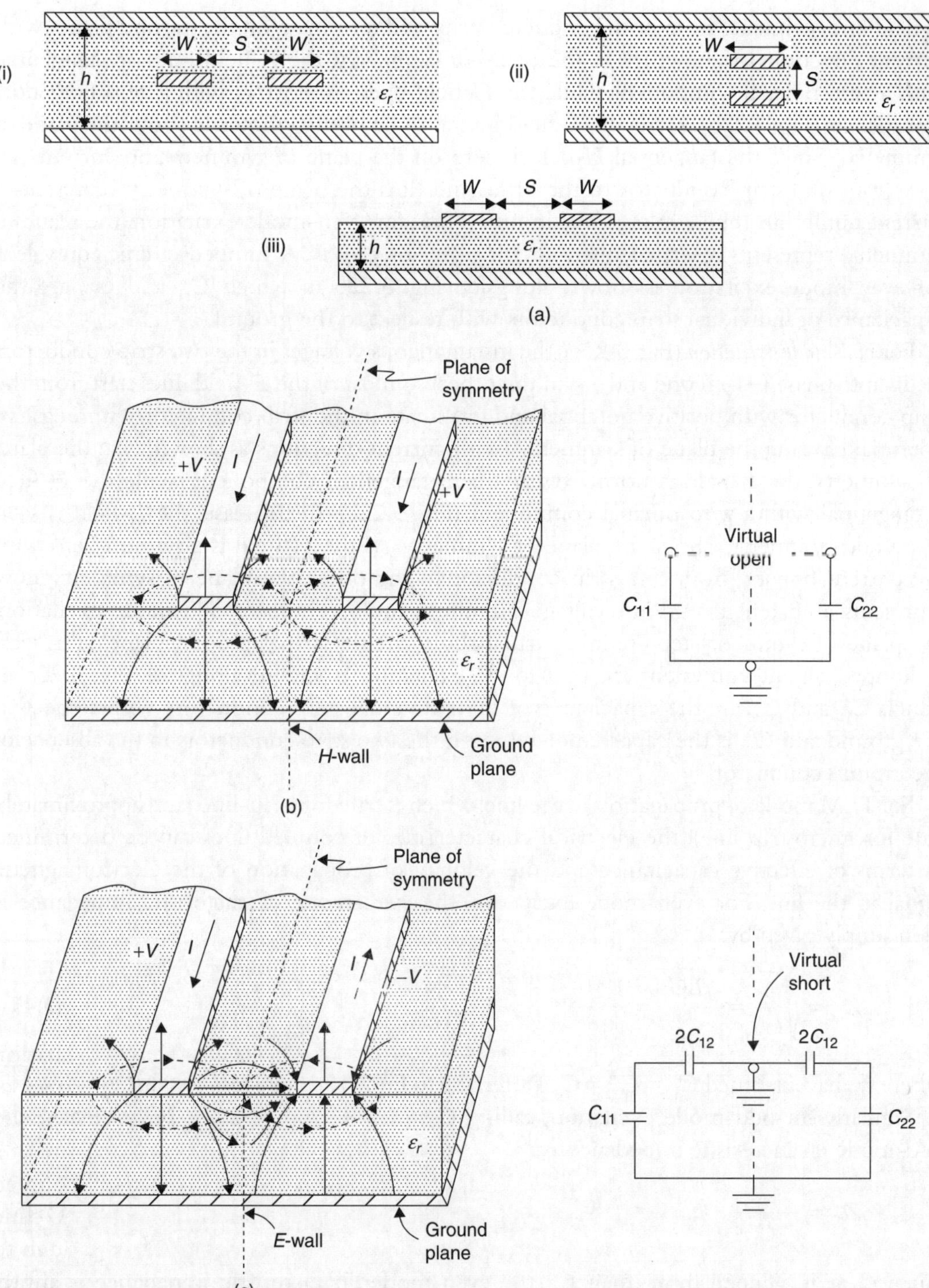

Fig. 3.22 (a) Coupled planar transmission line geometries (i) Edge-coupled strip-line (ii) Broadside-coupled strip-line (iii) Coupled microstrip, (b) Even-mode excitation in coupled microstrip line, (c) Odd-mode excitation in coupled microstrip line.

E-field lines originate from both conductor strips and terminate on the ground plane (with current directions as shown). On the plane of symmetry, the *E*-field lines are tangential with a zero normal component while the *H*-field lines are normal with a zero tangential component (Fig. 3.22b). Here, the *E*-field is seen to be even-symmetric about the plane of symmetry. Since the tangential *H*-field is zero on the plane of symmetry, no current can flow from one strip conductor to the other and thus the plane of symmetry represents a current minimum (null)/voltage maximum. Thus, for even-mode excitation, the plane of symmetry represents a virtual open-circuit plane or *H-wall*. A lumped circuit equivalent for even-mode excitation is shown alongside Fig. 3.22b in which C_{11} and C_{22} are the capacitance of individual strip conductors with respect to the ground.

For *odd-mode excitation* (Fig. 3.22c), the instantaneous voltages in the two strip conductors are in anti-phase ($+V$ in one and $-V$ in the other) so most of the *E*-field lines start from the strip conductor with positive potential and terminate in the strip conductor with negative potential crossing the plane of symmetry (with current directions as shown). On the plane of symmetry, the *E*-field is normal with a zero tangential component while the *H*-field is tangential with a zero normal component (Fig. 3.22c). In this case, the *E*-field is seen to be odd-symmetric about the plane of symmetry. As the *H*-field is now non-zero with the current flowing from one strip conductor to the other, the plane of symmetry now represents a current maximum/voltage minimum (null). Thus, for odd-mode excitation, the plane of symmetry represents a virtual short-circuit/virtual ground plane or *E-wall*. A lumped circuit equivalent for odd-mode excitation is shown alongside Fig. 3.22c in which C_{11} and C_{22} are the capacitance of the individual strip conductors with respect to the ground and C_{12} is the capacitance between the two strip conductors in the absence of the ground conductor.

For TEM mode of propagation in the line, which is true for strip-line (and approximately true for microstrip line), the electrical characteristics of coupled lines can be determined in terms of effective capacitance and the velocity of propagation of the electromagnetic signal in the line. For even-mode excitation, the even-mode characteristic impedance is then simply given by:

$$Z_{0e} = \sqrt{\frac{L}{C_e}} = \frac{\sqrt{LC_e}}{C_e} = \frac{1}{v_p C_e} \tag{3.45a}$$

where we have assumed $C_{11} = C_{22} = C_e$. Referring to Eq. 3.12d, for a loss-less line $v_p = 1/\sqrt{LC}$.

Similarly for odd-mode excitation, calling $C_o = C_{11} + 2C_{12} = C_{22} + 2C_{12}$ we have the odd-mode characteristic impedance as:

$$Z_{0o} = \frac{1}{v_p C_o} \tag{3.45b}$$

Since C_o is in general more than C_e, the even-mode characteristic impedance is always greater than the odd-mode characteristic impedance, that is $Z_{0e} > Z_{0o}$.

When the coupled line is operated in even or odd-mode, the respective impedances for each strip conductor with respect to the ground can be determined from Eqs 3.46a and 3.46b respectively, by evaluating the respective capacitance by conformal mapping technique if the mode of propagation is TEM (as in the case of a coupled strip-line). But for a quasi-TEM line, which is the case for a microstrip line, the capacitance needs to be determined numerically or by some quasi-static technique. Further, in the case of microstrip coupled lines, the phase velocity is different for the two modes of propagation caused by different field configurations of the modes in the air–dielectric interface. However, practical realisation uses compensating techniques like dielectric overlays and anisotropic substrates to make even and odd-mode phase velocities equal. Standard curves are available in text books [8, 9] for even and odd-mode impedances for different geometrical parameters of the coupled microstrip line that are useful for the design of various microwave components.

3.4.2 Variants of Microstrip Line

3.4.2.1 Inverted microstrip and trapped inverted microstrip

Some common variants of the basic microstrip line [8] are the inverted microstrip and trapped inverted microstrip. In the *inverted microstrip* (IM), the ground plane is air spaced underneath the strip conductor of the microstrip structure oriented in inverted form and unlike the conventional microstrip, it has no ground plane backing the dielectric substrate (Fig. 3.23a).

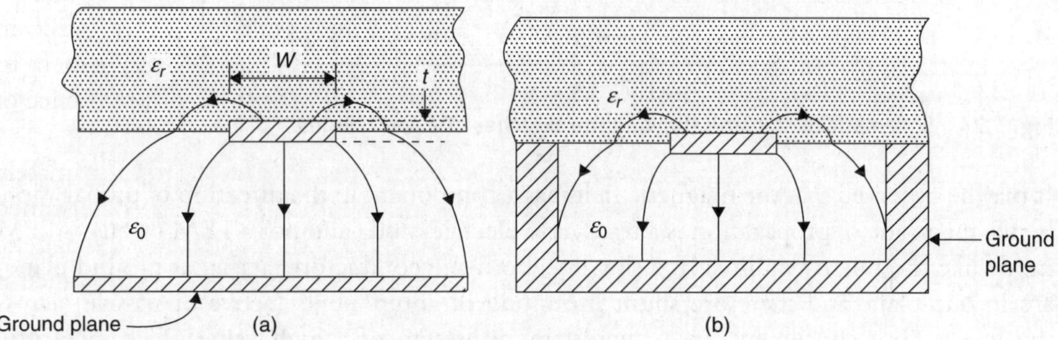

Fig. 3.23 (a) Inverted microstrip (b) Trapped inverted microstrip.

With such a structure, most of the field now lies within the air between the strip and the ground plane. This reduces $\varepsilon_{\text{reff}}$ and hence increasing the guide wavelength relative to that in a conventional microstrip operated at similar frequencies. In other words, it means that compared to a conventional microstrip, a wider line can be used for a prescribed characteristic impedance allowing better design tolerance at higher frequency. The *trapped inverted microstrip* (TIM) shown in Fig. 3.23b, is similar to the inverted microstrip except that now the ground plane is made into a slot or 'channel' shaped section effectively grounding and shielding three planes, away from the strip conductor. This reduces the possibility of edge radiations and provides for inherent suppression of some higher order modes which could otherwise propagate in the usual inverted microstrip. Both IM and TIM are potential candidates for applications at frequencies extending up to W-band (94 GHz), that is, for millimetre-wave applications. Other variants of the microstrip with relative advantages and disadvantages are discussed below.

3.4.2.2 Slot-line

Structurally, slot-line is a complementary version of the microstrip line. The schematic of a slot-line with the sketches for the electric and magnetic fields are shown in Fig. 3.24.

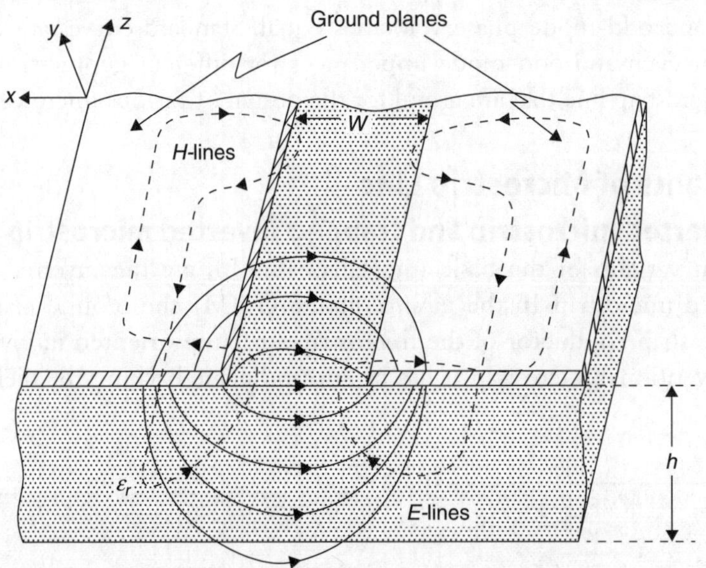

Fig. 3.24 Schematic of slot-line showing the sketches of *E* and *H* fields.

It may be observed that the magnetic field has a component in the direction of propagation; hence the mode of propagation is a transverse electric mode and not a TEM or quasi-TEM mode like the microstrip line. In a slot-line, both the conductors (acting as ground plane) are in one plane and therefore shunt mounting of components (active or passive) across the line is very convenient. In a microstrip, realisation of a high value of characteristic impedance requires a very narrow strip conductor width. On the contrary, high impedance in the case of a slot-line requires large slot widths. Since there is a technological limitation on the fabrication of very narrow conductor strips (a microstrip has a typical characteristic impedance of 20–120 Ω), higher characteristic impedances (50–300 Ω) can be easily realised in slot-lines. We can find a location (in the cross-section of the slot-line) where the magnetic field is circularly polarised (Fig. 3.24). This characteristic is useful for the design of several ferrite components, such as resonance isolator, phase shifter and so on. One disadvantage of a slot-line is that its *Q* factor is around 100, which is rather low compared to ~250 realisable with a microstrip line.

3.4.2.3 Coplanar-line

In a coplanar-line (also known as coplanar waveguide, CPW, any transmission line at microwave frequency, in generic terms, is a waveguide as it guides electromagnetic waves) all the conductors are situated on only one surface of the dielectric substrate. In this sense, a slot-line is a form of coplanar line. But a more useful form of coplanar line can be a coupled slot-line, in which a strip of thin metallic film is deposited on the surface of a dielectric slab with two ground planes running adjacent and parallel to the strip conductor on the same side (Fig. 3.25). A signal is applied to the centre strip conductor with respect to the two ground planes on the either side.

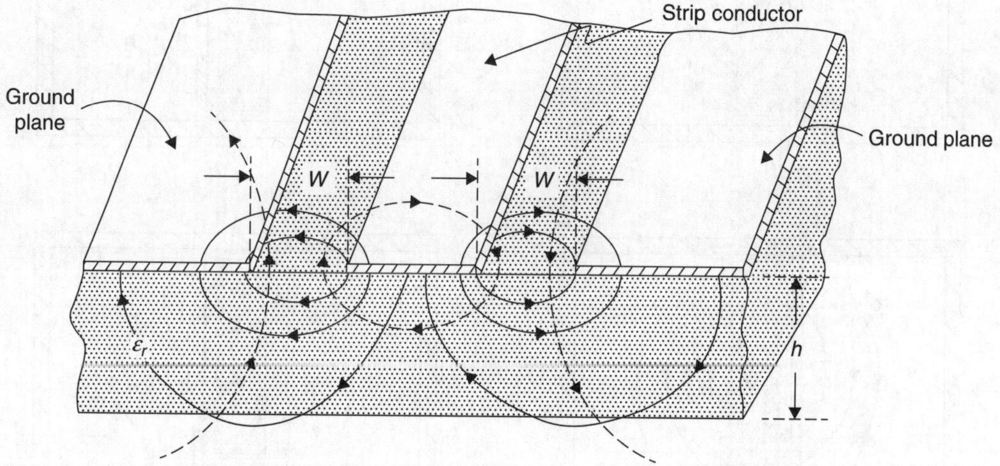

Fig. 3.25 Coplanar-line (coupled slot-line or coplanar waveguide, CPW) with two ground planes and a central strip conductor.

Unlike the slot-line, the dominant mode of propagation resembles a quasi-TEM. The magnetic field in the slots at the air dielectric interface is elliptically polarised, which is useful in the design of non-reciprocal ferrite devices. This geometry combines some of the advantages of microstrip and slot-line in a single structure. The microstrip geometry is convenient for series mounting of components—across a gap in the strip conductor, while slot-lines are well suited for shunt mounting of components across the slot. In a coplanar-line (coupled slot-line) both series and shunt mounted components can be easily incorporated. Impedances in the range of 40–150 Ω are realisable with a Q factor ~150. The coplanar structures – the slot-line and coplanar waveguide – tend to be excessively lossy at millimetre-wave frequencies because of their open configuration. Their utility thus far has been limited essentially to microwave frequencies. Enclosing these transmission lines in a rectangular metal shield can alleviate the problem of radiation. In fact, the shielded versions of slot-line and coplanar-line are known as unilateral fin-line and edge coupled fin-line, which we shall discuss next.

3.4.2.4 Fin-line

It is basically a slot-line/coplanar line inserted in the *E*-plane of a rectangular metal enclosure that gives the unilateral/edge-coupled fin-line (Figs 3.26a and 3.26b). Bilateral and antipodal configurations (Figs 3.26c and 3.26d) are also used with many variants to realise operational advantages.

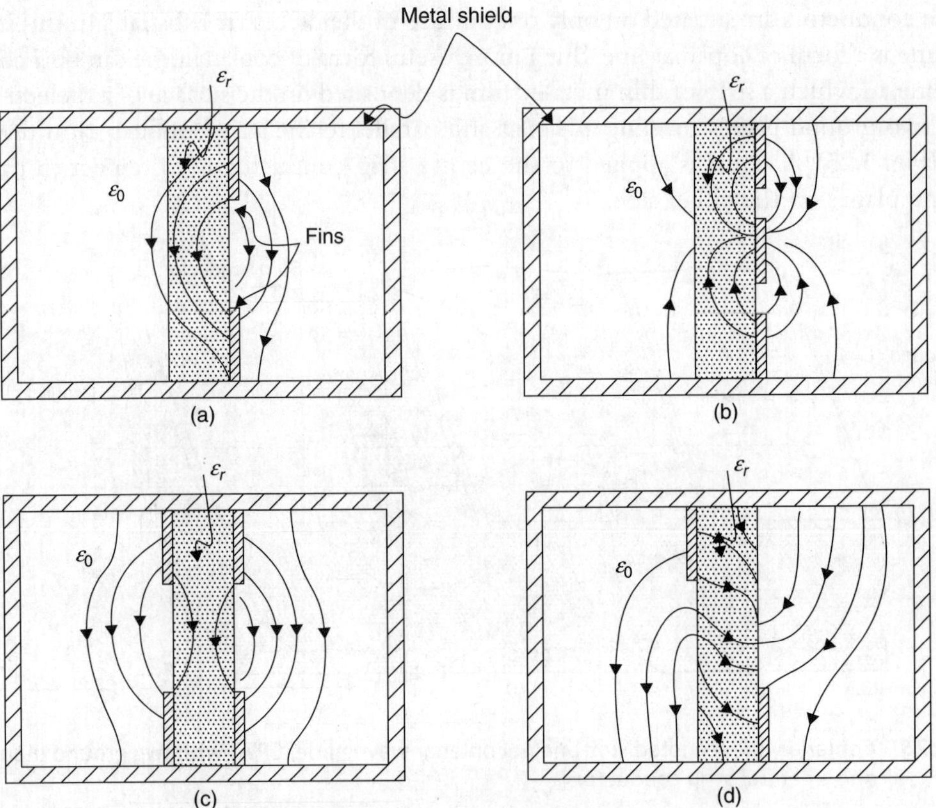

Fig. 3.26 Fin-lines (a) Unilateral (b) Edge-coupled (c) Bilateral (d) Antipodal.

The fin-line may be viewed as a ridged waveguide (details of the waveguide will be discussed in Chapter 4) with thin ridges backed by a dielectric substrate, or as a dielectric slab loaded waveguide with metal fins imprinted on the dielectric. The structure is considered quasi-planar because the entire circuit pattern, including the active devices, is incorporated on the planar surface of a dielectric substrate, while the design of the circuit also takes into account the effect of the waveguide housing. The rectangular metal enclosure reduces radiation losses and eliminates external electromagnetic interference. The structure is low-loss compared to a microstrip line and is a very useful transmission structure for millimetre-wave integrated circuits.

The mode of propagation in such a structure is a combination of TE and TM modes (called a hybrid mode), with a dominant TE the mode is designated as HE-mode and with a dominant TM it is designated as EH-mode. Because it can be viewed as a dielectric loaded/ridged waveguide the cut-off frequency is lowered for the propagation mode and thus larger bandwidth is realisable. The dimensions of the fin-line structure in the millimetre-wave frequency is compatible with beam-lead and chip devices and thus offers the potential for construction of passive and active integrated circuits in such transmission structures. Further, it permits easy transition to a standard rectangular waveguide and operates over the entire bandwidth of the waveguide.

3.4.2.5 Dielectric integrated guide

At higher millimetre-wave frequencies (>100 GHz) the wavelengths become comparable with cross-sectional dimensions of the dielectric slab and even fin-line poses operational problems. A useful transmission structure at these frequencies is the *dielectric integrated guide*, a basic form of which is the image guide (Fig. 3.27). Variants of the image guide are—trapped image guide, non-radiative dielectric guide, and so on, which take care of different losses incurred in the basic dielectric integrated guide.

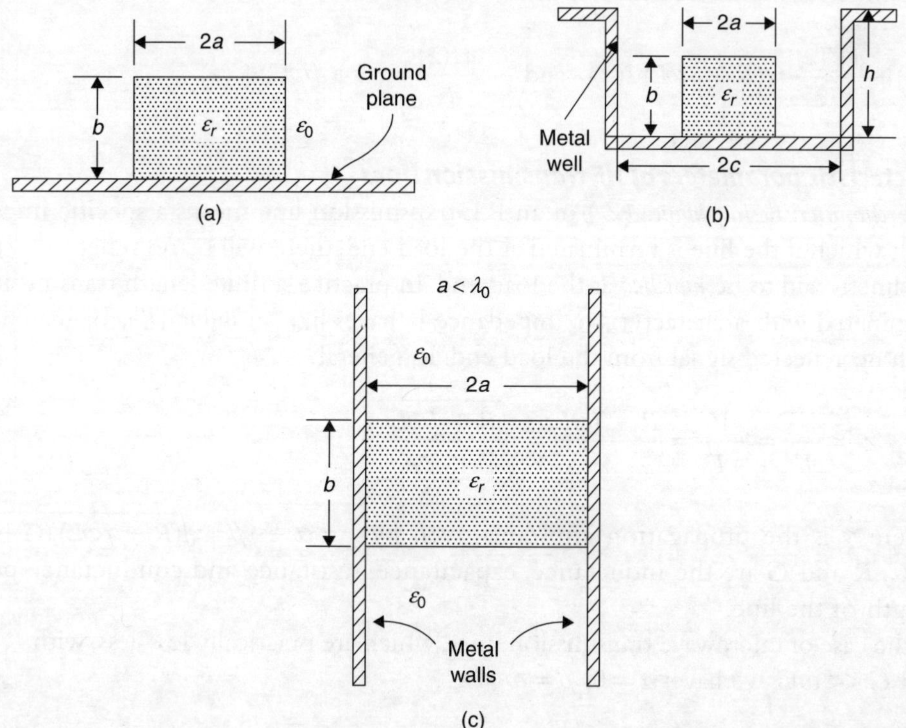

Fig. 3.27 Dielectric integrated guide (a) Image guide (b) Trapped image guide (c) Non-radiative dielectric guide.

While the principle of wave guiding is similar to the optical dielectric guides, configuration and fabrication techniques are distinctly different. An optical dielectric guide (commonly known as optical fibre) consists of an inner core made of silica glass surrounded by a concentric layer called cladding, also made of glass but of a slightly lower dielectric constant. They are formed using such techniques as sputtering, diffusion and ion-implantation, while in the image guide, the dielectric strip (alumina or boron nitride) can be grown on the metal ground plane by thick-film technology.

Although unloaded Q-factors of several thousand may be possible and operation above 100 GHz can be easily realised, many problems such as poor compatibility with active devices, mutual coupling and radiation from discontinuities and bends still remains to be solved for practical applications of such planar guiding structures.

RECAPITULATION

Transmission line equations

- Generalised transmission line equations (*Telegrapher's equations*) are given by:

$$-\frac{\partial v(z,t)}{\partial z} = Ri(z,t) + L\frac{\partial i(z,t)}{\partial t} \quad \text{and} \quad -\frac{\partial i(z,t)}{\partial z} = Gv(z,t) + C\frac{\partial v(z,t)}{\partial t}$$

- The time-harmonic transmission-line equations are:

$$-\frac{dV(z)}{dz} = (R + j\omega L)\,I(z) \quad \text{and} \quad -\frac{dI(z)}{dz} = (G + j\omega C)\,V(z)$$

Characteristic parameters of HF transmission lines

- The *characteristic impedance* (Z_0) of an RF transmission line means a specific impedance with which if the line is terminated at the load end, there will be no reflected wave and the line is said to be *matched* at the load end. In practice, a finite length transmission line terminated with a characteristic impedance behaves like an *infinite line* or matched line with no reflected signal from the load end. In general:

$$Z_0 = \frac{V^+}{I^+} = -\frac{V^-}{I^-} = \frac{R + j\omega L}{\gamma} = \sqrt{\frac{R + j\omega L}{G + j\omega C}}$$

where γ is the propagation constant given by $\gamma = \alpha + j\beta = \sqrt{(R + j\omega L)(G + j\omega C)}$; L, C, R and G are the inductance, capacitance, resistance and conductance per unit length of the line.

- In the case of microwave transmission lines, which are practically loss-less, with $R \ll \omega L$ and $G \ll \omega C$, we have: $\alpha = 0$, $\beta = \omega\sqrt{LC}$,

$$\text{whence:} \quad v_p = \frac{\omega}{\beta} = \frac{1}{\sqrt{LC}} = \frac{1}{\sqrt{\mu_0 \varepsilon_0}} = c \quad \text{and} \quad Z_0 = \sqrt{\frac{L}{C}}.$$

As the phase constant (β) is a linear function of frequency, a two-conductor microwave transmission line (coaxial line) will be *non-dispersive.* Further, a coaxial line at microwave frequency will also behave practically as a *distortionless* line as the phase velocity is independent of frequency.

Terminated HF transmission lines

- For terminated transmission line terminated with a load impedance Z_L:

$$Z_L = \frac{V(0)}{I(0)} = \frac{V^+ + V^-}{I^+ + I^-} = Z_0\frac{V^+ + V^-}{V^+ - V^-}$$

(the reference zero for the length of the line is considered to be the load plane).

- When $V^- = 0$, that is, there is no reflected wave and the line is matched, $Z_L = Z_0$. But for $Z_L \neq Z_0$ (for mismatched condition) there will be reflected waves (backward propagating waves, V^-) and also a transmitted wave. We define *load reflection coefficient* (Γ_L) or in general *reflection coefficient* (Γ) and transmission coefficient (T) as:

$$\Gamma = \frac{V^-}{V^+} = \frac{Z_L - Z_0}{Z_L + Z_0}; \qquad T = 1 + \Gamma = \frac{2Z_L}{Z_L + Z_0}$$

For a matched line $\Gamma = 0$, while for two extreme cases of mismatch: (i) short-circuited load ($Z_L = 0$), $\Gamma = -1$, there is voltage reversal ($V^+ = -V^-$) but there will be no current reversal ($I^+ = I^-$); (ii) open-circuited load ($Z_L = \infty$), $\Gamma = +1$, there will be no voltage reversal ($V^+ = V^-$) but with current reversal ($I^+ = -I^-$). Physically, this would mean that from a short-circuited load, the voltage wave will get reflected back to the generator side, while for an open-circuited load, the voltage wave will continue past the load against the generator direction and will give rise to a radiated signal caused by the fringing out of the electric field lines. However, in both cases, the magnitude of the maximum value of Γ is 1; thus $0 \leq |\Gamma| \leq 1$.

The transmission coefficient between two points in a circuit is often expressed in decibels when it is called the *insertion loss*, given by:

$$L_{\text{insertion}} = -20 \log |T| \text{ (in dB)}$$

- The *transmission line impedance equation* at higher RF/microwave frequency is given by:

$$Z(z) = Z_0 \frac{Z_L + j\, Z_0 \tan\beta z}{Z_0 + j\, Z_L \tan\beta z}$$

- A matched line is a *flat line* or *infinite line* and acts as a *non-resonant* line in which only a forward travelling wave is present. For a mismatched line there will be both the incident wave and the reflected wave and the two together will interfere on the line giving rise to the so called *standing wave* (having voltage/current maxima and minima, the maximum and minimum being separated by a quarter-wavelength) and acts as LC *resonant line* at RF. The ratio of the maximum of the standing wave pattern to the minimum is defined as the *standing-wave ratio*, designated by ρ which may also be expressed in terms of the reflection coefficient:

$$\rho = \frac{|V(z)|_{\max}}{|V(z)|_{\min}} = \frac{1 + |\Gamma_L|}{1 - |\Gamma_L|}$$

The bounds of standing wave ratio is given by $1 \leq \text{SWR} \leq \infty$. A high value of ρ on a line is undesirable as it results in large power loss.

The reflection coefficient and impedance at a voltage maximum/minimum of the standing wave pattern at any location on the line are given by:

$$\Gamma(z) = |\Gamma_L| \text{ and } Z_{\text{in}} = Z_0 \rho, \quad \text{at a voltage maximum}$$

$$\Gamma(z) = -|\Gamma_L| \text{ and } Z_{\text{in}} = Z_0/\rho, \quad \text{at a voltage minimum}$$

For pure resistive termination: (i) when $R_L > R_0$, $|\Gamma_L|$ is positive and SWR = R_L/R_0 (ii) when $R_L < R_0$, $|\Gamma_L|$ is negative and SWR is R_0/R_L. For pure reactive termination: (i) if $x = +1$ (inductance), $\Gamma_L = 1\angle 90°$ while (ii) if $x = -1$ (capacitance), $\Gamma_L = 1\angle -90°$ with SWR = ∞ (which is also the case for short and open-circuited load conditions).

- For a mismatched load, a portion of the available power is reflected back towards the source which is estimated in terms of *Return loss* (L_{return}) and the power dissipated in the load is estimated in terms of *Reflection loss* or the *Mismatch loss* ($L_{\text{reflection}}$) given by:

$$L_{\text{return}} = 20\log_{10}\left(\frac{1}{|\Gamma|}\right) \quad \text{and} \quad L_{\text{reflection}} = 10\log_{10}\frac{1}{(1-|\Gamma|^2)}$$

Transmission line as HF circuit elements

- For open-circuit termination, $Z_{\text{inoc}} = -jR_0 \cot\beta l$; thus, the input impedance of an open-circuited loss-less line is purely reactive and this impedance can be either capacitive or inductive depending on the length of the line.

  For short-circuit termination, $Z_{\text{insc}} = jR_0 \tan\beta l$; the input impedance of a short-circuited loss-less line can also be either purely inductive or capacitive, depending on the length of the line.

  By measuring the input impedance of line section under open and short conditions, characteristic impedance (Z_0)/ characteristic resistance (R_0) and the propagation constant (β) of the line can be determined using:

$$R_0 = \sqrt{Z_{\text{inoc}}\, Z_{\text{insc}}} = Z_0 \quad \text{and} \quad \beta = \frac{1}{l}\tan^{-1}\sqrt{\frac{Z_{\text{insc}}}{Z_{\text{inoc}}}}$$

- When the length of the line is an odd multiple of $\lambda/4$, $Z_{\text{in}} = \dfrac{R_0^2}{Z_L}$; thus a quarter-wave loss-less line transforms the load impedance to the input terminals as its inverse multiplied by the square of the characteristic resistance hence acting as a *quarter-wave impedance transformer*.

  When the length of a line is an integral multiple of $\lambda/2$, $Z_{\text{in}} = Z_L$; thus, a half wave loss-less line transfers the load impedance to the input terminals without change and hence acts as a *one to one impedance transformer*.

  The short and open-circuited transmission lines of quarter or half wave lengths are also very much in use as *stubs* for impedance matching.

Smith chart and its applications

- The *Smith chart* is a graphical means of solving transmission line problems and the chart represents a plot of normalised resistance (r) and normalised reactance (x) functions in the complex reflection coefficient ($\Gamma = r \pm jx$) plane. Most vector network analysers (useful for microwave circuit characterisation) have the Smith chart display as one of the primary screen options.
- Smith chart is bounded by $r = 0$ circle, where $|\Gamma| = 1$. All the r-circles pass through $r = \infty$ point, that is, through the open-circuit (o.c) point with the $r = 0$ circle having the largest radius (unity) and the r-circles become progressively smaller as r increases

from 0 (short-circuit, s.c point) towards ∞, ending at the o.c point. The x-circles become progressively smaller as $|x|$ increases from 0 towards ∞, ending at the o.c point and all the x-circles also appear to originate from the o.c point and are bounded by the $r = 0$ circle. The centre of the Smith chart (origin of the Γ-plane) represents $r = 1$ and $x = 0$, that is, a perfectly matched condition. The intersection of an r-circle and an x-circle defines a point that represents a normalised load impedance $z_L = r \pm jx$, which when multiplied by the characteristic impedance of the line gives the actual load impedance.

- The Smith chart has many applications in solving microwave circuit problems including the solution of stub matching problems and impedance measurement at microwave frequencies. In impedance measurement, if the load has an inductive reactance component then as per convention we need to move along the WTG (wavelength towards generator) direction when the load is replaced by the short; while if the load has a capacitive reactance component then we need to move along the WTL (wavelength towards load) direction.
- In the solution of some transmission line problems, it may be necessary at different stages of solving to use the impedance and admittance charts simultaneously; in such cases an impedance–admittance chart—Smith and Jones chart is used.
- When working with the design of oscillators and negative resistance amplifiers, the active load associated with such systems might make the magnitude of the reflection coefficient greater than unity. To accommodate the possibilities of $|\Gamma| > 1$, an open form of the Smith chart (called *compressed Smith chart*) is used, in which the standard chart happens to be only a part of the said compressed chart and for all practical purposes, a maximum value of $|\Gamma|$ is limited to 3.16.

Planar transmission lines

- Planar transmission lines are useful for MICs as this allows integration of both the passive and active devices via planar technology used for integrated circuit design. A *microstrip* is a practical variant of the basic planar transmission line, the strip-line. Some common variants of the microstrip are *inverted microstrip* and *trapped inverted microstrip* which are useful for millimeter-wave applications. Other variants of the microstrip are *slot-line* (which can provide larger characteristic impedance but Q value is low), *coplanar-line* (which combines some of the advantages of microstrip and slot-line in a single structure), *fin-line* (shielded versions of slot-line and coplanar line that helps to alleviate the radiation problem of the two structures and also low-loss compared to the microstrip and is thus more useful for millimeter-wave applications). At ultra-microwave frequencies (>100 GHz) a useful transmission structure is the *dielectric integrated guide*; the basic form of which is the image guide with variants such as the trapped image guide and non-radiative guide.
- The mode of propagation in a microstrip is quasi-TEM in the sense that the electromagnetic wave traverses in the air-dielectric composite, which is an inhomogeneous medium. But for all practical purposes, it is considered to be homogeneous by defining an effective relative permittivity ($\varepsilon_{\text{reff}}$), with characteristic impedance (Z_0) and effective wavelength (λ_{eff}) given by:

$$\varepsilon_{\text{reff}} = \frac{\varepsilon_r + 1}{2} + \frac{\varepsilon_r - 1}{2}\left\{\left(1 + 12\frac{h}{W}\right)^{-1/2} + 0.04\left(1 - \frac{W}{h}\right)^2\right\} \quad \text{for} \quad \left(\frac{W}{h}\right) \le 1$$

$$= \frac{\varepsilon_r + 1}{2} + \frac{\varepsilon_r - 1}{2}\left(1 + 12\frac{h}{W}\right)^{-1/2} \quad \text{for} \quad \left(\frac{W}{h}\right) \ge 1$$

$$Z_0 = \frac{60}{\sqrt{\varepsilon_{\text{reff}}}} \ln\left(\frac{8h}{W} + \frac{W}{4h}\right) \quad \text{for} \quad \left(\frac{W}{h}\right) \le 1$$

$$= \frac{120\,\pi}{\sqrt{\varepsilon_{\text{reff}}}\left\{\left(\frac{W}{h} + 1.393\right) + 0.667\ln\left(\frac{W}{h} + 1.444\right)\right\}} \quad \text{for} \quad \left(\frac{W}{h}\right) \ge 1$$

$$\lambda_{\text{eff}} = \frac{\lambda}{\sqrt{\varepsilon_r}}\left\{\frac{\varepsilon_r}{1 + 0.6\,(\varepsilon_r - 1)\,(W/h)^{0.0297}}\right\} \quad \text{for} \quad \left(\frac{W}{h}\right) < 0.6$$

$$= \frac{\lambda}{\sqrt{\varepsilon_r}}\left\{\frac{\varepsilon_r}{1 + 0.63\,(\varepsilon_r - 1)\,(W/h)^{0.1255}}\right\} \quad \text{for} \quad \left(\frac{W}{h}\right) > 0.6$$

Both the effective dielectric constant and the characteristic impedance vary with the geometry (W/h) of the microstrip line. Standard computed plots for Z_0 vs W/h and $\varepsilon_{\text{reff}}$ vs W/h for different parametric values of the dielectric constant (ε_r) of the substrate material are available in texts.

For design purposes, we need to have a relation that allows us to compute W/h ratios based on the given characteristic impedance Z_0 and the dielectric constant ε_r of the substrate material.

$$\frac{W}{h} = \frac{8e^A}{e^{2A} - 2} \quad \text{or} \quad \left(\frac{W}{h}\right) \le 2$$

$$= \frac{2}{\pi}\left[B - 1 - \ln(2B - 1) + \frac{\varepsilon_r - 1}{2\varepsilon_r}\left\{\ln(B - 1) + 0.39 - \frac{0.61}{\varepsilon_r}\right\}\right] \quad \text{for} \quad \left(\frac{W}{h}\right) \ge 2$$

where: $A = \dfrac{Z_0}{60}\sqrt{\dfrac{\varepsilon_r + 1}{2}} + \dfrac{\varepsilon_r - 1}{\varepsilon_r + 1}\left(0.23 + \dfrac{0.11}{\varepsilon_r}\right)$ and $B = \dfrac{377\,\pi}{2Z_0\sqrt{\varepsilon_r}}$

- Even and odd mode excitation in coupled transmission lines is of importance in the analysis of a number of microwave components like coupled-line coupler, short-slot hybrid, Wilkinson power divider/combiner and so on, including the design of coupled line filters. For even-mode excitation, the two strip conductors are at the same potential and carry equal currents in the same direction, the symmetry plane being the magnetic wall. For odd-mode excitation, the strip conductors are of opposite polarity and have oppositely directed currents; in this case, the symmetry plane represents an electric wall.

The expressions for even-mode characteristic impedance (Z_{0e}) and odd-mode characteristic impedance (Z_{0o}) are given below:

$$Z_{0e} = \frac{1}{v_p C_e} \quad \text{and} \quad Z_{0o} = \frac{1}{v_p C_o}$$

Review Questions

3.1 Why can't the twin-wire line be used at radio frequencies above 300 MHz?

3.2 Coaxial line is a better option than twin-wire line above 300 MHz but is again limited for operation up to 3 GHz—why?

3.3 Rectangular or circular waveguides can be used for microwave signals from 3–300 GHz, but are not suitable for MHz signals—why?

3.4 What do you understand by 'lumped parameter' and 'distributed parameter' network?

3.5 Mention the advantages and disadvantages of planar transmission lines.

3.6 What are the transmission line parameters? Discuss their physical relevance with reference to a coaxial line.

3.7 What are Telegrapher's equations? How do you derive time-harmonic transmission line equations from Telegrapher's equations?

3.8 Discuss the significance of the characteristic impedance of a transmission line.

3.9 At microwave frequencies, a coaxial line is non-dispersive and hence distortionless—explain.

3.10 What do you understand by load reflection coefficient? Discuss its significance with respect to a short-circuited and open-circuited load.

3.11 What are the minimum and maximum values of the reflection coefficient?

3.12 What is insertion loss? Is a high or low value of insertion loss desirable?

3.13 Voltage, current and impedance vary along the length of a RF transmission line—justify.

3.14 A mismatched (non-flat) RF transmission line supports standing waves—explain.

3.15 Discuss the characteristics of a standing wave in a mismatched RF transmission line.

3.16 What are the minimum and maximum values of the standing wave ratio?

3.17 What are the values of reflection coefficient and impedance at voltage minimum and maximum of a standing wave.

3.18 Discuss the nature of a standing wave and the reflection coefficient for different load conditions.

3.19 What do you understand by return loss and reflection loss with reference to a microwave load? Is a high or low value of return loss desirable?

3.20 Explain how short sections of open and short-circuited transmission lines can give distributed capacitance and inductance at microwave frequency.

3.21 Discuss how one can determine the characteristic resistance and propagation constant of a transmission line by measuring the input impedance under open and short conditions.

3.22 Discuss the principle of quarter-wave impedance transformation. How is one to one impedance transformation realised with a half-wave line?

3.23 What is the Smith chart? Mark the short-circuit point and open-circuit point on the Smith chart. The centre of the Smith chart represents matched load—explain.

3.24 The Smith chart is bounded by a $|\Gamma| = 1$ circle—explain. But to accommodate the situation $|\Gamma| > 1$ for oscillators and for negative resistance amplifiera, what type of Smith chart is used?

3.25 What do the WTG and WTL scales on the rim of the Smith chart represent? How are these scalings done?

3.26 What is the Smith and Jones chart?

3.27 For a lossy transmission line, the VSWR follows a spiralling path instead of a circle of constant radius—explain.

3.28 Name a few planar transmission lines useful for microwave integrated circuits. To realise high characteristic impedance, would you prefer a microstrip or a fin-line structure?

3.29 Discuss lucidly the quasi-TEM mode of propagation of electromagnetic waves in a microstrip line.

3.30 Explain in brief the concept of even and odd modes in coupled strip-line.

3.31 Name two variants of the microstrip line which are better suited at millimetre-wave frequency.

3.32 The slot-line is a complementary version of the microstrip with some advantages and disadvantages—discuss. What are the advantages of CPW?

3.33 What type of planar transmission line is used above 100 GHz?

Numerical Problems

3.1 In the design of a transmission line network, it is required to design a reactance of $j150$ as an element in the network. Show how this can be accomplished with (a) a loss-less, shorted transmission line with a characteristic impedance of 50 Ω at a wavelength of 1 m, (b) a loss-less open transmission line with a characteristic impedance of 75 Ω at a wavelength of 1 m. If a capacitive reactance of $-j150$ is need instead, what are the answers to (a) and (b)?

[**Ans:** (a) With a short-circuited transmission line of length $d_{sc} = (0.1987 + 0.5n)$ m
(b) With an open-circuited transmission line of length $d_{op} = (0.9262 + 0.5n)$ m
Answers for (a) and (b) for the second part of the problem are: $(0.8012 + 0.5n)$ m
and $(0.0737 + 0.5n)$ m respectively; where $n = 0, 1, 2, \ldots$]

3.2 Consider a loss-less transmission line of characteristic impedance R_0. A time harmonic voltage source of amplitude $V_g = 100$ V and an internal resistance $R_g = R_0 = 50\ \Omega$ is connected to the input terminals of the line, which is terminated with a load impedance $Z_L = 50 - j25$. Calculate the incident power P_{in}, reflection coefficient Γ, VSWR ρ, average power to the load P_L, the load current and load voltage I_L and V_L respectively.

[**Ans:** $P_{in} = 25$ W, $\Gamma = 0.243\ \angle -75°$, $\rho = 1.65$,
$P_L = 23.5$ W, $|I_L| = 0.96$ A $|V_L| = 54.1$ V]

3.3 A transmission line of characteristic impedance 100 Ω is terminated with a resistive load of 75 Ω. The voltage amplitude measured at the load has a magnitude of 100 V. Calculate (a) maximum and minimum voltage on the line (magnitude only), (b) location of the voltage minimum and maximum from the load end, (c) the minimum and maximum impedance on the line. Where does this occur? (d) Power transmitted to the load.

[**Ans:** (a) $V_{max} = 300$ V and $V_{min} = 100$ V (note $V_{min} = V_L = 100$ V, as $R_L < Z_0$); (b) The minima occurs at: $z = 0, 0.5\lambda, 1.0\lambda, 1.5\lambda \ldots$
while the maxima occurs at: $z = 0.25\lambda, 0.75\lambda, 1.25\lambda \ldots$
(c) $Z_{min} = 75\ \Omega$ and $Z_{max} = 225\ \Omega$; (d) $P_L = 200$ W]

3.4 A 3.4λ long transmission line having attenuation constant of 0.5 dB/λ and characteristic impedance of 50 Ω is connected to matched source that produces a signal of amplitude 20 V. Find (a) the power that is delivered by the source, (b) the power being delivered to the load, and (c) the amount of power that is lost in the line.

[**Ans:** $P_S = 981$ mW, $P_L = 663$ mW, $P_{loss} = 318$ mW]

3.5 In an experiment conducted on a 50 Ω loss-less transmission line terminated in an unknown load impedance, it is found that the standing wave ratio is 3.0. When the load is replaced by a short, the reference minima shifts 6 cm towards the load and the successive voltage minima are 30 cm apart. Using the Smith chart find (a) the load impedance, (b) the reflection coefficient of the load. Where does the first voltage minimum occur when the load is replaced by a short?

[**Ans:** $Z_L = (23.75 - j30)\ \Omega$, $\Gamma_L = 0.5\, e^{j252^0}$, when the load is replaced by a short,
the first voltage minimum occurs at 30 cm from the load]

3.6 A 70 Ω high frequency loss-less line is used at a frequency where $\lambda = 80$ cm with a load at $x = 0$ of $(140 + j91)$ Ω. Using the Smith chart find Γ,VSWR, distance to the first voltage maximum from the load, distance to the first voltage minimum from the load, the impedance at V_{max}, the impedance at V_{min}, the input impedance of a section of line that is 54 cm long.

[**Ans:** $\Gamma = 0.5 \angle 29°$; VSWR = 3.0; $d_{Vmax} = 0.04\lambda$ (towards generator) or 3.2 cm from the load, $Z_{max} = (210 + j0)$ Ω; $d_{Vmin} = 0.29\lambda = 23.2$ cm from load, $Z_{min} = (23.1 + j0)$ Ω, $Z_{in} = (39.2 - j49.7)$ Ω].

3.7 A loss-less transmission line of length 0.275λ and characteristic impedance 50 Ω is terminated in an impedance (120 + *j*100) Ω. Using the Smith chart find (a) the voltage reflection coefficient, (b) standing wave ratio, (c) input impedance, and (d) the location of a voltage maximum on the line.

[**Ans:** $\Gamma = 0.56\, e^{j24.5^0}$, $\rho = 3.7$, $Z_i = 12.5 - j7.0$, a voltage maximum occurs at 0.034λ from the load]

3.8 A microstrip line is designed at 1.0 GHz with a substrate material of thickness 0.031 inch having a dielectric constant $\varepsilon_r = 2.22$ to realise a characteristic impedance of 50 Ω. Calculate the strip width, the effective dielectric constant and the phase velocity of the propagating electromagnetic wave down the microstrip line.

[**Ans:** $W = 0.095$ inch, $\varepsilon_{reff} = 1.891$, $v_{pms} = 2.18 \times 10^{10}$ cm/s]

3.9 With the MATLAB program generate the curves given in Figs 3.21a 3.21b.

4

Guided Structures: Waveguides and Cavity Resonators

The inner vision of the soul is one that finds every medium transparent and none translucent

4.1 Introduction

It was mentioned in Chapter 3 that for frequencies above about 3 GHz, single conductor wave-guiding structures (waveguides) have reasonable practical dimensions, provide much lower loss and have substantially larger power handling capability. The possibility of electromagnetic wave propagation in hollow metallic tubes was theoretically predicted by Lord Rayleigh in 1897 (and used intuitively by Sir J C Bose of Calcutta, India, as a radiating tube in his millimetre-wave instrument designed during 1895–1897). The practical and well-understood usage of these waveguides received serious attention during World War II when their use with microwave radar and communication systems emerged as a crucially important application. In fact, waveguides and waveguide-based components for microwave applications were systematically studied and developed by Radiation Laboratory of MIT, US, during the 1940s. However, in spite of their many desirable properties, waveguides are often bulky at lower microwave frequencies (< 3 GHz) and hence their applicability in the higher end of the microwave spectrum (10–100 GHz) is more prevalent.

The simplest type of waveguide consists of two perfectly conducting plates, called parallel-plate waveguide (Fig. 4.1), between which electromagnetic waves of various types (known as 'modes') can be guided. The basic concept from which its idea originates was discussed in Section 2.7.2.2. In such a structure, electromagnetic waves are guided in the z direction by confining the wave energy in the x direction, while it has to be of infinite extent (theoretically) in the y direction. But this is not a practically suitable wave guiding structure, because for all practical purposes, the plates are always finite in extent in the y direction, with fringing fields at the edges. Hence electromagnetic energy will leak through the open sides of the guide and create undesirable stray couplings to other circuits and systems.

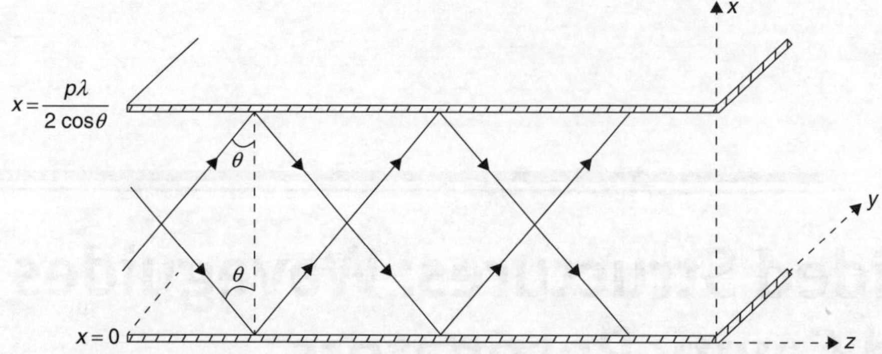

Fig. 4.1 Parallel-plate waveguide.

Thus, in practice, *waveguides* are hollow or dielectric-filled metallic tubes of rectangular or circular cross-section which confine the wave energy in both transverse directions (x and y directions). Further, enclosed metallic structures (which are essentially segments of waveguides, that is, truncated waveguides with closed end faces separated by integral multiples of half-wavelength) are designed to confine electromagnetic energy in all the three dimensions and are called *cavity resonators*. These are very important practical structures at microwave frequencies and are in fact used as resonant devices having a very high quality factor, Q.

In a single conductor, wave-guiding structure TEM mode cannot propagate (unlike two conductor wave-guiding structure, coaxial line) because such a mode would require either a centre conductor for allowing the axial conduction current to flow or an axial electric field for allowing the flow of an axial displacement current to satisfy Maxwell's magnetic curl equation. But a single conductor waveguide has no central conductor and TEM mode has no axial electric field, making TEM mode propagation impossible in the single conductor

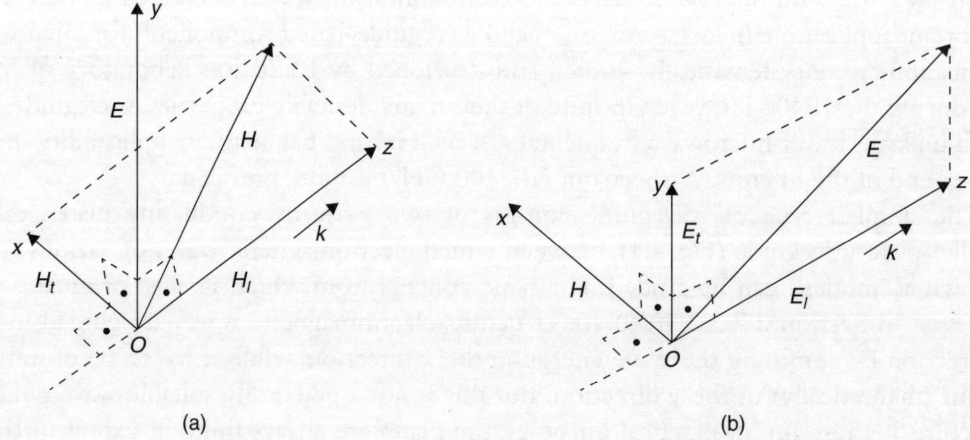

Fig. 4.2 (a) Transverse Electric (TE) mode or *H*-wave (b) Transverse Magnetic (TM) mode or *E*-wave.

wave-guiding structure. Thus, the mode of propagation in the single conductor waveguide has to be either transverse electric (TE) mode or *H*-wave, in which the electric field is always transverse while the magnetic field may have an additional component in the direction of propagation (Fig. 4.2a); or transverse magnetic (TM) mode or *E*-wave, in which the magnetic field is always transverse while the electric field may have an additional component in the direction of propagation (Fig. 4.2b).

However, these TE or TM modes propagating down the single conductor wave-guiding structure (waveguide) actually result from the superposition of TEM waves reflected back and forth from the metallic walls of the waveguide (as may be observed from Fig. 4.1).

4.2 Rectangular Waveguide

Guided wave analysis of the waveguide structure [11, 4] is based on the solution of Maxwell's equations with appropriate boundary conditions. For metallic waveguides, the boundary conditions on metal walls are given by $E_{\text{tangential}} = 0$ and $H_{\text{normal}} = 0$. The medium within the waveguide is normally source free ($J = \sigma E = 0$) and non-conducting ($\sigma = 0$). Under this condition for rectangular waveguide (Fig. 4.3) the Maxwell's curl equations (Eqs 2.5c and 2.5d) can be written in rectangular coordinate axes as:

$$[\nabla \times \vec{E} = -j\omega\mu\vec{H}] \qquad\qquad [\nabla \times \vec{H} = j\omega\mu\vec{E}]$$

$$\frac{\partial E_z}{\partial y} - \frac{\partial E_y}{\partial z} = -j\omega\mu H_x \qquad\qquad \frac{\partial H_z}{\partial y} - \frac{\partial H_y}{\partial z} = j\omega\varepsilon E_x$$

$$\frac{\partial E_x}{\partial z} - \frac{\partial E_z}{\partial x} = -j\omega\mu H_y \quad (4.1\text{a}) \qquad\qquad \frac{\partial H_x}{\partial z} - \frac{\partial H_z}{\partial x} = j\omega\varepsilon E_y \quad (4.1\text{b})$$

$$\frac{\partial E_y}{\partial x} - \frac{\partial E_x}{\partial y} = -j\omega\mu H_z \qquad\qquad \frac{\partial H_y}{\partial x} - \frac{\partial H_x}{\partial y} = j\omega\varepsilon E_z$$

While the wave equations in $\vec{E}$ and $\vec{H}$ (Eq. 2.7) are given by:

$$[\nabla^2 \vec{E}_s - \gamma^2 \vec{E}_s = 0] \text{ and } [\nabla^2 \vec{H}_s - \gamma^2 \vec{H}_s = 0], \text{ with } \gamma = \sqrt{(j\omega\varepsilon)(j\omega\mu)}$$

$$\frac{\partial^2 \vec{E}}{\partial x^2} + \frac{\partial^2 \vec{E}}{\partial y^2} + \frac{\partial^2 \vec{E}}{\partial z^2} = -\omega^2 \mu\varepsilon\vec{E} \tag{4.2a}$$

$$\frac{\partial^2 \vec{H}}{\partial x^2} + \frac{\partial^2 \vec{H}}{\partial y^2} + \frac{\partial^2 \vec{H}}{\partial z^2} = -\omega^2 \mu\varepsilon\vec{H} \tag{4.2b}$$

Let us assume that the wave propagates in the $+z$ direction with field variation of the form $e^{-\gamma_g z}$, where $\gamma_g = \alpha_g + j\beta_g$ is the complex propagation constant of the wave propagating down the waveguide. In general, the propgation constant (γ_g) in the guided structure will be different from the propgation constant (γ) in the unbounded medium. Note that γ_g

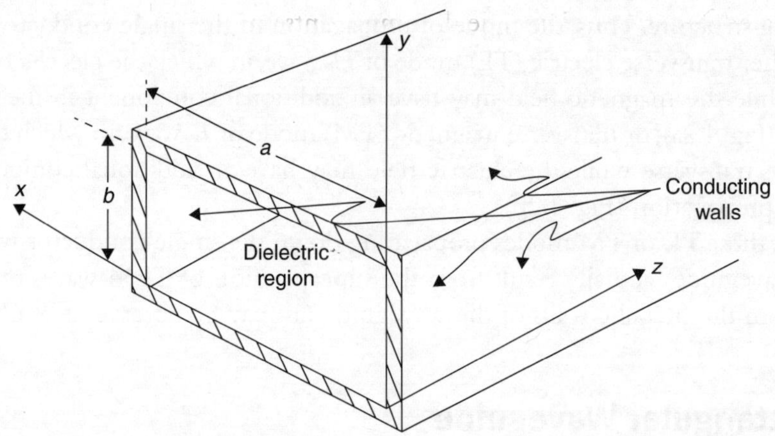

Fig. 4.3 Rectangular waveguide.

happens to be an imaginary number $j\beta_g$, if $\alpha_g = 0$, when the expressions (Eqs 4.2) represent a wave without attenuation—a *propagating wave.* On the other hand, if γ_g is real so that $\beta_g = 0$, there is no wave motion but only an exponential decrease in amplitude—an *evanescent wave* (dictionary meaning of *evanescent* is *tending to vanish like vapour*).

With the field variation assumed, we have $\frac{\partial}{\partial z} \rightarrow -\gamma_g$, and for a rectangular waveguide both x and y derivatives of the field will exist (as it is bounded with metallic walls in both x and y directions). Thus Eqs 4.1 can be written as:

$$\frac{\partial E_z}{\partial y} + \gamma_g E_y = -j\omega\mu H_x \qquad \frac{\partial H_z}{\partial y} + \gamma_g H_y = j\omega\varepsilon E_x$$

$$\frac{\partial E_z}{\partial x} + \gamma_g E_x = j\omega\mu H_y \quad (4.3a) \qquad \frac{\partial H_z}{\partial x} + \gamma_g H_x = -j\omega\varepsilon H_y \quad (4.3b)$$

$$\frac{\partial E_y}{\partial x} - \frac{\partial E_x}{\partial y} = -j\omega\mu H_z \qquad \frac{\partial H_y}{\partial x} - \frac{\partial H_x}{\partial y} = j\omega\varepsilon E_z$$

While the wave equations for E_z and H_z can be written as:

$$\frac{\partial^2 E_z}{\partial x^2} + \frac{\partial^2 E_z}{\partial y^2} + \gamma_g^2 E_z = -\omega^2 \mu\varepsilon E_z \quad (4.4a)$$

$$\frac{\partial^2 H_z}{\partial x^2} + \frac{\partial^2 H_z}{\partial y^2} + \gamma_g^2 H_Z = -\omega^2 \mu\varepsilon H_z \quad (4.4b)$$

The relationship among various field components that exist within the guide being represented by Eq. 4.3, may be expressed in terms of E_z and H_z with judicious manipulation of those equations. This will help us to understand more explicitly in analytical terms why the TEM mode cannot exist within single conductor guiding structures (waveguides).

The manipulated equations giving field components in terms of E_z and H_z are given below.

$$H_x = -\frac{\gamma_g}{k_c^2}\frac{\partial H_z}{\partial x} + j\frac{\omega\varepsilon}{k_c^2}\frac{\partial E_z}{\partial y} \tag{4.5a}$$

$$H_y = -\frac{\gamma_g}{k_c^2}\frac{\partial H_z}{\partial y} - j\frac{\omega\varepsilon}{k_c^2}\frac{\partial E_z}{\partial x} \tag{4.5b}$$

$$E_x = -\frac{\gamma_g}{k_c^2}\frac{\partial E_z}{\partial x} - j\frac{\omega\mu}{k_c^2}\frac{\partial H_z}{\partial y} \tag{4.5c}$$

$$E_y = -\frac{\gamma_g}{k_c^2}\frac{\partial E_z}{\partial y} + j\frac{\omega\mu}{k_c^2}\frac{\partial H_z}{\partial x} \tag{4.5d}$$

where $k_c^2 = \gamma_g^2 + k^2$ (4.5e)

in which, $k = \omega\sqrt{\mu\varepsilon}$ (Eq. 2.10).

It is clear from Eqs 4.5, that if E_z and H_z are simultaneously set to zero (which is the case for a TEM wave), then all fields within the waveguide will be nonexistent. Thus, the TEM mode of propagation is not possible in a waveguide; hence there must always be either an E_z or an H_z component. It is thus convention to talk of transverse electric (TE) mode (for which $E_z \equiv 0$) and transverse magnetic (TM) mode (for which $H_z \equiv 0$). TE and TM are US nomenclature, while in the literature some other terms are also used for TE and TM waves. In British nomenclature, TE waves are known as *H-waves* (as *H*-field has a component in the direction of propagation). The TM wave in British nomenclature is called the *E*-wave (as *E*-field has a component in the direction of propagation).

4.2.1 Transverse Electric (TE)-mode/*H*-wave

The wave equation for TE waves is given by Eq. 4.4b which can be re-written as:

$$\frac{\partial^2 H_z}{\partial x^2} + \frac{\partial^2 H_z}{\partial y^2} + (\gamma_g^2 + k^2)\, H_z = 0 \tag{4.6a}$$

When the wave equation for plane wave propagation in free space (for TEM wave, Eq. 2.7) is compared with Eq. 4.6, we find that the propagation constant for TE wave is given by:

$$\gamma_{g\mathrm{TE}} = \sqrt{k_c^2 - k^2} = \sqrt{k_c^2 - \omega^2\mu\varepsilon} \tag{4.6b}$$

(which follows from Eq. 4.5e and Eq. 2.8a).

It is worth noting that for TEM wave (which is the mode of propagation in free space and also in coaxial transmission line) the propagation constant $k = 0$, when $\omega = 0$, (that is, propagation starts just after DC and continues up to the frequency limited only by the attenuation of the signal caused by the losses in the dielectric, air for free space and the dielectric material between the two conductors of the coaxial line) while for TE wave in waveguide, the propagation constant $\gamma_{g\text{TE}} = 0$, when $k_c = \omega_c\sqrt{\mu\varepsilon}$. Zero propagation constant means no propagation. Thus, propagation starts after a *cut-off frequency* f_c, given by:

$$f_c = \frac{k_c}{2\pi\sqrt{\mu\varepsilon}} \tag{4.6c}$$

where k_c is called the *cut-off wave number.*

A question that naturally arises is, what happens when $\omega^2\mu\varepsilon < k_c^2$? For these values of k_c, the propagation constant $\gamma_{g\text{TE}}$ is real (Eq. 4.6b) which indicates that the signal will be non-propagating (since $\gamma = \alpha$, under this condition; with $\beta = 0$). It is seen subsequently that this non-propagation is not due to signal attenuation caused by any loss in the medium (in fact a high frequency waveguide system is loss-less) but physically it would mean *evanescence* of the wave.

However, when $\omega^2\mu\varepsilon > k_c^2$, the propagation constant given by Eq. 4.6 is imaginary and the wave propagates without attenuation (loss-less medium). This occurs at all frequencies above the cut-off frequency. Thus, a waveguide appears to behave as a high-pass filter (while the free-space and coaxial line acts as an all-pass filter which operates down to DC).

Next, let us take up the solution of wave equation (Eq. 4.6a) for TE modes which is a second order partial differential equation in two variables and can thus be solved by using separation of variables. For TE modes we write:

$$H_z(x, y, z) = H_z^0(x, y)\, e^{-\gamma_g z} \tag{4.7a}$$

To apply the method of separation of variables, let us assume solutions for $H_z^0(x, y)$ of the form:

$$H_z^0(x, y) = X(x)\,Y(y) \tag{4.7b}$$

On substituting this form of solution (Eq. 4.7b) for H_z in the wave equation (Eq. 4.6a) and using Eq. 4.5e, we have two ordinary differential equations—one only in X and the other only in Y:

$$\frac{1}{X}\frac{d^2X}{dx^2} + k_c^2 = A^2 \quad \text{and} \quad \frac{1}{Y}\frac{d^2Y}{dy^2} = -A^2 \tag{4.7c}$$

where A is a constant to be determined from boundary conditions. The general solutions of the second order differential equations (4.7c) are given by:

$$X = C_1 \cos(Bx) + C_2 \sin(Bx) \tag{4.7d}$$

$$Y = C_3 \cos(Ay) + C_4 \sin(Ay) \tag{4.7e}$$

where $B^2 = k_c^2 - A^2$. Substituting Eqs 4.7d and 4.7e in Eq. 4.7b we have:

$$H_z^0(x, y) = C_1C_3 \cos(Bx) \cos(Ay) + C_1C_4 \cos(Bx) \sin(Ay)$$
$$+ C_2C_3 \sin(Bx) \cos(Ay) + C_2C_4 \sin(Bx) \sin(Ay) \quad (4.7f)$$

Now, the boundary conditions to be satisfied by the electric and magnetic fields on the metallic walls of the rectangular waveguide (Fig. 4.3a) are $E_{tangential} = 0$ and $H_{normal} = 0$, which means:

On the side walls, that is, at $x = 0$ and $x = a$,

$$E_y = 0, \text{ and also } \frac{\partial H_z}{\partial x} = 0 \text{ (which follows from Eq. 4.5d)}$$

While, on the top and bottom walls at $y = 0$ and $y = b$,

$$E_x = 0, \text{ and also } \frac{\partial H_z}{\partial y} = 0 \text{ (which follows from Eq. 4.5c)}$$

Differentiating $H_z^0(x, y)$ given by Eq. 4.8c with respect to y and applying the boundary condition at $y = 0$ we have:

$$\frac{\partial H_z^0}{\partial y} = -0 + AC_1C_4 \cos(Bx) - 0 + AC_2C_4 \sin(Bx) = 0$$

It may be noted that for $A \neq 0$ we must have $C_4 = 0$ to make the above identity true (because, if we set $C_1 = C_2 = 0$, the field will be zero everywhere and so that is not considered).

Again, differentiating $H_z^0(x, y)$ with respect to x and applying the boundary condition at $x = 0$ we have:

$$\frac{\partial H_z^0}{\partial x} = BC_2C_3 \cos(Ay) = 0$$

With $B \neq 0$, We must have $C_2 = 0$ to make the above identity true because C_3 cannot be set to zero since the field will be zero everywhere.

We can thus write $H_z^0(x, y)$ as:

$$H_z^0(x, y) = C \cos(Bx) \cos(Ay) \quad (4.7g)$$

where $C = C_2C_4$ is simply a constant multiplier that is determined by the source that establishes the field. Note that the constants A and B are yet to be determined, for which we still have two more boundary conditions.

At $x = a$, we must have $\dfrac{\partial H_z^0(a, y)}{\partial x} = C \sin(Ba) \cos(Ay) = 0$

so for all $y, \rightarrow B = \dfrac{m\pi}{a}, m = 1, 2, 3, \cdots$

Similarly at $y = b$, we must have: $\dfrac{\partial H_z^0(x, b)}{\partial y} = C \cos(Bx) \sin(Ab) = 0$

so for all y, $\rightarrow A = \dfrac{n\pi}{b}$, $n = 1, 2, 3, \cdots$

These values of A and B for which the field solutions of the wave equation can satisfy the boundary conditions are known as *characteristic values* or *eigenvalues* (*eigen* is German for 'characteristic' or 'one's own'). It may be noted that physically m and n represent the number of half-cycle variations of the E-field (for TE modes)/H-field (for TM modes) in the x and y directions respectively. The eigenvalue solution of the wave equation with appropriate boundary conditions gives rise to different modes known as eigen modes (specific combinations of time-varying electric and magnetic fields) of propagation of electromagnetic waves down the waveguide.

The final expression for $H_z^0\,(x, y)$ is thus:

$$H_z^0\,(x, y) = C\cos\left(\frac{m\pi}{a}x\right)\cos\left(\frac{n\pi}{b}y\right) \tag{4.7h}$$

Noting that $E_z = 0$ for TE_{mn} modes, and assuming propagating modes so that $\gamma_g = j\beta_{gmn}$, the total set of field components in the phasor form for the propagating TE_{mn} modes can be written by using Eq. 4.7h along with Eqs 4.5a–4.5d.

$$H_z = C\cos\left(\frac{m\pi}{a}x\right)\cos\left(\frac{n\pi}{b}y\right)e^{-j\beta_{gmn}z}$$

$$H_x = -\frac{\gamma_g}{k_c^2}\frac{\partial H_z}{\partial x} = \frac{j\beta_{gmn}}{k_c^2}C\frac{m\pi}{a}\sin\left(\frac{m\pi}{a}x\right)\cos\left(\frac{n\pi}{b}y\right)e^{-j\beta_{gmn}z}$$

$$H_y = -\frac{\gamma_g}{k_c^2}\frac{\partial H_z}{\partial y} = \frac{j\beta_{gmn}}{k_c^2}C\frac{n\pi}{b}\cos\left(\frac{m\pi}{a}x\right)\sin\left(\frac{n\pi}{b}y\right)e^{-j\beta_{gmn}z} \tag{4.8}$$

$$E_x = -j\frac{\omega\mu}{k_c^2}\frac{\partial H_z}{\partial y} = \frac{j\omega\mu}{k_c^2}C\frac{n\pi}{b}\cos\left(\frac{m\pi}{a}x\right)\sin\left(\frac{n\pi}{b}y\right)e^{-j\beta_{gmn}z}$$

$$E_y = j\frac{\omega\mu}{k_c^2}\frac{\partial H_z}{\partial x} = -\frac{j\omega\mu}{k_c^2}C\frac{m\pi}{a}\sin\left(\frac{m\pi}{a}x\right)\cos\left(\frac{n\pi}{b}y\right)e^{-j\beta_{gmn}z}$$

The propagation properties of TE waves in a rectangular waveguide can be investigated by noting that $\gamma_{g\text{TE}} = \sqrt{k_c^2 - \omega^2\mu\varepsilon}$ (Eq. 4.6b) and $k_c^2 = A^2 + B^2$. Thus, we have the propagation constant as:

$$\gamma_{g\text{TE}} = \sqrt{\left(\frac{m\pi}{a}\right)^2 + \left(\frac{n\pi}{b}\right)^2 - \omega^2\mu\varepsilon} \tag{4.9a}$$

At low frequencies, when $\omega^2\mu\varepsilon$ is small, $\gamma_{g\text{TE}}$ is real; this implies that $\beta_{g\text{TE}} = 0$, that is, there is no phase shift and hence no wave motion along the waveguide. When $\omega^2\mu\varepsilon$ is greater than $[(m\pi/a)^2 + (n\pi/b)^2]$, $\gamma_{g\text{TE}}$ will be imaginary and may be written as $j\beta_{g\text{TE}}$, (since $\alpha = 0$, for perfectly conducting walls) exhibiting wave motion along the guide.

The frequency below which the wave propagation does not take place (when $\gamma_{g\text{TE}}$ is real) is the *cut-off frequency* and is given from Eq. 4.9a as:

$$\omega_{c\text{TE}_{mn}} = \frac{1}{\sqrt{\mu\varepsilon}}\sqrt{\left(\frac{m\pi}{a}\right)^2+\left(\frac{n\pi}{b}\right)^2} \quad \text{or} \quad f_{c\text{TE}_{mn}} = \frac{c}{2}\sqrt{\left(\frac{m}{a}\right)^2+\left(\frac{n}{b}\right)^2} \tag{4.9b}$$

Above the cut-off frequency, the phase constant can be written as:

$$\beta_{g\text{TE}_{mn}} = \sqrt{\omega^2\mu\varepsilon-\left(\frac{m\pi}{a}\right)^2+\left(\frac{n\pi}{b}\right)^2} = \beta\sqrt{1-\left(\frac{f_{c\text{TE}_{mn}}}{f}\right)^2} \tag{4.9c}$$

While the phase velocity of wave propagation as:

$$v_{p\text{TE}_{mn}} = \frac{\omega}{\beta_{g\text{TE}_{mn}}} = \frac{v_p}{\sqrt{1-\left(\frac{f_{c\text{TE}_{mn}}}{f}\right)^2}} \tag{4.9d}$$

where v_p $(=1/\sqrt{\mu\varepsilon})$.

The wavelength within the guide is called the guide-wavelength λ_g and is given by:

$$\lambda_{g\text{TE}_{mn}} = \frac{2\pi}{\beta_{g\text{TE}_{mn}}} = \frac{\lambda}{\sqrt{1-\left(\frac{f_{c\text{TE}_{mn}}}{f}\right)^2}} \tag{4.9e}$$

(where $\lambda = 2\pi/\beta$ is the free space wavelength).

Since m and n can have any integral value, multiple solutions for phase constant β are possible, giving rise to the possibility of different *modes of propagation* in a waveguide. From Eq. 4.9, it is observed that H_z exists even when both m and n are zero but all other field components are non-existent, thus the TE_{00} mode cannot exist. However the field components for the TE_{01} and TE_{10} modes are as follows:

(i) TE_{01} mode: $H_x = E_y = 0$

$$H_z = C\cos\left(\frac{\pi}{b}y\right)e^{-j\beta_{g01}z},\ E_x = \frac{j\omega\mu b}{\pi}C\sin\left(\frac{\pi}{b}y\right)e^{-j\beta_{g01}z},\ H_y = \frac{j\beta_{g01}b}{\pi}C\sin\left(\frac{\pi}{b}y\right)e^{-j\beta_{g01}z} \tag{4.10a}$$

(ii) TE_{10} mode: $E_x = H_y = 0$

$$H_z = C\cos\left(\frac{\pi}{a}x\right)e^{-j\beta_{g10}z},\ H_x = \frac{j\beta_{g10}a}{\pi}C\sin\left(\frac{\pi}{a}x\right)e^{-j\beta_{g10}z}$$

$$E_y = -\frac{j\omega\mu a}{\pi}C\sin\left(\frac{\pi}{a}x\right)e^{-j\beta_{g10}z} \tag{4.10b}$$

The presence of E_x and H_y field components for the TE_{01} mode indicate that power flow can take place in the z-direction; hence this mode can exist. Similarly the field components

E_y and H_x for the TE_{10} mode result in a power flow in the z-direction, suggesting that this is also a viable mode. Thus for the TE_{mn} mode, either m or n can be zero, but not both. If $a > b$, (which is unusually the case with a rectangular waveguide), then TE_{10} is the mode with a lower cut-off frequency compared to TE_{01}. Higher order modes possess higher cut-off frequencies in order to be propagated along a guide of a given dimension.

Finally, we can also calculate the wave impedance by taking the ratio of the transverse electric field and the transverse magnetic field in Eq. 4.9, this can be obtained both for frequencies below and above the cut-off frequency. Below the cut-off frequency, the propagation constant is real and is given by α_{gmn} while above the cut-off frequency it is $j\beta_{gmn}$, when we get:

Case I: For $f < f_c$,

$$Z_{wTE_{f<f_c}} = \frac{E_x}{H_y} = -\frac{E_y}{H_x} = j\frac{\omega\mu}{\alpha_{gmn}} \tag{4.11a}$$

Case II: For $f > f_c$,

$$Z_{wTE_{f>f_c}} = \frac{E_x}{H_y} = -\frac{E_y}{H_x} = \frac{\omega\mu}{\beta_{gTE_{mn}}} = \frac{\omega\mu}{\beta}\frac{1}{\sqrt{1-\left(\frac{f_{cTE_{mn}}}{f}\right)^2}} = \frac{\eta}{\sqrt{1-\left(\frac{f_{cTE_{mn}}}{f}\right)^2}} \tag{4.11b}$$

where Eqs 4.9c, 2.8a and 2.11 have been used.

It may be observed from Eq. 4.11b that in the propagation region of the waveguide, the wave impedance for the TE wave is frequency dependent and always larger than the wave impedance η for free space—for TEM waves except at $f \to \infty$, where the two are equal. At the cut-off, that is, when $f = f_{cTE}$, the wave impedance for TE wave propagation is infinite. This result gives a further physical insight regarding the cut-off. An infinite wave impedance at the cut-off means that for finite electric field intensity, the magnetic field intensity is trivially zero. The electromagnetic energy under this condition makes only a back and forth motion between the waveguide walls with no propagating signal along the guide. However, from Eq. 4.11a, it may be observed that wave impedance is purely inductive when the operating frequency is below the cut-off frequency and thus results in no average power flow. There is only energy storage in the waveguide due to reactive power while the wave is *evanescent*, that is, non-propagating.

4.2.2 Transverse Magnetic (TM)-mode/*E*-wave

The solution for TM waves can be carried out in a very similar way as discussed for TE waves, however, here we solve the wave equation for the E field, Eq. 4.4a, with boundary conditions: $E_x = 0$, at $y = 0$, $y = b$ and $E_y = 0$, at $x = 0$ and $x = a$. Noting that $H_z = 0$, for TM_{mn} modes, the total set of field components in the phasor form for the propagating TM_{mn} modes can be derived as:

$$E_z = C\sin\left(\frac{m\pi}{a}x\right)\sin\left(\frac{n\pi}{b}y\right)e^{-j\beta_{gmn}z}$$

$$E_x = -\frac{\gamma_g}{k_c^2}\frac{\partial E_z}{\partial x} = -\frac{j\beta_{gmn}}{k_c^2} C \frac{m\pi}{a} \cos\left(\frac{m\pi}{a}x\right) \sin\left(\frac{n\pi}{b}y\right) e^{-j\beta_{gmn}z}$$

$$E_y = -\frac{\gamma_g}{k_c^2}\frac{\partial E_z}{\partial y} = -\frac{j\beta_{gmn}}{k_c^2} C \frac{n\pi}{b} \sin\left(\frac{m\pi}{a}x\right) \cos\left(\frac{n\pi}{b}y\right) e^{-j\beta_{gmn}z} \quad (4.12)$$

$$H_x = j\frac{\omega\varepsilon}{k_c^2}\frac{\partial E_z}{\partial y} = \frac{j\omega\varepsilon}{k_c^2} C \frac{n\pi}{b} \sin\left(\frac{m\pi}{a}x\right) \cos\left(\frac{n\pi}{b}y\right) e^{-j\beta_{gmn}z}$$

$$H_y = -j\frac{\omega\varepsilon}{k_c^2}\frac{\partial E_z}{\partial x} = -\frac{j\omega\varepsilon}{k_c^2} C \frac{m\pi}{a} \cos\left(\frac{m\pi}{a}x\right) \sin\left(\frac{n\pi}{b}y\right) e^{-j\beta_{gmn}z}$$

As can be seen from Eq. 4.12, field components within the guide are non-zero only when both m and n are non-zero. Thus, the lowest cut-off frequency will occur for TM waves when $m = n = 1$, that is, for TM_{11} mode.

The cut-off frequency (f_{cTM}), phase constant (β_{gTM}), phase velocity (v_{pTM}) and the guide wavelength (λ_{gTM}) expressions are the same for TM_{mn} modes as for the TE_{mn} modes. The wave impedance for TM modes, namely Z_{wTM}, can be evaluated in the same manner as was done for TE waves (Eq. 4.12) and we will get:

Case I: For $f < f_c$,

$$Z_{wTM_{f<f_c}} = \frac{E_x}{H_y} = -\frac{E_y}{H_x} = -j\frac{\alpha_{gmn}}{\omega\varepsilon} \quad (4.13a)$$

Case II: For $f < f_c$,

$$Z_{wTM_{f>f_c}} = \frac{E_x}{H_y} = -\frac{E_y}{H_x} = \frac{\beta_{gTE_{mn}}}{\omega\varepsilon} = \frac{\beta}{\omega\mu}\sqrt{1-\left(\frac{f_{cTM_{mn}}}{f}\right)^2} = \eta\sqrt{1-\left(\frac{f_{cTM_{mn}}}{f}\right)^2} \quad (4.13b)$$

It may be concluded from Eq. 4.13a that for TM modes also, there is no average power flow in the waveguide for operating frequencies below the cut-off frequency as the wave impedance is now purely capacitive.

The plot of wave impedance as a function of frequency is shown in Fig. 4.4 for both TE and TM modes.

For both types of modes, the wave impedance above the cut-off frequency for the propagating wave is purely resistive thus allowing flow of power through the waveguide. However, for TE modes it starts from a very large value when the operating frequency is close to the cut-off frequency and gradually approaches the intrinsic impedance (η) of the unbounded medium while for TM modes it is always smaller than η. It is interesting to note that $\sqrt{Z_{wTE_{f>f_c}} \times Z_{wTM_{f>f_c}}} = \eta$.

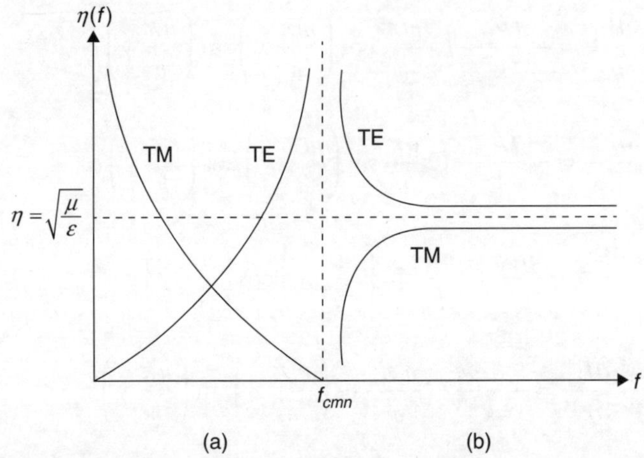

Fig. 4.4 Wave impedance of the TE and TM modes in the waveguide as a function of frequency (a) Below the cut-off (b) Above the cut-off.

4.2.3 TE-mode vs TM-mode and the Dominant Mode

Unlike the TM_{mn} modes (where neither m nor n can be zero), for TE_{mn} modes, it is possible to make either m or n (but not both) equal to zero without causing all the fields to vanish. Thus, a mode with a lower cut-off frequency for TE waves is possible rather than with TM waves. Normally, for a rectangular waveguide, m is taken along the broader dimension a, and n is along the narrower dimension b, with ($a > b$). Thus, TE_{10} mode (and not TE_{01}) has the lowest cut-off frequency and this mode is called the *dominant mode*. The modes with the same cut-off frequency are *degenerate modes* and in fact all corresponding TE_{mn} and TM_{mn} modes are degenerate modes (as both of them have the same cut-off frequency). Figure 4.5 shows some of the lower order TE/TM models possible in a rectangular waveguide with cut-off frequency relative to the dominant TE_{10} mode.

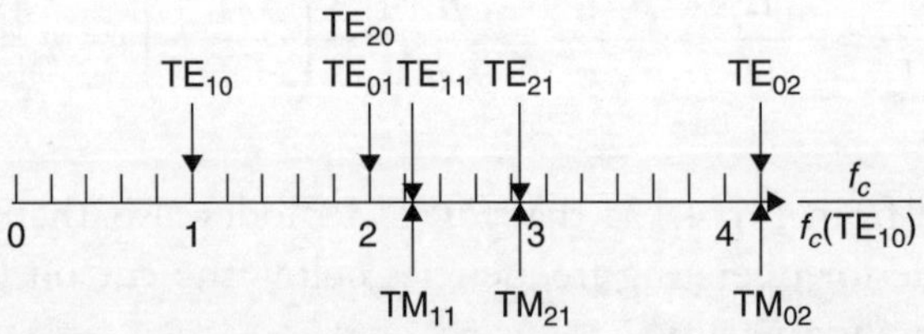

Fig. 4.5 Cut-off frequencies of the first few TE and TM modes in rectangular waveguide ($a = 2b$) relative to the cut-off frequency of the dominant (TE_{10}) mode.

The field components for TE_{10} mode has been given by Eq. 4.11b and other relevant parameters are given below.

$$\text{Cut-off frequency } f_{cTE_{10}} = \frac{c}{2}\sqrt{\left(\frac{1}{a}\right)^2} = \frac{c}{2a}, \text{ thus } \lambda_{cTE_{10}} = 2a$$

Propagation constant $\beta_{g\text{TE}_{10}} = \sqrt{\omega^2\mu\varepsilon - \left(\frac{\pi}{a}\right)^2} = \beta\sqrt{1-\left(\frac{f_{c\text{TE}_{10}}}{f}\right)^2}$

Phase velocity of wave propagation $v_{p\text{TE}_{10}} = \frac{\omega}{\beta_{g\text{TE}_{10}}} = \frac{v_p}{\sqrt{1-\left(\frac{f_{c\text{TE}_{10}}}{f}\right)^2}}$ (4.14)

Guide wavelength $\lambda_{g\text{TE}_{10}} = \frac{\lambda}{\sqrt{1-\left(\frac{f_{c\text{TE}_{10}}}{f}\right)^2}} = \frac{\lambda}{\sqrt{1-\left(\frac{\lambda}{2a}\right)^2}}$

Wave impedance★: $Z_{w\text{TE}_{10}} = \frac{\eta}{\sqrt{1-\left(\frac{f_{c\text{TE}_{10}}}{f}\right)^2}}$

★***Note***: Analogous to the transmission line, is the characteristic impedance (Z_0) of a rectangular waveguide. There are various ways of defining this characteristic impedance–voltage–current (*V–I*), power–current (*P–I*) and power–voltage (*P–V*); all of which depend on the guide dimensions in the same way but differ only by a constant and is given by [10, 11]:

$$Z_{0\text{TE}_{10}} = C\frac{b}{a}Z_{w\text{TE}_{10}}$$

where the constant $C = \pi/2$, with (*V–I*) definition, $\pi^2/8$ with (*P–I*) definition, and $C = 2$ with (*P–V*) definition. The choice of definition is arbitrary as long as the mode type and guide configuration remain fixed. Unless otherwise stated, a *modified* (*P–V*) definition is used in which the constant $C = 1$, when for air-filled rectangular waveguide $Z_{0\text{TE}_{10}} = 377\frac{b}{a}\frac{\lambda_g}{\lambda}$.

The significance of designing all practical waveguides for the dominant mode (except in some special situations) and some application view points in favour of using the dominant mode may be understood as follows. For wave propagation in a waveguide, the operating frequency f of the signal should be larger than the cut-off frequency f_c for that particular mode. It is known that higher order modes have higher value of cut-off frequency or smaller cut-off wavelength λ_c. Again $\lambda_c \,\alpha\, a$, the proportionality constant being smaller for higher order modes (typically $\lambda_{c\text{TE}_{10}} = 2a$ while $\lambda_{c\text{TM}_{11}} = 0.89a$). Thus, for a given operating frequency, a higher order mode will require a larger a dimension to fit the half-wavelength in the a dimension of the guide (generally, a is chosen to be $\approx \lambda/2$). In other words, the lower the order of the mode, the smaller will be the waveguide dimension. The dominant mode having lowest cut-off frequency will have the smallest a dimension (and hence the b dimension too as $b \approx 0.5a$ for practical design). Furthermore, a waveguide designed for the dominant mode has the advantage that it will not allow higher order modes to propagate via the guide as the half-waves for those modes cannot fit in the small a dimension of the guide; thus ensuring that no energy loss can occur through the spurious generation of other modes. A few more engineering considerations for the use of the TE_{10} dominant mode in a rectangular waveguide is that excitation of the mode is easier, attenuation of the signal while propagating down the guide in this mode is less and the E field has definite polarisation in one direction everywhere.

4.2.4 Field Patterns in a Rectangular Waveguide

The electric (E) and magnetic (H) field patterns in a rectangular waveguide in different TE and TM modes can be obtained from the field equations for such modes given by Eqs 4.8 and 4.12 respectively. The sketch of the field patterns for the dominant TE_{10} mode is given in Fig. 4.6 (a). The electric field (shown by full lines) is seen to be maximum at the centre of the broad wall of the guide for this mode and drops off sinusoidally to zero intensity at the narrow walls. The magnetic field (shown by dotted lines) is in the form of closed loops being parallel to the top and bottom broad walls and lies in the plane perpendicular to the electric field. A brief explanation of how we get the sketch of the electric and magnetic field lines in the xy plane (cross-sectional view), yz plane (side view) and xz plane (top view) may be given as follows. We observe from Eq. 4.10b that the TE_{10} mode has only three non-zero field components—H_z, H_x and E_y. In the xy plane say, when $z = 0$, both E_y and H_x vary as $\sin(\pi x/a)$ and are independent of y [Fig. 4.6a(i)]. In the yz plane, for example, at $x = a/2$, $\sin(\pi x/a) = 1$ and $\cos(\pi x/a) = 0$, when we only have E_y and H_x, both of which vary sinusoidally with βz, a sketch of E_y and H_x is shown in Fig. 4.6a(ii). The sketch in the xz plane shows all the three non-zero field components E_y, H_x and H_z. The slope of the H lines at $t = 0$ is governed by the following equation:

$$\left.\frac{H_x}{H_z}\right|_{TE_{10}} = \frac{\beta_{gTE_{10}}}{k_c^2}\left(\frac{\pi}{x}\right)\tan\left(\frac{\pi x}{a}\right)\tan\beta z$$

which can be used to draw the H lines in Fig. 4.6a(iii); these lines are independent of y.

The field patterns of other modes may also be derived analytically; however, with an intuitive approach those field patterns may be drawn without going into complicated analytical methods. For the TE_{20} mode, it is expected that the electric field will have two half-sinusoidal variations in the x-direction as may be observed from the cross-sectional view depicted in Fig. 4.6b(i). Accordingly, in the top view [Fig. 4.6b(iii)] two sets of magnetic loops are expected in the x-direction while the side view [Fig. 4.6b(ii)] seen from the $x = 0$ plane is similar to that for the TE_{10} mode (however, from the $x = a$ plane, both the magnetic and electric field directions will be reversed—not shown in the figure). In the TE_{11} mode, the electric field looks like a cobweb in the corners of the guide as indicated in the cross-sectional view [Fig. 4.6c(i)]. It may be observed that there is no half-wave change of electric intensity in both the x and y axes, with the electric intensity maximum in the centre of the waveguide. The magnetic field at any given cross-section is similar to those for the TE_{m0} modes; but it now varies along the y axis. For the TM_{11} mode, the electric field

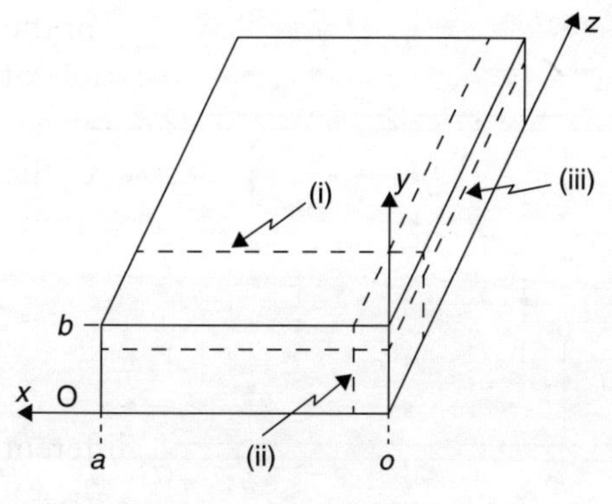

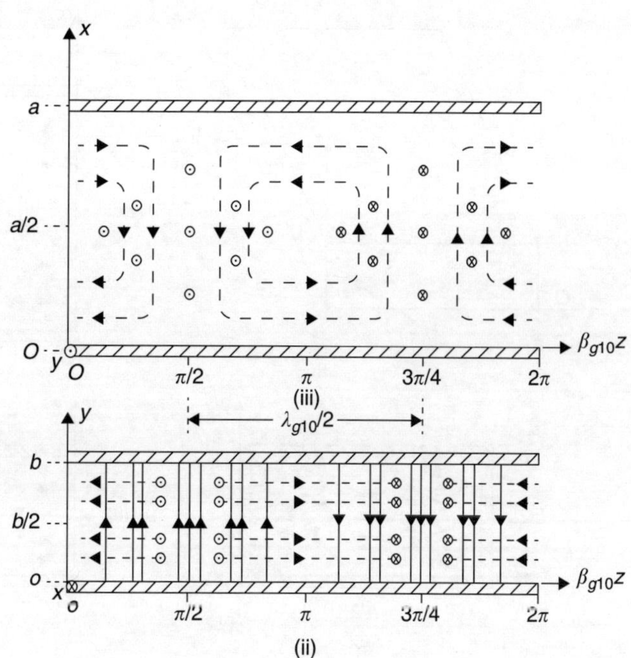

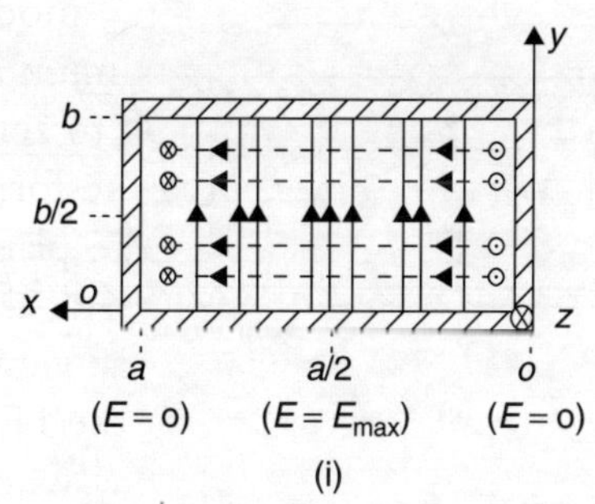

(ii) (image 3 lower part)

(a) $[TE_{10}]$ $\{\lambda_c = 2a\}$

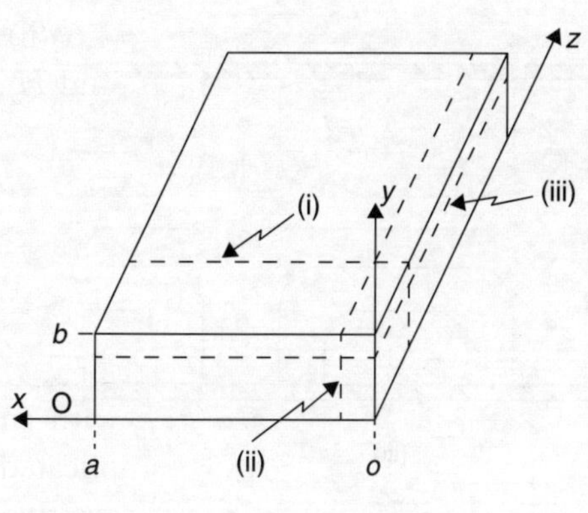

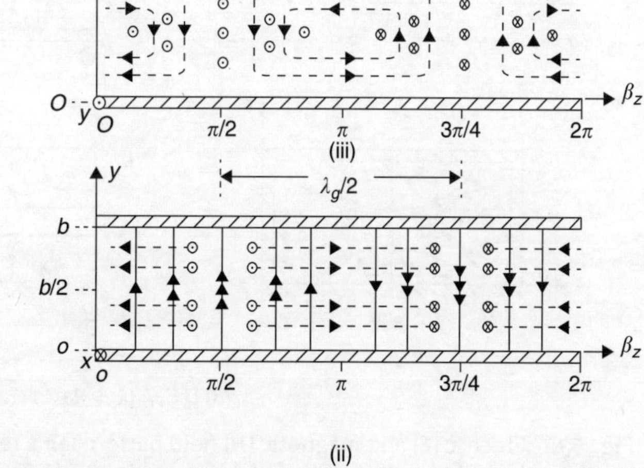

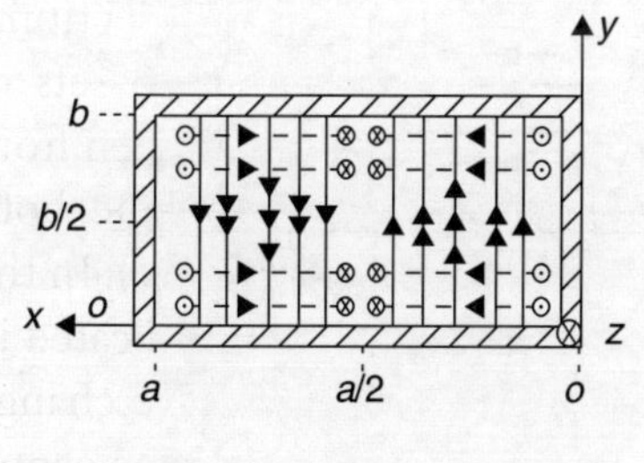

(i) (ii)

(b) $[TE_{20}]$ $\{\lambda_c = a\}$

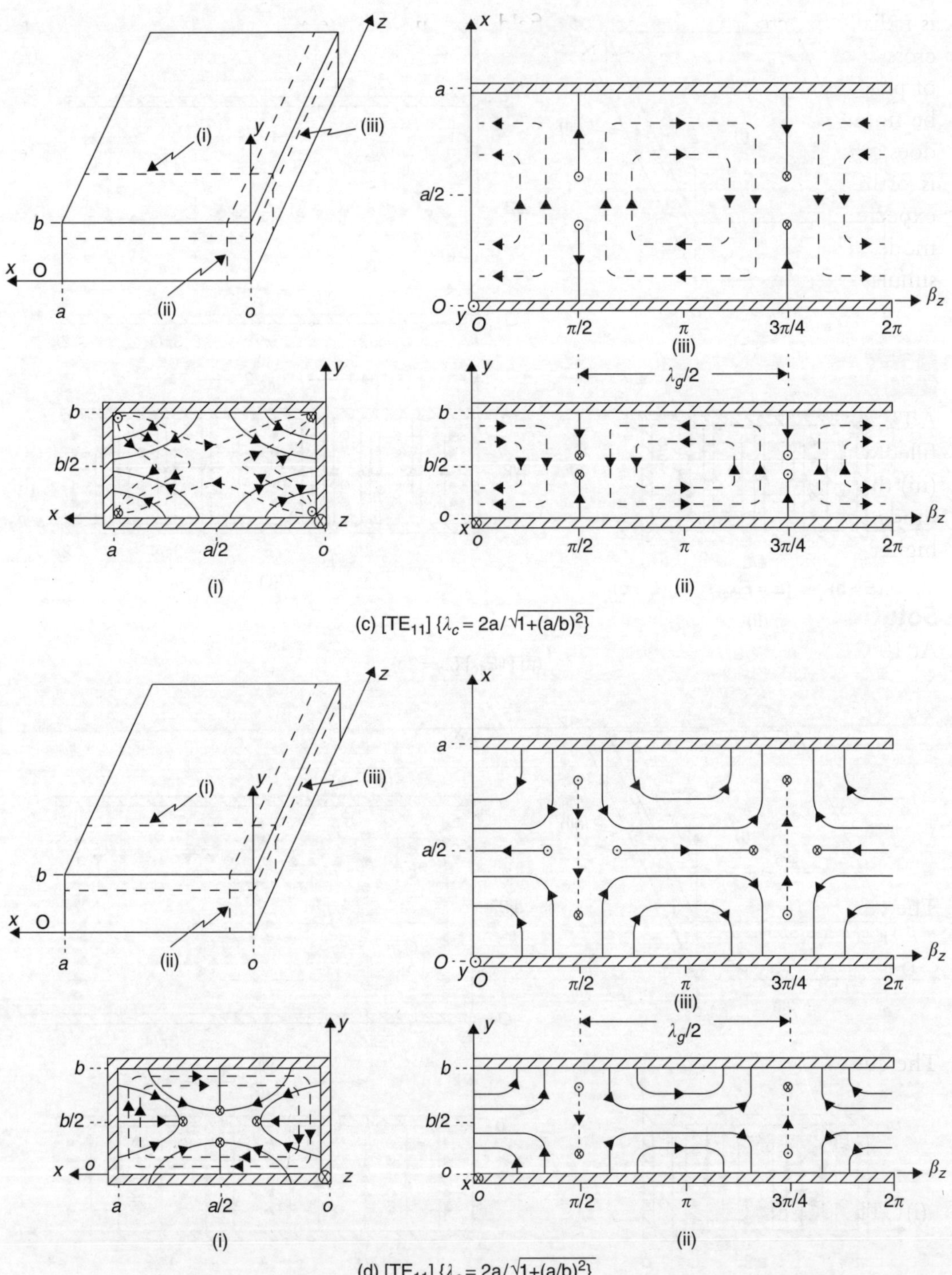

Fig. 4.6 Electric (*E*) and magnetic (*H*) field patterns in a rectangular waveguide [*E* with full lines (—), *H* with dashed lines (---), symbol dot ⊙ is for away from the paper and symbol cross ⊗ is into the paper. The field patterns are shown in different planes of cuts of the waveguide: (i) Cross-sectional view (ii) Side view (iii) Top view.

is radially directed and the magnetic field is annular in the x–y plane [Fig. 4.6d(i) for the cross-sectional view). As expected, the electric field now has a component in the direction of propagation, which the magnetic field possesses for TE modes. Also it may, in general, be noted for all modes of propagation that whenever the electric field touches a wall, it does so at right angles (that is strikes the metallic wall normally) while the magnetic field is orthogonal to the electric field and hence appears tangential to the metallic wall as expected from boundary conditions. The sketches for other non-dominant higher order modes may also be intuitively drawn or may be generated using a 3D electromagnetic field simulator (left to the students as an exercise).

EXAMPLE 4.1

A TE_{10} wave at 10.0 GHz propagates in a rectangular waveguide ($a = 2.5$ cm and $b = 1.25$ cm) filled with teflon having $\varepsilon_r = 2.1$. Determine (i) the phase constant, (ii) the guide wavelength, (iii) the phase velocity, (iv) the wave impedance, and (v) the percent bandwidth for single mode propagation in terms of the cut-off frequencies of the dominant mode and the first higher order mode.

Solution

At 10 GHz, the wavelength and phase velocity in unbounded teflon is:

$$\lambda_{\text{tef}} = \frac{c}{\sqrt{\varepsilon_r}}\frac{1}{f} = \frac{3\times 10^{10}}{\sqrt{2.1}\times 10\times 10^{9}} = 2.07 \text{ cm}$$

$$v_{p\text{tef}} = \frac{c}{\sqrt{\varepsilon_r}} = \frac{3\times 10^{10}}{\sqrt{2.1}} = 2.07\times 10^{10} \text{cm}$$

The cut-off frequency for TE_{10} mode is:

$$f_{c\text{TE}_{10}} = \frac{c}{2a} = \frac{3\times 10^{10}}{2\times 2.5} = 6.0 \text{ GHz}$$

The cut-off frequency for the next higher order mode TE_{20} is:

$$f_{c\text{TE}_{20}} = \frac{c}{a} = \frac{3\times 10^{10}}{2.5} = 12.0 \text{ GHz}$$

(i) The phase constant of the TE_{10} wave propagating down the teflon loaded guide is:

$$\beta_{g\text{TE}_{10}(\text{tef})} = \beta_{\text{tef}}\sqrt{1-\left(\frac{f_{c\text{TE}_{10}}}{f}\right)^2} = \frac{2\pi}{\lambda_{\text{tef}}}\sqrt{1-\left(\frac{6}{10}\right)} = \frac{2\times 3.142}{2.07}\times 0.632 = 1.96 \text{ rad/cm}$$

(ii) The guide wavelength:

$$\lambda_{g(\text{tef})} = \frac{\lambda_{\text{tef}}}{\sqrt{1-\left(\dfrac{f_{c\text{TE}_{10}}}{f}\right)^2}} = \frac{2.07}{0.632} = 3.275 \text{ cm}$$

(iii) The phase velocity:

$$v_{pg(\text{tef})} = \frac{v_{p\text{tef}}}{\sqrt{1-\left(\dfrac{f_{c\text{TE}_{10}}}{f}\right)^2}} = \frac{2.07\times 10^{10}}{0.632} = 3.275\times 10^{10} \text{ cm/s}$$

(iv) The wave impedance of the TE_{10} wave propagating down the teflon loaded guide:

$$Z_{w\text{TE}_{10}} = \frac{\eta_{\text{tef}}}{\sqrt{1-\left(\dfrac{f_{c\text{TE}_{10}}}{f}\right)^2}} = \frac{\eta_0/\sqrt{\varepsilon_r}}{0.632} = \frac{377/\sqrt{2.1}}{0.632} = 864\ \omega$$

(v) The per cent bandwidth for single mode propagation in terms of the cut-off frequencies of the dominant mode and the first higher order mode is given by:

$$\frac{(f_{c20}-f_{c10})}{(f_{c10}+f_{c20})/2}\times 100\% = \frac{12-6}{(6+12)}\times 200\% = 66.67\%$$

4.2.5 Wave Dispersion and Wave Velocities

It may be noted from Eq. 4.10c that the phase constant in a rectangular waveguide varies non-linearly with frequency. Thus, the phase velocity (Eq. 4.10d) is a function of frequency and it can be greater than the velocity of light and can even approach infinity as the operating frequency approaches the cut-off frequency of the mode. However, this is not the case for wave propagation in free space or in two-conductor transmission lines at microwave freqency, for example, in a coaxial line (where the phase constant is a linear function of frequency, $\beta = \omega\sqrt{\mu\varepsilon}$ and the phase velocity $v_p = \omega/\beta = 1/\sqrt{\mu\varepsilon} = c$, is independent of frequency). Thus, the waveguide is a *dispersive medium* (a medium in which phase velocity depends on frequency) even with no dispersive material loaded in it. Since all information-bearing signals consist of a band of frequencies, the waves of component frequencies will then travel with different phase velocities resulting in distortion in the signal wave while propagating via a waveguide.

The dispersion in the waveguide is different from conventional material dispersion, in which either the material media are lossy or exhibit properties such that ε and μ depend on frequency. But in a waveguide, since v_p is a function of frequency, any information-carrying signal (which is normally a modulated signal consisting of a band of frequencies) while travelling down the waveguide will get distorted or dispersed, as the waves of different frequencies will travel with different phase velocities. Thus, waveguide dispersion is not due to the nature of the material filling the waveguide but as a result of the boundary conditions of the waveguide structure imposing restrictions on the propagation characteristics, making phase velocity frequency dependant.

The *phase velocity* (v_p) is defined as the velocity of the constant phase points on the wave as the wave advances, that is, it is the speed of advance of the phase front of the wave containing the E and H fields. The phase velocity is applicable to a single frequency and does not transmit any information, hence, it is not the velocity with which the electromagnetic energy travels. The information is transmitted through a modulated carrier, and the speed of transmitting information is the speed of transmitting the information components in a modulated wave. A modulated wave is not a single frequency wave but group of waves (a wave-packet) with different frequencies, so the transmission velocity of the envelope of the wave-packet is called the *group velocity* (v_g). The group velocity is expressed as:

$$v_g = (d\beta/d\omega)^{-1} \tag{4.15a}$$

In a non-dispersive medium, $v_g = v_p$, while in a dispersive medium $v_g \neq v_p$. A medium is said to exhibit normal dispersion if $v_g < v_p$, and to exhibit anomalous dispersion if $v_g > v_p$. The waveguide as a dispersive medium, exhibits normal dispersion. The group velocity for TE_{mn}/TM_{mn} modes can be derived from Eq. 4.16a by using Eq. 4.10c and remembering that $\beta = \omega\sqrt{\mu\varepsilon}$ and $v_p = 1/\sqrt{\mu\varepsilon}$, when:

$$v_{g\text{TE/TM}} = v_p\sqrt{1-\left(\frac{f_{c(\text{TE/TM})}}{f}\right)^2} \tag{4.15b}$$

The variations of the phase and group velocities with operating frequency can thus be obtained from Eqs 4.10d and 4.16b respectively (Fig. 4.7).

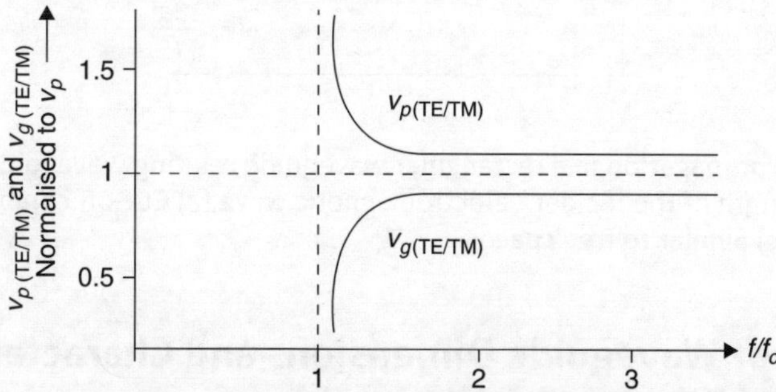

Fig. 4.7 Plot of phase velocity and group velocity in the waveguide.

Note that the group velocity $v_{g(\text{TE/TM})}$ is zero at the cut-off frequency, corresponding to the situation where the component uniform plane waves (TEM waves) constituting the waveguide modes (TE/TM modes) reflect back and forth between the two narrow walls of the waveguide with no propagation along the guide (Fig. 4.8a). As the frequency is increased above the cut-off frequency, the component TEM waves get obliquely reflected from the two narrow walls of the waveguide and results in the TE/TM waves as propagating modes in the waveguide (Fig. 4.8b). But when $f >> f_{c(\mathrm{TE/TM})}$, v_g approaches v_p asymptotically and the component TEM waves then just slide between the narrow walls of the waveguide (Fig. 4.8c).

The issues discussed in Figs 4.8a, b and c also follow from Eq. 2.44 when the separation between the two conducting plates is taken as equal to a (Fig. 4.1) and set (a) $\lambda = 2a/p$, making $\theta = 0°$; (b) $\lambda < 2a/p$, when θ has a finite value but less than 90°; and (c) $\lambda << 2a/p$, when $\theta = 90°$ respectively. It is interesting to note (Eqs 4.16b and 4.10d) that:

$$v_{p(\mathrm{TE/TM})} \times v_{g(\mathrm{TE/TM})} = v_p^2 \tag{4.15c}$$

where $v_p = 1/\sqrt{\mu\varepsilon}$. This interesting relationship between group and phase velocity is true only for propagating media for which (β_g vs ω) relationship is described by Eq. 4.10c. Note that the phase velocity of an electromagnetic wave can be more than the velocity of light but group velocity can never be greater than the velocity of light, otherwise causality will be violated as group velocity is the velocity with which the energy of the electromagnetic wave travels.

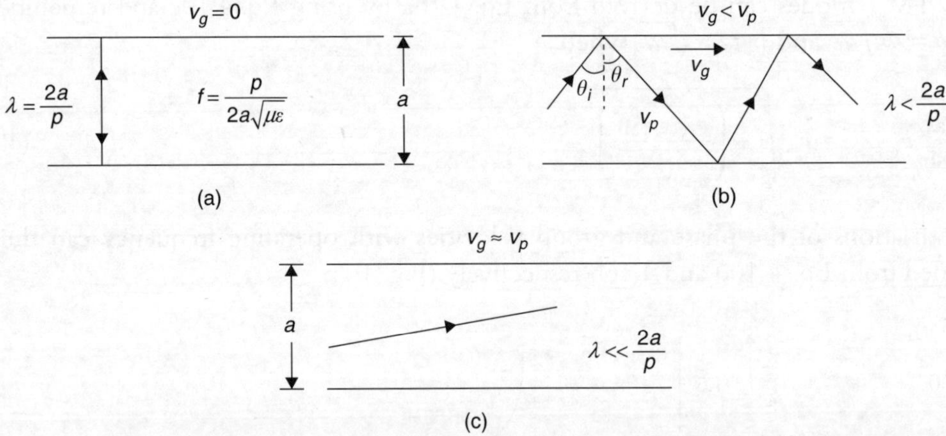

Fig. 4.8 Cut-off and propagation in a rectangular waveguide relating waveguide dimensions with the wavelength of the incident electromagnetic wave (a) Cut-off condition (b) Propagation condition (c) Similar to free space.

4.2.6 Choice of Waveguide Dimensions and Characteristics of Standard Rectangular Waveguides

In practice, the waveguide broad dimension (a) and narrow dimension (b) are so chosen that, only a single mode propagation can take place in the frequency band of operation. Multimode propagation is not desirable because each mode that could propagate has a different phase velocity and a different field configuration, causing energy distribution in undesired modes and inefficient energy available for the desired mode of propagation.

For $a = b$, the cut-off frequencies of TE_{10} and TE_{01} are equal (*a degenerate situation* where the modes differ only by rotation through a 90° angle), thus a square guide of this type is prone to undesirable mode conversion/coupling and not being used in practice for normal applications. Usually, a is chosen to be approximately equal to $2b$. For a given a at a particular frequency, too small a value for b increases attenuation (per unit transmitted power) and also reduces power handling capabilities of the guide (which will be seen shortly).

The value of a is chosen in terms of the wavelength of operation λ and necessarily we must have $\lambda/2 < a < \lambda$, to ensure the transmission of *only* the dominant TE_{10} mode. Choosing a too close to λ does not allow enough separation from the propagating mode with the cut-off frequency of the next higher order mode namely, TE_{20}, thus limiting the bandwidth of operation in the dominant mode. Again a value of a too close to $\lambda/2$ is not desirable since the wave impedance and the phase velocity vary too rapidly with frequency (Figs 4.4 and 4.7). Thus, a value of a of the order of 0.7λ is often chosen in practice. For example, at X-band (8.2–12.4 GHz), the outside measurements of a rectangular waveguide designated as EIA WR-90 by the Electronic Industry Association (***Note***: WR stands for Waveguide Rectangular), are 2.54 cm (1.0 inch) wide (a dimension) and 1.27 cm (0.5 inch) high (b dimension), and its inside measurements are 0.9 inch (2.286 cm) wide and 0.4 inch (1.016 cm) high; with wall thickness 0.5 inch (Table 4.1, which tabulates the characteristics of standard rectangular waveguides). It may be noted that WR-xxx refers to a waveguide for which the inside dimension of the width (a) of the waveguide is equal to xxx/1000 inches.

Table 4.1 Characteristics of standard rectangular waveguides

Waveguide designation		**Recommended frequency range for TE_{10} mode (GHz)**	**Physical dimensions of waveguide**				**Cut-off frequency for air-filled waveguide (GHz)**
			Inside in inches (cm)		**Outside in inches (cm)**		
US	**UK**		**Width**	**Height**	**Width**	**Height**	
WR 650	WG 6	1.14–1.73 (1.36)	6.500 (16.510)	3.250 (8.255)	6.660 (16.916)	3.410 (8.661)	0.907
WR 430	WG 8	1.72–2.61 (2.06)	4.300 (10.922)	2.150 (5.461)	4.460 (11.328)	2.310 (5.867)	1.372
WR 284	WG 10	2.60–3.95 (3.12)	2.840 (7.214)	1.340 (3.404)	3.000 (7.620)	1.5000 (3.810)	2.078
WR 187	WG 12	3.94–5.99 (4.73)	1.872 (4.755)	0.872 (2.215)	2.000 (5.080)	1.000 (2.540)	3.153
WR 137	WG 14	5.38–8.17 (6.46)	1.372 (3.485)	0.622 (1.580)	1.500 (3.810)	0.750 (1.905)	4.304
WR 90	WG 16	8.20–12.4 (9.84)	0.9000 (2.286)	0.4000 (1.016)	1.000 (2.540)	0.500 (1.270)	6.570
WR 62	WG 18	11.90–18.0 (14.2)	0.622 (1.580)	0.311 (0.790)	0.702 (1.783)	0.391 (0.993)	9.495

(*continued*)

Table 4.1 Continued

Waveguide designation US	Waveguide designation UK	Recommended frequency range for TE_{10} mode (GHz)	Inside in inches (cm) Width	Inside in inches (cm) Height	Outside in inches (cm) Width	Outside in inches (cm) Height	Cut-off frequency for air-filled waveguide (GHz)
WR 42	WG 20	17.60–26.7 (21.1)	0.420 (1.067)	0.170 (0.432)	0.500 (1.270)	0.250 (0.635)	14.088
WR 28	WG 22	26.40–40.0 (31.6)	0.280 (0.711)	0.140 (0.356)	0.360 (0.914)	0.220 (0.559)	21.184
WR 22	WG 23	32.90–50.1 (39.5)	0.224 (0.569)	0.112 (0.284)	0.304 (0.772)	0.192 (0.488)	26.461
WR 19	WG 24	40.00–60.0 (47.1)	0.188 (0.478)	0.0940 (0.239)	0.268 (0.681)	0.174 (0.442)	31.595
WR 15	WG 25	50.00–75.0 (59.9)	0.148 (0.376)	0.074 (0.188)	0.228 (0.579)	0.154 (0.391)	40.058
WR 12	WG 26	60.00–90.0 (72.6)	0.122 (0.310)	0.061 (0.155)	0.202 (0.513)	0.141 (0.358)	48.549
WR 8	WG 28	90.00–140 (111.0)	0.080 (0.203)	0.040 (0.102)	0.160 (0.406)	0.120 (0.305)	74.440
WR 7	WG 29	110.00–170 (136.3)	0.065 (0.165)	0.033 (0.084)	0.135 (0.343)	0.103 (0.262)	91.227
WR 4	WG 31	170.00–260 (206.0)	0.043 (0.109)	0.022 (0.056)	0.083 (0.211)	0.062 (0.157)	137.938

EXAMPLE 4.2

Design an air-filled X-band (8–12 GHz) rectangular waveguide such that the centre frequency of this band is at least 25% higher than the cut-off frequency of the TE_{10} mode and at least 25% lower than the next higher order mode, so that the dominant mode of propagation is TE_{10}.

Solution

For the TE_{10} mode, $f_{cTE_{10}} = c/2a$

According to first design criterion,

$$f = 10 \text{ GHz} \geq 1.25\ f_{cTE_{10}} = 1.25\ (c/2a)$$

From which we can find the lower bound of the dimension a as

$$a > 1.875 \text{ cm}$$

For the TE_{20} mode, $f_{cTE_{20}} = c/a$

Using the second design criterion, we have:

$$f = 10 \text{ GHz} < 0.75\ f_{cTE_{20}} = 0.8\ (c/a)$$

From which we can find the upper bound for the dimension a as:

$$a \leq 2.55 \text{ cm}$$

Thus, the dimension a is bounded by:

$$1.875 \text{ cm} \leq a \leq 2.55 \text{ cm}$$

For the TE_{01} mode, $f_{cTE_{01}} = c/2b$

Again using the second design criterion, we have:

$$f = 10 \text{ GHz} < 0.75\, f_{cTE_{01}} = 0.75\,(c/2b)$$

From which we can find the upper bound for the dimension of b as:

$$b \leq 1.25 \text{ cm } [= 0.5a \text{ (upper bound)}]$$

The design criteria do not set a lower bound of b. In fact, we do not need to calculate the lower bound of b as a larger b is desirable to increase the power handling capability, which is directly proportional to the cross-sectional area of the waveguide.

4.2.7 Launching of Modes in Rectangular Waveguides

In order to launch a particular mode in a waveguide, a type of probe or a loop is used that acts as a coupling antenna and launches the desired E and H fields in the guide for that mode. The location of the probe corresponds to the position of the desired maximum electric field. If two or more probes are required to launch a particular mode, care must be taken to ensure proper phasing between the currents in the antennae represented by the probes. This is realised by inserting an additional length of transmission line in one or more of the antenna feeders. The depth of insertion of the probe will determine the power it couples and by varying the depth of penetration of the probe within the guide, impedance-matching may also be achieved. A device that excites a given mode in the guide can also serve reciprocally to pick-up electromagnetic energy for that mode. The methods of excitation for various modes in rectangular waveguides are shown in Fig. 4.9.

For launching the TE_{10} mode in the rectangular waveguide, the coaxial feed line (acting as electric probe) is placed transverse to the broad wall and along the central line of this wall which corresponds to the location of the maximum of the E field for the mode (Figs 4.9a and 4.6a). The probe acting as antenna has its polarisation (orientation of the E field) parallel to the antenna itself and thus generates the electric field desired for the TE_{10} mode. The current in the coupling antenna sets up the magnetic field in the guide that lies in the plane parallel to the top and bottom broad wall of the waveguide. For the TE_{20} mode, two probes are used and the probes acting as antennae are fed 180° out of phase [by physically separating them by a distance of $(\lambda_g/2)_{TE_{20}}$ (Fig. 4.9b)]; so that out of phase current flowing through them will result in a zero E field at the centre as desired by this mode (Fig. 4.6b). For TM modes, the coaxial feed antenna has to be inserted into the waveguide in the longitudinal direction (that is, parallel with respect to the broad wall, since the E

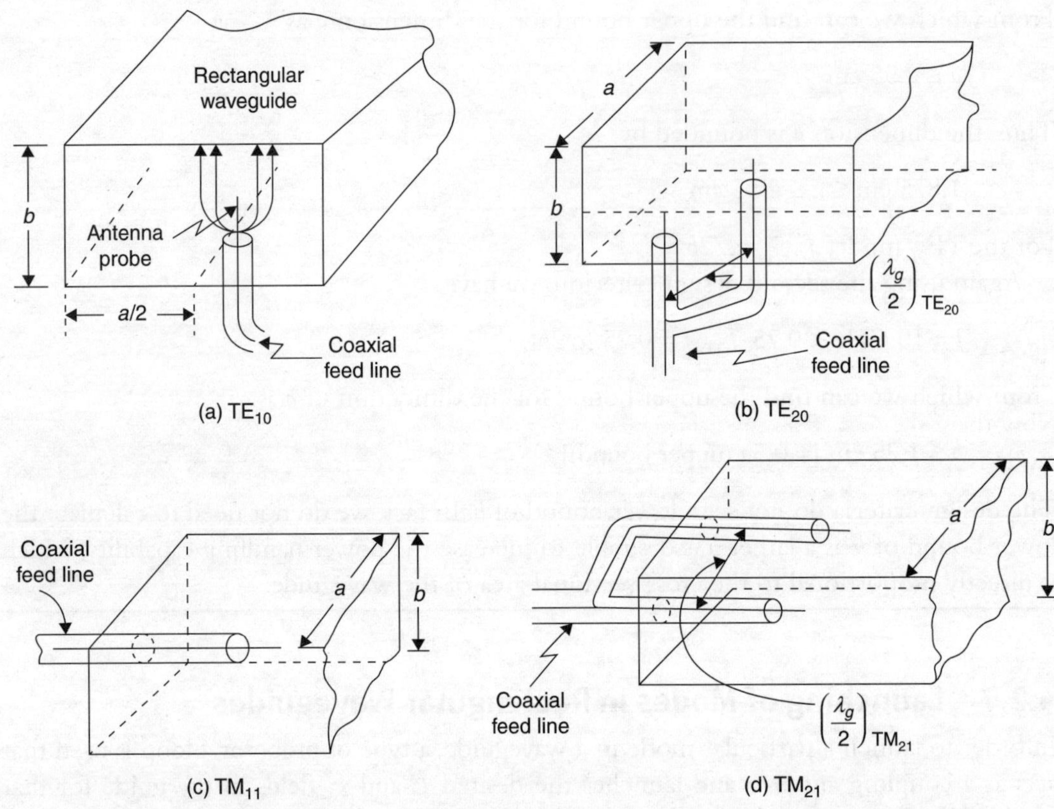

Fig. 4.9 Launching of different modes in the rectangular waveguide (a) TE_{10} mode (b) TE_{20} mode (c) TM_{11} mode (d) TM_{21} mode.

field of the TM mode has orthogonal orientation compared to the TE modes) as shown for TM_{11} and TM_{21} modes in Figs 4.9c and 4.9d respectively.

4.2.8 Power Handling Capacity of the Rectangular Waveguide

Many radar and communication systems require the transmission of very high microwave power densities through waveguides. The breakdown of the dielectric that fills the waveguide and the dimensions of the waveguide are usually the limiting factors in maximising the power-handling capability.

Let E_{br} be the breakdown electric field for the dielectric material inside the waveguide and E_0 be the maximum peak value of the electric field of the electromagnetic wave propagating down the waveguide. Then, in order to avoid breakdown of the dielectric inside the waveguide, we must have: $E_0 < E_{br}$. For a dominant TE_{10} mode in the rectangular waveguide, the largest electric field occurs along the centre line of the broad wall (Fig. 4.10), and we have from Eq. 4.10b:

$$E_0 = \left|E_y\right|_{\max} = \omega\mu a C/\pi$$

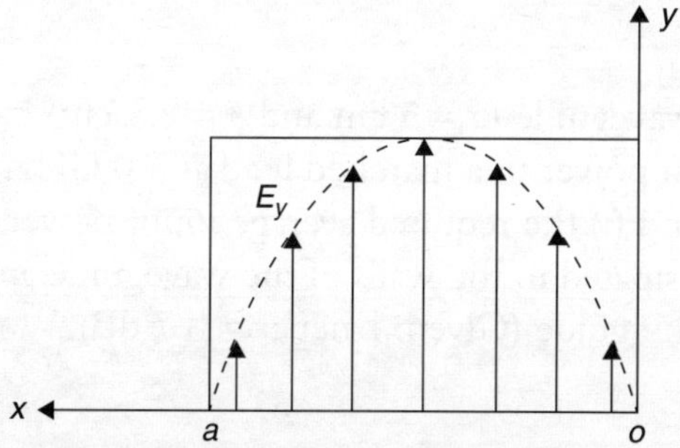

Fig. 4.10 Peak electric field distribution for the TE_{10} mode.

Now, the average value of electromagnetic power flowing down the waveguide for the TE_{10} mode is:

$$P_{av} = \frac{1}{2} Re \int_0^a \int_0^b (\vec{E} \times \vec{H}^\star) \cdot \vec{a}_z \, dx \, dy$$

$$= \frac{1}{2} Re \int_0^a \int_0^b (E_x H_y^\star - E_y H_x^\star) \, dx \, dy$$

$$= \frac{1}{2} Re \int_0^a \int_0^b -E_y (z=0) \cdot H_x^\star (z=0) \, dx \, dy \quad (\text{since } E_x = H_y = 0)$$

$$= \frac{1}{2} \int_0^a \int_0^b \left(\frac{\omega \mu a C}{\pi} \right) \left(\frac{\beta_{g10} a C}{\pi} \right) \sin^2 \left(\frac{\pi x}{a} \right) dx \, dy \ \ (\text{which follows from Eq. 4.10b})$$

$$= \frac{\omega \mu a^2 \beta_{g10} C^2 b}{2\pi^2} \int_0^a \sin^2 \left(\frac{\pi x}{a} \right) dx$$

$$= \frac{\omega \mu a^2 \beta_{g10} C^2 b}{2\pi^2} \left(\frac{a}{2} \right) \tag{4.16a}$$

Substituting for $C = \pi E_0 / \omega \mu a$ from above, we have P_{peak} in terms of the peak electric field as:

$$P_{\text{peak}} = 2P_{av} = \frac{E_0^2}{\eta} \frac{ab}{2} \sqrt{1 - \left(\frac{\lambda}{2a} \right)^2} \tag{4.16b}$$

It appears from Eq. 4.16 that the power handling capacity of the waveguide can be increased by choosing the dimensions a and b as large as possible. In fact, for a given value of a, P_{peak} is proportional to b, so that the power capacity may be increased by increasing b. However, for all practical purposes, since it is desirable to have the dominant TE_{10} mode as the only propagating mode, the upper bound of b is restricted for a given value of a.

EXAMPLE 4.3

An air-filled rectangular waveguide ($a = 5$ cm and $b = 2.5$ cm) having a length of 0.5 m is to deliver 2 kW maximum power to a matched load at 5.0 GHz. Assuming an attenuation constant of 0.05 dB/m, find (i) the required average input power to the waveguide, (ii) the total amount of power dissipated in the walls of the waveguide and (iii) maximum value of electric intensity within the guide (Given: 1 neper = 8.7 dB).

Solution

Given $a = 5$ cm and $b = 2.5$ cm, Thus, the cut-off frequency for the dominant mode is $c/2a = 3.0$ GHz while that for the next higher order mode (TE_{20}) is $c/a = 6.0$ GHz. The only propagating mode at 5 GHz is then the dominant mode ($f_c = 3.0$ GHz), because the next higher order mode has a cutoff frequency of 6.0 GHz > 5 GHz.

(i) $\alpha = 5 \times 10^{-2}$ dB/m $= 5.75 \times 10^{-3}$ np/m

Thus $P_{in} = P_{load} e^{2\alpha l} = 2.0 \times 10^3 e^{2(5.75 \times 10^{-3})} = 2 \times 10^3 e^{0.00575} = 2011.53$

(ii) Power dissipated $= P_{in} - P_{load} = (2011.53 - 2000)$ W $= 11.53$ W

(iii) From Eq. 4.16b, we have:

$$E_0 = \sqrt{\frac{4P_{av}\eta}{ab\left\{1-\left(\frac{\lambda}{2a}\right)^2\right\}^{1/2}}} = \sqrt{\frac{4 \times 2011.53 \times 377}{5 \times 2.5\left\{1-\left(\frac{6}{2 \times 5}\right)^2\right\}^{1/2}}} = 55.076 \text{ kV/m}$$

EXAMPLE 4.4

Determine the maximum electric field strength in an air-filled rectangular waveguide ($a = 2.5$ cm and $b = 1.0$ cm) when 200 kW of peak power is transmitted at 10 GHz. If a safety factor of 2.0 is specified, calculate the maximum power that can be transmitted through the guide at 15 GHz.

Solution

From Eq. 4.16b we can write:

$$E_0 = \sqrt{\frac{2P_{peak}\eta}{ab\left\{1-\left(\frac{\lambda}{2a}\right)^2\right\}^{1/2}}} = \sqrt{\frac{2 \times 200 \times 1000 \times 377}{2.5 \times 1.0\left\{1-\left(\frac{3}{2 \times 2.5}\right)^2\right\}^{1/2}}} = 868 \text{ kV/m}$$

When $f = 15$ GHz, $\lambda = 2.0$ cm. According to the problem, $E_0 = 30$ kV cm^{-1}/2.0 (since safety factor is 2.0). Thus from Eq. 4.16b we have:

$$P_{\text{peak}} = \frac{E_0^2}{\eta}\frac{ab}{2}\sqrt{1-\left(\frac{\lambda}{2a}\right)^2} = \left(\frac{3\times 10^3}{2}\right)\times\frac{1}{377}\times\frac{2.5\times 1.0}{2}\sqrt{1-\left(\frac{2}{2\times 2}\right)^2} = 0.646 \text{ MW}$$

4.2.9 Attenuation Due to Losses in the Rectangular Waveguide

So far, we have considered an ideal waveguide with perfectly conducting walls enclosing a perfect dielectric medium. But in practice, any imperfection in the conductivity of the wall will result in *conduction loss*. Also the dielectric medium inside the waveguide may have a non-zero conductivity resulting in *dielectric loss*. Thus, we can expect some power loss in a practical waveguide leading to attenuation of the fields as they propagate down the waveguide.

Case I: ***Attenuation due to conduction losses*** **(α_{cl})**

Attenuation due to conduction loss (α_{cl}) caused by the waveguide walls (being not perfect conductors) may be assumed to follow the usual signal attenuation in a lossy medium which is represented by a decaying field, namely, $e^{\alpha_{cl}z}$. Then, the average electromagnetic power P_{av} transmitted along the length of the waveguide (in the z-direction) is given by $P_{av} = P_0 e^{-2\alpha_{cl}z}$, P_0 being the power at $z = 0$. The rate of change of transmitted power along the longitudinal direction of the waveguide is then given by:

$$-\frac{\partial P_{av}}{\partial z} = 2\alpha_{cl} P_{av} \tag{4.17a}$$

The –ve sign signifies that the signal gets reduced (attenuated) as it propagates in the +ve z-direction. This reduction of power, that is, attenuation of the signal per unit length, must be equal to the power lost or dissipated per unit length of the waveguide. Therefore, from Eq. 4.17a:

$$\frac{\text{(Power lost per unit length)}}{\text{(Power transmitted)}} = \frac{2\alpha_{cl}P_{av}}{P_{av}} = 2\alpha_{cl}$$

Thus
$$\alpha_{cl} = \frac{\text{(Power lost per unit length)}}{2\text{(Power transmitted)}} \tag{4.17b}$$

To evaluate α_{cl}, we first need to determine the power lost per unit length of the waveguide which is caused by surface current flow in the imperfect conducting walls of the guide. Surface current can be determined by considering the dielectric medium inside the waveguide as region 1 and the conducting wall of the waveguide as region 2. The surface current is obtained from $\vec{J}_s = \hat{n}\times\vec{H}_1$, where $\hat{n}$ is the unit normal to the waveguide wall pointing into region 1 (with the assumption that the magnetic field in region 2, that is, in the metal is zero as the skin depth in region 2 is extremely small at microwave frequencies). We compute the power lost due to surface current in the waveguide walls located at $x = 0$ and $y = 0$ and, due to symmetry, doubling their sum we get the contribution from all four walls:

$$P_{\text{loss (Total)}} = 2[P_{\text{loss}}]_{x=0} + 2[P_{\text{loss}}]_{y=0} \tag{4.17c}$$

where $P_{\text{loss}} = \frac{1}{2} R_S I^2$, R_s $(= 1/\delta\sigma_{\text{wall}})$ being the surface resistance of the conducting walls of the waveguide and I is the amplitude of the total current on the surface of the wall. The surface current on waveguide walls for the TE_{10} mode may be written as (Eq. 4.10b):

$$\vec{J}_s(x=0) = \vec{a}_x \times \vec{a}_z \vec{H}_z = -\vec{a}_y \vec{H}_z = -\vec{a}_y C e^{-j\beta_{g10}z} = -\vec{a}_y C \cos\beta_{g10} z$$

$$\vec{J}_s(y=0) = \vec{a}_y \times (\vec{a}_x \vec{H}_x + \vec{a}_z \vec{H}_z)$$

$$= -\vec{a}_z \frac{j\beta_{g10} a C}{\pi} \sin\left(\frac{\pi x}{a}\right) e^{-j\beta_{g10}z} + \vec{a}_x C \cos\left(\frac{\pi x}{a}\right) e^{-j\beta_{g10}z}$$

$$= \vec{a}_x C \cos\left(\frac{\pi x}{a}\right) \cos\beta_{g10} z - \vec{a}_z \frac{\beta_{g10} a C}{\pi} \sin\left(\frac{\pi x}{a}\right) \sin\beta_{g10} z \tag{4.17d}$$

The surface currents on the inside walls at $x = 0$ and at $y = b$ are sketched in Fig. 4.11. The pattern is shown for a particular instant of time; as time advances, current flow changes in accordance with the magnetic field so that the entire pattern propagates down the waveguide.

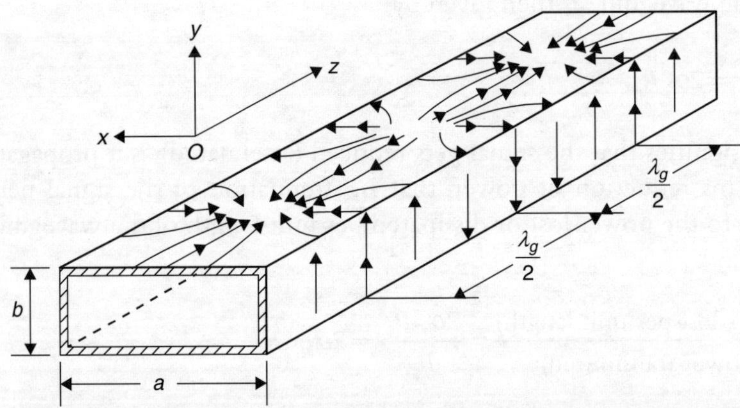

Fig. 4.11 **Surface current on the inside walls of the rectangular waveguide at $x = 0$ (narrow wall) and $y = b$ (broad wall).**

We can now evaluate the total power loss due to the four walls of the waveguide by using Eqs 4.17c and 4.17d as follows.

$$P_{\text{loss(Total)}} = 2\left\{\int_0^b \frac{R_s}{2} \left|\vec{J}_s(x=0)\right|^2 dy + \int_0^a \left|\vec{J}_s(y=0)\right|^2 dx\right\}$$

$$= C^2 R_s \left\{\int_0^b dy + \int_0^a \cos^2\left(\frac{\pi x}{a}\right) dx + \frac{\beta_{g10}^2 a^2}{\pi^2} \int_0^a \sin^2\left(\frac{\pi x}{a}\right) dx\right\}$$

$$= C^2 R_s \left(b + \frac{a}{2} + \frac{\beta_{g10}{}^2 a^3}{2\pi^2} \right)$$

$$= C^2 R_s \left\{ b + \frac{a}{2} \left(\frac{f}{f_{c\text{TE}_{10}}} \right)^2 \right\} \tag{4.17e}$$

where we have used the expressions for $\beta_{g\text{TE}_{10}}$ and $f_{c\text{TE}_{10}}$ from Eqs 4.14.

Note that for the TE_{10} mode of signal propagation in the rectangular waveguide, we derived in Eq. 4.16a, the average power transmitted via the waveguide. Thus, using Eqs 4.17e and 4.16a, we can finally evaluate α_{cl} from Eq. 4.17b as:

$$\alpha_{c\text{TE}_{10}} = \frac{R_s}{\eta} \frac{1 + (2b/a)(f_{c\text{TE}_{10}}/f)^2}{b\sqrt{1 - (f_{c\text{TE}_{10}}/f)^2}} \tag{4.17f}$$

For other TE_{mn} modes, it can be shown, with proper manipulation [4], that:

$$\alpha_{c\text{TE}_{mn}} = \frac{2R_s}{\eta} \frac{1}{b\sqrt{1 - (f_{c\text{TE}_{mn}}/f)^2}}$$

$$\times \left[\left(1 + \frac{b}{a}\right) \left(\frac{f_{c\text{TE}_{mn}}}{f} \right)^2 + \left\{ 1 - \left(\frac{f_{c\text{TE}_{mn}}}{f} \right)^2 \right\} \left\{ \frac{(b/a)\,[(b/a)\,m^2 + n^2]}{(b^2 m^2/a^2) + n^2} \right\} \right] \tag{4.17g}$$

Following a similar analysis it can be shown that for TM_{mn} modes:

$$\alpha_{c\text{TM}_{mn}} = \frac{2R_s}{\eta} \frac{1}{b\sqrt{1 - (f_{c\text{TM}_{mn}}/f)^2}} \times \frac{m^2 (b/a)^3 + n^2}{m^2 (b/a)^2 + n^2} \tag{4.17h}$$

The expressions of α_{cl} are different for TE_{mn} and TM_{mn} modes because of different current distribution. However, for all propagation modes, α_{cl} is directly proportional to R_s, which in turn, is inversely proportional to the conductivity of the wall σ_{wall}. Naturally, α_{cl} is zero (no power loss) if the guide walls are perfect conductors ($\sigma_{\text{wall}} \to \infty$).

The computed values of α_{cl} for a few TE_{mn} and TM_{mn} modes are shown in Fig. 4.12.

It may be observed that at the cut-off frequency of any particular mode (marked by an arrow along the frequency axis) attenuation approaches infinity, Fig. 4.12a (as expected from Eqs 4.17g and 4.17h). For all modes (TE/TM) the attenuation first decreases and then increases. Again, keeping the a/b ratio unchanged, if the size of the rectangular waveguide is increased then the attenuation decreases (Fig. 4.12b). This is because the resistive losses decrease as the waveguide becomes oversized. In fact, *oversized waveguides* are used at millimeter-wave frequencies if one wishes to push a large amount of power with reduced loss but with reasonable waveguide size. Further, resonators made with oversized waveguides

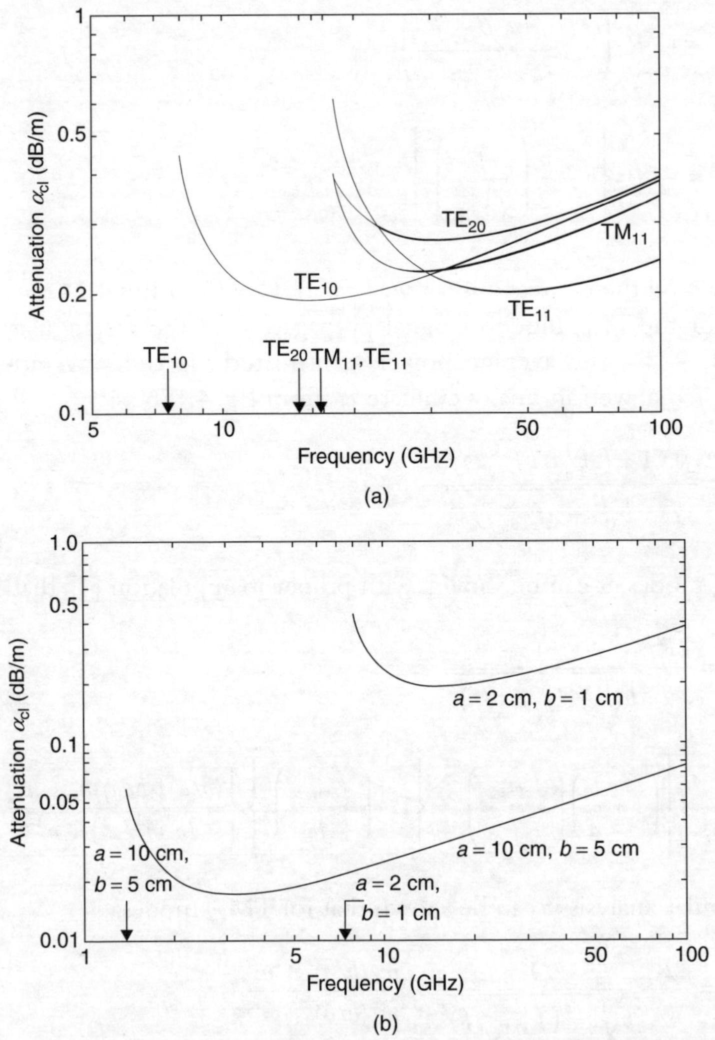

Fig. 4.12 (a) Computed values of attenuation in an air-filled rectangular waveguide ($a = 2$ cm and $b = 1$ cm) with $s = 5.8 \times 10^7$ S/m, (b) Comparison of attenuation in two waveguides having the same a/b ratio but different a and b dimensions with dominant mode (TE_{10}) propagation.

will have higher Q values which are of importance at millimeter-wave frequencies where the cavity size happens to be small even with an oversized waveguide.

***Case* II: *Attenuation due to dielectric losses* (α_{dl})**

Imperfections (non-zero conductivity) in the dielectric material between the conductors of a practical waveguide also lead to losses in the waveguide transmission system, resulting in dielectric loss. Such losses can be estimated by considering the permittivity of the dielectric filling to be complex (that is, $\varepsilon_c = \varepsilon' - j\varepsilon''$, in which loss due to non-zero conductivity is contained in ε''). Using Eq. 4.6b and replacing ε by ε_c, we have the complex propagation constant in the waveguide as:

$$\gamma_g = \alpha_{g(dl)} + j\beta_g = \sqrt{k_c^2 - \omega^2 \mu\varepsilon} = \sqrt{k_c^2 - \omega^2 \mu\varepsilon'(1 - j\tan\delta_c)}$$

$$= \sqrt{k_c^2 - \omega^2 \mu\varepsilon'} \left(1 - \frac{j\omega^2 \mu\varepsilon' \tan\delta_c}{\sqrt{k_c^2 - \omega^2 \mu\varepsilon'}}\right)^{1/2}$$

where $\tan\delta_c = \varepsilon''/\varepsilon'$, is the loss tangent of the dielectric. For most dielectrics, $\tan\delta_c << 1$, and so we can use the binomial expansion for the bracketed term on the right above, when we get:

$$\gamma_g = \sqrt{k_c^2 - \omega^2 \mu\varepsilon'} - \frac{j\omega^2 \mu\varepsilon' \tan\delta_c}{2\sqrt{k_c^2 - \omega^2 \mu\varepsilon'}}$$

$$= \frac{\omega^2 \mu\varepsilon' \tan\delta_c}{2\beta_g} + j\beta_g \tag{4.18a}$$

(since $\omega^2 \mu\varepsilon > k_c^2$, for a propagating wave)

where $j\beta_g = \sqrt{k_c^2 - \omega^{2\mu\varepsilon'}}$.

The real part of γ_g in Eq. 4.18a is the attenuation due to dielectric loss and is given by:

$$\alpha_{dl} = \frac{\omega^2 \mu\varepsilon' \tan\delta_c}{2\beta_g}$$

$$= \frac{\pi f \sqrt{\mu\varepsilon'}}{\sqrt{1 - (f_c/f)^2}} \tan\delta_c \tag{4.18b}$$

where we have used Eq. 4.9c.

EXAMPLE 4.5

A silver plated ($\sigma = 6.17 \times 10^7$ S-m^{-1}) WR-187 rectangular waveguide (a = 4.755 cm and b = 2.215 cm) filled with PTFE Teflon ($\varepsilon_r = 2.1, \mu_r = 1, \tan\delta = 1.5 \times 10^{-4}$) has to propagate the TE_{10} wave at 5.0 GHz. Determine the attenuation due to conduction and dielectric losses for the wave propagating down the dielectric loaded waveguide.

Solution

The surface resistance for the silver plated waveguide:

$$R_s = \sqrt{\frac{\pi f \mu_0}{\sigma_{\text{silver}}}} = \sqrt{\frac{3.142 \times 5 \times 10^9 \times 4 \times \pi \times 10^{-7}}{6.17 \times 10^7}} = 1.788 \times 10^{-2}$$

The wave impedance in the Teflon is:

$$\eta_{\text{Teflon}} = \frac{\eta_0}{\sqrt{\varepsilon_r}} = \frac{377}{\sqrt{2.1}} = 260.15\ \Omega$$

The cut-off frequency of the TE_{10} wave propagating down the Teflon loaded guide:

$$f_{c\text{TE}_{10}\,(\text{Teflon})} = \frac{v_p}{2a} = \frac{c}{2a\sqrt{\varepsilon_r}} = \frac{3\times10^{10}}{2\times4.755\times\sqrt{2.1}} = 2.176\ \text{GHz}$$

Thus, attenuation due to conduction loss from Eq. 14.17f is given by:

$$\alpha_{c\text{TE}_{10}} = \frac{1.788\times10^{-2}}{260.15}\,\frac{\{1+(2\times2.215/4.775)(2.176/6)^2\}}{2.215\sqrt{1-(2.176/6)^2}}$$

$$= 3.73\times10^{-5}\ \text{np/cm} = 0.0323\ \text{dB/m}$$

While the attenuation due to dielectric loss is given from Eq. 14.18b as:

$$\alpha_{d\text{TE}_{10}} = \frac{\pi f\sqrt{\varepsilon_r}}{c\sqrt{1-\left(f_{c\text{TE}_{10}}/f\right)^2}}\tan\delta_c = \frac{3.142\times5\times10^9\sqrt{2.1}\times1.5\times10^{-4}}{3\times10^{10}\times\sqrt{1-(2.176/6)^2}}$$

$$= 1.310\times10^{-4}\ \text{np/cm} = 0.1136\ \text{dB/m}$$

4.3 Circular Waveguide

The fundamental principles of analysis of circular waveguides are similar to those for the rectangular waveguide, but here we write the Maxwell's curl equations and the wave equations in cylindrical coordinates and solve the boundary value problem. A circular waveguide of radius a in the cylindrical coordinate system is shown in Fig. 4.13.

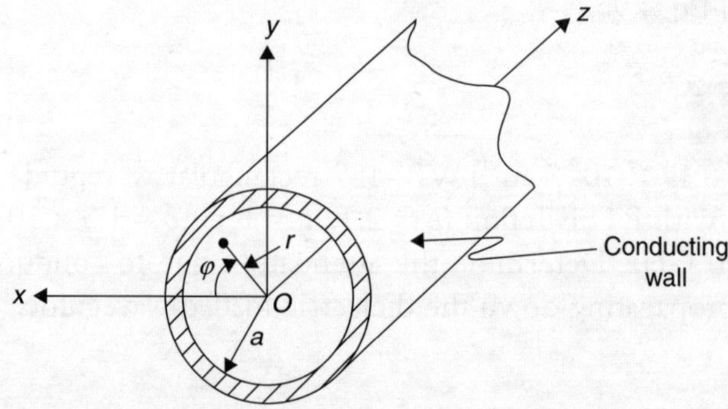

Fig. 4.13 Circular waveguide in cylindrical coordinate system.

We can write Maxwell's curl equations in the cylindrical coordinate system as:

$$\frac{1}{r}\cdot\frac{\partial E_z}{\partial\varphi} + \gamma_g E_\varphi = -j\omega\mu H_r$$

$$-\gamma_g E_r - \gamma_g H_r - \frac{\partial E_z}{\partial r} = -j\omega\mu H_\varphi$$

$$\frac{1}{r}\cdot\frac{\partial}{\partial r}(rE_\varphi) - \frac{1}{r}\cdot\frac{\partial E_r}{\partial \varphi} = -j\omega\mu H_z$$

$$\frac{1}{r}\cdot\frac{\partial H_z}{\partial \varphi} + \gamma_g H_\varphi = j\omega\varepsilon E_r \tag{4.19}$$

$$-\gamma_g H_r - \frac{\partial H_z}{\partial r} = j\omega\varepsilon E_\varphi$$

$$\frac{1}{r}\cdot\frac{\partial}{\partial r}(rH_\varphi) - \frac{1}{r}\cdot\frac{\partial H_r}{\partial \varphi} = j\omega\varepsilon E_z$$

where we have assumed propagation in the z-direction in a non-conducting medium and used $(\partial/\partial z) = -\gamma_g$.

The above equations may be combined for convenience to express the field components E_r, E_φ, H_r and H_φ in terms of E_z and H_z when we have:

$$k_c^2 H_r = j\frac{\omega\varepsilon}{r}\frac{\partial E_z}{\partial \varphi} - \gamma_g\frac{\partial H_z}{\partial r}$$

$$k_c^2 H_\varphi = -j\omega\varepsilon\frac{\partial E_z}{\partial r} - \frac{\gamma_g}{r}\frac{\partial H_z}{\partial \varphi} \tag{4.20}$$

$$k_c^2 E_r = -\gamma_g\frac{\partial E_z}{\partial r} - j\frac{\omega\mu}{r}\frac{\partial H_z}{\partial \varphi}$$

$$k_c^2 E_\varphi = -\frac{\gamma_g}{r}\frac{\partial E_z}{\partial \varphi} + j\omega\mu\frac{\partial H_z}{\partial r}$$

where $k_c^2 = \gamma_g^2 + k^2$.

The wave equation (Helmholtz equation or the Modal equation) for E_z in cylindrical coordinates may be written (using Eq. 2.8) as:

$$\frac{1}{r}\frac{\partial}{\partial r}\left(r\frac{\partial \vec{E}_z}{\partial r}\right) + \frac{1}{r^2}\frac{\partial^2 \vec{E}_z}{\partial \varphi^2} + \frac{\partial^2 \vec{E}_z}{\partial z^2} + k^2\vec{E}_z = 0$$

Remembering that $k^2 = \omega^2\mu\varepsilon$ and $(\partial/\partial z) = -\gamma_g$, we get:

$$\frac{\partial^2 \vec{E}_z}{\partial r^2} + \frac{1}{r}\frac{\partial \vec{E}_z}{\partial r} + \frac{1}{r^2}\frac{\partial^2 \vec{E}_z}{\partial \varphi^2} + (\gamma_g^2 + \omega^2\mu\varepsilon)\,\vec{E}_z = 0 \tag{4.21}$$

Proceeding in a similar manner to that followed in the case of the rectangular waveguide and noting that $E_z(r, \varphi, z) = E_z^0(r, \varphi)\,e^{-\gamma_g z}$, we can solve Eq. 4.21 using the method of separation of variables and assuming a solution for $E_z^0(r, \varphi)$ of the form $E_z^0(r, \varphi) = X(r)\,Y(\varphi)$. Substituting the expression for E_z in Eq. 4.21 and dividing by XY we get:

$$\frac{1}{X}\cdot\frac{d^2X}{dr^2} + \frac{1}{r\cdot X}\cdot\frac{dX}{dr} + k_c^2 = -\frac{1}{Yr^2}\cdot\frac{d^2Y}{d\varphi^2} \tag{4.22}$$

For the equality in Eq. 4.22 to hold for all values of r and φ, both sides must be equal to the same constant, which we take to be equal to n^2, where n is an integer. After some algebraic manipulation we thus have two separate algebraic equations, one in terms of $Y(\varphi)$ given by:

$$\frac{d^2Y}{d\varphi^2} + n^2Y = 0 \tag{4.23a}$$

and the other in terms of $X(r)$, that can be written as:

$$r^2\frac{d^2X}{dr^2} + r\frac{dX}{dr} + (k_c^2r^2 - n^2)\,X = 0 \tag{4.23b}$$

The general solution of Eq. 4.23a is of the form:

$$Y(\varphi) = C_1\cos(n\varphi) + C_2\sin(n\varphi) \tag{4.24a}$$

Equation 4.23b is known as Bessel's equation and has a general solution (with n being an integer):

$$X(r) = C_3J_n\,(k_cr) + C_4\,N_n\,(k_cr) \tag{4.24b}$$

where $J_n(.)$ and $N_n(.)$ are respectively the nth order Bessel functions of the first kind and the second kind (Fig. 4.14 for plots).

Since the function $N_n(.)$, also known as the Neumann function or Weber function, has a singularity (becomes infinite) at the origin ($r = 0$), such solutions cannot be used for any physical problem in which the origin is included. For example, the hollow and dielectric-filled waveguide problem, so that $C_4 = 0$. Thus the general solution for E_z is:

$$E_z^0(r,\varphi) = C_3J_n(k_cr)\,[C_1\cos(n\varphi) + C_2\sin(n\varphi)]\,e^{-\gamma_g z} \tag{4.25}$$

The relative amplitudes of C_1 and C_2 determine the orientation of the field in the guide, and for a circular guide with any particular value of n, the $\phi = 0$ axis can always be oriented to make either C_1 or C_2 equal to zero. Thus without any loss of generality we can write:

$$E_z^0(r,\phi) = C_nJ_n(k_cr)\cos(n\phi) \tag{4.26}$$

where C_n is a constant. The solution for $H_z^0(r,\phi)$ will have exactly the same form as $E_z^0(r,\phi)$ above. For the TM and TE waves, the remaining field components can be obtained by inserting $E_z^0(r,\phi)$ and $H_z^0(r,\phi)$ into Eq. 4.20.

4.3.1 TM and TE Modes in Circular Guides

Similar to the case of the rectangular waveguide, we divide the possible solutions for circular guides into transverse magnetic (TM) and transverse electric (TE) waves. For TM waves $H_z \equiv 0$ and the wave equation for E_z is used. The boundary condition $E_{\text{tangential}} = 0$ requires

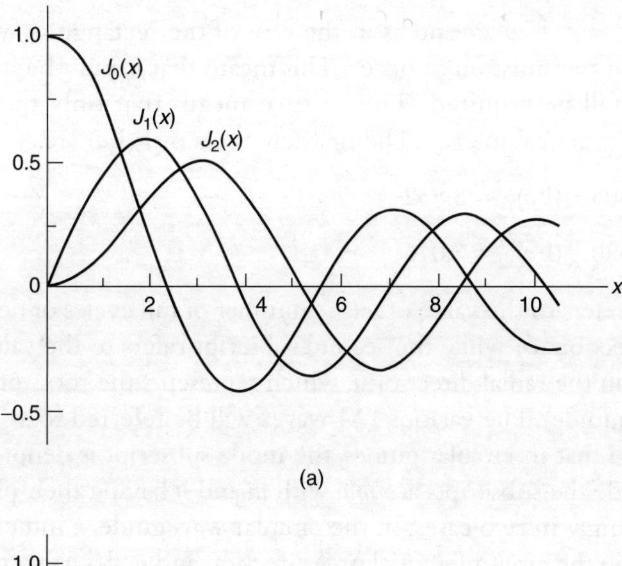

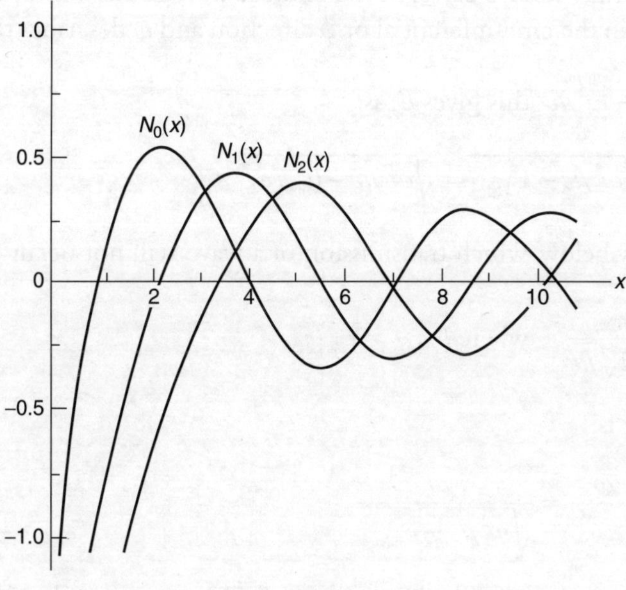

Fig. 4.14 Bessel functions (a) first kind (b) second kind.

that E_z must vanish at the surface of the guide, that is, $E_z^0(r, \phi) = 0$ at $r = a$. Thus from Eq. 4.26 we must have:

$$J_n(k_c a) = 0 \tag{4.27}$$

where a is the radius of the guide. Thus, we have a non-trivial solution of the boundary value problem at hand, as was obtained for the case of rectangular waveguide, only for discrete *eigenvalues* $k_c = (t_{nm}/a)$, where t_{nm} is the mth root of $J_n(x)$, that is, $J_n(t_{nm}) = 0$. The form of $J_n(x)$ for a few values of n is shown in Fig. 4.14 and Table 4.2 gives the first few values of t_{nm}. As before, $k_c^2 = \gamma_g^2 + \omega^2\mu\varepsilon$ and as in the case of the rectangular waveguides, k_c^2 must be less than $\omega^2\mu\varepsilon$ for transmission to occur. This means that k_c must be small, else extremely high frequencies will be required. This in turn means that only the first few roots of Eq. 4.27 will be of practical interest. The first few roots of $J_n(k_c a)$ are:

$$\begin{aligned} &(k_c a)_{01} = 2.405 \quad (k_c a)_{11} = 3.832 \\ &(k_c a)_{02} = 5.520 \quad (k_c a)_{12} = 7.016 \end{aligned} \tag{4.28}$$

The first subscript refers to the value of n (the number of full cycles of field variations in the circumferential direction, ϕ) while the second subscript refers to the value of m (the Bessel function variation in the radial direction r, which represents the roots of Bessel function in their order of magnitude). The various TM waves will be referred to as TM_{01}, TM_{12} and so on. It may be noted that in circular guides, the mode subscript is denoted as nm, while for the rectangular guide the subscripts are mn; with m and n having their physical significance mentioned accordingly in two cases. In the circular waveguide, n indicates the number of *full-cycle* variations in the circumferential or ϕ direction and m describes the radial variations of the field.

Since $\gamma_g = \sqrt{k_c^2 - \omega^2\mu\varepsilon}$, this gives β_g as:

$$\beta_{g\mathrm{TM}_{nm}} = \sqrt{\omega^2\mu\varepsilon - k_{c\mathrm{TM}_{nm}}^2} = \sqrt{\omega^2\mu\varepsilon - (t_{nm}/a)^2}$$

The *cut-off frequency* below which transmission of a wave will not occur is:

$$f_{c\mathrm{TM}_{nm}} = \frac{k_{c\mathrm{TM}_{nm}}}{2\pi\sqrt{\mu\varepsilon}}, \quad \text{where } k_{c\mathrm{TM}_{nm}} = t_{nm}/a$$

The phase velocity is:

$$v_{pg\mathrm{TM}_{nm}} = \frac{\omega}{\beta_{g\mathrm{TM}_{nm}}} = \frac{\omega}{\sqrt{\omega^2\mu\varepsilon - k_{c\mathrm{TM}_{nm}}^2}}$$

From Eq. 4.20, the various field components of TM waves can be derived in terms of E_z given by Eq. 4.26, when we obtain the expressions for TM_{nm} waves in the circular waveguide as:

$$\begin{aligned} E_Z &= C_n J_n\left(\frac{t_{nm}}{a} r\right)\cos(n\varphi)\, e^{-j\beta_{gnm}z} \\ E_r &= -\frac{\gamma_g}{k_c^2}\frac{\partial E_z}{\partial r} = -\frac{ja\beta_{gnm}}{t_{nm}} C_n J_n'\left(\frac{t_{nm}}{a} r\right)\cos(n\varphi)\, e^{-j\beta_{gnm}z} \\ E_\varphi &= -\frac{\gamma_g}{rk_c^2}\frac{\partial E_z}{\partial \varphi} = \frac{ja^2 n\beta_{gnm}}{rt_{nm}^2} C_n J_n\left(\frac{t_{nm}}{a} r\right)\sin(n\varphi)\, e^{-j\beta_{gnm}z} \end{aligned} \tag{4.29}$$

$$H_r = j\frac{\omega\varepsilon}{rk_c^2}\frac{\partial E_z}{\partial\varphi} = -\frac{j\omega\varepsilon n a^2}{rt_{nm}^2}C_n J_n\left(\frac{t_{nm}}{a}r\right)\sin(n\varphi)\,e^{-j\beta_{gnm}z}$$

$$H_\varphi = -\frac{j\omega\varepsilon}{k_c^2}\frac{\partial E_z}{\partial r} = -\frac{j\omega\varepsilon a}{t_{nm}}C_n J_n'\left(\frac{t_{nm}}{a}r\right)\cos(n\varphi)\,e^{-j\beta_{gnm}z}$$

$$H_z = 0$$

where $J_n'(x) = \frac{dJ_n(x)}{dx}$.

The solution for TE modes can be found in a similar manner, except for the fact that we now solve for $H_z^0(r, \phi)$ and remember that E_z is identically zero. We have from our previous discussion (refer to Eq. 4.26):

$$H^0{}_z(r, \phi) = C_n J_n(k_c r)\cos(n\phi) \tag{4.30}$$

The boundary condition $E_{\tan} = 0$ for TE wave means that the ϕ component of the electric field E_φ, which is tangential to the inner surface of the circular waveguide, at $r = a$ must vanish. However, with $E_z = 0$, we find from Eq. 4.20 that E_φ is proportional to $\partial H_z/\partial r$, and therefore to $J_n'(k_c r)$. Thus, for TE modes, the boundary conditions to be satisfied are $J_n'(k_c a) = 0$. If the roots of $J_n'(x)$ are defined as t_{nm}', so that $J_n'(t_{nm}') = 0$, where $t_{nm}' = 0$ is the mth root of J_{nm}', then k_{cnm} must have the value $k_{cnm} = t_{nm}'/a$. The first few values of t_{nm}' are given in Table 4.3.

Table 4.2 Values of t_{nm} [m^{th} roots of $J_n(x) = 0$] (for TM modes)

n	t_{n1}	t_{n2}	t_{n3}
0	2.405	5.520	8.654
1	3.832	7.016	10.174
2	5.135	8.417	11.620

Table 4.3 Values of t'_{nm} [m^{th} roots of $J'_n(x) = 0$] (for TE modes)

n	t'_{n1}	t'_{n2}	t'_{n3}
0	3.832	7.016	10.174
1	1.841	5.331	8.536
2	3.054	6.706	9.970

The first few roots of $J_n'(k_c a)$ are:

$$(k_c a)_{01} = 3.832 \qquad (k_c a)_{11} = 1.841$$

$$(k_c a)_{02} = 7.016 \qquad (k_c a)_{12} = 5.331 \tag{4.31}$$

The corresponding TE waves are referred to as TE_{01}, TE_{11} and so on.

The equations for β_g, f_c and v_{pg} are the same for the TM case except that $k_{cTE_{nm}}$ is now given by t_{nm}'/a, instead of t_{nm}/a. Then we have:

$$\beta_{gTE_{nm}} = \sqrt{\omega^2\mu\varepsilon - (t_{nm}'/a)^2}$$

$$f_{c\text{TE}_{nm}} = t'_{nm}/2\pi a\sqrt{\mu\varepsilon} \quad \text{or} \quad \lambda_{c\text{TE}_{nm}} = 2\pi a/t'_{nm} \tag{4.32}$$

$$v_{pg\text{TE}_{nm}} = \omega/\sqrt{\omega^2\mu\varepsilon - (t'_{nm}/a)^2}$$

The expressions for TE_{nm} waves in the circular waveguide can be obtained by substituting H_z from Eq. 4.30 into Eq. 4.20, when we get:

$$H_z = C_n J_n\left(\frac{t'_{nm}}{a}r\right)\cos(n\varphi)e^{-j\beta_{gnm}z}$$

$$H_r = -\frac{\gamma_g}{k_c^2}\frac{\partial H_z}{\partial r} = -\frac{ja\beta_{gnm}}{t'_{nm}}C_n J'_n\left(\frac{t'_{nm}}{a}r\right)\cos(n\varphi)e^{-j\beta_{gnm}z}$$

$$H_\varphi = -\frac{\gamma_g}{rk_c^2}\frac{\partial H_z}{\partial \varphi} = \frac{ja^2 n\beta_{gnm}}{rt'^2_{nm}}C_n J_n\left(\frac{t'_{nm}}{a}r\right)\sin(n\varphi)e^{-j\beta_{gnm}z} \tag{4.33}$$

$$E_r = -j\frac{\omega\mu}{rk_c^2}\frac{\partial H_z}{\partial \varphi} = \frac{j\omega\mu na^2}{rt'^2_{nm}}C_n J_n\left(\frac{t'_{nm}}{a}r\right)\sin(n\varphi)e^{-j\beta_{gnm}z}$$

$$E_\varphi = \frac{j\omega\mu}{k_c^2}\frac{\partial H_z}{\partial r} = -\frac{j\omega\mu a}{t'_{nm}}C_n J'_n\left(\frac{t'_{nm}}{a}r\right)\cos(n\varphi)e^{-j\beta_{gnm}z}$$

$$E_z = 0$$

It may be observed that TE_{11} (which corresponds to the lowest value of $t_{nm} = 1.814$) has the lowest cut-off frequency in the circular waveguide which is therefore the *dominant mode* in the circular waveguide (with $\lambda_c = 3.412a$).

Figure 4.15a shows some of the TE/TM modes possible in the circular waveguide with cut-off frequency relative to the dominant TE_{11} mode while Fig. 4.15 (b–e) shows the field pattern of some of the important lower order TE/TM modes including the dominant mode. The electric (E) and magnetic (H) field patterns follow the same general rules as far as the boundary conditions for E and H fields are concerned and the propagating wave has the same repetition rate of λ_g of the fields as was the case with the rectangular waveguide.

From Figs 4.15a and 4.5 it may be noted that the ratio of the cut-off frequency for TM_{01} (the next higher order mode) to the cut-off frequency for the TE_{11} mode (the dominant mode) in the circular waveguide is 2.405/1.841 = 1.3 [while for the rectangular waveguide, the ratio of the cut-off frequency of next higher order mode TE_{20} (c/a) to the dominant mode TE_{10} ($c/2a$) is equal to 2]. This means that the circular guide has a much narrower frequency range of operation for single mode propagation than the rectangular guide, which makes the rectangular waveguide more popular than the circular waveguide from the application point of view. Further, it may also be shown that for the same cut-off frequency of operation the cross-sectional area occupied by the circular waveguide in comparison to the cross-sectional area occupied by the rectangular waveguide is larger by a factor of more than 2 when both of them operate in their respective dominant modes.

It is interesting to note that the electric and magnetic field patterns of the dominant TE_{11} mode in the circular waveguide have a close similarity with those of the dominant TE_{10} mode in the rectangular waveguide. Thus, a gradual tapering from the rectangular to the circular cross-section launches the TE_{11} mode in the circular waveguide from the TE_{10} in the rectangular waveguide due to mode compatibility.

There is no TE_{10} mode in the circular waveguide but the TE_{01} mode exists. The TE_{01} in the circular waveguide has a special characteristic—it has no longitudinal surface current, the surface current (being perpendicular to the longitudinal magnetic field lines) is circumferential to the metallic wall of the waveguide which decreases with frequency. This makes the wall loss resistance (R_s) extremely low at microwave frequencies and thus microwave cavities designed with the TE_{01} mode in the circular waveguide can have very large Q ($= \omega L/R_s$) values which are useful in the design of wavemeters (cavity type microwave frequency meter) and echo boxes (used for radar performance characterisation).

It may further be noted that the TM_{01} mode in a circular waveguide possesses axial symmetry or rotational symmetry of the E-field which makes it useful in the design of

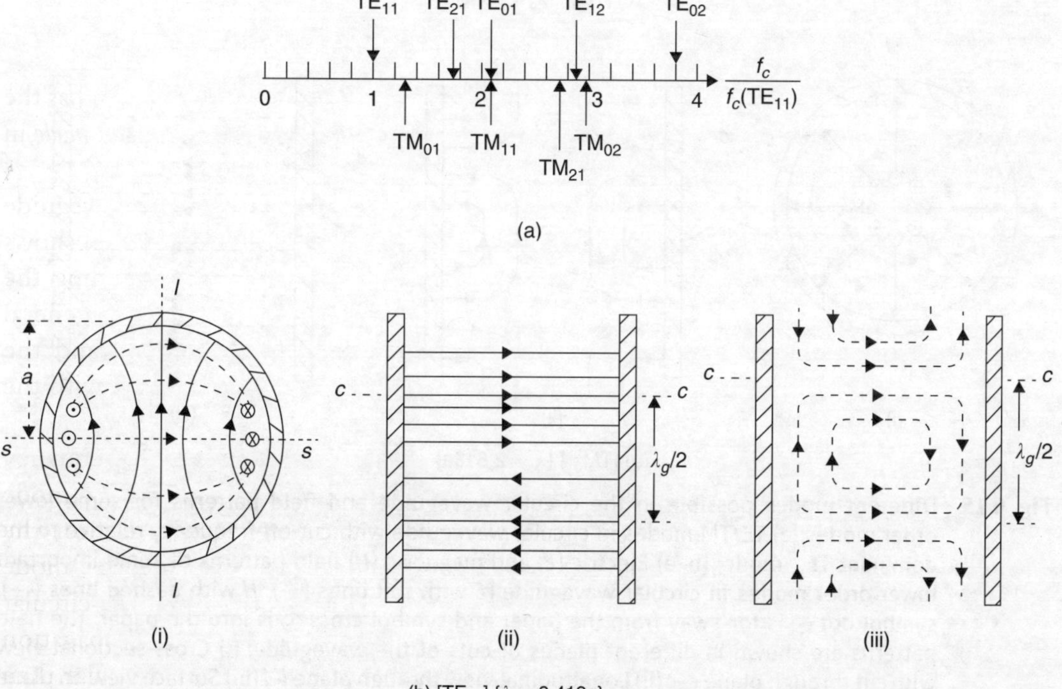

(b) [TE_{11}] {$\lambda_c = 3.412a$}

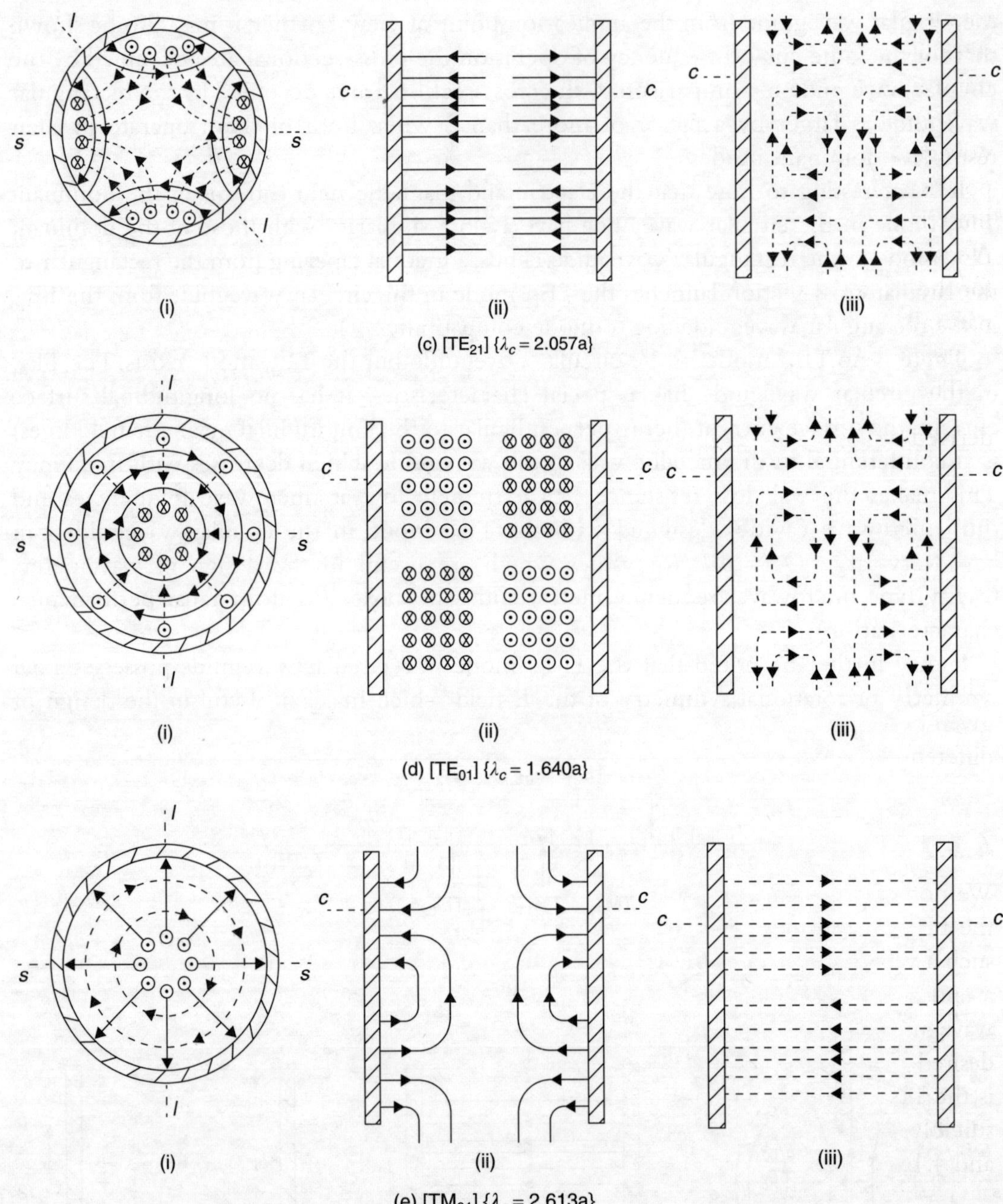

Fig. 4.15 Different modes possible in the circular waveguide and field patterns for some lower order modes (a) TE/TM modes in circular waveguide with cut-off frequency relative to the dominant TE_{11} mode; (b–e) Electric (*E*) and magnetic (*H*) field patterns of some important lower order modes in circular waveguide [*E* with full lines (—), *H* with dashed lines (---); symbol dot ⊙ is for away from the paper and symbol cross ⊗ is into the paper. The field patterns are shown in different planes of cuts of the waveguide: (i) Cross-sectional view with cut through plane *c–c* (ii) Longitudinal view through plane *l–l* (iii) Surface view on plane of cut through *s–s*.

rotary joint (Section 6.2.10) that is used between the transmitter and the mechanically steerable antenna in the surveillance radar system to maintain signal polarisation from the transmitter to the antenna unchanged.

It is interesting to note that the polarisation of the propagating wave in a rectangular waveguide is uniquely defined. However, there are certain applications that require dual polarisation capability; for example, a waveguide connected to a circularly-polarised antenna must be able to efficiently propagate both vertically and horizontally polarised waves. A square waveguide may provide such capability since with $b = a$, the cut-off frequencies for the TE_{10} and TE_{01} modes are the same. But, a circular waveguide with its field patterns naturally provides the most common type of a dual polarisation transmission line and is preferred over a rectangular waveguide.

The wave impedance for the TE and TM modes in the circular waveguide may be derived from $Z_w = \frac{E_r}{H_\varphi}$, when we have:

$$Z_{w(\text{TE})} \frac{\eta}{\sqrt{1-\left(\frac{f_{c\text{TE}_{nm}}}{f}\right)^2}}, \quad Z_{w(\text{TM})} = \eta\sqrt{1-\left(\frac{f_{c\text{TM}_{nm}}}{f}\right)^2} \tag{4.34}$$

These are seen to be exactly similar to the expressions obtained for the rectangular waveguide given by Eqs 4.11b and 4.13b respectively; only the corresponding cut-off frequencies are different.

4.3.2 Launching of Modes in Circular Waveguides

We noted earlier that for TE modes, $E_z = 0$ and for TM modes $H_z = 0$. Thus, to launch the TE_{11} mode in the circular waveguide, a coaxial line can be introduced in the guide in such a way that it excites the E field in the y-direction, having H field components in the x and z-directions (Figs 4.2a, 4.15 and 4.16a). To launch the TM_{01} mode in the circular waveguide, a coaxial line is connected as a longitudinal probe to the guide to excite the desired field (Figs 4.15 and 4.16b). Another important mode in the circular waveguide is the TE_{01} mode which may be excited by radiating slots in two rectangular waveguides suitably connected at diametrically opposite sides of the circular waveguide (Figs 4.15 and 4.16c).

4.3.3 Characteristics of Standard Circular Waveguides

The inner diameter of a circular waveguide is determined in terms of the frequency of the signal being transmitted. For example, at X-band frequencies (from 8–12 GHz), the inner diameter of a circular waveguide designated as EIA WC(94) (note WC stands for Waveguide Circular) by the Electronic Industry Association is 2.383 cm (0.938 inch). Table 4.4 tabulates the characteristics of the standard circular waveguides.

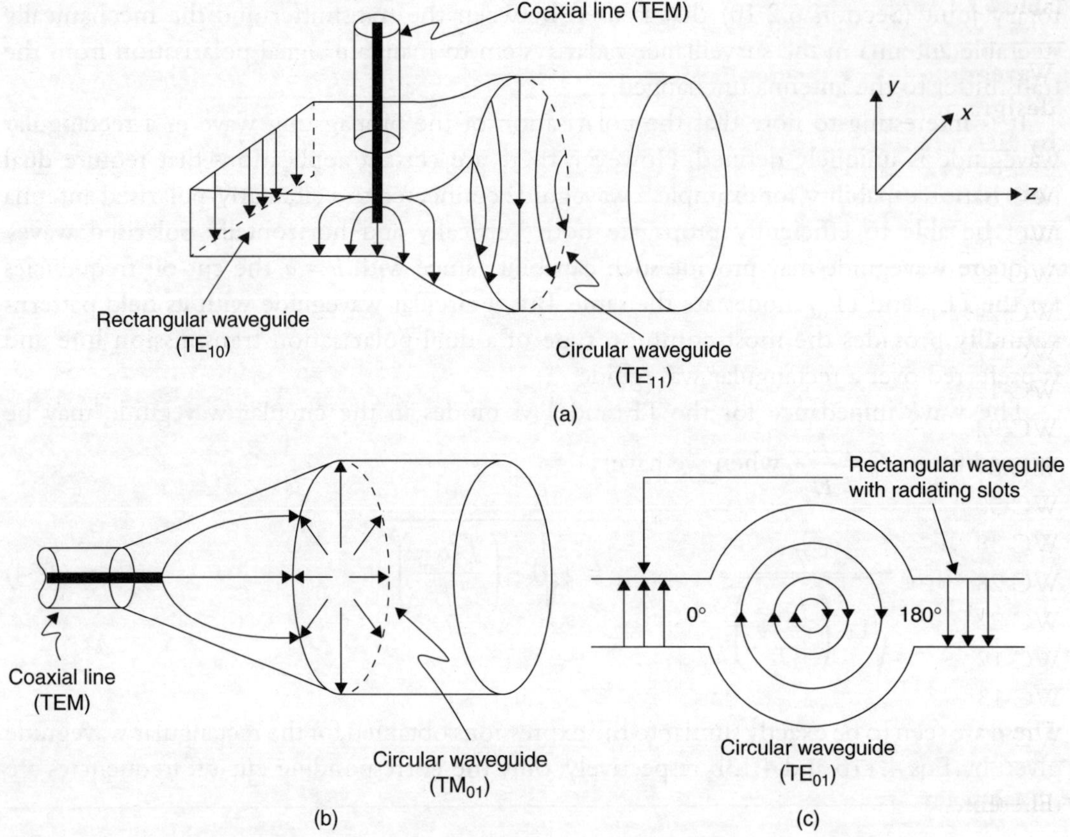

Fig. 4.16 Launching of different modes in the circular waveguide (a) TE_{11} mode (b) TM_{01} mode (c) TE_{01} mode.

4.3.4 Attenuation in Circular Waveguides

The attenuation produced by the imperfectly conducting walls of a circular waveguide can be calculated in the same manner as for rectangular waveguides and for the TE_{nm} mode, it is given by [4]:

$$\alpha_{c\mathrm{TE}_{nm}} = \frac{R_s}{a\eta} \frac{\left\{\left(\frac{f_{c\mathrm{TE}_{nm}}}{f}\right)^2 + \frac{n^2}{(t'_{nm})^2 - n^2}\right\}}{\left\{1 - \left(\frac{f_{c\mathrm{TE}_{nm}}}{f}\right)^2\right\}^{1/2}} \tag{4.35}$$

While for TM modes it is: $\alpha_{c\mathrm{TM}_{nm}} = \frac{R_s}{a\eta}\left\{1 - \left(\frac{f_{c\mathrm{TM}_{nm}}}{f}\right)^2\right\}^{-1/2}$ (4.36)

Computed values of α_{cl} for some of the lower order TE and TM modes are plotted as a function of frequency in Fig. 4.17a.

Table 4.4 Characteristics of standard circular waveguides

Waveguide designation by EIA★	Inside diameter (inch/cm)	Recommended frequency range for TE_{11} mode (GHz)	Cut-off frequency for air-filled waveguide (GHz)
WC 618	6.181 (15.700)	1.29–1.76	1.120
WC 451	4.511 (11.458)	1.76–2.42	1.534
WC 329	3.292 (8.362)	2.42–3.31	2.102
WC 240	2.403 (6.104)	3.31–4.54	2.880
WC 175	1.750 (4.445)	4.54–6.23	3.955
WC 128	1.281 (3.254)	6.21–8.51	5.402
WC 94	0.938 (2.383)	8.49–11.60	7.377
WC 69	0.688 (1.748)	11.60–15.90	10.057
WC 50	0.500 (1.270)	15.90–21.80	13.842
WC 38	0.375 (0.953)	21.20–29.10	18.446
WC 28	0.281 (0.714)	28.30–38.80	24.620
WC 22	0.219 (0.556)	36.40–49.80	31.617
WC 17	0.172 (0.437)	46.30–63.50	40.227
WC 13	0.125 (0.318)	63.50–87.20	55.280
WC 09	0.094 (0.239)	84.80–116.00	73.552

★ EIA (Electronic Industry Association, US) designation

The attenuation constant for TM waves is seen to decrease with frequency until $f = 3f_c$ and then increases slightly or remains constant for further increase of frequency. The same behaviour occurs for TE modes except for TE_{om} modes. For the latter, the attenuation constant decreases indefinitely with frequency. This is because, for such modes, wall currents are completely circumferential as the magnetic field is longitudinal everywhere. Field analysis shows that for a fixed power level, the circumferential currents decrease with frequency. This characteristic of the TE_{01} mode in circular waveguides appears to be its potential advantage as a transmission medium for wideband multi-channel system for long-distance communication at high frequencies, particularly at millimetre-wave frequencies. Although at one time this property of the TE_{01} mode made the circular waveguide designed with the TE_{01} mode attractive for long-haul communication, excitation of other modes like TE_{11} or TM_{01} with lower cut-off frequencies however, caused difficulties (causing significant loss of power in those undesired modes). In recent times, optical fibre transmission technology proved to have much greater potential with improved performance at a lower cost when compared with circular waveguide designed with TE_{01} mode of operation.

As in the case of the rectangular waveguide (Fig. 4.18b), it may be observed that the oversized circular waveguide also provides less attenuation due to reduced wall losses hence making it attractive for millimetre-wave applications when large amounts of power needs to be pushed through the guide with small attenuation being incurred.

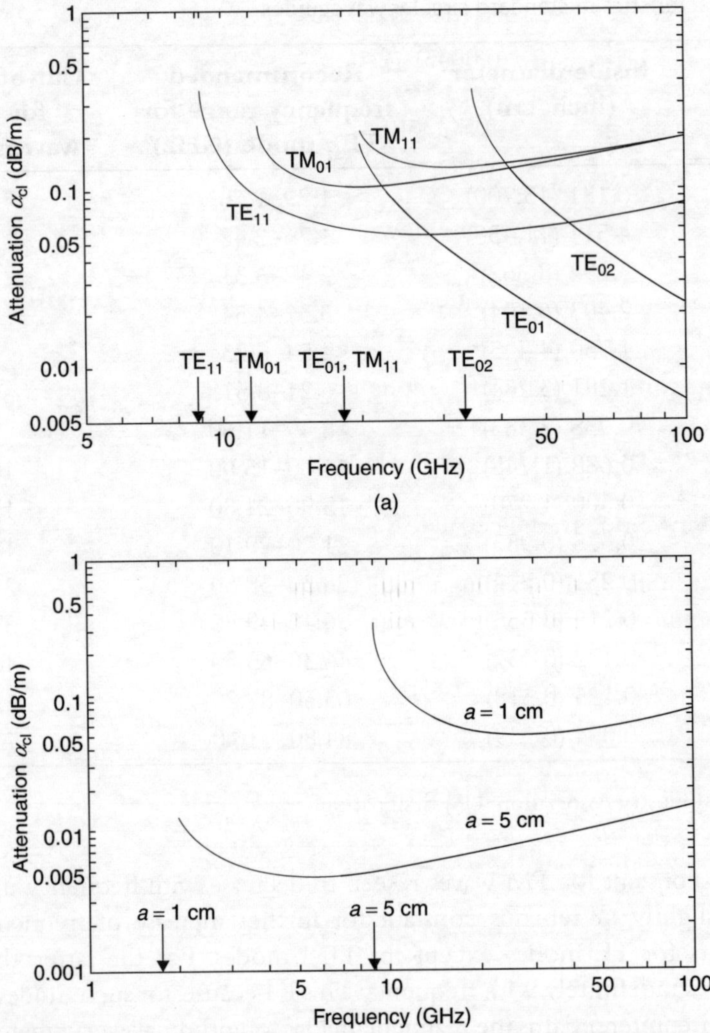

Fig. 4.17 (a) Computed values of attenuation in an air-filled circular waveguide ($a = 1$ cm) with $s = 5.8 \times 10^7$ S/m, (b) Comparison of attenuation in two waveguides having different a values with dominant mode (TE_{11}) propagation.

EXAMPLE 4.6

Determine the cut-off frequencies of the first two propagating modes of a circular waveguide having radius $a = 4.5$ mm and filled with a dielectric having dielectric constant $\varepsilon_r = 2.1$. If the guide is gold plated and the loss tangent of the dielectric is 0.001, calculate the attenuation in dB for a 60 cm length of guide operating at 15 GHz.

Solution

The first two propagation modes of a circular waveguide are TE_{11} and TM_{01} modes. The cut-off frequencies can be found using Eq. 4.32:

$$\text{TE}_{11}: f_c = \frac{t'_{11}c}{2\pi a\sqrt{\varepsilon_r}} = \frac{1.841(3\times10^8)}{2\pi(0.0045)\sqrt{2.1}} = 13.47\text{ GHz}$$

$$\text{TM}_{01}: f_c = \frac{t_{01}c}{2\pi a\sqrt{\varepsilon_r}} = \frac{2.405(3\times10^8)}{2\pi\,(0.0045)\sqrt{2.1}} = 17.60\text{ GHz}$$

So, only the TE_{11} mode can propagate down the waveguide at 15 GHz.

The wavenumber is $k = \omega\sqrt{\mu\varepsilon} = \dfrac{2\pi f\sqrt{\varepsilon_r}}{c} = \dfrac{2\pi(15\times10^9)\sqrt{2.1}}{3\times10^8} = 455.26\text{ m}^{-1}$

The propagation constant of the TE_{11} mode is

$$\beta_{g11} = \sqrt{k^2-(t'_{11}/a)^2} = \sqrt{(455.26)^2-(1.841/0.0045)^2} = 199.72\text{ m}^{-1}$$

Now, the attenuation due to dielectric loss is calculated from Eq. 4.18b:

$$\alpha_{dl} = \frac{\omega^2\mu\varepsilon'\tan\delta_\iota}{2\beta_{g11}} = \frac{(455.26)^2\,(0.001)}{2\times(199.72)} = 0.5188\text{ np/m}$$

The conductivity of gold is $\sigma = 4.10\times10^7$ S/m, so the surface resistance is:

$$R_s = \sqrt{\frac{\omega\mu_0}{2\sigma}} = \sqrt{\frac{2\pi\times15\times10^9\times4\pi\times10^{-7}}{2\times4.10\times10^7}} = 0.038\ \Omega$$

The wave impedance in the dielectric is:

$$\eta_{\text{Teflon}} = \frac{\eta_0}{\sqrt{\varepsilon_r}} = \frac{377}{\sqrt{2.1}} = 260.15\ \Omega$$

From Eq. 4.35, the attenuation due to metallic loss is:

$$\alpha_{c\text{TE}_{nm}} = \frac{0.038}{0.0045\times260.51}\,\frac{\left[\left(\dfrac{13.47}{15}\right)^2+\dfrac{1}{(1.841)^2-1}\right]}{\left[1-\left(\dfrac{13.47}{15}\right)^2\right]^{1/2}} = 0.09\text{ np/m}$$

So, the total attenuation factor is:

$$\alpha = \alpha_c + \alpha_d = 0.6088\text{ np/m}$$

We observe that the dielectric loss dominates. The attenuation in the 60 cm long guide is:

$$\text{Attenuation (dB)} = -20\log e^{-\alpha l} = -20\log e^{-(0.6088)(0.6)} = 3.17\text{ dB}$$

4.4 Cavity Resonators

The lumped circuit elements L and C are used at lower frequencies to design resonant circuits which can store electric and magnetic energies. However, at UHF and microwave frequencies, the utility of lumped elements for realising resonant behaviour becomes impractical due to radiation losses (as the dimensions of the lumped element becomes comparable to the wavelength of operation of the signal) and also due to high effective resistance (resulting from skin depth being extremely small at such high frequencies). To provide a resonant circuit at higher RF and microwave frequencies we prefer to use instead an enclosure (a cavity) completely surrounded by conducting walls. Such a shielded enclosure confines the electromagnetic fields inside and furnishes large areas for surface current flow, thus eliminating the problem of undesired radiation and high resistance effect. These enclosures have natural resonant frequencies and can provide very large quality factor ($Q = 5000$ to 6000 typical, going up to even 30,000) and are called *cavity resonators*.

A cavity resonator can be constructed simply by shorting the two ends of a segment of a waveguide (rectangular or circular) of suitable length (normally some multiple of half-wavelength). At each of these shorting ends, the incident wave will be completely reflected giving rise to standing waves within the guide. However, only those standing-wave patterns (*modes*) will exist that satisfy the boundary conditions at walls of the enclosed waveguide. The oscillating electric and magnetic fields get stored or trapped in the cavity for a particular mode, when it exists, only at a particular frequency called the *resonant frequency* of the resonator (determined by the resonator dimension) for operation in that particular mode.

Two common types of cavity resonators used at microwave frequencies are the *rectangular* and *cylindrical cavity resonators*. Other important types of resonators such as the *strip/disc resonator*, made of truncated microstrip (usually in the form of a disc) and *dielectric resonator*, typically consisting of a small disc- or cube-shaped dielectric, are also of practical importance in recent times. Strip and dielectric resonators can be easily incorporated into microwave and millimetre-wave integrated circuits and coupled to various forms of planar transmission lines. Again, at very short wavelengths, say in the sub-millimetre range, Q becomes too small for the usual type of cavities (since Q decreases as $f^{-1/2}$). A further difficulty is the miniature size of the cavity at wavelengths shorter than a few millimetres. To overcome these difficulties in the short-millimetre-wavelength region and throughout the sub-millimetre (infrared and optical) region, the optical Fabry–Perot has been adapted as a resonator.

A new type of resonator concept is emerging these days, known as *zeroth order resonator* (theoretically size independent), based on CRLH (composite right/left-handed) transmission line. Such resonators are metamaterial-inspired with possible application potential in designing miniaturised microwave components and antennae (details are given in Chapter 15).

4.4.1 Rectangular Cavity Resonator

The schematic of a rectangular cavity resonator is shown in Fig. 4.18, having dimensions a, b, l, such that $a > b$.

The cavity walls are metallic, and the material inside the cavity is typically a dielectric with constants μ and ε. As in the case for waveguide transmission, cavity modes can also be divided into two sets—TE and TM. However, specification of a unique signal propagation

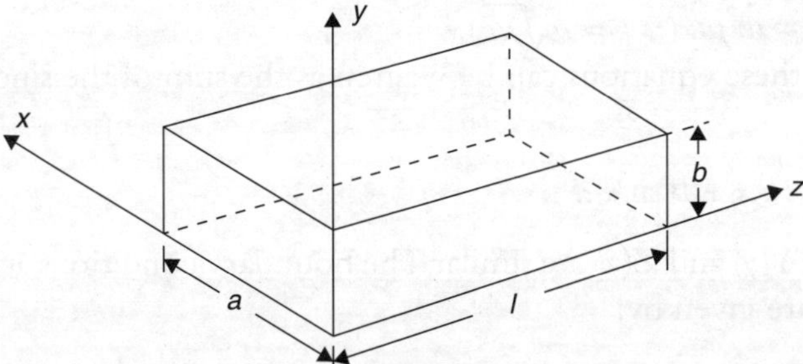

Fig. 4.18 Geometry of rectangular cavity resonator.

direction is required; here, we take it as the z-direction (otherwise a TE mode with respect to the z-axis could be a TM mode with respect to the y-axis). The existence of conducting walls at $z = 0$ and $z = l$ gives rise to multiple reflections and sets up standing waves in the resonator. A three-symbol (mnp) subscript is used to designate a TE or TM standing wave pattern in a cavity resonator. The indices m, n and p (similar to waveguides) indicate the number of half-sinusoid variations of the field along the x, y and z directions respectively.

The solution for the fields within the cavity may, in principle, be determined by extending the wave mode solution of waveguides by considering the cavity as a section of a standard waveguide with short-circuited terminations at $z = 0$ and $z = d$; with $d = n\lambda/2$ where $n = 1, 2, 3, \ldots$ However, to get a better insight, it will be worthwhile to begin with the wave equation and use the method of separation of variables to solve for the electric and magnetic fields that satisfy the boundary conditions of the cavity. [The reason why a waveguide closed at two ends with shorting plates behaves as a resonator when the distance between two shorting plates is $m\lambda_g/2$, where m is an integer, may be understood as follows. The round-trip phase difference for a wave travelling between two reflecting plates (short-circuit for microwave signal and reflecting mirrors for optical waves) is given by: $(2\pi/\lambda)2l$, where l is the separation between the two reflecting plates and λ is the wavelength of the electromagnetic signal ($2l$ being the round-trip path length). To realise resonance in the cavity of length l at the operating wavelength, the above mentioned round-trip phase difference should be equal to $2\pi m$. Thus from the equation $(2\pi/\lambda)2l = 2\pi m$ we get: $l = m\lambda/2$.]

TE$_{mnp}$ modes ($E_z \equiv 0, H_z \neq 0$): By equating the z components of Eq. 4.4b, the following equation is obtained:

$$\frac{\partial^2 H_z}{\partial x^2} + \frac{\partial^2 H_z}{\partial y^2} + \gamma_g^2 H_z = -\omega^2 \mu\varepsilon H_z \tag{4.37}$$

Using the technique of separation of variables to solve the above equation we can write:

$$H_z = X(x)\,Y(y)\,Z(z) \tag{4.38}$$

Substituting this in Eq. 4.37 above, we get:

$$\frac{1}{X}\frac{d^2X}{dx^2} = -k_x^2, \quad \frac{1}{Y}\frac{d^2Y}{dx^2} = -k_y^2 \quad \text{and} \quad \frac{1}{Z}\frac{d^2Z}{dx^2} = -k_z^2$$

where $k_x^2 + k_y^2 + k_z^2 = \omega^2\mu\varepsilon$ ($\because k = \omega\sqrt{\mu\varepsilon}$).

The solution to these equations can be written as the sum of the sine and cosine terms, when we have:

$$X(x) = A \cos k_x x + B \sin k_x x$$

The expressions for $Y(y)$ and $Z(z)$ are similar. The boundary conditions at the six conducting walls of the cavity are given by:

$$\frac{\partial \vec{H}_z}{\partial x} = 0, \text{ at } x = 0 \quad \text{and} \quad x = a, \quad \frac{\partial \vec{H}_z}{\partial y} = 0, \text{ at } y = 0 \quad \text{and} \quad y = b,$$

and $\quad \vec{H}_z = 0$, at $z = 0$ and $z = l$.

The application of these boundary conditions results in the expression for $\vec{H}_z$ from Eq. 4.38:

$$\vec{H}_z = H_0 \cos\left(\frac{m\pi}{a}\right) x \cos\left(\frac{n\pi}{b}\right) y \sin\left(\frac{p\pi}{l}\right) z \tag{4.39}$$

where $k_x = m\pi/a$, $k_y = n\pi/b$, $k_z = p\pi/l$ and m, n and p are positive integers. H_0 is a constant that is proportional to the signal level. Substituting the above expression for H_z into Eq. 4.37 yields:

$$\left\{\left(\frac{m\pi}{a}\right)^2 + \left(\frac{n\pi}{b}\right)^2 + \left(\frac{p\pi}{l}\right)^2 - \omega^2\mu\varepsilon\right\} \vec{H}_z = 0$$

If $H_z = 0$, all field components are zero and the solution is trivial. For the field to be non-zero, the following condition must hold:

$$\left(\frac{m\pi}{a}\right)^2 + \left(\frac{n\pi}{b}\right)^2 + \left(\frac{p\pi}{l}\right)^2 = \omega_r^2\mu\varepsilon$$

where $\omega_r = 2\pi f_r$, f_r being the resonant frequency of the rectangular cavity resonator. Solving for f_r we get:

$$f_r = \frac{1}{2\sqrt{\mu\varepsilon}} \sqrt{\left(\frac{m}{a}\right)^2 + \left(\frac{n}{b}\right)^2 + \left(\frac{p}{l}\right)^2} \tag{4.40}$$

Thus, an electromagnetic field can exist in the loss-less cavity only when the excitation frequency is exactly f_r, which clearly indicates a resonance behaviour. Since m, n and p can take on all integer values, the cavity has an infinite number of resonant modes, each with its own resonant frequency. For a rectangular cavity where b is conventionally assumed to be the smallest of the three dimensions, TE_{101} is the *primary mode*, that is, the mode having the *lowest* resonant frequency and this corresponds to the TE_{10} mode in a rectangular waveguide.

It is interesting to note that for a given cavity (having fixed a, b, l), the resonant frequencies are not harmonically related but consist of a set of irregularly spaced eigen values of the wave equation. This circumstance is quite unlike the case of a lumped resonant circuit (such as a parallel RLC circuit), which is resonant at $\omega_0 = \left(\sqrt{LC}\right)^{-1}$ as well as at $2\omega_0$, $3\omega_0$ and so on.

The electric and magnetic fields for TE_{mnp} mode in a rectangular resonator can be derived from Eq. 4.39 by considering the axial magnetic field $\vec{H}_z$ as the superposition of two oppositely travelling waves as:

$$\vec{H}_z = H_0 \cos k_x x \cos k_y y \left(\frac{e^{jk_z z} - e^{-jk_z z}}{2j} \right) = H_0 \cos k_x x \cos k_y y \sin k_z z \tag{4.41}$$

where the first exponential term represents a wave travelling in the $-z$ direction with a phase constant k_z and the second term represents a $+z$ directed wave. Using this form of $\vec{H}_z$ and the relations in Eq. 4.5, the remaining field components are given by:

$$E_x = -j \frac{\omega\mu}{k_c^2} \frac{\partial H_z}{\partial y} = j \frac{\omega_r \mu k_y}{k_c^2} H_0 \cos k_x x \sin k_y y \sin k_z z$$

$$E_y = j \frac{\omega\mu}{k_c^2} \frac{\partial H_z}{\partial x} = -j \frac{\omega_r \mu k_x}{k_c^2} H_0 \sin k_x x \cos k_y y \sin k_z z$$

$$E_z = 0 \tag{4.42}$$

$$H_x = -\frac{\gamma_g}{k_c^2} \frac{\partial H_z}{\partial x} = \frac{1}{k_c^2} \frac{\partial^2 H_z}{\partial x \partial z} = -\frac{k_x k_z}{k_c^2} H_0 \sin k_x x \cos k_y y \cos k_z z$$

$$H_y = -\frac{\gamma_g}{k_c^2} \frac{\partial H_z}{\partial y} = \frac{1}{k_c^2} \frac{\partial^2 H_z}{\partial y \partial z} = -\frac{k_y k_z}{k_c^2} H_0 \cos k_x x \sin k_y y \cos k_z z$$

where $k_x = m\pi/a$, $k_y = n\pi/b$, $k_z = p\pi/l$, and $k_c^2 = k_x^2 + k_y^2$; and note that $\frac{\partial}{\partial z} = -\gamma_g$.

These equations can be used to sketch the field patterns of the various cavity modes.

Subsequently, we will see that the primary mode (one with lowest resonant frequency) in a rectangular cavity resonator is the TE_{101} mode (with $m = 1$, $n = 0$ and $p = 1$) for which the field equations would simplify to:

$$H_z = H_0 \cos\left(\frac{\pi x}{a}\right) \sin\left(\frac{\pi z}{l}\right)$$

$$H_x = -\left(\frac{a}{l}\right) H_0 \sin\left(\frac{\pi x}{a}\right) \cos\left(\frac{\pi z}{l}\right)$$

$$E_y = -j \frac{\omega_r \mu a}{\pi} H_0 \sin\left(\frac{\pi x}{a}\right) \sin\left(\frac{\pi z}{l}\right) \tag{4.43a}$$

The field pattern for the TE_{101} cavity mode is shown in Fig. 4.19.

The resonant frequency for this mode is $f_{101} = \frac{1}{2\sqrt{\mu\varepsilon}} \sqrt{\frac{1}{a^2} + \frac{1}{l^2}}$ (4.43b)

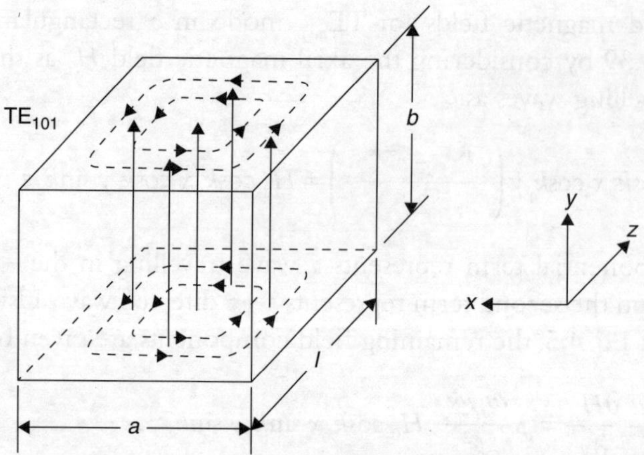

Fig. 4.19 Electric (—) and magnetic (---) field pattern of the TE_{101} mode in a rectangular cavity resonator.

Note that the field components have no variations in the y direction and one-half variations in the x and z directions. Further, it is apparent that the electric and magnetic fields are 90° out-of-phase. When the magnetic field is maximum, the electric field is zero and vice versa. This means that no real power flow occurs, the cavity merely stores the energy associated with the field. As in the case with an ordinary L–C circuit, the stored energy changes from magnetic to electric and back to magnetic every half cycle.

TM_{mnp} modes ($H_z \equiv 0, E_z \neq 0$): The wave equation to be solved for these modes is obtained by equating the z component of Eq. 4.4a:

$$\frac{\partial^2 E_z}{\partial x^2} + \frac{\partial^2 E_z}{\partial y^2} + \gamma_g^2 E_z = -\omega^2 \mu\varepsilon E_z \tag{4.44}$$

Again using the technique of separation of variables we can obtain:

$$\vec{E}_z = E_0 \cos\left(\frac{m\pi}{a}\right) x \sin\left(\frac{n\pi}{b}\right) y \cos\left(\frac{p\pi}{l}\right) z \tag{4.45}$$

where m, n and p represent the half wave periodicity of the field in the x, y and z directions respectively. The boundary conditions used to solve for $\vec{E}_z$ are:

$$\vec{E}_z = 0; \text{ at } x = 0, x = a, \quad \text{and} \quad \text{at } y = 0, y = b$$

while $\dfrac{\partial \vec{E}_z}{\partial z} = 0$; at $z = 0$ and $z = l$

Substituting Eq. 4.45 into Eq. 4.44 leads to the same resonance condition as for the TE modes. Thus, Eq. 4.40 is also valid for the TM modes in a rectangular cavity. The remaining field components are obtained by substituting the expression for E_z into Eq. 4.5. For the TM_{mnp} cavity modes they are:

$$E_x = -\frac{\gamma_g}{k_c^2}\frac{\partial E_z}{\partial x} = \frac{1}{k_c^2}\frac{\partial^2 E_z}{\partial x \partial z} = -\frac{k_x k_z}{k_c^2} E_0 \cos k_x x \sin k_y y \sin k_z z$$

$$E_y = -\frac{\gamma_g}{k_c^2}\frac{\partial E_z}{\partial y} = \frac{1}{k_c^2}\frac{\partial^2 E_z}{\partial y \partial z} = -\frac{k_y k_z}{k_c^2} E_0 \sin k_x x \cos k_y y \sin k_z z$$

$$E_z = E_0 \sin k_x x \sin k_y y \cos k_z z \tag{4.46}$$

$$H_x = j\frac{\omega\varepsilon}{k_c^2}\frac{\partial E_z}{\partial y} = j\omega_r\varepsilon\frac{k_y}{k_c^2} E_0 \sin k_x x \cos k_y y \cos k_z z$$

$$H_y = -j\frac{\omega\varepsilon}{k_c^2}\frac{\partial E_z}{\partial x} = -j\omega_r\varepsilon\frac{k_x}{k_c^2} E_0 \cos k_x x \sin k_y y \cos k_z z$$

$$H_z = 0$$

where $k_x = m\pi/a$, $k_y = n\pi/b$, $k_z = p\pi/l$, and $k_c^2 = k_x^2 + k_y^2$; and note that $\frac{\partial}{\partial z} = -\gamma_g$. The field patterns for the TM cavity modes may be deduced from the above relations.

It may be noted that the resonant frequency f_r of the TM modes are the same as the TE modes. Thus, for a given set of mode numbers, the TE and TM modes in a rectangular cavity resonator are *degenerate*—they resonate at the same frequency. For a rectangular cavity mode to exist, at least two of its integer values must be non-zero. For given set of *m*, *n* and *p* values, the existence of a mode in the cavity may be determined by substituting the values into the expressions for the field components. For instance, TM_{mn0} modes can exist in the cavity (Eq. 4.46) while TE_{mn0} modes cannot (Eq. 4.42). It may be observed from Eq. 4.40 that TE_{101} has the lowest resonant frequency and is thus the *primary mode* in a rectangular cavity resonator.

4.4.2 The Quality Factor (Q) of the TE_{101} Cavity Mode

The quality factor of a cavity resonator (which determines the frequency selectivity of the resonator) is, in general, a measure of the ability of the resonator to store energy in relation to time average power dissipation. The Q of a resonator is thus defined as:

$$Q = 2\pi\frac{\text{(Time average energy stored at resonance)}}{\text{(Energy dissipated in one period of the resonant frequency)}}$$

$$= \omega_0 \frac{\bar{W}_{\text{str}T}}{P_{\text{walls}T}} \tag{4.47}$$

Physically, $\bar{W}_{\text{str}T} = \bar{W}_e + \bar{W}_m$, where $\bar{W}_e$ and $\bar{W}_m$ are respectively the time average stored electric and magnetic energies. We can find $\bar{W}_e$ and $\bar{W}_m$ separately and add them together to find $\bar{W}_{\text{str}T}$ or can alternatively evaluate $\bar{W}_e$ or $\bar{W}_m$ and find $\bar{W}_{\text{str}T} = 2\bar{W}_e$ or $2\bar{W}_m$ since $\bar{W}_e = \bar{W}_m$. The physical meaning of $\bar{W}_e = \bar{W}_m$ is that over the course of one period, the stored energy in the cavity cycles between purely electric, partly electric and partly magnetic or purely magnetic storage, such that, on the average over one period, it is shared equally between the electric and magnetic forms.

We have
$$\bar{W}_e = \frac{1}{T_P}\int_0^{T_p} \frac{1}{2}\,\varepsilon E_0^2 \cos^2(\omega t - \beta z)\,dt$$

$$= \frac{1}{2}\,\varepsilon E_0^2\,\frac{1}{T_p}\int_0^{T_p}\left[\frac{1}{2} + \frac{1}{2}\cos\{2(\omega t - \beta z)\}\right]dt$$

$$= \frac{1}{4}\,\varepsilon E_0^2$$

Thus, the time-averaged stored electric energy in the cavity for the TE_{101} mode is given by:

$$\bar{W}_e = \frac{1}{4}\int_v \varepsilon\left|E_y\right|^2 dv$$

$$= \frac{\varepsilon}{4}\left(\frac{\omega_{101}\mu a}{\pi}\right)^2 H_0^2 \int_0^l \int_0^b \int_0^a \sin^2\left(\frac{\pi x}{a}\right)\sin^2\left(\frac{\pi z}{l}\right) dx\,dy\,dz \quad \text{(vide Eq. 4.43a)}$$

$$= \frac{\varepsilon\omega_{101}^2\mu^2 a^2}{4\pi^2} H_0^2\left(\frac{a}{2}\right) b\left(\frac{l}{2}\right)$$

$$= \frac{1}{4}\,\varepsilon\mu^2 a^3 b l f_{101} H_0^2 \quad \text{(vide Eq. 4.43b)}$$

$$= \frac{\mu}{16}\,abl\left(1 + \frac{a^2}{l^2}\right) H_0 \tag{4.48}$$

The total time average stored energy in the resonator is then:

$$\bar{W}_{\text{str}T} = 2\bar{W}_e = \frac{\mu H_0^2\,abl}{8}\left[\frac{a^2}{l^2} + 1\right] \tag{4.49}$$

Next, we need to determine the power dissipated in the six walls of the cavity resonator caused by the flow of surface current induced by the tangential magnetic field. Mathematically we can write:

$$P_{\text{wall}} = \frac{R_S}{2}\int_{\text{wall}} \left|H_{\tan}\right|^2 ds \tag{4.50}$$

where R_s is the surface resistance of the wall of the cavity material.

Now, for the top and bottom walls of the cavity:

$$P_{\text{wall}(t,b)} = 2\int_0^l \int_0^a \left(|H_z|^2 + |H_x|^2 \, dx \, dz \right) \tag{4.51a}$$

For right and left walls of the cavity:

$$P_{\text{wall}(r,l)} = 2\int_0^l \int_0^b \left(|H_z|^2 \, dy \, dz \right) \tag{4.51b}$$

For front and rear walls of the cavity:

$$P_{\text{wall}(f,r)} = 2\int_0^b \int_0^a \left(|H_x|^2 \, dx \, dy \right) \tag{4.51c}$$

where the contribution from the walls at $x = 0, y = 0$ and $z = 0$ has been doubled to account for the contributions from the walls at $x = a, y = b$ and $z = l$ respectively, which follows from the symmetry of the rectangular cavity. Substituting Eqs 4.51a–4.51c into Eq. 4.50 and H_x, H_z from Eq. 4.43a and after performing the integrations we get:

$$P_{\text{wall(Total)}} = \frac{R_s H_0^2}{2} \left\{ \left(\frac{al}{2} + \frac{a^3 b}{d} \right) + bl + \frac{a^3 b}{l^2} \right\} \tag{4.52}$$

Therefore, the quality factor Q_c (subscript c indicates that this quality factor is due to losses in the conducting walls only) can be obtained by using Eqs 4.43b, 4.49 and 4.52 and with some manipulations as:

$$Q_c = \frac{\pi\eta}{4R_s} \frac{\left[\left\{ 1 + \left(\frac{a}{l} \right)^2 \right\} \right]^{3/2}}{\left[1 + \left(\frac{a}{2b} \right) \left\{ 1 + \left(\frac{a}{l} \right)^2 + \left(\frac{2ba^2}{l^3} \right) \right\} \right]} \tag{4.53}$$

The Q_c of a cubical cavity ($a = b = l$) oscillating in the TE_{101} mode is $Q_c = 0.740\ (\eta/R_s)$. For an air-filled cavity, $\eta = 377\ \Omega$, with copper walls, $R_s = 2.60 \times 10^{-7} \sqrt{f_{101}}\ \Omega$ we have $Q_c = 1.07 \times 10^9 / \sqrt{f_{101}}$, with $a = c/\sqrt{2} f_{101}$ as the cubical dimension. For example, with a resonant frequency of 10 GHz, $a = 2.12$ cm, we have $Q_c = 10{,}700$. This high value of Q is typical of cavity resonators.

In addition to conduction losses (which determines Q_c), additional losses are expected in resonator cavities operating at microwave frequencies which arise because of the complex nature of the permittivity of the dielectric material filling the cavity at such frequencies. Thus, we have another Q factor, namely Q_d, which arises because of losses in dielectric material at microwave frequencies. Following the principle of dielectric heating discussed in Section 2.4.1, we can write the power dissipation in the dielectric filling the cavity as:

$$P_{\text{dielectric}} = \frac{1}{2}\omega\varepsilon'' \int_0^a \int_0^b \int_0^l |E_y|^2 \, dy \, dx \, dz$$

$$= \frac{\varepsilon''}{\varepsilon'}\,\omega_{101}\,\frac{\mu H_0^2 abd}{8}\left\{\left(\frac{a}{l}\right)^2 + 1\right\} \tag{4.54}$$

So that the quality factor due to dielectric losses is given by using Eqs 4.43b, 4.49 and 4.54 as:

$$Q_d = \omega_0 \frac{\bar{W}_{\text{str}T}}{P_{\text{dielectric}}} = \frac{\varepsilon'}{\varepsilon''} = \frac{1}{\tan\delta_c} \tag{4.55}$$

The total Q of the system is then given by $Q_T = \omega_0 \bar{W}_{\text{str}T}/(P_{\text{wall}} + P_{\text{dielctric}})$. Thus, for microwave cavities:

$$\frac{1}{Q_T} = \frac{1}{Q_d} + \frac{1}{Q_c} \tag{4.56}$$

It may be noted that the dielectric filling of the cavity substantially reduces the Q of the cavity, however, dielectric-filled cavities are smaller in size than air-filled ones, and the dielectric strength of plastics such as Teflon is substantially higher than that of air, thus allowing for greater energy storage capacity.

EXAMPLE 4.7

Consider an air-filled cubical cavity ($a = b = d$) designed to be resonant in the TE_{101} mode at 12 GHz with gold plated surfaces ($\sigma = 4.10 \times 10^7$ S-m^{-1}). Find the quality factor. If the cavity is polystyrene-filled, with $\varepsilon = \varepsilon_0$ (2.55 – j0.000842) and $\varepsilon_r = 2.56$. Find the total quality factor of the dielectric filled cavity.

Solution

From Eq. 4.43b, we have f_{101} for air-filled cubical cavity as:

$$f_{101} = \frac{1}{a\sqrt{2\mu_0\varepsilon_0}} = 12 \text{ GHz (Given)}$$

$$\text{whence } a_{\text{air}} = \frac{c}{\sqrt{2} f_{101}} = \frac{3\times 10^{10}}{\sqrt{2}\times 12\times 10^9} = 1.767 \text{ cm}$$

$$\text{Now } R_s = \frac{1}{\sigma\delta} = \frac{\sqrt{\pi f_{101}\mu\sigma}}{\sigma} = \sqrt{\frac{\pi\mu_0}{\sigma}}\sqrt{f_{101}} = \sqrt{\frac{\pi\times 4\pi\times 10^{-7}}{4.10\times 10^7}}\sqrt{12\times 10^9} = 3.39\times 10^{-2}\ \Omega\text{-cm}^{-1}$$

Thus the quality factor of the cavity from Eq. 4.53 is:

$$Q_{c(\text{air})(a=b=l)} = \frac{0.74\,\eta}{R_s} = \frac{0.74 \times 377}{3.39 \times 10^{-2}} = 8.229$$

When the cavity is dielectric-filled, λ will be reduced by a factor of $\sqrt{\varepsilon_r}$ while the resonant frequency will be increased by a factor of $\sqrt{\varepsilon_r}$. In order to keep the resonant frequency of the cavity unchanged, the $Q_{c(\text{air})}$ of the cavity will be decreased by a factor of $\sqrt{\varepsilon_r}$. Thus we have:

$$Q_{c\ (\text{dielectric})} = 8229/\sqrt{\varepsilon_r} = 5143$$

The dimension of the cavity will also be reduced by a factor of $\sqrt{\varepsilon_r}$:

$$a_{\text{dielectric}} = 1.767/\sqrt{2.56} = 1.104 \text{ cm}$$

The quality factor due to the dielectric loss is given by Eq. 4.55:

$$Q_d = \frac{\varepsilon'}{\varepsilon''} = \frac{2.55}{0.000842} = 3028$$

The total Q is then $Q_T = \dfrac{Q_c Q_d}{Q_c + Q_d} = 1905$

4.4.3 Cylindrical Cavity Resonator

A schematic of a cylindrical resonator of radius a and length l is shown in Fig. 4.20.

The field analysis that follows is similar to that of the rectangular resonator. The wave equations of the electric and magnetic fields, given in Eq. 2.8, may be written in cylindrical coordinates for the z component as:

$$\frac{1}{r}\frac{\partial}{\partial r}\left(r\frac{\partial \vec{E}_z}{\partial r}\right) + \frac{1}{r^2}\frac{\partial^2 \vec{E}_z}{\partial \varphi^2} + \frac{\partial^2 \vec{E}_z}{\partial z^2} = -\,\omega^2 \mu\varepsilon \vec{E}_z \qquad (4.57\text{a})$$

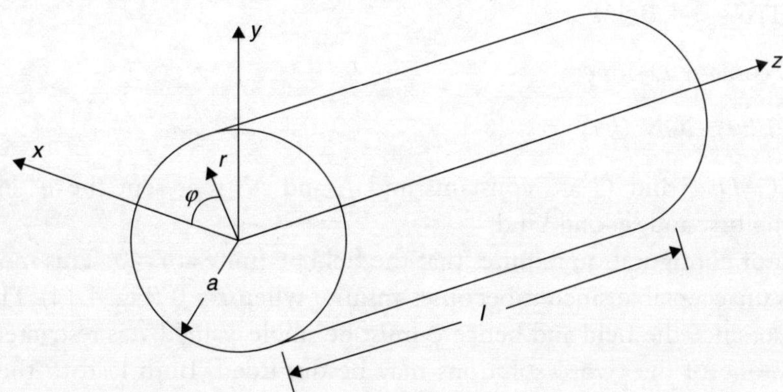

Fig. 4.20 Schematic of cylindrical cavity resonator.

$$\frac{1}{r}\frac{\partial}{\partial r}\left(r\frac{\partial \vec{H}_z}{\partial r}\right)+\frac{1}{r^2}\frac{\partial^2 \vec{H}_z}{\partial \varphi^2}+\frac{\partial^2 \vec{H}_z}{\partial z^2}=-\omega^2\mu\varepsilon\vec{H}_z \tag{4.57b}$$

TE$_{nmp}$ modes ($E_z = 0$, $H_z \neq 0$): To solve Eq. 4.57b, we again use the separation of variable technique. Let:

$$\vec{H}_z = \vec{R}(r)\,\vec{\phi}(\varphi)\,\vec{Z}(z) \tag{4.58}$$

Thus, substituting H_z from Eq. 4.58 to Eq. 4.57b, we have:

$$\frac{1}{r\vec{R}}\frac{\partial}{\partial r}\left(r\frac{\partial \vec{R}}{\partial r}\right)+\frac{1}{r^2\vec{\phi}}\frac{\partial^2\vec{\phi}}{\partial\varphi^2}+\frac{1}{\vec{Z}}\frac{\partial^2\vec{Z}}{\partial z^2}=-\omega^2\mu\varepsilon$$

Equating the first two terms to $-k_c^2$ and the third term to $-k_z^2$ results in the following equations:

$$\frac{r}{\vec{R}}\frac{\partial}{\partial r}\left(r\frac{\partial\vec{R}}{\partial r}\right)+k_c^2r^2=-\frac{1}{\vec{\phi}}\frac{\partial^2\vec{\phi}}{\partial\varphi^2} \tag{4.59a}$$

$$\frac{\partial^2\vec{Z}}{\partial z^2}+k_z^2\vec{Z}=0 \tag{4.59b}$$

where $k_z^2 + k_c^2 = \omega^2\mu\varepsilon$. Equating both sides of Eq. 4.59a to a constant (n^2) yields:

$$\frac{d^2\vec{\phi}}{d\varphi^2}+n^2\vec{\phi}=0 \tag{4.60a}$$

and

$$\frac{d^2\vec{R}}{dr^2}+\frac{1}{r}\frac{d\vec{R}}{dr}+\left(k_c^2-\frac{n^2}{r^2}\right)\vec{R}=0 \tag{4.60b}$$

Solutions to Eqs 4.59b, 4.60a and 4.60b are:

$$\begin{aligned}\vec{Z} &= A\cos k_z z + B\sin k_z z\\ \vec{\phi} &= C\cos n\varphi + D\sin n\varphi\\ \vec{R} &= FJ(k_c r)+GN_n(k_c r)\end{aligned} \tag{4.61}$$

where A, B, C, D, F and G are constants and J_n and N_n represent the nth order Bessel functions of the first and second kind.

The resonator configuration requires that the field be finite at $r = 0$. This means that the N_n solution is unacceptable since it becomes infinite when $r = 0$ (Fig. 4.14). Thus, G must equal zero. Also, since the field and hence ϕ must be single-valued, n is restricted to integer values. Finally, one of the two ϕ solutions may be discarded. Both lead to the same field patterns except that one is phase quadrature with the other in the ϕ direction. Setting $D = 0$ leaves only the $\cos n\phi$ term. Substituting Eq. 4.61 to Eq. 4.58 yields:

$$\vec{H}_z = CF \cos n\varphi \,(A \cos k_z z + B \sin k_z z)\, J_n(k_c r)$$

The boundary conditions at $z = 0$, and $z = l$ require that $\vec{H}_z = 0$. Therefore $A = 0$ and $k_z l = p\pi$, where p is a positive integer. The $\vec{H}_z$ solution reduces to:

$$\vec{H}_z = H_0 J_n(k_c r) \cos n\varphi \sin k_z z \tag{4.62}$$

where $k_z = p\pi/l$ and H_0 is an arbitrary amplitude constant. At $r = a$, the boundary condition requires that $d\vec{H}_z/dr = 0$, which means $J_n'(k_c a) = 0$. The prime denotes differentiation with respect to $k_c r$. The mth root of this equation is designated by $t_{nm}' = k_c a$. Several root values are listed in Table 4.3.

The resonance condition is obtained by substituting Eq. 4.62 into Eq. 4.57b, when we get:

$$\omega_r^2 \mu\varepsilon = k_c^2 + k_z^2 = \left(\frac{t_{nm}'}{a}\right)^2 + \left(\frac{p\pi}{l}\right)^2$$

Thus, the resonant frequency for TE_{nmp} modes in a cylindrical cavity resonator is given by:

$$f_{nmp} = \frac{c}{\sqrt{\mu_r \varepsilon_r}} \sqrt{\left(\frac{t_{nm}'}{\pi D}\right)^2 + \left(\frac{p}{2l}\right)^2} \tag{4.63}$$

where $D = 2a$. The integer p has the same meaning as in the rectangular cavity, namely, the number of half cycles in the z direction. On the other hand, n indicates the number of *full-cycle* variations in the circumferential or ϕ direction. Since the $J_n'(t_{nm}' r/a)$ function behaves like a damped sinusoid, m describes the radial variations of the field. By convention, n is the first of the three subscripts used to denote a cylindrical cavity mode.

The remaining field components can be derived from the relations in Eq. 4.20 by a procedure similar to that used for a rectangular cavity, when we have for the TE_{nmp} modes in a cylindrical resonator:

$$E_r = -j\frac{\omega\mu}{rk_c^2}\frac{\partial H_z}{\partial\varphi} = j\omega_r\mu\frac{nH_0}{rk_c^2} J_n(k_c r) \sin n\varphi \sin k_z z$$

$$E_\varphi = +j\frac{\omega\mu}{k_c^2}\frac{\partial H_z}{\partial r} = j\omega_r\mu\frac{H_0}{k_c} J_n'(k_c r) \cos n\varphi \sin k_z z$$

$$E_z = 0 \tag{4.64}$$

$$H_r = -\frac{\gamma_g}{k_c^2}\frac{\partial H_z}{\partial r} = \frac{1}{k_c^2}\frac{\partial H_z}{\partial r\partial z} = \frac{k_z}{k_c} H_0 J_n'(k_c r) \cos n\varphi \cos k_z z$$

$$H_\varphi = -\frac{\gamma_g}{rk_c^2}\frac{\partial H_z}{\partial\varphi} = \frac{1}{rk_c^2}\frac{\partial H_z}{\partial\varphi\partial z} = -\frac{nk_z}{rk_c^2} H_0 J_n(k_c r) \sin n\varphi \cos k_z z$$

$$H_z = H_0 J_n(k_c r) \cos n\varphi \sin k_z z$$

where $k_z = p\pi/l$, $k_c = t'_{nm}/a$, and J'_n is the derivative of J_n with respect to $k_c r$; and note that $\frac{\partial}{\partial z} = -\gamma_g$. The field patterns for the TE modes in a cylindrical cavity can be derived from Eqs 4.64 whose resonant frequencies are given by Eq. 4.63.

TM$_{nmp}$ modes ($H_z = 0$, $E_z \neq 0$): We may solve Eq. 4.57a for $\vec{E}_z$ by employing the same method used to obtain $\vec{H}_z$ for the TM modes. The boundary conditions in this case are $\vec{E}_z = 0$ at $r = a$ and $\partial\vec{E}_z/\partial z = 0$ at $z = 0$ and $z = l$. The remaining field components are obtained by substituting the resultant expression for $\vec{E}_z$ into Eq. 4.20. This yields the following field equations for the TM$_{nmp}$ modes in a cylindrical resonator cavity.

$$E_r = -\frac{\gamma_g}{k_c^2}\frac{\partial E_z}{\partial r} = \frac{1}{k_c^2}\frac{\partial E_z}{\partial r\partial z} = -\frac{k_z}{k_c}E_0 J'_n(k_c r)\cos n\varphi \sin k_z z$$

$$E_\varphi = -j\frac{\gamma_g}{rk_c^2}\frac{\partial E_z}{\partial \varphi} = \frac{1}{rk_c^2}\frac{\partial E_z}{\partial \varphi\partial z} = \frac{nk_z}{rk_c^2}E_0 J_n(k_c r)\sin n\varphi \sin k_z z$$

$$E_z = E_0 J_n(k_c r)\cos n\varphi \cos k_z z \tag{4.65}$$

$$H_r = \frac{j\omega\varepsilon}{rk_c^2}\frac{\partial E_z}{\partial \phi} = -j\omega_r\varepsilon\frac{nE_0}{rk_c^2}J_n(k_c r)\sin n\varphi \cos k_z z$$

$$H_\varphi = -j\frac{\omega\varepsilon}{k_c^2}\frac{\partial E_z}{\partial r} = -j\omega_r\varepsilon\frac{E_0}{k_c}J'_n(k_c r)\cos n\varphi \cos k_z z$$

$$H_z = 0$$

where $k_z = p\pi/l$, $k_c = t_{nm}/a$, and t_{nm} is the mth root of J_n (Table 4.2); and note that $\frac{\partial}{\partial z} = -\gamma_g$.

The field patterns for the TM modes in a cylindrical cavity can be derived from Eqs 4.65. The resonant frequencies of the modes are given by Eq. 4.63 *except* that t'_{nm} is replaced by t_{nm}. In most cases, these values are different when compared with those for the TE modes. As a result, most TE and TM modes in a cylindrical cavity resonator are *non-degenerate* (unlike those in rectangular cavity resonators), an exception being the TE$_{0mp}$ and the TM$_{1mp}$ modes. For $D/l < 0.99$, the *primary mode* in a cylindrical cavity is TE$_{111}$; otherwise it is the TM$_{010}$ mode.

4.4.4 The Quality Factor (*Q*) of TE$_{011}$ Cavity Mode

The TE$_{011}$ modes in a cylindrical cavity are of considerable engineering interest because of their high Q values and ease of tuning. Following the same procedure as that for the rectangular cavities, the Q_c of a cylindrical cavity for the TE$_{011}$ mode can be derived as [10]:

$$Q_{c(\text{TE}_{011})} = \frac{\lambda_r}{2\pi\delta}\frac{\left[(t'_{01})^2 + \left(\frac{\pi a}{l}\right)^2\right]^{3/2}}{(t'_{01})^2 + \frac{2a}{l}\left(\frac{\pi a}{l}\right)^2} \tag{4.66}$$

where λ_r is the wavelength at which the cavity resonates and δ is the skin depth at the resonant frequency. It may be observed that similar to the rectangular cavity, Q_c decreases with increasing frequency and when the cavity is dielectric filled, the overall unloaded cavity Q_T is obtained from Eq. 4.56.

EXAMPLE 4.8

Design a cylindrical cavity with length of the cavity equal to its diameter, with the TE_{011} mode resonating at 10 GHz. The cavity has silver-plated inner walls (σ for silver is 6.14×10^7 S-m^{-1}). Determine the Q of the cavity.

Solution

From Eq. 4.63 we have:

$$f_{011} = c\sqrt{\left(\frac{3.832}{2\pi a}\right)^2 + \left(\frac{1}{4a}\right)^2} = 10 \text{ GHz (given)}$$

This, when evaluated for a gives $a = 1.98$ cm (this can be compared with $a = 1.767$ cm, calculated for a rectangular cavity with TE_{101} mode as in Example 4.7).

The Q_c of this cavity may be calculated from Eq. 4.66 as:

$$Q_{c(TE_{011})} = \frac{\lambda_r}{2\pi\delta} \frac{\left\{(3.832)^2 + \left(\frac{\pi a}{2a}\right)^2\right\}^{3/2}}{(3.832)^2 + \frac{2a}{2a}\left(\frac{\pi a}{2a}\right)^2} \approx 30{,}792$$

where $\delta = \left(\sqrt{\pi f \mu_0 \sigma}\right)^{-1} = 6.42 \times 10^{-7}$ m and $t'_{nm} = 3.832$.

A very large Q value is possible with the TE_{011} mode in the cylindrical resonator that results in high selectivity possible with such cavities, thus making them very suitable for use in the design of frequency meter (wavemeter) at microwave frequencies. Further, note from the electromagnetic field distribution in a cylindrical resonator cavity with TE_{011} mode (Fig. 4.21) that the electric field is in the circumferential (ϕ) direction. The magnetic field is in the axial (z) direction near the side wall while it is in the radial (r) direction near the end walls of the cavity. Thus, the surface current flow is in the circumferential direction with no flow of such current between the side wall and the end wall of the cavity.

Therefore, there is no need for good electrical contact between the end wall and the side wall of the cavity. Thus, if one end of the cavity is closed by a movable metallic plunger that is not in contact with the side wall and is moved to change the length of the cavity (as is done in microwave wavemeter, Section 12.5.1), the TE_{011} mode can still be supported, since the current flow is not interrupted by any chance. However, a non-contacting moving plunger can suppress undesired modes like TE_{111} and TM_{010}, which are of lower resonant frequency but need current continuity between the end wall and the side wall. Based on the same reason of the requirement of current continuity, the mode TM_{111}, which has the same

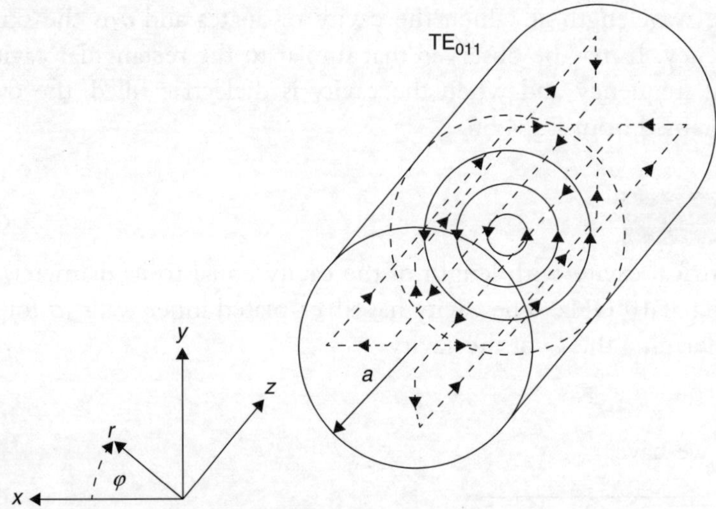

Fig. 4.21 Electric (—) and magnetic (---) field pattern of the TE_{011} mode in a cylindrical cavity resonator.

resonant frequency as the TE_{011} mode but lower Q, is suppressed due to non-contacting moving plunger of the wavemeter.

4.4.5 Mode Chart

In the design of cavity resonators, it is usually ensured that the cavity operates in a single mode. The *mode chart* is often used to check how many modes may exist in a cylindrical cavity resonator and also helps to know the frequency range over which a particular resonance can be tuned without interference from other modes.

The mode chart is drawn on the basis of the expression for the resonant wavelength of the TE_{nmp} and TM_{nmp} mode in a cylindrical waveguide which may be rewritten from Eq. 4.63 for an air-filled cavity with t'_{nm} and t_{nm} replaced by q_{nm}, when we have:

$$\lambda_{nmp} = \frac{1}{\sqrt{\left(\frac{q_{nm}}{2\pi a}\right)^2 + \left(\frac{p}{2l}\right)^2}}$$

On further modification, this may be written as:

$$(2af_{nmp})^2 = \left(\frac{cp}{2}\right)^2\left(\frac{2a}{l}\right)^2 + \left(\frac{cq_{nm}}{\pi}\right)^2 \tag{4.67}$$

So, there are linear relationships between $(2af_{nmp})^2$ and $(2a/l)^2$ with the slope of the curve being $(cp/2)^2$ and intercept $(cq_{nm}/\pi)^2$. These lines form the mode chart, are shown in Fig. 4.22.

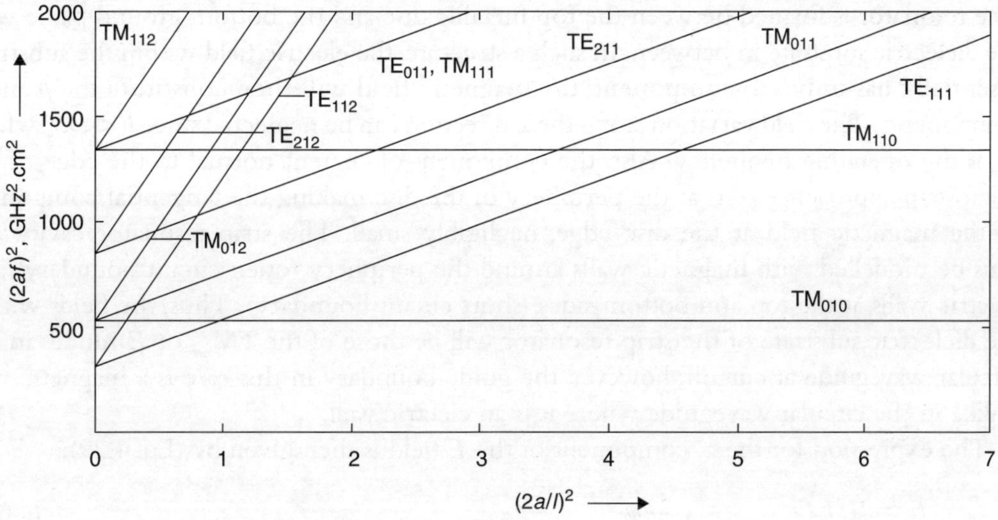

Fig. 4.22 Mode chart for cylindrical cavity resonator.

The mode chart is a useful chart for the design of the cylindrical cavity in the desired mode of operation. Because the examination of the chart gives an idea regarding the range of frequencies and the cavity dimensions $(2a/l)$, only a single mode can resonate (in the case of degenerate modes, however, two modes resonate at the same frequency). For example, we can think of a working rectangle for $(2a/l)$ in the range 2 to 3 when only the TE_{011} and TM_{111} modes can resonate in the frequency range corresponding to 1750 and 2000 GHz. If the TM_{111} mode is suppressed, then for the cavity dimensions obtained from the chart the resonator will operate in the TE_{011} mode without being affected by any other spurious resonances.

4.5 Strip/Disc Resonator

The strip or disc resonator designed around the microstrip (planar) transmission line is typically of circular disc type as shown in Fig. 4.23.

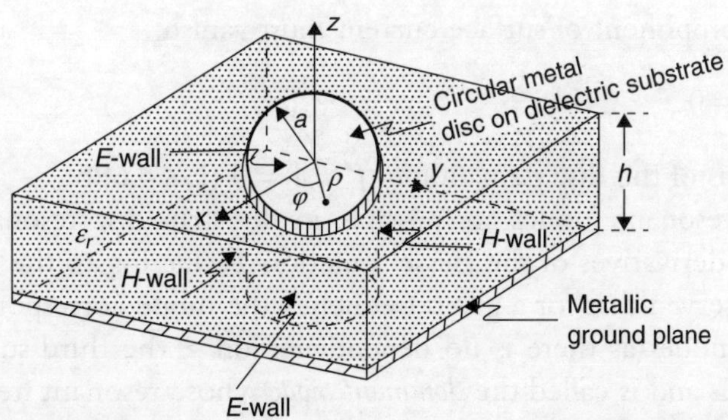

Fig. 4.23 Strip/disc resonator.

The resonator is formed between the top metallic disc and the bottom ground plane with the dielectric substrate in between. In such a structure, the electric field within the substrate essentially has only a z-component; the magnetic field will then constitute the ρ and ϕ components. The field variation along the z direction can be neglected since $h << \lambda_{op}$, where λ_{op} is the operating frequency. Also the component of current normal to the edge of the microstrip approaches zero at the periphery of the disc making the tangential component of the magnetic field at the disc edge, negligibly small. The strip resonator cavity can thus be modelled with magnetic walls around the periphery (open circuit boundary) and electric walls at the top and bottom sides (short circuit boundary). Thus, the fields within the dielectric substrate of the strip resonator will be those of the TM_{nm} or E-modes in the circular waveguide at cut-off; however, the guide boundary in this case is a magnetic wall unlike in the circular waveguide where it is an electric wall.

The expression for the z-component of the E field is then given by (Eq. 4.29):

$$E_z = E_0 J_n(k_{cnm(\text{TM})}\rho)\cos n\varphi \tag{4.68a}$$

The corresponding magnetic field components are (again following Eq. 4.29):

$$H_\rho = -\frac{jn}{\omega\mu\rho} E_0 J_n(k_{cnm(\text{TM})}\rho)\sin\varphi$$

$$H_\varphi = -\frac{jk_{cnm(TM)}}{\omega\mu} E_0 J'_n(k_{cnm(\text{TM})}\rho)\cos n\varphi \tag{4.68b}$$

The other field components are zero, that is:

$$E_r = E_j = H_z = 0 \tag{4.68c}$$

The surface current on the circular disc due to the magnetic field inside the cavity is given by:

$$J_{s(\text{disc})} = \hat{n}\times\vec{H} = \hat{\rho}H_\varphi - \hat{\varphi}H_\rho$$

where $\hat{\rho}$ and $\hat{\varphi}$ are unit vectors in the ρ and φ directions respectively. At the edge of the metallic disc, the component of surface current must vanish:

$$H_\varphi(\rho = a) = 0$$

where a is the radius of the disc demanding $J'_n(k_c a) = 0$ (Eq. 4.68b).

Thus, different resonant modes are possible in the strip/disc resonator corresponding to the zeros of the derivatives of the Bessel function. Referring to the roots of $J'_n(k_c a) = 0$ (Table 4.3), it is observed that for a given frequency the mode corresponding to $n = m = 1$, that is, the TM_{110} mode (as there is no dependence on z, the third subscript is zero) has the minimum radius and is called the *dominant mode* whose resonant frequency is given by equating $k_c a = 1.841$, when:

$$f_{110(\text{TM})} = \frac{1.841c}{2\pi\sqrt{\varepsilon_r}\,a} \tag{4.69}$$

where c is the velocity of light.

EXAMPLE 4.9

Determine the disc diameter of a strip resonator that resonates at 5.2 GHz and uses alumina having $\varepsilon_r = 9.8$ as the dielectric substrate.

Solution

Using Eq. 4.69 we have the radius of the strip resonator as:

$$a = \frac{1.841\,c}{2\pi\sqrt{\varepsilon_r}\,f_{110(\text{TM})}} = 0.5399 \text{ cm}$$

Thus, the disc diameter of the strip resonator is: $D = 2a = 10.79$ mm

4.6 Dielectric Resonator

With the advent of advanced ceramic technology in the 1970s, the use of low loss, temperature stable, high dielectric constant material to exhibit cavity property, that is, to confine electric and magnetic energy in the material, drew the attention of scientists and engineers towards developing a *dielectric resonator* (DR). It is basically a small cube or disc of dielectric material, typically barium tetratitanate and zirconium tetratitanate, with a dielectric constant (ε_r) in the range $10 < \varepsilon_r < 100$.

In the DR cavity, the waves are reflected from the interface between the dielectric and the air unlike the metal cavities where waves are reflected from metal walls. This implies that the conductor losses usually incurred in metal cavities is practically nil in the case of DR cavities. However, a high dielectric constant is desired for the DR to confine the electromagnetic field in the cavity and this also makes the size of the DR quite small (for disc type DR in which $D > L$, an approximate relation is $D = c/f_r\sqrt{\varepsilon_r}$, where D is the diameter of the disc type DR, L is its length, f_r is the resonant frequency of the DR and ε_r is the dielectric constant of the DR material). But, dielectric losses usually increase with increasing dielectric constant. The unloaded Q factor of a DR is decided practically by the losses in the dielectric and Q values up to several thousands can be achieved. Further, the temperature stability of the DR is comparable to that of invar. The DR can be used as a high performance miniaturised resonator at microwave and millimetre wave frequencies for the design of oscillators and filters in planar ciruits.

The dominant mode with lowest cut-off frequency in a disc-type ($L < D$) dielectric resonator (Fig. 4.24) is $TE_{01\delta}$, where $\delta = 2L/\lambda_g < 1$ is used to denote the z variation of the resonant mode (λ_g being the guide wavelength of the TE_{01} dielectric waveguide mode). It may be noted from the figure that for a distant observer, this mode appears as a magnetic dipole, so this mode is also known as the *magnetic dipole mode*.

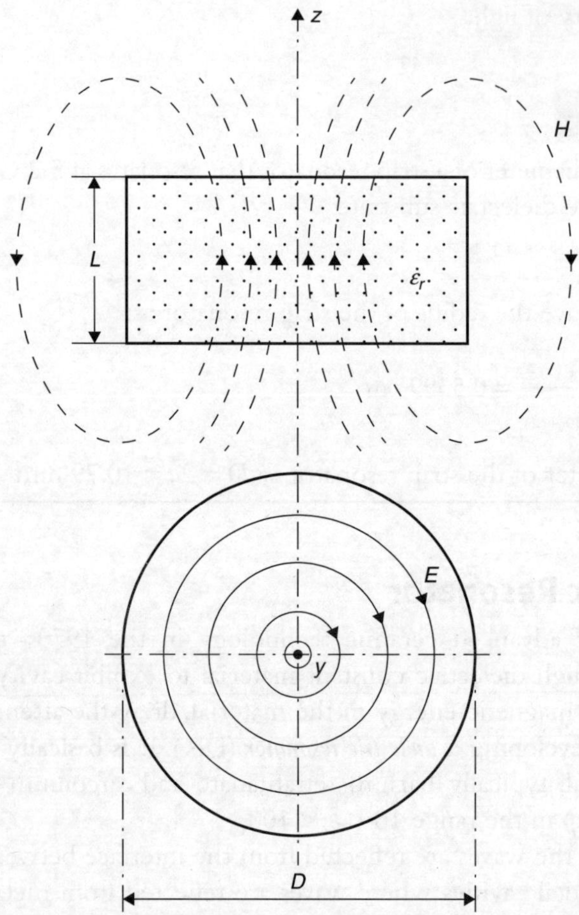

Fig. 4.24 Field lines of the resonant mode $TE_{01\delta}$ in an isolated dielectric resonator.

For a dielectric resonator, the accurate resonant frequency of its $TE_{01\delta}$ mode can be calculated only by complicated numerical procedures. However, for an approximate estimation of the resonant frequency of an isolated dielectric resonator, the following empirical formula may be used:

$$f_{01\delta} = \frac{68}{D\sqrt{\varepsilon_r}}\left(\frac{D}{2L} + 3.45\right) \tag{4.70}$$

where D and L are the diameter and length of the dielectric resonator, ε_r is the dielectric constant of the material of the resonator. In Eq. 4.70, the unit of frequency is GHz, and the unit for D and L is mm. The accuracy of this equation is about 2% in the range of $0.5 < (D/2L) < 2$ and $30 < \varepsilon_r < 50$.

EXAMPLE 4.10

Find the resonant frequency and approximate Q for the $TE_{01\delta}$ mode of a dielectric resonator made from barium tetratitanate, with $\varepsilon_r = 37$, and $\tan\delta_c = 0.0005$. The resonator dimensions are $D = 7.7$ mm and $L = 8.44$ mm.

Solution

From Eq. 4.70 we have $f_{01\delta} = \dfrac{68}{7.7\sqrt{37}}\left(\dfrac{7.7}{2\times 8.44} + 3.45\right) = 5.67$ GHz

The approximate Q due to dielectric loss (neglecting the fringing or leakage from the sides and ends of the dielectric resonator) from Eq. 4.55:

$$Q_d = \frac{1}{\tan\delta_c} = 2000$$

4.7 Coupling and Tuning of Microwave Resonators

The resonators we have discussed so far, be it the rectangular/cylindrical waveguide type, microstrip type or dielectric resonator type, need coupling to deliver power to or pick-up power from an external circuit. Also, tuning of the resonator cavity is required to effect any frequency change of the resonator, as in the case of a mechanically tuned oscillator/wavemeter design and so on.

In a rectangular waveguide type resonator, the aperture/iris coupling is very common in the transverse wall (Fig. 4.25a), used in oscillator/amplifier application to connect the cavity to the desired load. When a small aperture of diameter D is in the transverse wall of the guide, it perturbs the tangential magnetic field on this wall and thus causes inductive loading. The equivalent inductive susceptance represented by the aperture can be determined from the formula [7]:

$$jB = -j\frac{3ab}{D^3\beta} \tag{4.71}$$

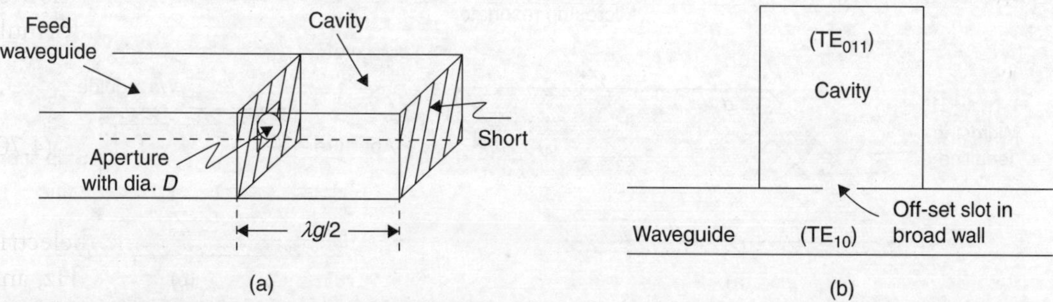

Fig. 4.25 (a) Aperture coupling in rectangular waveguide cavity (b) Slot coupling from rectangular to cylindrical waveguide cavity.

A cylindrical waveguide cavity resonator can be coupled to a rectangular waveguide via a slot (Fig. 4.25b) when the off-set radiating slot on the broad wall of the waveguide couples signals from the waveguide to the cavity as is used in the design of the wavemeter.

Apart from the aperture/slot coupling, power from cavities may also be coupled into or coupled out via (electric) probe coupling (Fig. 4.26a) or (magnetic) loop coupling (Fig. 4.26b) using a coaxial line (whose central conductor acts as the electric probe/ magnetic loop). The electric probe is set parallel to the electric field of the existing mode (TE_{011}) in the cavity while the magnetic loop is parallel to the magnetic field of the mode.

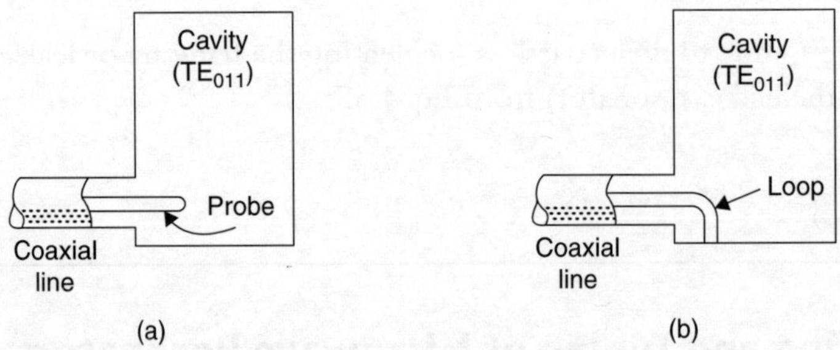

Fig. 4.26 (a) Electric probe coupling (b) Magnetic loop coupling.

Tuning of waveguide resonators can be carried out by inserting movable short-circuit plungers which can move at right angles to the axis of a cylindrical resonator (as is used in the wavemeter design, Fig. 12.24) or parallel to the transverse wall of a rectangular cavity (as is used in oscillator design for mechanical tuning, Fig. 9.44) thereby changing the length and hence the resonant frequency of the cavity. A metallic screw may also be inserted in a rectangular waveguide via the top broad wall cavity to tune the frequency of the oscillator (Fig. 7.26a).

Coupling of a microstrip resonator can be done by capacitive coupling (via the gap g) between a $\lambda/2$ open-circuited microstrip resonator and a microstrip feed line as shown in Fig. 4.27a. The fringing electric field at the gap between the open end of the resonator and the microstrip line results in capacitive coupling between the two. A strip line resonator may also couple the signal to a waveguide via aperture coupling (Fig. 4.27b).

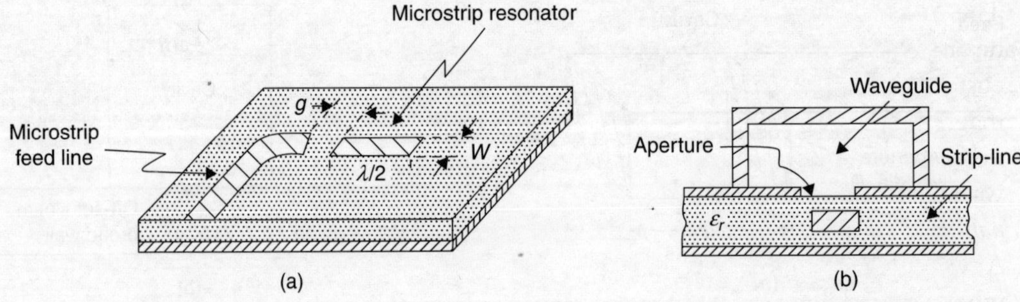

Fig. 4.27 (a) Microstrip resonator gap coupled to microstrip feed line (b) Strip-line cavity aperture coupled to waveguide.

In a dielectric resonator (DR), the $TE_{01\delta}$ mode can easily be coupled magnetically (with its fringing field) to a microstrip line (Fig. 4.28).

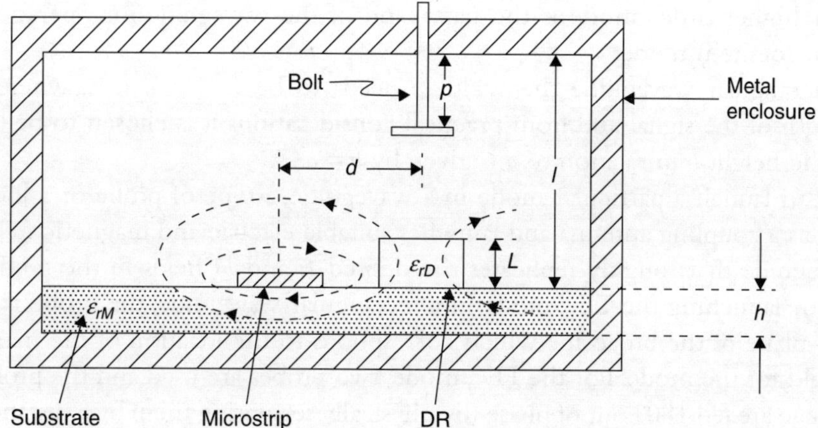

Fig. 4.28 Coupling between a dielectric resonator (DR) and a microstrip line with tuning by a bolt.

A DR resonator having a high dielectric constant (ε_{rD}) is placed on top of the substrate (having dielectric constant $\varepsilon_{rM} << \varepsilon_{rD}$) of the microstrip. The lateral distance d between DR and the conducting strip of the microstrip transmission line determines the amount of coupling between the two. Proper metallic shielding is required to minimise the radiation loss (and hence increase Q). It is found that the natural frequency of the resonator is determined not only by the size of the resonator but also by the spacing $(l - L)$. Thus, the resonator can be tuned mechanically by adjusting the depth of penetration p of the bolt inserted from the top cover of the metal shield. Electronic tuning may be realised with the change of bias controlled capacitance of a varactor diode mounted on the microstrip or by placing a photosensitive material (PSM) on the DR and perturbing the electromagnetic field around the DR with optically controlling the electric conductivity of the PSM.

Recapitulation

Rectangular waveguide

- The TEM mode of propagation is not possible in a waveguide, that is, there must always be an electric or magnetic field component along the propagation z-direction. It is thus customary to talk of transverse electric (TE) mode (for which $E_z \equiv 0$) and transverse magnetic (TM) mode (for which $H_z \equiv 0$). TE and TM are US nomenclature; in British nomenclature these are called respectively the H-wave and the E-wave. In rectangular waveguide 'mn' subscript is used for TE/TM modes where m and n respectively mean *half-cycle* variation of the field E/H in the x and y directions.
- A waveguide behaves as a high-pass filter with a cut-off frequency: $f_c = k_c/2\pi\sqrt{\mu\varepsilon}$, where k_c is called the *cut-off wave number*, while the free-space and coaxial line acts as an all-pass filter which *operates down to DC*. Practically all waveguides act actually as

band-pass filters in the sense that, out of the many possible modes, only one mode (in most cases preferably the *dominant mode*) is expected to propagate down a waveguide. The band of frequency over which the desired mode operates before the onset of the next higher order mode is the pass-band of the waveguide for operation in that particular (desired) mode.

- For a rectangular waveguide the width a needs to be $\lambda/2 < a < \lambda$, where λ is the wavelength of the signal and from practical considerations it is chosen to be 0.7λ. The waveguide height b in relation to a is given by $b = a/2$.
- In order to launch a particular mode in a waveguide, a type of probe or a loop is used that acts as a coupling antenna and launches suitable electric and magnetic fields within the waveguide that roughly replicates the desired E and H fields in the guide for that mode. For launching the TE_{10} mode in a rectangular waveguide, the probe is placed in the mid-plane of the broadside which corresponds to the location of the maximum of the E field for the mode. For the TE_{20} mode, two probes are used and the probes acting as antennae are fed 180° out of phase (by physically separating them by a distance of $\lambda/2$) as required by the field pattern to be launched, and so on.

Transverse electric (TE)-mode /H-wave ($E_z \equiv 0$)

- The frequency below which the wave propagation does not take place is the *cut-off frequency* and for TE mode in a rectangular waveguide, it is given by:

$$f_{c\text{TE}} = c/2\sqrt{(m/a)^2 + (n/b)^2}$$

where a and b are the width (along the x-direction) and height (along the y-direction) of the waveguide, m and n represent the number of half-cycle variations of the E-field in the x and y direction respectively.

- Above the cut-off frequency, the phase constant can be written as:

$$\beta_{g\text{TE}} = \sqrt{\omega^2\mu\varepsilon - (m\pi/a)^2 + (n\pi/b)^2} = \beta\sqrt{1-(f_{c\text{TE}}/f)^2}$$

While the phase velocity of wave propagation is:

$$v_{p\text{TE}} = \omega/\beta_{g\text{TE}} = v_p/\sqrt{1-(f_{c\text{TE}}/f)^2}$$

where v_p $(= 1/\sqrt{\mu\varepsilon})$ is the velocity of waves in free space.

The wavelength within the guide is called the *guide-wavelength* λ_g and is given by:

$$\lambda_{g\text{TE}} = 2\pi/\beta_{g\text{TE}} = \lambda/\sqrt{1-(f_{c\text{TE}}/f)^2}$$

(where $\lambda = 2\pi/\beta$ is the free space wavelength).

- Since m and n can have any integral value, multiple solutions for phase constant β are possible giving rise to the possibility of different *modes of propagation* in a waveguide. For the TE_{mn} mode, either m or n can be zero, but not both. Thus TE_{01} and also TE_{10} modes result in power flow in rectangular/square waveguide while the TE_{00} mode does not. If $a > b$, as is the case usually, then TE_{10} mode has the lowest cut-off frequency, however, in a square waveguide ($a = b$), both TE_{10} and TE_{01} have the same cut-off frequency, that is, they are the degenerate modes.

- Wave impedance in rectangular waveguide above cut-off frequency is given by:

$$Z_{w\text{TE}\,f>fc} = \frac{E_x}{H_y} = -\frac{E_y}{H_x} = \frac{\omega\mu}{\beta_{gmn}} = \frac{\omega\mu}{\beta}\frac{1}{\sqrt{1-(f_{c\text{TE}}/f)^2}} = \frac{\eta}{\sqrt{1-(f_{c\text{TE}}/f)^2}}$$

(However, the wave impedance is purely inductive when the operating frequency is below cut-off frequency and thus results in no average power flow in the *evanescent band*).

Transverse magnetic (TM)-mode/E-wave ($H_z \equiv 0$)

- The cut-off frequency ($f_{c\text{TM}}$), phase constant ($\beta_{g\text{TM}}$), phase velocity ($v_{p\text{TM}}$) and the guide wavelength ($\lambda_{g\text{TM}}$) expressions are the same for the TM_{mn} modes as for the TE_{mn} modes. The wave impedance above the cut-off for TM modes is given by:

$$Z_{w\text{TM}\,f>f_c} = \frac{E_x}{H_y} = -\frac{E_y}{H_x} = \frac{\beta_{gmn}}{\omega\varepsilon} = \frac{\beta}{\omega\mu}\sqrt{1-(f_{c\text{TM}}/f)^2} = \eta\sqrt{1-(f_{c\text{TM}}/f)^2}$$

(For the TM modes also there is no average power flow in the waveguide in the evanescent band since the wave impedance is now purely capacitive for operating frequency below the cut-off frequency).

TE-mode vs TM-mode and the dominant mode

- For a rectangular waveguide with $a > b$, all corresponding TE and TM modes are *degenerate modes*, that is, having the same cut-off frequency with TE_{10} and TM_{11} being the lowest possible modes in the two cases of which TE_{10} has the lowest cut-off frequency (largest cut-off wavelength) and hence is the *dominant mode*. The relevant parameters for the TE_{10} mode are:

Cut-off frequency $f_{c\text{TE}_{10}} = c/2\sqrt{(1/a)^2} = \dfrac{c}{2a}$, thus $\lambda_{c\text{TE}_{10}} = 2a$

Propagation constant $\beta_{g\text{TE}_{10}} = \sqrt{\omega^2\mu\varepsilon - (\pi/a)^2} = \beta\sqrt{1-(f_{c\text{TE}_{10}}/f)^2}$

Guide wavelength $\lambda_{g\text{TE}_{10}} = \lambda/\sqrt{1-(f_{c\text{TE}_{10}}/f)^2} = \lambda/\sqrt{1-(\lambda/2a)^2}$

Wave impedance $Z_{w\text{TE}_{10}} = \eta/\sqrt{1-(f_{c\text{TE}_{10}}/f)^2}$

Characteristic impedance $Z_{0\text{TE}_{10}}$ [modified $(P-V)$] $= 377\left(\dfrac{b}{a}\right)\dfrac{\lambda_{g\text{TE}_{10}}}{\lambda}$

Waveguide dispersion and wave velocities

- The waveguide is a *dispersive medium* (a medium in which the phase constant is a non-linear function of frequency and thus phase velocity depends on the frequency), unlike free space or two conductor transmission line (such as, a coaxial line), even though no dispersive material is loaded in it. Any information carrying signal (such as a modulated signal) will thus be dispersed or distorted while travelling down the waveguide.

  The group velocity (with which the information travels) for TE_{mn}/TM_{mn} modes is given by:

  $$v_{g\text{TE/TM}} = v_p\sqrt{1-(f_{c(\text{TE/TM})}/f)^2}, \quad \text{where } v_p = 1/\sqrt{\mu\varepsilon}$$

  In the propagating region, $v_{p(\text{TE/TM})} \times v_{g(\text{TE/TM})} = v_p^2$, where $v_{p\text{TE}} = v_p/\sqrt{1-(f_{c\text{TE}}/f)^2}$

Power handling capability and attenuation in a rectangular waveguide

- Peak electric field in a rectangular waveguide is given by:

  $$P_{\text{peak}} = 2P_{av} = \frac{E_0^2}{\eta}\frac{ab}{2}\sqrt{1-\left(\frac{\lambda}{2a}\right)^2}$$

  where E_0 is the maximum peak value of the electric field of the electromagnetic wave propagating down the waveguide.
- Attenuation due to conduction loss is given by:

  $$\alpha_{c\text{TE}_{10}} = \frac{R_s}{\eta}\frac{1+(2b/a)(f_{c\text{TE}_{10}}/f)^2}{b\sqrt{1-(f_{c\text{TE}_{10}}/f)^2}}$$

  where R_s is the surface resistance of the conducting walls of the waveguide and η is the wave impedance of the dielectric medium filling the waveguide.
- Attenuation due to dielectric loss is given by:

  $$\alpha_{d\text{TE}_{10}} = \frac{\pi f\sqrt{\mu\varepsilon'}}{\sqrt{1-(f_c/f)^2}}\tan\delta_c$$

  where $\tan\delta_c = \varepsilon''/\varepsilon'$, is the loss tangent of the dielectric.

Circular waveguide

- In circular guides, the mode subscript is denoted as *nm*, while for a rectangular guide, the subscripts are *mn* with *m* and *n* having their physical significance mentioned accordingly in two cases. In a circular waveguide, *n* indicates the number of *full-cycle* variations in the circumferential or φ direction and *m* describes the radial *r*, variations of the field.
- For TE and TM modes in a circular waveguide, the equations for β_{gnm}, f_c and v_{pg} are given by:

  $$\beta_{gnm} = \sqrt{\omega^2\mu\varepsilon - (k_{cnm})^2}$$

$$f_{cnm(\text{TE})} = k_{cnm}/2\pi\sqrt{\mu\varepsilon}$$

$$v_{pg} = \omega/\sqrt{\omega^2\mu\varepsilon - (k_{cnm})^2}$$

where $k_{cnm} = (t'_{nm}/a)$ for TE modes and (t_{nm}/a) for TM modes, with t'_{nm} being the mth root of $J'_n(x)$ and t_{nm} is the mth root of $J_n(x)$.

- The TE_{11} mode (which corresponds to the lowest value of t_{nm} = 1.814) has the lowest cut-off frequency in circular waveguide which is thus the *dominant mode* in a circular waveguide (with $\lambda_c = 3.412a$). The wave impedance for the TE and TM modes in a circular waveguide has similar expressions as for the rectangular waveguide, only the corresponding cut-off frequencies are different.
- In a circular waveguide, the ratio of the cut-off frequency for TM_{01} (the next higher order mode) to that of the TE_{11} mode (the dominant mode) is 2.405/1.841 = 1.3; while for rectangular waveguide, the ratio of the cut-off frequency of next higher order mode TE_{02} (c/a) to the dominant mode TE_{10} $(c/2a)$ is equal to 2. This means that the circular guide has a much narrower frequency range of operation for single mode propagation than the rectangular guide. Further, for the same cut-off frequency of operation, the area occupied by a circular waveguide in comparison to a rectangular waveguide is larger by a factor of more than 2 when both of them operate in their respective dominant modes. This makes the rectangular waveguide more popular than the circular guide from an application point of view. However, the TE_{01} mode and TM_{01} mode in the circular waveguide have special field characteristics and are used in the design of wavemeters and rotary joints respectively.

Attenuation in circular waveguides

- Attenuation suffered by TE_{nm} modes due to imperfectly conducting walls of a circular waveguide is given by:

$$\alpha_{c\text{TE}_{nm}} = \frac{R_s}{a\eta}\frac{\left\{\left(\dfrac{f_{c\text{TE}_{nm}}}{f}\right)^2 + \dfrac{n^2}{(t'_{nm})^2 - n^2}\right\}}{\left\{1-\left(\dfrac{f_{c\text{TE}_{nm}}}{f}\right)^2\right\}^{1/2}}$$

While those by TM_{nm} modes is,

$$\alpha_{c\text{TM}_{nm}} = \frac{R_s}{a\eta}\left\{1-\left(\frac{f_{c\text{TM}_{nm}}}{f}\right)^2\right\}^{-1/2}$$

where R_s is the surface resistance of the conducting walls of the waveguide and η is the wave impedance of the dielectric medium filling the waveguide.

Rectangular cavity resonator

- A three-symbol (*mnp*) subscript is used to designate a TE or TM standing wave pattern in a cavity resonator. The indices *m*, *n* and *p* (similar to waveguides) indicate the number of *half-sinusoid* variations of the field along the *x*, *y* and *z* directions respectively.
- Both TE_{mnp} and TM_{mnp} have the same expression for resonant frequency and thus for a given set of *m*, *n*, *p* the TE and TM modes are *degenerate*. In order for a rectangular cavity mode to exist, at least two of its integer values must be non-zero with non-vanishing field. The *primary mode* (that is, one with lowest resonant frequency) in a rectangular cavity resonator is the TE_{101} mode with resonant frequency:

$$f_{101} = \frac{1}{2\sqrt{\mu\varepsilon}}\sqrt{\frac{1}{a^2}+\frac{1}{l^2}}$$

- The quality factor of a rectangular cavity resonator due to losses in the conducting wall is given by:

$$Q_c = \frac{\pi\eta}{4R_s}\,\frac{\left\{1+\left(\frac{a}{l}\right)^2\right\}^{3/2}}{\left[1+\left(\frac{a}{2b}\right)\left\{1+\left(\frac{a}{l}\right)^2+\left(\frac{2ba^2}{l^3}\right)\right\}\right]}$$

- While due to losses in the dielectric filling the cavity is, $Q_d = \frac{1}{\tan\delta_c}$.
- Thus, the total Q is given by $\frac{1}{Q_T} = \frac{1}{Q_d} + \frac{1}{Q_c}$

Cylindrical cavity resonator

- In a cylindrical cavity resonator, mode designation is TE_{nmp} and TM_{nmp}; the integer *p* has the same meaning as in the rectangular cavity, namely, the number of *half-cycles* in the *z* direction while *n* indicates the number of *full-cycle* variations in the circumferential or φ direction and *m* describes the radial variations of the field.
- The resonant frequency for the TE_{nmp}/TM_{nmp} modes in a cylindrical cavity is given by:

$$f_r = \frac{c}{\sqrt{\mu_r\varepsilon_r}}\sqrt{\left(\frac{k_{cnm}}{\pi D}\right)^2+\left(\frac{p}{2l}\right)^2}$$

where $k_{cnm} = (t'_{nm}/a)$ for TE modes and (t_{nm}/a) for TM modes; with t'_{nm} being the mth root of $J'_n(x)$ and t_{nm} is the mth root of $J_n(x)$.
- The quality factor of a rectangular cavity resonator due to losses in the conducting wall is given by:

$$Q_{c(\text{TE}_{011})} = \frac{\lambda_r}{2\pi\delta} \frac{\left\{(t'_{01})^2 + \left(\frac{\pi a}{l}\right)\right\}^{3/2}}{(t'_{01})^2 + \frac{2a}{l}\left(\frac{\pi a}{l}\right)^2}$$

While due to dielectric filling and total Q follows similar ones as for rectangular cavity resonator.

Mode chart

- In the design of cavity resonators, we usually ensure that the cavity operates in a single mode. The *mode chart* is often used to check how many modes may exist in a cylindrical cavity and also helps to know the frequency range over which a particular resonance can be tuned without interference from other modes.

Strip/disc resonator

- The strip or disc resonator designed around a microstrip (planar) transmission line is typically the circular disc type in which the top metallic strip of the conventional microstrip line is truncated in the form of a disc. The resonator is formed between the top metallic disc and the bottom ground plane with the dielectric substrate in between.
- TM_{110} mode has the minimum radius and is called the *dominant mode* whose resonant frequency is given by:

$$f_{110(\text{TM})} = \frac{1.841c}{2\pi\sqrt{\varepsilon_r}\,a}$$

Dielectric resonator

- A small cube or disc (height L and diameter D) of low-loss, high dielectric constant (ε_r) material can be used as a microwave resonator, called a *dielectric resonator* (DR). The resonant frequency of the $TE_{01\delta}$ mode in a DR may be approximately estimated from:

$$f = \frac{68}{D\sqrt{\varepsilon_r}}\left(\frac{D}{2L} + 3.45\right) \quad (D > L,\ D \text{ and } L \text{ are in mm and } f \text{ is in GHz})$$

Coupling and tuning of microwave resonators

- In a rectangular waveguide type resonator, aperture/iris coupling is very common in a transverse wall, used in oscillator/amplifier application to connect the cavity to the desired load. The equivalent inductive susceptance represented by the aperture is given by, $jB = -j3ab/D^3\beta$. A cylindrical waveguide cavity resonator can be coupled to a rectangular waveguide via an off-slot as is used in the design of the wavemeter.
- Tuning of waveguide resonators can be done by inserting a movable short-circuit plunger at right angles to the axis of a cylindrical resonator or parallel to the transverse wall of a rectangular cavity thereby changing the length and hence the resonant frequency of the

cavity. A metallic screw may also be inserted in a rectangular cavity via the top broad wall to tune the frequency of the oscillator.

- Coupling of a microstrip resonator can be done by capacitive coupling between a $\lambda/2$ open-circuited microstrip resonator and a microstrip feed line. A strip line resonator may couple signals to a waveguide via aperture coupling.
- In a dielectric resonator (DR), the $TE_{01\delta}$ mode can easily be coupled magnetically to a microstrip line. Proper metallic shielding is required to minimise the radiation loss. The DR can be tuned by adjusting the depth of penetration of a bolt inserted from the top cover of the metal shield.

Review Questions

4.1 Why can't the TEM mode propagate in a single conductor transmission line, that is, in a waveguide?

4.2 The TE-mode (*H*-wave)/TM-mode (*E*-wave) propagate in waveguides: represent them pictorially.

4.3 Why is a parallel plate waveguide not a practical wave guiding structure?

4.4 How are cavity resonators made with truncated waveguides?

4.5 A coaxial line is an all-pass filter while a waveguide is a high pass filter—explain.

4.6 Discuss the significance of an evanescent band (stop band) and a propagation band (pass band) of a waveguide.

4.7 For the propagating TE_{mn}/TM_{mn} mode in a rectangular waveguide, what is the physical significance of the subscripts m and n?

4.8 What do you understand by the eigen value solution of the wave equation for a propagating wave in a waveguide?

4.9 Why can't the TE_{00} mode exist in a rectangular waveguide? TE_{10} is a preferable mode in a practical rectangular waveguide (with $a > b$) but not the TE_{01} mode, why?

4.10 In terms of wave impedance of the TE_{10} mode in a rectangular waveguide, how can you physically explain the nature of the propagating and evanescent waves?

4.11 What do you understand by the dominant mode in a waveguide? What are degenerate modes?

4.12 A waveguide is a dispersive medium even though no dispersive material is loaded in it—explain. How does it differ from conventional material dispersion?

4.13 Explain from a physical point of view the concepts of phase velocity and group velocity. Group velocity can never be greater than the velocity of light—discuss.

4.14 Group velocity is zero at the cut-off frequency causing no propagation of electromagnetic energy—explain.

4.15 Discuss how waveguide dimensions are chosen for a rectangular waveguide.

4.16 Discuss how different modes are launched in a rectangular waveguide.

4.17 Mention the factors that determine the maximum power handling capability of a rectangular waveguide.

4.18 What are the causes for attenuation of a signal travelling down a practical waveguide? Is it larger or smaller than the attenuation suffered by evanescent waves in the stop band of a waveguide?

4.19 What is an oversized waveguide? What are the advantages of using an oversized waveguide at millimetre wave frequency?

4.20 For propagating the TE_{nm}/TM_{nm} mode in a circular waveguide, what is the physical significance of the subscripts *n* and *m*?

4.21 Name the dominant mode in a circular waveguide. Compare the ratio of cut-off frequency of the dominant mode and the next higher order mode in circular and rectangular waveguides and comment on the result from a practical point of view.

4.22 Discuss the importance of TE_{01} and TM_{01} modes in a circular waveguide from the designer's point of view.

4.23 Discuss why a circular waveguide designed with TE_{01} may be useful as the transmission medium for long haul communication at millimetre-wave frequency?

4.24 What are cavity resonators? Name a few microwave cavity resonators. Explain why the length of a cavity resonator is an integral multiple of half-wavelength?

4.25 TE_{mnp} is a typical mode designation in a rectangular cavity resonator; what do the three subscripts *m*, *n* and *p* physically signify in the case of a rectangular cavity resonator?

4.26 TE_{101} is the primary mode in a rectangular cavity resonator—explain.

4.27 In a lumped resonant circuit, harmonic resonant modes are possible, while in a microwave resonator the resonant frequencies are not harmonically related—explain.

4.28 TE_{nmp} is a typical mode designation in a cylindrical cavity resonator; what do the three subscripts *n*, *m* and *p* physically signify in the case of a cylindrical cavity resonator?

4.29 In a wavemeter design, TE_{011} in a cylindrical waveguide is the preferred mode—explain. How does the design suppress the TM_{111} mode that has same resonant frequency as that for TE_{011}?

4.30 Discuss the significance of the mode chart in the design of cavity resonators.

4.31 Fields within the dielectric substrate of the strip resonator will be those of the TM_{nm} or *E*-modes in a circular waveguide at cut-off, but the guide boundary of the former is a magnetic wall instead of an electric wall—explain.

4.32 What is a dielectric resonator? A dielectric resonator supports 'magnetic dipole mode'—justify.

Numerical Problems

4.1 A waveguide is constructed so that the cross-section of the guide forms a triangle with sides of length a, a and $\sqrt{2}a$. The walls are perfect conductors and the inside of the guide has air as the dielectric. Determine the allowed modes of propagation and the cut-off frequency.

[**Ans:** Allowed mode is $TE_{n,n}$ with a cut-off frequency $f_n = \dfrac{nc}{\sqrt{2}a}$]

4.2 An air-filled rectangular waveguide operates at a frequency of 1 GHz and has $a = 5$ cm and $b = 2$ cm. Show that the TM_{21} is an evanescent mode at this frequency of operation. If the waveguide is instead excited at a frequency of 20 GHz, (a) calculate the cut-off frequency for the TM_{21} mode and show that it now represents the propagating wave, (b) determine the phase constant, phase velocity, group velocity, wavelength of propagation of the wave, (c) calculate the wave impedance of the TM_{21} mode and (d) find the average power transmitted by the waveguide if the amplitude of the applied electric field is 500 V/m.

[**Ans:** (a) $f_c = 9.6$ GHz, (b) $\beta_{21} = 367.47$ rad/m, $v_{p21} = 3.42 \times 10^8$ m/s, $v_{g21} = 2.63 \times 10^8$ m/s, $\lambda_{21} = 17$ mm, (c) $\eta_{TM21} = 330.42\ \Omega$ (d) $P_{21} = 315.6$ mW]

4.3 An air filled waveguide operates at 7 GHz. The dimensions of the waveguide are $a = 3$ cm, $b = 2$ cm. Calculate the maximum average power transmitted along the waveguide without causing a breakdown inside the waveguide at the dominant TE_{10} mode. The dielectric strength of air is 30 kV/cm. Use a safety factor of 10.

[**Ans:** 25.06 kW]

4.4 A parallel plate waveguide with the plates separated by a distance d is partially filled (along its length) with a dielectric slab of thickness d (with a dielectric constant 2.1). Assume that a TE_{10} mode propagates in the air-filled part of the guide at a frequency of $f = 3f_c$, where f_c is the cut-off frequency of the empty guide. Find (a) the reflection and transmission coefficients at the air–dielectric interface and (b) the standing wave ratio in the air and also in the dielectric.

[**Ans:** $|\Gamma| = 0.01588$, $T = 0.9841$; $SWR_{air} = 1.0322$, $SWR_{dielectric} = 1$]

4.5 Write a MATLAB program to generate the curves given in Figs 4.12a and 4.12b for a rectangular waveguide.

4.6 It is known that a circular waveguide occupies a larger area compared to a rectangular waveguide. Show that the ratio of the areas occupied by a circular waveguide to a rectangular waveguide is 2.18, when they are operated in their respective dominant modes having the same cut-off frequency.

4.7 Write a MATLAB program to generate the curves given in Figs 4.18a and 4.18b for a circular waveguide.

4.8 An air-filled rectangular cavity made of brass has dimensions: $a = 4$ cm, $b = 3$ cm and $d = 5$ cm. What is the dominant mode in the cavity? Calculate its resonant frequency. Find the Q of the cavity and the time-average stored electric and magnetic energies at the resonant frequency, assuming H_0 to be 0.1 A/m.

[**Ans:** Dominant mode is TE_{101}, $f_{101} = 4.8$ GHz, $Q_{101} = 6870$, $W_e = W_m = 7.7 \times 10^{-14}$ J]

4.9 List in ascending order, the resonant frequencies of the five lowest order modes possible in a cylindrical cavity resonator (air-filled) having radius of 2 cm and length 5 cm.

[**Ans:** TE_{111} (5.32 GHz), TM_{010} (5.74 GHz), TM_{011} (6.47 GHz), TE_{211} (7.88 GHz), TM_{111}/TE_{011} (9.62 GHz)]

4.10 Write a MATLAB program to generate the curves given in Fig. 4.23 for a mode chart of a cylindrical cavity resonator.

5 Microwave Network and Scattering Matrix

Thoughts get scattered in the material world;
while, in the spiritual world they are always in consolidation

5.1 Introduction

Devices and circuits in a network at low frequency are characterised in terms of Z (impedance), Y (admittance), or h (hybrid) parameters; the latter being often used to describe transistor characteristics. These parameters are defined in terms of voltage and current in the device/circuit. This is because at low frequency, the device/circuit dimensions are quite small compared to the wavelength, and thus the devices and circuits may be considered as '*lumped*' components for which voltages and currents can be defined uniquely at any point in the network. However, at microwave frequencies, the device/circuit dimensions are comparable to wavelength which makes them '*distributed*' components, that is, the phase of the signal changes from one point to another in the network. This calls for field analysis of microwave network. The voltages and currents are no longer measurable quantities (as they are difficult to measure) at microwave frequencies but are actually the derived quantities (voltage derived from electric field and current derived from magnetic field). The measurable quantities at microwave frequencies are the incident, reflected and transmitted power in the network (the reflected and transmitted powers being generated due to scattering of the incident signal at the microwave junction). Again, in view of the linearity of the field equations and most devices, the scattered wave amplitudes are linearly related to the incident wave amplitudes. The matrix describing this linear relationship between the reflected and transmitted waves with incident wave amplitude is called the *scattering matrix* and the microwave device/circuit is analysed in terms of the scattering parameters or *S-parameters*.

It would be seen shortly that unlike the Z, Y and ABCD parameters (the last one being used at low frequencies for cascaded networks), the S-parameters do not use short and open circuit terminations. From a practical view point this has significance, as it is difficult to obtain perfect open and short circuit termination at RF and microwave frequencies. Also, the impedance of short and open is frequency dependant at microwave frequencies and thus defining parameters to characterise microwave device/circuit should not use them. Further, the presence of short/open eventually causes complete reflection of signal from the load terminal leading to possible oscillation and might even cause damage to the source. But use of a matched load in defining S-parameters would cause absorption of the incident energy, thus causing no reflection back to the source.

Similar to the low frequency transmission matrix (the ABCD matrix) at microwave frequencies, the transmission or T matrix (otherwise known as the chain S-matrix) is used when dealing with multistage circuits or infinitely long periodic structures such as those used in circuits for travelling wave tubes.

Further, it may be noted that both S and T matrices are useful for linear circuits. But to analyse non-linear circuits and systems, a new class of parameters is defined, known as X-parameters. While S-parameters are useful for DC and small signal analysis in linear circuits, X-parameters are useful for large-signal conditions (which is usually the case with non-linear circuit analysis that we come across in harmonic mixing, non-linear reflection effects and so on). However, in the limit of small signal approximation, the X-parameters reduce to S-parameters. Calculations based on X-parameters are usually performed within a harmonic balance simulator environment.

S-parameters have many advantages over *Z*, *Y* and *h* parameters. They relate to familiar measurements such as gain, loss and reflection coefficient. The measured S-parameters of multiple devices can be cascaded to predict the overall system performance. If desired, *Z*, *Y* and ABCD-parameters can be derived from S-parameters and vice versa (vide Appendix I). From the RF designer's point of view, it is also very important that the S-parameters can be easily imported and used for circuit simulations in electronic-design automation (EDA) simulation tools. S-parameters are the shared language between simulation and measurement. Since the 1970s, the popularity of the S-parameters increased with the advent of the new generation network analysers which could perform S-parameter measurements with ease.

5.2 S-parameter Formulation of Microwave Network

Let us consider an n-port microwave junction (Fig. 5.1). Let a wave with an associated equivalent voltage V_1^+ be incident on the junction at terminal plane t_1. This will generate a reflected wave in line 1 given by $S_{11}V_1^+ = V_1^-$, where S_{11} is the reflection coefficient or scattering coefficient for line 1, with a wave incident at line 1. Waves will also be transmitted or scattered out to the remaining $(n - 1)$ lines from line 1 and will have amplitudes proportional to V_1^+. These amplitudes can be expressed as $V_i^- = S_{i1}V_1^+$ where $i = 2, 3, \ldots, n$, (S_{i1} is the transmission coefficient for a signal transmitted to i^{th} line from line 1).

When waves are incident in all the n lines, the wave voltage in a particular line has its own reflected wave voltage plus the scattered wave voltage contributions from all other $(n - 1)$ lines. Thus, in general, we can write:

$$V_1^- = S_{11}V_1^+ + S_{12}V_2^+ + S_{13}V_3^+ + \cdots + S_{1n}V_n^+$$

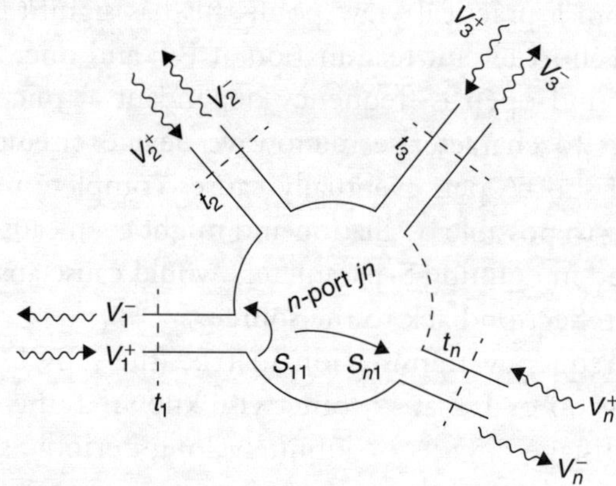

Fig. 5.1 An *n*-port microwave junction.

$$V_2^- = S_{21}V_1^+ + S_{22}V_2^+ + S_{23}V_3^+ + \cdots + S_{2n}V^+{}_n \tag{5.1}$$

$$\vdots \quad \vdots \quad \vdots \quad \vdots$$

$$V_n^- = S_{n1}V_1^+ + S_{n2}V_2^+ + S_{13}V_3^+ + \cdots + S_{nm}V_n^+$$

which can be written in matrix form as $[V^-] = [S][V^+]$, where both $[V^-]$ and $[V^+]$ are column matrices and S is an $n \times n$ matrix called scattering matrix given by:

$$\begin{bmatrix} V_1^- \\ V^-{}_2 \\ \cdot \\ \cdot \\ V_n^- \end{bmatrix} = \begin{bmatrix} S_{11} & S_{12} & S_{13} & \cdot & \cdot & S_{1n} \\ S_{21} & S_{22} & S_{23} & \cdot & \cdot & S_{2n} \\ \cdot & \cdot & \cdot & \cdot & \cdot & \cdot \\ \cdot & \cdot & \cdot & \cdot & \cdot & \cdot \\ S_{n1} & S_{n2} & S_{n3} & \cdot & \cdot & S_{nn} \end{bmatrix} \begin{bmatrix} V_1^+ \\ V_2^+ \\ \cdot \\ \cdot \\ V_n^+ \end{bmatrix} \tag{5.2}$$

The coefficients $S_{11}, S_{12}, \ldots S_{nn}$ are called scattering parameters (S-parameters) or scattering coefficients.

The scattering matrix definition so far was implicitly based on the assumptions that all ports have the same characteristic impedance. Though this may be the case in many practical situations, for general analysis we need to consider the case when each port has non-identical characteristic impedance. This will also bring out some significant meaning of the wave variables—the reflected plus transmitted wave variables (b_1, b_2) as dependent variables and the incident wave variables (a_1, a_2) as independent variables (Fig. 5.2 a two-port network); in terms of which we are going to define the S-parameters.

Taking each port's characteristic impedance into account, we define two normalised voltage waves:

$$a_i = \frac{V_i^+}{\sqrt{Z_{0i}}} \tag{5.3a}$$

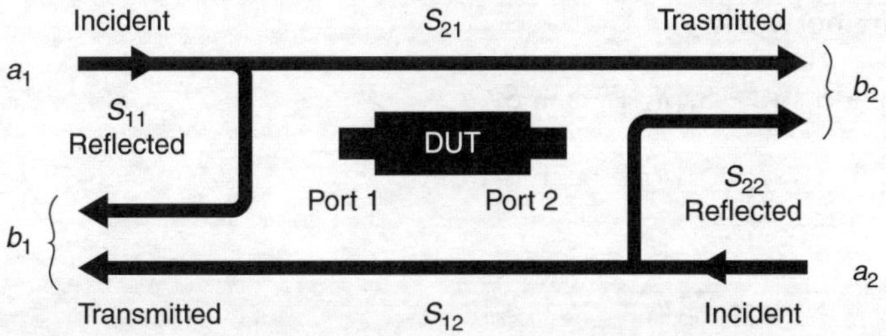

Fig. 5.2 A two-port microwave network showing the incident and reflected plus transmitted wave variables and the S-parameters.

$$b_i = \frac{V_i^-}{\sqrt{Z_{0i}}} \tag{5.3b}$$

where a_i and b_i represent the normalised incident and reflected plus transmitted voltages respectively at the ith port and Z_{0i} is the characteristic impedance of the ith port (which is a real number for a lossless-line).

Thus, the total voltage and current at each port can now be written as (Eq. 3.13):

$$V_i = V_i^+ + V_i^- = \sqrt{Z_{0i}}\,(a_i + b_i) \tag{5.4a}$$

$$I_i = I_i^+ - I_i^- = \frac{V_i^+}{Z_{0i}} - \frac{V_i^-}{Z_{0i}} = \frac{(a_i - b_i)}{\sqrt{Z_{0i}}} \tag{5.4b}$$

The average net power delivered to the ith port can now be expressed in terms of a_i and b_i (without consideration of the fact that different ports have different characteristic impedance) as:

$$P_i = \frac{1}{2}\,\text{Re}\,(V_i I_i^*) = \frac{1}{2}\,\text{Re}\left\{|a_i|^2 - |b_i|^2 + (a_i^* b_i - a_i b_i^*)\right\}$$
$$= \frac{1}{2}\left(|a_i|^2 - |b_i|^2\right) \tag{5.4c}$$

since $(a_i^* b_i - a_i b_i^*) = 0$, when we take the real part of it.

This equation is significant, as it clearly shows that the net power delivered to each port is equal to the normalised incident power, less by the normalised reflected power. By defining the wave variables in this way, it may be seen that the square of any wave variable becomes proportional to the average power in the wave, and these are now directly measurable quantities at microwave frequencies.

The generalised *S*-matrix can now be defined in terms of the normalised voltage wave variables *a* and *b* as:

$$[b] = [S][a] \tag{5.4d}$$

For a two-port network we may thus write:

$$b_1 = S_{11}\,a_1 + S_{12}\,a_2 \tag{5.4e}$$

$$b_1 = S_{21}\,a_1 + S_{22}\,a_2 \tag{5.4f}$$

And, for an n-port network:

$$\begin{aligned} b_1 &= S_{11}a_1 + S_{12}a_2 + S_{13}a_3 + \cdots + S_{1n}a_n \\ b_2 &= S_{21}a_1 + S_{22}a_2 + S_{23}a_3 + \cdots + S_{2n}a_n \\ \vdots\ & \quad\ \vdots \qquad\quad \vdots \qquad\quad \vdots \qquad\qquad\qquad \vdots \\ b_n &= S_{n1}a_1 + S_{n2}a_2 + S_{n3}a_3 + \cdots + S_{nn}a_n \end{aligned} \tag{5.4g}$$

where S_{11} and S_{22} are the reflection coefficients at ports 1 and 2 respectively and S_{12} and S_{21} are the transmission coefficients to port 1 from port 2 and to port 2 from port 1 respectively.

Now we try to understand the physical significance of the S-parameters. Referring to Eq. 5.4e, we find that:

$$S_{11} = \left.\frac{b_1}{a_1}\right|_{a_2=0} \tag{5.5a}$$

(this is the *input reflection coefficient*).

The variable a_2 is zero when there is no incident wave at port 2. The parameter S_{11} is, therefore, the reflection coefficient at port 1 when port 2 is terminated with a matched load.

Again from Eq. 5.4f we have:

$$S_{21} = \left.\frac{b_2}{a_1}\right|_{a_2=0} \tag{5.5b}$$

(this is the *forward transmission coefficient*).

This gives the ratio of a wave leaving port 2 for a wave entering port 1, that is, the transmission coefficient from port 1 to port 2 under matched condition at port 2 (for a lossy transmission line or waveguide, S_{21} should be less than 1, and it indicates the amount of attenuation).

The other two parameters can also be found from Eqs 5.4e and 5.4f as follows:

$$S_{22} = \left.\frac{b_2}{a_2}\right|_{a_1=0} \tag{5.5c}$$

(this is the *output reflection coefficient*).

$$S_{12} = \left.\frac{b_1}{a_2}\right|_{a_1=0} \tag{5.5d}$$

(this is the *reverse transmission coefficient*).

It may be observed that S_{22} is the reflection coefficient at port 2 with port 1 match terminated and that S_{12} gives the transmission coefficient from port 2 to port 1 under matched condition at port 1.

In general, the scattering parameters are complex numbers having both magnitude and phase.

EXAMPLE 5.1

Find the S-parameters of a waveguide component if the measured VSWR = 1.3 when the component is terminated with a matched load. It is also found that the power to the matched load is 60 mW for an input power of 100 mW. The same results are obtained when the component is reversed.

Solution

From Eq. 5.5a, $S_{11} = \left.\dfrac{b_1}{a_1}\right|_{a_2=0}$

which is the reflection coefficient at port 1 with no incident wave at port 2. The wave variable a_2 is zero since the component is terminated in the matched load.

We can find the reflection coefficient (Γ), using: $S_{11} = |\Gamma| = \dfrac{VSWR-1}{VSWR+1} = \dfrac{1.3-1}{1.3+1} = 0.13$

From Eq. 5.5b, $S_{21} = \left.\dfrac{b_2}{a_1}\right|_{a_2=0}$

Now, since power is proportional to the square of the wave variables a_1 and b_2,

$$\frac{|b_2|^2}{|a_1|^2} = \frac{P_2^-}{P_1^+} \quad \text{and} \quad S_{21} = \sqrt{\frac{P_2^-}{P_1^+}} = \sqrt{\frac{60\times10^{-3}}{100\times10^{-3}}} = 0.77$$

If the measured results are the same when the component is turned around, then $|S_{11}| = |S_{22}| = 0.13$ and $|S_{21}| = |S_{12}| = 0.77$. It may be noted that we have determined only the magnitude of the S-parameters. The angles (phase) depend on the reference position of the ports.

5.3 Properties of S-parameters

Depending upon the nature (reflection-less/reciprocal/loss-less) of the microwave junction, the S-parameters will satisfy certain conditions and these properties of the S-parameters are very useful in simplifying the S-matrix of a microwave junction and to draw useful inferences regarding the electrical properties of microwave passive circuit components and active devices.

Property I: Reflection-less junction: We have seen that the parameters S_{11} and S_{22} represent reflection coefficients at ports 1 and 2 under matched condition at ports 2 and 1 respectively. Therefore if the reflection coefficients are zero for a two-port network, then, $S_{11} = S_{22} = 0$. Most microwave components are designed to have small reflections (VSWR ~ 1.02). Thus, it is often justified to set the diagonal elements of their S-matrix to be zero when the S-matrix becomes a *skew-symmetric matrix*. For an ideal *n*-port with no reflection at any port under matched condition we thus have:

$$S_{ii} = 0, \quad i = 1, 2, 3 \ldots n, \quad \text{for matched junction with no reflection.}$$

Property II: Reciprocal junction: A reciprocal two-port junction has the same transmission characteristics in either direction, that is, the output of the two-port junction is the same for either port 1 or port 2 are used as input port. Thus, for a two-port reciprocal junction $S_{12} = S_{21}$. For an n-port junction, in general, $S_{ij} = S_{ji}$, $i \neq j$. This means that for a reciprocal microwave junction, $[S] = [S]^t$, where $[S]^t$ is the transpose of the matrix S; when the S-matrix becomes a *symmetrical matrix*.

Property III: Lossless junction: For a lossless junction, the total power leaving all the n-ports must equal to the sum of the incident power at all those n-ports. The mathematical statement of this power conservation condition is:

$$\sum_{n}(a_k a_k^{\star} - b_k b_k^{\star}) = 0 \quad \text{or} \quad \sum_{n} b_k b_k^{\star} = \sum_{n} a_k a_k^{\star} \tag{5.6}$$

This condition imposes a number of restrictions on the S-matrix coefficients which is discussed below. Following Eq. 5.4g, we may write:

$$b_k = \sum_{i=1}^{n} S_{ki} a_i \tag{5.7}$$

(where k is the parameter and i is the running variable). Substituting Eq. 5.7 in Eq. 5.6 we have:

$$\sum_{k=1}^{n} \left| \sum_{i=1}^{n} S_{ki} a_i \right|^2 = \sum_{k=1}^{n} a_k a_k^{\star} \tag{5.8}$$

Since a_ks are independent, we may choose all a_ks except a_i to be zero. Then, we get from Eq. 5.8 above:

$$\sum_{k=1}^{n} \left| S_{ki} a_i \right|^2 = a_i a_i^{\star} = |a_i|^2 \tag{5.9a}$$

or

$$\sum_{k=1}^{n} |S_{ki}|^2 = \sum_{k=1}^{n} S_{ki} S_{ki}^{\star} = 1 \tag{5.9b}$$

Since the index i is arbitrary, the above equation is valid for all values of i ($i = 1, 2, 3 \ldots n$). Equation 5.9 states that for a loss-less microwave network, the product of any column of the S-matrix with the complex conjugate of this same column equals unity. This is known as *unity property of S-matrix*.

In addition to the above, another constraint may be imposed on the S-parameters. By selecting all a_ks except a_s and a_r to be zero ($r \neq s$), Eq. 5.8 may be written as:

$$\sum_{k=1}^{n} \left| S_{ks} a_s + S_{kr} a_r \right|^2 = \sum_{k=1}^{n} (S_{ks} a_s + S_{kr} a_r)(S_{ks} a_s + S_{kr} a_r)^{\star} = |a_s|^2 + |a_r|^2 \tag{5.10}$$

Expanding the LHS and using Eq. 5.9a in a number of terms on the left and cancelling, we get:

$$\sum_{k=1}^{n} (S_{kr} S_{ks}^{\star} a_r a_s^{\star} + S_{ks} S_{kr}^{\star} a_s a_r^{\star}) = 0 \tag{5.11}$$

In view of the independent nature of a_s and a_r, we may choose in the first instance $a_s = a_r$, this leads to:

$$|a_s|^2 \sum_{k=1}^{n} (S_{kr}S_{ks}^{\star} + S_{ks}S_{kr}^{\star}) = 0 \tag{5.12a}$$

Instead, if we choose $a_s = ja_r$, with a_r real we obtain:

$$j\,|a_r|^2 \sum_{k=1}^{n} (-\,S_{kr}S_{ks}^{\star} + S_{ks}S_{kr}^{\star}) = 0 \tag{5.12b}$$

Since neither a_s nor a_r is zero, both Eqs 5.12a and b are satisfied only if:

$$\sum_{k=1}^{n} S_{kr}S_{ks}^{\star} = 0,\ r \neq s \tag{5.13}$$

This equation states that the product of any column of the S-matrix with the complex conjugate of any other column is zero. This is called the *zero property of S-matrix*.

A matrix with elements that satisfy the constraints stated in Eqs 5.9 and 5.13 is called the *unitary matrix* and thus a loss-less microwave junction satisfies the unitary property.

Property IV: Change of port position: The port position is the reference plane which determines the phase of the signal from that port. A change in the port position will effect the phase of the scattering coefficients. For analysis and practical design applications, sometimes it may be desirable to shift reference planes in order to make the angles of the S-parameters to be zero, that is, to make them real quantities (and not complex quantities) when it becomes physically realisable.

5.4 Signal Flow Graphs: Decomposition Rules and Meson's Rule

Signal flow graphs (SFG) are very useful for the analysis of microwave networks in terms of the incident wave (independent) variable *a* and the reflected plus transmitted wave (dependent) variable *b*. This graphical representation implicitly indicates the energy or power flow in the microwave network in terms of S-parameters. In fact, the signal flow graphs, through the use of graphical diagrams, provide a physical insight into the relationship between the cause (incident wave variables, *a*) and the effect (reflected plus transmitted wave variables, *b*) in the system. In essence, a signal flow graph is an alternative and yet a simpler method to the well-known 'block diagram' method in representing a complicated linear system. The advantage of the signal flow diagram method over the block diagram method is the availability of a flow-graph gain formula, which provides the relation between the system variables.

In a SFG, the wave variables are shown as *nodes* and S-parameters as *branches* connecting the nodes, the directions of branches are from *a* nodes to *b* nodes (Fig. 5.3; a two-port network). The nodes are shown by small circles and the branches have arrow on the line indicating the direction of signal flow and the letter near the arrow indicates the multiplication factor.

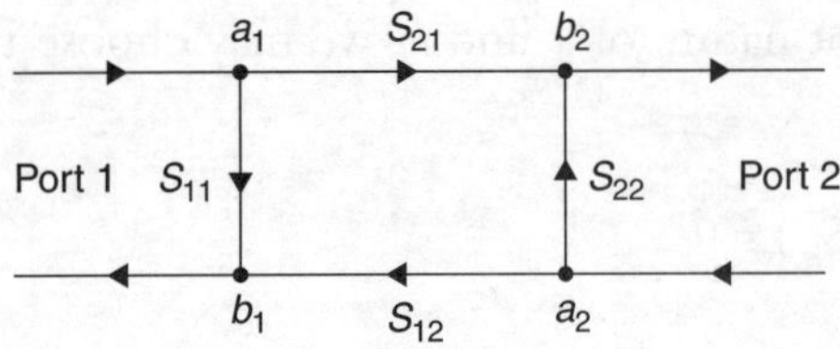

Fig. 5.3 Signal flow graph of a two-port network.

The signal flow graph gives an intuitive graphical illustration of the network behaviour. For example, a wave incident at port 1 at node a_1 will split, with a part going as reflected wave (S_{11}) generated at port 1 and as a transmitted wave (S_{21}) to port 2. Again at node b_2, a wave goes out through port 2 and if a load with non-zero reflection coefficient is connected at that port, then a reflected wave (S_{22}) will be generated at port 2 and a part of the signal from node a_2 will be transmitted (S_{12}) from port 2 to port 1.

Once a microwave network has been represented with a signal flow graph, the evaluation of the ratio of any combination of wave amplitudes at the input–output of the network can be done in two ways. One can use the *decomposition rules* or alternatively use *Meson's rule* of control theory. In the following sections we discuss the decomposition rules (without going into mathematical derivations) and also the Meson's rule in a nutshell. However, it may be noted that the use of the decomposition rule is much easier than Meson's rule, since for complicated graphs, the use of Meson's rule may be quite intricate.

Decomposition rules

Rule I (*Series or multiplication rule*): This rule states that two series branches (whose common node 2 has only one incoming and one outgoing wave) may be combined into a single branch from node 1 to node 3 with an effective transmission coefficient (multiplication factor) equal to the product of the transmission coefficient from node 1 to node 2 with that from node 2 to node 3 (Fig. 5.4a); mathematically, $S_{31} = S_{21}S_{32}$.

Rule II (*Shunt or addition rule*): This rule states that two or more parallel connecting branches (going from one node to another) may be replaced by a single branch with effective transmission coefficient equivalent to the sum of the transmission coefficients of the individual branches (Fig. 5.4b); mathematically, $S_{\text{eff}} = S_1 + S_2$.

Rule III (*Loop rule*): (*i*) *Self-loop*: When a node has a self-loop [a branch that begins and ends on the same node, Fig. 5.4c(*i*) of coefficient S, the self-loop can be eliminated by multiplying the transmission coefficients of the branches feeding that node by $1/(1 - S)$; thus: $S_{31} = S_{21}S_{32}/(1 - S_{22})$. (*ii*) *Feedback-loop*: A feedback loop [Fig. 5.4c(*ii*)] may be eliminated by dividing the input transmission coefficient by 1 minus the transmission factor around the loop which is the product $S_{23}S_{32}$ (by Rule I); thus: $S_{31} = S_{21}/(1 - S_{23}S_{32})$.

Rule IV (*Splitting rule*): Any node can be split into two separate nodes where each of the two nodes are connected only once to the incoming and outgoing nodes with the same coefficient (Fig. 5.4d).

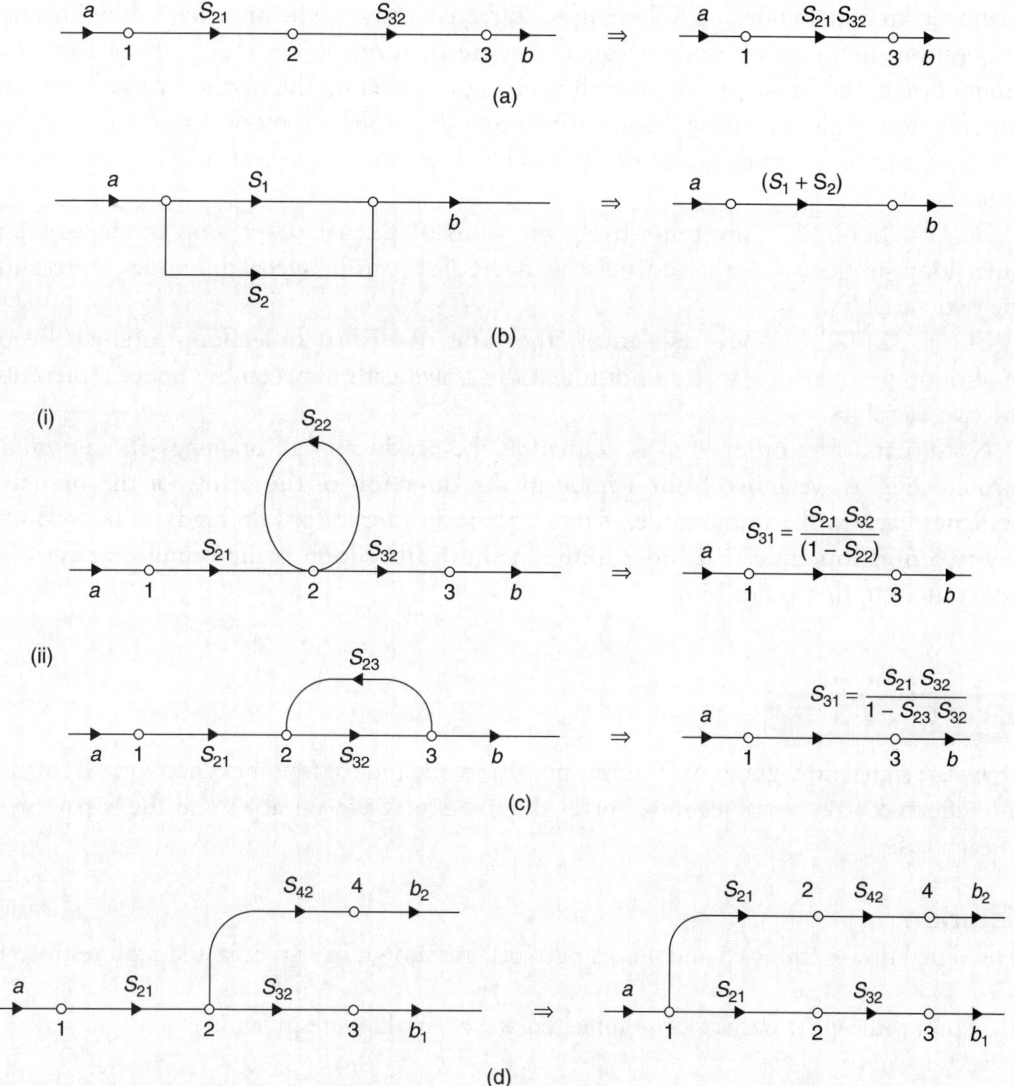

Fig. 5.4 Decomposition rule (a) Series rule (b) Shunt or addition rule (c) Loop rule: (i) Self-loop (ii)Feedback loop (d) Splitting rule.

Meson's Rule: It states that the transfer function, that is, the ratio of a dependent to independent variable is given by:

$$T = \frac{P_1\left\{1 - \sum L(1)^{(1)} + \sum L(2)^{(1)} - \cdots\right\} + P_2\left\{1 - \sum L(1)^{(2)} + \cdots\right\} + P_3\,(1 - \cdots)}{1 - \sum L(1) + \sum L(2) - \sum L(3) + \cdots} \tag{5.14}$$

where, P_1, P_2, P_3, … are the values of the various paths connecting the dependent and independent variables whose transfer function is to be determined. A path is a branch starting at an initial node and following the directed arrow of the branches to the final node in question such that no node is touched more than once. The value of the path is the product of all the branch values (coefficients) covered along the particular path. It is to be remembered while traversing along a path no node is touched more than once.

$\Sigma L(1), \Sigma L(2), \ldots$ are the sums of all *first order loop* products, *second order loop* products, … respectively.

$\Sigma L(1)^{(1)}, \Sigma L(2)^{(1)}, \ldots$ are respectively the sums of all first order loop products, second order loop products, …, that do not touch the *first* path between the nodes representing the two variables.

$\Sigma L(1)^{(2)}, \Sigma L(2)^{(2)}, \ldots$ are respectively the sums of all first order loop products, second order loop products, …, that do not touch the *second* path between the nodes representing the two variables.

Note that a first order loop is defined as the product of all branches encountered in a round trip as we move from a node in the direction of the arrow of the branch or branches back to the initial node. While a second order loop is defined as the product of any two non-touching first order loops. A third order loop is the product of any three non-touching first order loops.

EXAMPLE 5.2

Draw the signal flow graph of (a) three-port network and (b) four-port network. Determine the reflected wave variables in terms of the incident wave variables and the S-parameters in each case.

Solution

The signal flow graph of 3 and 4 port network are shown in Figs 5.5a and 5.5b respectively.

1. For a three-port network the reflected wave variables are given by:

$$b_1 = S_{11}a_1 + S_{12}a_2 + S_{13}a_3$$

$$b_2 = S_{21}a_1 + S_{22}a_2 + S_{23}a_3$$

$$b_3 = S_{31}a_1 + S_{32}a_2 + S_{33}a_3$$

2. For a four-port network the reflected wave variables are given by:

$$b_1 = S_{11}a_1 + S_{12}a_2 + S_{13}a_3 + S_{14}a_4$$

$$b_2 = S_{21}a_1 + S_{22}a_2 + S_{23}a_3 + S_{24}a_4$$

$$b_3 = S_{31}a_1 + S_{32}a_2 + S_{33}a_3 + S_{34}a_4$$

$$b_4 = S_{41}a_1 + S_{42}a_2 + S_{43}a_3 + S_{44}a_4$$

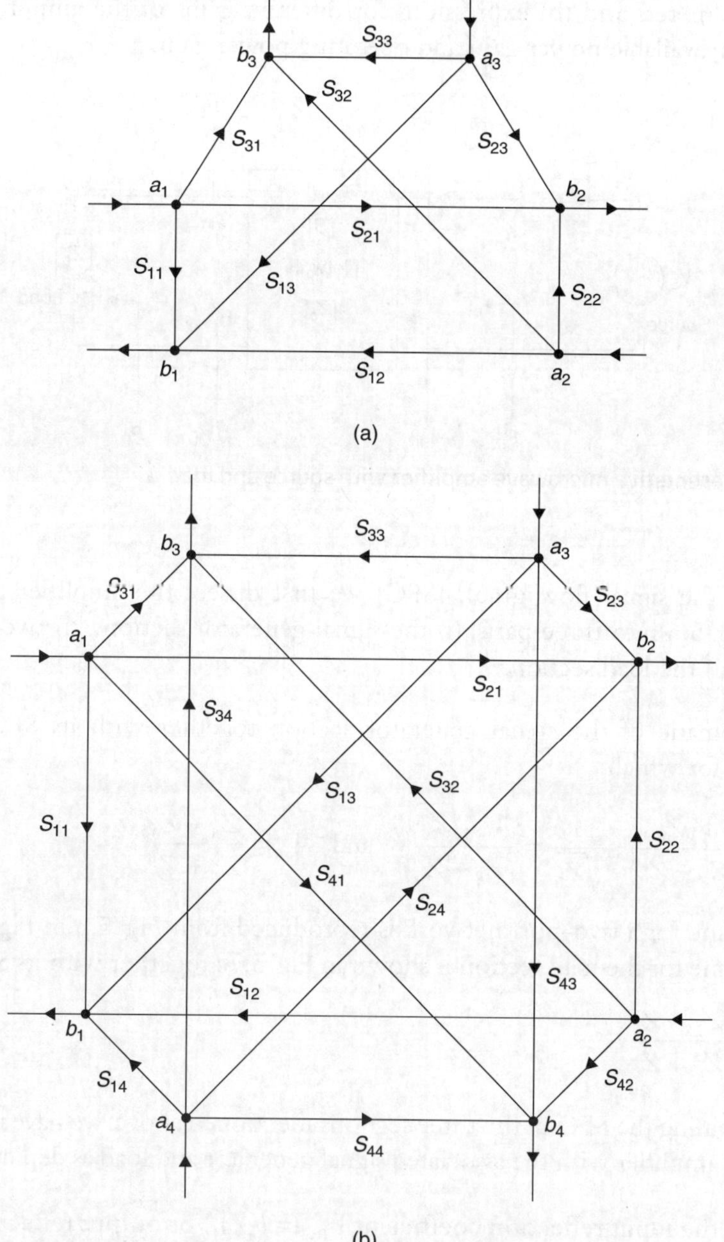

Fig. 5.5 Signal flow graph of (a) 3-port network (b) 4-port network.

EXAMPLE 5.3

A two-port network that represents a microwave amplifier is shown in Fig. 5.6. It has a load and a source at the output and input ends respectively. Determine its signal flow graph and obtain the (a) input reflection coefficient (Γ_{IN}) and output reflection coefficient (Γ_{OUT}) in terms of S-parameters and (b) expressions for different gains of the amplifier such as the transducer gain, available power gain and operating power gain.

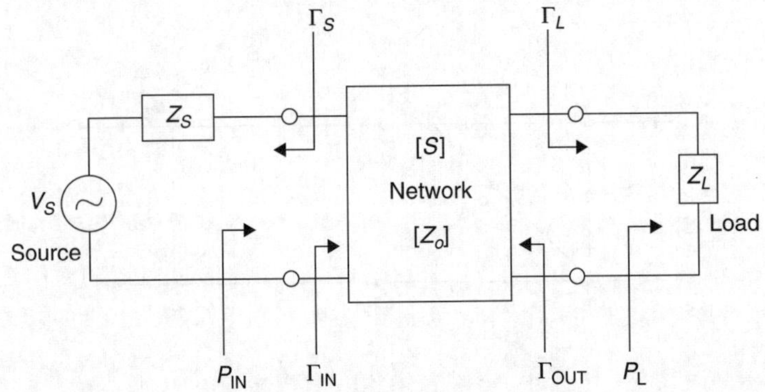

Fig. 5.6 A representative microwave amplifier with source and load.

Solution

To determine the signal flow graph (SFG) we first dissect the amplifier network block diagram of Fig. 5.6 into three parts: (i) the signal generator section, (ii) two-port network section, and (iii) the load section.

(i) The schematic of the signal generator section together with its SFG is shown in Fig. 5.7a, for which:

$$b_s = \frac{V_S Z_0}{(Z_S + Z_0)} \cdot \frac{1}{\sqrt{Z_0}} = \frac{V_S \sqrt{Z_0}}{(Z_S + Z_0)} \quad \text{and} \quad \Gamma_S = \frac{Z_S - Z_0}{Z_S + Z_0}$$

(ii) The scheme for a two-port network is reproduced from Fig. 5.3 in Fig. 5.7b.

(iii) The scheme for the load section is shown in Fig. 5.7c together with its SFG, for which:

$$\Gamma_L = \frac{Z_L - Z_0}{Z_L + Z_0}$$

On combining the SFG of the three sections mentioned above, we have the SFG for the two-port amplifier with the associated signal generator and load as depicted in Fig. 5.7d.

(a) To obtain the input reflection coefficient Γ_{IN} (= b_1/a_1) or output reflection coefficient Γ_{OUT} (= b_2/a_2), we may use either the decomposition rule (method I) or Meson's rule (method II).

Method I: To find the input reflection coefficient $\Gamma_{IN} = b_1/a_1$, we redraw the necessary part of the signal flow graph of Fig. 5.7d in Fig. 5.7e. Using decomposition rule in steps (vide Fig. 5.7e (i–iv) we can derive the expression for Γ_{IN}, given by:

$$\Gamma_{IN} = S_{11} + \frac{\Gamma_L S_{12} S_{21}}{1 - S_{22}\Gamma_L} = \frac{S_{11} - \Delta\Gamma_L}{1 - S_{22}\Gamma_L}$$

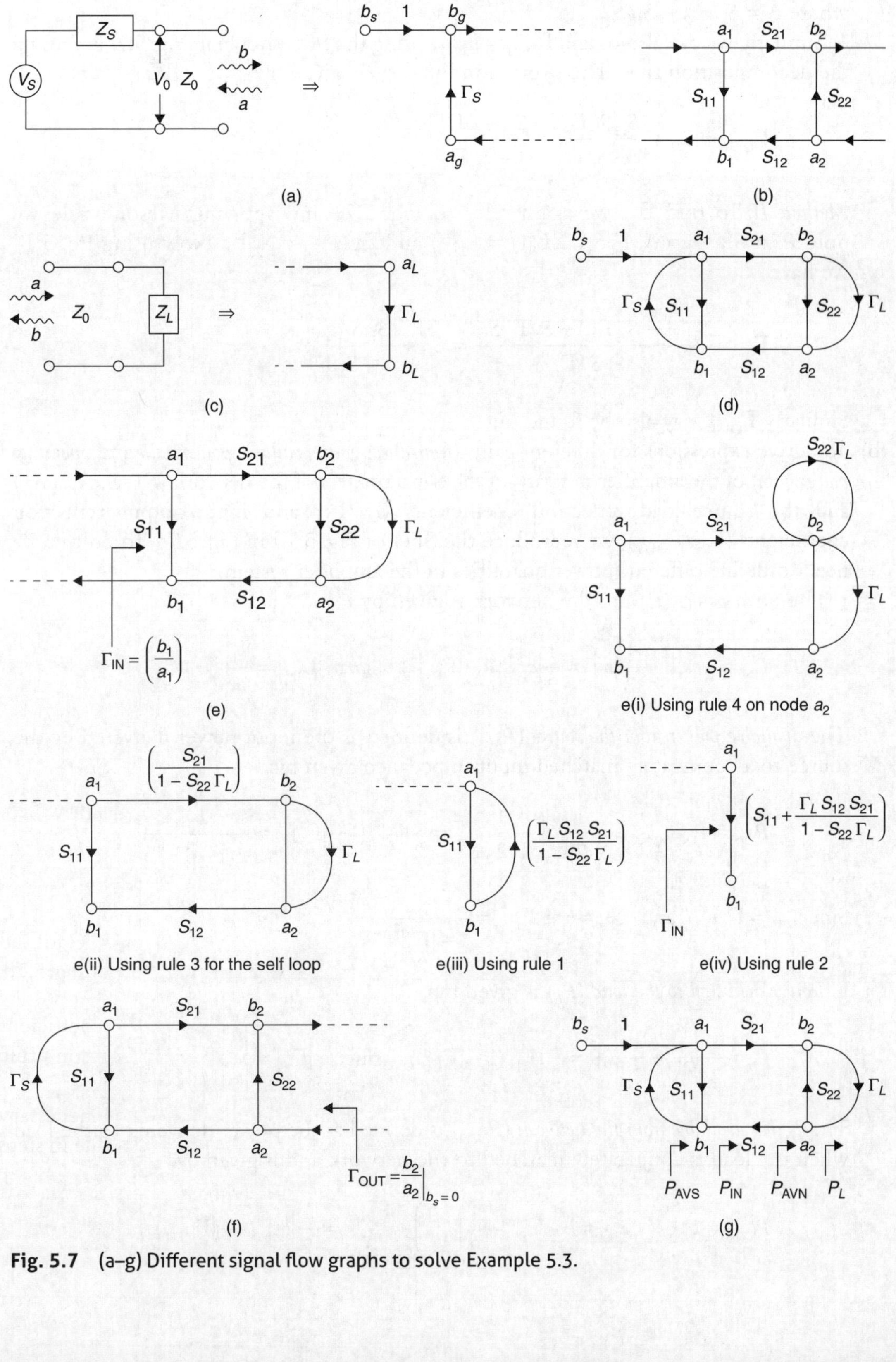

(a)

(b)

(c)

(d)

$\Gamma_{IN} = \left(\frac{b_1}{a_1}\right)$

(e)

e(i) Using rule 4 on node a_2

e(ii) Using rule 3 for the self loop

e(iii) Using rule 1

e(iv) Using rule 2

$\Gamma_{OUT} = \frac{b_2}{a_2}\Big|_{b_s=0}$

(f)

(g)

Fig. 5.7 (a–g) Different signal flow graphs to solve Example 5.3.

where $\Delta = S_{11}S_{22} - S_{12}S_{21}$.

Similarly we can also obtain $\Gamma_{OUT} = b_2/a_2$, using the SFG shown in Fig. 5.7f and using the decomposition rule. The expression for Γ_{OUT} is given by:

$$\Gamma_{OUT} = S_{22} + \frac{S_{12}S_{21}\Gamma_S}{1 - S_{11}\Gamma_S} = \frac{S_{22} - \Delta\Gamma_S}{1 - S_{11}\Gamma_S}$$

Method II: To find Γ_{IN} we again refer to Fig. 5.7e and applying Meson's rule we note $P_1 = S_{11}$, $P_2 = S_{21}\Gamma_L S_{12}$, $\Sigma L(1) = S_{22}\Gamma_L$ and $\Sigma L(1)^{(1)} = S_{22}\Gamma_L$. Now using Eq. 5.14 we have:

$$\Gamma_{IN} = \frac{S_{11}(1 - S_{22}\Gamma_L) + S_{21}\Gamma_L S_{12}}{1 - S_{22}\Gamma_L} = S_{11} + \frac{S_{12}S_{21}\Gamma_L}{1 - S_{22}\Gamma_L}$$

Similarly Γ_{OUT} may also be found out.

(b) To derive expressions for different gains (*transducer gain*, *available power gain* and *operating power gain*) of the amplifier in terms of the S-parameters of the network (S_{11}, S_{22}, S_{21}, S_{12}) and the source/load reflection coefficients (Γ_S/Γ_L) and input/output reflection coefficients (Γ_{IN}/Γ_{OUT}), we reproduce the SFG of Fig. 5.7d in Fig. 5.7g. To do this we need to define different power quantities in the amplifier system.

The *input power* (P_{IN}) to the network is given by:

$$P_{IN} = \frac{1}{2}|a_1|^2 - \frac{1}{2}|b_1|^2 = \frac{1}{2}|a_1|^2(1 - |\Gamma_{IN}|^2), \quad \text{since } |\Gamma_{IN}| = \frac{|b_1|}{|a_1|}$$

The *available power from the source* (P_{AVS}) is defined as the input power delivered by the source to a conjugately matched input impedance given by:

$$P_{AVS} = P_{IN}\big|_{\Gamma_{IN} = \Gamma_S^*} = \frac{1}{2}|b_g|^2 - \frac{1}{2}|a_g|^2 = \frac{1}{2}|b_g|^2(1 - |\Gamma_S|^2) = \frac{b_s}{2(1 - |\Gamma_S|^2)}$$

$$\text{Since, } a_g = \Gamma_S^* b_g \quad \text{and} \quad b_g = b_s + \Gamma_S a_g = \frac{b_s}{1 - |\Gamma_S|^2}$$

Power delivered to the load (P_L) is given by:

$$P_L = \frac{1}{2}|b_2|^2 - \frac{1}{2}|a_2|^2 = \frac{1}{2}|b_2|^2(1 - |\Gamma_L|^2), \quad \text{since } |\Gamma_L| = \frac{|b_2|}{|a_2|}$$

The *power available from the network* (P_{AVN}) is defined as the power delivered to the load when the load is conjugately matched to the network and is given by:

$$P_{AVN} = P_L\big|_{\Gamma_L = \Gamma_{OUT}^*} = \left(\frac{1}{2}|b_2|^2 - \frac{1}{2}|a_1|^2\right)_{\Gamma_L = \Gamma_{OUT}^*} = \frac{1}{2}|b_2|^2(1 - |\Gamma_{OUT}|^2)$$

Now the *transducer power gain* (G_T) is defined as the ratio of P_L to P_{AVS}, when:

$$G_T = \frac{|b_2|^2}{|b_s|^2}(1-|\Gamma_L|^2)(1-|\Gamma_S|^2)$$

The value of b_2/b_s may be determined from Mason's rule referring to Fig. 5.7g. It may be seen that there are three first order loops having values $S_{11}\Gamma_S$, $S_{22}\Gamma_L$ and $S_{21}\Gamma_L S_{12}\Gamma_S$ while there is only one second order loop having a value $S_{11}\Gamma_S S_{22}\Gamma_L$. Thus, using Meson's rule we can write:

$$\frac{b_2}{b_s} = \frac{S_{21}}{1 - S_{11}\Gamma_S - S_{22}\Gamma_L - S_{21}\Gamma_L S_{12}\Gamma_S + S_{11}\Gamma_S S_{22}\Gamma_L}$$

Substituting this value of b_2/b_s in the expression for G_T we get:

$$G_I = \frac{|S_{21}|^2(1-|\Gamma_S|^2)(1-|\Gamma_L|^2)}{|(1-S_{11}\Gamma_S)(1-S_{22}\Gamma_L) - S_{12}S_{21}\Gamma_S\Gamma_L|^2}$$

This may also be expressed as:

$$G_T = \frac{1-|\Gamma_S|^2}{|1-\Gamma_{IN}\Gamma_S|^2}|S_{21}|^2\frac{1-|\Gamma_L|^2}{|1-S_{22}\Gamma_L|^2}$$

Or, we might write it in yet another way:

$$G_T = \frac{1-|\Gamma_S|^2}{|1-S_{11}\Gamma_S|^2}|S_{21}|^2\frac{1-|\Gamma_L|^2}{|1-\Gamma_{OUT}\Gamma_L|^2}$$

Next, the *available power gain* (G_{AV}) is defined as the ratio of P_{AVN} to P_{AVS}, when:

$$G_{AV} = \frac{|b_2|^2}{|b_s|^2}(1-|\Gamma_S|^2)(1-|\Gamma_{OUT}|^2)$$

The expression for b_2/b_s with $\Gamma_L = \Gamma^{*}_{OUT}$, after some manipulation becomes:

$$\frac{b_2}{b_s} = \frac{S_{21}}{(1-S_{11}\Gamma_S)(1-|\Gamma_{OUT}|^2)}$$

When G_{AV} can be expressed as:

$$G_{AV} = \frac{1-|\Gamma_S|^2}{|1-S_{11}\Gamma_S|^2}|S_{21}|^2\frac{1}{1-|\Gamma_{OUT}|^2}$$

Finally, the *operating power gain* or simply *power gain* (G_P) is defined as the ratio of P_L to P_{IN}, when:

$$G_P = \frac{|b_2/b_s|^2}{|a_1/b_s|^2}\left(\frac{1-|\Gamma_L|^2}{1-|\Gamma_{IN}|^2}\right)$$

Using Meson's rule, we obtain:

$$\frac{a_1}{b_s} = \frac{1-S_{22}\Gamma_L}{1-S_{11}\Gamma_S-S_{22}\Gamma_L-S_{21}\Gamma_L S_{12}\Gamma_S+S_{11}\Gamma_S S_{22}\Gamma_L}$$

Thus, G_P can be written as:

$$G_P = \frac{1}{1-|\Gamma_{IN}|^2}|S_{21}|^2\frac{1-|\Gamma_L|^2}{|1-S_{22}\Gamma_L|^2}$$

Recapitulation

S-parameter formulation of microwave network

- Microwave network behaviour is characterised with a scattering matrix based on the linear relationship between the reflected and transmitted waves with incident wave amplitude and the microwave device/circuit is analysed in terms of the scattering parameters or *S-parameters*. To analyse the non-linear behaviour of a device/circuit under large-signal condition at microwave frequency, X-parameters are used which in the limit of small-signal approximation reduces to S-parameters.
- At microwave frequencies, current and voltage are not measurable quantities, instead the measurable quantity is microwave power, the wave variables *a* (incident wave variable) and *b* (reflected wave variable) are so defined that they represent average power in the wave. [$a_i = V_i^+/\sqrt{Z_{0i}}, b_i = V_i^-/\sqrt{Z_{0i}}$, where a_i and b_i represent the normalised incident and reflected voltages respectively at the ith port and Z_{0i} is the characteristic impedance of the ith port (which is a real number for loss-less line).]
- The physical meaning of four S-parameters of a microwave network is $S_{11} \rightarrow$ input reflection coefficient, $S_{22} \rightarrow$ output reflection coefficient, $S_{21} \rightarrow$ forward transmission coefficient and $S_{12} \rightarrow$ reverse transmission coefficient. In general, the scattering parameters are complex numbers having both magnitude and phase.

Properties of S-parameters

- (i) The S-matrix of a reflection-less junction is a *skew-symmetric matrix* (all the diagonal elements of the matrix are zero), (ii) for a reciprocal junction S-matrix is a *symmetrical matrix*, (iii) in case of loss-less microwave junction (which is always a practical situation) the S-matrix satisfies both the *unity property* and *zero property*; thus the S-matrix of a loss-less microwave junction is said to satisfy the *unitary property*, (iv) change of port position is done to make S-parameters real quantities in order to be physically realisable.

Signal flow graphs

- The signal flow graphs (SFG) are very useful for the analysis of microwave networks in terms of incident wave (independent) variable *a* and the reflected wave (dependent) variable *b*. This graphical representation implicitly indicates the energy or power flow in the microwave network in terms of S-parameters.
- Once a microwave network has been represented with signal flow graph, the evaluation of the ratio of any combination of wave amplitudes at the input–output of the network can be done in two ways. One can use either the *decomposition rules* or alternatively use the *Meson's rule* of control theory.
- *Decomposition rules: Rule I* (*Series or multiplication rule*): $S_{31} = S_{21}S_{32}$ (two series branches S_{21} and S_{32} are combined to give an effective transmission coefficient, S_{31}). *Rule II* (*Shunt or addition rule*): $S_{\text{eff}} = S_1 + S_2$ (two or more parallel branches may be replaced by a single branch with effective transmission coefficient equal to the sum of transmission coefficient of individual branches). *Rule III* (*Loop rule*): (*a*) *Self-loop*: $S_{31} = S_{21}S_{32}/(1 - S_{22})$, (*b*) *Feedback-loop*: $S_{31} = S_{21}/(1 - S_{23}S_{32})$. *Rule IV* (*Splitting rule*): Any node can be split into two separate nodes where each of the two nodes are connected only once to the incoming and outgoing nodes with the same coefficient.
- *Meson's rule:* It states that the transfer function, that is, the ratio of a dependent to independent variable is given by:

$$T = \frac{P_1\left\{1 - \sum L(1)^{(1)} + \sum L(2)^{(1)} - \cdots\right\} + P_2\left\{1 - \sum L(1)^{(1)} + \cdots\right\} + P_3\,(1 - \cdots)}{1 - \sum L(1) + \sum L(2) - \sum L(3) + \cdots}$$

where, P_1, P_2, P_3, ... are the values of the various paths connecting the dependent and independent variables whose transfer function is to be determined. Other symbols have their meaning as discussed in the text.

REVIEW QUESTIONS

5.1 What are S-parameters? Why can't *Z*, *Y* or *h*-parameters used for low-frequency device–circuit analysis be used to analyse microwave devices/circuits? Instead, why are S-parameters used?

5.2 What are X-parameters? When you need to use X-parameters instead of S-parameters in microwave device/circuit analysis?

5.3 The square of any wave variable used in defining S-parameters is proportional to the average power in the wave—justify.

5.4 Discuss the physical significance of the four S-parameters used to characterise the microwave circuit performance.

5.5 The S-matrix of a reflection-less microwave junction is a skew-symmetric matrix—justify.

5.6 The S-matrix of a reciprocal microwave junction is a symmetric matrix—justify.

5.7 Show that the S-matrix of a loss-less microwave junction is a unitary matrix.

5.8 Change of port position is sometimes required to make S-parameters real—explain.

5.9 What is the significance of a signal flow graph for microwave networks? Name the rules used to determine the ratio of wave variables at input–output of microwave networks.

5.10 State and explain the decomposition rule and Meson rule used to characterise microwave networks represented by signal flow graphs.

5.11 State the formulae for transducer gain, available power gain and operating power gain of a microwave amplifier in terms of S-parameters and different reflection coefficients.

6 Microwave Passive Circuit Components

Passivate all misadventure of vice with the divine spirit of virtue

6.1 Introduction

A microwave communication or radar system consists of both active devices and passive circuit components. Initial developments of microwave passive components primarily at the UHF band were based on the coaxial transmission line. However, the development of microwave passive circuit components using waveguides goes back to the extensive research work of MIT Radiation Laboratory, US in the 1940s [1]. They developed *E* and *H*-plane waveguide *T*-junctions, various types of couplers like Bethe hole and Schwinger coupler, multi-hole directional coupler, the waveguide magic-T and so on. From the 1960s onwards, with the advent of planar transmission lines such as strip-line and microstrip line, many of those couplers were re-invented leading to the development of the Wilkinson power divider/combiner, branch line qudrature hybrid and coupled line directional coupler. In recent times, the development of metamaterial-inspired microwave passive components with planar circuits has been of great research interest. This has the potential of miniaturisation and efficient operation of couplers, phase shifters and so on. Note that all these passive circuit components are constructed from suitable lengths of transmission lines/waveguides based on distributed-circuit parameters and are thus frequency dependent. However, with advanced photolithographic and thin-film technology, lumped-circuit elements are also now possible at microwave frequencies with dimensions being very small compared to the operating wavelength and are thus practically frequency-independent allowing the design of wide-band circuits. Lumped-circuit elements help in miniaturisation and are useful in hybrid microwave integrated circuits (HMICs) and also in monolithic microwave integrated circuits (MMICs). In this chapter, we will discuss in detail, the waveguide and planar transmission line based passive components. A brief discussion on lumped-circuit elements at microwave frequencies and distributed-circuit elements using microstrip transmission line are also included.

6.2 Waveguide and Planar Transmission Line Based Components

Most of the microwave passive components used in modern microwave system design are either waveguide-based or planar transmission line based (Table 6.1). Let us now discuss the

Table 6.1 Waveguide and planar transmission line based components

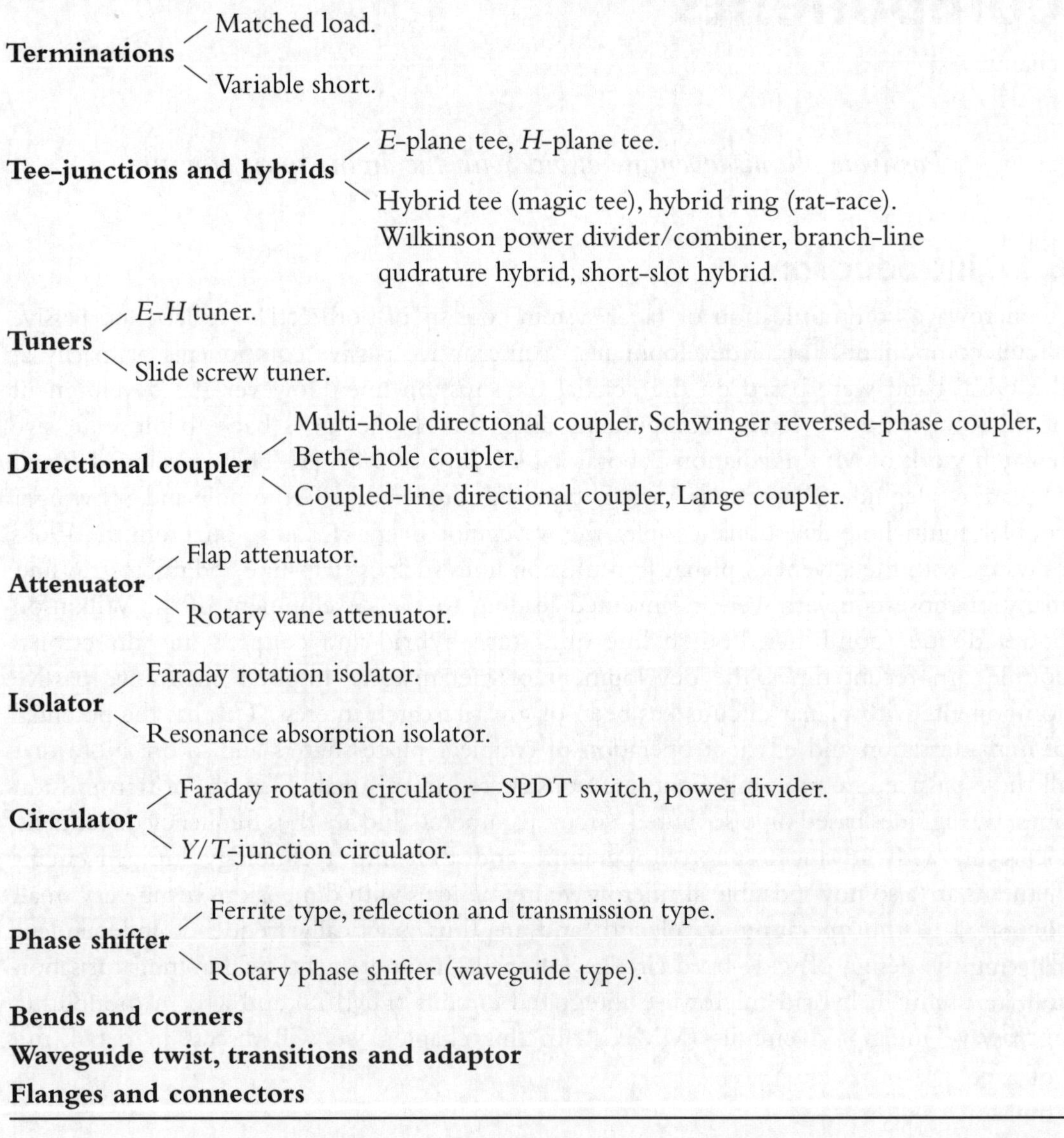

Component	Types
Terminations	Matched load.
	Variable short.
Tee-junctions and hybrids	*E*-plane tee, *H*-plane tee.
	Hybrid tee (magic tee), hybrid ring (rat-race). Wilkinson power divider/combiner, branch-line qudrature hybrid, short-slot hybrid.
Tuners	*E-H* tuner.
	Slide screw tuner.
Directional coupler	Multi-hole directional coupler, Schwinger reversed-phase coupler, Bethe-hole coupler.
	Coupled-line directional coupler, Lange coupler.
Attenuator	Flap attenuator.
	Rotary vane attenuator.
Isolator	Faraday rotation isolator.
	Resonance absorption isolator.
Circulator	Faraday rotation circulator—SPDT switch, power divider.
	Y/*T*-junction circulator.
Phase shifter	Ferrite type, reflection and transmission type.
	Rotary phase shifter (waveguide type).
Bends and corners	
Waveguide twist, transitions and adaptor	
Flanges and connectors	

physical principle of their operation and applications including the design of some of the components. The S-parameter characterisation of some important components will also be discussed at the end of this chapter.

6.2.1 Terminations

Terminations have only an input port and are thus in effect, one port. The *matched load* provides a termination that absorbs all the incident power (the power is being absorbed or dissipated in a lossy material) without any reflection (because of this it is also called a *dissipative termination* or *reflectionless termination*) thereby terminating the line or guide in its characteristic impedance. The *variable short* is a termination that produces an adjustable reactive load and being a short circuited load (with arrangement for variation of the position of the short circuit) it reflects all the incident power. The phase of the reflected power is varied by varying the position of the short circuit, and this is equivalent to changing the reactance of the termination. The terminations are extensively used when measuring impedance or scattering parameters, in a directional coupler as a matched load, in mechanical tuning of oscillators and in an *E–H* tuner as variable short.

The *matched load* consists of a wedge or pyramid shaped structure (made of highly dissipative ceramic material such as bakelite or graphite coated with carbon film) securely fixed inside a rectangular waveguide (Fig. 6.1). The length of the wedge or pyramid structure is approximately $2\lambda_g$ and the pointed end is oriented towards the incoming signal to keep the reflection of the incoming signal to a minimum (VSWR $\approx$ 1.02).

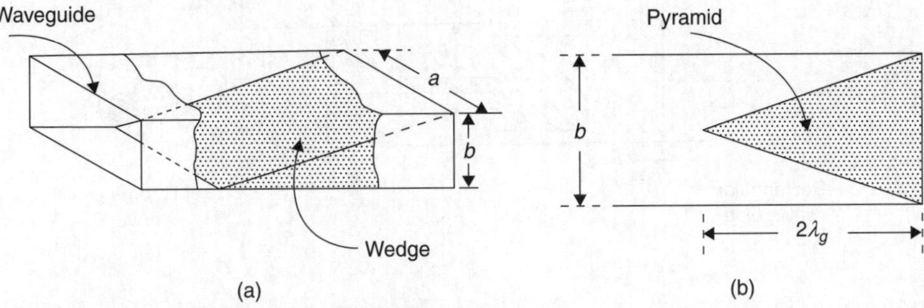

Fig. 6.1 Matched load (a) With wedge-shaped lossy dielctric in waveguide (b) With pyramid-shaped lossy dielectric in waveguide (side view).

The simplest form of a *variable short* may be a sliding block of copper or any other good conductor which has a snug fit in a rectangular waveguide. But this simple short circuit is not a good microwave short as it might cause the signal to leak past the block making the reflection coefficient less than 1. A practical form of a microwave short circuit is the choke type variable short (Fig. 6.2) that uses the impedance transformation properties of $\lambda/4$ transformer (Fig. 6.2a). Impedance Z_s which is very small (say 1 to 2 ohms, that is, that of a copper/brass block) when viewed through a two section quarter-wave impedance transformer the impedance Z_{seen} works out to be:

$$Z_{seen} = \left(\frac{Z_1}{Z_2}\right)^2 \cdot Z_s \tag{6.1a}$$

If we choose $Z_2 > Z_1$, then Z_{seen} will be a better short circuit than Z_s by a factor of $(Z_1/Z_2)^2$.

Practical realisation of this type of variable short can be done as follows. With reference to Fig. 6.2b, the sections A and B may be considered to represent coaxial structures for which characteristic impedance is known to be proportional to (b/a) where a and b are the dimensions of the inner and outer conductors of the coaxial structure. Then, for section A, characteristic impedance $Z_1 \; \alpha \; b/(b-2b_1)$. Thus, for a fixed b, to obtain a small Z_1, we need to make $(b-2b_1)$ large, which effectively means we have to decrease b_1. However, it is to be noted that b_1 can be made as small as possible being consistent with the requirement that the front section of the sliding structure must not touch the upper and lower waveguide walls. Next, for section B, characteristic impedance $Z_2 \; \alpha \; b/(b-2b_2)$. Thus, for a fixed b, to obtain a large Z_2, we need to make $(b-2b_2)$ small, which effectively means we have to increase b_2. Note that

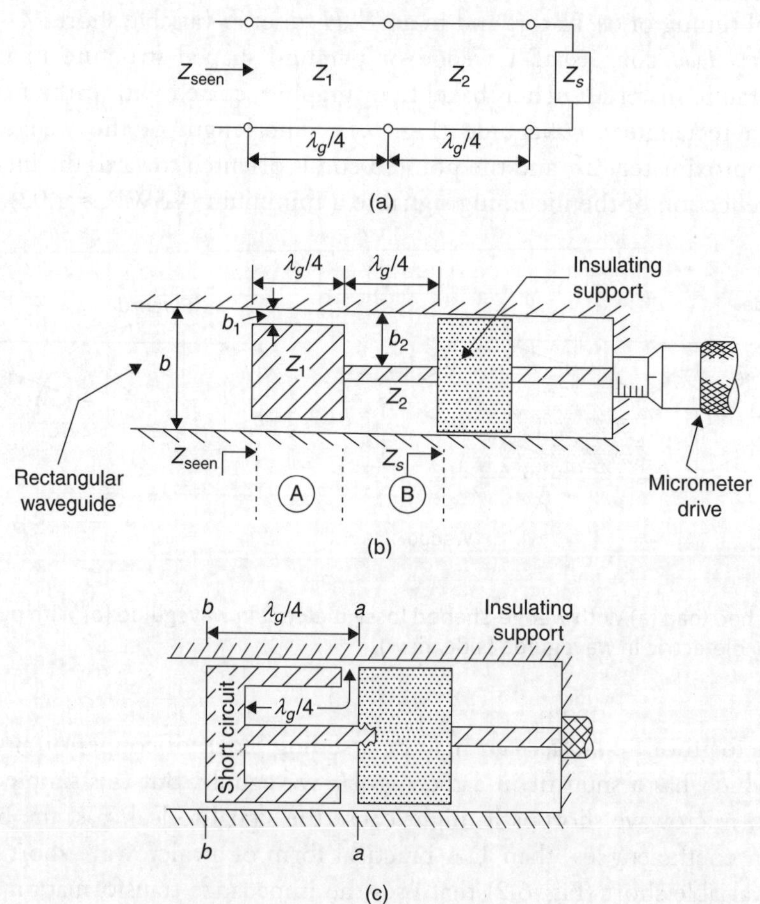

Fig. 6.2 Choke-type variable short (a) Two section $\lambda/4$ impedance transformation (b) Practical variable short (c) Alternative design of variable short.

b_2 can be made as large as possible consistent with maintaining the mechanical strength of the plunger. Thus, Eq. 6.1a may be rewritten in terms of the physical dimensions of the *low–high* (as Z_1 represents low impedance and Z_2 represents high impedance) short structure as:

$$Z_{\text{seen}} = \left(\frac{b_1}{b_2}\right)^2 Z_s \tag{6.1b}$$

By good mechanical design, b_2 can be made about 10 times bigger than b_1 and hence an improvement in performance by a factor of 10^{-2} for the short may be realised with choke type variable short over the non-choke type, that is, the simple copper block type variable short.

An alternative design of a variable short is shown in Fig. 6.2c. This consists of two section folded quarter-wave transformers. The inner line section ($\approx \lambda_g/4$) transforms the short circuit impedance to an ideal open circuit at the plane *aa*, while the outer line section ($\approx \lambda_g/4$) transforms the open circuit impedance at *aa* into an ideal short circuit impedance at the plane *bb* (refer to Fig. 3.6). Thus, the incoming signal to the plane *bb* finds a perfect short circuit.

6.2.2 Tee-Junctions and Hybrids

Tee junctions are used in waveguide systems either to combine two RF signals of opposite phase (use *E-plane tee*) or of same phase (use *H-plane tee*) or to split a signal into two signals (normally having 3 dB power division) having opposite phase or same phase. The *E–H tee* or *hybrid tee* (also known as *magic tee*) is used in the design of a balanced mixer found in the front end of microwave receivers. The *hybrid ring* or *rat-race* is used for the same purpose as the magic tee but is a convenient structure for coaxial line and planar circuits.

An *E-plane tee* (Fig. 6.3a) is a waveguide junction in which the axis of its side arm is parallel to the plane of the *E* field in the main guide (in the collinear arms). A signal fed at the *E*-arm is coupled to the collinear arms which are equal in magnitude (and equal to half that is fed into the *E*-arm; as 3 dB power division takes place at the junction, J_n) but opposite in phase. In other words, the signal fed in the *E*-arm generates odd-symmetric fields in the junction J_n (Fig. 6.3b). Similarly, two signals fed at the two collinear arms will produce a signal at the *E*-arm which is the difference of the signals at the two ports of the collinear arms. Electrically, the *E*-arm is in series with the collinear arms and hence this type of tee junction is also known as a *series tee*.

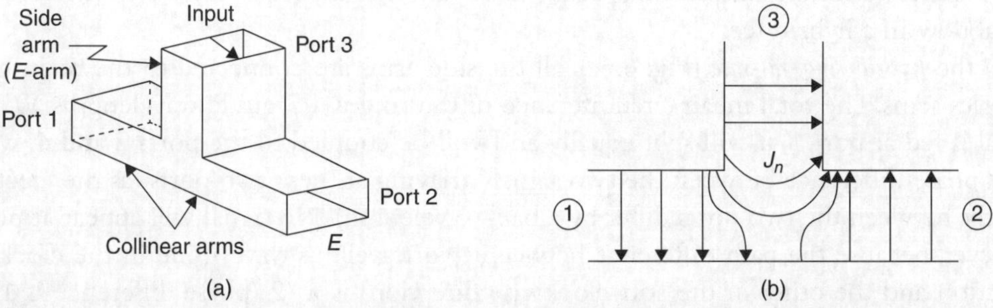

Fig. 6.3 *E*-plane tee or series tee (a) Structure (b) *E*-field with odd-symmetry at the junction J_n.

In a *H-plane tee* (Fig. 6.4a) the side arm is shunting the *E*-field (the *H*-arm is in shunt with the collinear arms and thus is also known as *shunt tee*) or it is parallel to the *H*-field plane in the main guide forming the collinear arms. A signal fed at the *H*-arm will produce signals of equal magnitude (with 3-dB power division at the junction, J_n) and of same phase. This means that the signal fed in the *H*-arm generates an even-symmetric field in the junction J_n (Fig. 6.4b). Similarly, two signals fed at the two collinear arms will be in phase and additive at the *H*-arm.

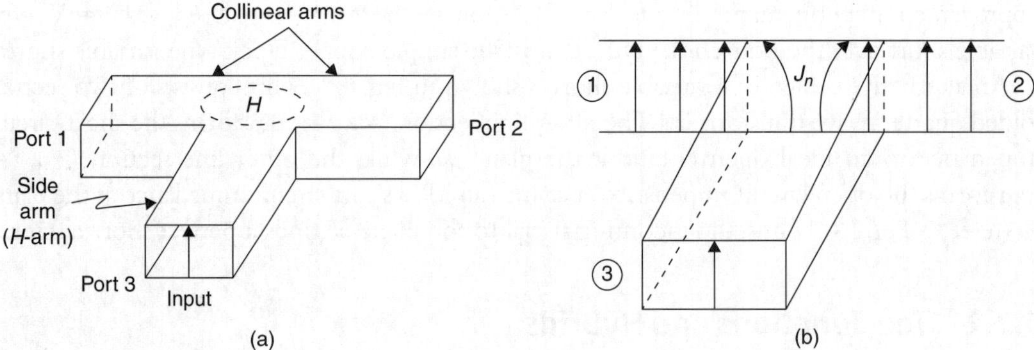

Fig. 6.4 *H*-plane tee or shunt tee (a) Structure (b) *E*-field with even-symmetry at the junction J_n.

The *hybrid tee/magic tee* (Fig. 6.5a) is a four-port microwave junction. A signal fed at the *H*-arm (shunt arm) will experience a 3-dB power division at the junction (with even symmetric field, Fig. 6.5b) and will produce signals of equal magnitude and of same phase at the collinear arms, but no signal will be coupled at the *E*-arm (if the collinear arms are identically terminated). Similarly a signal fed at the *E*-arm (series arm) will couple signals of equal magnitude (with 3-dB power division at the junction) but of opposite phase (180° phase difference) at the collinear arms (because of the odd-symmetric field generated at the junction, Fig. 6.5c), but with no signal coupling to the *H*-arm (if the collinear arms are identically terminated). Thus, with identical impedance termination of the collinear arms, a signal fed at the *E*-arm will not couple any signal to the *H*-arm (that is, they are perfectly cross-coupled) and vice-versa (this is the *magic*; unlike flow of water expected in such a configuration of metallic pipe junction). However, a signal fed at the *E*-arm will couple that extent of signal to the *H*-arm and vice versa, by which the terminating impedances in the two collinear arms are mismatched; again this is another aspect of the *magic* of microwave signal flow in a hybrid tee.

In the *hybrid ring/rat-race* (Fig. 6.6a), all the side arms are connected to the main guide as series arms. The total mean circumference of the ring is 1.5 guide wavelengths. When a signal is fed at port 1, it will split equally and will be coupled to the ports 2 and 4, with a 180° phase difference between the two signals arriving at these two ports (as the electrical lengths between the two ports differ by a half-wavelength). No signal will appear at port 3, however, because the path difference between two travelling waves (one in the clockwise direction and the other in the anti-clockwise direction) is $\lambda_g/2$ (phase difference is 180°), thus causing cancellation of signals at this port. Similarly, if a signal is fed at port 3, it will

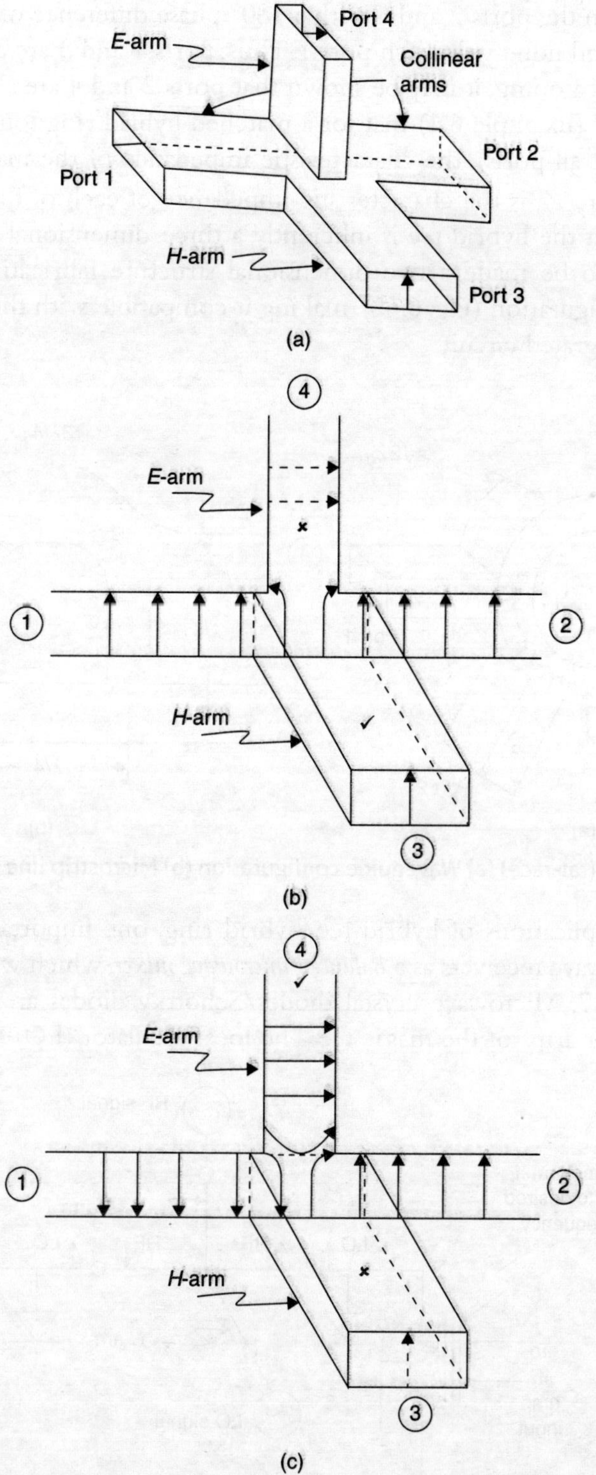

Fig. 6.5 Hybrid tee or magic tee (a) Structure (b) *E*-field pattern with input in *H*-arm (c) *E*-field pattern with input in *E*-arm.

split equally between the ports 2 and 4 with a 180° phase difference of the signals arriving at these two ports and none will reach port 1. Thus, ports 1 and 3 are decoupled (isolated) ports. With similar reasoning, it may be shown that ports 2 and 4 are also decoupled ports. It can be established (Example 6.1) that for a matched hybrid ring (one having practically negligible VSWR at all ports), the characteristic impedance of the main guide should be equal to $\sqrt{2}Z_0$, where Z_0 is the characteristic impedance of each of the series (side) arms. It may be noted that the hybrid tee is inherently a three-dimensional structure, while the hybrid ring can also be made a two-dimensional structure fabricating the structure in microstrip line configuration (Fig. 6.6b), making it compatible with the planar technology of a microwave integrated circuit.

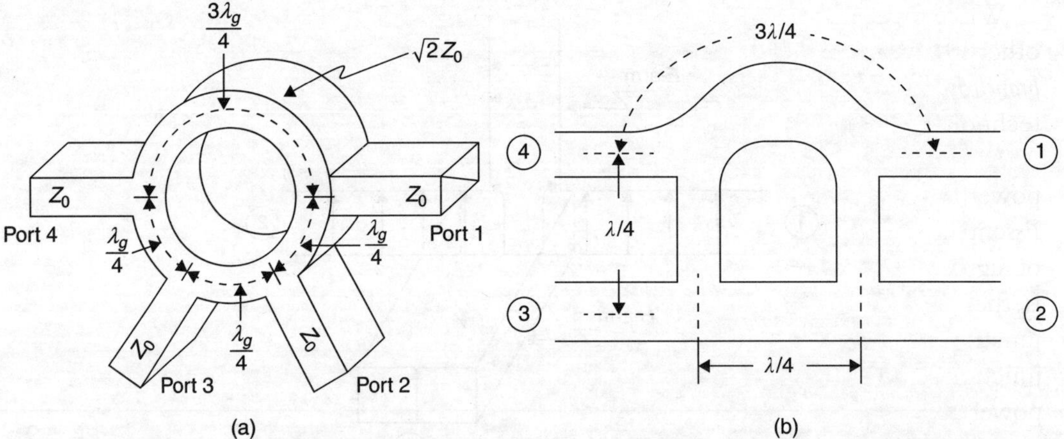

Fig. 6.6 Hybrid ring (rat-race) (a) Waveguide configuration (b) Microstrip line configuration.

There are many applications of hybrid tee/hybrid ring, one important use being in the front-end of microwave receivers as a *balanced microwave mixer*, which will be discussed with reference to Fig. 6.7. Microwave crystal diodes/Schottky diodes are located at the two ends of the collinear arms of the magic tee. The local oscillator (LO) signal is fed into the

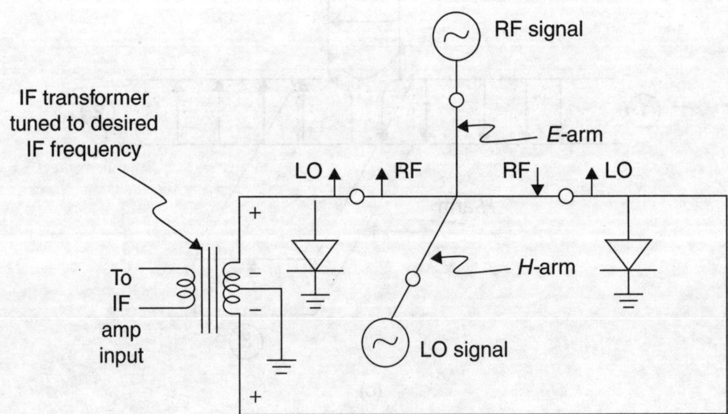

Fig. 6.7 Balanced microwave mixer using hybrid tee.

H-plane arm and will arrive at the diodes in phase. The input microwave (RF) signal is fed into the *E*-plane arm and will arrive at the two diodes in anti-phase. The IF signals (the difference frequency signal produced by the mixing of the LO and RF signals) generated in the two diodes will be 180° out of phase. However, the LO noise will be in phase at the diodes. By feeding the diode outputs into a balanced IF input (push–pull arrangement), the local oscillator noise is seen to be cancelled out but the IF signals add-up in phase giving full strength of IF signal to the IF amplifier input. The use of such a balanced microwave mixer using magic tee and push–pull arrangement at the IF port helps to separate the RF and LO ports so that LO noise is not unduly radiated through the antenna at the RF port (as *E* and *H* arm are cross-coupled with identical termination of the collinear arms) and also noise of the LO is not allowed to reach the IF port.

While on the discussion about tee junctions and hybrids, it is worth learning about two other important microwave passive components—the *Wilkinson power divider/combiner* and *branch line quadrature hybrid*, normally designed with strip-line/microstrip line using planar technology.

A *Wilkinson power divider/combiner* is a three-port network that divides the input power at port 1 equally between the two output ports 2 and 3 (which are also in phase). Essentially it acts as a 3 dB power divider or coupler. Reversing the direction of flow of signal, that is, feeding inputs to ports 2 and 3 will result in a signal at port 1 which is the phasor sum of the input signals at ports 2 and 3; thus making the Wilkinson junction act also as a power combiner (which is in fact true for any 3 dB power divider junction, such as *E*-, *H*-plane tee and so on). A schematic diagram of a 3 dB Wilkinson power divider is shown in Fig. 6.8. It may be noted that the characteristic impedance of each of the $\lambda/4$ line should be $Z_{0(\lambda/4)} = \sqrt{Z_0 \cdot 2Z_0} = \sqrt{2}Z_0$; as these two matched $\lambda/4$ lines appear in shunt so their combined input impedance (each $2Z_0$) must result in Z_0 of the input line. Further, it is assumed that the output ports 2 and 3 are perfectly matched, which seldom occurs. To take care of mismatch at any of the output ports, a resistor $R = 2Z_0$ is connected across the output arm; thus, when a mismatch occurs at one of the output ports, the reflected signal will get split equally between the $\lambda/4$ line and the resistor

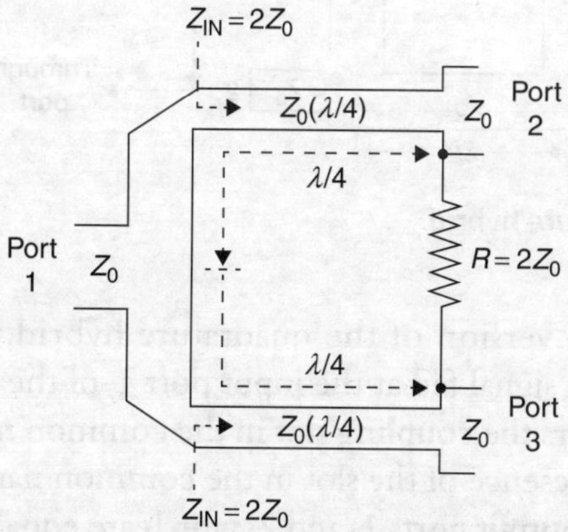

Fig. 6.8 Wilkinson power divider/combiner.

R [as the resistor $R = 2Z_0$ and input impedance $(2Z_0)$ of each $\lambda/4$ line are equal]. Under this condition, the two signals reaching the input port will be in phase while to the other output port will see phase cancellation. The Wilkinson power divider/combiners are easily constructed in strip-line (and not in waveguides) because the impedance of various lines (which are different) can be easily adjusted by controlling the strip width (the characteristic impedance of a strip-line is inversely proportional to the strip width).

The hybrid ring discussed above is a four-port network with 180° phase difference between the signals emerging from the two output ports (which is also true for hybrid tee when signal is fed in the E-arm). While the *branch-line quadrature hybrid* (Fig. 6.9) is another type of four-port microwave network in which the phase difference of the signals arriving at the two output ports is 90°. In this coupler, a signal fed at port 1 is evenly divided into ports 2 and 3, with the signal at the coupled port 3 is advanced in phase by 90° with respect to the signal at the through port 2 (as shown); while the signals arriving at port 4 are in phase opposition and thus cancel out making port 4 an isolated port. The coupler behaves identically when power is applied at any other port. For instance, if a signal is applied at port 2, it is split evenly between ports 1 and 4 with no power being coupled to port 3. For 3 dB power coupling between the two output ports of the coupler with the phase relationship of the signals as mentioned above, the characteristic impedance of the different line sections should be as follows: $Z_{01} = \sqrt{Z_0 \times \dfrac{Z_0}{2}} = Z_0/\sqrt{2}$ (as it has to transform the Z_0 impedance of the line at input port 1 to the parallel combination of two output lines of impedance Z_0 each) and $Z_{02} = Z_0$ as shown in Fig. 6.9. Branch line hybrid demanding different characteristic impedance of the lines is also easily realisable in a strip-line. These couplers are particularly useful in applications where strong coupling is desired.

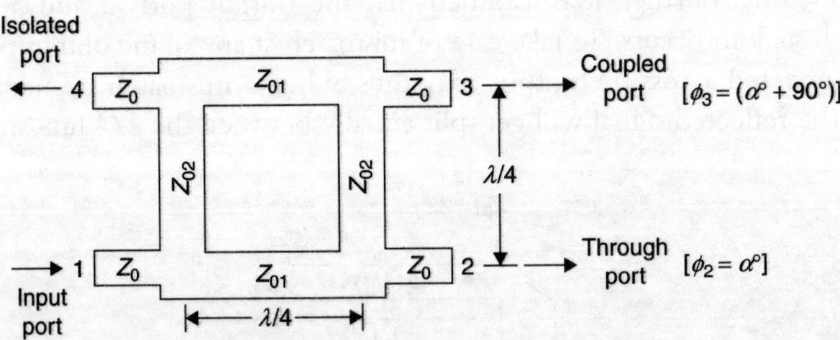

Fig. 6.9 Branch-line quadrature hybrid.

There is also a waveguide version of the quadrature hybrid, known as *short-slot hybrid* (Fig. 6.10). In this hybrid, a signal fed at the input port 1_i of the main guide proceeds along that guide until it encounters the coupling slot in the common narrow wall of the main and the auxiliary guides. The presence of the slot in the common narrow wall of the two guides produce two signals at the output ports 1_o and 2_o which are equal in magnitude (3 dB power division). The signal at the output port 2_o leads by 90° corresponding to the signal at input port 1_o, with all ports perfectly matched. The reason for this 3 dB quadrature coupling at the two output ports may be understood in terms of even- and odd-mode coupling [10]. In fact, every time the travelling wave propagating down either of the guide faces the slot, it advances in phase by 90°.

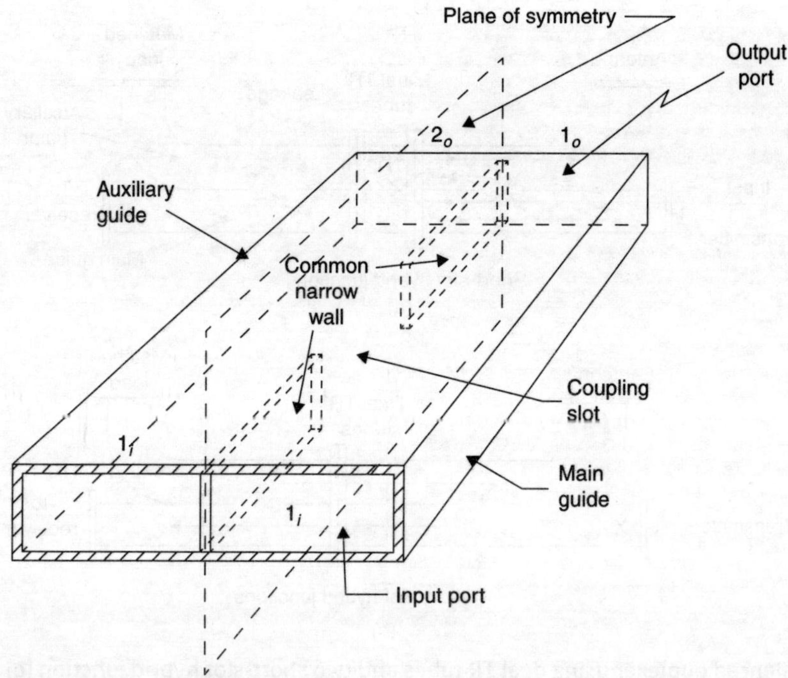

Fig. 6.10 Schematic of short-slot hybrid.

The short-slot type of quadrature hybrid is used to design a *balanced duplexer* that is required for a monostatic radar system where a single antenna is used for both transmission and reception. The scheme for a balanced duplexer using a short-slot hybrid is shown in Fig. 6.11. The microwave signal from the transmitter travelling down the main guide splits at the slot of the hybrid producing two signals; one at the main guide and the other in the auxiliary guide, the latter signal being advanced in phase by 90°. The transmit–receive (TR) tube (which is a gas discharge device designed to breakdown and ionise quickly with an incident RF power of high magnitude and de-ionise quickly after the power is removed; while with a weak signal input, it allows the signal to pass through) in both guides is fired by this signal which acts as a short circuit across the guide and reflects the signals without causing any relative phase shift between them. However, the reflected signal in the auxiliary guide from TR2 while travelling back to the input port gets advanced in phase by another 90° on facing the slot and thus the two signals, one from TR1 (with no phase change) and another from TR2 (with 180° phase opposition) arrives at the transmitter terminal of input port and thus no signal is fed back to the transmitter terminal. On the other hand, the two signals reflected from TR1 and TR2 are in phase and get reinforced at the antenna terminal of the input port in the transmit mode. On reception through an antenna, the signal level is insufficient to fire the TR tubes, and the signals in the two guides pass through the TR tubes unhindered and recombine in phase at the receiver terminal, with little loss of energy. In the transmission mode, if there is any leakage signal at the terminal 2_o of the auxiliary guide, then it gets absorbed in the matched load.

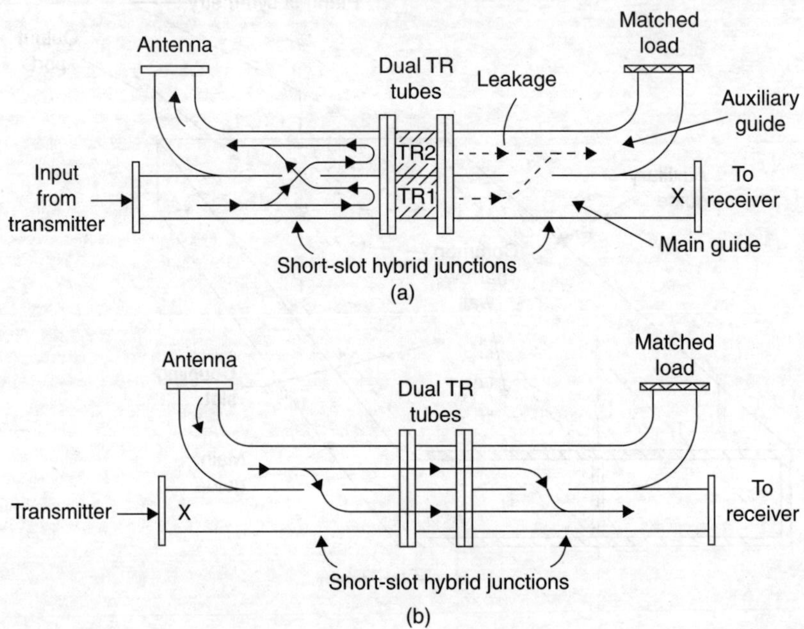

Fig. 6.11 Balanced duplexer using dual TR tubes and two short-slot hybrid junction (a) Signal paths in transmit mode (b) Signal paths in receive mode.

EXAMPLE 6.1

The unfolded hybrid ring circuit is shown in Fig. 6.12. Remembering that the side arms are in series with the main guide, show that the impedance of the main line forming the ring should be made equal to $\sqrt{2}Z_0$, where Z_0 is the characteristic impedance of each of the side arms.

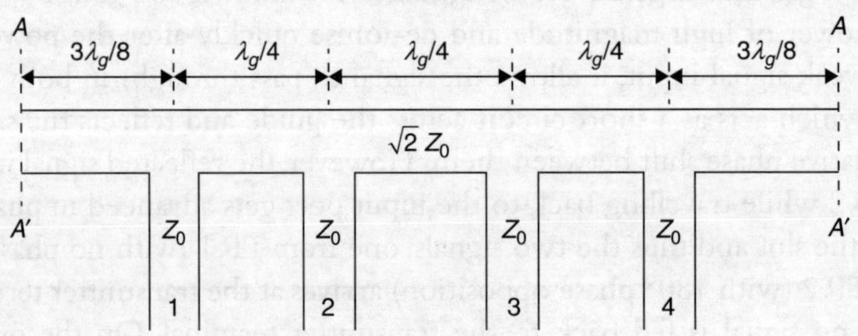

Fig. 6.12 Unfolded hybrid-ring circuit (refer to Fig. 6.6).

Solution

Relative to port 1, port 2 and port 4 are spaced at odd multiples of $\lambda/4$. Port 3 is spaced at an even multiple of $\lambda/4$ from port 1. At the design frequency of the ring, a $\lambda/2$ transformer is effectively interposed between port 1 and port 3. This means that the load presented to port 3 will appear at port 1 (as the $\lambda/2$ transformer is a one to one transformer). Now, since the side arms are in series, the total impedance seen by the ring at port 1 will be $Z_0 + Z_0 = 2Z_0$. In order to maintain a match in the system, the impedance inverting action of $\lambda/4$ transformer is noted between port 1 to port 2 and 4. For this, it is evident that the impedance of the main line forming the ring should be made equal to $Z_{ring} = \sqrt{2Z_0 \times Z_0} = \sqrt{2}Z_0$.

6.2.3 Tuners

Tuners are used to tune out impedance mismatches existing in a waveguide system.

The *E–H tuner* consists of a series arm (*H*-arm) and a shunt arm (*E*-arm) that are mounted upon a main waveguide (Fig. 6.13a). Each arm is terminated with a variable (adjustable) short circuit. This basically resembles a series-shunt stub impedance matching scheme of the transmission line (Chapter 7) in the case of a waveguide system. VSWR values up to 20 can be reduced to 1.02 with proper adjustments of movable shorts in the *E* and *H* arms.

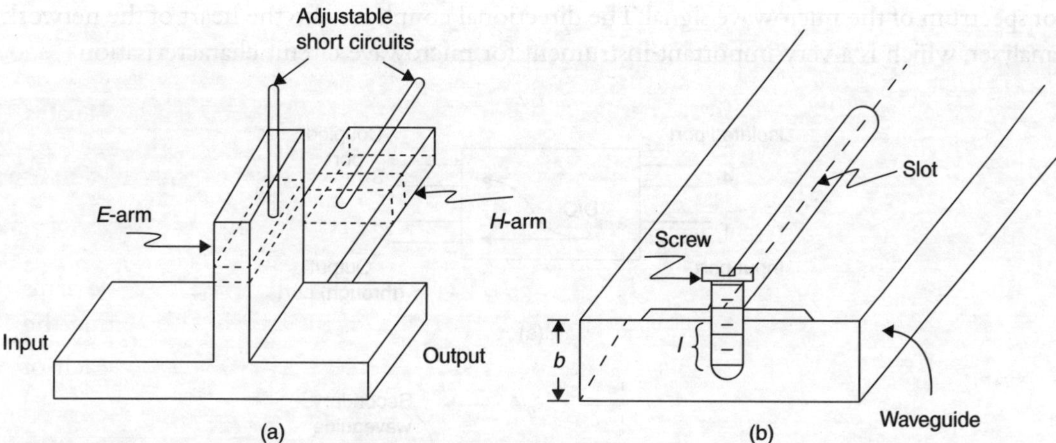

Fig. 6.13 Tuners (a) *E–H* tuner (b) Slide screw tuner.

The *slide screw tuner* consists of a slotted section of a rectangular waveguide in which a metallic screw can be inserted and whose depth of penetration (l) and longitudinal position can be adjusted (Fig. 6.13b). The slot on the broad wall of the waveguide is cut along the central line of the broad wall (where the strength of the *E*-field is maximum) and which represents a non-radiating slot (as it does not cut across the current lines, Fig. 4.11); and the slot length is approximately a guide wavelength. The variation of the depth of penetration of the screw and the longitudinal position of the screw inside the waveguide introduce extra shunt susceptance at the plane of the screw and thereby tunes out undesired impedance mismatch. By controlling the depth of penetration of the screw ($<\lambda_g/4$), a shunt capacitive susceptance can be introduced, Section 7.4 (thus tuning out inductive mismatches), while the longitudinal movement of the penetrated screw by half-wavelength (corresponding to the initial position obtained for realising capacitive susceptance) will change the corresponding shunt capacitive susceptance being introduced to an equivalent inductive susceptance (which may be used to tune out capacitive mismatches). Only by changing the depth of penetration of the screw inside the waveguide, inductive susceptance is not possible to realise (as the depth of penetration of more than $\lambda_g/4$ is required to do so, Section 7.4). But the narrow dimension of a waveguide is $\approx \lambda_g/4$ from waveguide design consideration, hence the depth of penetration of screw inside the waveguide cannot be made more than $\lambda_g/4$; thus the need for longitudinal movement of the screw in a slide screw tuner.

6.2.4 Directional Coupler

It is a four-port microwave passive component (having an input port, output/through port, coupled port and isolated port) used for sampling out a known amount of power along the coupled port while the rest of the power goes along the output (through) port and none of the signal is expected to go to the isolated port (Fig. 6.14a, shows the schematic of a waveguide directional coupler). The sampled signal may be used for monitoring the frequency or spectrum of the microwave signal. The directional coupler is also the heart of the network analyser, which is a very important instrument for microwave circuit characterisation.

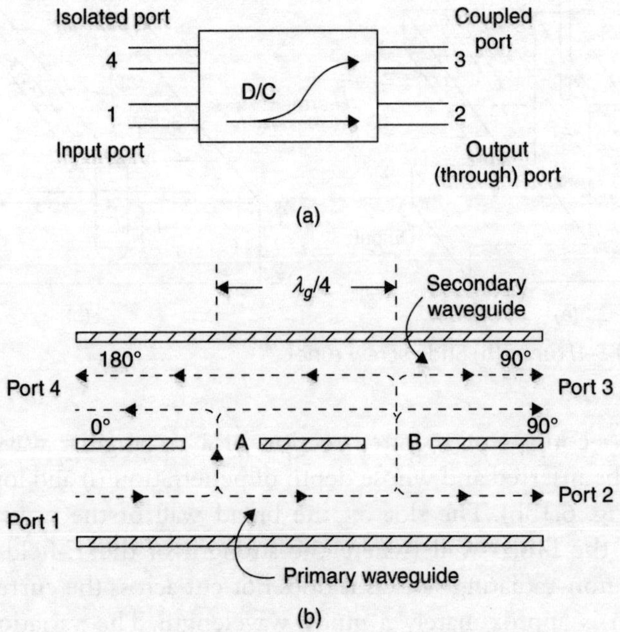

Fig. 6.14 Directional coupler (a) Schematic (b) Two-hole directional coupler.

The most common type of directional coupler is a *multi-hole directional coupler*. This consists of two waveguides (a primary waveguide and a secondary waveguide) sharing a common broad wall. There are a number of holes in the common wall to provide coupling from the primary waveguide to the secondary waveguide. The principle of operation of this type of coupler may be understood considering two holes (Fig. 6.14b). Let a signal enter at port 1 in the primary waveguide; this will be coupled to the secondary waveguide through the holes A and B in the common broad wall of the two waveguides (the holes are located quarter wavelength apart in the longitudinal direction and are offset from the central line on the broad wall of the guides so that they perform as coupling holes acting as radiating slots). The field coupled from the primary waveguide via the holes in the common broad wall excites waves in the secondary waveguide. Each of the holes excites two waves; one travelling to the left and the other travelling towards the right. Because of the quarter wavelength spacing between the two holes, the two waves travelling towards the left are out of phase (that is, differs in phase by 180°) and therefore cancel each other. On the other hand, the two waves travelling towards the right are in phase (each 90° phase advanced) and thus reinforce each other. Thus, an output appears at port 3 and no power is ideally coupled to port 4. The amount of power coupled to port 3 depends on the size and location of the holes in the common broad wall. It may be noted that the main part of the input signal entering port 1 will go out of port 2 in the primary waveguide.

The two ports 1 and 2 in the primary waveguide are respectively the input and output (through) ports while the two other ports in the secondary waveguide are the coupled port (port 3) and the isolated port (port 4). The isolated port is always terminated with a matched load.

The characteristics of a directional coupler are expressed in terms of the *coupling factor* (*C*) and *directivity* (*D*). The coupling factor is a measure of the fraction of power (P_{o3}) coupled out to a specific port, the coupled port in the secondary waveguide, for an incident wave power P_{i1} at the input port of the primary waveguide. *C* is defined as:

$$C = 10\log_{10}\left(\frac{P_{i1}}{P_{o3}}\right) = -10\log_{10}\left(\frac{P_{o3}}{P_{i1}}\right) = -20\log_{10}|S_{31}| \tag{6.2a}$$

For practical directional couplers, the typical value for coupling factor may be 10, 20 or 30 dB.

The directivity is a measure of how well the forward travelling wave (the incident wave) in the primary waveguide couples only to a specific port (to the coupled port and not to the isolated port) of the secondary waveguide under a matched condition. *D* is defined as:

$$D = 10\log_{10}\left(\frac{P_{o3}}{P_{o4}}\right) = 20\log_{10}\left|\frac{S_{31}}{S_{41}}\right| \tag{6.2b}$$

where P_{o3} and P_{o4} are the output powers at port 3 (the coupled port) and port 4 (isolated port) respectively, for an input power P_{i1} at input port 1. Ideally P_{o4} should be zero and thus *D* should be infinite; but for practical directional couplers *D* is of the order of 20 to 30 dB.

The directivity may also be interpreted as the difference between the *isolation* $10\log_{10}(P_{i1}/P_{04})$ and the coupling factor:

$$D = 10\log_{10}\left(\frac{P_{i1}}{P_{o4}}\right) - 10\log_{10}\left(\frac{P_{i1}}{P_{o3}}\right) = 10\log_{10}\left(\frac{P_{o3}}{P_{o4}}\right) \tag{6.2c}$$

which is precisely that given by Eq. 6.2b.

The two-hole directional coupler discussed above works satisfactorily only at a single frequency as its operation depends on the physical separation (quarter wavelength) of the holes A and B. As in the case of the impedance transformer (which will be discussed in Chapter 7), the bandwidth of a directional coupler can also be increased by using multiple pairs of coupling holes and can realise maximally flat/equal ripple design [9]. In addition to the multi-hole directional coupler, there are other types of waveguide couplers such as the Schwinger reversed-phase coupler and the Bethe-hole coupler (Figs 6.15a and 6.15b).

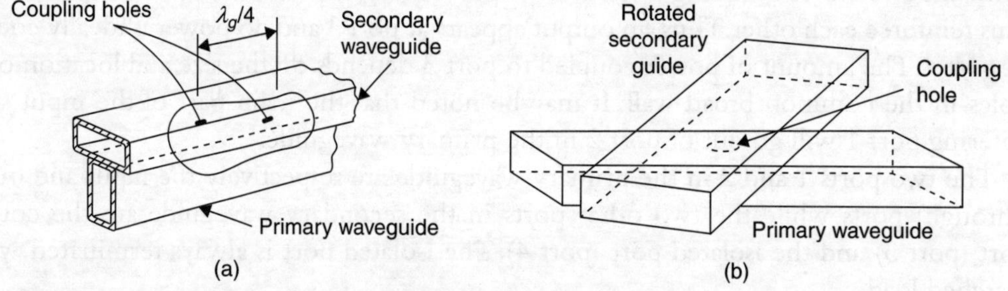

Fig. 6.15 (a) Schwinger reversed-phase coupler (b) Bethe-hole coupler.

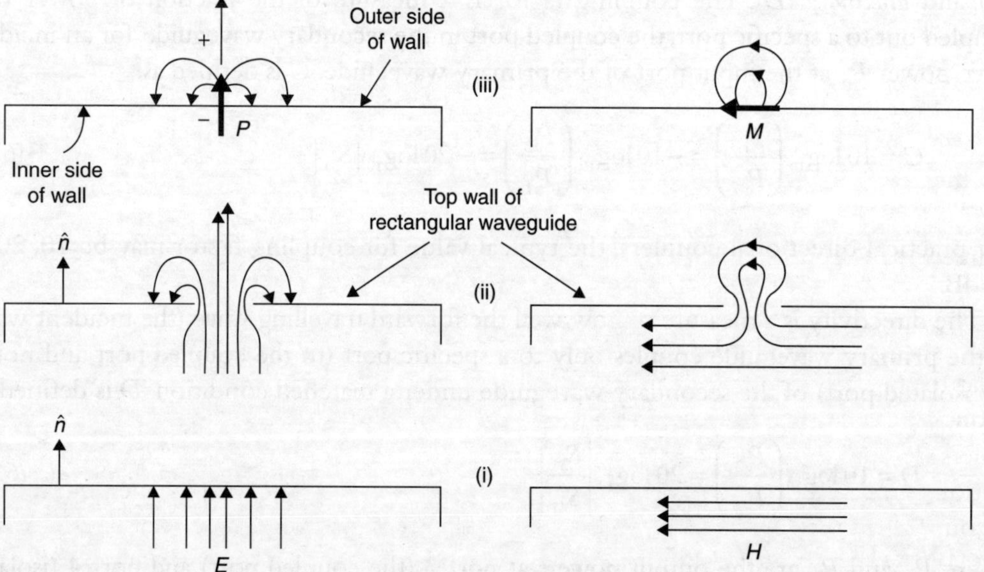

Fig. 6.16 Aperture on the broad wall of a waveguide with equivalence of the aperture to the electric and magnetic dipoles.

The analysis of all types of directional couplers mentioned above is primarily based on the following physical concept. To a first approximation, a small hole (aperture) in a conducting wall (say the broad wall of a rectangular waveguide) couples the electric and magnetic field from the inner side of the wall to the outer side (Fig. 6.16) in such a way that it is equivalent to an electric dipole normal to the aperture and having a strength proportional to the normal component of the exciting electric field, plus a magnetic dipole in the plane of the hole and having a strength proportional to the exciting tangential magnetic field. The constants of proportionality depend on the hole (aperture) size, shape and offset of the hole from the central line on the broad wall of the guide.

EXAMPLE 6.2

A directional coupler has coupling factor of 10 dB and a directivity of 30 dB. If the power in the isolated port is 40 μW, find the power in the input port and also in the through port. What is the insertion loss of the coupler?

Solution

Given $C = 10\log_{10}\left(\dfrac{P_{i1}}{P_{o3}}\right) = 10 \text{ dB, thus } \left(\dfrac{P_{i1}}{P_{o3}}\right) = \text{antilog } (1) = 10$ (i)

$$D = 10\log_{10}\left(\frac{P_{o3}}{P_{o4}}\right) = 30 \text{ dB, thus } \left(\frac{P_{o3}}{P_{o4}}\right) = \text{antilog } (3) = 1000 \quad \text{(ii)}$$

Hence from (i) and (ii) we have $(P_{i1}/P_{o4}) = 10{,}000$.

When $P_{i1} = 10{,}000 \times 40 \times 10^{-6} = 0.4$ W; the power in the input port (1)

From (i), $P_{o3} = \dfrac{0.4}{10} = 0.04$ W

Thus, the power in the through port (2) is $P_{o2} = P_{i1} - (P_{o3} + P_{o4}) = 0.36$ W

The insertion loss of the coupler is given by $IL = 10\log_{10}\left(\dfrac{P_{o2}}{P_{i1}}\right) = -0.457$ dB

Similar to the waveguide directional coupler discussed above, in strip-line or in microstrip version, we come across the *coupled-line directional coupler* (Fig. 6.17a). Note that waveguide couplers are *forward type couplers* as the coupled power in the secondary guide flows in the same direction as the input power in the primary guide (Fig. 6.14a) while in strip-line/microstrip line couplers, the coupled energy travels in the opposite direction of the input energy flow (Fig. 6.17a) and hence are also known as *backward couplers*. The coupled line coupler uses the concept of even and odd mode coupling of coupled transmission line (Section 3.4.1.1). If $C_v = (V_3/V_1)$ is the voltage coupling factor, then it can be derived [9] that the coupled line directional coupler has even and odd mode impedances $Z_{0e} = Z_0\sqrt{(1+C_v)/(1-C_v)}$ and $Z_{0o} = Z_0\sqrt{(1-C_v)/(1+C_v)}$; thus $Z_{0e} > Z_{0o}$, where Z_0 is the characteristic impedance of the microstrip line. Using standard curves for Z_{0e} and Z_{0o} plotted for different values of W/h with S/h as parameters [8] or standard formulations available in literature [12] the design of such microstrip based coupled-line couplers can be done.

However, this type of coupler is suitable for weak couplings, typically 20 dB or so. A tight coupling would require extreme closeness of the lines which may be impractical and would also demand a combination of even and odd mode characteristic impedances that may be non-realisable. One way to increase the coupling between edge-coupled lines is to use several lines parallel to each other, so that the fringing fields at both edges of a line contribute to the coupling. A practical version of such a coupled-line coupler is the *Lange coupler* (Fig. 6. 17b). Here, four coupled lines are used with interconnections to provide tight coupling; and thus the geometry is also known as an interdigitated structure. This coupler can easily achieve a 3 dB coupling ratio. The design takes care to compensate for the unequal even and odd mode phase velocities, which also improves the bandwidth. There is a 90° phase difference between the output lines (ports 2 and 3), so the Lange coupler is a type of quadrature hybrid coupler similar to a branch line quadrature hybrid or branch-line

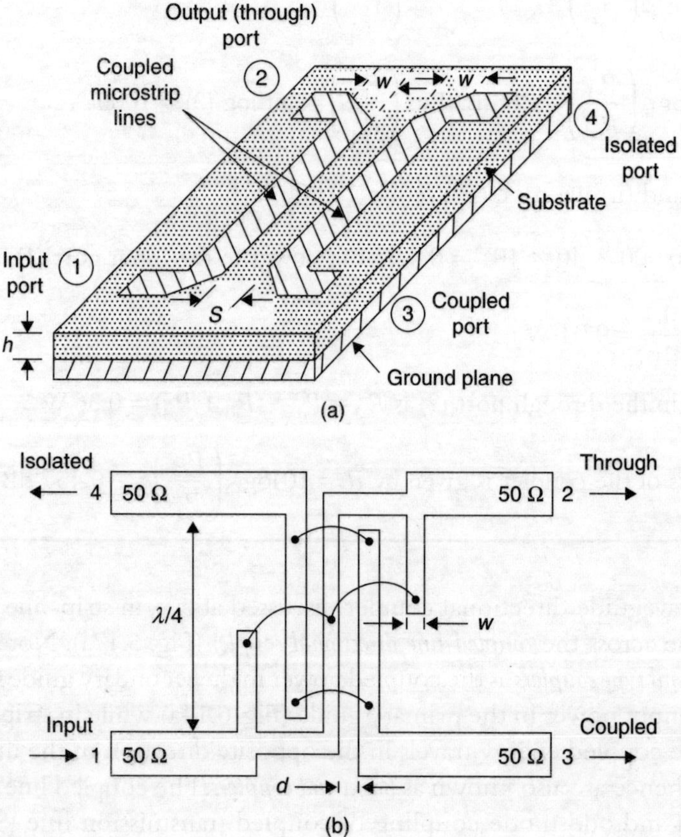

Fig. 6.17 Microstrip layout of (a) Coupled-line directional coupler (b) Lange coupler.

coupler and Wilkinson's power divider. However from a designer's point of view, the Lange coupler occupies less space compared to the other two structures and thus lends itself to a very compact design. Lange couplers are used extensively as power combiner/splitter in the design of microwave amplifiers, mixers and modulators.

EXAMPLE 6.3

Design and characterise a coupled microstrip line based directional coupler at 10 GHz having a coupling factor of 20 dB. The designed coupler is expected to have a directivity and return loss close to 20 dB. In addition to the analytical approach for design, use Electronic Design Automation (EDA) simulation tool for optimised design and performance characterisation.

Solution

With the given coupling factor, the voltage coupling factor is $C_v = 10^{-20/20} = 10^{-1} = 0.1$.

The even and odd mode characteristic impedances are then given by:

$$Z_{0e} = Z_0\sqrt{\frac{(1+C_v)}{(1-C_v)}} = 55.27\ \Omega \quad \text{and} \quad Z_{0o} = Z_0\sqrt{\frac{(1-C_v)}{(1+C_v)}} = 45.22\ \Omega$$

Following Akhtarzad et al [12] synthesis process, we can evaluate $(W/h)_{es}$ and $(W/h)_{os}$ from the expression of Z_0 with W/h of the conventional microstrip line (Eq. 3.44) in terms of the even and odd mode impedances for a single microstrip line, namely, Z_{0es} $(= Z_{0e}/2)$ and Z_{0os} $(= Z_{0o}/2)$, when we get $(W/h)_{es} = 2.3$ and $(w/h)_{os} = 3.55$.

Then the values of S/h and W/h ratios for the coupled lines are determined from the following expressions [12]:

$$\frac{S}{h} = \frac{2}{\pi}\cosh^{-1}\left[\frac{\cosh\left\{\frac{\pi}{2}(W/h)_{es}\right\} + \cosh\left\{\frac{\pi}{2}(W/h)_{os}\right\} - 2}{\cosh\left\{\frac{\pi}{2}(W/h)_{os}\right\} - \cosh\left\{\frac{\pi}{2}(W/h)_{es}\right\}}\right]$$

$$(W/h)_{es} = \frac{2}{\pi}\cosh^{-1}\left(\frac{2d-g+1}{g+1}\right) \quad \text{and} \quad (W/h)_{os} = \frac{2}{\pi}\cosh^{-1}\left(\frac{2d-g-1}{g-1}\right)$$

$$g = \cosh\left(\frac{\pi}{2}\cdot\frac{S}{h}\right) \quad \text{and} \quad d = \cosh\left(\frac{\pi W}{h} + \frac{\pi S}{2h}\right)$$

The values of S/h and W/h can be evaluated either by analytical calculation or by using MATLAB program (Appendix III for MATLAB program) and are given by 0.49 and 1.18 respectively. Hence, the calculated values of S and W are respectively 0.1862 mm and 0.4484 mm (since substrate height is 0.38 mm). The analytically calculated length of the coupler is typically a quarter wavelength, that is, L (= 2.95 mm) Eq. (3.43).

Table 6.2 Optimised design and performance data

Design parameters	Value	Simulation parameters	Simulated value @ 10 GHz
Frequency	10 GHz	Return loss	17.85 dB
Relative permittivity of the substrate (ε_r)	9.6	Insertion loss	0.11 dB
Substrate height (h)	0.38 mm	Coupling	20.67 dB
Conductor thickness (t)	0.017 mm	Isolation	37.75 dB
50 Ω Line width (W_1)	0.38 mm	Directivity	17.08 dB
Coupled line width (W_2)	0.64 mm		
Separation (S)	0.1 mm		
Length (L)	0.63 mm		

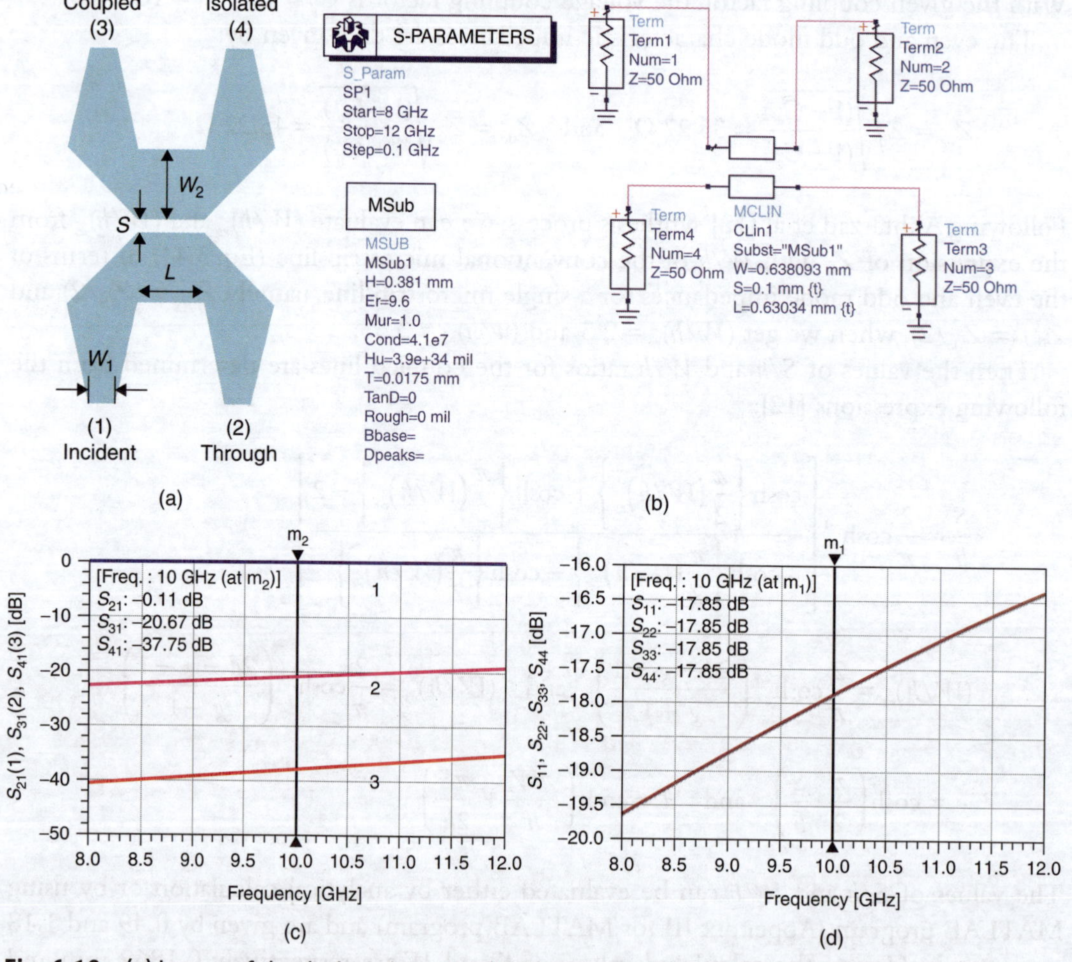

Fig. 6.18 (a) Layout of the designed coupler (b) Screen-shot of schematic design in EDA simulator (c) Plot of insertion loss (S_{21}), coupling factor (S_{31}) and isolation (S_{41}) vs frequency (d) Plot of return loss (S_{11}, S_{22}, S_{33} and S_{44}) at all the four ports vs frequency.

It may be noted that when the coupler is simulated in the schematic of the EDA simulator target specification, particularly the directivity, was not achievable. Thus the design optimisation was done by tailoring the dimensions W (= W_2), S and L. The final design data and the performance data after optimisation is given in Table 6.2. The layout of the directional coupler, screen-shot of schematic design in the EDA simulator and the performance characteristics (coupling factor S_{31}, isolation S_{41}, insertion loss S_{21} including the return loss at all the four ports – S_{11}, S_{22}, S_{33} and S_{44} – spanning the X-band (8–12 GHz) frequency range as obtained in simulation is shown in Fig. 6.18. The directivity of the coupler can be evaluated in terms of isolation and coupling factor as was indicated in Eq. 6.2c.

6.2.5 Attenuator

It is used to insert a known amount of reduction of output signal power with respect to the input power in a waveguide transmission system. A *flap attenuator* is commonly used, but for precession work, a *rotary vane attenuator* is used.

The *flap-type attenuator* consists of a lossy resistive card or flap mounted on two thin metallic rods and the card can be moved parallel to the side wall of a rectangular waveguide (Fig. 6.19). If the card is placed more towards the centre, more attenuation can be obtained (as it intercepts, parallel to its surface, larger strength of the electric field for the TE_{10} mode in a rectangular waveguide). The shape of the card is made such that minimum reflection is caused by the edges of the card facing input and output ends of the guide. To minimise reflections from mounting rods, they are oriented perpendicular to the electric field and are placed $\lambda_g/2$ apart so that the reflection from one, if any, will tend to cancel those from the other. Fixed attenuators of flap-type are typically 3, 6 or 10 dB, and variable ones are 0–20 dB.

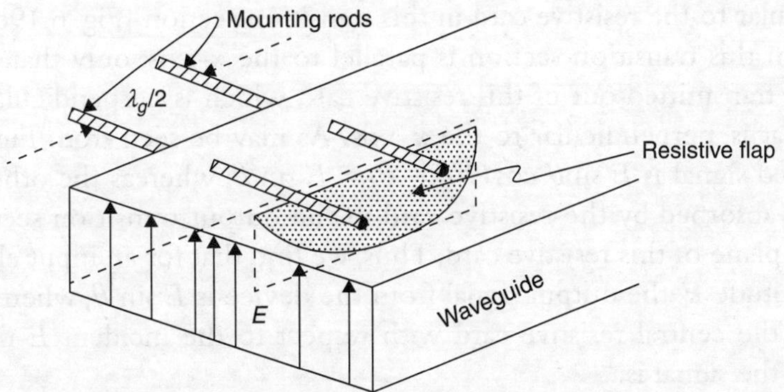

Fig. 6.19 Schematic of a flap-type waveguide attenuator.

Perhaps the most satisfactory precision microwave attenuator developed to date is the *rotary vane attenuator*. The schematic of this attenuator is shown in Fig. 6.20a. The basic parts of this device consist of two rectangular to circular waveguide tapered sections together with an intermediate section of circular waveguide. A very thin tapered resistive card (in the shape of bat-wing) is placed at the transition sections at the input and output ends of the device; oriented parallel to the broad wall of the rectangular guide in the input and output ends. Another resistive card assembly is placed in the intermediate circular waveguide section which is rotatable and can be oriented at an angle θ with the normal to the axis of the circular waveguide. The resistive cards are made of lossy dielectric material.

The principle of operation of the attenuator may be understood as follows. The input TE_{10} signal in the rectangular waveguide is transformed into the TE_{11} mode in a circular waveguide due to field compatibility in the two guides. The presence of the resistive card in the input transition section does not affect the input signal that enters the circular waveguide as the *E*-field of the signal is at right angles to the plane of this resistive card. Since the resistive card in the central section is rotatable, its orientation say θ, relative to the incident TE_{11} mode signal, can be varied. If it can thus be established that the attenuation of the signal is proportional to θ, the amount by which the incident signal is attenuated is then adjustable by changing θ. How the attenuation (A) of the signal is proportional to the angle of rotation (θ) of the central resistive card may be understood as follows.

With reference to Fig. 6.20b, let the central resistive card be oriented at an angle θ relative to the direction of the electric field polarisation E of the TE_{11} mode in a circular waveguide. The TE_{11} mode polarised in the y-direction may be decomposed into the sum of two TE_{11} modes polarised along the u and v directions (as shown). That component of the E field which is parallel to the central resistive card ($u = E\cos\theta$) will be absorbed by the lossy dielectric material of the card [Fig. 6.20c(i)] whereas the portion which is polarised perpendicular to the central resistive card ($v = E\sin\theta$) will be transmitted out of this resistive card. However, upon entering the output transition section, the transmitted signal from the central resistive card, $E\sin\theta$, is again not perpendicular to the resistive card in this transition section [Fig. 6.19c(ii)]. Since the resistive card in this transition section is parallel to the x-axis, only that component of $E\sin\theta$ will be transmitted out of this resistive card which is perpendicular to the plane of the card, that is, perpendicular to the x-axis. As may be seen from Fig. 6.20c(ii), the final transmitted signal is $E\sin\theta\cos(90° - \theta) = E\sin^2\theta$; whereas the other component, $E\sin\theta\cos\theta$, is absorbed by the resistive card in the output transition section since it is parallel to the plane of this resistive card. Thus, we find that for an input electromagnetic signal of magnitude E, the output signal from the device is $E\sin^2\theta$, where θ is the angle of rotation of the central resistive card with respect to the incident E-field. Then the attenuation of the signal is:

$$A_{dB} = -20\log_{10}\left(\frac{E\sin^2\theta}{E}\right) = -40\log_{10}(\sin\theta) \qquad (6.3)$$

The S matrix of this attenuator is given by:

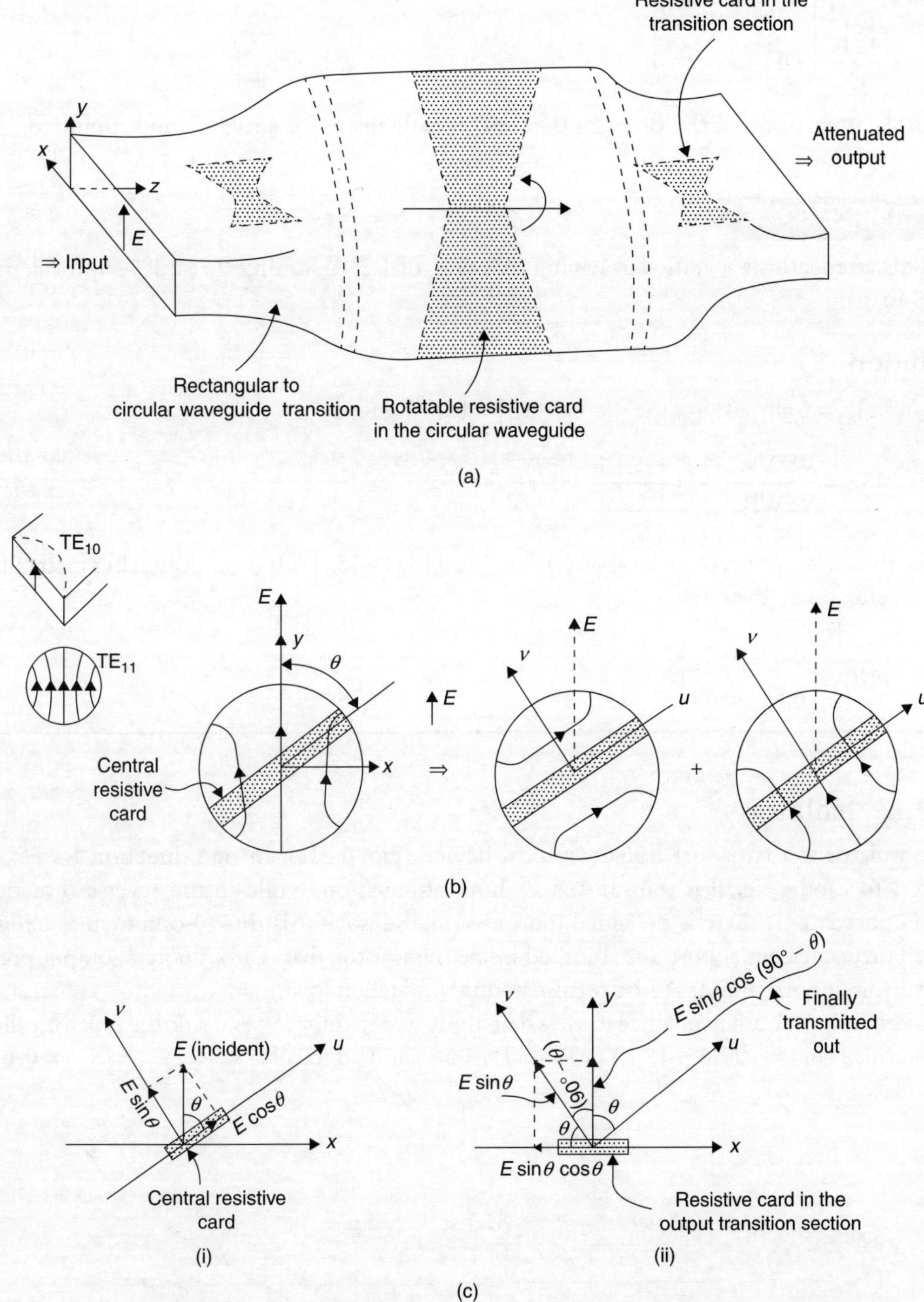

Fig. 6.20 Rotary vane attenuator (a) Schematic structure (b) TE_{11} mode being decomposed in two orthogonal directions (c) (i) Signal at the output of the central resistive card (ii) Signal at the output of the attenuator.

$$[S] = \begin{bmatrix} 0 & \sin^2\theta \\ \sin^2\theta & 0 \end{bmatrix}$$

Variable attenuation of the order of 0–50 dB is realisable with a rotary vane attenuator.

EXAMPLE 6.4

A 5 dB attenuator is specified as having a VSWR of 1.2. Assuming that it is reciprocal, find its *S*-matrix.

Solution

$$A_{dB} = 5 = -20\log_{10}|S_{21}| \quad \text{or} \quad |S_{21}| = 10^{-0.25} = 0.56$$

Now, $$S_{11} = \frac{\text{VSWR}-1}{\text{VSWR}+1} = \frac{1.2-1}{1.2+1} = 0.09$$

Since the attenuator is reciprocal $|S_{12}| = |S_{21}|$ and $|S_{22}| = |S_{11}|$. Thus the scattering matrix for this attenuator is given by:

$$[S] = \begin{bmatrix} 0.09 & 0.56 \\ 0.56 & 0.09 \end{bmatrix}$$

6.2.6 Isolator

The isolator is a two-port non-reciprocal device (Fig. 6.21a). In one direction (say, from port 1 to 2), the signal is transmitted without attenuation, while in the reverse direction (from port 2 to 1) there is no signal transmission; the isolator is thus also known as *uniline*. Any power incident at port 2 is absorbed inside the isolator, that is, from port 2 (output port), port 1 (input port) appears to be terminated in a matched load.

Ideally $|S_{21}| = 1$, but in practice it is less than unity due to finite losses inside the isolator, called the *insertion loss* $(= -20\log_{10}|S_{21}|)$ or *forward attenuation*. Also, ideally $|S_{12}| = |S_{11}| = |S_{22}| = 0$, but

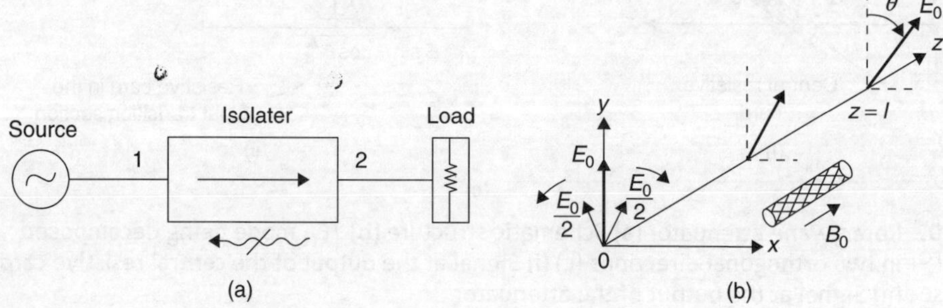

Fig. 6.21 (a) Schematic for isolator (b) Faraday rotation in ferrite.

in practice $|S_{12}|$ is greater than zero due to some imperfect absorption of the signal in the reverse direction, that is, due to non-ideal *isolation* or *reverse attenuation* ($-20\log_{10}|S_{12}|$). Ideally, insertion loss should be zero and isolation infinite, but the practical values of insertion loss and isolation are 0.6–1.0 dB and 30–40 dB respectively.

Isolators are generally used to improve the frequency stability of microwave oscillators. At microwave frequencies, the load is generally reactive in nature, so any load mismatch will cause frequency drift of the oscillator if the oscillator is not isolated from the load. Thus, an isolator should always be placed (with the arrow direction from source to load) between the source and the load in a microwave system to ensure good frequency stability of the oscillator.

EXAMPLE 6.5

Find the scattering matrix of a matched isolator with 1.0 dB insertion loss and 30 dB isolation. Neglect any reflection.

Solution

$$\text{Insertion loss} = 1.0 = -20\log_{10}|S_{21}| \quad \text{or} \quad |S_{21}| = 10^{-0.05} = 0.89$$

$$\text{Isolation} = -20\log_{10}|S_{12}| \quad \text{or} \quad |S_{12}| = 10^{-1.5} = 0.03$$

Since we can neglect reflection, $|S_{22}| = |S_{11}| = 0$

Thus $[S] = \begin{bmatrix} 0 & 0.03 \\ 0.89 & 0 \end{bmatrix}$

The isolator can be constructed by using the non-reciprocal property of the rotation of plane of polarisation of an electromagnetic signal propagating down a magnetised ferrite medium called *Faraday rotation*. The principle of Faraday rotation in magnetised ferrite medium is explained below.

Let us consider a loss-less ferrite material medium with a static magnetic field $\vec{B}_0$ applied along the z-direction (Fig. 6.21b). Let a plane TEM wave that is linearly polarised along the y-axis at $z = 0$ be propagating in the positive z-direction. This linearly polarised wave may be decomposed into the sum of a left and right circularly polarised wave which is the natural mode of propagation in the direction of $\vec{B}_0$ (as shown). With detailed analysis (Appendix II) it may be shown that these two circularly polarised waves have different effective permeabilities, μ^+ and μ^-, and hence different propagation constants β^+ and β^- (since $\beta = \omega\sqrt{\varepsilon\mu}$) and hence eventually have different phase velocities v_p^+ and v_p^- (since $v_p = \omega/\beta$) while propagating through the magnetised ferrite medium. As a result, the plane of polarisation of the input electromagnetic signal gets rotated (by an angle θ, clockwise) while propagating down the ferrite medium and this is known as Faraday rotation in magnetised ferrite—analogous to a similar phenomenon in optics. The amount of rotation of the plane of polarisation of the output wave (which is again a linearly polarised wave) with respect to the input wave polarisation depends on the length and composition of the ferrite material and also on the strength of the DC magnetic field.

The Faraday rotation is a non-reciprocal effect in the sense that if the direction of propagation is reversed (keeping the magnetic field direction unchanged), that is, if we consider the propagation of the wave back to the plane $z = 0$, the plane of polarisation continues to rotate in the same direction and thus the original polarisation direction is not restored; instead the wave will come back at $z = 0$ after being rotated at an angle 2θ (clockwise) with respect to the x-axis.

The schematic arrangement of an isolator using Faraday rotation in magnetised ferrite – *Faraday rotation isolator* – is shown in Fig. 6.22. The main part of the isolator is built in a circular waveguide which is tapered at the input and output ends to the rectangular waveguide sections. As shown in the figure, the orientation of the output rectangular waveguide section is physically at 45° clockwise with respect to the input rectangular waveguide. A ferrite rod (thin with both ends being pencil shaped) is placed along the axis of the circular waveguide region of the assembly. The length and composition of the ferrite material and the magnitude of the static magnetic field is selected such that the E field vector of the electromagnetic wave (at the desired operating frequency) travelling along the ferrite (in either direction) gets rotated by 45°. In addition to the ferrite rod two resistive vanes (made of lossy dielectric material) are also placed suitably; one of the resistive vanes is located inside the tapered transition section in the input side while the other one is located inside the tapered transition section in the output side. The resistive vanes are placed parallel to the broad walls of the corresponding rectangular waveguide sections at the input and output ends. Edges of the vanes looking towards the ferrite rod are tapered as shown (to minimise undue reflections). In practice, the ferrite rod and the two resistive vanes are supported mechanically by a low-loss dielectric material (commonly polyfoam), not shown in the figure. The static magnetic field is obtained by a permanent magnet or solenoid placed on the outside surface around the circular waveguide.

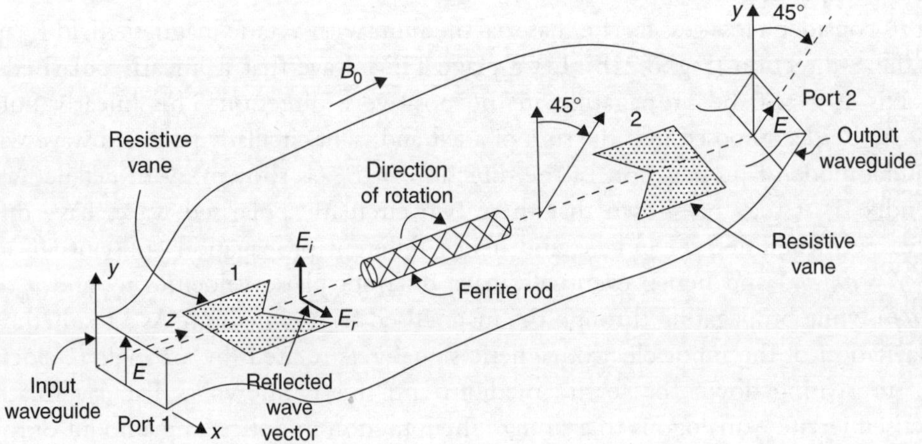

Fig. 6.22 Faraday rotation isolator.

The principle of operation of this isolator may be understood as follows. An electromagnetic wave incident at port 1 with electric field direction shown for the TE_{10} mode in a rectangular waveguide would excite a TE_{11} mode in a circular waveguide. The E field of this TE_{11} mode is parallel to the E field of the incident wave and is thus perpendicular to the resistive vane numbered 1 (near the input end). Therefore, the power incident at port 1 is not being affected by the presence of the resistive vane near this end. This wave next encounters the magnetised ferrite rod, and as it travels down the rod, the plane of polarisation of the incident electric field, that is, the direction of the E field gets rotated by 45° clockwise (looking into the direction of propagation, that is, from input to output). After this rotation, the E field vector becomes perpendicular to the vane numbered 2 (near the output end) and therefore unaffected by its presence. Since the physical orientation of the port 2 is fixed at 45° clockwise with respect to port 1, the wave coming out of the vane numbered 2 propagates down the rectangular waveguide section at the output unaffected. Thus, the input power is transmitted to the output practically un-attenuated. Next, let us consider a wave incident to the device from port 2. In this case, the E field direction is such that this wave is unaffected by the resistive vane numbered 2 (being perpendicular to the plane of this vane). While travelling down the magnetisd ferrite rod, it encounters the static magnetic field (B_0) in the reverse direction (to that for the wave from port 1 to port 2). Therefore, the E vector rotates counter-clockwise (now looking into the direction of propagation from 2 to 1). The rotation of the plane of the E field is again in the direction marked in the figure. However, after this rotation, the E field lines become parallel to the resistive vane numbered 1 and the power is absorbed by this vane. The edge of the vane is tapered to reduce any possible reflection towards port 2. Even if such a reflection does take place, the E field of the reflected wave becomes parallel to the resistive vane numbered 2 also; and residual power, if any, gets dissipated in this vane. Thus, the electromagnetic power travelling from port 2 towards port 1 is not transmitted but gets totally absorbed inside the isolator.

The Faraday rotation isolator is suitable for low and medium power applications at microwave frequency, say up to 2 kW peak power. However, at high peak power levels, non-linear effects in ferrite make such isolators unsuitable, when the *resonance absorption isolator* is used. The principle of operation of such an isolator is based on the non-reciprocal absorption property of circularly polarised waves at and around the ferrite resonance frequency of a magnetised ferrite (Fig. 6.23a). It is seen that the attenuation constant α^+ for right circularly polarised (clockwise polarised or positive polarisation) wave is very high while α^- for left circularly polarised (anticlockwise polarised or negative polarisation) wave is very small when the operating frequency is in the vicinity of ferrite resonance frequency (ω_o) [7].

This non-reciprocal absorption in ferrite at ω_o is used in the design of resonance absorption isolator using a left-circularly polarised wave (as α^- gives low attenuation) for transmission in the forward direction and right circularly polarised wave for transmission in the reverse direction. The power flow in the reverse direction is absorbed in the ferrite (as α^+ is very high) and dissipated via the waveguide wall. To design such a device, the ferrite material has to be placed inside the waveguide at a location where it can see a circularly polarised wave (Fig. 6.23b). On studying the field distribution for the TE_{10} mode in a rectangular waveguide (Eq. 4.11b), it is seen that (i) the E field has only one component

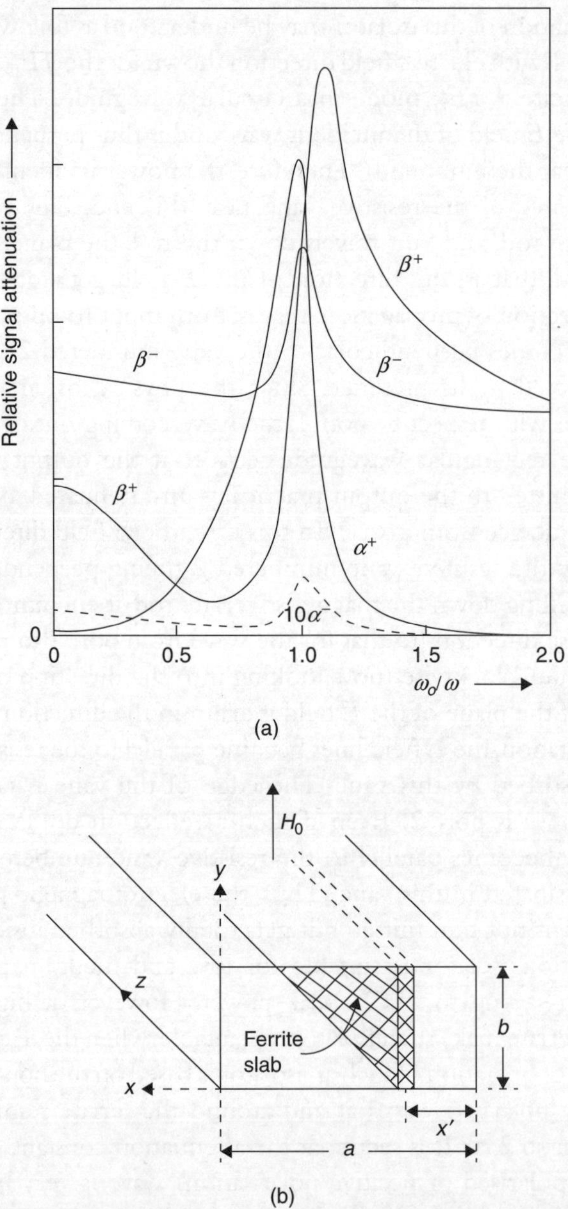

Fig. 6.23 (a) Absorption of electromagnetic signal by magnetised ferrite near ferrite resonance frequency (ω_o) (b) Rectangular waveguide loaded with ferrite slab.

E_y and cannot be circularly polarised at any location while (ii) the H field has two normal components H_x and H_z which are 90° out of phase. Thus, at any point where the two components of the H field are equal in magnitude, we have a circularly polarised magnetic field. This occurs when $x' = (a/\pi) \tan^{-1}(\pi/a\beta)$, where x' is the offset distance of the ferrite inside the waveguide from the narrow wall of the waveguide, a is the broad dimension of the waveguide and β is the propagation constant. Keeping this in mind, the ferrite is positioned inside the waveguide as shown in Fig. 6.22b, where the static magnetic field H_0 is also shown. In the range $0 < x' < a/2$, where the ferrite is placed inside the waveguide, the ratio of H_x to H_z is given by $H_x/H_z = \mp j$ (it can also be shown that in the range $x' > a/2$, $H_x/H_z = \pm j$). With respect to the y-axis, this solution corresponds to a left circularly polarised wave (used in the case of forward transmission with the least possible attenuation or absorption of signal) for propagation in the +z direction and a right circularly polarised wwwave (used in the case of reverse transmission with the largest possible attenuation of the signal) for propagation in the $-z$ direction.

By proper design, a forward loss of 0.5 dB/inch at 10 GHz and a reverse loss of 6–10 dB/inch or even more can be realised. Since the ferrite is in thermal contact with the metal wall of the waveguide, it acts as a heat-sink to dissipate the heat generated inside the isolator caused by reverse signal loss in the ferrite and the dielectric.

6.2.7 Circulator

A circulator is a multi-port waveguide junction in which the signal can flow from the nth port to the $(n + 1)^{\text{th}}$ port in one direction (Fig. 6.24a). Although there is no restriction in the number of ports, 4 and 3 port circulators are in common use. A 4-port circulator (like isolator) also uses the non-reciprocal property of Faraday rotation in magnetised ferrite medium. The methodology used for the design of a Faraday rotation isolator may be extended for the design of 4-port *Faraday rotation circulator*. The basic configuration is similar to the isolator, however, the resistive vanes in the input and output transition sections are not present in the circulator structure. Instead, arrangements are made to couple out signals through two extra ports that would otherwise have been dissipated in the resistive vanes in the case of the isolator. The two extra ports in the case of the circulator are oriented radially outwards (with respect to the circular waveguide) and are attached to the circular waveguide section as depicted in Fig. 6.24b. Angular positions of these two ports (numbered 3 and 4) with respect to ports 1 and 2 are shown in Fig. 6.24c, which shows that the ports 1, 2, 3 and 4 are oriented such that the E field can couple to these ports successively in numerical order after going through a rotation of 45° in the clockwise direction in each step.

The principle of operation of a 4-port Faraday rotation circulator may be understood as follows. For a signal incident at port 1, the E vector is in the *aa* direction. It does not get coupled to port 3 as ports 1 and 3 are cross-coupled (90° with each other). When the wave travels down the magnetised ferrite, the direction of the E field vector gets rotated by 45° clockwise to get oriented in the *bb* direction. Thus, the power incident at port 1, appears at port 2 (it is not coupled to port 4 as ports 2 and 4 are cross-coupled). By similar arguments we can see that the signal incident at port 2 will be coupled to port 3 and so on, in a cyclic order (in this case, in the clockwise direction). A 4-port Faraday rotation circulator may be easily converted into a 3-port circulator by terminating one of the ports by a perfect short circuit.

A 4-port circulator can be used as a *single-pole-double throw* (*SPDT*) *switch*. The switching action can be obtained just by reversing the direction of the magnetic field B_0. This

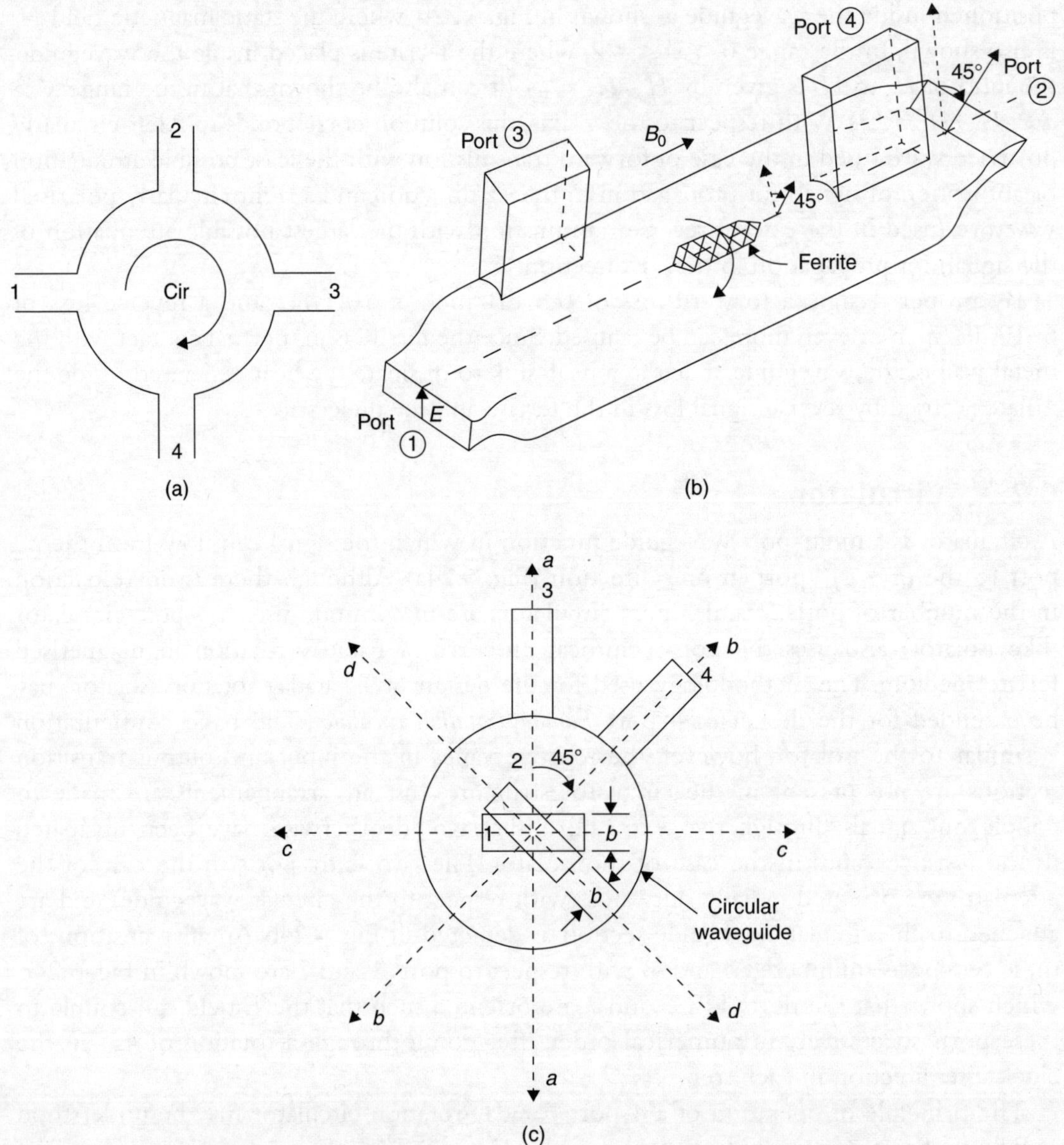

Fig. 6.24 Circulator (a) Schematic (b) Physical structure (c) Relative position of ports with respect to each other.

changes the rotation of the E field vector from the clockwise to the anti-clockwise direction and couples the power to port 4 (*dd* direction) instead of the port 2 (*bb* direction); Fig. 6.23c. Thus, port 1 can be connected to port 2 or to port 4 by changing the direction of current in the electromagnet used for generating the magnetic field to which the ferrite rod is subjected. The 4-port circulator may also be used as a *power divider*. We have just seen that for a magnetic field in the $+z$ direction, the power incident at port 1 gets coupled to port 2, while for a magnetic field in the opposite direction ($-z$) the power is transmitted to port 4. If the magnetic field is absent, the direction of the E field vector remains unchanged (hypothetically) and is half way between that required for coupling to ports 2 and 4. This is expected to result in half of the power getting coupled to port 2 and the remaining half to port 4. It is thus possible to have any desired fraction of incident power to appear at port 2 and the other half at port 4 by adjusting the magnetic field such that the E field vector rotates in between *bb* and *dd* directions. The components of the E field vector along *bb* and *dd* determine the amount of power coupled to port 2 and port 4, respectively. For a variable power divider of this type, port 3 is terminated with a matched load.

A more commonly used circulator for both waveguide and strip-line configuration is a Y/T-*junction circulator.* It consists of a symmetrical distribution of ferrite material at the junction of three rectangular waveguides (Fig. 6.25a) or strip-line (Fig. 6.25b) ports having a 120° angle between the ports (the port arrangement in the present case resemble the English letter *Y*; it may also be arranged resembling the letter *T*). The ferrite material is usually in the form of a cylindrical rod or disc or in the form of a prism and the DC magnetic field (H_0) is applied along the axis of the cylinder/disc as shown.

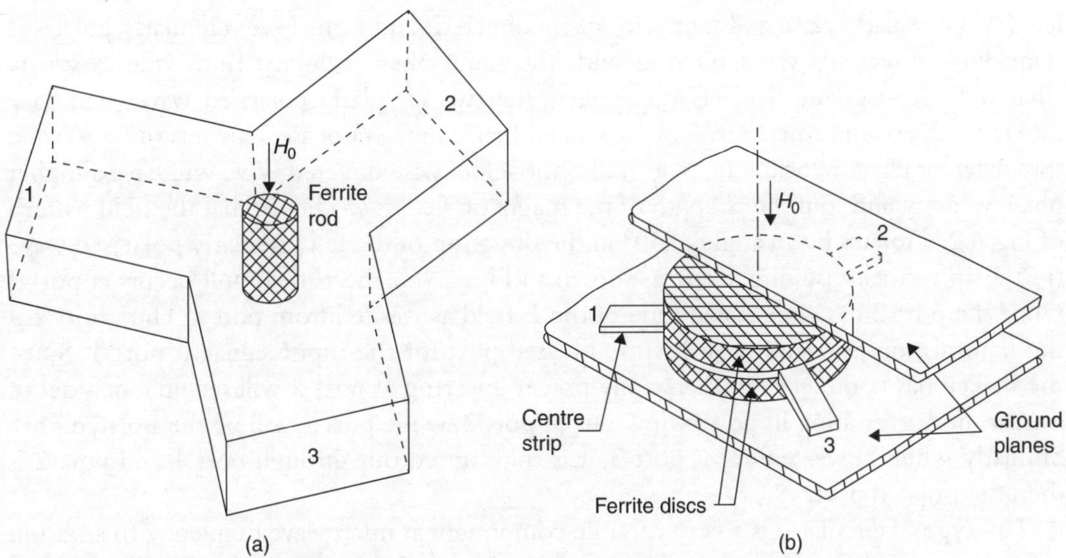

Fig. 6.25 Y/T-junction circulator (a) Using rectangular waveguide (b) Using strip-line.

The resonance of the ferrite cylinder/disc/prism is the key feature in the operation of this type of circulator. The lowest frequency resonance mode is the dipolar mode, for which the E field vectors are perpendicular to the plane of the disc and the H field vectors lie parallel to the plane of the disc. This configuration mode when excited from port 1, is illustrated in Fig. 6.26a for the un-magnetised case. Thus, ports 2 and 3 will see voltages that are 180° out of phase with the input voltage and approximately half the amplitude of the input voltage at port 1.

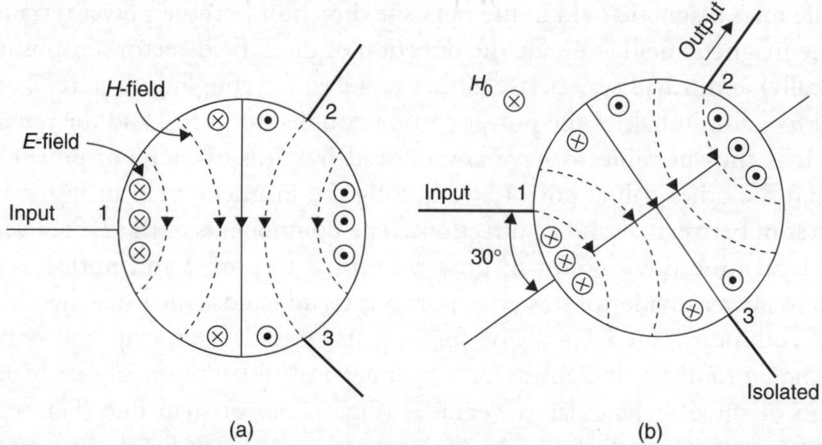

Fig. 6.26 Dipolar mode showing E (⊗ and ⊙) and H (– – –) fields (a) In the absence of the magnetic field H_0 (b) In the presence of the magnetic field H_0.

The field pattern in Fig. 6.26a may be considered to be due to the interference of a left (–ve) circularly polarised (anti-clockwise directed) and right (+ve) circularly polarised (clockwise directed) waves moving with the same phase velocity. But, when a steady (that is, DC) magnetic field (H_0) is applied, the two circularly polarised waves will have different effective permeabilities ($\mu^+ > \mu^-$) and hence different phase constants ($\beta^+ > \beta^-$) and thus different phase velocities ($v_p^+ < v_p^-$). The anti-clockwise directed wave will have a higher phase velocity and so if the strength of the magnetic field is so chosen that the field pattern of Fig. 6.26a rotates by an angle of 30° in the direction of the left circularly polarised wave (in the anti-clockwise direction), as shown in Fig. 6.26b, the E field null occurs at port 3 while the port 2 has same magnitude of the E field as was fed from port 1. Thus, port 2 is the transmission port and port 3 is the isolated port for the input signal at port 1. Since the device has complete symmetry, any power entering at port 2 will set-up a new set of similar modes resulting in power flow out of port 3, while port 1 will be the isolated port. Similarly when power enters at port 3, it is transmitted out through port 1 and port 2 is then the isolated port.

This type of circulator is a very versatile component at microwave frequency. In addition to its use as a circulator, it can also be used as an isolator (by terminating one of the ports in matched load) or as a switch (by reversing the direction of the magnetic field).

Circulators find many other applications in microwave systems. A 3-port circulator can be used to separate the incident and reflected signals in a reflection type amplifier when port 2 of the circulator is terminated with a negative resistance device like a tunnel diode, IMPATT or Gunn diode (Fig. 6.27a). The 3-port circulator can also be used as a diplexer (Fig. 6.27b) in the radar trans-receiver system for isolating the transmitted and received signals, where a single antenna is used both for transmission and receiving the echo signal (as in the monostatic radar). Also, in some cases, two circulators are connected in cascade making a 6-port device with a transmitter (Tx), antenna and receiver connected to ports 1, 2 and 3 respectively with matched load (ML) at ports 4, 5 and 6 to improve isolation between the transmitter and receiver (Fig. 6.27c).

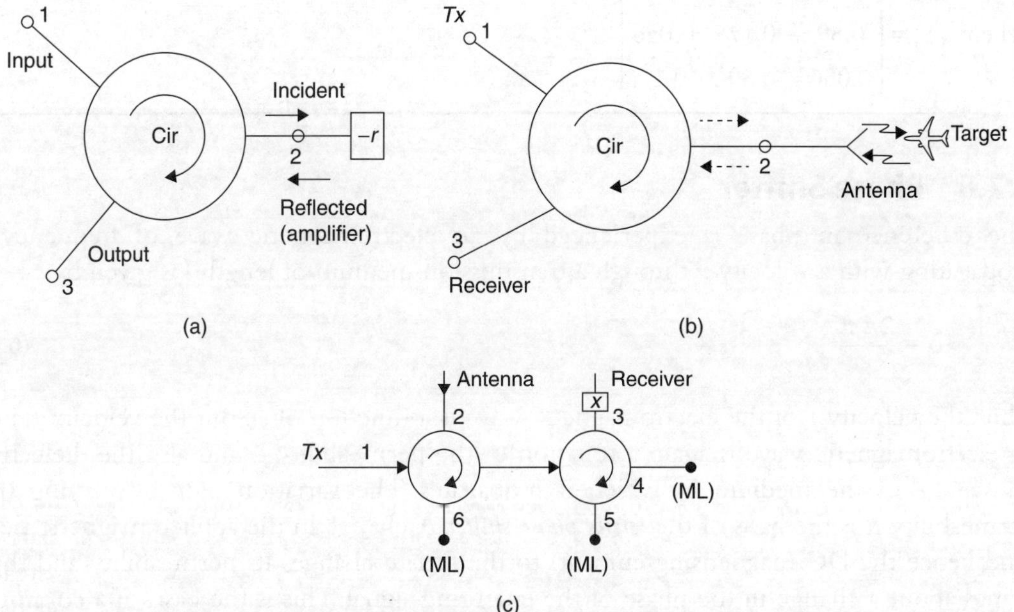

Fig. 6.27 Applications of circulator in (a) Reflection type amplifier with negative resistance diode (b) Diplexer for monostatic radar (c) Cascaded circulators for better isolation between transmitter and receiver.

EXAMPLE 6.6

The specifications of a three-port circulator are given below. Insertion loss = 1 dB, isolation = 25 dB and VSWR = 1.4. Characterise the circulator by its S-parameters.

Solution

$$\text{Insertion loss (dB)} = -20\log_{10}|S_{21}| \text{ that is, } 1 = -20\log_{10}|S_{21}|$$

Or $|S_{21}| = 10^{-1/20} = 0.89$

Therefore for the circulator: $|S_{21}| = |S_{32}| = |S_{13}| = 0.89$

$$|S_{12}| = 10^{-25/20} = 0.056$$

$$\text{Isolation (dB)} = -20\log_{10}|S_{12}| \text{ that is, } 25 = -20\log_{10}|S_{12}|$$

Therefore for the circulator: $|S_{12}| = |S_{23}| = |S_{31}| = 0.056$

The principal diagonal elements are determined by: $S_{11} = |\Gamma| = \dfrac{\text{VSWR} - 1}{\text{VSWR} + 1} = \dfrac{1.4 - 1}{1.4 + 1} = 0.17$

Therefore: $|S_{11}| = |S_{22}| = |S_{33}| = 0.17$

When $[S] = \begin{bmatrix} 0.17 & 0.056 & 0.89 \\ 0.89 & 0.17 & 0.056 \\ 0.056 & 0.89 & 0.17 \end{bmatrix}$

6.2.8 Phase Shifter

The difference in phase φ experienced by an electromagnetic wave of frequency f propagating with a velocity v through a transmission medium of length l is given by:

$$\phi = \frac{2\pi fl}{v} \tag{6.4}$$

Thus, the velocity v of the electromagnetic wave is a function of φ. But the velocity (v) of an electromagnetic wave is again a function of the permeability μ and also the dielectric constant ε of the medium in which it propagates. The variation of v by varying the permeability μ is the basis of the *ferrite phase shifter*. A change in the applied magnetic field (and hence the DC magnetising current) to the ferrite changes its permeability and thus brings about a change in the phase of the emerging signal. This is the basis of a common type of ferrite phase shifter known as a Reggia–Spencer phase shifter.

Realisation of phase change with variation in the length (l) of the transmission line is used in the design of *reflection and transmission type phase shifter*.

The *reflection type phase shifter* may be built around a waveguide circulator or with a branch line quadrature hybrid in a planar version. As shown in Fig. 6.28a, the output from port 2 of the circulator is fed through a transmission line of characteristic impedance Z_0 to the voltage variable capacitance (X_c) of a varactor diode. The reflection coefficient Γ is given by:

$$\Gamma = \frac{jX_c - Z_0}{jX_c + Z_0} \tag{6.5a}$$

with reflection phase change: $\phi = -2\tan^{-1}\left(\frac{X_c}{Z_0}\right)$ (6.5b)

If the value of X_c is changed from X_1 to X_2 by varying the reverse bias voltage to the varactor diode that will change the phase of the output signal at port 3 of the circulator continuously from φ_1 to φ_2 with respect to the input signal fed at port 1. This type of phase shifter is known as an *analogue phase shifter*.

Referring to Fig. 6.28b, it may be observed that a branch-line quadrature hybrid (also known as 3 dB hybrid coupler) along with two PIN diodes (acting as SPST switches) and reflecting shorts as termination can make another type of reflection phase shifter at microwave frequencies. The quadrature hybrid has the property that a signal fed at port 1 is divided equally between the output ports 2 and 3 with no signal at port 4 (Section 6.2.2).

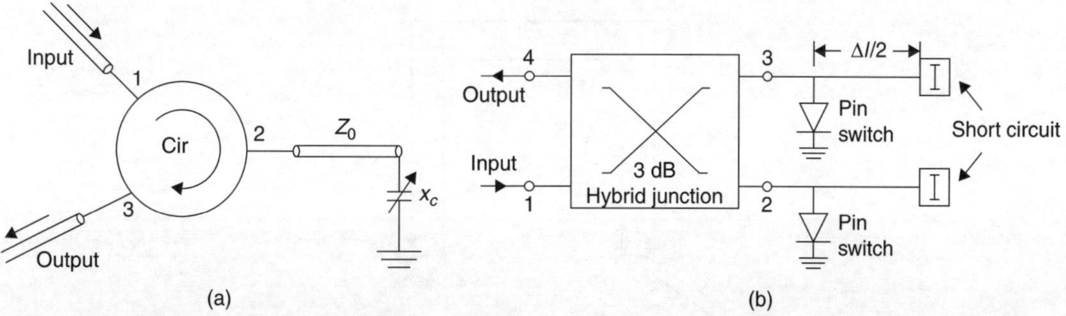

Fig. 6.28 Reflection type phase shifter (a) Using varactor diode (b) Using branch line quadrature hybrid.

With the PIN diodes ON (forward biased or closed) the signal will get reflected from the diodes while being OFF (reverse biased or open) the signal will return from the short circuit terminations. The signals at port 2 and 3, after being reflected from either the diode switch or the short circuited termination combine at port 4; while none reaching at port 1. The difference in the two way path length Δl with the diode switches ON and OFF gives rise to the phase shift of $\Delta\phi = \beta.\Delta l$. This is a *digital type phase shifter*, cascading N such hybrids will form an N-bit phase shifter.

A basic type of *transmission type phase shifter* is a switched line phase shifter which is a time delay circuit in which the phase is obtained by switching between two transmission line sections of different lengths, l_1 and l_2, by means of switches (PIN diode switches are commonly used); Fig. 6.29. Depending upon the ON (forward biased or closed) or OFF (reverse biased or open) condition of the switch pairs (S_1 and S_2) or (S_3 and S_4), the signal will follow the path lengths l_1 or l_2 and hence the phase shift incurred in these two switching conditions is $\Delta\phi = \beta(l_2 - l_1)$. This is also a digital type phase shifter. The use of $\lambda/4$ line sections provide isolation between the ON path and OFF path thereby preventing undue resonance if the length of either line becomes a half-wavelength at the frequency of operation.

In addition to the above mentioned techniques, a microwave phase shifter may also be designed in a waveguide system using quarter-wave and half-wave plates to construct a *rotary phase shifter* which is a precision phase shifter and is used in microwave measurement laboratories as the calibration standard. The construction of this type of phase shifter resembles the rotary vane attenuator discussed earlier; however, in this case the central rotatable resistive card of the attenuator is replaced by a rotatable half-wave plate while the two resistive cards in the input and output transition sections of the attenuator are replaced by quarter-wave plates. The quarter-wave plate introduces a 90° differential phase shift and in addition converts a linearly polarised TE_{11} mode wave in a circular waveguide into a circularly polarised mode wave and vice versa; while the half-wave plate introduces a 180° differential phase shift and in addition produces a phase shift equal to twice the angle θ through which it is rotated.

The essential parts of the phase shifter are three circular waveguide sections, two of which are placed at the input and output ends (both being fixed) and the third central one

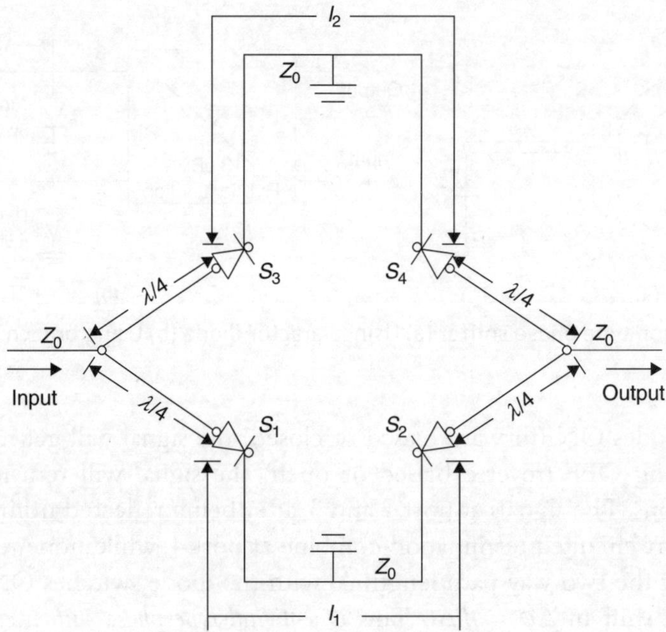

Fig. 6.29 Transmission type phase shifter.

that is rotatable. The fixed sections are quarter-wave plates and the central rotatable one is the half-wave plate (Fig. 6.30a). In the discussion that follows, the quarter-wave and half-wave plates are dielectric loaded circular waveguides; besides dielectric slabs, circular guides may also be loaded with metallic fins or rods to produce 90° and 180° differential phase shift.

The principle of operation of the phase shifter may be understood as follows [10]. The linearly polarised (vertical polarisation) input electromagnetic signal can be thought to be decomposed into two components—one along the dielectric slab (u-direction) and the other, perpendicular to the plane of the slab (v-direction) of the quarter-wave plate in the input side. The component along the dielectric slab will propagate with a reduced phase velocity (v_p) and thus the propagation constant $\beta_u > \beta_v$ (since $\beta = \omega/v_p$). By choosing the length (l) of the dielectric slab properly we may make $(\beta_u l - \beta_v l) = \pi/2$. Since $v_{pu} < v_{pv}$, the wave emerging from the quarter-wave plate in the input side will be a clockwise rotating circularly polarised wave with its vector-phasor components as $(E/\sqrt{2})\ \angle(-\beta_v l - \pi/2)$ in the u-direction and $(E/\sqrt{2})\ \angle(-\beta_v l)$ in the v-direction (Fig. 6.30b).

When $\theta = 0°$, that is, the central rotatable section is with zero rotation, all the three dielectric slabs in the wave plates are in line and in this case the components along the u-direction and the perpendicular v-direction to all the dielectric slabs arrive at the output of the phase shifter delayed by $(4\beta_v l + 2\pi)$ and $4\beta_v l$ respectively. The two components being in phase (2π radian phase difference being equivalent to zero phase difference), the output is a vertically polarised wave given by $E\angle(-4\beta_v l)$. The orientation of the output transition (circular to rectangular waveguide) is such that the wave is delivered to the output waveguide without any reflection or loss.

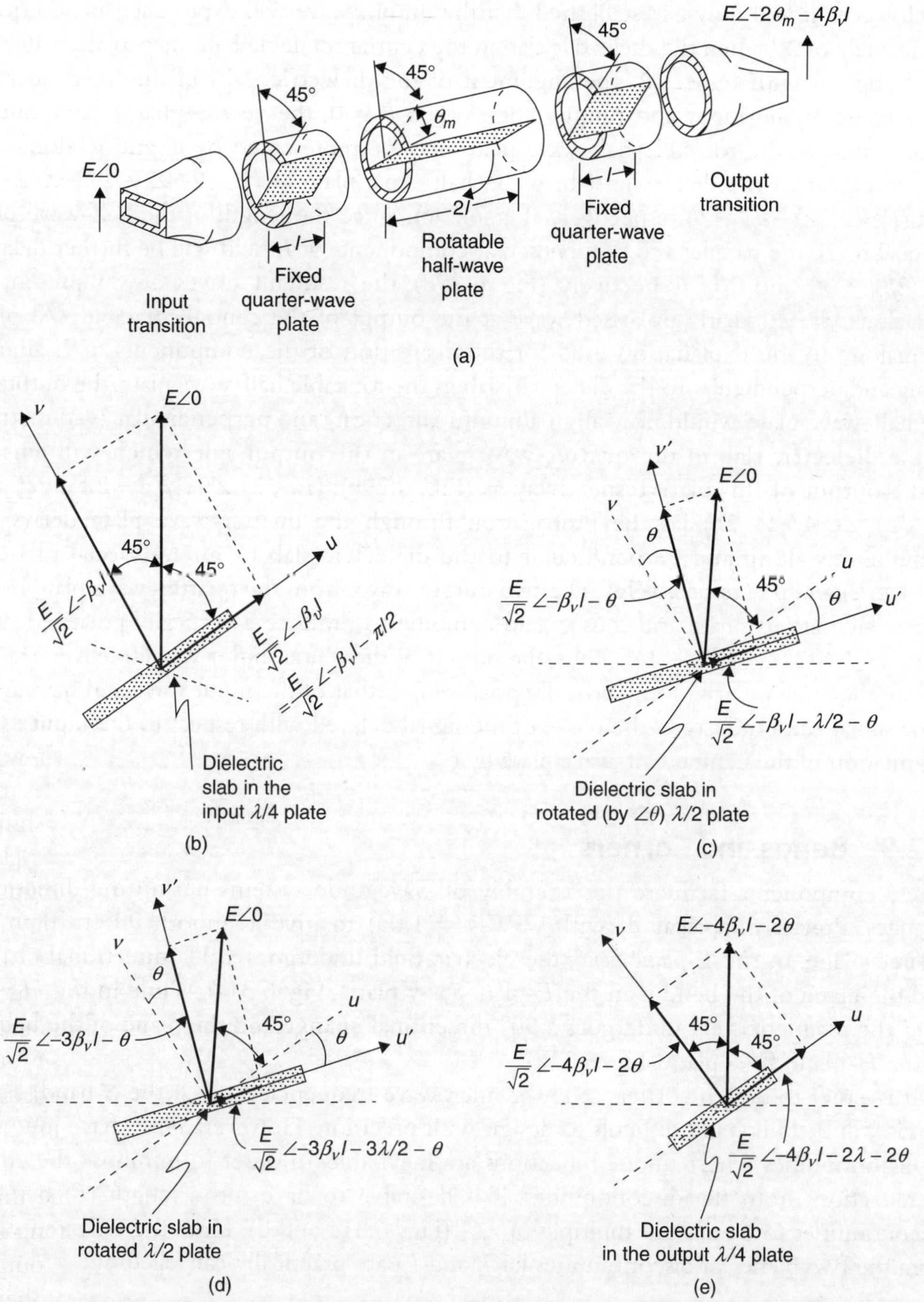

Fig. 6.30 Rotary phase shifter (a) Schematic (b) Vector phase components of the *E*-field at the output of $\lambda/4$ plate in the input side (c) Vector phase components of the *E*-field at the input of rotatable $\lambda/2$ plate when rotated by an angle θ (d) Vector phase components of the *E*-field at the output of rotatable $\lambda/2$ plate (e) Vector phase components of the *E*-field at the output of the $\lambda/4$ plate in the output side.

However, it will now be established that the output wave will experience an additional phase delay of 2θ when the dielectric slab in the central rotatable half-wave plate is aligned at an angle θ with respect to the alignment of the dielectric slabs in the fixed quarter-wave plates at the input and output sides. When $\theta \neq 0$, the vector–phasor components at the input of the rotatable half-wave plate may be represented by u' and v' along and perpendicular to the dielectric slab in the half-wave plate as $(E/\sqrt{2})\ \angle(-\beta_v l - \pi/2 - \theta)$ and $(E/\sqrt{2})\ \angle(-\beta_v l - \theta)$ respectively (Fig. 6.29c). Since the length of the half-wave plate is equal to $2l$, the parallel and perpendicular components of E field will be further delayed by $(2\beta_v l + \pi)$ and $2\beta_v l$ respectively (Fig. 6.29d); the resultant waves now represent an anti-clockwise circularly polarised wave at the output of the central rotatable $\lambda/2$ plate. In analogy to the explanation used for the derivation of the components of E aligned along and perpendicular to the dielectric slab in the rotatable half-wave plate, the output of the half-wave plate would now align along (u-direction) and perpendicular (v-direction) to the dielectric slab in the quarter-wave plate in the output side which will emerge at the output of this plate respectively as (Fig. 6.29e): $(E/\sqrt{2})\ \angle(-4\beta_v l - 2\pi - 2\theta)$ and $(E/\sqrt{2})\ \angle(-4\beta_v l - 2\theta)$ [as the propagation through the quarter-wave plate delays the components along and perpendicular to the dielectric slab by an additional phase of $(\beta_v l + \pi/2)$ and $\beta_v l$, respectively]. The two output waves from the quarter-wave plate in the output side are in phase and thus would combine to produce a vertically polarised wave of value: $E/\sqrt{2}\ \angle(-4\beta_v l - 2\theta)$. Thus, the output of the phase shifter would deliver a signal having the same polarisation (vertically polarised) as that of the input wave and having the same signal amplitude, only the phase of the signal delayed with respect to the input by 2θ for rotation of the central half-wave plate by θ.

6.2.9 Bends and Corners

These components facilitate the assembly of waveguide systems permitting directional changes. Precisely fabricated (with VSWR = 1.05) to give a smooth 90° turn in the desired plane. In the *E-plane bend*, the electric field undergoes a 90° directional change and the plane of the bend is in the E-plane (x–y plane, Fig. 6.31a). While in the *H-plane bend*, the magnetic field undergoes a 90° directional change and the plane of the bend is in the H-plane (x–z plane, Fig. 6.31b).

It is easier to design corners at lower microwave frequencies (say at the S-band) as the bends will be bulky and difficult to design with precision. However, *corners* have junctions of discontinuities where undue reflections are inevitable. In order to minimise the effects of reflections from the discontinuities, it is desirable to have mean length (L) between discontinuities to be an odd multiple of $\lambda/4$ (Figs 6.31c and d). Thus, the reflected waves from the two planes of discontinuities (at J_1 and J_2) are practically cancelled out.

6.2.10 Waveguide Twist, Transitions and Adaptor

The *waveguide twist* is used whenever a 90° (or 45°) change in the plane of polarisation is required (Fig. 6.32a). A long enough ($\approx 2\lambda_g$) rectangular waveguide is gradually twisted mechanically to minimise reflections (VSWR < 1.05). At the output end of a Faraday rotation isolator we need to use a 45° waveguide twist to connect it to the main waveguide line.

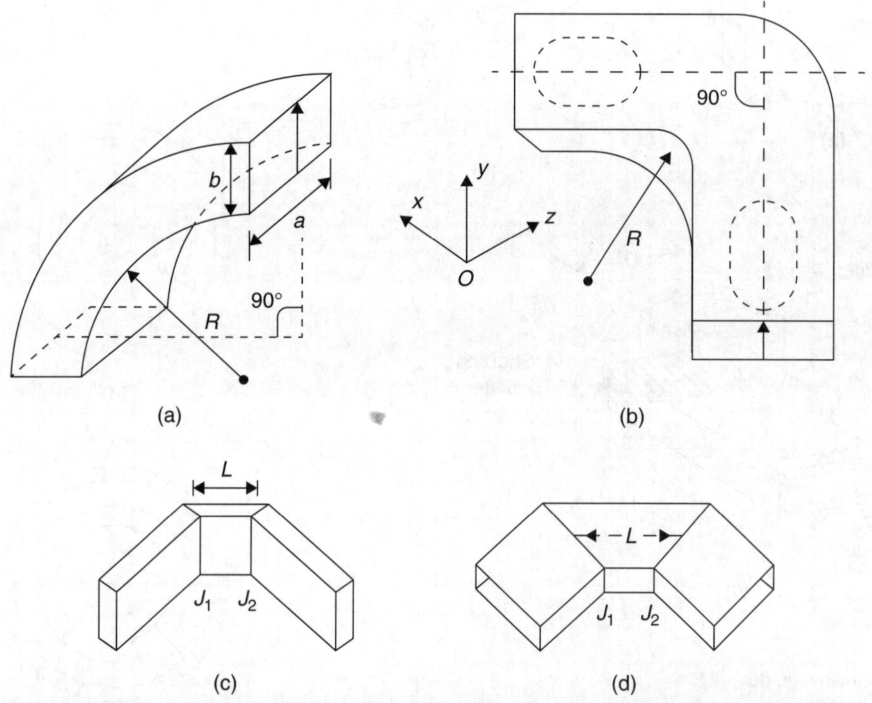

Fig. 6.31 Bends (a) *E*-plane bend (b) *H*-plane bend (c) *E*-plane corner (d) *H*-plane corner.

The *transitions* between waveguides of different size, shape and/or modes are also required in microwave system design. The most commonly used transition is one where we need a dominant mode transmission from rectangular to circular waveguide (Fig. 6.32b); which is used at the input and output ends of a rotary vane attenuator and rotary phase shifter. Examples of the use of transition from the dominant mode in one waveguide type (rectangular) to a non-dominant (higher order) mode in another type of waveguide (circular) are also there. In the case of design of rotary joints that are used for mechanical beam steering of surveillance radar antenna, we need to use the TM_{01} mode in the circular waveguide (note that TE_{11} is the dominant mode in a circular waveguide) as the *E*-field for this mode has axial symmetry, Fig. 4.15 (which does not distort the transmission characteristics of the antenna while rotating). Thus, here we need transition from the TE_{10} in a rectangular waveguide to the TM_{01} in a circular waveguide. The scheme for this is shown in Fig. 6.32c(i) along with a rotary joint in Fig. 6.32c(ii). The orientations of the two types of waveguides help in smooth transition of the magnetic field patterns (and hence the electric field patterns) of the two modes as shown. The diameter D of the circular waveguide should be such that its operating frequency is above the cut-off frequency of TM_{01} mode but should be below the cut-off frequency of the next higher order mode. However, having done this, the non-dominant higher order mode (TM_{01}) will

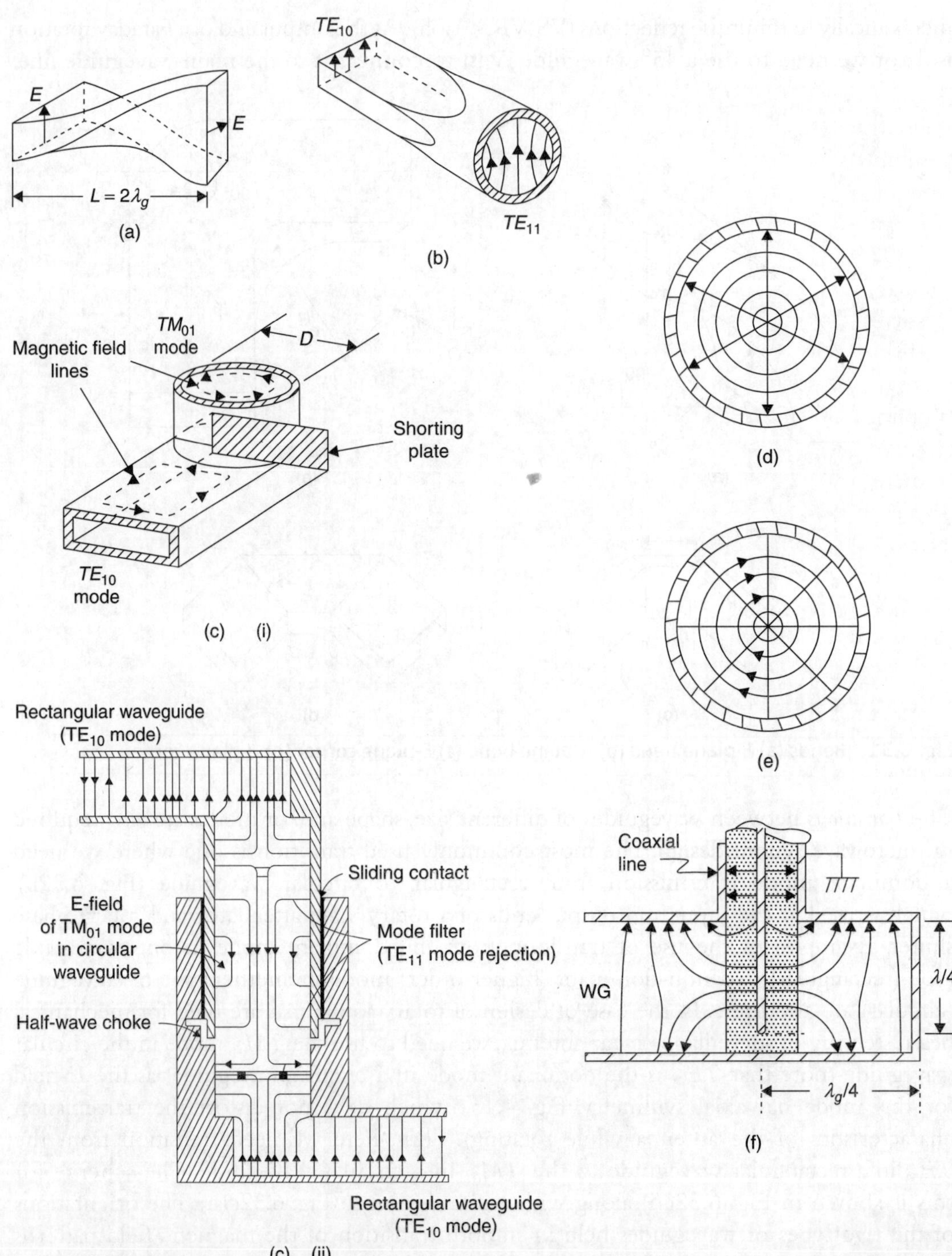

Fig. 6.32 (a) 90° waveguide twist (b) Rectangular to circular waveguide transition (c)(i) TE_{10} mode in rectangular waveguide to TM_{01} mode in circular waveguide transition (ii) Rotary joint (d) Mode filter for TM_{01} mode transmission (e) Mode filter for TE_{01} mode transmission (f) Coaxial to waveguide adapter.

be corrupted by the dominant mode (TE_{11} mode) in a circular waveguide that has a cut-off frequency lower than that of the TM_{01} mode. Thus, to suppress the TE_{11} undesired mode (being automatically launched in this case), we need to use a *mode filter*. A simple way to realise this mode filtering is to use thin metallic rings embedded in a low-loss dielectric disc (resembling a thin Swiss-roll) which is placed inside the circular waveguide (a cross-sectional view of which is shown in Fig. 6.32d). Another example of the requirement of transition from the TE_{10} in a rectangular waveguide to TE_{01} in a circular waveguide is in realising very high Q $(= \omega L/R_s)$ of the cylindrical cavity which uses the TE_{011} mode, as only the TE_{01} in a circular waveguide has circumferential surface current lines on the waveguide walls making R_s small. Such high Q cylindrical cavities are used in the design of wavemeters (used for the measurement of microwave frequencies) Section 12.5.1, and also in the design of echo boxes (used in the performance study of radar systems). This is normally done by slot or iris coupling of the signal from a rectangular waveguide to a circular waveguide, Section 4.7. However, when the TE_{01} in a circular waveguide is used, the corrupting TE_{11} would be there, which should be filtered out to obtain the pure TE_{01} mode in a circular waveguide. This mode filtering is now realised with the help of a mode filter which looks like the one shown in Fig. 6.32c, with the E field of the TE_{11} mode being parallel to the metallic wires while the E field of the TE_{01} mode will be perpendicular to the metallic wires at every location. Arrow headed lines in Figs 6.32d and 6.32e represent the E-field for TM_{01} and TE_{01} modes respectively in a circular waveguide. For E-field pattern of TE_{11} mode in a circular waveguide, refer Fig. 4.15.

In most of the microwave test benches, we need a *waveguide-to-coaxial adapter*. It employs a short section of a rectangular waveguide (WG) coupled to a coaxial line with an N-type connector (Fig. 6.32f). The inner conductor of a coaxial line with its dielectric cladding is inserted via the central line on the top of the broad wall of a rectangular waveguide while the outer conductor of the coaxial line is soldered to the waveguide wall. The coaxial line acts as a $\lambda/4$ pickup antenna. One end of the waveguide is closed by a short circuit placed at $\lambda_g/4$ from the plane of the coaxial probe and the input signal enters the waveguide through the other end (Fig. 6.32f). This arrangement smoothly adapts the TEM mode of the coaxial line to the TE_{10} mode in the rectangular waveguide.

6.2.11 Flanges and Connectors

While joining two waveguides, WG1 and WG2 *flanges* are used. There are two types of flanges: plane-to-plane flange and plane-to-choke flange. In a plane-to-plane flange (Fig. 6.33a) two plane flanges are bolted together so that the mating surfaces make good ohmic contact; particularly at the junction points J_1 and J_2 which should act as a perfect short circuit, that is, the longitudinal current lines on the broad wall of the waveguides should be continuous. If the mating surfaces are flat and clean, reflections at junctions can be neglected for low power and medium power applications with VSWR down to 1.05 or so. But for high power applications (as in high power radar) plane-to-plane flanges are not advisable to use as undesired leakage of microwave signal at the junctions J_1 and J_2 may lead to unsafe health hazards.

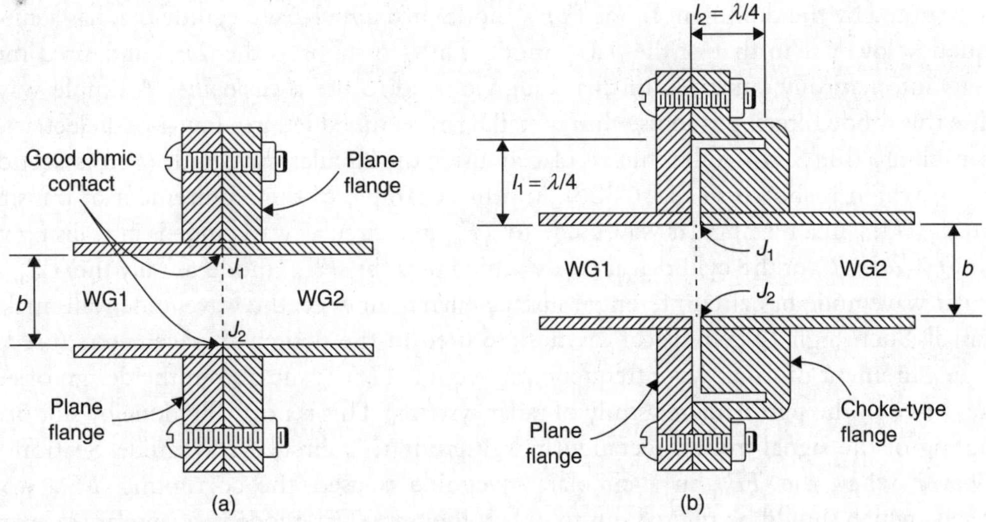

Fig. 6.33 Waveguide flanges (a) Plane-to-plane type flange (b) Plane-to-choke-type flange.

An effective connection between waveguides, particularly in high power applications, involves the use of one plane flange and one choke-type flange (Fig. 6.33b). In such flange joints, though the junctions J_1 and J_2 represent DC open, from a microwave point of view, they are short circuit junctions as the groove of length $l_1 + l_2 = \lambda/2$, represents a short circuited half wave line making the junction points J_1 and J_2 with zero impedance. In other words, the two waveguide walls at J_1 and J_2 points will appear continuous at microwave frequency thus ensuring the continuity of longitudinal RF surface currents on the broad wall of the two waveguides at the junction points J_1 and J_2; which are otherwise DC open.

Connectors are inevitably useful components in microwave systems. *BNCs* (Bayonet Notch Connector) are useful RF connectors for coaxial cables. These connectors are made to match the characteristic impedance of a cable at either 50 Ω or 75 Ω and are normally useful up to 3 GHz. *N-type* connectors are used at higher microwave frequencies, typically up to 20 GHz and are also available in 50 Ω or 75 Ω versions. *SMA*s (Sub-miniature version A) are precision coaxial connectors generally designed for a 50 Ω system and can be used from DC to 26 GHz; commonly used at 18–26 GHz. But at and above 26 GHz, that is, in millimetre-wave frequencies, K-connectors (that principally uses air dielectrics) are used to cover from 26–110 GHz (3.5 mm at 26 GHz, 2.9 mm at 40 GHz, 2.4 mm at 50 GHz, 1.85 mm at 67 GHz and 1.0 mm at 110 GHz). Sample photographs of different types of connectors used at microwave and millimetre-wave frequencies are shown in Fig. 6.34.

6.3 Lumped-circuit Elements at Microwave Frequency

At microwave frequencies we cannot use conventional low frequency lumped-circuit *R*, *L*, *C* elements as the dimensions of such circuit elements become comparable to the wavelength of operation. Thus, commonly used circuit elements at microwave frequencies are realised

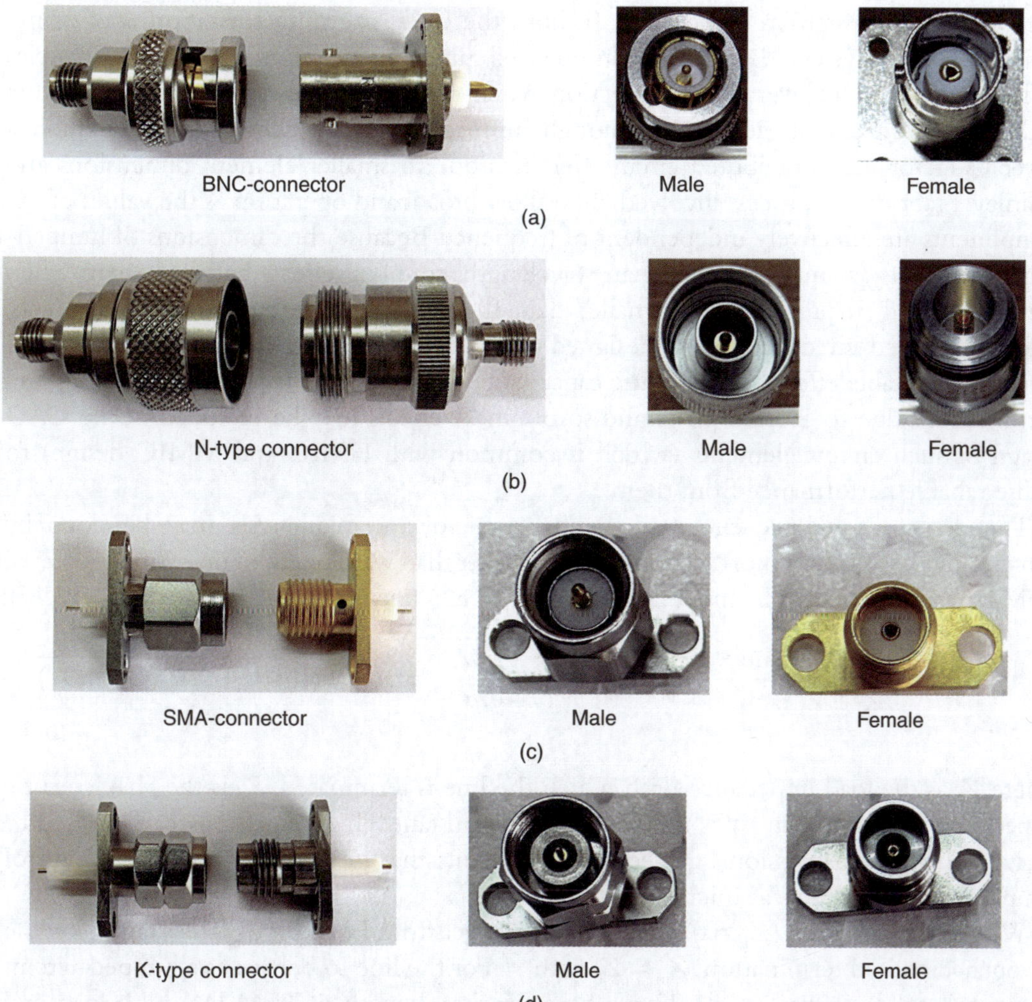

Fig. 6.34 Microwave and millimetre-wave connectors (a) BNC connector (b) N-type connector (c) SMA connector (d) K-type connector.

with distributed circuits, that is, some length of transmission line with suitable termination (Section 3.2.4). In fact, the lumped-circuit elements for operation at microwave frequency should have its size across any dimension less than at least one tenth of the operating wavelength. In other words, the lumped-circuit elements at any frequency should be electrically small so that the current distribution in the element is practically uniform (that is, exhibiting negligible phase shift). These types of circuit elements were not feasible at microwave frequencies in the past, however, with the advent of advanced photolithographic and thick and thin-film technology of today, it is now possible to design lumped-circuit elements at microwave frequencies up to 30 GHz and with limited values even up to 60 GHz.

The use of lumped-circuit elements at microwave frequencies has many advantages as far as the monolithic microwave integrated circuit (MMIC) and hybrid microwave integrated circuit (HMIC) design are concerned. In both the cases, size miniaturisation is of prime concern and subsystem design with lumped-circuit elements at microwave frequencies definitely leads to an overall size reduction when compared with the same design done with distributed circuit elements. Although lumped-circuit elements typically exhibit a lower Q factor than distributed circuit elements due to smaller element dimensions and multilevel fabrication process involved, they allow broadband operation as the values of the components are effectively independent of frequency. Because the dimensions of lumped-circuit elements are much smaller than wavelength, coupling effects between them when they are placed in proximity are smaller than those of the distributed circuit elements. Though lumped-circuit elements designed at microwave frequencies might suffer from certain undesirable effects like parasitic capacitance, spurious resonance, fringing fields and perturbations due to ground plane and so on, these effects may be included in the CAD design of such circuit elements (which is common with HMIC and MMIC design) to realise reliable performance from them.

The design of lumped-circuit elements at microwave frequencies may be done by considering them as very short sections (much smaller than wavelength of operation, $l << \lambda$) of TEM transmission lines. The input impedance of a TEM line of length l is given by (Eq. 3.17):

$$Z_{in} = Z_0 \frac{Z_L + Z_0 \tanh \gamma l}{Z_0 + Z_L \tanh \gamma l}, \text{ with } Z_0 = \sqrt{\frac{R + j\omega L}{G + j\omega C}} \quad \text{and} \quad \gamma = \sqrt{(R + j\omega L)(G + j\omega C)} \tag{6.6a}$$

where Z_L is the load impedance with which the line is terminated, Z_0 is the characteristic impedance of the line and γ is the propagation constant. These equations can be used to derive equivalent expressions and equivalent circuits that give insight into the design of lumped-circuit elements at microwave frequencies.

With short-circuited ($Z_L = 0$) termination, we get from Eq. 6.6a, $Z_{in} = Z_0 \tanh \gamma l$ while for open-circuited termination, $Z_{in} = Z_0 \coth \gamma l$. For the line to behave as a lumped-circuit element ($l << \lambda$), that is, ($\gamma l << 1$), we have for short length of TEM line with load end short-circuited (with first order approximation):

$$Z_{in} = Z_0 \gamma l = (R + j\omega L)l \tag{6.6b}$$

While with load end open-circuited (with second order approximation):

$$Z_{in} = Z_o \left(\frac{1}{\gamma l} + \frac{\gamma l}{3} \right) = \frac{G}{(\omega C)^2 l} + \frac{1}{j\omega C l} + \frac{Rl}{3} \tag{6.6c}$$

where we have considered $G << \omega C$ and $\omega L << R$, and have used the series expansions:

$$\tanh x = x - \frac{1}{3} x + ... \quad \text{and} \quad \coth x = x^{-1} + \frac{x}{3} - .....$$

It may thus be observed from Eq. 6.6b that the two terminal circuit elements will behave as an inductor in series with a resistor. When the thickness of conductors making the TEM line are much greater than the skin depth at the operating frequency (which is a practical situation at microwave frequency), conductors may be considered to be loss-less ($R = 0$), and hence a lumped inductor is obtained at microwave frequencies. It may be noted that in order to realise such an inductor, it is not necessary to have both the conductors of the transmission line, instead a metallic strip/wire in the form of loop or spiral will suffice (Fig. 6.35). Again, from Eq. 6.6c, it may be observed that the two-terminal circuit element behaves as a capacitor with a series resistance $[G/(\omega C)^2 l + Rl/3]$; while if we consider $G = 0$ it would represent a resistor with a capacitance. A single gap or inter-digital set of gaps in

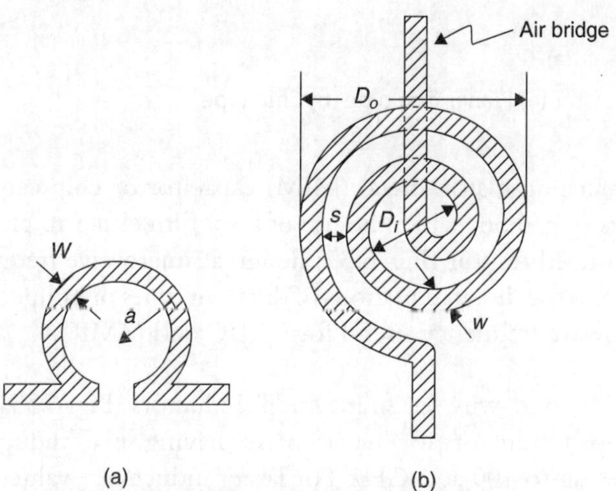

Fig. 6.35 Lumped planar inductor (a) Round wire (loop type) (b) Spiral type.

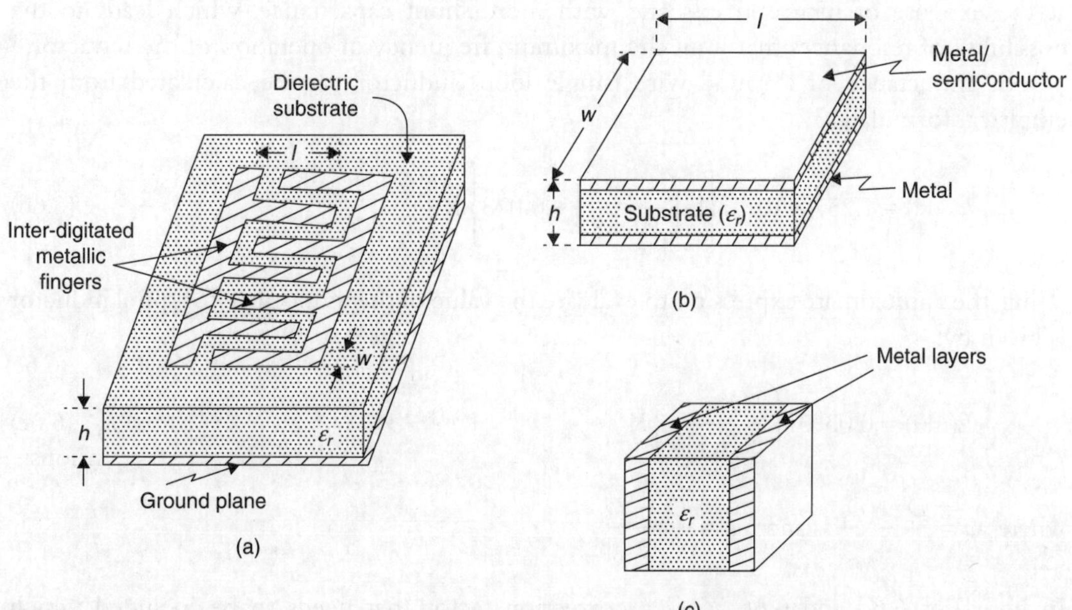

Fig. 6.36 Lumped capacitor (a) Inter-digital type (b) MIM/MIS type (c) Chip type.

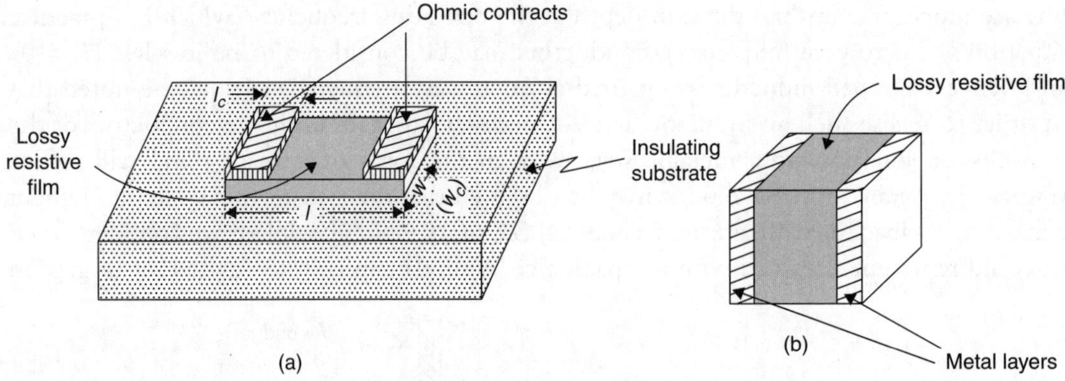

Fig. 6.37 Lumped resistor (a) Planar film type (b) Chip type.

transmission lines, metal–insulator–metal (MIM) capacitor or chip capacitor are the types of lumped capacitors (Fig. 6.36) while the use of lossy film either in planar from or in chip form is used for a lumped resistor (Fig. 6.37) design at microwave frequencies.

Next, we will discuss the design equations of different types of lumped inductor, capacitor and resistor at microwave frequency useful for HMIC and MMIC.

Lumped inductor: Round-wire (or single-loop) inductors (Fig. 6.35a) are used as series-connected type lumped inductors for hybrid MICs having small inductance values (0.5 to 4 nH) with Q values up to 100 at 2 GHz. For larger inductance values (10 nH or higher), a spiral inductor with air bridge (for connecting the centre of the spiral to the ground plane) (Fig. 6.35b) is used for MMIC applications. Larger inductance values are obtained at the expense of increased loss and with more shunt capacitance, which leads to the possibility of resonances that limit the maximum frequency of operation of the inductor.

The inductance of a round-wire (single-loop) inductor can be calculated from the empirical formula [13]:

$$L(\text{nH}) = 1.257 \times 10^{-3} a \left\{ \ln\left(\frac{a}{W+t}\right) + 0.078 \right\} K_g \tag{6.6d}$$

While the approximate expression to evaluate the value of the inductance for spiral inductor is given by:

$$L(\text{nH}) = 0.03937 \left(\frac{a^2 n^2}{8a + 11c}\right) K_g \tag{6.6e}$$

where $a = \dfrac{D_o + D_i}{4}$ and $c = \dfrac{D_o - D_i}{2}$

In both Eqs 6.6d and 6.6e, K_g is a correction factor that needs to be included which accounts for the presence of ground plane and is given by:

$$K_g = 1.4 + 0.217 \ln\left(\frac{W}{5t}\right) \qquad 5 < \frac{W}{t}, < 100 \qquad \text{(round wire)}$$

$$= 1 + 0.333\left(1 + \frac{S}{W}\right)^{-1.7} \qquad \text{(spiral)} \qquad (6.6f)$$

where t is the thickness of the conductor in either case.

Lumped capacitor: The inter-digital capacitor (Fig. 6.36a) consists of an inter-digital type metallisation pattern printed on a dielectric substrate and backed by a metallic ground plane. The inter-digital capacitor relies on the fringing capacitance between the long common-edge areas of the metal fingers, which are separated by just a few microns. These capacitors are usually formed in MMIC using the same interconnect metallisation as used for the transmission lines. This fringing capacitance is fairly low, so these capacitors are only able to reach capacitance values of a few pF. The approximate value of capacitance of an inter-digital capacitor structure can be calculated from the empirical formula [13]:

$$C = (\varepsilon_r + 1)l\{(N-3)A_1 + A_2\} \text{ pF} \qquad (6.6g)$$

with $A_1 = 4.409 \tanh\left\{0.55\left(\frac{h}{w}\right)^{0.45}\right\} 10^{-6} \text{ pF/}\mu\text{m}$

and $A_2 = 9.92 \tanh\left\{0.52\left(\frac{h}{w}\right)^{0.5}\right\} 10^{-6} \text{ pF/}\mu\text{m}$

where l, w and h represent the finger length (usually $\leq \lambda/4$), finger width and the height of the substrate respectively, and N is the number of fingers of the inter-digital capacitor.

Greater capacitance values of up to 25 pF can be obtained either with metal–insulator–metal (MIM) capacitor used in MMIC or chip capacitor for hybrid MIC. The MIM capacitor (Fig. 6.36b) consists of two metallic layers (of area lw) between which an insulating oxide layer (silicon dioxide) of thickness h is sandwiched. There are also MIS (metal–insulator–semiconductor) capacitors in which one terminal of the capacitor is made of a highly doped semiconductor instead of metal. The capacitance of MIM/MIS capacitor can be calculated from:

$$C = \varepsilon_0 \varepsilon_r \frac{lw}{h} \qquad (6.6h)$$

The chip capacitor (Fig. 6.36c) is formed by sandwiching a high permittivity dielectric (titanium oxide, a ceramic) between two metal layers. This is basically a modified MIM.

Larger values of capacitance (compared to an inter-digital type capacitor) can be realised with MIM/MIS or chip capacitor because they are constructed of two relatively large plates of metal separated by a much smaller distance, and the gap in between is also filled with an insulating dielectric material, which further increases the capacitance.

Lumped resistor: The resistor elements realised as lumped components that may be used in MMIC and hybrid MIC are respectively of planar film type or chip type (Figs 6.37a and 6.37b). In the planar film type resistor, a thin resistive film of lossy material (like nichrome, tantalum nitride or doped semiconductor material) is deposited on an insulating material (usually a ceramic material—alumina or quartz) having ohmic contacts. The total resistance of a planar film resistor is the sum of resistive film and the resistance of the two ohmic contacts and is given by [13]:

$$R = R_s \frac{l}{W} + 2R_{sc}\frac{l_c}{W_c} \tag{6.6i}$$

where R_s, l, W are respectively the sheet resistance (ρ/t, ρ the bulk resistivity and t the thickness), length and width of the film and and R_{sc}, l_c, W_c are the corresponding quantities for the ohmic contacts.

Thin-film and thick-film hybrid technologies have been used to manufacture lumped element chip resistors for low power and high power applications. In both cases, an alumina substrate is used for low power applications and beryllia or aluminium nitride substrate is used for high power applications.The lossy film which is deposited on the substrate material is nichrome or tantalum nitride for thin-film technology while various compositions of ruthenium oxide paste are used for thick-film technology.

Resistance values in the range 50 to 500 ohms are typically realisable, however, low resistance values are hard to obtain.

Sometimes lumped circuit realisation of resonators are required for HMIC and MMIC design. This can be done by combining the planar printed inductors and capacitors as shown in Figs 6.38a and 6.38b for series resonant circuit and parallel resonant circuit which are made up of a round wire (or single-loop) inductor and inter-digital capacitor.

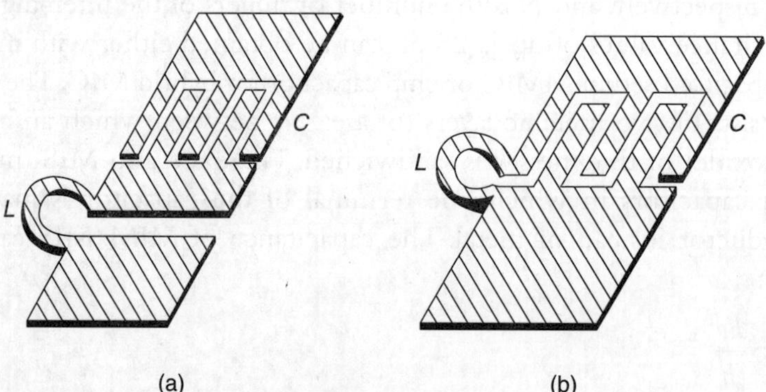

Fig. 6.38 Resonators with lumped circuit elements (a) Series resonator (b) Parallel resonator.

6.4 Distributed-Circuit Elements Using Microstrip Transmission Line

We need to design distributed-circuit elements such as distributed inductor (either in series or in parallel with the host transmission line) or distributed capacitor (in general in shunt as for by-pass capacitor) with microstrip transmission line in the design of planar microwave filters, input/output matching networks for amplifiers and so on. How such distributed inductors (*L*) and capacitors (*C*) can be realised with short sections of conventional transmission line or waveguides was discussed in Chapter 3. However, in MIC design they are to be designed with planar transmission line, especially, microstrip line. The design of such distributed-circuit elements (*L, C*) at microwave frequency using microstrip line will be discussed below.

The distributed-circuit elements are also realised with short section of microstrip line ($l < \lambda_{eff}/10$, where λ_{eff} is the wavelength of the propagating wave in microstrip line given by Eq. 3.43) typically the electrical length is chosen to be $\theta = \beta l = \pi/6$, that is, $l = \lambda_{eff}/12$;

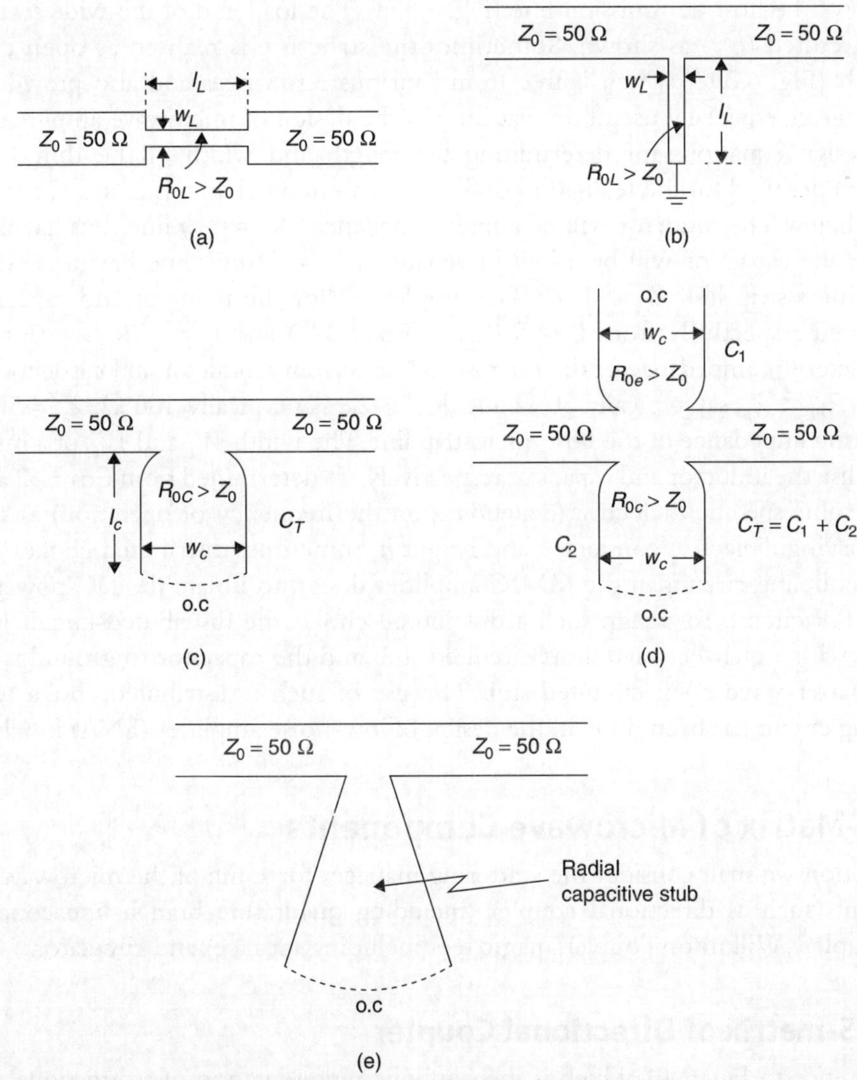

Fig. 6.39 Realisation of distributed inductor and capacitor with microstrip line (a) Series inductor (b) Shunt inductor (c) Shunt capacitor (d) Two shunt capacitor equivalent to single capacitor of (c) (e) Radial stub type capacitor.

otherwise the equivalent component effect will depart more and more from the ideal lumped component being realised with distributed circuit. The distributed series inductor is realised by a thin section of microstrip line being sandwiched between two 50 Ω host transmission lines (Fig. 6.39a) while shunt inductance is realised with a thin section of the line connected between two 50 Ω host line in which the other end of the thin line is grounded (Fig. 6.39b) through a via connection to the ground plane of the printed circuit board (PCB) on which the component is formed. The distributed shunt capacitor is realised with a wide line connected by two host lines of characteristic impedance 50 Ω on either side of it (Fig. 6.39c). The capacitor section is attached to the 50 Ω host lines with small tapered section as indicated in the figure to improve impedance transition. A further improvement is possible by splitting the capacitor into two parts and placing it on both sides of the host transmission line (Fig. 6.39d). The load end of the wide section line is open circuited (o.c) as shown. Sometimes the structure is realised as open circuited radial stub (Fig. 6.39e) which is free from spurious resonances and also provides broad band operation; especially useful on bias lines in the design of microwave amplifiers.

The design equations for determining the length and width of the thin and wide microstrip lines used for the realisation of distributed-circuit elements L and C respectively are given below. The inductor will be a high impedance ($X_L = \omega L$) line thus having a thin line while the capacitor will be a low impedance ($X_C = 1/\omega C$) line having a wide line; since for microstrip line $Z_{ch}\ \alpha\ 1/W$. The line length for the inductor and capacitor will be calculated respectively from $l_L = fLR_{0L}\lambda_{\text{eff}}$ (Eq. 3.32b) and $l_C = fCR_{0C}\lambda_{\text{eff}}$ (Eq. 3.31b). The characteristic impedance of the microstrip line realising inductor and capacitor may be chosen as $R_{0L} < Z_0$, say typically 30 Ω and $R_{0C} > Z_0$, say typically 100 Ω; $Z_0 = 50$ Ω, the characteristic impedance of the host microstrip line. The width W_L and W_C of a microstrip line to realise the inductor and capacitor respectively are determined from Eqs 3.42 and 3.41 choosing some specific dielectric (depending on the frequency of operation) as substrate material having dielectric constant ε_r and height h. Sometimes distributed chokes are used as DC decoupling circuits in the MMIC amplifier design to isolate the DC power supply from the RF circuit. To design such a distributed choke, the distributed-circuit inductor is realised with a quarter-wave short circuited stub and the capacitor to ground is realised with a quarter-wave open circuited stub. The use of such a distributed choke type DC decoupling circuit has been done in the design of low-noise amplifier (LNA) in Chapter 9.

6.5 S-Matrix of Microwave Components

In this section we shall consider the scattering matrices for some of the microwave passive components such as directional coupler (including quadrature branch line coupler and Lange coupler), Wilkinson coupler, magic tee (including rat race) and circulator.

6.5.1 S-matrix of Directional Coupler

With reference to Fig. 6.40, which is the symbolic representation of a directional coupler the generalised scattering matrix of this 4-port device is:

$$[S] = \begin{bmatrix} S_{11} & S_{12} & S_{13} & S_{14} \\ S_{21} & S_{22} & S_{23} & S_{24} \\ S_{31} & S_{32} & S_{33} & S_{34} \\ S_{41} & S_{42} & S_{43} & S_{44} \end{bmatrix} \tag{6.7a}$$

Fig. 6.40 Symbolic representation of directional coupler.

Using electromagnetic properties of the directional coupler and the properties of S-parameters, the S-matrix of the directional coupler can be exhibited in a simpler and elegant form.

In a directional coupler, all the 4 ports are matched. Thus, all the ports are reflection-less. Hence, we can put all the diagonal elements of the S-matrix given in Eq. 6.7a as zero ($S_{ii} = 0$) when we get:

$$[S] = \begin{bmatrix} 0 & S_{12} & S_{13} & S_{14} \\ S_{21} & 0 & S_{23} & S_{24} \\ S_{31} & S_{32} & 0 & S_{34} \\ S_{41} & S_{42} & S_{43} & 0 \end{bmatrix} \tag{6.7b}$$

A signal fed at port 1 will not reach at port 4 and similarly a signal fed at port 2 will not reach at port 3 (for a waveguide directional coupler), thus:

$$S_{41} = S_{14} = S_{32} = S_{23} = 0 \tag{6.7c}$$

when Eq. 6.7b can be written as:

$$[S] = \begin{bmatrix} 0 & S_{12} & S_{13} & 0 \\ S_{12} & 0 & 0 & S_{24} \\ S_{13} & 0 & 0 & S_{34} \\ 0 & S_{24} & S_{34} & 0 \end{bmatrix} \tag{6.7d}$$

In the above, we have used the reciprocal nature of directional coupler ($S_{ij} = S_{ji}$).

Next, we apply the loss-less property of the directional coupler at microwave frequency to simplify further the S-matrix given by Eq. 6.7d. Using the unity property of the S-matrix we can write from Eq. 6.7d:

$$|S_{12}|^2 + |S_{13}|^2 = 1 \tag{6.7e}$$

$$|S_{12}|^2 + |S_{24}|^2 = 1 \tag{6.7f}$$

$$|S_{13}|^2 + |S_{34}|^2 = 1 \tag{6.7g}$$

$$|S_{24}|^2 + |S_{34}|^2 = 1 \tag{6.7h}$$

While using zero property of the S-matrix, we get from Eq. 6.7d:

$$S_{12}S_{24}^{\star} + S_{13}S_{34}^{\star} = 0 \tag{6.7i}$$

Now from Eqs 6.7e and 6.7f: $|S_{13}| = |S_{24}|$ (6.7j)

While from Eqs 6.7f and 6.7h: $|S_{12}| = |S_{34}|$ (6.7k)

We may choose the reference planes in the respective ports (that is, 1 with respect to 2 and 3 with respect to 4) in such a way that S_{12} and S_{34} become real quantities, hence we may write:

$$S_{12} = S_{34} = \alpha \tag{6.7l}$$

where α is a real quantity and physically α^2 represents the power to the through port, that is, the transmission factor.

Applying the result of Eq. 6.7l in Eq. 6.7i, we get:

$$\alpha(S_{24}^{\star} + S_{13}) = 0$$

or, $S_{24}^{\star} + S_{13} = 0 \quad (\text{since } \alpha \neq 0)$ (6.7m)

Thus $S_{13} = -S_{24}^{\star} = -j\beta$ (6.7n)

where β is a real quantity and physically β^2 represents the coupling factor. The negative sign preceding β indicates that the signal flow from 1 to 3 or from 2 to 4 basically represents the attenuation of the signal (coupling a fraction of the input signal to the coupled port). Substituting Eqs 6.7l and 6.7n in Eq. 6.7d, we have:

$$[S] = \begin{bmatrix} 0 & \alpha & j\beta & 0 \\ \alpha & 0 & 0 & j\beta \\ j\beta & 0 & 0 & \alpha \\ 0 & j\beta & \alpha & 0 \end{bmatrix} \tag{6.7o}$$

From Eqs 6.7l and 6.7n and any of the Eqs 6.7e to 6.7h it can be seen that:

$$\alpha^2 + \beta^2 = 1 \tag{6.7p}$$

This is also justified from the principle of conservation of power. With simple calculations, for a 30 dB directional coupler, the coupling factor $\beta^2 = 0.001$ and transmission factor $\alpha^2 = 0.9994$.

There is an alternative way to present the S-matrix of the directional coupler. Since the transmission factor α^2 and the coupling factor β^2 represent the average power of the electromagnetic signal while the S-parameters are expressed in equivalent voltage we can replace α and β by $1/\sqrt{2}$ in Eq. 6.7o, when we have:

$$[S] = \frac{1}{\sqrt{2}}\begin{bmatrix} 0 & 1 & j & 0 \\ 1 & 0 & 0 & j \\ j & 0 & 0 & 1 \\ 0 & j & 1 & 0 \end{bmatrix} \tag{6.7q}$$

It may further be noted that in some texts, ports 2 and 4 are interchanged when compared with the port numbering used in this text. Then, the S-matrix of the directional coupler comes out as:

$$[S] = \frac{1}{\sqrt{2}}\begin{bmatrix} 0 & j & 1 & 0 \\ j & 0 & 0 & 1 \\ 1 & 0 & 0 & j \\ 0 & 1 & j & 0 \end{bmatrix} \tag{6.7r}$$

The directional coupler we have been discussing so far is a 90° hybrid and is similar to the quadrature (90°) branch line coupler/hybrid and also the Lange coupler. All these couplers are known as symmetrical couplers. The S-matrix for all these type of qudrature coupler/hybrids will be represented by Eq. 6.7q or Eq. 6.7r, depending on the port numbering as mentioned above. But, for other hybrid structures such as magic tee and rat-race which are 180° hybrids (also known as the anti-symmetrical coupler) the S-matrix is different from the above and is derived below.

The Wilkinson coupler (used as power divider/combiner) is a 3-port RF device whose S-matrix can directly be written from its physical principle of operation and is given by:

$$[S] = \frac{j}{\sqrt{2}}\begin{bmatrix} 0 & 1 & 1 \\ 1 & 0 & 0 \\ 1 & 0 & 0 \end{bmatrix} \tag{6.7s}$$

6.5.2 S-matrix of Magic Tee

With reference to Fig. 6.41, which is the symbolic representation of a magic tee, we find that it is also a four-port device and hence its general S-matrix is similar to that given in Eq. 6.7a.

We can apply the electromagnetic properties of a magic tee and the properties of the S-parameters to get the S-matrix of the magic tee in a simple and elegant form.

If ports 1 and 2 are identically terminated, then a signal at port 3 reaches port 1 and 2 with equal magnitude and at same phase, and hence:

$$S_{13} = S_{23} \tag{6.8a}$$

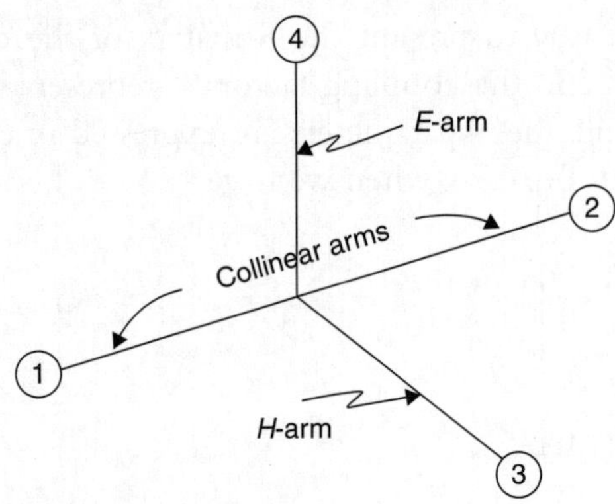

Fig. 6.41 Symbolic representation of a magic tee.

Also, a signal fed at port 4 will be coupled to port 1 and 2 with equal magnitude but with opposite phase, thus:

$$S_{14} = -S_{24} \tag{6.8b}$$

The magic tee is reciprocal in the sense that two signals of equal magnitude and of same phase fed at ports 1 and 2 will add and couple the signal to port 3, also two signals of equal magnitude and opposite phase fed at ports 1 and 2 will add and couple the signal to port 4. So, for a reciprocal junction, we can apply the symmetry property of the S-matrix, that is, $S_{ij} = S_{ji}$, $i \neq j$, thus:

$$\begin{aligned} S_{21} &= S_{12} \qquad & S_{32} &= S_{23} \\ S_{31} &= S_{13} \qquad & S_{42} &= S_{24} \\ S_{41} &= S_{14} \qquad & S_{43} &= S_{34} \end{aligned} \tag{6.8c}$$

Further, with identical terminations at ports 1 and 2, a signal fed at port 3 will not be coupled to port 4 and vice versa (as these ports are de-coupled ports), thus:

$$S_{34} = S_{43} = 0 \tag{6.8d}$$

Substituting the results obtained in Eqs 6.8a–d in Eq. 6.7a we get:

$$[S] = \begin{bmatrix} S_{11} & S_{12} & S_{13} & S_{14} \\ S_{12} & S_{22} & S_{13} & -S_{14} \\ S_{13} & S_{13} & S_{33} & 0 \\ S_{14} & -S_{14} & 0 & S_{44} \end{bmatrix} \tag{6.8e}$$

Matching elements that do not destroy the symmetry of the junction may be placed in the E-plane arm and the H-plane arm, so as to make:

$$S_{44} = S_{33} = 0 \tag{6.8f}$$

Using this, the S-matrix becomes:

$$[S] = \begin{bmatrix} S_{11} & S_{12} & S_{13} & S_{14} \\ S_{12} & S_{22} & S_{13} & -S_{14} \\ S_{13} & S_{13} & 0 & 0 \\ S_{14} & -S_{14} & 0 & 0 \end{bmatrix} \tag{6.8g}$$

Now we use the unitary property of the *S*-matrix (assuming the structure to be loss-less). Putting the sum of the products of each term of the first column with its conjugate equal to unity (applying the unity property) we get:

$$|S_{11}|^2 + |S_{12}|^2 + |S_{13}|^2 + |S_{14}|^2 = 1 \tag{6.8h}$$

And similarly the expression from the second column of Eq. 6.8g is:

$$|S_{12}|^2 + |S_{22}|^2 + |S_{13}|^2 + |S_{14}|^2 = 1 \tag{6.8i}$$

Subtracting Eq. 6.8i from Eq. 6.8h we get:

$$|S_{11}|^2 - |S_{22}|^2 = 0 \quad \text{or} \quad |S_{11}| = |S_{22}| = 0 \tag{6.8j}$$

Again from columns 3 and 4 of Eq. 6.8g, using unity property:

$$2|S_{13}|^2 = 1 \quad \text{or} \quad |S_{13}| = \frac{1}{\sqrt{2}}, \quad \text{and} \quad 2|S_{14}|^2 = 1 \quad \text{or} \quad |S_{14}| = \frac{1}{\sqrt{2}} \tag{6.8k}$$

Using the relation 6.8k in 6.8h:

$$|S_{11}|^2 + |S_{12}|^2 + \left(\frac{1}{2}\right) + \left(\frac{1}{2}\right) = 1 \quad \text{or} \quad |S_{11}|^2 + |S_{12}|^2 = 0 \tag{6.8l}$$

This sum can equal zero only when both S_{11} and S_{12} are individually zero:

$$S_{11} = S_{12} = 0 \tag{6.8m}$$

Thus, from Eq. 6.8j:

$$S_{22} = 0 \tag{6.8n}$$

Substituting the results from Eqs 6.8m and 6.8n in Eq. 6.8g, we have:

$$[S] = \begin{bmatrix} 0 & 0 & S_{13} & S_{14} \\ 0 & 0 & S_{13} & -S_{14} \\ S_{13} & S_{13} & 0 & 0 \\ S_{14} & -S_{14} & 0 & 0 \end{bmatrix} \tag{6.8o}$$

By proper choice of terminal planes in the *E* and *H*-arms, S_{13} and S_{14} can be made real quantities. Thus, the scattering matrix of a magic tee may be exhibited in the form:

$$[S] = \frac{1}{\sqrt{2}} \begin{bmatrix} 0 & 0 & 1 & 1 \\ 0 & 0 & 1 & -1 \\ 1 & 1 & 0 & 0 \\ 1 & -1 & 0 & 0 \end{bmatrix} \tag{6.8p}$$

EXAMPLE 6.7

2 mW and 4 mW of microwave power are fed to the two ports of the collinear arms of a magic tee with all its ports match terminated. Using the *S*-matrix, calculate the power output from all other ports.

Solution

For a magic tee with input at ports 1 and 2, we have the following matrix equation:

$$\begin{bmatrix} b_1 \\ b_2 \\ b_3 \\ b_4 \end{bmatrix} = \frac{1}{\sqrt{2}} \begin{bmatrix} 0 & 0 & 1 & 1 \\ 0 & 0 & 1 & -1 \\ 1 & 1 & 0 & 0 \\ 1 & -1 & 0 & 0 \end{bmatrix} \begin{bmatrix} a_1 \\ a_2 \\ a_3 \\ a_4 \end{bmatrix}$$

Carrying out the matrix multiplication:

$$b_1 = 0, \quad b_2 = 0, \quad b_3 = \frac{1}{\sqrt{2}}(a_1 + a_2), \quad b_4 = \frac{1}{\sqrt{2}}(a_1 - a_2)$$

Since power is proportional to the wave variable squared:

$$P_{i1} = 0.5a_1^2 = 2 \text{ mW}, \quad \text{thus } a_1 = 2 \text{ mW}$$

$$P_{i2} = 0.5a_2^2 = 4 \text{ mW}, \quad \text{thus } a_2 = 2\sqrt{2} \text{ mW}$$

When $b_3 = \frac{1}{\sqrt{2}}(2 + 2\sqrt{2}) = 3.414 \text{ mW}, \quad b_4 = \frac{1}{\sqrt{2}}(2 - 2\sqrt{2}) = -0.5857 \text{ mW}$

Therefore: $P_{03} = \frac{b_3^2}{2} = \frac{(3.414)^2}{2} = 5.827 \text{ mW}$ and $P_{04} = \frac{b_4^2}{2} = \frac{(-0.5857)^2}{2} = 0.1715 \text{ mW}$

EXAMPLE 6.8

Two magic tees are connected back to back and 4 mW of microwave power is fed through port 1 of one of the magic tee. Find the power output from all the remaining ports.

Solution

Let the input of 4 mW be applied through port 1 of the first magic tee on the left. Thus, with reference to Fig. 6.42, we observe that $P_{i1} = 4 \text{ mW} = a_1^2/2$, when $a_1 = 2\sqrt{2}$.

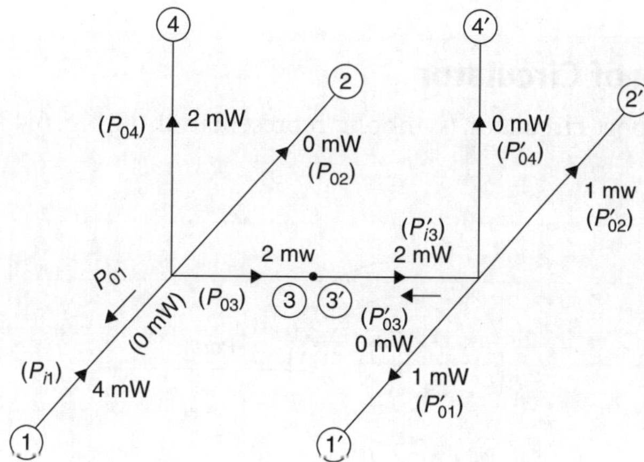

Fig. 6.42 Two magic tees connected back to back (Example 6.8).

The *S*-matrix for the first magic tee can be written as:

$$\begin{bmatrix} b_1 \\ b_2 \\ b_3 \\ b_4 \end{bmatrix} = \frac{1}{\sqrt{2}} \begin{bmatrix} 0 & 0 & 1 & 1 \\ 0 & 0 & 1 & -1 \\ 1 & 1 & 0 & 0 \\ 1 & -1 & 0 & 0 \end{bmatrix} \begin{bmatrix} a_1 \\ 0 \\ 0 \\ 0 \end{bmatrix}$$

Carrying out the matrix multiplication we get:

$$b_1 = 0, \quad b_2 = 0, \quad b_3 = \frac{a_1}{\sqrt{2}}, \quad b_4 = \frac{a_1}{\sqrt{2}}$$

Substituting the value of a_1 from the above we have $b_3 = b_4 = 2$

$$P_{o3} = P_{o4} = 0.5\left|b_{3,4}\right|^2 = 0.5(2)^2 = 2 \text{ mW} \quad \text{and} \quad P_{o1} = P_{o2} = 0.5\left|b_{1,2}\right|^2 = 0$$

Now $P_{03} = 2$ mW is input P'_{i3} to the second magic tee at port 3′, thus the S-matrix equation for the second magic tee is given by:

$$\begin{bmatrix} b'_1 \\ b'_2 \\ b'_3 \\ b'_4 \end{bmatrix} = \frac{1}{\sqrt{2}} \begin{bmatrix} 0 & 0 & 1 & 1 \\ 0 & 0 & 1 & -1 \\ 1 & 1 & 0 & 0 \\ 1 & -1 & 0 & 0 \end{bmatrix} \begin{bmatrix} 0 \\ 0 \\ a'_3 \\ 0 \end{bmatrix}$$

From which $b_1' = \frac{a_3'}{\sqrt{2}}, b_2' = \frac{a_3'}{\sqrt{2}}$ and $b'_3 = 0, b'_4 = 0$

Thus, we get $P'_{o1} = 0.5|b_1'|^2 = 1.0$ mW, $P'_{o2} = 0.5|b_2'|^2 = 1.0$ mW, $P'_{o3} = P'_{o4} = 0$

6.5.3 S-matrix of Circulator

Let us consider a 3-port circulator (symbolic representation in Fig. 6.43).

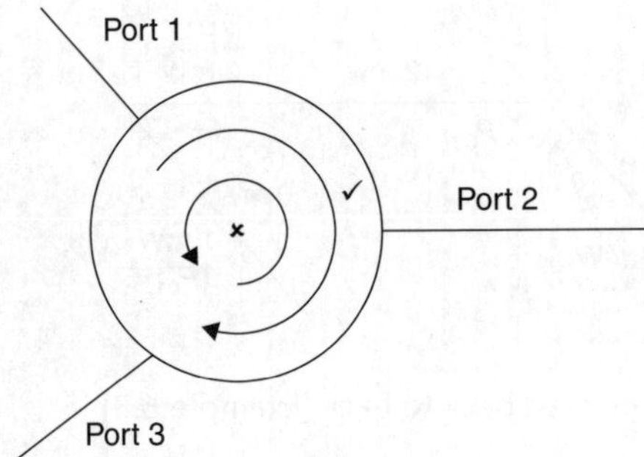

Fig. 6.43 Symbolic representation of 3-port circulator.

The generalised S-matrix for a 3-port device is:

$$[S] = \begin{bmatrix} S_{11} & S_{12} & S_{13} \\ S_{21} & S_{22} & S_{23} \\ S_{31} & S_{32} & S_{33} \end{bmatrix} \tag{6.9a}$$

If all the three ports are perfectly matched, that is, there is no reflection at the respective ports then:

$$S_{11} = S_{22} = S_{33} = 0 \tag{6.9b}$$

The device is non-reciprocal, thus referring to Fig. 6.43:

$$S_{12} = 0, S_{23} = 0, S_{31} = 0 \tag{6.9c}$$

Also, $$S_{12} \neq S_{21}, S_{13} \neq S_{31}, S_{23} \neq S_{32} \tag{6.9d}$$

Thus, applying 6.9b, 6.9c and 6.9d to Eq. 6.9a, we get:

$$[S] = \begin{bmatrix} 0 & 0 & S_{13} \\ S_{21} & 0 & 0 \\ 0 & S_{32} & 0 \end{bmatrix} \tag{6.9e}$$

Assuming the circulator to be loss-less we get from unity property:

$$|S_{21}|^2 = |S_{32}|^2 = |S_{13}|^2 = 1 \tag{6.9f}$$

Thus from matrix 6.9e, we have the *S*-matrix for a perfect 3-port circulator to be:

$$[S] = \begin{bmatrix} 0 & 0 & 1 \\ 1 & 0 & 0 \\ 0 & 1 & 0 \end{bmatrix} \tag{6.9g}$$

EXAMPLE 6.9

Prove that any loss-less, matched, *non-reciprocal* three port microwave junction is a perfect circulator. Also show that it is impossible to construct a perfectly matched, loss-less, *reciprocal* 3-port junction.

Solution

A perfectly matched 3-port junction has the scattering matrix of the form:

$$[S] = \begin{bmatrix} 0 & S_{12} & S_{13} \\ S_{21} & 0 & S_{23} \\ S_{31} & S_{32} & 0 \end{bmatrix} \tag{Ia}$$

(Since it is perfectly matched at all ports, there will be no self-reflection at the respective ports. Thus: $S_{11} = S_{22} = S_{33} = 0$.)

For a non-reciprocal junction the scattering matrix is no longer symmetrical, that is, $S_{ij} \neq S_{ji}$. However, if the junction is loss-less, conservation of power requires that the S-matrix is unitary. Applying unitary condition to Eq. (Ia) we have:

$$|S_{21}|^2 + |S_{31}|^2 = 1 \tag{Ib}$$

$$|S_{12}|^2 + |S_{32}|^2 = 1 \tag{Ic}$$

$$|S_{13}|^2 + |S_{23}|^2 = 1 \tag{Id}$$

(where the unity property is used.)

$$S_{31}S_{32}^{\star} = S_{21}S_{23}^{\star} = S_{12}S_{13}^{\star} = 0 \tag{Ie}$$

(where the zero property is used.)

Let us assume that $|S_{21}| \neq 0$, that is, there is transmission from port 1 to port 2. Then from Eq. Ie, $|S_{23}| = 0$, which means from Eq. Id, $|S_{13}| = 1$ and from Eq. Ie, $|S_{12}| = 0$. Thus, from Eq. Ic, $|S_{32}| = 1$ which means we have from Eq. Ie, $|S_{31}| = 0$. So we have:

$$|S_{21}| = |S_{13}| = |S_{32}| = 1 \tag{If}$$

and, $$|S_{23}| = |S_{12}| = |S_{31}| = 0 \tag{Ig}$$

Thus, there is perfect transmission from port 1 to port 2, from port 2 to port 3 and from port 3 to port 1. There is zero transmission in any other direction. Thus, the resultant scattering matrix of any matched, loss-less, *non-reciprocal* three port junction must have the form:

$$[S] = \begin{bmatrix} 0 & 0 & 1 \\ 1 & 0 & 0 \\ 0 & 1 & 0 \end{bmatrix} \tag{Ih}$$

(Since the choice of terminal planes will make S_{13}, S_{21} and S_{32} real.)

If the above analysis is applied to a loss-less, perfectly matched, *reciprocal* three-port junction, the reciprocity would require $S_{13} = S_{21} = S_{32} = 0$; since $S_{23} = S_{12} = S_{31} = 0$ from Eq. Ig.

The resultant scattering matrix would then vanish. The conclusion to be drawn from this result is that *it is impossible to construct a perfectly matched, loss-less, reciprocal three-port junction.*

RECAPITULATION

Waveguide and planar transmission line based components

- Distributed passive components at microwave frequencies use a conventional transmission line/waveguide/planar transmission line as the distributed circuit element to design passive components.
- *Terminations:* A *matched load* is one that absorbs all the incident power in a highly dissipative ceramic material (like bakelite or graphite coated with carbon film), thereby terminating the guide or the line in its characteristic impedance). A *variable short* produces an adjustable reactive load and being a short circuited load, it reflects all the incident power. The phase of the reflected power is varied by varying the position of the short circuit. A practical form uses two sections of quarter wave impedance transformers having low (Z_1) and high (Z_2) characteristic impedances when we have $Z_{seen} = (Z_1/Z_2)^2 Z_s$ thus, Z_{seen} will be a better short circuit than Z_s by a factor $(Z_1/Z_2)^2$.
- *Tee-junctions and hybrids: E-plane tee* is a waveguide junction in which a signal fed at the *E*-arm is coupled to the collinear arms which are equal in magnitude but opposite in phase, it is called a shunt tee. In a *H-plane tee*, a signal fed at the *H*-arm will produce signals of equal magnitude and of same phase, it is called a series tee. The *hybrid tee/magic tee* is a four-port microwave junction. With identical termination (impedance) of the collinear arms, a signal fed at the *E*-arm will not couple any signal to the *H*-arm and vice versa, however, a signal fed at the *E*-arm will couple that extent of signal to the *H*-arm by which the terminating impedances in the two collinear arms are mismatched. The *hybrid ring* or *rat-race* is used for the same purpose as the magic tee but is a convenient structure for coaxial line and planar circuits. In the hybrid ring, all the side arms are connected

to the main guide as series arms and the characteristic impedance of the main guide should be equal to $\sqrt{2}Z_0$, where Z_0 is the characteristic impedance of each of the series/side arms. Both hybrid tee/hybrid ring are 180° hybrids. There are many applications of hybrid tee/hybrid ring, one important use being in the front-end of microwave receivers as balanced microwave mixers. A *Wilkinson power divider/combiner* is a three-port network that divides the input power equally between the two output ports which are in phase; while two in-phase signals fed at the output ports will combine their power giving the phasor sum of those signals at the input port thus acting as power combiner (which is in fact true for any 3 dB power divider junction, such as the *E*, *H*-plane tee). Wilkinson power divider/combiners demand lines of different characteristic impedances and are thus easily constructed in strip-line (and not in waveguides) because the impedance of various lines can be easily adjusted by controlling the strip width. A *branch line quadrature hybrid* is another type of four-port microwave network in which the phase difference of the signals arriving at the two output ports is 90° while the fourth port is an isolated port. A branch line coupler that demands lines of different characteristic impedances is also easily realisable in strip-line. There is also a waveguide version of the quadrature hybrid known as a *short-slot hybrid*, it is used to design a *balanced duplexer* that is necessary for monostatic radar system where a single antenna is used for both transmission and reception.

- *Tuners:* Tuners are used to tune out impedance mismatches existing in a waveguide system. *E–H tuner* consists of a main waveguide upon which a series arm (*H*-arm) and a shunt arm (*E*-arm) are mounted. It resembles the series-shunt stub impedance matching scheme of the transmission line. A *slide screw tuner* consists of a slotted section of the waveguide in which a metallic screw can be inserted from the top broad wall via a non-radiating slot and whose depth of penetration and longitudinal position can be adjusted by moving it with the help of a carriage through a longitudinal slot half wavelength long in order to introduce respectively the capacitive and inductive susceptance for impedance matching.
- *Directional coupler:* It is a four-port microwave passive component (having an input port, output port, coupled port and isolated port) used for sampling out a known amount of power along the coupled port while the rest of the power goes along the output port and none of the signal is expected to go via the isolated port. The characteristics of a directional coupler are expressed in terms of the *coupling factor* (*C*) and *directivity* (*D*). $C = 10\log_{10}\left(P_{i1}/P_{o3}\right) = -20\log_{10}|S_{31}|$ and $D = 10\log_{10}\left(P_{o3}/P_{o4}\right) = 20\log_{10}|S_{31}/S_{41}|$. For a practical directional coupler, *C* is 10, 20 or 30 dB while *D* is of the order of 20–30 dB. The most common type of waveguide directional coupler is a *multi-hole directional coupler*, in which two waveguides share a common broad wall which contains a pair of slots whose size and offset from the central line on the broad wall determine the coupling factor. In the microstrip version, the *coupled line directional coupler* is used, which functions on the basis of even and odd-mode coupling in a coupled microstrip line. A modified version of the coupled line coupler with strong coupling is the *Lange coupler*.
- *Attenuator:* It is used to insert a known amount of reduction of output signal power with respect to the input in a waveguide transmission system. A *flap attenuator* is commonly

used, but for precession work, a *rotary vane attenuator* is used. In a flap attenuator, a resistive card of lossy dielectric material of proper shape is moved parallel to the narrow wall of the waveguide from the wall towards the centre to effect attenuation of the signal. Fixed attenuators of the flap type are typically 3, 6, 10 dB, and variable ones are 0–20 dB. In a rotary vane attenuator, a lossy restive card is moved about the axis of a circular waveguide and the angle of rotation (θ) of the resistive card with respect to the polarisation of the input electromagnetic signal give the attenuation A_{dB} of the signal given by $A_{dB} = -40 \log_{10} (\sin\theta)$. Variable attenuation of the order of 0–50 dB is realisable with a rotary vane attenuator.

- *Isolator:* The isolator is a two port non-reciprocal device where the signal is transmitted in one direction without attenuation, while in the reverse direction there is no signal transmission; thus it is also known as an *uniline*. Ideally for an isolator, insertion loss ($= -20\log_{10}|S_{21}|$) should be zero and isolation ($-20\log_{10}|S_{12}|$) infinite, but the practical values of insertion loss and isolation are 0.6–1.0 dB and 30–40 dB respectively. A *Faraday rotation isolator* is constructed by using the non-reciprocal property of the rotation of the plane of polarisation of the electromagnetic signal propagating down a magnetised ferrite medium, known as *Faraday rotation*. The Faraday rotation isolator is suitable for low and medium power applications at microwave frequency, say up to 2 kW peak power. However, at high peak power levels, non-linear effects in ferrite make such isolators unsuitable, in which case the *resonance absorption isolator* is used. The principle of operation of such an isolator is based on the non-reciprocal absorption property of circularly polarised waves at and around the resonant frequency of a magnetised ferrite.
- *Circulator:* A circulator is a multi-port waveguide junction in which the signal can flow from the nth port to the $(n + 1)^{th}$ port in one direction. A 4-port circulator (like isolator) also uses the non-reciprocal property of Faraday rotation in magnetised ferrite medium. A 4-port circulator can be used as a *single-pole-double throw (SPDT) switch*. The switching action can be obtained just by reversing the direction of the magnetic field B_0. The 4-port circulator may also be used as a *power divider*; this is achieved by adjusting the magnetic field such that the E field vector rotates around its mean position for a no-field situation. A more commonly used circulator for both waveguide and strip-line configuration is a *Y/T-junction circulator*. It consists of a symmetrical distribution of ferrite material at the junction of three waveguide or strip-line ports having a 120° angle between the ports. The ferrite material is usually in the form of a cylindrical disc or rod, or in the form of a prism and the DC magnetic field is applied along the axis of the cylinder. The resonance of the ferrite cylinder/disc/prism is the key feature in the operation of this type of circulator. This type of circulator is a very versatile component at microwave frequency. In addition to its use as a circulator, it can also be used as an isolator (by terminating one of the ports in matched load) or as switch (by reversing the direction of the magnetic field).
- *Phase shifter:* The difference in phase ϕ experienced by an electromagnetic wave of frequency f propagating with a velocity v through a transmission medium of length l is given by $\phi = 2\pi fl/v$. The variation of v of a travelling electromagnetic signal by varying the permeability μ of the ferrite medium when it is subjected to a magnetic field is the basis of the *ferrite phase shifter*. Realisation of phase change with variation of the length (l) of

transmission line is used in the design of *reflection and transmission type phase shifter*; where the varactor diode with its voltage variable capacitance and the PIN diode as switch are used as active devices and those may be analogue and digital type phase shifters. In addition to the above mentioned techniques, the microwave phase shifter may also be designed using quarter-wave and half-wave plates to construct the *rotary phase shifter* which is a precision phase shifter.

- *Bends and corners:* These components facilitate the assembly of waveguide systems permitting directional changes. In the *E-plane bend*, the electric field undergoes a 90° directional change and the plane of the bend is in the *E*-plane; while in the *H-plane bend*, the magnetic field undergoes a 90° directional change and the plane of the bend is in the *H*-plane. It is easier to design *corners* at lower microwave frequencies (say at the *S*-band) as the bends will be bulky and difficult to design with precision.
- *Waveguide twist, transitions and adaptor:* The *waveguide twist* is used whenever a 90° (or 45°) change in the plane of polarisation is required. A long enough ($\approx 2\lambda_g$) rectangular waveguide is gradually twisted mechanically to minimise reflections (VSWR < 1.05). The *transitions* between waveguides of different size, shape and/or modes are also used in microwave systems. Rectangular to waveguide transition is required in the rotary vane attenuator and rotary phase shifter. Transition from TE_{10} in a rectangular waveguide to the TM_{01} in a circular waveguide is required in the rotary joint used for mechanical beam steering of surveillance radar antenna. Also transition from the TE_{10} in a rectangular waveguide to TE_{01} in a circular waveguide is required in the design of a wavemeter. In most of the microwave test benches, we need a *waveguide-to-coaxial adaptor*; here the TEM mode of the coaxial line is smoothly adapted in the rectangular waveguide as the TE_{10} mode.
- *Flanges and connectors: Flanges* are used to join two waveguides. There are two types of flanges—plane-to-plane flange and plane-to-choke flange. If the mating surfaces are flat and clean for plane-to-plane flange, reflections at junctions can be neglected for low power and medium power applications. However, for high power applications, plane-to-choke flange is used that ensures the continuity of longitudinal RF surface currents on the broad wall of the two waveguides at the junction points which are otherwise DC open. *Connectors* are inevitably useful components in microwave systems. BNC connector is useful for frequencies below 3 GHz, *Type N* connectors are used at higher microwave frequencies, typically up to 20 GHz. *SMA* connectors are commonly used in the range 18–26 GHz. But above 26 GHz, especially in millimetre-wave frequencies, K-type connectors are used to cover from 26–110 GHz.

Passive lumped-circuit elements at microwave frequencies

- Passive components such as inductors, capacitors and resistors to be used in hybrid MIC/MMIC design are sometimes preferred to be of lumped-element type which are usually less than one tenth of a wavelength, hence effectively independent of frequency thus allowing broadband design. These components work satisfactorily up to 20 GHz and with limited values even up to 60 GHz.

- The inductance of a *round-wire (single loop) inductor* is given by the empirical formula: $L(\text{nH}) = 1.257 \times 10^{-3} a \left[\ln\left(\frac{a}{W+t} \right) + 0.078 \right] K_g$.

  Inductance for a *spiral inductor* is given by: $L(\text{nH}) = 0.03937 \left[\frac{a^2 n^2}{8a + 11c} \right] K_g$; where $a = \frac{D_o + D_i}{4}$ and $c = \frac{D_o - D_i}{2}$; K_g is a correction factor that needs to be included which accounts for the presence of ground plane and is given by:

$$K_g = 1.4 + 0.217 \ln\left(\frac{W}{5t} \right) \qquad 5 < \frac{W}{t}, < 100 \qquad \text{(round wire)}$$

$$= 1 + 0.333 \left(1 + \frac{S}{W} \right)^{-1.7} \qquad \text{(spiral)}$$

  where t is the thickness of the conductor in either case.

- Capacitance for *inter-digital capacitor* can be calculated from the empirical formula:

$$C = (\varepsilon_r - 1) l \{ (N-3) A_1 + A_2 \} \text{ pF}$$

  with $A_1 = 4.409 \tanh\left\{ 0.55 \left(\frac{h}{w} \right)^{0.45} \right\} 10^{-6}$ pF/μm

  and $A_2 = 9.92 \tanh\left\{ 0.52 \left(\frac{h}{w} \right)^{0.5} \right\} 10^{-6}$ pF/μm

  where l, w and h represent the finger length (usually $\leq \lambda/4$), finger width and the height of the substrate respectively, and N is the number of fingers of the inter-digital capacitor.

  The capacitance of the *MIM/MIS capacitor* can be calculated from $C = \varepsilon_0 \varepsilon_r (lw/h)$, where lw is the area of the metallic layers and h is the thickness of the oxide layer.

- The resistor elements realised as lumped components that may be used in MMIC and hybrid MIC are respectively of planar film type or chip type. Resistance values in the range 50 to 500 ohms are typically realisable, however, low resistance values are hard to obtain.
- Lumped circuit realisation of resonators (either as series resonant or parallel resonant circuit) required for HMIC and MMIC design can be done by combining the planar printed inductors and capacitors (made up of a single turn loop inductor and inter-digital capacitor).

Distributed-circuit elements using microstrip transmission line

- Distributed-circuit elements such as distributed inductor (either in series or in parallel with the host transmission line) or distributed capacitor (in general in shunt as for by-pass capacitor) with microstrip transmission line is required in the design of planar microwave filters, input/output matching networks for amplifiers and so on.

- The distributed circuit elements are also realised with short section of microstrip line ($l < \lambda_{eff}/10$, where λ_{eff} is the wavelength of the propagating wave in microstrip line) typically the electrical length is chosen to be $\theta = \beta l = \pi/6$, that is, $l = \lambda_{eff}/12$. The distributed series inductor is realised by a thin section of microstrip line being sandwiched between two 50 Ω host transmission lines while shunt inductance is realised with a thin section of the line connected between two 50 Ω host line in which the other end of the thin line is grounded through a via to the ground plane of the PCB on which the component is formed. The distributed shunt capacitor is realised with a wide line connected by two host lines of characteristic impedance 50 Ω on either side of it. The load end of the wide section line is open circuited. Sometimes the structure is realised as an open circuited radial stub which is free from spurious resonances and also provides broad band operation; especially useful on bias lines in the design of microwave amplifiers.

S-matrix of microwave components

- The S-matrix of *directional coupler* is given by $[S] = \frac{1}{\sqrt{2}}\begin{bmatrix} 0 & 1 & j & 0 \\ 1 & 0 & 0 & j \\ j & 0 & 0 & 1 \\ 0 & j & 1 & 0 \end{bmatrix}$

- The S-matrix of *magic tee* is given by $[S] = \frac{1}{\sqrt{2}}\begin{bmatrix} 0 & 0 & 1 & 1 \\ 0 & 0 & 1 & -1 \\ 1 & 1 & 0 & 0 \\ 1 & -1 & 0 & 0 \end{bmatrix}$

- The S-matrix of a *3-port circulator* is given by $[S] = \begin{bmatrix} 0 & 0 & 1 \\ 1 & 0 & 0 \\ 0 & 1 & 0 \end{bmatrix}$

Review Questions

6.1 Matched load terminates the transmission line (waveguide) in its characteristic impedance—justify. Name a typical lossy dielectric used to design a matched load.

6.2 Variable short represents an adjustable reactive load—justify. Discuss the design principle of the choke-type variable short at microwave frequency.

6.3 Why is an E–H tee termed as a magic tee? Discuss how it can be used in the design of a balanced microwave mixer.

6.4 Show that the main guide forming the hybrid ring should have a characteristic impedance of $\sqrt{2}Z_0$, where Z_0 is the characteristic impedance of each of the series arms.

6.5 Discuss the principle of operation of Wilkinson's power divider/combiner. It is amenable to design in strip-line—discuss.

6.6 In a quadrature hybrid (branch line or short-slot type) the phase difference between two output ports is 90°—discuss. Explain the principle of operation of balanced duplexer using a short-slot hybrid.

6.7 Discuss the principle of slide-screw tuner used for impedance matching in a waveguide system. Why does the screw need to be moved longitudinally to introduce the matching inductive susceptance?

6.8 What do you understand by 'coupling factor' and 'directivity' of a directional coupler? Mention a few uses of directional coupler.

6.9 Lange coupler is a practical type of coupled line coupler—discuss.

6.10 Discuss the principle of flap attenuator. When would you prefer to use rotary vane attenuator?

6.11 What happens to the power reflected from the load in a uniline (isolator)? Why should an isolator always be placed between a signal source and the load in a microwave system?

6.12 Define the terms 'insertion loss' and 'isolation' in the case of an isolator; what are the practical values for these quantities?

6.13 Faraday rotation in a magnetised ferrite medium is non-reciprocal in nature—explain.

6.14 Discuss the principle on which the resonance absorption isolator operates.

6.15 What is a circulator? How can you convert a 4-port circulator to an SPDT microwave switch and as a power divider?

6.16 Explain the principle of operation of a *Y*/*T* junction circulator. How can you convert this circulator to an isolator?

6.17 Discuss the application of a circulator in the design of a reflection type amplifier and diplexer.

6.18 Discuss the principle of a ferrite phase shifter and the reflection and transmission type phase shifter.

6.19 Discuss the principle of operation of a rotary phase shifter using a rotatable half-wave plate and two quarter-wave plates.

6.20 Discuss the use of bends and corners in configuring a microwave system.

6.21 At the output end of a Faraday rotation isolator, we need to use a 45° waveguide twist to connect it to the main waveguide line—explain why.

6.22 Discuss how mode transition from the TE_{10} in a rectangular waveguide to the TM_{01} in a circular waveguide is realised. Explain how the undesired TE_{11} mode is filtered out in this case.

6.23 Discuss how mode transition from the TE_{10} in a rectangular waveguide to the TE_{01} in a circular waveguide is realised. Explain how the undesired TE_{11} mode is filtered out in this case.

6.24 Discuss how a waveguide to coaxial adapter is designed.

6.25 Discuss the principle of operation of a plane-to-choke flange used in high power radar systems.

6.26 Name a few common types of microwave connectors mentioning their operating frequency ranges.

6.27 What are the distributed and lumped passive components at microwave frequencies? Discuss their relative advantages and disadvantages.

6.28 Discuss how a lumped inductor, capacitor, resonator and resistor are realised at microwave frequency.

6.29 Discuss how distributed-element inductors and capacitors at microwave frequencies are realised with a microstrip line. What is the advantage of a radial stub capacitor?

6.30 State the S-matrix for directional coupler, magic tee and 3-port circulator.

7 Impedance Matching in Transmission Lines and Waveguides

If your actions match your words—
Know, that is a test of your genuinity

7.1 Introduction

Impedance matching is a very important issue related to the design of a microwave system. When a transmission line is connected to a generator, a load, or another transmission line, the impedances may, in general, be mismatched. This results in undue reflection at the location of the mismatched impedances and hence a standing wave gets set up on the line, which is definitely undesirable. It is thus often necessary to match a transmission line to a load or to generator, for the purpose of eliminating such undesirable standing waves on the line. Similarly, if a discontinuity exists, such as the connection of unmatched line sections, it is often necessary to eliminate this mismatch before the lines can be connected to each other. The results of mismatch on a high frequency line can be disastrous, as a large amount of reactive power may travel along the line which can easily damage circuitry, especially generators. From a common visual experience, a very long (as if a never-ending) narrow-gauge/broad-gauge railway track gives the feel of a matched line system (an infinite line from which no reflected signal comes back from the load side) while the junction of narrow-gauge to broad-gauge railway track or vice-versa (both are of finite length with a discontinuity between the two) gives the visual feeling of a mismatched line.

A transmission line or waveguide is said to be matched to a load if the load impedance is equal to the characteristic impedance of the line. Similarly, if the line impedance is equal to the generator impedance, the two are matched. To match a load to a line (or a generator for that matter), a matching network is connected between the line and the load. There are two types of impedance matching networks that are particularly useful at microwave frequencies. One matching technique, known as *stub matching*, employs short-circuited (or open circuited) sections of transmission line (a stub) as reactive elements, which may be connected either across (in parallel) or along (in series) the parent transmission line at a suitable distance from the load. The stub matching is applicable where the need is for narrow band matching (as the length of the stub is frequency dependent). The other method of impedance matching is based on the principle of impedance transformation. The theory of small reflection at the junction of multiple quarter-wave impedance transformers is used to design broadband impedance transformers (since the single quarter-wave transformer is narrow band as its length is frequency dependent). The *stepped* and the *tapered impedance transformers* are commonly used for impedance matching at microwave frequencies. Incorporation of an inductive/capacitive window (iris) or introduction of a tuning screw in a waveguide is also used as a means for impedance matching in certain applications for a microwave circuit design. Further, a simple yet practical technique is to use an L-section network formed with a lumped component/planar distributed line reactive elements for impedance matching; used for input/output matching in the design of microwave amplifiers.

7.2 Stub Matching

The idea of *stub matching* is to connect open or short circuited *stub* (sections of transmission lines) either in shunt (in parallel) or in series with the parent transmission line. From a theoretical point of view there is no difference between the open or short circuited stub, and we know that for a given susceptance or reactance, there is only a difference in lengths ($\lambda/4$) of an open or short circuited stub. However, from an engineering point of view, open circuited stubs are easier to fabricate in planar transmission lines as in microstrips and strip-lines (since a via hole through the substrate to the ground plane is not required). However, for a coaxial line and waveguide, short circuited stubs are usually preferred, because the cross sectional area of such an open circuited line may be large enough (electrically) to radiate (from its open end), in which case the stub is no longer purely reactive. Further, shunt stubs are commonly used in coaxial transmission lines as it is difficult to construct a series stub in such transmission line structures.

Stub matching (or tuning, as it effectively tunes out impedance mismatches) may be realised either with one piece of stub or with two/three pieces of stubs. The idea of stub matching relies on the fact that the impedance of the stub and/or its location on the parent transmission line is so chosen that the combined impedance of the parent line and the stub (or stubs) equals the characteristic impedance of the parent line. In single stub matching, the stub is placed at a suitable distance from the load and the susceptance (for shunt stub) or reactance (for series stub) presented by the stub cancels out the susceptance/reactance of the loaded line at the location of the stub thereby terminating the parent transmission line in its characteristic impedance. A disadvantage with single stub matching is that every load requires a new stub position. Thus, if matching is desired for a variable impedance load, then single stub matching is difficult to use in practice. The use of two stubs spaced by a fixed amount and located at a fixed distance from the load may be used to overcome this disadvantage. However, a double stub matching system of this type may not be able to match all possible values of load admittance. A triple stub matching arrangement can overcome this problem of double stub tuner, in fact, a triple stub tuner may be considered to be two double stub tuners in series.

One can seek analytical solutions for stub matching problems, which are usually more accurate. However, Smith chart solutions are fast, intuitive and accurate enough in practice and for this reason Smith chart solutions are routinely used by microwave engineers. We shall discuss both the analytical and Smith chart methods for solution of stub matching problems.

7.2.1 Single Stub Matching

In *single stub matching* (Fig. 7.1), the two adjustable parameters are, the distance d from the load to the stub position, and the stub length l which determines the value of susceptance/reactance provided by the shunt/series stub. For the shunt stub matching (Fig. 7.1a), the basic idea is to select d so that the normalised input admittance of the line y_{inB}, seen looking into the line from B–B′ located at a distance d from the load is of the form $y_{\text{inB}} = 1 + jb_{\text{B}}$. Then, the stub length l is so chosen that it represents a susceptance of $y_{\text{SB}} = -jb_{\text{B}}$, resulting in a matched condition at the stub location ($y_{\text{inB}} + y_{\text{SB}} = 1 + jb_{\text{B}} - jb_{\text{B}} = 1$). For the series stub case (Fig. 7.1b), the distance d is selected so that the normalised input impedance z_{in} of the line, seen looking into the line at a distance d from the load is of the form $z_{\text{inB}} = 1 + jx_{\text{B}}$. Then, the stub length l is chosen to exhibit a stub reactance of $z_{\text{SB}} = -jx_{\text{B}}$, resulting in a matched condition ($z_{\text{inB}} + z_{\text{SB}} = 1 + jx_{\text{B}} - jx_{\text{B}} = 1$). Let us now carry out an analysis of both the shunt and series stub matching using a single stub (open or short circuited) to derive analytic expressions for d and l which are the design parameters for a given transmission line and load. It may be noted that single stub matching (and in fact any stub matching) cannot be used to match a purely complex load (a load with no real part) to any line which has real characteristic impedance (a microwave transmission line) as the stub always exhibits a

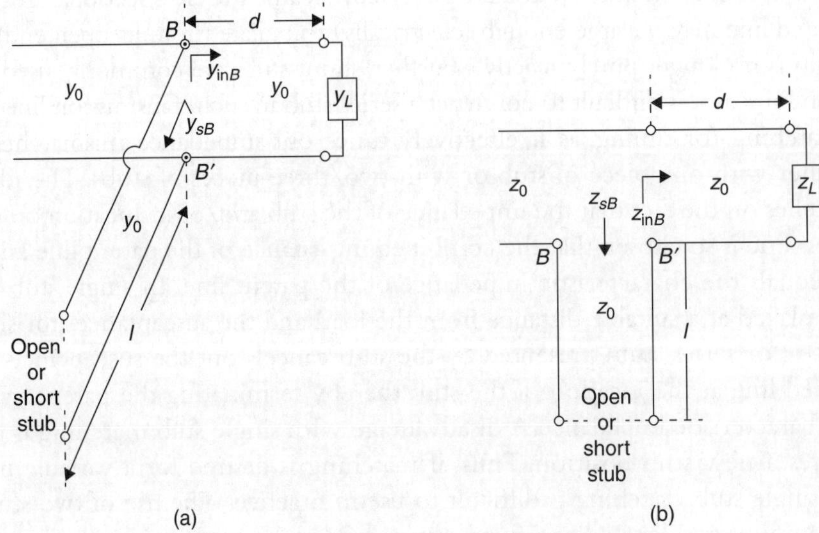

Fig. 7.1 Single stub matching (a) With shunt stub (b) With series stub.

purely imaginary reactance/susceptance that tunes out jx_B/jb_B making $z_{in}/y_{in} = 0$, leading to a perfect mismatch (!).

Case I Shunt stub matching: The analytical derivation for shunt type single stub matching can be carried out as follows [14,9]. Let the normalised load impedance terminating the transmission line be $z_L = 1/y_L = r_L + jx_L$ (Fig. 7.1a). Then the impedance of the line seen from a distance d from the load (looking towards the load) at the stub position B–B′ is given from Eq. 3.17b as:

$$z_B = \frac{(r_L + jx_L) + jt}{1 + j(r_L + jx_L)t} \tag{7.1}$$

$$\text{where,} \quad t = \tan \beta d \tag{7.2}$$

The normalised admittance is then:

$$y_B = \frac{1}{z_B} = g_B + jb_B \tag{7.3}$$

$$\text{where,} \quad g_B = \frac{r_L(1+t^2)}{r_L^2 + (x_L + t)^2} \tag{7.4}$$

$$b_B = \frac{r_L^2 t - (1 - x_L t)(x_L + t)}{r_L^2 + (x_L + t)^2} \tag{7.5}$$

To find expression for d, we resort to the matched condition at the stub position, which demands that the susceptance b_B due to the loaded line at the stub position B–B′ is cancelled out by the stub susceptance b_{SB} ($b_{SB} = -b_B$) and the normalised admittance across the line at the stub position then becomes equal to unity, $y_B = g_B = 1$. Using this condition for matching in Eq. 7.4, we obtain:

$$(r_L - 1)t^2 - 2x_L t + (r_L - r_L^2 - x_L^2) = 0 \tag{7.6}$$

Solving Eq. 7.6 for t we get:

$$t = \frac{x_L \pm \sqrt{r_L\{(1 - r_L^2) + x_L^2\}}}{(r_L - 1)}, \quad \text{for } r_L \neq 1 \tag{7.7}$$

While for $r_L = 1$, $t = -x_L/2$. Thus, from Eq. 7.2, the two principal solutions for d are:

$$d = \frac{\lambda}{2\pi} \tan^{-1}(t), \quad \text{for } t \geq 0 \tag{7.8a}$$

$$= \frac{\lambda}{2\pi} \{\pi + \tan^{-1}(t)\}, \quad \text{for } t < 0 \tag{7.8b}$$

To find the required stub length l, we use the value of t as given by Eq. 7.7 in Eq. 7.5 to find the stub susceptance, $b_{SB} = -b_B$; when we can write for an open circuited stub:

$$l_{oc} = \frac{\lambda}{2\pi}\tan^{-1}(b_{SB}) = \frac{\lambda}{2\pi}\tan^{-1}(-b_B), \quad b_B \geq 0 \tag{7.9a}$$

$$= \frac{\lambda}{2\pi}\{\pi + \tan^{-1}(b_{SB})\} = \frac{\lambda}{2\pi}\{\pi + \tan^{-1}(-b_B)\}, \quad b_B < 0 \tag{7.9b}$$

For a short circuited stub:

$$l_{sc} = \frac{\lambda}{2\pi}\tan^{-1}\left(-\frac{1}{b_{SB}}\right) = \frac{\lambda}{2\pi}\tan^{-1}\left(\frac{1}{b_B}\right), \quad b_B \geq 0 \tag{7.10a}$$

$$= \frac{\lambda}{2\pi}\left\{\pi + \tan^{-1}\left(-\frac{1}{b_{SB}}\right)\right\} = \frac{\lambda}{2\pi}\left\{\pi + \tan^{-1}\left(\frac{1}{b_B}\right)\right\}, \quad b_B < 0 \tag{7.10b}$$

Thus, for a given load impedance, both d and l values can be calculated using the above equations. If the lengths given by Eqs 7.9 and 7.10 happen to be negative, then $\lambda/2$ can be added to give a positive result.

Case II Series stub matching: Similar analysis as that done for the shunt stub may be carried out for series stub matching as well (Fig. 7.1b) and the expressions for the design parameters d and l may be evaluated from the following expressions. The location of the stub from load position d, is given by Eq. 7.8 with:

$$t = \frac{1}{(g_L - 1)}\left[b_L \pm \sqrt{g_L\{(1 - g_L)^2 + b_L^2\}}\right], \quad \text{for } g_L \neq 1 \tag{7.11}$$

while for $g_L = 1$, $t = -b_L/2$. Again, the length of the stub for open circuited and short circuited stub is given by the following expressions.

For open circuited stub:

$$l_{oc} = \frac{\lambda}{2\pi}\tan^{-1}\left(\frac{1}{x_{SB}}\right) = \frac{\lambda}{2\pi}\tan^{-1}\left(-\frac{1}{x_B}\right), \quad x_B \geq 0 \tag{7.12a}$$

$$= \frac{\lambda}{2\pi}\left\{\pi + \tan^{-1}\left(\frac{1}{x_{SB}}\right)\right\} = \frac{\lambda}{2\pi}\left\{\pi + \tan^{-1}\left(-\frac{1}{x_B}\right)\right\}, \quad x_B < 0 \tag{7.12b}$$

And for short circuited stub:

$$l_{sc} = \frac{\lambda}{2\pi}\tan^{-1}(-x_{SB}) = \frac{\lambda}{2\pi}\tan^{-1}(x_B), \quad x_B \geq 0 \tag{7.13a}$$

$$= \frac{\lambda}{2\pi}\{\pi + \tan^{-1}(-x_{SB})\} = \frac{\lambda}{2\pi}\{\pi + \tan^{-1}(x_B)\}, \quad x_B < 0 \tag{7.13b}$$

where, $x_B = \dfrac{g_L^2 t - (1 - b_L t)(b_L + t)}{g_L^2 + (b_L + t)^2}$ (7.14)

If the length given by Eqs 7.12 or 7.13 is negative, $\lambda/2$ can be added to give a positive result.

7.2.2 Double Stub Matching

In the *double stub matching* circuit (Fig. 7.2a), the load may be at an arbitrary distance from the first stub. Although this is more representative of a practical situation, the circuit of Fig. 7.2b, where the load y_L' has been transformed back to the position of the first stub, is easier to deal with and does not lose any generality.

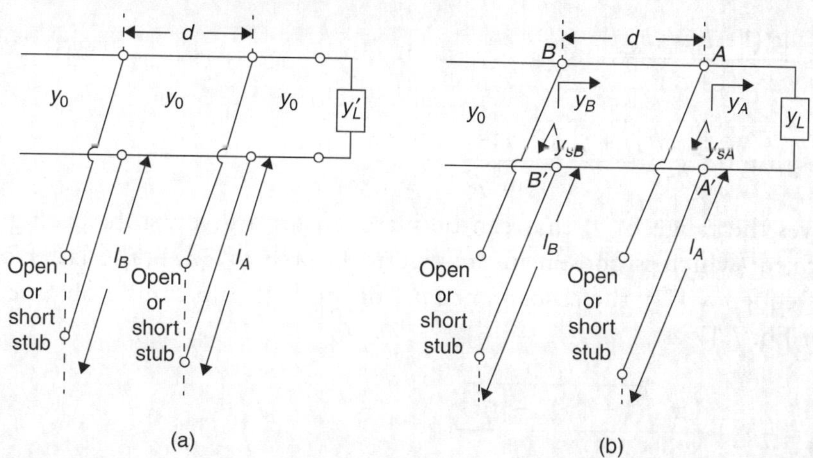

Fig. 7.2 Double stub matching (a) Representative practical situation (b) Transformed load for ease of analysis.

The stubs shown in Fig. 7.2 are shunt stubs, which are usually easier to implement in practice than series stubs; the latter could be used just as well, in principle. In either case, the stubs can be open circuited or short circuited.

The analytical derivation for evaluating the design parameters of the double stub matching/tuner is given below [9].

The normalised admittance to the left of the first stub located at A A′ is given by:

$$y_A = g_L + j(b_L + b_{SA}) \tag{7.15}$$

where, $y_L = g_L + jb_L$ is the normalised load admittance and b_{SA} is the normalised susceptance of the first stub. After transforming through a length d of the transmission line, the normalised admittance to the right of the location of the second stub B–B′ is:

$$y_B = \frac{g_L + j(b_L + b_{SA} + t)}{1 + jt(g_L + jb_L + jb_{SA})} \tag{7.16}$$

where $t = \tan\beta d$. At this point, the matching condition demands that the real part of y_B must equal unity, which leads to the equation:

$$g_L^2 - g_L\frac{(1+t^2)}{t^2} + \frac{(1-b_Lt-b_{SA}t)^2}{t^2} = 0 \tag{7.17}$$

Solving for g_L gives:

$$g_L = \frac{(1+t^2)}{2t^2}\left\{1 \pm \sqrt{1-\frac{4t^2(1-b_Lt-b_{SA}t)^2}{(1+t^2)^2}}\right\} \tag{7.18}$$

Since g_L is real, the quantities within the square root must be non-negative and so:

$$0 \le \frac{4t^2(1-b_Lt-b_{SA}t)^2}{(1+t^2)^2} \le 1 \tag{7.19}$$

This implies that: $0 \le g_L \le \dfrac{1+t^2}{t^2} = \dfrac{1}{\sin^2\beta d}$

This gives the range of g_L that can be matched for a given stub spacing, d. After d has been fixed (which is chosen in practice to be $\lambda/8$ or $3\lambda/8$, the latter one is used at higher frequency [7]); the susceptance b_{SA} of the first stub can be obtained from the solution of Eq. 7.17 as:

$$b_{SA} = -b_L + \frac{1 \pm \sqrt{(1+t^2)g_L - g_L^2t^2}}{t} \tag{7.20}$$

While the normalised susceptance of the second stub can be obtained from the imaginary part of Eq. 7.16 when it is equated to $-jb_{SB}$, which yields:

$$b_{SB} = \frac{\pm\sqrt{g_L(1+t^2) - g_L^2t^2} - g_L}{g_Lt} \tag{7.21}$$

The upper and lower signs in Eqs 7.20 and 7.21 correspond to the same solutions. The stub lengths for open and short circuited stubs are:

For open circuited stub:

$$l_{oc} = \frac{\lambda}{2\pi}\tan^{-1}(b) \tag{7.22}$$

For short circuited stub:

$$l_{sc} = \frac{\lambda}{2\pi}\tan^{-1}\left(-\frac{1}{b}\right) \tag{7.23}$$

where $b = b_{SA}$ or b_{SB} as appropriate.

The analytical solution of the stub matching problem discussed so far may be used to obtain very accurate results by solving the analytical equations through computer programs (Numerical Problem 7.4), however, such accuracies are seldom required in actual problem solutions and thus stub matching problem may be solved more easily by using the Smith chart.

7.2.3 Smith Chart Technique for Solving Stub Matching Problems

EXAMPLE 7.1

A loss-less line of characteristic impedance $Z_0 = 50\ \Omega$ is to be matched to a load impedance $Z_L = 50/[2 + j(2 + \sqrt{3})]\ \Omega$ by means of a short circuited shunt stub. The characteristic impedance of the stub is 100 Ω. Determine the stub position and the stub length so that a match is obtained between the line and the load. Use Smith chart to solve the problem.

Solution

Step 1: Shunt stub problems can be conveniently solved by computing the normalised load admittance and entering it in the Smith chart used as an admittance chart. From the given data, the normalised load admittance is:

$$y_L = \frac{Z_0}{Z_L} = 50\,\frac{\{2 + j(2 + \sqrt{3})\}}{50} = 2 + j3.732$$

Step 2: Next, draw an admittance circle through the point y_L which intersects the normalised unit conductance circle ($g = 1$ circle) at two points y' and y'', where $y' = 1.0 - j2.70$ and $y'' = 1.0 + j2.70$ (Fig. 7.3). Each one of these provides one possible solution.

Step 3(a): First choose y'. Now move from the load point (y_L) on the admittance circle *towards the generator* (WTG scale) to reach this point y'. As seen from Fig. 7.3, the lines extended from the centre of the chart through y_L and y' intersect the WTG scale at 0.217λ and 0.30λ respectively. Thus, the distance d_1 of the stub position from the load plane (Fig. 7.4) is obtained from Fig. 7.3 as:

$$d_1 = (0.30 - 0.217)\lambda = 0.083\lambda$$

Step 3(b): At the junction corresponding to the point 1–1′ of the line (Fig. 7.4), the condition for impedance matching at the junction with the stub and the load requires that:

$$Y_{11} = Y' + Y_s, \quad \text{or} \quad y_{11} Y_0 = y' Y_0 + y_s Y_{0s}$$

where Y_s is the susceptance that a stub of length l will contribute at 1–1′ plane and Y' is the line admittance. Y_0 and Y_{0s} are the characteristic impedance of the line and the stub respectively.

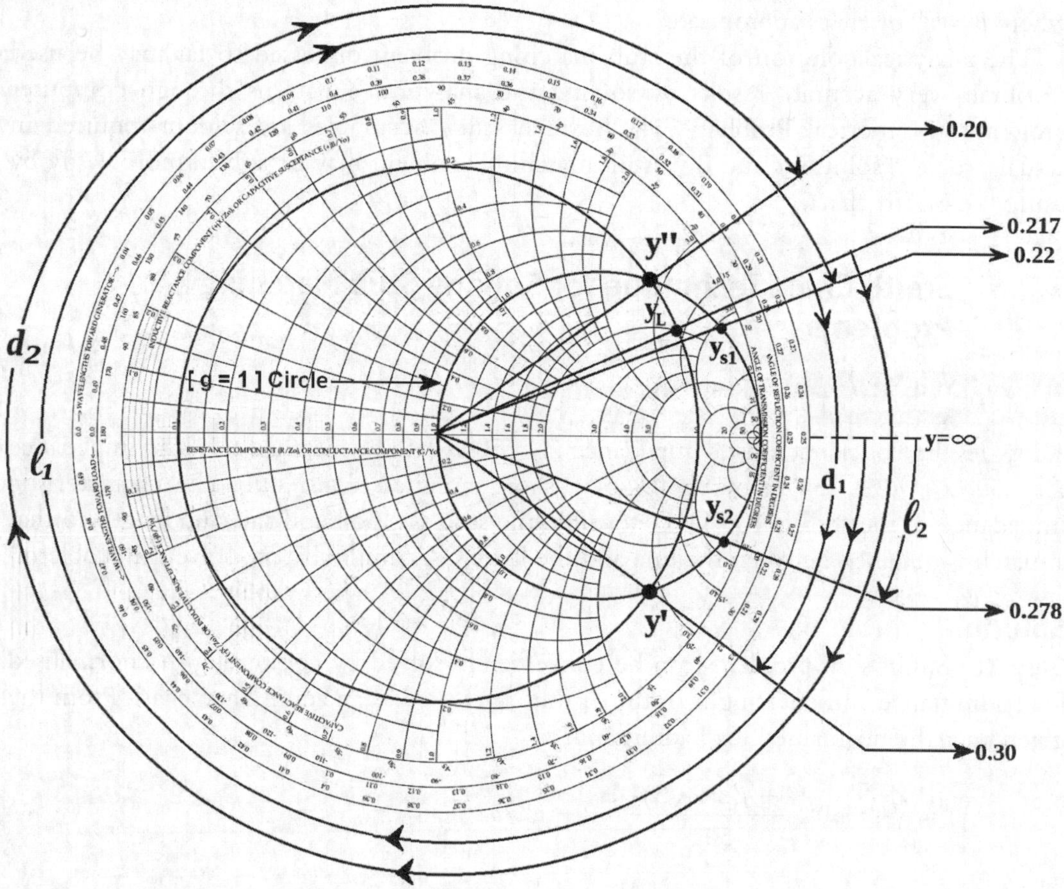

Fig. 7.3 Smith chart solution of single (shunt) stub matching problem (Example 7.1).

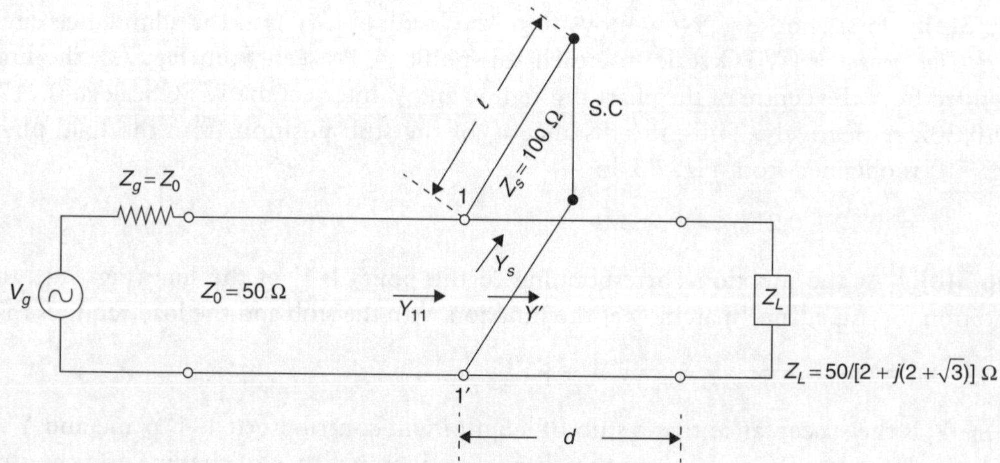

Fig. 7.4 Scheme for short circuited single stub matching (Example 7.1).

Thus $y_s = (y_{11} - y')Y_0/Y_{0s} = [1.0 - (1.0 - j2.7)]100/50 = j5.40$, which is the normalised stub susceptence (capacitive) for matching at the 1–1′ junction.

[**Note:** Since the characteristic impedance of the stub is different from that of the line, y_s is different from $-b'$ (since $y' = g - jb'$). But if the characteristic impedances were equal, then $y_s = -b'$, as the stub is added to cancel the susceptance of the loaded line at the 1–1′ plane.]

Step 3(c): To obtain the length of the stub, we start at $y = \infty$ ($z = 0$, the short circuit point which corresponds to the right side of the real axis in the admittance chart, as shown in Fig. 7.3). Since the stub contributes a susceptance of $+j5.40$, we enter this on the Smith chart and starting at $y = \infty$ we move along the WTG circle until we reach the point $y_{s1} = +j5.40$. Lines extended from the centre of the chart through $y = \infty$ and $y_{s1} = +j5.40$ are drawn that intersect the WTG circle at 0.250λ and 0.22λ respectively. Thus, the stub length is:

$$l_1 = 0.25\lambda + 0.22\lambda = 0.47\lambda$$

Step 4: If we had chosen y'', then $y_{s2} = -j5.40$ (the required normalised stub susceptance is inductive) and it may be seen from Fig. 7.3 that on moving from the load point (y_L) on the admittance circle *towards the generator* (WTG scale) we reach the point y'' by covering a distance of d_2 given by:

$$d_2 = 0.50 - (0.217 - 0.20) = 0.50 - 0.017 = 0.483\lambda$$

In this case, the stub length is obtained by starting at $y = \infty$ and moving along the WTG circle until we reach the point $y_{s2} = -j5.40$, when we get:

$$l_2 = 0.278 - 0.25 = 0.028\lambda$$

[**Note:** Out of the two possible solutions, it is advisable to choose the one that has the stub position closest to the load (in this case choose the solution with d_1). This helps to improve bandwidth of the match and also reduces the possibility of losses that might otherwise be caused by large standing wave ratio on the line between the stub and the load.]

EXAMPLE 7.2

A transmission line of characteristic impedance $Z_0 = 75\ \Omega$ is to be matched to a load impedance $Z_L = 15 + j60\ \Omega$ by means of a short circuited series stub having same characteristic impedance as the line. Determine the stub position and the stub length so that a match is obtained between the line and the load. Use Smith chart to solve the problem.

Solution

Step 1: Series stub problems are convenient to solve by computing the normalised load impedance and entering it in the Smith chart used as an impedance chart. From the given data, normalised load impedance is $z_L = Z_L/Z_0 = 15 + j60/75 = 0.2 + j0.8$.

Step 2: Next, draw an impedance circle through the point z_L which intersects the normalised unit resistance circle ($r = 1$ circle) at two points z' and z'', where $z' = 1.0 + j2.55$ and $z'' = 1.0 - j2.55$ (Fig. 7.5). Each one of these provides one possible solution.

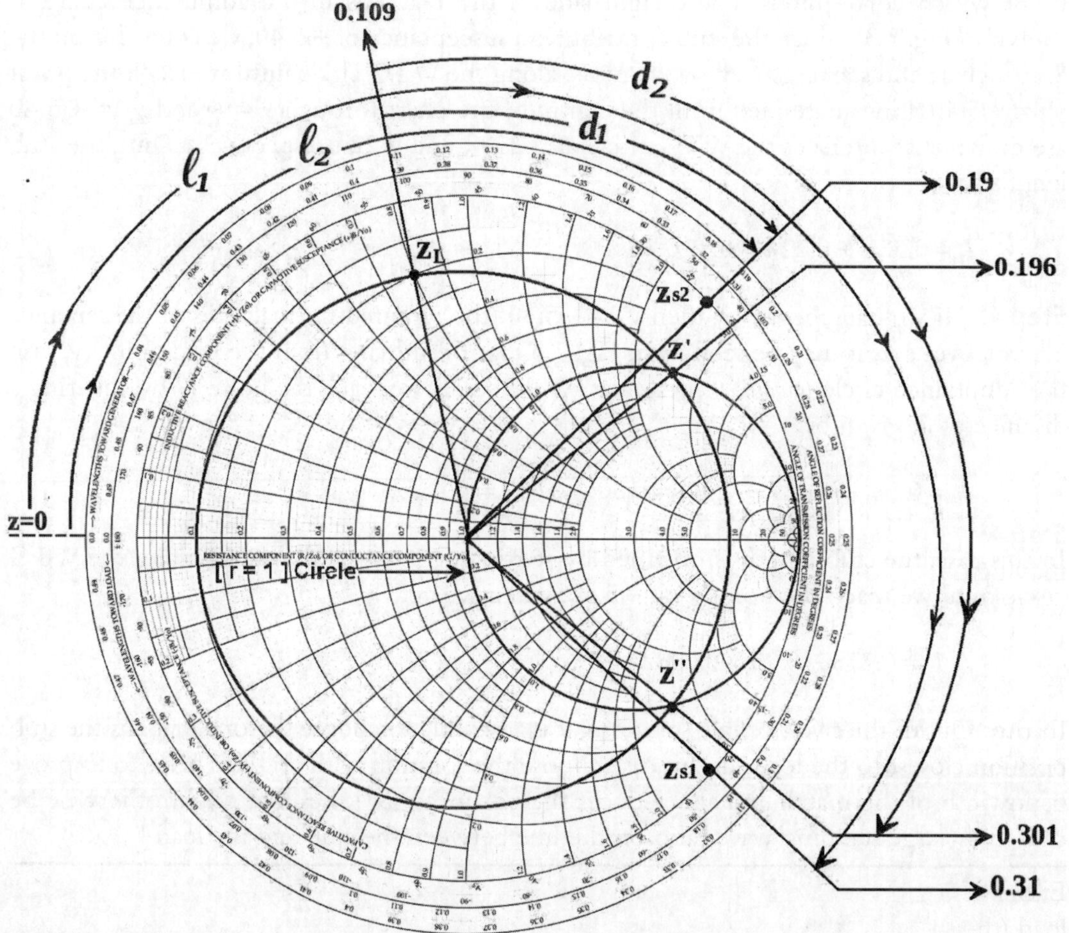

Fig. 7.5 Smith chart solution of single (series) stub matching problem (Example 7.2).

Step 3(a): First choose z'. Now move from the load point (z_L) on the impedance circle *towards the generator* (WTG scale) to reach this point z'. As seen from Fig. 7.5, the lines extended from the centre of the chart through z_L and z' intersect the WTG scale at 0.109λ

and 0.196λ respectively. Thus, the distance d_1 of the stub position from the load plane (Fig. 7.6) is obtained from Fig. 7.5 as:

$$d_1 = (0.190 - 0.109)\lambda = 0.081\lambda$$

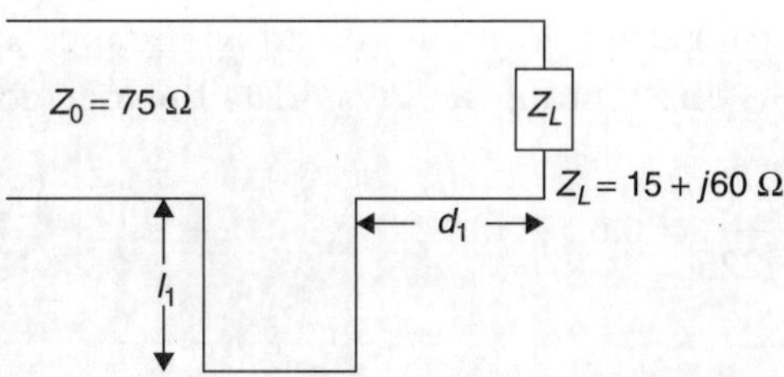

Fig. 7.6 Scheme for short circuited series stub matching (Example 7.2).

Step 3(b): The normalised line reactance at z' is $j2.55$. The stub length must be such that its normalised input impedance is $-j2.55$. This required impedance is marked as point z_{s1} in the Smith chart. Now, starting at $z = 0$ (short circuit point, which corresponds to the left side of the real axis in the impedance chart, Fig. 7.5) we move along the WTG circle until we reach the point $z_{s1} = -j2.55$. Lines extended from the centre of the chart through $z = 0$ and $z_{s1} = -j2.55$ are drawn that intersect the WTG circle at 0.00λ and 0.31λ respectively. Thus, the stub length is:

$$l_1 = 0.31\lambda - 0.00\lambda = 0.31\lambda$$

Step 4: If we had chosen z'', then $z_{s2} = +j2.55$ and it may be seen from Fig. 7.5 that on moving from the load point (z_L) on the impedance circle *towards the generator* (WTG scale) we reach the point z'' by covering a distance of d_2 given by:

$$d_2 = 0.31\lambda - 0.109\lambda = 0.301\lambda$$

In this case, the stub length is obtained by starting at $z = 0$ and moving along the WTG circle until we reach the point $z_{s2} = +j2.55$, when we get:

$$l_2 = 0.190\lambda - 0.00\lambda = 0.190\lambda$$

Either solution is correct. However, from a practical point of view, the closest stub to the load (the first solution) needs to be chosen.

EXAMPLE 7.3

A transmission line of characteristic impedance $Z_0 = 300\ \Omega$ is to be matched to a load impedance $Z_L = 150 + j225\ \Omega$ by means of a double stub (short circuited shunt stubs) matching network such that the two stubs are 0.1λ apart. The load impedance is 0.2λ from the first stub. Determine the stub lengths so that a match is obtained between the line and the load. Use Smith chart to solve the problem.

Solution

Step 1: At first we determine the normalised load admittance and enter it in the Smith chart used as an admittance chart. From the given data the load admittance is:

$$y_L = \frac{Z_0}{Z_L} = \frac{300}{150 + j225} = 0.62 - j0.92$$

[This could also been done first by locating the normalised load impedance $z_L = Z_L/Z_0 = 0.5 + j0.75$ in the Smith chart used as an impedance chart and then finding the diametrically opposite point on the Smith chart, which gives $y_L = 0.62 - j0.92$.]

Step 2: The load admittance at the location of stub 1 (the new load admittance) is found by moving from point y_L towards the generator a distance of 0.2λ (Fig. 7.7), this brings us to the point $y_L' = 0.36 + j0.425$ (Fig. 7.8). [When we move the diametrically opposite point to y_L', we arrive at $z_L' = 1.25 - j1.42$, which is the normalised impedance of the (new) load with the load impedance being shifted to the location of the stub 1.]

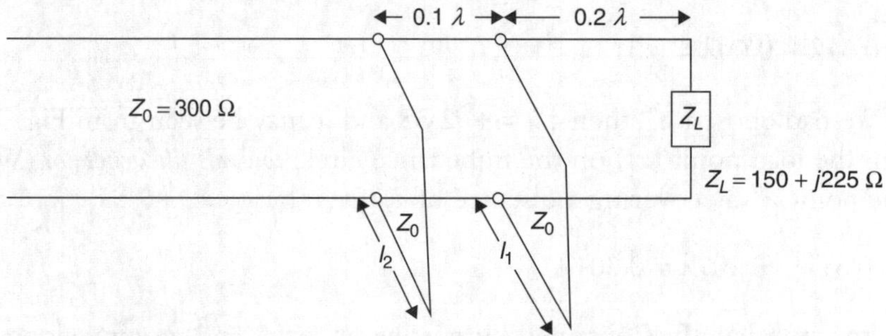

Fig. 7.7 Scheme for double stub (short circuited) matching (Example 7.3).

Step 3: At this point, this admittance y_L' is matched with stub 2 (forgetting for the moment stub 1) acting as a short circuited single stub and then shift by 0.1λ towards the load to reach the position of stub 1 and make the final match. To do this, we first draw the unit conductance circle g_2 for stub 2 and then moving 0.1λ towards the load, draw the unit conductance circle g_1 for stub 1 as shown in Fig. 7.8.

Step 4: The solution process starts with stub 1 at the load. The stub's impedance is purely imaginary and thus, it can only change the susceptance of the combined stub and load while the conductance remains the same. To find the combined admittance on the unit circle for stub 1, we start from the load point y_L and move on the constant conductance circle (with

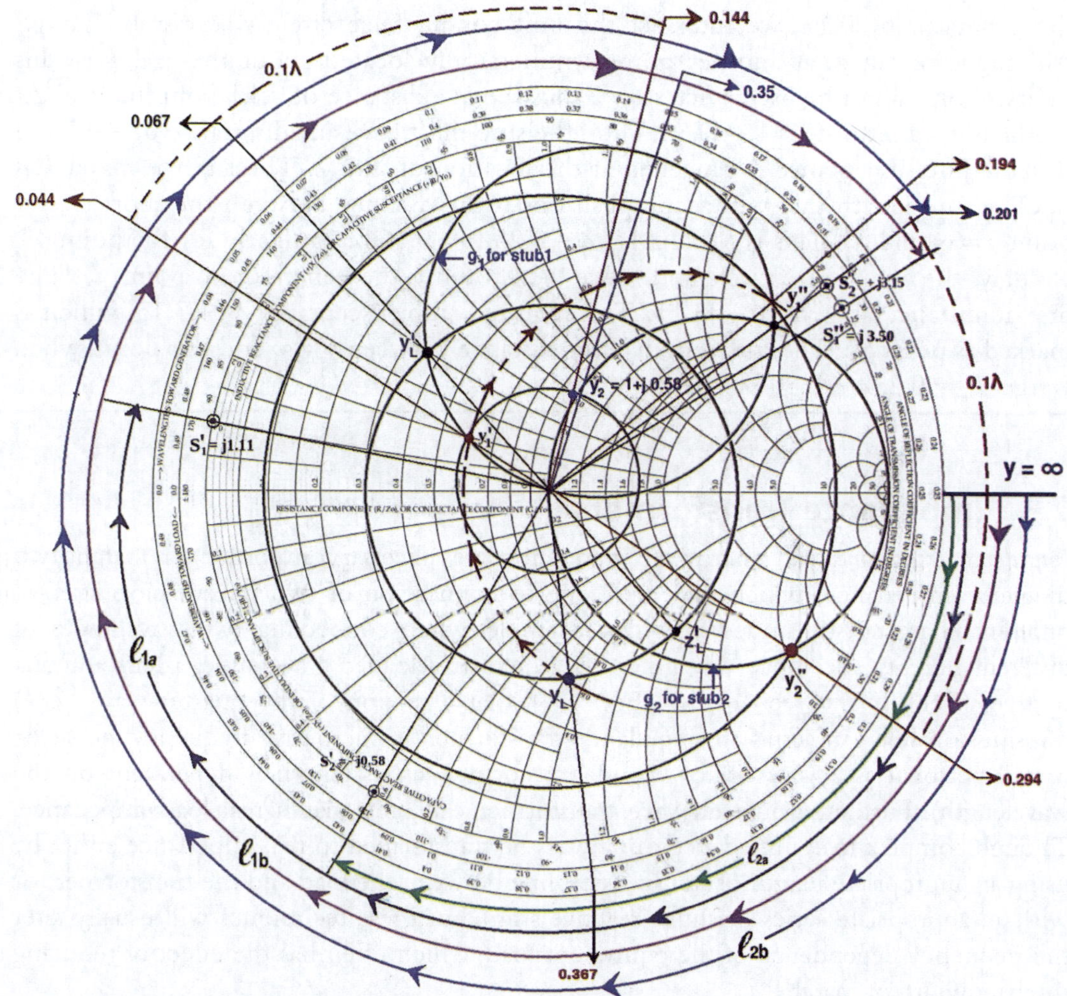

Fig. 7.8 Smith chart solution of double (shunt) stub matching problem (Example 7.3).

$g_L = 0.62$) which cuts the g_1 circle at two points—y_1' and y_1''. The admittances at y_1' and y_1'' are respectively $0.62 + j0.19$ and $0.62 + j0.2.58$. Thus, the susceptance contributed by stub 1 is $y_{1a}' = y_1' - y_L = (0.62 + j0.19) - (0.62 - j0.92) = j1.11$ and $y_{1b}'' = y_1'' - y_L = (0.62 + j0.2.58) - (0.62 - j0.92) = j3.50$. These two values are shown as S_1' and S_1'' in Fig. 7.8. The possible stub lengths are found by moving from $y = \infty$ point (the short circuit point in the admittance chart) *towards the generator* (WTG) circle, to the points S_1' and S_1'', when we get $l_{1a} = 0.266$ and $l_{1b} = 0.456\lambda$.

Step 5: Now we consider the admittances at y_1' and y_1'' as the new load admittances. From here on, we treat the problem as a single stub matching for each of these admittances and the distance between the load and stub 2 being known and equal to 0.1λ. We draw the admittance circle through y_1', and starting at y_1' as we move towards the generator by 0.1λ, we intersect the unit conductance circle g_2 for stub 2 at y_2'. Although we cut g_2 at another point symmetrically located about the real axis, this intersection cannot be used since stub 2 must be at a distance of 0.1λ from the load. At y_2', the line admittance is $1 + j0.58$. Thus, the stub must have an admittance of $-j0.58$ so that the line susceptance is cancelled at the location of stub 2. The latter is marked as S_2'. The stub length that will accomplish this is the distance between the short circuit point $y = \infty$ and S_2'. This is $l_{2a} = (0.416\lambda - 0.25\lambda) = 0.166\lambda$. Similarly for the point y_1'', we draw the admittance circle and move 0.1λ towards the generator, to point y_2''. The line admittance at y_2'' is $1 - j3.15$. The required stub susceptance is $+j3.15$, which is marked as point S_2''. The stub length is the distance between $y = \infty$ and this point, when we get $l_{2b} = 0.25\lambda + 0.201\lambda = 0.451\lambda$.

7.3 Impedance Transformers

Impedance transformers at microwave frequency are necessary in order to match two dissimilar impedances which may arise due to connection of two transmission lines of different characteristic impedances (for example, when connecting two waveguides of different heights, an empty waveguide to a dielectric loaded waveguide, a lens antenna to free space). It was seen earlier (Eq. 3.34), that a section of the quarter-wave ($\lambda/4$) transmission line can act as an impedance transformer at microwave frequency; however, it will perform as a very narrow band transformer (as its length is dependent on the wavelength). Further, a quarter-wave transformer can only match real load impedance. Though complex load impedance can always be transformed to real impedance either by using an appropriate length of transmission line between the load and the transformer (or with an appropriate series or shunt reactive stub), but these techniques will usually alter the frequency dependence of the equivalent load, which often has the effect of reducing the bandwidth of match.

In order to realise impedance matching over a broad band of frequencies, a multi-section quarter-wave transformer is used. This is done in principle by cascading multiple sections of quarter-wave lines having progressively increasing/decreasing characteristic impedance. The analysis and design of such an impedance transformer is done on the basis of the *theory of small reflections* in which the discontinuity (small impedance mismatch) between the junctions of two consecutive quarter-wave sections is considered to exhibit a 'small reflection' (with VSWR, of the order of say, 1.05). The theory of small reflection also finds application in the design of other microwave devices, such as directional couplers, filters and antenna arrays.

7.3.1 Frequency Response of Quarter-Wave Transformer

A single section quarter-wave matching transformer is shown in Fig. 7.9, the characteristic impedance of the matching section being $Z_{ch} = \sqrt{Z_1 Z_L}$. This matching transformer is capable of perfect impedance matching at a single frequency f_0 and at all the odd harmonics for which the length becomes equal to an odd multiple of a quarter wavelength ($\lambda_0/4$). The bandwidth of such a transformer can be determined by considering the electrical length $\beta l = \theta$ at any other frequency $f \neq f_0$ and finding the variation of the reflection coefficient (Γ) with frequency.

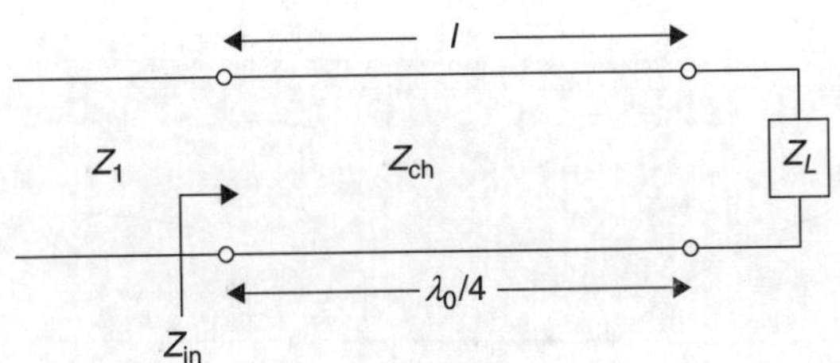

Fig. 7.9 A quarter-wave impedance matching transformer.

At any frequency, the input impedance seen looking into the matching section is given by (Eq. 3.17b):

$$Z_{\text{in}} = Z_{ch} \frac{Z_L + jZ_{ch} \tan\theta}{Z_{ch} + jZ_L \tan\theta} \tag{7.24}$$

Thus, the reflection coefficient is:

$$\Gamma = \frac{Z_{\text{in}} - Z_1}{Z_{\text{in}} + Z_1}$$

$$= \frac{Z_L - Z_1}{Z_L + Z_1 + j2 \tan\beta l \sqrt{Z_1 Z_L}} \tag{7.25}$$

The latter form is obtained by using $Z_{ch} = \sqrt{Z_1 Z_L}$.

The magnitude of Γ is readily evaluated and is given by:

$$|\Gamma| = \frac{|Z_L - Z_1|}{\left\{(Z_L + Z_1)^2 + 4\tan^2 \beta l Z_1 Z_L\right\}^{1/2}}$$

$$= \frac{1}{\left[1 + \left\{4Z_1 Z_L/(Z_L - Z_1)^2\right\} \sec^2\theta\right]^{1/2}} \tag{7.26}$$

Now, if we assume that the frequency is near the design frequency f_0, then $l \approx \lambda_0/4$ and $\theta \approx \pi/2$. Then, $\sec^2\theta >> 1$, and Eq. 7.26 simplifies to:

$$|\Gamma| \approx \frac{|Z_L - Z_1|}{2\sqrt{Z_1 Z_L}} |\cos\theta| \quad \text{for } \theta \text{ near } \pi/2 \tag{7.27}$$

The plot of Eq. 7.27 representing the variation of $|\Gamma|$ with θ ($= \beta l = f\pi/2f_0$) is essentially the frequency response for a single-section quarter-wave transformer. As may be noted from Fig. 7.10, the variation of $|\Gamma|$ vs θ is periodic in nature because of the periodic variation of input impedance with frequency.

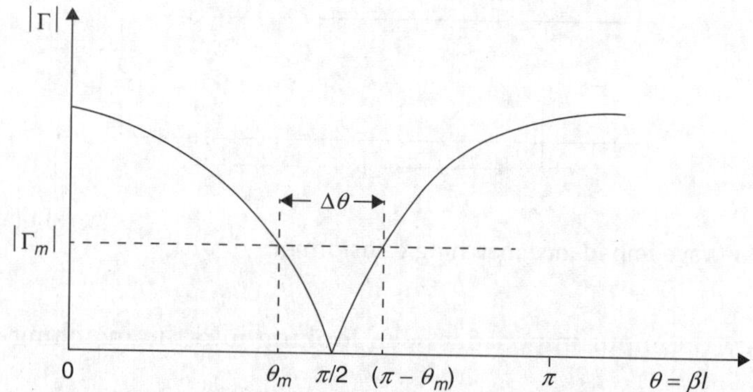

Fig. 7.10 Frequency response of a single section quarter-wave transformer.

The bandwidth of the transformer is defined for the range of $|\Gamma|$ having a maximum tolerable limit $|\Gamma_m|$ around the mid-band frequency f_0, where $|\Gamma| = 0$ and $\theta = \pi/2$. The value of θ at the edge of the useful pass-band may be found from Eq. 7.26 by equating $|\Gamma|$ to $|\Gamma_m|$, when:

$$\theta_m = \cos^{-1}\left|\frac{2|\Gamma_m|\sqrt{Z_1 Z_L}}{(Z_L - Z_1)\sqrt{1-|\Gamma_m|^2}}\right| \tag{7.28}$$

Thus, the transformer bandwidth is given by:

$$\Delta f = 2(f_0 - f_m) = 2\left(f_0 - \frac{2f_0}{\pi}\theta_m\right)$$

While the fractional bandwidth is:

$$\frac{\Delta f}{f_0} = 2 - \frac{4}{\pi}\cos^{-1}\left|\frac{2|\Gamma_m|\sqrt{Z_1 Z_L}}{(Z_L - Z_1)\sqrt{1-|\Gamma_m|^2}}\right| \tag{7.29}$$

A plot of the reflection coefficient magnitude versus normalised frequency is shown in Fig. 7.11 for various mismatched loads and it may be observed that larger bandwidth is realisable with smaller impedance ratio (Z_L/Z_1) between the two impedances Z_L and Z_1

to be matched. In other words the bandwidth of single section quarter wave transformer increases as the impedances to be matched (Z_L and Z_1 in the present case) becomes closer to each other (for less mismatched load).

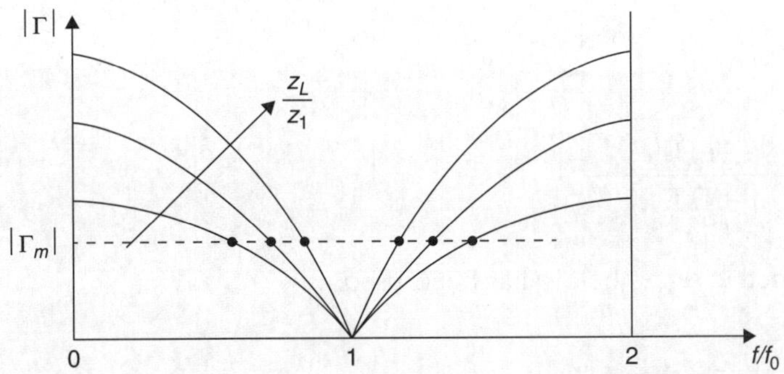

Fig. 7.11 Effect of load mismatches on the bandwidth of single section quarter-wave transformer.

The above analysis assumed that the impedances Z_1 and Z_{ch} are frequency independent, which is a good approximation for TEM transmission lines. However, for a waveguide system, the wave impedance varies with frequency (Eqs 4.11 and 4.13) (unlike Z_{ch} of a TEM line) and this complicates the analysis considerably. Another factor being neglected in the above analysis (which is true for both transmission lines and waveguides) is the presence of shunt susceptance which appears at the junctions of the transformer ends, due to sudden change of dimensions of the lines with different impedances. This can, however, often be compensated for by making small adjustments in the length of the matching section.

7.3.2 Theory of Small Reflections

The overall reflection coefficient at the input of any structure having multiple discontinuities/junctions (causing reflection) may be evaluated by considering all those multiple reflections taking place at different discontinuities (junctions) of the structure. It will be seen that for small reflections, the total reflection coefficient can be obtained only from the first order reflections.

Let us first consider a single-section quarter-wave transformer having two discontinuities at which the reflection coefficient (Γ) and transmission coefficient (T) are indicated by appropriate suffices in Γ and T (Fig. 7.12a). These are expressed by:

$$\Gamma_1 = \frac{Z_2 - Z_1}{Z_2 + Z_1}, \quad \Gamma_2 = -\Gamma_1, \quad \Gamma_3 = \frac{Z_3 - Z_2}{Z_3 - Z_2} \tag{7.30}$$

$$T_{21} = 1 + \Gamma_1, \quad T_{12} = 1 + \Gamma_{21} = 1 - \Gamma_1 \tag{7.31}$$

Let us assume that a unit amplitude wave is incident at the input. Inspection of Fig. 7.12b shows that the net reflected wave amplitude due to multiple partial reflections is:

$$\Gamma = \Gamma_1 + T_{21}T_{12}\Gamma_3 e^{-j2\theta} + T_{12}T_{21}\Gamma_3^2\Gamma_2 e^{-j4\theta} + \cdots$$

$$= \Gamma_1 + T_{12}T_{21}\Gamma_3 e^{-j2\theta}\sum_{n=0}^{\infty}\Gamma_2^n\Gamma_3^n e^{-j2n\theta}$$

$$= \Gamma_1 + \frac{T_{12}T_{21}\Gamma_3 e^{-j2\theta}}{1-\Gamma_2\Gamma_3 e^{-j2\theta}} \tag{7.32}$$

where the geometric series below has been used.

$$\sum_{n=0}^{\infty} x^n = \frac{1}{1-x}, \quad \text{for } |x|<1$$

Using $\Gamma_2 = -\Gamma_1$, $T_{21} = 1 + \Gamma_1$ and $T_{12} = 1 - \Gamma_1$ from Eqs 7.30 and 7.31, Eq. 7.32 may be expressed as:

$$\Gamma = \frac{\Gamma_1 + \Gamma_3 e^{-j2\theta}}{1+\Gamma_1\Gamma_3 e^{-j2\theta}} \tag{7.33}$$

Assuming small reflections, $|\Gamma_1|$ and $|\Gamma_3| << 1$, we can approximate Eq. 7.33 as:

$$\Gamma \approx \Gamma_1 + \Gamma_3 e^{-j2\theta} \tag{7.34}$$

This result states that, the total reflection is dominated by the reflection from the initial discontinuity between Z_1 and Z_2, that is, Γ_1 and the first reflection from the discontinuity between Z_2 and Z_L, that is, $\Gamma_3 e^{-2j\theta}$. The $e^{-2j\theta}$ term accounts for the phase delay when the incident wave travels up and down the line. Thus, for small reflections, the total reflection coefficient can be obtained only from the first order reflection. For example, with $\Gamma_1 = \Gamma_3 \leq 0.2$, the maximum error is only 4%.

Stepped impedance transformer: Next, let us consider the problem of matching two impedances Z_1 and Z_{n+1} using $(n - 1) = N$ sections, that is, n steps of the quarter-wave transformers (note that number of sections N is always less than the number of steps by 1; in Fig. 7.12a we had single section and a two step transformer). These n steps will generate n reflection coefficients $\Gamma_1, \Gamma_2, \ldots, \Gamma_n$. Because of the discontinuities at the junctions, each component wave will undergo multiple reflections back and forth, between any two successive steps. Therefore, for $(n - 1)$ sections, the total reflection coefficient Γ_T at any frequency f, can be written as:

$$\Gamma_T = \Gamma_1 + \Gamma_2 e^{-j2\theta} + \Gamma_3 e^{-j4\theta} + \cdots \Gamma_n e^{-j2(n-1)\theta} \tag{7.35}$$

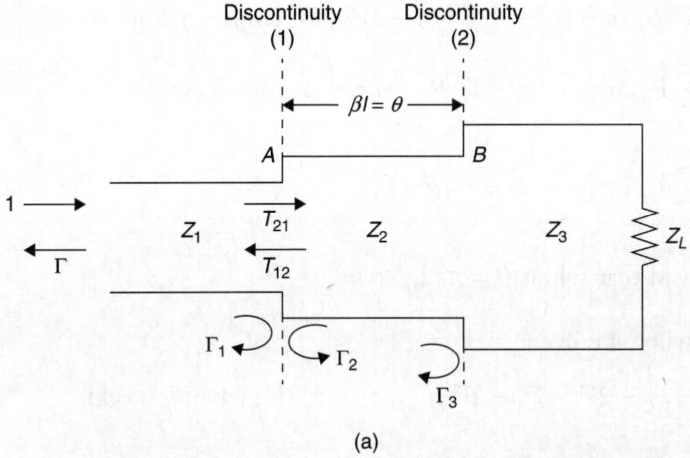

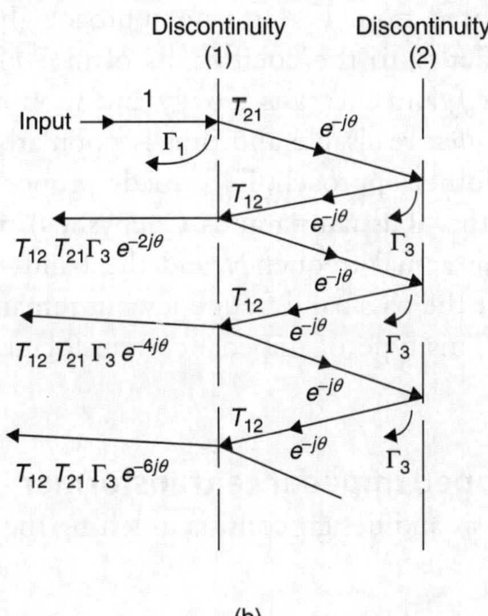

Fig. 7.12 (a) Single section quarter-wave transformer with two discontinuities (b) Multiple partial reflections of waves at the two discontinuities (reflecting surfaces).

where $\Gamma_i = (Z_{i+1} - Z_i)\,(Z_{i+1} - Z_i)$, the reflection coefficient at the junction between the $(i + 1)^{\text{th}}$ and i^{th} sections for a small reflection.

Assuming that the transformer is symmetrical with respect to a transverse reference plane at the centre (this however does not mean that the Z_ns are symmetrical):

$$\Gamma_1 = \Gamma_n, \Gamma_2 = \Gamma_{n-1}, \Gamma_3 = \Gamma_{n-2} \cdots \tag{7.36}$$

Then,

$$\Gamma_T = 2\{\Gamma_1 \cos(n-1)\theta + \Gamma_2 \cos(n-3)\theta + \Gamma_3 \cos(n-5)\theta + \cdots\} \tag{7.37}$$

$$= 2\sum_{k=1}^{m} \Gamma_k \cos\{n-(2k-1)\}\theta; \quad m = n/2 \quad \text{(for } n \text{ even)} \tag{7.38}$$

$$= 2\sum_{k=1}^{m} \Gamma_k \cos\{n-(2k-1)\}\theta + \Gamma_{(n+1)/2}; \quad m = (n-1)/2 \quad \text{(for } n \text{ odd)} \tag{7.39}$$

It may be observed that when $\theta = \pi/2$,

$$\Gamma_T = 0; \text{ when } n \text{ is even} \tag{7.40}$$

$$= \Gamma_{(n+1)/2} - 2\Gamma_{(n-1)/2} + 2\Gamma_{(n-3)/2} + \ldots \pm 2\Gamma_1; \text{ when } n \text{ is odd} \tag{7.41}$$

By properly choosing the reflection coefficient Γ_k, $k = 1, 2, \ldots n$, a variety of pass band characteristics of a multi-section transformer can be realised. The impedances Z_ks for various sections are calculated from Γ_ks. In one approach the Γ_ks are set proportional to binomial coefficients, that is, to the coefficients of $(a + b)^{n-1}$. In this approach, the magnitude of Γ_T is zero at f_0 and increases slowly and monotonically on either side of f_0. The curve at f_0 is the *flattest* realisable and thus is popularly known as *maximally flat/binomial* transformer. In another approach, Γ_T is made proportional to the Tchebyscheff (an alternative spelling of this Russian name is Chebyshev) polynomial $T_{n-1}(x)$, where x is a function of both the actual frequency and the bandwidth. In this approach, Γ_T oscillates in magnitude over the pass band frequency but remains less (in magnitude) than a prescribed ripple value. This type of impedance transformer is known as *equal ripple/Tchebyscheff* transformer.

7.3.2.1 Binomial stepped impedance transformer

Binomial transformer analysis and design consists in letting the Γ_ks be proportional to the coefficients of $(a + b)^{n-1}$:

$$\Gamma_k = \Gamma_1 C_{k-1}^{n-1} \tag{7.42}$$

The reflection coefficient Γ_T can now be expressed as:

$$\Gamma_T = 2\Gamma_1 \sum_{k=1}^{m} C_{k-1}^{n-1} \cos\{n-(2k-1)\}\theta; \quad m = n/2, \text{ for } n \text{ even} \tag{7.43}$$

$$= 2\Gamma_1 \sum_{k=1}^{m} C_{k-1}^{n-1} \cos\{n-(2k-1)\}\theta + C_{(n-1)/2}^{n-1}; \quad m = (n-1)/2, \text{ for } n \text{ odd} \tag{7.44}$$

Expanding $\cos p\theta$ in terms of $\cos\theta$, where $p = n - (2k - 1)$, we get:

$$\cos p\theta = 2\cos\,(p-1)\theta\,\cos\theta - \cos\,(p-2)\theta \tag{7.45}$$

It may be verified that, when the expansion of cos $p\theta$ in terms of cosθ is substituted into the expression of Γ above, only one term involving $\cos^{(n-1)}\theta$ with coefficient $2^{(n-2)}$ does not cancel out, that is:

$$\Gamma = 2\,\Gamma_1\, 2^{(n-2)} \cos^{(n-1)}\theta = \Gamma_1\, 2^{(n-1)} \cos^{(n-1)}\theta \tag{7.46}$$

Since all the derivatives of Γ at $f = f_0$, where $\theta = \pi/2$ up to the $(n-2)^{\text{th}}$ term vanishes, the binomial distribution provides a *maximally flat* response as shown in Fig. 7.13.

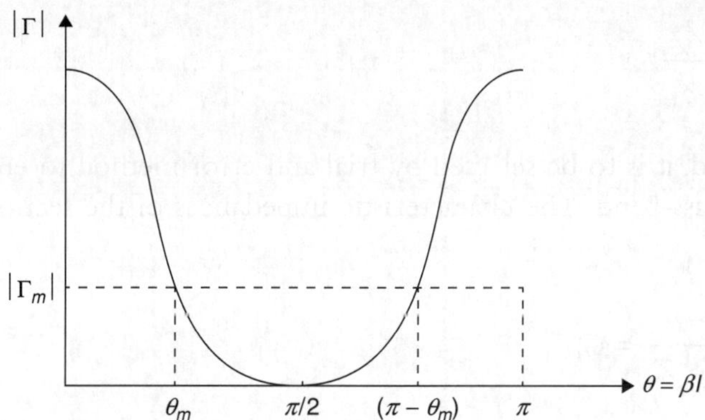

Fig. 7.13 Maximally-flat response.

In the design of the *binomial transformer*, the problem is stated in terms of a tolerable (maximum) reflection coefficient $|\Gamma| = |\Gamma_m|$ over a specified band of frequencies (pass-band) from f_1 to f_2 with $f_0 = (f_1 - f_2)/2$ realisable with n number of sections.

The maximum value of Γ_1 that occurs at the band edges may then be computed from the following equation.

$$|\Gamma_1| \le \frac{|\Gamma_m|}{2^{n-1}\cos^{n-1}\theta_1} \quad \text{and} \quad |\Gamma_2| \le \frac{|\Gamma_m|}{2^{n-1}\cos^{n-1}\theta_2} \tag{7.47}$$

where θ_1 and θ_2 respectively correspond to the total phase shift per section at f_1 and f_2, which are respectively, the low end and the high end of the frequency band. It may be noted that for symmetry of the response curve, $\cos\theta = |\cos\theta_2|$, or $\theta_2 = \pi - \theta_1$.

Now, let us put together some of the design formulae for the *binomial impedance transformer* [15,16]. Let r_w be the ratio of the wavelengths at the band edge frequencies f_1 and f_2. Then:

$$r_w = \frac{\lambda_{g1}}{\lambda_{g2}} = \frac{\theta_2}{\theta_1} = \frac{\pi - \theta_1}{\theta_1} = \frac{\theta_2}{\pi - \theta_2} \tag{7.48}$$

$$\theta_1 = \frac{\pi}{1 - r_w} = \frac{2\pi l}{\lambda_{g1}} = \frac{\pi\lambda_{g0}}{2\lambda_{g1}} \tag{7.49}$$

$$\theta_2 = \frac{\pi r_w}{1 + r_w} = \frac{2\pi l}{\lambda_{g2}} = \frac{\pi\lambda_{g0}}{2\lambda_{g2}} \tag{7.50}$$

$$l = \frac{\lambda_{g0}}{4} = \frac{\theta_1\lambda_{g1}}{2\pi} = \frac{\theta_2\lambda_{g2}}{2\pi} \tag{7.51}$$

$$\lambda_{g0} = \frac{2\lambda_{g1}\lambda_{g2}}{\lambda_{g1} + \lambda_{g2}} \tag{7.52}$$

If n is not specified, it is to be selected by trial and error method to ensure that $|\Gamma| \le |\Gamma_m|$ over the desired pass-band. The characteristic impedances of the sections are determined from the relation:

$$\Gamma_k = \frac{Z_{k+1} - Z_k}{Z_{k+1} + Z_k} = C_{k-1}^{n-1} \tag{7.53}$$

where $Z_{(i+1)}$ is the characteristic impedance of the ith section and Γ_i is the reflection coefficient at the input of the ith section. Further, if the reflections are small (in all practical designs, $\Gamma_k \le 0.35$ when error is less than 4%) we can write:

$$\ln\left(\frac{Z_{k+1}}{Z_k}\right) = 2\left\{\frac{Z_{k+1} - Z_k}{Z_{k+1} + Z_k} + \frac{1}{3}\left(\frac{Z_{k+1} - Z_k}{Z_{k+1} + Z_k}\right)^3 + \cdots\right\}$$

$$= 2\left(\Gamma_k + \frac{1}{3}\Gamma_k^3\right)$$

$$\approx 2\Gamma_k$$

$$\text{Thus,}\quad \Gamma_k = \frac{1}{2}\ln\frac{Z_{k+1}}{Z_k} = C_{k-1}^{n-1}\frac{1}{2}\ln\frac{Z_2}{Z_1} \tag{7.54}$$

$$\text{when,}\quad \ln\left(\frac{Z_{k+1}}{Z_k}\right) = C_{k-1}^{n-1}\ln\frac{Z_2}{Z_1} \tag{7.55}$$

Therefore, the impedances can be computed from the corresponding reflection coefficients:

$$\Gamma_1 = \frac{1}{2}\ln\left(\frac{Z_2}{Z_1}\right),\quad Z_2 = Z_1\{\text{anti}\ \ln(2\Gamma_1)\}$$

$$\Gamma_2 = \frac{1}{2}\ln\left(\frac{Z_3}{Z_2}\right), \quad Z_3 = Z_2\{\text{anti } \ln(2\Gamma_2)\}$$

$$\vdots \qquad \vdots \qquad \vdots \qquad \vdots$$

$$\Gamma_k = \frac{1}{2}\ln\left(\frac{Z_{k+1}}{Z_k}\right), \quad Z_{k+1} = Z_k\{\text{anti } \ln(2\Gamma_k)\} \tag{7.56}$$

By adding up to n terms:

$$\Gamma_T = \Gamma_1 + \Gamma_2 + \cdots + \Gamma_n = \frac{1}{2}\ln\left(\frac{Z_{n+1}}{Z_1}\right) \tag{7.57}$$

This gives the total reflection coefficient of n steps or $(n-1) = N$ section transformer.

EXAMPLE 7.4

Design a two-section binomial impedance transformer to match a 225 Ω load (typical characteristic impedance of an X-band rectangular waveguide) to a reduced height waveguide having a characteristic impedance of 75 Ω at 10 GHz. What bandwidth is obtained for a maximum tolerable reflection coefficient of 0.05? Evaluate the design dimensions of the transformer and characterise the transformer by optimising the design using a 3D electromagnetic field simulator.

Solution

Given that the number of sections, $N = 2$. Thus number of steps, $n = 2 + 1 = 3$.

Here, $Z_1 = 75\ \Omega, Z_{n+1} = Z_4 = 225\ \Omega, \Gamma_m = 0.05, f_0 = 10$ GHz ($\lambda_0 = 3.0$ cm).

Since $n = 3$, we have from Eq. 7.57:

$$\Gamma_T = \Gamma_1 + \Gamma_2 + \Gamma_3 = \frac{1}{2}\ln\left(\frac{Z_4}{Z_1}\right) = \frac{1}{2}\ln 3 = \frac{1.099}{2} = 0.549$$

As the transformer is symmetrical, $\Gamma_1 = \Gamma_3$ and $\Gamma_2 = \Gamma_1 C^{(3-1)}_{(2-1)} = 2\Gamma_1$ (Eq. 7.42)

Therefore, $\Gamma_1 + 2\Gamma_1 + \Gamma_1 = 0.549$, or $\Gamma_1 = 0.137$ and $\Gamma_2 = 0.274$

Now, since $\Gamma_k = \frac{1}{2}\ln\left(\frac{Z_{k+1}}{Z_k}\right)$; $Z_{k+1} = Z_k\{\text{anti } \ln(2\Gamma_k)\}$

So, $Z_2 = Z_1\{\text{anti } \ln(2\Gamma_1)\} = 75\{\text{anti } \ln(2 \times 0.137)\} = 98.64\ \Omega$

$Z_3 = Z_2\{\text{anti } \ln(2\Gamma_2)\} = 98.64\{\text{anti } \ln(2 \times 0.274)\} = 170.63\ \Omega$

Length of each section is, $\frac{\lambda_g}{4} = 0.993$ cm (since $\lambda_0 = 3$ cm and $\lambda_g = 3.975$ cm)

Now, $\Gamma_m = \Gamma_1 2^{n-1} \cos^{n-1} \theta_1 = \Gamma_1 2^{n-1} \cos^{n-1} \theta_2$ (Eq. 7.47)

Thus, $0.05 = 0.137 \times 2^2 \times \cos^2 \theta_1$ or, $\cos^2 \theta_1 = 0.091$; when, $\cos\theta_1 = 0.302$

with, $\theta_1 = 71.33°$ and, $\theta_2 = (180 - \theta_1) = 108.67°$

Therefore, $\theta = (\theta_2 - \theta_1) = (108.67° - 71.33°) = 37.34°$

When: $r_w = \frac{\theta_2}{\theta_1} = \frac{108.67}{71.33} = 1.523$

The bandwidth of the matching transformer may be calculated approximately by using the TEM line equation as:

$$\Delta f = \left(\frac{2f_0}{\pi}\right)\left(\frac{\theta_2}{\theta_1}\right) = \left(\frac{2 \times 10}{180}\right)(37.34) = 4.148 \text{ GHz}$$

The design dimensions of the transformer may be evaluated as follows. From Fig. 7.14, the physical length of each section of the transformer is $\lambda_g/4$ (electrical length $\theta = 90°$). The step heights are given by $s_1 = (b_2 - b_1)$, $s_2 = (b_3 - b_2)$ and $s_3 = (b_4 - b_3)$; where b is given by $b = \frac{Z_0 a}{377}\sqrt{1 - \left(\frac{\lambda_0}{2a}\right)^2}$ (which follows from the expression for characteristic impedance of rectangular waveguide, Section 4.2.3).

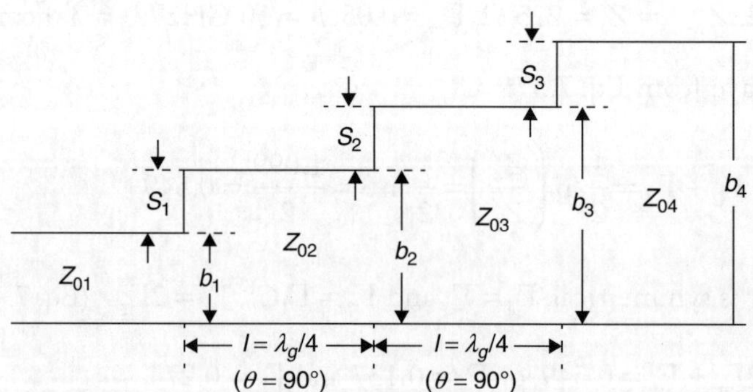

Fig. 7.14 Schematic of stepped impedance transformer (Example 7.4 and 7.6) [$n = 3$ (step), $N = 2$ (section)].

For an X-band waveguide $a = 0.9'' = 2.286$ cm and with $\lambda_0 = 3$ cm, $\lambda_g = 3.975$ cm. Here $Z_{01} = 75\ \Omega$, $Z_{02} = 98.64\ \Omega$, $Z_{03} = 170.625\ \Omega$, $Z_{04} = 225\ \Omega$. It may be calculated that $b_1 = 0.3431$ cm,

$b_2 = 0.4513$ cm, $b_3 = 0.7807$ cm, $b_4 = 1.0295$ cm. Thus, $s_1 = 0.1082$ cm, $s_2 = 0.3294$ cm and $s_3 = 0.2488$ cm.

The impedance transformer designed through the synthesis process (discussed above) is simulated in a 3D electromagnetic (EM) field simulator to realise an optimised design with little trimming of the physical dimensions of the transformer. The optimised simulated structure of the transformer along with different dimensions and a comparison of the simulation result and synthesis result are shown in Fig. 7.15a–7.15c.

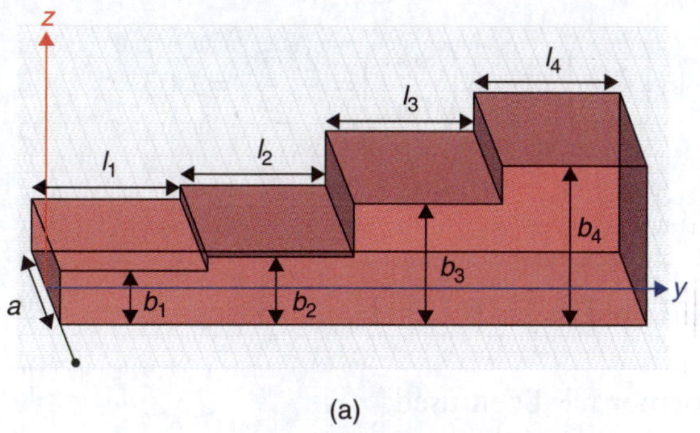

(a)

Parameters	Synthesis dimensions	Simulation dimensions
a	22.86	22.86
b_1, (l_1)	3.431, (9.5)	3.431, (9.5)
b_2, (l_2)	4.513, (9.5)	4.50, (9.4)
b_3, (l_3)	7.807, (9.5)	7.80, (9.1)
b_4, (l_4)	10.295, (9.5)	1.0295, (9.5)

(b)

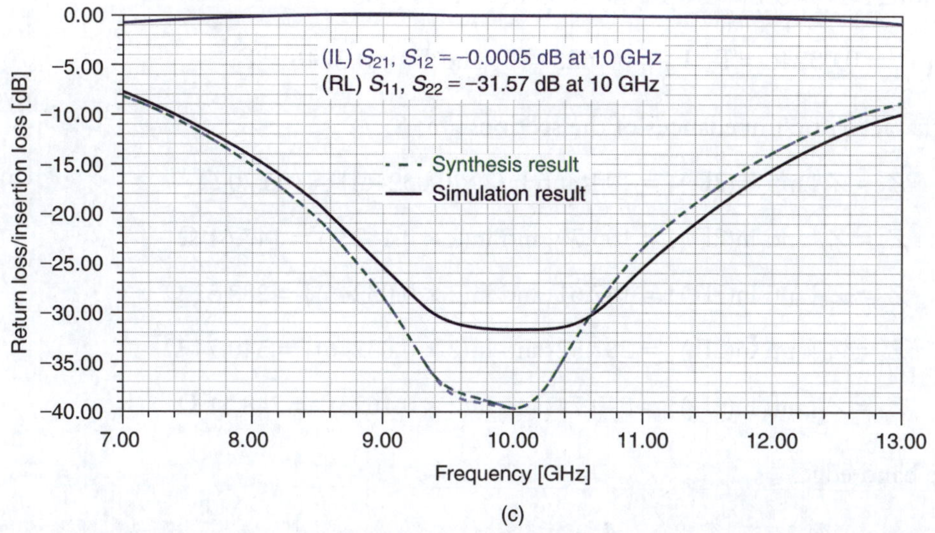

(c)

Fig. 7.15 3D EM field simulator based design of a binomial transformer (a) Structure (b) Design dimensions (mm) (c) Comparison of the simulation result and synthesis result.

EXAMPLE 7.5

Design a five-section quarter-wave stepped impedance transformer at 10 GHz to match a 100 Ω line to an 800 Ω load. Assume binomial distribution pass-band response of the reflection coefficients for the band edge frequency ratio of 1.75. Find the maximum VSWR at the input of the transformer.

Solution

Given that the number of sections $N = 5$. Thus number of steps $n = 5 + 1 = 6$.

$$Z_1 = 100\ \Omega,\ Z_{n+1} = Z_7 = 800\ \Omega$$

Assuming small reflection at each step:

$$\Gamma_T = \Gamma_1 + \Gamma_2 + \Gamma_3 + \Gamma_4 + \Gamma_5 + \Gamma_6 = \frac{1}{2}\ln\left(\frac{Z_7}{Z_1}\right) = \frac{1}{2}\ln\left(\frac{800}{100}\right) = \frac{2.079}{2} = 1.039$$

For binomial distribution: $\Gamma_k = \Gamma_1 C^{(n-1)}_{(k-1)} = \Gamma_1 C^5_{k-1}$

Thus, $\Gamma_1 = \Gamma_1 C^5_0 = \Gamma_6; \Gamma_2 = \Gamma_1 C^5_1 = 5\Gamma_1 = \Gamma_5;\quad \Gamma_3 = \Gamma_1 C^5_2 = 10\Gamma_1 = \Gamma_4$

where, the symmetry of the binomial transformer has been used.

Therefore, $2(\Gamma_1 + 5\,\Gamma_1 + 10\,\Gamma_1) = 1.039$ or $32\Gamma_1 = 1.039$

When $\Gamma_1 = 0.03246 = \Gamma_6, \Gamma_2 = 0.1623 = \Gamma_5$ and $\Gamma_3 = 0.3246 = \Gamma_4$

The characteristic impedances of the sections are:

$$Z_2 = Z_1\{\text{anti}\ \ln(2\Gamma_1)\} = 100\{\text{anti}\ \ln(2 \times 0.03246)\} = 106.70\ \Omega$$

$$Z_3 = Z_2\{\text{anti}\ \ln(2\Gamma_2)\} = 106.70\{\text{anti}\ \ln(2 \times 0.1623)\} = 147.61\ \Omega$$

$$Z_4 = Z_3\{\text{anti}\ \ln(2\Gamma_3)\} = 147.61\{\text{anti}\ \ln(2 \times 0.03246)\} = 282.52\ \Omega$$

$$Z_5 = Z_4\{\text{anti}\ \ln(2\Gamma_4)\} = 282.52\{\text{anti}\ \ln(2 \times 0.03246)\} = 540.74\ \Omega$$

$$Z_6 = Z_5\{\text{anti}\ \ln(2\Gamma_5)\} = 540.74\{\text{anti}\ \ln(2 \times 0.1623)\} = 748.11\ \Omega$$

At the band edge,

$$\theta_1 = \frac{\pi}{1 + r_w} = \frac{\pi}{1 + 1.75} = \frac{\pi}{2.75} = 1.142 \text{ radians} = 65.\,44° \text{ (since } r_w = 1.75\text{, given)}$$

Therefore, $\cos\theta_1 = 0.4156$ and $\cos^5\,\theta_1 = 0.01239$

Now, $\Gamma_m = \Gamma_1\,2^{n-1}\,\cos^{n-1}\,\theta_1 = 0.03246 \times 2^5 \times \cos^5\,\theta_1 = 0.012869$

$$\text{Thus maximum VSWR} = \frac{1 + |\Gamma_m|}{1 - |\Gamma_m|} = \frac{1 + 0.012869}{1 - 0.012869} = 1.02607$$

[**Note:** For a practical stepped impedance transformer design, more than 2 to 3 steps are not required because the impedance ratios met with are normally 3 to 5 and hardly as large as 8 in the present case. However, if that happens, then stepped impedance transformation is used to take care of matching impedance ratio up to 4 or 5 and further impedance transformation is done by other novel techniques (discussed for coaxial-waveguide cavity in chapter 9)].

7.3.2.2 Tchebyscheff stepped impedance transformer

One may also let $\Gamma_k s$ be proportional to $T_{n-1}(x)$, where $T_n(x)$ is the Tchebyscheff polynomial of order n given by:

$$T_n(x) = \cos(n \cos^{-1} x); \qquad -1 < x < 1$$

$$T_n(x) = \cosh(n \cosh^{-1} x); \quad x < -1, x > 1 \tag{7.58a}$$

While x is a parameter for change of variable, which is a function of frequency and is given by $x = \cos\theta / \cos\theta_1 = \cos\theta / |\cos\theta_2|$. Instead of maximally flat pass-band characteristics of the binomial transformer, the Tchebyscheff transformer exhibits equal ripple characteristics in which the maximum tolerable reflection coefficient $|\Gamma_m|$ vanishes at as many different frequencies in the pass-band as there are transformer sections (N). In contrast with the binomial matching transformer, the Tchebyscheff transformer optimises bandwidth at the expense of the pass-band ripple. If such a pass-band characteristic can be tolerated, the Tchebyscheff transformer can provide a substantially wider bandwidth compared to the binomial transformer, for a given number of sections.

The nth order Tchebyscheff polynomial is a polynomial of degree n, and is denoted by $T_n(x)$. The first four Tchebyscheff polynomials are:

$$T_1(x) = x$$

$$T_2(x) = 2x^2 - 1$$

$$T_3(x) = 4x^3 - 3x \tag{7.58b}$$

$$T_4(x) = 8x^4 - 8x^2 + 1$$

Higher order polynomials can be found by using the following recurrence formula:

$$T_k(x) = 2 \times T_{k-1}(x) - T_{k-2}(x) \tag{7.58c}$$

The first four Tchebyscheff polynomials are plotted in Fig. 7.16 from which the following important properties of Tchebyscheff polynomials may be noted.

(i) For $-1 \le x \le 1, |T_k(x) \le 1|$, that is, all the polynomials are limited within the range (1, −1), which means that the Tchebyscheff polynomials oscillate between ±1 and hence exhibit equal ripple characteristics that is mapped in the pass-band of the transformer.

(ii) For $|x| > 1$, $|T_k(x) \geq 1|$ which falls outside the pass-band and $|T_k(x)|$ increases faster with x as k increases.

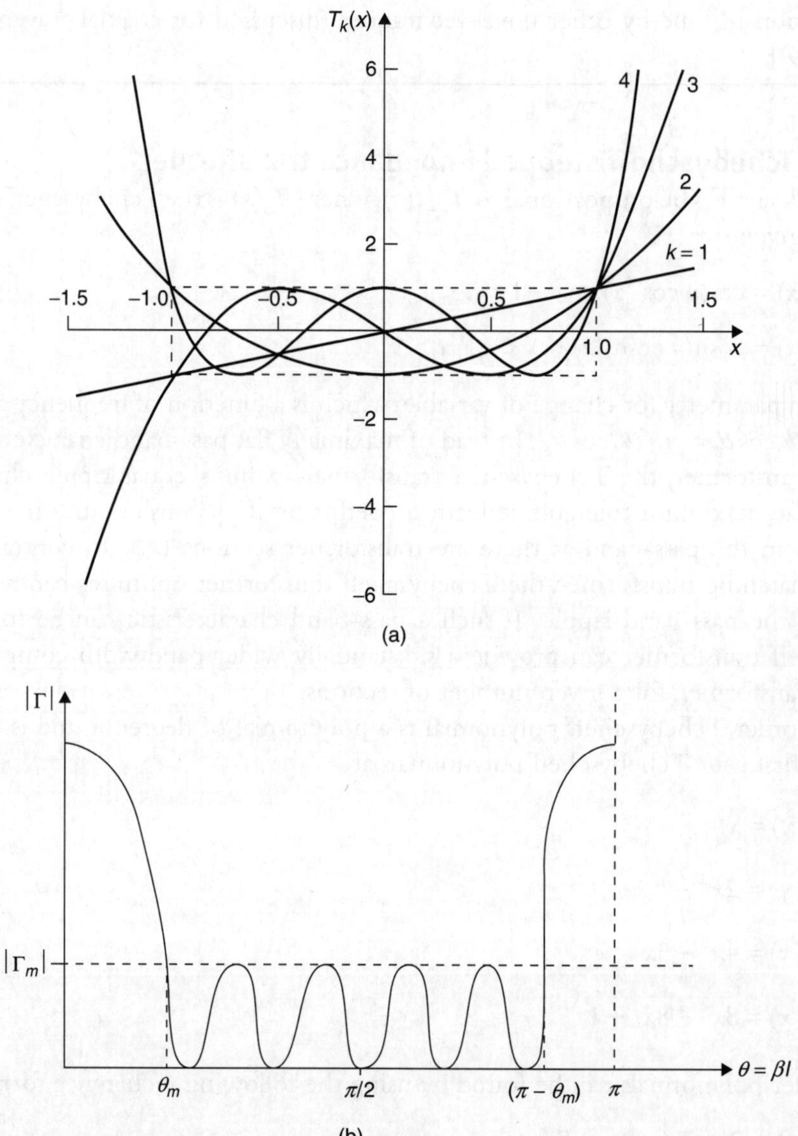

Fig. 7.16 (a) Tchebyscheff polynomials of different order (here k up to 4) (b) Equal-ripple characteristics for a five-section transformer.

For the design of a Tchebyscheff impedance matching transformer, on specifying a maximum tolerable reflection coefficient $|\Gamma_m|$, we can write with reference to a transverse plane at the centre by using the total reflection coefficient expressions given by Eqs 7.38 and 7.39:

$$\Gamma_T = |\Gamma_m| T_{n-1}(x) = 2\sum_{k=1}^{m} \Gamma_k \cos\{n-(2k-1)\}\theta;\ m = n/2 \text{ (for } n \text{ even)} \tag{7.59}$$

$$= 2\sum_{k=1}^{m} \Gamma_k \cos\{n-(2k-1)\}\theta + \Gamma_{(n+1)/2};\ m = (n-1)/2 \text{ (for } n \text{ odd)} \tag{7.60}$$

This ensures that $|\Gamma_m T_{n-1}(x)| \le |\Gamma_m|$, since $|T_{n-1}(x)| \le 1$. The $\Gamma_k s$ are obtained by expanding the right hand side of Eqs 7.59 and 7.60 in terms of powers of $\cos\theta$ and substituting $x\cos\theta_1$ for $\cos\theta$ and equating the coefficient of like powers of x with those on the left hand side. The Tchebyscheff transformer gives the largest bandwidth for a specified $|\Gamma_m|$ and number of sections and for a given bandwidth gives the lowest ripple $|\Gamma_m|$.

EXAMPLE 7.6

Design a two-section Tchebyscheff impedance transformer to match a 225 Ω load to a 75 Ω line at 10 GHz. The maximum VSWR allowed is 1.1. Find the corresponding bandwidth. Evaluate the step dimensions of the transformer. Using 3D electromagnetic field simulator, characterise the designed transformer.

Solution

Given that the number of sections $N = 2$. Thus number of steps $n = 2 + 1 = 3$.

$$Z_1 = 75\ \Omega,\ Z_{n+1} = Z_4 = 225\ \Omega$$

$$|\Gamma_m| = \frac{1.1-1}{1.1+1} = \frac{0.1}{2.1} = 0.0476 \text{ (since maximum VSWR is specified as 1.1)}$$

Total reflection coefficient for symmetrical transformer (n odd) is:

$$\Gamma = |\Gamma_m| T_{n-1}(x) = 0.0476\ T_2(x) = 2\ (\Gamma_1 \cos 2\theta) + \Gamma_2 \quad \text{(Eq. 7.60)}$$

or,
$$0.0476\ (2x^2 - 1) = 2\ \{\Gamma_1\ (2\cos^2\theta - 1)\} + \Gamma_2$$
$$= 2\ \{\Gamma_1\ (2x^2 \cos^2\theta_1 - 1)\} + \Gamma_2$$

Equating the coefficients of x^2, x and so on,

$$2 \times 0.0476 = 4\ \Gamma_1 \cos^2\theta_1 \text{ or, } \Gamma_1 = 2 \times \frac{0.0476}{4} \cos^2\theta_1$$

and
$$-0.0476 = -2\ \Gamma_1 + \Gamma_2 \text{ or, } \Gamma_2 = 2\ \Gamma_1 - 0.0476$$

Now at zero frequency, $\theta\ (= \beta l) = 0$, and we can write for general reflection coefficient:

$$\left(\frac{Z_4 - Z_1}{Z_4 + Z_1}\right) = \Gamma_m \Gamma_{n-1}(\sec\theta_1) \quad \left(\text{since } x = \frac{\cos\theta}{\cos\theta_1} = \sec\theta_1\right)$$

or, $$\left(\frac{225-75}{225+75}\right)=0.0476\,T_2(\sec\theta_1)$$

or, $$\frac{150}{300}=0.0476\,(2\sec^2\theta_1-1)$$

or, $\sec^2\theta_1 = 5.752$, when $\cos^2\theta_1 = 0.1738$ and $\cos\theta_1 = 0.4168$ with $\theta_1 = 65.367°$

$$= 0.36315\pi \text{ radians}$$

Therefore, $$\Gamma_1=\left(\frac{2\times 0.0476}{4\times 0.1738}\right)=0.1369$$

$$\Gamma_2 = 2\,\Gamma_1 - 0.0476 = 2\times 0.136 - 0.0476 = 0.22627$$

Now, $$\Gamma_1=\left(\frac{Z_2-Z_1}{Z_2+Z_1}\right),\ \text{and } \Gamma_2=\left(\frac{Z_3-Z_2}{Z_3+Z_2}\right)$$

When, $$Z_2=\frac{(1+\Gamma_1)}{(1-\Gamma_1)}Z_1=\frac{1+0.1369}{1-0.1369}\times 75=98.79\ \Omega$$

$$Z_3=\frac{(1+\Gamma_2)}{(1-\Gamma_2)}Z_2=\frac{1+0.22627}{1-0.22627}\times 98.79=156.57\ \Omega$$

The fractional bandwidth $$\frac{\Delta f}{f_0}=\frac{4}{\pi}\left(\frac{\pi}{2}-\theta_1\right)$$

$$=\frac{4}{\pi}\left(\frac{\pi}{2}-0.36315\pi\right)$$

or, $$\frac{\Delta f}{f_0}=0.5474$$

Thus the bandwidth is $\Delta f = 0.5474 \times 10$ GHz $= 5.4744$ GHz

[**Note:** When compared with the result of Example 7.4, it is observed that the Tchebyscheff transformer is capable of providing a larger bandwidth than the binomial transformer].

The step dimensions of the transformer can be calculated using the necessary formula and the data values as appropriate (following Example 7.4) and in this design we obtain $b_1 = 0.3431$ cm, $b_2 = 0.45193$ cm, $b_3 = 0.7162$ cm, $b_4 = 1.0295$ cm and hence $s_1 = 0.10883$ cm, $s_2 = 0.26427$ cm and $s_3 = 0.3133$ cm (where $Z_{01} = 75\ \Omega$, $Z_{02} = 98.79\ \Omega$, $Z_{03} = 156.57\ \Omega$, $Z_{04} = 225\ \Omega$. The schematic of the transformer designed with the synthesis process is similar to that shown in Fig. 7.14.

The impedance transformer designed through the synthesis process (discussed above) is simulated in a 3D electromagnetic field simulator to realise the optimised design.

The optimised simulated structure of the transformer along with different dimensions and a comparison of the simulation result and synthesis result are shown in Fig. 7.17a–7.17c.

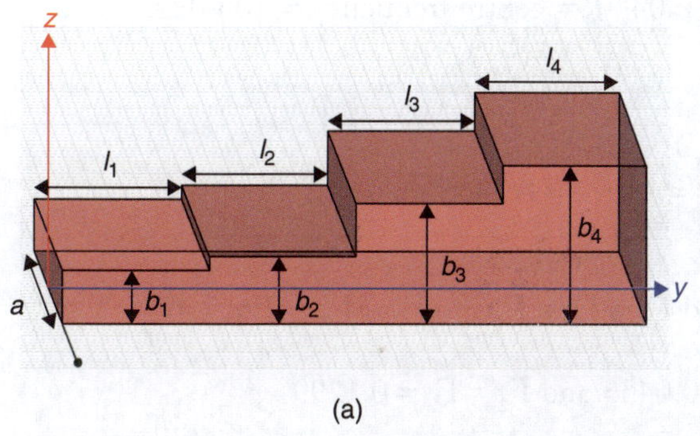

(a)

Parameters	Synthesis dimensions	Simulation dimensions
a	22.86	22.86
b_1, (l_1)	3.431, (9.5)	3.431, (9.5)
b_2, (l_2)	4.519, (9.5)	4.80, (9.5)
b_3, (l_3)	7.162, (9.5)	7.3, (9.5)
b_4, (l_4)	10.295, (9.5)	10.00, (9.5)

(b)

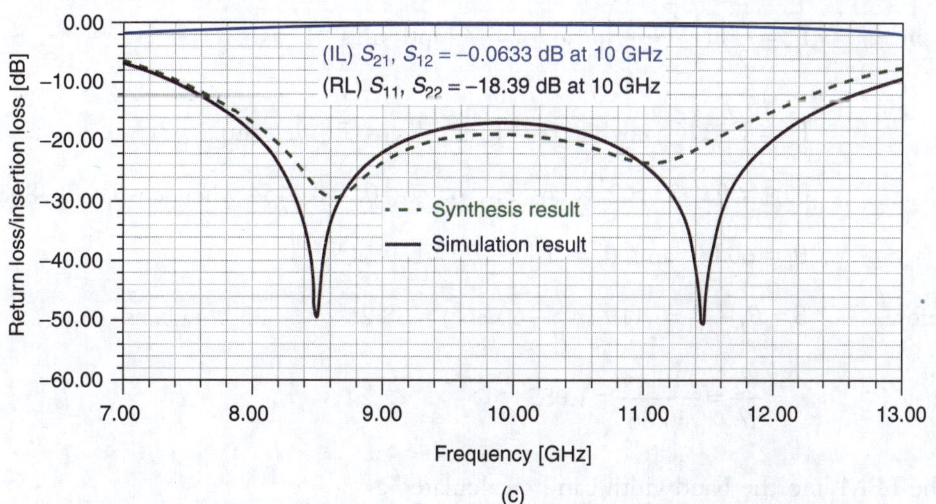

(c)

Fig. 7.17 3D EM field simulator based design of Tchebyscheff transformer (a) Structure (b) Design dimensions (c) Comparison of the simulation result and synthesis result.

EXAMPLE 7.7

Design a 3-section binomial and also a Tchebyscheff transformer at 10 GHz with the following data: impedances to be matched are 100 Ω to 50 Ω, allowable maximum reflection coefficient $\Gamma_m = 0.04$. Show that the bandwidth realisable with the Tchebyscheff transformer is larger than that with the binomial transformer but with a penalty of non-zero ripple in the pass-band of the Tchebyscheff transformer.

Repeat the design for 1, 2 and 4 section transformer in each case and with the plot of reflection coefficient magnitude vs frequency show that greater bandwidth is realisable using more number of sections for both the impedance transformers.

Solution

Binomial transformer: Given that the number of sections $N = 3$. Thus number of steps $n = 3 + 1 = 4$.

Here, $Z_1 = 50\ \Omega$, $Z_{n+1} = Z_5 = 100\ \Omega$, $\Gamma_m = 0.04$, f_0 = centre frequency = 10 GHz.

Since $n = 4$, we have from Eq. 7.57:

$$\Gamma_T = \Gamma_1 + \Gamma_2 + \Gamma_3 + \Gamma_4 = \frac{1}{2}\ln\left(\frac{Z_5}{Z_1}\right) = \frac{1}{2}\ln 2 = 0.3465$$

As the transformer is symmetrical, $\Gamma_1 = \Gamma_4$ and $\Gamma_2 = \Gamma_3 = \Gamma_1 C_{(3-1)}^{(4-1)} = 3\Gamma_1$ (Eq. 7.47)

Therefore, $2\Gamma_1 + 6\Gamma_1 = 0.3465$, or $\Gamma_1 = 0.0433$ and $\Gamma_2 = \Gamma_3 = 0.1299$

Now since: $\Gamma_k = \dfrac{1}{2}\ln\left(\dfrac{Z_{k+1}}{Z_k}\right)$; $Z_{k+1} = Z_k\{\text{anti ln}(2\Gamma_k)\}$

Now, $\Gamma_m = \Gamma_1 2^{(n-1)} \cos^{(n-1)}\theta_1 = \Gamma_1 2^{(n-1)} \cos^{(n-1)}\theta_2$ (Eq. 7.47)

Thus, $0.04 = 0.0433 \times 2^3 \times \cos^3\theta_1$ or, $\cos^3\theta_1 = 0.1154$ when, $\cos\theta_1 = 0.4868$

with $\theta_1 = 60.86°$ and, $\theta_2 = (180 - \theta_1) = 119.14°$

Therefore, $\theta = \theta_2 - \theta_1 = (119.14° - 60.86°) = 58.28° = 0.3237\pi$ radians

When $r_w = \dfrac{\theta_2}{\theta_1} = \dfrac{119.14}{60.86} = 1.957$

For the TEM line, the bandwidth can be calculated as:

$$\Delta f = \left(\frac{2f_0}{\pi}\right)\left(\frac{\theta_2}{\theta_1}\right) = (2 \times 0.32327)f_0 = 0.6475 f_0\text{GHz} = 6.475\ \text{GHz}$$

Tchebyscheff transformer: Given that the number of sections $N = 3$. Thus, number of steps $n = 3 + 1 = 4$.

Here, $Z_1 = 50\ \Omega$, $Z_{n+1} = Z_5 = 100\ \Omega$, $\Gamma_m = 0.04$, f_0 = centre frequency = 10 GHz.

Total reflection coefficient for symmetrical transformer (n even) is:

$$\Gamma = |\Gamma_m| T_{n-1}(x) = 0.04\ T_3(x) = 2\,(\Gamma_1 \cos 3\theta + \Gamma_2 \cos\theta) \quad \text{(Eq. 7.59)}$$

or, $0.04\,(4x^3 - 3x) = 2\,\{\Gamma_1\,(4\cos^3\theta - 3\cos\theta) + \Gamma_2\cos\theta\}$

$= 2\,\{\Gamma_1\,(4x^3\cos^3\theta_1 - 3x\cos\theta_1) + \Gamma_2 x\cos\theta_1\}$

Equating the coefficients of x^4, x^2 and so on,

$$4 \times 0.04 = \Gamma_1 \cos^3 \theta_1 \times 8; \text{ or } \Gamma_1 = \frac{0.04}{2} \times \cos^3 \theta_1$$

and $\quad (-3\,\Gamma_1 + \Gamma_2) \times 2 \cos\theta_1 = -3 \times 0.04$

Now at zero frequency, $\theta = 0$, and we can write for general reflection coefficient:

$$\left(\frac{Z_5 - Z_1}{Z_5 + Z_1}\right) = \Gamma_m T_{n-1}(\sec\theta_1)\left(\text{since } x = \frac{\cos\theta}{\cos\theta_1} = \sin\theta_1\right)$$

or, $\quad \left(\dfrac{100 - 50}{100 + 50}\right) = 0.04\, T_3(\sec\theta_1)$

or, $\quad \dfrac{50}{150} = 0.04\,(4\sec\theta^3\,\theta_1 - 3\sec\theta_1)$

or, $\quad \sec\,\theta_1 = 1.47$, when $\cos\theta_1 = 0.68$ with $\theta_1 = 47.156° = 0.2619\pi$ radians

The fractional bandwidth $\dfrac{\Delta f}{f_0} = \dfrac{4}{\pi}(0.5 - 0.2619)\pi$

or, $\quad \dfrac{\Delta f}{f_0} = 0.9524$

Thus, the bandwidth is $\Delta f = 0.95 f_0$ GHz = 9.524 GHz

Therefore, the bandwidth obtained with a Tchebyscheff transformer (Δf = 9.524 GHz) is greater than that realisable with a binomial transformer (Δf = 6.475 GHz); however, for

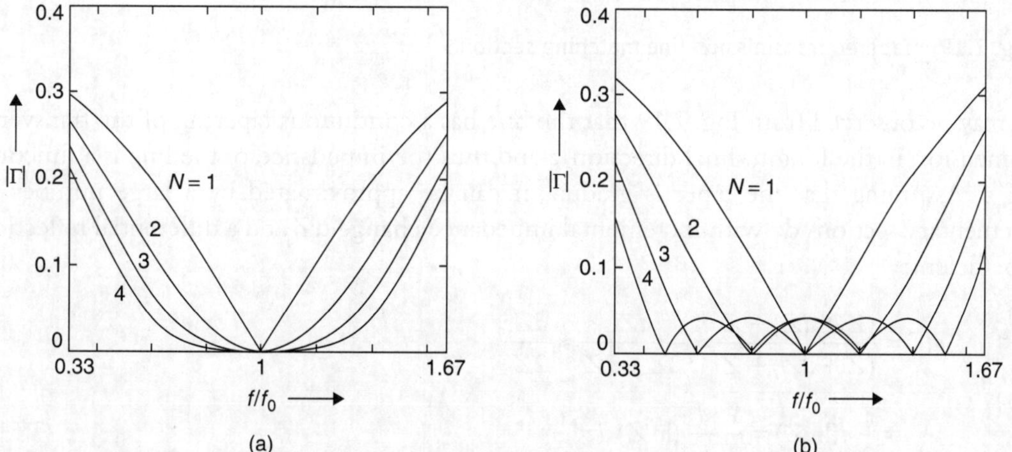

Fig. 7.18 Reflection coefficient magnitude vs frequency for different of number of sections of the matching transformer (a) Binomial type (b) Tchebyscheff type.

a Tchebyscheff transformer this increase of bandwidth is obtained with the penalty of tolerating pass-band ripple.

The plots of reflection coefficient magnitude vs frequency for the binomial transformer and Tchebyscheff transformer are shown in Fig. 7.18a and 7.18b respectively which clearly indicates that greater bandwidth is realisable using more number of sections for both the impedance transformers. It may also be observed that for a given maximum reflection coefficient and for a given number of sections the Tchebyscheff transformer can provide larger bandwidth than the binomial transformer. It may be observed that for the Tchebyscheff transformer, the number of dips in the ripple is equal to the number of sections (N) and one less than the number of steps n ($= N+1$), that is, the order of the filter.

7.3.2.3 Tapered impedance matching transformers

In the previous section we have seen how two transmission lines of different characteristic impedances can be matched by interposing a number of discrete steps to change the impedance level. As the number of discrete sections increases, the step changes in characteristic impedance between the sections become smaller, and we practically approach a continuously tapered impedance matching transformer as shown in Fig. 7.19. By changing the type of taper, we can obtain different pass-band characteristics.

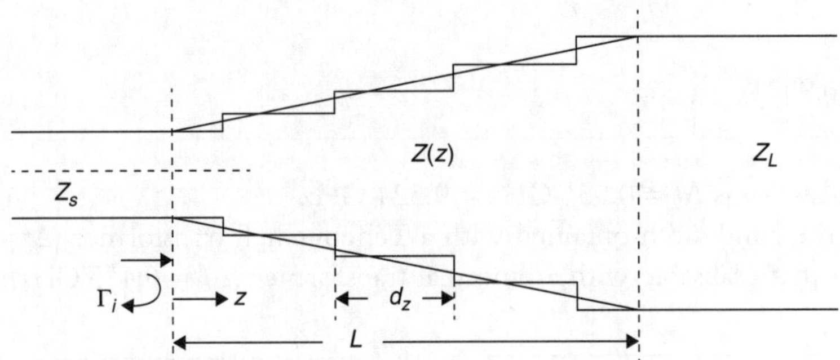

Fig. 7.19 Tapered transmission line matching section.

It may be observed from Fig. 7.19 that the line has a continuous tapering of the transverse dimension in the longitudinal direction z and thus the impedance of the line is a function of z. Assuming that the taper is gradual, it can be approximated by a large number of elementary sections dz with incremental impedance change dZ and a differential reflection coefficient:

$$d\Gamma = \frac{(Z + dZ) - Z}{(Z + dZ) + Z} = \frac{dZ}{2Z + dZ} \approx \frac{dZ}{2Z}$$

$$= \frac{1}{2}\, d(\ln Z) = \frac{1}{2}\frac{d}{dz}(\ln Z)\, dz \tag{7.61a}$$

since d$Z << Z$.

The input reflection coefficient (the total reflection coefficient at $z = 0$) for a tapered line of length L can now be found (following the theory of small reflections) by summing all partial reflections with their appropriate phase shifts as:

$$\Gamma_i = \frac{1}{2}\int_0^L \frac{d}{dz}(\ln Z)\, dz\, e^{-j2\beta z} \tag{7.61b}$$

Thus if $Z(z)$ is known, Γ_i can be found as a function of frequency. Alternatively, for a given response characteristic of Γ_i, $Z(z)$ can be determined. The latter approach is difficult and is generally avoided in practical design. Now we shall consider three special cases of $Z(z)$ impedance tapers and evaluate the resulting responses.

Case I: Exponential taper: For exponential taper, $Z(z)$ varies exponentially. We can write:

$$Z(z) = Z_0 e^{pz}, \quad \text{for } 0 < z < L \tag{7.62a}$$

where the exponential taper control parameter p can be evaluated from:

$$p = \frac{1}{L}\ln\left(\frac{Z_L}{Z_0}\right) \tag{7.62b}$$

In this impedance matching, Z_0 is matched to Z_L using exponential impedance matching transformer of length L (inset in Fig. 7.20).

Now following Eq. 7.61b and using Eqs 7.62a and b, we have:

$$\Gamma_i = \frac{1}{2}\int_0^L e^{-j2\beta z}\frac{d}{dz}(\ln e^{pz})\, dz = \frac{\ln(Z_L/Z_0)}{2l}\int_0^L e^{-j2\beta z}\, dz = \frac{\ln(Z_L/Z_0)}{2}\left(\frac{\sin\beta L}{\beta L}\right)e^{-j\beta l} \tag{7.62c}$$

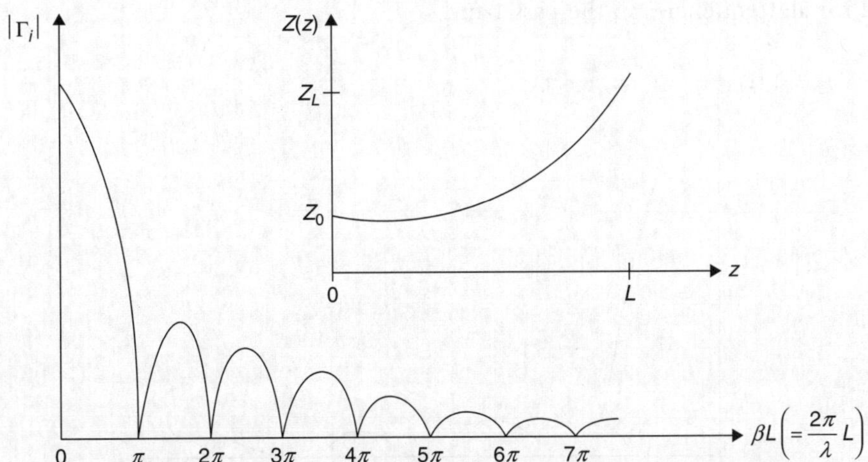

Fig. 7.20 Exponential taper with reflection coefficient magnitude response (impedance response is in the inset).

In this derivation, it is assumed that that $\beta = 2\pi/\lambda$, is independent of z (considering the TEM case).

The magnitude of the reflection coefficient in Eq. 7.62c has been sketched in Fig. 7.20. It may be noted that the peaks in $|\Gamma_i|$ decrease with increasing length, as expected, and that the length should be greater than $\lambda/2$ ($\beta L > \pi$) to minimise the mismatch at low frequencies.

Case II: Taper with triangular distribution: For a triangular taper, $Z(z)$ is given by:

$$Z(z) = Z_0 e^{2(z/L)^2 \ln(Z_L/Z_0)} \qquad 0 \le z \le L/2$$

$$= Z_0 e^{(4z/L - 2z^2/L^2 - 1)(\ln Z_L/Z_0)} \qquad L/2 \le z \le L \tag{7.63a}$$

Using Eq. 7.63a, we can write:

$$\frac{d(\ln Z/Z_0)}{dz} = (4z/L^2)\ln(Z_L/Z_0) \qquad 0 \le z \le L/2$$

$$= \{4/(L - 4z)/L^2\}\ln(Z_L/Z_0) \quad L/2 \le z \le L \tag{7.63b}$$

Substituting Eq. 7.63a into Eq. 7.61b and integrating it gives:

$$\Gamma_i = \frac{1}{2} e^{-j\beta l} \ln\left(\frac{Z_L}{Z_0}\right)\left\{\frac{\sin(\beta L/2)}{(\beta L/2)}\right\}^2 \tag{7.63c}$$

The magnitude of $|\Gamma_i|$ has been plotted in Fig. 7.21. It may be noted that compared to the exponential taper, the percentage of the minor lobe amplitude to that of the major lobe is much smaller for this taper. Though this small value of reflection coefficient occurs for a taper length of twice that required for the exponential taper (first null for triangular taper occurs at $\beta L = 2\pi$, whereas for the exponential taper it occurs at $\beta L = \pi$), this taper is however preferred in many practical situations as much smaller values of $|\Gamma_i|$ can be obtained for all frequencies in the pass-band.

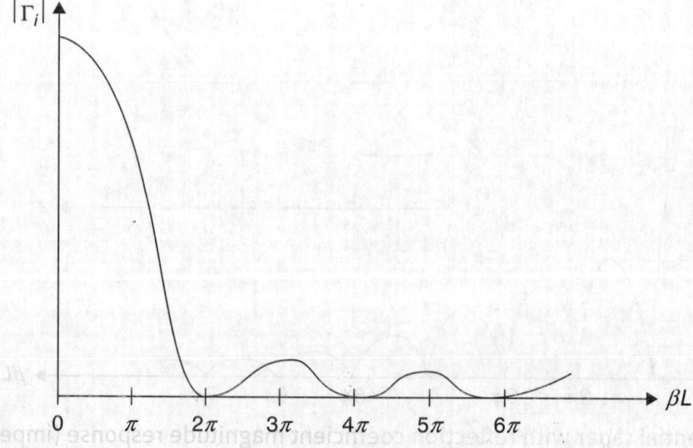

Fig. 7.21 Triangular taper with reflection coefficient magnitude response.

Case III: Klopfenstein taper: Looking into the infinite number of possibilities (at least theoretically) of designing a tapered impedance transformer by choosing the impedance function $Z(z)$ with relative advantages and disadvantages as far as the performance characteristics and size of the transformer is concerned, a question naturally arises whether we can design a transformer that gives minimum reflection in the pass-band yet with a minimum size—whether an optimised design is possible. This end has been realised with the *Klopfenstein taper* transformer design in which shortest length between the two impedances to be matched with continuous tapering sections can be realised with the lowest possible reflection coefficient over the desired pass-band. In other words, from a designer's point of view, with a given maximum tolerable reflection coefficient specification in the pass-band, the length of the matching transformer section is shortest for the *Klopfenstein taper* when compared with other taper distribution functions like exponential/triangular taper.

With a pre-designated minimum reflection coefficient (maximum allowable ripple) in the pass-band Γ_m, taper length L, input impedance Z_0, and load impedance Z_L, the logarithm of the characteristic impedance variation for the Klopfenstein taper is given by [17]:

$$\ln Z(z) = 0.5 \ln Z_0 Z_L + \Gamma_m A^2 \varphi\left(\frac{2z}{L} - 1, A\right) \quad 0 \le z \le L \tag{7.64a}$$

where $\Gamma_m = \dfrac{\Gamma_0}{\cosh A}$, $A = \cosh^{-1}\left(\dfrac{\Gamma_0}{\Gamma_m}\right)$ and $\Gamma_0 = 0.5 \ln\left(\dfrac{Z_L}{Z_0}\right)$

The function $\varphi(z, A)$ is defined as:

$$\varphi(z, A) = -\varphi(-x, A) = \int_0^z \frac{I_1(A\sqrt{1-y^2})}{A\sqrt{1-y^2}}\, dy \quad \text{for } |z| \le 1 \tag{7.64b}$$

where I_1 is the modified Bessel function which has the following special values:

$$\varphi(0, A) = 0,\ \varphi(z, 0) = 0.5z \text{ and } \varphi(1, A) = \frac{(\cosh A - 1)}{A^2}$$

The resulting input reflection coefficient is given by:

$$\Gamma_i = \Gamma_0 e^{-j\beta L} \frac{\cos\sqrt{(\beta L)^2 - A^2}}{\cosh A} \quad \text{for } \beta L > A$$

$$= \Gamma_0 e^{-j\beta L} \frac{\cos\sqrt{A^2 - (\beta L)^2}}{\cosh A} \quad \text{for } \beta L < A \tag{7.64c}$$

The comparative performance characteristics and design of exponential taper, triangular taper and Klopfenstein taper will be clear in the solved example given below.

EXAMPLE 7.8

Design an exponential taper, triangular taper and a Klopfenstein taper in a waveguide system with maximum allowable ripple in the pass-band to be $\Gamma_m = 0.025$ to match a reduced height waveguide with characteristic impedance 58 Ω to a full height waveguide having characteristic impedance 222 Ω (typical for an X-band waveguide with $\lambda_0 = 3$ cm). Show the plot of impedance variation with normalised length of the matching waveguide and the variation of the resulting reflection coefficient magnitude with the electrical length βl of the guide. Design all three types of tapered transformers with 3D electromagnetic field simulators including performance characterisation.

Solution

Exponential taper: From Eq. 7.62a the impedance variation is given by:

$$Z(z) = Z_0 e^{pz} \text{ for } 0 < z < L, \text{ with } p = \frac{1}{L}\ln\left(\frac{Z_L}{Z_0}\right) = 1.342/L$$

The reflection coefficient response is given by Eq. 7.62c as:

$$|\Gamma_i(\theta)| = \frac{\ln(Z_L/Z_0)}{2}\left(\frac{\sin\beta L}{\beta L}\right)$$

Triangular taper: From Eq. 7.63a, the impedance variation is given by:

$$Z(z) = Z_0 e^{2(z/L)^2 \ln(Z_L/Z_0)} \qquad 0 \le z \le L/2$$

$$= Z_0 e^{(4z/L - 2z^2/L^2 - 1)(\ln Z_L/Z_0)} \qquad L/2 \le z \le L$$

with $Z_0 = 58$ Ω and $Z_L = 222$ Ω. The reflection coefficient response is given by Eq. 7.63c as:

$$|\Gamma_i(\theta)| = \frac{1}{2}\ln\left(\frac{Z_L}{Z_0}\right)\left\{\frac{\sin(\beta L/2)}{(\beta L/2)}\right\}^2$$

Klopfenstein taper: We have in this case:

$$\Gamma_0 = 0.5\ln\left(\frac{Z_L}{Z_0}\right) = 0.671 \text{ and } A = \cosh^{-1}\left(\frac{\Gamma_0}{\Gamma_m}\right) = \cosh^{-1}\left(\frac{0.671}{0.025}\right) = 3.98$$

The impedance taper characteristics need to be evaluated numerically using Eq. 7.64a. The reflection coefficient response is determined from Eq. 7.64c:

$$\Gamma_i(\theta) = \Gamma_0 e^{-j\beta L}\frac{\cos\sqrt{(\beta L)^2 - A^2}}{\cosh A}$$

The pass-band for *Klopfenstein taper* is defined as $\beta L > A = 3.98 = 1.267\pi$.

The MATLAB plotted curves (Appendix III for program) showing impedance variation $Z(z)$ with z/L and resulting reflection coefficient (magnitude) $|\Gamma|$ with L for all the three types of taper transformers are shown in Fig. 7.22a and 7.22b respectively. It is observed that the Klopfenstein taper gives the desired response with $|\Gamma| \le \Gamma_m = 0.025$ for electrical length $\beta L = 1.3\pi$ ($\approx 1.267\pi$), which is shorter than either the triangular or exponential taper responses. The response of Klopfenstein taper in the pass-band resembles the equal-ripple characteristics similar to Tchebyscheff type impedance matching transformer.

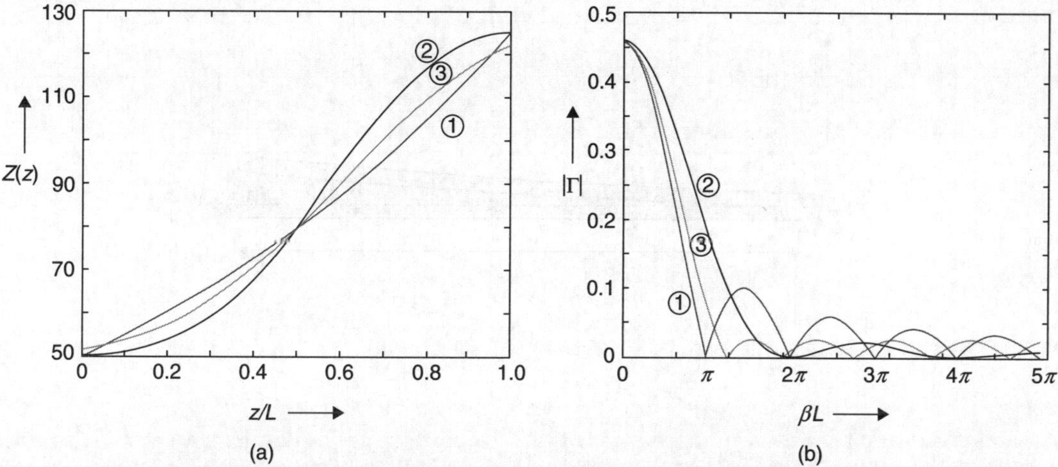

Fig. 7.22 (a) Impedance variation with normalised taper length for (1) Exponential, (2) Triangular, and (3) Klopfenstein taper (b) Reflection coefficient (magnitude) variation with frequency for the three types of tapers.

The curve in Fig. 7.22a may be used to design transformers to evaluate their physical dimensions using Table 7.1 for the three types of tapered transformers.

Table 7.1 Data for drawing the exponential, triangular and Klopfenstein taper structures

Parameters (*a*, *b* and *L* in mm; *Z* in Ω)	**Exponential**	**Triangular**	**Klopfenstein**
a	22.86	22.86	22.86
L	126.3	74.5	23.68
$b_1, L_1\ (Z_1)$	2.65, 0.0 (58)	2.65, 0.0 (58)	2.65, 0.0 (58)
$b_2, L_2\ (Z_2)$	3.67, 30.618 (80.31)	3.10, 18.06 (67.91)	3.10, 5.741 (67.83)
$b_3, L_3\ (Z_3)$	5.15, 62.512 (112.7)	5.12, 36.87 (112)	5.13, 11.72 (112.2)
$b_4, L_4\ (Z_4)$	7.32, 95.682 (160.3)	8.56, 55.68 (187.1)	8.54, 17.7 (186.8)
$b_5, L_5\ (Z_5)$	10.16, 126.3 (222)	10.16, 74.5 (222)	10.16, 23.68 (222)

The table of data given in Table 7.1 is used to draw the transformer contours in a 3D electromagnetic field simulator (Fig. 7.23a) and to characterise their performance in terms of the simulated return loss and insertion loss (Fig. 7.23b). The values of different b and L dimensions are obtained from the table of data generated while drawing Fig. 7.22a with MATLAB.

From Fig. 7.23b, it may be observed that for an exponential taper, the return loss variation repeats every 0.95 GHz which is expected to take place every π radians (equivalent to every 1.18 GHz) if the plot was $|\Gamma|$ vs βl (Figs 7.20 and 7.22b). But for a triangular taper this repetition is seen to be almost at double that frequency as expected.

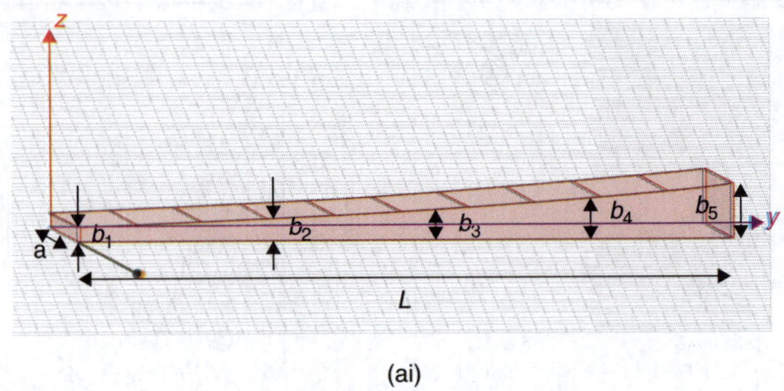

(ai)

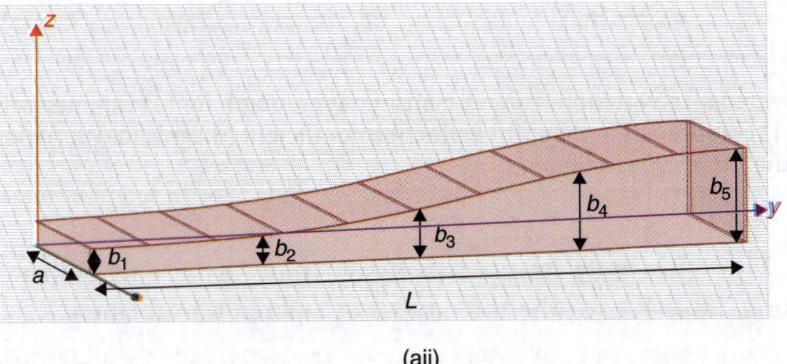

(aii)

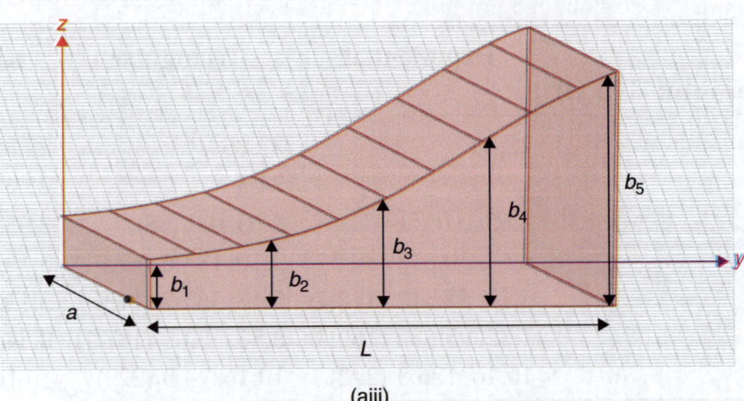

(aiii)

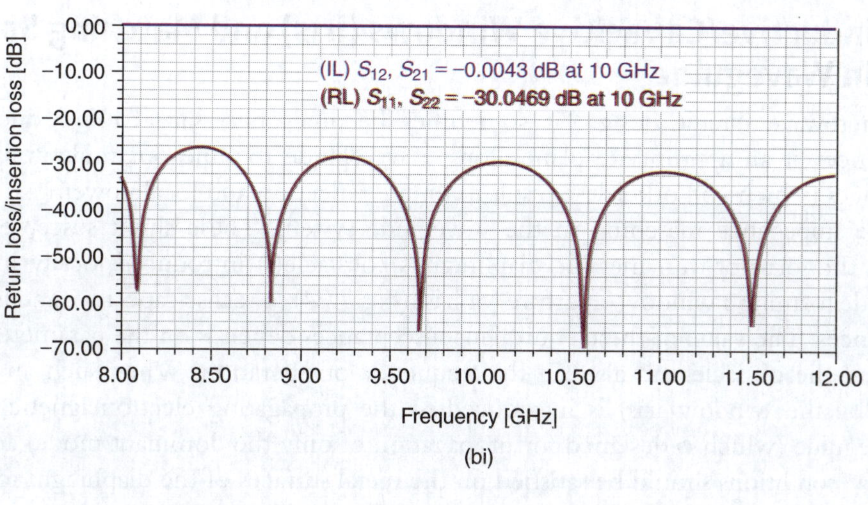

(bi)

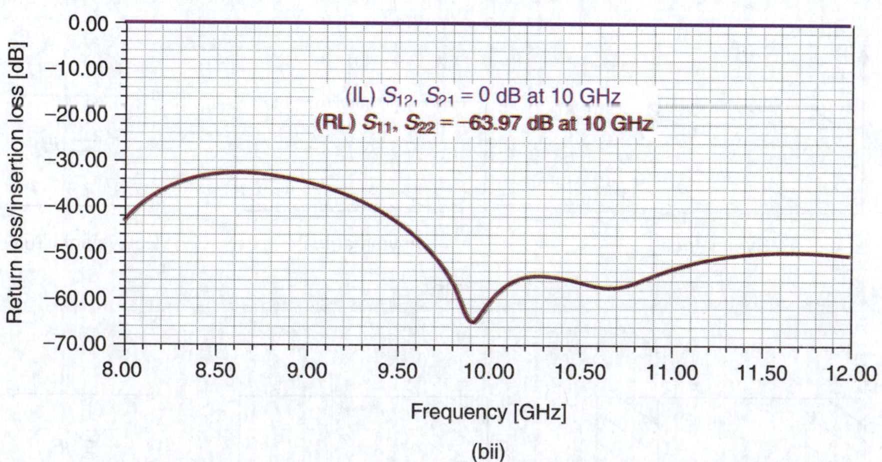

(bii)

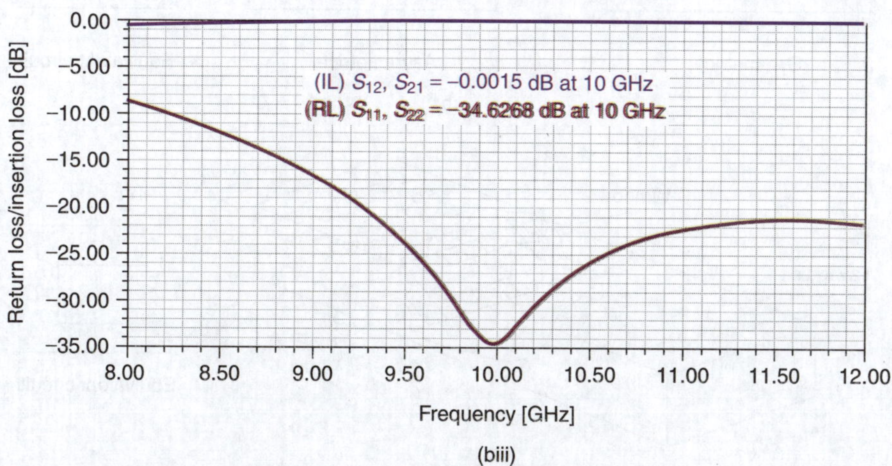

(biii)

Fig. 7.23 (a) Transformer contours in 3D electromagnetic field simulator and (b) Performance characteristics of (i) Exponential taper (ii) Triangular taper (iii) Klopfenstein taper.

7.4 Inductive/Capacitive Windows (iris) and Matching Screws in Waveguide

In a microwave circuit design, it sometimes becomes necessary to introduce shunt succeptances at an appropriate plane along a waveguide to tune out undesired reactive impedances (which will otherwise result in undue reflection and local power loss) in order to realise impedance matching in the waveguide system. These shunt susceptences are often in the form of thin metallic diaphragm(s)/obstacle(s) introduced locally inside the waveguide forming a window/aperture (or iris) (Fig. 7.24) to realise capacitive and inductive susceptances. The value of these equivalent susceptances depends on the parameters of the guide and the obstacle and also on the frequency of operation. When such an obstacle (caused by the window/iris) is introduced to the propagating electromagnetic wave in the waveguide (which is designed for propagation of only the dominant mode) additional boundary conditions should be satisfied on the metal surfaces of the diaphragm for higher

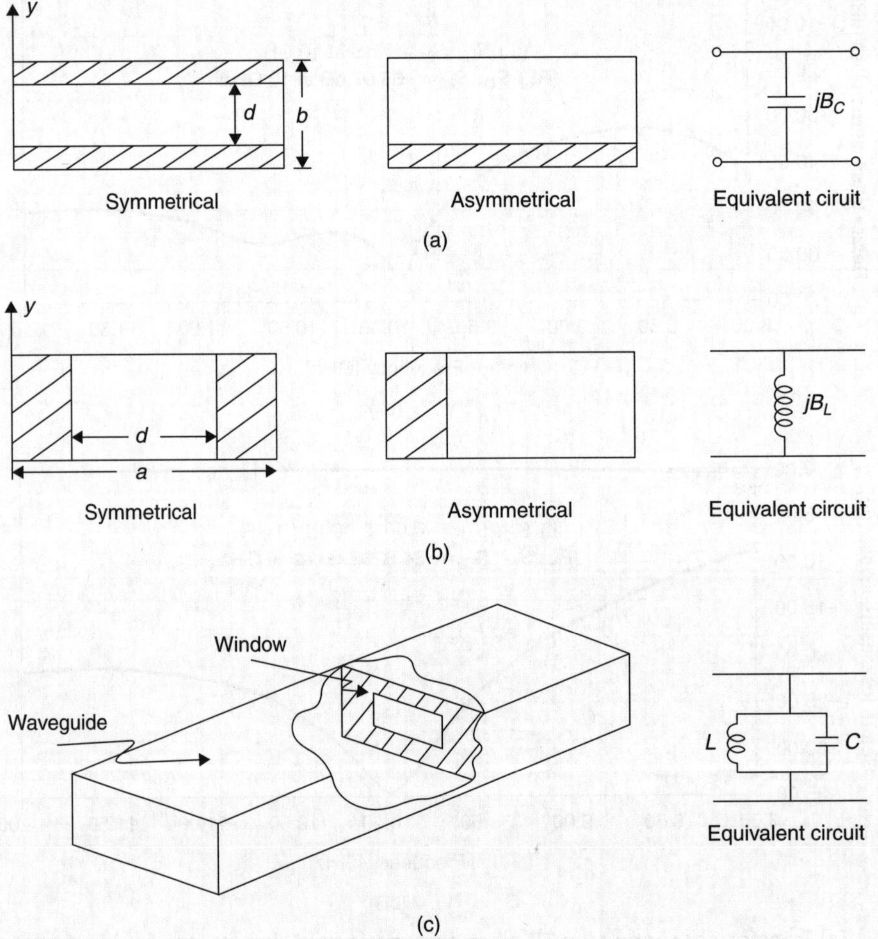

Fig. 7.24 Window/iris in waveguide as susceptance (a) Capacitive window (b) Inductive window (c) Resonant window.

order modes. The higher order modes are non-propagating evanescent modes in such iris loaded waveguides and are trapped/localised in the iris. Analytical characterisation of the electromagnetic situation of the presence of the window/iris may be done by *modal analysis* but that is quite involved. However, qualitative understanding is simple and approximate formulae are quite useful for design.

The horizontal diaphragm (both symmetrical and asymmetrical, Fig. 7.24a) extending from one narrow wall of the waveguide to the other narrow wall (in which the edges of the diaphragm are perpendicular to the electric plane in the guide), provides capacitive susceptance across the waveguide and hence such an obstacle is known as *capacitive iris.* This is because the presence of the diaphragm effectively reduces the b dimension of the waveguide along the y direction (which is the direction of the electric field of the dominant TE_{10} mode) thereby increasing the electric energy storage at the location of the diaphragm causing capacitive loading. The shunt capacitive susceptance across the waveguide is directly proportional to the size of the opening. An approximate expression for the normalised capacitive susceptance in the case of symmetrical capacitive window is given by [7]:

$$b_{Cd} = \frac{B_{Cd}}{Y_{10}} = \frac{4b}{\lambda_g} \ln\{\mathrm{cosec}(\pi d/2b)\} \tag{7.65}$$

where Y_{10} is the reciprocal of $Z_{wTE_{10}}$ given by Eq. 4.14 and λ_g is the guide wavelength given by Eq. 4.14. A plot of $(b_{Cd}\lambda_g/4b)$ vs (d/b) with different parametric values of (b/λ_g) is shown in Fig. 7.25a which includes in the inset, the diagram of a symmetrical capacitive window with its equivalent circuit. It may be observed that for $d = b$, the shunt capacitive susceptance happens to be practically zero and this is expected as it then represents the case of an open circuit (no presence of diaphragm). While for $d = 0$, that is, for $d/b = 0$, the shunt capacitive susceptance approaches infinity as is expected from a short circuit (as the diaphragm then fills the whole cross-section of the guide). This approximate formula is accurate to within 5% in the usual range of operating frequencies.

The vertical diaphragm (both symmetrical and asymmetrical, Fig. 7.24b), extending from one broad wall of the waveguide to the other broad wall (in which the edges of the diaphragm are perpendicular to the magnetic plane in the guide), provides inductive susceptance across the waveguide and hence such an obstacle is known as *inductive iris.* This is because now the presence of the conducting diaphragms in the y direction provides additional current paths through them causing an increase in the magnetic energy storage at the location of the diaphragm(s) resulting in inductive loading. The shunt inductive susceptance across the waveguide is directly proportional to the size of the opening. An approximate expression for the normalised inductive susceptance in the case of symmetrical inductive window is given by:

$$b_{Ld} = \frac{B_{Ld}}{Y_{10}} = \frac{\lambda_g}{a} \cot^2(\pi d/2a) \tag{7.66}$$

The plot of $(b_{Ld}\, a/\lambda_g)$ vs (d/a) with different parametric values of (a/λ_g) is shown in Fig. 7.25b which includes in the inset, the diagram of a symmetrical inductive window

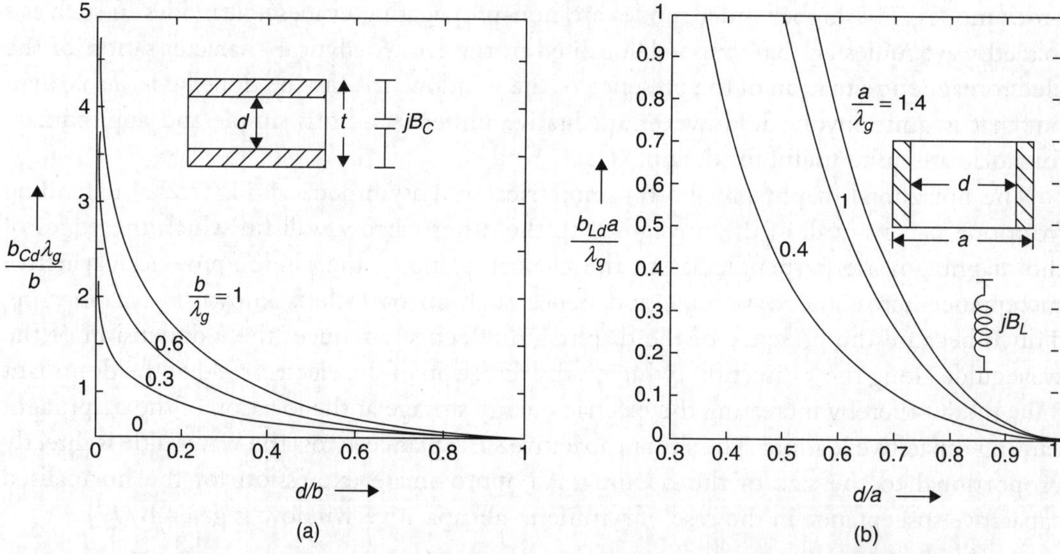

Fig. 7.25 (a) Plot of $(b_{Cd}\lambda_g/4b)$ vs (d/b) with different parametric values of (b/λ_g) for symmetrical capacitive window (shown in the inset), (b) Plot of $(b_{Ld}a/\lambda_g)$ vs (d/a) with different parametric values of (a/λ_g) for symmetrical inductive window (shown in the inset).

with its equivalent circuit. The cases for $d = a$ and $d = 0$, represent similar situations of open and short circuit in terms of the shunt inductive susceptance as was observed with shunt capacitive susceptance mentioned above.

The iris shown in Fig. 7.24c includes in the same structure the capacitive and inductive effect (have diaphragm portions across both the magnetic and electric planes in the waveguide) and may thus be represented by an equivalent shunt resonant circuit. At the resonant frequency, the iris acts as a high shunt resistance. Above or below resonance, the iris acts as a capacitive or inductive suceptance. Such a *resonant iris* can be used as a mode filter which may be understood as follows. If the waveguide is made oversized either to allow multimode propagation demanded by some specific application or to realise low attenuation for the dominant mode, then the presence of the resonant iris/window will allow the dominant mode to pass through it but prevent the higher order modes. This happens because the waveguide dimension at the location of the window is suitable for the dominant mode (as this mode demands smallest possible *a* dimension, Section 4.2.3) to be tapped and passed onto the other side of the window while for higher order modes the window's a dimension is too small to allow them to propagate on the other side of the window making them evanescent beyond the window location. Thus, *a* resonant window/iris (Fig. 7.24c) acts as a *mode filter*, allowing the dominant mode to pass in the output direction and filtering out the higher order modes.

Because the irises are by nature with fixed dimensions with no scope for adjustment of their size in a particular design, they are normally used to tune out fixed amount of mismatches. However, insertion of a metallic screw (which is readily adjustable) into a waveguide through one of the broad walls can also provide capacitive/inductive susceptances across the cross-sectional plane of a waveguide (Fig. 7.26a). The nature of susceptance being presented by the variable depth screw at the location of its insertion may be understood with simple physical explanation. The tuning screw inserted in the waveguide behaves much like a dipole antenna, if due consideration is accorded to its image caused by the lower broad wall of the waveguide (as the inside of waveguide walls normally have mirror finish with silver or gold plating). A dipole antenna is known to behave as an open circuited transmission line (with current and voltage pattern as shown in Fig. 7.26b) having an input impedance $Z_{in} = -jZ_0 \cot\beta l$. We may now consider two cases of the depth of penetration (l) of the screw. Case I: here $l < \lambda/4$, say $l = \lambda/8$, when it turns out that $Y_{in} = +jY_0$ (capacitive susceptance) and case II: here $l > \lambda/4$, say $l = 3\lambda/8$, when it turns out that $Y_{in} = -jY_0$ (inductive susceptance). It may be noted that in Section 6.2.3, it was indicated that for a slide screw tuner only capacitive susceptance can be realised with the depth of penetration of the screw ($l < \lambda/4$) in the waveguide (since the b dimension of a rectangular waveguide

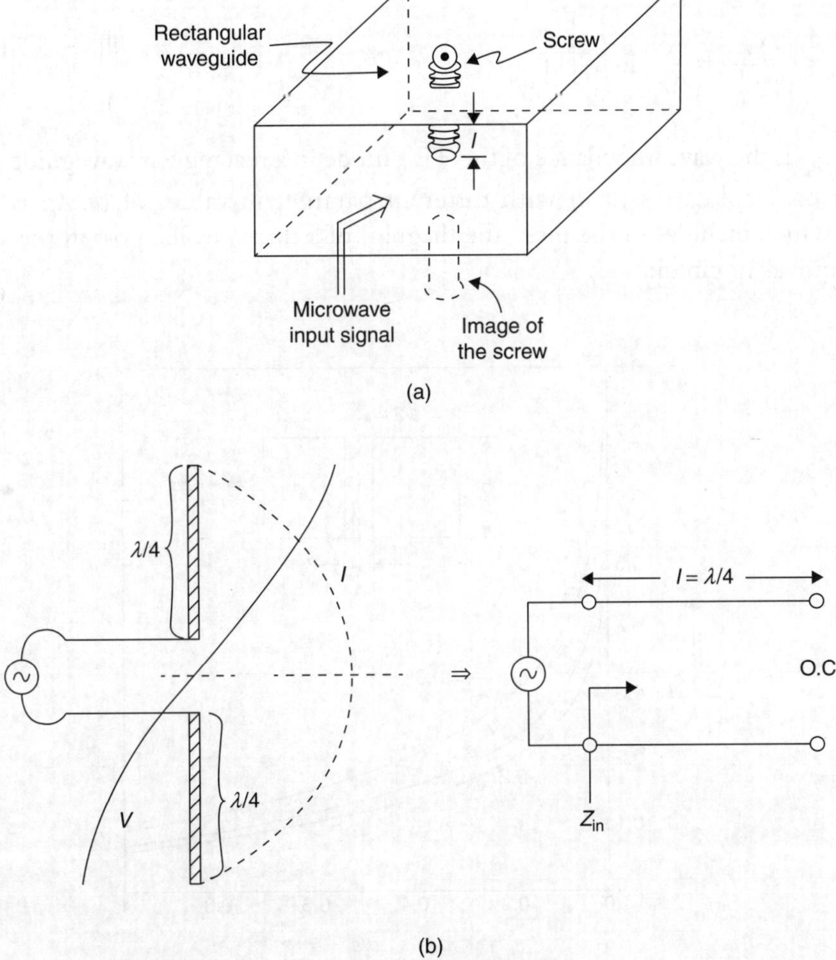

Fig. 7.26 (a) Metallic screw inserted in a rectangular waveguide (b) Equivalence with a dipole antenna and hence an open circuited transmission line.

is approximately a quarter wavelength). Thus, from a practical consideration, the inductive susceptance could not be realised with depth of penetration of the screw in the waveguide and was realised with longitudinal movement of the screw along the length of the guide.

In addition to tuning out impedance mismatches in waveguide circuits, tuning screws of this sort are also used for fine mechanical tuning of the frequency of microwave oscillator (with variable depth capacitive post/screw). No simple approximate formula is available for the capacitive susceptance presented by the variable depth capacitive screw but is determined in terms of the equivalent T-network parameters for the structure (which is beyond the scope of present discussion). Further, a thin metallic post in the waveguide behaves as an equivalent inductance (shown in the inset of Fig. 7.27) which is used in designing bias-post mounted microwave oscillators with semiconductor diodes such as IMPATT/Gunn (detailed discussion is given in Section 9.7). For a thin inductive post, however, the approximate formula for the inductive reactance presented by the metallic post of diameter d is given by [7]:

$$x_L = \frac{X_L}{Z_{w\mathrm{TE}_{10}}} = \frac{a}{2\lambda_g} \ln\left(\frac{2a}{d}\right) \tag{7.67}$$

where $Z_{w\mathrm{TE}_{10}}$ is the wave impedance of the TE_{10} mode in a rectangular waveguide (Eq. 4.14).

A plot of $(2x_L\lambda_g/a)$ vs (d/a) with different parametric values of (a/λ_g) is shown in Fig. 7.27 which includes in the inset, the diagram of a thin metallic post in the waveguide with its equivalent circuit.

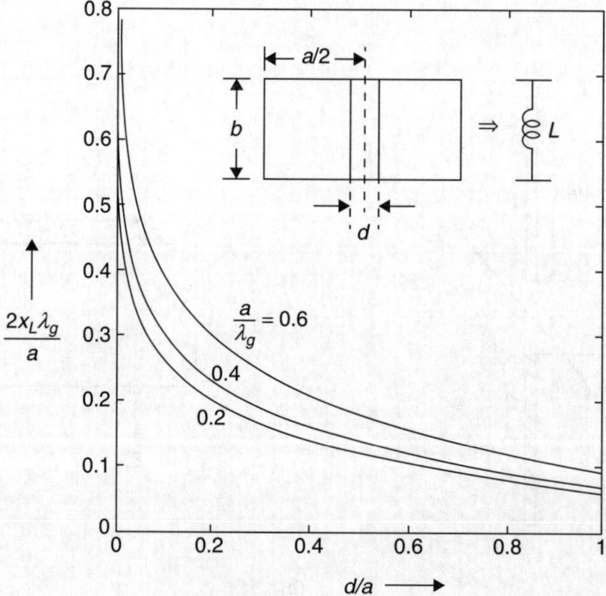

Fig. 7.27 Plot of $(2x_L\lambda_g/a)$ vs (d/a) with different parametric values of (a/λ_g) for a thin metallic post in the waveguide (shown in the inset).

7.5 Impedance Matching with Lumped Component/ Planar Distributed Line Reactive Elements

A simple yet practically useful technique of impedance matching in certain applications in microwave circuit design is to use either a lumped component inductor/capacitor or planar distributed line reactive elements (inductance/capacitance). In fact, in the design of microwave amplifiers, the source and load impedances should be matched respectively with the input and output impedance of the active device and this is done by such reactive element impedance matching. When two reactive elements are used to realise impedance matching, the network is called an L-section network, while three-element matching networks are either T or Π-section type. The names are descriptive of the configuration (L, T, Π) formed by the reactive elements. Lumped-element inductance and capacitance (Section 6.3) may be used typically up to 20 GHz and with limited values up to 60 GHz. However, for MMIC applications, a planar distributed line (microstrip line) is also used to realise the reactive elements. Let us now discuss the L-section matching technique with the Z–Y Smith chart, that is, *Smith and Jones chart.*

With two different reactive elements L and C, there are 2^2 (four) possible configurations for the L-section network (Fig. 7.28).

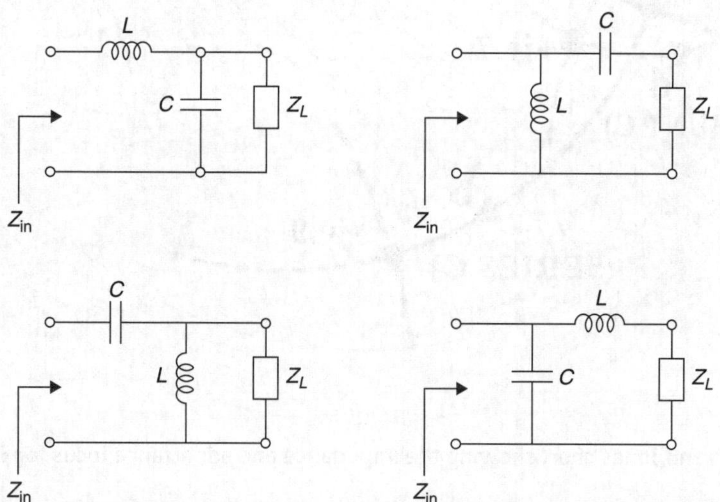

Fig. 7.28 Four possible configurations for L-section network for impedance matching.

We shall now discuss with the help of some examples how the Smith and Jones chart (Z–Y Smith chart) can be used to design L-section matching networks.

EXAMPLE 7.9

Illustrate with the use of a Z–Y Smith chart, the effect of adding a series/shunt inductance/ capacitance to impedance/admittance in the following cases:

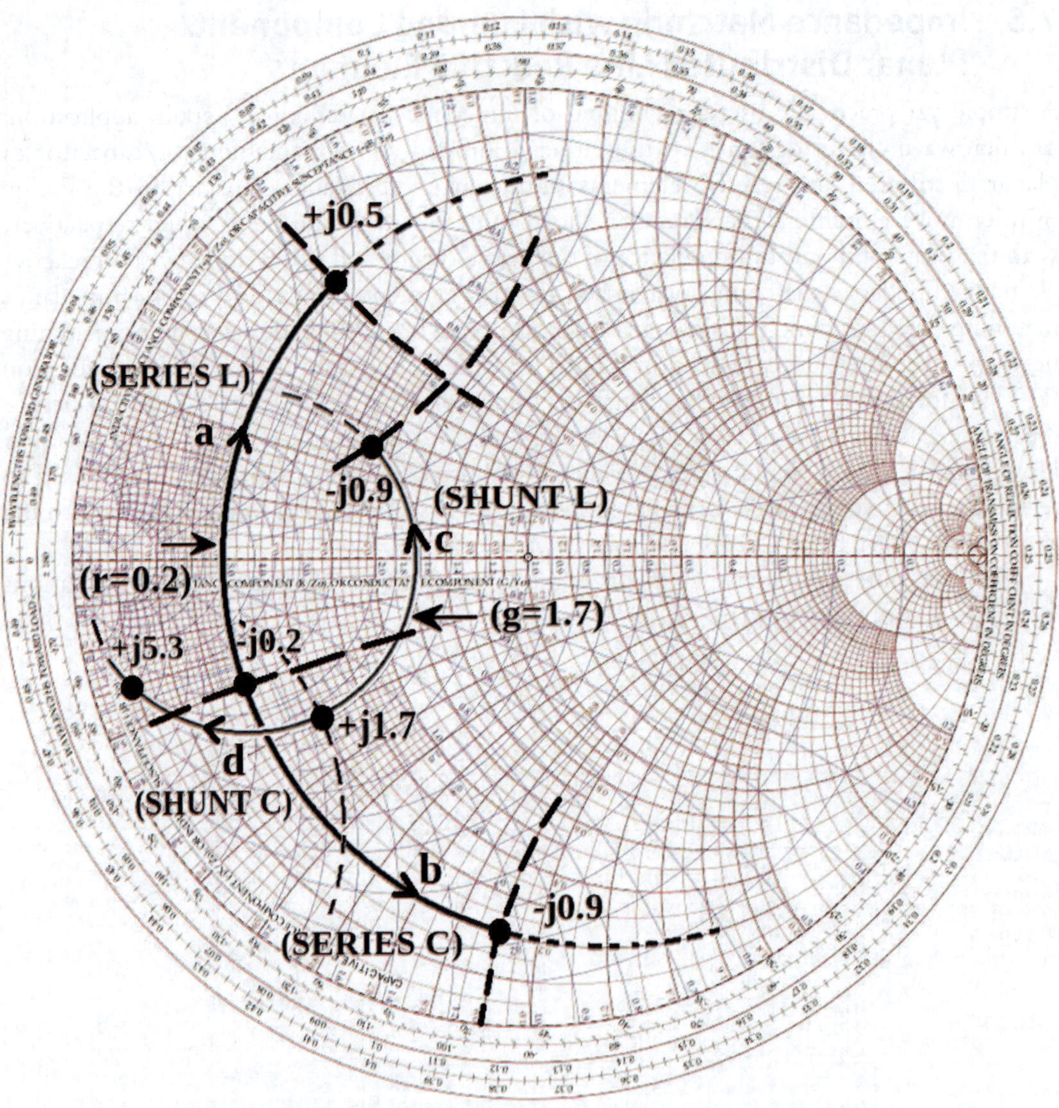

Fig. 7.29 Smith and Jones chart showing the impedance and admittance locus for Example 7.10.

(i) A series inductor L ($z_L = +j0.7$) to an impedance $z = (0.2 - j0.2)$.
(ii) A series capacitor C ($z_C = -j0.7$) to an impedance $z = (0.2 - j0.2)$.
(iii) A shunt inductor L ($y_L = -j2.6$) to an admittance $y = (1.7 + j1.7)$.
(iv) A shunt capacitor C ($y_C = +j3.6$) to an admittance $y = (1.7 + j1.7)$.

Solution

(i) When a *series inductor* with a normalised reactance of $+j0.7$ is added to an impedance $(0.2 - j0.2)$, we need to move in a constant-resistance circle in the Smith chart starting from the initial reactance value of $-j0.2$ to the final value of $+j0.5$ which makes a *clockwise rotation* as shown by curve 'a' in Fig. 7.29.

(ii) When a *series capacitor* with a normalised reactance of $-j0.7$ is added to an impedance $(0.2 - j0.2)$, we need to move in a constant-resistance circle in the Smith chart starting from the initial reactance value of $-j0.2$ to the final value of $-j0.9$ which makes an *anticlockwise rotation* as shown by curve 'b' in Fig. 7.29.

(iii) When a *shunt inductor* with a normalised susceptance of $-j2.6$ is added to an admittance $(1.7 + j1.7)$, we need to move in a constant-conductance circle in the Smith chart starting from the initial susceptance value of $+j1.7$ to the final value of $-j0.9$ which makes an *anticlockwise rotation* as shown by curve 'c' in Fig. 7.29.

(iv) When a *shunt capacitor* with a normalised susceptance of $+j3.6$ is added to an admittance $(1.7 + j1.7)$ we need to move in a constant-conductance circle in the Smith chart starting from the initial susceptance value of $+j1.7$ to the final value of $+j5.3$ which makes a *clockwise rotation* as shown by curve 'd' in Fig. 7.29.

EXAMPLE 7.10

Design two matching networks at 1.0 GHz, one with series L and shunt C while the other one with series C and shunt L, to match a load impedance of $(15 + j15)\ \Omega$. The characteristic impedance of the line to be used is 75 Ω. Use a Z–Y Smith chart to solve the problem.

Solution

The normalised load impedance is $z_L \dfrac{Z_L}{Z_o} = \dfrac{15 + j15}{75} = (0.2 + j0.2)$ which is first located in the Smith chart marked by point A (Fig. 7.30).

Matching with series L and shunt C: When we add a *series* L with normalised inductive reactance of $+j2$, say, we need to move along the constant-resistance circle of 0.2 in a *clockwise direction* to reach an impedance of $(0.2 + j0.4)$ which is incidentally an admittance of $(1 - j2)$ located at point B (Fig. 7.30). Now, the final matching of the load would need a *shunt capacitive* susceptance of $+j2$ (capacitive reactance of $1/j2 = -j5$), moving along the unit constant-conductance circle in *clockwise direction* to reach the origin of the Smith chart, which is the location for perfect matching. Thus, with a series inductance $L = 0.2 \times 75/2\pi \times 10^9 = 2.38$ nH and shunt capacitance $C = 1/0.5 \times 75 \times 2\pi \times 10^9 = 4.24$ pF the load impedance can be matched with the 75 Ω line—the matching network is shown in Fig. 7.31a.

Matching with series C and shunt L: However, when we add a *series* C with normalised capacitive reactance of $-j0.6$, say, we need to move along the constant-resistance circle of 0.2 in an *anticlockwise direction* to reach at impedance of $(0.2 - j0.4)$ which is

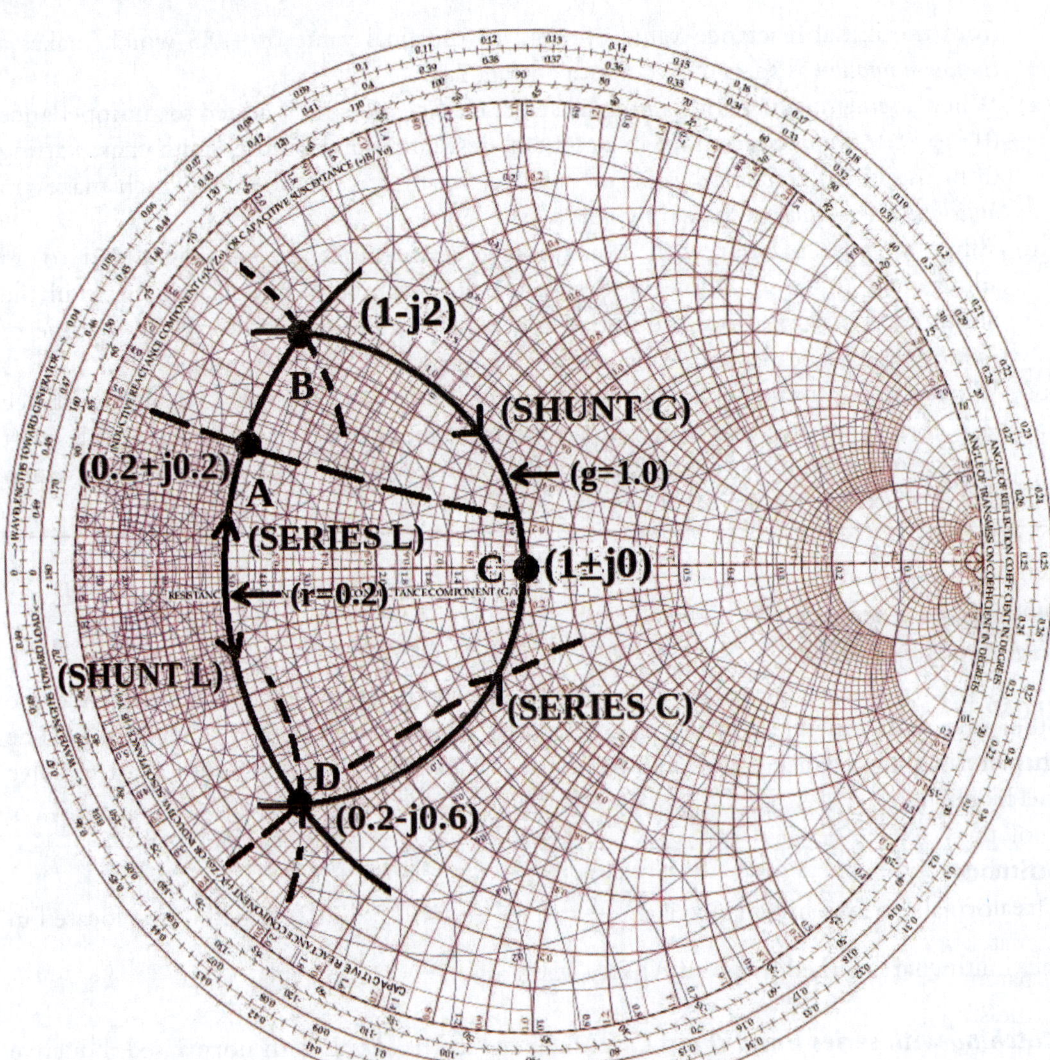

Fig. 7.30 Smith and Jones chart showing the impedance and admittance locus of example 7.11.

incidentally an admittance of (1 + j2) located at point D (Fig. 7.30). Now the final matching of the load would need a *shunt inductive* susceptance of $-j2$ (inductive reactance of $1/-j2 = +j0.5$), moving along the unit constant conductance circle in an *anticlockwise direction* to reach the origin of the Smith chart which is the location of perfect matching. Thus, with a series capacitance $C = 1/0.6 \times 75 \times 2\pi \times 10^9 = 3.54$ pF and shunt inductance $L = 0.5 \times 75/2\pi \times 10^9 = 5.97$ nH the load impedance can be matched with the 75 Ω lin (Fig. 7.31b).

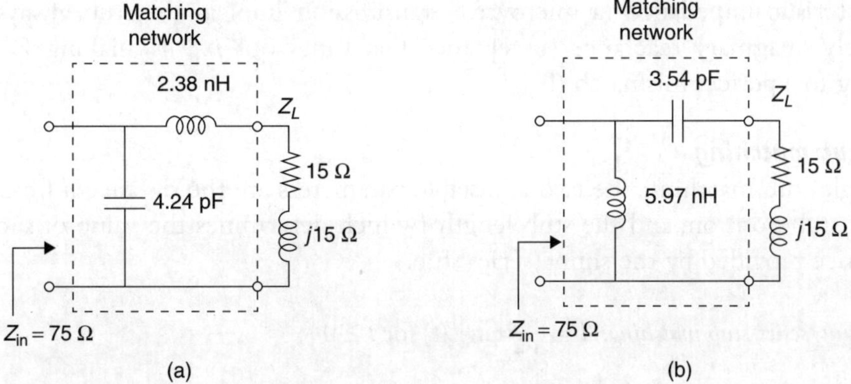

Fig. 7.31 Matching network for Example 7.11 (a) With series L and shunt C (b) With series C and shunt L.

(**Note:** In both cases of matching, the aim of choosing the initial value of the first reactive element was to move finally into a unit conductance circle to effect the final matching with the second reactive element.)

RECAPITULATION

Stub matching

- The idea of stub matching is to connect open or short circuited 'stub' (sections of transmission lines) either in shunt or in series with the parent transmission line. Stub matching (or tuning, as it effectively tunes out the impedance mismatches) may be realised either with one piece of stub or with two/three pieces of stubs. The idea of stub matching relies on the fact that the impedance of the stub and/or location of it on the parent transmission line is so chosen that the combined impedance of the parent line and the stub (or stubs) equals the characteristic impedance of the parent line.
- In single stub matching, the stub is placed at a suitable distance from the load and the suceptance (for shunt stub) or reactance (for series stub) presented by the stub cancels out the load suceptance at the location of the stub thereby terminating the parent transmission line in its characteristic impedance. A disadvantage with single stub matching is that every load requires a new stub position. Thus, if matching is desired for a variable impedance load, then single stub matching is difficult to use. The use of two stubs spaced by a fixed amount and located at a fixed distance from the load may be used to overcome this disadvantage. However, a double stub matching system of this type may not be able to match all possible values of load admittance. A triple stub matching arrangement can overcome this problem of double stub tuner, in fact, a triple stub tuner may be considered to be two double stub tuners in series.
- It may be noted that single stub matching (in fact, any stub matching) cannot be used to match a purely complex load (a load with no real part) to any line which has real

characteristic impedance (a microwave transmission line) as the stub always exhibits a purely imaginary reactance/susceptance that tunes out jx_B/jb_B making $Z_{in}/Y_{in} = 0$, leading to a perfect mismatch (!)

Single stub matching

- In single stub matching, the two adjustable parameters are the distance d from the load to the stub position, and the stub length l which determines the value of suceptance/reactance provided by the shunt/series stub.

For shunt/series stub matching: $d = \dfrac{\lambda}{2\pi}\tan^{-1}(t)$ for $t \geq 0$

$$= \frac{\lambda}{2\pi}\{\pi + \tan^{-1}(t)\} \text{ for } t < 0$$

where $t = \dfrac{1}{(r_L - 1)}\left[x_L \pm \sqrt{r_L[(1-r_L)^2 + x_L^2\}}\right]$ for $r_L \neq 1$ and $t = -\dfrac{x_L}{2}$, for $r_L = 1$ [for shunt stub]

$$t = \frac{1}{(g_L - 1)}\left[b_L \pm \sqrt{g_L\{(1-g_L)^2 + b_L^2\}}\right] \text{ for } g_L \neq 1 \text{ and } t = -\frac{b_L}{2}, \text{ for } g_L = 1 \text{ [for series stub]}$$

Shunt stub: When an *open circuited shunt stub* is used:

$$l_{oc} = \frac{\lambda}{2\pi}\tan^{-1}(b_{SB}) = \frac{\lambda}{2\pi}\tan^{-1}(-b_B), \quad b_B \geq 0$$

$$= \frac{\lambda}{2\pi}\{\pi + \tan^{-1}(b_{SB})\} = \frac{\lambda}{2\pi}\{\pi + \tan^{-1}(-b_B)\}, \quad b_B < 0$$

While, if a *short circuited shunt stub* is used:

$$l_{sc} = \frac{\lambda}{2\pi}\tan^{-1}\left(-\frac{1}{b_{SB}}\right) = \frac{\lambda}{2\pi}\tan^{-1}\left(\frac{1}{b_B}\right), \quad b_B \geq 0$$

$$= \frac{\lambda}{2\pi}\left\{\pi + \tan^{-1}\left(-\frac{1}{b_{SB}}\right)\right\} = \frac{\lambda}{2\pi}\left\{\pi + \tan^{-1}\left(\frac{1}{b_B}\right)\right\}, \quad b_B < 0$$

(where b_B is the susceptance of the loaded line looking into the load from the stub position and b_{SB} is the stub susceptance).

Series stub: When an *open circuited series stub* is used:

$$l_{oc} = \frac{\lambda}{2\pi}\tan^{-1}\left(\frac{1}{x_{SB}}\right) = \frac{\lambda}{2\pi}\tan^{-1}\left(-\frac{1}{x_B}\right), \quad x_B \geq 0$$

$$= \frac{\lambda}{2\pi}\left\{\pi + \tan^{-1}\left(\frac{1}{x_{SB}}\right)\right\} = \frac{\lambda}{2\pi}\left\{\pi + \tan^{-1}\left(-\frac{1}{x_B}\right)\right\}, \quad x_B < 0$$

While, if a *short circuited series stub* is used:

$$l_{sc} = \frac{\lambda}{2\pi}\tan^{-1}(-x_{SB}) = \frac{\lambda}{2\pi}\tan^{-1}(x_B), \quad x_B \geq 0$$

$$= \frac{\lambda}{2\pi}\{\pi + \tan^{-1}(-x_{SB})\} = \frac{\lambda}{2\pi}\{\pi + \tan^{-1}(x_B)\}, \quad x_B < 0$$

(where x_B is the reactance of the loaded line looking into the load from the stub position and x_{SB} is the stub reactance).

Thus, for a given load impedance, both d and l values can be calculated using the above equations.

Double stub matching

- In double stub matching, the first stub is located at the load plane and the second stub is at a distance d from the first stub towards the generator. In practice, d is chosen to be $\lambda/8$ ($3\lambda/8$ at higher frequency).

  The stub lengths for open and short circuited shunt stubs are:

  For an open circuited stub $l_{oc} = \frac{\lambda}{2\pi}\tan^{-1}(b)$ and

  for a short circuited stub $l_{sc} = \frac{\lambda}{2\pi}\tan^{-1}\left(-\frac{1}{b}\right)$

  where $b = b_{SA}$, and b_{SB} is the stub susceptance of the first stub and the second stub respectively given by:

$$b_{SA} = -b_L + \frac{1 \pm \sqrt{(1+t^2)g_L - g_L^2 t^2}}{t} \text{ and } b_{SB} = \frac{\pm\sqrt{g_L(1+t^2) - g_L^2 t^2} - g_L}{g_L t}$$

  with $g_L = \frac{1+t^2}{2t^2}\left\{1 \pm \sqrt{1 - \frac{4t^2(1 - b_L t - b_{SA} t)^2}{(1+t^2)^2}}\right\}, \quad (t = \tan\beta d)$

  The analytical solution of stub matching problem may be used to obtain very accurate results by solving the analytical equations through computer programs, however, such accuracy is seldom required in actual problem solutions and thus stub matching problems may be solved more easily by using the Smith chart.

Impedance transformers

- A quarter-wave ($\lambda/4$) transmission line can act as an impedance transformer at microwave frequency. However, it will perform as a very narrow band transformer (as its length is dependent on wavelength). Further, a quarter-wave transformer can only match real load impedance.

- In order to realise impedance matching over a broad band of frequencies, a multi-section quarter-wave transformer is used. This is done in principle by cascading multiple sections of quarter-wave lines having progressively increasing/decreasing characteristic impedance. The analysis and design of such an impedance transformer is done on the basis of the *theory of small reflections*.
- Frequency response for a single-section quarter-wave transformer is given by:

$$|\Gamma| \approx \frac{|Z_L - Z_1|}{\sqrt{2Z_1Z_L}}|\cos\theta|, \text{ where } \theta = \beta l = f\pi/2f_0$$

The fractional bandwidth is:

$$\frac{\Delta f}{f_0} = 2 - \frac{4}{\pi}\cos^{-1}\left|\frac{2|\Gamma_m|\sqrt{Z_1Z_L}}{(Z_L - Z_1)\sqrt{1-|\Gamma_m|^2}}\right|$$

Larger bandwidth is realisable with smaller impedance ratio (Z_L/Z_1) between the two impedances Z_L and Z_I to be matched.

- The overall reflection coefficient at the input of any structure having multiple discontinuities (causing reflection) may be evaluated by considering all those multiple reflections taking place at different discontinuities (junctions) of the structure. For *small reflections*, the total reflection coefficient can be obtained only from the first order reflections.

$$\Gamma \approx \Gamma_1 + \Gamma_3 e^{-j2\theta}$$

Thus, the total reflection (Γ) is dominated by the reflection from the initial discontinuity between Z_1 and Z_2, that is, Γ_1 and the first reflection from the discontinuity between Z_2 and Z_L, that is, $\Gamma_3 e^{-2j\theta}$.

- Depending on the choice of the reflection coefficients Γ_k, $k = 1, 2, \ldots n$, for a *stepped impedance transformer*, a variety of pass band characteristics of a multi-section transformer can be realised; the common ones being *maximally flat* or *binomial* transformer and *Tchebyscheff* or *equal ripple* transformer.
- As the number of discrete sections increases, the step changes in characteristic impedance between the sections become smaller, and we practically approach a continuously *tapered impedance matching transformer*. Common ones are exponential taper, triangular taper and Klopfenstein taper.

Inductive/capacitive windows (iris) and matching screws in waveguide

- The horizontal diaphragm (both symmetrical and asymmetrical) extending from one narrow wall of the waveguide to the other narrow wall provides capacitive susceptance across the waveguide and hence such an obstacle is known as a *capacitive iris*. An approximate expression for the normalised capacitive susceptance in case of symmetrical capacitive window is given by:

$$b_{Cd} = \frac{B_{Cd}}{Y_{10}} = \frac{4b}{\lambda_g} \ln\left\{\operatorname{cosec}\left(\frac{\pi d}{2b}\right)\right\}$$

where Y_{10} is the reciprocal of wave impedance $Z_{w\text{TE}10}$ and λ_g is the guide wavelength.

- The vertical diaphragm (both symmetrical and asymmetrical) extending from one broad wall of the waveguide to the other broad wall provides inductive susceptance across the waveguide and hence such an obstacle is known as an *inductive iris*. An approximate expression for the normalised inductive susceptance in the case of a symmetrical inductive window is given by:

$$b_{Ld} = \frac{B_{Ld}}{Y_{10}} = \frac{\lambda_g}{a} \cot^2\left(\frac{\pi d}{2a}\right)$$

- The same structure that includes both the capacitive and inductive irises together is a *resonant iris* that can be used as a *mode filter* to allow dominant mode to travel down the guide unaffected while the non-dominant modes are filtered out (suppressed).
- A metallic screw inserted in a waveguide can provide a capacitive succceptance across the waveguide if the depth of penetration of the screw is less than a quarter wavelength while an inductive suceptance is provided if the depth of penetration is more than a quarter wavelength. This is used for impedance matching to tune out impedance mismatches in waveguide systems.
- A thin metallic post in a waveguide behaves as an equivalent inductance which is used in designing a bias-post mounted microwave oscillator with semiconductor diodes like IMPATT/Gunn. The approximate formula for the inductive reactance presented by the metallic post of diameter d is given by:

$$x_L = \frac{X_L}{Z_{w\text{TE}_{10}}} = \frac{a}{2\lambda_g} \ln\left(\frac{2a}{d}\right)$$

where $Z_{w\text{TE}_{10}}$ is the wave impedance of the TE_{10} mode in a rectangular waveguide.

Impedance matching with lumped component/planar distributed line reactive elements

- A simple yet practically useful technique of impedance matching in certain applications in a microwave circuit design is to use either a lumped component inductor/capacitor or planar distributed line reactive elements (inductance/capacitance). In fact, in the design of microwave amplifiers, the source and load impedances are to be matched respectively with the input and output impedance of the active device and this is done by such reactive element impedance matching. When two reactive elements are used to realise the impedance matching, the network is called L-section network, while three-element matching networks are either T or Π-section type.

Review Questions

7.1 Why is impedance matching in transmission line/waveguide system essential at microwave frequencies?

7.2 What does impedance matching imply? Mention a few techniques used for realising impedance matching at microwave frequencies.

7.3 Briefly explain the concept of stub matching. Why can't lines with real characteristic impedance (microwave transmission lines) be matched to a purely reactive load with stub matching?

7.4 Explain the concept of single stub matching. Discuss the limitations of single stub matching and how they are taken care of with double stub/triple stub matching?

7.5 Show that for quarter-wave impedance transformer, larger bandwidth is realisable with smaller impedance ratio (Z_L/Z_1) between the two impedances Z_L and Z_I to be matched.

7.6 Discuss the characteristics of *binomial* or *maximally flat* transformer and *Tchebyscheff* or *equal ripple* transformer for microwave impedance matching.

7.7 Compare the performance characteristics of exponential taper, triangular taper and Klopfenstein taper impedance transformer.

7.8 Discuss how the horizontal diaphragm extending from one narrow wall of the waveguide to the other wall represents an equivalent capacitive susceptance.

7.9 Discuss how the vertical diaphragm extending from one broad wall of the waveguide to the other broad wall represents an equivalent inductive susceptance.

7.10 Discuss how the nature of susceptance presented by a variable depth screw at the location of its insertion in a rectangular waveguide changes with the depth of penetration.

7.11 Discuss the effect of adding series/shunt inductance/capacitance to impedance/admittance in LC matching using the Z–Y Smith chart.

Numerical Problems

7.1 The single stub method is used to match a load impedance (37.5 + j37.5) Ω to a 75 Ω transmission line. (a) Find the required length and position of a short-circuited stub made of a section of same 75 Ω line (b) repeat part (a) assuming that the short circuited stub is made of a section of a line that has a characteristic impedance of 50 Ω.

[**Ans:** (a) $d_1 = 0.25\lambda$, $l_1 = 0.312\lambda$, $d_2 = 0.0745\lambda$, $l_2 = 0.0875\lambda$
(b) $d_1 = 0.209\lambda$, $l_1 = 0.4225\lambda$, $d_2 = 0.027\lambda$, $l_2 = 0.078\lambda$]

7.2 A high frequency loss-less 70 Ω line, with $\varepsilon_r = 2.1$, is terminated in $Z_R = 50\angle 30°$ at 320 MHz. The load is to be matched with a shorted section of 50 Ω line, with $\varepsilon_r = 2.3$, connected in parallel; the stub must be at least 5 cm from the load. If such matching is possible, find the distance from the load and the length of the stub.

[**Ans:** distance from load = 21 cm and stub length = 11.1 cm]

7.3 A 70 Ω double stub tuner is used to match a load Y_R = (4.76 + j1.43) mS at 600 MHz to a 70 Ω loss-less air dielectric line. The first stub is located at the load and the separation between the stubs is 10 cm. Find the shorted-stub lengths for matched condition.

[**Ans:** For the first solution: 0.109λ = 5.45 cm and for the second solution: 0.417λ = 20.85 cm]

7.4 Solve Examples 7.1 and 7.3 with the analytical approach of stub matching as detailed in Sections 7.2.1 and 7.2.2 writing suitable MATLAB programs.

7.5 Design a 3-section binomial transformer to match a 125 Ω load to a 75 Ω line at 10 GHz. The maximum VSWR that can be tolerated is 1.15. Calculate the realisable bandwidth.

[**Ans:** $n = 4, \Gamma_T = 0.255, \Delta f = 9.18$ GHz]

7.6 Design a 4-section Tchebyscheff impedance transformer to match a 200 Ω load to a 50 Ω line at 12 GHz. The maximum VSWR allowed is 1.2. Find the corresponding bandwidth.

[**Ans:** $n = 5, \Delta f = 14.80$ GHz]

7.7 Write suitable MATLAB program to generate the curves shown in Figs 7.18(a) and 7.18(b) with reference to Example 7.7.

7.8 Design an exponential taper, triangular taper and a Klopfenstein taper with the maximum allowable ripple in the pass-band as $\Gamma_m = 0.025$ to match a 50 Ω line to 125 Ω load. Show the plot of impedance variation with normalised length of the line and the variation of the resulting reflection coefficient magnitude with the electrical length βl of the line. Write suitable MATLAB programs to generate the curves.

7.9 Write a MATLAB program to generate the graphs shown in Figs 7.20(a), 7.20(b) and Fig. 7.21(b) and discuss their physical implications.

8 Microwave Filters

We should respond to our conscience
which acts as a filter to lead our path towards goodness

8.1 Introduction

A filter is a two-port network which is ideally required to provide perfect transmission for all frequencies in certain *pass-band* regions and to exhibit infinite attenuation in the *stop-band* region. Such ideal characteristics cannot be realised in practice, and the goal of the filter designer is to approximate the ideal requirements to within an acceptable tolerance. Filters are usually categorised by their frequency response characteristics—low-pass, high-pass, band-pass and band-stop/band-reject (Fig. 8.1). The application of filters can be found virtually in any type of communication, radar and test/measurement systems.

At low frequencies, inductors and capacitors have simple frequency characteristics and a very general and complete synthesis procedure has been developed for the design of filters at low frequency using them. However, filter design at microwave frequencies (where distributed parameter elements must be used) is much more complicated due to complex frequency behaviour of microwave circuit elements. In spite of these added complications at microwave frequencies, a number of useful techniques have been developed for the design of microwave filters. The design of a narrow band filter (the term *narrowband* is generally used to indicate bandwidths of less than 10%, while *wideband* usually denotes bandwidths in excess of 40%) at microwave frequencies is particularly straight forward since many microwave elements will have frequency characteristics essentially like those of an ideal inductive or capacitive reactance over a limited frequency range. In such cases, a low frequency prototype filter may be used as a model. The microwave filter is realised by replacing all inductors and capacitors by suitable microwave circuit elements (like lengths of transmission line sections suitably terminated at the load end) that have similar frequency characteristics over the frequency range of interest. During the past few years, lumped inductors and capacitors have been fabricated and used at microwave frequencies (but with limited values and up to

60 GHz or so). If lumped elements are used for the filter circuit at microwave frequencies, the microwave filter design becomes identical to the low frequency filter design.

There are two low frequency filter synthesis techniques in common use—*image-parameter* method (and variation thereof, such as the *constant-k* and *m-derived* filters) and *insertion-loss* method. The image-parameter method provides a filter design having the required pass-band and stop-band characteristics, but does not specify the exact frequency characteristics over each region. This method thus suffers from the shortcoming that a good deal of cut-and-dried procedures must be followed (often must be iterated several times) to achieve the desired results. But the insertion-loss method begins with a complete specification of physically realisable frequency characteristics, and from this, a suitable filter network can be easily synthesised at microwave frequencies.

The microwave filter theory began with the pioneering work by Darlington, Fano, Lawson, and Richards just before World War II. The image parameter method of filter design was developed in the late 1930s and was useful for low frequency filters in radio and telephony. In the early 1950s, G Matthaei, L Young, E Jones, S Cohn and others of the Stanford Research Institute, became very active in microwave filter and coupler development.

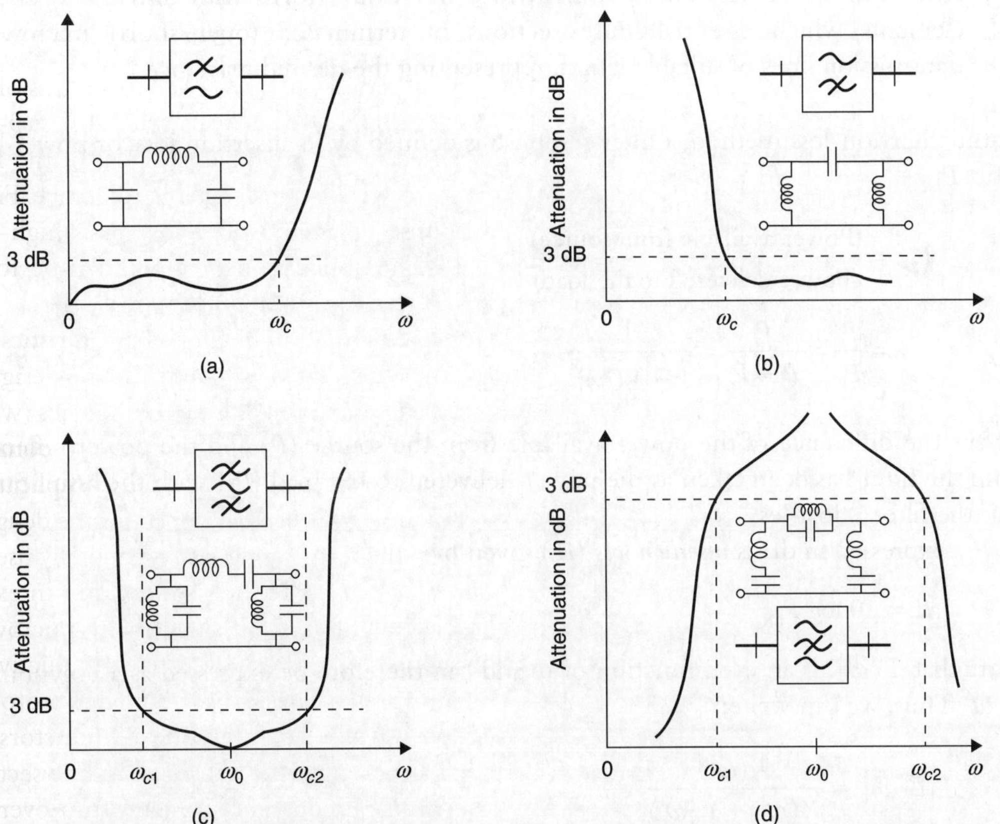

Fig. 8.1 Filter characteristics with equivalent electrical circuit (a) Low-pass filter (b) High-pass filter (c) Band-pass filter (d) Band-stop filter.

A voluminous handbook on filters and couplers resulted from their work and is still a valuable reference [18]. Most microwave filter designs these days is done with computer-aided design (CAD) packages based on Electronic Design Automation (EDA) tool.

8.2 Insertion-loss Method for Low-pass Prototype Design

A perfect filter should have zero insertion loss in the pass-band, infinite attenuation in the stop-band and a linear phase response (to avoid signal distortion) in the pass-band. However, such filters do not exist in practice. Thus, some compromises must be made and this constitutes the art of filter design.

The design of a microwave filter with insertion-loss method is done in following steps:

(i) A low-pass prototype filter is first designed in terms of normalised impedance and frequency with desired pass-band characteristics.
(ii) This prototype network is then transformed to the required filter type (low-pass, high-pass, band-pass or band-stop) with proper scaling of the prototype circuit elements in terms of impedance and frequency.
(iii) Finally, the filter network is realised in a microwave form with distributed circuit elements which essentially use sections of terminated (open/short) microwave transmission lines of suitable length representing the desired reactance.

In the insertion-loss method, a filter response is defined by its insertion loss, or power loss ratio, P_{LR}:

$$P_{LR} = \frac{\text{(Power available from source)}}{\text{(Power delivered to the load)}}$$

$$= \frac{P_i}{P_L} = \frac{P_i}{P_i - P_r} = \frac{1}{1 - |\Gamma(\omega)|^2} \tag{8.1a}$$

where, the difference of the power available from the source (P_i) and the power reflected from the filter has been taken as the power delivered to the load (P_L) with the assumption that the filter is loss-less.

P_{LR} expressed in dB is *insertion loss* (*IL*) given by:

$$IL = 10 \log P_{LR} \tag{8.1b}$$

Note that $|\Gamma(\omega)|^2$ is an even function of ω and can therefore be expressed as a polynomial in ω^2. Thus, we can write:

$$|\Gamma(\omega)|^2 = \frac{X(\omega^2)}{X(\omega^2) + Y(\omega^2)} \tag{8.2}$$

Substituting this in Eq. 8.1a we have,

$$P_{LR} = 1 + \frac{X(\omega^2)}{Y(\omega^2)} \tag{8.3}$$

where X and Y are real polynomials in ω^2. Thus, for a filter to be physically realisable its power loss ratio must be of the form depicted in Eq. 8.3.

In principle, there can be an unlimited number of physical networks for which the power loss ratio can be specified but most of them would be too complex to realise in practice. The most useful power loss ratios for a microwave filter design are those that give a *maximally-flat* pass-band response and those that give an *equal-ripple* response in the pass-band. Such pass-band response characteristics correspond to those of the binomial (maximally flat) and Tchebyscheff (equal ripple) multi-section quarter-wave transformers discussed in Chapter 7.

8.2.1 Case I Maximally-flat/Butterworth Filter

The power loss ratio for a maximally-flat/Butterworth filter is specified by:

$$P_{LR} = 1 + A_p^2(\Omega)^{2n} \tag{8.4}$$

where $\Omega = (\omega/\omega_c)$ represents the normalised frequency variable of the prototype filter and n is the order of the filter, ω and ω_c are the running frequency variable and cut-off frequency of the actual filter. From Fig. 8.2a, it may be observed that the pass-band extends from $\Omega = 0$ ($\omega = 0$) to $\Omega = \Omega_p$ ($\omega = \omega_c$) with the maximum value of P_{LR} in the pass-band being $(1 + A_p^2)$, where A_p determines the attenuation tolerance in the pass-band. Normally, A_p is chosen to be equal to 1 when the band-edge power loss ratio $(1 + A_p^2)$ happens to be equal to -3 dB (Fig. 8.2a). For $\Omega > 1$, the attenuation increases without bound with the increase of Ω and the rate of increase is more (sharper) with order n of the filter, while for $\Omega < 1$, P_{LR} shows progressive closeness to the Ω axis as n increases. For $\Omega >> 1$ ($\omega >> \omega_c$), $P_{LR} \approx A_p^2(\omega/\omega_c)$; which indicates that the insertion loss incurred by the filter far-off the cut-off frequency increases at the rate of $6n$ dB/octave (or $20n$ dB/decade).

This type of filter is known as a maximally flat filter because its power loss ratio given by Eq. 8.4a has the maximum number $(2n - 1)$ of zero derivatives at $\Omega = 0$, approximating the ideal low pass filter response at $\Omega = 0$ but with gradual deviation from that as Ω approaches the cut-off frequency Ω_p (Fig. 8.2a).

8.2.2 Case II Equal-ripple/Tchebyscheff Filter

The power loss ratio of equal-ripple/Tchebyscheff filter (that has some tolerable ripples of equal amplitude unlike the maximally-flat one) is represented by:

$$P_{LR} = 1 + A_p^2 T_n^2(\Omega) \tag{8.5}$$

where $T_n(\Omega)$ is the Tchebyscheff polynomial of first kind having order n (vide Section 7.3.2.2), which is given below:

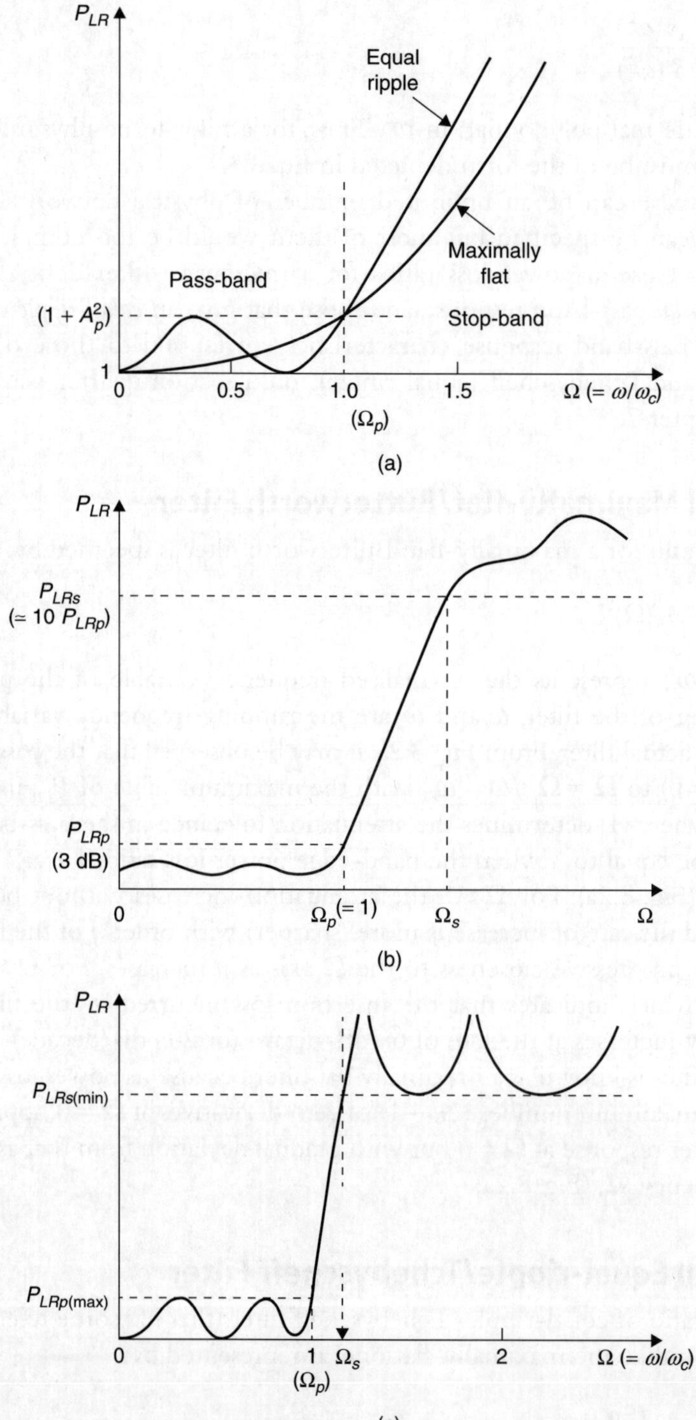

Fig. 8.2 (a) Maximally-flat and equal-ripple low-pass filter response (b) Low-pas filter characteristics showing the pass-band edge frequency (Ω_p) and stop-band edge frequency (Ω_s) and corresponding power loss ratios ($P_{LRp,\,s}$) (c) Elliptic function low-pass filter.

$$T_n(\Omega) = \cos(n\cos^{-1}\Omega) \quad \text{for} \quad |\Omega| \le 1$$
$$= \cosh(n\cosh^{-1}\Omega) \quad \text{for} \quad |\Omega| \ge 1 \tag{8.6}$$

The pass-band will have equal ripples of amplitude $(1 + A_p^2)$ (Fig. 8.2a), since $T_n(x)$ varies between ± 1, for $x \le 1$. For large values of x, $T_n(x) \approx \frac{1}{2}(2x)^n$, so that for $\Omega >> 1$ ($\omega >> \omega_c$), $P_{LR} \approx (A_p^2/4)\,[2(\omega/\omega_c)^{2n}]$; that is, insertion loss incurred by this filter also increases at the rate of $6n$ dB/octave (or $20n$ dB/decade) far-off the cut-off frequency but it is $2^{(2n-2)}$ times larger than that for a maximally-flat filter. Thus, the equal-ripple filter will have a sharper rate of rise above cut-off with better stop-band characteristics (approximately 'brick-wall' characteristics outside the pass-band) compared to a maximally-flat filter, though with a penalty of ripple within the pass-band. However, for a specified rate of increase of the power loss ratio beyond the cut-off frequency, the equal-ripple characteristics can provide smallest pass-band tolerance (A_p). In this sense the equal-ripple filter represents an optimum design.

It may further be observed that for a given insertion loss, the equal-ripple filter needs fewer elements (number of elements is equal to the order n of the filter, as will be seen shortly) than a maximally-flat filter which may be established as follows. Let the power loss ratio at the pass-band edge frequency Ω_p (= 1) be P_{LRp} while the same at stop-band edge frequency Ω_s be P_{LRs} (Fig. 8.2b). Denoting insertion loss at Ω_p and Ω_s respectively as $IL_p = 10 \log P_{LRp}$ and $IL_s = \log P_{LRs}$ it can be derived from Eq. 8.4 that for a maximally flat filter:

$$n \ge \frac{\log\{(P_{LRs} - 1)/(P_{LRp} - 1)\}^{1/2}}{\log(\Omega_s/\Omega_p)} \tag{8.7a}$$

while for an equal-ripple filter it can be derived by using Eq. 8.5 and the expression for $T_n(x)$ from Section 7.3.2.2, that:

$$n \ge \frac{\cosh^{-1}\{(P_{LRs} - 1)/(P_{LRp} - 1)\}^{1/2}}{\cosh^{-1}(\Omega_s/\Omega_p)} \tag{8.7b}$$

From a comparison of the two expressions given by Eqs 8.5a and 8.5b, it may be concluded that given the same requirements for the insertion-loss characteristics in the stop-band (which determines the roll-off of the filter), equal-ripple filter requires fewer elements (lower order of the filter) compared to the maximally-flat filter. In other words, with the same order of the filter, equal-ripple filter will have a better roll-off beyond cut-off.

8.2.3 Case III Elliptic Filter

The maximally-flat and equal-ripple responses both have monotonically increasing attenuation in the stop-band. In many applications it is required to specify a minimum stop-band attenuation, when a better cut-off rate can be obtained. Such filters are called elliptic function filters and have equal-ripple response in the pass-band as well as in the stop-band (Fig. 8.2c). The maximum attenuation in the pass-band $P_{LRp\ (\max)}$ as well as the minimum attenuation in the stop-band $P_{LRs\ (\min)}$ are specified. The synthesis for the elliptic function filter is quite involved and beyond the scope of this text; however, details are available in suitable reference books [19].

Low-pass prototype filter design technique: The low-pass filter prototype design is done in terms of normalised frequency and impedance. This normalisation simplifies the design of filters for arbitrary frequency, impedance and type of the filter. The low-pass prototypes are then scaled to the desired frequency and impedance.

Let the low-pass prototype be a ladder network of capacitance and inductance (Fig. 8.3). The load impedance is chosen to be equal to unity (1 Ω) and the source impedance R_s (Fig. 8.3a) and $G_s = 1/R_s$ (Fig. 8.3b) are chosen to be unity. The element values are denoted by g_k, and are same for both the circuits; it is capacitance in farads (for k odd) and inductance in henries (for k even) in Fig. 8.3a, while it is inductance in henries (for k odd) and capacitance in farads (for k even) in Fig. 8.3b.

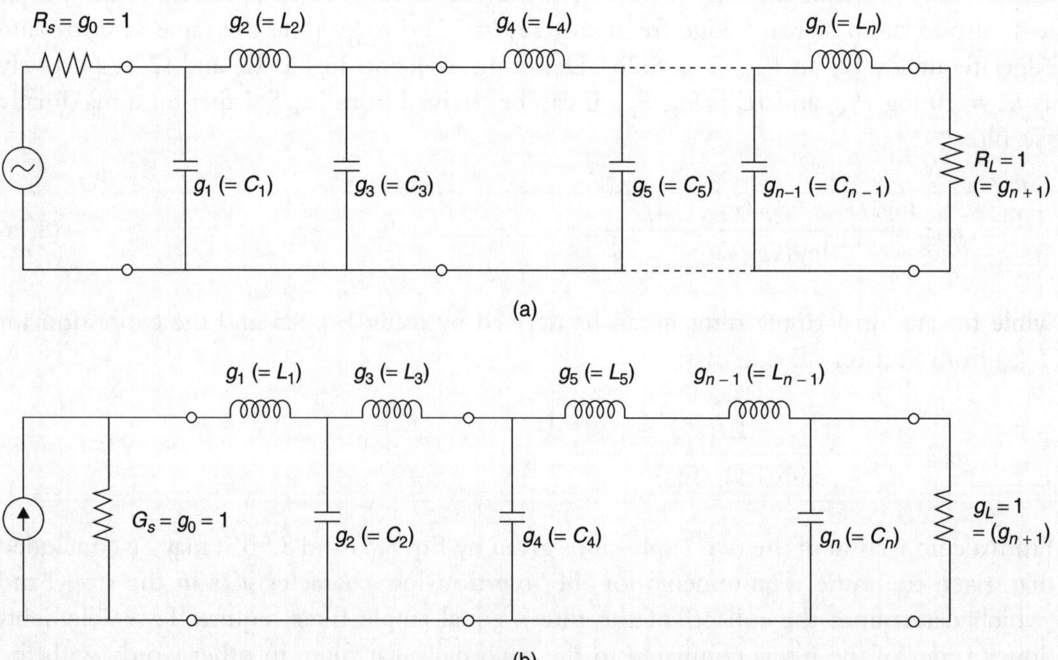

Fig. 8.3 Low-pass prototype filter network (a) Ladder network with $g_k = C_k$ (when k odd), $g_k = L_k$ (when k even) (b) Dual of (a) with $g_k = L_k$ (when k odd), $g_k = C_k$ (when k even).

The element values of the prototype low-pass filter with equal-ripple and maximally-flat responses for the normalised value of cut-off angular frequency (Ω = 1) can be calculated from Bennett's formula [15].

For an equal-ripple low-pass prototype filter:

$$g_0 = 1 \text{ for all } n \quad \text{and} \quad g_1 = \frac{2a_1}{\gamma}$$

$$g_k = \frac{4a_{k-1}a_k}{b_{k-1}g_{k-1}} \quad \text{with} \quad k = 2, 3, 4 \ldots n$$

$$g_{n+1} = 1, \quad \text{for} \quad n \text{ odd}$$

$$= \coth^2\left(\frac{\beta}{4}\right), \quad \text{for } n \text{ even} \tag{8.8}$$

where $a_k = \sin\dfrac{2k-1}{2n}\pi$, $b_k = \gamma^2 + \sin^2\dfrac{k\pi}{n}$, $\beta = \ln\left(\coth\dfrac{A_{pm}}{17.37}\right)$, $\gamma = \sinh\left(\dfrac{\beta}{2n}\right)$

with $k = 2, 3, \ldots n$.

[**Note:** A_{pm} is the maximum attenuation in dB in the pass-band (the maximum ripple height). In other words, $IL_{\text{at } \omega = \omega c}$, that is, $10 \log_{10} (1 + A_p^2)$ since at $\omega = \omega_c$, $T_n(x) = 1$ (Eq. 7.58a)].

For maximally-flat low-pass filter:

$$g_0 = 1 = g_{n+1}, \text{ for all } n$$

$$g_k = 2 \sin\left\{\frac{(2k-1)\pi}{2n}\right\}, k = 1, 2, \ldots n \tag{8.9}$$

This analytical approach is used to evaluate the element values for a prototype low-pass filter. Such tables are available in the standard text [18], however, they are also given in Table 8.1 and Table 8.2 for equal-ripple and maximally-flat cases, respectively. The tables have been generated using a MATLAB program (the flow chart and program has been included in Appendix III).

Table 8.1 Values of g_k for different order (n) of Tchebyscheff filter with maximum tolerable ripple height $A_{pm} = IL_{\text{at } \omega = \omega_c} = 10 \log_{10}(1 + A_p^2)$ is varied as parameter

(a) A_{pm} = 0.1dB

n	g_1	g_2	g_3	g_4	g_5	g_6	g_7	g_8	g_9	g_{10}	g_{11}
1	0.3052	1.0000									
2	0.8430	0.6220	1.3554								
3	1.0315	1.1474	1.0315	1.0000							
4	1.1088	1.3061	1.7703	0.8180	1.3554						
5	1.1468	1.3712	1.9750	1.3712	1.1468	1.0000					
6	1.1681	1.4039	2.0562	1.5170	1.9029	0.8618	1.3554				
7	1.1811	1.4228	2.0966	1.5733	2.0966	1.4288	1.1811	1.0000			
8	1.1897	1.4346	2.1199	1.6010	2.1699	1.5640	1.9444	0.8778	1.3554		
9	1.1956	1.4425	2.1345	1.6167	2.2053	1.6167	2.1345	1.4425	1.1956	1.0000	
10	1.1999	1.4481	2.1444	1.62651	2.2253	1.6418	2.2046	1.5821	1.96281	0.8853	1.3554

(*continued*)

Table 8.1 Continued

(b) A_{pm} = 0.2 dB

n	g_1	g_2	g_3	g_4	g_5	g_6	g_7	g_8	g_9	g_{10}	g_{11}
1	0.4342	1.0000									
2	1.0378	0.6745	1.5386								
3	1.2275	1.1525	1.2275	1.0000							
4	1.3028	1.2844	1.9761	0.8468	1.5386						
5	1.3394	1.3370	2.1660	1.3370	1.3394	1.0000					
6	1.3598	1.3632	2.2394	1.4555	2.0974	0.8838	1.5386				
7	1.3722	1.3781	2.2756	1.5001	202756	1.3781	1.3722	1.0000			
8	1.3804	1.3875	2.2963	1.5217	2.3413	1.4925	2.1349	0.8972	1.5386		
9	1.3860	1.3938	2.3093	1.5340	2.3728	1.5340	2.3093	1.3938	1.3860	1.0000	
10	1.3901	1.3983	2.3181	1.5417	2.3904	1.5536	2.3720	1.5066	2.514	0.9034	1.5386

(c) A_{pm} = 0.3 dB

n	g_1	g_2	g_3	g_4	g_5	g_6	g_7	g_8	g_9	g_{10}	g_{11}
1	0.5349	1.000									
2	1.1800	0.6966	1.6967								
3	1.3706	1.1383	1.3729	1.000							
4	1.4450	1.2540	2.1270	0.8541	1.6967						
5	1.4810	1.2995	2.3090	1.2999	1.4852	1.000					
6	1.5009	1.3221	2.3784	1.4026	2.2428	0.8881	1.6967				
7	1.5130	1.3349	2.4124	1.4407	2.4126	1.3355	1.5191	1.000			
8	1.5210	1.3429	2.4318	1.4591	2.4744	1.4342	2.2786	0.9010	1.6967		
9	1.5264	1.3483	2.4440	1.4695	2.5039	1.4695	2.4443	1.3491	1.5342	1.000	
10	1.5304	1.3521	2.4521	1.4760	2.5203	1.4862	2.5032	1.4462	2.2944	0.9076	1.6967

(d) A_{pm} = 0.4 dB

n	g_1	g_2	g_3	g_4	g_5	g_6	g_7	g_8	g_9	g_{10}	g_{11}
1	0.6213	1.0000									
2	1.2983	0.7054	1.8435								
3	1.4902	1.1184	1.4926	1.0000							
4	1.5642	1.2228	2.2534	0.8509	1.8435						
5	1.5999	1.2635	2.4308	1.2639	1.6042	1.0000					
6	1.6195	1.2836	2.4978	1.3558	2.3658	0.8818	1.8435				
7	1.6315	1.2950	2.5306	1.3896	2.5308	1.2956	1.6377	1.0000			
8	1.6394	1.3022	2.5492	1.4059	2.5904	1.3838	2.4003	0.8936	1.8435		
9	1.6447	1.3069	2.5609	1.4151	2.6186	1.4151	2.5613	1.3077	1.6528	1.0000	
10	1.6486	1.3103	2.5687	1.4208	2.6343	1.4298	2.6178	1.3944	2.4156	0.8997	1.8435

(continued)

Table 8.1 Continued

(e) $A_{pm} = 0.5$ dB

n	g_1	g_2	g_3	g_4	g_5	g_6	g_7	g_8	g_9	g_{10}	g_{11}
1	0.6986	1.0000									
2	1.4029	0.7071	1.9841								
3	1.5963	1.0967	1.5963	1.0000							
4	1.6703	1.1926	2.3661	0.8419	1.9841						
5	1.7058	1.2296	2.5408	1.2296	1.7058	1.0000					
6	1.7254	1.2479	2.6064	1.3137	2.4758	0.8696	1.9841				
7	1.7372	1.2583	2.6381	1.3444	2.6381	1.2583	1.7372	1.0000			
8	1.7451	1.2647	2.6564	1.3590	2.6964	1.3389	2.5093	0.8796	1.9841		
9	1.7504	1.2690	2.6678	1.3673	2.7239	1.3673	2.6678	1.2690	1.7504	1.0000	
10	1.7543	1.2721	2.6754	1.3725	2.7392	1.3806	2.7231	1.3485	2.5239	0.8842	1.9841

Table 8.2 Values of g_k for different order (n) of Butterworth (maximally-flat) low-pass prototype filter for $IL = 3$ dB when $\Omega\ (= \omega/\omega_c) = \Omega_p = 1$

n	g_1	g_2	g_3	g_4	g_5	g_6	g_7	g_8	g_9	g_{10}	g_{11}
1	2.000	1									
2	1.413	1.4159	1								
3	0.999	2.0000	1.0023	1							
4	0.765	1.8473	1.8485	0.7679	1						
5	0.617	1.6175	2.0000	1.6193	0.6206	1					
6	0.517	1.4136	1.9315	1.9323	1.4159	0.5205	1				
7	0.444	1.2464	1.8014	2.0000	1.8028	1.2489	0.4479	1			
8	0.390	1.1107	1.6624	1.9613	1.9619	1.6642	1.1133	0.3931	1		
9	0.347	0.9995	1.5315	1.8790	2.0000	1.8801	1.5336	1.0023	0.350	1	
10	0.312	0.9076	1.4137	1.7815	1.9752	1.9756	1.7830	1.4159	0.910	0.315	1

8.3 Filter Transformations from Prototype

The low-pass filter prototype of the previous section was a normalised design (with $R_L = R_s = 1$ and $\Omega = 1$ at the edge of the pass-band). However, the prototype design may be used as the basis for the design of actual low-pass, high-pass, band-pass and band-stop filters at any desired impedance and frequency. To do this, we need to scale the prototype circuit elements in terms of frequency and impedance, the procedure for which is given below.

Frequency transformation/mapping basically transforms or maps a response (Tchebyscheff/Butterworth) of the low-pass prototype in the frequency domain Ω (Fig. 8.2a) into the frequency domain ω of an actual/practical filter in which it is expressed. The frequency transformation will have an effect on all reactive components but will have no effect on the resistive elements of the actual filter.

The impedance scaling is required to remove the $g_0 = 1$ normalisation of the source impedance, as a practical filter might need to work for any desired value of source impedance, say $Z_s = Z_0$. For impedance scaling, a prototype reactive element in the series branch (appearing in the impedance branch) needs to be multiplied by the impedance scaling factor I_t while a prototype reactive element in the shunt branch (appearing in the admittance branch) needs to be divided by the scaling factor I_t. Thus, with reference to Fig. 8.3a, the series inductances (L_k) and shunt capacitances (C_k) of the prototype will be represented by:

$$L_k' = I_t L_k \quad \text{and} \quad C_k' = \frac{C}{I_t} \tag{8.10}$$

where $I_t = Z_0/g_0$.

The impedance scaling accounts for the actual impedance of the practical filter and since it is independent of frequency, the same transformation/scaling holds for any type of filter.

Let us now discuss transformation from low-pass prototype to actual low-pass, high-pass, band-pass and band-stop filters.

8.3.1 Prototype to Actual Low-pass Transformation

For frequency scaling from low-pass prototype to practical low-pass filter, the latter with a cut-off frequency ω_c in the angular frequency axis ω, we need to change $\Omega_p = (\omega/\omega_c) = 1$, to the actual band-edge cut-off frequency ω_c. This demands the frequency transformation from Ω to Ω', where $\Omega' = \left(\dfrac{\Omega_p}{\omega_c}\right)\omega$, for the prototype reactive elements, allowing us to write the following equations: $j\Omega' L = j\left(\dfrac{\Omega_p}{\omega_c}\right)\omega L_k = j\omega L_k'$ and $j\Omega' C_k = j\left(\dfrac{\Omega_p}{\omega_c}\right)\omega C_k = j\omega C_k'$, whence:

$$L_k' = \frac{\Omega_p L_k}{\omega_c}, \quad C_k' = \frac{\Omega_p C_k}{\omega_c} \tag{8.11}$$

When both impedance and frequency scaling are required the results, Eqs 8.10 and 8.11, can be combined to give:

$$L_k' = \left(\frac{\Omega_p}{\omega_c}\right) L_k I_t \tag{8.12a}$$

$$C_k' = \left(\frac{\Omega_p}{\omega_c}\right)\frac{C_k}{I_t} \tag{8.12b}$$

The effect of the frequency transformation observed here basically represents a stretching out of the original low-pass prototype filter pass-band which shifts the cut-off from unity

($\Omega_p = 1$, for prototype) to $\omega_c > 1$ (as desired by the actual low-pass filter pass-band) in the frequency scale (compare Fig. 8.4b with 8.4a).

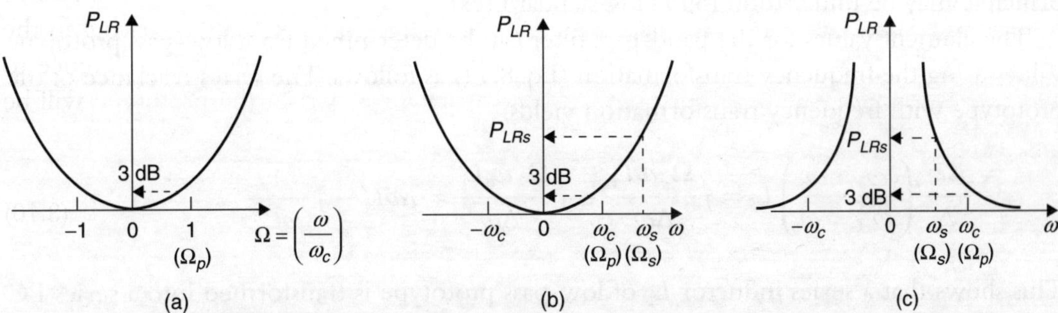

Fig. 8.4 Filter response (a) Low-pass prototype (b) Actual low-pass with frequency scaling (c) High-pass with frequency scaling.

8.3.2 Low-pass Prototype to High-pass Transformation

A conversion from low-pass to high-pass demands that in the frequency scale, $\omega = 0$ should map to $\omega = \pm\infty$ and vice versa (Fig. 8.4c) and inductance and capacitance of low-pass configuration be interchanged in the high-pass configuration. To implement this, we need to use the frequency transformation: $\Omega' = -\left(\frac{\Omega_p}{\omega}\right)\omega_c$ to get the actual high-pass filter elements. The negative sign converts inductors/capacitors of the low-pass prototype to capacitors/inductors in the actual high-pass filter. Having done this frequency transformation and including the impedance transformation, the actual high-pass filter elements can be derived as:

$$C_k' = \left(\frac{1}{\omega_c \Omega_p}\right)\frac{1}{L_k I_t} \quad \text{and} \quad L_k' = \left(\frac{1}{\omega_c \Omega_p}\right)\frac{I_t}{C_k} \tag{8.13}$$

This shows that the low-pass prototype to actual high-pass transformation effectively replaces all inductances L_k by capacitances C_k', and the C_k must be replaced by inductances L_k'.

8.3.3 Low-pass to Band-pass and Band-stop Transformations

The low-pass prototype filter design may also be transformed to exhibit band-pass and band-stop responses (Figs 8.5a and 8.5b). The low-pass response can be converted to a band-pass response by means of the following frequency transformation (remembering that band-pass is a combination of low-pass and high-pass characteristics):

$$\Omega' = \frac{\Omega_p}{\Delta}\left(\frac{\omega}{\omega_0} - \frac{\omega_0}{\omega}\right) \tag{8.14}$$

where $\Delta = (\omega_{c2} - \omega_{c1})/\omega_0$ is the fractional bandwidth of the pass-band with ω_{c1} and ω_{c2} denoting the angular frequency at the edges of the pass-band (they correspond to $\Omega = \pm 1$ in the low-pass prototype). The mid-band frequency ω_0 is given by $\sqrt{\omega_{c1}\omega_{c2}}$. The mapping principle may be understood following standard texts [7, 9].

The element values for the band-pass filter can be determined from low-pass prototype values using the frequency transformation (Eq. 8.14) as follows. The series reactance of the prototype with frequency transformation yields:

$$\frac{j\Omega_p}{\Delta}\left(\frac{\omega}{\omega_0} - \frac{\omega_0}{\omega}\right)L_k = j\,\frac{\Omega_p \omega L_k}{\Delta\omega_0} - j\,\frac{\Omega_p \omega_0 L_k}{\Delta\omega} = j\omega L_k' - j\,\frac{1}{\omega C_k'}$$

This shows that a series inductor L_k of low-pass prototype is transformed into a series LC circuit in band-pass filter with element values:

$$L'_{kse} = \left(\frac{\Omega_p}{\Delta\omega_0}\right) I_t L_k \quad \text{and} \quad C'_{kse} = \left(\frac{\Delta}{\Omega_p \omega_0}\right)\frac{1}{I_t L_k} \tag{8.15}$$

(where impedance transformation is also included).

Similarly, the shunt susceptance of the prototype with frequency transformation yields:

$$\frac{j\Omega_p}{\Delta}\left(\frac{\omega}{\omega_0} - \frac{\omega_0}{\omega}\right)C_k = j\,\frac{\Omega_p \omega C_k}{\Delta\omega_0} - j\,\frac{\Omega_p \omega_0 C_k}{\Delta\omega} = j\omega C_k' - j\,\frac{1}{\omega L_k'}$$

This shows that a shunt capacitor C_k of low-pass prototype is transformed to a shunt LC circuit in band-pass filter with element values (including the impedance transformation):

$$L'_{ksh} = \left(\frac{\Delta}{\Omega_p \omega_0}\right)\frac{I_t}{C_k} \quad \text{and} \quad C'_{ksh} = \left(\frac{\Omega_p}{\Delta\omega_0}\right)\frac{C_k}{I_t} \tag{8.16}$$

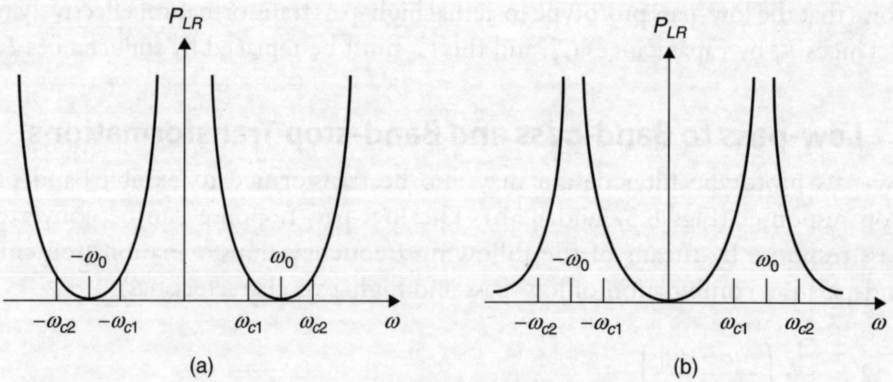

Fig. 8.5 Filter response with frequency scaling (a) Band-pass (b) Band-stop.

Table 8.3 Summary of prototype to actual filter transformation

$[\Delta = (\omega_{c2} - \omega_{c1})/\omega_0;\ \Omega_p = 1\ ;\ I_t = Z_0/g_0\ (\text{with } g_0 = 1)]$

Low-pass prototype	Actual low-pass	High-pass	Band-pass	Band-stop
L_k	L'_k	C'_k	L'_{kse} C'_{kse}	L'_{ksh} C'_{ksh}
$g_k = L_k$	$L'_k = \left(\frac{\Omega_p}{\omega_c}\right) L_k I_t$	$C'_k = \left(\frac{1}{\omega_c \Omega_p}\right)\frac{1}{L_k I_t}$	$L'_{kse} = \left(\frac{\Omega_p}{\Delta\omega_o}\right) I_t L_k$ $C'_{kse} = \left(\frac{\Delta}{\Omega_p \omega_o}\right)\frac{1}{I_t L_k}$	$L'_{ksh} = \left(\frac{\Omega_p \Delta}{\omega_o}\right) I_t L_k$ $C'_{ksh} = \left(\frac{1}{\Omega_p \Delta\omega_o}\right)\frac{1}{I_t L_k}$
C_k	C'_k	L'_k	L'_{ksh} C'_{ksh}	L'_{kse} C'_{kse}
$g_k = C_k$	$C'_k = \left(\frac{\Omega_p}{\omega_c}\right)\frac{C_k}{I_t}$	$L'_k = \left(\frac{1}{\omega_c \Omega_p}\right)\frac{I_t}{C_k}$	$L'_{ksh} = \left(\frac{\Delta}{\Omega_p \omega_o}\right)\frac{I_t}{C_k}$ $C'_{ksh} = \left(\frac{\Omega_p}{\Delta\omega_o}\right)\frac{C_k}{I_t}$	$L'_{kse} = \left(\frac{1}{\Omega_p \Delta\omega_o}\right)\frac{I_t}{C_k}$ $C'_{kse} = \left(\frac{\Omega_p \Delta}{\omega_o}\right)\frac{C_k}{I_t}$

The low-pass filter elements are thus converted to series resonant circuits (low impedance at resonance) in the series arms, and parallel resonant circuits (high impedance at resonance) in the shunt arms. However, both series and parallel resonator elements have the same resonant frequency, ω_0.

To realise the band-stop response (Fig. 8.6b) we need to use inverse transformation:

$$\Omega' = \frac{\Omega_p}{\Delta}\left(\frac{\omega}{\omega_0} - \frac{\omega_0}{\omega}\right)^{-1} \tag{8.17}$$

where, Δ and ω_0 have a similar meaning as mentioned above. Then the series inductors of low-pass prototype are converted to shunt LC circuits in the band-stop filter having element values:

$$L'_{ksh} = \left(\frac{\Omega_p \Delta}{\omega_0}\right) I_t L_k \quad \text{and} \quad C'_{ksh} = \left(\frac{1}{\Omega_p \Delta \omega_0}\right)\frac{1}{I_t L_k} \tag{8.18}$$

The shunt capacitors of the low-pass prototype are converted to series LC circuits in the band-stop filter having element values:

$$L'_{kse} = \left(\frac{1}{\Omega_p \Delta \omega_0}\right)\frac{I_t}{C_k} \quad \text{and} \quad C'_{kse} = \left(\frac{\Omega_p \Delta}{\omega_0}\right)\frac{C_k}{I_t} \tag{8.19}$$

The element transformations from low-pass prototype to actual low-pass, high-pass, band-pass and band-stop filter (with both frequency and impedance scaling) are summarised in Table 8.3.

8.4 Microwave Implementation

So far we have discussed lumped-element filters that generally work well at low frequencies but have limited scope to be used directly at microwave frequencies. Though lumped components may be implemented at microwave frequencies, they are available only up to a certain frequency and with a limited range of values. Thus, microwave filters are mostly designed with distributed component inductors and capacitors. In addition, at microwave frequencies, the distance between filter components sometimes becomes negligibly small. To convert lumped elements to distributed components with transmission line sections Richard's transformation is used, while Kuroda's identities can be used to separate filter elements by using transmission line sections. Such additional transmission line sections normally do not affect the filter response, however, it is possible to design microwave filters that take advantage of these sections to improve the filter response.

8.4.1 Richard's Transformation

This transformation was introduced by P Richard to synthesise an LC network using open and short circuited transmission line sections or stubs. Here, an inductance can be replaced by a short circuited stub of length l and characteristic impedance Z_k and a capacitance can be replaced by an open circuited stub of length l and characteristic impedance Y_k. The new reactance and susceptance functions now become:

$$jX_k' = jZ_k \tan\beta l = jZ_k \tan\left(\frac{\omega l}{c}\right) = jZ_k \tan\theta = jg_k \tan\theta \tag{8.20a}$$

$$jB_k' = jY_k \tan\theta = jg_k \tan\theta \tag{8.20b}$$

And the power loss ratio for the filter is given by:

$$P_{LR} = 1 + P(\tan^2\theta) \tag{8.20c}$$

The frequency transformation $\Omega' = \tan\theta$ maps the whole Ω' axis periodically into intervals of length π in the θ domain. Hence the inductors and capacitors of a lumped element filter can be replaced with short-circuited and open-circuited stubs as illustrated in Fig. 8.6. The reason why the length of the line should be $\lambda/8$ at ω_c is that for a short circuited transmission line when length of the line is made $\lambda/8$, the characteristic impedance of the line represents inductive impedance at ω_c while an open circuited transmission line of length $\lambda/8$ represents a capacitive impedance at ω_c. This equivalence of transmission line length representing lumped inductance and capacitance by short circuited and open circuited lines respectively of length $\lambda/8$ is valid not only at ω_c but also holds approximately good for any ω far off the cut-off frequency. This is because, at $\omega = \omega_c$, for short circuited stub with $l = \lambda/8$, $\omega_c L = Z_0$; while for $\omega << \omega_c$, when $\tan(\pi/4) \approx \pi/4$, $\omega L = 0.8\, Z_0$.

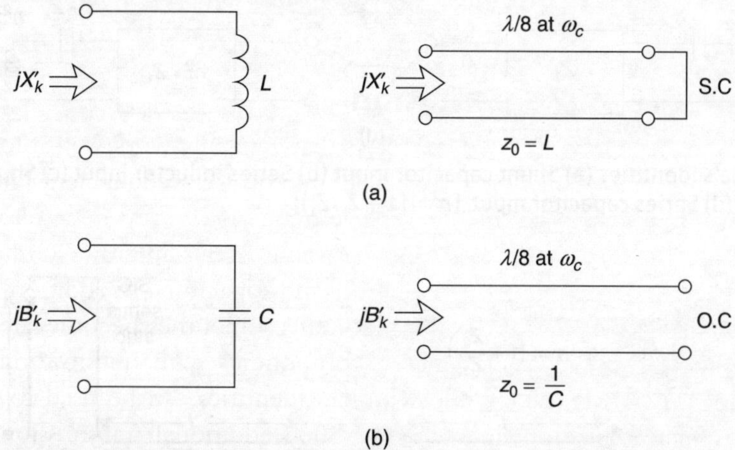

Fig. 8.6 Richard's transformation (a) For an ideal inductor to a short-circuited stub (b) For a capacitor to an open-circuited stub.

8.4.2 Kuroda's Identities

Kuroda's identities (transforms) are used to convert a section of transmission line with a shorted series stub into an electrically equivalent section of transmission line with an open parallel stub and vice versa. As a result, an identical ABCD-parameter matrix is produced that performs the same function. These identities may also be used to separate physically the transmission line stubs and also to change impractical characteristic impedances into more realisable ones. To be precise, these identities are used to achieve a more practical microwave filter implementation.

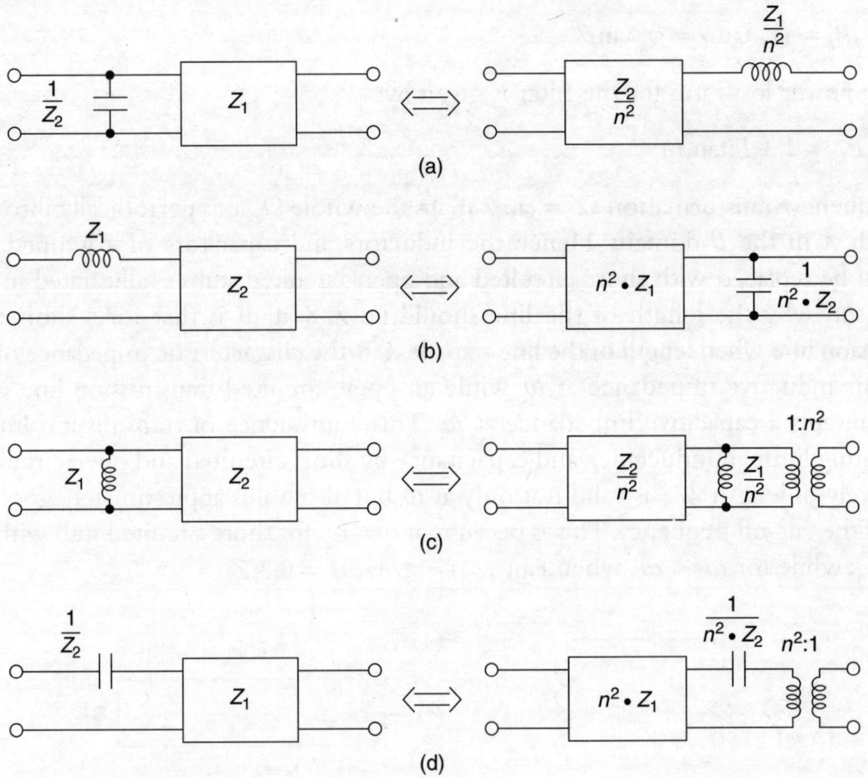

Fig. 8.7 Kuroda's identities (a) Shunt capacitor input (b) Series inductor input (c) Shunt inductor input (d) Series capacitor input, $[n^2 = (1 + Z_2/Z_1)]$.

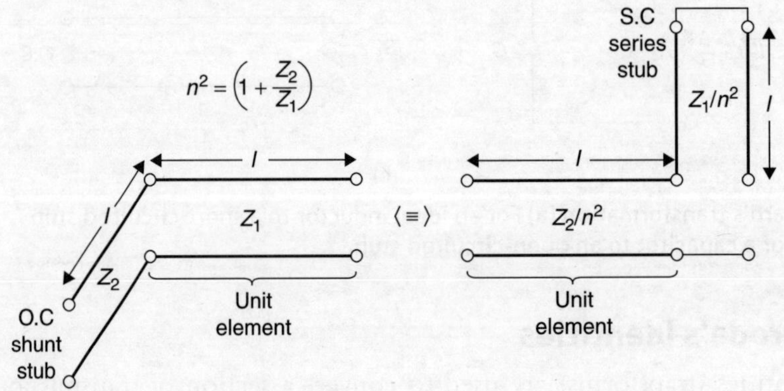

Fig. 8.8 Equivalent circuit illustrating Kuroda's identity (a) and (b) in Fig. 8.7.

Four such network equivalences are there (Fig. 8.7); the equivalence is only valid in the S-plane defined as:

$$S = j\Omega' = j \tan \beta l = j \tan \theta \tag{8.21}$$

Each box in Fig. 8.7 represents a unit element or a transmission line of indicated characteristic impedance and having length $\lambda/8$ at $\omega = \omega_c$. The inductors and capacitors represent short-circuited and open-circuited stubs, respectively. The two circuits of identity (a) in Fig. 8.7 can be redrawn as in Fig. 8.8, which are equivalent.

EXAMPLE 8.1

Design a third order low-pass filter in a 50 Ω microstrip transmission line system with substrate material having $\varepsilon_r = 10$ and substrate height equal to 0.65 mm. The cut-off frequency of the filter is desired to be at 1.8 GHz. Assume 0.5 dB equal-ripple pass-band characteristics.

Solution

Given the order of the filter (n) = 3 (odd); $g_{n+1} = g_4 = g_0 = 1.0$

From Table 8.1(e), the normalised low-pass prototype element values are given by:

$$g_1 = 1.5963\ (= L_1),\quad g_2 = 1.0967\ (= C_2), g_3 = 1.5963\ (= L_3).$$

The corresponding lumped-element circuit is shown in Fig. 8.9a.

Now we use Richard's transformations to convert the series inductors in the form of series stubs, and shunt capacitors in the form of shunt stubs (Fig. 8.9b). According to Eq. 8.20, the normalised characteristic impedance of a series stub (inductor) is $z_{0se} = L$, and the normalised the characteristic impedance of a shunt stub (capacitor) is $z_{0sh} = 1/C$. The length of each stub will be $\lambda/8$ at ω_c.

As the series stubs of Fig. 8.9b would be very difficult to implement in microstrip form, we will use one of Kuroda's identities to convert these to shunt stubs. First we add unit elements at either end of the filter (Fig. 8.9c). These redundant elements do not affect the filter performance since they are matched to the source and load ($z_0 = 1 = R = R_L$). Now we can apply Kuroda's identity (b) (Fig. 8.7) to both ends of the filter to convert series elements to shunt elements. In both cases we have:

$$n^2 = 1 + \frac{Z_2}{Z_1} = 1 + \frac{1}{1.5963} = 1.626 \text{ (since } Z_1 = z_{0se} = 1.5963\ \Omega \text{ and } Z_2 = 1\ \Omega)$$

The result is shown in Fig. 8.9d, which follows from Fig. 8.9c where we use $Z_2/n^2 = 1$ (when $Z_2 = 1.626$) and $Z_1/n^2 = z_{0se}$ (1.5963) (when $Z_1 = 2.596$) on the series elements to convert it to shunt elements.

Finally, we need to scale the circuit in terms of impedance and frequency, which simply involves multiplying the normalised characteristic impedances by 50 Ω and choosing the line and stub lengths to be $\lambda/8$ at 1.8 GHz ($l = 2.083$ cm). The final circuit is shown

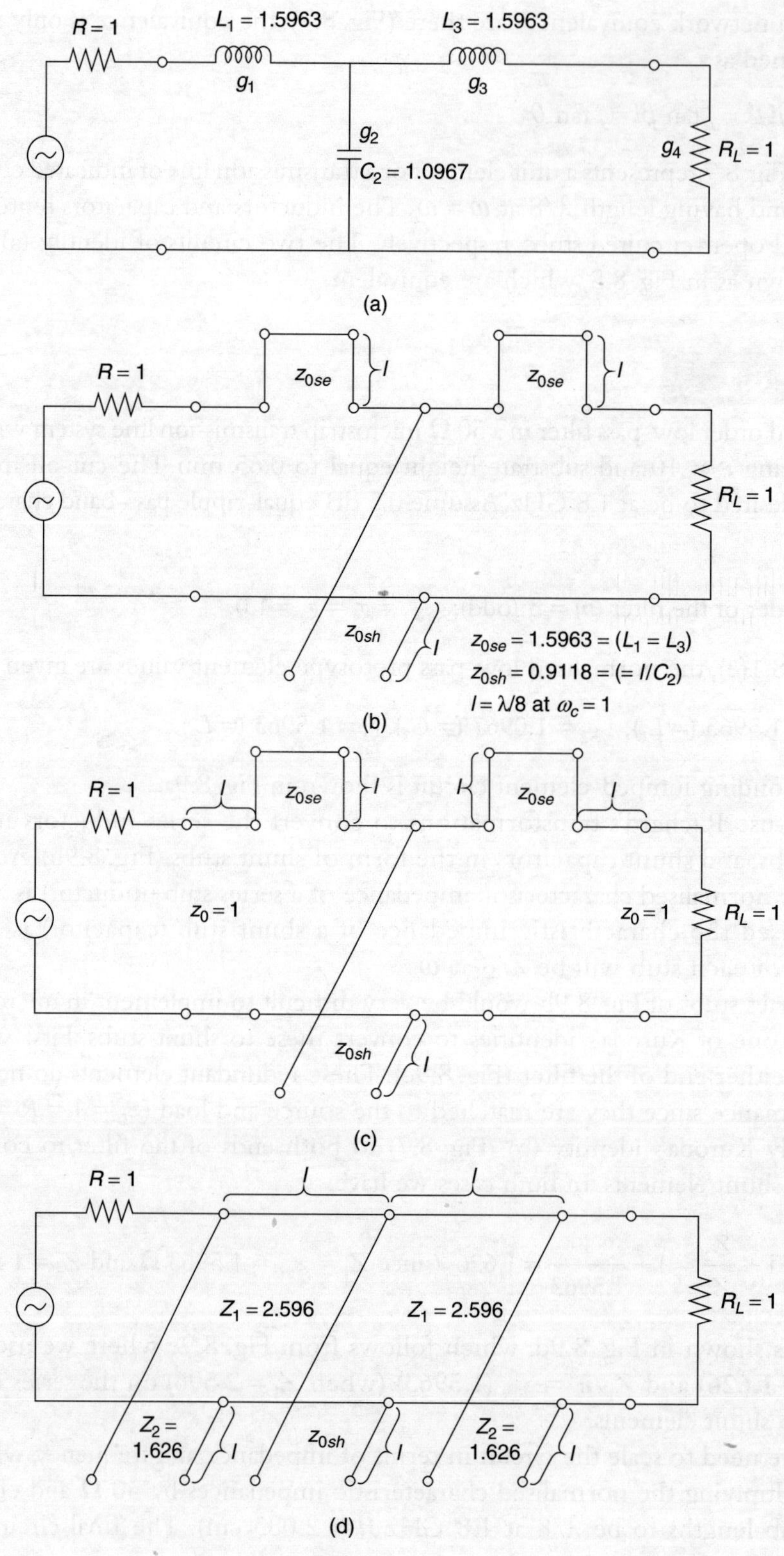

R = 1
L1 = 1.5963
g1
L3 = 1.5963
g3
g2
C2 = 1.0967
g4
RL = 1
(a)
R = 1
z0se
l
z0se
l
RL = 1
z0sh
l
z0se = 1.5963 = (L1 = L3)
z0sh = 0.9118 = (= l/C2)
l = λ/8 at ωc = 1
(b)
R = 1
l
z0se
l
l
z0se
l
z0 = 1
z0 = 1
RL = 1
z0sh
l
(c)
R = 1
l
l
RL = 1
Z1 = 2.596
Z1 = 2.596
Z2 =
1.626
l
z0sh
l
Z2 =
1.626
l
(d)

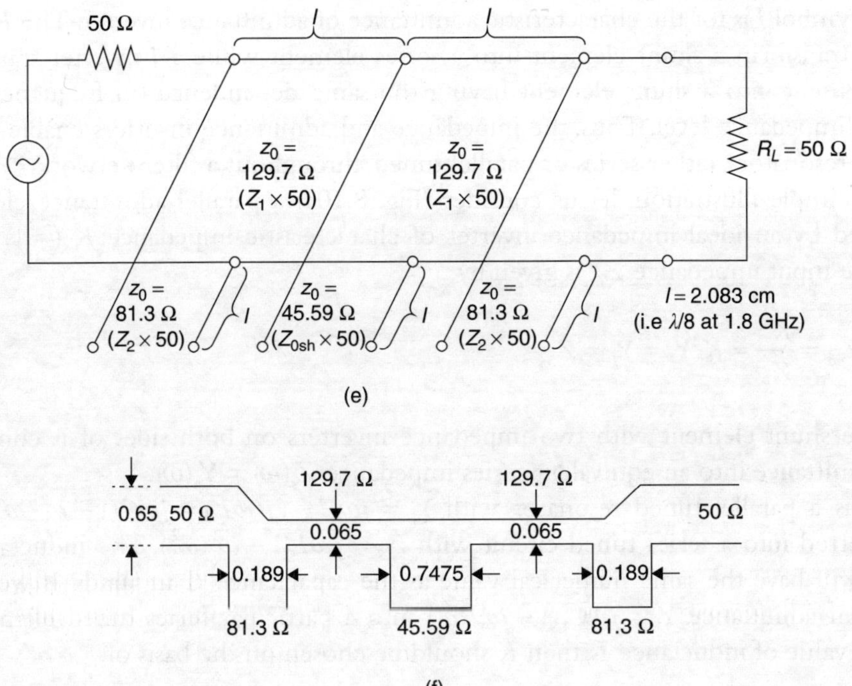

Fig. 8.9 (a) Lumped-element low-pass filter prototype (b) Richard's transformation applied to convert *L* and *C* to series and shunt stubs (c) Unit elements added to the ends of the filter (d) Application of Kuroda's identity (e) After impedance and frequency scaling (f) Microstrip implementation of the filter (all dimensions in mm).

Fig. 8.9e, with the microstrip layout in Fig. 8.9f. It may be observed that lines with smaller characteristic impedances have larger width as the characteristic impedance of a microstrip line is inversely proportional to the strip width (Fig. 3.21a).

8.4.3 Impedance and Admittance Inverters

At microwave frequencies, it is desirable from a fabrication point of view, to use either all series or all shunt elements to implement filter networks. If a combination of both series and shunt elements occur (as in Example 8.1), then it is advisable to convert the series elements to shunt elements and vice-versa. Kuroda's identity can be used to realise such conversions; however, another practical approach is to use impedance (K) or admittance (J) inverters. Such inverters are especially useful for the design of band-pass and band-stop filters at microwave frequency.

The impedance and admittance inverters, in essence, are ideal quarter-wave transformers as a load connected at one end of such a transformer is seen to be inverted with respect to the characteristic impedance squared at the input (Eq. 3.34). There is no basic difference between the inverting properties of impedance/admittance inverters; the only distinction is that the symbol K is used to denote the characteristic impedance of impedance inverter and the symbol J is for the characteristic admittance of admittance inverter. The K inverter helps to transform a shunt element into a series element while a J inverter transforms a series element into a shunt element having the same dependence on frequency and an arbitrary impedance level. Thus, the impedance and admittance inverters enable us to use identical resonators, either series or parallel tuned, throughout a filter network.

For a simple illustration, let us consider (Fig. 8.10a) a parallel admittance element Y_p connected by an ideal impedance inverter of characteristic impedance K (= 1) on both sides. The input impedance Z_{in} is given by:

$$Z_{in} = \frac{K^2}{Z_P} = K^2 Y_P = Y_P = Z_S \tag{8.22}$$

Thus, the shunt element with two impedance inverters on both sides of it converts the shunt admittance into an equivalent series impedance $Z_s(\omega) = Y_p(\omega)$.

If Y_p is a parallel tuned resonator with $Y_P = j\omega C - (j/\omega L) = j\omega C(1 - \omega_0^2/\omega^2)$, it will be converted into a series tuned circuit with $Z_S = j\omega L(1 - \omega_0^2/\omega^2)$. The inductance L in henries will have the same numerical value as the capacitance C in farads. If we want to convert an admittance $Y_1 = j\omega C_1(1 - \omega_0^2/\omega^2)$ into a particular series tuned element with arbitrary value of inductance L, then K should be chosen on the basis of:

$$K^2 j\omega C_1\left(1 - \frac{\omega_0^2}{\omega^2}\right) = j\omega L\left(1 - \frac{\omega_0^2}{\omega^2}\right) \tag{8.23}$$

or $$K = \sqrt{\frac{L}{C_1}}$$

Thus, we can transform a shunt element into a series element having the same dependence on frequency but with an arbitrary impedance level.

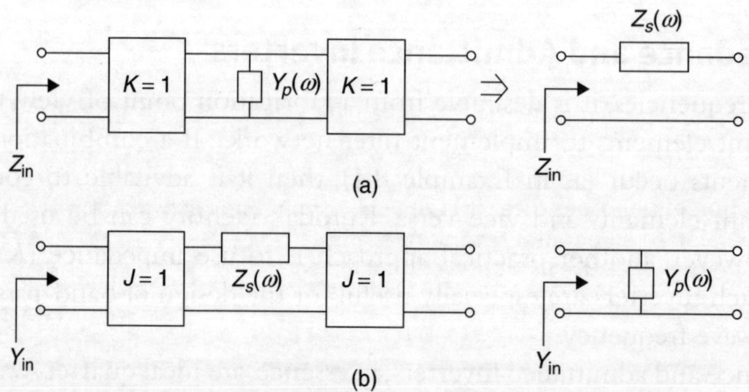

Fig. 8.10 (a) Impedance (K) inverter (b) Admittance (J) inverter.

In Fig. 8.10b is shown a series element $Z_s(\omega)$ with the admittance inverter having a characteristic admittance J (= 1) connected on both sides. The input admittance will be:

$$Y_{\text{in}} = \frac{J^2}{Y_S} = J^2 Z_S = Y_P \tag{8.24}$$

Thus, the series element with the two admittance inverters on either side will appear from the input terminals to be equivalent to a shunt admittance.

We can use inverters to convert the circuit shown in Fig. 8.11a into one with two identical series tuned resonators (using K inverters) or one with two identical parallel tuned resonators (using J inverters).

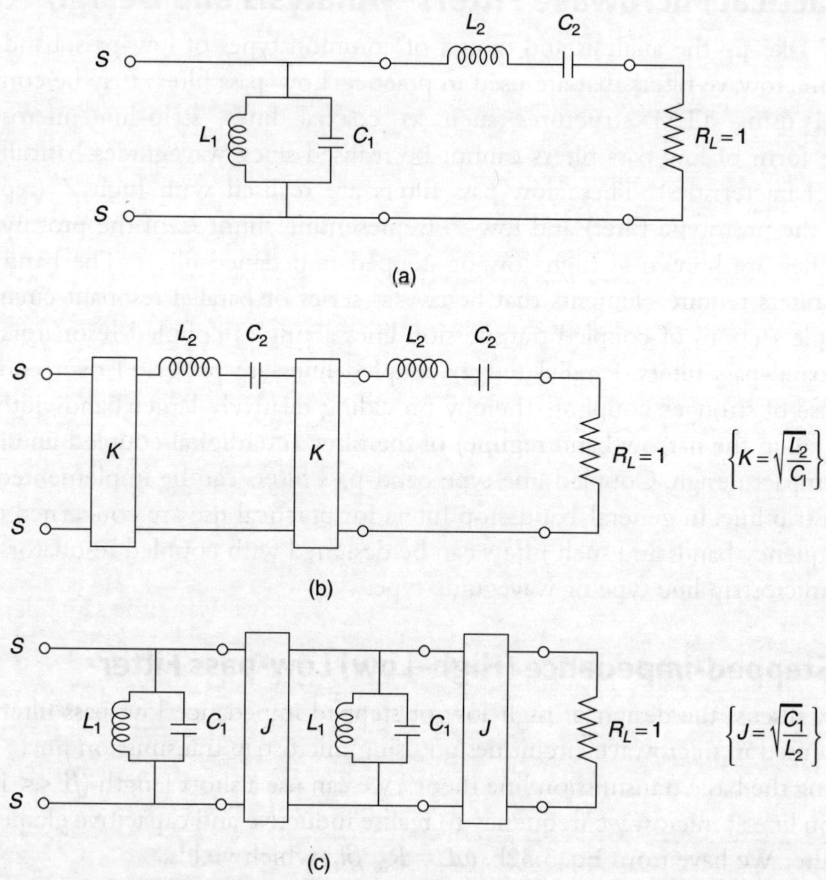

Fig. 8.11 (a) A section of band-pas filter (b) Conversion to all series resonator by *K*-inverter (c) Conversion to all parallel resonator by *J*-inverter.

For Fig. 8.11b we choose K so that:

$$K^2 j\omega C_1\left(1-\frac{\omega_0^2}{\omega^2}\right) = j\omega L_2\left(1-\frac{\omega_0^2}{\omega^2}\right) \tag{8.25}$$

That is, $K = \sqrt{\dfrac{L_2}{C_1}}$

While for Fig. 8.11c, J is chosen so that:

$$J^2 j\omega L_2\left(1-\frac{\omega_0^2}{\omega^2}\right)=j\omega C_1\left(1-\frac{\omega_0^2}{\omega^2}\right) \tag{8.26}$$

That is, $J=\sqrt{\dfrac{C_1}{L_2}}$

8.5 Practical Microwave Filters—Analysis and Design

Let us now take-up the analysis and design of common types of low-pass, band-pass and band-stop microwave filters that are used in practice. Low-pass filters may be conveniently constructed using TEM structures such as coaxial lines, strip-line/microstrip line (waveguide form of low pass filters cannot be realised since waveguides basically exhibit high pass characteristics). These low-pass filters are realised with high-Z (representing series L of the prototype filter) and low-Z (representing shunt C of the prototype filter). Popularly, they are known as high–low or stepped-impedance filters. The band-pass and band-stop filters require elements that behave as series or parallel resonant circuits (Table 8.3). Multiple sections of coupled transmission lines acting as coupled resonators are used to design band-pass filters. Parallel (edge) coupled lines are preferred over end coupled lines because of stronger coupling, thereby providing relatively larger bandwidth (though it will still fall in the narrow band regime) of the filter. Interdigital coupled lines, however, provide compact design. Coupled line type band-pass filters can be implemented in strip-line/microstrip line. In general, band-stop filters for practical use are concerned only with narrow frequency bands and such filters can be designed with coupled resonators either of strip-line/microstrip line type or waveguide type.

8.5.1 Stepped-impedance (High–Low) Low-pass Filter

Let us now discuss the design of high-low or stepped impedance low-pass filter which is commonly used in microwave circuit design using microstrip transmission line.

Following the basic transmission line theory, we can use a short length ($\beta l << 1$) loss-less transmission line at microwave frequency to realise inductive and capacitive elements of the low-pass filter. We have from Eq. 3.32b, $\omega L = R_{0L}\,\beta l_L$, which yields:

$$l_L=\frac{f_c L\lambda}{R_{0L}} \quad \text{or} \quad R_{0L}=\frac{f_c L\lambda}{l_L} \tag{8.27a}$$

While, from Eq. 3.31b, $1/\omega C = R_{0C}/\beta l_C$, which yields:

$$l_C=f_c C\lambda R_{0C} \quad \text{or} \quad R_{0C}=\frac{l_C}{f_c C\lambda} \tag{8.27b}$$

where f_c is the cut-off frequency of the filter. It may be noted that L and C in Eqs 8.27a and 8.27b are the total inductance and capacitance of the line and not the per unit length of the line. The line providing inductive impedance will represent a line having high characteristic impedance while the line providing capacitive impedance will represent a line having small characteristic impedance.

EXAMPLE 8.2

Design a stepped impedance low-pass filter having a Tchebyscheff response with a 0.3 dB ripple and a cut-off frequency of 2.4 GHz. It is desired to have at least 35 dB insertion loss at 4.0 GHz in the stop band. Realise the filter in microstrip transmission line having a dielectric substrate of thickness 0.65 mm with $\varepsilon_r = 10$ at 2.4 GHz. Assume 50 Ω terminations at both load and source ends and the highest practical line impedance are 120 Ω and the lowest being 25 Ω.

Solution

This problem will be solved both by analytical synthesis process and also using Electronic Design Automation (EDA) simulation tool.

Method I: Analytical synthesis process

For the given ripple magnitude, 0.3 dB $(A_{pm}) = 10 \log_{10}(1 + A_p^2)$, thus we have $A_p^2 = 0.07152$.

The power-loss ratio of an actual low-pass filter having equal-ripple, that is, Tchebyscheff response is given by:

$$P_{LR} = 1 + A_p^2 T_n^2(\Omega') \quad \text{with} \quad \Omega' = \frac{\omega_s}{\omega_c} \text{ (Fig. 8.5b), (since } \Omega_p = 1)$$

Now, $T_n(\Omega') = \cosh(n \cosh^{-1} \Omega')$ where $\Omega' = \omega_s/\omega_c = 4/2.4 = 1.67 > 1$ (as per the given data).

Substituting the respective values ($P_{LRs} = 35$ dB and $\Omega' = 1.67$) in the expression for P_{LR} and solving for n, we get $n = 5.48$, which can be taken approximately as 5.

[**Note:** $[1 + 0.0715 \cosh^2(n \cosh^{-1} 1.67)] = 10^{35/10}$ or $[\cosh(n \cosh^{-1} 1.67)] = [(10^{3.5} - 1)/0.715]^{1/2} = 210.27$ or $(n \cosh^{-1} 1.67)] = \cosh^{-1}(210.27) = 6.04$, whence $n = 6.04/\cosh^{-1} 1.67 = 5.48 \approx 5$]. Therefore, a five-section low-pass filter is required.

Since n is odd, $g_{n+1} = g_6 = g_0 = 1.0$ while $g_1 = 1.4810 = g_5$, $g_2 = 1.299 = g_4$, $g_3 = 2.3090$ (which follows from Table 8.1c). The low-pass prototype filter is shown in Fig. 8.12a.

Next we use impedance and frequency scaling to find the inductor and capacitor values of the prototype (referring to Eqs 8.12a and 8.12b), where $R_L = R_s = Z_0$ (say) = 50 Ω and $\omega_c = 2\pi \times 2.4$ GHz, when $C_1 = C_5 = 1.965$ pF, $C_3 = 3.0639$ pF and $L_2 = L_4 = 4.3093$ nH.

[**Note:** $\Omega_p = 1$ and $I_t = Z_0/g_0 = Z_0$ since $g_0 = 1$].

These lumped elements can be realised with sections of microstrip lines with length and width having corresponding suffixes as shown in Fig. 8.12b. The capacitive elements will be the ones with the lowest line impedance, $Z_{0C} = 25$ Ω, while the inductive elements

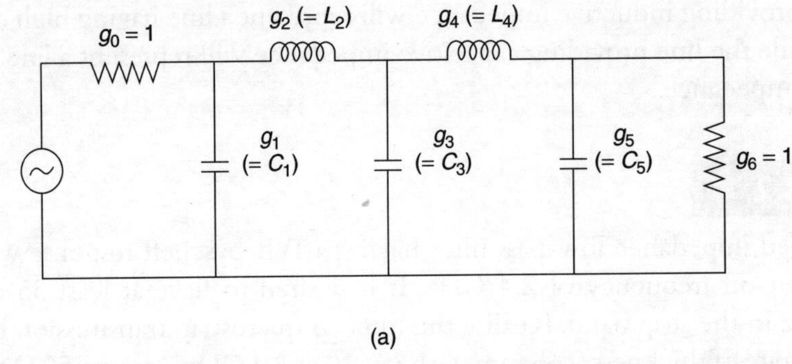

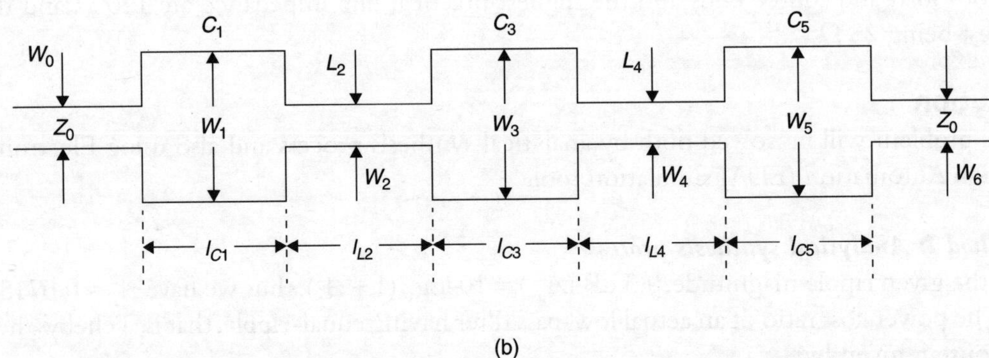

Fig. 8.12 (a) Low-pass prototype corresponding to Example 8.2 (b) Microstrip implementation scheme.

will have the highest line impedance, $Z_{0L} = 120\ \Omega$, and the impedance on the load and source side being $Z_0 = 50\ \Omega$. As the filter will be constituted with low and high impedance sections, they are also known as low–high–low filters. Note that the low impedance section will have a larger strip width while the high impedance section will have smaller strip width (since, for a microstrip line, $Z_0\ \alpha\ 1/W$). From Fig. 3.21, we clearly find that for $Z_{0C} = 25\ \Omega$, $W/h > 1$ and for $Z_{0L} = 120\ \Omega$, $W/h < 1$.

The length of each section will be calculated using the following formulae:

$$L_{C1} = l_{C5} = f_c \lambda_{gC} R_{0C} C_1 \quad \text{and} \quad l_{C3} = f_c \lambda_{gC} R_{0C} C_3. \text{ While, } l_{L2} = l_{L4} = f_c L_2 \frac{\lambda g_L}{R_{0L}}.$$

In the above, the value of λ_g will be calculated from $\lambda_g = \lambda_0 / \sqrt{\varepsilon_{\text{reff}}}$; where (Eq. 3.41):

$$\varepsilon_{\text{reff}} = \frac{\varepsilon_r + 1}{2} + \frac{\varepsilon_r - 1}{2} \left\{ \left(1 + 12\frac{h}{W}\right)^{-1/2} + 0.04\left(1 - \frac{W}{h}\right)^2 \right\} \quad \text{for} \quad \left(\frac{W}{h}\right) \le 1$$

$$= \frac{\varepsilon_r + 1}{2} + \frac{\varepsilon_r - 1}{2} \left(1 + 12\frac{h}{W}\right)^{-1/2} \quad \text{for} \quad \left(\frac{W}{h}\right) > 1$$

The values for W/h can be calculated from Eq. 3.44:

$$\frac{W}{h} = \frac{8e^A}{e^{2A} - 2} \quad \text{for} \quad \left(\frac{W}{h}\right) \le 2$$

$$= \frac{2}{\pi}\left[B - 1 - \ln(2B-1) + \frac{\varepsilon_r - 1}{2\varepsilon_r}\left\{\ln(B-1) + 0.39 - \frac{0.61}{\varepsilon_r}\right\}\right] \quad \text{for} \quad \left(\frac{W}{h}\right) \ge 2$$

$$\text{where} \quad A = \frac{Z_0}{60}\sqrt{\frac{\varepsilon_r + 1}{2}} + \frac{\varepsilon_r - 1}{\varepsilon_r + 1}\left(0.23 + \frac{0.11}{\varepsilon_r}\right) \quad \text{and} \quad B = \frac{377\pi}{2Z_0\sqrt{\varepsilon_r}}$$

The expression for Z_0 for microstrip line being given by Eq. 3.42:

$$Z_0 = \frac{60}{\sqrt{\varepsilon_{\text{reff}}}}\ln\left(\frac{8h}{W} + \frac{W}{4h}\right) \quad \text{for} \quad \left(\frac{W}{h}\right) \le 1$$

$$= \frac{120\pi}{\sqrt{\varepsilon_{\text{reff}}}\left\{\left(\frac{W}{h} + 1.393\right) + 0.667\ln\left(\frac{W}{h} + 1.444\right)\right\}} \quad \text{for} \left(\frac{W}{h}\right) > 1$$

Necessary calculations (either done analytically or by writing MATLAB program) yield:

$$W_1 = W_5 = W_3 = 2.00347 \text{ mm}, \; W_2 = W_4 = 0.0392 \text{ mm}$$

While, $W_0 = W_6 = 0.6216$ mm (Fig. 3.12a, with $Z_0 = 50\ \Omega$ and $\varepsilon_r = 10$)

The cut-off frequency of the filter being at 2.4 GHz, so $\lambda_0 = 12.5$ cm, when:

$$\varepsilon_{\text{reff}} = 5.977 \quad \text{for} \left(\frac{W}{h}\right) \le 1 \quad \text{and} \quad \varepsilon_{\text{reff}} = 7.534 \quad \text{for} \left(\frac{W}{h}\right) > 1.$$

Thus, $\lambda_{gL} = 5.1128$ cm and $\lambda_{gC} = 4.5539$ cm.

Whence, $l_{C1} = l_{C5} = 6.028$ mm, $l_{L2} = l_{L4} = 4.406$ mm and $l_{C3} = 9.399$ mm.

[The MATLAB program is given in Appendix III].

Method II: Using the EDA simulation tool

The characteristics of a stepped impedance low-pass filter obtained with the Electronic Design Automation (EDA) simulation tool is depicted in Fig. 8.13a.

The curve 1 is obtained with ideal (loss-less having infinite Q value) lumped components as evaluated from analytical synthesis. However, all practical circuit elements will have a finite loss (finite Q value). Curve 2 is obtained with a Q value of 169 (say) that shows some finite insertion loss in the pass-band of the filter, but the filter response otherwise is similar to the

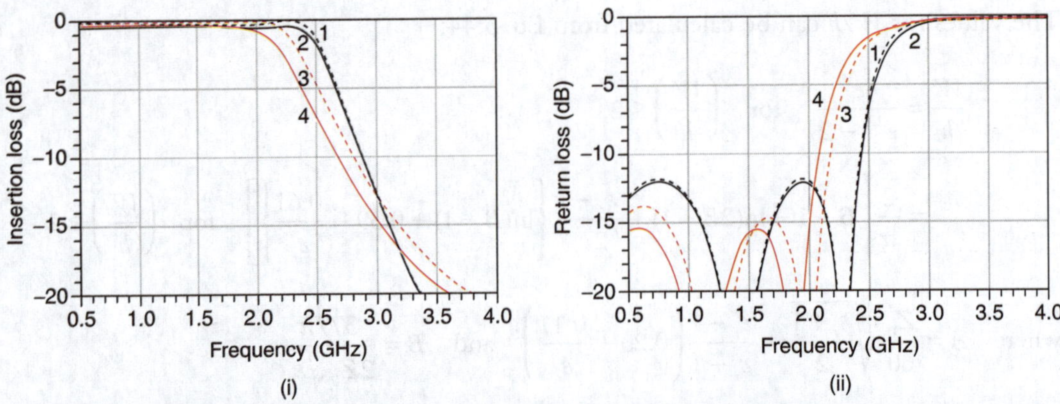

Fig. 8.13(a) Stepped impedance low-pass filter (LPF) characteristics obtained with EDA simulation tool (i) Transmission characteristics (ii) Reflection characteristics.

ideal filter since no parasitic effect has been considered so far. The distributed realisation using the microstrip configuration may introduce additional losses arising from surface wave propagation in the quasi-TEM structure (as indicated by curve 3 in Fig. 8.13a, obtained with a circuit simulator). The complete electromagnetic (EM) simulation (using method of moments) takes care of extra parasitics due to the cross-coupling between elements along with above mentioned non-idealities. This is manifested in the form of a reduced bandwidth and changed filter skirt in the result (curve 4 in Fig. 8.13a). The return loss is seen to be nearly –15 dB down from 0 dB reference (in fact 10 dB down is fine, however, for a good design –20 dB down is desirable, which is done by optimisation). The screenshot

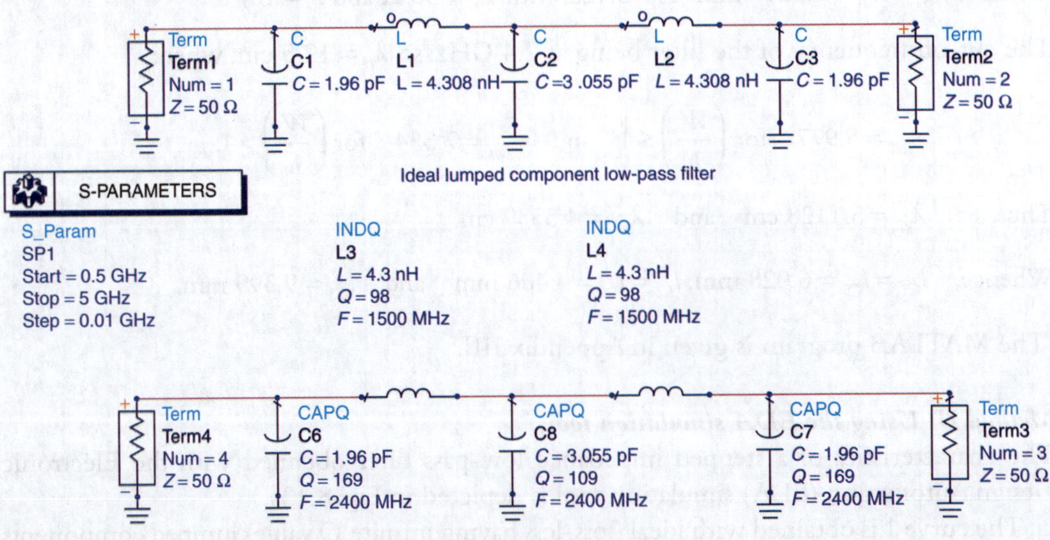

Fig. 8.13(b) Screenshot of the schematic for ideal lumped component and practical lumped component LPF.

of the schematic for an ideal lumped component and practical lumped component LPF is shown in Fig. 8.13b; microstrip realisation with a circuit simulator is shown in Fig. 8.13c, while the layout of electromagnetic (EM) simulation of the stepped impedance LPF in EDA simulation tool is shown in Fig. 8.13d.

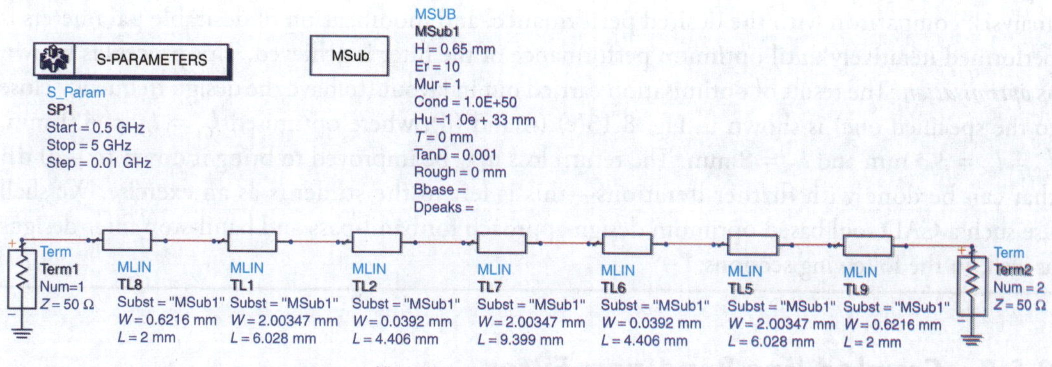

Fig. 8.13(c) Screenshot of microstrip realisation (with circuit simulator) of stepped impedance LPF.

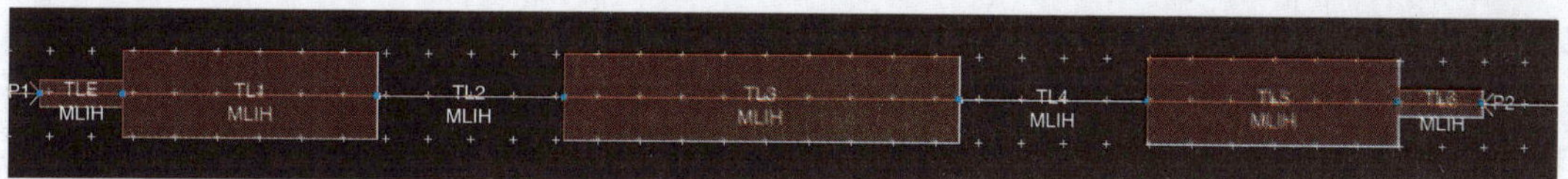

Fig. 8.13 (d) Layout of EM simulation for stepped impedance LPF corresponding to Example 8.2.

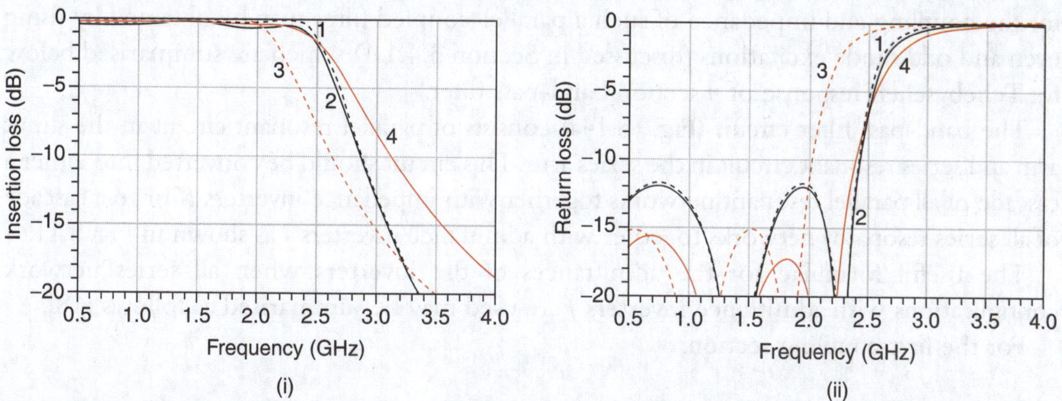

Fig. 8.13(e) Stepped impedance low-pass filter (LPF) characteristics obtained after optimisation with the EDA simulation tool (i) Transmission characteristics (ii) Reflection characteristics.

Note: A few observations may be made with respect to the design of the stepped impedance filter either with the analytical synthesis method or using the computer-aided-design CAD based EDA simulation tool. As per the problem, the design data demands $W_2 = W_4 = 0.0392$ mm which is a critically small dimension from a fabrication point of view. Also, microstrip realisation either with a circuit simulator or an EM simulation shows that the cut-off frequency is shifted from the design specification. For design optimisation of a particular filter, one often starts with a given set of filter specifications and an initial design is realised through analytical synthesis. Filter performance is then evaluated with the initial design data obtained above, either with a circuit simulator or with EM simulation. If the results fail to satisfy the desired specifications/realisable dimensions, the desirable (optimisation) parameters of the filter are altered in a systematic manner. The sequence of filter analysis, comparison with the desired performance, and modification of desirable parameters is performed iteratively until optimum performance of the filter is achieved. This process is known as *optimisation*. The result of optimisation carried out in layout (to have the design frequency close to the specified one) is shown in Fig. 8.13(e) (i) and (ii); where optimised $l_{C1} = l_{C5} = 4.9$ mm, $l_{L2} = l_{L4} = 3.3$ mm and $l_{C3} = 8$ mm. The return loss may be improved to bring it down to –20 dB that can be done with further iterations—this is left to the students as an exercise. We shall use such a CAD tool based optimum design approach for band-pass and band-stop filter designs as well, in the following sections.

8.5.2 Coupled-line Band-pass Filter

The band-pass filter circuit can be derived from low-pass prototype (Fig. 8.3a and Table 8.3) which is shown in Fig. 18.14a. It consists of series and parallel resonant circuits. Thus, the basic design unit for such a filter will be resonators which need to be multiple coupled structures. When strip-line/microstrip line is used for the design, the resonators can be realised with a parallel-coupled microstrip line as shown in Fig. 8.14b [8]. The resonator elements are half-wavelength long at the centre frequency and open at both ends and maximum coupling occurs between the resonators when the overlapping regions between the resonators, that is, l_1, l_2, l_3, ... l_n is quarter-wave ($= \lambda_g/4$) long. The detailed derivation for the coupling and impedance of such a parallel-coupled filter may be obtained by using even and odd mode excitations (discussed in Section 3.4.1.1) which are summarised below for Tchebyscheff response of n section band-pass filter.

The band-pass filter circuit (Fig. 18.14a) consists of parallel resonant circuit in the shunt arm and series resonant circuit in the series arm. This circuit should be converted into either a cascade of all parallel resonant networks together with impedance inverters K or to a cascade of all series resonator networks together with admittance inverters J as shown in Fig. 8.11.

The useful formulae for the admittances of the inverters when all series network configurations with admittance inverters J are used may be summarised as follows [18]:

For the first coupling section:

$$\frac{J_{10}}{Y_0} = \left(\frac{\pi\Delta}{2g_0g_1}\right)^{1/2} \tag{8.28}$$

For the final coupling section:

$$\frac{J_{n+1,n}}{Y_0} = \left(\frac{\pi\Delta}{2g_ng_{n+1}}\right)^{1/2} \tag{8.29}$$

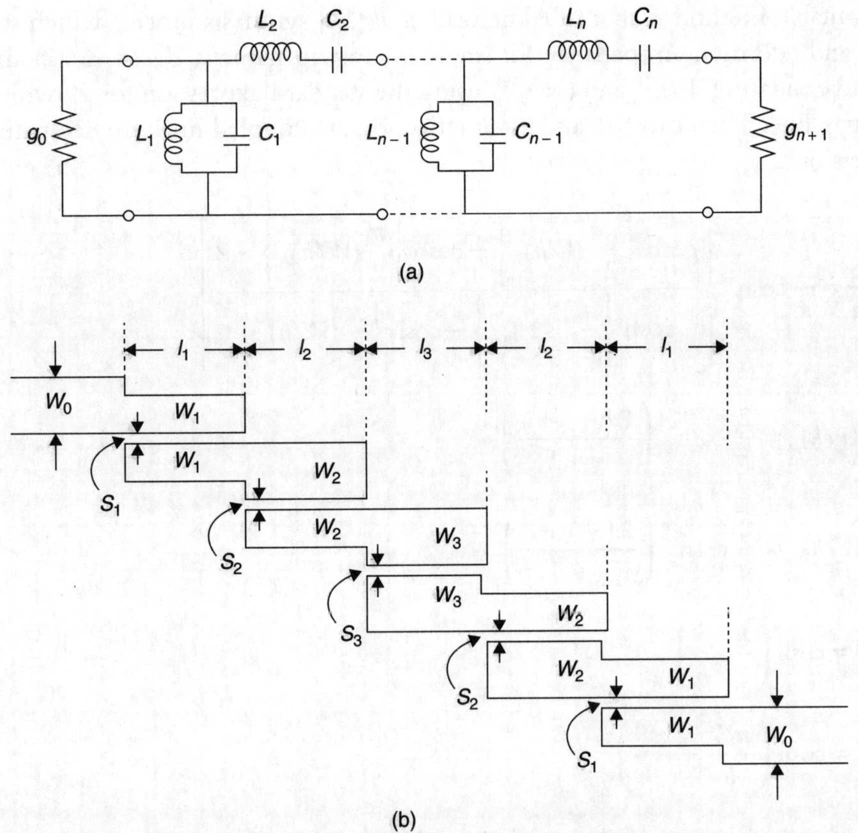

Fig. 8.14 (a) Band-pass filter circuit derived from low-pass prototype (b) General microstrip configuration for a 5-section parallel-coupled band-pass filter.

For the intermediate coupling section:

$$\frac{J_{k+1,k}}{Y_0} = \left(\frac{\pi\Delta}{2\omega_c' g_n \sqrt{g_k g_{k+1}}} \right)^{1/2}, \quad k = 1 \text{ to } (n-1) \tag{8.30}$$

where $\Delta = (\omega_{c2} - \omega_{c1})/\omega_0$ is the fractional bandwidth.

In the microstrip filter, the line dimensions are determined from the knowledge of even and odd mode coupled line impedances which are given in terms of admittances of the inverters.

$$(Z_{0e})_{k+1,k} = Z_0(1 + J_{k+1,k}Z_0 + J_{k+1,k}^2 Z_0^2) \tag{8.31a}$$

and $$(Z_{0o})_{k+1,k} = Z_0(1 - J_{k+1,k}Z_0 + J_{k+1,k}^2 Z_0^2) \tag{8.31b}$$

where Z_0 is the characteristic impedance of the input/output lines of the filter.

Several workers described the variation of Z_{0o} and Z_{0e} as functions of W/h, S/h and ε_r to evaluate the microstrip widths (W) and separations (S) of the parallel coupled resonant lines. A suitable method is that of Akhtarzad *et al* [12] synthesis process which starts with the even and odd mode impedance for single microstrip, namely, Z_{0es} (= $Z_{0e}/2$) and Z_{0os} (= $Z_{0o}/2$) and evaluates $(W/h)_{es}$ and $(W/h)_{os}$ using the standard expression for Z_0 with W/h for a microstrip line. Then the S/h and W/h ratios for the coupled lines are determined from the expressions:

$$\frac{S}{h} = \frac{2}{\pi}\cosh^{-1}\left[\frac{\cosh\left\{\frac{\pi}{2}(W/h)_{es}\right\} + \cosh\left\{\frac{\pi}{2}(W/h)_{os}\right\} - 2}{\cosh\left\{\frac{\pi}{2}(W/h)_{os}\right\} - \cosh\left\{\frac{\pi}{2}(W/h)_{es}\right\}}\right] \tag{8.32a}$$

$$(W/h)_{se} = \frac{2}{\pi}\cosh^{-1}\left(\frac{2d - g + 1}{g + 1}\right) \tag{8.32b}$$

$$(W/h)_{so} = \frac{2}{\pi}\cosh^{-1}\left(\frac{2d - g - 1}{g - 1}\right) \tag{8.32c}$$

where $$g = \cosh\left(\frac{\pi}{2}\cdot\frac{S}{h}\right) \tag{8.32d}$$

and $$d = \cosh\left(\frac{\pi W}{h} + \frac{\pi S}{2h}\right) \tag{8.32e}$$

Using Eq. 8.32a, we can calculate S/h when $(W/h)_{es}$ and $(W/h)_{os}$ are known. Finally, using either Eq. 8.32b or 8.32c, we can determine W/h.

The physical lengths of the resonators and the coupled regions are respectively half and quarter wavelengths long. The coupled region length is determined in terms of the mid-band wavelength $\lambda_{gm} = (\lambda_{go} + \lambda_{ge})/2$, where λ_{go} and λ_{ge} are the odd-mode and even-mode guide wavelengths respectively, being calculated from:

$$\lambda_{go} = \frac{300}{f\,(\text{GHz})}\frac{Z_{0o}}{Z_{0o\,\text{air}}}\ \text{mm}, \quad \lambda_{ge} = \frac{300}{f\,(\text{GHz})}\frac{Z_{0e}}{Z_{0e\,\text{air}}}\ \text{mm} \tag{8.33}$$

where, $Z_{0e\,\text{air}}$ and $Z_{0o\,\text{air}}$ are the even and odd mode characteristic impedances of the air-spaced coupled lines which can be calculated in terms of the corresponding characteristic impedance of the single line, namely, $Z_{0es\,\text{air}}$, $Z_{0os\,\text{air}}$ using $Z_{0e\,\text{air}} = 2Z_{0es\,\text{air}}$ and $Z_{0o\,\text{air}} = 2Z_{0os\,\text{air}}$. The values for $Z_{0es\,\text{air}}$, $Z_{0os\,\text{air}}$ can be calculated by using the standard formula for Z_0 vs W/h (with $\varepsilon_r = 1$) (Fig. 3.21a).

EXAMPLE 8.3

Design a parallel-coupled microstrip band-pass filter having Tchebyscheff characteristics with pass-band ripple of 0.2 dB between the pass-band edge frequencies of 2.27 GHz and

2.53 GHz. It is desired to have at least 30 dB insertion loss at 2.65 GHz in the stop-band. The filter impedance is 50 Ω. Consider a substrate of dielectric constant $\varepsilon_r = 10$ and a height $h = 0.65$ mm.

Solution

This problem will be solved both by analytical synthesis process and also by using the EDA simulation tool.

Method I: Analytical synthesis process

Step 1: At first, we determine the number of sections for the filter from the specified attenuation characteristics. The centre frequency: $f_0 = \sqrt{f_{c1} f_{c2}} = \sqrt{2.27 \times 2.53} = 2.4$ GHz and the fractional bandwidth $\Delta = (f_{c2} - f_{c1})/f_0 = (2.53 - 2.27)/2.4 = 0.108$.

With the given equal-ripple magnitude (A_{pm}) as 10 $\log_{10}(1 + A_p^2) = 0.2$ dB, we have $A_p^2 = 0.047$. The power-loss ratio of the band-pass filter having equal-ripple, that is, Tchebyscheff response, is given by:

$$P_{LR} = 1 + A_p^2 T_n^2(\Omega') \text{ where } T_n(x) = \cosh\,(n \cosh^{-1} \Omega')$$

$$\text{Here, } \Omega' = \frac{\Omega_p \omega_0}{\omega_{c2} - \omega_{c1}} \left(\frac{\omega_s}{\omega_0} - \frac{\omega_0}{\omega_s} \right) = \frac{2.4}{2.53 - 2.27} \left(\frac{2.65}{2.4} - \frac{2.4}{2.65} \right) = 1.83 > 1 \text{ (with } \Omega_p = 1)$$

Substituting the respective values ($P_{LRs} = 30$ dB and $\Omega' = 1.83$) in the expression for P_{LR} and solving for n, we get $n = 4.679$, which can be taken approximately as 5. Therefore, a five-section band-pass filter is required.

Step 2: Now we need to determine the values of the prototype elements to realise the specifications.

For 0.2 dB ripple with $n = 5$, the low-pass prototype elements are: $g_1 = 1.3394 = g_5, g_2 = 1.337 = g_4$ and $g_3 = 2.166$ (Table 8.1b), while $g_0 = g_6 = 1.0$ (since n is odd).

Step 3: Next, we determine the inverter values (say we use the J inverter, that is, with cascaded series resonators). As per symmetry, for the first and final coupling we have:

$$\frac{J_{10}}{Y_0} = \frac{J_{6,5}}{Y_0} = \sqrt{\frac{\pi\Delta}{2 g_0 g_1}} = \sqrt{\frac{\pi\Delta}{2 g_5 g_6}} = \sqrt{\frac{\pi \times 0.108}{2 \times 1 \times 1.3394}} = 0.356$$

For the intermediate coupling section ($k = 1$ to 2)

$$\frac{J_{2,1}}{Y_0} = \frac{\pi\Delta}{2\sqrt{g_1 g_2}} = \frac{J_{5,4}}{Y_0} = \frac{\pi \times 0.180}{2\sqrt{(1.3394 \times 1.337)}} = 0.127$$

$$\frac{J_{3,2}}{Y_0} = \frac{\pi\Delta}{2\sqrt{g_2 g_3}} = \frac{J_{4,3}}{Y_0} = \frac{\pi \times 0.180}{2\sqrt{(1.337 \times 2.166)}} = 0.0997$$

Step 4: Then we have to obtain the even and odd mode coupled line characteristic impedances Z_{0e} and Z_{0o} (refer to Eqs 8.31a and 8.31b).

Looking into the symmetry, only $k = 0$ to 2 are required to be computed, when $k = 3$ and $k = 4$, these will respectively take the values of $k = 1$ and $k = 0$. Following Eqs 8.31a and 8.31b we have:

For $k = 0$, $(Z_{0e})_{1,0} = Z_0(1 + J_{1,0}Z_0 + J_{1,0}^2 Z_0^2) = 50\ \{1 + 0.356 + (0.356)^2\} = 74.14\ \Omega$

$$(Z_{0o})_{1,0} = Z_0(1 - J_{1,0}Z_0 + J_{1,0}^2 Z_0^2) = 50\ \{1 - 0.356 + (0.356)^2\} = 38.54\ \Omega$$

For $k = 1$, $(Z_{0e})_{2,1} = Z_0(1 + J_{2,1}Z_0 + J_{2,1}^2 Z_0^2) = 50\ \{1 + 0.127 + (0.127)^2\} = 57.16\ \Omega$

$$(Z_{0o})_{2,1} = Z_0(1 - J_{2,1}Z_0 + J_{2,1}^2 Z_0^2) = 50\ \{1 - 0.127 + (0.127)^2\} = 44.46\ \Omega$$

For $k = 2$, $(Z_{0e})_{3,2} = Z_0(1 + J_{3,2}Z_0 + J_{3,2}^2 Z_0^2) = 50\ \{1 + 0.0997 + (0.0997)^2\} = 55.48\ \Omega$

$$(Z_{0o})_{3,2} = Z_0(1 + J_{3,2}Z_0 + J_{3,2}^2 Z_0^2) = 50\ \{1 - 0.0997 + (0.0997)^2\} = 45.51\ \Omega$$

Step 5: Next, we need to determine the microstrip widths (W) and separations (S) of the parallel coupled resonant lines. To do this, we use the Akhtarzad et al synthesis approach when the even and odd mode impedances for a single microstrip for different values of k are given by:

For $k = 0$, $(Z_{0es})_{1,0} = \dfrac{(Z_{0e})_{1,0}}{2} = \dfrac{74.14}{2} = 37.07\ \Omega$

$$(Z_{0os})_{1,0} = \frac{(Z_{0o})_{1,0}}{2} = \frac{38.54}{2} = 19.27\ \Omega$$

For $k = 1$, $(Z_{0es})_{2,1} = \dfrac{57.1}{2} = 28.58\ \Omega$, $(Z_{0os})_{2,1} = \dfrac{44.46}{2} = 22.23\ \Omega$

For $k = 2$, $(Z_{0es})_{3,2} = \dfrac{55.48}{2} = 27.74\ \Omega$, $(Z_{0os})_{3,2} = \dfrac{45.51}{2} = 22.75\ \Omega$

Then, using Z_0 (in terms of W/h) given by Eq. 3.42, and inserting the value of $\varepsilon_{\text{reff}}$ in it from Eq. 3.41, we get:

$$Z_0 = \frac{60\ln\left(\dfrac{8h}{W} + \dfrac{W}{4h}\right)}{\left[\dfrac{\varepsilon_r + 1}{2} + \dfrac{\varepsilon_r - 1}{2}\left\{\left(1 + \dfrac{12h}{W}\right)^{-1/2} + 0.04\left(1 - \dfrac{W}{h}\right)^2\right\}\right]^{1/2}} \quad \text{for } \left(\frac{W}{h}\right) \le 1$$

$$= \frac{120\pi}{\left[\dfrac{\varepsilon_r + 1}{2} + \dfrac{\varepsilon_r - 1}{2}\left(1 + \dfrac{12h}{W}\right)^{-1/2}\right]^{1/2} \left\{\dfrac{W}{h} + 1.4 + 0.667\ln\left(\dfrac{W}{h} + 1.444\right)\right\}} \quad \text{for } \left(\frac{W}{h}\right) > 1$$

Now, for every Z_{0es} and Z_{0os}, we calculate the corresponding $(W/h)_{es}$ and $(W/h)_{os}$ from the above equation; then we get:

$$\left\{\left(\frac{W}{h}\right)_{es}\right\}_{1,0} = 1.67, \left\{\left(\frac{W}{h}\right)_{es}\right\}_{2,1} = 2.512, \left\{\left(\frac{W}{h}\right)_{es}\right\}_{3,2} = 2.63$$

$$\left\{\left(\frac{W}{h}\right)_{os}\right\}_{1,0} = 4.41, \left\{\left(\frac{W}{h}\right)_{os}\right\}_{2,1} = 3.66, \left\{\left(\frac{W}{h}\right)_{os}\right\}_{3,2} = 3.52$$

Next, the corresponding S/h values can be calculated from Eq. 8.32a when we have:

For $\quad k = 0, \dfrac{S}{h} = 0.125, k = 1, \dfrac{S}{h} = 0.422, k = 2, \dfrac{S}{h} = 0.500$

Then, the W/h ratio for the coupled line can be calculated from any of the two relations given in Eqs 8.32b and 8.32c which yields:

For $\quad k = 0, \dfrac{W}{h} = 0.945, k = 1, \dfrac{W}{h} = 1.157, k = 2, \dfrac{W}{h} = 1.17$

Step 6: Finally, we calculate (following Eq. 8.33) the physical lengths of the resonators and the coupled regions which are respectively half and quarter wavelengths long. The final design data for the resonators is given in Table 8.4 (***Note***: $W_0 = 0.65$ mm, with $Z_0 = 50\ \Omega$ and $\varepsilon_r = 10$, Fig. 3.21a). These data values given in the table may be obtained either with analytical calculations or may be evaluated by writing a MATLAB program (for MATLAB program, refer to Appendix III).

Table 8.4 Design data of the resonators

Coupled lines	W/h	W (mm)	S/h	S (mm)	$l = \lambda_{gm}/4$ (mm)
$k = 0$	0.945	0.61425 (W_1)	0.125	0.08125 (S_1)	11.214 (l_1)
$k = 1$	1.157	0.75205 (W_2)	0.422	0.2743 (S_2)	11.391 (l_2)
$k = 2$	1.17	0.7605 (W_3)	0.5006	0.32539 (S_3)	11.3945 (l_3)
$k = 3$	1.157	0.75205 (W_2)	0.422	0.2743 (S_2)	11.391 (l_2)
$k = 4$	0.945	0.61425 (W_1)	0.125	0.08125 (S_1)	11.214 (l_1)

Method II: Using EDA simulation tool

The EDA tool generated BPF filter characteristics with microstrip realisation (curve 1 is obtained with circuit simulator) and with complete EM simulation (curve 2) is shown in Fig. 8.15a. The layout for coupled microstrip BPF obtained with EM simulation is shown in Fig. 8.15b and the screenshot of optimisation in layout is depicted in Fig. 8.15c.

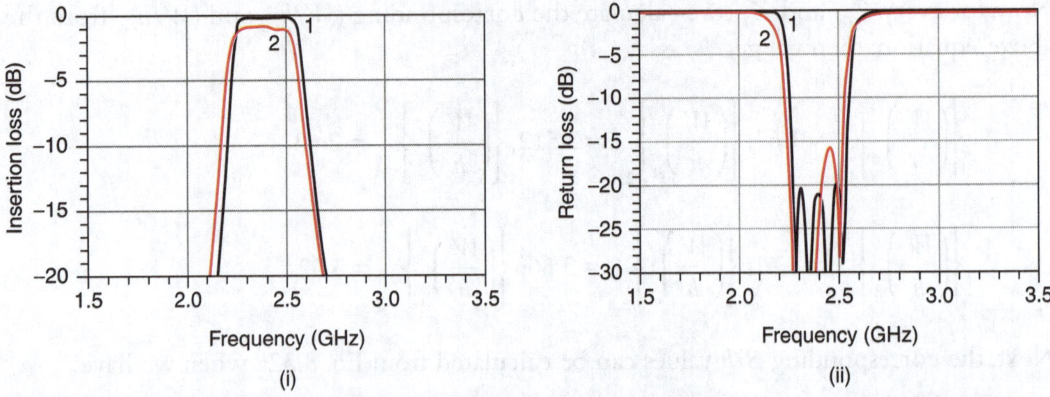

Fig. 8.15 (a) Coupled-line microstrip band-pass filter (BPF) characteristics obtained with EDA simulation tool (i) Transmission characteristics (ii) Reflection characteristics.

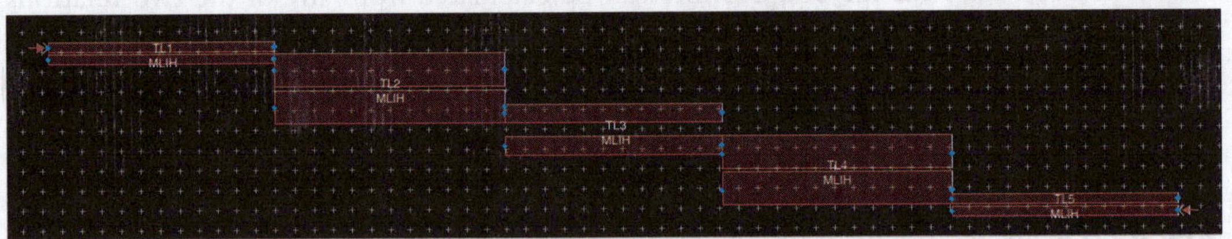

Fig. 8.15 (b) Layout of EM simulation for coupled microstrip BPF corresponding to Example 8.3.

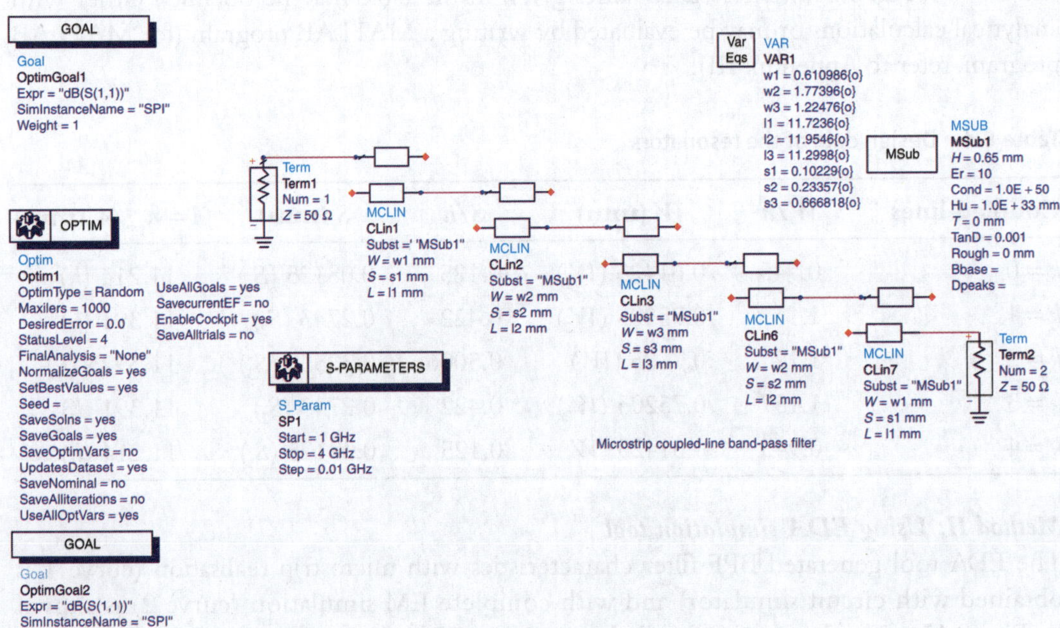

Fig. 8.15 (c) Screenshot of BPF design optimisation in layout.

The optimised design data of the EM simulated BPF (Fig. 8.15b) is given in Table 8.5 (Fig. 8.14b for *W*, *S* and *l*).

Table 8.5 Optimised design data of the EM simulated BPF

W (mm)	*S* (mm)	*l* (mm)
$W_1 = 0.61$	$S_1 = 0.102$	$l_1 = 11.72$
$W_2 = 0.77$	$S_2 = 0.23$	$l_2 = 11.95$
$W_3 = 1.22$	$S_3 = 0.67$	$l_3 = 11.3$

8.5.3 Coupled-resonator Band-stop Filter

The band-stop filter circuit can be derived from a low-pass prototype (Fig. 8.3 and Table 8.3) which is shown in Fig. 8.16a. It consists of parallel and series resonant circuits and hence resembles the case of a band-pass filter. Here also, we need to use either all series or all parallel resonators with the application of J/K inverter concept. When this is done, the lumped element ladder network of Fig. 8.16a may be equivalently replaced by all series resonators with quarter-wave coupling network (Fig. 8.16b) or with all shunt resonators with quarter-wave resonators (Fig. 8.16c). The design procedure for this type of filter using a microstrip line is summarised below [18].

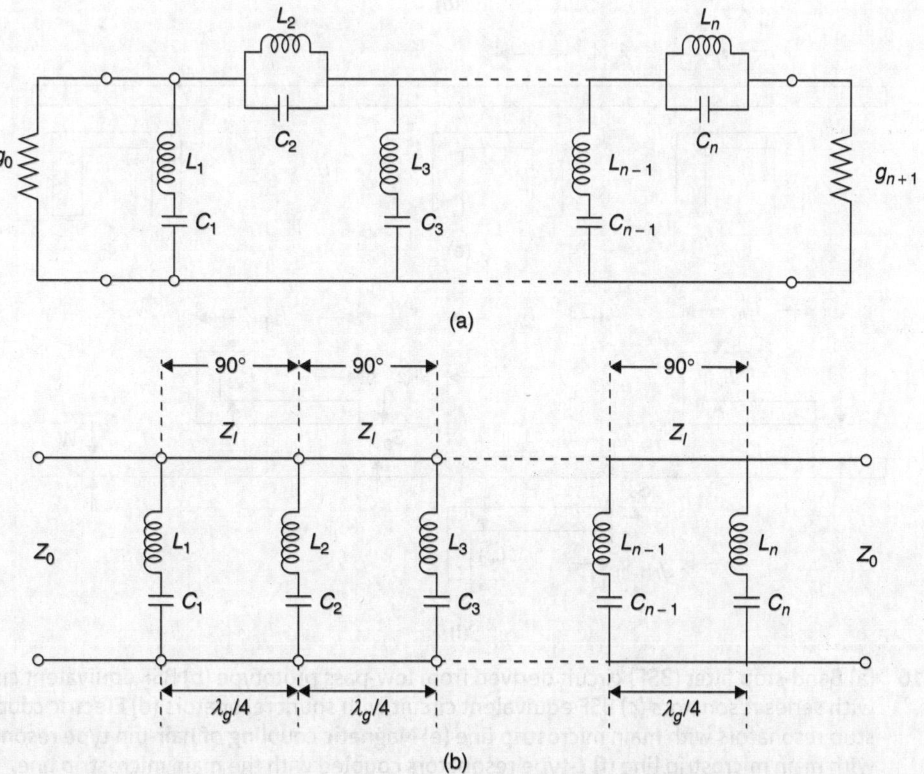

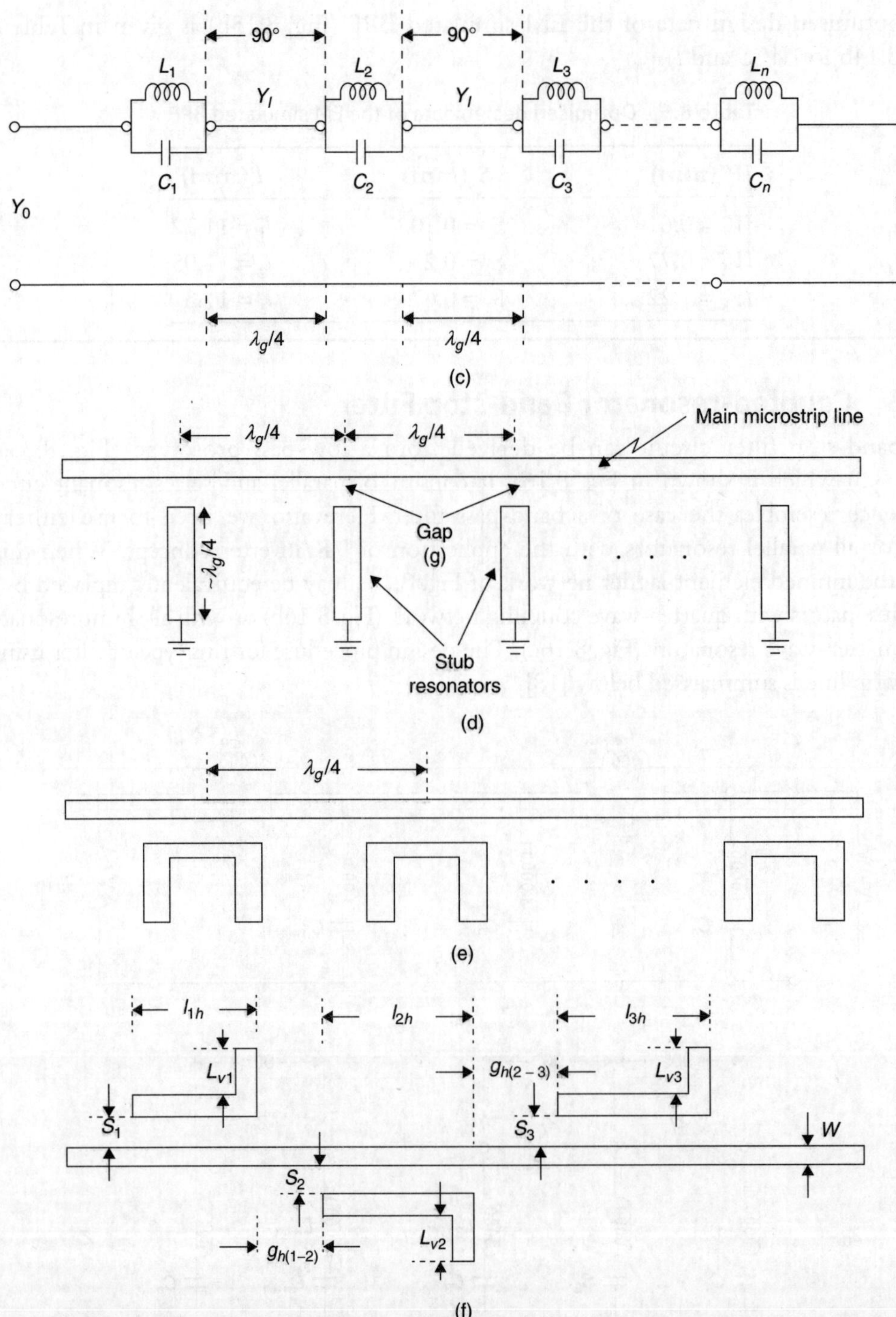

Fig. 8.16 (a) Band-stop filter (BSF) circuit derived from low-pass prototype (b) BSF equivalent circuit with series resonators (c) BSF equivalent circuit with shunt resonators (d) Electric coupling of stub resonators with main microstrip line (e) Magnetic coupling of hair-pin type resonators with main microstrip line (f) *L*-type resonators coupled with the main microstrip line.

Let Z_0 be the characteristic impedance of the main microstrip line and Z_I be the characteristic impedance of the quarter-wave coupling networks (of the inverters) then:

$$Z_I = Z_0, \quad \text{for } n \text{ odd} \quad \text{and} \quad Z_c = Z_I \left(\frac{1}{g_0 g_{n+1}} \right)^{1/2}, \quad \text{for } n \text{ even}$$

The filter design is carried out by defining *reactance slope parameters*: $x = \left. \frac{\omega_0}{2} \frac{dX}{d\omega} \right|_{\omega = \omega_0}$, with $X = (\omega L - 1/\omega C)$ of the resonators given by:

$$\frac{x_1}{Z_0} = \frac{1}{g_0 g_1 \Delta}, \quad \left. \frac{x_k}{Z_0} \right|_{k(\text{even})} = \left(\frac{Z_I}{Z_0} \right)^2 \frac{g_0}{g_k \Delta}, \quad \left. \frac{x_k}{Z_0} \right|_{k(\text{odd}), \neq 1} = \left(\frac{Z_I}{Z_0} \right)^2 \frac{1}{g_0 g_k \Delta}$$

where $\Delta = (\omega_{c2} - \omega_{c1})/\omega_0$, is the fractional bandwidth.

In strip-line/microstrip realisation, the inductances in series/parallel resonant circuits can be realised with a stub (less than quarter-wave long if realised with a short-circuited stub and less than half-wavelength long if realised with an open-circuited stub) and the capacitance by means of the gap (g) between the main microstrip line and the stub (Fig. 8.16d). The coupling between the main microstrip line and the resonators may be electrical (Fig. 8.16d) or may also be magnetic (with hair-pin type resonator, Fig. 8.16e). For practical purposes (as will be seen while solving Example 8.4), to realise significant capacitance through the gap (g), L-type resonators (that combine both electrical and magnetic coupling) may be used (Fig. 8.16f) [19]. Anyway, when electric coupling (Fig. 8.16d) is used, the length of the stub is determined from: $l_k = \varphi_l / \beta_{\text{eff}} = \varphi_l \lambda_{\text{eff}} / 2\pi$, where φ_l is the phase delay incurred by the signal in propagating via the line. Evaluation of φ_l is done by using the formula $F(\varphi_l) = \varphi_l \sec^2 \varphi_l + \tan \varphi_l$, which is generally evaluated numerically and a table for $F(\varphi_l)$ vs φ_l is given in Table 8.6. If a short circuited stub is used, then $F(\varphi_l) = 2x_k / Z_{c\,(\text{stub})}$ but if an open circuited stub is used then $F(\varphi_l) = x_k / Z_{c\,(\text{stub})}$, where $Z_{c\,(\text{stub})}$ is the characteristic impedance of the line forming the stub. The series gap capacitance is determined from $C_{gk} = 1/(\omega_0 Z_{c(\text{stub})} \tan \varphi_l)$.

The design of such filters using waveguides is available in advanced references [18].

Table 8.6 Values of $F(\varphi_l)$ for different values of φ_l (in degrees) obtained by numerical solution of equation $F(\varphi_l) = \varphi_l \sec^2 \varphi_l + \tan \varphi_l$

φ_l	$F(\varphi_l)$	φ_l	$F(\varphi_l)$	φ_l	$F(\varphi_l)$	φ_l	$F(\varphi_l)$
15	0.5485	35	1.6106	55	4.3459	75	23.2721
16	0.5890	36	1.6865	56	4.6082	76	26.6739
17	0.6302	37	1.7660	57	4.8935	77	30.8879
18	0.6722	38	1.8493	58	5.2051	78	36.1958
19	0.7153	39	1.9368	59	5.5462	79	43.0132
20	0.7593	40	2.0288	60	5.9208	80	51.9730

(continued)

Table 8.6 Continued

φ_l	$F(\varphi_l)$	φ_l	$F(\varphi_l)$	φ_l	$F(\varphi_l)$	φ_l	$F(\varphi_l)$
21	0.8044	41	2.1256	61	6.3336	81	64.0787
22	0.8507	42	2.2277	62	6.7903	82	80.9982
23	0.8982	43	2.3356	63	7.2974	83	105.6711
24	0.9471	44	2.4498	64	7.8628	84	143.6788
25	0.9975	45	2.5708	65	8.4961	85	207.7040
26	1.0495	46	2.6993	66	9.2088		
27	1.1031	47	2.8360	67	10.0151		
28	1.1586	48	2.9817	68	10.9322		
29	1.2160	49	3.1373	69	11.9819		
30	1.2755	50	3.3038	70	13.2301		
31	1.3372	51	3.4824	71	14.5948		
32	1.4014	52	3.6743	72	16.2921		
33	1.4683	53	3.8810	73	18.1752		
34	1.5378	54	4.1043	74	20.4861		

[PROGRAM]:

```
for i=15:75
    p(i,1)=i/57.296;
y(i,1)=p(i,1)*(sec(p(i,1)))^2+tan(p(i,1));
```

EXAMPLE 8.4

Design a strip-line based band-stop filter with Tchebyscheff response having a frequency of infinite attenuation at 2.4 GHz, fractional bandwidth 0.05 (narrow band type), pass band ripple of 0.1 dB with 20 dB minimum attenuation occurring at 2% of the centre frequency. Given the characteristic impedance of the strip-line is 50 Ω with dielectric substrate of $\varepsilon_r = 10$ and height $h = 0.65$ mm. Show calculations using an open-circuited stub and also short-circuited stub lines having nominal characteristic impedance of 70 Ω to realise the inductance of the series resonators.

Solution

This problem will be solved both by analytical synthesis process and also by using Electronic Design Automation (EDA) simulation tool.

Method I: Analytical synthesis process

Step 1: At first we determine the number of sections for the filter.

With the given pass-band ripple magnitude (A_{pm}) as $10 \log_{10}(1 + A_p^2) = 0.1$ dB, we have $A_p^2 = 0.023$.

The power-loss ratio of the band-stop filter having equal-ripple (Tchebyscheff response) is given by:

$$P_{LR} = 1 + A_p^2 T_n^2(\Omega') \text{ where } T_n(x) = \cosh(n \cosh^{-1} \Omega')$$

Here, $$\Omega' = \frac{\Omega_p \omega_0}{\omega_{c2} - \omega_{c1}} \left(\frac{\omega_s}{\omega_0} - \frac{\omega_0}{\omega_s} \right)^{-1} = 0.05 \left(\frac{2.424}{2.4} - \frac{2.4}{2.424} \right)^{-1} = 2.5 > 1 \text{ (with } \Omega_p = 1)$$

[since the frequency at 20 dB attenuation is $f_s = (2.4 \text{ GHz} + 0.5 \times 2\% \text{ of } 2.4 \text{ GHz}) = 2.424$ GHz) and the fractional bandwidth $\Delta = (\omega_{c2} - \omega_{c1})/\omega_0 = 0.05$]

Substituting the respective values ($P_{LRs} = 20$ dB and $\Omega' = 2.5$) in the expression for P_{LR} and solving for n we get $n = 3.108$, which can be taken approximately as 3. Therefore a three-section band-stop filter is required.

Step 2: Now we need to determine the values of the prototype elements to realise the specifications.

For a 0.1 dB ripple with $n = 3$, the low-pass prototype elements are $g_1 = 1.0315 = g_3$ and $g_2 = 1.1474$ (Table 8.1a), while $g_0 = g_4 = 1.0$.

Step 3: We can now proceed for microwave realisation of the filter using strip-line circuit. Here $n = 3$ (odd), thus the impedance of the quarter-wave long coupling lines will be $Z_I = Z_0$. The reactance slope parameters of the series resonators are thus given by:

$$\frac{x_1}{Z_0} = \frac{1}{g_0 g_1 \Delta} = \frac{x_3}{Z_0} = \frac{1}{1 \times 1.0315 \times 0.05} = 19.32$$

$$\frac{x_2}{Z_0} = \left(\frac{Z_I}{Z_0} \right)^2 \frac{g_0}{g_k \Delta} = \frac{1}{1.1474 \times 0.05} = 17.43 \text{ (since } Z_I = Z_0)$$

Therefore $x_1 = x_3 = 19.32 \times 50 = 966\ \Omega$ and $x_2 = 17.43 \times 50 = 871.5\ \Omega$

Step 4: Now, if we use *open-circuited stub* lines (with characteristic impedance of the line forming the stub $Z_{c\,(\text{stub})} = 70\ \Omega$) to realise the inductance of the series resonators, then:

$$F(\varphi_{l1}) = F(\varphi_{l3}) = \varphi_{l1} \sec^2 \varphi_{l1} + \tan \varphi_{l1} = \frac{x_1}{Z_{c(\text{stub})}} = \frac{966}{70} = 13.8$$

$$F(\varphi_{l2}) = \varphi_{l2} \sec^2 \varphi_{l2} + \tan \varphi_{l2} = \frac{x_2}{Z_{c(\text{stub})}} = \frac{871.5}{70} = 12.45$$

From the $F(\varphi_l)$ vs φ_l values (Table 8.6) we find:

$$\varphi_{l1} = \varphi_{l3} = 70.5° = 1.2309 \text{ radians}, \ \varphi_{l2} = 69.5° = 1.2134 \text{ radians}$$

Thus, the stub lengths are given by:

$$l_1 = l_3 = \frac{\varphi_{l1,3}\lambda_{\text{eff}}}{2\pi} = \frac{1.2309 \times 3.972}{2\pi} = 0.7781 \text{ cm}$$

$$l_2 = \frac{\varphi_{l2}\lambda_{\text{eff}}}{2\pi} = \frac{1.2134 \times 3.972}{2\pi} = 0.767 \text{ cm}$$

(since $\lambda_{\text{eff}} = \lambda_0/\sqrt{\varepsilon_r} = (30/2.4)/\sqrt{9.9} = 3.972$ cm and $\lambda_{\text{eff}}/2 = 1.986$ cm).

Step 5: If we use *short-circuited stub* lines to realise the inductance of the series resonator, then:

$$F(\varphi_{l1}) = F(\varphi_{l3}) = \varphi_{l1}\sec^2\varphi_{l1} + \tan\varphi_{l1} = \frac{2x_1}{Z_{c(\text{stub})}} = \frac{2 \times 966}{70} = 27.6$$

$$F(\varphi_{l2}) = \varphi_{l2}\sec^2\varphi_{l2} + \tan\varphi_{l2} = \frac{2x_2}{Z_{c(\text{stub})}} = \frac{2 \times 871.5}{70} = 24.9$$

From the $F(\varphi_l)$ vs φ_l values (Table 8.6) we find:

$$\varphi_{l1} = \varphi_{l3} = 76.2° = 1.3304 \text{ radians}, \varphi_{l2} = 75.5° = 1.3182 \text{ radians}$$

Thus, the stub lengths are given by:

$$l_1 = l_3 = \frac{\varphi_{l1,3}\lambda_{\text{eff}}}{2\pi} = \frac{1.3304 \times 3.972}{2\pi} = 0.8410 \text{ cm}$$

$$l_2 = \frac{\varphi_{l2}\lambda_{\text{eff}}}{2\pi} = \frac{1.3182 \times 3.972}{2\pi} = 0.8333 \text{ cm}$$

Step 6: The gap capacitances can be calculated from $C_{gk} = 1/(\omega_0 Z_{c(\text{stub})} \tan\varphi_l)$, where $Z_{c(\text{stub})} = 70\ \Omega$, assumed to be equal for all the stubs. Thus, when *open-circuited stubs* are used,

$$C_{g1,3(o.c)} = \left(\frac{1}{2\pi \times 2.4 \times 10^9 \times 70 \times \tan 70.5^0}\right) = 0.335 \text{ pF}$$

and $$C_{g2(o.c)} = \left(\frac{1}{2\pi \times 2.4 \times 10^9 \times 70 \times \tan 69.5^0}\right) = 0.354 \text{ pF}$$

If a *short-circuited stub* is used:

$$C_{g1,3(s.c)} = \left(\frac{1}{2\pi \times 2.4 \times 10^9 \times 70 \times \tan 76.2^0}\right) = 0.232 \text{ pF}$$

and $$C_{g2(o.c)} = \left(\frac{1}{2\pi \times 2.4 \times 10^9 \times 70 \times \tan 75.5^0}\right) = 0.245 \text{ pF}$$

Step 7: Finally, we need to determine the line widths of the strip line transmission line to complete our design. For the main line, characteristic impedance is 50 Ω and for the resonator line it is 70 Ω; while the dielectric constant of the substrate material for the strip-line is $\varepsilon_r = 10$ in either case. From Fig. 3.21a, we can calculate the strip-width as follows.

$$\text{For 50 } \Omega \text{ line, } \frac{W}{h} = 1.0, \text{ thus } W = 1.0 \times 0.65 = 0.65 \text{ mm}$$

$$\text{For 70 } \Omega \text{ line, } \frac{W}{h} = 0.49, \text{ thus } W = 0.49 \times 0.65 = 0.3185 \text{ mm}$$

Method II: Using EDA simulation tool

The microstrip realisation of the coupled-resonator band-stop filter (with circuit simulator of the EDA simulation tool) using the dimensions obtained through analytical synthesis is shown in Fig. 18.17a (curve 1), which does not depict any band-stop characteristics whatsoever. This is because the realisable gap capacitance between the host microstrip line and individual stub resonators is very small. However, the use of the calculated capacitance values corresponding to the gap (g) shows significant band-stop performance when simulated in the circuit simulator of the EDA tool (curve 2 in Fig. 8.17a). Curve 3

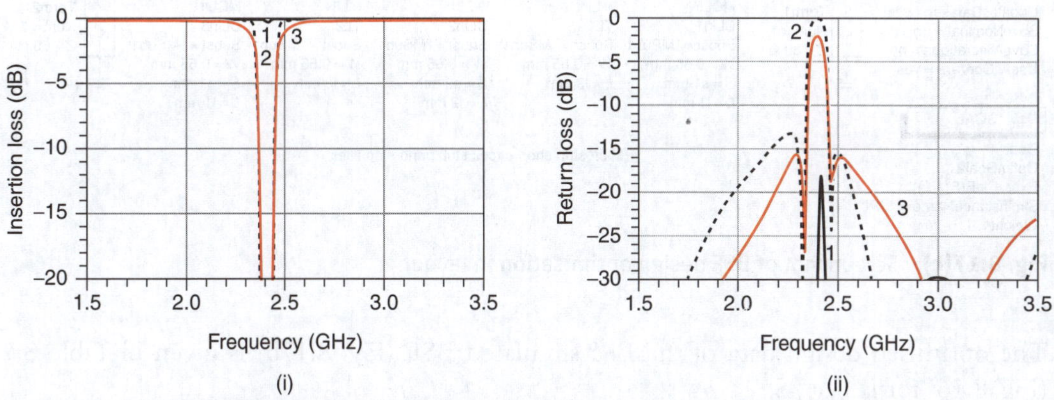

Fig. 18.17(a) Microstrip band-stop filter (BSF) characteristics (i) Transmission characteristics (ii) Reflection characteristics.

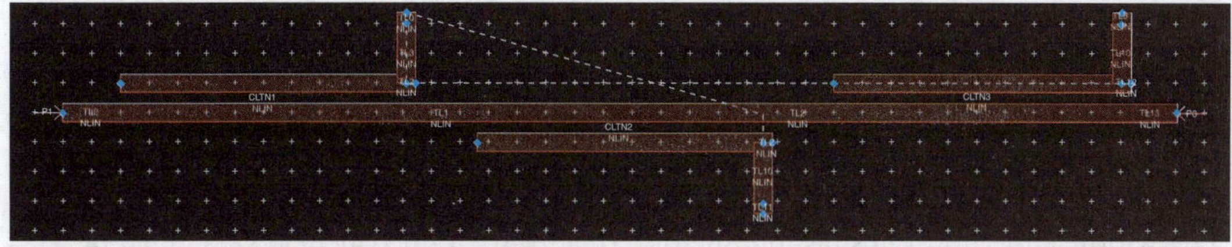

Fig. 18.17(b) Layout of EM simulation with L-shaped resonator for coupled-resonator BSF corresponding to Example 8.4.

in Fig. 8.17a obtained with EM simulation of the filter structure when the resonators are of L-type (that provides much larger capacitance between the host microstrip line and the resonators) shows pretty good band-stop characteristics when compared with conventional stub resonators (curve 1 in Fig. 8.17a) as is used in the analytical synthesis. The layout of EM simulation with L-type resonators is shown in Fig. 8.17b together with the screenshot of the optimisation in layout in Fig. 8.17c.

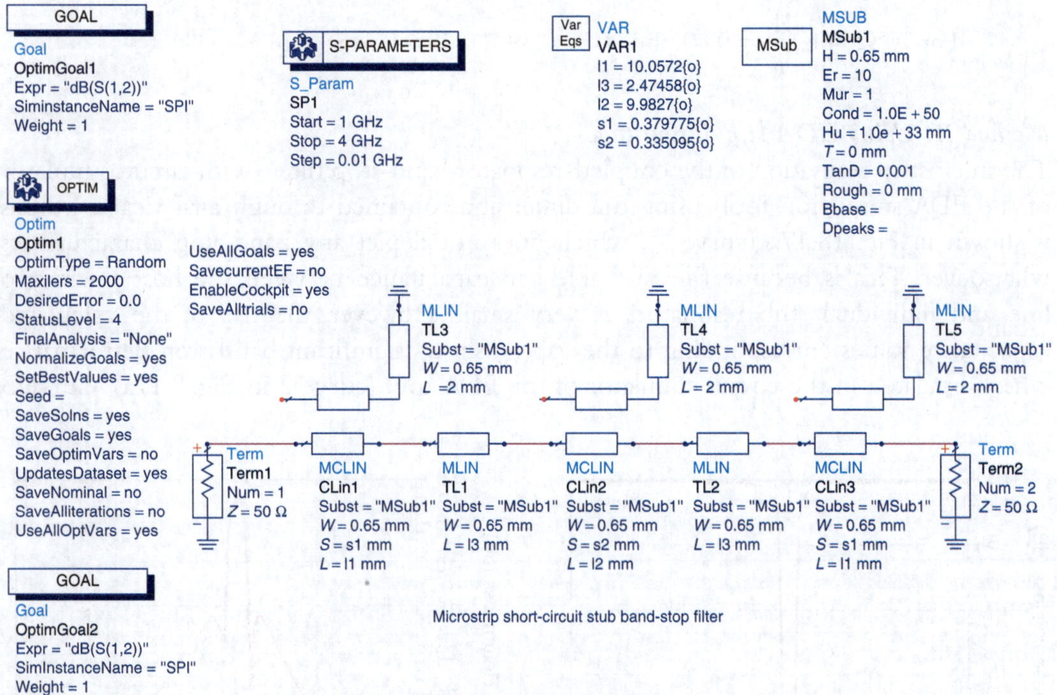

Fig. 8.17(c) Screenshot of BSF design optimisation in layout.

The optimised design data of the EM simulated BSF (Fig. 8.17b) is given in Table 8.7 (Fig. 8.16f for l_h, l_v, g_h, S).

Table 8.7 Optimised design data of the EM simulated BSF

l_h (mm)	l_v (mm)	g_h (mm)	S (mm)
$l_{h1} = 10 = l_{h3}$	$l_{v1} = l_{v2} = l_{v3} = 2.5$	$g_{h(1-2)} = 2.47$	$S_1 = 0.38 = S_3$
$l_{h2} = 9.98$		$g_{h(2-3)} = 2.47$	$S_2 = 0.335$

The width of the host transmission line of characteristic impedance 50 Ω is given by $W =$ 0.65 mm (Fig. 3.21a, with $Z_0 = 50$ Ω and $\varepsilon_r = 10$).

RECAPITULATION

Insertion-loss method for low-pass prototype design

- The design of microwave filter with insertion-loss method is done in the following steps:
  (i) A low-pass prototype filter is first designed in terms of normalised impedance and frequency with desired pass-band characteristics.
  (ii) This prototype network is then transformed to the required filter type (low-pass, high-pass, band-pass and band-stop) with proper scaling of the prototype circuit elements in terms of impedance and frequency.
  (iii) Finally, the filter network is realised in microwave form with distributed circuit elements which essentially use sections of suitably terminated (open or short) microwave transmission lines of suitable length representing the desired reactance.
- The power loss ratio for a maximally-flat filter is specified by $P_{LR} = 1 + A_p^2(\Omega)^{2n}$.
- The power loss ratio of an equal-ripple filter is represented by $P_{LR} = 1 + A_p^2 T_n^2(\Omega)$.
- The equal-ripple filter will have a sharper rate of rise above cut-off with better stop-band characteristics (approximately 'brick-wall' characteristics outside the pass-band) compared to the maximally-flat filter, though with a penalty of ripple within the pass-band. However, for a specified rate of increase of the power loss ratio beyond the cut-off frequency, the equal-ripple characteristics can provide the smallest pass-band tolerance. In this sense, the equal-ripple filter represents an optimum design. Further, for a given insertion loss, the equal-ripple filter needs fewer elements than the maximally-flat filter.

Low-pass prototype filter design technique

- The low-pass filter prototype design is done in terms of normalised frequency and impedance. This normalisation simplifies the design of filters for arbitrary frequency, impedance and type of the filter. The low-pass prototypes are then scaled to the desired frequency and impedance.
- The element values of the prototype low-pass filter with *equal-ripple* and *maximally-flat* responses for the normalised value of cut-off angular frequency can be calculated from Bennett's formula.

  For an equal-ripple low-pass prototype filter:

$$g_0 = 1, \text{ for all } n$$

$$g_1 = \frac{2a_1}{\gamma}$$

$$g_k = \frac{4a_{k-1}a_k}{b_{k-1}g_{k-1}} \text{ with } k = 2, 3, 4 \ldots n$$

$$g_{n+1} = 1, \text{ for } n \text{ odd}$$

$$= \coth^2\left(\frac{\beta}{4}\right), \text{ for } n \text{ even}$$

where $a_k = \sin\frac{2k-1}{2n}\pi, b_k = \gamma^2 + \sin^2\frac{k\pi}{n}, \beta = \ln\left(\coth\frac{A_{pm}}{17.37}\right), \gamma = \sinh\left(\frac{\beta}{2n}\right)$

with $k = 2, 3, \ldots n$.

For a maximally-flat low-pass filter:

$$g_0 = 1 = g_{n+1}, \text{ for all } n$$

$$g_k = 2\sin\left\{\frac{(2k-1)\pi}{2n}\right\}, k = 1, 2, \ldots n$$

This analytical approach is used to evaluate the element values for a prototype low-pass filter.

Filter transformations from prototype

- Prototype design may be used as the basis for design of actual low-pass, high-pass, band-pass and band-stop filters at any desired impedance and frequency. To do this we need to scale the prototype circuit elements in terms of impedance and frequency.

  The impedance scaling is required to remove the $g_0 = 1$ normalisation of the source impedance as a practical filter might need to work for any desired value of source impedance, say $Z_s = Z_0$. With impedance sacling, the series inductances (L_k) and shunt capacitances (C_k) of the prototype will be represented by: $L'_k = I_t L_k$ and $C'_k = C_k/I_t$, where $I_t = Z_0/g_0$. Impedance transformation is frequency independent and thus the same transformation/scaling holds for any type of filter. However, frequency scaling is different for different type of filters.
- ***Prototype to actual low-pass transformation***: This demands the frequency transformation from Ω to Ω', where $\Omega' = (\Omega_p/\omega_c)\omega$, of the prototype reactive elements, when the actual low-pass elements are $L'_k = (\Omega_p/\omega_c)L_k I_t$ and $C'_k = (\Omega_p/\omega_c)(C_k/I_t)$.
- ***Low-pass prototype to high-pass transformation***: Here, the frequency transformation demands Ω to be replaced by Ω', where $\Omega' = -(\Omega_p/\omega)\omega_c$, to get the actual high-pass filter elements as: $C'_k = (1/\omega_c\Omega_p)(1/L_k I_t)$ and $L'_k = (\omega_c\Omega_p)(I_t/C_k)$.
- ***Low-pass to band-pass transformation***: Replace Ω to Ω', where, $\Omega' = (\Omega_p/\Delta)[(\omega/\omega_0) - (\omega_0/\omega)]$ to get band-pass filter elements. Here, the series inductor L_k of low-pass prototype is transformed into a series LC circuit in band-pass filter with element values: $L'_{kse} = (\Omega_p/\Delta\omega_0)I_t L_k$ and $C'_{kse} = (\Delta/\Omega_p\omega_0)(1/I_t L_k)$ while a shunt capacitor C_k of low-pass prototype is transformed to a shunt LC circuit in band-pass filter with element values $L'_{ksh} = (\Delta/\Omega_p\omega_0)(I_t/C_k)$ and $C'_{ksh} = (\Omega_p/\Delta\omega_0)(C_k/I_t)$ where $\Delta = (\omega_{c2} - \omega_{c1})/\omega_0$ is the fractional bandwidth of the pass-band with ω_{c1} and ω_{c2} denoting the angular frequency at the edges of the pass-band.
- ***Low-pass to band-stop transformation***: Here we replace Ω to Ω', where $\Omega' = (\Omega_p/\Delta)[(\omega/\omega_0) - (\omega_0/\omega)]^{-1}$ to get band-stop filter elements. Here, series inductors of the low-pass prototype are converted to parallel LC circuits in band-stop filter having element values $L'_{ksh} = (\Omega_p\Delta/\omega_0)\, I_t L_k$ and $C'_{ksh} = (1/\Omega_p\Delta\omega_0)(1/I_t L_k)$ while shunt capacitors of the low-pass prototype are converted to series LC circuits in band-stop filter having

element values $L'_{kse} = (1/\Omega_p \Delta\omega_0)(I_t/C_k)$ and $C'_{kse} = (\Omega_p \Delta/\omega_0)(C_k/I_t)$, where Δ and ω_0 have a similar meaning as in the case of a band-pass filter.

Microwave implementation

- Microwave filters are mostly designed with distributed component inductors and capacitors. In addition, at microwave frequencies, the distances between filter components sometimes become negligibly small. To convert lumped elements to distributed components with transmission line sections *Richard's transformation* is used, while *Kuroda's identities* can be used to separate filter elements by using transmission line sections. Such additional transmission line sections do not affect the filter response. However, it is possible to design microwave filters that take advantage of these sections to improve the filter response.
- During the past few years, lumped inductors and capacitors have been fabricated and used at microwave frequencies (but with limited values and upto 60 GHz or so). If lumped elements are used for the filter circuit at microwave frequencies, the microwave filter design becomes identical to the low frequency filter design.

Impedance and admittance inverters

- At microwave frequencies it is desirable, from a fabrication point of view, to use either all series or all shunt elements to implement filter networks. Kuroda's identity can be used to realise such conversions; however, another practical approach is to use impedance (*K*) or admittance (*J*) inverters. The impedance and admittance inverters, in essence, are ideal quarter-wave transformers. The *K* inverter helps to transform a shunt element into a series element while a *J* inverter transforms a series element into a shunt element having the same dependence on frequency and an arbitrary impedance level. Thus, the impedance and admittance inverters enable us to use identical resonators, either series or parallel tuned, throughout a filter network.

Review Questions

8.1 Name four common types of filters. Mention the typical bandwidths for narrow band and broadband filters.

8.2 Why is the insertion-loss method commonly used as the basis for the design of a microwave filter?

8.3 Enumerate the steps to be followed to design a microwave filter using the insertion loss method.

8.4 Discuss the advantages and disadvantages of maximally-flat and equal-ripple filters in terms of insertion loss and performance in the stop-band. What is elliptic function filter?

8.5 Discuss the analytical algorithm for evaluating the element values for a prototype low-pass filter.

8.6 Give in a tabular form, the element values of actual low-pass, high-pass, band-pass and band-stop filter in the series and shunt arms with respect to the prototype element.

8.7 Discuss the usefulness of Richard's transformation and Kuroda's identities for microwave implementation of lumped-element filter.

8.8 What are K and J inverters? With typical examples show how a K inverter helps to transform a shunt element into a series element while a J inverter transforms a series element into a shunt element.

8.9 Discuss how a practical low-pass filter can be designed at microwave frequencies using stepped impedance or low-high impedance concept in microstrip transmission line structure.

8.10 Discuss the principle of coupled-line band-pass filter design at microwave frequencies.

8.11 Discuss the principle of the coupled-resonator band-stop filter design at microwave frequencies.

Numerical Problems

8.1 Design a third order low-pass filter in a 50 Ω microstrip transmission line system with the substrate material having $\varepsilon_r = 4.3$ and substrate height equal to 31 mil (1 mil = 10^{-3} inch). The cut-off frequency of the filter is desired to be at 5.2 GHz. Assume 0.4 dB equal-ripple pass-band characteristics.

[**Ans:** Hints: $n = 1.671$ and $l = 6.94$ cm]

8.2 Design a stepped impedance low-pass filter having a Tchebyscheff response with a 0.1 dB ripple and a cut-off frequency of 5.2 GHz. It is desired to have at least 30 dB insertion loss at 7.2 GHz in the stop-band. Realise the filter in microstrip transmission line having a dielectric substrate of thickness 31 mil with $\varepsilon_r = 4.3$ at 5.2 GHz. Assume 50 Ω terminations at both load and source ends and the highest practical line impedance are 110 Ω and the lowest being 30 Ω. Evaluate the result by writing a MATLAB program.

[**Ans:**

A_p^2	n	C_1 (pF)	L_2 (nH)	C_3 (pF)	L_4 (nH)
0.02329299	7	0.67908	2.18765	1.28405	2.40890
C_5 (pF)	L_6 (nH)	C_7 (pF)	W_1 (mm)	W_2 (mm)	W_3 (mm)
1.28405	2.18765	6.7908	3.27978	0.27269	3.27978
W_4 (mm)	W_5 (mm)	W_6 (mm)	W_7 (mm)	$W_0 = W_8$ (mm)	ε_{reff} for $(W/h) \le 1$
0.27269	3.27978	0.27269	3.27978	1.53142	2.95454
ε_{reff} for $(W/h) > 1$	λ_0 (cm)	λ_{gC} (cm)	λ_{gL} (m)	l_{C1} (mm)	l_{L2} (mm)
3.48756	5.769	3.089	3.356	3.56	3.47
l_{C3} (mm)	l_{L4} (mm)	l_{C5} (mm)	l_{L6} (mm)	l_{C7} (mm)	
6.72	3.82	6.72	3.47	3.56	

Using the above synthesis data, characterise the filter with the help of EDA simulation tool.]

8.3 Design a parallel-coupled microstrip band-pass filter having Tchebyscheff characteristics with pass-band ripple of 0.3 dB between the pass-band edge frequencies of 5.32 GHz and 5.08 GHZ. It is desired to have at least 25 dB insertion loss at 5.85 GHz in the stop-band. The filter impedance is 50 Ω. Consider a substrate of dielectric constant $\varepsilon_r = 4.3$ and height $h = 31$ mils.

[**Ans:** Hints: $f_0 = 5.2$ GHz, $\Delta = 0.046$, $A_p^2 = 0.0715$, 2-section

$$\left\{\left(\frac{W}{h}\right)_{es}\right\}_{1,0} = 3.23, \left\{\left(\frac{W}{h}\right)_{es}\right\}_{2,1} = 3.43, \left\{\left(\frac{W}{h}\right)_{es}\right\}_{3,2} = 3.73$$

$$\left\{\left(\frac{W}{h}\right)_{os}\right\}_{1,0} = 6.56, \left\{\left(\frac{W}{h}\right)_{os}\right\}_{2,1} = 7.13, \left\{\left(\frac{W}{h}\right)_{os}\right\}_{3,2} = 6.885$$

Coupled lines	W/h	W(mm)	S/h	S(mm)	$l = \lambda_{gm}/4$ (mm)
$k = 0$	1.11	0.87	0.093	0.073	6.88
$k = 1$	0.99	0.78	0.07	0.055	7.355
$k = 2$	1.14	0.898	0.107	0.084	7.27

]

8.4 Design a strip-line based band-stop filter with Tchebyscheff response having a frequency of infinite attenuation at 5.2 GHz, fractional bandwidth 0.04 (narrow band type), pass-band ripple of 0.2 dB having 25 dB minimum attenuation occurring at 3% of the centre frequency. Given the characteristic impedance of the strip-line is 50 Ω with dielectric substrate of $\varepsilon_r = 4.3$ and height $h = 31$ mils. Show calculations using an open-circuited stub and also short-circuited stub lines having nominal characteristic impedance of 80 Ω to realise the inductance of the series resonators.

[**Ans:** Hints: $A_p^2 = 0.047$, 6-section

If we use *open-circuited stub* lines (with characteristic impedance of the line forming the stub $Z_{c\,(\text{stub})} = 80\ \Omega$) to realise the inductance of the series resonators then, stub lengths are given by:

$$l_1 = l_6 = 0.584 \text{ cm}, l_2 = l_5 = 0.576 \text{ cm}, l_3 = l_4 = 0.575 \text{ cm}, l_7 = 0.582 \text{ cm}.$$

If we use *short-circuited stub* lines to realise the inductance of series resonator then, stub lengths are given by:

$$l_1 = l_6 = 0.619 \text{ cm}, l_2 = l_5 = 0.592 \text{ cm}, l_3 = l_4 = 0.587 \text{ cm}, l_7 = 0.613 \text{ cm}.$$

The gap capacitances:

When an *open-circuited stub* is used:

$$C_{g1,6(o.c)} = 0.098 \text{ pF}, C_{g2,5(o.c)} = 0.105 \text{ pF}, C_{g3,4(o.c)} = 0.106 \text{ pF},$$
$$\text{and } C_{g7(o.c)} = 0.1 \text{ pF.}$$

If a *short-circuited stub* is used:

$$C_{g1,6(s.c)} = 0.065 \text{ pF}, C_{g2,5(s.c)} = 0.091 \text{ pF}, C_{g3,4(s.c)} = 0.095 \text{ pF},$$
$$\text{and } C_{g7(s.c)} = 0.071 \text{ pF.}$$

The width of the main microstrip line (Z_0 = 50 Ω) and resonator lines ($Z_{0(stub)} = 80\ \Omega$) are given by:

$$\text{W}_{\text{main line}} = 2.1 \times 0.707 = 1.48 \text{ mm}; \text{W}_{\text{stub line}} = 0.44 \times 0.707 = 0.31 \text{ mm}]$$

9 Microwave Active Devices: Oscillators, Amplifiers and Power Combiners

Yes, when you surpass the limits you have seen so far
—that is when your true journey begins

9.1 Introduction

Microwave active devices, required primarily for signal generation and amplification, are available both as tube and solid-state (semiconductor) devices depending on the frequency and power level of the microwave signal required. The power levels obtainable from solid-state sources are low compared to that from tube sources. Thus, for high power applications, tubes are preferred. However, power combining techniques in solid-state sources are making them compete strongly with medium power tubes. Solid-state sources generally require low power supply voltage for operation and do not need an electron gun or focussing magnetic field and are thus very compact and light weight compared to the tube devices. These characteristics are particularly useful for air/space borne and military applications where light weight and small size devices and circuits are in great demand in addition to their quality performance.

At frequencies above about 1.0 GHz, the performance of conventional tubes such as triodes, tetrodes and pentodes are impaired by the parasitic reactance arising out of inter-electrode capacitance and the inductance due to the lead wires. Further, the transit-time of electrons from the cathode to anode becomes comparable with the period of the microwave signal; this also impairs the performance of the conventional tubes at microwave frequencies. An altogether new principle utilising the transit-time effect in favour of the operation of microwave tubes led to the development of linear-beam tubes (such as the klystron and

TWT) and crossed-field tubes (such as the magnetron and crossed field amplifier) for microwave signal generation and amplification. Yet another technique is used to generate high power millimeter-wave signals which utilises transverse interaction between the electromagnetic signal and the electron beam, as is done in a gyrotron, ubitron or peniotron. These tubes are also known as fast-wave electron tubes.

Solid-state (semiconductor) devices such as the *p–n* junction diode, bipolar junction transistor (BJT) and field effect transistor (FET), that are very popular at lower RF applications, suffer from high frequency limitations. The semiconductor–semiconductor *p–n* junction diode is thus replaced by a metal–semiconductor junction diode or Schottky junction diode to get over the problem of minority carrier storage of the *p–n* junction diode for operation at microwave frequencies. Another microwave solid-state diode is the tunnel diode. This uses the tunnelling phenomenon and is useful as a low gain pre-amplifier in space applications. Though the frequency range of operation of a BJT can be extended to some extent for high frequency operations, the type of FET useful at microwave frequencies is the metal–semiconductor field effect transistor (MESFET) which is a useful microwave amplifier, especially in a planar design using monolithic microwave integrated circuits (MMIC). In addition to these homojunction three-terminal devices (which use only one type of semiconductor Si/GaAs), nowadays heterojunction three-terminal devices (that use two or more semiconductors), such as heterojunction bipolar transistor (HBT) and high electron mobility transistor (HEMT), are becoming very useful. Compared to homojunction microwave BJT, the HBT is capable of operation at higher millimeter wave frequency spectrum while compared to MESFET. The HEMT is a low-noise device with higher switching speed and can be used to design amplifiers at higher millimeter-wave frequency. Further, some new classes of microwave solid-state diodes have also become very important from an application point of view for quite some time (typically beyond 30 GHz). For example, the gunn diode (a type of transferred electron device, TED) and the IMPATT diode (a type of avalanche transit time, ATT device) used for the design of microwave/millimeter-wave oscillators and power combiners.

Apart from signal generation and amplification, microwave active devices are also used for the design of detectors, mixers, multipliers, phase shifters, switches, attenuators, tuners and so on. For these applications some of the tube devices were used in the past. Microwave solid-state devices like Schottky diode, varactor diode, PIN diode and so on, find great use in modern microwave systems. Different types of microwave tubes and solid-state devices with principal applications are tabulated in Table 9.1 and Table 9.2 respectively.

9.2 Microwave Tubes

Out of the different microwave tubes mentioned in Table 9.1, linear beam tubes like the klystron (both as amplifier and oscillator), travelling wave tube (TWT) as amplifier, crossed field tube like magnetron as CW and pulsed oscillator, and fast-wave electron tube like gyrotron as high power millimeter-wave signal generator will be discussed under this section.

Table 9.1 Microwave tube devices

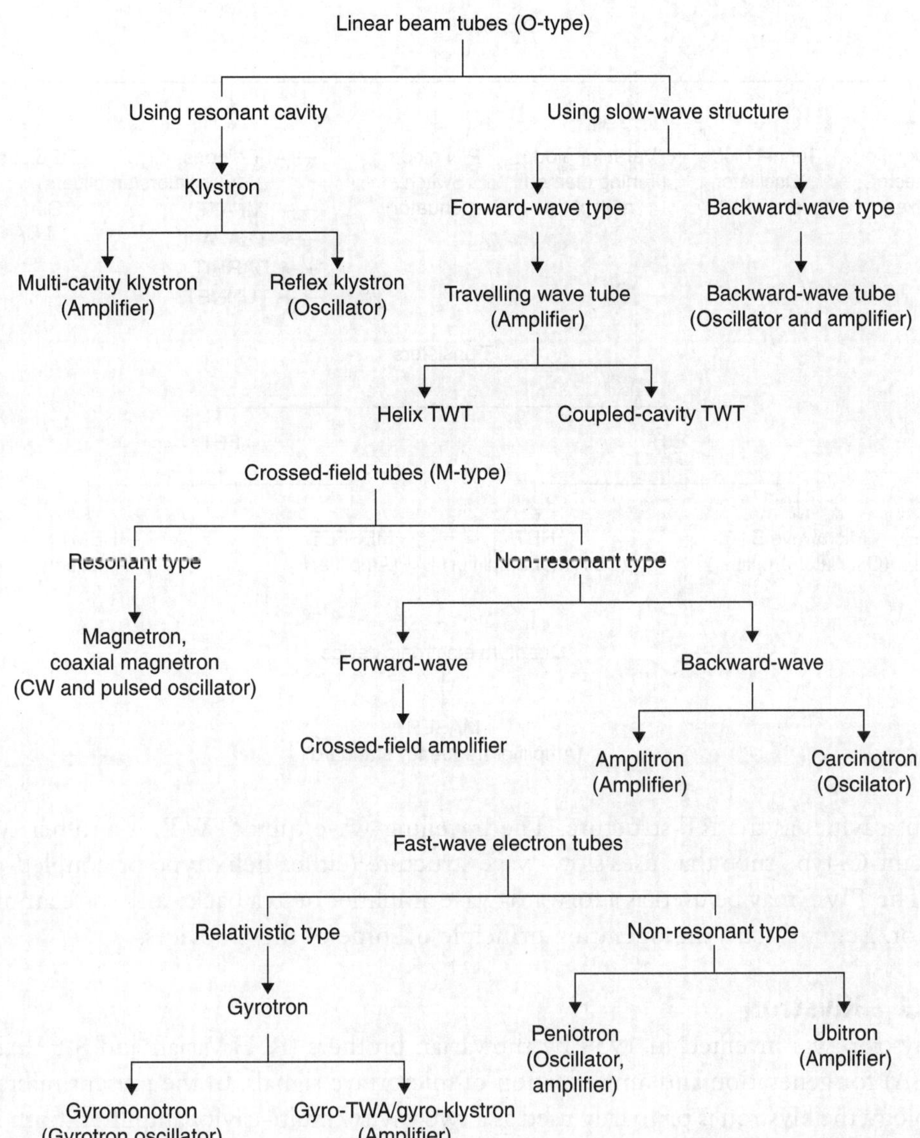

9.2.1 Linear Beam Tubes

In a linear beam tube, the magnetic field is applied in the axial direction of the tube and this merely takes part in focussing the electron stream (to form a long cylindrical collinear electron beam) without taking part in the interaction of the electron beam with the RF field. They are known as O-type tubes from the French acronym TPO (tubes à propagation des ondes) meaning the original type of tubes. The class of O-type tube started with klystron, multi-cavity klystron as amplifier and reflex klystron as oscillator, which use

Table 9.2 Microwave solid-state devices

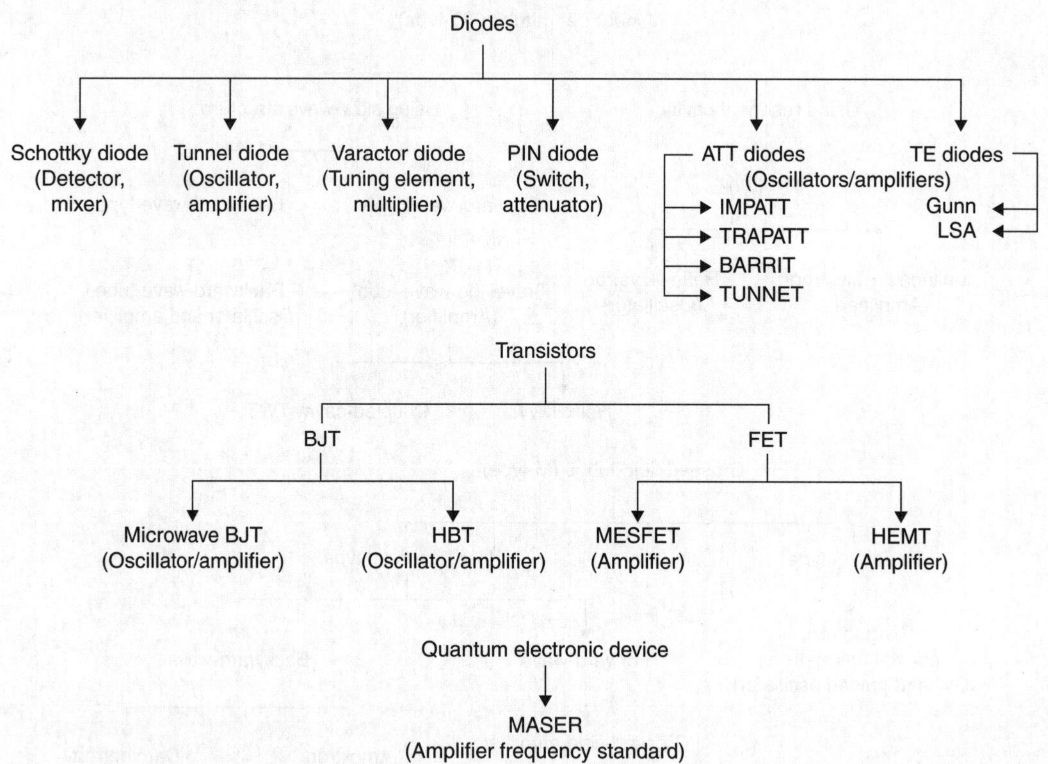

resonant cavities as the RF structure. The travelling-wave tube (TWT) is another type of important O-type tube that uses slow-wave structure (either helix type or coupled-cavity type). The TWT may be used as a forward-wave amplifier or as a backward-wave amplifier/oscillator. Let us discuss the operating principle of some of these devices.

9.2.1.1 Klystron

The klystron was invented in 1939 by the Varian brothers (R H Varian and S F Varian of the USA) for generation and amplification of microwave signals. In the present microwave technology, the klystron is primarily used as a two cavity/multi-cavity amplifier (both in the CW and pulsed mode) in microwave transmitters of UHF TV, long range radar, troposcatter link, in the earth stations of satellite communication, in linear particle accelerators and so on. The frequency range of application varies from 500 MHz to 94 GHz, with a gain of 30–40 dB at UHF and 60–70 dB in the higher microwave and millimeter-wave frequencies. Because of its high noise figure of 30–35 dB, the klystron amplifier is not used in the receiver stage of microwave systems. Klystron as oscillator, the reflex klystron, is a single cavity device with a power output of the order of 100–500 mW at the X-band.

Two cavity/multi-cavity klystron amplifier

Principle of operation: The principle of operation of a two-cavity/multi-cavity klystron amplifier may be understood as follows. The electrons emitted from the cathode are focussed and accelerated by an electron gun (the Pierce gun) to form a long cylindrical electron beam before the beam enters the RF interaction region 'A' (Fig. 9.1a). As the electron beam of the klystron passes the input resonant cavity (called the buncher) the velocity of electrons gets modulated by the input microwave signal (to be amplified) applied to the buncher. This velocity modulated electron beam, while traversing down the drift space (which is an RF field-free zone and this space is quite long and the transit-time of the electron in this region is put to use for the operation of the tube) results in density modulated or bunched beams. An output resonant cavity (called the catcher) placed at a suitable location along the electron path in the drift space, helps to extract the RF power from the density modulated electron beam and delivers the amplified power to the load.

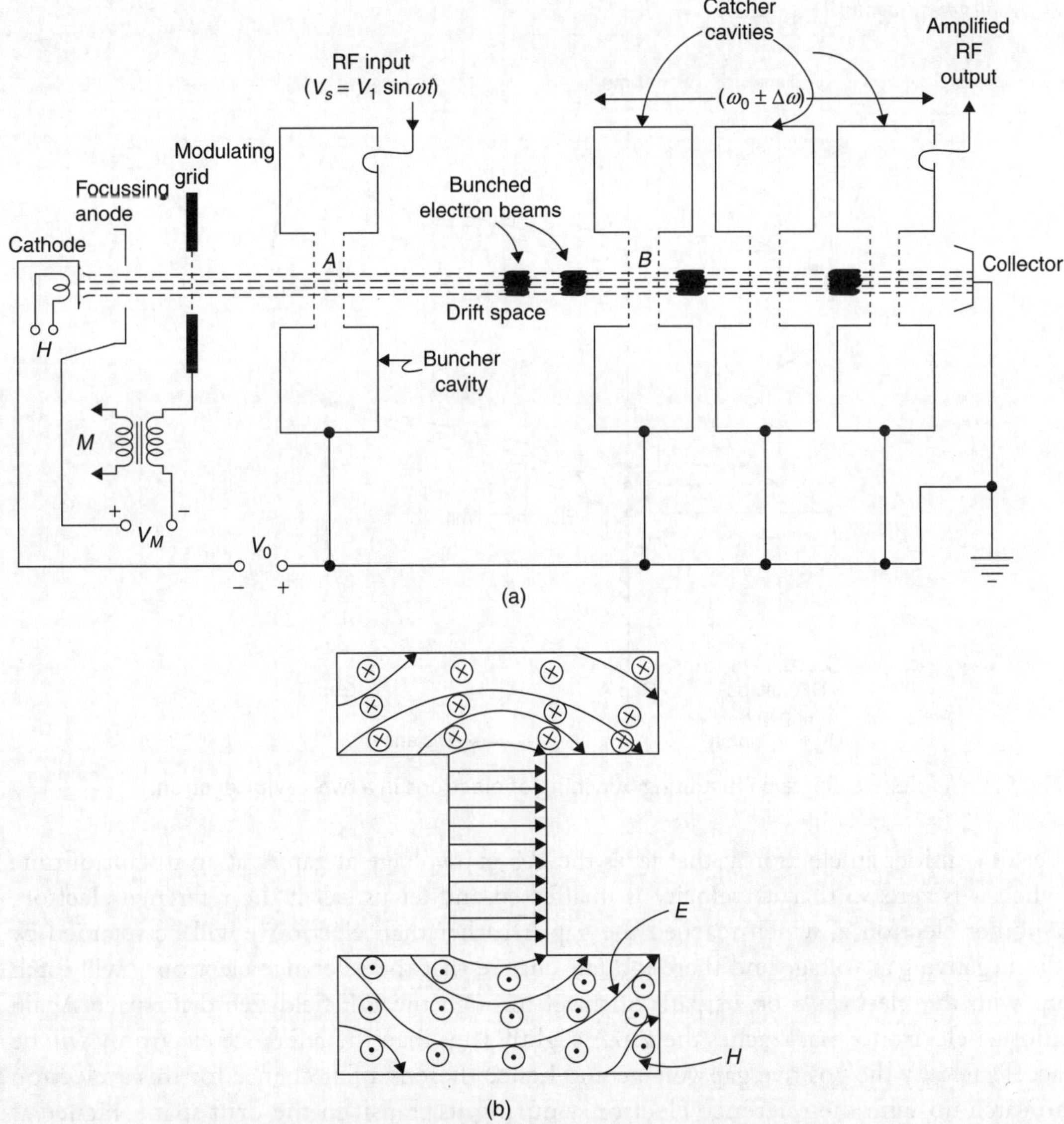

Fig. 9.1 (a) Schematic of two-cavity klystron amplifier (b) Re-entrant cavity.

The buncher and catcher cavities are made 're-entrant' (Fig. 9.1b), so that the electric field across the gap of these cavities is strong enough to interact with the electron beam passing through these cavity gaps. The electric field component of the electromagnetic field in the cavity is parallel to the electron beam axis. Thus, the electrons passing the resonator cavity see an accelerating field for half a cycle of the RF field and a retarding field for the other half time period. Consequently some of the electrons in the electron beam will be accelerated and some others are retarded. This phenomenon is called *velocity modulation* of the electron beam. As those electrons move further from the buncher cavity along the RF field-free drift space, the velocity modulated electrons cause *bunching* of electrons. The bunching effect converts the velocity modulation into what is known as the *density modulation* of the electron beam. This may be illustrated by a distance–time diagram, known as *Applegate diagram* (Fig. 9.2).

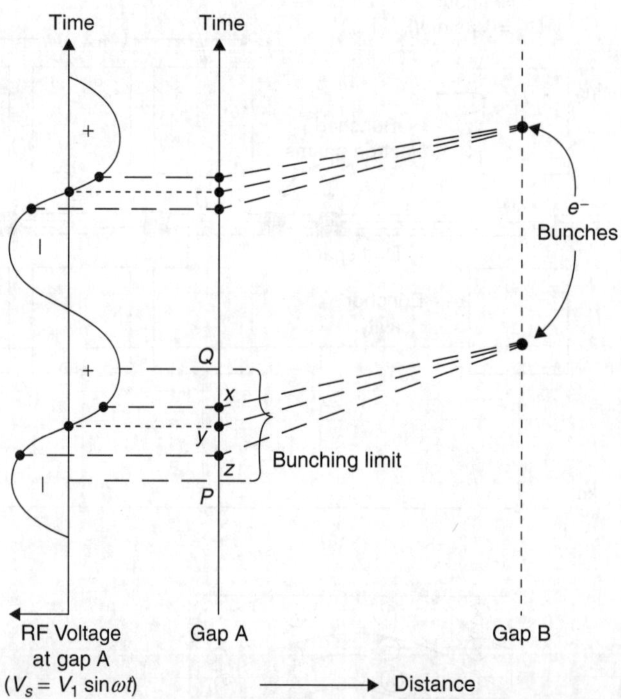

Fig. 9.2 Applegate diagram illustrating bunching of electrons in a two-cavity klystron.

Let us consider an electron y, that faces the RF gap voltage at gap A, at an instant of time when it is zero, so that its velocity is unaffected and let us call it the reference electron. Another electron z, which reached the gap A earlier than electron y will be retarded by the negative gap voltage and there is a fair chance that the reference electron y will catch up with the electron z on its path of travel through the RF field-free drift space. Again another electron x that reaches the gap A slightly later than the reference electron y will be accelerated by the positive gap voltage and hence there is a fair chance for this x electron to catch up with the reference electron y during its transit in the drift space. Hence at a distance d (say) from the buncher cavity in the drift space, they are *bunched* together. The bunch then moves with the average velocity of the electron beam. As the beam progresses further down the drift space, the bunching becomes more and more complete as more and more of the faster electrons catch-up with the bunches ahead. The catcher or the output cavity is placed at an optimum distance d_{opt}, from the buncher cavity where the bunching is maximum. The expression for d_{opt} is given by [16]:

$$d_{opt} = \frac{3.682\, v_0 V_0}{\omega \beta_i V_1} \tag{9.1}$$

where, v_0 is the electron beam velocity, V_0 is the beam voltage applied between the cathode and the collector, ω is the angular frequency of the input microwave signal to be amplified, β_i is the beam coupling coefficient, and V_1 is the amplitude of the input microwave signal applied to the buncher cavity.

The expression in Eq. 9.1 may be derived analytically; however, the parameter dependence of d_{opt} may also be understood physically as follows. The inverse proportionality of d_{opt} with V_1 and ω is due to the fact that when the input signal amplitude is weak, bunching will be more complete after a longer time of travel of the electrons via the drift space, while, if the frequency of input signal is high, then bunching limits are small and hence bunch formation takes place over a small time period and thus over a smaller length of drift space. The direct proportionality of d_{opt} with beam voltage or electron beam velocity is justified because when the electrons are accelerated with higher beam voltage, they will fly past longer length along the tube per unit time (the limit of this effect will also be controlled by debunching effect to be discussed soon).

It may be noted from the Applegate diagram that bunching occurs once per cycle of the input signal, centring on the reference electron y, that is, the bunching limit is PQ. The electrons arriving at gap A slightly after the second limit Q cannot be accelerated sufficiently to catch the reference electron, and the reference electron is unable to catch any electron passing gap A just before the first limit P. Bunches thus arrive at the catcher gap B once per cycle of the RF signal applied to the buncher cavity. Hence, the density of the electrons in the gap B varies cyclically with time which is same as the time variation of the input RF signal applied to the buncher. This periodic arrival of electron bunches at gap B results in an RF current (it may be noted that the electrons appear at gap A at a uniform rate while at B, there is a rate of change of the arrival of electrons in the form of bunches at the frequency of the input RF signal). If an alternating field can be made to exist in the output cavity (catcher cavity) and if the bunches of electrons pass through it at an instant of time in the cycle at which the AC field in the catcher cavity has a retarding value, the electrons will be slowed down by the field and the kinetic energy from the electron beam will be delivered to the RF field of the catcher cavity resonator. Since a little RF power applied to the buncher cavity results in enhanced beam current resulting in the catcher cavity, considerable power amplification is realised.

The requirement for the existence of a retarding RF field across the catcher cavity need not be realised with an external excitation, rather, the electron beam passing through the catcher cavity results in this field across the cavity by *beam coupling* which may be understood as follows. From Fig. 9.3, we can see the catcher cavity at two instants of time for the entry and exit of the electron bunch through it. We know that a point charge in the space between the grids of the re-entrant cavity would induce upon the inner surfaces of the grids a total charge equal in magnitude to the point charge, but of opposite polarity, the ratio of induced charges on the two grids being inversely proportional to the ratio of distances of the point charge from the respective grids. Now, when the first bunch of electrons enter the catcher cavity, their movement from the vicinity of grid 1 to grid 2 reduces the positive charge being induced on the outside of grid 1, as is evident from the diagram depicting the entry and exit instants of the electron through the catcher cavity (Fig. 9.3). This eventually means that the next bunch of electrons appearing cyclically with input RF frequency at the catcher cavity will see a retarding field (one going from higher positive potential to lower positive potential value, that is, with a negative gradient) at its entry instant to the catcher cavity. Thus, the electron bunch will be retarded in velocity and would impart some of its KE to the RF field (the retarding field in the catcher cavity) causing exchange of energy from the electron beam to the RF field of the catcher cavity.

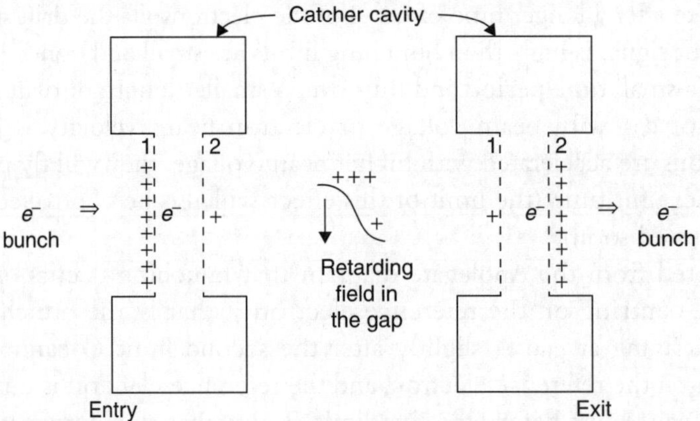

Fig. 9.3 Schematic to explain the beam coupling in the catcher cavity of a klystron.

EXAMPLE 9.1

A double resonator klystron is to give maximum power output at a fundamental frequency of 6 GHz. The accelerating voltage is 1000 volts and the amplitude of the input microwave signal is 3 volts. Assuming that the beam coupling coefficient is 1, what is the optimum length of the drift space?

Solution

The electron beam velocity v_0 is given by:

$$v_0 = \sqrt{2eV_0/m} = 0.593 \times 10^6 \sqrt{V_0} \text{ m/s} = 18.75 \times 10^8 \text{ cm/s, as } V_0 = 1000 \text{ V}$$

$$\omega = 2\pi f = 2 \times 3.142 \times 6 \times 10^9 = 37.7 \times 10^9 \text{ radians}$$

Given $V_1 = 3$ volts and $\beta_i = 1$.

Thus, from Eq. 9.1,

$$d_{\text{opt}} = \frac{3.682\, v_0 V_0}{\omega \beta_i V_1} = \frac{3.682 \times 18.75 \times 10^8 \times 1000}{37.7 \times 10^9 \times 1 \times 3} = 61 \text{ cm}$$

Practical considerations: Let us now discuss some of the practical considerations related to the operation of two/multi-cavity klystron amplifiers. The usual drift length of a klystron may be approximately 60 cm or more (as has been seen in the above example). Thus, mutual repulsion of electrons during its motion along the drift space might cause *debunching* of electrons. To counteract this undesired effect of debunching, an axial magnetic field is applied externally to the klystron tube by means of an electromagnet. The magnetic field helps to confine the electrons to a relatively long thin beam exerting a radially directed force to the electron beam in the tube. Water cooling is required for such electromagnets and also at the anode which collects the bombarding electron beam after its passage through the catcher cavity. The size and weight of the klystron is primarily contributed by the electromagnet, cooling arrangement and the large power supply (of the order of kilovolts) for the beam voltage applied between the cathode and the anode.

The out-of-phase electrons that arrive at the catcher cavity at random between bunches give rise to high noise figure in a two-cavity klystron amplifier. In a practical system, more than two cavities (4–6) are used to increase the gain of the klystron amplifier and to reduce the noise figure to some extent (though even after that, the noise figure is typically 25–30 dB). The multi-cavity klystron also helps to improve the bandwidth of operation of the amplifier by stagger tuning; three catcher cavities are tuned to frequencies $(f_0 - \Delta f)$, f_0 and $(f_0 + \Delta f)$ for a 4-stage klystron amplifier. However, even with stagger tuning, the amplifier bandwidth is never more than 8% to 10% as resonant cavities are involved.

A common method for pulsing the beam of a klystron amplifier is with an electrode in the electron gun that controls the klystron beam current. In this method, a negative bias is applied to the modulating grid (also known as shadow grid) which cuts off the electron bunch to reach the buncher cavity. The presence of the modulating grid also helps to realise amplitude modulation of the klystron, if desired.

Reflex klystron oscillator: The klystron amplifier discussed above can be used as an oscillator if a fraction of the output power is fed back to the input cavity with proper phase and maintenance of the proper loop gain. However, in this configuration, if the oscillator frequency needs to be changed, the resonant frequency of each cavity and the phase shift must be readjusted for necessary positive feedback for oscillation to take place at a new frequency. But, with the advent of the CW magnetron with better power output and higher efficiency, the conversion of the klystron amplifier to operate as microwave oscillator has never been used in practical applications. The form of klystron oscillator that has found wide applications as low to medium power microwave oscillator is the *reflex klystron* which is a single-cavity klystron with a different principle of operation when compared with two-cavity/multi cavity klystron amplifier. The reflex klystron oscillator has been used for a long time in the laboratory for microwave measurements, as local oscillator in microwave receivers, in airborne Doppler radar and so on. However, in the present microwave systems, its competitor, the Gunn oscillator with similar or better performance features, is fast replacing it in many microwave system applications.

Principle of operation: The reflex klystron also has an electron gun but of much smaller size; however, it does not need any external magnetic field for beam focussing as its drift space is much smaller compared to the two cavity/multi cavity klystron amplifier. Here, the electron beam injected from the cathode is accelerated towards the RF cavity by the beam voltage V_a and focussed by the focussing anode (Fig. 9.4).

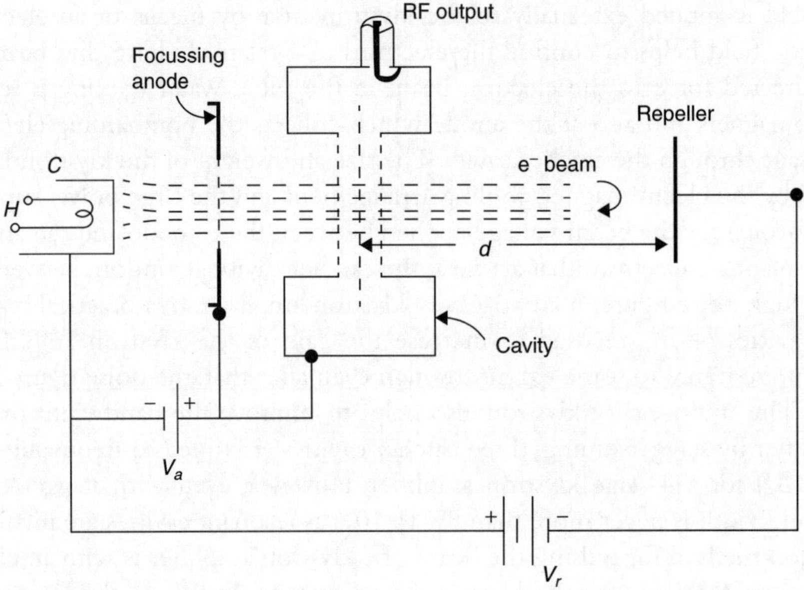

Fig. 9.4 Schematic of reflex klystron.

The cavity gap voltage (which originates as initial stimulus as in any oscillator from the noise or switching transient) velocity modulates the electron beam. Then some of the electrons are accelerated by the gap field (facing the positive half cycle) and some others are retarded by the gap field (falling in the negative half cycle) while some others are not affected at all as they fall in the time corresponding to the mean line of the gap field. But none of these electrons can reach the *repeller* which is given a negative potential. Hence, the electrons are returned to the cavity in the form of bunches (as explained in Fig. 9.5). The bunches can be made to form once per cycle of the RF gap field by choosing the distance of the repeller from the cavity and by adjusting the negative voltage applied to it at a particular frequency of operation. Further, a little adjustment of the repeller voltage would lead to the situation when the bunches can be made to return to the cavity during the retarding phase of the RF gap field. This will result in the exchange of kinetic energy

(KE) from the electron beam to the RF gap field leading to the growth of the oscillation amplitude in the cavity. The electrons are finally collected by the walls of the cavity. The conversion efficiency of the microwave oscillator relates to the conversion of the DC power supply signal via the electron beam to the generated microwave signal. Stable oscillations build up in the cavity when the KE delivered to the cavity from the electron beam gets over the losses in the cavity and the load. The signal from the RF cavity can be coupled out to the load via magnetic loop coupling (Fig. 9.5).

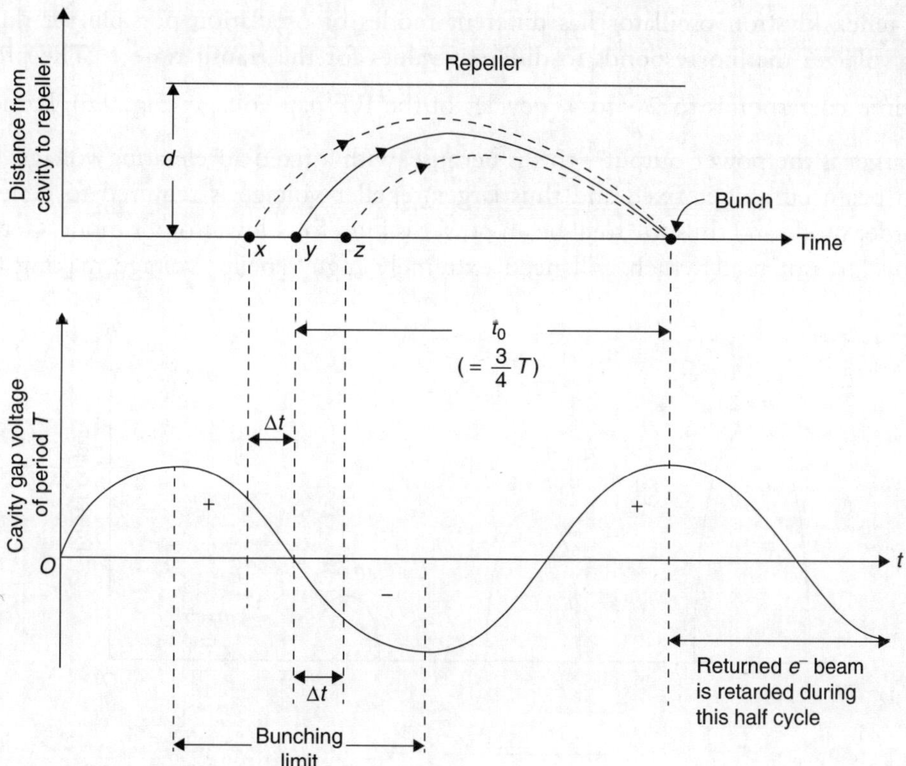

Fig. 9.5 Bunching mechanism in reflex klystron.

The bunch formation and subsequent oscillation mechanism in the reflex klystron oscillator may be understood as follows (Fig. 9.5). Let us consider an electron y which passes through the resonator gap when the voltage across the gap is zero (changing from +ve to –ve). This electron, termed as the reference electron, returns to the resonator at a time t_0 later. Another electron x that passes through the gap Δt time earlier (with reference to the y electron) will possess higher initial velocity compared to the reference electron y (since it falls in the +ve half cycle of the RF gap field). Thus, the electron x would travel further towards the repeller and hence a longer time would be required for it to come back to the resonator. It would thus be caught up by the reference electron y. Similarly, the electron z that passes the resonator gap Δt time after the reference electron y, starts with a smaller initial velocity (since it falls in the –ve half cycle of the RF gap field) and thus take less time to come back to the resonator. Thus, the electron z ultimately catches up with the electrons y and x forming an electron *bunch* around the reference electron y. The electron bunches thus formed will deliver power to the resonator if they pass through the resonator at an instant of time when the field across it can retard the bunch (during the half cycle when it is changing from +ve to –ve; in other words, having negative slope of the field). It may be observed from Fig. 9.5, that this happens when $t_0 = \left(n + \frac{3}{4}T\right)$, where $n = 0, 1, 2, \ldots$ and T is the period of the RF voltage across the resonator gap.

The reflex klystron oscillator has different modes of oscillation possible for different repeller voltages that corresponds to different values for the transit time t_0. The practical transit time corresponds to $2\frac{3}{4}$ to $4\frac{3}{4}$ cycles of the RF gap voltage (Fig. 9.6). Earlier the mode, larger is the power output—this is because, with a fixed accelerating voltage V_a, the electron beam current is fixed and thus larger repeller voltage is required to generate a lower order mode and thus the total beam power is high for a lower order mode. Generally, n = 0 or 1 is not used which will need extremely high repeller voltage making power

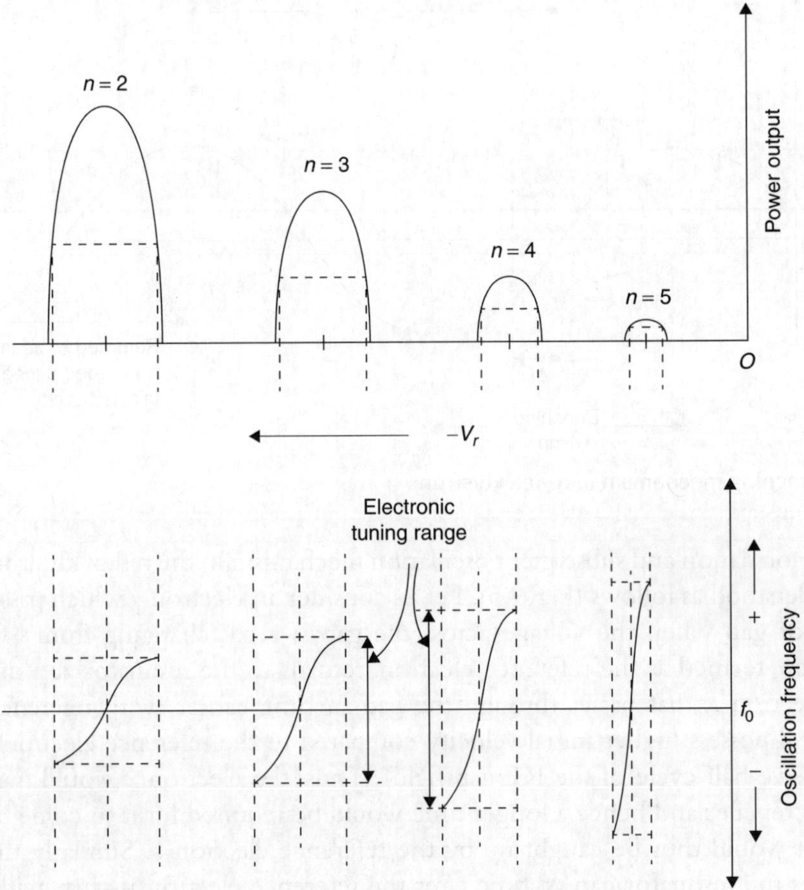

Fig. 9.6 Oscillation modes and frequency tuning of reflex klystron oscillator.

dissipation large and hence lowering the oscillator efficiency. Higher values of n (> 4) are not used since gain mechanism would be insufficient to get over the losses in the system.

Frequency tuning characteristics: Frequency tuning characteristics of a reflex klystron oscillator (Fig. 9.6) may be understood as follows. If the electron bunch (which on exchanging its KE to the RF gap field gives rise to the RF current) returns to the cavity at a time little shorter than $t_0 = \left(n + \frac{3}{4}T\right)$, the RF current would lag behind the gap RF voltage and hence an inductive impedance or capacitive susceptance is presented to the cavity, lowering the oscillation frequency of a particular mode. This situation demands a higher negative repeller voltage compared to the resonant condition for that particular mode of oscillation. On the other hand, if the repeller voltage is set at a slightly lower negative voltage for a particular mode, then capacitive impedance or inductive susceptance would raise the oscillation frequency above the resonant value for that particular mode of oscillation. Thus, if the time varying bias modulation voltage is superimposed on the DC repeller voltage of the reflex klystron, then electronic tuning of the oscillator can be obtained. This is useful for frequency modulation and automatic frequency control in the klystron oscillator.

A point of caution for the operation of a reflex klystron is that the repeller voltage should always be applied before the cathode supply voltage. Otherwise, accelerated electrons from the cathode might bombard the repeller electrode and destroy it or produce over heating.

9.2.1.2 Travelling-wave tube

The travelling wave tube (TWT) is a very broadband (octave bandwidth, $f_2/f_1 = 2$, where $\Delta f = f_2 - f_1$) and moderately low noise (noise figure of the order of 8–10 dB) microwave amplifier with reasonably high gain (50–60 dB). It is extensively used as the front-end RF amplifier in the ground station of space communication receivers. Also due to its high reliability, that is, having large mean time before failure (MTBF), the TWT is used in space borne systems. It finds application as a power amplifier in very high power radar transmitters used in master oscillator power amplifier (MOPA) configuration. A popular use of TWT in military radar systems is for radar jamming [in electronic counter measure (ECM)]. In this application, the TWT is fed with a broadband noise source and its output is transmitted to confuse enemy radar.

The inventors of TWT, Kompfner of the UK and Pierce of the USA (shortly after World War II they invented it independently) were dissatisfied with the brief interaction between the electron beam and the RF field in the klystron (interaction occurring only at the discrete gaps of a relatively few resonant cavities, which limits its bandwidth) and so they invented a special technique by which a continuous interaction can take place between the electron beam and the RF field all along the length of the tube over a large number of cycles of the RF signal to be amplified, thereby providing an enormous bandwidth.

Principle of operation: In a TWT (Fig. 9.7), the microwave signal which, in general, travels with the velocity of light, is slowed down by a slow wave structure (which may

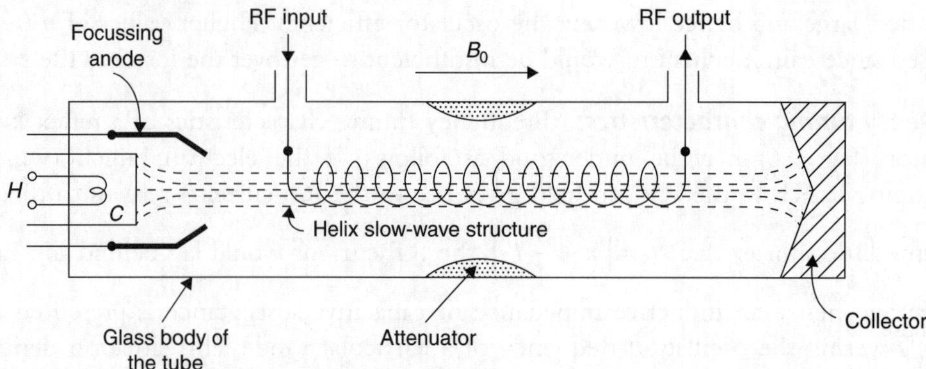

Fig. 9.7 Schematic of travelling wave tube (TWT).

be a helix or a coupled-cavity structure). This slowed RF field (with velocity ≈ 10% of the velocity of light), while travelling from input to output of the tube, interacts continuously with the electron beam over the total length of the slow wave structure and over a large number of cycles of the input microwave signal to be amplified. Due to this close interaction of the electron beam and the RF field, a broadband amplified microwave signal is obtained at the output of the tube.

The interaction of the electron beam with the slowed RF field existing in the slow-wave structure and the consequent amplification of the microwave signal can be understood as follows. When the electron beam enters the helix structure, an interaction takes place between the axial electric field of the slowed electromagnetic field (which is a travelling wave and not a standing wave) and the moving electrons. The electric (E) field of the time varying slow wave (axial) RF signal causes a velocity modulation of otherwise uniform velocity electrons passing through the helix axis. The velocity modulation tends to bunch the electrons at regular intervals of one wavelength (Fig. 9.8a). If the velocity of electrons (v_e) caused by the DC beam voltage is slightly greater than the axial velocity, that is, the phase velocity (v_p) of the slowed RF wave, then the bunches would face the decelerating RF field and will be retarded and hence impart some of its KE to the RF field. This continues all along the length of the slow wave structure and over a large number of cycles of the RF field. Therefore, the bunches form more and more completely and consequently delivering more and more of their KE to the RF field thereby growing or amplifying the RF field exponentially from input to output (Fig. 9.8b) and finally the retarded electrons are collected by the anode.

It is found that the helix supports a slow wave which propagates with its electric field $\vec{E}$ along the axial direction of the helix with a phase velocity $v_p = c \sin\psi$, where ψ is the pitch angle and c is the velocity of light [7]. From Fig. 9.9a we can write, $\sin\psi = p/2\pi a$, where p and a are respectively the pitch and radius of the helix. Finally we have the expression for v_p as:

$$v_p = \frac{c}{n\pi D} \tag{9.2}$$

where n is the number of turns per unit length of the helix and D is the helix diameter.

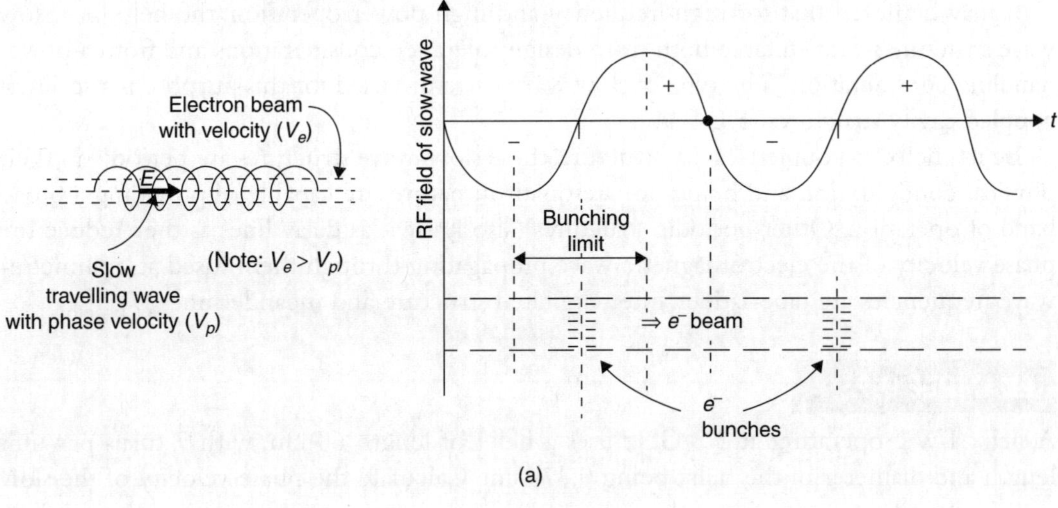

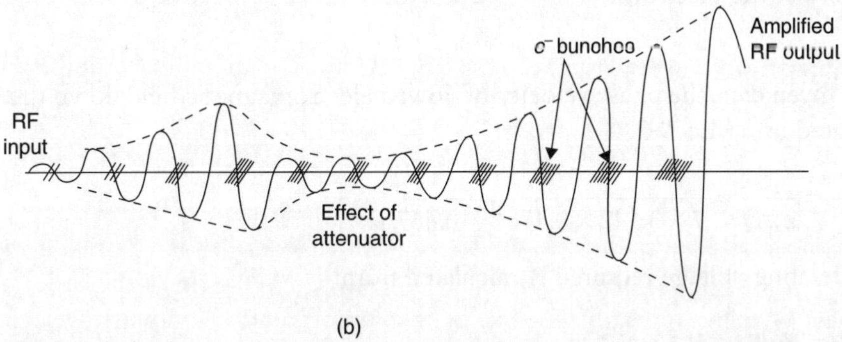

Fig. 9.8 (a) Bunch formation in TWT (b) Growth of RF field showing the effect of the presence of attenuator.

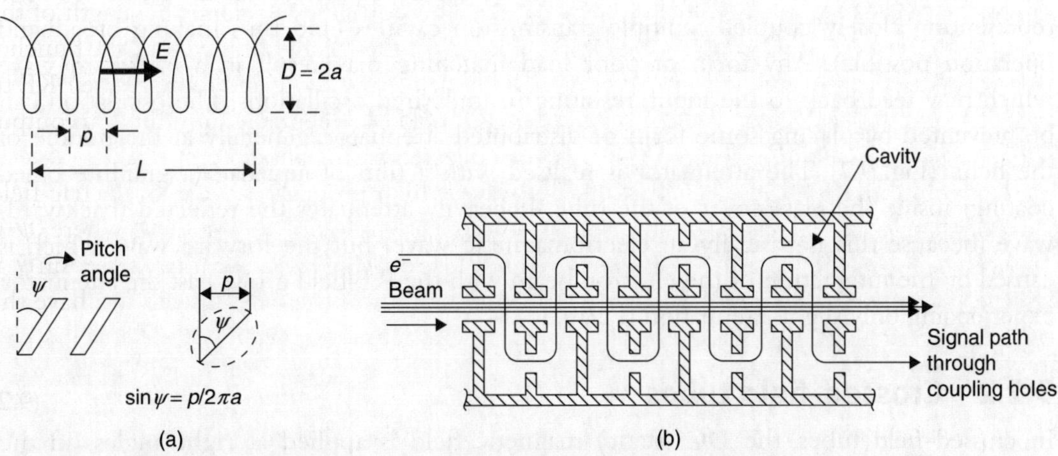

Fig. 9.9 (a) Helix as slow-wave structure (b) Coupled-cavity structure.

It may be noted that for high frequency and high power operation, the helix as a slow wave structure is not suitable both from design tolerance considerations and from a power handling consideration. The type of slow wave structure used for this purpose is a series of coupled cavity structures (Fig. 9.9b).

Be it a helix or coupled-cavity structure, these slow-wave structures are periodic in their physical configuration and being non-resonant in nature, are capable of providing a broad band of operation. Other periodic structures (also known as delay lines as they reduce the phase velocity of the electromagnetic wave propagating through them) used at millimetre-wave frequencies are tape-ladder, inter-digitated structure and meander lines [7].

EXAMPLE 9.2

A helix TWT operating at 3.3 GHz uses a helix of length 60 cm, with 7 turns per unit length and diameter of the helix being 0.476 cm. Calculate the phase velocity of the slow wave that will propagate down the helix and the accelerating voltage required to produce this velocity of the electrons.

Solution

With the given data, the phase velocity of slowed electromagnetic field down the helix may be calculated using Eq. 9.2:

$$v_p = \frac{c}{n\pi D} = \frac{c}{7 \times 3.142 \times 0.476} = \frac{c}{10.467}$$

The accelerating voltage required is calculated from:

$$v_p = \frac{v_p^2 m_e}{2e} = \frac{0.5 \times c^2 \times m_e}{(10.467)^2 \times e} = 2.335 \text{ kV (since } \frac{e}{m_e} = 1759 \times 10^8 \text{ Ckg}^{-1})$$

Prevention of oscillation: TWT is a fairly high gain amplifier; with the helix structure representing closely coupled multiple transmission cavities (and thus making broadband operation possible). Any form of poor load matching may result in a backward wave which may feed back to the input, resulting in undesired oscillations. The oscillation can be prevented by placing some form of distributed attenuator, generally at the middle of the helix (Fig. 9.7). The attenuator is realised with a film of aquadag (a graphite based coating) inside the glass cover of the tube. It heavily attenuates the reflected (backward) wave (because this is basically an electromagnetic wave) but the forward wave which is caused by the interaction of the electron beam with the RF field grows past the attenuator, experiencing only slight attenuation (Fig. 9.8b).

9.2.2 Crossed-field tubes

In crossed-field tubes, the DC (static) magnetic field is applied at right angles (in the perpendicular direction) to the DC electric field—crossed to each other; both these externally applied DC magnetic and DC electric fields take part in the interaction of the electron beam with the RF field for operation of these tubes. They are also known as M-type tubes after the French acronym TPOM (tubes à propagation des ondes à champs magnétique) meaning tubes for propagation of waves in a magnetic field. The primary crossed-field tube is the magnetron that uses coupled resonant cavities as an RF structure with its varied forms as coaxial and inverted magnetrons. Pulsed magnetrons are primarily used as microwave oscillators in radar transmitting systems while continuous wave (CW) magnetrons are used in microwave heating. The use of a non-resonant RF structure in a crossed-field device results in forward and backward crossed field amplifiers (CFA) which are designed for specific practical use. For example, the backward wave crossed field amplifier developed by Raytheon Company in 1960 (trade name amplitron) is a high power, high gain and high efficiency microwave tube, and it has many applications in airborne radar systems and space borne communication systems.

9.2.2.1 Magnetron

The cavity type (or travelling wave) magnetron was developed by Randall and Boot of the UK at around the same time when the klystron was developed in the USA. Magnetrons are primarily used for generating pulsed signals for radar transmitters. Maximum available peak powers are 2 MW at X-band and 8 kW at W-band with a typical duty cycle of 0.01. CW magnetrons are also available and are used for industrial heating and also in our household cooking appliance–the microwave oven; at the heart of which is a CW magnetron (typically, frequency: 2.45 GHz, CW power: 500W and efficiency is 40% to 50%).

Principle of operation: The mechanism of generation of microwave signals in a magnetron involves the interaction between the electromagnetic fields existing in a system of coupled microwave cavities and the electron path resulting from their motion in static (DC) electric and magnetic fields oriented at right angles to each other (crossed-field). The pulsed RF signal is produced with the magnetron by switching the magnetron ON and OFF by a pulser.

The cavity magnetron or the travelling wave magnetron oscillator, is a diode, usually of cylindrical construction (Fig. 9.10). It employs a radial DC electric field, an axial DC magnetic field and an anode structure with vane type or hole and slot type re-entrant cavities. The cathode (in the form of a cylindrical rod made of barium strontium oxide for copious thermionic emission) is surrounded by the anode with coupled cavities. The output is taken from one of the cavities, by means of a coaxial line or waveguide, depending on the power and frequency.

The electron beam in a cylindrical magnetron under the effect of the crossed electric and magnetic field describe an epi-cycloidal path around the cathode (Fig. 9.11a). This motion of the electron path is represented by the superposition of a slow rotation around the cathode (R_0 rotation) and a faster looping on the surface of the cathode (r_0 rotation). The magnetic field strength can be suitably set so that the electron path just grazes past the anode surface [Fig. 9.11b(ii)], giving the cut-off condition (called the Hull cut-off condition, after E W Hull who first proposed the idea of a crossed-field device). The cut-off DC magnetic field is given by [16]:

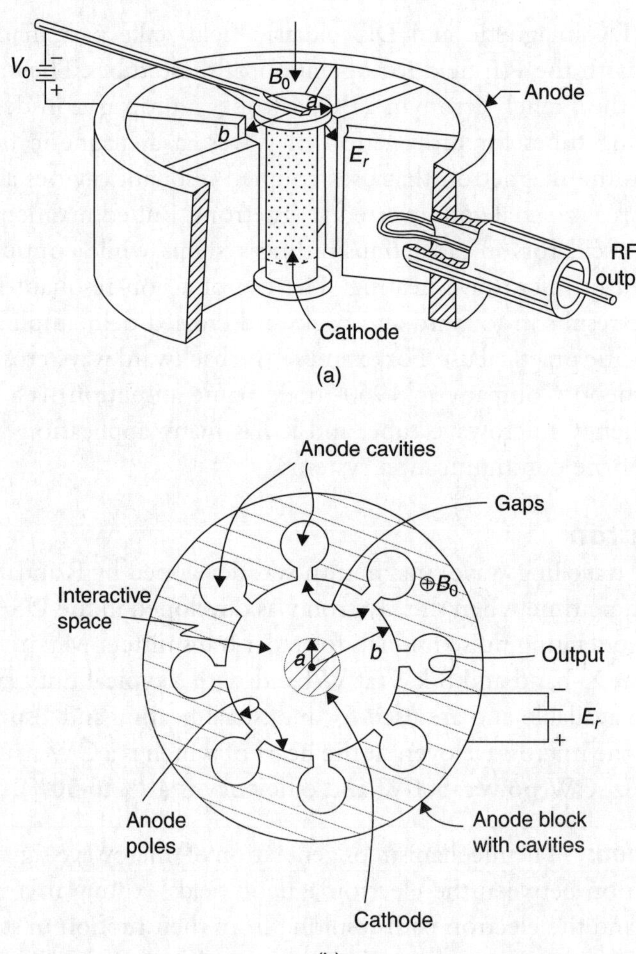

Fig. 9.10 **Cavity magnetron oscillator (a) Cylindrical magnetron with vane-type anode cavities (b) Cross-sectional view with hole and slot type cavities.**

$$B_{0c} = \frac{(8V_a m/e)^{0.5}}{b\left(1 - \dfrac{a^2}{b^2}\right)} \tag{9.3}$$

where a is the cathode radius, b is the radius of the anode block (Fig. 9.10) and V_a is the applied positive voltage to the anode with respect to the cathode. In practical operation of the device, the applied DC magnetic field B_0 is chosen to be greater than B_{0c} [Fig. 9.11b(iii)].

The equation of electron trajectory (the epi-cyloidal path) around the cathode in a cylindrical magnetron can be derived by solving the problem in the cylindrical coordinate system; however, it is worthwhile to understand the nature of the electron trajectory in a planar magnetron and extend that physical picture in a cylindrical system as follows. It may be derived [20] that in a planar magnetron (Fig. 9.12a) the equations of motion of the electron from the cathode to the anode are governed by:

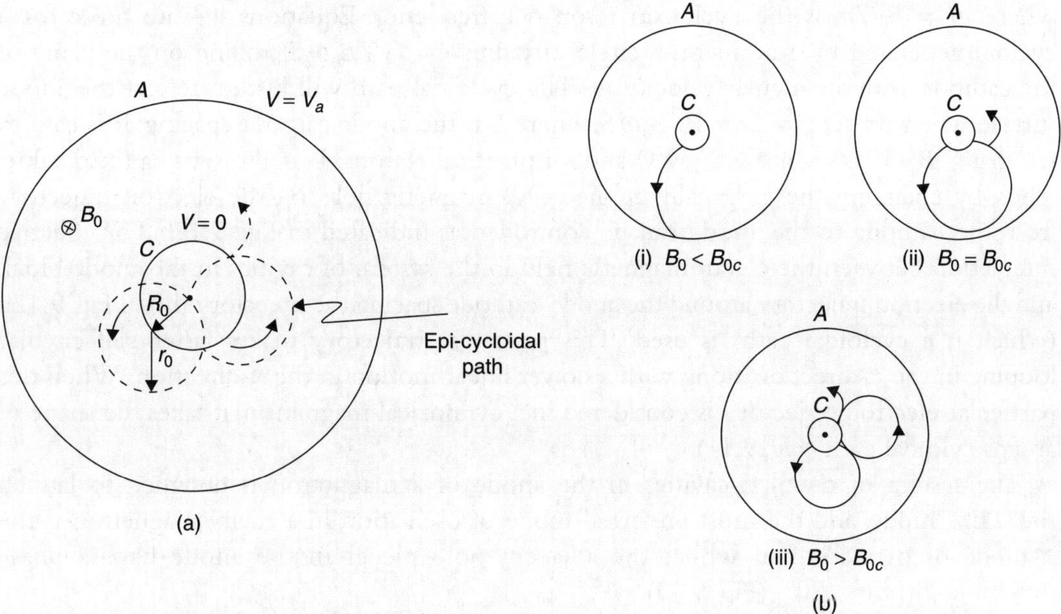

Fig. 9.11 (a) Epi-cycloidal path of the electron around the cathode of a cylindrical magnetron (b) Applied magnetic field (B_0) with respect to the cut-off field (B_{0c}) for control of epi-cycloidal path around the cathode.

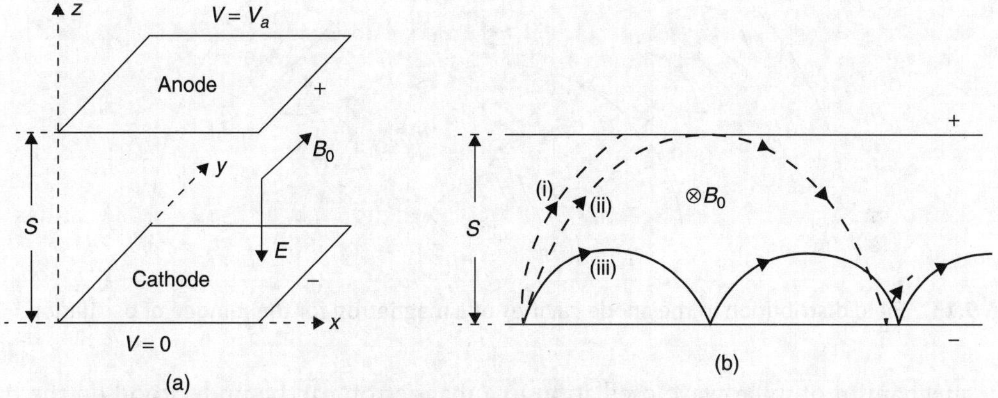

Fig. 9.12 (a) Schematic of a planar magnetron (b) Electron trajectory in a planar magnetron with different strengths of the applied magnetic field: (ii) With $B_0 = B_{0c}$, (i) With $B_0 < B_{0c}$ and (iii) With $B_0 > B_{0c}$; B_0 is the applied DC magnetic field and B_{0c} is the cut-off magnetic field.

$$
\begin{aligned}
x &= \frac{V_a}{B_0 \omega_c S} (\omega_c t - \sin \omega_c t) \\
y &= 0 \\
z &= \frac{V_a}{B_0 \omega_c S} (1 - \cos \omega_c t)
\end{aligned}
\tag{9.4}
$$

where $\omega_c = B_0 e/m$ is the cyclotron resonance frequency. Equations 9.4 are those for a cycloid generated by a point on a circle of radius $r = (V_a/B_0\omega_c S)$ rolling on the plane of the cathode with an angular velocity ω_c. The cycloidal path will just graze past the anode surface when we set $S = 2r = V_a/B_0\omega_c S$, where S is the anode cathode spacing. It is easy to see that $S^2B_0^2/V_a$ is a constant (say k) and for practical reasons, V_a is also set to a fixed value. Thus, by changing the externally applied DC magnetic field B_0, the electron trajectory from the cathode to the anode can be controlled as indicated in Fig. 9.12b. For effective interaction between the electromagnetic field in the system of cavities in the anode block and the electron trajectory around the anode-cathode spacing the trajectory (iii) in Fig. 9.12b (which is a cycloidal path) is used. This particular trajectory makes faster half-circular looping in the z-direction along with a slower linear motion in the x-direction. When this particular electron trajectory is considered in a cylindrical magnetron, it takes the shape of an epi-cyloidal path (Fig. 9.11).

The system of coupled cavities in the anode of a magnetron is designed to launch the TE_{110} mode, and the most preferred mode of oscillation in a cavity magnetron is the π-mode of oscillation in which the adjacent pole pieces in the anode have a phase difference of π (= 180°) (Fig. 9.13).

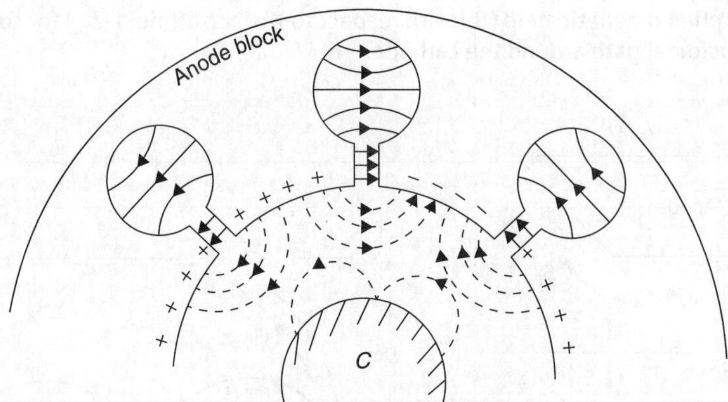

Fig. 9.13 Field distribution in the anode cavities of a magnetron for the π-mode of oscillation.

The mechanism of microwave oscillation in a magnetron can be understood on the basis of the interaction of the epi-cycloidal electron paths with the RF gap fields in the coupled microwave cavities and subsequent exchange of energy from the electron beam to the RF field in the cavities (Fig. 9.14).

Depending on the instant of emission of electrons from the cathode surface with respect to the RF gap field, there can be *unfavoured* and *favoured* electrons (Fig. 9.14). If an electron starts from the cathode at an instant in the cycle of the RF gap field when it is increasing from –ve to +ve (as for the '*a*' electron), the field helps to pull the electron towards the anode on its outward flight and pushes it towards the cathode in its return flight because then the RF gap field is changing from +ve to –ve. Thus, the '*a*' type electron is accelerated all along its path and hence takes energy from the RF gap field and returns quickly to the cathode or may even hit the anode on its first outward flight. This type of electron is known as an *unfavoured* electron as it extracts energy from the field existing in the resonator system. On the other hand, if an electron leaves the cathode surface at an instant of the RF gap field when the field changes from +ve to –ve (as for the '*b*' electron), the field pushes the electron towards the cathode on its outward flight while on its return flight it sees the field changing from –ve to +ve, thereby pulling the electron towards the anode on its return flight. The '*b*' type electron is thus retarded all along its path of travel from the cathode to the anode, delivering energy to the resonator system over a number of cycles of the RF gap field. This type of electron is thus known as a *favoured* electron. The same number of each kind of electron, '*a*' and '*b*', leaving the cathode surface in unit time is on an average, equally probable. Thus, as a whole, the resonator system is supplied with energy (which is the difference of energy delivered by the favoured electrons and the energy extracted by the unfavoured electrons) from the electron beam due to interaction of the RF gap field and the epi-cycloidal electron motion under crossed DC electric and magnetic fields.

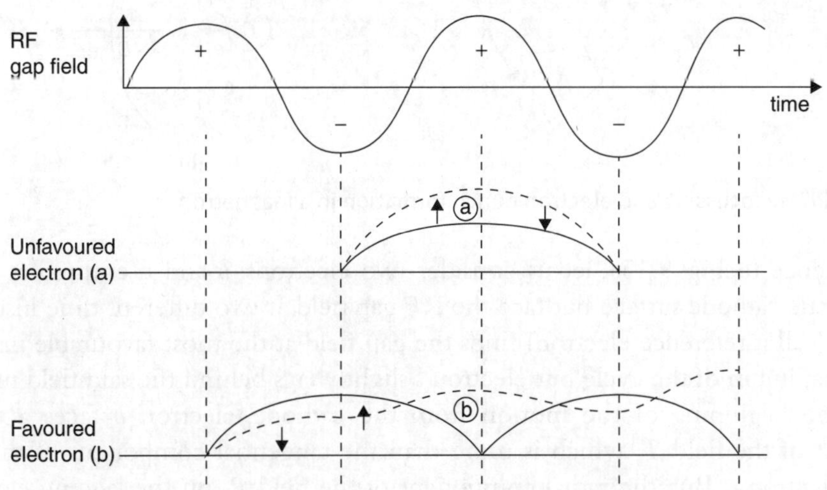

Fig. 9.14 Interaction of the electron path with the RF gap field across a cavity and exchange of energy from electron beam to the RF field in cavity magnetron.

The original stimulus for oscillation, that is, the initial RF gap field in the cavity system, originates from the switching transient which subsequently builds up with the energy supplied from the DC power supply, through the electron beam via the interaction of the electron beam with RF gap field (hence the efficiency of the oscillator is to be understood as the DC to RF conversion efficiency). The oscillation builds up until it overcomes the losses in the resonator system and the load coupled to it, when sustained oscillation at a frequency determined by the cavity dimensions takes place.

It is understood that the so called favoured electrons are responsible for energy exchange from the DC power supply via the electron beam to the RF field of the coupled anode cavity system in a magnetron. However, the unfavoured electrons are seen to extract some energy from the RF field and have thus earned the negative name—*unfavoured*. The question that naturally arises is, is it of no use at all even in some other perspective? In nature, nothing is futile—the unfavoured electrons return to the cathode surface with a very high velocity and tend to heat it due to this bombardment. This is known as *back heating* and is useful in the sense that it is often possible to shut-off the filament supply in a magnetron after a few minutes and just rely on the back heating to maintain the correct cathode temperature.

The favoured electrons deliver energy to the anode cavity system through electron bunches and the *bunching* of the favoured electrons in the magnetron takes place due to *phase focussing*, which may be understood with reference to Fig. 9.15.

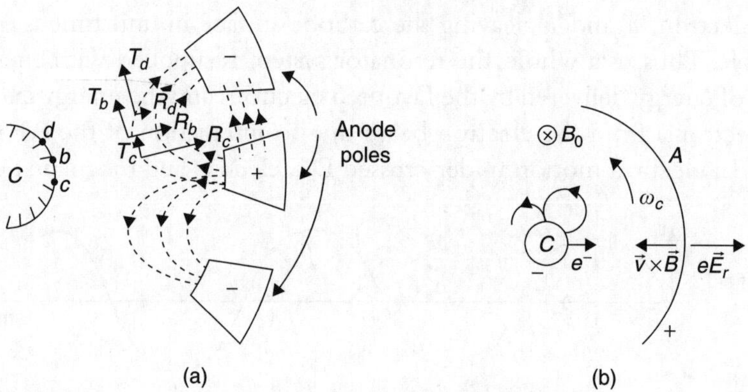

Fig. 9.15 Phase focussing and electron bunch formation in a magnetron.

With reference to Fig. 9.15a, let us consider two electrons, *b* and *c*, emitted at the same time from the cathode surface but face the RF gap field at two different time instants. The electron *b* (call it reference electron) finds the gap field at the most favourable instant, that is, at the maximum of the cycle but electron *c* slightly lags behind the said field maximum. Now, at the beginning of the motion from the cathode, electron *b* faces a tangential component of the field T_b which is more than the tangential component of the field T_c faced by electron *c*. But, the radial component of the field R_c on the lagging electron *c* is more than the radial component of the field R_b on the electron *b*. As this happens, the DC magnetic field exerts a stronger force on electron *c* tending to bend it back to the cathode and also accelerating it somewhat in the anti-clockwise direction (Fig. 9.15b). In turn, the lagging electron *c* gets a good chance to catch up with the reference electron *b*. In a similar way, a leading electron *d* is retarded and also gets caught-up by the *b* electron. Thus, the electrons *c* and *d*, which were respectively lagging and leading in phase with respect to the reference electron *b*, while being emitted from the cathode are brought to form a *bunch* in the anode–cathode spacing by *phase focussing*.

The result of this bunch formation by phase focussing is the *wheel-spoke* situation of the electron cloud in the anode–cathode space (Fig. 9.16). These spokes of the electron bunches form continuously and rotate in synchronism with the RF gap field, delivering energy to the cavity system. Hence, the descriptive name of a cavity magnetron is *travelling-wave magnetron*; this has nothing to do with the travelling wave tube (TWT) discussed in the previous section.

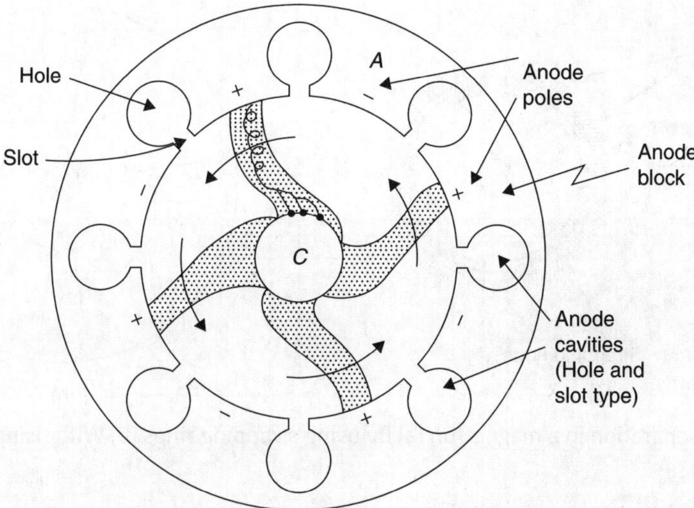

Fig. 9.16 Wheel and spoke situation of the electron cloud in the *anode–cathode* spacing of a travelling wave magnetron.

EXAMPLE 9.3

A cylindrical magnetron operating in the frequency range 2.72–2.89 GHz has the following parameters: $a = 0.119$ inch, $b = 0.315$ inch and $V_a = 20$ kV. Calculate the cut-off magnetic flux density for the magnetron. Given $e/m = 1.76 \times 10^{11}$ Ckg^{-1}.

Solution

With the given data, we can calculate B_{oc} by using Eq. 9.3:

$$B_{oc} = \frac{(8V_a m/e)^{0.5}}{b\left(1-\dfrac{a^2}{b^2}\right)} = \frac{(8\times 20\times 10^3/1.76\times 10^{11})^{0.5}}{0.8\left\{1-\left(\dfrac{0.302}{0.80}\right)^2\right\}\times 10^{-2}} = \frac{9.534\times 10^{-4}}{0.686\times 10^{-2}} = 13.90\times 10^{-2}\ \text{tesla}$$

Practical Considerations

(i) Mode separation: Because a magnetron has multiple coupled cavities, several closely spaced resonances are possible giving rise to *mode jumping*. This might result in oscillation at reduced power and at a wrong frequency. At microwave frequencies, the use of *strapping rings*, and at millimeter-wave frequencies, the use of the *rising-sun resonator* system can solve the problem of mode jumping.

In the strapping ring technique, two heavy gauge wires connecting alternate anode poles are used (Fig. 9.17a). Thus, in the π-mode, since the adjacent anode poles are 180° out of phase, there is no potential drop along the strapping rings. Therefore, capacitive loading of the cavities is caused by the strapping rings, thereby decreasing the resonant frequency of the π-mode of oscillation. In other modes (different from π) there is a potential drop along the

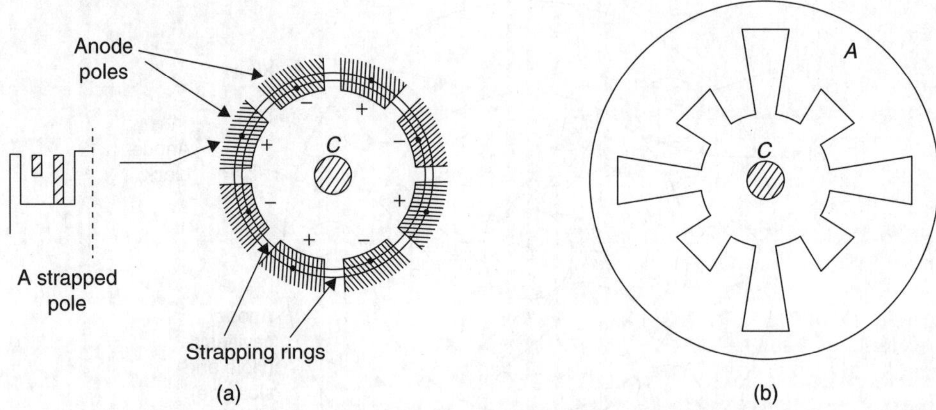

Fig. 9.17 Mode separation in a magnetron (a) By using strapping rings (b) With rising-sun resonators.

strapping ring and hence a current flows in the strapping rings cause inductive loading of the cavity system by the strapping rings thereby increasing the resonant frequency of the non-π modes. Thus, the π and non-π modes of oscillations are separated in the requency spectrum reducing the possibility of mode jumping.

At millimeter-wave frequencies, the losses in the strapping rings become high (as strap length becomes comparable to wavelength) and the strap construction becomes difficult mechanically. Further, at high frequencies, it is desirable to have a large number of cavities to ensure an effective RF field to interact with the epicycloidal electron path around the cathode. To meet all these requirements and with the basic aim of mode separation, the *rising-sun* construction (Fig. 9.17b) for the anode structure is required at millimeter-wave frequencies.

In the rising-sun resonator technique, the two sizes of resonators taken individually have widely different resonant frequencies (as they are of different lengths, normally 2:1). When the resonators are coupled together, the π-modes of both sets of resonators couple strongly making it predominant and its frequency lies intermediate between the resonant frequencies of the two sets of resonators. For other modes (non-π modes) the coupling is very weak and the frequency of these modes remains in the locations of the resonant frequency of the individual resonators. Thus, the π-mode is effectively separated from the other undesired modes.

(ii) Frequency tuning: Capacitive tuning of a magnetron can be done by bringing a metallic ring in the space between the anode and cathode where the RF electric field is high. This is known as *cookie-cutter* tuning in which tuning rings are spaced between the strapping rings (Fig. 9.18a). Inductive tuning of a magnetron can be done by inserting conducting rods in the hole part of the hole and slot cavities in the anode, where the RF magnetic field is high (Fig. 9.18b). This is known as '*crown-of-thorns*' tuning.

The combined capacitive and inductive tuning in a magnetron can result in a very broad band of tuning (1.5 : 1) which can be made very quickly using a bellow type piston operated by the servo mechanism. The microprocessor control of such system can provide

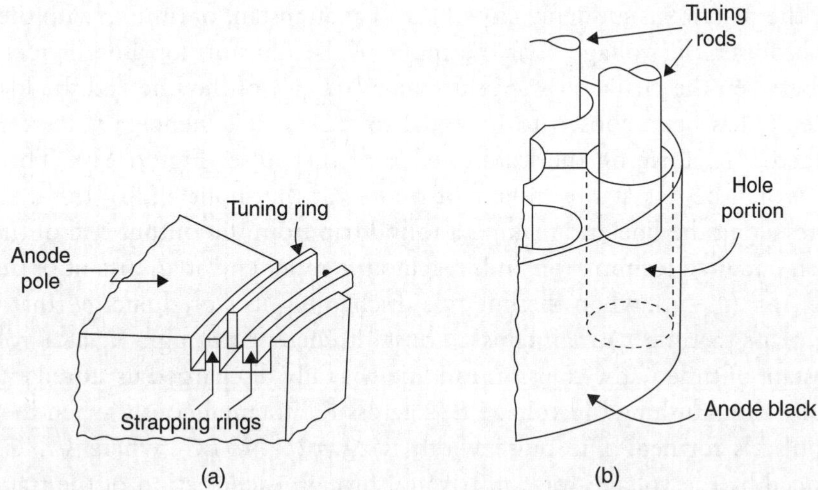

Fig. 9.18 Frequency tuning in a magnetron (a) Cookie-cutter tuning (b) Crown-of-thorns tuning.

a frequency agile or dither tuning of the magnetron which is very useful as noise anti-jamming technique, that is, in electronic counter counter measures (ECCM) which are important in modern electronic warfare.

9.2.2.2 Pulsing of magnetron with pulse forming network (PFN)

Microwave signal generation by a magnetron, discussed so far, produces continuous wave (CW) RF signals. But, for a pulse radar system where a magnetron is used exclusively as power oscillator, the CW magnetron oscillator needs to be pulsed in order to provide a pulsed RF signal. The pulser/modulator/keyer used for this purpose is normally designed to generate a pulse signal of kilovolt amplitude with typical pulse width of a microsecond and pulse repetition time of a millisecond. This unipolar (DC) pulse signal is applied between the anode and the cathode of the magnetron tube and hence is said to perform anode modulation, which needs to carry the full beam power of the tube [in contrast to the klystron and TWT where pulsing is done by applying the modulation signal to the control/shadow grid (grid modulation) and needs to carry only a fraction of the full beam power]. There are various techniques such as, line-type modulator, that uses a pulse forming network (PFN) and hard tube pulser to generate such high voltage DC pulses of which the former is commonly used for fixed pulse width (PW) and pulse repetition frequency (PRF) while the latter is used when the need for changing PW and PRF arises.

A simple scheme for line type or line-pulsing modulator is shown in Fig. 9.19a. It is well known that a transmission line is able to store magnetic energy and electric energy respectively in its distributed inductance and capacitance. When the switch S (an electronic switch, typically a hydrogen thyratron or a silicon controlled rectifier, whose 'ON' and 'OFF' is controlled by the trigger signal from the synchroniser unit in the radar system that also triggers the time base of the radar display unit) is open (OFF), the transmission line is charged slowly (controlled by the charging inductance L_c) to the DC supply voltage E.

Suppose the switch S is suddenly closed (ON) at an instant of time t_1, with the trigger from the synchroniser, the voltage E at the input of the transmission line divides immediately equally between the characteristic impedance ($\sqrt{L/C}$) of the line and the load impedance R_L (since R_L has been chosen to be equal to $\sqrt{L/C}$) and hence a voltage $E/2$ suddenly appears across R_L forming the leading edge of the pulse (Fig. 9.19b). The discharge of the line which begins at the instant of time t_1 at the input end of the transmission line propagates along the line and making a round trip from the output end of the line (which is an open circuited termination end) reaches the input end at an instant of time t_2. During the time $\Delta t = (t_2 - t_1)$, when the line was discharging its stored energy (that was stored in the line prior to time t_1) maintains a constant current through R_L at a voltage of $E/2$. At an instant of time t_2, the transmission line has fully discharged its stored energy through R_L thereby extinguishing the voltage $E/2$ across R_L instantaneously when the trailing edge of the pulse is formed. The pulse width is $\Delta t = 2 \times N\sqrt{LC}$, where $\sqrt{L/C}$ is the delay experienced by the voltage wave in passing through each section of the transmission line, N is the number of sections and the factor 2 appears because of the round trip of the signal via the transmission line. Thus, a rectangular pulse of width $2N\sqrt{LC}$ appears across the load (in practice the load is nothing but the magnetron tube itself). The switch S which is trigger controlled (from the synchroniser) is made 'ON' for a short period (preferably equal to the pulse width) and it is then automatically OFF. Again during the 'OFF' period of the switch (which for the RF tube is the inter-pulse period) the line is again charged, then again with a trigger the switch is closed (ON) and the next pulse is generated. In this way a high voltage pulse train can be generated with this line type pulser.

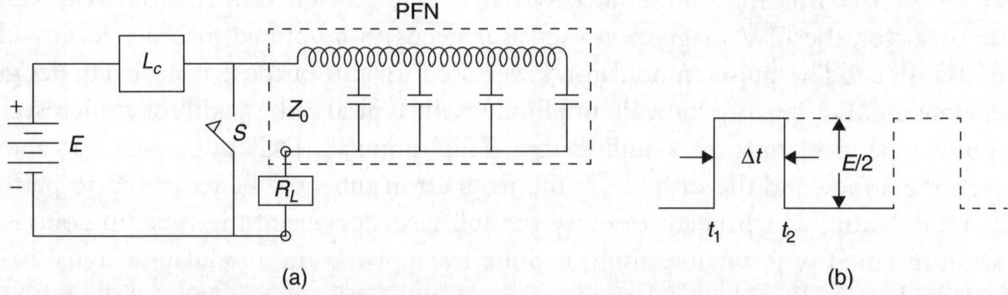

Fig. 9.19 **(a) Schematic of line-pulsing modulator (b) Generated pulse waveform across R_L.**

A normal RF transmission line cannot be used for this purpose. To generate a pulse having pulse width of, say, microsecond order (which is typical with pulse radar requirement), the normal RF transmission line length when calculated would be $l = c \times \Delta t/2 = 3 \times 10^8 \times 1 \times 10^6/2 = 150$ m (!) which is impractical. Thus, an artificial transmission line realised with lumped circuit elements in T, π or L configuration is used in cascade to simulate the RF transmission line. These artificial transmission lines are known as pulse forming network (PFN) or lumped constant delay lines (Fig. 9.19a). It might consist of an air-core inductance with taps along its length to which is attached capacitances to ground. The PFN are often enclosed in metal boxes with each of the circuit elements immersed in oil to give compact high voltage insulation. The networks are labelled with the value of their pulse length, characteristic resistance and safe operating voltage.

9.2.2.3 Coaxial magnetron

It was observed that in a conventional magnetron, the problem of mode jumping may be solved using strapping rings or with rising-sun resonators. However, another technique is used these days in which no strapping ring is required and also anode cavity length need not be of two different sizes but a stabilising cavity operating at TE_{011} mode is placed co-axially with the main magnetron structure (Fig. 9.20). This permits improvements in power output, efficiency, stability, tunabilty and life of operation of magnetron tube.

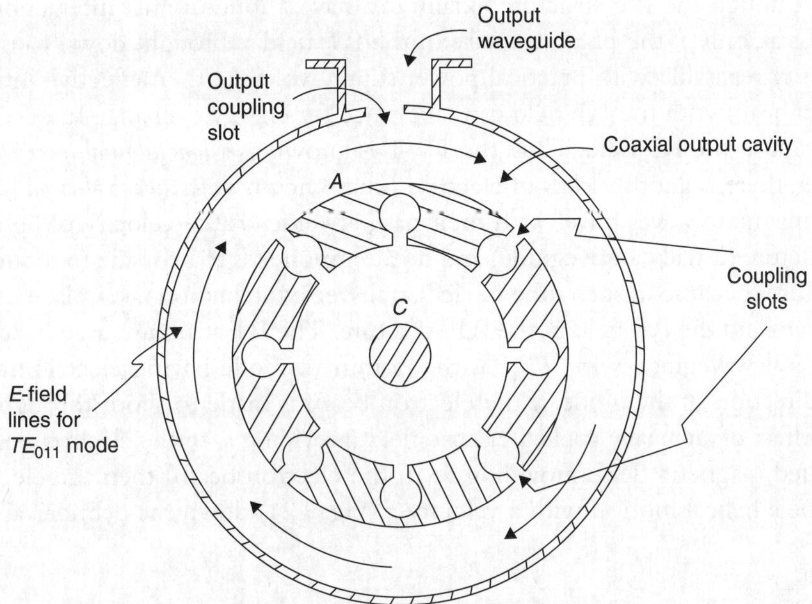

Fig. 9.20 Schematic of a coaxial magnetron.

In a coaxial magnetron, the TE_{011} mode in the stabilising cavity has the electric field in the circumferential, that is, the φ direction. Thus, if the alternate anode cavities have slots for coupling the signal to the stabilising cavity, then those anode cavities will be in phase as desired for the usual π-mode of oscillation in the magnetron oscillator. Since most of the RF energy is now stored in the stabilising cavity rather than in the resonator region, stable and broadband tuning of the magnetron oscillator may be realised with a non-contacting plunger inserted in the stabilising cavity (as the TE_{011} mode in a cylindrical cavity resonator does not need current continuity between the side wall and end wall of the cavity, Section. 4.4.4). Further, a large number of resonator cavities may now be used in the anode structure to increase the power output of the magnetron oscillator without being concerned with the complications in realising mode separation, since the job of mode separation is taken care of by the stabilising cavity. The larger structure of the coaxial magnetron permits a less conservative design and hence extends the operating life of the tube by a factor of more than ten compared to the conventional magnetron.

9.2.3 Fast-Wave Electron Tubes

It was observed while discussing the travelling wave tube (TWT), that the phase velocity of the electromagnetic wave propagating down the RF structure (helix or coupled cavity) was brought down to a practically realisable electron beam velocity so that the electron beam can effectively interact with the RF field (thus, the TWT is also known as the *slow-wave electron tube*). This was required in the TWT because the electric field of the RF signal is along the axial direction of the helix (or coupled cavities) and thus, an electron beam shot through the RF structure axially can have a longitudinal interaction with the RF field. Thus, only if the phase velocity of the RF field is brought down to the electron beam velocity realisable with practical power supply voltages, then effective interaction of the electron beam with the RF field can take place for energy exchange to occur from the electron beam to the RF field. Thus, the TWT is known as a *longitudinal interaction device*.

However, there is another class of electron tubes known as *fast-wave electron tubes,* where the electromagnetic waves travel with their natural velocity (the velocity of light) through the RF structure (usually a waveguide) and no attempt needs to be made to reduce its phase velocity. This is because in such tube devices, transverse interaction takes place between the electron beam and the RF field in the RF structure. The RF structure used in such devices is a cylindrical waveguide with TE_{11} as the dominant mode having electric field lines in the radial direction of the guide. If the electron beam is made to shoot through the guide under the effect of an axially applied electric field (providing $v_{\parallel}$ to the electron beam) and an axially applied magnetic field (providing $v_{\perp}$ to the electron beam) then the electron beam will describe a helical motion with a velocity v (Fig. 9.21) down the cylindrical guide.

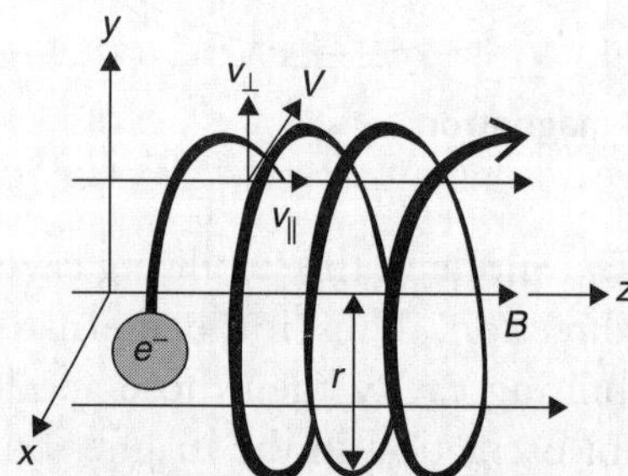

Fig. 9.21 Helical trajectory of an electron subjected simultaneously to an axial electric field and an axial magnetic field.

The electrons are then said to describe a cyclotron orbit having a radius r with cyclotron resonance frequency f_c around the helical path given by:

$$r = \frac{v_{\parallel}}{2\pi f_{op}} \tag{9.5a}$$

and

$$f_c = \frac{eB}{2\pi m_e} \tag{9.5b}$$

where $v_\parallel$ is the axial velocity of the electron beam imparted by the beam voltage V_0 applied between the cathode and the anode of the tube and is given by:

$$v_\parallel = \sqrt{\frac{2eV_0}{m}} \tag{9.5c}$$

The electric field of the dominant mode (TE_{11}) in the waveguide (which is a cylindrical guide) will then interact with the transverse component ($v_\perp$) of the electron beam in the guide and results in an effective (transverse) interaction of electron beam and the RF field. Thus, in such *transverse interaction devices*, no slowing down of the electromagnetic field is required. Thus, these devices, in contrast to the longitudinal interaction slow-wave electron tubes, are known as fast-wave electron tubes (as the electromagnetic waves travel with their natural velocity). In recent years, many fast-wave electron tubes such as gyrotron, peniotron and ubitron have been developed for high power microwave, millimeter-wave and even terahertz signal generation and amplification.

9.2.3.1 Gyrotron

The Gyrotron was invented by M I Petelin and I M Orlova of the USSR in the late 1960s. It is a very promising high power and high efficiency source with the operating frequency ranging from 20–250 GHz. To date, a CW power of 220 kW at 30 GHz and a pulsed power of 1 MW at 100 GHz have been achieved. The efficiency is up to 50%. Potential applications of the gyrotron include radar, communication and plasma heating of controlled nuclear fusion devices. Apart from this, the US military is developing a gyrotron based non-lethal crowd control system named *Active Denial System* which will use high power microwave signals to heat the skin of the mob (crowd) it is directed at, to an unpleasant level.

Principle of operation: As mentioned above, in a transverse interaction fast-wave electron device such as the gyrotron, electrons describe a helical motion within the cylindrical waveguide caused by an axial magnetic field (which gives it a cyclotron orbit), and an axial electric field (caused by the so called beam voltage) that helps to shoot the electrons through the waveguide from the cathode to the anode (Fig. 9.22). Initially, the phase of the electrons in their cyclotron orbits are random, but the phase bunching of electrons necessary for the transfer of energy from the electron beam to the microwave field inside the waveguide occurs because of the dependence of the effective relativistic-electron mass on energy. The bunching process in the gyrotron depends on a relativistic effect known as *Cyclotron Resonance Maser* (CM) interaction between the electromagnetic wave and the electron beam. It may however be noted that unlike the conventional maser, it is not dependent on material property in any way. Further, the speed of electron in a gyrotron is slightly relativistic, that is comparable to, but not close to, the velocity of light unlike a free electron laser.

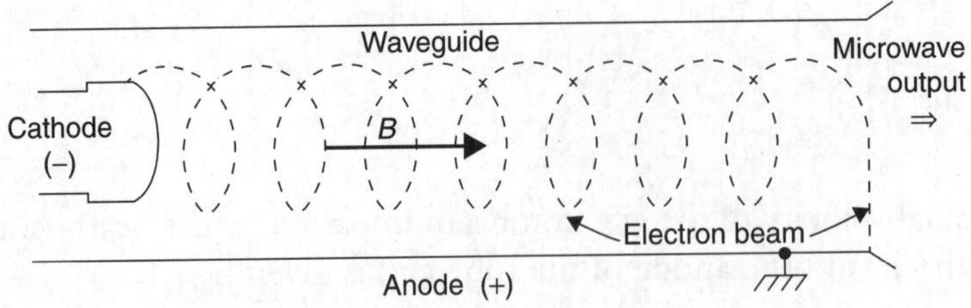

Fig. 9.22 Schematic of a gyrotron tube.

The physical mechanism of interaction of electrons with the electromagnetic field taking place within the gyrotron device may be understood as follows [21]. Those electrons interacting with the electric field of the dominant mode of the initial microwave stimulus existing in the waveguide of the gyrotron that absorb energy from the field (accelerating electrons) become heavier, rotate more slowly and recede in phase (lagging phase) while those electrons that deliver energy to the field (decelerating electrons) become lighter, rotate faster, and advance ahead in phase (leading phase) (Fig. 9.23a). This velocity modulation of electrons in the interaction space results in phase bunching of electrons. If the bunched electrons and the microwave electric field in the waveguide are synchronised in such a way that the bunched electrons are kept decelerated (Fig. 9.23b), the lost kinetic energy of the bunched electrons is converted to microwave potential energy and the initial stimulus of the microwave field in the waveguide of the gyrotron tube gets stronger. This is how microwave oscillation/amplification is accomplished in a gyrotron tube.

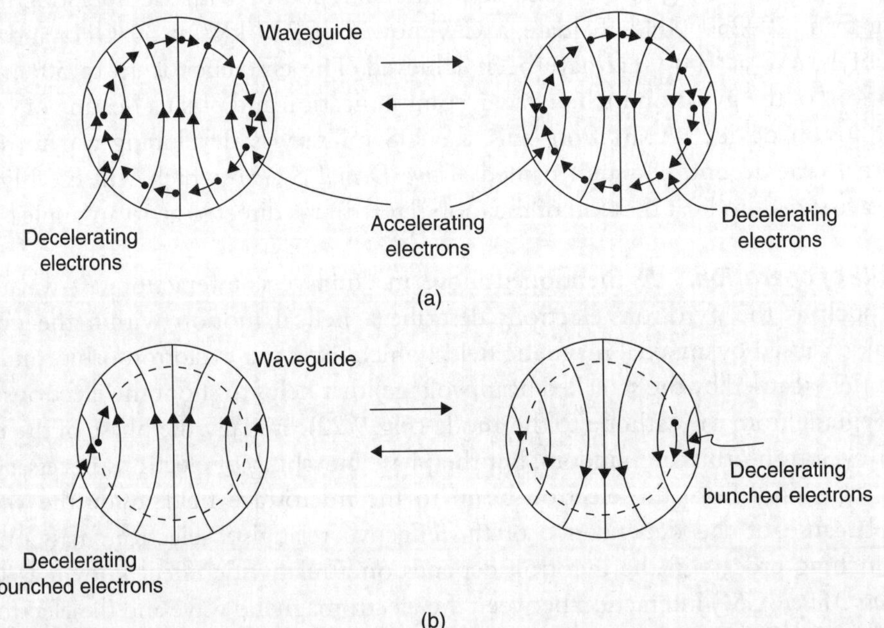

Fig. 9.23 Interaction of the RF field and electron beam in a gyrotron (a) Velocity modulation of a helical electron beam interacting with the RF field in the waveguide (b) Bunching and energy delivery.

Unlike the klystron, magnetron or TWT, a gyrotron tube is not designed with resonant cavities—here, the electron beam interacts with the electromagnetic field modes in a wave-guiding structure. The resonant cavity structure of the klystron and magnetron, and the coupled transmission cavities in the form of a slow-wave structure in the TWT, limits the possibility of realising high power operation of these devices at higher frequencies as the resonant cavity dimension is inversely proportional to the operating frequency. Thus, the fundamental size restriction of conventional microwave tubes is not present in the case of fast-wave electron devices such as the gyrotron and hence this class of device can easily produce kilowatt level power at higher microwave frequencies.

It was mentioned that the electron bunch formation due to velocity modulation of electrons by the cyclotron resonance maser interaction between the electron beam and the RF field had to be synchronised with the RF field to make the energy exchange from the electron beam to RF field possible. This can happen when the cyclotron resonance frequency ($eB/2\pi m_e$), Eq. 9.5b, becomes equal to the operating frequency (f_{op}) of the device. Thus, the magnetic flux density required to operate a gyrotron at a desired frequency is given by:

$$B = \frac{2\pi m_e f_{op}}{e} \tag{9.6}$$

EXAMPLE 9.4

A 240 GHz gyrotron oscillator with an output power of 3.5 kW operates with a beam voltage of 30 kV. Calculate the radius of the cyclotron orbit that the electron beam describes and the magnetic flux density applied axially to the gyrotron tube.

Solution

The axial velocity of the electrons resulting from the beam voltage is given by:

$$v_{\|} = \sqrt{\frac{2eV_0}{m}} = \sqrt{2 \times 1759 \times 10^8 \times 30 \times 10^3} = 10.27 \times 10^7 \text{ ms}$$

Thus, radius of the cyclotron orbit is $r = \dfrac{v_{\|}}{2\pi f_{op}} = \dfrac{10.27 \times 10^7}{2 \times 3.142 \times 240 \times 10^9} = 6.8 \times 10^{-5}$ m

The magnetic flux density need to be applied axially is given by:

$$B = \frac{2\pi m_e f_{op}}{e} = 2 \times 3.142 \times 240 \times 10^9 / 1759 \times 10^8 = 8.5 \text{ telsa}$$

We note from the above example that the radius of the cyclotron orbit for a single helix orbit discussed so far, is quite small compared to the diameter of the cylindrical waveguide (which is approximately a half-wavelength). This will cause a very weak interaction between the electron beam and the RF field in the guide. However, in actual design, the electrons are made to follow a coiled (also known as double coil) helical path (Fig. 9.24) which allows effective interaction between the electron beam and the RF field.

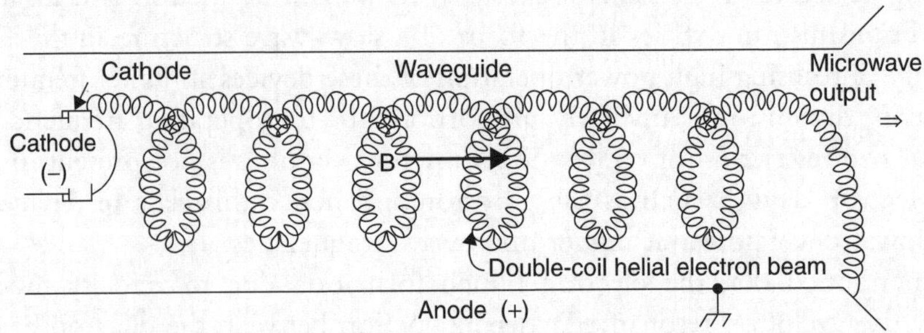

Fig. 9.24 Double coil helical electron beam for effective interaction of an electron beam with the RF field in the waveguide.

In addition to the gyrotron, other fast-wave electron tubes have also been developed—peniotorn, ubitron. The peniotron has many similarities with the gyrotron, it however differs from a gyrotron in the sense that it operates on non-relativistic mass change process. An ubitron uses a linear circuit (like a rectangular waveguide) and a periodic electron trajectory (using undulator as in the free electron laser). Thus, like a gyrotron, its interaction area is large (may be 100 times larger than that of a TWT at the same frequency) and hence much greater power output is possible with it.

All these fast-wave electron tube devices are capable of generating/amplifying microwave signals with a large amount of power at higher microwave frequencies and hence they are opening up new application possibilities at higher microwave frequencies demanding large amounts of microwave power.

9.3 Microwave Solid-State (Semiconductor) Devices

The list of different microwave solid-state (semiconductor) devices and their important applications have been shown in Table 9.2. In this section we will first discuss the principle of operation of the Schottky diode, tunnel diode and varactor diode. The PIN diode along with its various applications at microwave frequencies will also be discussed. Next, the principle of operation of Gunn and IMPATT diodes including their equivalent circuit will be discussed as these diodes are useful in the design of microwave and millimeter-wave oscillator, amplifier and power combiner. Among different microwave transistors, BJT and HBT will be discussed in brief while the principle of operation of MESFET and HEMT device along with their equivalent circuit will be discussed; these field-effect transistors are useful for the design of microwave amplifiers.

9.3.1 Microwave Solid-State Diodes

Microwave solid-state diodes are either semiconductor–semiconductor contact diodes (such as tunnel diode, PIN diode, varactor diode, IMPATT diode) or metal–semiconductor contact diode (Schottky diode) or bulk semiconductor diode (Gunn diode). The semiconductor materials normally used are elemental semiconductors like silicon (Si) and germanium (Ge) and compound semiconductors like gallium arsenide (GaAs), indium phosphide (InP), silicon carbide (SiC) and gallium nitride (GaN). The compound semiconductors are found to be preferable for high frequency design as they have high electron mobility and also for high power applications. The microwave diodes such as Schottky diode, varactor diode and PIN diode are primarily used for signal detection and mixing, frequency multiplication and tuning and also for switching and other signal control applications while microwave diodes such as tunnel diode, Gunn diode and IMPATT diode are used for signal generation and amplification.

9.3.1.1 Schottky diode, tunnel diode and varactor diode

Schottky diode: The Schottky diode is basically a metal–semiconductor contact diode unlike a *p–n* junction diode which is a semiconductor (*p*-type)-semiconductor (*n*-type) junction diode. The fundamental concept of transport phenomena in metal–vacuum contact given by the German physicist W H Schottky was found to form the basis of understanding the carrier transport in metal–semiconductor contacts and hence such diodes have been named after Schottky. The Schottky diode is a rectifying diode like a *p–n* junction diode but it is a majority carrier diode (unlike a *p–n* junction diode which has both majority and minority carriers) and thus Schottky diodes do not suffer from the minority carrier storage problem generally faced with the *p–n* junction diode, allowing its operation even up to 100 GHz. Schottky diodes are extensively used in microwave detectors, mixers and so on. Apart from the application in two terminal devices as diode, the Schottky junction is also used in three-terminal devices like metal semiconductor field effect transistors (MESFET) where the gate–source contact is a Schottky junction. The Schottky rectifying diode is also known as a *hot electron diode* since the work function of the semiconductor (normally *n*-type is used because of high mobility of electrons) is made *smaller* than that of the metal (thus the Fermi level in the metal is at a lower energy level compared to that in the semiconductor) causing flow of electrons having higher energy in the semiconductor (as if it is at a higher temperature) to the metal in which electrons are at lower energy level. In this connection, it may be mentioned that metal semiconductor contacts also arise in the case of *ohmic contact*, which is also a Schottky contact (but non-rectifying one), to be made to all sorts of semiconductor devices for interfacing with external circuitry. In such cases, the work function in a semiconductor is *larger* than that of the metal being chosen making practically no potential barrier at the contact and hence the I–V characteristics become linear, that is, the metal–semiconductor contact is ohmic.

We can now discuss the principle of operation of an *n*-type Schottky diode in terms of the energy band diagram. The structure (i) and symbol (ii) of Schottky diode is shown in Fig. 9.25a where the metal (Al) contact with the lightly doped *n*-type semiconductor forms the Schottky rectifying contact while the metal with heavily doped *n*-type semiconductor (n^+) forms the ohmic contact. For the Schottky rectifying contact diode, the metal is chosen with a larger work function ($q\varphi_m$) compared to the *n*-type semiconductor ($q\varphi_{sc}$) (Fig. 9.25b) and hence the energy band diagram for such a Schottky junction while in contact will be as shown in Fig. 9.25c [where $\chi = (E_{ref} - E_c)/q$ is the semiconductor's electron affinity]. Since the work function of the metal is larger than that of the semiconductor, the Fermi level of electrons in the semiconductor will be at a higher energy level (E_{Fsc}) compared to that of the electrons in the metal (E_{Fm}). Thus, when the metal and the semiconductor are in contact, electrons flow from the semiconductor to the metal (aligning the Fermi levels at equilibrium), leaving behind ionised donors in the semiconductor (forming a space–charge region) and resulting in a thin sheet of negative charge on the metal side of the contact. This forms the potential hill (with a contact potential of $q\varphi_{Bs}$) at the junction and the potential barrier φ_{Bs} prevents further migration of electrons in equilibrium from the

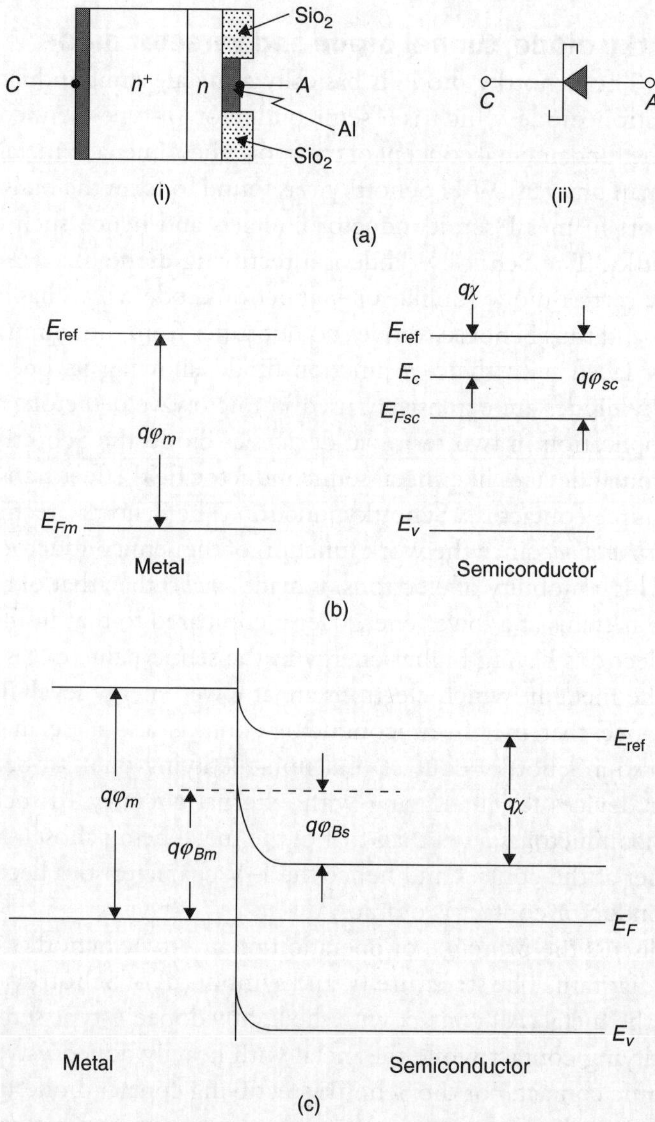

Fig. 9.25 Schottky diode (a) (i) Structure (ii) Circuit symbol (b) Metal and semiconductor energy band diagram before contact (c) Schottky junction band diagram.

semiconductor to the metal. The electrons on the metal side face a potential barrier φ_{Bm}, which is known as the *Schottky barrier.*

The equilibrium potential difference φ_{Bs} is controllable by the bias to the diode, which can be decreased with forward bias voltage and increased with reverse bias voltage. When forward bias V_a is applied (with the metal at a positive potential and the semiconductor at a negative potential), the potential barrier on the semiconductor side is decreased from φ_{Bs} to $\varphi'_{Bs} = (\varphi_{Bs} - V_a)$ (Fig. 9.26a), allowing the flow of electrons above the equilibrium value from the semiconductor to the metal. However, the flow of electrons from the metal to the semiconductor is not affected by the bias as the Schottky barrier remains unchanged. Thus, there is a net electron flow from the semiconductor to the metal. If the Schottky junction is reverse biased, the potential barrier in the semiconductor is increased from φ_{Bs} to $\varphi''_{Bs} = (\varphi_{Bs} + V_a)$ (Fig. 9.26b), causing a reduced flow of electrons from the semiconductor to the metal compared to the equilibrium value. However, the flow of electrons from the metal to the semiconductor remains unchanged as before. The I–V characteristics of the diode are given by $I = I_0(e^{qV_a/kT} - 1)$, which is similar to the diode equation for a *p–n* junction diode. However, in this case the reverse saturation current I_0 depends on the electron flow from the metal to the semiconductor across the barrier φ_{Bm} which is unaffected by the applied bias voltage.

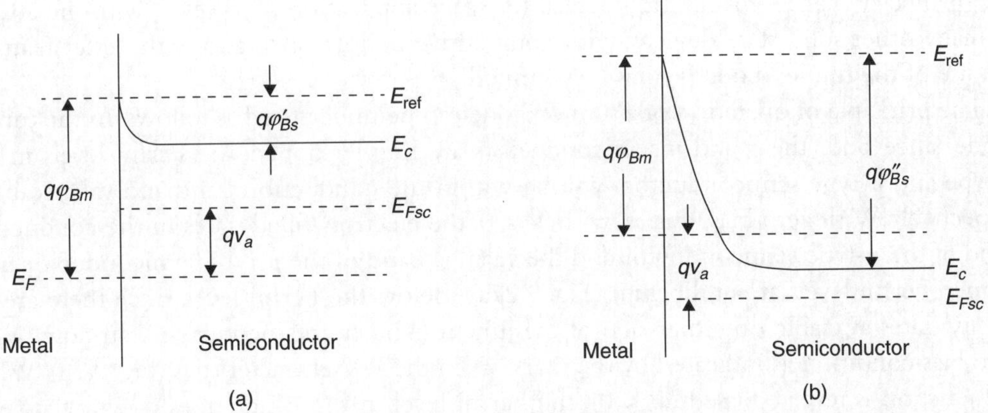

Fig. 9.26 Energy band diagram of an *n*-type Schottky diode with (a) Forward bias (b) Reverse bias.

A question that naturally arises is, why is the Schottky diode a better choice than a *p–n* junction diode as far as the microwave frequency of operation is concerned? The answer is very simple—as the Schottky contact diode is a majority carrier device, its speed of operation is no longer dependent on extracting minority carrier charges via diffusion or recombination as in a *p–n* junction diode. Thus, the switching speed of a Schottky diode can be of the order of picoseconds, which is an order of magnitude faster than that of a *p–n* junction diode. Minority carrier dependence of a *p–n* junction diode makes it temperature dependent which is not the case with a Schottky diode and hence the noise performance of a Schottky diode is also better than a *p–n* junction diode. However, a Schottky diode has a higher reverse current density (100 to 1000 times) compared to a *p–n* junction diode.

Tunnel diode: The tunnel diode was a landmark invention by L Esaki in 1957 for generation and amplification of microwave signals with a solid-state device that uses a quantum mechanical phenomenon called *tunnelling*. Basically, it is a *p–n* junction diode. However, both the *p* and *n* semiconductors are heavily doped (may be 1000 times more compared to an ordinary *p–n* junction diode) so that the Fermi levels in the *n* and *p* semiconductors are aligned in such a way that the conduction band electron states on the *n*-side get more or less aligned with valence band hole states on the *p*-side. This heavy doping thus results in a junction that has a depletion layer (space–charge region) which is so thin (even less than 10 nm) that tunnelling of carriers can occur via the junction. The thinness of the junction naturally makes the transit time of carriers via the junction very small, thus permitting the operation of the device in microwave and millimeter-wave frequencies. The tunnel diode offers *negative differential resistance* in part of their operating range and was therefore found to be an excitingly important device to design solid-state oscillators and amplifiers when it was first introduced. In fact, the receiver front end of the Intelsat V satellite, used in 1977, was a microstrip tunnel diode amplifier operating in the 14 to 15.5 GHz band (which was found to give a better performance at that frequency compared to the transistor based front-end amplifier of that time). Some major advantages of the tunnel diode amplifier are that power supply requirement is very modest (typically, a few milliamperes at 10 V), it has a low noise figure (typically 5–6 dB at 15 GHz) and is immune to the ambient radiation encountered in interplanetary space—making it practical for space applications. However, with the advent of many other solid-state devices, with competitive performance and with better features, the use of the tunnel diode is now very limited.

The principle of operation of a tunnel diode can be understood as follows. In the tunnel diode, since both the *p* and *n* semiconductors are heavily doped, the Fermi levels in the *n*-type and *p*-type semiconductors will be within the conduction band and valance band respectively. With zero bias, that is, with $V = 0$, the electron-filled states in the conduction band of the *n*-type semiconductor and the valance band of the *p*-type semiconductor have aligned Fermi levels at equilibrium (Fig. 9.27a). Below the Fermi level, since there are no empty states available on either side of the junction, no tunneling current can flow under zero bias condition as indicated by the mark 'O' in the I–V characteristics (Fig. 9.28). When the junction is reverse biased ($V < 0$), the Fermi level in the *n*-side goes down relative to that in the *p*-side and hence the electrons in the valance band of the *p*-side find empty states in the conduction band of the *n*-side. They can tunnel from the *p*-side to the *n*-side via the thin junction (Fig. 9.27b). With increasing reverse bias, a larger number of electrons from the *p*-side can find empty states on the *n*-side thus increasing the tunnel current (as indicated by the portion OR in Fig. 9.28). When a small forward bias is applied to the junction, the Fermi level of the *n*-side is raised (with respect to that on the *p*-side) and the electrons from the *n*-side can now tunnel through the junction to the *p*-side (Fig. 9.27c). This happens because the empty states in the *p*-side valence band are at the same energy level as the occupied states in the conduction band of the *n*-side. Tunnelling in the other direction (that is, from the *p*-side to the *n*-side) cannot occur since the valence band electrons on the *p*-side are now opposite to the forbidden energy gap (E_g) on the *n*-side. The tunnelling current increases with increased forward bias until a maximum or peak is reached

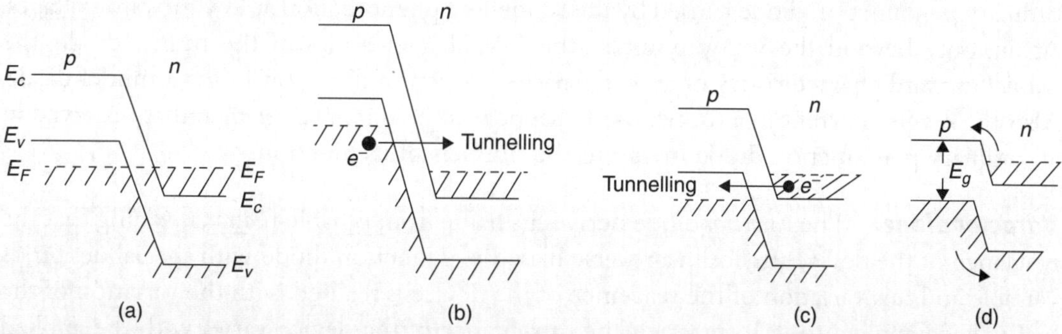

Fig. 9.27 Energy band diagram of the tunnel diode under different input bias conditions. (a) Zero bias (b) Reverse bias (c) Forward bias (d) High forward bias beyond valley point.

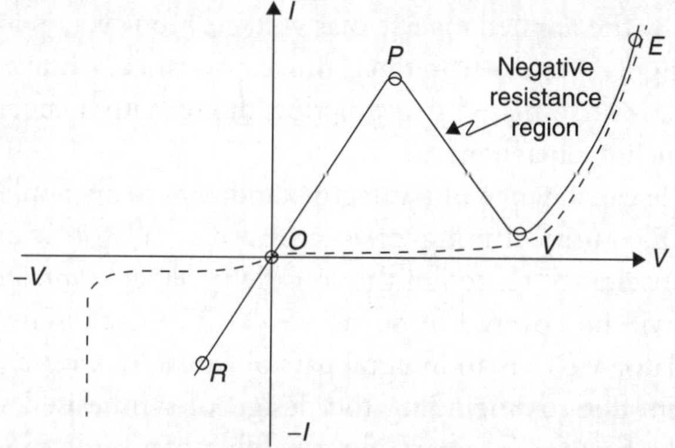

Fig. 9.28 The current–voltage (I–V) characteristics of the tunnel diode.

(as indicated by point P in the I–V characteristics). Beyond this point when the applied forward bias is increased further, the energy level of the *n*-side is raised further allowing fewer *n*-side free electrons to find empty states on the *p*-side, in the process decreasing the tunnelling current (Fig. 9.28). This will continue with increasing forward bias resulting in the drop of forward current via the diode and hence the diode duels in the negative resistance region of its I–V characteristics. Finally, a forward bias is reached at which there are no conduction band electrons of *n*-side opposite to the valence band vacant states in the *p*-side and the possibility of tunnelling stops. Under this forward bias condition of the diode, it is said that the I–V characteristics of the tunnel diode has reached the *valley point* indicated by the point V in Fig. 9.28. If the forward bias is increased beyond this limit, normal diffusion current begins to flow as in case of an ordinary *p–n* junction diode (Fig. 9.27d) and the forward current increases exponentially with voltage as indicated by curve VE in the I–V characteristics.

A few observations may be made with respect to the I–V characteristics of the tunnel diode when compared with the I–V characteristics of an ordinary *p–n* junction diode (shown by dotted curve in Fig. 9.28). In addition to the negative resistance region in its I–V characteristics, the tunnel diode exhibits a much larger forward current compared to ordinary *p–n* junction diode caused by the tunnelling phenomenon at low forward voltages. Again, only beyond the valley point in the I–V characteristics of the tunnel diode, the *normal* forward characteristics of an ordinary *p–n* junction diode set in for a tunnel diode. Also, the reverse current in a tunnel diode happens to be much larger than that observed in an ordinary *p–n* junction diode owing to the thinness of its junction.

Varactor diode: The *varactror* diode derives its name from variable reactor, meaning that the reactance of the device (basically a reverse biased *p–n* junction diode with special design) is variable and this variation of the reactance of the device is realised with the variation of the bias voltage applied to it. In practice, the capacitance of the device varies with the applied bias voltage and hence it's basically a voltage variable capacitance and also known as *varicap*. In fact, any reverse biased *p–n* junction diode exhibits a capacitance (as the depletion region acts as a dielectric medium while the anode and the cathode act as plates of the capacitor) which is controllable by the applied reverse bias voltage. However, in a normal *p–n* junction diode, the device designer prefers to reduce this capacitance while in the varactor diode design, it is basically a reverse biased *p–n* junction diode with a special design to increase this reverse bias dependant capacitance.

The voltage variable capacitance of a varactor diode has many applications in microwave engineering design. Electronic tuning of microwave oscillators can be done with this voltage dependent capacitance to design the so called voltage controlled oscillator (VCO), the details of which will be covered in Section 9.4.1.3 in the discussion on the varactor-tuned IMPATT oscillator.VCO is an integral part of phase-lock loop (PLL) used in various communication system designs including the design of synthesised sources at microwave frequency. Utilising the harmonic generation capability of a varactor (caused by non-linear variation of capacitance with voltage) frequency multipliers can be designed. Such frequency multipliers are used to generate higher microwave and millimeter-wave frequency signal by multiplying the frequency of a stable lower frequency signal source; otherwise signal generation at such high microwave and millimeter-wave frequencies may be expensive or may not have the desired signal quality. A varactor diode is also used as a pump source in parametric amplifier design, where reactive mixing of the varactor diode is used to develop a negative resistance that results in low-noise microwave signal amplification.

A varactor diode exhibits a variable capacitance with the change of applied reverse bias (V_{rb}) to the diode (Fig. 9.29a) according to the formula:

$$C_j(V_{rb}) = C_0\left(\frac{1+V_{rb}}{\varphi}\right)^{-\gamma} \tag{9.7a}$$

where C_0 is the junction capacitance for zero bias voltage, V_{rb} is the magnitude of the reverse bias voltage, φ is the barrier potential and γ is a constant which depends on the nature of the *p–n* junction (0.5 for abrupt junction and 0.3 for linearly graded junction). The equivalent circuit of a packaged varactor diode is given in Fig. 9.29b where, $C_j(V_{rb})$ is the variable junction capacitance of the diode, R_S is the series resistance of the diode, C_P and L_P are the capacitance and inductance due to the package parasitics.

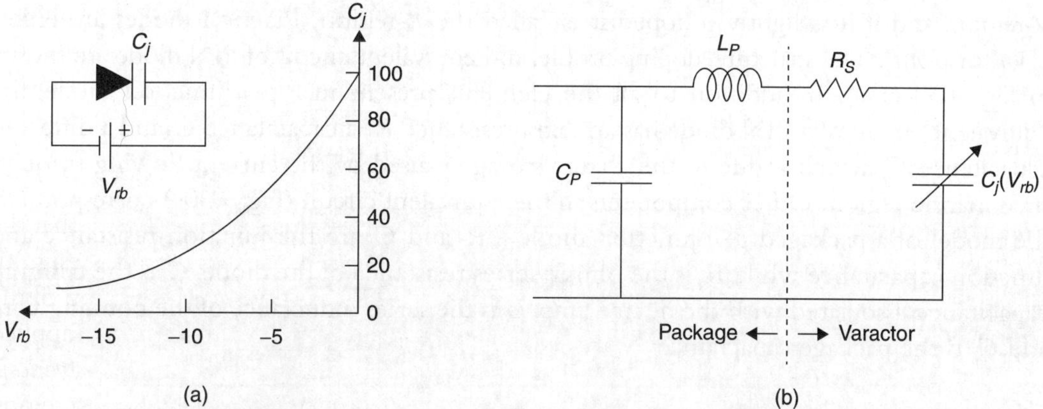

Fig. 9.29 **(a) Voltage variable capacitance of the varactor diode along with symbolic representation of the diode (in the inset) (b) Equivalent circuit of a packaged varactor diode.**

When an AC voltage $v = V \sin \omega t$ is applied to the reverse biased diode, the diode capacitance varies as:

$$C_j = C_0 \left\{1 + \frac{(V_{rb} + V \sin \omega t)}{\varphi}\right\}^{-\gamma}$$

$$= C_0 - C_1 \sin \omega t + C_2 \sin \omega t - C_3 \sin \omega t + \cdots$$

The current due to the varying capacitance will be:

$$i(t) = I_1 \cos \omega t + I_2 \cos 2\omega t + I_3 \cos 3\omega t + \cdots \quad (9.7b)$$

This shows that the non-linear property of the varactor diode can be used to realise harmonic generation, that is, to design a frequency multiplier where the resonant circuit tunes to the desired harmonic frequency. Multiplying factors greater than 4 are not used with ordinary non-linear capacitance varactors because of the significant decrease of the efficiency and output power which would result. However, the so called step-recovery varactor diodes are now available for high-factor (may be of the order of 12) frequency multiplication.

Step recovery diodes, also known as charge-storage diodes or snap-off varactors, are silicon epitaxial diffused junction diodes, which store charge when they are conducting with a forward bias. When the reverse bias is applied, the diode very briefly discharges the stored energy in the form of a sharp pulse very rich in harmonics. Thus, the design of multipliers with a large multiplication factor can be done with step recovery diodes and they are also used as comb generators.

9.3.1.2 PIN diode and its applications

The structure of a PIN diode consists of heavily doped *p* and *n* regions separated by the *i*-region (intrinsic region) of high resistivity. The typical *i*-region thickness is 10 to 100 μm. In practice, the *i*-region of the diode has a light *n*-type dopant when it is called the

ν-region, and if it is lightly p doped it is called the π-region. Practical diodes are either $p^+\nu n^+$ or $p^+\pi n^+$. The structure, doping profile, and equivalent circuit of PIN diode are shown in Fig. 9.30 (a–c). In addition to all the elements present in a p–n junction diode, the equivalent circuit of a PIN diode has an extra resistance R_i, a capacitance C_i and a diffusion capacitance C_d which is due to the charge storage caused by the current flowing through the intrinsic region. Other components in the equivalent circuit (Fig. 9.30c) correspond to the model of a packaged p–n junction diode—R_j and C_j are the junction resistance and junction capacitance while R_s is the ohmic series resistance of the diode, C_f is the fringing capacitance associated with the device chip, L_s is the series inductance of the bonding wire and C_p is the package capacitance.

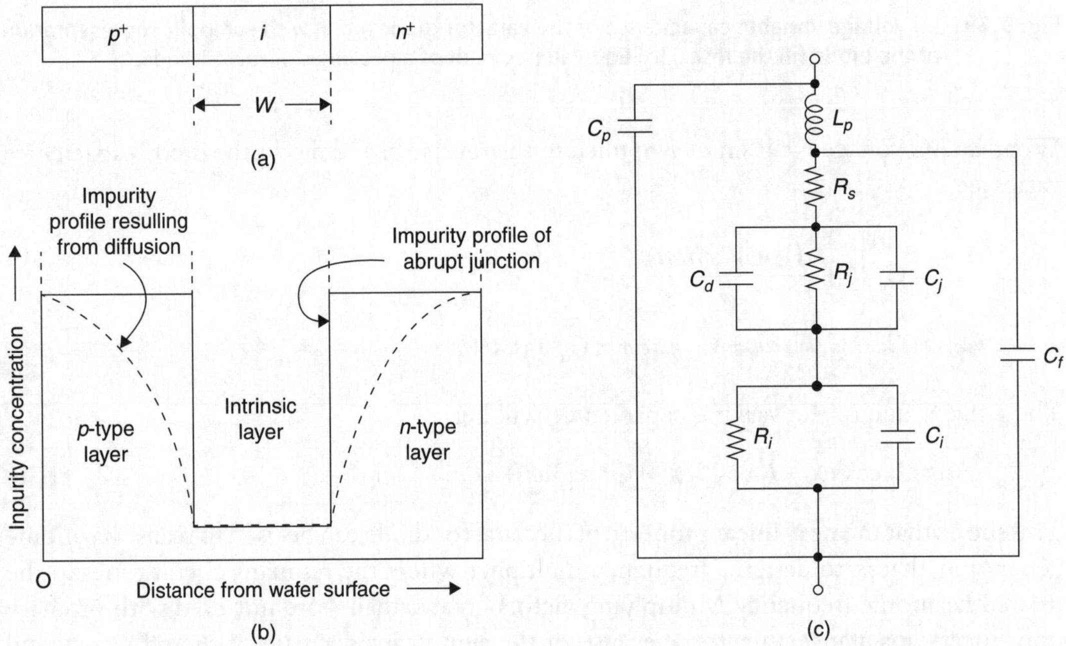

Fig. 9.30 PIN diode (a) Structure (b) Doping profile (c) Equivalent circuit of packaged PIN diode.

Principle of operation: At zero bias, two space charge regions are formed in the p and n-regions adjacent to the i-region because of the diffusion of holes and electrons across the junctions. Thus we have a region of fixed negative charge in the p-region and a region of fixed positive charge in the n-region with equal charges in the two regions. When a reverse bias is applied, the space charge regions in the p and n-regions become wider.

When a forward bias is applied to the diode, electrons and holes are injected into the i-region from the n and p-regions respectively. With the injection of carriers into the i-region, the carrier concentration in the i-region increases above the equilibrium level (however, carrier concentration diminishes with depth into the i-region due to recombination). If the carrier lifetimes in the i-region are relatively long and the i-region is not too thick, the i-region becomes flooded with carriers at a reasonable forward bias level. When this happens, the diode exhibits a low resistance. To realise this characteristic of a PIN diode, it is essential that the lifetime of carriers in the *i*-region be greater than the time period at the operating frequency. For this reason, the PIN diode cannot be operated at lower radio frequencies. Typically, for a carrier lifetime of around 100 ns, the lowest frequency of operation of the PIN diode is 10 MHz.

The effect of controlling the carrier density and hence the resistance of the *i*-layer by varying the forward bias is known as *conductivity modulation.* This phenomenon plays a key role in the use of PIN diodes in microwave active attenuators, modulators, limiters and automatic levelling circuits. It may be recalled that the ordinary *p–n* junction diodes have relatively thin depletion layers and do not exhibit the phenomenon of conductivity modulation as seen in PIN diodes.

Applications of PIN diode

PIN attenuator: We have seen that the resistance of the intrinsic region of a forward biased PIN diode is a strong function of the bias current (the two are inversely proportional). The relation is given by $R_i = CI_{fb}^{-x}$, where I_{fb} is the forward bias current in mA and the value of the constant C, typically 26, depends on the width of the *i*-region, average mobility and lifetime of carriers and x is typically 0.87. This property of the PIN diode, acting as a bias controlled resistor, is used in the design of the current controlled variable attenuator or active attenuator at microwave frequencies.

The most commonly used configuration of the PIN attenuator is π (pi) configuration (Fig. 9.31a) in which the circuit elements R_P and R_S of a conventional π attenuator are forward biased PIN diodes. From the theory of the π-type resistive attenuator it is known that the attenuation ratio A is given by:

$$A = \left(\frac{E_1}{E_2}\right) = \left(\frac{R_P + Z_0}{R_P - Z_0}\right) \tag{9.8}$$

where $R_P = Z_0\left(\dfrac{A+1}{A-1}\right)$ and $R_S = \dfrac{A^2-1}{2A}$.

EXAMPLE 9.5

A 3 dB attenuator is to be designed with PIN diodes for a 50 Ω system. Calculate the values of the resistors required for this purpose and the corresponding bias current to be fed to the PIN diodes which form the π network.

Solution

Here, $Z_0 = 50\ \Omega$, $A = 2$. Thus $R_P = 150\ \Omega$ and $R_S = 37.5\ \Omega$.

Using the relation $R_i = 26\ I_{fb}^{-0.87}$, the current through the PIN diodes representing R_P and R_S may be calculated as $I_{fbP} = 130\ \mu\text{A}$ and $I_{fbS} = 650\ \mu\text{A}$.

PIN based ALC: The PIN attenuator can be used in a very simple way to design an automatic levelling circuit (ALC). The PIN attenuator is basically a bias controlled resistor (BCR) having $R \alpha I^{-x}$ and also $A \alpha R^{-1}$. Thus, if we configure a circuit like that shown in Fig. 9.31b it is easy to understand that the drooping in the RF current of the RF source can be automatically levelled by this simple circuit arrangement.

PIN modulator: This is a three port device, one of the ports is for feeding the modulating signal and the other two ports are RF ports, to one of which the signal to be modulated is applied. Heart of the PIN modulator is the bias controlled PIN attenuator (BCR, Fig. 9.31c). The RF output will vary according to the bias current (sine wave, square wave or pulse signal) applied to the PIN diodes via the bias port. As the input impedance of the π attenuator does not change with the value of attenuation, the frequency or the output power of the input microwave source does not get distorted by the modulator, that is, the source always sees a matched load irrespective of the attenuation value.

PIN limiter: A PIN limiter is often used to protect a sensitive receiver from a nearby high power transmitter. A sudden application of high RF energy will inject electrons and holes into the intrinsic region of the diode causing it to have a low resistance, that is, to establish a short circuit path. This causes the incident high power to be reflected back from this short circuit established by the incident signal, and this type of limiter is known as self-bias limiter. A typical PIN limiter is shown in Fig. 9.31d. Two PIN diodes are connected in shunt across a transmission line and the diodes are spaced $\lambda/4$ apart so that the individual reflections cancel out and a low VSWR is obtained. The choke shown provides a return path for the self-bias.

PIN switches: The PIN diode has a very high resistance when in reverse bias and a very low resistance when in the forward bias. Typical values are greater than 10 MΩ and less than 1 Ω, respectively. Since these values are much higher or much lower than commonly used transmission line impedances, the PIN diode behaves as an open-circuited and short-circuited element under two bias conditions, respectively. For switching applications, PIN diodes may be mounted in the transmission line in two possible configurations (series or shunt) to realise single pole single throw (SPST) microwave switch [Fig. 9.31e (i) and (ii)]. A PIN diode may also be configured to realise a single pole double throw (SPDT) switch (Fig. 9.31f). Two PIN diodes A and B are used and the distance between the diode location and the junction point J are quarter wavelength each at centre frequency. When the diode A is forward biased with diode B reverse biased, the impedance looking from J to the diode A is infinite, while looking to diode B is zero and thus transmission takes place from port 1 to port 3. Similarly when diode B is forward biased and diode A is reverse biased, transmission takes place form port 1 to port 2.

PIN phase shifter: The PIN phase shifter has been discussed in Section 6.2.8.

9.3.1.3 Gunn and IMPATT diodes

Gunn diode: Gunn diode is a bulk-effect device (unlike IMPATT and PIN diode which are junction devices) and hence in a strict sense of the word, it cannot be called a diode;

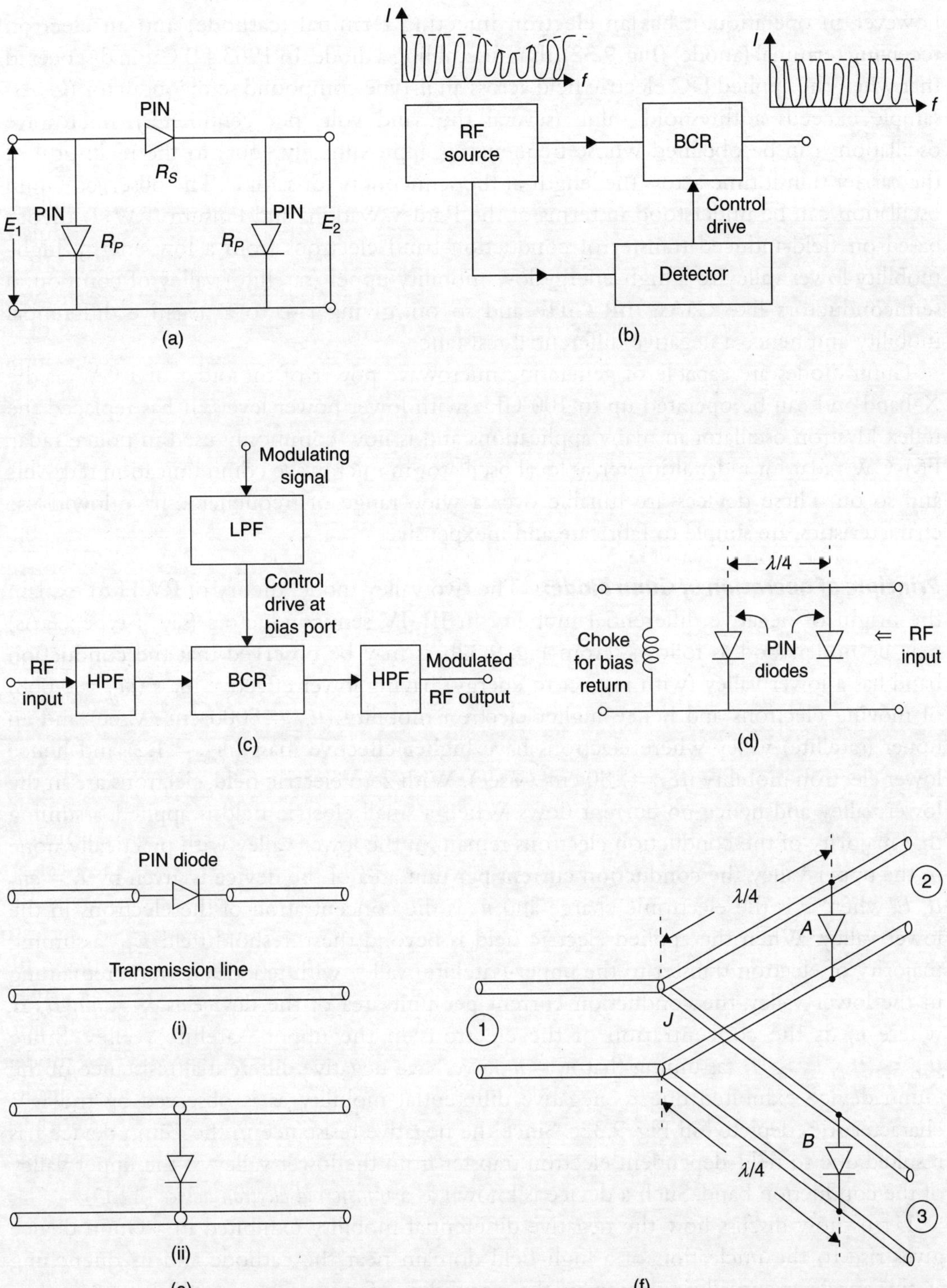

Fig. 9.31 (a) PIN attenuator (π-type) (b) ALC with PIN attenuator (c) PIN modulator (d) PIN limiter (e) SPST PIN switches (i) Series switch (ii) Shunt switch (f) SPDT PIN switch.

however, in operation, it has an electron injecting terminal (cathode) and an electron receiving terminal (anode) (Fig. 9.32), thus resembling a diode. In 1963, J B Gunn discovered that when the applied DC electric field across an n-type compound semiconductor (GaAs) sample exceeds a threshold value (several thousand volts per centimeter), microwave oscillations can be obtained whose frequency is approximately equal to the reciprocal of the carrier transit time across the length of the semiconductor sample. The observed Gunn oscillation can be understood in terms of the Ridley, Watkins and Hilsum (RWH) theory based on field-induced transfer of conduction-band electrons from a low-energy, high-mobility lower valley to a high-energy, low-mobility upper (satellite) valley of compound semiconductors like GaAs, InP, CdTe and so on, giving rise to a negative differential mobility and hence a negative differential resistance.

Gunn diodes are capable of generating microwave power of the order of 1.0 W at the X-band and can be operated up to 100 GHz with lower power levels. It has replaced the reflex klystron oscillator in many applications and is now commonly used in police radar, FM-CW radar for radar altimeter, as local oscillator in microwave communication receivers and so on. These devices are tunable over a wide range of frequencies, have low noise characteristics, are simple to fabricate and inexpensive.

Principle of operation of Gunn diode: The two valley model theory of RWH to explain the origin of negative differential mobility in III–IV semiconductors (say n-type GaAs) may be understood as follows. From Fig. 9.32b, it may be observed that the conduction band has a lower valley (with respect to energy) having lower effective mass ($m_{eL} = 0.068$) of moving electrons and hence higher electron mobility ($\mu_{eL} = 8000$ cm^2/V.sec) and an upper (satellite) valley where electrons have higher effective mass ($m_{eU} = 1.2$) and hence lower electron mobility ($\mu_{eU} = 180$ cm^2/V.sec). With zero electric field, electrons are in the lower valley and hence no current flows. When a small electric field is applied, assuming that majority of the conduction electrons remain in the lower valley with practically none in the upper valley, the conduction current per unit area of the device is given by $I_L = qn_L \mu_{eL}E$, where q is the electronic charge and n_L is the concentration of the electrons in the lower valley. When the applied electric field is beyond the threshold field E_{Th}, assuming majority of electron transfer to the upper (satellite) valley with practically none remaining in the lower valley, the conduction current per unit area of the device is $I_U = qn_U \mu_{eU}E$, where n_U is the concentration of the electrons in the upper (satellite) valley. Since $\mu_{eU} << \mu_{eL}$, $I_U >> I_L$ (assuming that $n_L = n_U$), we have negative differential resistance of the Gunn device exhibited due to negative differential mobility as is observed by the I–V characteristics depicted in Fig. 9.32c. Since the negative resistance in the Gunn device has resulted due to field-dependent electron transfer from the lower valley to the upper valley of the conduction band. Such a device is known as a *transferred electron device* (TED).

Let us now discuss how the negative differential mobility exhibited in a Gunn device gives rise to the nucleation of a high-field domain near the cathode and its quenching at the anode eventually resulting in the possibility of microwave oscillations in such a device. The Gunn device may be so fabricated that its resistance is quite high at the ohmic contact on the left-hand side of the n-active region (at $x = 0$) which is connected to the

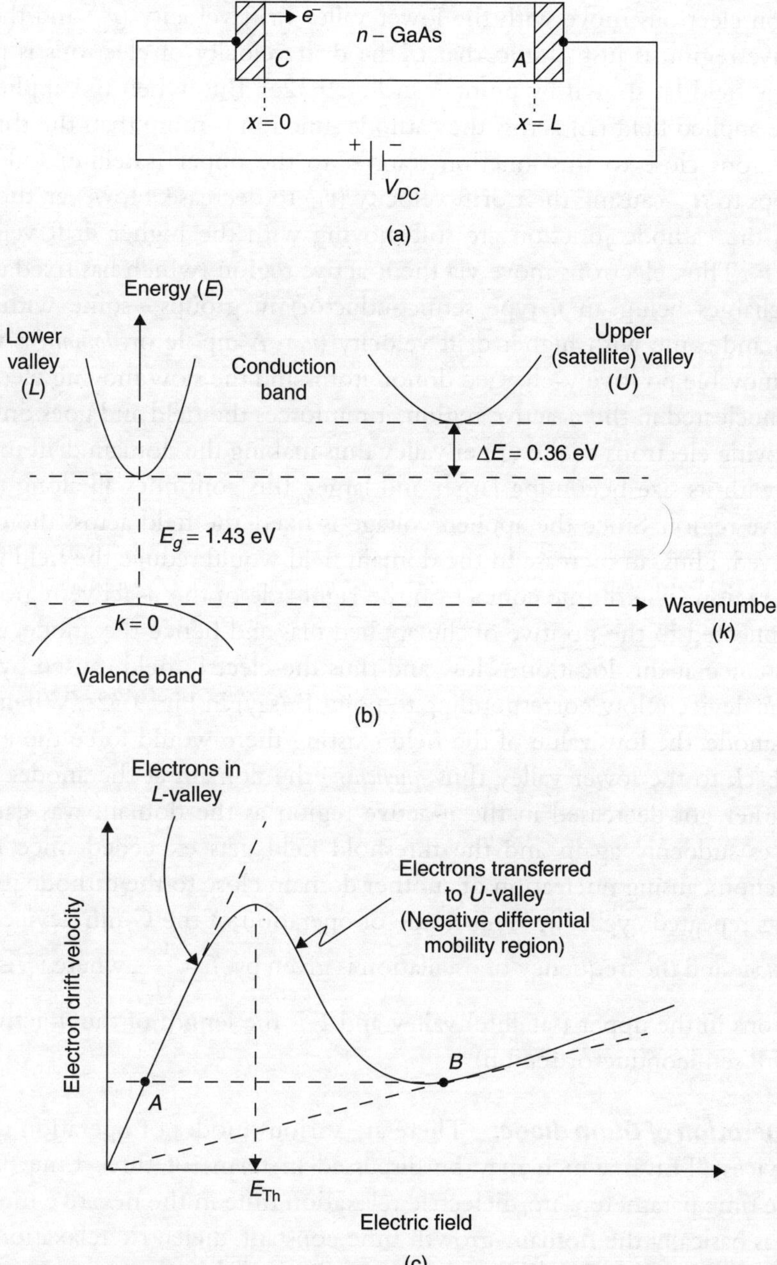

Fig. 9.32 Gunn diode (a) Structure (b) Two valley model for *n*-type GaAs (c) Negative differential mobility (negative resistance) in Gunn diode.

negative of the applied bias and hence termed as the cathode (Fig. 9.32a). Thus, the applied voltage will cause a comparatively higher field to appear across the cathode junction and the high-field domain caused by negative differential mobility (with $E > E_{Th}$) is thus said to originate (*nucleate*) near the cathode junction. Initially, when the applied voltage is low, all conduction electrons move with the lower valley drift velocity (v_L) and the behaviour of the n-active region is just ohmic, that is, the drift velocity of electrons is proportional to the electric field (as shown by point A in Fig. 9.32c). But, when the applied voltage is such that the applied field (E) across the cathode junction is more than the threshold field E_{Th}, the electrons close to this junction transfer to the upper (satellite) valley and their mobility drops to μ_{eU} causing their drift velocity (v_U) to decrease. However, those electrons not close to the cathode junction are still moving with the higher drift velocity of the lower valley v_L. Thus, electrons move via the n-active region (which has fixed donor atoms as positive charges being an n type semiconductor) in groups—some with lower drift velocity (v_U) and some with higher drift velocity (v_L). A dipole or *domain* is thus formed with the immovable positively-charged donor atoms and the slow moving electrons. Once a domain is nucleated in the n-active region, it reinforces the field and goes on transferring the faster moving electrons to the upper valley thus making the domain drift in the positive x-direction with its size becoming larger and larger; this continues all along the length L of the n-active region. Since the applied voltage is fixed, the field across the device needs to be conserved. Thus, an increase in the domain field would reduce the field in the rest of the n-active region. The ohmic contact on the right side of the n-active region (at $x = L$), which is connected to the positive of the applied bias and hence the anode, is made such that the resistance at this location is low and thus the electric field caused by the applied voltage to the device is low, corresponding to point B (say) in Fig. 9.32c. When the domain reaches the anode, the low value of the field existing there would force the carriers to be transferred back to the lower valley, thus *quenching* the domain at the anode. The electric field that earlier got decreased in the n-active region as the domain was gaining in size now increases suddenly again and the threshold field gets exceeded once more in the cathode junction causing nucleation of another domain close to the cathode junction; such a process gets repeated cyclically. This mode of operation of the Gunn device is known as *transit-time mode* and the frequency of oscillation is given by $f_o = \frac{v_U}{L}$, where v_U is the velocity of the electrons in the upper (satellite) valley and L is the length of the n-active region (of the GaAs/InP semiconductor region).

Modes of operation of Gunn diode: There are various modes of operation of transferred electron devices (TEDs) which may be discussed in terms of three time parameters in TEDs. These time parameters are, dielectric relaxation time in the negative mobility region (T_D) which is basically the domain-growth time constant, dielectric relaxation time in the positive mobility region (T_d) and the transit time (T_t), which is the time taken for a mature domain to travel from the cathode to the anode.

(i) *Transit time mode*: This mode exists when the device is operated in a low Q resonator at a bias voltage just above the threshold voltage. Stable domains are formed. The period of oscillation is the time taken for a domain to drift from its point of nucleation (at the cathode contact) to the point of quenching (at the anode). As soon as the domain is collected at the anode, the voltage swings over the threshold level and the cycle repeats.

(ii) *Quenched and delayed domain mode*: Let a TED be placed in a high Q cavity and biased in the negative mobility region. The total voltage $V(t)$ across the device is the sum of the bias voltage V_b and the instantaneous value of the RF voltage $V_1 \cos(\omega_0 t - \varphi)$, where ω_0 is the resonant frequency of the cavity, corresponding to a time period $T_0 = 2\pi/\omega_0$. If V_1 is comparable to V_b, the total voltage $V(t)$ can swing below the threshold voltage during parts of the cycle and can either cause the domain to collapse or prevent the nucleation of a new domain. The following two modes are then possible.

(a) Quenched domain mode ($T_D < T_0 < T_t$): Since T_0 is greater than T_D, a domain that nucleates at the cathode when the voltage swings over the threshold value has sufficient time to mature. However, since T_0 is less than T_t, the voltage will swing below the threshold voltage V_T. When the voltage swings below the domain sustaining voltage V_s (which is less than V_T) the domain will be quenched even before it reaches the anode.

(b) Delayed domain mode ($T_D < T_t < T_0$): When T_t is less than T_0, a domain can reach the anode before the voltage falls below V_s and the next domain will not be nucleated until the voltage rises above the threshold value. The resonator therefore 'delays' the formation of the domain and this mode is called inhibited or delayed domain mode.

Thus, by choosing the resonant circuit for quenched domain mode and the delayed domain mode, the Gunn oscillator can be made to operate both below and above the intrinsic transit time frequency.

(c) Limited space-charge accumulation (LSA) mode: The three modes mentioned above, are transit-time limited modes, that is, they are dependent on the physical dimension (thickness) of the device. For such devices, the power-impedance product falls off as $1/f^2$. This is because, as the device operates at a fixed value of the electric field, the voltage across the device decreases linearly with increasing design frequency. For a higher frequency of operation, the device thickness and hence the device capacitance decrease, leading to a decrease in the impedance of the device. But, for the LSA mode, which does not depend on the transit time, the power-impedance product will not fall off as $1/f^2$ and the device designed to operate in this mode has the potential advantage of possible operation at higher frequency and can deliver a relatively higher power output.

In the LSA mode, the device is mounted in a resonator cavity tuned to a frequency $f_0 = 1/T_0$, which is several times above the transit time frequency. The device is biased well above the threshold voltage. As the voltage swings above the threshold value, the space charge fluctuation starts building up at the cathode. Since $T_0 < T_D$, the total voltage swings below the threshold before a domain can form. Again, as $T_d < T_0$, the accumulated space charge is drained in a very small fraction of the RF cycle. The device therefore spends most of the RF cycle in the negative mobility regime, and the space charge is not allowed to build up. Hence the name *limited space-charge accumulation mode.*

IMPATT diode: Unlike the Gunn diode, the IMPATT diode uses a semiconductor junction to realise negative resistance; however, similar to the Gunn diode, the transit time of carriers from the cathode to anode is again utilised in this device though with a different operating principle. In 1958, W T Read proposed a reverse biased p^+nin^+ diode structure consisting of an avalanche region (localised at the p^+–n interface where a very high electric field develops) as the carrier injection mechanism and subsequent drifting of the charges via a constant field drift region to realise dynamic negative resistance at microwave frequency. The mode of oscillation conceived by Read has been called the IMPATT (for IMPact Avalanche Transit Time) mode. The Read structure is difficult to fabricate; however, in 1965 Johnson, De Loach and Cohen showed that any reverse biased p–n junction diode is capable of realising IMPATT phenomena under suitable operating conditions. The IMPATT diode is also known as an avalanche transit time (ATT) device as it combines avalanche multiplication of carriers in the high field region and subsequent transit of carriers via the constant field drift region of the device. Other modes of operation in which the ATT class of devices operate are TRAPATT (for TRApped-Plasma Avalanche-Triggered Transit), BARITT

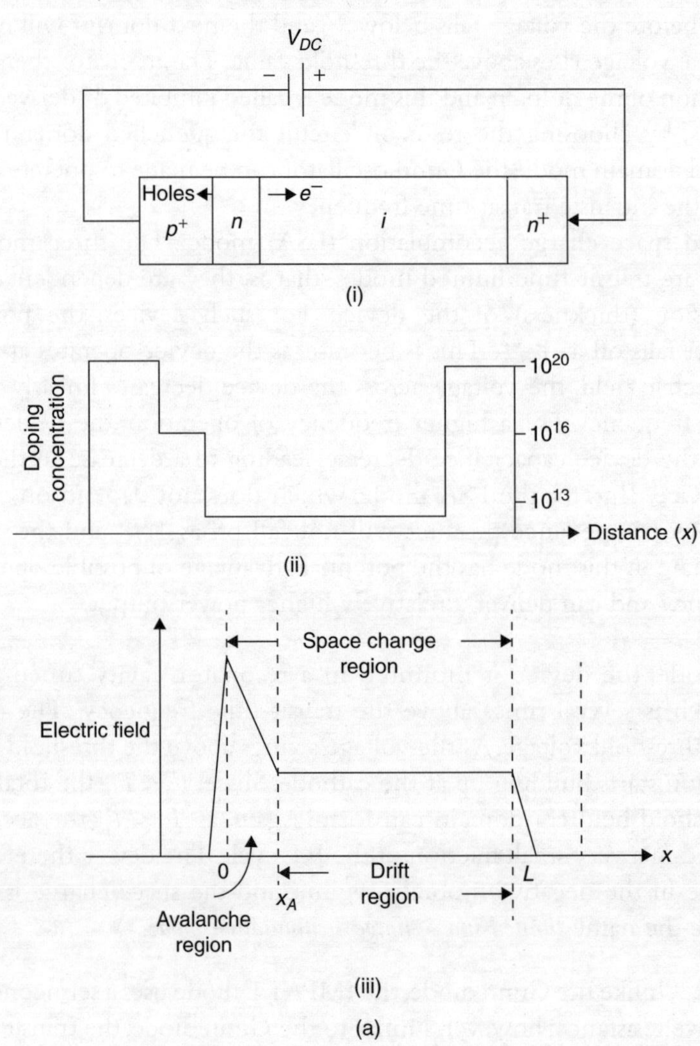

(a)

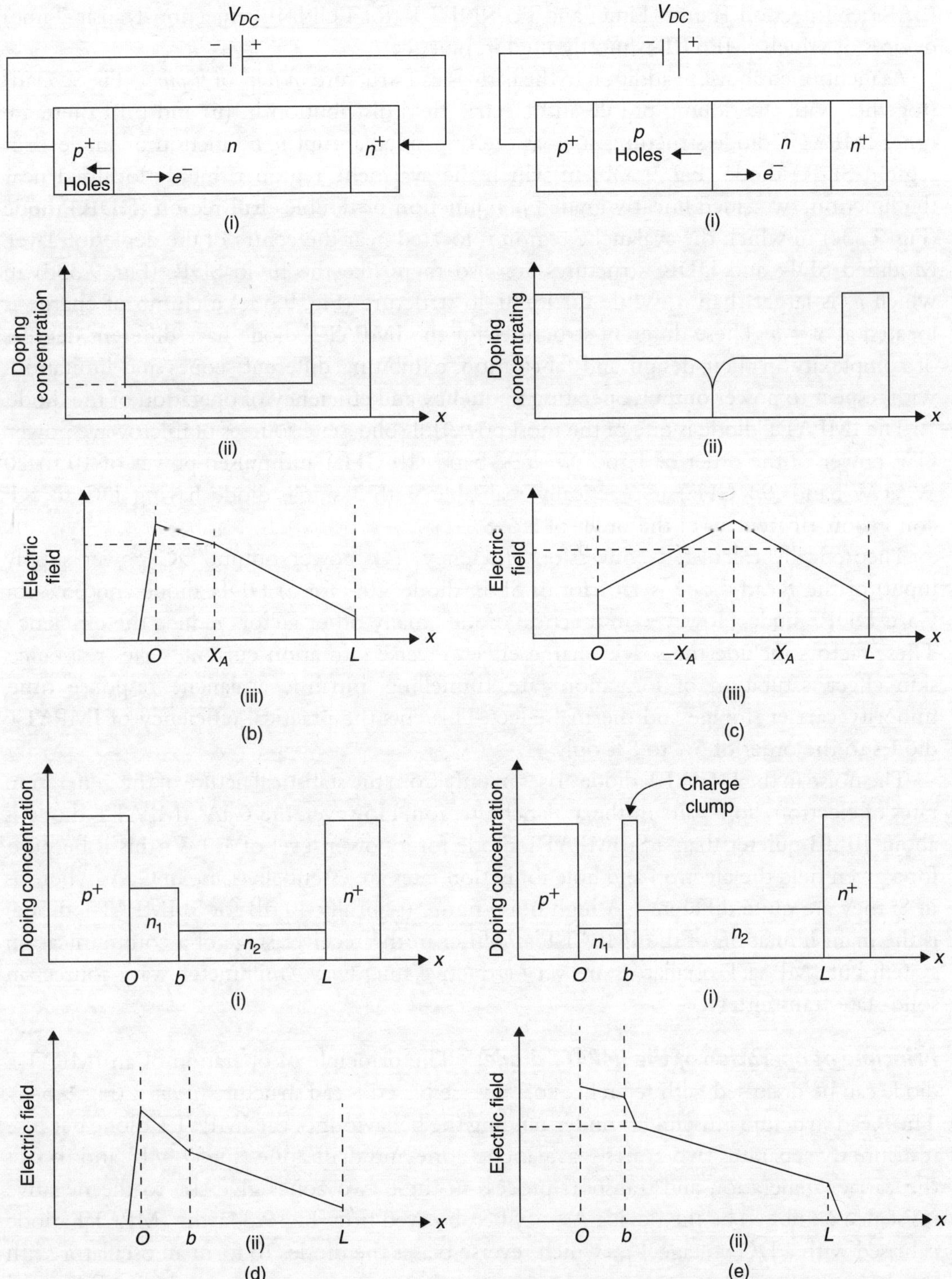

Fig. 9.33 IMPATT diodes (a) Read diode (i) Structure (ii) Doping profile (iii) Electric field distribution (b) SDR diode (c) DDR diode (d) hi–lo diode (e) lo–hi–lo diode.

(BARrier-Injection Transit-Time) and TUNNETT (for TUNNEl-injection Transit-Time) devices of which IMPATT is mostly used in practice.

As mentioned above, in addition to the basic Read structure (p^+nin^+ or n^+pip^+), [Fig. 9.33a(i) (together with the doping profile and electric field distribution in (ii) and (iii)] there are other IMPATT diode structures such as the one-sided abrupt *p–n* junction or single-drift region (SDR) diode (Fig. 9.33b) in which the avalanche region is highly localised near the junction, two sided linearly graded *p–n* junction or double-drift region (DDR) diode (Fig. 9.33c) in which the avalanche region is located near the centre of the depletion layer. Modified SDR and DDR structures are also there like the hi–lo SDR (Fig. 9.33d) in which n_1 is larger than n_2 while for lo–hi–lo structure (Fig. 9.33e) a clump of charge is located at $x = b$. These different structures for the IMPATT diode have different degrees of complexity in their design and fabrication, exhibiting different scopes and limitations with respect to power output, operating frequency and efficiency of operation of the diode.

The IMPATT diode is one of the most powerful solid-state sources of microwave power. CW power of the order of 1 to 3 W at X-band (10 GHz) and pulsed power of 10 to 20 W at W-band (94 GHz) is practically realisable with a single diode having DC to RF conversion efficiencies of the order of 5%–8%.

Theoretically calculated conversion efficiency (RF power output/DC power supply input) of the Read diode is 15% for Si SDR diode, 20% for Si DDR diode and 35% for GaAs DDR diode. However, in practical diodes, many other factors reduce the efficiency. These factors include the space–charge effect, reverse saturation current, series resistance, skin effect, saturation of ionisation rate, tunnelling, intrinsic avalanche response time, minority-carrier storage and thermal effect. This cuts the practical efficiency of IMPATT diodes to the order of 5% to 8% only.

The noise in the IMPATT diode arises mainly from the statistical nature of the generation rates of electron–hole pairs in the avalanche region. However, the GaAs IMPATT diode is about 10 dB quieter than a Si IMPATT diode for a power level of 1.0 W (this is because, for a given field the electron and hole ionisation rates are essentially same in GaAs where as in Si they are quite different). A high noise figure (typically 40 dB for Si IMPATT diode) is the main limitation of the IMPATT amplifier in the receiver stage of a communication system but IMPATT oscillators are very attractive microwave/millimeter-wave sources in solid-state transmitters.

Principle of operation of the IMPATT diode: The principle of operation of an IMPATT diode can be discussed with reference to a reverse biased Read structure (n^+pip^+) (Fig. 9.34a). The Read structure is useful for understanding the behaviour of the IMPATT diode because it distinctly separates two zones—avalanche zone and drift zone (Fig. 9.34b) and in fact the carrier generation and transport process in these two zones give rise to the negative resistance exhibited by the device. As may be observed from Fig. 9.34a, the IMPATT diode is biased with a DC voltage V_{DC} which reverse biases the diode. To form an oscillator with the device it is necessary to mount the device in a resonant cavity across which a RF signal (Fig. 9.34c) may be maintained (originating from switching transient as initial stimulus) at a frequency determined by the cavity dimensions. Thus, operation of the IMPATT device may be understood on the basis of biasing it with a DC voltage V_{DC} superimposed by an AC voltage V_{AC} (due to the RF signal in the cavity). Until the total applied voltage ($V_{DC} + V_{AC}$) exceeds the breakdown voltage V_{br}, the IMPATT diode operates as an ordinary diode.

If the diode is biased with a reverse bias voltage $V_{DC} = V_{br}$, during the positive half cycle of the RF signal the diode will be driven beyond the breakdown threshold voltage V_{br} and during the negative half cycle of the RF signal it will get biased below the breakdown threshold voltage. Hence, during the positive half cycle of the RF signal the avalanche ionisation process will set in at the n^+–p junction (in the avalanche zone, Fig. 9.34b) due to a very high electric field [of the order of E_0 (> E_{br}) = 2×10^5V/m for a Si diode] caused by the effective voltage ($V_{DC} + V_{AC}$) across the diode. The avalanche phenomenon has an exponential increasing character and thus the avalanche current I_{AV} reaches its maximum when the RF voltage is at $\omega t = \pi$ (Fig. 9.34d). Thus, it may be observed that the avalanche current I_{AV} lags in phase with respect to the RF voltage by 90°. During the negative half cycle of the RF voltage, the total electric field caused by the effective voltage ($V_{DC} - V_{AC}$) across the diode is below the critical value E_0 and the avalanche process decays. The delayed current pulse I_{AV} generated in the avalanche zone of the device during the positive half cycle of the RF voltage is next injected into the drift zone. In this way, the avalanche region acts as an injector of carriers producing a pulse of current at each RF cycle, which lags the applied RF voltage by 90°.

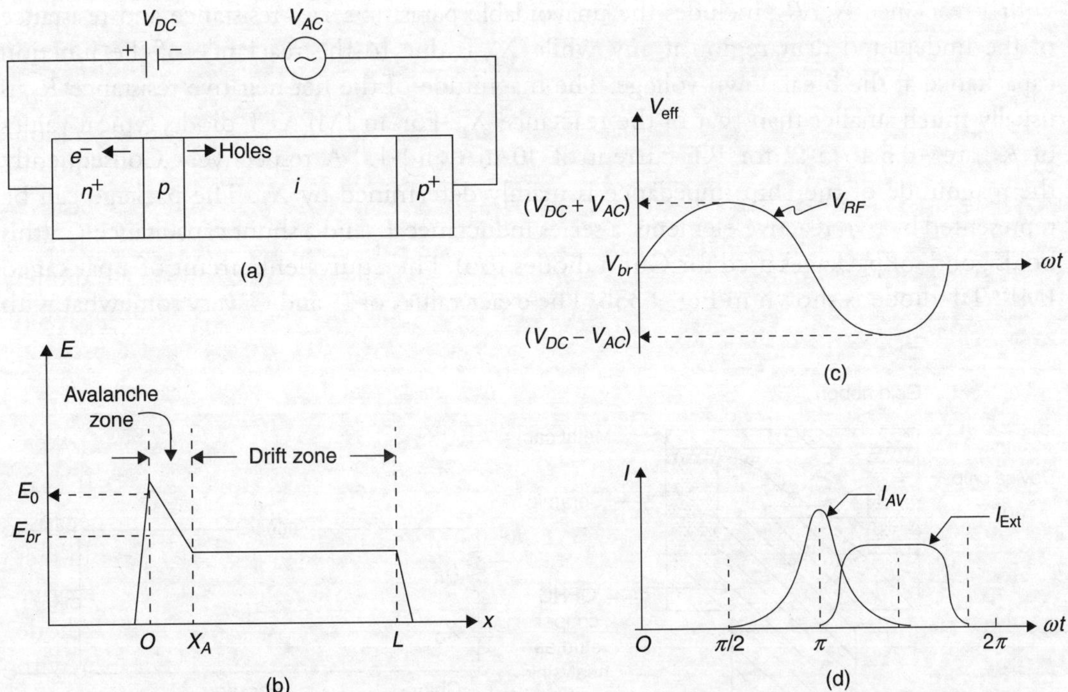

Fig. 9.34 (a) Reverse biased Read diode with AC voltage across it (b) Avalanche and drift zones of Read diode (c) RF voltage giving AC signal across the diode (d) Avalanche current and induced current in the external circuit.

The electric field due to the reverse bias voltage V_{br} is large enough to sweep the generated carriers out of the p-region (the direction of drift being dependent on each carrier's charge polarity). The electrons are immediately swept into the n^+ region and do not contribute any further phase lag to the current. But the holes drifting across the intrinsic region (also known as drift region) with saturated drift velocity v_s will cause an additional phase lag of the current (I_{Ext}) in the external circuit depending on the length L of the drift space. If L ($= \tau v_s$, where τ is the transit time of the carriers, that is, the holes across the drift region) is so chosen that an additional phase lag of 90° is realised, then the current in the external circuit (I_{Ext}) is delayed in total by 180° relative to the applied RF voltage (Fig. 9.34d). The IMPATT device thus exhibits a negative resistance (since a device possesses negative resistance when the AC current generated lags the applied voltage by a phase angle between 90° and 270°). Ideally $\omega\tau = \pi$ determines the operation frequency of the diode as an oscillator when $f_{op} = 1/2\tau = v_d/2L$; but for all practical designs, for maximum realisable power from the device, $f_{op} = 0.7v_d/2L$ is used. Typically, with $v_s = 10^7$ cm/s, $f_{op} = 30$ GHz can be realised with the device having a drift zone length $L = 1.23$ μm.

Equivalent circuit of packaged IMPATT diode: Practical IMPATT diodes are fabricated with integral heat-sink (Fig. 9.35a), in which the base of the package is oxygen-free high conductivity (OFHC) copper acting as heat-sink (for high a power device, however, it is made of diamond). The equivalent circuit of the packaged diode consists of the diode chip and the package parasitics. The diode chip is represented by negative resistance R_D in series with a reactance X_D. R_D includes the unavoidable parasitic series resistance and resistance of the undepleted drift region if any, while X_D is due to the reactance of the junction capacitance at the breakdown voltage. The magnitude of the net negative resistance R_D is usually much smaller than that of the reactance X_D. For an IMPATT diode, typical values of R_D are –0.5 to 3 Ω for RF current of 40 mA and 1.0 A respectively. Consequently, the magnitude of the chip impedance is mainly determined by X_D. The package can be represented by two reactive elements, a series inductance L_p and a shunt capacitance C_p (this holds good for packages used for Gunn diodes too). The equivalent circuit of a packaged IMPATT diode is shown in Fig. 9.35b. The exact values of L_p and C_p vary somewhat with

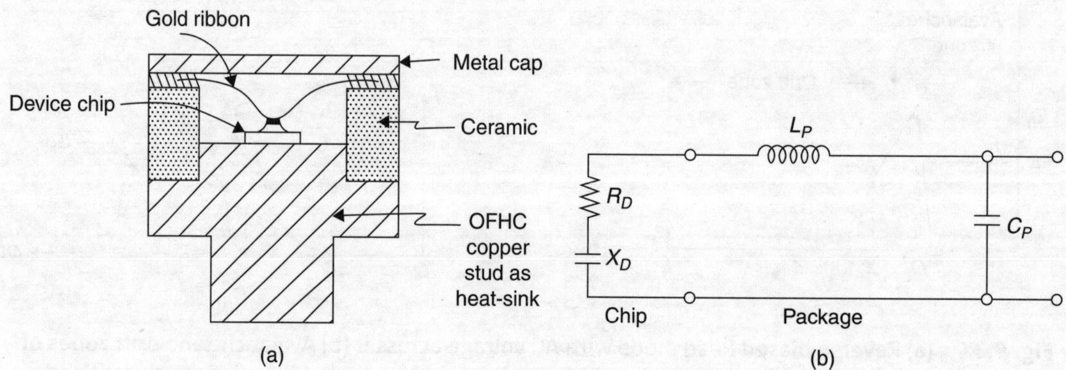

Fig. 9.35 (a) IMPATT device with integral heat-sink (b) Equivalent circuit of packaged IMPATT diode.

the package style and frequency of operation. Typical values for an X-band IMPATT diode are $L_p = 0.6$ nH and $C_p = 0.3$ pF, respectively while for a Ka-band device these values are typically $L_p = 2.5$ nH and $C_p = 15$ pF.

The packaged IMPATT/Gunn diodes (along with their engineering drawings showing dimensions) used for CW and pulsed operation are shown in Fig. 9.36. High power CW diode has a threaded bottom structure in order to have a firm fitting with the heat-sink of the mounting structure of the microwave cavity so that good heat dissipation can be realised. However, the pulsed diode is non-threaded since it operates with a small duty cycle (0.01% to 0.03%). Thus, its requirement for average power dissipation is much less compared to a CW diode.

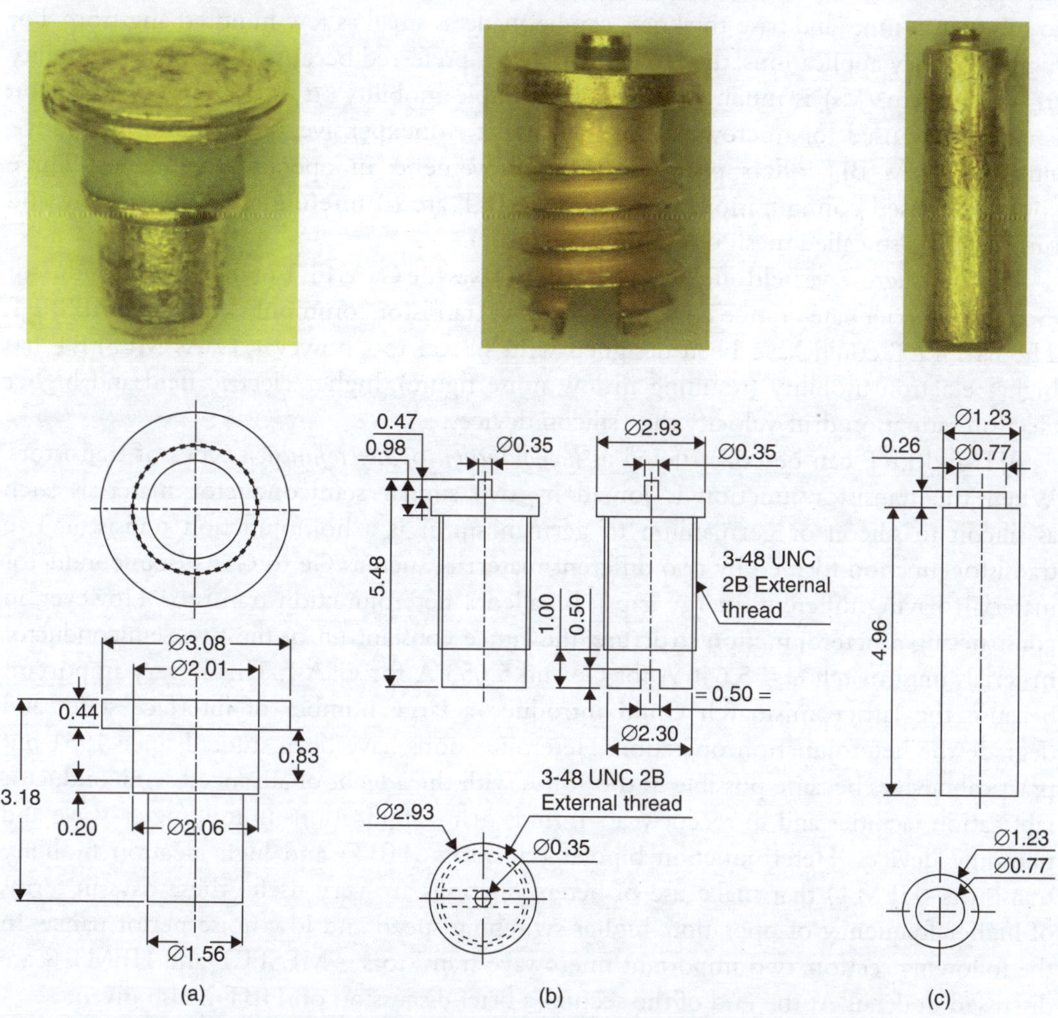

Fig. 9.36 Photograph of packaged IMPATT and Gunn diodes with engineering sketches (all dimensions in mm) (a) Low power CW IMPATT diode at X-band (b) High power CW Gunn diode at Ka-band (c) Pulsed IMPATT diode at W-band (*Courtesy*: University of Calcutta and SAMEER Kolkata centre).

9.3.2 Microwave Transistors

Since its invention in 1949, the transistor (***trans**fer res**istor***), in which the *voltage* on one terminal *is transferred to control the current* through the other two terminals, was confined only to lower RF frequency applications. However, with the advent of many technological breakthroughs in semiconductor processing technology such as alloy junction, grown junction, epitaxial, lithography, dry etching and in other related areas, high frequency transistors (above 1 GHz) have been made possible subsequently for commercial use.

For microwave transistors, a control on the active area dimensions and package parasitics is crucial and these should be considerably reduced. For a bipolar junction transistor (BJT), the emitter stripe width and base thickness are critical parameters. In lithography, with the advanced diffusion and ion implantation techniques, the emitter stripe width can be reduced to submicron units and base thickness can be made as small as few hundred ångström. For high frequency applications, the *n–p–n* structure is preferred because the electron mobility (μ_n = 1500 cm^2/V.s) is much higher than the hole mobility (μ_p = 450 cm^2/V.s). Silicon is commonly used for microwave BJT because it is inexpensive, durable and integrative, although GaAs BJT offers prospects for improvements in operating frequency. Three commonly used configurations of microwave BJT are (i) interdigitated, (ii) overlay, and (iii) matrix (also called mesh or emitter grid type).

The first microwave field effect transistor (FET) was the GaAs FET or metal semiconductor (Schottky barrier gate-source contact) field effect transistor commonly known as MESFET. The MESFET could have been designed with silicon too, however, GaAs MESFET has higher electron mobility (resulting in low noise figure), higher electric field and higher electron saturation drift velocity than silicon devices.

BJT and FET can be constructed as *homojunction* or *heterojunction* types of transistors. When the transistor junction is joined by two similar semiconductor materials such as silicon to silicon or germanium to germanium, it is a homojunction transistor. The transistor junction formed by two different materials, such as Ge to GaAs (semiconductor materials having different energy gaps), is called a heterojunction transistor. However, in constructing a heterojunction structure, the lattice constant (*a*) of the two semiconductor materials must match (a = 5.646 Å for Ge and 5.653 Å for GaAs). This is very important because the lattice mismatch could introduce a large number of interface states and degrade the heterojunction operation. Heterojunctions have been studied since 1951 but practical devices became possible in the 1980s with the advent of advanced semiconductor fabrication facilities and in recent years it finds many applications in millimeter-wave and photonic devices. Heterojunction bipolar transistors (HBT) and high electron mobility transistors (HEMT) that make use of heterojunctions are very useful these days in terms of higher frequency of operation, higher switching speed and low noise performance. In the following section, two important microwave transistors – MESFET and HEMT – are discussed in detail. At the end of the section a brief discussion on HBT is also given.

9.3.2.1 MESFET and HEMT devices

Metal Semiconductor Field Effect transistor (MESFET) and High Electron Mobility Transistor (HEMT) fall in the class of field effect transistors (FET). The field effect transistors are unipolar, majority carrier devices (unlike BJTs that are bipolar devices in which both majority and minority carriers take part in the device operation). This makes the FET devices faster than BJTs. It may be noted that there are further basic differences between BJT and FET devices both from a physical and an operational point of view. From a physical point of view in case of BJTs, the applied bias voltages to the emitter and collector with respect to the base directly controls the carrier flow (the current) from emitter to the collector while in the FETs, the applied voltage to the gate of the device manifests as a field across the gate terminals that controls the resistivity of the channel and hence the carrier current from source to drain. However, from an operational point of view, the BJTs are known as current controlled devices since the base current acts as the controlling parameter to change the V_{CE} vs I_C characteristics, while FETs are known as voltage controlled devices as the gate voltage acts as the controlling parameter to change the V_{DS} vs I_D characteristics.

Compared to the junction field effect transistor (JFET), the MESFET has better high frequency performance since it uses the Schottky barrier gate-source junction (unlike JFET where the gate-source junction is a *p–n* junction, limiting the high frequency performance of JFET). The HEMT device has better noise performance with high switching speed and can operate at much higher frequencies compared to MESFET because it can realise much higher electron mobility (may be double compared to MESFET) using heterostructure in which the carrier electrons move as a 2D electron gas via an un-doped channel and are thus separated from their parent donors reducing scattering by ionised impurities.

Various types of FET devices principally depend on how the gate terminal is isolated from the channel so that practically no gate current can flow. For example, in JFET, a *p–n* junction is used in reverse bias condition that isolates the gate, while in MESFET, the gate forms a Schottky barrier with the semiconductor of the channel for gate isolation. In the HEMT device, a large band-gap semiconductor is used to isolate the gate from the channel. In this connection, one should not get confused between MESFET and MOSFET (metal-oxide field-effect transistor) or IGFET (insulated gate field-effect transistor) in which an oxide layer is used to isolate or insulate the channel from the gate.

In the following section we will discuss the principle of operation of both MESFET and HEMT devices followed by the equivalent circuit of a packaged HEMT/MESFET device.

Metal semiconductor field effect transistor (MESFET): The MESFET uses a metallic gate (unlike JFET that uses a *p*-type semiconductor as gate material), thus, the gate-source junction in MESFET is a Schottky contact. Therefore, MESFET is also known as a Schottky barrier FET. The semiconductor material in which the channel of MESFET is formed may be either Si or GaAs; preferably the latter, since GaAs has higher saturation drift velocity compared to Si and electron mobility in GaAs is also higher compared to that in Si. Thus, MESFETs are also commonly known as GaAs FET. The use of GaAs also allows the device operation at higher temperature and at higher power levels when compared with Si based MESFET. Further, *n*-type epitaxial GaAs is used in MESFET since electrons as majority carriers have higher mobility (compared to holes if *p*-type were used) thus making it a low-noise and high speed device. MESFETs are used as active devices for designing low-noise as well as power amplifiers including their applications as switches, attenuators and so on. For high frequency operation the gate length should be very small; typically, with a 1 μm gate length, GaAS FET can operate up to 32 GHz. With the advent of micron and sub-micron fabrication facility, MESFET has become a very important low-noise (noise figure down to even 2 dB at 10 GHz) amplifier in microwave integrated circuit (MIC) and monolithic microwave integrated circuit (MMIC) system design.

Principle of operation of MESFET: The schematic of a MESFET using *n*-channel GaAs is shown in Fig. 9.37a. The substrate is a semi-insulating GaAs (typically of 100 μm thickness) which is basically chromium (Cr) doped GaAs having a resistivity as large as 10^7–10^8 Ω-cm. On this non-conducting substrate material, a thin active layer (having thickness 0.15 to 0.30 μm) of *n*-type GaAs (which is sulphur/tin doped GaAs) is grown either through ion-implantation of donor atoms into the substrate material or by epitaxial growth of doped material to form the *n*-type channel region of the field-effect transistor. Using the photolithographic process, the active layer is etched and applied with the Schottky gate (G) metal (aluminum, Al) and source (S) and drain (D) ohmic contacts with Au-Ge. Sometimes n^+ implant (as shown) is also used for better ohmic contact and to realise reduced source and drain parasitic series resistance improving its high frequency performance. Often a high resistivity GaAs epitaxial layer of thickness 3 μm, known as buffer layer, is grown between the *n*-type GaAs active layer and the semi-insulating substrate material.

The current–voltage (I–V) characteristics of MESFET, the V_{DS} vs I_D curve for various V_{GS} values (Fig. 9.37b) may be understood as follows. From Fig. 9.37a, we see, the source end is at a negative potential with respect to the drain (supply voltage V_{DS}) so that electrons travel from the source to the drain via the *n*-channel zone. The gate is held at negative potential with respect to the source (with the gate-source voltage V_{GS}) and thus the Schottky junction formed between the gate and the *n*-channel is reverse biased. When V_{DS} is small with $V_{GS} = 0$, a drain current I_D flows from the source to the drain, following ohms law, with the *n*-channel acting as a resistor. The channel resistance changes when a negative voltage is applied at the gate terminal because of the formation of positive immobile charges in the *n*-channel region (and hence an electric field E_x in the *x*-direction) that have resulted due to the reverse biased Schottky junction. Thus, the conducting channel region narrows down (decreasing its conductivity) with increasing negative voltage to the gate. For a particular negative voltage to the gate, the channel completely gets closed with the space–charge region due to immobile charges (whose conductivity is zero) and drain current I_D happens to be zero—this is known as *pinch-off* condition.

Now, let us see what happens when V_{DS} is increased beyond its small value mentioned above, keeping $V_{GS} = 0$. Under this condition, with increasing V_{DS}, the drain current I_D increases. The ohmic voltage drop in the channel resistance due to this drain current (I_D), with the direction indicated, will have a gradient from drain to the source making larger reverse bias to be applied to the Schottky junction close to the drain location than that at the source location. Thus, the depletion region intrudes more into the conducting channel close to the drain with decreasing gradient when approaching close to the source (Fig. 9.37a). Eventually a voltage V_{DS} is reached at which the channel is finally pinched-off. This voltage is not very sharply defined in I–V characteristics (Fig. 9.37b) when the current I_D begins to level off and approach a constant value. It is important to note that pinch-off does not

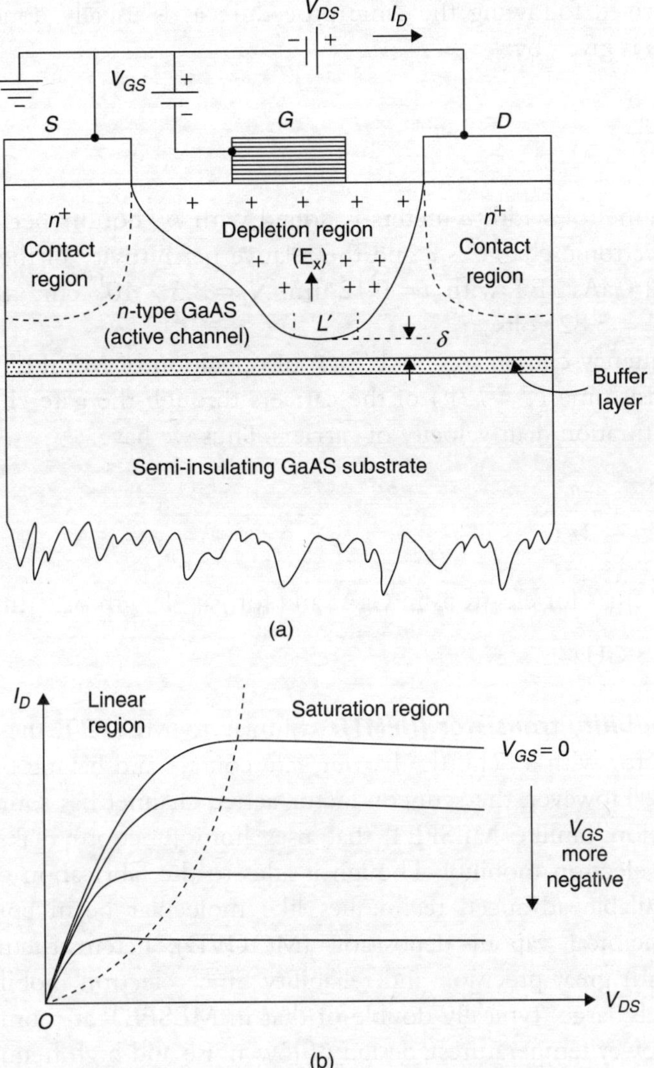

Fig. 9.37 MESFET device (a) Cross-sectional view (b) I–V characteristics.

mean that the channel is completely closed making I_D zero. If this had happened, then the ohmic voltage drop that causes the necessary back bias to maintain the space charge would extinguish! In fact a small minimum channel width (δ) exists close to the drain location at pinch-off. Further increase of V_{DS} beyond pinch-off would cause an increase of E_x in an adjacent channel section towards the source thereby increasing L' (the effective channel length for channel width of δ, Fig. 9.37a) with increasing V_{DS}, while δ remains at a fixed value keeping I_D constant with increasing V_{DS} beyond pinch-off.

If now a gate voltage $V_{GS} > 0$ is applied with the polarity indicated to provide additional reverse bias, pinch-off will occur for smaller values of V_{DS}, and the maximum drain current will be smaller (Fig. 9.37b).

It may be derived following the same procedure as is usually done with JFET that pinch-off voltage is given by:

$$V_p = \frac{qN_d a^2}{2\varepsilon_0 \varepsilon_r} \tag{9.9}$$

where the n-channel of width a meters is doped with N_d donors per cubic meter. The magnitude of electronic charge is q and the relative permittivity of the channel material is ε_r. For a typical GaAs FET with $a = 0.12$ μm, $N_d = 8.5 \times 10^{17}$ cm^{-3}, $\varepsilon_r = 13.2$, it can be calculated that $V_P = 5.82$ volts.

The high frequency cut-off ($f_{c\max} = 1/2\pi\tau$) of operation of MESFET is determined in terms of the transit time ($\tau = L/v_s$) of the carriers through the gate channel of width L, where v_s is the saturation drift velocity of carriers. Thus we have:

$$f_{c\max} = \frac{1}{2\pi t} = \frac{v_s}{2\pi L} \tag{9.10}$$

with $v_s = 2 \times 10^7$ cm/s for electrons in GaAs and for gate length of 1 μm, we find that

$$f_{c\max} = 31.8 \text{ GHz}$$

High electron mobility transistor (HEMT): Similar to MESFET, the HEMT is also a field-effect transistor with a Schottky barrier gate contact and having ohmic contacts for source and drain. However, the semiconductor active channel has some novelty realised with heterojunction (unlike MESFET that uses homojunction) making it possible to realise very high electron mobility. Though it adds to the fabrication complexities, with the presently available advanced techniques like molecular beam epitaxy (MBE) and metal-organic chemical vapour deposition (MOCVD), heterostructure fabrication is made possible with great precision and reliability. Since electron mobility in an HEMT device is extremely large (typically double of that in MESFET at room temperature and much higher at lower temperatures), design of low-noise and high frequency microwave/millimeter-wave amplifiers have become commonplace with HEMT devices especially in monolithic microwave integrated circuit (MMIC). Though the specific mode of carrier transport used in HEMTs was first investigated in 1969, it was not until 1980 that the first experimental device was operational. Nowadays, HEMT devices are used in a wide range of RF design applications including cellular telecommunications, direct broadcast receivers (DBS), electronic warfare radar, radio astronomy and any RF design application that requires a combination of low noise and very high frequency performance.

Principle of operation of HEMT: The cross-sectional view of an HEMT device (Fig. 9.38a) is seen to have many similarities with MESFET as far as the semi-insulating substrate and the source, drain, gate contacts are concerned. But, to realise superior transport properties for the carriers (realisation of high electron mobility), heterostructure is used to form a special active channel where electrons travel as a two-dimensional electron gas (2-DEG) being unhindered by the donor atoms (that normally reduces the electron drift velocity and hence mobility in homojunction devices due to scattering). The heterostructure is formed between a high band-gap doped semiconductor, typically an *n*-type AlGaAs, and a low band-gap undoped semiconductor, commonly GaAs (and not Si as Si has lower electron mobility than GaAs). Such heterostructure forms a triangular shaped quantum

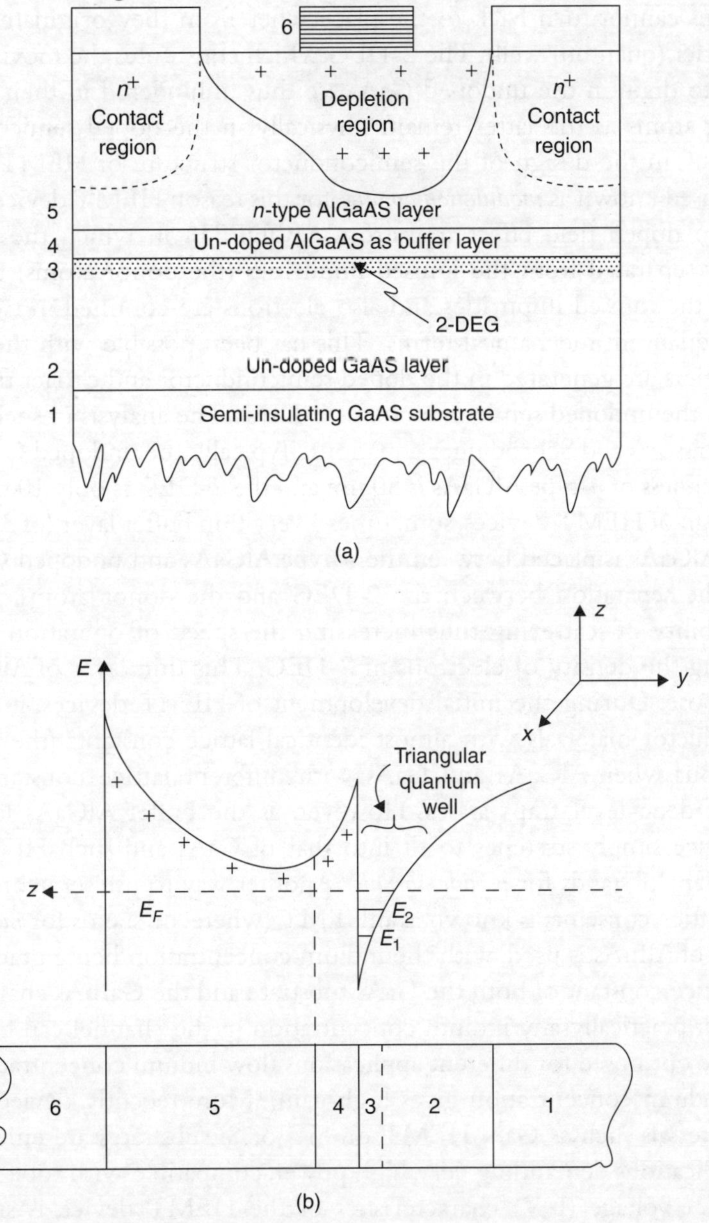

Fig. 9.38 HEMT device (a) Cross-sectional view (b) Band-bending leading to triangular quantum well at the GaAs/n-type AlGaAs heterojunction used in HEMT.

well (can be thought to be the inverse of the potential barrier we commonly come across in homojunctions) where the electrons are confined in the direction of growth of the structure (z-direction) in quantum energy states (Fig. 9.38b) and can move freely in the x and y directions (parallel to the interface) thus forming the so called 2-DEG (Fig. 9.38a) that moves from source to drain as a sheet of electron stream. This happens because of high electron affinity of GaAs which gets flooded with electrons from the n-type AlGaAs doped (normally with Si) semiconductor due to quantum mechanical tunnelling effect and these electrons cannot turn back to the place wherefrom they originated because of the potential barrier (quantum well). The 2-DEG which constitutes the moving electron sheet from source to drain in the undoped GaAs are thus unhindered in their motion by their parent donor atoms as the latter remain physically in the doped semiconductor (n-type AlGaAs). Thus, in the design of the semiconductor structure of HEMT devices, a novel technique is used known as *modulation doping* (for this reason HEMT devices are also known as modulation doped field-effect transistors, MODFET) in which the carrier electrons are physically separated from the ionised impurities (the donor atoms) thereby reducing scattering by the ionised impurities and also electrons are confined in two dimensions as a 2-DEG in quantum mechanical terms. This has been possible with the heterostructure in which carriers are generated in the doped semiconductor and carrier transport is finally manifested in the undoped semiconductor as 2-DEG. The analysis of such devices is done in terms of Poisson's and Schrödinger equations. Typically, the undoped GaAs is 1 μm thick while the thickness of n-type AlGaAs is 50 nm and the 2-DEG is only 10 nm (100 Å) thick.

In the design of HEMT devices, sometimes a very thin buffer layer (of 3–6 nm thickness) of undoped AlGaAs is placed between the n-type AlGaAs and undoped GaAs (Fig. 9.38a) to increase the separation between the 2-DEG and the donor atoms. This reduces the further possibility of scattering thus increasing the speed of operation, however, at the cost of cutting the density of electrons in 2-DEG. This thin layer of AlGaAs also serves another purpose. During the initial development of HEMT devices, it was preferred to use semiconductor materials with almost identical lattice constants (the spacing between the atoms). But when AlGaAs and GaAs with different lattice constants are used, the possibility of deep-level traps arises. However, as the buffer AlGaAs layer is very thin its crystal lattice simply stretches to fit into that of GaAs and such HEMT is known as pHEMT (where 'p' stands for *pseudomorphic*). Another way to use semiconductor materials of different lattice constants is known as mHEMT (where 'm' stands for *metamorphic*). Here, a buffer layer of AlInAs is used, with the indium concentration being graded so that it can match the lattice constant of both the GaAs substrate and the GaInAs channel. This has the advantage that practically any indium concentration in the channel can be realised, so the devices can be optimised for different applications (low indium concentration provides low noise, high indium concentration gives high gain). More recently, devices based on large band-gap materials such as GaN HEMT on SiC or Si substrates are emerging for power amplifier applications demanding very high power, compatible with tube based amplifiers.

The current voltage (I–V) characteristics of the HEMT device is similar to that of MESFET (Fig. 9.37b). For low values of source to drain bias, a current flows from drain to source through the 2-DEG and this current saturates at higher values of drain to source voltage similar to the case of MESFET. The 2-DEG density is controllable by the gate bias voltage. A negative gate bias will reduce the 2-DEG concentration thereby decreasing the source to drain current (I_{DS}) while a positive gate bias voltage will increase it.

Equivalent circuit of HEMT/MESFET device: The equivalent circuit model of HEMT/MESFET device has some *intrinsic elements* (related to the inherent device characteristics) and *extrinsic elements* (related to the parasitics of the packaged device). The intrinsic elements of the device are bias dependent and hence considered to be non-linear while the extrinsic elements are considered to be constant as they are not dependent on the bias applied to the device. Such a small-signal model of the device is very useful in the design of amplifiers as they can link the measured S-parameters and the electrical processes occurring within the device. Each element in the equivalent circuit (Fig. 9.39a) represents a lumped element

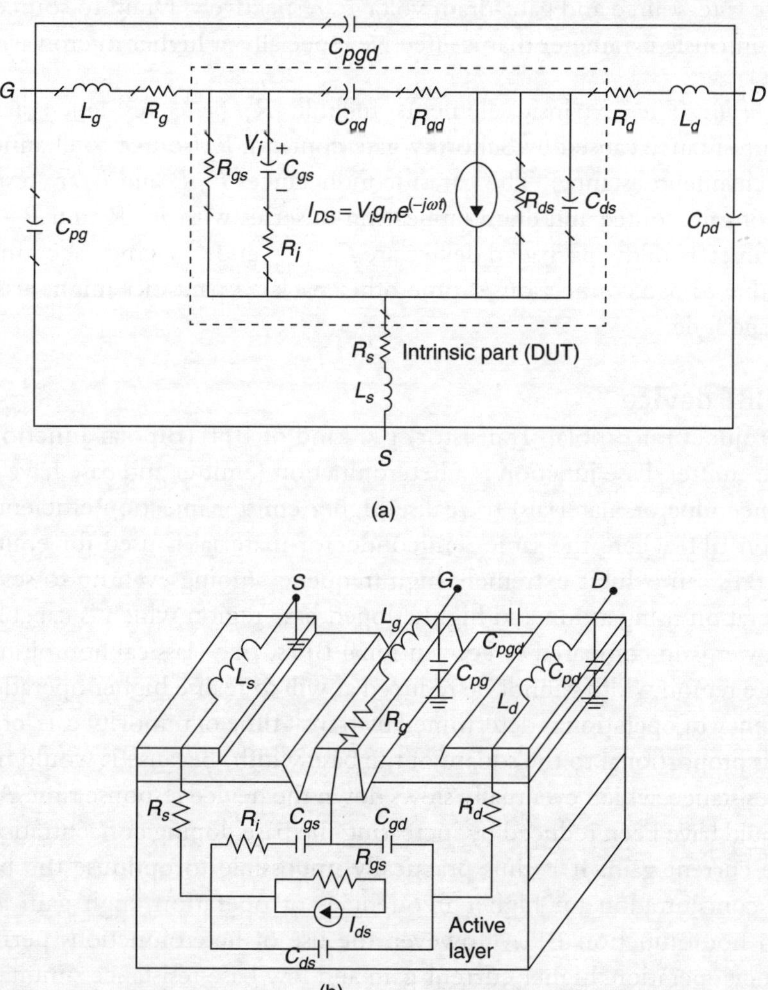

Fig. 9.39 **(a) Equivalent circuit of packaged HEMT/MESFET device (b) Physical origin of the lumped elements of equivalent circuit.**

approximation to some aspects of the device physics whose origin in the packaged device is shown in Fig. 9.39b. The values of these intrinsic and extrinsic elements depend on the structure, material, channel type and dimensions of the Schottky barrier gate. Large values of the extrinsic resistances will seriously decrease the power gain and efficiency and increase the noise figure of the amplifier. The individual circuit elements of the device model are discussed in brief in the following.

Intrinsic elements: The most important intrinsic element is the *transconductance* $g_m = dI_{DS}/dV_{DS}$ (for constant V_{GS}); it is directly proportional to gate width and the gain mechanism of the device is controlled by this parameter. The input (channel) resistance R_i and drain to source resistance R_{ds}, gate to source resistance R_{gs}, and gate to drain resistance R_{gd} are also intrinsic parameters of the device. Other intrinsic device parameters are gate-to-source capacitance C_{gs} and gate-to-drain capacitance C_{gd}, which are due to charge depletion with respect to the gate-source and gate-drain voltages respectively. Drain to source capacitance C_{ds} is also an intrinsic parameter that is effective especially at higher microwave frequencies.

Extrinsic elements: The extrinsic elements include R_g (gate resistance arising out of metallisation resistance caused by Schottky gate contact), R_s (source to channel resistance), R_d (drain to channel resistance). The parasitic inductances L_g, L_s and L_d are respectively due to the leads of gate, source and drain (appearing in series with R_g, R_s and R_d). Some more extrinsic parameters of the packaged device are C_{pgd}, C_{pg} and C_{pd} which account for parasitic capacitance due to pad connections. Some other package parasitics might arise depending on the package style.

9.3.2.2 HBT device

HBT (Heterojunction Bipolar Transistor) is a kind of BJT (Bipolar Junction Transistor) in which the emitter–base junction is a heterojunction (emitter and base have two different types of semiconductor materials) to realise higher emitter injection efficiency unlike the homojunction BJT where the same semiconductor material is used for emitter, base and collector. HBTs can exhibit extremely high frequency (going even up to several hundred GHz) of operation using a thin and highly doped base region which is capable of realising very high power gain compared to conventional BJTs. In a classical homojunction bipolar transistor, base region width cannot be reduced at will to realise higher operation frequency (as the frequency of operation is determined by transit time of minority carriers through the base which is proportional to the square of the base width) because it would mean increase of the base resistance, which eventually slows down the device response time. Although base resistance could have been reduced by increasing the base doping concentration, this would decrease the current gain. It is thus practically impossible to optimise the base thickness and doping concentration for higher frequencies of operation, high gain and low base resistance in homojunction BJT. However, the use of heterojunctions permit improved high frequency operation, higher current gain and low base resistance simultaneously.

In HBT, a wider band gap semiconductor material (AlGaAs) is used for *n*-type emitter and a comparatively narrow band gap material (GaAs) is used for *p*-type base; the collector being an n-type GaAs material. The base is a heavily doped thin p^+ GaAs layer with a concentration greater than 5×10^{18} cm^{-3} in which beryllium (Be) or carbon (C) is used as the dopant material. The collector doping concentration may be of the order of 3×10^{16} cm^{-3}. The collector and base material are the same in most HBTs, although in some cases, a wide band gap collector material is used to improve the collector base breakdown voltage. A heavily doped n^+ GaAs layer with a concentration of the order of 4×10^{18} cm^{-3} is used as ohmic contact to emitter and collector (Fig. 9.40). AuBe/Pd/Au is a typical base contact metallisation and AuGe/Ni/Au is often used for the emitter and collector contacts. The HBT structure is fabricated/grown on a semi-insulating GaAs substrate.

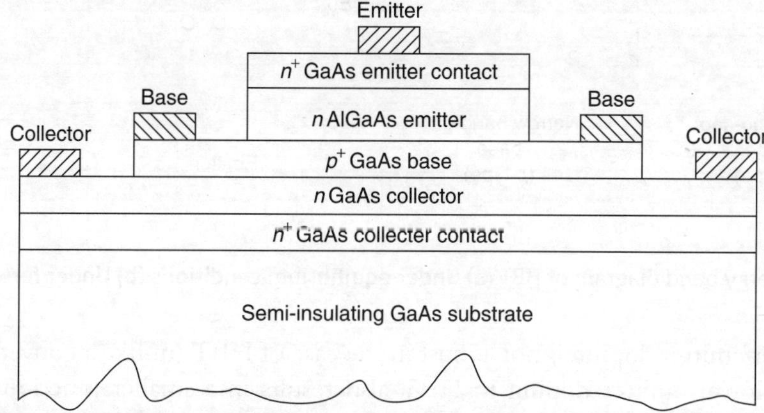

Fig. 9.40 Cross-sectional schematic of an HBT device structure.

From Fig. 9.41, it may be observed that when a forward bias is applied to the emitter–base junction, holes need to overcome a much larger potential barrier than the electrons. As a result the hole current (I_p) injected into the emitter region is much smaller than the electron current (I_n) injected into the base region even if the doping in the base region is higher than that of the emitter region. This has a significant consequence relating to the superiority of performance and the design of HBT over BJT. The figure of merit of an *npn* bipolar transistor is determined in terms of the emitter injection efficiency (which is the ratio of electron current injected from the emitter into the base and the hole current injected from the base into the emitter) that in effect also determines the current gain ratio. In a conventional homojunction BJT, high emitter injection efficiency and high current gain ratio can be realised with higher emitter doping together with lighter base doping. But in HBT, the above mentioned strong asymmetry of carrier injection in the emitter–base heterojunction provides the realisation of very high injection efficiency and hence the high current gain without restricting the doping concentration of base and emitter as demanded in the case of a conventional BJT. The use of a heterojunction thus allows the realisation of high current gain even with high doping concentration of the base region. A thin, highly doped base simultaneously satisfies the requirements for a low transit time for electrons and a low base resistance giving HBT an excellent high frequency performance.

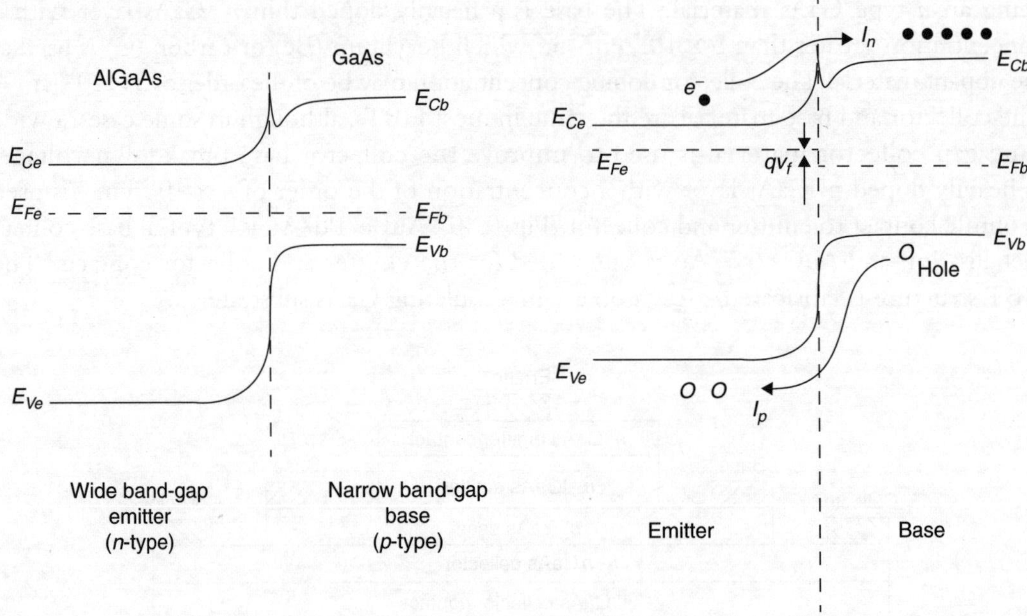

Fig. 9.41 Energy band diagram of HBT (a) Under equilibrium conditions (b) Under forward bias (V_f).

Since a higher emitter doping is not a must in the case of HBT (unlike a conventional BJT), the use of a lower emitter doping in HBT also results in a smaller junction capacitance leading to faster HBT operation.

9.4 Microwave Solid-State Oscillators, Amplifiers and Power Combiners

Microwave solid-state oscillators and amplifiers can be designed with both two-terminal active devices (diodes) such as tunnel diodes, Gunn diodes and IMPATT diodes and with three-terminal active devices (transistors) such as high frequency BJT/HBT, MESFET and HEMT devices. In this section, we will discuss the fundamental theoretical background for the design of oscillator/amplifier with diodes, oscillator/amplifier configurations with Gunn/IMPATT diodes and typical IMPATT oscillator design using 3D electromagnetic field simulator. This will be followed by the discussion on the fundamental theoretical background of amplifier/oscillator design with transistors at microwave frequencies and the design of low-noise amplifier (LNA) with HEMT device using electronic design automation (EDA) simulation tool and a brief discussion on the design approach of the transistor oscillator with dielectric resonator (DR). Finally, a concise discussion on power combiners with IMPATT/Gunn diode will also be included.

9.4.1 Oscillator/Amplifier with Two-Terminal Device (Diode)

Diodes (two-terminal devices) like tunnel diodes, Gunn diodes and IMPATT diodes (all of which are negative resistance devices) are used for the design of oscillators and amplifiers at microwave frequencies. Tunnel diodes are used for the design of low gain amplifiers while Gunn diodes are used for the design of low to medium power oscillators and IMPATT diodes are used for the design of medium and high power microwave oscillators (both CW and pulsed type). IMPATT diodes are also used in the design of injection-locked amplifiers. In this section we will first discuss the fundamental theoretical background of such oscillators and amplifiers followed by different oscillator/amplifier configurations used for IMPATT/Gunn diode based design and finally design examples will be given to illustrate the methodology for the design of IMPATT oscillator/varactor tuned IMPATT oscillator with 3D electromagnetic field simulator.

9.4.1.1 Fundamental theoretical background of microwave diode oscillator/amplifier

With respect to an IMPATT diode (though the general principle is true for any negative resistance device), the simplest equivalent circuit has been seen to be a negative resistance (R_D) in series with a capacitive reactance (X_D) (Section 9.3.1.3). The circuit part of the oscillator/amplifier represented by Z_C includes the impedance matching circuit (that matches the device having impedance Z_D with the load R_L) and the diode package parasitics and is represented by a positive resistance R_C in series with an inductive reactance X_C. In other words, the driving point impedance of the device is given by, $Z_D = (-R_D - jX_D)$, where the magnitude of $R_D > 0$) and Z_D is a function of frequency (f), DC current ($I_{d.c}$), RF current (I_{RF}) and temperature (T). The circuit impedance looking into the device terminals is given by $Z_C = (R_C + jX_C)$, where Z_C is a function of frequency (f) only. A generalised oscillator/amplifier scheme is shown in Fig. 9.42a and the condition for stable oscillation is:

$$|R_D| = |R_C| \quad \text{and} \quad |X_D| = |X_C| \tag{9.11}$$

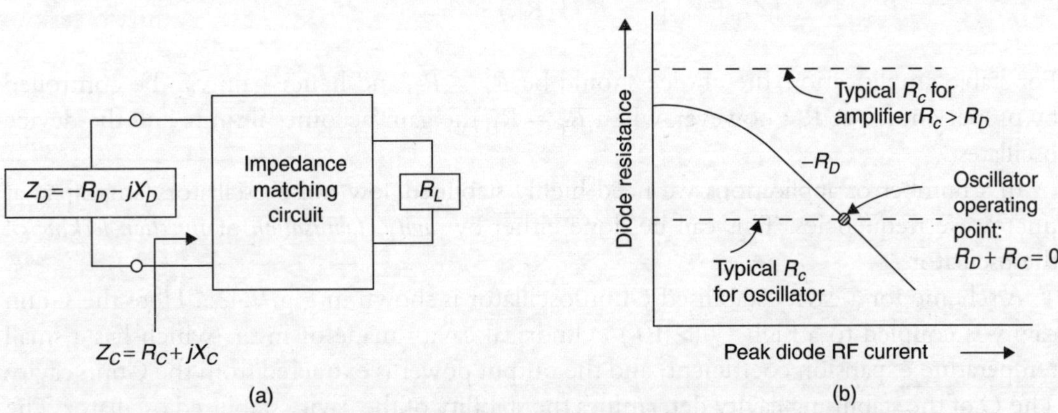

Fig. 9.42 (a) Generalised oscillator/amplifier scheme (b) Variation of the magnitude of negative resistance of diode with RF signal amplitude.

The stability of the oscillator is ensured by the net zero resistance of the circuit–diode combination and the frequency of oscillation is determined by the resonant frequency of the system consisting of device and the circuit; the total reactance of which is zero at resonance. In general, the two conditions in Eq. 9.11 will not be satisfied simultaneously in all conditions, since the negative resistance of the diode will not in general be equal to the circuit resistance at the resonant frequency. However, stable oscillation is realised in the following way. The oscillation is started, as usual by a random noise fluctuation caused by the switching transient, which will grow if the total resistance of the device-circuit combination system is negative (for positive resistance there will be decay of the signal). This is possible when with proper circuit design the magnitude of R_D is made to be larger than R_C. Now, the inherent property of the IMPATT diode (or any negative resistance device) is that the magnitude of the negative resistance of the diode decrease as the RF signal amplitude generated in the device increases (Fig. 9.42b). Thus, with device-circuit interaction, the oscillation continues to grow for several cycles until the magnitude of the diode negative resistance decreases to the value of the positive resistance of the circuit, when stable oscillation is achieved. Thus, at oscillation frequency f_0,

$$R_C(f_0) = R_D(f_0, I_0, I_{RF}, T) \tag{9.12a}$$

$$X_C(f_0) = -X_D(f_0, I_0, I_{RF}, T) \tag{9.12b}$$

The output power of the oscillator is governed by Eq. 9.12a while the oscillation frequency is determined from Eq. 9.12b.

The *reflection type amplifier* is a common configuration for signal amplification with a two-terminal negative resistance device (as mentioned in Section 6.2.7). A circulator is used to separate the input signal P_i from the amplified output power P_o. The power gain (G_P) of the amplifier is given by:

$$G_p = \frac{P_o}{P_i} = |\Gamma|^2 = \left|\frac{Z_D - Z_C}{Z_D + Z_C}\right|^2 = \frac{(R_C + R_D)^2}{(R_C - R_D)^2} \quad \text{(assuming } X_D = X_C\text{)} \tag{9.12c}$$

For stable amplification, the choice should be $R_C > R_D$ and hence gain can be controlled by proper choice of R_C; however, when $R_C = R_D$ the gain becomes infinite and the device oscillates.

In a number of applications we need highly stabilised, low noise oscillators/amplifiers at microwave frequencies. This can be done either by *cavity stabilisation* or *injection-locking* of the oscillator.

A scheme for a cavity stabilised Gunn oscillator is shown in Fig. 9.43a. Here the Gunn cavity is coupled to a high Q (> 10^4) cylindrical cavity made of invar (which has a small temperature expansion coefficient) and the output power is extracted from the Gunn cavity. The Q of the stabilising cavity determines the stability of the cavity stabilised oscillator. The degree of frequency stability is measured by *stabilisation factor* S given by:

$$S = \frac{Q_{LS}}{Q_{LUS}} \tag{9.12d}$$

A measure of FM (frequency modulation) noise reduction is given by $20\log_{10}S$, where Q_{LS} and Q_{LUS} are respectively the loaded Q factor of the stabilised and unstabilised oscillator. Substantial FM noise reduction (> 20 dB) can be readily achieved using cavity stabilisation, however, tunabilty of such an oscillator is limited to a very narrow band.

The scheme for injection-locked oscillator/amplifier is shown in Fig. 9.43b. Here an input signal (also known as locking signal which is a highly stable signal) of frequency f_i and having power P_i is injected through a circulator into a free-running oscillator whose frequency is f_o with power output P_o. By bringing f_i close to f_o, the free-running oscillator can be injection-locked by the externally applied (input) signal at frequency f_i when all the power P_o of the free-running oscillator appears at f_i and assumes the stability of the injected signal over a locking range of 2Δf, which is given by:

$$2\Delta f = \frac{2f_o}{Q_e}\left(\frac{P_o}{P_i}\right)^{-1/2} \tag{9.12e}$$

where Q_e is the external Q of the oscillator circuit.

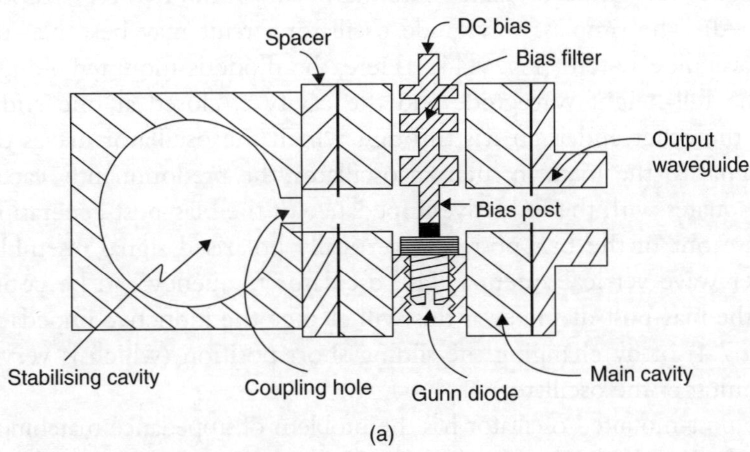

(a)

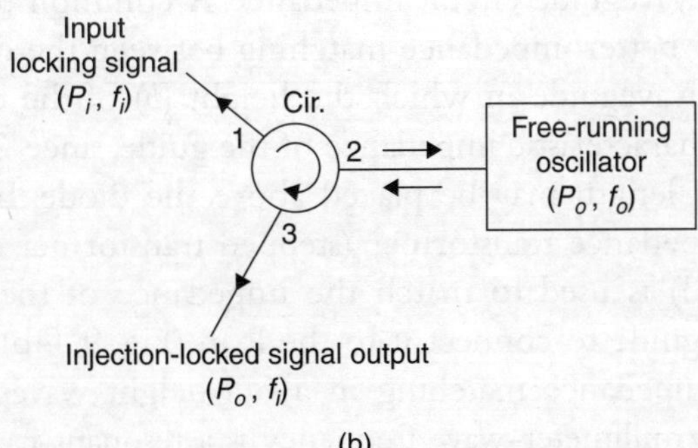

(b)

Fig. 9.43 (a) Scheme for cavity stabilised Gunn oscillator (b) Scheme for injection-locked oscillator/amplifier.

This phenomenon is helpful to reduce the noise of an oscillator by locking it to an external low-noise source. Its use as an amplifier is also important since within the locking range $2\Delta f$, a small signal input at f_i manifests as a large output signal at f_i. The gain of such an injection-locked amplifier is higher compared to the usual stable-mode amplifiers and is given by:

$$G_p = \frac{P_o}{P_i} = \frac{4}{Q_e^2}\left(\frac{f_o}{2\Delta f}\right)^2 \quad (9.12f)$$

Sophisticated phase-lock loop (PLL) techniques are also available which can get over the limitations of cavity stabilisation and injection locking techniques. The cavity stabilisation is limited in tunability and demands tight dimensional constraints at higher microwave frequencies while injection locking needs a highly stable reference oscillator and lower Q_e for large locking bandwidth worsening the FM noise performance of the locked oscillator.

9.4.1.2 Oscillator/amplifier configurations with IMPATT/Gunn diodes

Commonly used waveguide oscillator circuits of an Gunn/IMPATT diode are shown in Fig. 9.44(b–d). The simplest waveguide oscillator circuit may be a bias-post mounted full-height waveguide system (Fig. 9.44a). Here, the diode is mounted along with a bias-post inside the full-height waveguide and the cavity is closed at one end by a sliding short and on the other end is an iris through which the oscillator drives the generated microwave signal to the load. In such an oscillator, the predominantly capacitive device impedance resonates with the inductive impedance of the bias-post generating oscillations which emanate out of the bias-post as a vertically polarised signal resembling radiation from a quarter-wave vertical antenna. The oscillator frequency can be controlled either by changing the bias-post diameter as this will change the inductive impedance presented by it (Section 7.4) or by changing the sliding short position (which is very effective for mechanical tuning of the oscillator).

The simple post-mounted oscillator has the problem of impedance matching via the bias-post between the IMPATT/Gunn device (for X-band diode the magnitude of R_D is of the order of 1 Ω) and the waveguide circuit impedance. A common technique at microwave frequencies for realising better impedance matching between the device and the circuit is to use reduced-height waveguide in which the height (b) of the device may be reduced [thereby reducing the characteristic impedance of the guide, since $Z_0 \, \alpha \, b$, Eq. 4.14 (Note)] and a post of a smaller length may be placed above the diode. In such a configuration of the oscillator, an impedance transformer (stepped transformer or tapered transformer, Sections 7.3.2.1–7.3.2.3) is used to match the impedance of the reduced-height guide to the full-height waveguide to connect it to the load (Fig. 9.44b). Another technique to realise device–circuit impedance matching in a full-height waveguide for narrow band operation at microwave/millimeter-wave frequency is a resonant-cap structure (Fig. 9.44c). The resonant-cap is a circular metal disc attached to the end of a bias-post which is inserted through the top broad wall of the waveguide. The cap forms a localised radial-line cavity around the diode, with the radius of the cap approximately a quarter wavelength. The frequency of the oscillator is not easily tuned in this type of configuration, since the dimensions of the radial-line cavity (which has a very high Q value) set the frequency of oscillation quite precisely and the realisable bandwidth of the oscillator is typically 1% to

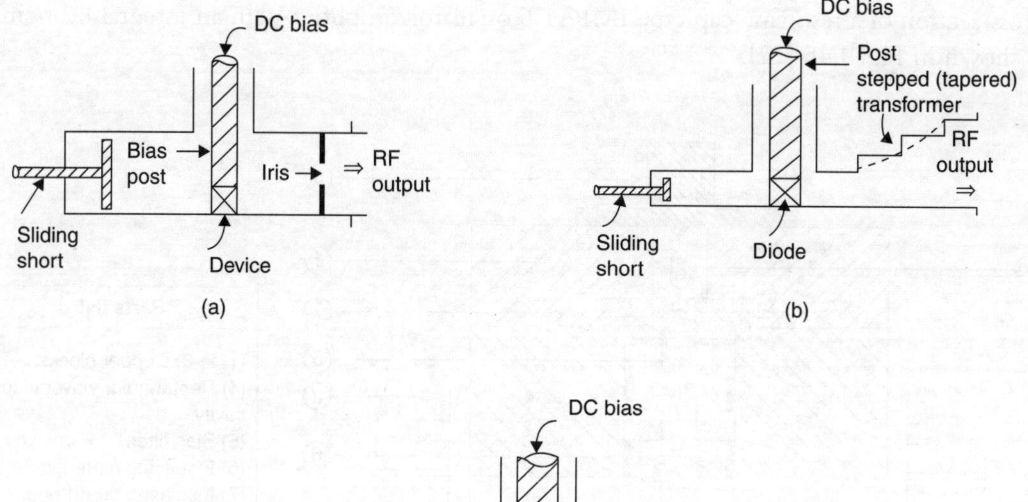

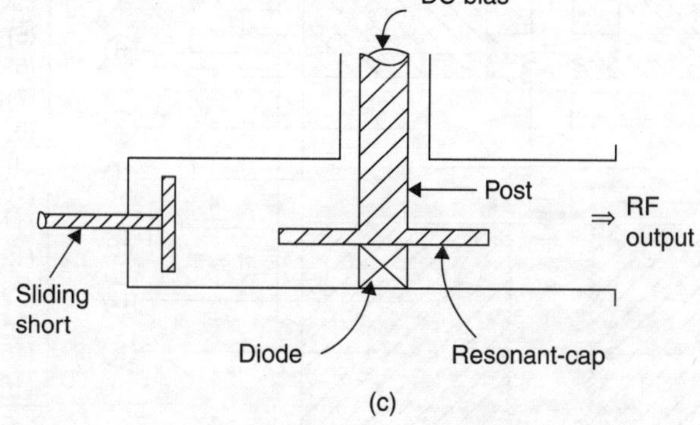

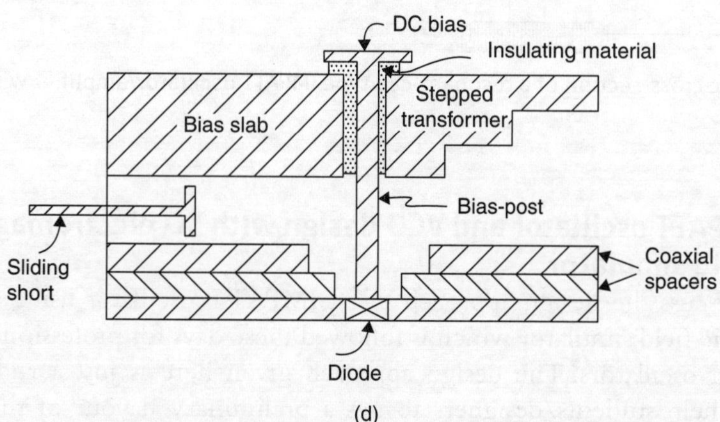

Fig. 9.44 Oscillator/amplifier configurations with IMPATT/Gunn diodes (a) Bias-post type (b) Reduced-height waveguide type (c) Resonate-cap type (d) Coaxial-waveguide type.

2%. Another technique useful at millimeter-wave frequencies for broad-band operation is coaxial-waveguide configuration in reduced-height waveguide (Fig. 9.44d). In a coaxial-waveguide cavity, the impedance of the millimeter-waveguide is first reduced with a stepped impedance transformer and then using coaxial spacers as shown. The power and bandwidth of the coaxial-waveguide type oscillator is controllable by changing the hole-diameter of the coaxial spacers, that is, it's coupling with the IMPATT device. Schematic cross-section of a resonant-cap type IMPATT oscillator/amplifier with an integral heat-sink is shown in Fig. 9.45 [22].

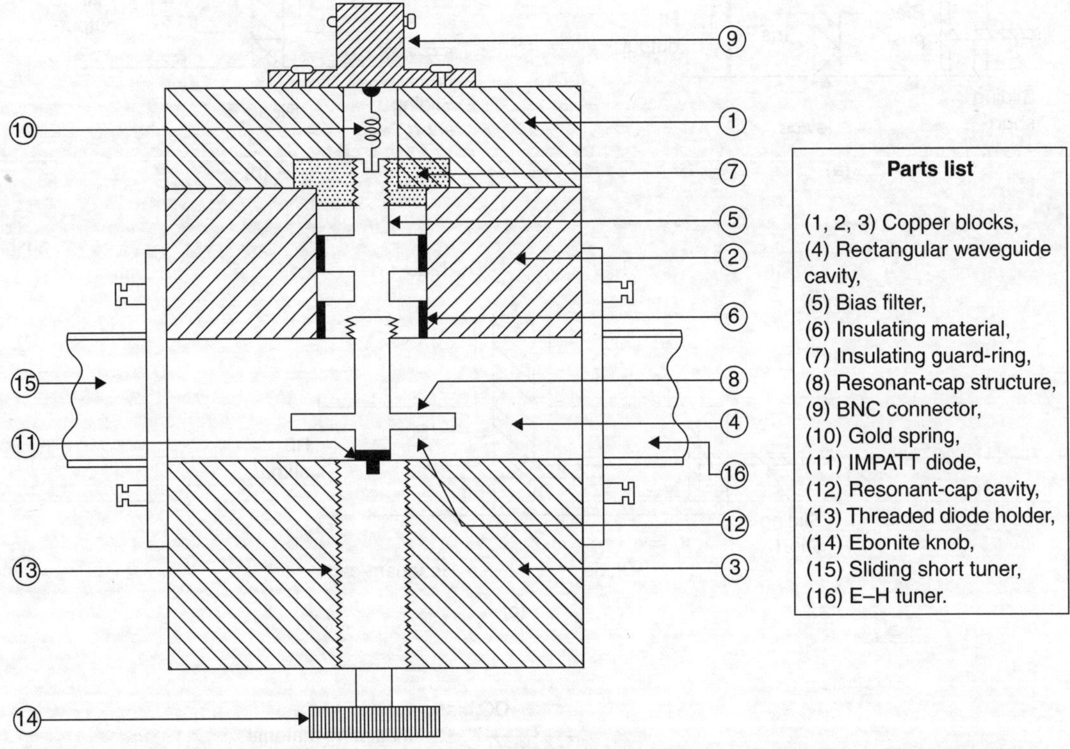

9.45 Schematic cross-section of a resonant-cap type IMPATT oscillator/amplifier with integral heat-sink.

9.4.1.3 IMPATT oscillator and VCO design with 3D electromagnetic field simulator

Let us now discuss the design approach of an IMPATT oscillator using a standard 3D electromagnetic field simulator which is followed these days for professional engineering design of such oscillators. The design approach given here is just a rudimentary one which might help students/designers to get a preliminary flavour of the design steps while designing such oscillators. A design approach for electronic tuning of such a single diode IMPATT oscillator with a varactor diode will also be given which is commonly used as a voltage controlled oscillator (VCO) at microwave frequencies (Section 9.3.1.1, varactor diode).

EXAMPLE 9.6

A microwave oscillator has to be designed at 10 GHz using X-band (8–12) GHz IMPATT diode and three types of circuit configurations—bias-post in full-height waveguide, bias-post in reduced-height waveguide and resonant-cap in full-height waveguide. Design (i) the necessary bias filter and characterise it (ii) the oscillator on the basis of device-circuit interaction in 3D electromagnetic field simulator and compare the performance characteristics of three oscillator configurations in terms of device-circuit matching (iii) the impedance transformer necessary to match the reduced-height waveguide to the full-height waveguide in case of reduced-height waveguide cavity oscillator.

Solution

(i) *Bias-filter design and its characterisation:* The bias-filter (Fig. 9.46a) is a low–high–low (stepped impedance) coaxial-type low-pass filter that is used to allow the DC power supply voltage (from the DC port) to bias the active device (IMPATT diode) and prevents the RF signal generated by the oscillator at the RF port from reaching the DC power supply. Thus, the design of this filter should on one hand be done to restrict the maximum frequency of operation to f_{max} = 8–12 GHz (X-band), typically say 10 GHz for TEM mode (the principal mode) of propagation via the coaxial line filter (Fig. 9.46a) and the cut-off frequency of the filter should be less than 0.5 f_{max}.

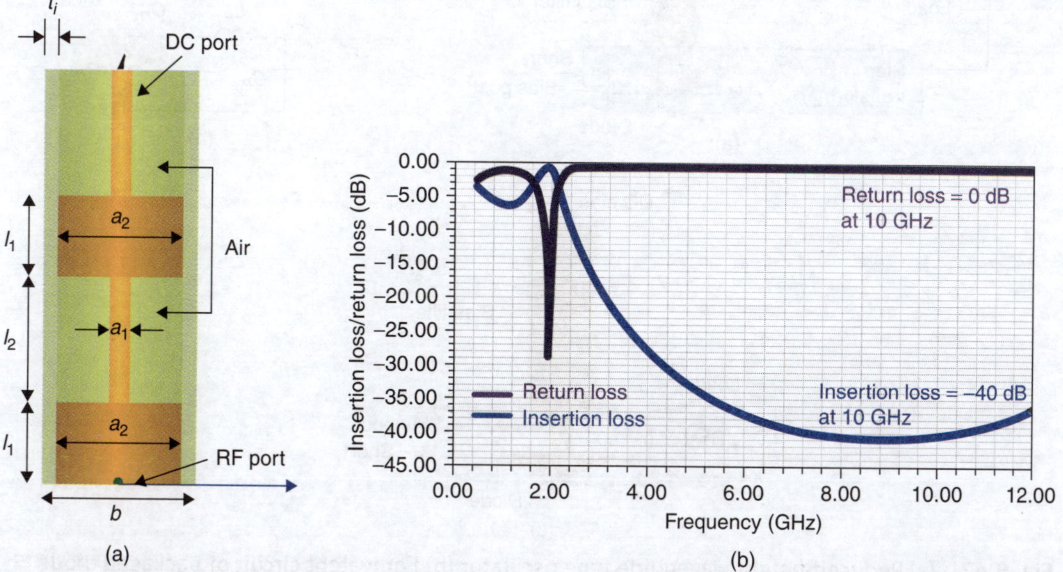

Fig. 9.46 Low–high–low coaxial type low-pass filter (a) Schematic (b) Simulation characteristics.

The condition to be met for the TEM mode propagation via the coaxial line without allowing higher order modes to propagate can be derived from field analysis as $(a_2 + b) < 2c/\pi f_{max}$ (Fig. 9.46a). With $f_{max} = 10$ GHz ($\lambda_{max} = 3$ cm), $(a_2 + b) = 1.91$ cm, and to satisfy the condition mentioned above we may chose $(a_2 + b)$ to be 10% less than the calculated value, that is, we choose $(a_2 + b) = 17.2$ mm. The insulating material (Teflon with $\varepsilon_r = 2.5$) is of thickness $t_i = 0.85$ mm when we have $(b - a_2) = 2t_i = 1.7$ mm. Thus, we find $b = 9.45$ mm, when $a_2 = 7.75$ mm. The Z_{low} represented by this section of length l_1 $(= \lambda_{max}/4\sqrt{\varepsilon_r}) = 0.474$ cm (acting as coaxial line) can be calculated from $Z_0 = \frac{138}{\sqrt{\varepsilon_r}} \log_{10}\left(\frac{D}{d}\right)$ (Table 3.2), when we have $Z_{low} = 7.505\ \Omega$. The high impedance Z_{high} to be represented by the section of the low-pass filter having length l_2 $(= \lambda_{max}/4) = 0.75$ cm and having diameter a_1 is determined by choosing $Z_{high} = 10Z_{low}$. Thus, with $Z_{high} = 75.05\ \Omega$ we can calculate $a_1 = 2.701$ mm. With these dimensions of b, a_1 and a_2 we may calculate $C = \frac{2\pi\varepsilon_0\varepsilon_r}{\ln(D/d)}$ for a line of length l_1, and $L = \frac{\mu_0}{2\pi} \ln \frac{D}{d}$ for a line of length l_2 (the formula for C and L are per unit length of the coaxial line (Table 3.1). When this is done we get $C = 341.28 \times 10^{-14}$ F and $L = 187.5 \times 10^{-11}$ H and the cut-off frequency $f_c = 1/\pi\sqrt{2LC} = 2.81$ GHz. It may be observed from the simulation characteristics of the filter (Fig. 9.46b) that around 2 GHz, there is a dip in the return loss and a peak of insertion loss and above that frequency the return loss and insertion loss sharply approaches to lower and higher values respectively and becomes 0 dB and –40 dB at and around 10 GHz.

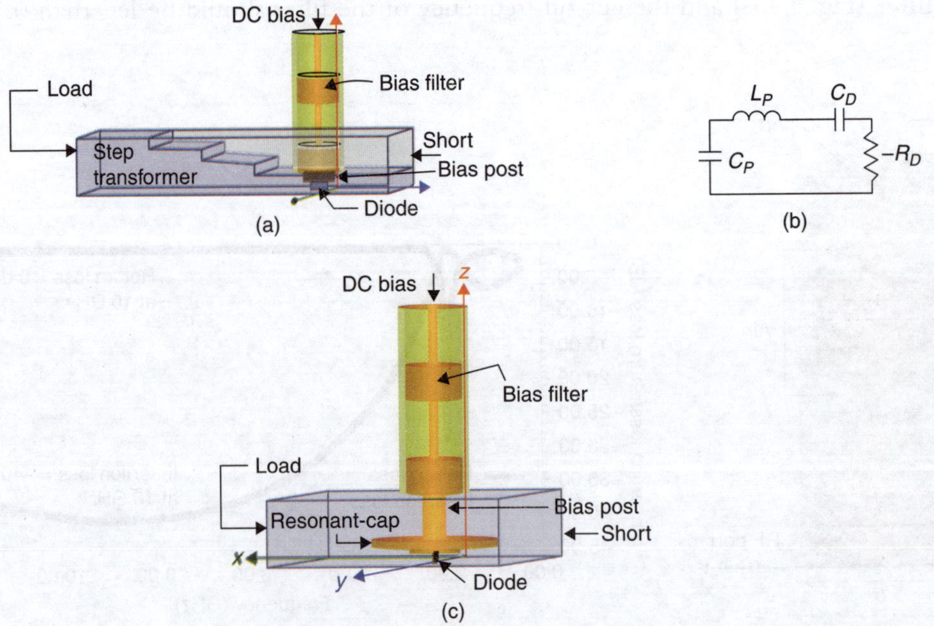

Fig. 9.47 (a) Reduced-height waveguide type oscillator (b) Equivalent circuit of packaged diode (c) Resonant-cap type oscillator.

(ii) *3D electromagnetic field simulator based oscillator design approach and characterisation*: The oscillator system with a bias-filter, bias-post in reduced-height waveguide cavity and integral post cum resonant-cap cavity in full-height waveguide along with the diode has been shown in Fig. 9.47a and 9.47c respectively (use of only bias-post in full-height waveguide could not realise device-circuit impedance match and was thus not used). The IMPATT device equivalent circuit is also shown in Fig. 9.47b with different diode parameters: $R_D = -0.7\ \Omega$, $C_D = 0.2$ pF, $L_P = 0.6$ nH and $C_P = 0.3$ pF. The oscillator system has been designed in a 3D electromagnetic field simulator considering device-circuit interaction to get the resonance at the desired freqency in which the effects of bias-post/resonant-cap cavity, short (which is normally set at $\lambda_g/4$) and the active device has been included. The bias-post diameter in case of post in reduced-height wavegide is chosen to be 6.0 mm and the resonant-cap diameter is 1.98 cm (with post above it with dia = 4 mm). The simulation result (Fig. 9.48) indicates that the resonance is around 10.3 GHz with a return loss of −11.8 dB for reduced-height waveguide type oscillator and for resonant-cap type oscillator the resonant freqency and return

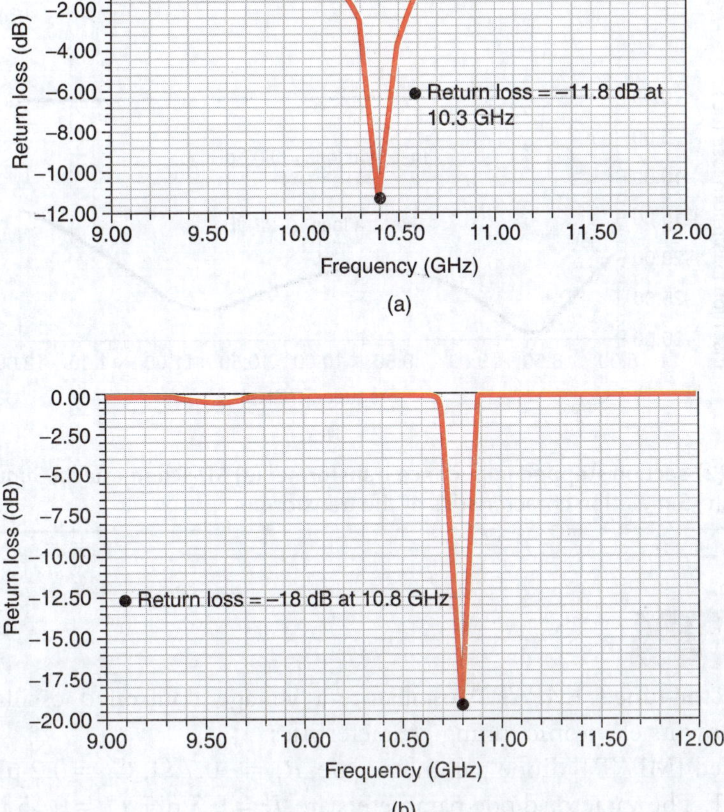

Fig. 9.48 Plot of frequency vs return loss of oscillator system (a) For diode embedded in reduced-height waveguide cavity (b) For diode embedded in resonant-cap cavity.

loss are seen to be at 10.8 GHz and –18 dB respectively. It may be observed that the resonant-cap type oscillator provides a better impedance match due to device–circuit interaction. However, both the oscillators can be further optimised for impedance matching and the freqency of resonance (this is left to the students as an exercise).

(iii) The design of two-section stepped impedance transformer required in the reduced-height cavity type oscillator (Fig. 9.47a) may be done by following the procedure given in Sections 7.3.2.1 and 7.3.2.2. The transformer structure with design data of simulation and the simulation characteristics of the designed transformer are given in Fig. 9.49.

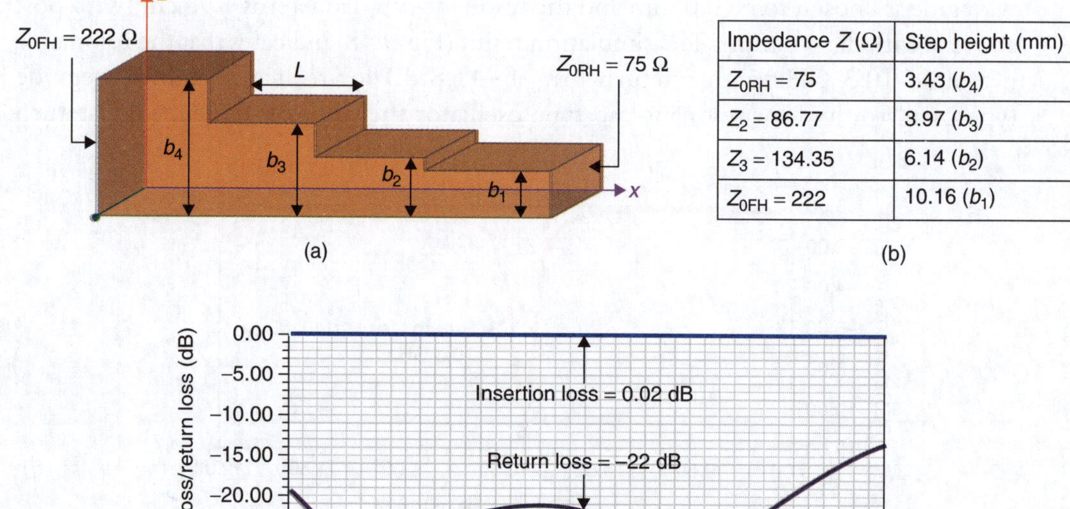

Impedance Z (Ω)	Step height (mm)
$Z_{0RH} = 75$	3.43 (b_4)
$Z_2 = 86.77$	3.97 (b_3)
$Z_3 = 134.35$	6.14 (b_2)
$Z_{0FH} = 222$	10.16 (b_1)

Fig. 9.49 Two (2)-section stepped impedance transformer (a) 3D schematic (b) Simulation design data (c) Simulation chracteristics of the transformer.

EXAMPLE 9.7

Design a varactor-tuned IMPATT oscillator (a voltage controlled oscillator, VCO) at X-band and show its electronic tuning characteristics.

The packaged IMPATT diode parameters are, $R_D = -0.7\ \Omega$, $C_D = 0.2$ pF, $L_P = 0.6$ nH and $C_P = 0.3$ pF. The varactor diode parameters are, $L_p = 0.3$ nH, $C_p = 0.15$ pF, $R_S = 1.2\ \Omega$, $C_j(V_{rb}) = 3$ pF @ 0V and $C_j(V_{rb}) = 0.2$ pF @ 20V. The bias voltage for a varactor diode can be changed from 0 volt to 20 volt.

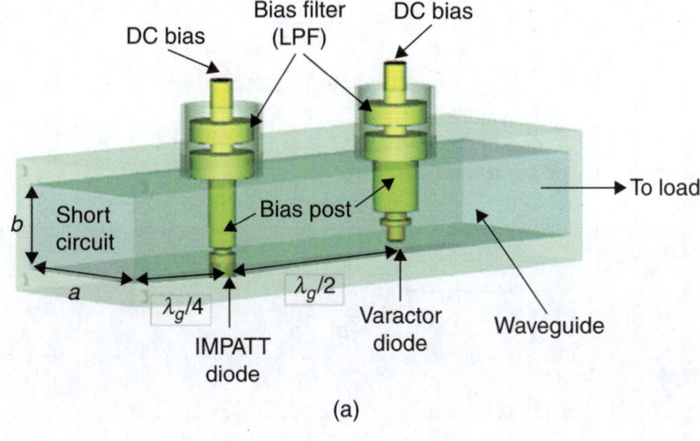

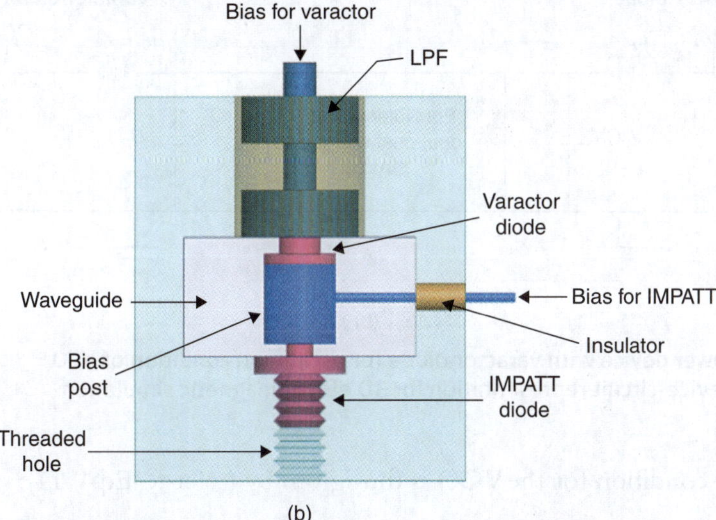

Fig. 9.50 Varactor-tuned oscillator using (a) Staggered configuration (b) Antipodal configuration.

Solution

The varactor tuned oscillator using IMPATT/Gunn diode may be of two types—parallel tuned (staggered) configuration and series tuned (antipodal) configuration (Fig. 9.50). In staggered configuration the distance between the two posts is $\lambda_g/2$ and the distance between post and short circuited plane is $\lambda_g/4$. But in the antipodal configuration, the electrical length between the IMPATT and varactor is zero. Thus, wideband VCOs are generally designed with antipodal configuration. Our design in this problem will be with the parallel tuned configuration.

The oscillation condition for a varactor tuned IMPATT oscillator can be written with reference to Fig. 9.51a, where the block marked as equivalent one port power device is the IMPATT oscillator itself consisting of the bias-posts in the waveguide and the IMPATT diode, while the varactor diode is represented by its junction capacitance $C_j\,(V_{rb})$ and series resistance R_S.

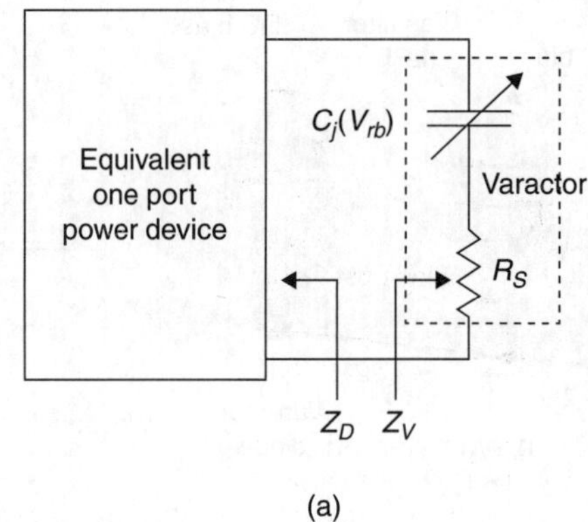

(a)

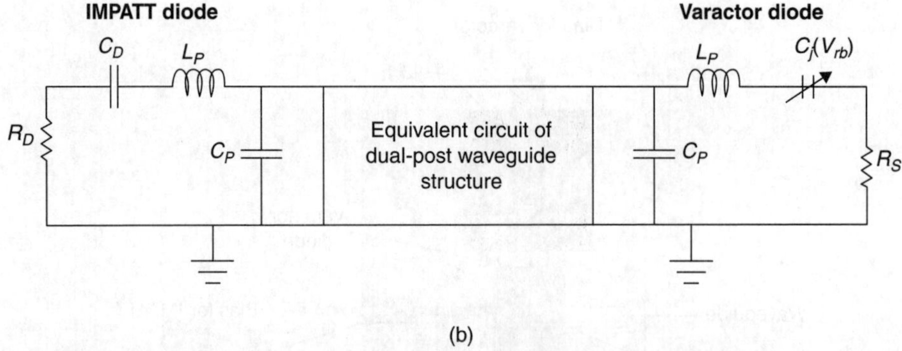

(b)

Fig. 9.51 (a) Power device with varactor diode for oscillation condition of VCO (b) Device-circuit representation for 3D electromagnetic simulation.

The oscillation condition for the VCO is thus given by (refer to Eq. 9.11):

$$R_S + R_D(\omega, I) = 0 \quad \text{and} \quad -\frac{1}{\omega C_j(V_{rb})} + X_D(\omega, I) = 0 \tag{9.13}$$

To design the VCO in a 3D electromagnetic field simulator, we refer to Fig. 9.51b, where the equivalent circuit of dual-post waveguide structure in conjunction with the equivalent packaged device (both IMPATT and varactor) impedances are shown. The device–circuit interaction is optimised in the simulator to realise good impedance matching between the device and the circuit. It may be observed from Fig. 9.52a that with zero varactor bias voltage ($V_{rb} = 0$) the oscillation frequency of VCO is at 10.09 GHz with good device–circuit impedance matching (return loss –22 dB). The varactor bias voltage change causes the change of its capacitance and hence the frequency change of the VCO and with varactor reverse bias set to $V_{rb} = -20$ volts, the oscillation frequency is at 10.50 GHz (with return loss of –20 dB) (Fig. 9.52b). The electronic tuning of the IMPATT oscillator with the change of varactor bias voltage of the VCO is seen to be almost linear with a frequency change of 410 MHz for varactor bias variation from 0–20 volts (Fig. 9.52c).

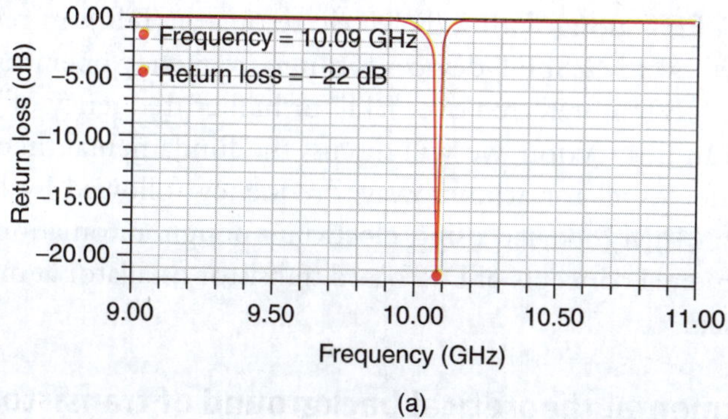

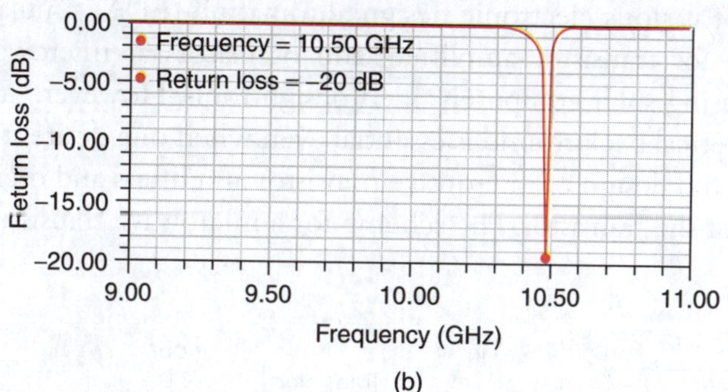

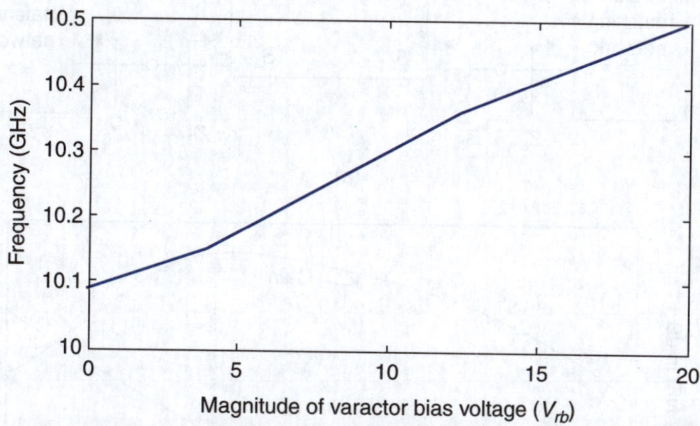

Fig. 9.52 Varactor-tuned IMPATT oscillator (VCO) characteristics (a) With zero varactor bias (V_{rb} = 0 V) (b) With V_{rb} = 20 V (c) VCO tuning characteristics.

9.4.2 Amplifier/Oscillator with Three-Terminal Microwave Device (Transistor)

Transistors (three-terminal device) like MESFET, HEMT and BJT are used for the design of amplifiers at microwave frequencies while transistor-based microwave oscillators may be designed with BJT/FET along with a dielectric resonator (DR). In recent times, HEMT devices are preferred over MESFET devices for high frequency operation with better noise performance while HBT is preferred over BJT for higher frequency of operation and with higher power gain. In this section, we will discuss the fundamental theoretical background of transistor amplifier/oscillator at microwave frequency followed by low-noise amplifier (LNA) design with HEMT device using electronic design automation (EDA) simulation tool. Finally, fundamental concepts of DR based transistor oscillator at microwave frequency will also be discussed.

9.4.2.1 Fundamental theoretical background of transistor amplifier/oscillator

With the advent of various electronic design automation (EDA) simulation tools, modern design procedures for transistor amplifiers and oscillators at microwave frequencies are now carried out using such computer-aided design tools. However, understanding of the fundamental theoretical background is essential even when one desires to apply such design tools. The analysis and design of microwave transistor amplifiers and oscillators are based on the S-parameters of the transistor. The scheme for a microwave transistor amplifier is given in Fig. 9.53a.

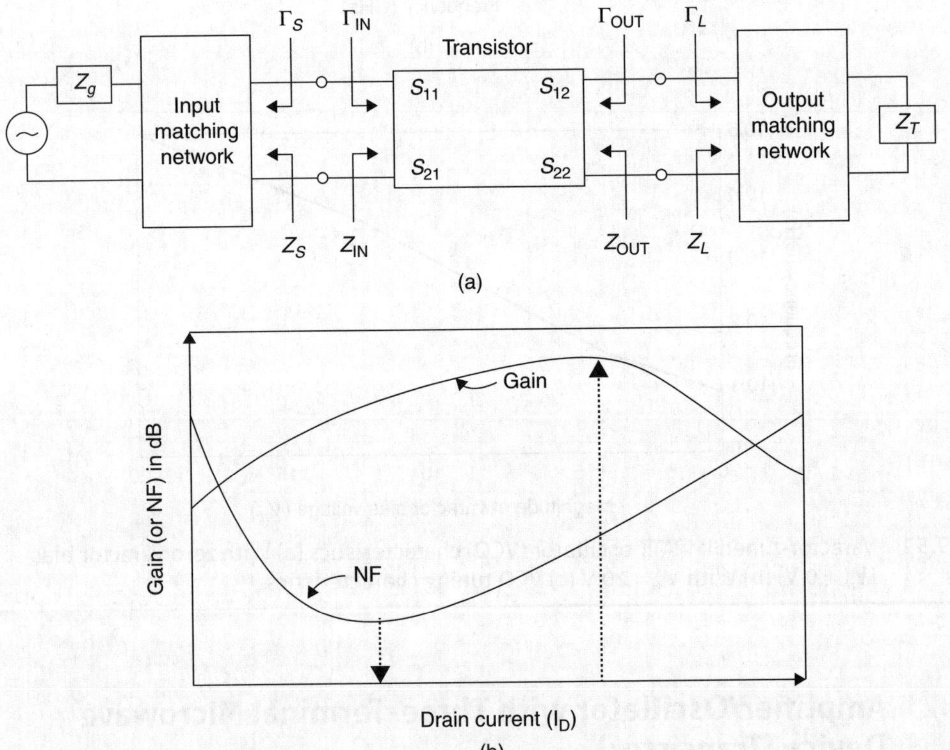

Fig. 9.53 (a) Schematic for a microwave transistor amplifier (b) Noise figure and associated gain as a function of drain current.

In the amplifier configuration, the power available from the source is delivered to the input of the transistor through the input matching network. This is designed to transform the generator impedance Z_g (usually 50 Ω) to the source impedance Z_S (as seen looking from the active device terminals towards the generator). The active device together with the matched generator is expected to deliver maximum power to the load. To realise this, an output matching network is used to transform the terminating impedance Z_T (usually 50 Ω) to the load impedance Z_L. These matching networks are designed to realise the required stability, gain, noise performance and bandwidth of the amplifier. Matching networks may be L, Π, or T-type of which L-section networks (Section 7.5) are simple to design and practically very common.

The amplifiers may be designed for *unilateral* case when $|S_{12}| = 0$, (the transmitted signal from load to the source is neglected) or for *bilateral* case when $|S_{12}| \neq 0$ (effect of load on source needs to be considered).

From Fig. 9.53b, we can see that realisation of maximum gain and low noise figure from an amplifier are at odds. For realising maximum gain, we need to make conjugate impedance match (in terms of impedance/reflection coefficient) at the input and output $Z_S = Z^{\star}_{IN}$ and $Z_L = Z^{\star}_{OUT}$. For multistage amplification (a requirement in almost all practical situations) we have from the well known Friss' formula, $F_n = F_1 + \dfrac{F_2 - 1}{G_1} + \dfrac{F_3 - 1}{G_1G_2} + ... + \dfrac{F_n - 1}{G_1G_2...G_{n-1}}$. This indicates that the overall noise figure of a multistage amplifier can be kept low if the noise figure (F_1) of the first stage is made small and its gain (G_1) is made moderately high (need not be the highest realisable gain). Thus, a common practice for a multistage amplifier design is to make the first stage a low-noise amplifier compromising the gain of the amplifier to some extent, while the subsequent stages are designed to have maximum possible gain with little concern for minimising the noise figure (as the noise figure of subsequent stages adds very little to the overall noise figure). Thus, in the design of a low-noise amplifier (LNA) impedance matching at the input need not be a conjugate match, rather Z_S is made equal to $Z_{opt} (\neq Z^{\star}_{IN})$. This might result in a little loss of gain of the amplifier (due to a little impedance mismatch), however, low-noise performance can be realised with the amplifier so designed.

The transistor oscillator design approach at microwave frequencies is similar to that of the amplifiers except that the circuit is purposely designed to be unstable. It will be seen shortly that for amplifier we need $K > 1$ while for oscillator $K < 1$, where K is known as the *stability factor*. The operating principle of three-terminal (transistor) oscillator is similar to that for the two-terminal (diode) oscillator discussed in Section 9.4.1.1.

Now, let us discuss the formulations and graphs necessary to understand the principle of operation of a microwave amplifier and its design based on S-parameters of the active device.

Amplifier gain: Many *power gain* equations are derived in terms of the S-parameters and the reflection coefficient at the input and output of microwave amplifiers. While discussing S-parameters in Chapter 5, we discussed how to derive different gain expressions in

Example 5.3, namely, *transducer power gain*, *available power gain*, and *operating power gain* or simply *power gain*. These are repeated here for convenience.

Transducer power gain (G_T) is given by:

$$G_T = \frac{1-|\Gamma_S|^2}{|1-\Gamma_{IN}\Gamma_S|^2}|S_{21}|^2\frac{1-|\Gamma_L|^2}{|1-S_{22}\Gamma_L|^2} \tag{9.14a}$$

Or alternatively as:

$$G_T = \frac{1-|\Gamma_S|^2}{|1-S_{11}\Gamma_S|^2}|S_{21}|^2\frac{1-|\Gamma_L|^2}{|1-\Gamma_{OUT}\Gamma_L|^2} \tag{9.14b}$$

$$\text{where } \Gamma_{IN} = S_{11} + \frac{\Gamma_L S_{12}S_{21}}{1-S_{22}\Gamma_L} \text{ and } \Gamma_{OUT} = S_{22} + \frac{S_{12}S_{21}\Gamma_S}{1-S_{11}\Gamma_S} \tag{9.14c}$$

Some special cases of transducer power gain may also be mentioned.

(i) When both input and output networks are matched to source impedance and load impedance then we have $\Gamma_S = \Gamma_L = 0$; hence, the gain is *matched transducer power gain* given by:

$$G_{TM} = |S_{21}|^2 \tag{9.14d}$$

(ii) If it is assumed that the amplifier is unilateral, that is, $|S_{12}| = 0$, then $\Gamma_{IN} = S_{11}$ and $\Gamma_{OUT} = S_{22}$ and thus *unilateral transducer power gain* is given by:

$$G_{TU} = \frac{1-|\Gamma_S|^2}{|1-S_{11}\Gamma_S|^2}|S_{21}|^2\frac{1-|\Gamma_L|^2}{|1-S_{22}\Gamma_L|^2} \tag{9.14e}$$

(iii) One can optimise the input reflection coefficient Γ_S and output reflection coefficient Γ_L making $\Gamma_S = S_{11}^*$ and $\Gamma_L = S_{22}^*$ to realise *maximum unilateral transducer power gain* given by:

$$G_{TU_{MAX}} = \frac{|S_{21}|^2}{(1-|S_{11}|^2)(1-|S_{22}|^2)} \tag{9.14f}$$

Available power gain (G_{AV}) is given by:

$$G_{AV} = \frac{1-|\Gamma_S|^2}{|1-S_{11}\Gamma_S|^2}|S_{21}|^2\frac{1}{1-|\Gamma_{OUT}|^2} \tag{9.14g}$$

Operating power gain or simply *power gain* (G_P) is given by:

$$G_P = \frac{1}{1-|\Gamma_{IN}|^2}|S_{21}|^2\frac{1-|\Gamma_L|^2}{|1-S_{22}\Gamma_L|^2} \tag{9.14h}$$

However, it may be mentioned that, the gain of an amplifier we talk about is the operating power gain or simply the power gain.

Stability considerations for amplifier: The design of an amplifier should ensure its *stability* in the sense that it does not oscillate either unconditionally or conditionally. The amplifier stability depends on the S-parameters of the active device (transistor), the matching networks and the terminations. The two port amplifier network (Fig. 9.53a) is said to be unconditionally stable at a given frequency if the real part of the input and output impedances of the active device are positive ($Re\ [Z_{IN}] > 0$ and $Re\ [Z_{OUT}] > 0$) for all passive load and source impedances. If the two port network is not unconditionally stable, it is potentially unstable or conditionally stable when the real part of input and output impedances are negative ($Re\ [Z_{IN}] < 0$ and $Re\ [Z_{OUT}] < 0$) for some passive load and source impedances at a specific frequency.

Thus, for an unconditionally stable case at a given frequency, we can write in terms of reflection coefficient:

$$|\Gamma_S| \le 1; |\Gamma_L| \le 1 \tag{9.15a}$$

$$|\Gamma_{IN}| = \left|S_{11} + \frac{S_{12}S_{21}\Gamma_L}{1-S_{22}\Gamma_L}\right| < 1 \tag{9.15b}$$

$$|\Gamma_{OUT}| = \left|S_{22} + \frac{S_{12}S_{21}\Gamma_S}{1-S_{11}\Gamma_S}\right| < 1 \tag{9.15c}$$

While, for a potentially unstable (or conditionally stable) case, we have in terms of reflection coefficient:

$$|\Gamma_{IN}| > 1 \quad \text{or} \quad |\Gamma_{OUT}| > 1 \tag{9.16}$$

The conditions for amplifier stability can be derived on the basis of graphical analysis of Eqs 9.15b and 9.15c as follows. When we set $|\Gamma_{IN}|$ and $|\Gamma_{OUT}|$ equal to unity, a boundary will be obtained beyond which the amplifier is unstable. It can be shown that solving for Γ_L from the solution of equation $|\Gamma_{IN}| = 1$ yields circles in the Γ_L plane known as *output stability circle* while solving for Γ_S from the solution of equation $|\Gamma_{OUT}| = 1$ yields circles in the Γ_S plane known as *input stability circle* [23]. The equations representing these circles are given by:

$$\left|\Gamma_L - \frac{(S_{22}-\Delta S_{11}^*)^*}{|S_{22}|^2-|\Delta|^2}\right| = \left|\frac{S_{12}S_{21}}{|S_{22}|^2-|\Delta|^2}\right| \tag{9.17a}$$ [For output stability circle]

$$\left|\Gamma_S - \frac{(S_{11} - \Delta S_{22}^{\star})^{\star}}{|S_{11}|^2 - |\Delta|^2}\right| = \left|\frac{S_{12}S_{21}}{|S_{11}|^2 - |\Delta|^2}\right| \qquad (9.17b) \quad \text{[For input stability circle]}$$

where $\Delta = S_{11}S_{22} - S_{12}S_{21}$ (9.17c)

The radii and centres of output stability and input stability circles are given by:

With $|\Gamma_{IN}| = 1$, the *output stability circle* in the Γ_L plane has:

$$\text{Radius } (r_L) = \left|\frac{S_{12}S_{21}}{|S_{22}|^2 - |\Delta|^2}\right| \qquad (9.18a)$$

$$\text{Centre } (c_L) = \frac{(S_{22} - \Delta S_{11}^{\star})^{\star}}{|S_{22}|^2 - |\Delta|^2} = \frac{C_L^{\star}}{|S_{22}|^2 - |\Delta|^2} \qquad (9.18b)$$

where $C_L^{\star} = (S_{22} - \Delta S_{11}^{\star})^{\star}$.

With $|\Gamma_{OUT}| = 1$, the *input stability circle* in the Γ_S plane has:

$$\text{Radius } (r_S) = \left|\frac{S_{12}S_{21}}{|S_{11}|^2 - |\Delta|^2}\right| \qquad (9.19a)$$

$$\text{Centre } (c_S) = \frac{(S_{11} - \Delta S_{22}^{\star})^{\star}}{|S_{11}|^2 - |\Delta|^2} = \frac{C_S^{\star}}{|S_{11}|^2 - |\Delta^2|} \qquad (9.19b)$$

where $C_S^{\star} = (S_{11} - \Delta S_{22}^{\star})^{\star}$.

Knowing the S-parameters of the active device at a specified frequency the stability circles can be plotted on the Smith chart (using Eqs 9.18 and 9.19) to investigate the stable and unstable regions of operation of the amplifier in the Γ_L plane and Γ_S plane.

The stability circles are shown in Fig. 9.54 when $|\Gamma_{IN}| = 1$ and $|\Gamma_{OUT}| = 1$; on one side of the stability circle boundary, in the Γ_L plane, we will have $|\Gamma_{IN}| < 1$ and on the other side $|\Gamma_{IN}| > 1$. Similarly, in the Γ_S plane on one side of the stability circle boundary, we will have $|\Gamma_{OUT}| < 1$ and on the other side $|\Gamma_{OUT}| > 1$. It may be noted that whenever $|\Gamma_{IN}|$ or $|\Gamma_{OUT}| < 1$, we have stable operation of the amplifier, while for the case $|\Gamma_{IN}|$ or $|\Gamma_{OUT}| > 1$, unstable operation of the amplifier results.

To identify these zones of stable and unstable operation on the Smith chart we consider a typical example. With $Z_L = Z_0, \Gamma_L = 0$ and hence $|\Gamma_{IN}| = |S_{11}|$ (Eq. 9.14c). Now with $|S_{11}| < 1$, $|\Gamma_{IN}| < 1$, we have a stable region of operation while with $|S_{11}| > 1$, $|\Gamma_{IN}| > 1$, we have an unstable region of operation (Fig. 9.55) in which the shaded region indicates the stable region of operation). Similar zones of stable and unstable regions of operation on the Smith chart in the Γ_S plane are shown in Fig. 9.56.

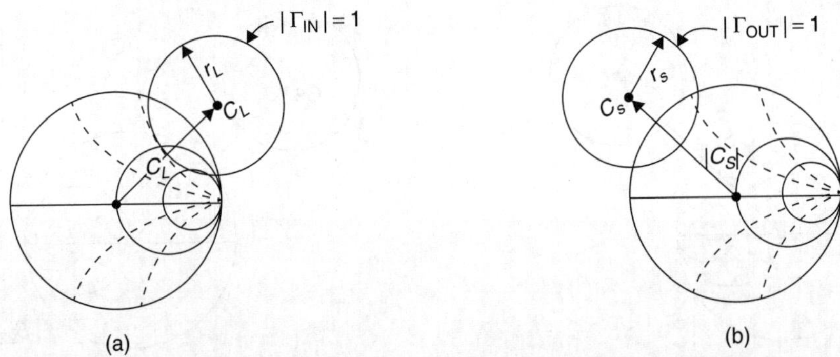

Fig. 9.54 Stability circle construction in the Smith chart (a) Plot in Γ_L plane (b) Plot in Γ_S plane.

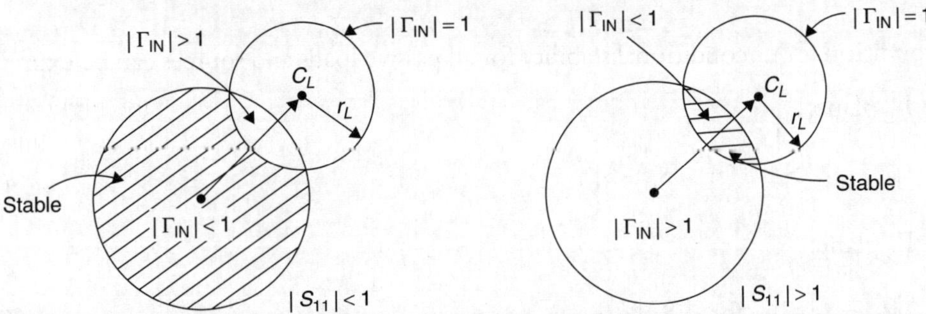

Fig. 9.55 Smith chart showing the stable and unstable regions in the Γ_L plane for conditional stability (potentially unstable case). [$|\Gamma_{IN}| = |S_{11}|$, for $\Gamma_L = 0$].

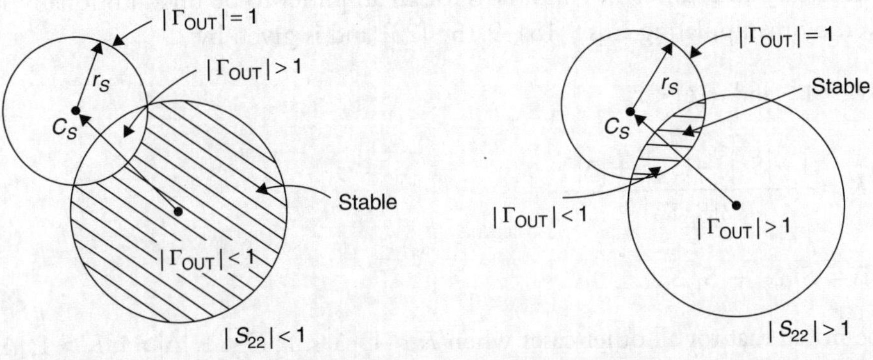

Fig. 9.56 Smith chart showing the stable and unstable regions in the Γ_S plane for conditional stability (potentially unstable case). [$|\Gamma_{OUT}| = |S_{22}|$, for $\Gamma_S = 0$].

The above discussion relates to the potentially stable (conditionally stable) case of amplifier from which it is clear that for unconditionally stable case the stability circle should be outside the Smith chart (Fig. 9.57).

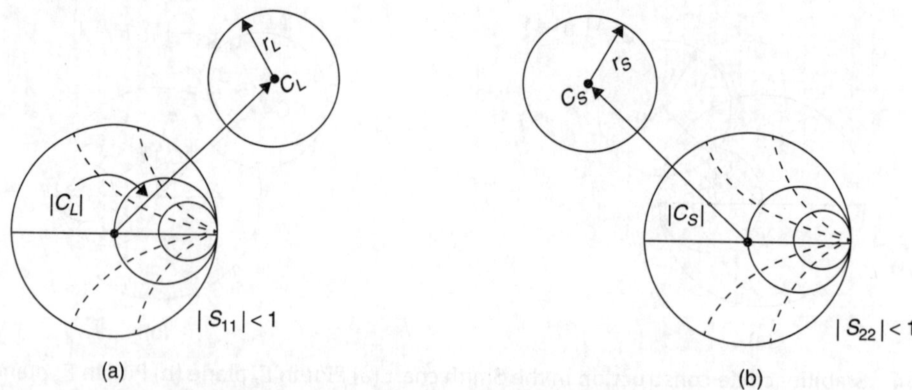

Fig. 9.57 Stability circles with respect to the Smith chart for unconditional stability (a) Plot in Γ_L plane (b) Plot in Γ_S plane.

The condition for unconditional stability for all passive loads and sources can be expressed as:

In the Γ_L plane:

$$\left||c_L| - r_L\right| > 1 \quad \text{for} \quad |S_{11}| < 1 \tag{9.20a}$$

In the Γ_S plane:

$$\left||c_S| - r_S\right| > 1 \quad \text{for} \quad |S_{22}| < 1 \tag{9.20b}$$

If either $|S_{11}| > 1$ or $|S_{22}| > 1$, the network cannot be unconditionally stable since the termination $\Gamma_L = 0$ or $\Gamma_S = 0$ will make $|\Gamma_{IN}| > 1$ or $|\Gamma_{OUT}| > 1$ (Eq. 9.14c).

The necessary and sufficient conditions for an amplifier to be unconditionally stable can be derived by manipulating Eqs 9.15a–9.15c [23] and is given by:

$$K > 1 \quad \text{and} \quad |\Delta| < 1 \tag{9.21a}$$

where, $$K = \frac{1 - |S_{11}|^2 - |S_{22}|^2 + |\Delta|^2}{2|S_{12}S_{21}|} \tag{9.21b}$$

and $$\Delta = S_{11}S_{22} - S_{12}S_{21} \tag{9.21c}$$

It may be noted that for all other cases when $K < 1, |\Delta| < 1$; $K < 1, |\Delta| > 1$; $K > 1, |\Delta| > 1$; the amplifier is potentially unstable or conditionally stable. Manufacturers do not fabricate transistors with $K > 1, |\Delta| > 1$; however, certain terminations or feedback network added to the transistor can sometimes create a potentially unstable amplifier with $K > 1, |\Delta| > 1$.

EXAMPLE 9.8

The scattering parameters of two different microwave transistors are given below:

(TR_1) $S_{11} = 0.752\angle{-138°}$, $S_{12} = 0.092\angle 8.3°$, $S_{21} = 1.52\angle 40.5°$, $S_{22} = 0.5\angle{-100°}$

(TR_2) $S_{11} = 0.775\angle{-165°}$, $S_{12} = 0.030\angle 34°$, $S_{21} = 6.01\angle 90°$, $S_{22} = 0.360\angle{-35°}$

(i) Determine the nature of stability of amplifiers made with these transistors.
(ii) For the potentially unstable amplifier draw the input and output stability circles on the Smith chart.

Solution

(i) To know the nature of stability of the amplifiers designed with the above mentioned two transistors we need to know the values of Δ and K.

When the design is done with TR_1 we have:

$$\Delta = S_{11}S_{22} - S_{12}S_{21}$$
$$= (0.752\angle{-138°} \times 0.5\angle{-100°}) - (0.092\angle{-8.3°} \times 1.52\angle{-40.5°})$$
$$= (0.376\angle{-238°} - 0.139\angle 48.8°)$$
$$- (0.199 - j0.3188) - (0.0915 + j0.10458)$$
$$= 0.1075 - j0.42338$$

Thus $|\Delta| = \sqrt{(0.1075)^2 + (0.42338)^2} = 0.4368$

And $\varphi = \tan^{-1}\left(\dfrac{-0.42338}{0.1075}\right) = \angle{-75.75°}$

Hence $\Delta = 0.4368\angle{-75.75°}$ with $|\Delta| = 0.4368$

Now $K = \dfrac{1 - |S_{11}|^2 - |S_{22}|^2 + |\Delta|^2}{2|S_{12}S_{21}|} = \dfrac{1 - 0.5655 - 0.25 + 0.19079}{0.27968} = 1.341$

Thus for amplifier design with TR_1 will make $K > 1$ and $|\Delta| < 1$; making the designed amplifier unconditionally stable.

When the design is done with TR_2 we have:

$$\Delta = S_{11}S_{22} - S_{12}S_{21}$$
$$= (0.775\angle{-165°} \times 0.360\angle{-35°}) - (0.03\angle{-34°} \times 6.01\angle{-90°})$$
$$= (0.279\angle{-200°} - 0.1803\angle 124°$$
$$= (0.262 - j0.0954) - (0.1008 - j0.1494)$$
$$= 0.1612 + j0.054$$

Thus $|\Delta| = \sqrt{(0.1612)^2 + (0.054)^2} = 0.1697$

And $\varphi = \tan^{-1}\left(\dfrac{0.054}{0.1612}\right) = \angle 17.64°$

Hence $\Delta = 0.1697\angle -17.64°$ with $|\Delta| = 0.1697$

Now $K = \dfrac{1 - |S_{11}|^2 - |S_{22}|^2 + |\Delta|^2}{2|S_{12}S_{21}|} = \dfrac{1 - 0.6006 - 0.1296 + 0.02879}{0.3606} = 0.8280$

Thus, for amplifier design with TR_2 will make $K > 1$ and $|\Delta| < 1$, making the designed amplifier conditionally stable (potentially unstable).

(ii) The transistor TR_2 is found to be potentially unstable. Thus for this transistor we need to draw the input and output stability circles by evaluating the centre and radius of the circles using necessary equations as follows.

From Eq. 9.43 we have for output stability circle:

$$C_L^{\star} = (S_{22} - \Delta S_{11}^{\star})^{\star} = S_{22}^{\star} - \Delta^{\star} S_{11} = 0.360\angle 35° - (0.1697\angle -17.64° \times 0.775\angle -165°)$$

$$= 0.360\angle 35° - 0.1309\angle -182.64° = (0.2948 + j0.2064) - (0.13076 - j0.00602)$$

$$= 0.16204 + j0.2124$$

$$|C_L^{\star}| = \sqrt{(0.16204)^2 + (0.2124)^2} = 0.2670$$

$$\varphi = \tan^{-1}\left(\frac{0.2124}{0.16204}\right) = \angle 52.65°$$

Thus: $c_L = \dfrac{C_L^{\star}}{|S_{22}|^2 - |\Delta|^2} = \dfrac{0.2670\angle 52.65°}{0.1296 - 0.0287} = 2.646\angle 52.65°$

And $r_L = \left|\dfrac{S_{12}S_{21}}{|S_{22}|^2 - |\Delta|^2}\right| = \dfrac{0.030 \times 6.01}{(0.360)^2 - (0.1697)^2} = 1.786$

Again, from Eq. 9.44, we have for input stability circle:

$$C_S^{\star} = (S_{11} - \Delta S_{22}^{\star})^{\star} = S_{11}^{\star} - \Delta^{\star} S_{22} = 0.775\angle 165° - (0.1697\angle -17.64° \times 0.360\angle -35°)$$

$$= 0.775\angle 165° - 0.061\angle -52.64° = (0.7485 + j0.2005) - (0.0370 - j0.0484)$$

$$= 0.7115 + j0.2489$$

$$|C_S^{\star}| = \sqrt{(0.7115)^2 + (0.2489)^2} = 0.7537$$

$$\varphi = \tan^{-1}\left(\frac{0.2489}{0.7115}\right) = \angle 19.27°$$

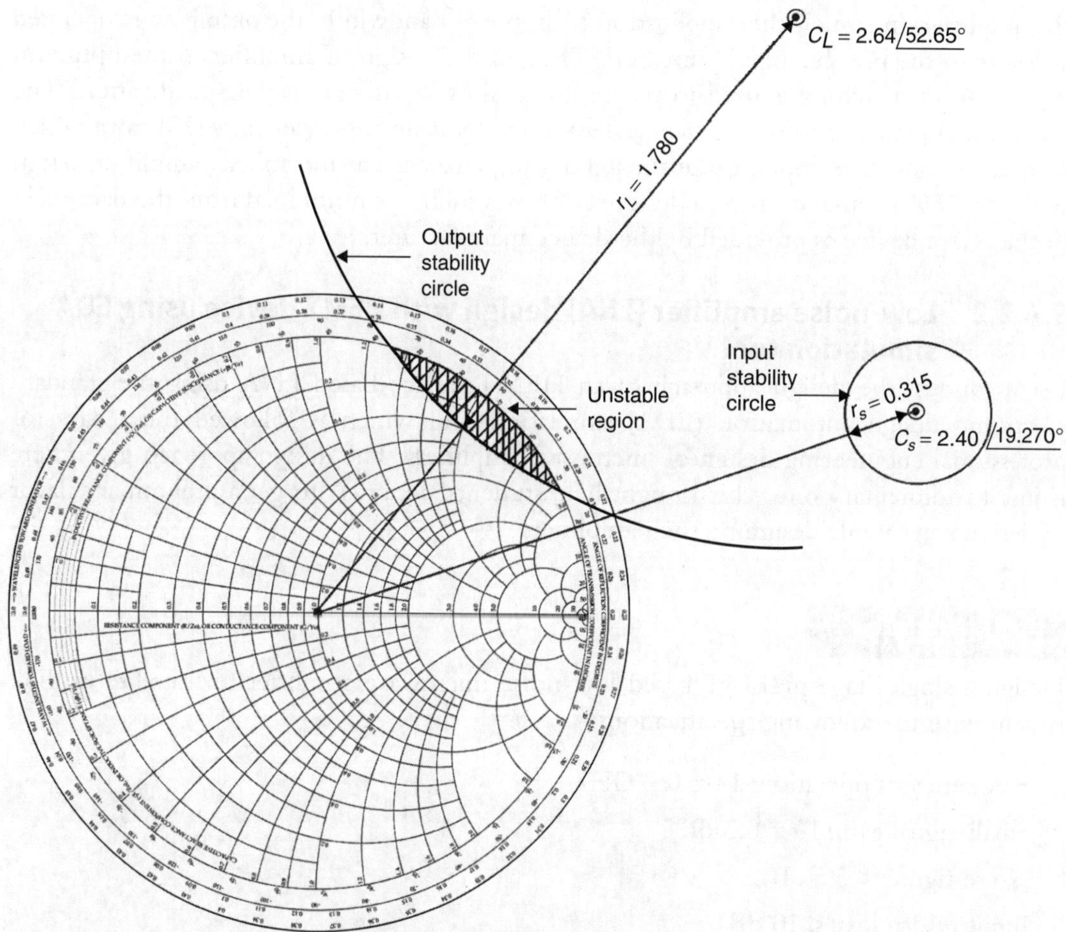

Fig. 9.58 Smith chart plot showing the input and output stability circles for Example 9.7.

Thus: $$c_S = \frac{C_S^*}{|S_{11}|^2 - |\Delta|^2} = \frac{0.7537\angle 19.27°}{0.6006 - 0.287} = 2.40\angle 19.27°$$

And $$r_S = \left|\frac{S_{12}S_{21}}{|S_{11}|^2 - |\Delta|^2}\right| = \frac{0.030 \times 6.01}{(0.775)^2 - (0.1697)^2} = 0.315$$

A Smith chart plot showing the input and output stability circles are given in Fig. 9.58.

A few considerations for microwave amplifier design may be discussed briefly in the following. In the design of microwave amplifiers, it is not possible to realise maximum gain and minimum noise figure at the same time as the two requirements demand different input/output impedance matching. In fact, in the design of low-noise amplifier (LNA), a compromise is done to realise the desired noise figure at the cost of trading-off the gain of the amplifier. In some other application to improve bandwidth, the amplifier is operated at less than the possible maximum gain. Thus, in the design of amplifiers some optimum input/output matching is used to realise the goal of the design as per specification. The analytical approach is to use *constant gain circle* and *constant noise figure circle* [23] from which to find the optimum input/output matching impedance. But in modern amplifier design with the EDA simulation tool, a designer chooses such optimum load from the data sheet of the active device as provided by the device manufacturer.

9.4.2.2 Low-noise amplifier (LNA) design with HEMT device using EDA simulation tool

Let us discuss the design approach of an HEMT device based LNA using the standard electronic design automation (EDA) simulation tool which is followed these days for professional engineering design of microwave amplifiers. The design approach given here is just a rudimentary one which might help students/designers to get a preliminary flavor of design steps while designing such amplifiers.

EXAMPLE 9.9

Design a single-stage pHEMT based low-noise microwave amplifier for satellite up-link system with the following specifications:

Frequency of operation: 14 ± 0.2 GHz

Small-signal gain: 10 ± 0.5 dB

Noise figure: < 3.5 dB

Input return loss: < 10 dB

Output return loss: < 15 dB

Solution

The design of the amplifier can be done in the following steps using EDA simulation tool.

Step 1: *Selection of the device:* The selection of the pHEMT device is done on the basis of the cut-off freqency of the device, maximum availbale gain and the noise figure desired from the amplifier. The device manufacturer provides the small-signal model of the device that contains the values of the intrinisc and extrinsic elements of the packaged device (Fig. 9.59). The designer is required to rebuild the model using the given circuit elements in the circuit simulator prior to the initiation of the design.

The device manufacturers may also provide the S2P file for the device which is the 2-port S-parameter file that contains the raw 2-port S-parameter data and noise data of the device over a given operating freqency range and for a particular bias point.

Step 2: *Biasing:* The biasing of a FET requires positive voltage at the drain terminal ($+V_{DS}$) and a negative voltage on the gate ($-V_{GS}$), the source is usually kept grounded. The

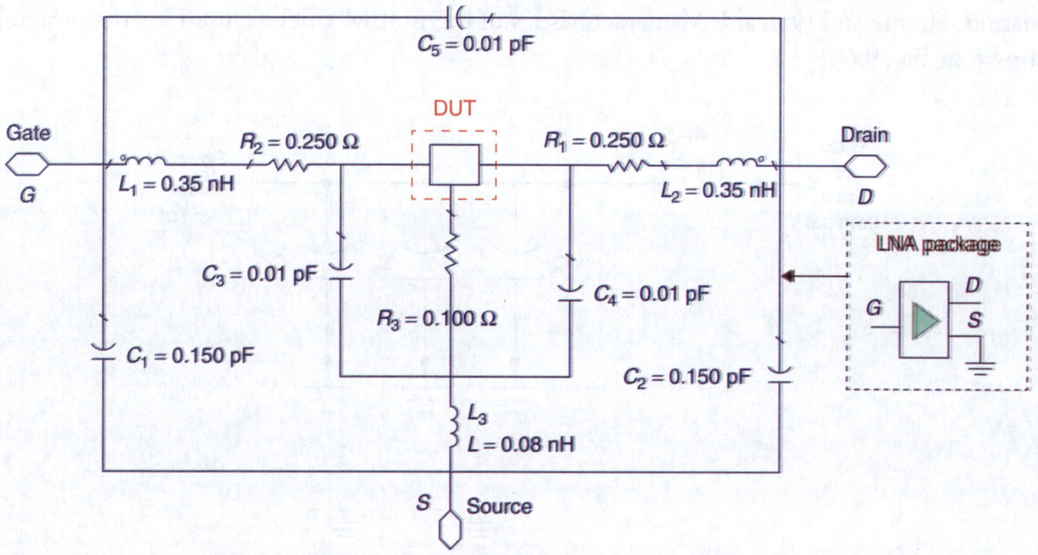

Fig. 9.59 Small-signal model (equivalent circuit) of packaged pHEMT device.

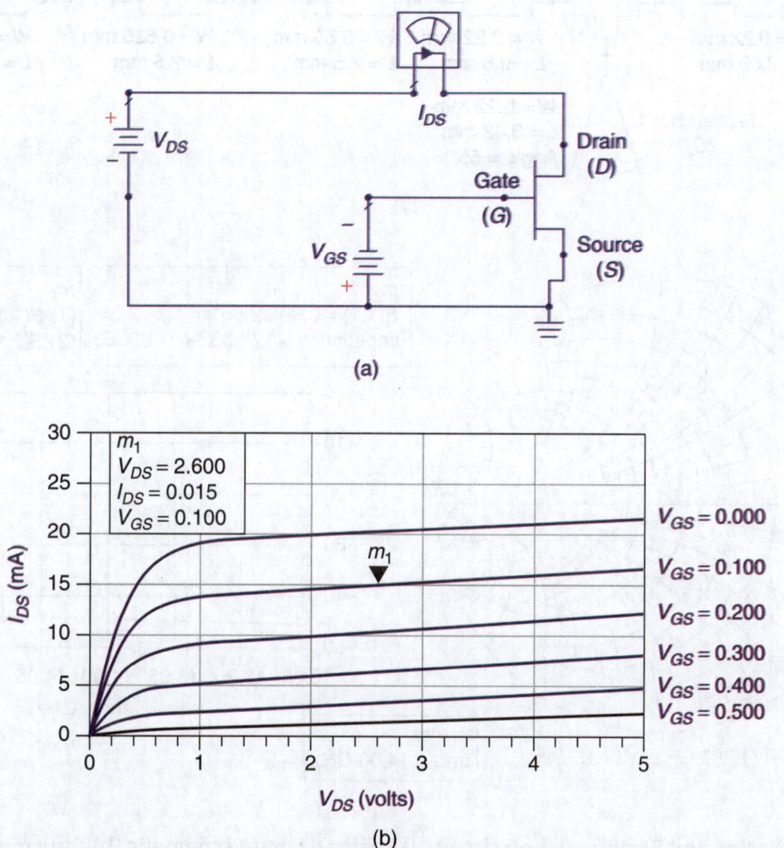

Fig. 9.60 (a) Biasing scheme and (b) I–V curves for pHEMT device used in the design.

biasing scheme and typical I–V characteristics of the pHEMT device used for the design is shown in Fig. 9.60.

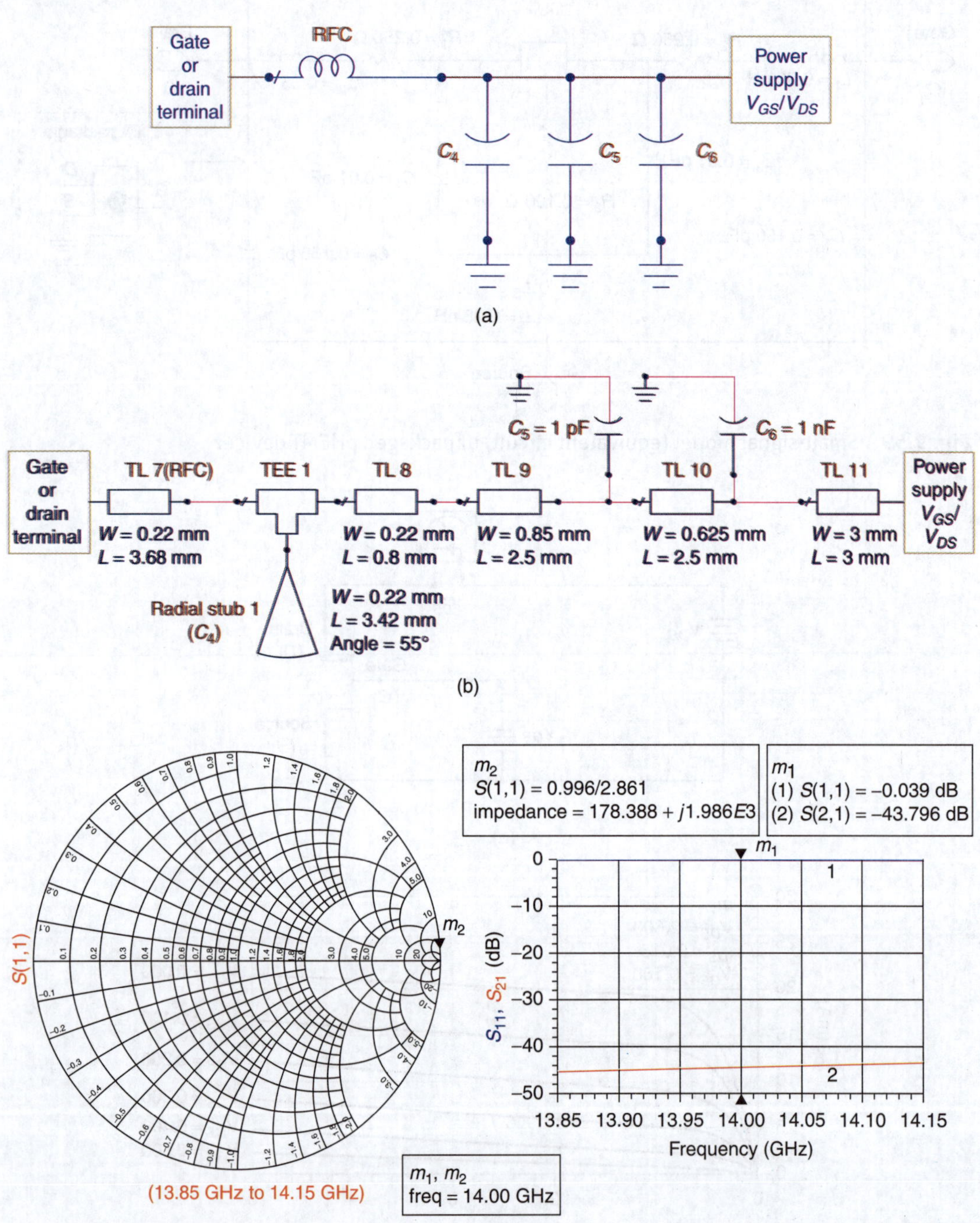

Fig. 9.61 Biasing and RF de-coupling circuit (a) Lumped circuit schematic (b) Transmission line realisation (c) Simulation result of bias de-coupling circuit.

Bias-decoupling circuit is an essential part of amplifier design to prevent the RF signal at the amplifier from entering into the path of the DC supply. To realise this, an inductor acting as RF choke and a few by-pass capacitors to ground are used (Fig. 9.61a). The combination basically acts as band-reject filter at the RF freqency. The inductive choke may be either a surface mount inductor which provides high impedance to RF signal, or a short-circuited transmission line stub that behaves as open circit to the RF signal. The capacitor may be either a surface mount capacitor or an open circuit stub that behaves as a short circuit for the RF signal. The lumped components are used from the component library of the simulation system and the distributed components are designed with 10 mil (0.025 mm) thick Rogers 5880 substrate for microstrip based transmission line stub design. The bias decoupling circuit used is shown in Fig. 9.61b, while the simulation result shown in Fig. 9.61c indicates that S_{21} is below −43 dB (ensuring complete isolation of RF port from the DC port) and S_{11} is 0.999 (indicating complete refelection of the signal at the RF port). A stabilising resistor is usually introduced in the decoupling section for the gate terminal

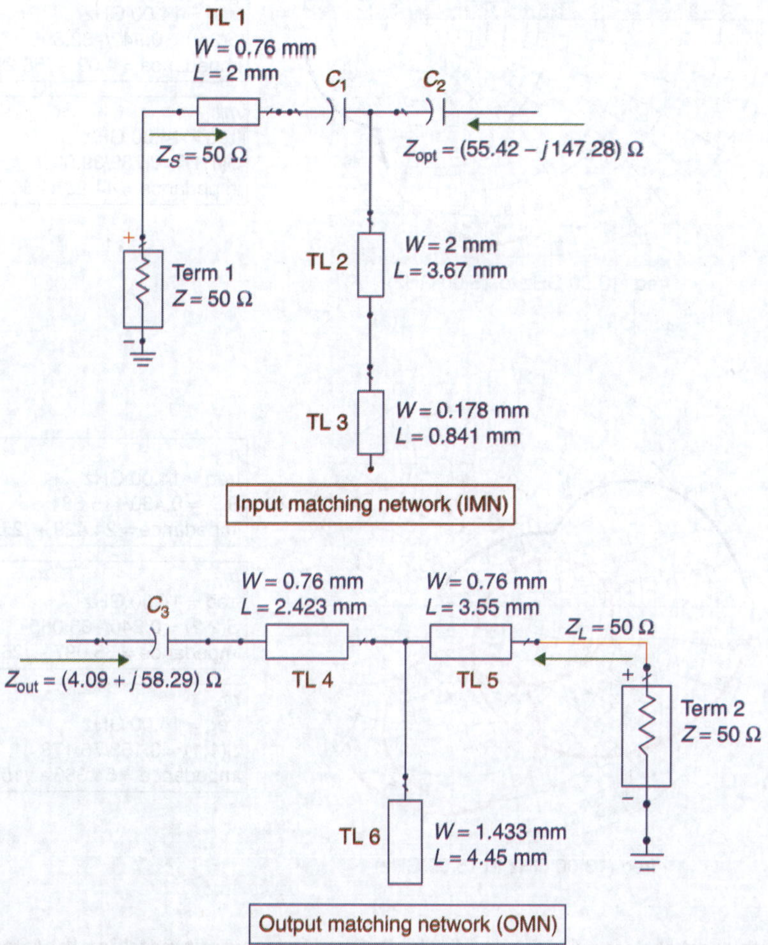

Fig. 9.62 Input and output matching sections.

that ensures small-signal stabilty and prevents any possible oscillation to build-up at low freqencies.

Step 3: *Input and output maching circuits:* The impedance matching of the active device at the input port and output port is an important issue in the design of microwave amplifier to realise the desired gain and noise performance. The device input impedance (Z_{IN}) is matched with the source impedance (Z_S) which is usually 50 Ω. For low-noise amplifier design, (may be at the cost off trading off gain to some extent) the Z_{IN} used is known as

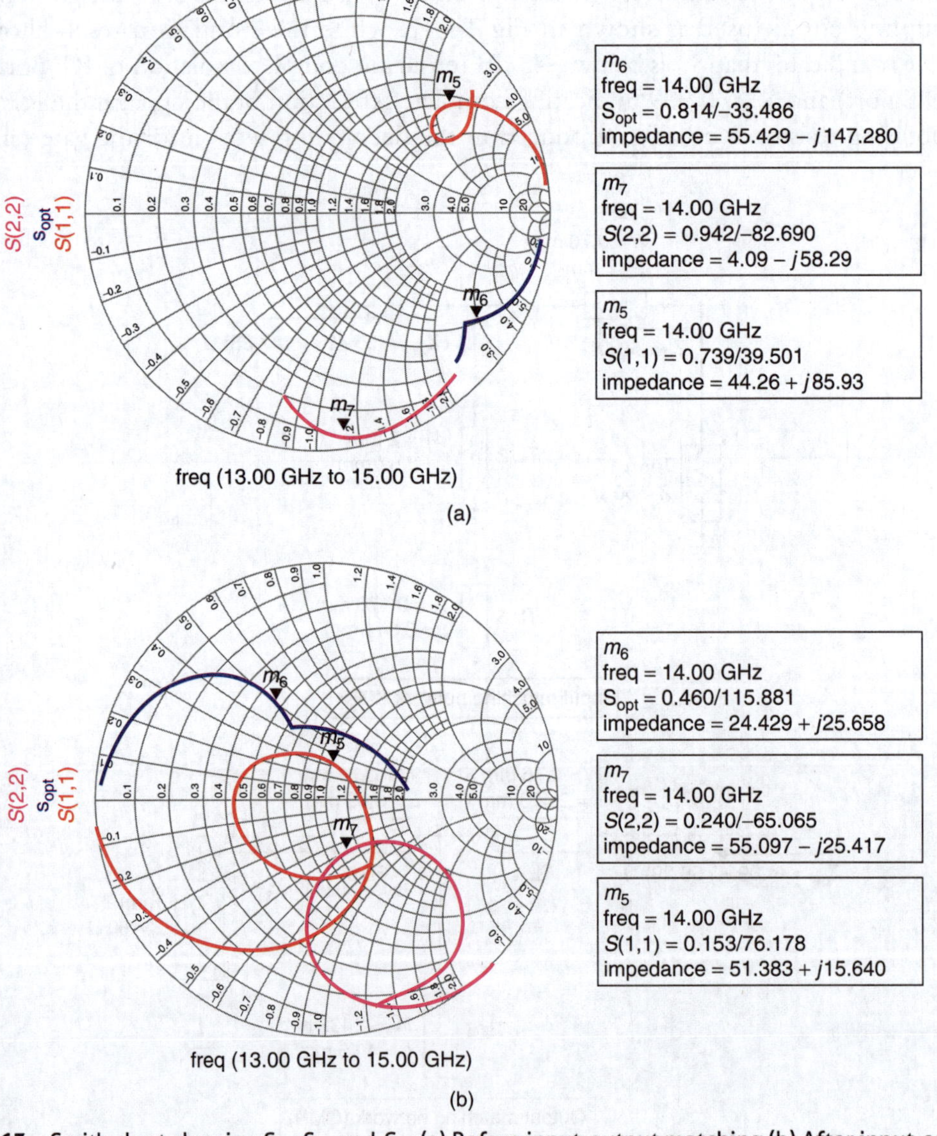

Fig. 9.63 Smith chart showing S_{11}, S_{22} and S_{opt} (a) Before input-output matching (b) After input-output matching.

optimum Z_{IN} and is indicated as Z_{opt}. The device output impedance Z_{OUT} is matched with the load impedance (Z_L) that is again usually 50 Ω. In the present design. Z_{IN}, Z_{OUT} and Z_{opt} are respectively given by (44.26 – j85.93) Ω, (4.09 + j 58.29) Ω and (55.42 – j147.28) Ω, which are obtained from the 2-port S-parameter data supplied by the device manufacturer. For conventional amplifier design where gain has to be maximised, a conjugate match has to be done, that is, $Z_S = Z_{IN}^*$ and $Z_L = Z_{OUT}^*$ (for a practical bilateral case, $S_{12} \neq 0$). However, for low-noise amplifier design, the input matching is done with $Z_S = Z_{opt}$, sacrificing a little bit of the gain of the amplifier. $L/\Pi/T$-section impedance matching technique (Section 7.5) has been used for impedance matching both at the input and output of the device. The inductance is realised with short-circuited stub and capacitor with open-circuited stub in microstrip transmission line. The matching sections are shown in Fig. 9.62. Initial matching of the input impedance is done between Z_{opt} and Z_S, then the matching is slightly detuned to achieve good input match (in terms of S_{11}) at the cost of increasing the noise figure within permissible limits. DC blocking capacitors are also included as part of the matching circuitry. The values of S_{11}, S_{22} and S_{opt} before and after impedance matching are shown in Fig. 9.63. After incorporating above impedance matching sections the curves for S_{11}, S_{22} and S_{opt} are observed to approach the 50 Ω point, however, S_{11} and S_{opt} are seen to be largely seperated from each other. Thus, the input matching has not been able to bring them efficiently to 50 Ω and a trade-off between optimum noise figure and good input match (for realising maximum gain) has been resorted to, for LNA design.

Step 4: *Schematic/circuit-level design:* The circuit schematic of a single stage LNA can be built by combining diffrent blocks of Fig. 9.64 discussed in earlier steps. The initial amplifier

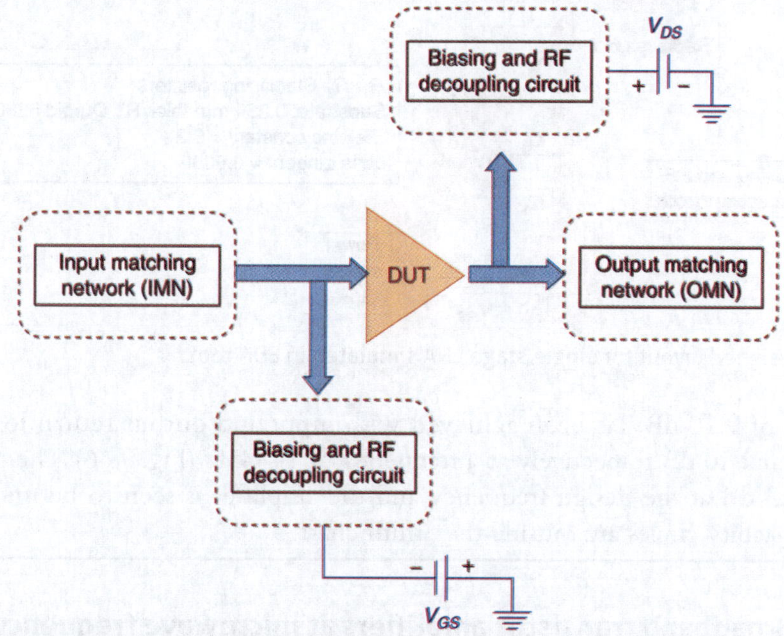

Fig. 9.64 Block diagram of one-stage amplifier.

design is done using the *circuit simulator* of the EDA simulation tool. However, to include the parasitic effects, the system has to be finally optimised in *layout* using the method of moments (MoM) solver that performs the electromagnetic (EM) simulation.

Step 5: *Layout design and optimisation:* The final optimised layout developed for a single stage LNA is shown in Fig. 9.65 together with electromagnetic (EM) simulation result for gain, noise figure and stability of the amplifier in Fig. 9.66.

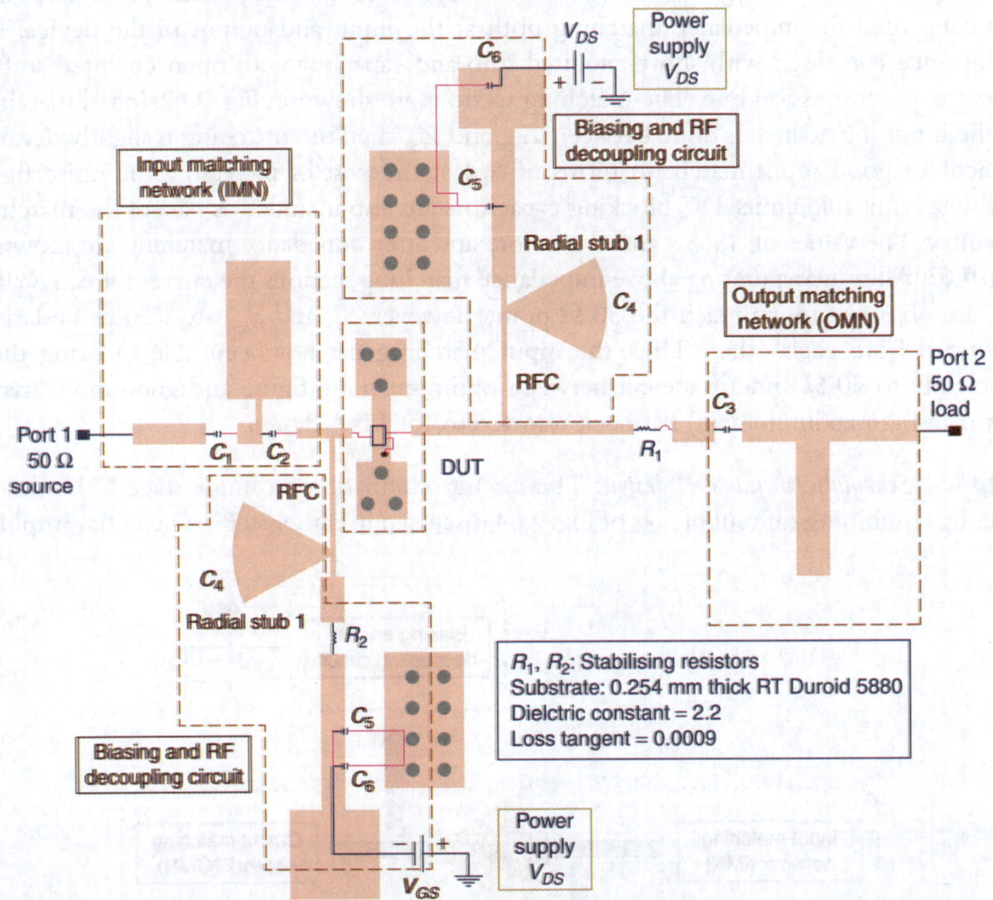

Fig. 9.65 Designed layout for single stage LNA simulated in EDA tool.

A final gain of 9.10 dB has been achieved with input and output return loss of better than 14 dB and 20 dB respectively at a frequency of 14 GHz (Fig. 9.66). The noise figure achived is 3.3 dB at the design frequency and the amplifier is seen to be unconditinally stable as the stabity circles are outside the Smith chart.

9.4.2.3 Broadband transistor amplifiers at microwave frequency

The transistor amplifier theory and design discussed above used either conjugate match for realisation of maximum gain or a little trade-off of gain for low-noise performance.

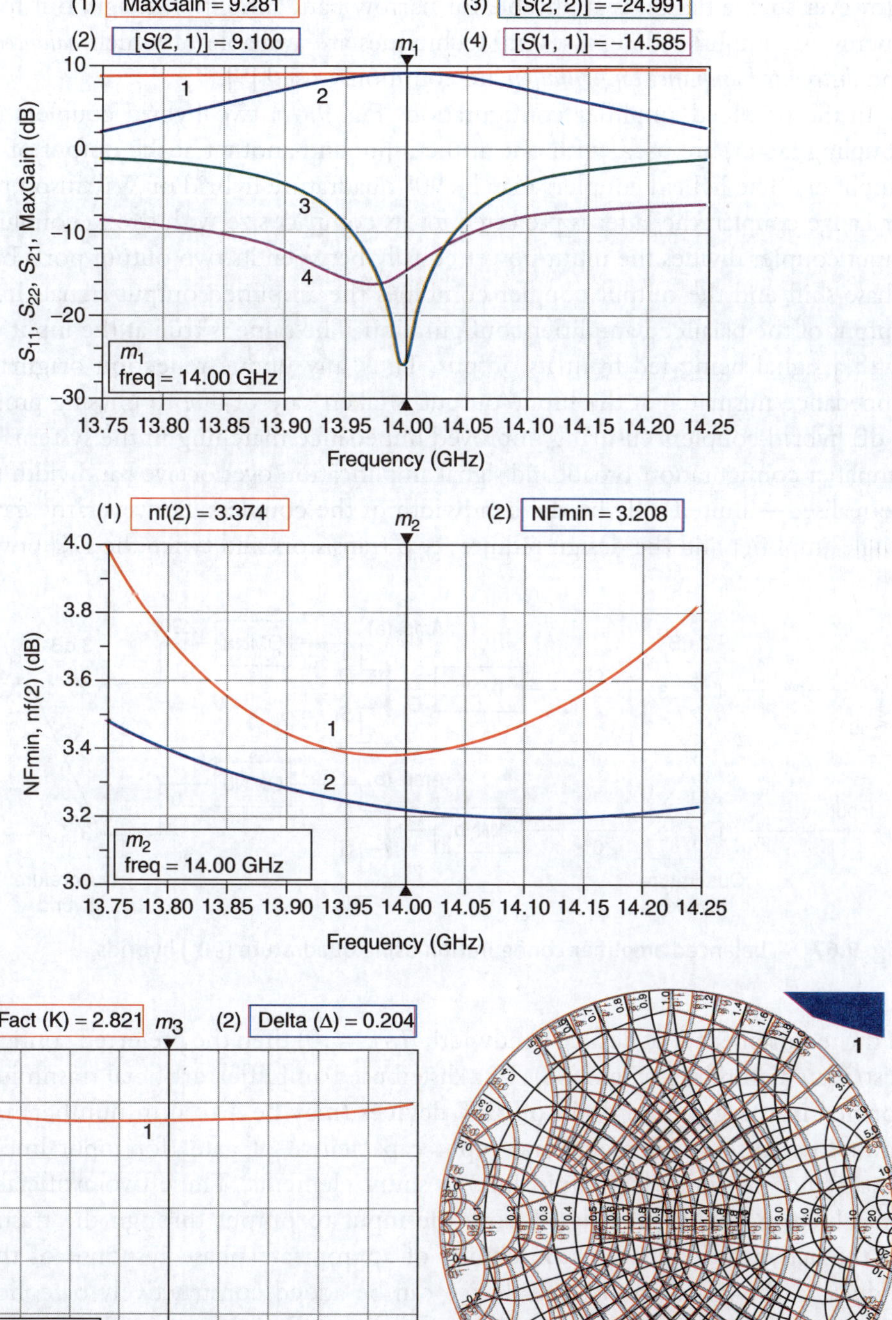

Fig. 9.66 EM-simulated result optimised in layout showing the gain, noise figure and stabilty of the designed amplifier.

However such a design is applicable for narrow band amplifier design. But for broad band microwave amplifier design, several techniques are available of which *balanced configuration* and *distributed amplifier configuration* are commonly used [9].

In the balanced amplifier configuration (Fig. 9.67) two hybrid couplers (having 3 dB coupling factor) are used, with one at the input and another at the output of two identical amplifiers. The hybrid couplers may be 90° quadrature hybrid or Wilkinson power divider or Lange coupler; the latter is preferred for its compact size with strong coupling. The 3 dB input coupler divides the input power equally between its two output ports but with a 90° phase shift and the output coupler combines the amplified output signals in phase at the output of the balanced amplifier configuration. The same is true at the input of the system with a signal being fed from its output. Thus, any undue reflection originating from an impedance mismatch at the input/output is taken care of due to phasing properties of the 3 dB hybrid couplers ensuring improved impedance matching in the system. With such an amplifier configuration, broadband signal amplification over octave bandwidth ($f_2/f_1 = 2$) can be realised—limited only by the bandwidth of the couplers. However, the gain is that of a single amplifier and the design requires two transistors and twice the DC power.

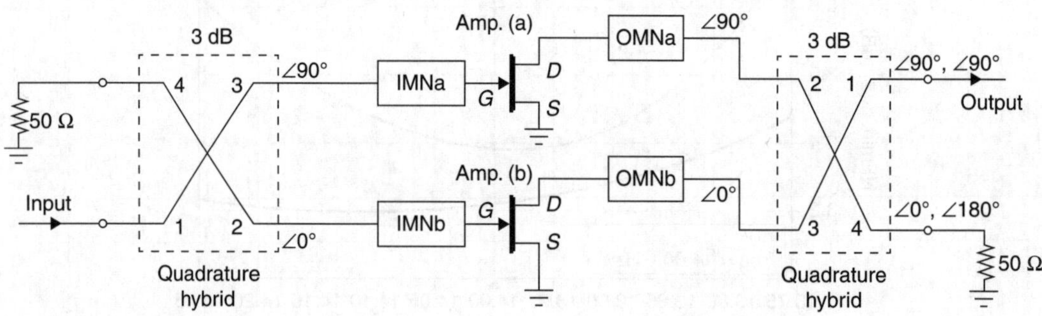

Fig. 9.67 A balanced amplifier configuration using quadrature (90°) hybrids.

If the requirement is of decade bandwidth ($f_2/f_1 = 10$) then the preferred configuration is the distributed amplifier (Fig. 9.68). In a distributed amplifier, artificial trasnmission lines are formed in the gate and drain of FET devices (may be 4 to 6 in number) using lumped inductors as series elements and input capacitance of gate (for input line) and output capacitance of drain (for output line) as shunt elements. These two artificial transmission lines help in guiding the signal from the input to output through the trasnconductance of the active devices. With the choice of appropriate phase response of the connected inductors, the signals from each device can be added constructively or coherently in the direction of the output giving an amplified signal. Broad bandwidth is realised with such amplifiers because in such a configuartion, unlike the conventional cascaded amplifiers, the transconductances of active devices can be added without increasing the parasitic capacitance associated with gate and drain. In cascaded amplifiers, the parasitic capacitance C_p contributed by C_{gd} and C_{gs} adds up for multiple active devices while in a distributed amplifier, only the effective capacitance of a single stage is seen by the amplifier making gain bandwidth product (GBW) rise instead of remaining constant. The non-additive effect of parasitic capacitance occurs by 'distributing' the parasitic capacitance in the artificial transmission lines. The original concept of distributed amplifier was proposed in 1936 by Percival to overcome the traditional GBW limit of vacuum tube amplifiers.

In addition to the large (decade) bandwidth of such an amplifier configuration, the input and output VSWR are very low, thereby allowing easy cascading of several amplifier units. However, the increase in the number of amplifier stages does not monotonically add to the overall gain. This is because the extra attenuation of the artificial transmission lines, contributed by parasitic resistance of the active devices R_i and R_{ds}, competes with the extara gain of the additional stages. The optimum number of active devices (say HEMT) in a distrbuted amplifier is given by $N_{opt} = \ln \frac{(A_d/A_g)}{(A_d - A_g)}$, where A_g and A_d are the attenuation per section in neper of gate and drain transmission lines. Though the distributed amplifier is capable of providing a large bandwidth, the overall circuit becomes large and does not give as much gain as a cascaded amplifier with the same number of stages.

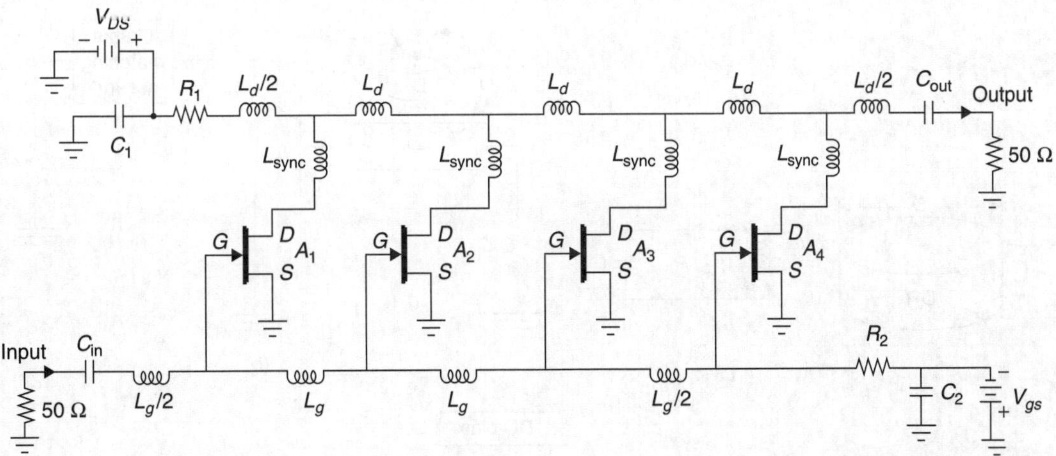

Fig. 9.68 Configuration of a 4-stage distributed amplifier ($L_d \rightarrow$ *inductance per section along drain line*, $L_g \rightarrow$ *inductance per section along gate line*, $L_{sync} \rightarrow$ *phase synchronisation inductance per section*).

9.4.2.4 Dielectric resonator (DR) based transistor oscillator

In Section 4.6, we discussed the basics of the dielectric resonator (DR) operating in the fundamental $TE_{01\delta}$ mode and in Section 4.7, coupling of DR with microstrip line was discussed with means for tuning the resonant frequency of DR. High Q value of DR along with its precise temperature stability can be used in conjunction with an active device (BJT/MESFET/HEMT) to design highly stable oscillators at microwave frequency, known as *dielectric resonator oscillator* (DRO), the oscillation frequency being determined by the DR used. In such an oscillator, the DR acts as a resonating element which can be modelled electrically as an *L*, *C*, *R* network (Fig. 9.69a) and connected to a suitable port of the active device.

A commonly used DR material for such oscillator design is barium titanate with a Q value (9000 @ 10 GHz) and temperature coefficient of frequency is typically ± 8 ppm/°C.

Some practical circuit configurations of DRO are shown in Fig. 9.69. In Fig. 9.69b, the DR acts as a series feedback element. The DR mounted in a microstrip line is connected to the terminating port and the DR-transistor combination produces a negative resistance at the load port. In Fig. 9.69c, the DR acts as a parallel feedback element. The parallel feedback arrangement can provide a wide band tuning, however, this configuration is difficult to construct as it needs coupling of two transmission lines.

Commercial electronic design automation (EDA) simulation tool can be used to design a DRO. Such a tool has programmes that can model typical housing used in the coupling of a DR to a microstrip line. The DR manufacturer specifies the radius and height of the DR for a particular resonant frequency along with its Q value. Knowing the dielectric constant and height of the microstrip line on which the DR is mounted and setting *mode* parameter as 0 (for $TE_{01\delta}$) the DRO design can be done with EDA simulation tool [23].

Mechanical tuning of DRO using a tuning screw was discussed in Section 4.7 in which a tuning bandwidth of 0.1% to 1% can be realised. Electrical tuning of DRO can

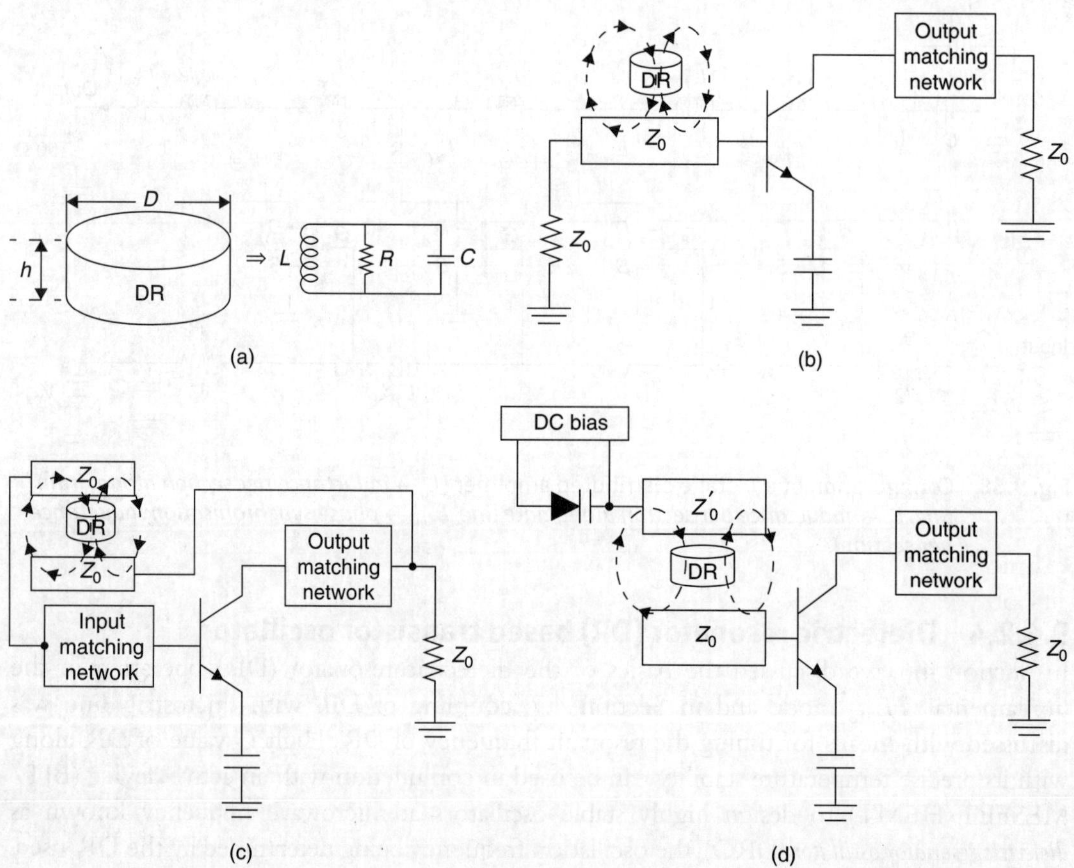

Fig. 9.69 (a) Dielectric resonator (DR) with equivalent circuit (b) DR as series-feedback element (c) DR as shunt-feedback element (d) Electronic tuning of DRO using varactor diode.

be realised with a varactor diode (Fig. 9.69d). Here the varactor is coupled to the DR resulting in two coupled circuits. Since a varactor is a voltage variable capacitance, the DRO can be tuned over 1%–2% of the resonant frequency of the DRO by varying the reverse bias voltage to the varactor.

9.4.3 Power Combiners with IMPATT/Gunn Diodes

It is known that the power output (P_O) realisable from a device decreases rapidly with the inverse square of the frequency of operation (f_{op}): ($P_O \, \alpha \, f_{op}^{-2}$). Thus, to achieve a high power output at microwave and millimeter-wave frequencies with solid-state devices like Gunn and IMPATT, power combiners are needed to combine the power from several individual devices. This is the demand of the day since almost all radar transmitters for airborne and space applications including military applications prefer to use solid-state power sources as one of their stages in power amplification. A microwave power combiner, in principle, desires to add the microwave power of several devices (as in device-level combining) or of several single-device oscillators (as in circuit-level combining) as efficiently as possible. Device-level combining is accomplished by clustering the devices in a region whose extent is small compared with a wavelength and is thus limited in the number of devices that can be efficiently combined. However, circuit-level combining is not limited to placing the diodes in an area small compared to a wavelength. Circuit-level combiners are more practical and are, in general, of two categories—those which combine the output of N devices in a single stage (known as N-way in single stage) and those which combine the output of N devices in multiple stages (known as N-way in multiple stage). The latter category is simpler and more widely used; however, the former category has the possibility of higher power combining efficiency since power generated does not have to pass through several stages of combining. The performance of a power combiner is characterised in terms of *power combining efficiency* defined as $\eta_c = P_{OT}/N.\, P_{OS}$, where P_{OT} is the net power at the output of the combiner using N number of active devices and P_{OS} is the power output expected from single device oscillator using the same devices as in the combiner. Typical power combining efficiency of a well designed power combiner may be around 90%.

The basic N-way in the single stage power combiner designed with resonant cavity combining structure was due to Kurokawa and Magalhaes [24] in 1971. Here the sum of output powers from a number of devices is obtained by coupling their outputs to a single resonator (Fig. 9.70). It basically represents pairs of double coaxial entry waveguide systems in which each coaxial line of a pair is located close to the side wall of the rectangular waveguide where the magnetic field maxima exist. The active device is mounted at one end of the coaxial line and the central conductor of the line helps to feed the necessary DC bias to the device. Since along the propagation direction of the rectangular waveguide the magnetic field maxima repeats at every $\lambda_g/2$, the coaxial line pairs mounted in the E-plane of the guide are spaced by $\lambda_g/2$ in the z-direction (Fig. 9.70). The combiner output is taken via an iris from one end of the guide (say at $z = l$) and the other end (at $z = 0$ plane) the guide is closed by a sliding short.

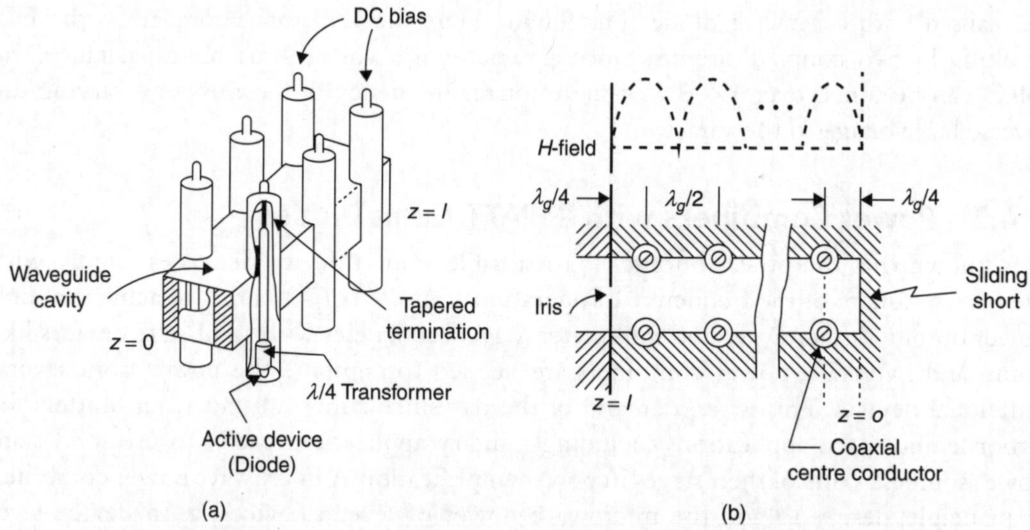

Fig. 9.70 Rectangular waveguide resonant cavity combiner (Kurokawa–Magalhaes type) (a) Schematic view of the combiner (b) Sectional view of the combiner as seen from the top.

Another simple, compact yet elegant duo-diode power combiner can be a twin-cap combiner in which IMPATT (or Gunn) diodes embedded in radial-line cavities combine their power at the electric field maximum of the waveguide in which the twin-cap structure is mounted [25] (Fig. 9.71). The twin-cap structure constitutes coupled radial-line cavities resembling a couplet of the conventional resonant-cap cavity used in the design of single-diode oscillator. This duo-diode configuration may be extended similar to the scheme used by Kurokawa and Magalhaes [24] to design a multi-diode power combiner.

N-way in multiple stages may be realised in many ways. One common approach is to use short-slot quadrature hybrid (in waveguide version) or branch-line qudrature (90°) hybrid (in microstrip version, discussed in Section 6.2.2). The power combiner scheme (in Fig. 9.72) in which four quadrature hybrids are used with a hybrid coupling adapter and four IMPATT sources (operating in reflection mode amplifier) whose combined power will be delivered in the output load. The operation of the power combiner may be understood as follows. The input power fed to the port 1 of the qudrature hybrid marked QH I feeds power to the through port 2 and coupled port 3 with the signal phase 90° and 180°, while no signal is ideally coupled to the isolated port 4. The signal from port 2 of QH I enters the input port 1 of second quadrature hybrid QH II and reaches the IMPATT amplifiers marked A1 and A2 with signal phase 180° and 270° respectively. Assuming that these signals are reflected without experiencing any phase change from the amplifiers will reinforce (being of same phase) the signals at port 4 of QH II but the reflected signals reaching back from the two amplifiers at port 1 of QH II will cancel out (being of phase opposition). Similarly the signal from port 3 of QH I after entering the port 1 of third qudrature hybrid QH III will reinforce signals reflected from the IMPATT amplifiers marked A3 and A4 and will not deliver back any signal to the port 1 of QH III (as shown). Finally the signals from

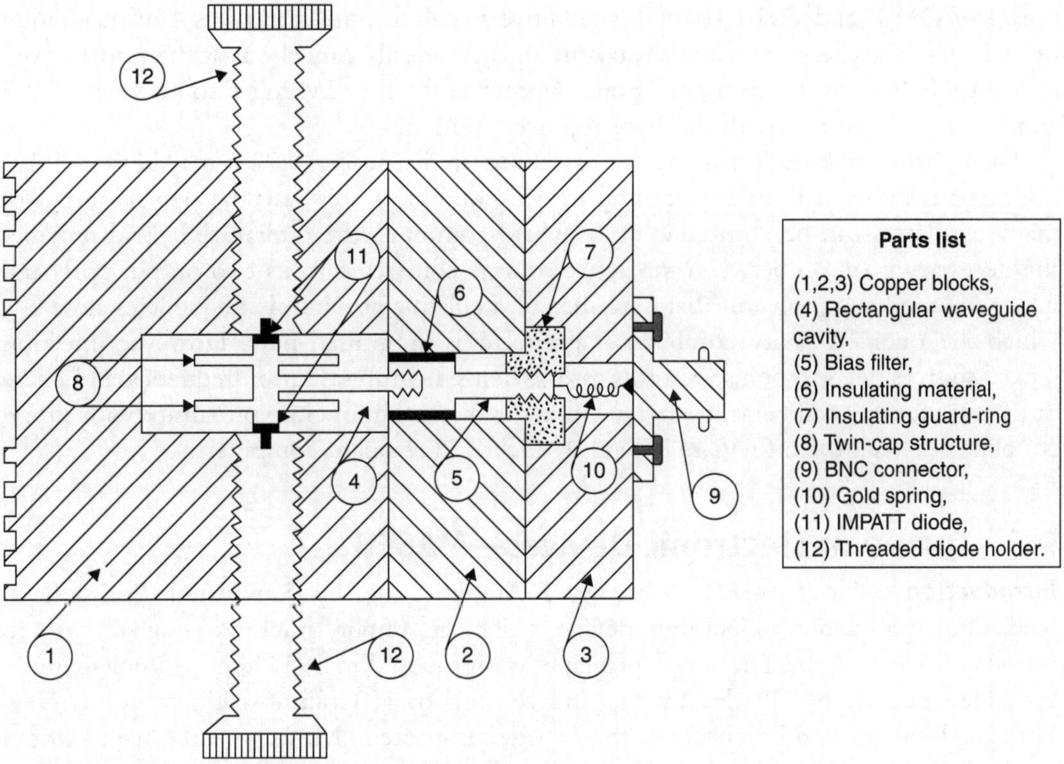

Parts list

(1,2,3) Copper blocks,
(4) Rectangular waveguide cavity,
(5) Bias filter,
(6) Insulating material,
(7) Insulating guard-ring
(8) Twin-cap structure,
(9) BNC connector,
(10) Gold spring,
(11) IMPATT diode,
(12) Threaded diode holder.

Fig. 9.71 Twin-cap IMPATT power combiner.

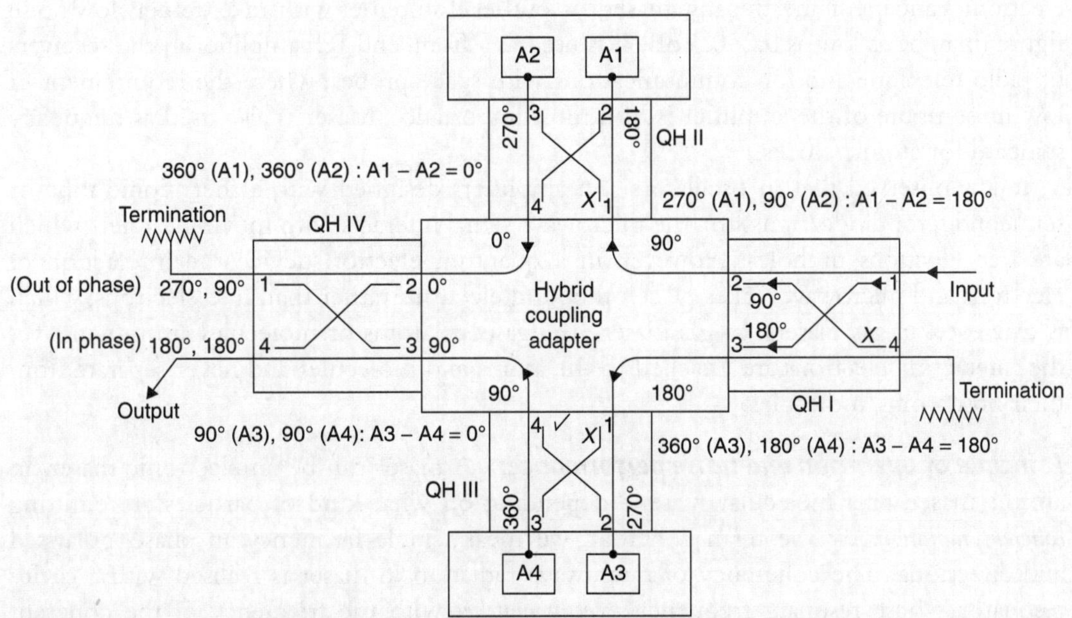

Fig. 9.72 Power combiner using quadrature hybrids.

port 4 of QH II and QH III will give additive signals at port marked as 4 of quadrature hybrid QH IV (where the output is taken) and will ideally cancel out at the port marked as 1 of QH IV. Thus the combined power output of the four IMPATT sources A1, A2, A3 and A4 can be obtained with this hybrid power combiner.

The hybrid combiner is suitable for cascading (in fact in Fig. 9.72 it may be noted that the input is received from a preceding hybrid combiner); this way, the power output of many oscillators can be combined. In a hybrid combiner, the qudrature hybrid provides the necessary isolation between the active sources connected to its two output ports and hence the modeing problem (that is, chance of oscillating at multiple frequencies close by), which the resonant cavity combiner is prone to, is minimised here. However, injection locked sources need to be used in this case (when synchronisation of frequency and phase of the two sources can be assured), but since the diodes are in close proximity with strong coupling between them they are inherently locked in resonant combiners.

9.5 Quantum Electronic Devices—MASER

Introduction: The MASER (Microwave Amplification by Stimulated Emission of Radiation) is a quantum electronic device which can amplify microwave signals with an *extremely low noise figure*. The maser principle was proposed in 1954 almost simultaneously but independently by Charles Townes and co-workers at Columbia University in New York and by Basov and Prochorov at the Lebedev Institute in Moscow—all three of whom received the Nobel Prize for their contribution to science in 1964. The shorter-wavelength counterpart of maser is LASER (Light Amplification by Stimulated Emission of Radiation); both maser and laser have many important applications in basic science, communication electronics and medicine. Because maser is a low-level amplifier with an extremely low noise figure (may be as low as 0.2–0.3 dB) it is used as a front-end RF amplifier in the receivers of radio telescope and for communications with space probes, where the requirement of low noise figure of the amplifier is of utmost importance. Maser is also used as frequency standard for atomic clocks.

It was observed that in oscillators and amplifiers designed with a thermionic tube or semiconductor diode/transistor, the microwave signal interacts with moving carriers which are free electrons or holes. However, in a quantum electron device the interaction of electrons and microwaves takes place on bound electrons rather than free electrons, which is true even in gas masers. In gas masers, the gaseous atoms or molecules are moving, yet the interacting electrons are attached to the atoms and molecules and never separate from their host atoms/molecules.

Principle of operation and noise performance: A maser can be an electronic maser, an atomic maser, or a molecular maser—depending on what kind of particles are emitting *coherent radiation*. By the term coherent, we mean, single-frequency, in-phase, polarised and directional. The coherency of microwave radiation in maser is realised with a cavity resonator whose resonant frequency *exactly matches* with the frequency of the coherent radiation obtained by *stimulated emission* caused by quantum transition of the active particles in the maser material. Inside the cavity resonator, there is a standing wave of microwave signals in which all electromagnetic fields are in phase or change in unison with respect to time and have coherency with the said stimulated radiation. This is because the cavity energised by the radiation of quantum transition has its frequency locked to the frequency of quantum transition (f_{QT}):

$$f_{QT} = \frac{E_{HS} - E_{LS}}{h} \tag{9.22}$$

where E_{HS} and E_{LS} are respectively the energy of the particle at higher energy level and lower energy level of the allowable quantum transition states in the material used for designing the maser and h is Plank's constant.

In an ionically bonded gas like ammonia or a paramagnetic crystal like ruby, it is possible to artificially create '*density inversion*' by *pumping* or feeding energy through a pump oscillator (whose frequency is equal to the value given by Eq. 9.22) whereby molecular/electron density of high energy state molecules/electrons becomes greater than the molecular/electron density of low energy state molecules/electrons. This is known as *population inversion* caused by the pump signal. Particles forced to higher energy states always have a tendency to make a downward transition (dictated by the law of thermodynamics), provided such an energy state is quantum mechanically allowable. When a particle just *absorbs* energy and is excited to a higher energy state, it will return to the lower energy state *of its own accord* in a very short span of time (10^{-8} s) with *spontaneous emission* of energy. But when the above mentioned population inversion is caused (forced) by pumping, the particles are kept in a *meta-stable* higher energy state from which the lifetime of emission is typically 10^{-3} s. For *stimulated emission* to take place in maser (or in laser) a photon of frequency f_{QT} (the input signal) is fired at the atoms of the maser material already in a higher energy state with a pump signal. The atom thus absorbing this photon quickly emits two photons (having the same frequency f_{QT}) that generates four photons which eventually produce eight photons and so on and the chain process goes on until all of them finally get back to the allowable lower energy state in the lifetime of emission from the meta-stable state. The *stimulated emission* of radiation thus caused by a huge number of photons results in a signal output at f_{QT} which has a much higher power compared to the small pump (stimulation) power. Just as an example, if 10^{12} photons are fed into the system and 10^{15} photons are emitted at signal frequency in one second, the power gain of amplification is 30 dB. This is the fundamental concept of *m*icrowave *a*mplification by *s*timulated *e*mission of *r*adiation (*maser*).

The noise figure of the maser amplifier is extremely small because there is no resistance involved and no electron stream to produce shot noise. Further the possibility of any thermal noise is eliminated as in most applications the maser is cryogenically refrigerated to 2 K, the temperature of liquid helium. The only source of possible noise is quantum mechanical noise. The quantum mechanical noise arises from any spontaneous emission that might take place in addition to the stimulated emission in maser. The spontaneous emission occurs with random phase from one atom to the next giving rise to the said quantum mechanical noise. But, for stimulated (or induced) emission, all atoms emit radiation that is in phase and coherent with the stimulating radiating field. Cutting the pump power down will cut the quantum mechanical noise though at the same time this will reduce the available gain of the amplifier. However, the gain and noise are traded off depending upon the optimum performance desired for particular application.

Maser as frequency standard: It may be observed from Eq. 9.22 that the maser signal frequency (f_{QT}) is determined in terms of the energy gap between E_{HS} and E_{LS}. There are characteristic values for ($E_{HS} - E_{LS}$) for each maser material—ammonia (23.870 GHz), cesium (9.192631830 GHz), sodium (1.772 GHz) and hydrogen (1.420 GHz). The characteristic frequencies are so accurate that they are used as frequency standards and time standards, as in the *atomic clock*. The atomic clock built at Harvard University in 1960 has a cumulative error which could cause it to be incorrect by 1 second after more than 30,000 years! A simple scheme for maser controlled standard frequency generator is shown in Fig. 9.73. Here, a small portion of the output signal from the voltage controlled oscillator (VCO) is sampled through the coupler and fed into the maser. When the VCO frequency coincides with the characteristic frequency of the maser, the signal is amplified by the maser and the output is detected, amplified and fed into a tracking circuit that controls the power supply voltage of the VCO to maintain the frequency coincidence of VCO and maser amplifier. The output thus obtained is a standard frequency signal which can be used as frequency standard determined by the characteristic frequency of the maser material.

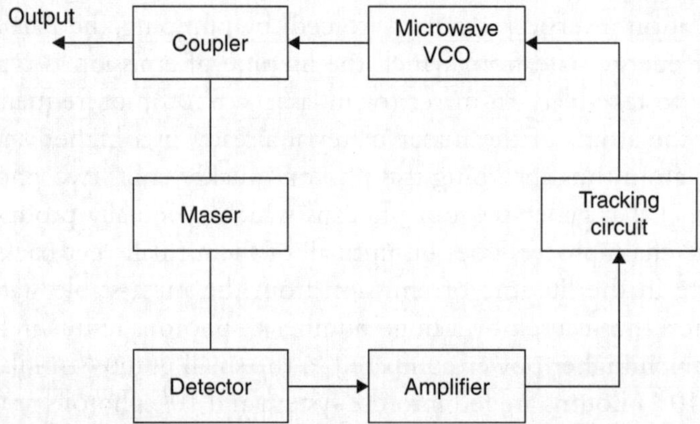

Fig. 9.73 Block diagram of a maser controlled standard frequency generator.

Maser as amplifier: Frequency standard masers are in general gas masers. In fact, the first atomic clock was built with an ammonia gas maser (while hydrogen and cesium featured predominantly among the materials used subsequently) since they all have well defined single line frequency spectra. However, gas masers suffer from the disadvantage of yielding amplifiers that work only at a single frequency whose bandwidth is very narrow (typically for ammonia gas maser, the bandwidth is only 3 kHz with an operating frequency close to 24 GHz). Solid-state masers, either cavity type or travelling-wave type, are used for amplifier application with the latter being capable of providing useful bandwidth for the maser amplifier. Solid-sate maser is based on paramagnetic crystals such as ruby. A ruby crystal is a crystal of Al_2O_3 with a small amount of Cr in it. The diagram of a ruby maser is shown in Fig. 9.74. As may be observed from the figure, the pump power (sufficient to produce significant density inversion) is fed to the cavity resonator through a coupling loop and the pump frequency is chosen in terms of Eq. 9.22. The energy difference ($E_{HS} - E_{LS}$) in the ruby crystal is maintained in the microwave frequency region by controlling the externally applied magnetic field. The signal power is fed to the ruby crystal placed inside the cavity resonator via port 1 of the circulator. Thus, the cavity resonator is a dual mode resonator, resonating at pump frequency and also at the signal frequency in different modes—with the two modes orthogonal to each other. The two modes interact with each other only in the ruby crystal when the maser action is activated. The amplified signal is taken out of the system via the port 3 of the circulator. In a practical maser, the maser is placed in a double dewar which contains liquid helium and liquid nitrogen. The liquid helium (LHe3) cools the maser as low as 2 K. To achieve this low temperature and to prevent evaporation and boiling loss of liquid helium, the maser is also cooled by liquid nitrogen (LN2) at 77 K.

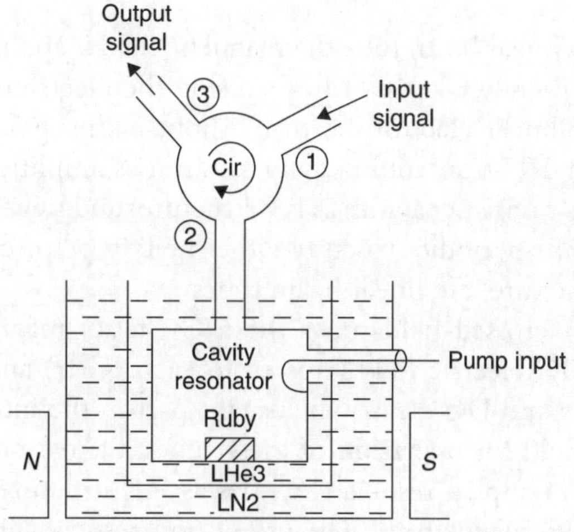

Fig. 9.74 Schematic of a cavity-type cooled ruby maser.

The cavity maser discussed above is also a narrow-band type amplifier (may be maximum of 10% bandwidth) as the bandwidth is controlled primarily by the cavity resonance characteristics. The solution of the problem is to use some form of travelling-wave (TW) structure in which the ruby crystal is mounted. Here the input signal fed at the input port of the slow-wave structure grows via the slow-wave structure at the expense of the pump signal (thus providing broad bandwidth) while the output is taken at the output of the slow-wave structure. It may be noted that the circulator is not needed here, as was used in the case of cavity type maser, for isolating the input and output signals. A typical TW maser operating at 1.6 GHz may have a gain of 25 dB with a bandwidth of 25 MHz and a typical noise figure of 0.3 dB.

Recapitulation

Microwave active devices

- Microwave active devices required primarily for signal generation and amplification are available both as tube and solid-state devices depending on the frequency and power level of the microwave signal required. Commonly used tubes for this purpose are linear-beam tubes (like klystron and TWT) and crossed-field tubes (like magnetron and CFA) and fast-wave electron tubes (like gyrotron, peneotron). Commonly used solid-state devices for signal generation and amplification at microwave frequency are IMPATT, Gunn and tunnel diodes and MESFET, HEMT and HBT transistors. The power levels obtainable from solid-state sources are low compared to that from tube sources. Thus for high power applications, tubes are preferred. However, power combining techniques in solid-state sources are making them a strong competitor of medium power tubes.
- Apart from signal generation and amplification, microwave active devices are also used for the design of detectors, mixers, multipliers, phase shifters, switches, attenuators, tuners and so on. For these applications microwave solid-state devices such as Schottky diode, varactor diode and PIN diode find great importance in microwave systems.

Microwave tubes

- *Linear beam tubes*: In a linear beam tube the magnetic field is applied in the axial direction of the tube and this merely takes part in focussing the electron stream, that is, to form a long cylindrical collinear electron beam (without taking part in the electron beam RF field interaction). Klystron: multi-cavity klystron as amplifier and reflex klystron as oscillator (which uses resonant cavities as RF structure) and travelling-wave tube (TWT) (that uses non-resonant periodic structure either helix type or coupled-cavity type that acts as slow-wave structure) are linear beam tubes.
- *Crossed field tubes*: In crossed-field tubes the DC (static) magnetic field is applied at right angles to the DC electric field (crossed to each other) and both these externally applied DC magnetic and DC electric fields take part in the interaction of the electron beam with the RF field for operation of these tubes. Primary crossed-field tube is the magnetron (that uses coupled resonant cavities as RF structure) with its varied forms as coaxial and inverted magnetrons. The use of non-resonant RF structure in crossed-field device results in forward and backward crossed field amplifiers (CFA) which are designed for specific practical use.
- *Fast-wave electron tubes*: In TWT, the phase velocity of an electromagnetic wave is brought down to nearly 10% of the velocity of light to make effective interaction (along the length of the tube, longitudinal interaction) with the electron beam being accelerated with practically realisable DC power supply—in such a tube the electromagnetic wave is slowed down, so it is called a slow-wave electron tube. In a fast-wave electron tube like a gyrotron, the electromagnetic wave travelling with its natural velocity (velocity of light) interacts with the electron beam moving in a helical path within a waveguide (under the combined effect of longitudinally applied electric and magnetic fields) having a transverse component of velocity (that is, transverse interaction takes place between the two). Since the electromagnetic wave need not be slowed down to interact with the

electron beam, in contrast to the slow-wave electron tube these devices are known as fast-wave electron tubes.

- *Two cavity/multi-cavity klystron amplifier*: Brief principle of operation of two cavity/multi cavity klystron amplifier is as follows: the electrons emitted from the cathode are focussed and accelerated by an electron gun to form a long cylindrical electron beam, before the beam enters the RF interaction region. As the electron beam of the klystron passes the input resonant cavity (*buncher*) the velocity of the electrons gets modulated by the input microwave signal (to be amplified) applied to the buncher. This velocity modulated electron beam while traversing down the drift space results in density modulated or bunched beams. An output resonant cavity (*catcher*) placed at a suitable location along the electron path in the drift space, helps to extract the RF power from the density modulated electron beam and delivers the amplified power to the load.

  The *velocity modulation* of the electron beam and consequent formation of density modulated electron beam (*bunches*) can be understood in terms of the *Applegate diagram*. The catcher is placed at an optimum distance d_{opt} from the buncher cavity where the bunching is maximum. The expression for d_{opt} is given by $d_{opt} = 3.682\ v_0 V_0/\omega\beta_i V_1$.

  The retarding field in the catcher cavity results from *beam coupling* which helps to retard the electron beam appearing once per cycle at the catcher cavity and results in energy transfer from the electron beam to the RF field and hence consequent amplification of the RF signal applied to the buncher cavity.

  The multi-cavity klystron helps to improve the gain of the amplifier and also the bandwidth of operation by stagger tuning. Three catcher cavities are tuned to frequencies: $(f_0 - \Delta f)$, f_0 and $(f_0 + \Delta f)$ for a 4-stage klystron amplifier. However, even with stagger tuning, the amplifier bandwidth is never more than 8%–10% as resonant cavities are involved.

- *Reflex klystron oscillator*: The form of the klystron oscillator that has found wide application as low to medium power microwave oscillator is the *reflex klystron* which is a single cavity klystron with a different principle of operation when compared to two-cavity/multi-cavity klystron amplifiers. The principle of operation of this oscillator is as follows. The cavity gap voltage (which originates as an initial stimulus as in any oscillator from the noise or switching transient) velocity modulates the electron beam which when returned by the repeller forms bunches which can be made to reach the cavity once per cycle during the retarding phase of the RF gap field in the cavity. This will result in the exchange of kinetic energy (KE) from the electron beam to the RF gap field growing the oscillation amplitude in the cavity. Stable oscillations build up in the cavity when the KE delivered to the cavity from the electron beam gets over the losses in the cavity and the load.

- *Travelling-wave tube*: The principle of operation of the TWT is as follows. The microwave signal which, in general, travels with the velocity of light is slowed down by a slow wave structure (which may be a helix or a coupled-cavity structure). This slowed RF field (with velocity ≈ 10% of the velocity of light), while travelling from input to output of the tube, interacts continuously with the electron beam over the total length of the slow wave structure and over a large number of cycles of the input microwave signal to be amplified. Due to this close interaction of the electron beam and the RF field, a broadband amplified microwave signal is obtained at the output of the tube.

The travelling wave tube is a very broad band amplifier providing even octave bandwidth ($f_2/f_1 = 2$, where $\Delta f = f_2 - f_1$) and moderately low noise (noise figure of the order of 8 to 10 dB) microwave amplifier with reasonably high gain (50 to 60 dB).

- *Magnetron*: The mechanism of generation of microwave signals in a magnetron involves the interaction between the electromagnetic fields existing in a system of coupled microwave cavities (TE_{011} for cylindrical cavities) and the electron path (epi-cycloidal path for cylindrical magnetron) resulting from their motion in static (DC) electric and magnetic fields oriented at right angles to each other (crossed field). The pulsed RF signal is produced with the magnetron by switching the magnetron ON and OFF by a pulser.

  Hull cut-off condition for cylindrical cavity magnetron is given by $B_{0c} = (8\,V_a m/e)^{0.5}/b[1-(a/b)^2]$.

  Bunch formation in a magnetron takes place with *phase focussing*. The result of bunch formation by phase focussing is the *wheel-spoke* situation of the electron cloud in the anode–cathode space which rotates continuously in a synchronised manner with the RF gap field in the coupled cavity system, delivering energy to the cavity system. Hence the descriptive name of cavity magnetron is the *travelling wave magnetron* [this has nothing to do with the travelling wave tube (TWT)].

  The possibility of mode jumping in a magnetron can be solved by using *strapping rings* at microwave frequencies and by using *rising-sun resonators* at millimeter-wave frequencies. Capacitive tuning in a magnetron can be done by *cookie-cutter* tuning while inductive tuning is done with *crown of thorns* tuning. Pulsing of a magnetron can be done with line pulsing modulator that uses a pulse forming network (PFN).

  In a coaxial magnetron, a stabilising cavity operating at TE_{011} mode is placed coaxially with the main magnetron structure. This permits improvements in power output, efficiency, stability, tunabilty and life of operation of the magnetron tube.

- *Gyrotron*: In this tube, electrons describe a helical motion within a cylindrical waveguide caused by an axial magnetic field (which gives it a cyclotron orbit) and an axial electric field (caused by the so called beam voltage) is used that shoot the electrons through the waveguide from the cathode to anode. Initially the phase of the electrons in their cyclotron orbits are random, but the phase bunching of electrons necessary for transfer of energy from the electron beam to the microwave field inside the waveguide occurs because of the dependence of the effective relativistic-electron mass on energy. The bunching process in a gyrotron depends on a relativistic effect known as *Cyclotron Resonance Maser* (CM) interaction between the electromagnetic wave and the electron beam. The bunched electrons and the microwave electric field in the waveguide are synchronised in such a way that the bunched electrons are kept decelerated, the lost kinetic energy of the bunched electrons is converted to microwave potential energy and the initial stimulus of the microwave field in the waveguide of the gyrotron tube gets stronger. This is how microwave oscillation/amplification is accomplished in a gyrotron tube. The synchronisation mentioned above is realised by making the cyclotron resonance frequency ($eB/2\pi m_e$) equal to the operating frequency (f_{op}) of the device. Thus, the magnetic flux density required to operate a gyrotron at a desired frequency is given by $B = 2\pi m_e f_{op}/e$.

Unlike the klystron, magnetron and TWT, a gyrotron tube is not designed with resonant cavities. Here the electron beam interacts with the electromagnetic field modes in a waveguiding structure. The resonant cavity structure of the klystron and magnetron and the coupled transmission cavities in the form of slow-wave structure in TWT limits the possibility of realising high power operation of these devices at a higher frequency as the resonant cavity dimension is inversely proportional to the operating frequency. Thus, the fundamental size restriction of conventional microwave tubes is not present in the case of fast-wave electron devices like gyrotron and hence this class of device can easily produce kilowatt level power at higher microwave frequencies.

Microwave solid-state (semiconductor) devices (diodes and transistors)

- *Schottky diode*: It is a metal-semiconductor contact diode unlike the *p–n* junction diode which is a semiconductor (*p*-type)-semiconductor (*n*-type) junction diode. The Schottky diode is a rectifying diode like the *p–n* junction diode but it is a majority carrier diode (unlike the *p–n* junction diode which has both majority and minority carriers) and thus Schottky diodes do not suffer from a minority carrier storage problem normally faced with the *p–n* junction diode, allowing its operation even up to 100 GHz.
- *Tunnel diode:* It is basically a *p–n* junction diode but both *p* and *n* semiconductors are heavily doped so that Fermi levels in *p* and *n* semiconductors are aligned in such a way that the conduction band electron states on the *n*-side get more or less aligned with valence band hole states on the *p*-side and thus resulting in a junction which has a depletion layer that is so thin (even less than 10 nm) that *tunnelling* of carriers can occur via the junction. The thinness of the junction naturally makes transit time of carriers via the junction very small hence permitting the operation of the device in microwave and millimetre-wave frequencies. The tunnel diode offers *negative differential resistance* in part of their operating range and is used to design solid-state oscillators and amplifiers.
- *Varactor diode:* It is a reverse biased *p–n* junction diode with a special design that exhibits a large variation of capacitance with the variation of the applied reverse bias voltage unlike a normal *p–n* junction diode where the aim is to reduce this capacitance change. Voltage variable capacitance of a varactor diode can be used in the design of VCO which is an integral part of PLL used in communication systems design. Utilising the harmonic generation capability of varactor (caused by non-linear variation of capacitance with voltage) frequency multipliers can be designed. A varactor diode exhibits a variable capacitance with the change of applied reverse bias (V_{rb}) to the diode according to the formula $C_j(V_{rb}) = C_0\,(1 + V_{rb}/\varphi)^{-\gamma}$.
- *PIN diode and its applications:* The structure of the PIN diode consists of heavily doped *p* and *n* regions separated by an *i*-region (intrinsic region) of high resistivity. The effect of controlling the carrier density and hence the resistance of the *i*-layer, by varying the forward bias in a PIN diode is known as *conductivity modulation*. To realise this characteristic of the PIN diode, it is essential that the lifetime of carriers in the *i*-region be greater than the time period at the operating frequency. For this reason, the PIN diode cannot be operated at lower radio frequencies. Typically, for a carrier lifetime of around 100 ns, the lowest frequency of operation of the PIN diode is 10 MHz. This phenomenon

of conductivity modulation plays a key role in the use of PIN diodes in the design of microwave active attenuators, modulators, limiters and automatic levelling circuits.

- *Gunn diode*: It is a bulk-effect device based on field-induced transfer of conduction-band electrons from a low-energy, high-mobility lower valley to a high-energy, low-mobility upper (satellite) valley of compound semiconductors like GaAs, InP and CdTe, giving rise to a negative differential mobility and hence a negative differential resistance. Gunn diodes are capable of generating microwave power of the order of 0.5 to 1.0 W at the X-band and can be operated up to 100 GHz with a lower power level. It has replaced the reflex klystron oscillator in many applications and is now commonly used in police radar, FM-CW radar for radio altimeter, as local oscillator in microwave communication receivers and so on. There are various modes of operation of transferred electron devices (TED): *transit time mode*, *quenched and delayed domain mode*, and *limited space-charge accumulation* (LSA) *mode*.
- *IMAPTT diode*: Unlike the Gunn diode, the IMPATT (IMPact Avalanche Transit Time) diode uses a semiconductor junction to realise negative resistance; however, similar to the Gunn diode, the transit time of carriers from the cathode to anode is again utilised in this device though with a different operating principle. The IMPATT diode is also known as avalanche transit time (ATT) device as it combines avalanche multiplication of carriers in the high-field avalanche region and subsequent transit of carriers via the constant field drift region of the device. Other modes of operation in which the ATT class of devices operate are TRAPATT (TRApped-Plasma Avalanche-Triggered Transit), BARITT (BARrier-Injection Transit-Time) and TUNNETT (TUNNEl-injection Transit-Time) devices of which IMPATT is mostly used in practice. The IMPATT diode is one of the most powerful solid-state sources of microwave power. CW power of the order of 1 to 3W at the X-band (10 GHz) and pulsed power of 10 to 20 W at W-band (94 GHz) is practically realisable with DC to RF conversion efficiencies of the order of 5%–8%. High noise figure (typically 40 dB for Si IMPATT diode) is the main limitation of the IMPATT amplifier in the receiver stage of a communication system but IMPATT oscillators are very attractive microwave/millimetre-wave sources in solid-state transmitters.
- *Metal semiconductor field effect transistor* (MESFET): Compared to the junction field effect transistor (JFET) the MESFET has better high frequency performance since it uses the Schottky barrier gate-source junction (unlike JFET where the gate-source junction is a *p–n* junction, limiting the high frequency performance of JFET). MESFETs are used as active devices for designing both low-noise as well as power amplifiers including their applications as switches, attenuators and so on. The high frequency cut-off of MESFET is determined in terms of the transit time of the carriers through the gate channel of width L and is given by $f_{cmax} = v_s/2\pi L$, where v_s is the saturation drift velocity of carriers.
- *High electron mobility transistor (HEMT)*: Similar to the MESFET, the HEMT is also a field-effect transistor with a Schottky barrier gate contact and having ohmic contacts for the source and drain. However, the semiconductor active channel has some novelty realised with a heterojunction (unlike MESFET that uses a homojunction) making it possible to realise very high electron mobility. To realise superior transport properties for the carriers (realisation of high electron mobility), the heterostructure formed between an undoped

GaAs and *n*-type AlGaAs is used to form a special active channel where electrons travel as a two-dimensional electron gas (2-DEG) in the undoped GaAs being unhindered by the donor atoms (that normally reduces the electron drift velocity and hence mobility in homojunction devices due to scattering). Nowadays, HEMT devices are used in a wide range of RF design applications including cellular telecommunications, direct broadcast receivers (DBS), electronic warfare radar, radio astronomy and any RF design application that requires a combination of low-noise and very high frequency performance.

- *Hetrojunction bipolar transistor (HBT):* HBTs can exhibit extremely high frequency (going even up to several hundred GHz) of operation and capable of realising a higher power gain compared to conventional BJTs using a thin and highly doped base region. In HBT, a wider band gap semiconductor material (AlGaAs) is used for *n*-type emitter and a comparatively narrow band gap material (GaAs) is used for *p*-type base, the collector being an *n*-type GaAs material.

Microwave solid-state oscillators, amplifiers and power combiners

- Microwave solid-state oscillators and amplifiers can be designed with both two-terminal active devices (diodes) like tunnel diode, Gunn diode and IMPATT diode as well as with three-terminal active devices (transistors) like high frequency BJT/HBT, MESFET and HEMT devices. To achieve a high power output at microwave and millimeter-wave frequency with solid-state devices like Gunn and IMPATT, power combiners are needed to combine the power from several individual devices either in device level or in circuit level.
- *Fundamental theoretical background of microwave diode oscillator/amplifier*: Microwave diodes like tunnel diode, Gunn diode and IMPATT diode (all of which are negative resistance devices) are commonly used for the design of oscillators and amplifiers at microwave frequencies. The device–circuit interaction gives rise to stable oscillations at a frequency f_0 governed by: $R_C(f_0) = R_D(f_0, I_0, I_{RF}, T)$ and $X_C(f_0) = -X_D(f_0, I_0, I_{RF}, T)$, where the driving point impedance of the device is given by $Z_D = (-R_D - jX_D)$ while the circuit impedance looking into the device terminals is given by $Z_C = (R_C + jX_C)$.

  The power gain (G_p) of the reflection mode amplifier is given by:

$$G_p = \frac{(R_C + R_D)^2}{(R_C - R_D)^2}$$

  In a number of applications we need highly stabilised, low noise oscillators/amplifiers at microwave frequencies. This can be done either by *cavity stabilisation* or *injection-locking* of the oscillator.
- *Oscillator/amplifier configurations with IMPATT/Gunn diodes*: Commonly used oscillator/amplifier configurations are bias-post type with the cavity formed in full-height waveguide, reduced-height waveguide type cavity to realise better impedance matching between the device and the circuit, and resonant-cap type cavity in a full-height waveguide. In a resonant-cap type oscillator, the resonant-cap acts as a quarter-wave radial-line impedance transformer between the device and the circuit and provides even

better impedance matching between the device and the circuit, however, this is a narrow band oscillator.

- *Amplifier/oscillator with three-terminal microwave device (transistor)*: Microwave transistors like MESFET, HEMT and BJT are used for the design of amplifiers at microwave frequencies while transistor based microwave oscillators may be designed with BJT/FET along with a dielectric resonator (DR).

The necessary and sufficient conditions for an amplifier to be unconditionally stable are given by:

$$K > 1 \quad \text{and} \quad |\Delta| < 1$$

$$\text{where, } K = \frac{1 - |S_{11}|^2 - |S_{22}|^2 + |\Delta|^2}{2|S_{12}S_{21}|} \quad \text{and} \quad \Delta = S_{11}S_{22} - S_{12}S_{21}$$

- For all other cases when $K < 1$, $|\Delta| < 1$; $K < 1$, $|\Delta| > 1$; $K > 1$, $|\Delta| > 1$; the amplifier is potentially unstable or conditionally stable.

The transistor oscillator design approach at microwave frequencies is similar to that of the amplifier's except that the circuit is purposely designed to be unstable. It is seen that for an amplifier we need $K > 1$ while for an oscillator $K < 1$, where K is known as the stability factor.

Many *power gain* equations are derived in terms of the S-parameters and the reflection coefficient at the input and output of microwave amplifiers. Different gain expressions—*transducer power gain* (G_T), *available power gain* (G_{AV}) and *operating power gain* or simply *power gain* (G_P) are given by:

$$G_T = \frac{1 - |\Gamma_S|^2}{|1 - \Gamma_{IN}\Gamma_S|^2} |S_{21}|^2 \frac{1 - |\Gamma_L|^2}{|1 - S_{22}\Gamma_L|^2} \qquad G_{AV} = \frac{1 - |\Gamma_S|^2}{|1 - S_{11}\Gamma_S|^2} |S_{21}|^2 \frac{1}{1 - |\Gamma_{OUT}|^2}$$

$$G_P = \frac{1}{1 - |\Gamma_{IN}|^2} |S_{21}|^2 \frac{1 - |\Gamma_L|^2}{|1 - S_{22}\Gamma_L|^2}$$

$$\text{where, } \Gamma_{IN} = S_{11} + \frac{\Gamma_L S_{12} S_{21}}{1 - S_{22}\Gamma_L} \quad \text{and} \quad \Gamma_{OUT} = S_{22} + \frac{S_{12}S_{21}\Gamma_S}{1 - S_{11}\Gamma_S}$$

For broad band microwave amplifier design, several techniques are avilable of which balanced configuration and distributed amplifier configuration are commonly used. Balanced configuration can realise octave bandwidth while the distributed amplifier configuration can give even decade bandwidth.

- *Dielectric resonator (DR) based transistor oscillator:* High Q value of DR along with its precise temperature stability can be used in conjunction with an active device (BJT/MESFET/HEMT) to design highly stable oscillators at microwave frequencies, known as dielectric resonator oscillators (DRO), the oscillation frequency being determined by the DR used.

In such an oscillator, the DR acts as a resonating element which can be modelled electrically as an *L*, *C*, *R* network and connected to a suitable port of the active device. DRO can be designed with DR either as the series feedback element or as the parallel feedback element. The parallel feedback arrangement can provide a wide band tuning, however, this configuration is difficult to construct as it needs coupling of two transmission lines.

- *Power combiners with IMPATT/Gunn diodes*: The performance of a power combiner is characterised in terms of *power combining efficiency* defined as $\eta_c = P_{OT}/N.\,P_{OS}$, where P_{OT} is the net power at the output of the combiner using *N* number of active devices and P_{OS} is the power output expected from single device oscillator using the same devices as in the combiner. Typical power combining efficiency of a well-designed power combiner may be 90% or so.
- A microwave power combiner, in principle, desires to add the microwave powers of several devices (as in device level combining) or of several single-device oscillators (as in circuit level combining) as efficiently as possible. The circuit level combiners are more practical and are, in general, of two categories; those which combine the output of *N* devices in a single stage (known as *N*-way in single stage) and those which combine the output of *N* devices in multiple stages (known as *N*-way in multiple stage). The latter category is simpler and more widely used; however, the former category has the possibility of higher power combining efficiency since power generated does not have to pass through several stages of combining.
- The basic *N*-way in a single stage power combiner designed with resonant cavity combining structure was due to Kurokawa and Magalhaes. Here the sum of output powers from a number of devices is obtained by coupling their outputs to a single resonator and the combing of power takes place with the devices mounted at magnetic field maxima. Another simple, compact yet elegant duo-diode power combiner can be a twin-cap combiner in which IMPATT (or Gunn) diodes embedded in radial-line cavities combine their power at the electric field maximum of the waveguide in which the twin-cap structure is mounted. *N*-way in multiple stages may be realised in many ways, one common approach is to use short-slot quadrature hybrid (in waveguide version) or branch-line qudrature (90°) hybrid (in microstrip version).
- The hybrid combiner is suitable for cascading so that the power output of many oscillators can be combined. In a hybrid combiner, the qudrature hybrid provides necessary isolation between the active sources connected to its two output ports and hence the modeing problem which the resonant cavity combiner is prone to is minimised here. However, injection locked sources need to be used in this case (when synchronisation of frequency and phase of the two sources can be assured) but in resonant combiners since the diodes are in close proximity with strong coupling between them they are inherently locked.

Quantum electronic devices—MASER

- The MASER (Microwave Amplification by Stimulated Emission of Radiation) is a quantum electronic device which can amplify microwave signals with an *extremely low noise figure*. Therefore, it is used as a front-end RF amplifier in the receiver of radio telescopes and for communications with space probes—where the requirement of low noise figure of the amplifier is of utmost importance. Maser also has an important application as a frequency standard for atomic clocks.

- A maser can be an electronic maser, an atomic maser, or a molecular maser—depending on what kind of particles emit *coherent radiation*. Frequency standard masers are in general gas masers. However, gas masers suffer from the disadvantage of yielding amplifiers that work only at a single frequency whose bandwidth is very narrow (typically for ammonia gas maser bandwidth is only 3 kHz with operating frequency close to 24 GHz). Solid-state maser – either cavity type or travelling-wave type – are used for amplifier applications with the latter being capable of providing useful bandwidth for the maser amplifier. Solid-sate maser is based on paramagnetic crystals such as ruby.

Review Questions

9.1 Why are solid-state sources preferred over tube sources at microwave frequencies for air/space borne and military applications especially for low and medium power applications?

9.2 Why can't conventional tube devices be used at microwave frequencies above about 1 GHz? Name a few tube devices that are practically used at microwave frequencies.

9.3 Why can't a *p–n* junction diode be used at microwave frequencies? Instead, a Schottky barrier diode is used.

9.4 Name a few microwave solid-state devices that are used as amplifiers and oscillators at microwave and millimetre-wave frequencies.

9.5 Name a few solid-state devices that are used for the design of detectors, mixers, multipliers, phase shifters, switches, attenuators and tuners at microwave frequencies.

9.6 What are linear beam tubes and crossed-field tubes? Name a few of them.

9.7 Discuss the principle of microwave amplification in a two-cavity klystron. How is the bandwidth of the amplifier improved with stagger tuning in a multi-cavity klystron?

9.8 Discuss the principle of the reflex klystron as a microwave oscillator.

9.9 The TWT is a low noise and broad-band microwave amplifier compared to a klystron—explain why and how. What types of slow-wave structures are used in the design of a TWT?

9.10 Discuss how the interaction of the epi-cycloidal electron trajectory with the RF field in a system of coupled anode cavities generates microwave signals in a magnetron. Why is a cavity magnetron also known as a travelling-wave magnetron?

9.11 Discuss how strapping rings and rising-sun resonator structures help to realise mode separation in microwave and millimetre-wave magnetrons.

9.12 Discuss the techniques of frequency tuning in a magnetron.

9.13 Discuss the principle of pulsing of a magnetron with a pulse forming network. Why can't a conventional transmission line be used for this purpose?

9.14 What is a coaxial magnetron? What are its advantages over a conventional magnetron?

9.15 What are first-wave electron tubes? How does this type of tube differ from a slow-wave electron tube like the TWT?

9.16 Discuss the principle of operation of a gyrotron. Why is a gyrotron capable of generating a much higher power at millimetre-wave frequencies compared to all other electron tubes that use resonant cavities for their design?

9.17 Name a few solid-state diodes and mention their applications.

9.18 Why is a Schottky diode and not a *p–n* junction diode be used at microwave frequencies? When is a Schottky contact rectifying and when is it ohmic?

9.19 Discuss the principle of operation of a Schottky contact diode in terms of the energy band diagram.

9.20 What is a tunnel diode? How can it be used for signal generation and amplification at microwave frequencies?

9.21 Discuss the principle of realising negative differential resistance with a tunnel diode.

9.22 What is a varactor diode? Discuss how the voltage variable capacitance of a varactor can be used for harmonic generation. What is a snap-off varactor?

9.23 What do you understand by conductivity modulation in a PIN diode? A PIN diode is a useful microwave device but cannot be used at lower RF—explain.

9.24 Discuss the principle of a PIN modulator and a PIN limiter. Explain how a PIN attenuator can be used to design an automatic levelling circuit.

9.25 A Gunn device is a bulk semiconductor device but it acts as a diode—explain.

9.26 What is negative differential mobility? Discuss the principle of operation of a Gunn diode in terms of negative differential mobility.

9.27 What are the different modes of operation realisable with a Gunn diode?

9.28 What is an IMPATT diode? Name a few more devices of the ATT class. Why is IMPATT a noisy device?

9.29 Discuss the principle of operation of an IMPATT diode and explain the origin of negative resistance in the operation of such a device.

9.30 Give a brief idea of the packaged IMPATT/Gunn diode.

9.31 Mention the salient features of microwave transistors. Name a few microwave transistors that are used in recent times for high frequency and low noise applications.

9.32 Why is BJT called a current controlled device while MESFET/HEMT are called voltage controlled devices? Why do MESFET/HEMT fall in the class of field effect transistors?

9.33 In what respect is MESFET better than JFET, and HEMT better than MESFET as microwave devices?

9.34 Discuss the structure and principle of operation of a MESFET device.

9.35 Discuss the structure and principle of operation of a HEMT device. What is 2D electron gas (2-DEG)? Explain the concept of realising high electron mobility in HEMT.

9.36 What are the pHEMT and mHEMT devices?

9.37 With respect to the equivalent circuit of HEMT/MESFET, explain the physical origin of intrinsic and extrinsic parameters of a packaged device.

9.38 Discuss the structure and principle of operation of a HBT device.

9.39 Discuss the fundamental theoretical background for a microwave diode oscillator/amplifier. What is an injection-locked oscillator/amplifier?

9.40 Discuss some typical oscillator/amplifier configurations with an IMPATT/Gunn diode.

9.41 Discuss the technique of electronic tuning of IMPATT/ Gunn oscillator with a varactor diode to design VCOs.

9.42 Define different amplifier gains we come across in the case of microwave amplifiers and give their expressions in terms of S-parameters.

9.43 What do you understand by conjugate matching in the design of microwave amplifiers? What are the unilateral and bilateral cases? What is the speciality in matching as far as LNA is considered?

9.44 Mention the necessary and sufficient conditions for the unconditional stability of a microwave amplifier. When is it potentially unstable (or conditionally stable)? What should be the stability factor of a microwave oscillator?

9.45 Discuss balanced amplifier and distributed amplifier techniques used for the design of broad-band transistor amplifiers at microwave frequencies.

9.46 Discuss the principle of a DR based transistor oscillator.

9.47 What do you understand by device level combining and circuit level combining of microwave power? Mention the techniques for circuit level microwave power combiners. Define power combining efficiency.

9.48 Discuss the basic *N*-way in a single stage power combiner designed with a resonant cavity combining structure. Also discuss the *N*-way in a multiple stage combiner using a quadrature hybrid.

9.49 Discuss the principle of operation of a MASER. Give a scheme for frequency standard based on maser.

9.50 Why can't a gas maser be used for amplification of microwave signals with maser? Discuss the scheme for a ruby laser. What is travelling wave maser?

10 Microwave Antennae and Wave Propagation

Something that emerges apparently from invisibility
—creativity

10.1 Introduction

In the microwave frequency range, the wavelength of the electromagnetic signal becomes comparable to or even smaller than the size of the antenna and thus diffraction phenomena are dominant as in optical systems (such as reflectors, lenses). Microwave antennae can be made relatively high gain compared to any lower frequency RF antenna since its aperture size (D) with respect to wavelength (λ) of operation can be made equal to or larger than unity, $D/\lambda \geq 1$. This is absolutely necessary as power available from microwave active devices is much smaller than that available with lower frequency RF devices (as $P_{\text{available}} \; \alpha \, 1/f$). Thus effective radiated power from the antenna (which is proportional to the product of antenna gain and power available from transmitter) remains practically the same whether operating at lower frequency RF or at microwave frequency.

At lower radio frequency (RF) we use separate transmitting and receiving antennae (for example, a vertical mast antenna is used for FM transmission while a whip antenna is common for FM reception, turnstile antenna is used for TV transmission while a Yagi–Uda antenna is used for TV reception and so forth). But at microwave frequencies, the same type of antenna can be used for both transmission and reception. For example, a parabolic dish antenna is used for both transmission and reception in point-to-point microwave communication and satellite communication. In fact, in a monostatic pulsed radar operating at microwave frequency, the same (single) antenna is used for both transmission and reception. Wire antennae of various forms (loop antenna, rhombic antenna) are extensively used at lower RF applications, while at microwave frequencies, wire antennae are rare. However, a variant of the wire antenna known as the helical antenna (which can selectively generate right-handed or left-handed circularly polarised electromagnetic signals) is used in some microwave applications.

At present many different forms of microwave antennae are used whose principle of operation is more or less similar (dependent on diffraction principle) and differ significantly from antennae operating at lower RF. The differences between individual microwave antennae lie in the design features, feeding methods and so on. In general, there are five types of conventional microwave antennae—horn, reflector, slot, lens and surface-wave (dielectric rod) antenna. Each of these types has several sub-types with a large number of variants.

Apart form these conventional diffraction-type microwave antennae, many civil and military applications demand planar (printed) antennae such as microstrip patch antenna and its variants which are useful for low power applications. In recent times, dielectric resonator antenna is also gaining importance for millimeter-wave applications.

In general, microwave signal propagation is classified as space wave propagation, commonly known as line-of-sight (LOS) propagation. Also, the microwave signal is scattered by the troposphere making troposcatter microwave communication possible in difficult geographical terrain. In subtropical regions the phenomenon of super-refraction allows the propagation of microwave signals through atmospheric ducts making long range microwave signal propagation possible by duct propagation over the sea or ocean surface. Further, microwave signal propagation around the globe is done these days popularly and elegantly via communication satellites.

10.2 Principle of Operation and Performance Parameters of Conventional Microwave Antennae

10.2.1 Horn Antenna

A waveguide by itself is an inefficient radiator because the field fringing out of the mouth (the open end) of the waveguide finds that the waveguide impedance is poorly matched with the free space into which it fringes out. However, this impedance matching can be improved if the mouth of the waveguide leading to free space is flared either in the E or in the H plane of the waveguide. A waveguide having a progressively increasing cross-section (flared) in the E or H plane and terminating in a radiating aperture is known as a *horn*.

For a rectangular waveguide with a dominant TE_{10} mode, the E field is in the y direction, H is in the x and z-direction and the wave propagates in the z-direction (Fig. 4.6). Thus a widening or flaring out of the guide in the E-plane (y-direction) of a rectangular waveguide gives a sectoral E-plane horn (Fig. 10.1a). Physically, this means that the broad walls of the waveguide are flared in the plane (x–y plane) parallel to the electric field vector. While a widening in the H-plane (x-direction) gives a sectoral H-plane horn (Fig. 10.1b)—here the narrow dimension of the waveguide walls are flared in the plane (x–z plane) parallel to the magnetic field vector. A combination of the two produces a pyramidal horn (Fig. 10.1c). When the mouth of a circular waveguide is flared out we get a conical horn (Fig. 10.1d).

The process of radiation from a horn antenna may be understood as follows. It has a source S (may be the open end of the inner conductor of a coaxial cable or a dipole

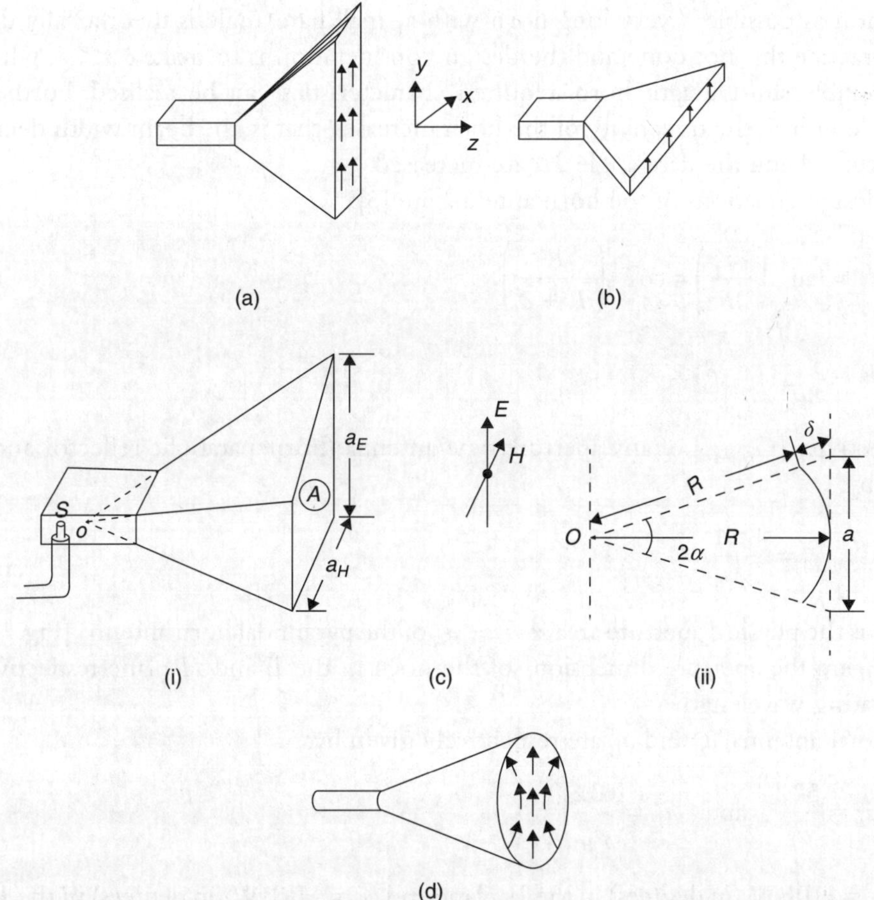

Fig. 10.1 **Horn antennae of different types (a) Sectoral *E*-Plane horn (b) Sectoral *H*-plane horn (c) Pyramidal horn (d) Conical horn.**

antenna) which feeds the signal to the waveguide part of the horn antenna and an aperture *A* bounded by the edges of the horn (Fig. 10.1c). The electromagnetic waves radiated from the source *S* propagate first over the normal waveguide portion, then pass to the horn and from the aperture of the latter enter external (free) space. The propagation of the field has a standing wave nature between the opposite walls (of the waveguide) and a travelling wave nature in the direction from the apex *O* of the horn to the aperture *A*. Beyond the limits of the aperture of the horn, the propagation of the electromagnetic wave is in accordance with the laws of diffraction. The characteristics of the radiation fields of the horn antenna are determined by the dimensional parameters of the horn—the length *R*, aperture *A* ($a_E \times a_H$) and apex angle (flare angle) 2α [longitudinal section of the horn in Fig. 10.1c(ii)].

Due to flaring, the spherical waves emanating from the aperture of the horn antenna will find a phase difference ($\varphi = 2\pi\delta/\lambda$) between the waves at the edges of the horn and at the centre. When this phase difference is π, the two waves will interfere destructively causing a decrease in gain and an increase in side lobe levels. Thus, to obtain as uniform an aperture distribution as possible, a very long horn with a small flare angle is theoretically desirable. But in practice this not done and the design consideration is to make $\delta << \lambda$, when with a considerably short length horn, required characteristics can be realised. Further, for a constant length R, the directivity of the horn increases, that is, the beam width decreases as the aperture A and the flare angle 2α are increased.

The design equations of the horn antenna are [6]:

$$\alpha = \tan^{-1}\left(\frac{a}{2R}\right) = \cos^{-1}\left(\frac{R}{R+\delta}\right)$$

$$R = \frac{a^2}{8\delta}, \qquad (\delta << R) \tag{10.1}$$

The directivity [$G_{D(\max)}$] of any aperture type antenna (horn, parabolic reflector and so on) is given by:

$$G_{D(\max)} = \frac{4\pi A}{\lambda^2} \tag{10.2}$$

where A is the physical aperture area $A = a_E \cdot a_H$ of the pyramidal horn antenna [Fig. 10.1c(i)], a_E and a_H are the aperture dimensions of the horn in the E and H-plane respectively, λ is the operating wavelength.

For horn antenna, a_E and a_H are respectively given by:

$$a_E = \frac{52\lambda}{\theta_E} \quad \text{and} \quad a_H = \frac{63\lambda}{\varphi_H} \tag{10.3}$$

where θ_E = HPBW (in degrees) in the E-plane and φ_H = HPBW (in degrees) in the H-plane.

Thus the directivity of pyramidal horn antenna (or of any aperture-type antenna) is given by:

$$G_{D(\max)} = \frac{41{,}257}{\theta_E \cdot \varphi_H} \tag{10.4a}$$

While the power gain (G) is given by (Eq. 2.56):

$$G = \eta_a G_{D(\max)} = \frac{28{,}879}{\theta_E \cdot \varphi_H} \tag{10.4b}$$

[since antenna aperture efficiency $\eta_a = (A_e/A) = 0.7$; A_e is the effective aperture of the horn antenna. The significance of η_a will be discussed in connection with reflector antenna in the next sub-section].

EXAMPLE 10.1

A pyramidal horn antenna has HPBW in the E and H-planes respectively as 20° and 30°. Determine the directivity and the physical dimensions of the aperture of the antenna if the operating frequency is 10 GHz. What will be the effective aperture of the antenna if the antenna aperture efficiency is 0.7?

Solution

Directivity $G_{D(\max)} = \dfrac{41,257}{\theta_E \cdot \varphi_H} = 68.76°$ (since $\theta_E = 20°$ and $\varphi_H = 30°$)

Thus, $10 \log_{10}[G_{D(\max)}] = 18.37$ dB

Now $a_E = \dfrac{52\lambda}{\theta_E} = 2.6\,\lambda = 7.80$ cm and $a_H = \dfrac{63\lambda}{\varphi_H} = 2.1\,\lambda = 6.3$ cm (since $\lambda = 3$ cm)

The physical aperture of the antenna is $A = a_E \cdot a_H = (2.6 \times 2.1)\lambda^2 = 5.46\,\lambda^2$

Since $\eta_a = 0.7$, the effective aperture of the antenna is $A_e = 0.7\,A = 3.822\,\lambda^2 = 34.40$ cm^2

The physical aperture area A of the antenna may also be obtained from:

$$A = \left(\frac{G_{D(\max)}}{4\pi}\right)\lambda^2 = \left(\frac{68.76}{4\pi}\right)\lambda^2 = 5.47\lambda^2$$

The horn antenna, both sectoral and pyramidal, is commonly used as primary feed for a parabolic reflector and its variants. The conical horn is also used in conjunction with a lens antenna at millimeter-wave frequencies to realise a practically approximated radiating point source. Sometimes the inner walls of the horn antenna are corrugated (Fig. 10.2a), which improves the edge diffraction and pattern symmetry and also reduces cross polarisation (less E field in the H plane). The corrugations in the horn walls, act as quarter-wave chokes that reduce the E field to a very low value at the horn edges for all polarisations, thus preventing waves from diffracting around the edges of the horn antenna. Instead of corrugations, sometimes a rolled section is attached to the outside of the aperture edge of a horn (Fig. 10.2b) that leads to significant improvement in the pattern, impedance and bandwidth characteristics.

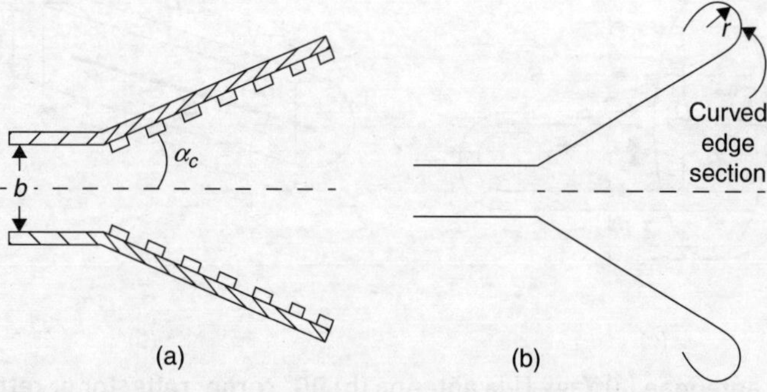

Fig. 10.2 Special horn antennae (a) Corrugated horn (b) Aperture matched horn.

10.2.2 Reflector Antenna

Reflectors may be used to modify the radiation pattern of a radiating element (as in rod reflector and corner reflector) or to act as an antenna by itself (secondary reflector) being fed by a primary feed antenna (as in a parabolic antenna). In a Yagi–Uda antenna (Fig. 10.3a), a reflector element (in the form of a rod) is placed behind the driven element which helps to reduce the backward radiation of the driven element (normally a folded half-wave dipole) and also provides substantial gain in the forward direction. While a corner reflector (realised with two plane reflecting surfaces oriented at an angle $\alpha < 180°$) is capable of producing a sharper radiation pattern compared to a flat-sheet reflector ($\alpha = 180°$). In fact a corner reflector with $\alpha = 90°$, also known as a *retroreflector*, is used as a passive target for radar and communication applications to return the signal exactly in the same direction from which it is coming (Fig. 10.3b). This knowledge is used in the design of most defense ships with minimum sharp corners to reduce the chances of their detection by enemy radar, which is basically a means of electronic counter measure (ECM). However, most commonly used reflector antennae are the *parabolic reflectors* (paraboloid of

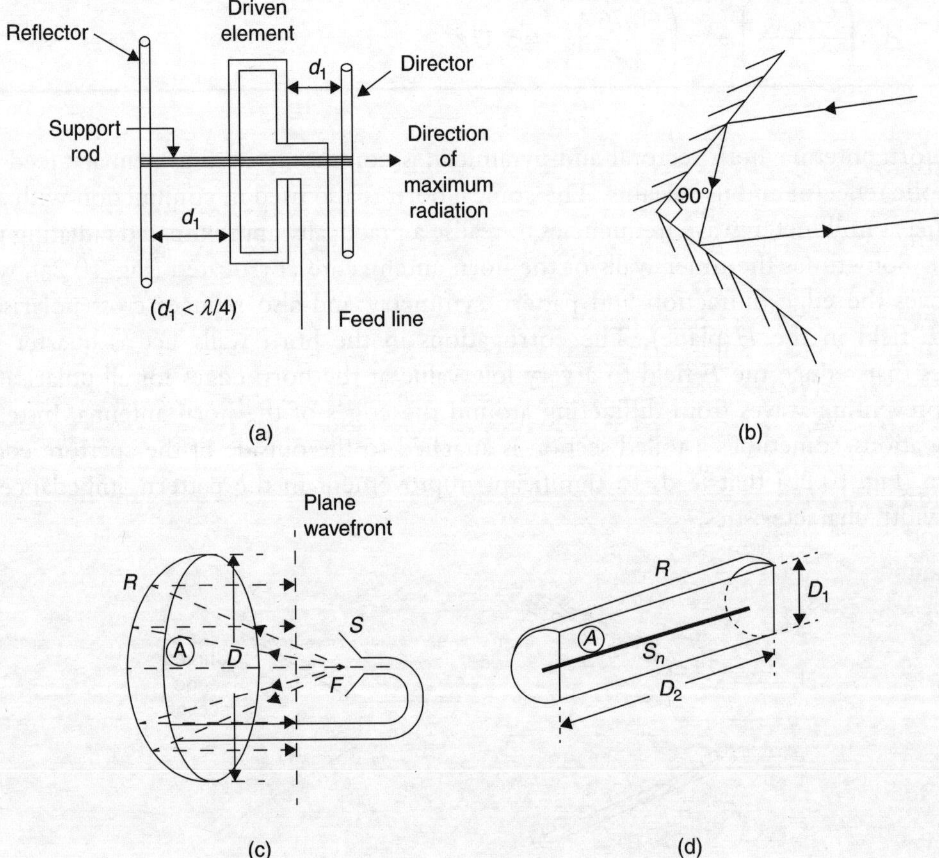

Fig. 10.3 Reflector antennae (a) Yagi–Uda antenna (b) 90° corner reflector or retroreflector antenna (c) Parabolic reflector antenna (d) Parabolic cylinder antenna.

revolution and parabolic cylinder, Figs 10.3c and 10.3d) which act as secondary reflectors. In this type of antenna there are two elements—the exciter or primary feed (*S*) and the reflector proper which is called the secondary reflector (*R*). In the case of paraboloid of revolution, also known as parabolic antenna (Fig. 10.3c), the primary feed is a point source located at the focus *F*, establishing an electromagnetic radiation field in the form of spherical waves. While, in the case of a parabolic cylinder antenna, the exciter is a line of sources (S_n) located along the focal antenna line which establishes an electromagnetic radiation field in the form of a cylindrical wave. The primary feed commonly used is a horn or dipole antenna. Frequently, to reduce wind resistance, the reflector surface is not made continuous but is in the form of a grid having a grid gap *g*. The girded structure acts as a continuous structure to the incident electromagnetic signal of wavelength λ when $g << \lambda$, from effective medium theory.

The process of radiation from the reflector antenna may be understood as follows. The electromagnetic wave radiated by the exciter in both cases arrives at the surface of the reflector, and after being reflected from it is eventually directed back towards the exciter in the form of a plane wave. Emerging into the free space, the reflected plane wave takes on the character of a spherical wave at relatively large distances from the antenna, in accordance with the law of diffraction, having a substantially higher directivity than the primary radiation of the exciter.

The aperture *A* of the reflector is the plane passing through the edges of the reflector. For the paraboloid of revolution the aperture is, in general, a circle of diameter *D* while for the parabolic cylinder it is rectangle with sides D_1 and D_2. In a number of practical cases, various forms of truncated paraboloids of revolution are employed where the aperture may take the form of a rectangle or a circle with excised segments.

The parabolic reflector antenna has a very sharp directional pattern with *pencil-beam* like main lobe and a number of low amplitude side lobes. The half power beam width of radiation of such an antenna is approximately given by:

$$\varphi = \frac{70\lambda}{D} \tag{10.5}$$

where *D* is the aperture diameter and λ is the operating wavelength and the above equation is good for all practical cases where $D >> \lambda$. The power gain of a parabolic dish antenna or parabolic reflector (which are synonymously used for paraboloid of revolution) can be derived as:

$$G \approx 6\left(\frac{D}{\lambda}\right)^2 \tag{10.6}$$

[This follows from $G = \dfrac{4\pi A_e}{\lambda^2} = \eta_a \dfrac{4\pi A}{\lambda^2} = \eta_a \dfrac{4\pi(\pi D^2)}{4\lambda^2}$; with $\eta_a = 0.6$ or 0.7]

where A_e is the effective aperture of the parabolic dish antenna, *A* is the physical area of the antenna aperture and η_a is the antenna efficiency of aperture type antenna. The term η_a is actually the product of several terms. η the radiation efficiency, η_i the aperture illumination efficiency (how well the aperture is utilised for collimating the radiated energy), η_1, η_2, η_3, include all other effects like spillover losses, phase error losses due to surface errors, aperture blockage (if any), depolarisation losses and others. For all practical purposes, taking into consideration the characteristics of horn/dipole as primary feed, η_a varies from 0.5 to 0.7; typically it is 0.60.

The parabolic reflector antenna is by far the most common type of microwave antenna used to generate a sharp *pencil-beam* with a number of applications in microwave communication and in tracking radar applications. However, shaped *fan-beam,* a common type being cosecant squared pattern is required for airport surveillance radar. For deep space communication and satellite tracking where low-noise amplifiers are to be used as the front-end of the receiver, a special type of Cassegrain feed parabolic antenna is used. With a pill-box reflector antenna (again a variant of parabolic reflector) one can obtain antenna beam pattern which is very narrow horizontally but not so directive vertically. In the ship to ship radar, we need such antenna patterns as the azimuth directivity should be excellent for this application, but elevation selectivity is not of practical importance as another ship is bound to be on the surface of the sea.

10.2.2.1 Feed mechanism for parabolic antennae

The feed mechanism of a parabolic reflector is determined in terms of the application it is used for. For all normal applications, the primary feed antenna is a horn or dipole placed at the focus of the parabolic reflector. Care is taken so that the feed antenna and its reflector do not obstruct any reasonable amount of reflection from the reflector. This depends on the choice of the relative size of the primary antenna and the reflector. The finite size of the primary antenna also influences the beam width of the antennae using parabolic reflectors. Not being a true point source, the feed antenna does not fit exactly at the focus and this aberration effect broadens the main lobe to some extent with side lobes being reinforced.

Another feed method, the *Cassegrain feed* (Fig. 10.4a), uses a hyperbolic subreflector as a secondary reflector. One of the foci of the hyperboloid coincides with the focus of the parabolic reflector, resulting in the action shown (for transmission). The rays emitted from the feed horn are reflected from the hyperboloid reflector and finally collimated rays are produced by the main parabolic reflector. The Cassegrain feed is used when low-noise receivers are to be placed very close to the primary feed, in order to keep the losses in the waveguide feeding the antenna low. The proximity of the receiver to the primary feed could also be done by placing the receiver at the focus, but this is impossible due to large size of the low-noise receiver (whose ancillary equipments such as the cooling apparatus are large though the actual RF amplifier may be relatively small in size).

The hyperbolic subreflector of the Cassegrain antenna appears to lead to a possible aperture blocking as well. Aperture blocking in this case may be reduced by using a large parabolic reflector compared to the hyperbolic subreflector, which is expensive. Another way may be to decrease the size of the subreflector, which requires that the

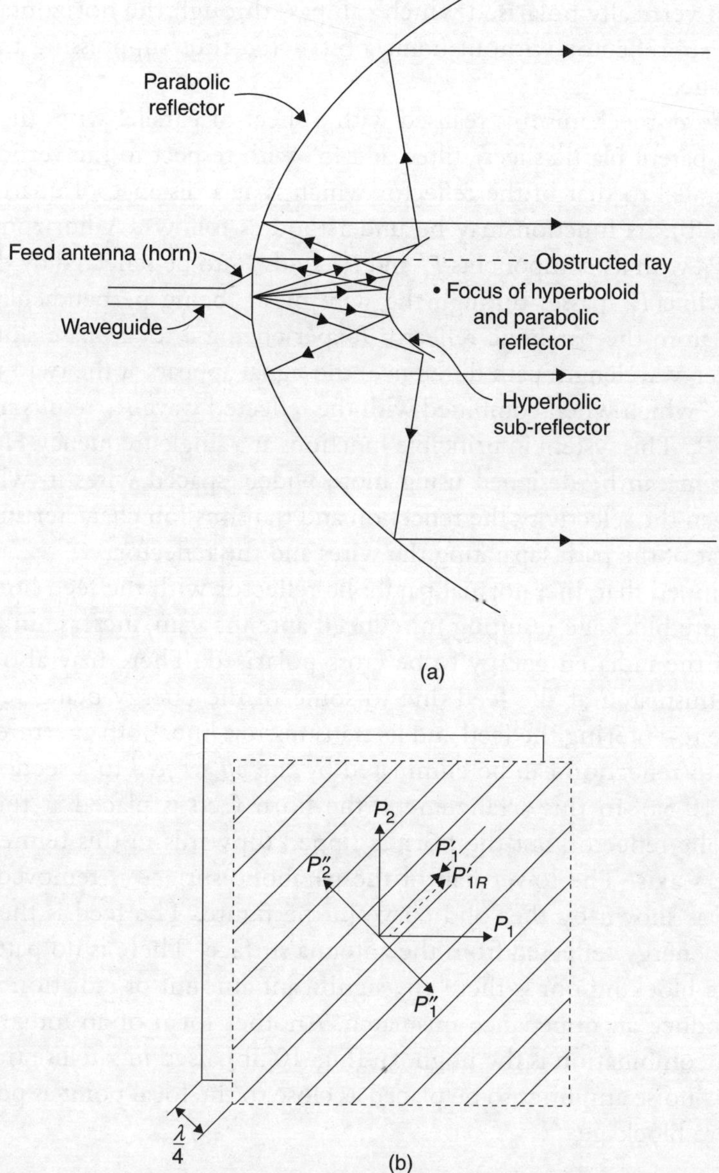

Fig. 10.4 (a) Cassegrain feed reflector antenna (b) Twist reflector to get rid of aperture blockage.

feed be made more directive or moved closer to the subreflector in order to minimise the possible spillover from the subreflector. But, the practical system uses a novel technique in which the primary feed (horn) radiates horizontally polarised waves which after reflection from the hyperbolic subreflector (consisting of horizontal grid of wires) reflects horizontally polarised waves to the parabolic reflector. At the parabolic reflector, the horizontally polarised wave received from the subreflector is rotated 90° by a special mechanism called a *twist reflector* and the wave reflected out of the parabolic reflector is then vertically polarised which can pass through the horizontal wire grid of the hyperbolic subreflector without being obstructed thus suppressing the problem of aperture blockage.

The *twist reflector* mechanism is realised with a sheet of parallel wires (held in place by means of a transparent plastic sheet), tilted at ±45° with respect to the vertical and located on a surface parallel to that of the reflector which is at a distance of quarter wavelength from it (Fig.10.4b). Its function may be understood as follows. A horizontally polarised incident wave P_1, with its components P_1' and P_1'' finds P_1' to be reflected by the wire (being parallel to it) while P_1'' passes through the wire mesh (being perpendicular to it). After being reflected from the parabolic reflector (experiencing a total phase shift of π, that is, twice the quarter-wavelength path difference) the signal appears at the twist reflector wire-mesh plane as P_2''which when combined with the reflected wave P_{1R}' results in the vertically polarised wave P_2. This system in principle functions at a single frequency. However, a large bandwidth system can be designed using more widely spaced wires in which there is a trade-off between the selectivity, the reflection and transmission characteristics of the sheet of wires and that of the path separating the wires and the reflector.

It was mentioned that, in a normal parabolic reflector with the feed horn at the focus results in aperture blockage resulting in reduced antenna gain, increase in side lobes, and causes some of the radiated energy to be cross-polarised. There may also be a problem of impedance mismatch at the feed due to some of the energy being reflected by the antenna surface re-entering the feed and its transmission line. Both aperture blockage and mismatch due to reflection can be eliminated by the *off-set feed* to a section of parabolic reflector (Fig. 10.5a). In this arrangement, the horn feed is placed at the focus of the original parabolic reflector, but the horn is tipped (upwards in this figure) with respect to the parabola's axis. The lower half of the parabolic surface is removed, leaving only the upper half as shown by the solid curve in the figure. The feed is therefore outside the path of the energy reflected from the antenna surface. There is no pattern distortion due to aperture blocking nor is there any significant amount of radiation intercepted by the feed to produce an impedance mismatch. Another form of an integrated parabolic reflector–horn combination is the hoghorn (Fig. 10.5b), used in satellite tracking systems requiring a low-noise amplifier to be placed as close to the focal point as possible without causing aperture blockage.

10.2.2.2 Fan-beam with parabolic antenna

In addition to *pencil-beam* (which is used for point to point communication and secret or covert communication and for tracking radar) realisable with a parabolic reflector antenna, *fan-beam* (antenna beam pattern with unequal azimuth and elevation beam widths) is also required in many applications. A common application where a specially shaped fan-beam is required is in airport surveillance radar (ASR). Such an antenna pattern is known as a *cosecant-squared beam pattern* which is realisable with a Csc^2 antenna. In ship-to-ship radars, a fan-beam is required that should have sharp azimuth pattern while the elevation pattern is of no importance. Such a fan-beam can be obtained with a *pill-box antenna*.

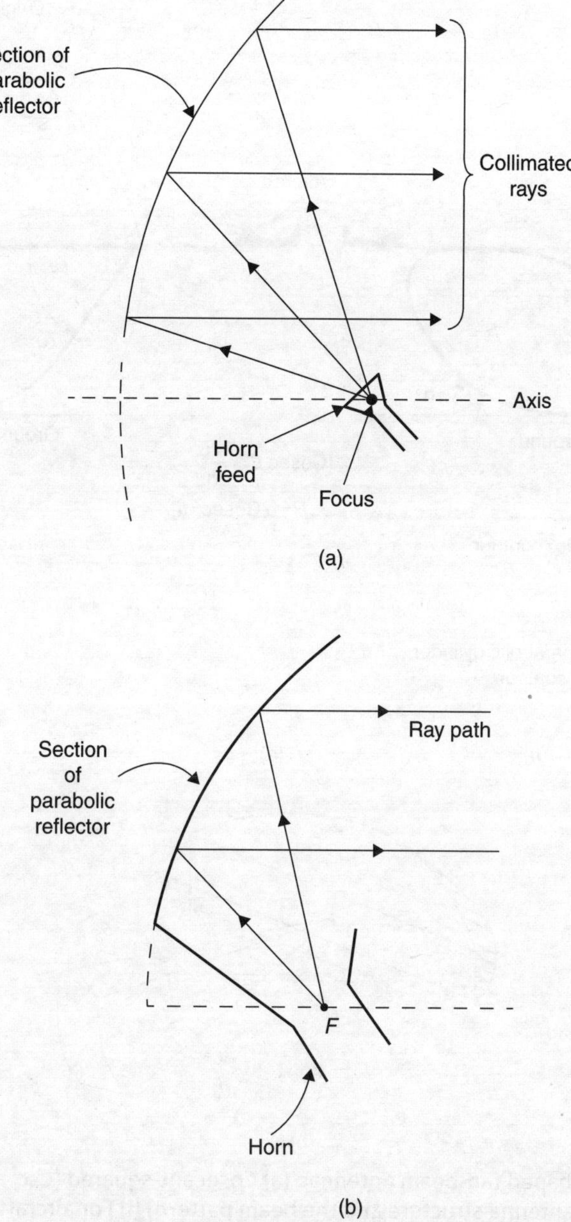

Fig. 10.5 Variants of parabolic reflector–horn combination (a) Off-set feed parabolic reflector (b) Hoghorn.

The cosecant-squared (Csc^2) antenna is a shaped parabolic cylinder antenna that provides a special type of fan-beam required either for the search of aircraft targets in the sky by the airport surveillance radar or for ground mapping by the radar installed in the aircraft (Fig. 10.6a). The power density pattern of such an antenna is proportional to cosec$^2\theta$, where θ is the elevation angle of the beam with respect to the target to be detected by

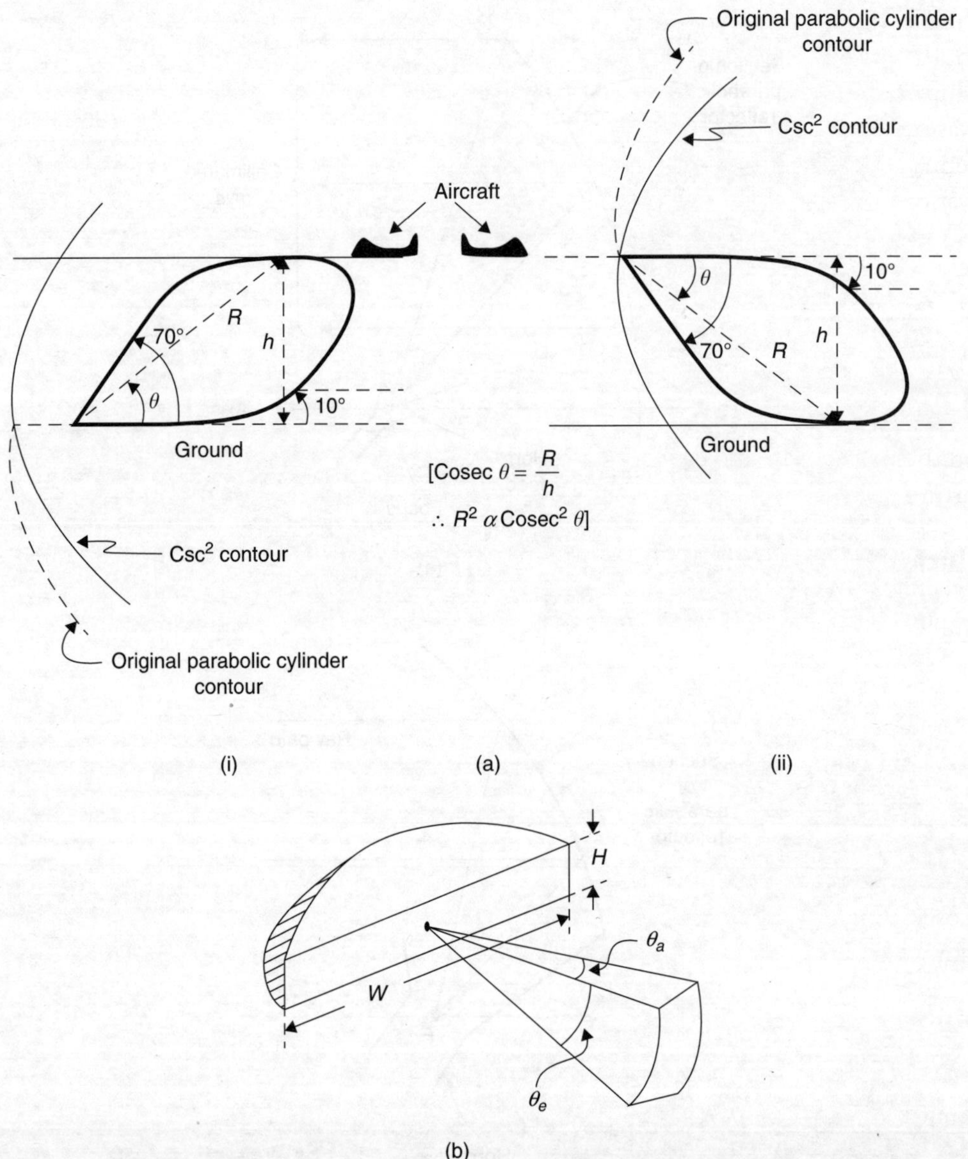

Fig. 10.6 Specially shaped fan-beam antennae (a) Cosecant-squared (Csc²) antenna (with vertical section of antenna structure and the beam pattern) (i) For aircraft surveillance (ii) For ground mapping (b) Pill-box antenna with typical fan-shaped beam with large elevation beam width θ_e (with poor elevation resolution) and small azimuth beam width θ_a (with good azimuth resolution).

the surveillance radar or the ground below the aircraft to be mapped by the radar in the aircraft. For such applications, very good azimuth resolution (to search it in 360° azimuth) is desired. Thus, the beam needs to be of very small azimuth beam width while elevation resolution is not strictly desired. In fact, an elevation resolution of only up to 30° is enough (as the height of the target will be measured with a height measuring instrument, normally a radio altimeter). Thus, the elevation beam width is considerably large (may be 5 to 10 times that of the azimuth beam width). The pill-box antenna (Fig. 10.6b) used in ship-to-ship radar also needs a very sharp azimuth pattern and practically no need for elevation resolution as both the ships are on the same sea surface. For Csc^2 antenna, pill-box antenna and also for parabolic cylinder antenna, the general formulae for gain, azimuth and elevation beam width are given by:

$$G = \eta_a \frac{4\pi WH}{\lambda^2}, \theta_a = 0.88\frac{\lambda}{W} \text{ (radian)}, \theta_e = 0.88\frac{\lambda}{H} \text{ (radian)} \quad (10.7)$$

EXAMPLE 10.2

Find the power gain of a 3 m diameter parabolic reflector antenna at 4 GHz assuming 55% aperture efficiency. What is the half power beam width of the antenna?

Solution

The power gain of the antenna is given by $P_g = \eta_a \left(\frac{\pi D}{\lambda}\right)^2 = 8685$

The half power beam width is calculated from $\varphi = \frac{70\lambda}{D} = 1.75°$

EXAMPLE 10.3

A cosecant squared antenna is 16 ft wide and 4.8 ft high. Calculate its azimuth and elevation beam width and also the directivity of the antenna. Take the operating frequency as 2.8 GHz.

Solution

Given: $W = 16$ ft $= 16 \times 30.48$ cm $= 487.68$ cm, $H = 4.8$ ft $= 4.8 \times 30.48$ cm $= 146.30$ cm. Thus, azimuth beam width $\theta_a = 0.88\frac{\lambda}{W} = 0.88 \times \frac{10.71}{487.68} = 0.0193$ radians $= 1.107°$ and elevation beam width $\theta_e = 0.88\frac{\lambda}{H} = 0.88 \times \frac{10.71}{146.30} = 0.064$ radians $= 3.69°$ [Eq. 10.7].

$$\lambda = \frac{3 \times 10^{10}}{2.8 \times 10^9} = 10.71 \text{ cm.}$$

The directivity of the antenna is given by:

$G_{D(\max)} = \frac{41,257}{\theta_E \cdot \varphi_H} = \frac{41,257}{(3.69° \times 1.107°)} = 40.04$ dBi, that is, approximately 40 dB above an isotropic antenna.

10.2.3 Slot Antenna

A slot cut into a metal plane is found to behave as a radiator being complementary to a dipole antenna (Fig. 10.7a) in terms of the well known Babinet's principle. They have similar radiated fields, however, the polarisation of the radiated signal is orthogonal to each other. For example, a horizontal slot will produce a vertically polarised radiation pattern, while a horizontal dipole will produce a horizontally polarised radiation pattern. A flat sheet with $\lambda/2$ slot radiates equally on both sides of the sheet, but unidirectional radiation

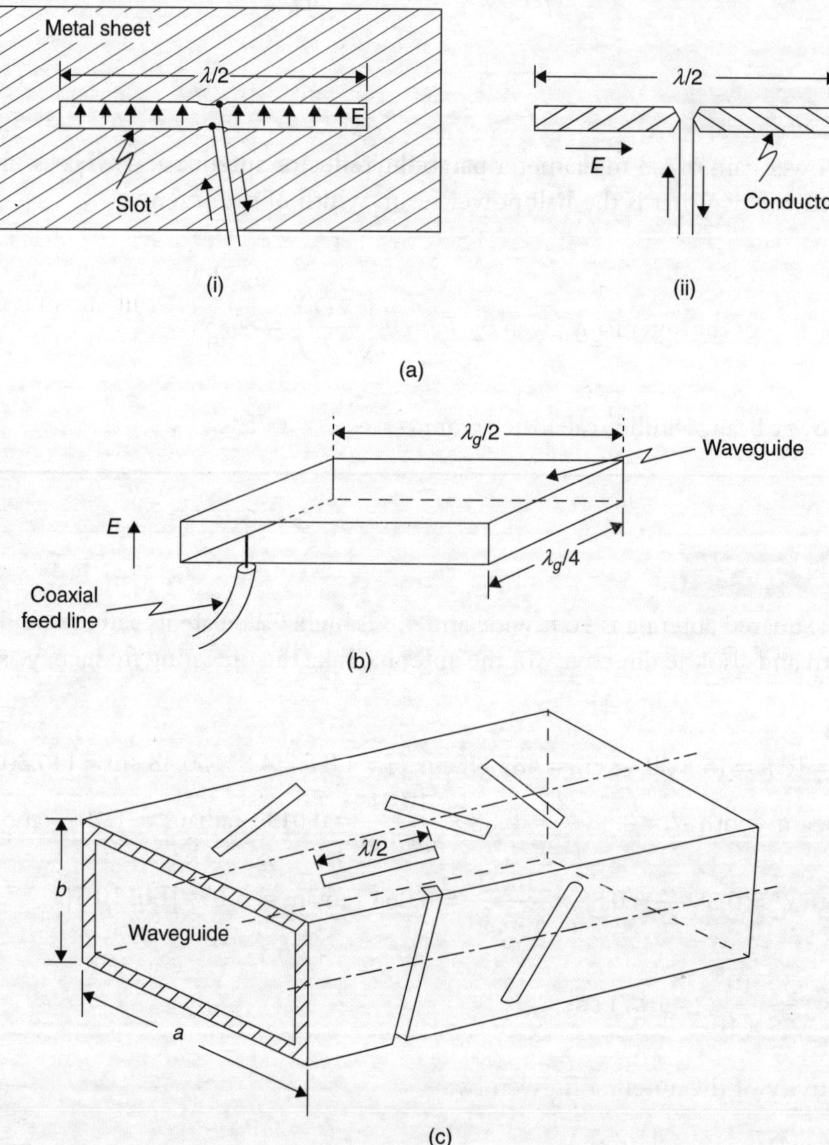

Fig. 10.7 Slot antennae (a) (i) Slot in a metal plane and (ii) Its complementary dipole (b) Cavity-backed slot antenna (c) Possible radiating slots cut into the rectangular waveguide walls.

may be obtained by terminating the slot by a cavity in the form of a parallelopiped having a depth of $\lambda_g/4$ (Fig. 10.7b). The quarter-wave section of the waveguide thus formed reflects infinite impedance to the slot, thus causing radiation to occur from only one side and also doubles the radiation resistance (the action being similar to a reflector). The radiation from only one side of a large flat sheet may in practice be realised if we cut thin slots on the walls of a waveguide, with the slot length being oriented properly with respect to the surface current lines on the waveguide wall. Slot antennae are useful in many applications, especially where flush mounting is required, for example, in high-speed aircraft. Usually such antennas are placed on the fuselage or the wing of the aeroplane. An antenna made of an array of slots on a cylinder has a special feature in that, it can produce omnidirectional radiation pattern in the horizontal plane with horizontal polarisation and is of considerable importance for television broadcasting.

A slot in a conducting sheet may be conveniently energised with a coaxial transmission line with the outer conductor of the coaxial line being bonded to the metal sheet. However, a centre-fed slot is not suitable from an impedance matching point of view as the terminal resistance at the centre of a resonant $\lambda/2$ slot in a large sheet is about 500 Ω while the characteristic impedance of a standard coaxial line is 50/75 Ω. Thus, the appropriate feed point is determined and off-centre feeding of slot is done (Fig. 10.7b). For a waveguide fed slot antenna, slots of length $\lambda/2$ are cut into the waveguide wall (with length/width of the slot >> 1, that is, the slots should be very narrow) which interrupt the flow of surface current thereby coupling the power from the waveguide's modal field into free space. For surface current distribution on the waves of rectangular waveguide, Fig. 4.11 may be referred to. Possible radiating slots cut in the walls of a rectangular waveguide are shown in Fig. 10.7c. Slots cut parallel to the surface current vector (along the central line on the broad wall of a waveguide) will be non-radiating slots (such slots are used in the design of microwave instruments such as standing wave detectors, slide screw tuners and so on). The strength of the radiation field of a slot antenna will depend on the position of the slot (which determines how the surface current lines are perturbed by the presence of the slot) and the number of slots (thus, slot array antennae are normally used to realise significant directivity). For the operation of a waveguide slot antenna, the input signal is launched into the waveguide via a coaxial probe that propagates down the waveguide and on reaching the slot emerges into external space and propagates thereafter according to the law of diffraction.

Since the slot and dipole antennae are complementary structures (as established by Booker by extending Babinet's principle in optics) it can be derived [11] that the slot and dipole impedances are related by $Z_s Z_d = \eta_0/4$; where Z_s and Z_d are respectively the slot and dipole input impedances and η_0 (= 377 Ω) is the free space impedance. Thus, the slot input impedance is given by:

$$Z_s = \frac{\eta_0^2}{4}\frac{(R_d - jX_d)}{(R_d^2 + X_d^2)} \tag{10.8}$$

where $Z_d = (R_d + jX_d)$, is the dipole input impedance. If the width of the dipole is very small, $w << \lambda$ (which will in fact be true for the complementary slot antenna too), then

$Z_d = (73 + j42.5)\ \Omega$, at the centre of a thin dipole (it may be noted that $R_d \approx R_{rad}$, since $R_{rad} >> R_{loss}$, Section 2.8.2). Thus the impedance at the centre of the slot (corresponding to the complementary dipole) calculated from Eq. 10.8 is $Z_s = (363 - j211)\ \Omega$. It may be noted that an inductive dipole will correspond to a capacitive slot and vice versa. A dipole which is greater than a half-wave length is known to be inductive (being an open-circuited transmission line with a length of more than $\lambda/2$, Fig. 3.6a) and hence a slot longer than a half-wave length is capacitive (being complementary to the dipole); similarly a slot shorter than a half-wavelength is inductive in nature.

10.2.4 Lens Antenna

The lens antenna resembles a conventional optical lens that has a source of radiation (the object) and the lens itself. The radiator may be a weakly directional antenna acting as a point source being placed at the focal point of the lens. Frequently the lenses are placed in the aperture of a conical horn (Fig. 10.8a), in which case the aperture of the horn should be large and the length small. This ensures that practically all the energy radiated by the horn is directed towards the lens and the radiator acts as a point source *S* located at the centre of the spherical wave and coinciding with the focus of the lens in this situation. The diverging spherical wave fronts from the horn would be collimated by the lens forming a plane wave front. The collimation (the formation of the parallel rays) results due to the refraction mechanism of the lens.

Lenses and reflectors are both used as collimating elements in microwave antennae. However, it may be noted that reflectors have only one degree of freedom—the reflector surface. It has no internal loss and chromatic aberration, is relatively easy to support and can be perforated (can be a grid structure) to reduce weight and wind loads. Lenses have, in general, more than one degree of freedom—the surface of the lens and index of refraction. Though lens antennae have no aperture blockage by the feed, they have internal and surface reflection losses and must be edge supported and are relatively heavy and bulky.

Apart from being used in conjunction with a horn antenna, the lens antenna may be used as an antenna in its own right, especially at millimeter-wave frequencies where low-loss dielectric lens antennae can be made competitive in weight and performance when compared with reflector antennae. In a *dielectric lens antenna* or *delay lens antenna*, the electrical path length of the electromagnetic signal is increased by the lens medium (made either of polystyrene or lucite with $\varepsilon_r = 2.56$ and $n = 1.6$) and thus the wave is retarded by the lens (Fig. 10.8b). To reduce the bulk/weight of the dielectric lens antenna, *zoning* is used [Fig. 10.8c (i) and (ii)]. Here the lens is divided into several circular zones such that the electrical path length between adjacent zones differs by an integral multiple of wavelength (the phase shift along the ray FAB and along FCD will differ by $2\pi n$ radians) so that the phase front is not disturbed. However, this produces ring-like regions with a perturbed phase front falling in the transition region between zones leading to aperture blockage. An alternative way of zoning of a dielectric lens antenna is where the dielectric material is removed from the refracting surface [Fig. 10.8c (ii)]. A further problem of the dielectric lens antenna is caused due to the reflection at the dielectric–air interface. To reduce this reflection, a

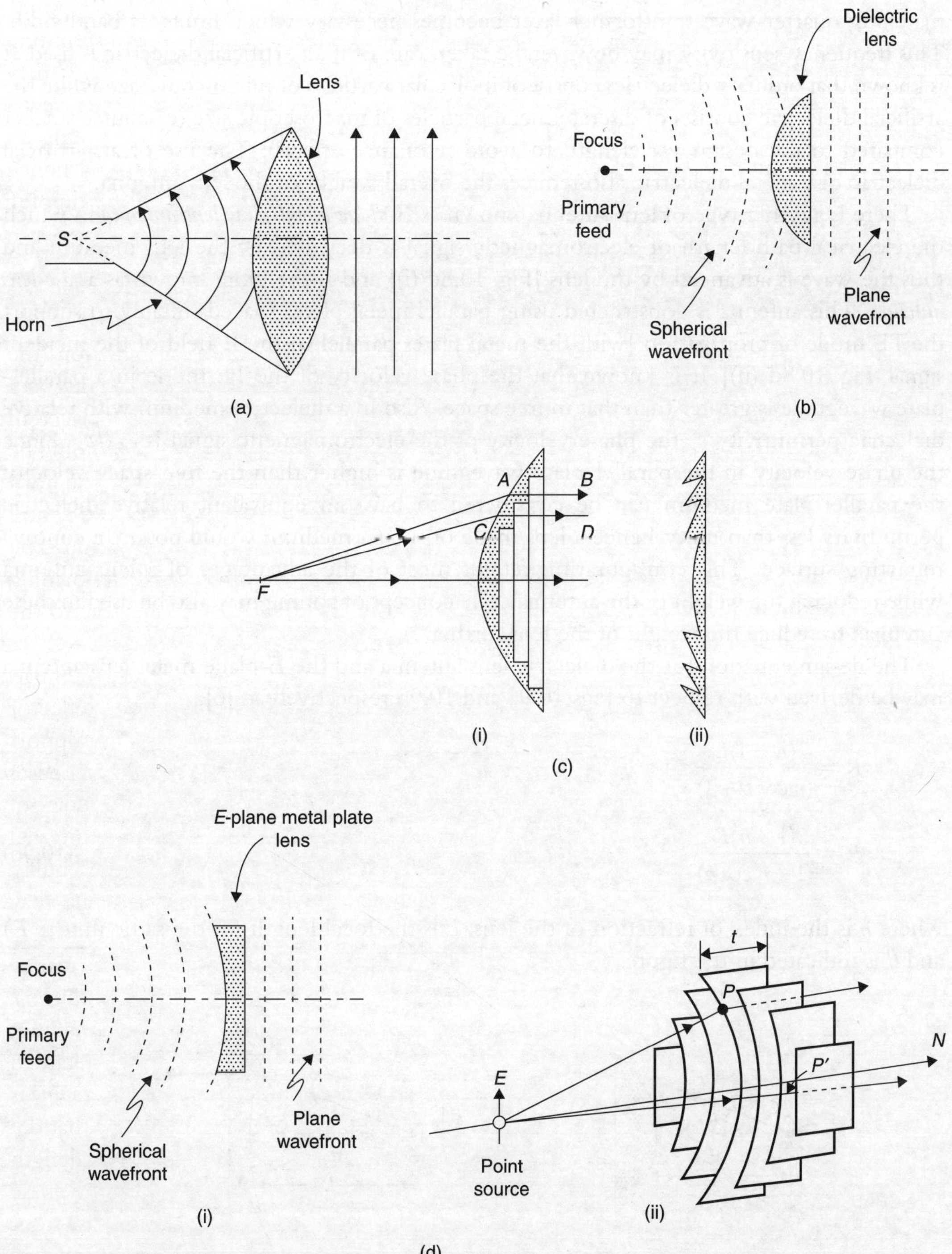

Fig. 10.8 Lens antennae (a) Horn-lens antenna (b) Dielectric lens (delay lens) antenna (c) Two methods of zoning in dielectric lens antennae (i) Zoning the non-refracting surface (ii) Zoning the refracting surface (d) E-plane metal lens antenna (i) Schematic (ii) Practical construction.

matching quarter-wave transformer layer becomes necessary which limits its bandwidth. This frequency sensitivity may, however, be taken care of if an artificial dielectric is used. It is known that ordinary dielectrics consist of molecular particles of microscopic size while the artificial dielectric consists of discrete metal particles of macroscopic size (of course smaller compared to the design wavelength to avoid resonance effects). The use of an artificial dielectric or a metal dielectric also reduces the overall weight of the lens antenna.

There is another type of lens antenna known as *E-plane metal-plate lens antenna* in which the electrical path length of electromagnetic signal is decreased by the lens medium and thus the wave is advanced by the lens [Fig. 10.8d (i)] and is therefore known as a *fast-lens antenna*. This antenna is constructed using parallel metal plates spaced suitably to support the TE mode of propagation [with the metal plates parallel to the E-field of the incident signal, Fig. 10.8d (ii)]. It is known that the phase velocity of the TE mode in a parallel-plate waveguide is greater than that in free space. Also in a dielectric medium with relative dielectric permittivity ε_r, the phase velocity of the electromagnetic signal is $c/\sqrt{\varepsilon_r}$. Since the phase velocity in the parallel-plate waveguide is higher than the free-space velocity, the parallel plate medium can be considered to have an equivalent relative dielectric permittivity less than unity, hence a lens made of such a medium would possess a concave refracting surface. This configuration retains most of the advantages of a lens antenna while reducing the weight of the antenna. The concept of zoning may also be used in these antennae to reduce the weight of the lens further.

The design equation for the dielectric lens antenna and the *E*-plane metal lens antenna may be derived with respect to Figs 10.9a and 10.9b respectively as [6]:

$$R = \frac{(n-1)L}{(n\cos\theta - 1)} \tag{10.9a}$$

$$R = \frac{(1-n)L}{(1-n\cos\theta)} \tag{10.9b}$$

where n is the index of refraction of the lens, L is the focal length (in the same unit as R) and θ is indicated in the figure.

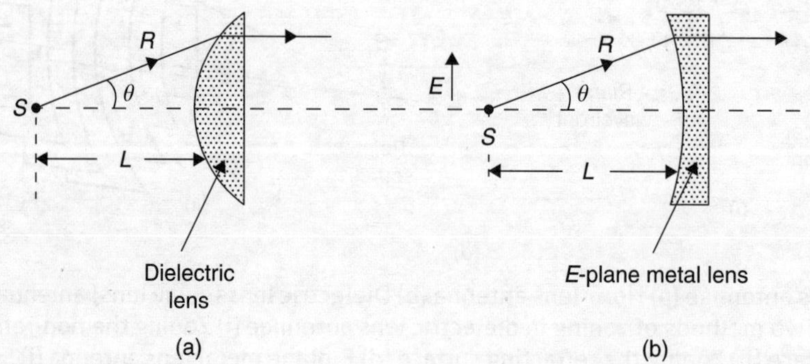

Fig. 10.9 Dielectric lens and *E*-plane metal lens antennae with geometric parameters indicated.

10.2.5 Surface Wave (Dielectric Rod) Antenna

At millimeter-wave frequencies, dielectric rod type surface wave antennae provide significant performance advantages and are a low cost alternative to conventional antenna designs such as Yagi–Uda and horn antennae, which are often more difficult to manufacture at these frequencies. Not surprisingly, the dielectric rod antenna is also nature's favourite choice when it comes to nanometer-wave applications—the retina of the human eye is an array of more than 100 million dielectric (rod or cone type) antennae.

Surface waves are unwanted byproducts in many antenna designs that degrade the performance of the antenna. However, if surface waves are properly controlled, they can be utilised to build promising antennae—as is the case with dielectric rods and other types of surface wave antennae. The dielectric rod type surface wave antenna has three parts—a dielectric rod *R* of length *l* (typically $6\lambda_0$, where λ_0 is the free-space wavelength), a metal base *B* (fixed on one of the ends of the dielectric rod and forming a section of waveguide filled with the dielectric), and the primary source of electromagnetic radiation *S* (fixed in the part of the rod shielded by the base, Fig. 10.10a). With waveguide excitation, the TE wave is initially excited in a hollow waveguide via a coaxial line which arrives at the base *B* filled with the dielectric. In the absence of the dielectric rod, the wave excited in the base would emerge from it similar to that from an open ended waveguide according to the laws of diffraction. The presence of the dielectric rod constrains the wave excited in the base to propagate further along the dielectric. This propagation is by surface wave which occurs in the boundary separating the dielectric rod and the external medium (air) and is a kind

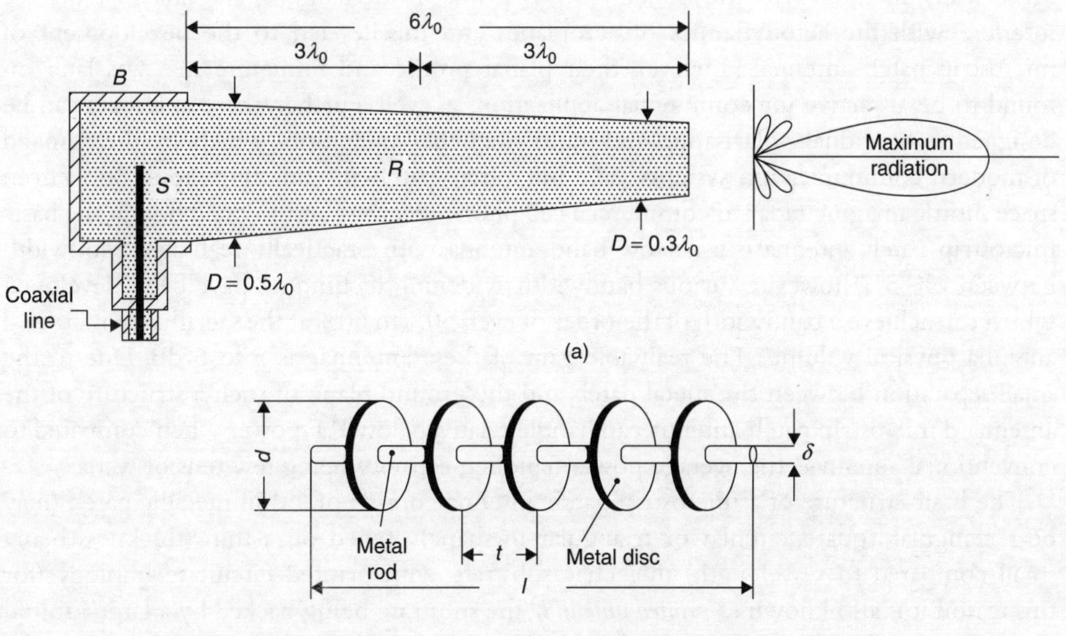

Fig. 10.10 Surface-wave antenna (a) Dielectric rod type (b) Metal rod and metal disc (artificial dielectric) type.

of wave that propagates parallel to the interface and decays vertically to it. Since a surface wave radiates only at discontinuities, the total pattern of this antenna (normally end-fire) is formed by the interference between the feed and terminal patterns. This type of antenna is also known as *polyrod antenna* as the dielectric used for the rod is made of polystyrene. The ratio of power guided outside the rod (radiated power as surface wave) to the power guided along the rod is both functions of the rod diameter (D) and the dielectric constant (ε_r). In such an antenna, the choice of polystyrene ($\varepsilon_r = 2.56$) and $D \leq \lambda/2$ results in good radiation efficiency of the antenna. Using $D = 0.5\lambda$ at the butt to 0.3λ at the end, that is with a tapered dielectric rod, reduced minor lobes can be realised.

The dielectric material instead of being a single rod of natural dielectric material (as used in dielectric rod type surface wave antenna) could alternatively be an artificial one. For example, a metal rod with metal discs fixed to it (Fig. 10.10b), which is another type of surface wave antenna.

Relatively infrequent use of surface wave antennae in the past (though this type of antenna concept was developed in the 1950s) was due to lack of adequate design and analysis tools. However, in recent times the availability of advanced simulation tools like FDTD (Finite Difference Time Domain) to solve electromagnetic field problems with arbitrary shaped media has made possible the design of such antennae to any blue print demanded by the designer.

10.3 Microstrip Patch Antenna

Historically, in the 1970s the military needs primarily demanding antenna that do not interfere with the aerodynamics of aeroplanes and missiles led to the development of microstrip patch antenna. However, their planar profile and miniature size was later on found to be attractive for commercial applications as well. Further, these antennae can be designed to have dual-polarisation and multi-frequency operations which are the demand of modern communication systems. The microstrip patch antenna finds application from space shuttle imaging radar to commercial cell phone receivers and so on. Though the basic microstrip patch antenna is a narrow band antenna with practically realisable bandwidth between 2%–5%, however, various bandwidth-widening techniques have been developed which can achieve a bandwidth of the order of even 50% to 60% at the sacrifice of increased antenna physical volume. The realisable gain of these antennae is 5 to 6 dB. Due to the small separation between the metal patch and the ground plane of such a structure of the antenna, a microstrip patch antenna can handle relatively low RF power when compared to conventional antennae; the average power handling capacity being few tens of watts.

The basic structure of a microstrip patch antenna consists of a thin metallic patch (may be rectangular, square, circular or triangular in shape) etched on a thin (thickness being small compared to wavelength) dielectric substrate with printed circuit technology (for this reason it is also known as *printed antenna*), the substrate being backed by a large ground plane (Fig. 10.11a). The dielectric substrate is preferably of low dielectric constant may be 2 to 3 (commonly used one is PTFE, polytetrafluoroethylene) for good radiation efficiency (as lower the dielectric constant, less is the confinement of electromagnetic signal in the substrate dielectric material allowing larger fringing field and higher radiation capability).

The operation of a microstrip patch antenna may be understood with reference to Fig. 10.11a, which shows a rectangular patch of length L and width W printed on a dielectric substrate of thickness/height h (which is basically a truncated microstrip of length L and width W). The region between the patch and the ground plane may be treated as a cavity that is surrounded by magnetic walls around the periphery (open boundary) and by electric walls at the top and bottom sides (short circuit boundary); as was discussed in Section 4.5. Since thin substrates are used, the field inside the cavity is uniform along the thickness/height of the substrate. The length L of the patch is chosen to be approximately half guide-wavelength ($\lambda_{geff}/2$, λ_{geff} being the effective guide wavelength of the microstrip line whose truncated version is the patch antenna) and the cavity mode is of TM type (unlike the quasi-TEM in microstrip line). With the fundamental TM_{10} mode in the open boundary cavity (for a rectangular patch), the electric field varies one $\lambda_{geff}/2$ along the length (L), and there is no variation along the width (W) of the patch, Fig. 10.11b. The field is seen to be ideally zero near the patch centre $x = 0$, gradually increasing on either side of the centre (with phase opposition) and becomes maximum near the patch edges $x = \pm L/2$ [Fig. 10.11b(i)], causing a pair-wise cancellation of the electric field on either side of the patch centre along L. Thus the lengthwise edges 3 and 4 act as non-radiating edges. The fringing electric field across the truncated edges W of the microstrip has a non-vanishing electric field E_x [Fig. 10.11b (ii)] which are in phase, making W the radiating edge (edges 1 and 2). The radiated field (caused by the fringing fields along the edge W indicated by radiating slots in the figure) finally comes out from the open microstrip resonator in the broadside direction z, which is the direction of the Poynting vector for the corresponding electric and magnetic fields along E_x and H_y respectively. Excitation of other modes such as, $TM_{01,}$ TM_{20} and so on, would lead to different radiation patterns. It may be noted that L is chosen in terms of the resonator length ($\lambda_{geff}/2$) while choice of a larger value of W would increase the radiated field and also the directivity of the antenna. However, too large a value of W might excite higher order modes. Since the operation of such an antenna depends on the resonator length L being approximately $\lambda_{geff}/2$, it is a narrow band antenna.

For an efficient transfer of power from the transmission line feeding input power to the patch antenna, the input impedance of the patch and the characteristic impedance of the transmission line have to be matched. It is observed that the impedance seen by the transmission line when attached to the radiating edge is very high [as voltage is finite and current is zero, Fig. 10.11b (ii)] and also the impedance decreases as one moves towards the centre of the patch. Therefore, depending on the characteristic impedance of the transmission line, an appropriate point on the patch has to be chosen as the feed point.

A microstrip patch antenna can be fed in a number of ways. The coaxial probe is by far the most common feeding method (Fig. 10.12a). The coaxial probe usually has a characteristic impedance of 50 Ω, thus the location of the probe should be at the 50 Ω point (which normally demands off-set feeding) of the patch to ensure impedance matching for optimum antenna performance. In practice, the centre conductor of the coaxial line is

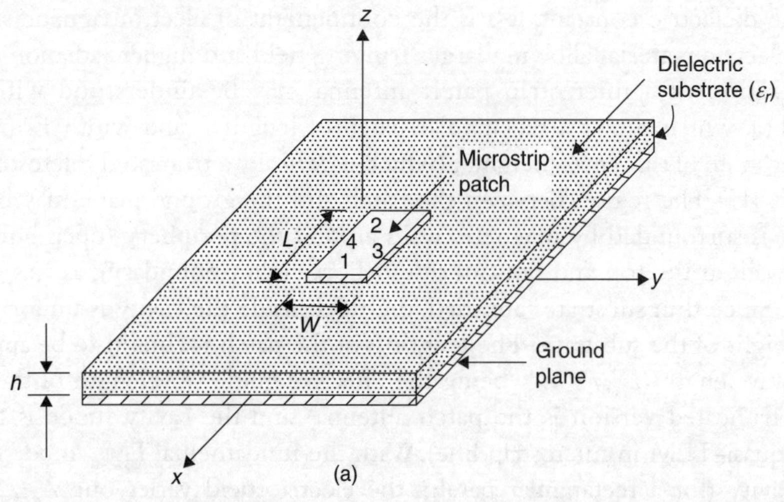

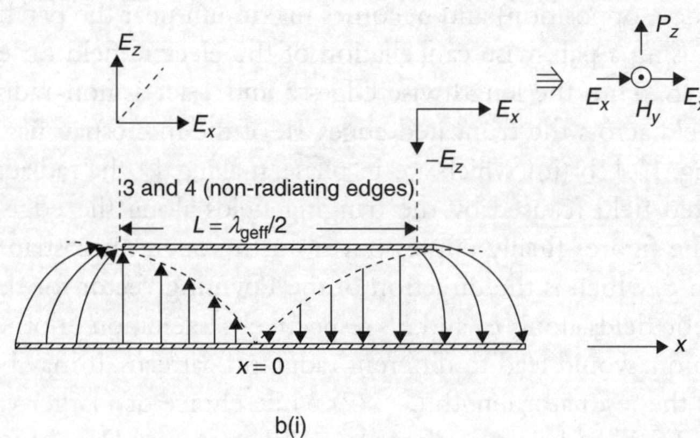

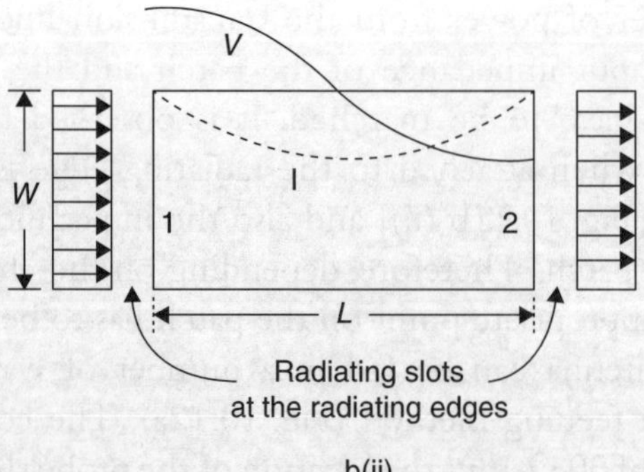

Fig. 10.11 Microstrip patch antenna (a) Basic structure of the antenna showing radiating and non-radiating edges (b) (i) Electric field variation along non-radiating edges (ii) Fringing electric field across the radiating edges.

soldered to the patch which needs a hole to be drilled through the substrate material and the connector protrudes outside the bottom ground plane, so the true planar nature of the antenna is not maintained. A patch excited by a microstrip line feed is shown in Fig. 10.12b. This feed arrangement has the advantage that it can be etched on the same substrate, so the total structure remains planar. However, the microstrip line needs to be matched with the patch impedance with a quarter wave impedance transformer. Further, radiation from the feed line leads to an increase of cross-polar level; and in the millimeter-wave range the size of the feed line becomes comparable to the patch size causing increased undesired radiation. Other feeding methods like an aperture-coupled feed and a proximity feed are also used depending on the need of the application [26].

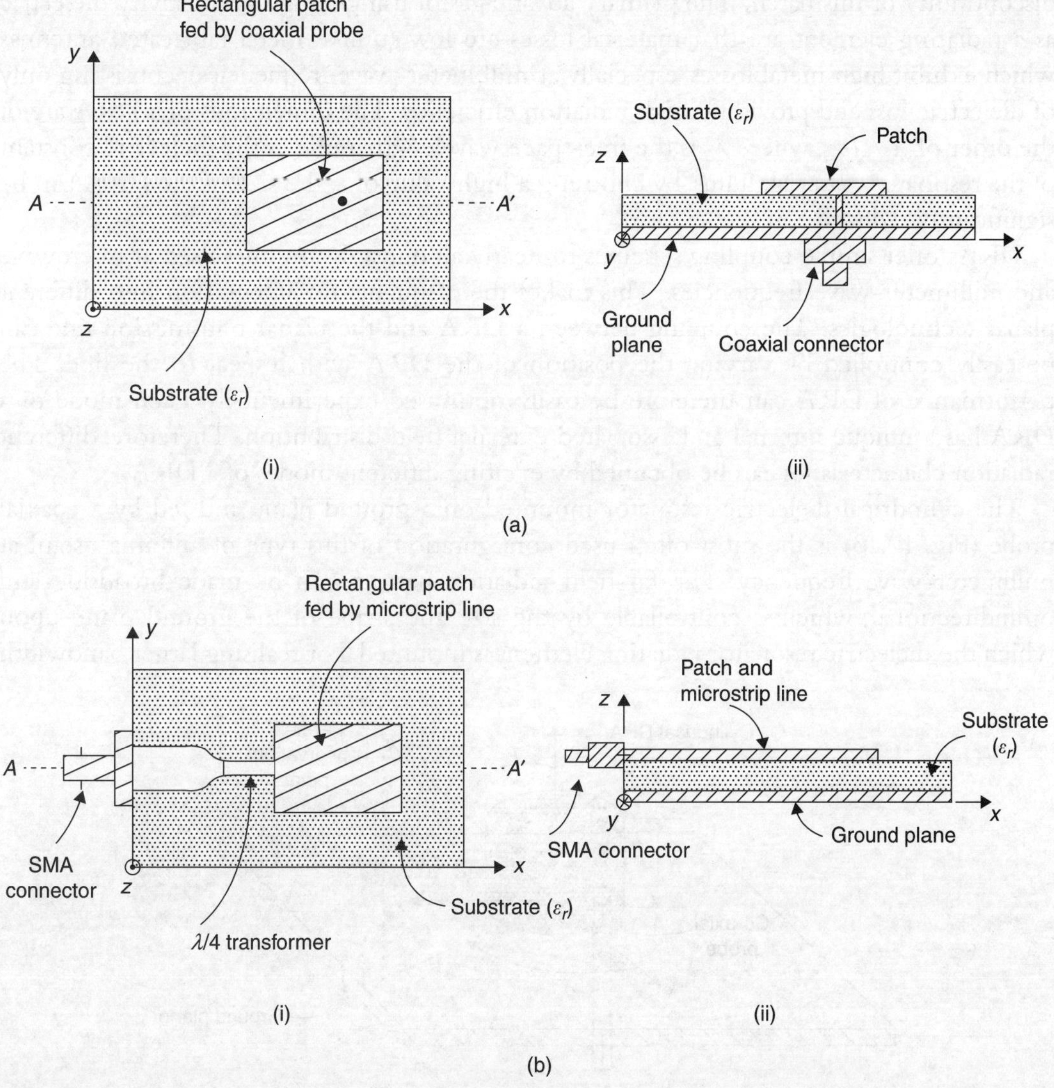

Fig. 10.12 Feeding of microstrip patch antenna (a) Coaxial feed (b) Microstrip feed.

10.4 Dielectric Resonator Antenna

In recent years, the dielectric resonator antenna (DRA) has emerged as a new and viable alternative to conventional low-gain elements such as dipoles, monopoles and microstrip patches. A DRA consists of a block of ceramic material (which may be of various shapes—hemispherical, cylindrical and rectangular; preferably cylindrical), performing as a dielectric resonator, being backed by a metallic ground plane (Fig. 10.13). The incident electromagnetic signal fed into the dielectric resonator of the DRA bounces back and forth between the resonator walls (top and bottom ones), forming standing waves. The vertical walls of the resonator allow radiation to take place as the dielectric material of the resonator (typically with a dielectric constant of 40 or so) and the air (with dielectric constant nearly unity) surrounding it has a steep gradient of dielectric constant leading to discontinuity or mismatch. The primary advantages of using a high permittivity dielectric as a radiating element are that material losses are low (unlike metal fabricated antennas which exhibit high metal losses, especially at millimeter-wave frequencies), consisting only of dielectric loss and providing high radiation efficiency. The dimensions of a DRA are of the order of $\lambda_0/\sqrt{\varepsilon_r}$, where λ_0 is the free-space wavelength and ε_r is the dielectric constant of the resonator material. Thus, by choosing a high value of ε_r the size of the DRA can be significantly reduced.

DRAs offer simple coupling schemes to nearly all transmission lines used at microwave and millimeter-wave frequencies. This makes them suitable for integration into different planar technologies. The coupling between a DRA and the planar transmission line can be easily controlled by varying the position of the DRA with respect to the line. The performance of DRA can therefore be easily optimised experimentally. Each mode of a DRA has a unique internal and associated external field distribution. Therefore, different radiation characteristics can be obtained by exciting different modes of a DRA.

The cylindrical dielectric resonator mounted on a ground plane and fed by a coaxial probe (Fig. 10.13) is the most often used configuration of this type of antenna useful at millimeter-wave frequency. The far-field radiation pattern can be made broadside and omnidirectional, which is controllable by the size and shape of the ground plane upon which the dielectric resonator radiating element is mounted. For realising larger bandwidth

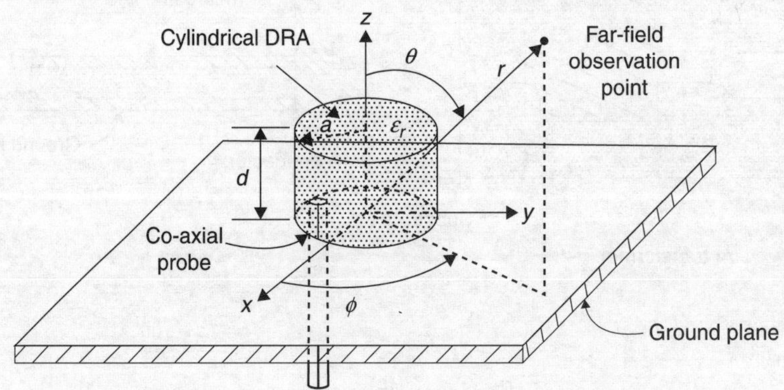

Fig. 10.13 Geometry of a cylindrical DRA with a coaxial feed arrangement.

of the antenna, a lower value of ε_r is desirable. The TM_{110} mode is used in a cylindrical dielectric resonator antenna, as this mode is the fundamental mode of a cylindrical resonator radiating element having lowest resonant frequency, given by:

$$f_{TM_{110}} = \frac{1}{2\pi a\sqrt{\mu_0 \varepsilon_r}}\sqrt{(1.841)^2 + \left(\frac{\pi a}{2d}\right)^2} \tag{10.10}$$

where a is the radius of the cylinder and d its height.

10.5 Principle of Microwave Signal Propagation

Microwave signal propagation from a transmitting antenna to the receiving antenna can take place via a *space wave* (generally along the *line of sight*) which is limited to a maximum coverage distance of around 100 km due to the curvature of the earth. However, satellite microwave communication using a transponder satellite placed in a geostationary orbit for global coverage is not limited by the earth's curvature. For terrestrial microwave communication, atmospheric conditions (temperature, pressure and humidity) in some geographical terrains makes *beyond-the-horizon communication* possible (covering even 1000 km) with atmospheric duct propagation and troposcatter propagation.

10.5.1 Line-of-Sight (LOS) Propagation

At radio frequencies above about 30 MHz, the ionosphere is not able to cause multiple refraction and hence ultimate reflection of the electromagnetic energy any further in the form of a *sky wave* (which is used for short-wave radio broadcast using HF range of radiowave spectrum) back to the earth, while the *ground wave* or *surface wave* (which is used for medium wave radio broadcast using a MF range of radiowave spectrum) signal attenuates to a negligible amplitude in a kilometer or so. Useful radiowave signal propagation can, however, be achieved at VHF, UHF and microwave frequencies by means of the *space wave* travelling along the Line-of-Sight (LOS) between the elevated transmitting and receiving antennas in normal atmospheric conditions.

When the distance between the transmitting and receiving antennae is considerably greater than the heights of the antennae (even for elevated ones), the received signal at the receiver antenna is due to the direct ray from the transmitter and the ground reflected ray (Fig. 10.14a); both might have equal amplitude but, in general, differing in phase. The field strength due to the space wave at the receiving antenna, assuming flat earth, is given by:

$$E = \frac{4\pi h_t' h_r'}{\lambda d} E_0 \tag{10.11}$$

where E_0 is the field strength due to a direct ray at unit distance and depends on the transmitter power and the directivity of the transmitting antenna, heights of transmitting and receiving antennae placed at a distance d apart. h_t' and h_r' are the effective heights of the transmitting and receiving antennae from the supposed flat earth.

The LOS propagation is limited by the curvature of the earth, that is, the direct waves following the optical horizon get blocked by the curvature of the earth to some limited range. However, the varying density of the atmosphere and diffraction around the curvature of the earth allows a horizon to the space wave called the *radio horizon* (Fig. 10.14b) which is about four-thirds (4/3) of the optical horizon. The radio horizon of an antenna is given, with good approximation, by the empirical formula $d_t = 4\sqrt{h_t}$, where d_t and h_t are respectively the distance from the transmitting antenna (in km) and the height (in m) of the transmitting antenna above the ground. The same formula naturally applies to the receiving antenna. Thus, the total distance from the transmitting to receiving sites in terms of the radio horizon will be given by:

$$d = d_t + d_r = 4\sqrt{h_t} + 4\sqrt{h_r} \tag{10.12}$$

A simple calculation shows that for a transmitting antenna at a height of 225 m above ground level, the radio horizon is 60 km. If the receiving antenna is 16 m above ground

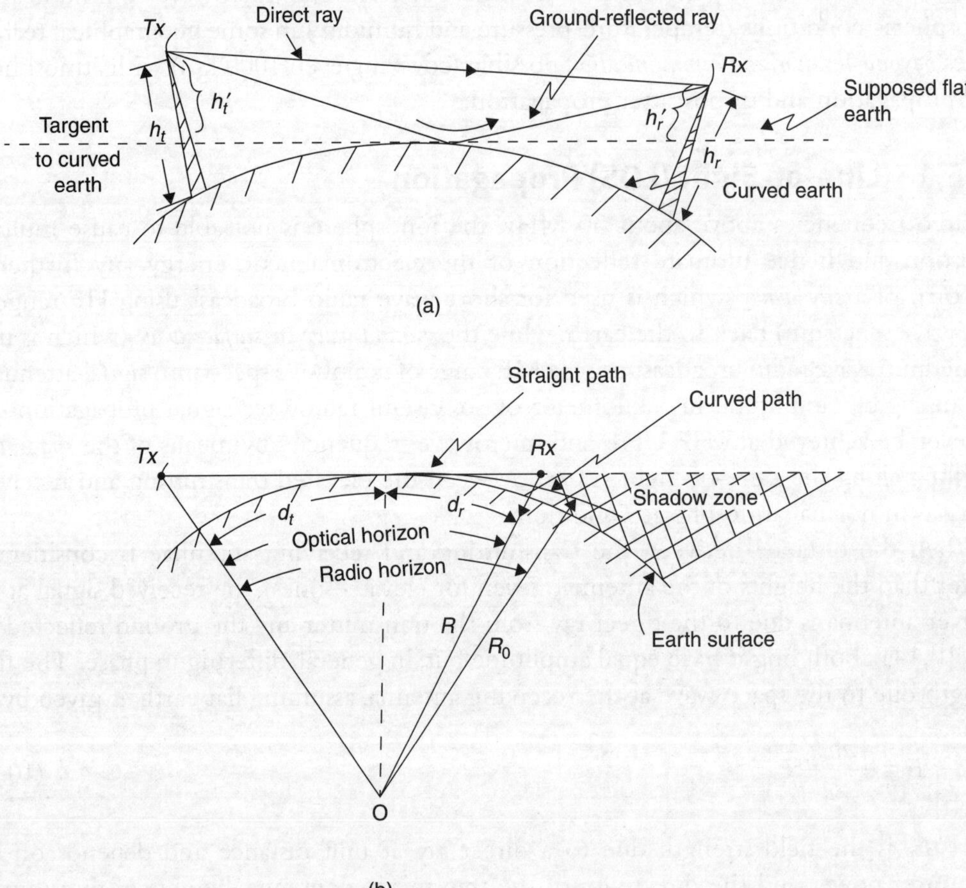

Fig. 10.14 (a) LOS propagation of space wave (b) Optical horizon and radio horizon in LOS propagation of radiowaves.

level, the total distance is increased to 76 km. A greater distance between transmitter and receiver may be obtained by locating them on hill tops, but links longer than 100 km are hardly ever used in commercial communications.

10.5.2 Atmospheric Duct Propagation

The effects of the atmosphere on space wave propagation are of significant importance. The earth's atmosphere extends all the way up to a height of 600 km. The atmosphere is divided into several layers—troposphere, stratosphere, mesosphere and ionosphere. The propagation of a space wave is mostly affected by the atmospheric conditions in the troposphere that extends from the surface of the earth to a height of 6 km at the poles and 18 km at the equator. Typical value of refractive index of air is $n = 1.000289$, which is a function of the atmospheric conditions and its value decreases as a function of height (h) above the surface of the earth. A quantity, *refractivity* $N = (n - 1) \times 10^6$, is defined to characterise the wave propagation mode in standard and anomalous atmospheric conditions in terms of the refractivity gradient dN/dh. At the surface of the earth, the refractivity is equal to 289. For the standard atmosphere, the refractivity falls off linearly up to a height of 2 km above the surface of the earth and is expressed by the relation $N = 289 - 39h$, where h is in km. A plot of N vs h (Fig. 10.15a) depicts the regions of sub-refraction, super-refraction and ducting of radiowave signals for different dN/dh values depending on the atmospheric conditions. If the slope of the refractivity is less than that of the standard atmosphere, $dN/dh < -39/\text{km}$, the wave is refracted more than that in a standard atmosphere and is known as *super-refraction*. In such cases, the radius of curvature of the radiowave path can be calculated (for low grazing angle) in terms of the refractivity gradient with an approximate formula $r = 10^6/(-dN/dh)$. For $dN/dh = -157/\text{km}$, $r = 6370$ km (approximately equal to the radius of the earth), and thus a horizontally transmitted wave would travel parallel to the surface of the curved earth. If $dN/dh < -157/\text{km}$, the radius of curvature of the ray is smaller than the radius of the earth and the radiowave then touches the surface of the earth and gets reflected from the earth's surface giving rise to *atmospheric duct propagation* (Fig. 10.15b).

Under certain atmospheric conditions called the *temperature inversion*, more likely in the sub-tropical region than the temperate zone, microwave signals can propagate by *duct propagation* to a long distance reaching even to 1000 km, caused by super-refraction in the atmospheric duct.

Temperature inversion may be caused when a layer of warm air is trapped above cooler air, such as air over the surface of water in the sea or ocean. This is an abnormal atmospheric condition because, in the standard atmosphere the air temperature decreases with height. The refractive index will then decrease far more rapidly with height than in the standard atmosphere. The rapid reduction of refractive index will do to microwave what the slower reduction of refractive index of standard atmosphere, of ionised layer, does to HF waves; complete bending down thus takes place. Microwave signals are then continuously refracted in the duct and reflected by the ground (Fig.10.15b), propagating around the curvature of the earth making beyond-the-horizon microwave signal propagation possible to a long distance over sea or ocean.

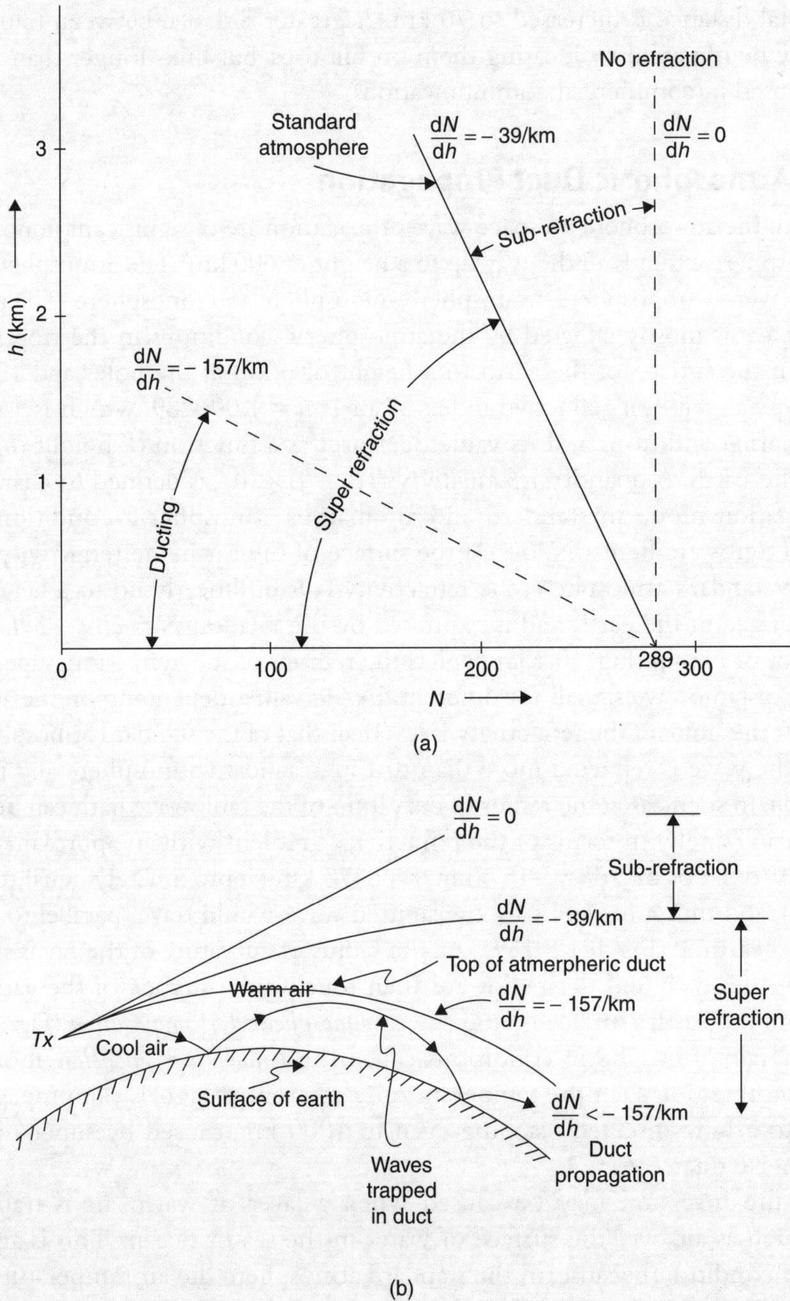

Fig. 10.15 (a) Plot of N vs h depicting the regions of sub-refraction, super-refraction and ducting of radiowave signal for different dN/dh values (b) Atmospheric duct propagation.

10.5.3 Tropsperic Propagation

The curvature of the earth limits the maximum distance (to the extent permitted by the radio horizon) of signal propagation from the transmitting antenna to the receiving antenna by the LOS propagation mode. However, tropospheric scattering phenomena can be used to establish a communication link over a long distance much beyond the radio horizon. The troposphere can scatter electromagnetic signals due to its inhomogeneous nature attributed to the refractive index changes and turbulence caused due to sudden changes in temperature or humidity or due to the presence of dust particles in the tropospheric part of the atmosphere.

Scattering of energy in a radiowave occurs in the troposphere for a wide range of frequencies centred around 900, 2000 and 5000 MHz. The amount of scattered power depends on the magnitude of the variations in the refractive index and on the distance over which these variations occur. Transmission over a troposcatter link covering a distance of the order of 750 km or more between transmitter and receiver is possible.

In troposcatter link, the transmitter and receiver antenna beams share a common volume of the atmosphere, called the scatter volume (Fig. 10.16), which is just few kilometers from the surface of the earth. The scattering volume has several regions of irregularities which have different shapes and sizes, and positions of which are changing randomly. Thus, the signal at the receiver is the superposition of scattered signals from these irregularities, creating a multi-path environment. That is, there exists more than one path through which the signals from the transmitter can reach the receiver with different transmission delays. The signals received in a torposcatter link thus undergo *fading* (received signal strength changes as a function of time).

The troposcatter link is often used to provide long distance telephone and other communication over rough or inaccessible terrain where microwave link or coaxial cable is difficult or impossible to use.

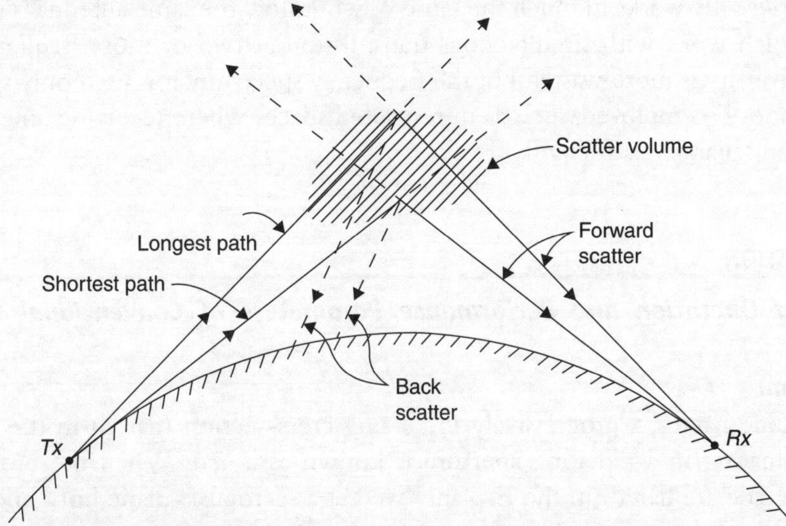

Fig. 10.16 Troposcatter propagation.

10.5.4 Fading and its Mitigation

Fading is the fluctuation of radio signal strength observed at the input of a receiver. This is caused by the interference of two or more signals which left the source at the same time but arrive at the destination by different paths and hence at different time phases. Thus, at any given instant of time, the total signal at the receiver is the vector sum of the separate components, which have random relative phases. The receiver thus sees a weighted sum of delayed replicas of the transmitted signal from these multi-paths, interfering with each other constructively or destructively, depending on the phase of the replica signals. This may develop due to meteorological variations resulting in anomalous propagation conditions, causing severe degradation in the performance of the radio system. Fading is more likely with smaller wavelengths (at higher frequencies), because greater phase shift will be produced at shorter wavelengths.

The mitigation for the fading of signal at microwave receivers, wherever it occurs, can be done with *diversity reception*. There are two forms—space diversity and frequency diversity.

AGC (Automatic Gain Control, used in the IF section) in a RF/microwave receiver helps greatly to minimise some effects of fading, however, it cannot help when the signal fades into the noise level. Diversity reception system makes use of the fact that although fading may be severe at some instant of time, some frequency, and at some point on the earth, it is extremely unlikely that signals at different points or different frequencies will fade simultaneously.

In *space diversity*, two or more receiving antennae are employed, separated by nine or more wavelengths. There are as many receivers as antennas, and arrangements are made to ensure that the AGC from the receiver with strongest signal at the moment cuts off the other receivers. Thus, only the signal from the strongest receiver is passed to the common output stage.

Frequency diversity works in much the same way, but now the same antenna is used for the receivers, which work with simultaneous transmission at two or more frequencies. Since frequency diversity is more wasteful of the frequency spectrum, it is used only where space diversity cannot be employed, such as in restricted spaces where receiving antennae could not have been separated sufficiently.

Recapitulation

Principle of Operation and Performance Parameters of Conventional Microwave Antennae

Horn antenna

- A waveguide having a progressively increasing cross-section (flared) in the E/H plane and terminating in a radiating aperture is known as a *horn*. When the broad walls of the waveguide are flared (in the E-plane) we get a sectoral E-plane horn and when the narrow walls of the waveguide walls are flared (in the H-plane) we get a sectoral H-plane horn while a combination of the two produces a pyramidal horn. A conical horn results from the flaring of the mouth of a circular waveguide.
- Half-power beam widths (HPBW) of pyramidal horns are $\theta_E = 52\ \lambda_o/a_E$, $\phi_H = 63\ \lambda_o/a_H$. The directivity $[G_{D(\max)}]$ and power gain $[G = \eta_a G_{D(\max)}]$ of pyramidal horn are given by $G_{D(\max)} = 41{,}257/\theta_E \cdot \phi_H$ and $G = 28{,}879/\theta_E \cdot \phi_H$ respectively, where $\eta_a \approx 0.7$.

- The horn antenna, both sectoral and pyramidal, is commonly used as primary feed for a parabolic reflector and its variants. The conical horn is also used in conjunction with a lens antenna at millimeter-wave frequencies to realise practically approximated radiating point source.
- Sometimes the inner walls of the horn antenna are corrugated which improves the edge diffraction and pattern symmetry and also reduces cross polarisation.

Reflector antenna

- Most commonly used reflector antennae are the parabolic reflectors (paraboloid of revolution and parabolic cylinder) which act as secondary reflectors having a horn or dipole antenna as the primary feed. In an array antenna (Yagi–Uda type), reflector rods are also used to suppress backward radiation and to improve forward gain. Retroreflectors (90° corner reflectors) are used as passive targets for radar and communication applications to return the signal exactly in the same direction from which it is coming. This has defense applications as an electronic counter measure.
- Frequently, to reduce wind resistance, the reflector (parabolic dish/cylinder) surface is not made continuous but is in the form of a grid with a grid gap g. The structure practically acts as a continuous structure to the incident electromagnetic signal of wave length λ when $g << \lambda$, from effective medium theory.
- Half power beam width of a pencil beam radiation pattern of parabolic dish antenna is given by $\varphi = 70\lambda/D$, while the power gain is approximately given by $G \approx 6\ (D/\lambda)^2$.

Feed mechanism for parabolic antennae

- For all normal applications, the primary feed antenna is a horn or dipole placed at the focus of the parabolic reflector. Not being a true point source, the feed antenna does not fit exactly at the focus and this aberration effect broadens the main lobe by some extent with the side lobes being reinforced.
- Another feed method is the *Cassegrain feed* as that uses a hyperbolic subreflector as a secondary reflector. One of the foci of the hyperboloid coincides with the focus of the parabolic reflector. The hyperbolic subreflector has the possibility of aperture blocking which may be solved by radiating horizontally polarised waves by the primary feed (horn) which after reflection from the hyperbolic subreflector (consisting of a horizontal grid of wires) reflects the horizontally polarised wave to the parabolic reflector. At the parabolic reflector, the horizontally polarised wave received from the subreflector is rotated 90° by a special mechanism called *twist reflector* and thus the wave reflected out of the parabolic reflector is then vertically polarised and can pass through the horizontal wire grid of the hyperbolic subreflector without being obstructed thus suppressing the problem of aperture blockage.
- A normal parabolic reflector with the feed at the focus results in aperture blockage resulting in reduced antenna gain, increase in side lobes, and causes some of the radiated energy to be cross polarised. These problems may be solved by using special paraboloid-horn combinations like off-set feed and *hoghorn* antennae.

- In addition to a *pencil-beam* (which is used for point to point communication and secret communication) realisable with a parabolic reflector antenna, *fan-beams* (antenna beam pattern with unequal azimuth and elevation beam width) are also required in many applications. A common application where a specially shaped fan-beam (a cosecant-squared antenna beam) is required is in airport surveillance radar (ASR). For such a fan-beam, azimuth beam width is narrow while the elevation beam pattern is sufficiently wide (may be ten times that of the azimuth beam width).
- The pill-box antenna used in ship-to-ship radar also needs a very sharp azimuth pattern and practically no need for elevation resolution as both the ships are on the same sea surface. For Csc^2 antenna, pill-box antenna and also for parabolic cylinder antenna, the general formulae for gain, azimuth and elevation beam width are given by:

$$G = \eta_a \frac{4\pi WH}{\lambda^2}, \theta_a = 0.88\frac{\lambda}{W}\text{(radian)}, \theta_e = 0.88\frac{\lambda}{H}\text{(radian)}$$

Slot antenna

- A slot cut into a metal plane behaves as a radiator being complementary to a dipole antenna. A horizontal dipole will produce a horizontally polarised radiation pattern while a horizontal slot will produce a vertically polarised radiation pattern.
- The radiation from only one side of a large flat sheet may in practice be realised if we cut thin slots on the walls of a waveguide, with the slot length being oriented properly with respect to the surface current lines on the waveguide wall. Slot antennae are useful in many applications, especially where flush mounting is required as, for example, in high-speed aircraft.
- A slot in a conducting sheet may be conveniently energised with a coaxial transmission line with the outer conductor of the coaxial line being bonded to the metal sheet. However, a centre-fed slot is not suitable from an impedance matching point of view as the terminal resistance at the centre of a resonant $\lambda/2$ slot in a large sheet is about 500 Ω while the characteristic impedance of standard coaxial line is 50/75 Ω. Thus, the appropriate feed point is determined and off-centre feeding of slot is done.
- Slot and dipole antennae are complementary structures with their impedances related by $Z_s Z_d = \eta_0/4$; where Z_s and Z_d are respectively the slot and diode impedances and η_0 (= 377 Ω) is the free space impedance.
- A dipole which is greater than a half-wavelength is known to be inductive and hence a slot longer than half-wavelength is capacitive (being complementary to the dipole); similarly a slot shorter than a half-wavelength is inductive in nature.

Lens antenna

- Frequently the lenses are placed in the aperture of a conical horn; when the diverging spherical wave fronts from the horn would be collimated by the lens forming a plane wave front.
- Lenses and reflectors are both used as collimating elements in microwave antennae. Though lens antenna has no aperture blockage by the feed, they have internal and surface reflection losses and must be edge supported and are relatively heavy and bulky.

- Apart from being used in conjunction with a horn antenna, the lens antenna may be used as an antenna in its own right, especially at millimeter-wave frequency where low-loss dielectric lens antenna can be made competitive in weight and performance compared with reflector antenna.
- In a *dielectric lens antenna* or *delay lens antenna*, the electrical path length of the electromagnetic signal is increased by the lens medium and thus the wave is retarded by the lens. To reduce the bulk/weight of dielectric lens antenna, *zoning* is used—the lens is divided into several circular zones such that the electrical path length between adjacent zones differs by an integral multiple of the wavelength so that the phase front is not disturbed. An alternative way of zoning of a dielectric lens antenna is where dielectric material is removed from the refracting surface. Normally an artificial dielectric or metal–dielectric structure is used to have better performance and to reduce size and weight.
- There is another type of lens antenna known as an *E-plane metal-plate lens antenna* in which the electrical path length of electromagnetic signal is decreased by the lens medium and thus the wave is advanced by the lens and is known as a *fast lens antenna*. This antenna is constructed using parallel metal plates spaced suitably to support the TE mode of propagation.

Surface wave (dielectric rod) antenna

- A dielectric rod antenna is a type of surface wave antenna which is also known as a polyrod antenna as the dielectric used for the rod is made of polystyrene. In such an antenna, the choice of polystyrene ($\varepsilon_r = 2.56$) and $D \leq \lambda/2$ results in good radiation efficiency of the antenna, with end-fire pattern. Using $D = 0.5\lambda$ at the butt to 0.3λ at the end, with tapered dielectric rod, reduced minor lobes can be realised.

Microstrip patch antenna

- The basic structure of a microstrip patch antenna consists of a thin metallic patch etched on a thin (thickness being small compared to wavelength) dielectric substrate (commonly PTFE) with printed circuit technology; the substrate being backed by a large ground plane. This is a narrow band (2% to 5% bandwidth) and low gain (realisable gain being 5 to 6 dB) antenna.
- For efficient transfer of power from the transmission line feeding input power to the patch antenna, we need to match input impedance of the patch to the characteristic impedance of the transmission line. Four types of feeding techniques are used—coaxial feed, microstrip feed, aperture-coupled feed and proximity feed; of which coaxial feed is common.

Dielectric resonator antenna

- A dielectric resonator antenna (DRA) consists of a block of ceramic material performing as a dielectric resonator which is backed by a metallic ground plane. Using a high value dielectric constant ceramic material (typically 40 or so) makes it an efficient radiator.
- The dimension of a DRA is of the order of $\lambda_0/\sqrt{\varepsilon_r}$, where λ_0 is the free-space wavelength and ε_r is the dielectric constant of the resonator material. Thus, by choosing a high value of ε_r the size of the DRA can be significantly reduced.

- The cylindrical dielectric resonator mounted on a ground plane and fed by a coaxial probe is the most often used configuration of this type of antenna useful at millimeter-wave frequency.

Principle of Microwave Signal Propagation

Line-of-Sight (LOS) propagation

- At radio frequencies above about 30 MHz, the ionosphere is not able to cause reflection of the electromagnetic energy any further in the form of a *sky wave* back to the earth, while the *ground wave* or *surface wave* signal attenuates to a negligible amplitude in a kilometre or so. Useful radiowave signal propagation achieved at VHF, UHF and microwave frequencies is by means of the *space wave* travelling along the LOS between the elevated transmitting and receiving antennae in normal atmospheric conditions.
- The LOS propagation is limited by the curvature of the earth, that is, the direct waves following the optical horizon gets blocked by the curvature of the earth to a limited range. However, the varying density of the atmosphere and because of diffraction around the curvature of the earth allows a horizon to the space wave called the *radio horizon* which is about four-thirds (4/3) of the optical horizon.
- The total distance from the transmitting to receiving sites in terms of radio horizon is given by $d = d_t + d_r = 4\sqrt{h_t} + 4\sqrt{h_r}$.

Atmospheric duct propagation

- Under certain atmospheric conditions called *temperature inversion*, more likely in the sub-tropical region than the temperate zone, microwave signals can propagate by *duct propagation* to a long distance reaching even 1000 km, caused by super-refraction in the atmospheric duct.
- Temperature inversion may be caused when a layer of warm air is trapped above cooler air, such as air over the surface of water in the sea or ocean. This is an abnormal atmospheric condition because in the standard atmosphere, the air temperature decreases with height. The refractive index will then decrease far more rapidly with height than standard atmosphere. The rapid reduction of the refractive index will do to microwave what the slower reduction of refractive index of standard atmosphere of the ionised layer does to HF waves—complete bending down thus takes place. Microwave signals are then continuously refracted in the duct and reflected by the ground propagating around the curvature of earth making beyond-the-horizon microwave signal propagation possible to a long distance over the sea or ocean.

Tropospheric propagation

- Scattering of energy in a radiowave occurs in the troposphere for a wide range of frequencies centred around 900, 2000 and 5000 MHz. The amount of scattered power depends on the magnitude of the variations in the refractive index and on the distance over which these variations occur. Transmission over a troposcatter link over a distance of the order of 750 km or more between transmitter and receiver is possible.

- The troposcatter link is often used to provide long distance telephone and other communication, as an alternative to microwave link or coaxial cable over rough or inaccessible terrain.

Fading and its mitigation

- Fading is the fluctuation of radio signal strength observed at the input of a receiver caused by the interference of two or more signals which left the source at the same time but arrive at the destination by different paths and hence at different time phases. The receiver thus sees a weighted sum of delayed replicas of the transmitted signal from these multi-paths, interfering with each other either constructively or destructively, depending on the phase of the replica signals. This may develop due to meteorological variations resulting in anomalous propagation condition causing severe degradation in the radio system performance. Fading is more likely with smaller wavelengths (at higher frequencies), because greater phase shift will be produced at shorter wavelength.
- The mitigation for the fading of signal at microwave receivers, wherever it occurs, can be done with *diversity reception*. There are two forms—space diversity and frequency diversity.
- In *space diversity*, two or more receiving antennae are employed, separated by nine or more wavelengths. There are as many receivers as antennae and arrangements are made to ensure that the AGC from the receiver with strongest signal at the moment cuts off the other receivers. Thus only the signal from the strongest receiver is passed to the common output stage.
- *Frequency diversity* works in much the same way, but now the same antenna is used for the receivers, which work with simultaneous transmission at two or more frequencies. Since frequency diversity is more wasteful of the frequency spectrum, it is used only where space diversity cannot be employed, such as in restricted spaces where receiving antennae could not have been separated sufficiently.

Review Questions

10.1 What are the specialties of a microwave antenna when compared with lower frequency RF antennae?

10.2 Name a few conventional microwave antennae. What type of wire antenna is used at microwave frequencies?

10.3 Printed antenna and dielectric resonator antenna are becoming popular these days—why?

10.4 Name a few propagation modes for microwave signal communication.

10.5 Horns being flared waveguides are better radiators—explain. Name a few types of horn antenna.

10.6 Discuss the principle of operation of a horn antenna and mention some of its design considerations.

10.7 Write the expressions for half power beam width in the E and H plane of a horn antenna.

10.8 Mention some of the applications of horn antennae.

10.9 Why is the inner wall of a horn antenna sometimes corrugated?

10.10 Mention some of the common types of reflectors used either in array antenna or as an antenna itself apart from the parabolic dish antenna that is used as a secondary reflector.

10.11 Discuss the principle of operation of a parabolic dish antenna and parabolic cylinder antenna.

10.12 Practical parabolic dish or parabolic cylinder antennae are made of a girded structure—discuss.

10.13 Write the expressions for half power beam width and power gain of a parabolic dish antenna.

10.14 Mention the problems that may arise for the primary feed (horn) not being a point focus in all practical cases.

10.15 Explain why *Cassegrain feed* method needs to be used when low noise receivers are to be placed very close to the primary feed.

10.16 Discuss the principle of *Cassegrain feed* mechanism.

10.17 Explain the principle of *twist reflector* mechanism used at the parabolic reflector of *Cassegrain feed* parabolic antenna.

10.18 What are the *off-set feed* parabolic reflector and Hoghorn antenna? What are their advantages?

10.19 What is a *fan-beam*? Why is the cosecant-squared antenna beam pattern used for airport surveillance radar (ASR)?

10.20 A slot cut into a metal plane is complementary to a dipole antenna—explain.

10.21 Discuss how thin slots cut on a waveguide wall can act as a radiating slot or a non-radiating slot. Mention the use of the non-radiating slot in a waveguide.

10.22 Mention some typical applications of the slot antenna.

10.23 Why is off-centred feeding preferred over centre-feeding in the case of a slot antenna?

10.24 The impedance of a slot of length of more than half-wavelength is capacitive in nature—explain.

10.25 Explain how a lens placed in the mouth of a conical horn antenna produces a plane wave front.

10.26 Lens and reflector both are used as collimating elements in microwave antennae—compare their relative advantages and disadvantages.

10.27 What is a dielectric lens antenna? Discuss the importance of zoning in such antennae. Why is an artificial dielectric or metal dielectric preferred over a natural dielectric?

10.28 What is an *E-plane metal-plate lens antenna*? Why are these known as *fast lens antennae*?

10.29 What are surface wave antennae? Mention the conditions to be satisfied by dielectric rod type surface wave antennae to improve their radiation efficiency.

10.30 What are microstrip patch antennae? Mention their advantages and disadvantages when compared with other conventional microwave antennae.

10.31 Explain the radiation mechanism of a microstrip patch antenna. Why is a microstrip patch antenna a narrow band and low gain antenna? Mention the typical bandwidth and realisable gain of such antennae.

10.32 Mention the common feed mechanisms used for a microstrip patch antenna. Discuss the relative advantages and disadvantages of coaxial feed and microstrip feed used for microstrip patch antennae.

10.33 What are dielectric resonator antennae (DRA)? Discuss their radiation mechanism.

10.34 Discuss the advantages of using a high value of dielectric constant for the ceramic material used as dielectric resonator of DRA.

10.35 What is the mode of propagation of microwave signals in a normal atmosphere? Name a few techniques of beyond-the-horizon communication of microwave signals.

10.36 What is radio horizon? How is it related to optical horizon? What is the distance from the transmitting to receiving antenna in terms of transmitting and receiving antenna heights considering radio horizon for radiowave propagation?

10.37 Explain the principle of atmospheric duct propagation caused by temperature inversion in sub-tropical zones making microwave signal propagation beyond the horizon through atmospheric ducts.

10.38 Explain the principle of establishing communication links beyond the horizon using troposphere.

10.39 What do you understand by 'fading of radiowave signal'? Why is the effect of fading more severe at higher radio frequencies such as at microwave frequencies?

10.40 What is diversity reception? Discuss how the effect of fading can be mitigated by using space diversity/frequency diversity.

11 Radar Principle and Systems

Mind is the finest intelligent radar (biological!) ever been possible

11.1 Introduction

Radar and microwave were almost synonymous in the days of World War II when a pressing need for the development of RADAR (RAdio Detection And Ranging) system to detect and find the range of enemy aircraft with radio signals at microwave frequencies triggered the real development of microwave engineering, especially in the MIT Radiation Lab, USA. Truly, necessity is the mother of invention! In recent times, we come across several similar terms—SONAR (SOund Navigation And Ranging), SODAR (SOnic Detection And Ranging), LIDAR (LIght Detection And Ranging). Sonar is basically an acoustic means of detection and range determination of under-water targets such as submarines, on the basis of the received target echo signal. The acoustic frequencies used in sonar systems vary from very low (infrasonic) to extremely high (ultrasonic). Sodar is basically a type of upward looking in-air sonar which is a meteorological instrument used as a wind profiler to measure the scattering of sound waves by atmospheric turbulence. Acoustic means of location in air was used much before the introduction of radar and was known as an echo-sounder. Lidar involves remote sensing technology that measures distance and characteristics of the target by illuminating it with laser signal and processing the reflected light signal with varied applications covering geology, seismology and atmospheric physics. However, in military terms, 'lidar' means a 'sensor' which is utilised for the purposes of detection, identification and the precise measurement of the range of a target. The information generated by the sensor is used in targetting, and computerised command and control of automated weaponry.

In nature, bats use a technique similar to radio navigation/radar that enables them to fly about in the dark. They emit supersonic cries and listen to any echo received from the object/obstacle in their path and automatically process the data to determine the presence of the obstacle and its distance (d) in terms of the elapsed time Δt and the velocity (v) of the supersonic signal being sent by them ($d = v\Delta t/2$). Before these inventions, people also used a sonic form of radar when, for example, if one was lost in the fog at sea of a cliff-bound shore, one could shout (through a megaphone) and wait for the echo signal to be received from the cliff—giving the approximate direction and distance of the coast. The principle of pulse ranging (which, in general, is the heart of modern microwave radar systems) was first used, much before the radar technique came into being, for measuring the height of the ionosphere using an instrument known as an ionosond (IONOsphere SOuNDing). The idea was to send a short burst of radio signal in the HF region from the transmitter vertically upward and the reflected signal, the echo, was picked up by the receiver and the height of the ionospheric layer was calculated from $h = c\Delta t/2$, where c is the velocity of light.

Prior to and during the early development of microwave radar, various Radio Aids to Navigation (RAN) were developed for wireless means of navigating aircraft and ships; especially using radio frequencies in the HF and VHF range of the electromagnetic spectrum. Among the preliminary forms of RANs are DF (direction finder) for ships and DME (distance measuring equipment) used both for ships and aircraft. More sophisticated forms of RANs are VOR (very high frequency omni range) or Doppler VOR, ILS (instrument landing system), TACAN (tactical air navigation), hyperbolic navigation and so on. VOR is a receiving unit (operating in the frequency band 108–117.95 MHz) installed in the aircraft to determine their own position (bearing) and stay on course by receiving radio signals transmitted by a network of fixed ground radio transmitters installed in airports. ILS is a ground-based instrument approach system that provides precision lateral and vertical guidance to an aircraft approaching and landing on a runway, using a combination of radio signals (in the frequency range 108–112 MHz), especially helpful during reduced visibility in the night and also at the time of fog, rain or blowing snow and, in many cases, high-intensity lighting arrays are also used to enable a safe landing of the aircraft. TACAN is a navigation system used by the US military aircraft that provides both bearing and distance (slant-range) to a ground or ship-borne station. It is a more accurate version of the VOR/DME system that normally provides bearing and range information for civil aviation. Hyperbolic navigation of early days (Omega, Decca systems; after the names of the companies) is replaced these days by satellite navigation systems like the Global Positioning System (GPS). Though modern radio aids to navigation in some applications operate at microwave frequencies (5–15 GHz), for example, MLS (microwave landing system), the principle of radio aids to navigation is different from those of radar and is thus not within the scope of this chapter.

Radar can be either monostatic (which uses a single antenna for both transmission and reception of echo signal) or bistatic (where two different antennae are used for transmission and reception). Also we can have either a primary radar (passive radar, where the radar target re-radiates the signal caused by electromagnetic scattering and the echo signal thus generated is processed in the radar receiver to get the range and bearing of the target) or a secondary radar (active radar or radar beacon in which the target actually possesses a receiver–transmitter system that gives rise to the echo signal). Pulse radar is commonly used since the inception of the earliest form of radar for locating the target and finding its range. Pulse-Doppler and MTI radar are variations in the system design of basic pulse radar to realise moving target indication and elimination of clutter (echo from fixed targets like hill, mountain, tall buildings). Continuous wave (CW) radar can provide the speed of the target, however, FM-CW [frequency modulated (FM) CW] radar can provide both the velocity and distance of the target (as is used in a radio altimeter for measuring the height of an aircraft from the ground below). Apart from these, specialised radars like surveillance radar (for early warning in military applications and target search in civilian applications), tracking radar, synthetic aperture radar (SAR), pulse compression radar are used for special purposes.

11.2 Radar Range Performance Equation

11.2.1 Introduction

The radar range performance equation relates the maximum achievable range by a radar system to the characteristics of the transmitter, receiver, antenna, target and the environment. It is useful not just as a means of determining the maximum distance to which a target can be detected by the radar but can serve both as a tool for understanding the radar operation technically and as a basis for radar system design. In the following subsections we will derive the radar range equation for primary (or passive) radar and secondary (or active) radar (beacon radar) including the effect of receiver noise. Let us also discuss the physical reason for the generation of radar echo signal which is practically re-radiated from the target and which is the basis of operation of primary radar systems.

11.2.2 Primary Radar Range Equation

Let P_t be the (peak) power from the pulse radar transmitter (Tx). Then, power density at the location of the target (Fig. 11.1), if the transmitting antenna is assumed to be isotropic (with power gain $G_t = 1$), is given by $P_t/4\pi R^2$; where $4\pi R^2$ is the area of an imaginary sphere of radius R (R being the distance/range of the target from the transmitter).

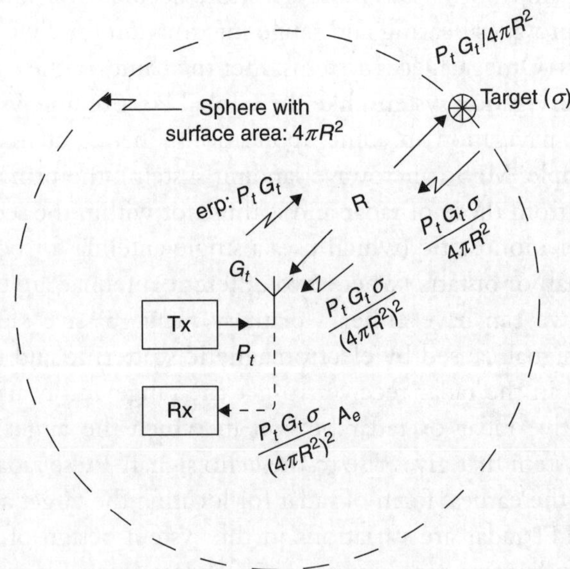

Fig. 11.1 Pictorial representation of different quantities of power at the radar and the target location to derive radar range performance equation.

However, all radar transmitting antennae are directive in nature with finite power gain ($G_t > 1$). Thus, the effective radiated power (erp) from the radar transmitter with its associated antenna is $P_t G_t$. The target intercepts a portion of the incident electromagnetic signal from the transmitter and re-radiates (the physical reason will be discussed in Section 11.2.4) in various directions. The measure of the amount of incident power intercepted by the target and re-radiated back into the direction of the radar is known as radar cross-section (RCS) of the target which is denoted by σ and has the unit of area. Thus the power density of the echo signal received at the receiver (Rx) of the radar system is given by:

$$P_r = \frac{P_t}{4\pi R^2} G_t \sigma \frac{1}{4\pi R^2} A_e \tag{11.1}$$

where, A_e is the capture area of the receiving antenna of the radar. The maximum radar range R_{max} is the distance at and beyond which the target cannot be detected. It occurs when the received echo signal power P_r equals just the minimum detectable signal S_{min} by the radar receiver, when from Eq. 11.1 we have:

$$R_{max} = \left\{ \frac{P_t G_t \sigma A_e}{(4\pi)^2 S_{min}} \right\}^{1/4} = \left\{ \frac{P_t G^2 \lambda^2 \sigma}{(4\pi)^3 S_{min}} \right\}^{1/4} = \left(\frac{P_t A_e^2 \sigma}{4\pi \lambda^2 S_{min}} \right)^{1/4} \tag{11.2a}$$

where, $A_e = G_r \lambda^2/4\pi$ (Eq. 2.47), and we assume $G_t = G_r = G$ for monostatic radar (pulse radar being mostly/always monostatic).

11.2.2.1 Comments from range equation

This is the fundamental form of radar range equation for *primary* (*passive*) *radar*. The radar range equation can be used to get an idea of the effect of changing different parameters on radar performance and also some of the trade-offs required in designing a radar system. A few observations from Eq. 11.2a in connection with maximum range realisable with primary radar are, (i) $R_{max} \alpha P_t^{1/4}$, in order to double the maximum range, the transmitter power has to be increased sixteen fold—this is an impractical way to increase the maximum range of a primary radar; (ii) $R_{max} \alpha (1/S_{min})^{1/4}$, thus, a decrease in minimum detectable signal S_{min} (by increasing the receiver sensitivity) will have the same effect as raising the transmitter power—appears to be a very attractive alternative. However, if S_{min} is made very low, the receiver will be very sensitive even to weak signals—which, in case of jamming and interference might lead to overloading of the receiver front end. Thus, an optimum compromise between the increase of transmitter power and improvement of receiver sensitivity needs to be done from application point of view. For example, in the case of military radar which might face ECM (electronic counter measure), that is jamming, there might be a need to employ a receiver with modest sensitivity and large transmitter power which may not be cost effective but required to guard against a possibility of getting jammed by enemy target. (iii) $R_{max} \alpha A_e^{1/2}$ (α D) and $\lambda^{-1/2}$ (since $A_e = \eta_a A$, where $\eta_a \approx 0.60$ for a parabolic dish antenna and $A = \pi D^2/4$, D is the aperture diameter of the parabolic dish antenna). Thus, the most practical way to double the maximum range of the radar appears to be by doubling the aperture diameter of the antenna or by reducing the wavelength or increasing the operating frequency of the radar system. But, an increase in D (keeping λ constant) or decrease in λ (keeping D constant) would eventually mean (Eq. 10.5) that the half-power beam width (φ) of the antenna is reduced. This might be advantageous for tracking radar where discrimination between close targets is desired; however for search or surveillance radar this is a disadvantage since this will take a longer time to sweep a certain area of the sky and might lose sight of some fast moving targets.

All the parameters of the radar range equation given by Eq. 11.2a are to some extent under the control of the radar designer, except for the RCS (σ) of the target. In practice, however, the simple radar range equation does not predict the range performance of actual radar equipment to a satisfactory degree of accuracy. The predicted value of radar range is usually too optimistic and in the worst case the realisable maximum range might be half of that predicted (!). This may be due to a number of reasons. The equation does not consider losses due to meteorological conditions along the propagation path of the radar signal which, in general, will be statistical in nature. Two more quantities $S_{\min}$ and σ are also statistical in nature. Further, most of the radar will have scanning/surveillance facility and this also has an effect on the range performance. The effect of scanning on range performance and the effect of loss will be discussed in Section 11.6 and a more realistic expression for range performance will be derived.

11.2.3 Effect of Noise on Range Equation

The inclusion of receiver noise modifies the radar range equation which can be derived in two ways. In the first approach, we evaluate the minimum signal power at the input of the receiver (S_i) for which the signal to noise ratio at the receiver output (S_o/N_o) will be minimum, which needs to be fed to the input of the IF amplifier of the receiver system. For the threshold detection, in the first approach we set S_i to be equal to $S_{\min}$ in Eq. 11.2a. In the second approach, we find the equivalent noise power at the input of the receiver $N_{eq(ir)}$ caused by a noisy receiver and for threshold detection we equate $N_{eq(ir)}$ with $S_{\min}$ in Eq. 11.2a.

To make necessary derivations for the two cases, we start with the basic formula of noise figure $F_r = (S_i/N_i)/(S_o/N_o)$, where S_i and N_i are the available signal and noise power at the input of the receiver and S_o and N_o are the available signal and noise power at the output of the receiver. It is known that $N_i = k_B T_0 B_r$ is the available thermal noise at the input of the receiver having bandwidth B_r and T_0 (= 290 K) is the ambient temperature in which the receiver is kept while k_B is the Boltzmann's constant. The expression for F_r may also be expressed as $F_r = N_o/k_B T_0 B_r G_a$, where N_o is the available noise power at the output of the receiver, G_a (= S_o/S_i) is the available gain of the receiver. Thus, we can write in the first case for threshold detection:

$$S_i = S_{\min} = k_B T_0 B_r F_r \left(\frac{S_o}{N_o}\right)_{\min} \tag{11.2b}$$

The expression for noise figure may also be manipulated as $(F_r - 1) = N_{eq(ir)}/k_B T_0 B_r$, from which we can write in the second case for threshold detection:

$$N_{eq(ir)} = S_{\min} = (F_r - 1)\, k_B T_0 B_r \tag{11.2c}$$

Substituting 11.2b/11.2c in Eq. 11.2a, we can write the modified radar range equation when the effect of receiver noise is included as:

$$R_{\max} = \left\{ \frac{P_t G^2 \lambda^2 \sigma}{(4\pi)^3 F_r N_i (S_o/N_o)_{\min}} \right\}^{1/4} = \left\{ \frac{P_t G^2 \lambda^2 \sigma}{(4\pi)^3 N_i (F_r - 1)} \right\}^{1/4} \tag{11.2d}$$

11.2.4 Physical Reason of Re-radiation of Signal from a Radar Target (The Generation of Echo Signal in Primary Radar)

The process that occurs at a radar target to generate the echo signal from a passive target of a primary (passive) radar system when irradiated by the radar transmitter signal may be understood as follows. The target, in general, may have conductivity, permittivity and permeability which differ from those of the surroundings representing a discontinuity to the incident signal that falls on the target. This results in an induced alternating current at the frequency of incident signal that will flow in the target material. These currents may only be displacement current if the target happens to be a pure dielectric; however, be it a displacement current or conduction current, this time-varying current in the target material will give rise to a radiated electromagnetic field from the target which is the so called *echo signal* emanating out of the target being irradiated by the radar transmitter signal. If any dimension of the target parallel to the *E*-field of the incident signal received from the transmitter happens to be a multiple of a half-wavelength of the signal frequency, the induced current will be large and thus a powerful echo signal will be generated from the target. It is apparent that the currents induced in various parts of a complex target, such as an aircraft, will be of different amplitudes and phases and these will change rapidly with different aspects of the target so that the calculation of the total re-radiated field is virtually impossible. Of course, elementary consideration shows that a large flat area, such as the side of a ship in the broadside-on position will give a very large echo while the swept wing fighter plane in the head-on position will give a small echo signal.

11.2.5 Secondary (Beacon) Radar Range Equation

In certain applications, a *secondary* (*active*) *radar* is used which is known as a *RAdar BeaCON* (*RACON*). This is a small radar set consisting of a receiver, a separate transmitter and an antenna (which is often omnidirectional). When another radar (called *interrogator*) sends a coded set of pulses (normally Morse code) to interrogate the beacon; the beacon radar (called *transponder*) responds by sending back its specific pulse code. It may be noted that RACON does not transmit pulses continuously like a search or tracking radar, but only responds to correct interrogation. In military applications, such a RACON system is used for IFF (Identification of Friend or Foe) and in civilian applications aircrafts carry them to identify themselves to the ATC (Air Traffic Control) radar installed at the airport. The presence of a RACON on a target increases enormously the distance over which a target can be detected, ranged and tracked. Let us now derive the range equation for interrogator–transponder system (Fig. 11.2) used in an active radar system using RACON.

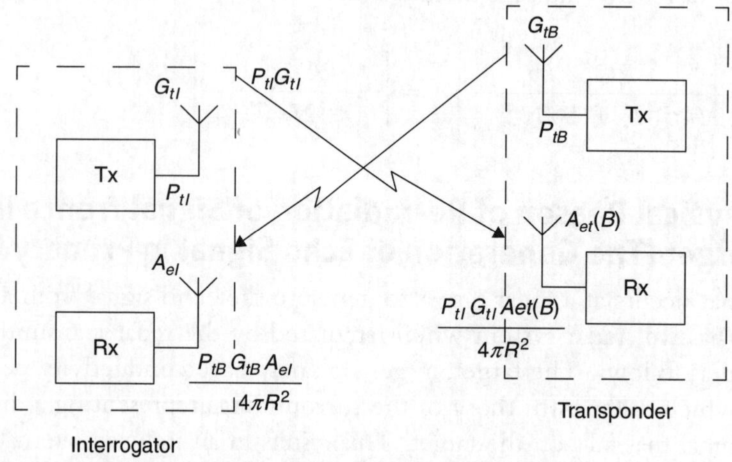

Fig. 11.2 Interrogator–transponder system for radar beacon (RACON).

Let P_{tI} be the power transmitted from the interrogating radar, G_{tI} the gain of its transmitting antenna and $A_{et(B)}$ the capture area of the transponder (beacon) receiving antenna, then the power intercepted by the beacon receiving antenna is $P_{tI}G_{tI}A_{et(B)}/4\pi R^2$. If $P_{\min, B}$ is the minimum power receivable by the beacon, the maximum range for *interrogation link* will be given by:

$$r_{\max, I} = \sqrt{\frac{P_{tI}A_{eI}A_{et(B)}}{\lambda^2 k_B T_0 B_r (F_{rB} - 1)}} \tag{11.3a}$$

where we have used $G_{tI} = 4\pi A_{eI}/\lambda^2$ and $P_{\min, B} = N_{r, B} = k_B T_0 B_r (F_{rB} - 1)$, and A_{eI} is the capture area of the interrogator receiving antenna. It has been assumed that the bandwidth and antenna noise temperature of the beacon are the same as those of the interrogating radar.

Applying similar reasoning it may be derived that the maximum range for a *reply link* from the beacon is given by:

$$r_{\max, R} = \sqrt{\frac{P_{tB}A_{et(B)}A_{eI}}{\lambda^2 k_B T_0 B_r (F_{rI} - 1)}} \tag{11.3b}$$

To calculate the maximum theoretical range for such active tracking, both Eqs 11.3a and 11.3b are solved and the lower of the two values obtained is used for range determination. Otherwise, one of the links (having shorter range) will be lost.

Now a significant comment can be made by comparing Eqs 11.2 and 11.3. The maximum range for a primary radar is proportional to the fourth root of the peak transmitted power while for a secondary radar it is proportional to the square root of the peak power. Thus, with the same transmitter power of the radar, a target having a RACON has almost double the range coverage compared to the passive radar.

While working with radar, a common unit used for range (distance) measurement is *nautical mile* (based on the circumference of the earth and is approximately one minute of arc measured along any meridian) and the velocity of the target in *knots* which is one nautical mile per hour. Conversion formulae are, 1 nautical mile (nmi) = 1.852 km = 1.1508 miles and 1 knot = 1.15 miles/h or 1.85 km/h. These are the units used especially by navigators in the shipping and aviation industries.

EXAMPLE 11.1

An S-band air surveillance radar (ASR) operating at 2.2 GHz uses a peak pulse power of 1.5 MW that must have maximum range coverage of 90 nmi (166.68 km) for aircrafts having a radar cross-section of 1.5 m^2. If the minimum receivable power of the radar receiver is 0.5 × 10^{-12} W, calculate the diameter (D) of the parabolic dish antenna which can detect the target (assume antenna aperture efficiency η_a to be 0.6).

Solution

We have from Eq. 11.2, $R_{max} = \left[\dfrac{P_t G^2 \lambda^2 \sigma}{(4\pi)^3 S_{min}}\right]^{1/4}$ and we know $G = 4\pi\eta_a A/\lambda^2 = \eta_a(\pi D/\lambda)^2$

(since $A_e = \eta_a A$ and $A = \pi D^2/4$). Here $\lambda = 3 \times 10^8/2.2 \times 10^9 = 0.136$ m, whence we have:

$$D = \frac{R_{max}(4)^{3/4} S_{min}^{1/4} \lambda^2}{P_t^{1/4} \sigma^{1/4} \eta_a^{1/2} \pi^{1/4}} = \frac{166.68 \times 10^3 \times (4)^{3/4} \times (0.5 \times 10^{-12})^{1/4} \times (0.136)^{1/2}}{(1.5 \times 10^6)^{1/4} \times (1.5)^{1/4} \times (0.6)^{1/2} \times (\pi)^{1/4}} = 3.66 \text{ m}$$

EXAMPLE 11.2

A bistatic radar (shown below) operates at 2.5 GHz with the directive gain of transmitting and receiving antenna as 35 dB and 21 dB respectively. To obtain a return power of 12 pW at the receiver, calculate the minimum power to be transmitted from the transmitter.

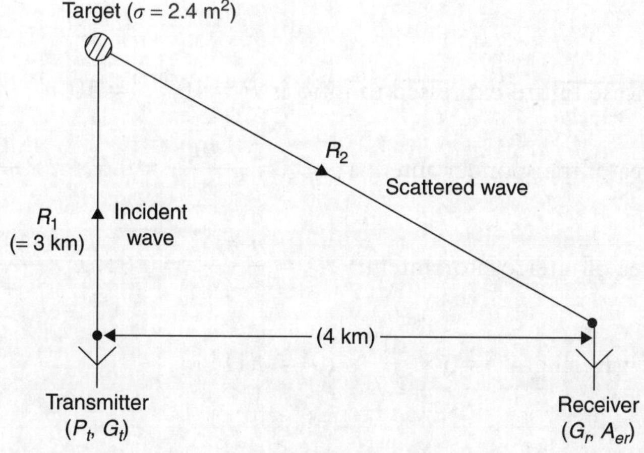

Solution

The power received at the target location is $P_{\text{traget}} = \dfrac{P_t G_t}{4\pi R_1^2}$.

The received power at the receiver is:

$$P_r = P_{\text{target}} \frac{\sigma}{4\pi R_2^2} A_{er} = \frac{P_t G_t}{4\pi R_1^2} \frac{\sigma}{4\pi R_2^2} \frac{G_r \lambda^2}{4\pi} = \frac{G_t G_r}{4\pi} \left(\frac{\lambda}{4\pi R_1 R_2} \right)^2 \sigma P_t$$

Given: $G_t = 35$ dB $= 10^{35/10} = 3162.3$, $G_r = 21$ dB $= 10^{21/10} = 125.9$, $R_1 = 3$ km, $R_2 \sqrt{3^2 + 4^2}$ $= 5$ km

$$\lambda = \frac{3 \times 10^8}{2.5 \times 10^9} = 0.12 \text{ m}, \; P_r = 12 \text{ pW}, \; \sigma = 2.4 \text{ m}^2$$

$$\text{Thus } 12 \times 10^{-12} = \frac{3162.3 \times 125.9}{4\pi} \left(\frac{0.12}{4\pi \times 3 \times 10^3 \times 5 \times 10^3} \right)^2 \times 2.4 \times P_t$$

Therefore, $P_t = 389$ W

EXAMPLE 11.3

Beacon radar operating at 2.7 GHz used in a space mission to track the space vehicle has the following specifications for the interrogator and the transponder:

Interrogator		*Transponder*	
Transmitter peak power	0.6 MW	Transmitter peak power	60 W
Antenna diameter	50 m	Antenna diameter	0.8 m
Noise figure of the receiver	1.2	Noise figure of the receiver	10 dB
Receiver bandwidth	6 kHz	Receiver bandwidth	6 kHz

Calculate the maximum detection range of this radar.
(Given: ambient temperature = 17°C, Boltzmann's constant $k_B = 1.38 \times 10^{-23}$ J/K)

Solution

The transponder noise figure expressed in ratio is $F_t = 10^{10/10} = 10$

$$\text{Capture area of transponder antenna } A_{et(B)} = \frac{\eta_a \pi D_t^2}{4} = 0.6 \times \pi \times \frac{(0.8)^2}{4} = 0.30 \text{ m}^2$$

$$\text{Capture area of interrogator antenna } A_{eI} = \frac{\eta_a \pi D_I^2}{4} = 0.6 \times \pi \times \frac{(50)^2}{4} = 1178 \text{ m}^2$$

$$\text{Operating wavelength } \lambda = 3 \times \frac{10^8}{2.7} \times 10^9 = 0.111 \text{ m}$$

Thus, the maximum range corresponding to the transponder link from Eq. 11.3b is:

$$r_{\max, R} = \sqrt{\frac{P_{tB} A_{et(B)} A_{eI}}{\lambda^2 k_B T_0 B_r (F_{rI} - 1)}} = \sqrt{\frac{6 \times 10 \times 0.30 \times 1178}{(0.11)^2 \times 1.38 \times 10^{-23} \times 290 \times 6 \times 10^3 \times (1.2 - 1)}}$$

$$= 6 \times 10^8 \text{ km}$$

while the maximum range corresponding to the interrogation link is:

$$r_{\max, I} = \sqrt{\frac{P_{tI} A_{eI} A_{et(B)}}{\lambda^2 k_B T_0 B_r (F_{rB} - 1)}} = \sqrt{\frac{0.6 \times 10^6 \times 1178 \times 0.30}{(0.11)^2 \times 1.38 \times 10^{-23} \times 290 \times 6 \times 10^3 \times (10 - 1)}}$$

$$= 9 \times 10^9 \text{ km}$$

Since the reply link due to the transponder beacon has the lower of the two values, for the maximum range, we need to take 600 million kilometres as the maximum range for the beacon radar. However, various losses like system loss, clutter and other factors contributed by nature will decrease this maximum possible range may be by a factor of 10. But by increasing the transponder antenna diameter and the transmitter power, the maximum range may be increased to more than 1,200 million kilometres and tracking of space vehicle to Jupiter and even to Saturn may be done using such beacon radar.

11.3 Pulse and CW Radar

11.3.1 Introduction

In a pulse radar only 'bursts' of RF signal (in which the RF signal from the RF oscillator acts as the 'carrier signal' and pulse signal acts as the 'modulating signal' giving the envelope of the pulsed RF signal) is sent from the transmitter over a very short period of time (may be microseconds) towards the target. In continuous wave (CW) radar, the RF signal sent towards the target is present at all instants of time (Fig. 11.3).

In pulse radar, the transmission is done during the 'ON' period (τ) of the pulse and the reception of echo signal takes place during 'OFF' period, that is, the interval between pulses (during inter-pulse period, $T = 1/f_r$) and hence the time elapsed (Δt) between the instant of transmission and the instant of reception is precisely known which helps to determine the range of the target from $R = c.\Delta t/2$, where c is the velocity of light with which the RF signal propagates. Most of the high power radars are pulse radar. But for many applications, especially looking into cost effectiveness and sometimes looking into technical requirement, CW (unmodulated and also modulated-CW) radar is used. Unmodulated CW (purely CW) radar cannot measure the range of the target because the instant of transmission and instant of reception cannot be segregated since the transmitter is 'ON' for all instants of time and so we cannot talk of a received echo for a transmitted signal at any particular instant of time. However, measurement of relative velocity of a moving target, as in police speed monitors,

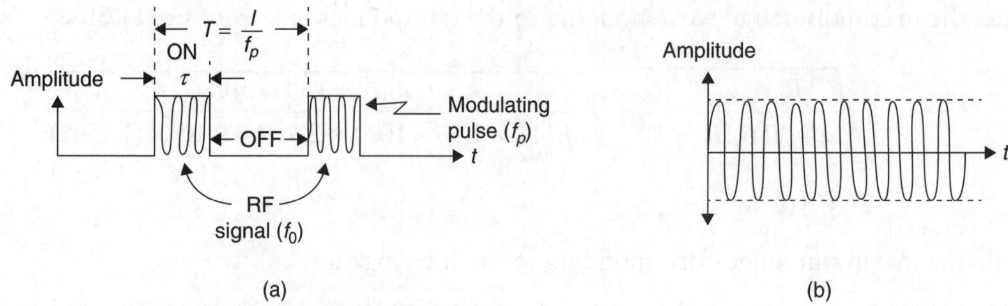

Fig. 11.3 (a) Pulsed RF signal (b) CW RF signal.

in speedometers for collision avoidance in very high speed trains and so on, where large transmitter power is not required, such purely CW radar can perform the job at a very low cost. It may be noted that the principal advantage of CW radar technique of measurement of velocity of target over other non-radar techniques (like tachometer) is that there need not be any physical contact with the object whose speed is being measured. Anyway, unlike the unmodulated CW radar the modulated CW radar say FM–CW radar can measure range (in fact it is widely used as radio altimeter—the height measuring equipment in aircraft) in which the CW carrier from the transmitter is frequency modulated (FM) and the timing mark is the changing frequency of the modulating signal. It may be noted that pulse radar can use a single antenna for both transmission and reception (because when transmission is done no reception of echo signal takes place and vice versa) and hence are monostatic in nature while in the unmodulated CW radar, transmission and reception are not segregated so transmitting and receiving antennae, in general, need to be different and thus are bistatic in nature.

The advantage of pulse radar is obviously its capability to transmit very high (of the order of megawatt) peak power so that long range detection can be done keeping the average power low (since $P_{av} = P_{\text{peak}} \times Duty\ cycle$, as duty cycle is typically 0.001). However, pulse radar has limitations of the so called maximum unambiguous range ($R_{\text{unamb}} = c/2f_p$) and the minimum range resolution ($\Delta R = c\tau/2$); the details of which will be discussed shortly. The disadvantage of CW radar in general, is due to its inability to generate large CW power; thus CW radar is mostly limited to low power short range applications. But, CW radar can, in principle, operate ideally down to zero range which is a very important aspect of CW radar in case of radio altimeters. Further, it will be seen shortly that bandwidth requirement of pulse radar is much larger compared to CW radar. Even though CW radar has specific advantages for some applications, most airborne radars are pulse radar. This is chiefly because pulse radar does not suffer from leakage/self-interference/self-jamming which the CW radar may be. In general, this is caused by transmitter interference with the reception of echo signal. In a radio altimeter (using FM–CW radar) leakage signal is, however, avoided with transmitting and receiving antennae being placed physically apart in the aircraft fuselage and the received echo is shifted in frequency with respect to the transmitted signal realised with frequency modulation of the transmitted signal. But for airborne applications where space limitation is stringent, monostatic radar configuration is preferred making pulse radar useful for most applications.

11.3.2 Maximum Unambiguous Range and Minimum Range Resolution of Pulse Radar

The *maximum unambiguous range* for pulse radar relates to the pulse repetition frequency (PRF) f_p of the pulse train—the reason for which may be understood as follows. Once the transmitted pulse is emitted by the radar, a sufficient length of time must elapse to allow any echo signals to return and be detected before the next pulse is transmitted. Therefore, the rate at which the pulses are transmitted is determined by the longest range at which targets are expected. If the PRF is too high (if pulses are sent in quick succession), echo signals from some targets (especially the distant ones) might arrive after the transmission of the next pulse, there might be some ambiguity in measuring the range. Such an echo (known as second/multiple time around echo or second-trace echo) would make the target appear at a much shorter range than it actually is and could be misleading. The range beyond which the targets appear as second time around echo is called the maximum unambiguous range and is given by $R_{\text{unamb}} = c/2f_p$. Thus, for maximum unambiguous range we need low PRF. But this conflicts with the fact that in radar the returned echo signal is integrated with a view to increase the probability of distinction between the target echo signal and possible noise. Integration builds-up the target echo signal (being repetitive/periodic in nature) but significantly cancels out the noise signal (being random in nature). The greater the number of pulses received/collected from the target—higher the PRF, better is the signal to noise ratio as far as the detection probability of the target is concerned. Thus a compromise has to be made by a radar designer between these two conflicting requirements depending upon the particular application. For example, in the case of ATC surveillance, radar 1000 pulses per second (pps) is desired (for better detection probability) while for satellite tracking, radar as low as 30 pps is used (for larger R_{unamb}).

The *minimum range resolution* (ΔR) of pulse radar is dependent on the pulse width (τ), $\Delta R = c\tau/2$. This is because, a pulse of 1 μs duration will extend 300 m along the direction of propagation and if there happen to be two targets within that 300 m, for a time span of 1 s they will be simultaneously 'illuminated' by the pulse, and at the receiver, there may appear to be one echo only. Thus, to get range discrimination among targets, the shortest possible pulse width (τ) must be used. If $\tau = 1$ μs, then range resolution obtainable in range measurement is 150 m. This is the minimum range of the radar, that is, no echo will be received from targets closer than 150 m if 1 μs pulse is used. In this case also a conflicting requirement arises. In the range equation $R_{\max}$ α P_t/B_r, but B_r α $1/\tau$ (which will be seen shortly) hence $R_{\max}$ α $P_t.\tau$ (pulse energy). This indicates that for a fixed P_t, the use of wide pulses would give larger maximum range (however, it may be noted that an increase of PW keeping PRF unchanged has the effect of increasing the duty cycle of the pulsed RF signal making average power high thus requiring high power rating of components). In fact, it may be noted that an increase in P_t for realising larger range is not practical as that would demand larger size and power consumption of components making them costlier. In fact it is the pulse energy that is a more fundamental issue and so increase of pulse width is more realistic to get larger maximum range. But this conflicts with the realisable minimum range resolution (which demands decrease of pulse width). A special type of signal processing technique used in a special kind of radar known as *pulse compression radar* which resolves this conflicting requirement and in which high range resolution of short pulses and large pulse energy of long pulses can be taken care of simultaneously (Section 11.8.2).

11.3.3 Bandwidth Requirement of Pulse and CW Radar

A brief discussion on the bandwidth requirement of pulse and CW radar would be befitting at this stage. Ideally, an infinite bandwidth is required to take care of the energy content of a single pulse from a pulse train (Fig. 11.4).

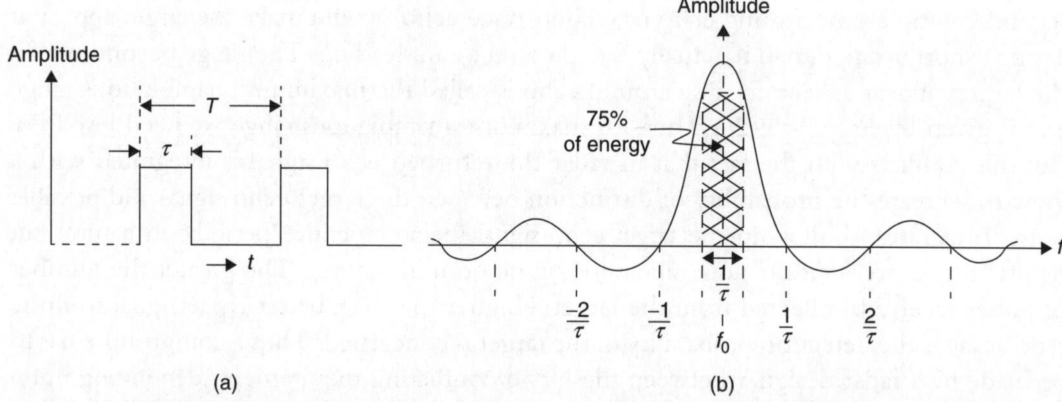

Fig. 11.4 (a) A pulse train (in time domain) (b) Frequency domain representation of a single pulse from the pulse train.

It may be observed from Fig. 11.4b that a great majority of the energy is contained between the first zeros on either side of f_0 corresponding to a bandwidth of $2/\tau$. In fact 75% of the energy is contained in half this bandwidth between $f_0 + \frac{1}{2\tau}$ and $f_0 - \frac{1}{2\tau}$. Thus, a practical bandwidth that will preserve the pulse shape and hence the information content more or less faithfully is $\approx \frac{1}{\tau}$, that is, $B_r = \frac{1}{\tau}$; a rule of thumb for pulse radar. If τ is of the order of μs, the bandwidth is of the order of MHz. But it will be seen that the bandwidth required for CW radar is much smaller than the pulse radar. For unmodulated CW it is known that the spectral density function for cos $\omega_0 t$ and sin $\omega_0 t$ exists only at ω_0, that is, ideally an impulse function. But, in practice, a pure sinusoid does not exist and hence a little bandwidth of a few Hz/hundred Hz is required. For FM–CW the bandwidth requirement is given by the well known Carson's rule of communication which states that the required bandwidth (B) for a frequency modulated RF signal is given by $B = 2(f_\Delta + \alpha f_m)$ where f_Δ is the maximum frequency deviation (which is 75 kHz for commercial FM according to FCC regulation) and $\alpha = 1$ (or 2 for worst case practical design) and W is the modulation frequency. If $f_m = 10$ kHz then the maximum desired value of B_T is of the order of 200 kHz. Thus, the bandwidth requirement for the receiver of a CW radar is much smaller than the pulse radar.

EXAMPLE 11.4

A pulse radar transmitter has maximum peak power capability of 10 MW and uses a duty cycle of 0.001 for the pulse signal. If PRF is 300 Hz, what is the range resolution of the radar? Also, calculate the maximum unambiguous range and a suitable bandwidth of this radar.

Solution

We know that duty cycle = Pulse width (τ) × PRF. Thus $\tau = \dfrac{0.001}{300} = 0.33\ \mu\text{s}$

Range resolution is given by $\Delta R = \dfrac{c\,\tau}{2} = 3 \times 10^8 \times 0.33 \times \dfrac{10^{-6}}{2} = 49.95\ \text{m}$

Maximum unambiguous range $R_{\text{unamb}} = \dfrac{c}{2f_p} = 3 \times \dfrac{10^8}{2} \times 300 = 500\ \text{km}$

Suitable bandwidth of the radar is $B_t = \dfrac{1}{\tau} = \dfrac{10^6}{0.33} = 3.03\ \text{MHz}$

11.3.4 System Block Diagram of Pulse Radar

The system block diagram of a pulse radar is given in Fig. 11.5 while a typical assembled system is shown in Fig. 11.6.

A cycle of operation of pulse radar begins with the 'triggering' of the *modulator/pulser/keyer* (may be line type, using PFN as discussed in Section 9.2.2.2) that sends out a high power high voltage pulse (with a typical pulse duration of 1 μs and PRF of the order of ms) to the *RF oscillator* (which may be a magnetron). The output of the magnetron is then pulsed RF, that is, the RF carrier from the magnetron being pulse modulated by the pulse voltage from the modulator. The trigger signal from the modulator comes from the *synchroniser* that also triggers the time-base signal of the CRT of the *plan-position indicator (PPI)*. The pulsed RF signal (with say MW peak power level) then travels through the *duplexer/diplexer* and down the feeder waveguide to the antenna system. The duplexer may be of balanced type (Section 6.2.2) and the diplexer may be of circulator type (Section 6.2.7). The function of duplexer/diplexer is to allow the high power pulsed RF signal from the transmitter to reach the antenna and during this transmission process (during the presence of the pulse signal) none of the transmitted signals allowed to leak to the receiver (the receiver is properly isolated/cut-off from main microwave circuit during the transmit cycle, otherwise the sensitive receiver will be damaged).

During the reception cycle (which takes place during the inter-pulse period of the pulse train) the received echo signal is routed to the front-end of the receiver (which may be a mixer in general) preventing any leakage of that weak signal to the transmitter section (otherwise loss of sensitivity will be incurred). The *antenna* used in the radar system is generally some form of shaped paraboloid (for example, cosecant squared antenna in

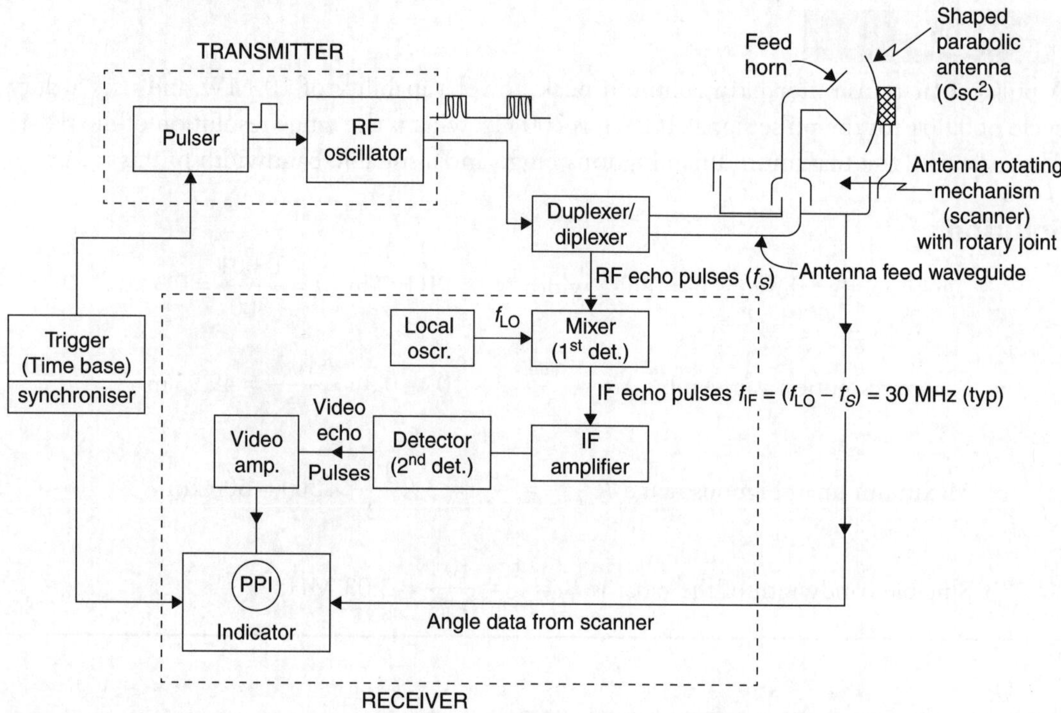

Fig. 11.5 Pulse radar block diagram showing transmitter, receiver, duplexer, rotating antenna and synchroniser.

case of surveillance radar, Section 10.2.2.2) acting as a secondary reflector with horn (or sometimes dipole) as primary feed. Polarisation normally used for surveillance radar is horizontal (as man made noise is generally vertically polarised) and also sometimes circular polarisation is used (to avoid undue clutter due to rain drops). The antenna beam has to be rotated in some fashion or the other for surveillance/tracking—antenna beam steering is needed in radar. For mechanical beam steering, a *rotary joint* (Section 6.2.10) is used between the rotating antenna and the waveguide feeder from the transmitter so that signal polarisation is not affected due to antenna rotation. For electronic beam steering, a *phased array antenna* is used (when rotary joint is not required since the antenna need not be rotated mechanically) which can realise high speed scanning by electronic means and can generate complex tracking patterns with computer processing and are thus very useful for military applications. The echo signals are picked up by the same antenna (as pulse radars are mostly monostatic) and sent down the waveguide feeder line to the front-end of the receiver. The receiver front-end may be a low-noise RF amplifier say a TWT or even a MASER amplifier for satellite tracing (where long distance is involved and no chance of excessive clutter or ECM) or may be just the mixer (in case of military radar where there is chance of jamming and also in the case of air and sea search radar where there is chance of heavy clutter; in either case the low-noise RF amplifier will be overloaded due to saturation). The *mixer* is a non-linear device in which the echo signal (f_s) and the local oscillator signal

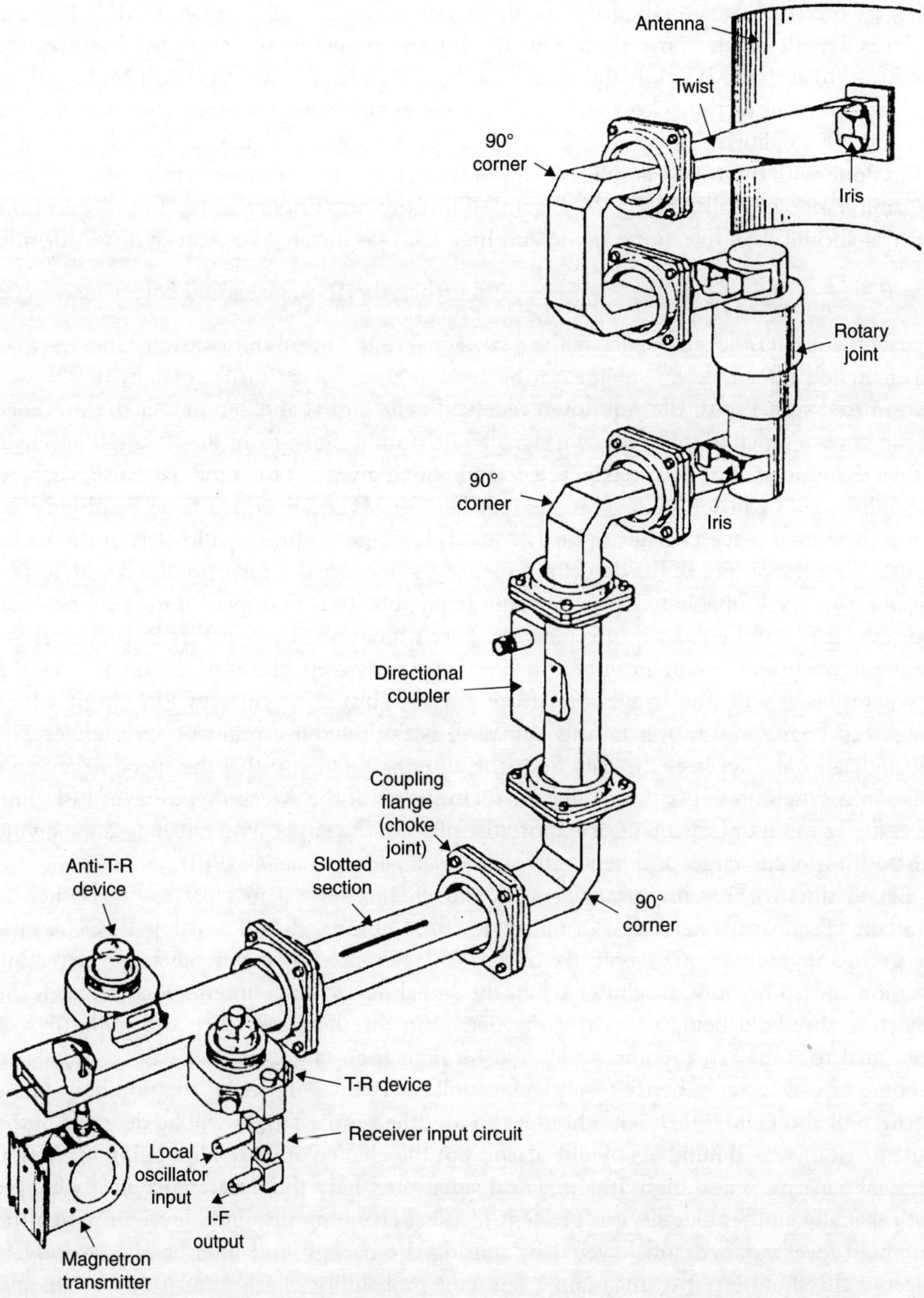

Fig. 11.6 A typical pulse radar system with a magnetron as pulsed RF source, rotary joint for mechanical beam steering, shaped parabolic antenna for fan-beam required for surveillance and other associated microwave passive components.

(f_{LO}) are mixed or heterodyned and the difference signal $f_{IF} = (f_{LO} - f_s) = 30$ MHz (typical) is derived with down conversion. The IF amplifier contains the same intelligence (the modulation) as in the RF, only the carrier has been translated down to IF (30 MHz) so that the amplification of the echo signal can be done at the HF level using conventional and low cost HF amplifiers. The IF stage is the place where most of the gain occurs (~100 dB) and this is done with multistage amplifiers (as single stage amplifier cannot provide the required dynamic range to realise such a high gain). Thus the noise figure of the front-end of the receiver should have low-noise figure and high gain (as dictated receiver by Friss' formula $F_n = F_1 + \frac{F_2 - 1}{G_1} + \frac{F_3 - 1}{G_1 G_2} + \ldots + \frac{F_n - 1}{G_1 G_2 \ldots G_{n-1}}$). Further, the IF stage should have *matched filter* characteristics in order to maximise the peak signal to the mean noise power ratio. In radar, this matched filter receiver design can be used as one can correlate (match) the known transmitted signal with the unknown received echo signal and can detect the presence of the known signal in the unknown signal. The output signal from the IF amplifier chain is then demodulated or detected by the *detector* which gives out the modulation signal; here, the pulse signal whose DC level is the rectified 30 MHz (IF) signal. The bandwidth of this pulse signal is seen earlier to be $1/\tau$ in MHz range (as pulse width τ is of the order of μs). Thus the output of the detector which is a video signal is then amplified with a *video amplifier* to a level suitable to drive the cathode ray tube (CRT) display of the plan position indicator (PPI) of the radar receiver system. It was mentioned earlier that PPI receives the timing signal from the synchroniser which initiates the sweep, that is, time base of the CRT is synchronised with the triggering of the pulser. Thus, the centre of PPI (from where the sweep begins and moves radially outward) is the reference point or zero range. The PPI indicator also receives the data from the antenna rotator so that the speed of antenna rotation is synchronised with the angular rotation (ϕ) of the sweeping beam on PPI. This gives the azimuth or bearing information (the plan) of the target with radial distance giving the position of the target and hence the name *plan position indicator* (PPI).

Let us discuss some important features and circuits related to radar receivers such as constant false alarm rate (CFAR), sensitivity time control (STC) and log-FTC. A false alarm (the appearance of a target like blip in PPI) is caused by erroneous target detection decision caused by noise or clutter when the signal due to such 'interference' exceeds the detection threshold being chosen. If the detection threshold chosen is too high, then as the signal to noise ratio required will also be high then targets with a weak echo signal become non-detectable, hence fewer targets will be detected; however, the number of false alarms will also be low. If the threshold is too low, then more targets will be detected at the expense of increased numbers of false alarms but the chance of weak blips being masked in large false alarms is also high. In a practical situation where the interference level changes both spatially and temporally, it is prudent to use a changing threshold level, in which the threshold level is raised and lowered by sensing the background interference (adaptively selects a threshold level) to maintain a constant probability of false alarm or *constant false alarm rate* (CAFR), making optimum detection probability possible.

Sensitivity time control STC, also known as *swept gain attenuators*, are special circuits used to adjust the receiver amplifier gain with time during a pulse repetition period, that is, during a single sweep. STC is commonly used in the IF section of the radar receivers to control the gain of such amplifiers with the help of voltage controlled active attenuators designed with PIN diodes to attenuate the strong echo signals received from nearby ground clutter. This is required because fixed gain amplifiers would otherwise get saturated by such strong echo signals driving the amplifier beyond its dynamic range. STC is required as an additional aid because conventional clutter rejection techniques while rejecting strong ground clutter would generally reduce the overall sensitivity of the radar receivers.

Fast time constant FTC, is a differentiator circuit with RC time constant chosen equal to the pulse width of the transmitted pulse which allows the passage of short duration desired echo signals received from the targets; however, they strongly attenuate the large clutter patches due to sea/rain clutter as these signals are comparatively much larger than the time constant of the FTC. The FTC is normally preceded by logarithmic amplifier in the IF section so that the display in the CRT of PPI can have large dynamic range.

11.4 Radar Cross-section of Targets and Radar Clutter

11.4.1 Radar Cross-section of Targets

In radar studies, a 'target' refers to any object which on receiving the incident RF signal from the radar transmitter re-radiates (reflects/scatters) the signal which is detected at the receiver of the detecting radar as an echo signal. The target intercepts a portion of the incident RF signal depending on its shape and its aspect sensitivity (the direction from which it is viewed by the detecting radar; for sphere there is no aspect sensitivity, since the sphere is a sphere no matter from what direction it is viewed from, but all practical complex targets such as aircraft and ships are aspect sensitive) and re-radiates it in various directions. The measure of the incident power intercepted by the target and re-radiated back in the direction of the detecting radar is known as the *radar cross-section of the target*, denoted by σ and has the unit of area. It is a characteristic of a particular target constituent (dependent on its material content) and is a measure of its aspect as seen by the detecting radar. Table 11.1 gives a brief list of radar cross-section of different types of radar targets at microwave frequencies. From Eq. 11.1, while deriving the range equation and denoting P_i $(= P_t G_t / 4\pi R^2)$ as the power density of the incident power at the location of the target and denoting P_e as the power scattered back to the radar as echo signal we can write $P_e = \dfrac{P_i}{4\pi R^2} \cdot \sigma$, and thus:

$$\sigma = \frac{P_e R^2}{P_i / 4\pi} = 4\pi \frac{P_e R^2}{P_i} \tag{11.4a}$$

Since the solid angle of an object that is very far away is roughly proportional to the ratio of its area to square of the distance; thus, for the echo signal falling over unit area at the receiver of the detecting radar, $1/R^2$ represents unit solid angle. Thus, radar cross-section of a target may be defined as 4π times the ratio of the power per unit solid angle scattered back towards the receiver of the detecting radar to the power density of the signal transmitted from the radar that falls on the target as incident signal. Since power density is proportional to the magnitude of the field strength squared, we can formally define radar cross-section of a target as:

$$\sigma = \lim_{R \to \infty} 4\pi R^2 \left| \frac{E_e}{E_i} \right|^2 \tag{11.4b}$$

Note that the target should be at sufficient distance ($R \to \infty$) to permit the incident electromagnetic field at the target to be approximated by plane waves propagating in the direction of the target away from the transmitter of the radar.

Table 11.1 Radar cross-section of targets at microwave frequencies

Item	**σ (sq. m)**
Jumbo jet	100
Large fighter plane	6
Small fighter plane	2
Cabin cruiser	10
Small pleasure boat	2
Automobile	100
Bicycle	2
Man	1
Bird	0.01
Insect	10^{-5}

In principle, the radar cross-section of targets can be determined in terms of the calculated scattered field by solving Maxwell's equations with proper boundary conditions. The general problem of scattering of plane waves by the most simple shape (a sphere) is available in classical electromagnetic theory books. The plot of $\sigma/\pi a^2$ vs $2\pi a/\lambda$ is shown in Fig. 11.7 [27], where σ is the scatter cross-section for a metal sphere of radius a and λ is the operating wavelength. There are three major zones in this characteristic depending on the target size in relation to the operating wavelength. The zone where $a >> \lambda$ (which is the case with almost all practical radar targets like aircraft and ships), $\sigma/\pi a^2 \approx 1$ and hence the radar cross-section of the target approaches a value of πa^2 (representing the projected area of a sphere, the area of a circular region of radius a). The other extreme zone corresponds to $a << \lambda$; for raindrops and cloud particles the Rayleigh scattering mechanism of RF wave dominates in this zone. Thus, weather radar used for meteorological early warning need to use much higher frequency (millimetre-waves) compared to the usual radar frequencies. The radar cross-section of rain droplets may be calculated from $\sigma_r = 1.96 \times 10^4 a^6/\lambda^4$, which follows from the dipole moment of a sphere immersed in an electromagnetic field (the dielectric constant of the material of the sphere being $\varepsilon_{rs} >> 1$, for rain drops).

Though the wave equation solution provides an academic understanding of the scattering mechanism related to radar cross-section of targets, no practical radar target

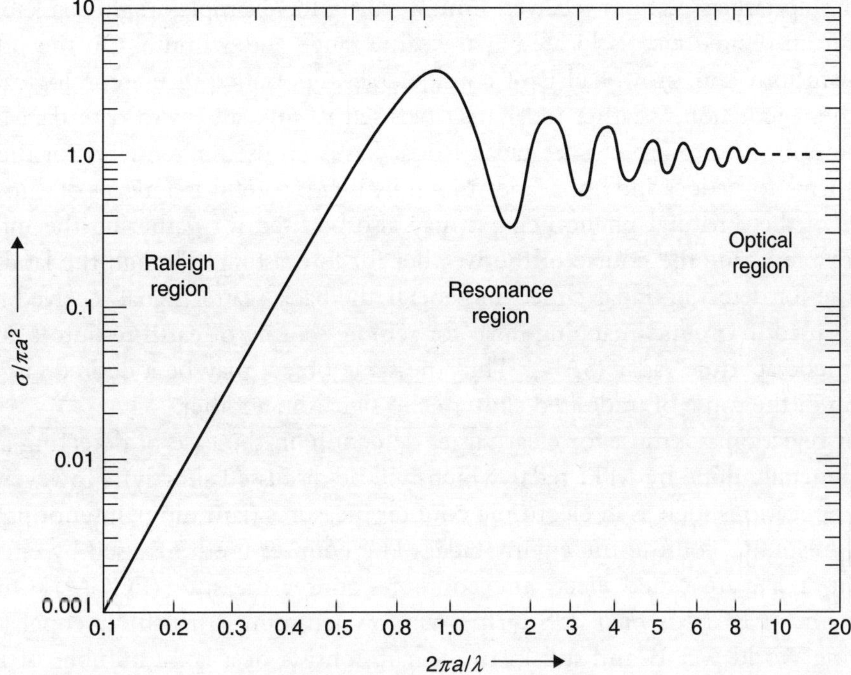

Fig. 11.7 Radar cross-section of a target (sphere) in relation to the sphere size normalised with respect to the wavelength of operation.

fits such a solution. Thus, for actual complex targets, computational techniques like the method of moments (MoM) that solves integral equations governing the distribution of induced electromagnetic fields on the surface of the target may be used. However, the method is limited by computer memory and execution time depending on the complexity of the target contour. In fact, for all practical purposes, radar engineers prefer echo signal measurement of scale model or actual full-size model target to find its radar cross-section.

11.4.2 Radar Clutter

Depending on the type of application, a target happens to be a wanted target or an unwanted target. The radar echo received from an unwanted target for a particular application is known as *radar clutter* (confusing noise) because such echoes can 'clutter' the display (PPI) of the radar receiver and make the detection of a wanted target very difficult. Echoes from land, sea and weather, in general, are considered to be harmful to radar while detecting an aircraft or ship. Because mountain, hills and tall buildings would act as stationary (extended) clutter and would make aircraft detection difficult (however, the use of MTI radar helps to distinguish moving targets from such stationary clutter, the latter being seen in the PPI display as black patches). A radio altimeter (a type of FM–CW radar) desires echo signals from land/sea below to measure the height of the aircraft. Again, whether the radar detection of meteorological objects like storm cloud, rain, snow or hail is a blessing or a curse depends upon view point of the application. Weather echoes are a nuisance to the radar operator whose job is to detect aircraft or ship targets. Echoes received from a storm, for example, might mask or confuse the echo from desired targets located at the same range and azimuth. On the other hand, radar return from rain, snow, hail is of considerable importance in meteorological research and weather prediction. Weather radar may be used to give an up-to-date data pattern of precipitation in the area around the radar. Radar has been used extensively for the study of thunderstorms, tornadoes and hurricanes. Not only is radar useful as a means of studying basic properties of these natural phenomena, it may also be used for gathering the information needed for predicting the course of the weather for forecasting. Though the land clutter is bothersome for detecting radar targets in general, the back-scatter signal received from land as target is used in ground-mapping radar for remote sensing of earth resources and also in most synthetic aperture radar (SAR). Thus the same object may be a desired target in one application or the cause of undesired clutter echo signals in another.

Clutter rejection is a must for clear target detection, in the case of detecting an aircraft or ship (generally done by MTI radar which will be discussed shortly). However, in some military applications such as in electronic counter measures (jamming), intentional artificial clutter is created to confuse the enemy radar. The counter methods used by enemy radar to combat ECM are called electronic counter–counter measure (ECCM). One of the earliest forms of ECM is *chaff* (US term)/window jamming (British)/dueppel (German) used during World War II and still in use. Chaff consists of a large number of half-wave ($\lambda/2$) dipole reflectors, usually in the form of metallic foil strips packaged as a bundle and released by an aircraft which acts as a reflecting cloud to the detecting radar. Chaff may be used either to deceive or confuse. Spot chaff is the name usually associated with the deception role, while corridor chaff is a confusion counter measure. The spot chaff, dropped as individual bundles, deceive the detecting radar with additional target-like signals received from the chaff, while corridor chaff is produced by aircraft continuously releasing chaff to form a long corridor like cloud thereby masking the desired target from the detecting radar, much like a smoke screen. However, with advanced MTI (moving target indicator) radar, relatively slow moving chaff is automatically discriminated from the actual target. A *decoy* is another means of ECM. It is a small aircraft-like vehicle made to appear as a realistic target to the radar. If sufficient numbers of decoys are present, the enemy radar system can be misled and bomber aircraft can safely perform their job. Since decoys can be made to closely resemble real targets and since a decoy might conceivably carry a bomb too, one defence strategy is to destroy all unfriendly targets, including known decoys.

Other methods of passive ECM are to reduce the radar cross-section of the target to be detected by the detecting radar. One way to reduce radar cross-section of the target is to design it with least possible flat or gently curved surfaces—the bomber aircrafts are shaped with cone–sphere. Radar cross-section of a target can also be reduced by using electromagnetic absorbent materials. Interference absorber uses quarter-wavelength thick coating of RF absorbing material (may be carbonyl iron) on the bomber aircraft surface so that energy reflected from the front surface of the absorbing material cancels (by destructive interference, due to 180° phase shift) the signal which enters the coating material and is reflected back from the inner surface (from the body of the aircraft). It is inherently a narrow band technique since it is dependent on the wavelength of the incident signal.

Another way to reduce the radar cross-section of a bomber aircraft may be the use of a continuous gradation from one absorber to another which attenuates the incident signal at different absorbing materials and finally extinguishes it. This technique demands a much thicker layer of absorbing material than the destructive interference absorber; however, the advantage is that it is a broad band technique. In recent years, research is going on with a new type of stealth technology with metamaterials which can make the target 'invisible' by cloaking it with a metamaterial cloak.

It may appear that different types of clutter originating from the land or sea may be characterised analytically by considering the clutter object as an electromagnetic scatterer. But as the clutter echoes are highly variable in nature, it is often a pragmatic approach to describe them by probability density functions or probability distribution. The *Raleigh distribution* assumes that there are a large number of randomly located independent scatterers on the clutter surface that is being illuminated by the radar signal; in other words, it is used to characterise relatively uniform clutter. However, this is not a suitable representation of clutter when the resolution cell size and grazing angle are small. In such cases *log–normal distribution* is used in which the echo power expressed in dB is Gaussian. Another distribution of importance is the *Weibull distribution* that can be made to fit clutter characterisation that lies between the Raleigh and log–normal. The Raleigh is actually a special case of Weibull while with suitable selection of the distribution's parameters, the Weibull can be made to approach the log–normal. Other distribution functions may also be used as per the clutter characteristics.

11.5 CW and FM–CW radar

11.5.1 Introduction

It was mentioned earlier that unmodulated CW radar can measure only the relative velocity of targets and not its range while FM–CW radar is capable of measuring the range of the target too. This is because the frequency modulation of the transmitted signal in FM–CW radar results in a time delay between the received echo and the transmitted signal hence allowing range determination in terms of equivalent frequency shift. However, it may be noted that in case of both unmodulated and modulated CW radar, Doppler shift resulting due to relative motion of the target and the radar need to be accounted for. In fact, in unmodulated CW radar, this Doppler shift is used to measure the relative velocity of the target with respect to the radar. However, in FM–CW radar, the frequency shift incurred in the echo signal due to Doppler and that due to the frequency modulation of the transmitter signal need to be processed and interpreted properly to get the unambiguous range measurement. Details of all these issues will be discussed now; however, Doppler effect with respect to radar will be discussed first.

11.5.2 Doppler Effect in Radar

The apparent (observed) frequency of sound/electromagnetic wave to an observer depends on the relative radial motion of the source and the observer; this was due to Christian Doppler in 1842—known as *Doppler effect*.

Let a transmitter transmit at a frequency f_t. If the stationary observer receives it, he/she would note f_t wave crests (or troughs) per second. Suppose the observer situated on a platform is approaching the source (transmitter) with a relative radial velocity v_r, he/she now encounters more than f_t crests per second. In fact the number observed under this condition is given by $f_t + f_d' = f_t[1 + (v_r/c)]$, with $v_r < 10c$. Thus, $f_d' = f_t v_r/c$, where f_d' is the Doppler frequency shift. If the platform of the observer recedes with respect to the transmitter then f_d' would be –ve. In radar involving a moving target, the signal undergoes a Doppler shift when impinged upon the target. The target eventually becomes the 'source' of the reflected wave so that for the echo signal to be received from the target at the radar site during the round trip we now have a moving source and stationary observer (the radar receiver). The two are still approaching each other, and so the Doppler shift occurs once again making the overall Doppler shift double. Thus, the Doppler frequency for an approaching target in radar is given by $f_d = 2f_d' = 2f_t v_r/c = 2v_r/\lambda$. For example, if v_r = 200 mph, λ = 3 cm, f_d = 5.96 kHz (in the audio range).

The Doppler effect is observed only for radial motion or if the motion has a radial component of velocity, but it is zero for tangential motion. In other words, no Doppler shift will be observed if the target moves across the field of view of the radar (Fig. 11.8), that is, if the target moves along TT having tangential motion (which is across the field of view of the radar antenna beam) then $v_r = 0$ making $f_d = 0$. But if the target moves in any other path as indicated by v then a radial component of velocity v_r exists resulting in a Doppler frequency shift.

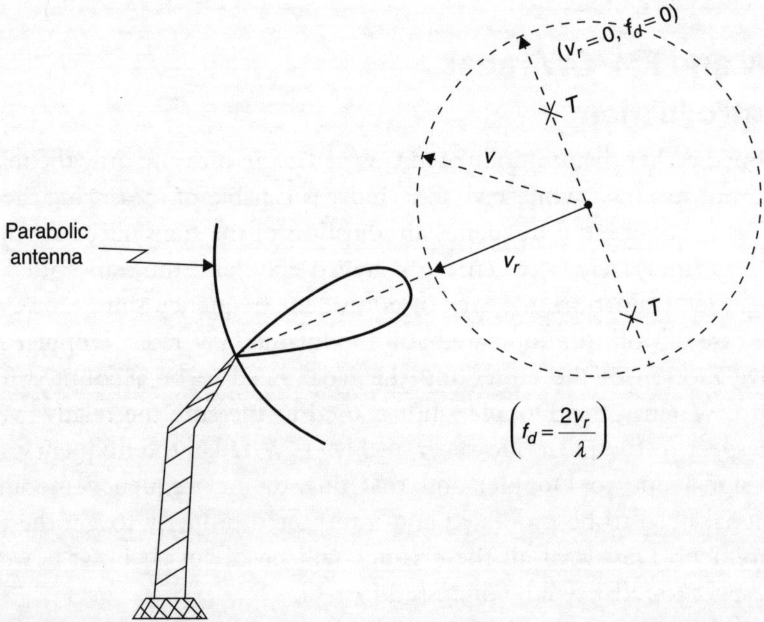

Fig. 11.8 Doppler effect in radar.

EXAMPLE 11.5

A police radar using CW microwave signal at 10 GHz from a Gunn source measures a Doppler frequency of 2.350 kHz from a car receding from the stationary police vehicle on the highway, in a 100 km/h speed limit zone. Has the car exceeded the speed limit? If so, what should the police officer do?

Solution

$$\text{Wavelength of CW microwave signal is } \lambda = 3 \times \frac{10^8}{10} \times 10^9 = 0.03 \text{ m}$$

Radial velocity of the car with respect to the police radar is $v_r = f_d\ \lambda/2 = 2350 \times 0.03/2 = 32.25$ m/s $= 32.25 \times 10^{-3} \times (60)^2$ km/h $= 126.9$ km/h. Thus, the car has exceeded the speed limit considerably and hence the police officer will give chase.

11.5.3 Principle of FM–CW Radar

In FM–CW radar, the transmitter frequency is changed periodically (changed as a function of time in a known manner, normally frequency modulated) by a saw-tooth or sinusoidal signal (usually linear saw-tooth in the form of a triangular signal is used). Then, at any instant of time, the echo signal received from a target will differ in frequency from the frequency of the signal being transmitted, by an amount equal to the change in transmitter frequency which has occurred during the time that has elapsed between the transmission of a wave and the reception of the echo at the radar receiver. The dashed line in Fig. 11.9a refers to the echo signal which is delayed by a time $T = 2R/c$ with respect to the transmitted signal. If the echo signal is heterodyned (mixed) with a portion of the transmitter signal in a non-linear element such as a diode, a beat-note (difference frequency) f_b will be produced. It is clear that greater the target distance, greater will be the beat-note frequency, and hence its magnitude is then a direct measure of the range of the target.

The beat-note frequency is given by $f_b = T \cdot \frac{df}{dt}$, where df/dt is the rate of change of transmitter frequency. Referring to Fig. 11.9a we have $\frac{df}{dt} = \frac{\Delta f/2}{1/4 f_m} = \Delta f \cdot 2 f_m$ and thus:

$$f_b = \frac{4R \cdot \Delta f \cdot f_m}{c} \tag{11.5a}$$

Hence the range of the target is measurable by knowing f_b given by Eq. 11.5a.

In the above deduction, it was implicit that the target is stationary and also there is no relative motion of the target with respect to the radar. Otherwise a Doppler frequency shift (f_d) will be superimposed on the FM range beat-note (f_r) (Fig. 11.9b), whose processing needs to be different; else, erroneous range measurement would result (this will be discussed in more detail at the end of this section). Further, if there is more than one target at two different ranges then the mixer output will contain more than one difference signal or beat-frequency; hence the processing of mixer output for range information needs to be done accordingly.

In fact, FM–CW radar is mostly applied for stationary, single target (ground/sea below) range measurement from the aircraft when it is called a *radio altimeter* (operating in 4.2 to 4.4 GHz) to measure the height (~10 km, maximum) of the aircraft from the surface of the earth. Since a large back-scatter signal is received from the vast ground/sea below (having large radar cross-section) and only short ranges are involved, the radio altimeter thus needs low power transmitter and low gain antenna. The power requirement of altimeter is of the order of 1 watt and may be obtained from a klystron/Gunn source. The transmitting and receiving antennae may be slot type and mounted on opposite sides of the fuselage of the aircraft to reduce the possibility of signal coupling. Since the relative motion between the aircraft and the ground is small, the effect of Doppler frequency shift may usually be neglected at larger heights.

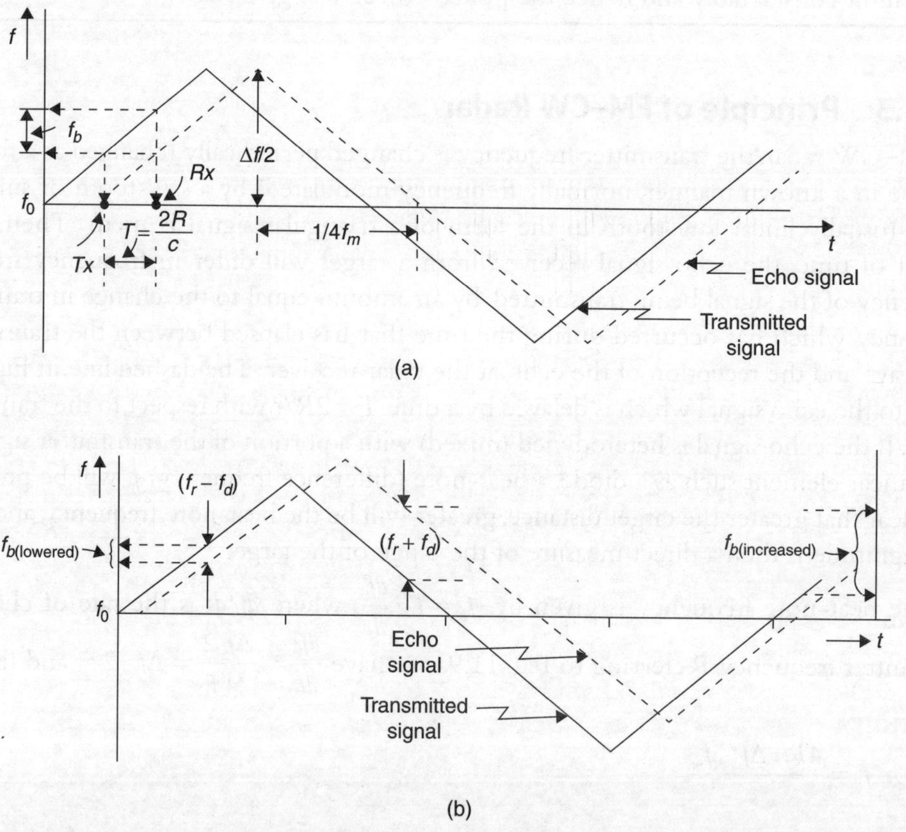

Fig. 11.9 Frequency–time relationship in FM–CW radar (a) For stationary targets (b) When there is relative motion between the target and the radar.

The absolute accuracy of the radar altimeter is usually more important at low altitudes (when the aircraft is about to land) than at high altitudes (when it is flying at nearly constant height over ground or sea). Errors of a few meters might not be significant when cruising at altitudes of 10 km but are of great importance if the altimeter is part of a radio (optically blind) landing system. The minimum beat-frequency that can be detected in the cycle counter (measuring the beat-frequency) is equal to f_m since $f_b = nf_m$ which follows from Eq. 11.5a considering $4R \cdot \Delta f/c = n$, (where $n = 1, 2, 3, \ldots$). Thus, if we equate f_b to

f_m in Eq. 11.5a, we will get the expression for minimum range that can be measured by the altimeter. Its *fixed-error* is:

$$R_{\min} = \frac{c}{4\Delta f} = \frac{75}{\Delta f(\text{MHz})}\ \text{m} \tag{11.5b}$$

Though the minimum range is not directly proportional to the carrier frequency used, a larger frequency sweep is realisable with higher carrier frequency thus making greater accuracy possible with the radio altimeter during touchdown of the aircraft if it happens to be the radio aid for blind landing of aircraft.

Here, we assumed that the target is stationary (which is normally the case as the relative velocity of the high flying target and the ground below is practically negligible). However, if there is a relative velocity between the two, then in addition to the FM beat-note (f_b) the Doppler frequency shift (f_d) will also be present and the frequency time plot will change. For an approaching target, the frequency difference between the transmitted signal and the received echo over the positive slopes decreases making the beat-note $f_{b(\text{lowerd})} = (f_r - f_d)$ while over the negative slope it increases making beat-note $f_{b(\text{increased})} = (f_r + f_d)$, Fig. 11.9b. The range frequency f_r may be obtained by measuring the average beat-frequency, $f_r = 0.5[f_{b(\text{increased})} + f_{b(\text{lowerd})}]$, while the Doppler frequency is determined from $f_d = 0.5\ [f_{b(\text{increased})} - f_{b(\text{lowerd})}]$. In this discussion we implicitly assumed that $f_r > f_d$. On the other hand, if $f_r < f_d$ (as may be the case with high speed targets at short ranges), the role of averaging and evaluation of difference frequency reverses, that is, averaging will give the Doppler frequency and the difference will give the range frequency.

EXAMPLE 11.6

An FM–CW radar operates at a carrier frequency of 4.2 GHz. The modulation frequency has a maximum sweep of 6% of the carrier. What is the fixed-error of this radio altimeter?

Solution

The fixed error is given by $R_{\min(\text{fixed})} = \dfrac{75}{\Delta f(\text{MHz})}$ (Eq. 11.5b)

Here $\Delta f = 6 \times 4.2 \times \dfrac{10^9}{100} = 252$ MHz

Thus, fixed-error of this radio altimeter is $R_{\min(\text{fixed})} = \dfrac{75}{252} = 29.76$ cm

11.5.4 System Block Diagram of CW and FM–CW Radar

It was mentioned that an unmodulated CW radar is capable of measuring only the relative velocity of a target with applications in speed monitors for automobile traffic control by the police and also in speedometers for collision avoidance in very high speed trains. A simple block diagram of a CW radar is given in Fig. 11.10a, which is a homodyne type receiver—in which no intermediate frequency (IF) is generated (however, in heterodyne/super-heterodyne receivers, there is always an IF). The Doppler frequency signal (f_d) appears as a beat-frequency at the output of the detector (mixer) which is fed with the return echo signal ($f_0 \pm f_d$) and mixes with the leakage signal from the transmitter at f_0 via the circulator.

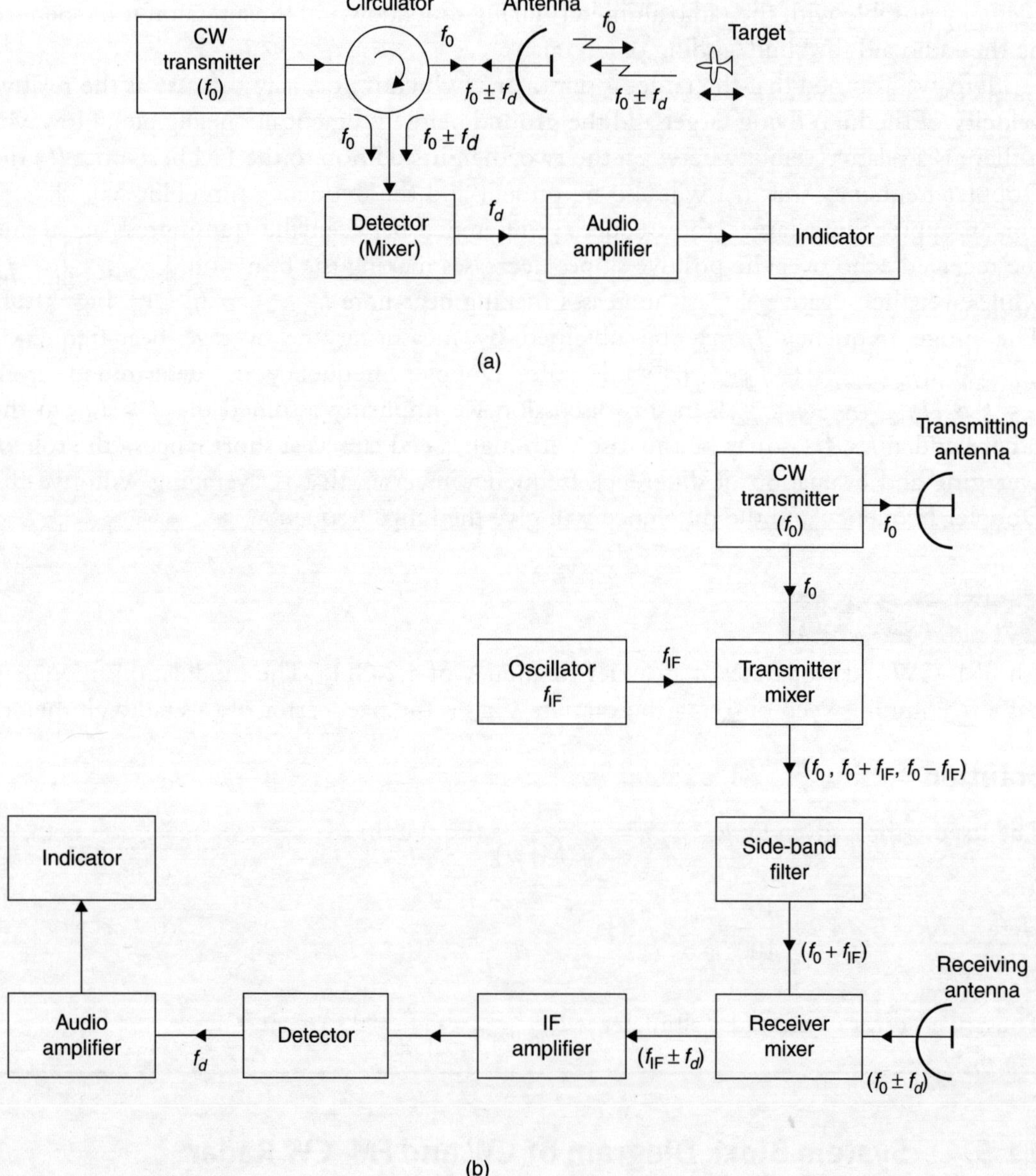

Fig. 11.10 Block diagram of a CW radar system (a) Simple homodyne type (b) Sophisticated super-heterodyne type.

The sensitivity of such a low cost receiver is poor because the Doppler frequency signal in most cases is in the audio range and at such frequencies the detector (mixer) suffers from shot-noise/flicker noise (modulation noise) which is inversely proportional to frequency.

Thus, for applications where a highly sensitive receiver is desired, this simple receiver is not useful and a super-heterodyne receiver is used where signal processing is done with a non-zero IF thus rendering flicker noise negligible compared to the usual receiver noise (thermal noise). The block diagram of such a CW radar receiver is shown in Fig. 11.10b. In this scheme, separate antennae are used for transmission and reception to improve isolation. Instead of the usual local oscillator (LO) used in the super-heterodyne receiver, the LO signal is derived in this receiver by mixing a portion of the transmitter signal with another oscillator whose frequency is equal to that of the receiver IF. Since the output of this transmitter mixer will have two sidebands on either side of the carrier plus higher harmonics, a narrow band filter is used to select one of the side bands to act as the reference signal. Extra receiver sensitivity (may be of the order of 30 dB improvement over the simple homodyne scheme discussed above) is realised here because the output of the receiver mixer is now at IF (which may be typically 30 MHz) at which the shot noise of the mixer diode is practically nil.

A simple block diagram of an FM–CW radar system is shown in Fig. 11.11. The FM transmitter is frequency modulated by a saw-tooth modulator (in fact any other modulating signal could also be used in principle). A portion of the transmitter signal is also used as a reference signal to the receiver mixer to produce the beat-frequency. The beat-frequency is amplified and limited to remove any possible amplitude fluctuations. The frequency of the amplitude-limited beat note is measured with a cycle counting frequency counter calibrated in distance. This simple arrangement is basically a homodyne type receiver. Thus, for better sensitivity super-heterodyne type receiver needs to be used as mentioned for an unmodulated CW radar; the scheme for which is available in standard text books on radar [27].

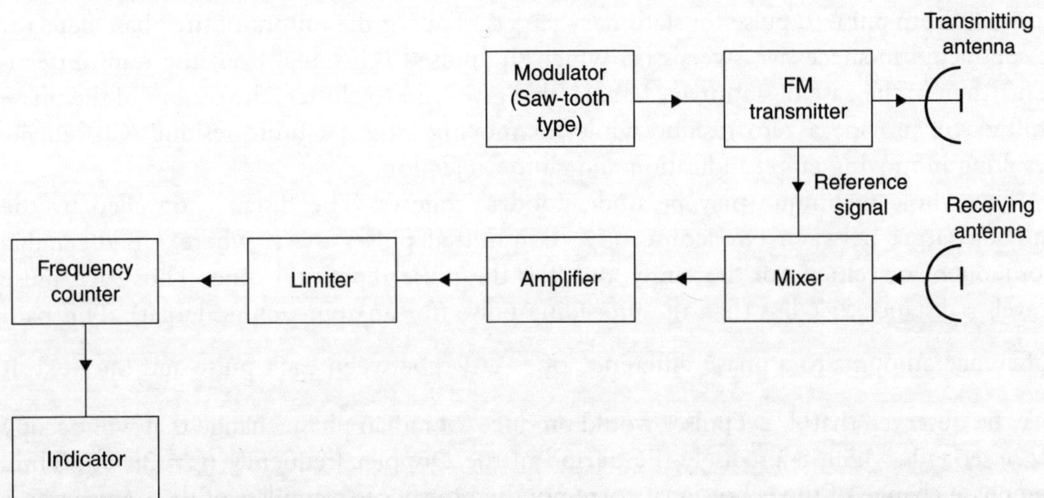

Fig. 11.11 Block diagram of a simple FM–CW radar system.

11.6 Pulse Doppler and MTI Radar

11.6.1 Introduction

In a conventional pulse radar, the Doppler shift that occurs due to relative motion of the target and the radar, may also be used to measure the velocity of the target. Then it is called *pulse Doppler radar*. More importantly, pulse Doppler radar with a modified arrangement is used to differentiate between the moving target and the clutter (in fact for clutter rejection) in which version it is known as a *moving target indicator* (MTI) radar. The MTI usually operates with ambiguous Doppler measurement (with the so called blind speed) but with unambiguous range measurement (with no second time around/second-trace echo), while the opposite is generally the case with pulse Doppler radar. Its PRF is usually high enough (unlike in the case of MTI where the PRF is chosen to be low for maximum unambiguous range) to operate with unambiguous Doppler (no blind speed) but at the expense of range ambiguity.

MTI is very much necessary in high quality air-surveillance radar to detect the target in the presence of clutter (that may be caused by stationary objects such as mountains, hills, buildings and trees) and is primarily used for clutter rejection (since blind speed need not be a concern in civilian applications). But for military applications, fighter aircrafts might intentionally fly at blind speed to evade detection by victim radar; hence pulse Doppler radar with high PRF need to be used for ranging and finding the velocity of the target. However, multiple PRF MTI can take care of both unambiguous range and avoidance of blind speed.

11.6.2 MTI Radar Principle

The basic principle of MTI radar depends on the Doppler effect in a pulse radar system in which phase shift occurs in the echo signal from one pulse to the next (at a rate proportional to the Doppler frequency) for moving targets while phase shift of the echo signal remains constant from pulse to pulse for stationary targets. Thus, if the output of the phase detector is subtracted in successive sweeps (in which the pulsed RF signal from the transmitter is sent), then in the case of stationary targets (that gives rise to clutter) the output of the phase shifter will provide a zero residue while for moving targets a finite residue will remain; resulting in moving target indication and clutter rejection.

The whole technique may be understood as follows. The distance travelled by the moving target between two consecutive transmitted pulses is vT, where v is the radial component of velocity of the target and T is the pulse repetition time. Thus each pulse travels a distance $2vT$ less than the preceding pulse (for an approaching target). This path difference amounts to a phase difference of $\frac{2\pi}{\lambda} \cdot 2vt$, between each pulse and the next. It may be observed that $\lambda/2vt$ pulses would produce 2π radian phase change that will occupy $\lambda/2v$ seconds which is precisely the period of the Doppler frequency ($f_d = 2v_r/\lambda$). Thus, the phase change of the echo signal corresponding to successive pulses of the transmitter is proportional to the Doppler shift.

The above mentioned phase change between one pulse and the next is detected at the output of a phase detector (usually a Foster Seely discriminator) which gives a proportional amplitude variation of the echo pulses. The echo returned from a stationary target exhibits a constant phase change from successive pulses (in successive sweeps of the pulse trigger of the transmitter) while for that received from a moving target, this phase change keeps on changing from one pulse to the next (Fig. 11.12). Thus, if we subtract the output of the phase sensitive detector for each pulse from the preceding one, by delaying the earlier output by a time equal to the pulse repetition period (pulse interval) 1/PRF, then the echo signals from the stationary target will cancel out due to subtraction (being at DC level and of equal strength in terms of phase change) while the echo signals from moving targets will leave an uncancelled residue after subtraction (being AC in nature varying at Doppler

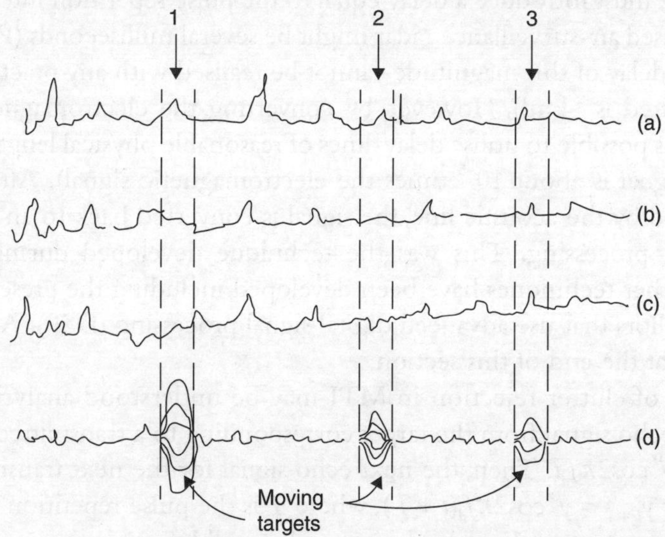

Fig. 11.12 An MTI radar A-scope display showing echo amplitude as a function of time at successive sweeps (a–c) and the superposition of many sweeps (d) showing up a 'butterfly' video signal at the location of moving targets (marked by arrows at 1, 2, 3).

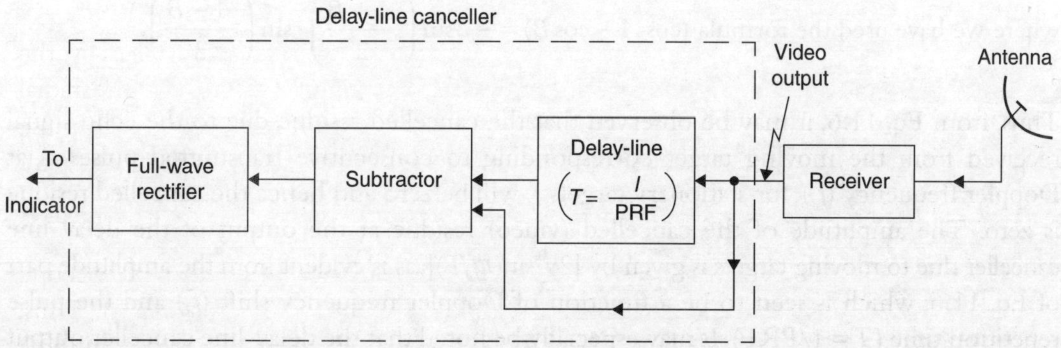

Fig. 11.13 Delay-line canceller with MTI receiver and antenna.

frequency with unequal strength from pulse to pulse due to changing phase resulting from target motion). Thus, the output of the subtractor feeding the display (A-scope or PPI) will show up only the moving targets that look like a 'butterfly' in an A-scope (Fig. 11.12).

The signal processing of this clutter rejection at the receiver of the MTI radar may be done by using the so called *delay-line canceller/discriminator* (Fig. 11.13).

Here, the video portion of the receiver is divided into two channels. One is a normal video channel while the other video channel experiences a time delay equal to one pulse repetition period (1/PRF). The outputs from the two channels are subtracted from one another when stationary targets leave no residue while the moving target has an uncancelled residue caused by pulse to pulse cancellation. The output of the subtractor is again a bipolar video as was the input. Thus, its output is converted to a unipolar video with the full-wave rectifier before this video signal can modulate the intensity of the display. We have observed that the delay-line must introduce a delay equal to the pulse repetition interval, which for a typical ground-based air-surveillance radar might be several milliseconds (PRF being of the order of kHz). A delay of this magnitude cannot be realised with any practical transmission line (length required is ~km!). However, by converting the electromagnetic signal to an acoustic signal, it is possible to utilise delay-lines of reasonable physical length (since velocity of the acoustic signal is about 10^{-5} times the electromagnetic signal). After the necessary delay is introduced by the acoustic line, the signal is converted back to an electromagnetic signal for further processing. This was the technique developed during World War II, however, many other techniques have been developed including the presently used digital delay-line chancellors that use advanced digital signal processing (DSP). A scheme for that will be discussed at the end of this section.

The principle of clutter rejection in MTI may be understood analytically as follows. Let the received echo signal from the target corresponding to a transmitted pulse sent at an instant t be $\gamma_k = \gamma^0 \cos 2\pi f_d t$. Then, the next echo signal for the next transmitted pulse will be represented by $\gamma_{k+1} = \gamma^0 \cos 2\pi f_d (t + T)$, where T is the pulse repetition time. Hence the video signal at the output of the delay line canceller will be:

$$\gamma_{dlc} = (\gamma_k - \gamma_{k+1}) = 2\gamma^0 \sin(\pi f_d T) \sin 2\pi f_d \left(t + \frac{T}{2} \right) \tag{11.6}$$

where we have used the formula $(\cos A - \cos B) = -2 \sin\left(\frac{A+B}{2}\right) \cdot \sin\left(\frac{A-B}{2}\right)$.

Thus, from Eq. 11.6, it may be observed that the cancelled residue due to the echo signal received from the moving target corresponding to consecutive transmitted pulses is at Doppler frequency (f_d), for stationary targets f_d will be zero and hence the cancelled residue is zero. The amplitude of the cancelled (video) residue at the output of the delay-line canceller due to moving targets is given by $[2\gamma^0 \sin(\pi f_d T)]$, as is evident from the amplitude part of Eq. 11.6; which is seen to be a function of Doppler frequency shift (f_d) and the pulse repetition time ($T = 1/\text{PRF}$). It may especially be noted that the delay-line canceller output will also be zero if the argument ($\pi f_d T$) in the amplitude factor of Eq. 11.6 is $\pi, 2\pi, \ldots$ (even if $f_d \neq 0$), that is, when $\pi f_d T = n\pi$ or when $v_n = \dfrac{n\lambda f_p}{2}$; (since $f_d = 2v_r/\lambda$, and $T = 1/f_p$). Thus, even if the target is moving, its velocity corresponding to v_n given by the above equation leaves a zero residue at the output of the delay line canceller and hence v_n is known as the *blind speed*; that is, for target motion at these speeds, the target is not discernable by the MTI radar as a moving target (hence also known as *eclipsing effect*). More about blind speed will be discussed shortly.

11.6.3 MTI Radar System Block Diagram

In MTI radar, the change of phase of the echo signal that occurs between consecutive transmitted pulses is detected by a phase detector which must be supplied with a phase reference (should be that of the transmitted signal, so that phase detection takes place in coherence with the phase of the transmitted signal) with respect to which it has to determine the phase change of the echo signal. This phase reference signal from the transmitted signal originates differently in two configurations of transmitter used in MTI radar systems—one being the power oscillator (POS) type and the other being the master oscillator power amplifier (MOPA) type. MTI using POS type transmitter (available from

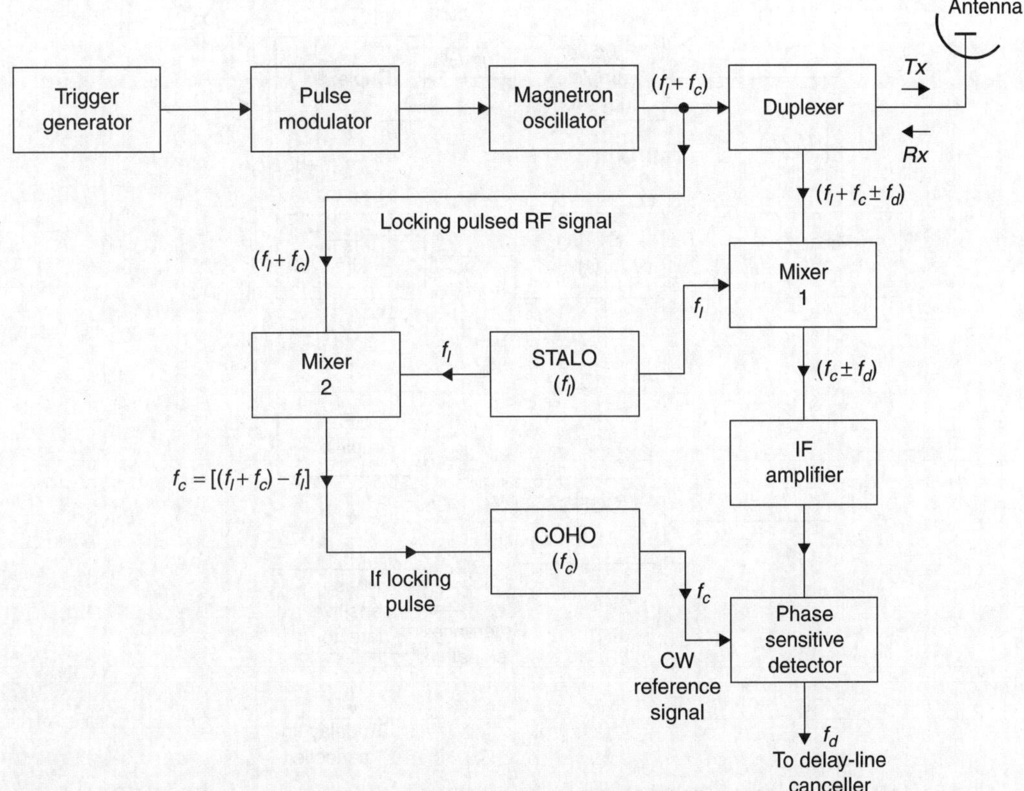

Fig. 11.14 MTI radar system block diagram using power oscillator (POS) type transmitter.

its inception and used even today in conventional radar for aircraft detection) uses a pulsed magnetron power oscillator (Fig. 11.14) to get the transmitted signal while in an MOPA type transmitter (used in high power radar for satellite tracking) the transmitted signal is obtained as a sum of the output of two stable oscillators STALO (STAble Local Oscillator) and COHO (COHerent Oscillator) which is eventually amplified with power amplifiers like klystron/TWT (which are pulsed just with a shadow grid, Fig. 11.15) to get the final transmitted signal. A magnetron is a free running RF oscillator giving CW signals (being pulsed with line type modulator), therefore, its original phase reference in the RF is lost after pulsing. Thus, in POS type transmitter the phase reference of transmitter to be given to the phase detector should be replenished in every pulse. To do this, it may be observed from Fig. 11.14 that a portion of the transmitted signal is mixed with the STALO output to produce an IF (intermediate frequency) beat-signal whose phase is directly related to the phase of the transmitter. This IF pulse is applied to the COHO causing the phase of the COHO CW oscillator to 'lock' in step with the phase of the IF reference pulse. The

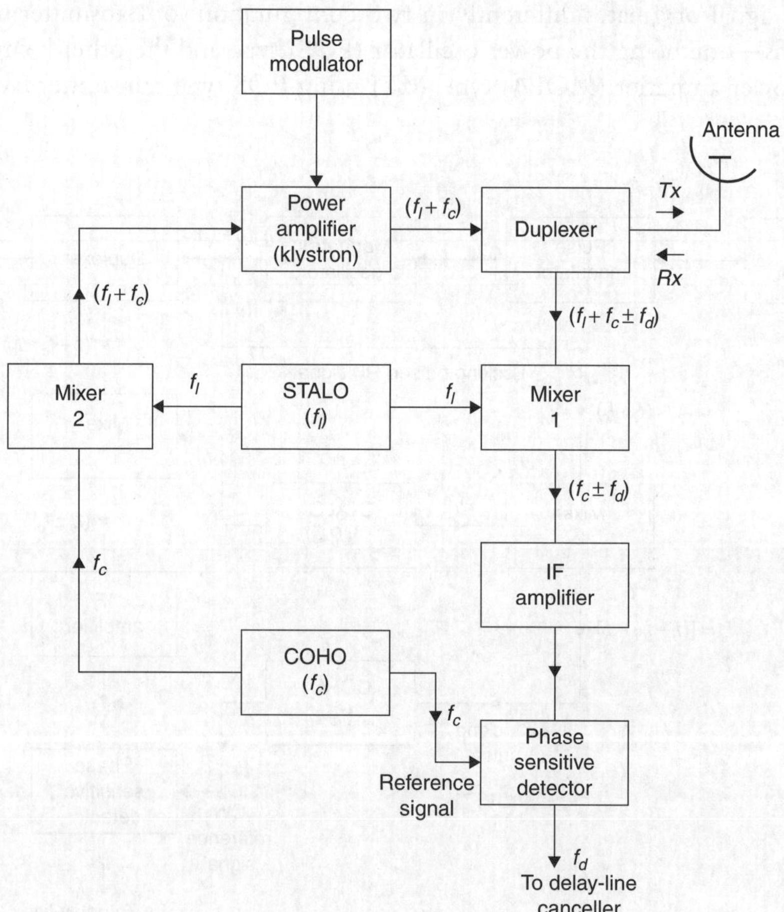

Fig. 11.15 MTI radar system block diagram using master oscillator power amplifier (MOPA) type transmitter.

phase of the COHO is then related to the phase of the transmitted pulse and may be used as reference signal for the echo received from that particular transmitted pulse. Upon the next transmission, another IF locking pulse is generated to re-lock the phase of the COHO until the next locking pulse comes along and this continues. However, from Fig. 11.15 it may be observed that in the case of a MOPA type transmitter, the COHO directly gives the phase reference of the transmitted signal to the phase detector (as the transmitter signal originates as the sum of a COHO and STALO signal). Although the phase of the STALO influences the phase of the transmitted signal, any STALO phase shift is cancelled on reception because the STALO, that is a part of the transmitted signal in this scheme, also acts as the local oscillator in the receiver.

Digital MTI (DMTI): As with any communication system, in the radar system too digital signal processing is preferred over analogue signal processing due to its many advantages. The IF signal that feeds the phase sensitive detector in the MTI receiver may be processed either with analogue delay-line canceller (as discussed earlier) or with a digital delay-line canceller; the latter system has become common in modern radar systems, thus digital MTI (DMTI) is very popular in modern MTI systems. A simple block diagram of a digital MTI processor is shown in Fig. 11.16. The output of the IF amplifier is passed through two channels—one is the in-phase channel (I-channel) in which the phase detector is driven with the COHO signal directly while the second one is the quadrature channel (Q-channel) in which the phase detector is driven with the COHO signal being phase shifted by 90°. Thus, the output from the phase detectors in the two channels will differ in phase by 90° (will be in phase quadrature). The output from the phase detectors in the two channels (*I* and *Q*) are then processed through the standard blocks of sample and hold (where the video signal is sampled at a rate to obtain one or more samples within each range resolution cell), analogue to digital (A/D) converter (where the voltage samples are converted into a series of digital words), digital memory (where the digital words are stored for one pulse repetition period, the digital storage media in the memory may be shift registers), and finally the subtractor [in which output from digital memory are subtracted in successive sweeps corresponding to the pulse repetition time (PRT)] of the transmitted pulses. The digital signals of the *I* and Q-channels are processed as $\sqrt{I^2 + Q^2}$ and converted to analogue signal by digital to analogue (D/A) converter and this unipolar video is ready for display in A-scope or PPI display.

Blind phase and its mitigation with I–Q channel of DMTI: The advantage of using the quadrature channel in DMTI is that it helps to eliminate the problem of *blind phases* that have every possibility of occurrence in MTI radar systems. Blind phase reduces the receiver sensitivity and hence it is desirable to eliminate this even with analogue signal processing. However, it is seldom done with analogue delay line canceller because of complexity involved with the addition of a second channel of analogue delay line. The blind phases are different from the blind speeds as the latter occur when the echo pulses appear at the same point in the consecutive Doppler cycles, that is, $\Delta\varphi = 2\pi$ (Fig. 11.17a). However, from Figs. 11.17b and 11.17c, it may be observed that for the I-cannel and Q-channel

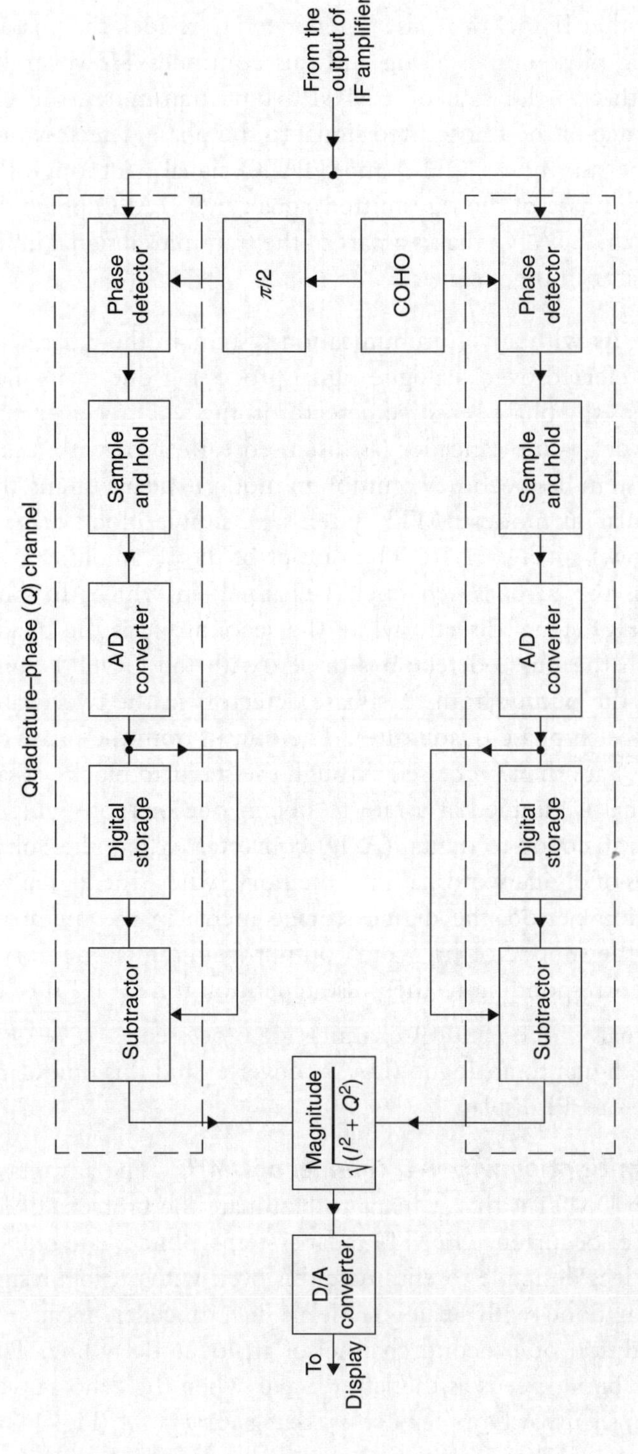

Fig. 11.16 Block diagram of DMTI signal processing scheme using (I–Q) channels.

respectively, the difference of signal amplitudes $(a_2 - a_1)$ and $(a_4 - a_3)$ are zero while $(b_2 - b_1)$ and $(b_4 - b_3)$ are non-zero; similarly $(a_3 - a_2)$ and $(a_5 - a_4)$ are non-zero while $(b_3 - b_2)$ and $(b_5 - b_4)$ are zero. Thus, the use of the Q-channel takes care of the signal components lost in the I-channel and vice versa. Use of only one channel (as is done in analogue signal processing of MTI radar) would thus result in some loss of echo signal strength when the received echo pulses are in the time instants of the Doppler cycle shown in Figs. 11.17b and 11.17c. The phase of the pulse train relative to that of the Doppler signal mentioned above is a typical case to illustrate the effect of the so called blind phase, however, with other phase and frequency relationship there is still a loss of signal with a single channel MTI. An extreme case can occur when the blind phase results in a complete loss of signal if f_d = PRF/2; the phase relationship between the two is such that the echo pulses lie on the zeros of the

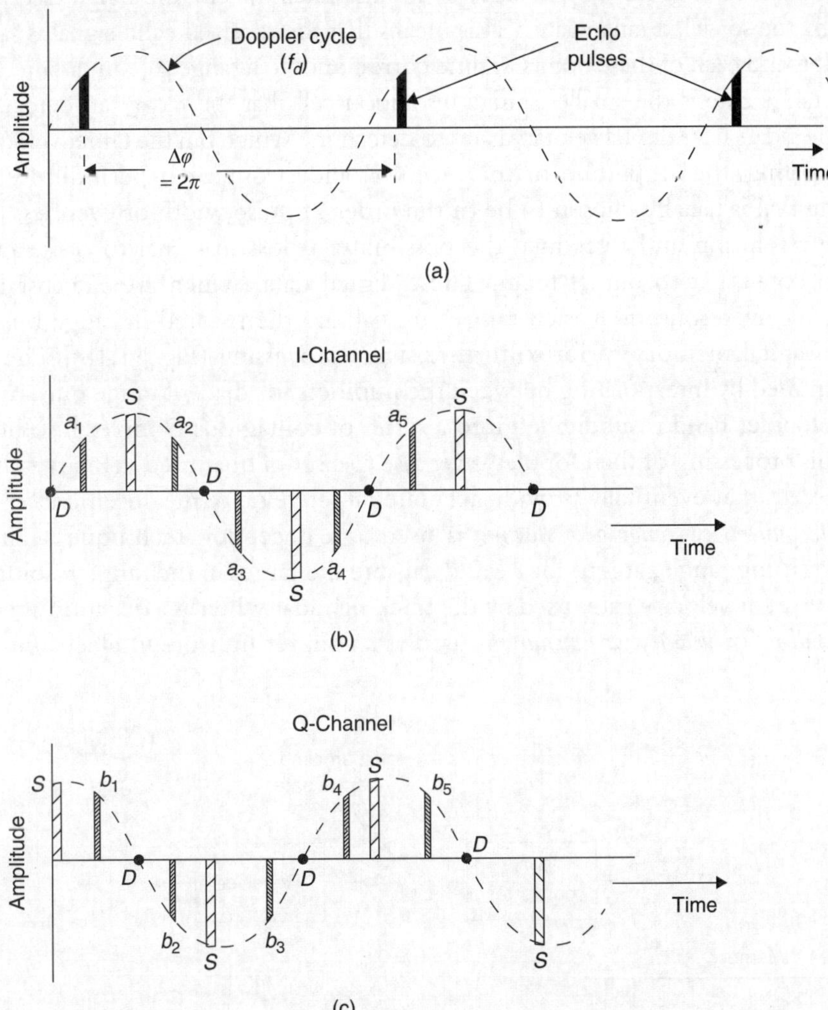

Fig. 11.17 Blind speed and blind phase in MTI radar (a) Occurrence of blind speed (b) Effect of blind phase in I-channel and (c) In Q-channel.

Doppler frequency sine wave (as shown by big dots D at the zero crossing of the Doppler cycle). This is not the condition of blind speed, nevertheless there is total loss of echo signal. However, if the phase relationship is shifted by 90°, as is done with the Q-channel in DMTI signal processing, then all the echo pulses occur in the peaks of the Doppler frequency sine wave of Q-channel (for those lost in I-cannel) and in I-channel for those lost in Q-channel, as shown by *S* marks in Figs. 11.17b and 11.17c. Thus, the combined I–Q channel scheme used in DMTI solves the problem of signal loss that would otherwise occur due to blind phases in conventional single channel MTI using analogue signal processing.

Range gate and velocity gate: In this connection it is worthwhile discussing the principle of the *range gate* and *velocity gate* that are used to process the echo signal (by digital technique) at the receiver of a radar system for accurate range determination and velocity determination of targets. While detecting/tracking a target by a pulse radar system, the range discrimination is regulated by the so called range gates. This means that the received echo signal is sampled at given time slots and each of these points in time corresponds to a range gate in distance. In other words, each range gate is the smallest range resolution cell that the radar can detect. Just like range gate, the radars have digital velocity gates to determine which bin the target velocity fits in.

In range gating, the amplitude of the receiver's video output is periodically sampled (sampling interval is usually chosen to be of the order of pulse width or even less to realise better range resolution and to reduce the possibility of loss in signal to noise ratio) and quantised to convert it to digital form. These digital data (which are the instantaneous levels of the receiver's output at each range interval) are then stored in range bins (which are basically digital memory) prior to further signal processing (Fig. 11.18). The range is finally determined by interpolating between the numbers in adjacent range bins. In velocity gating, the Doppler band is subdivided into a series of contagious Doppler resolution cells (or bins). This processing of the Doppler gives rise to a set of quantised velocity cells similar to the range cells that eventually provide accurate target velocity measurement.

Range gate pull-off or *range gate stealing* is used as a deception technique against pulse tracking radar using range gates to locate and capture the target in the range resolution cell. The same is true for velocity gates used by the tracking radar when as a deception technique, *velocity gate pull-off* or *velocity gate stealing* is used as a counter measure in electronic warfare.

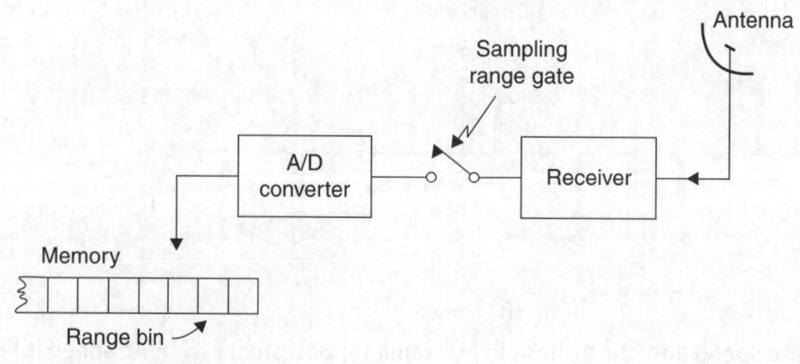

Fig. 11.18 Scheme for range gating.

MTI from a moving target: When the radar itself is in motion, as is the case when the radar is mounted in a ship or in an aircraft, the detection of a moving target in the presence of clutter is more difficult compared to when the radar is stationary. The Doppler frequency shift of the clutter is no longer at DC (not of constant level) but it varies with the speed of the radar platform, the direction of the antenna with respect to the clutter in the azimuth and elevation. Thus, the overall Doppler is changed (now it is the sum of the Doppler due to the moving target and the Doppler due to the relative velocity of clutter with respect to the radar platform, which is the clutter Doppler shift). Thus there is the need for compensation for the clutter Doppler shift. This can be done by changing the frequency of the COHO to compensate for the clutter Doppler shift. A tuneable oscillator is coupled to the COHO and its frequency tuneability is made proportional to the clutter Doppler frequency shift. Alternatively, an extra phase shifter can be placed in the delay-line and by adjusting its phase in sympathy with the phase change corresponding to the clutter Doppler shift the compensation for the clutter Doppler shift can be taken care of.

11.6.4 Blind Speed and Multiple Staggered PRF to Mitigate the Effect of Blind Speed

While operating an MTI radar we come across the so called *blind speed*, which was seen when the target happens to have a velocity whose radial component results in a phase shift of exactly 2π radians between successive pulses, this is the same as having no phase shift at all. The target thus appears stationary, and the echo received from it is cancelled by the MTI action. A radial velocity corresponding to this situation is known as blind speed, as is any integral multiple. If the target moves a distance of half wavelength ($\lambda/2$) between successive pulses ($T = 1/\text{PRF}$) then it can be readily seen that the phase shift of echo signal will be $\frac{2\pi}{\lambda}\cdot 2(vT = \lambda/2)$, that is, 2π. Thus the speed corresponding to this radial velocity of the target (or at any integral multiple of it) for a given operating frequency of the radar and for a given PRF of the pulse signal is the *blind speed* of the target and is given by:

$$v_{bn} = \frac{n\lambda}{2}\cdot\frac{1}{T} = \text{PRF}\cdot\frac{n\lambda}{2} \tag{11.7}$$

EXAMPLE 11.7

An MTI radar operates at 3 GHz with a PRF of 500 Hz. Calculate the three lowest blind speeds of the radar in knots. Comment on the results thus obtained.

Solution

Here, the operating wavelength $\lambda = 3\times\frac{10^8}{3}\times 10^9 = 0.1\text{ m}$

Thus, the three lowest blind speeds are (Eq. 11.7):

$$v_{b1} = \text{PRF} \cdot \frac{\lambda}{2} = 500 \times \frac{0.1}{2} \text{ m/s} = 25 \text{ m/s}$$

$$= 25 \times 10^{-3} \times 60 \times 60 \text{ km/h} = 90 \text{ km/h}$$

$$= 48.59 \text{ nmi/h or } 48.9 \text{ knots}$$

$$v_{b2} = \text{PRF} \cdot \frac{2\lambda}{2} = 2 \times 500 \times \frac{0.1}{2} \text{ m/s}$$

$$= 50 \text{ m/s} = 50 \times 10^{-3} \times 60 \times 60 \text{ km/h} = 180 \text{ km/h}$$

$$= 97 \text{ nmi/h or } 97 \text{ knots}$$

$$v_{b3} = \text{PRF} \cdot \frac{3\lambda}{2} = 3 \times 500 \times \frac{0.1}{2} \text{ m/s}$$

$$= 75 \text{ m/s} = 75 \times 10^{-3} \times 60 \times 60 \text{ km/h} = 270 \text{ km/h}$$

$$= 145.7 \text{ nmi/h or } 145.7 \text{ knots}$$

For commercial air navigation, blind speed is not of much concern as the problem of blind speed does not persist after a few pulse time elapses. Also a wideband receiver with microprocessor on board the aircraft can analyse the transmitted frequency and the PRF and can adjust the radial velocity accordingly to come out from the blind speed patch. However, for defense applications, a fighter aircraft might fly intentionally with blind speed as a counter measure to evade the radar detection. In such cases, the problem may be addressed by resorting to *multiple or staggered* MTI discussed below.

From Eq. 11.7, we observe that if the first blind speed is to be greater than the maximum radial velocity expected from the target, the product λf_p must be large. It demands that the MTI radar should operate at longer wavelength (at lower frequency) or with high PRF, or both. But on practical grounds, there are usually constraints other than blind speeds which determine the wavelength/frequency of operation of the radar and PRF to be used. Therefore, blind speeds may not be easily avoided. Lower radar frequencies have the disadvantages that antenna beam width, for a given antenna size, are wider than at higher frequencies and would not be satisfactory in applications where angular accuracy and angular resolution is important (both for surveillance and tracking of targets). High PRF results in range ambiguity ($R_{\text{unamb}} = c/2f_p$).

The effect of blind speeds can be significantly reduced, without incurring range ambiguity, if the MTI radar is operated with more than one PRF and the echo signal is processed accordingly. Here the MTI radar is needed to time-share its PRF between two or more values. The PRF may be switched in every other scan or may be altered in every other pulse. When the switching from one PRF to another is pulse to pulse, it is known as *staggered MTI*. However, this extension of the first blind speed is at the cost of decreased sensitivity and has the problem of uncancelled clutter residue caused by second time around clutter echo.

11.7 Surveillance and Tracking Radar

11.7.1 Introduction

The first and foremost function of radar is to search for targets through surveillance, that is, to obtain long range information on all targets in a certain volume of sky or area of sea or land. In airports, two such radar are used—one is airport surveillance radar (ASR) that typically operates in the S-band with a coverage of minimum 60 nmi to search for targets in the vicinity of the airport, and the other is air-route surveillance radar (ARSR) that operates at the L band with a typical coverage range of 200 nmi that provides coverage between airports. Such radar surveillance in search of aircraft targets are normally carried out with a rotating antenna covering 360° in azimuth and elevation up to 30° typically with a *fan-beam* realisable with cosecant squared (Csc^2) antenna (Section 10.2.2.2). The target position map (in range and azimuth) thus obtained can directly be correlated with a geographical map or chart of the area in a PPI. Thus the surveillance radar provides information regarding the coordinates where the target is expected and on the basis of which the tracking radar is positioned or 'put-on' to the target to track it (for military application) or to guide it (for air-navigation). The job of the tracking radar is then to 'acquire' the target by performing a limited search with a *pencil-beam* in the area of the designated target coordinates. The tracking radar continuously determines the target path which may be processed to determine the future position of the target. A tracking radar data is often used to measure the relative velocity of the target in terms of the rate of change of range instead of using Doppler shift (the latter technique being normally used in MTI radar).

A surveillance radar using fan-beam for search of targets can also provide tracking information in terms of scanning data which may be processed to predict the future position of targets. This type of track-while-scan was used in the early stages of radar development for ground control approach (GCA) system for aircraft landing. However, the modern system prefers to use separate surveillance radar (using a fan-beam) for getting initial information about the target coordinates (such a radar is also known as an acquisition radar) and tracking of the designated target then takes place continuously with a pencil beam. There are various techniques of target tracking like sequential lobbing, conical scan and monopulse tracking the details of which will be discussed in Section 11.7.3.

11.7.2 Effect of Scanning on Range Performance

In a surveillance or search radar, the radar antenna beam is required to search a specified volume of space within a specified time. Let Ω be the angular region (which might represent a region with 360° in azimuth and 30° in elevation as is usual with cosecant squared antenna used in surveillance radar) to be searched/scanned in the scan time t_s. If Ω_0 ($= \theta_a \times \theta_e$) is the solid angular beam width (typically 1.5° to 2°) of the scanning antenna and t_0 ($= nT_p = n/f_p$) is the time on target (where n number of pulses that hit the target in scan time t_s), then the scan time is $t_s = t_0(\Omega/\Omega_0)$, when:

$$\frac{t_s}{\Omega} = \frac{n}{f_p \Omega_0} = \frac{1}{6\omega_s} \tag{11.8a}$$

[Here, we have assumed that the antenna scan rate is ω_s rpm when the time on target/*look time* or *dwell time* (T_D) of the antenna is $(\Omega_0/360) \times (60/\omega_s) = \Omega_0/6\omega_s$. The look time/dwell time is the length of time that the antenna beam is 'looking at the target' in each antenna swing/beam scan. The number of pulses (n) that hit the target during the dwell time is known as *hits per scan* and is given by $\Omega_0 f_p/6\omega_s$. The hits per scan is a number that tells us how many echo pulses are received at the radar receiver from a specified target during every swing of the radar antenna].

Again, the peak power from the pulse radar transmitter can be written [remembering that $P_{\text{peak}} = P_{av}/\text{DC}$, where DC = pulse width (τ) × PRF (f_p)] as:

$$P_{\text{peak}(t)} = \frac{P_{av}}{\tau \cdot f_p} \tag{11.8b}$$

Note that many pulses that hit the target and return from it as echo signals in each scan by the antenna are summed up (integrated in the receiver) to improve the signal to noise ratio at the output of the detector for the echo signal received from the target. This integration of pulses improves the signal to noise ratio as the pulses are periodic in nature while the noise is random in nature. The integration of pulses is done after the signal is detected by the detector (usually designated as second detector). This post-detection is simple to implement (compared to pre-detection which needs to be done at the IF output before this second detection process), however, includes the noise being generated in the detection process. Detection efficiency may be defined as:

$$E_i(n) = \frac{(S_o/N_o)_n}{n \cdot (S_i/N_i)_1} \tag{11.8c}$$

where $(S_o/N_o)_n$ is the post-detection signal to noise ratio at the output of the second detector when n number of pulses are integrated while $(S_i/N_i)_1$ is the signal to noise ratio at the input of the second detector corresponding to a single pulse. For post-detection integration process, $E_i(n) \approx 0.8$. The integration efficiency in the detection process is seen to be less than one because there is some loss in the integration process caused by the non-linear action of the second detector.

Substituting Eqs 11.8a, 11.8b and 11.8c in the range Eq. 11.2 and remembering that in the equation, $S_{\min} = k_B T_0 B_r F_r (S_i/N_i)_1$, we get the range equation for surveillance radar as:

$$R^4_{\max} = \frac{P_{av} A_e \sigma E_i(n)}{4\pi k_B T_0 F_r (S_o/N_o)_n L_s} \cdot \frac{t_s}{\Omega} \tag{11.8d}$$

where we have used $G = 4\pi/\Omega_0$, and $B\tau = 1$ for a matched filter design generally used in the radar receiver, L_s represents the system loss which is always more than 1.

EXAMPLE 11.8

An airport surveillance radar (ASR) has the following technical specifications:

Frequency of operation 2.7–2.9 GHz
Transmitter peak power 500 kW
Pulse width 1.1 μs
Pulse repetition frequency 960 Hz
Csc^2 antenna size 16 ft wide (*W*) and 4.8 ft high (*H*)
Azimuth and elevation search volume 360° × 30°
Antenna rotation rate 12.5 rpm
Noise figure of the receiver 4.0 dB
Post-detection signal to noise ratio 6.0 dB
Detection efficiency 80%
System losses 10 dB

Calculate the detection range of this radar. (Given: Radar cross-section of the target is 30 m^2 and the antenna aperture efficiency is 0.6.)

Also show that the number of pulses hitting the target in the present case results in an echo signal strength improvement by 17 dB due to pulse integration taking place at the receiver.

Solution

At first we make the following calculations:

$$P_{av} = P_{\text{peak}(t)} \times \text{PW} \times \text{PRF} = 500 \times 10^3 \times 1.1 \times 10^{-6} \times 960 = 528\ \text{W}$$

$$A_e = \eta_a \times A = 0.6 \times 16 \times 4.8\ \text{ft}^2 = 46.08 \times (0.3048)^2\ \text{m}^2 = 4.28\ \text{m}^2$$

$$F_r = 10^{4/10} = 2.51, \left(\frac{S_o}{N_o}\right)_n = 10^{6/10} = 3.98, L_s = 10^{10/10} = 10$$

Since $\omega_s = 12.5$ rpm, $\dfrac{t_s}{\Omega} = \dfrac{1}{6\omega_s} = \dfrac{1}{6 \times 12.5} = 0.0133$

Substituting all the relevant values in Eq. 11.8d we have:

$$R_{\max}^4 = \frac{528 \times 4.28 \times 30 \times 0.8 \times 0.0133}{4\pi \times 1.38 \times 10^{-23} \times 290 \times 2.51 \times 3.98 \times 10} = 1.435 \times 10^{20}\ \text{m}$$

Thus $R_{\max} = 109.46\ \text{km} = 59\ \text{nmi}$

Now, the number of pulses hitting the target is given by Eq. 11.8a as:

$$n = \frac{\Omega_0 f_p}{6\omega_s} = \frac{4.08 \times 960}{6 \times 12.5} \approx 52$$

[where, for the given Csc^2 antenna, it can be calculated that azimuth beam width $\theta_a = 1.107°$ and elevation beam width $\theta_e = 3.69°$ (Example 10.3). Thus, $\Omega_0 = \theta_a \times \theta_e = 4.08°$]

Hence, 52 pulses that hit the target during successive pulse periods would make the effective echo strength 52 times stronger or expressed in terms of dB—17 dB.

11.7.3 Tracking Radar Principle and Techniques

Once the target has been 'acquired', its coordinate designation in terms of range, bearing (both azimuth and elevation), and Doppler are obtained with a surveillance radar, then the present and future course of movement of the target need to be known for guiding the target (as is the case with aviation application) or putting the weapon on the target (as in defence application) and this is done by tracking radar. In all the tracking radar techniques, an angle-error signal is generated by comparing the target position at that moment (as sensed by the antenna beam) with respect to a reference direction (normally the antenna boresight axis). This angle-error signal actuates a servomechanism system to position the movable antenna to make the angular error zero. When the angle-error becomes zero, the target is located along the antenna boresight axis/reference line and the target is said to be 'tracked'. The range and the relative velocity of the target (which is the rate of change of range) can also be continuously tracked in a tracking radar system.

Tracking radar systems may be of different types depending on the application requirements. Single target trackers (STT) continuously track a single target at a relatively rapid data rate (typically 10 observations per second) as in a military guided-missile weapon control radar. Automatic detection and track (ADT) is used in modern air surveillance both for civil and military applications. Its data rate is lower than STT but it can simultaneously track a large number of targets. Tracking in ADT is done open loop which means that the antenna position is not controlled by the processed tracking data as it is done in the case of STT. Phased array radar tracking (PART) combines the rapid data rate of STT and also has the ability of ADT to hold many targets in track with the help of electronically steerable phased array radar. Multiple targets are tracked on a time-shared basis under computer control since the beam of an electronically scanned array can be very rapidly switched from one angular direction to another, may even be in a few micorseconds. Track while scan (TWS) was used in the past for air-defence radars, aircraft landing radars, and in some airborne intercept radars to hold multiple targets; however, it has only moderate data rate.

The tracking radar technique uses two general principles to generate the angle-error signal. In one approach (*lobe switching and conical scan*), the formation of angular error signal requires the processing of a number of echo signal pulses during each sweep of the transmitted pulse train. The shortcoming of this approach is that the angle-error signal has every possibility of being sensitive to the amplitude fluctuations in the received echo signals resulting from random variations in the target cross-section. But, this pulse to pulse amplitude fluctuation of the echo signal will have no effect on tracking accuracy (in the generation of angle-error signal) if the angular error measurement is done on the basis of a single pulse. Thus, the modern approach is to use *monopulse* (as it uses single echo pulse for angle error signal generation) tracking methodology. It will be observed shortly that from an operational point of view, the conventional lobe switching and conical scan of tracking utilises one antenna beam on a time shared or sequential basis (it is a single channel tracker) while in monopulse tracking, more than one antenna beam is used simultaneously (it is a multi-channel, on reception, tracker). Monopulse tracing radar is used in air and naval defence systems for guiding surface-to-air missiles and also in ships for surface-to-air fire control systems.

Lobe-switching and conical scan techniques: In the *lobe switching* technique, the magnitude and direction of the angular error signal is obtained by switching the antenna beams LOBE A and LOBE B alternately between two positions 1 and 2 [Fig. 11.19a(i)]. If the target is along the equisignal direction (the boresight axis of the antenna), centred between the lobes of the antenna in two switched positions, the received echo will be of same strength for both lobes. Any offset of the target from the boresight axis will result in a stronger echo corresponding to one lobe (as per the diagram for LOBE B) when compared with that received corresponding to the other lobe (LOBE A), Fig. 11.19a(ii). The difference of the strength of the echo for the two switched positions is a measure of the magnitude of the angular error. The sign of the difference determines the direction in which the antenna needs to be rotated in order to align its boresight axis with the direction of the target. These two switching positions is good for tracking either in azimuth or in elevation; however, for two dimensional tracking (tracking in both azimuth and elevation), two more switching positions are required in the orthogonal coordinate. Thus, for two dimensional tracking with lobe switching, a cluster of four feed horns illuminating a single reflector antenna can provide successive lobe switching required for right–left and up–down sector tracking. In fact, a fifth feed horn will also be required at the central position of the cluster of five horn feeds for signal transmission.

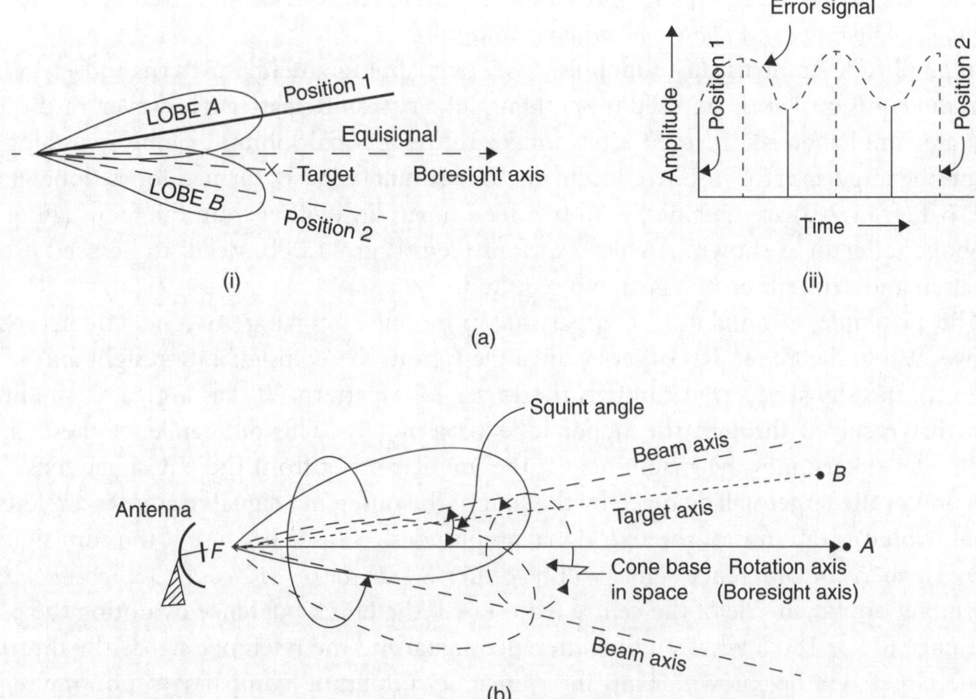

Fig. 11.19 (a) (i) Lobe switching (ii) Nature of error signal amplitude with time (b) Conical scan.

The *conical scan* may be thought as a logical extension of lobe switching in which the antenna beam off-set around a circular contour (Fig. 11.19b) can be realised either by tilting the feed with the reflector fixed or by keeping the feed horn fixed and rotating the reflector dish antenna (a typical conical scan rotating speed may be 30 rotations per second, which must be much smaller than PRF). As the conical scan pattern generated in space can be imagined to be the base of a cone with its apex at the focus F of the reflector, it is called conical scan. When the line of sight of the target (target axis) with the target at A coincides with the rotation axis of the beam (the boresight axis of the antenna) the echo signal strength is zero (there is no modulation of the echo signal) and hence the target is said to be tracked. However, a modulated echo signal (at a frequency equal to the rotation frequency of the beam) is observed if the target is offset from the boresight axis (say at position B). The amplitude of the echo signal modulation is dependent on the shape of the antenna pattern, the angle between the beam rotation axis and the beam axis at the moment (called the squint angle), and the angle between the line of sight of the target (target axis) and the boresight axis. The phase of the echo signal modulation is dependent on the target axis and the beam rotation axis. This conical scan modulation is extracted from the echo signal and used to actuate a two dimensional servomechanism system to position the antenna on the target. The system block diagram for a conical scan tracing radar is available in standard text books on radar [27].

Monopulse techniques: There are in general, two monopulse radar techniques depending on the manner the angle-error information is extracted from the received echo signal—*amplitude comparison* and *phase comparison* monopulse.

In amplitude comparison monopulse, two overlapping antenna patterns (offset by $\pm\theta_0$ from the equisignal axis) are used (resembling lobe switching pattern; however, in this case they are simultaneously formed lobes unlike the sequential lobing in lobe switching) to obtain the angular error in one coordinate. The simultaneously formed lobes (lobe A and lobe B, Fig. 11.20a) are obtained with two feed horns little offset from the focus (F) of the parabolic reflector as shown. A cluster of four feeds (Fig. 11.20b) needs to be used if both elevation and azimuth error signals are required.

The principle of amplitude comparison monopulse tracking may be understood as follows. When the target T is offset by an angle θ from the equisignal (boresight axis of the antenna), the signal received through the lower lobe pattern (B) has a greater amplitude than that received through the upper lobe pattern (A). The difference of these signal amplitudes determines the magnitude of the angular offset from the eqiusignal axis. The direction of the target will be given by the sign of the difference signal relative to a reference signal (which is the sum of the two signal amplitudes). This is because if the sum signal is $K_s \cos \omega t$ then the difference signal will be either $K_d \cos \omega t$ or $-K_s \cos \omega t$ $[K_d \cos(\pi + \omega t)]$, depending on which side of the centre (equisignal) the target is. Hence by noting the phase difference (0° or 180°) between the difference signal and the reference signal, the direction of the target will be known. Thus in essence, in amplitude comparison monopulse, the transmitter feeds all the four horns so that a sum signal is transmitted. On reception, proper signal routing provides the following three signals—the sum signal ($\Sigma = A + B + C + D$), the horizontal (azimuth) difference signal $[\Delta Az = (A + B) - (C + D)]$, and the vertical (elevation) difference signal $[\Delta El = (A + C) - (B + D)]$.

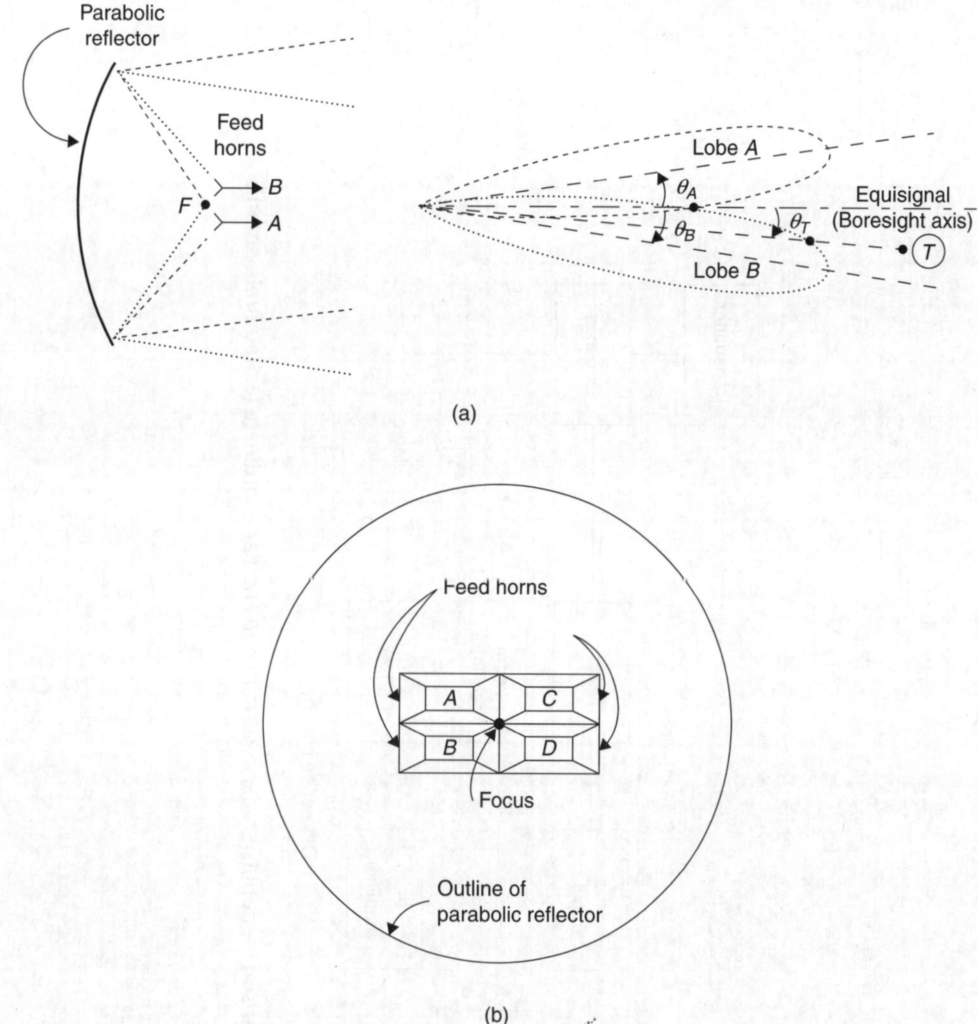

Fig. 11.20 (a) Parabolic reflector antenna with feed horns and the simultaneous lobe patterns (b) Parabolic reflector with feed horns for both azimuth and elevation tracking in amplitude comparison monopulse technique.

The block diagram of amplitude comparison monopulse tracking radar is given in Fig. 11.21 for a particular angular coordinate (azimuth/elevation). The two adjacent antenna feeds (placed row-wise for vertical/elevation tracking or placed column-wise for horizontal/ azimuth tracking) are connected to two collinear arms (1 and 2) of a magic tee. The magic tee will provide a sum signal in the *H*-arm and a difference signal in the *E*-arm corresponding to the signals being fed at the collinear arms. The transmitter is connected via the duplexer to the *H*-arm of the magic tee so that the sum signal is transmitted. On reception, the output from the *H*-arm of magic tee is connected via the duplexer to the sum channel from which the range information is extracted. The output from the *E*-arm of the magic tee is taken to the difference channel to extract the angle-error information

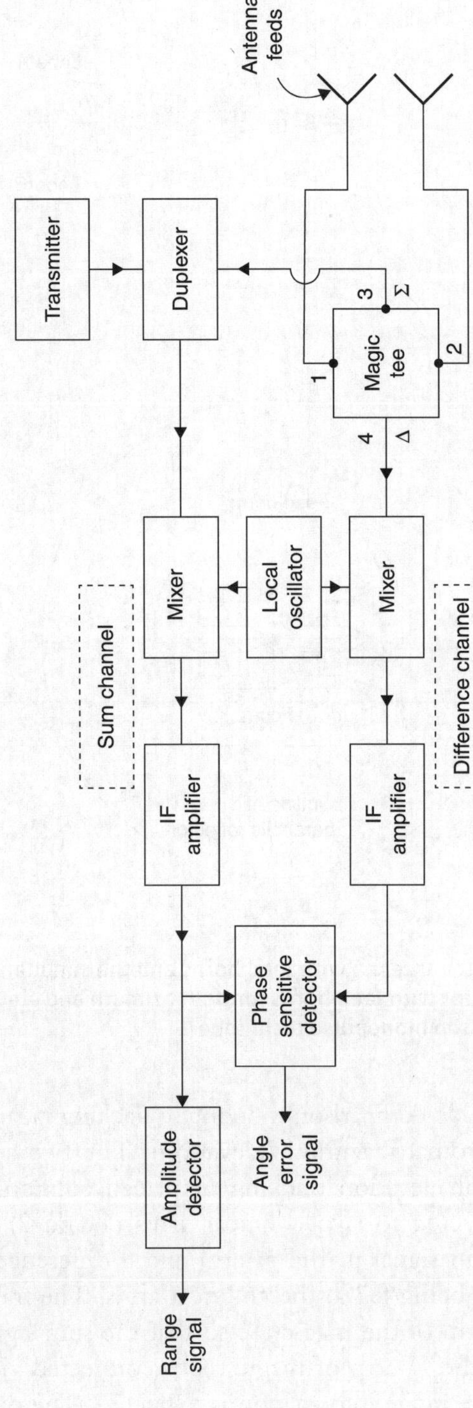

Fig. 11.21 Block diagram of amplitude comparison monopulse radar system in one angular coordinate (azimuth/elevation).

(both its magnitude and phase). The output of the phase-sensitive detector provides this angle-error information whose magnitude is proportional to the amount of angular error and whose sign is proportional to the direction. Both the sum and difference channel signal processing is done with super-heterodyne receiver. The block diagram of a monopulse radar with a provision for extracting angle-error signal in both azimuth and elevation is a little more complicated and may be found in standard reference books on radar [27].

Unlike in amplitude comparison monpolulse, two separate antennae placed adjacent to each other are used in *phase comparison monopulse* that sends out the lobe patterns A and B with the boresight axis of the antennae making same angle θ with respect to the target when un-tracked (Fig. 11.22a). Thus, the return echo signal received from the target corresponding to illumination by both the antenna lobe patterns will have the same amplitude irrespective of the value of the angle θ. However, the angular error (θ) that exists between the target

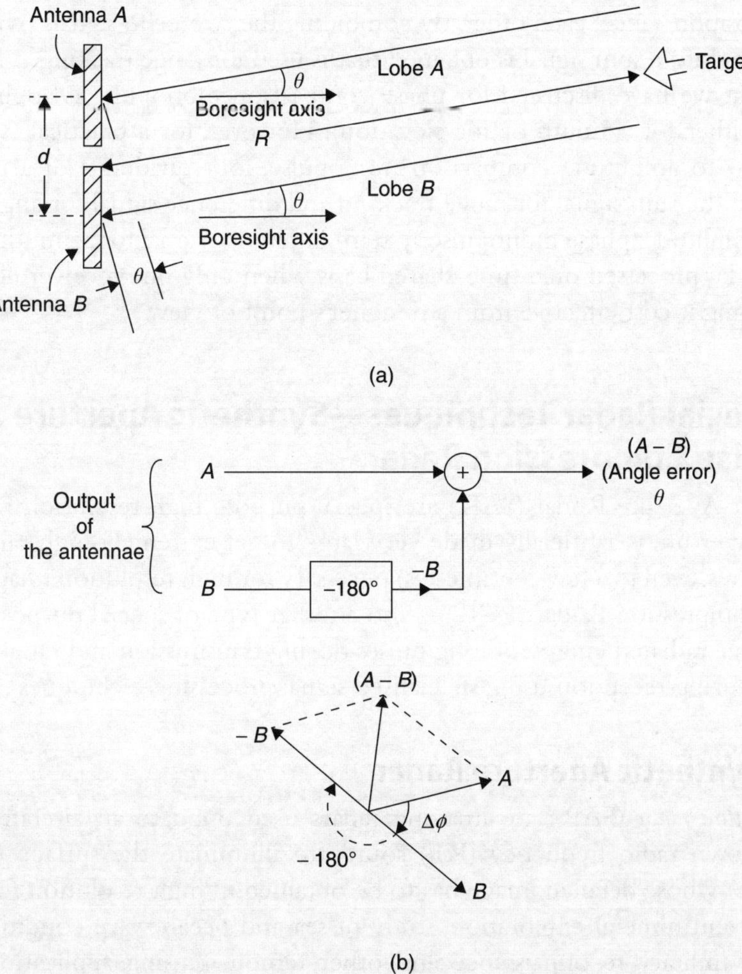

Fig. 11.22 (a) Phase comparison monopulse antenna with the beams directed to the target (b) Extraction of angle error signal.

and the boresight axis of the antenna beam patterns will give rise to a path difference (d) between the two echo signals and hence a phase difference of $\Delta\varphi = 2\pi d/\lambda \approx 2\pi\theta/\lambda$ (since, for all practical situations $d << R$; θ is thus very small). Hence a comparison of the phase of the two received echo signals is the estimate of the angle error θ. The antenna used in phase comparison monopulse system may be a planar array antenna in which the two antennae producing lobes A and B mentioned above are basically represented by the two halves of the array.

The angle-error signal is extracted from the two received echo signals that differ in phase (as mentioned above) if the target is not being tracked (if $\theta \neq 0$) by introducing an external phase shift of 180° to one of the echo signals and then summing up the two (Fig. 11.22b). When angular error exists, the phase difference of the two echo signals results in a proportional difference signal which is used to actuate the servomechanism to align the antenna beam in such a way that θ becomes zero and the target is thus tracked. When the target is tracked, the said difference signal becomes zero as the two echo signals of the same amplitude and in phase opposition cancel each other. By combining the two echo signals (without the said external phase shift) a sum signal is obtained that is used for range tracking.

So far what we have discussed for phase comparison monopulse is applicable for one coordinate (either for azimuth or for elevation). However, for a practical two coordinate system, similar to amplitude comparison monopulse, four quadrants of a planar array is used to obtain the sum signal for range tracking and difference signal for angle tracking. In either case (amplitude/phase monopulse system) difference signal for azimuth and elevation tracking may be processed on a time-shared basis when only one receiver channel will be required making it cost effective from a designer's point of view.

11.8 Special Radar Techniques—Synthetic Aperture and Pulse Compression Radar

The Synthetic Aperture Radar (SAR) are special purpose high resolution radar in which the antenna aperture is artificially made very large to get extremely high range resolution (down to metres, even to a few centimetres) necessary for high resolution imaging purposes. The Pulse Compression Radar (PCR) is also another type of special purpose radar which can realise large radiated energy of long pulses during transmission and range resolution of short pulses during reception using specialised signal processing techniques.

11.8.1 Synthetic Aperture Radar

Synthetic aperture radar (SAR) are imaging radars used onboard an aircraft or spacecraft that uses its own radio frequency (RF) source to illuminate the surface (generally, the earth's surface) whose detailed image has to be obtained in high resolution for the purpose of geological and mineral exploration, study of sea and ocean wave conditions including oil slicks, for military reconnaissance and other remote sensing applications. The radio frequency used in such radar operation is generally of centimetre wavelength where atmospheric absorption is negligible. Earth surface imaging is also being done with optical signals or infra-red signals (which is expected to provide much better image resolution as the operating wavelength is comparatively lower than the RF signal). However, atmospheric absorption at those frequencies and adverse weather conditions limit their use in many applications. In fact, the radar image obtained with RF signals is quite different from the images acquired with visible or infra-red frequencies. Because of these differences, radar and optical/IR imaging data can be complementary to one another as they offer different perspectives of the earth's surface providing different information contents.

To improve the image resolution of such microwave/RF radar, the antenna aperture is increased enormously with a special *artificial* technique and hence the name SAR; as if the aperture is *synthetically* generated. The use of RF naturally permits mapping underneath the earth's surface (as RF can penetrate the earth's surface, lower the RF frequency larger is the depth of penetration) which is required for mineral exploration. It is interesting to note that satellite based SAR imaging helped the rescue team to rescue a child lying alive underneath debris for more than 4 days during the earthquake in Nepal in February 2015.

The SAR is a sophisticated type of high resolution imaging radar using radio frequency in which a large antenna aperture is synthetically/artificially generated. Similar RF imaging may also be done using real aperture radar (RAR) having a real aperture D but with poor image resolution (Fig. 11.23a, points A and B are resolvable by SAR which is not at all possible with RAR). This is because the antenna beam width (θ) is inversely proportional to its aperture dimension (D) for a give frequency of operation (since $\theta = k\lambda/D$, Section 10.2.2) making $\theta_{\text{SAR}} << \theta_{\text{RAR}}$. With RAR, to realise improved image resolution, either larger antenna aperture or shorter wavelength/higher frequency of operation needs to be used. Larger antenna has practical physical limitations [for example, to realise 25 m along-track/cross-range resolution at 100 km distance from the target using 25 cm wavelength RF signal an RAR would need 1 km (!) aperture diameter antenna] while the use of shorter wavelength (that is, higher frequency) RF generally leads to higher cloud and atmospheric attenuation, reducing the all-weather capability of the RF based imaging radar and also capable of providing less depth of penetration of RF signal into the earths' surface. To overcome the limitations of RAR a very large antenna aperture may be *synthesised artificially* with SAR thereby realising improved cross-range resolution even with centimetre wavelength RF signal; however, it requires special recording and processing of backscatter echo signal data.

In SAR, a radar with a given real antenna aperture whose radiated beam (having a beam width θ_a) spreads along the y-direction and shot towards the range (x) direction, is carried in a moving aircraft or spacecraft (Fig. 11.23b). As a result of this, a large antenna aperture is effectively generated 'synthetically' in the y-direction. As the aircraft fly along the y-direction with a velocity v, an echo signal is received by it corresponding to each transmitted pulse (marked by × in the figure). If such echoes received at every location of the aircraft differing by a distance $\Delta d = vT_p = v/f_p$ (where f_p is the pulse repetition frequency of the transmitted pulses) in its path of travel (y-direction) are summed up for say, n number of transmitted pulses, then the resulting effect would be similar to a linear array antenna whose length is the distance travelled by the aircraft during those n transmitted pulses. The element spacing of such an antenna with a synthesised aperture is Δd. It may be noted that this synthetic antenna aperture is a sequentially generated one in contrast to

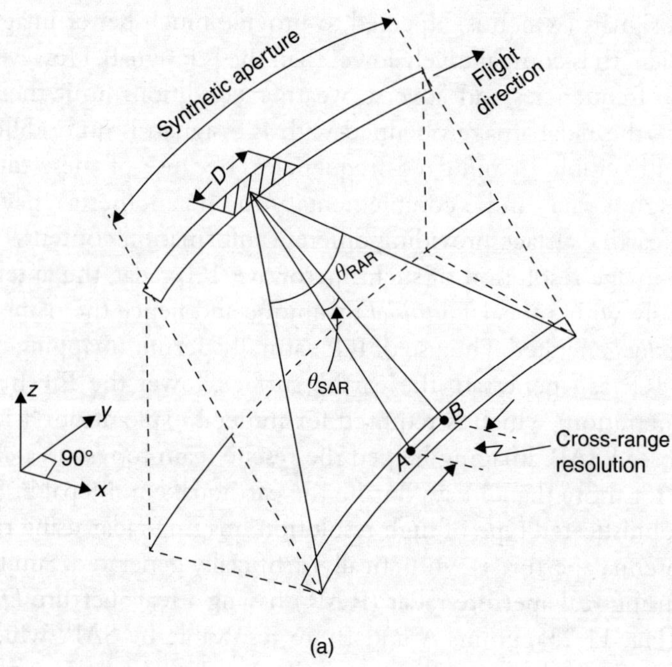
Synthetic aperture
Flight direction
D
θ_{RAR}
θ_{SAR}
z
y
90°
x
B
A
Cross-range resolution

(a)

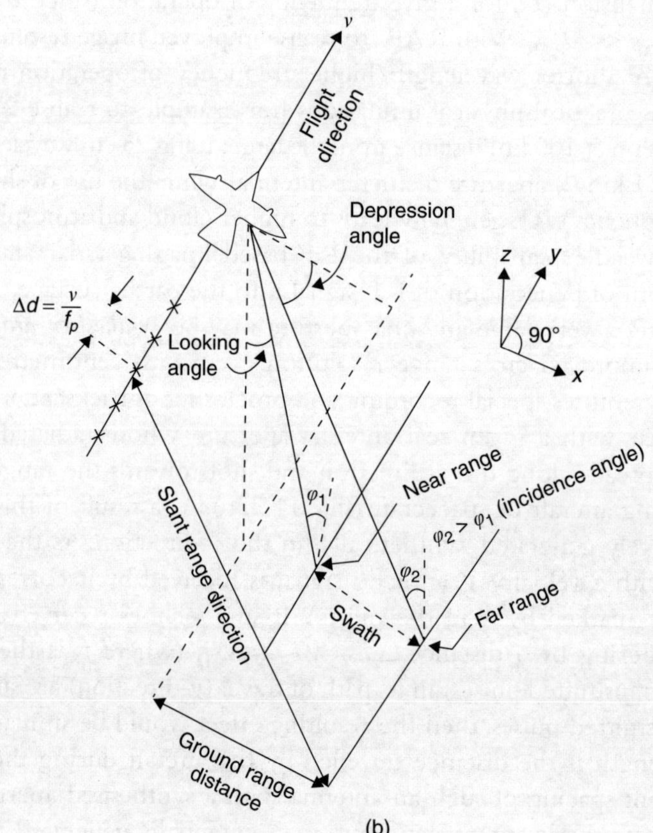
v
Flight direction
Depression angle
$\Delta d = \frac{v}{f_p}$
Looking angle
z
y
90°
x
Near range
φ_1
$\varphi_2 > \varphi_1$ (incidence angle)
φ_2
Slant range direction
Swath
Far range
Ground range distance

(b)

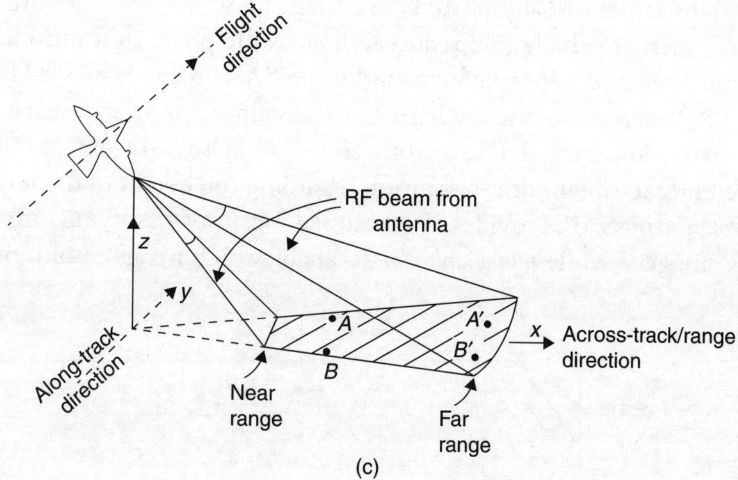

Fig. 11.23 (a) Concept of RAR and SAR (b) SAR principle and the viewing geometry of imaging radar (c) Along-track spatial resolution of RAR.

the simultaneously generated equivalent aperture of a real array antenna. SAR/RAR is also sometimes called side-looking airborne radar (SLAR) as the antenna beam looks sidewise (across-track/range direction) corresponding to its synthetically generated aperture in the direction of motion of the aircraft/spacecraft. Current microwave/millimetre-wave SAR can provide range resolution down to a metre or even down to a few centimetres.

SAR has certain advantages over RAR as far as cross-range/along-track spatial resolution is concerned. With RAR, the along-track resolution is determined by the angular width of the radiated RF signal and the slant range distance (Fig. 11.23c). At far ranges, as the radar illumination beam has to travel a longer distance from the sensor (located at the aircraft/spacecraft) the along-track resolution decreases (becomes coarser); thus the targets A and B in near range are seen to be resolvable but the targets A′ and B′, that have the same separation as A and B, is not resolvable being in the far range. However in SAR, the targets at far range, where the illuminating beam is widest, are illuminated for a longer period of time compared to the targets at the near range. The expanding beam width, combined with the increased time for which a target is within the beam as ground range increases, balance each other such that the resolution remains constant across the entire swath. Thus SAR can realise good along-track resolution with a reasonable size antenna operating at relatively longer wavelength and the range resolution is also independent of the slant range.

Aperture synthesis in SAR can be done by viewing the object in various ways giving rise to varieties of SAR techniques like swath (or strip) mapping SAR, spotlight SAR, inverse SAR imaging, interferometric SAR, polarimetric SAR and tomographic SAR. Spotlight SAR provides a higher image resolution compared to swath mapping SAR but at the expense of coverage area. *Swath mapping* SAR is also known as search SAR because it is useful for imaging large areas at relatively coarse resolution. *Spotlight* SAR on the other hand is used to obtain a relatively fine resolution image of a known location or target of interest (Fig. 11.24). Both swath mapping and spotlight SAR use a moving platform (the aircraft) and the target (usually earth) being imaged is stationary. They are not useful for targets in motion such as vehicles, ships and aircraft, as the phase shift introduced by target motion is not included and hence defocussing of the SAR image is inevitable. For moving targets, *inverse* SAR *imaging* is used where the imaging radar is stationary. *Interferometric* SAR involves two SAR images taken from antennae at slightly different locations and compared coherently to obtain fine-resolution regarding the height of the terrain or targets in the image. *Polarimetric* SAR provides an additional means of classifying different areas and objects in SAR imagery while *tomographic* SAR enables high image resolution.

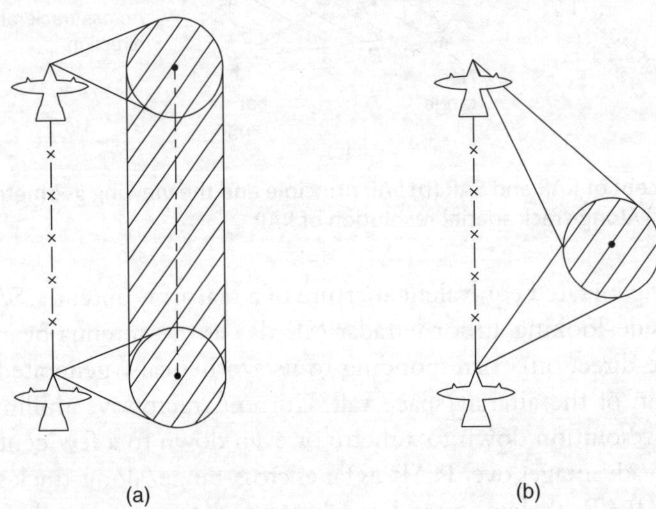

Fig. 11.24 SAR mapping modes (a) Swath (strip) mapping (b) Spotlight mapping.

11.8.2 Pulse Compression Radar

A *pulse compression radar* (PCR) provides the large radiated energy ($E_r = P_{peak} \times \tau$) capability of long duration pulse (for a given peak power) during transmission while realising the small range resolution of a short duration pulse during reception in the same radar system using some sort of signal processing. It is known that long or broad pulses have a large energy content, thus the signal-to-noise ratio of the received echo is higher but its bandwidth is small (being inversely proportional to pulse width) and hence posses poor range resolution ($\Delta R = c/2B_r$). But a short or narrow pulse has a large bandwidth and hence has a good range resolution but has small energy content. Thus, PCR is an optimised performance radar providing the advantages of both the long and short duration pulses simultaneously in the same system by using a judicious signal processing technique.

The PCR accomplishes its task by employing some form of frequency or phase modulation of the transmitted RF signal contained within the long duration pulse. Commonly used modulations are—linear frequency modulation (FM) and phase-coded pulse. In linear FM pulse compression, the transmitted signals envelope is a train of long duration pulses with the RF carrier contained in it being frequency modulated. Upon reception, the FM pulsed

RF echo signal is passed through a delay-line (known as pulse compression filter) whose delay time is a function of frequency (related to the FM characteristics of the transmitted carrier) thereby compressing the pulse. Thus the transmission is with long duration pulse energy while the received signal is processed so that the intelligence is retrieved from a compressed narrow pulse. Hence a good signal-to-noise ratio and good range resolution is simultaneously realised with the PCR radar. In the phase-coded pulse, the long duration transmitter pulse having a duration T is divided into N sub-pulses each of width τ. The phase of each sub-pulse may be 0 or π chosen either at random or following a certain code sequence (commonly used code is a PN code). The output of the receiver matched filter will be a compressed pulse of width τ with an amplitude N times greater than that of the long transmitted pulse. Different pulse compression schemes have different advantages depending on the application point of view; however, the linear FM pulse compression scheme is the most widely used one.

A scheme for linear frequency modulation (LMF) pulse compression radar described and patented by R H Dicke is given in Fig. 11.25. The transmitted signal is a frequency modulated (FM) waveform usually generated at low power and amplified by a power amplifier. The transmitted rectangular pulse waveform along with the up-chirp signal frequency modulating the RF transmitted signal are shown in Fig. 11.26a and 11.26b respectively; while the FM waveform of the RF transmitted signal is shown in 11.26c. On reception, the received echo of the transmitted FM signal is passed through the pulse compression filter (a delay-line whose velocity of propagation is proportional to frequency). The function of this filter is to speed up the higher frequencies at the trailing edge of the pulse while slowing down the lower frequencies at the leading edge of the pulse thereby

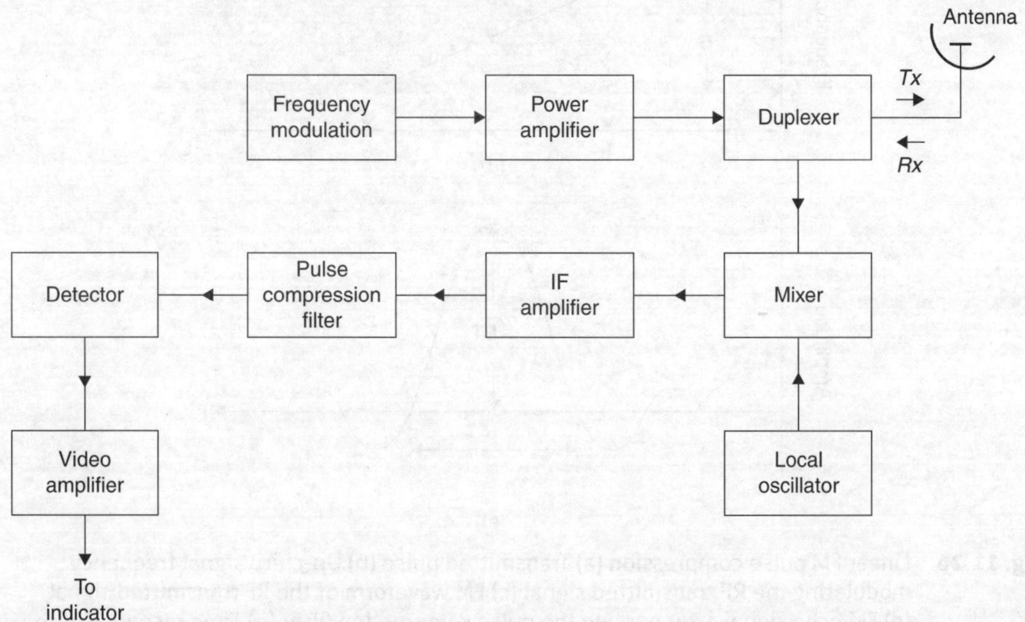

Fig. 11.25 Block diagram of an FM pulse compression radar.

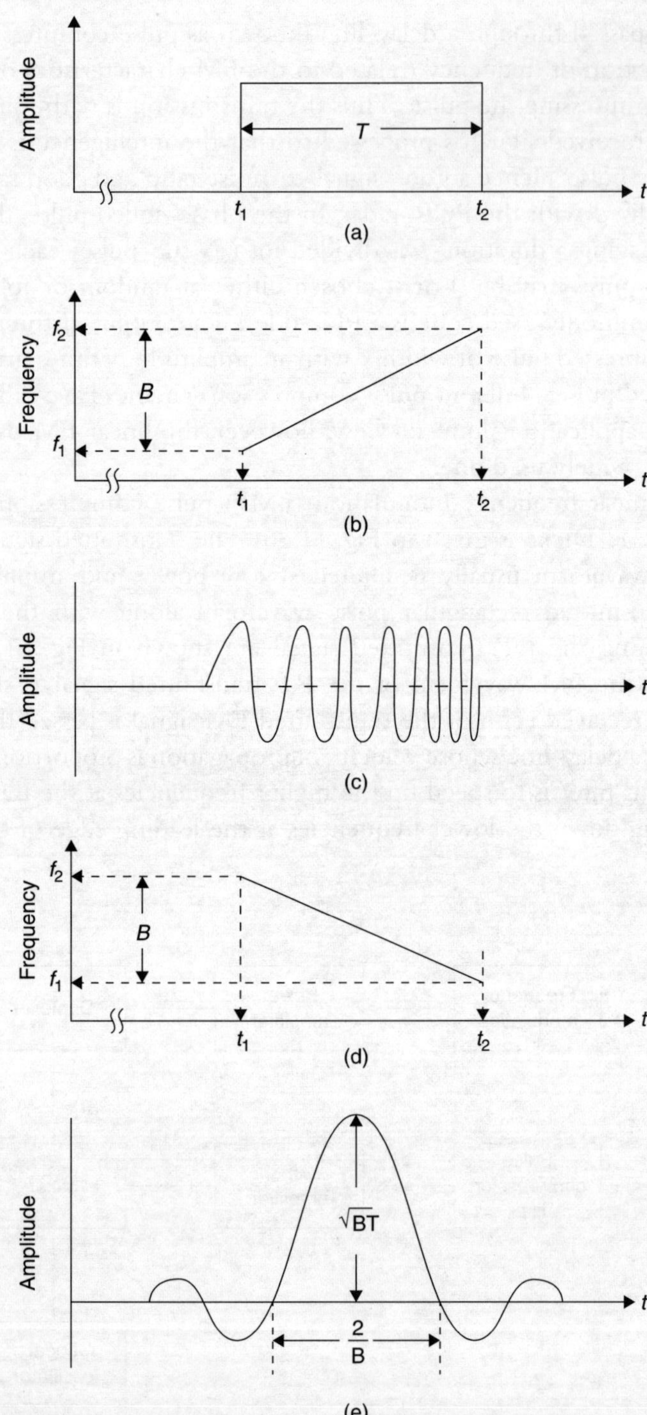

Fig. 11.26 Linear FM pulse compression (a) Transmitted pulse (b) Up-chirp signal frequency modulating the RF transmitted signal (c) FM waveform of the RF transmitted signal (d) FM echo signal after passing the pulse compression filter (e) Time response of the pulse compression filter.

compressing the signal to a bandwidth $1/B$, where $B = (f_2 - f_1)$ (Fig. 11.26d); the time response of the pulse compression filter is shown in Fig. 11.26e. The pulse compression filter is designed as a matched filter (as is usual with any radar receiver) and hence its output envelope is the autocorrelation function of the input. In the present case the output is proportional to $\sin(\pi BT)/(\pi BT)$ and the peak power of the pulse is increased by the pulse compression ratio $BT = T/\tau$ after passing through the filter.

Recapitulation

Radar range performance equation

Primary radar range equation

- The radar range performance equation relates the maximum achievable range by a radar system to the characteristics of the transmitter, receiver, antenna, target and the environment. The primary radar range equation is given by:

$$R_{max} = \left[\frac{P_t G_t \sigma A_e}{(4\pi)^2 S_{min}}\right]^{1/4} = \left[\frac{P_t G^2 \lambda^2 \sigma}{(4\pi)^3 S_{min}}\right]^{1/4} = \left[\frac{P_t A_e^2 \sigma}{4\pi \lambda^2 S_{min}}\right]^{1/4}$$

- **Note:** (i) $R_{max} \propto P_t^{1/4}$: To double the maximum range, the transmitter power needs to be increased sixteen fold—this is an impractical way to increase the maximum range of a primary radar. (ii) $R_{max} \propto (1/S_{min})^{1/4}$: A decrease in the minimum detectable signal S_{min} (by increasing the receiver sensitivity) will have the same effect as raising the transmitter power—appears to be very attractive alternative, however, it is application dependent. To guard against *jamming* it is better to employ a receiver with modest sensitivity and large transmitter power. (iii) $R_{max} \propto A_e^{1/2}$ ($\propto D$) and $\lambda^{-1/2}$: Thus, the most practical way to double the maximum range of the radar appears to be by doubling the diameter of the antenna or by reducing the wavelength or increasing the operating frequency of the radar system, however, the choice is again application dependent (tracking/surveillance).
- When the receiver noise is included, the radar range equation is modified to:

$$R_{max} = \left[\frac{P_t G^2 \lambda^2 \sigma}{(4\pi)^3 F_r N_i (S_o/N_o)_{min}}\right]^{1/4} = \left[\frac{P_t G^2 \lambda^2 \sigma}{(4\pi)^3 N_i (F_r - 1)}\right]^{1/4}$$

Secondary (beacon) radar range equation

- Maximum range for *interrogation link* is $r_{max,r} = \sqrt{\dfrac{P_{tI} A_{eI} A_{eB}}{\lambda^2 k_B T_0 B_r (F_{rB} - 1)}}$
- Maximum range for *reply link* from the beacon is $r_{max,R} = \sqrt{\dfrac{P_{tB} A_{eB} A_{eI}}{\lambda^2 k_B T_0 B_r (F_{rI} - 1)}}$
- The lower of the two values thus obtained is used for range determination, otherwise one of the links (having sorter range) will be lost.

Pulse and CW radar

- Unmodulated CW (purely CW) radar cannot measure the range of the target because the instant of transmission and instant of reception cannot be segregated since the transmitter is 'ON' for all instants of time and so we cannot talk of a received echo for a transmitted signal at any particular instant of time. However, measurement of the relative velocity of a moving target – as in speed monitors for automobile traffic control by the police and in speedometers for collision avoidance in very high speed trains – where large transmitter power is not required, such purely CW radar can perform the job at a very low cost.
- Unlike the unmodulated CW radar, the modulated CW radar, say the FM–CW radar, can measure range (in fact it is widely used as a radio altimeter in aircraft) in which the CW carrier is frequency modulated (FM) and the timing mark is the changing frequency of the modulating signal.
- The advantage of pulse radar is obviously its capability to transmit very high (of the order of megawatt) peak power so that long range detection can be done keeping the average power low (since $P_{av} = P_{\text{peak}} \times Duty\ cycle$, as duty cycle is typically 0.001). However, pulse radar has limitations of the so called maximum unambiguous range ($R_{\text{unamb}} = c/2f_p$) and the minimum range resolution ($\Delta R = c\tau/2$). But, CW radar can, in principle, operate down to zero range which is a very important aspect of CW radar in case of radio altimeters.
- Bandwidth requirement of the pulse radar is much larger compared to the CW radar. For pulse radar, the receiver bandwidth required is $B_r = \frac{1}{\tau}$ (if τ is of the order of μs, the bandwidth is of the order of MHz). For unmodulated CW, ideally an impulse function and bandwidth of a few Hz/hundred Hz is required. For FM–CW, the bandwidth requirement is given by $B = 2(f_\Delta + \alpha f_m)$. If $f_m = 10$ kHz then maximum desired value of B is of the order of 200 kHz. Thus, the bandwidth requirement for receiver of CW radar is much smaller than the pulse radar.

Radar cross-section of targets and radar clutter

- The measure of the incident power intercepted by the target and re-radiated back in the direction of the detecting radar is known as the *radar cross-section of the target*, denoted by σ and has the units of area. It is a characteristic of a particular target constituent (dependent on its material content) and is a measure of its aspect; the size as seen by the detecting radar.

$$\sigma = \lim_{R \to \infty} 4\pi R^2 \left|\frac{E_e}{E_i}\right|^2$$

- Depending on the type of application, a target happens to be a wanted target or an unwanted target. The radar echo received from an unwanted target for a particular application is known as *radar clutter* (confusing noise) because such echo can 'clutter' the display (PPI) of the radar receiver and make the detection of the wanted target very difficult.
- From the application point of view, clutter rejection is a must for clear target detection. For example, in the case of detecting an aircraft or ship (which is generally done by MTI radar).

However in some military applications, for example, in ECM, intentional artificial clutter is created (with chaff/decoy) to confuse the enemy radar. The counter methods used by enemy radar to combat ECM are called electronic counter–counter measure (ECCM).

- Clutter echoes are highly variable in nature and are described by probability density functions or probability distributions such as *Raleigh distribution*, *log-normal distribution* and *Weibull distribution*.

Doppler effect in radar and FM–CW radar

- In case of radar, the Doppler shift is doubled due to relative motion of the target and the radar caused by roundtrip of the incident and the echo signal when $f_d = 2f_d' = 2f_t v_r/c = 2v_r/\lambda$.

  The Doppler effect is observed only for radial motion or if the motion has a radial component of velocity, but it is zero for tangential motion.
- In FM–CW radar the transmitter frequency is changed periodically by frequency modulating the RF signal with a saw-tooth signal and the received echo that is delayed with respect to the transmitted signal is heterodyned with the transmitted signal to get the beat-note f_b which gives a measure of the target distance $f_b = \dfrac{4R \cdot \Delta f \cdot f_m}{c}$.
- Common use of the FM–CW radar is in the radio altimeter (operating in 4.2 to 4.4 GHz) to measure the height (~10 km, maximum) of the aircraft from the surface of the earth. When the radio altimeter used for blind landing of aircraft we talk of a fixed-error given by $R_{\min} = \dfrac{c}{4\Delta f} = \dfrac{75}{\Delta f(\text{MHz})}$ m.

Pulse Doppler and MTI radar

- Pulse Doppler radar usually operates with high enough PRF in order to operate with unambiguous Doppler (no blind speed) but at the expense of range ambiguity while MTI is also a type of pulse Doppler radar but operates with ambiguous Doppler measurement (with the so called blind speed) but with unambiguous range measurement utilising low enough PRF. Pulse Doppler radar is used primarily for defence applications where an aircraft might intentionally fly with blind speed to evade detection by enemy radar. But MTI is primarily used for clutter rejection for civilian applications where the target is not expected to move with blind speed. The use of multiple PRF MTI can take care of both unambiguous range and avoidance of blind speed.
- The basic principle of MTI radar depends on the Doppler effect in a pulse radar system in which phase shift occurs in the echo signal from one pulse to the next (at a rate proportional to the Doppler frequency) for moving targets while phase shift of the echo signal remains constant from pulse to pulse for stationary targets. Thus, if the output of the phase detector is subtracted in successive sweeps (in which the pulsed RF signal from transmitter is sent) then in the case of stationary targets (that gives rise to clutter) the output of the phase shifter will provide a zero residue while for moving targets a finite residue will remain. This results in moving target indication and clutter rejection. The output of the subtractor feeding the display (A-scope or PPI) will show up only the moving targets that looks like a 'butterfly'. The signal processing of this clutter rejection

at the receiver of the MTI radar may be done by using the so called *delay-line canceller/discriminator.*

- Digital MTI (DMTI) that uses combined I–Q channel scheme and digital signal processing solves the problem of signal loss that would otherwise occur due to blind phases in conventional single channel MTI using analogue signal processing (where the use of I–Q channels is difficult to implement).
- While operating an MTI radar, we come across *blind speed*; this occurs when the target happens to have a velocity whose radial component results in a phase shift of exactly 2π radians between successive pulses. This is the same as having no phase shift at all and thus the target appears stationary and echoes received from it are cancelled by the MTI action. A radial velocity corresponding to this situation is known as blind speed (v_{bn}), as are any integral multiples.

$$v_{bn} = \frac{n\lambda}{2} \cdot \frac{1}{T} = PRF \cdot \frac{n\lambda}{2}$$

- The effect of blind speeds can be significantly reduced, without incurring range ambiguity, if the MTI radar is operated with more than one PRF and the echo signal is processed accordingly.

Surveillance and tracking radar

- Surveillance radar is used to obtain long range information on all targets in a certain volume of sky or area of sea or land. In airports two such radar are used—one is airport surveillance radar (ASR) that typically operates in the S-band and has a coverage of minimum 60 nmi to search for targets in the vicinity of the airport, and the other is the air-route surveillance radar (ARSR) that operates at the L band and with a typical coverage range of 200 nmi that provides coverage between airports. The surveillance antenna used for this purpose is a fan-shaped beam antenna, the cosecant squared (Csc^2) antenna.
- The job of a tracking radar is to 'acquire' the target by performing a limited search with a *pencil beam* in the area of the designated target coordinates (provided by the surveillance radar). The tracking radar continuously determines the target path which may be processed to determine the future position of the target (either to guide the aircraft to make a safe landing or to direct a weapon on the target for defence applications). A tracking radar data is often used to measure the relative velocity of the target in terms of the rate of change of range instead of using Doppler shift (the latter technique being normally used in MTI radar).
- The range performance equation is significantly modified when the effect of antenna scanning searching the target is included and is given by:

$$R_{\max}^4 = \frac{P_{av} A_e \sigma E_i(n)}{4\pi k_B T_0 F_r (S_o/N_o)_n L_s} \cdot \frac{t_s}{\Omega}$$

- The tracking radar technique uses two general principles to generate the angle-error signal. In one approach (*lobe switching and conical scan*), the formation of an angular error signal requires the processing of a number of echo signal pulses during each sweep of the transmitted pulse train. The shortcoming of this approach is that the angle-error signal has every possibility of being sensitive to the amplitude fluctuations in the received echo signals resulting from random variations in the target cross-section. But, this pulse to pulse amplitude fluctuation of the echo signal will have no effect on tracking accuracy (in the generation of angle-error signal) if the angular error measurement is done on the basis of a single pulse. Thus, the modern approach is to use *monopulse* (as it uses a single echo pulse for angle-error signal generation) tracking methodology.

- There are in general, two monopulse radar techniques depending on the manner the angle-error information is extracted from the received echo signal—*amplitude comparison* and *phase comparison* monopulse.

Synthetic aperture and pulse compression radar

- The synthetic aperture radar (SAR) are special purpose high resolution radar in which the antenna aperture is made very large artificially (synthetically), to get extremely high range resolution (down to metres even to a few centimetres) necessary for high resolution imaging purposes. The PCR is also another type of special purpose radar which can realise large radiated energy of long pulses during transmission and range resolution of short pulses during reception using specialised signal processing techniques.
- Aperture synthesis in SAR can be done by viewing the object in various ways giving rise to varieties of SAR techniques like swath (or strip) mapping SAR, spotlight SAR, inverse SAR imaging, interferometric SAR, polarimetric SAR and tomographic SAR.
- The PCR accomplishes its task by employing some form of frequency or phase modulation of the transmitted RF signal contained within the long duration pulse. Commonly used modulations are—linear frequency modulation (FM) and phase coded pulse.

REVIEW QUESTIONS

11.1 What is RADAR? What are SONAR, SODAR and LIDAR? Mention their applications.

11.2 Name a few radio aids to navigation and mention their uses.

11.3 From the radar range equation, comment on what are the trade-offs of different design parameters to improve the maximum range of a radar.

11.4 Discuss how the range equation of radar is modified due to the presence of receiver noise.

11.5 Discuss the physical reason of re-radiation of signal from a radar target, that is, the generation of echo signal in the primary radar.

11.6 What is radar beacon (RACON)? Give a simple sketch of the interrogator–transponder system used in RACON.

11.7 Why can't range measurement of the target be done with unmodulated CW radar? Explain how FM–CW and pulse radar solve this problem.

11.8 What do you understand by monostatic and bistatic radar? Why is unmodulated CW radar, in general, bistatic in nature?

11.9 Pulse radar has the limitations of maximum unambiguous range and minimum range resolution—explain.

11.10 Discuss the bandwidth requirement of pulse and CW radar.

11.11 Explain the operation of a pulse radar with a suitable block diagram.

11.12 Discuss the terms CFAR, STC and FTC with reference to a radar receiver.

11.13 What do you understand by radar cross-section (RCS) of targets? How is the RCS of a target determined?

11.14 What is radar clutter? In ECM, intentional artificial clutter is created to confuse the enemy radar—discuss a few such techniques. Name a few statistical techniques to estimate land/sea clutter.

11.15 Discuss the Doppler effect in radar.

11.16 Explain the principle of operation of FM–CW radar. What is fixed error of such radar? Higher carrier frequency reduces the fixed error—explain.

11.17 'MTI is pulse Doppler radar', but in a strict sense, there is a subtle difference between MTI and pulse Doppler radar—explain.

11.18 Discuss the principle of MTI radar. How does it realise clutter rejection and depict the presence of a moving target in the background of clutter?

11.19 Explain with a block diagram the principle of delay-line cancellor/discriminator used in moving target indication. What is eclipsing effect and blind speed?

11.20 Discuss with the necessary block diagram, the principle of operation of MTI system using (a) POS type transmitter (b) MOPA type transmitter.

11.21 What is DMTI? Explain the principle of operation of a DMTI system with the help of a block diagram. Discuss how the problem of blind phase is solved with the DMTI scheme.

11.22 Explain the principle of range gate and velocity gate for accurate range and velocity measurement of radar targets.

11.23 Discuss the principle of occurrence of blind speed in MITI radar. Explain how the occurrence of blind speed can be mitigated by the use of multiple PRF.

11.24 What is surveillance radar? Name two common types of such radar used in airports indicating their coverage range and purpose. Discuss how the effect of scanning modifies the radar range performance equation.

11.25 What is tracking radar? Name a few tracking radar sytems used in practice.

11.26 Discuss the principle of lobe-switching and conical scan for generating angle-error signal in tracking radar. What is monolpulse tracking?

11.27 Discuss the principle of amplitude monopulse technique. Explain the scheme for amplitude monolpulse tracking radar for a particular angular coordinate with the help of a block diagram.

11.28 What is phase comparison monopluse? Discuss how angular error is derived in such systems when the target has not yet been tracked?

11.29 What is synthetic aperture radar (SAR)? How does it realise a much better cross-range resolution compared to a real aperture radar? Name a few modes of aperture synthesis in SAR.

11.30 What is pulse compression radar (PCR)? Explain the principle of operation of linear frequency modulation (LFM) pulse compression radar system with a block diagram. What is phase coding PCR?

NUMERICAL PROBLEMS

11.1 A surveillance radar operates at 3 GHz and uses a 1.5 m diameter dish antenna for both transmitting and receiving. If the mean transmitting power is 10 kW and the noise level at the input of the receiver is –130 dBW, calculate the maximum range at which an aircraft of cross-section 1.5 m^2 can reliably be detected. Assume 6 dB loss and a signal to noise ratio (SNR) of 12 dB.

[**Ans:** 15.6 km]

11.2 A satellite tracking radar operating at 2.5 GHz has the following specifications for the interrogator and the transponder:

Interrogator		*Transponder*	
Transmitter peak power	0.5 MW	Transmitter peak power	50 W
Antenna diameter	64 m	Antenna diameter	1 m
Noise figure of the receiver	1.1	Noise figure of the receiver	13 dB
Receiver bandwidth	5 kHz	Receiver bandwidth	5 kHz

Calculate the maximum detection range of this radar.

(Given: Ambient temperature = 17°C, Boltzmann constant $k_B = 1.38 \times 10^{-23}$ J/K)

[**Ans:** Maximum detection range of this beacon radar is 1,030 million km]

11.3 A pulse radar operates with a peak power of 1 MW having a pulse width of 1 μs and pulse repetition period of 1 ms. Find the duty cycle, average power and energy of the pulse. What are the range resolution, maximum unambiguous range and bandwidth of this radar?

[**Ans:** DC = 0.001, P_{av} = 1 kW, E_p = 1 J, ΔR = 150 m, R_{unamb} = 150 km, B_r = 1 MHz]

11.4 A 5 GHz police radar measures a Doppler frequency of 927 Hz from a car approaching the stationary police patrol vehicle, in an 80 km/h speed limit zone. What is the velocity of the car with respect to the police radar? Has the car exceeded the specified speed limit?

[**Ans:** v = 100 km/h, yes the car has exceeded the specified speed limit.]

11.5 It is desired that a pulse radar have maximum unambiguous range of 200 nmi and no blind speed of target less than 270 knots. Calculate the highest frequency of operation of the radar. Given: 1 nmi = 1.852 km and 1 knot = 1.85 km/h.

[**Ans:** Highest radar operating frequency: 437.4 MHz.]

11.6 (a) A pulsed radar has a pulse width (PW) of 1 μs, what is the extent (in m) of the pulse energy in space in the range coordinate? (b) What should be the pulse repetition frequency of a radar in order to achieve a maximum unambiguous range of 80 nmi? (c) What is the duty cycle of a radar that has PW of (a) and PRF of (b)? (d) A radar operating at 2.6 GHz uses a Csc^2 antenna that has rectangular aperture of 20 ft width and 6 ft height. Find the azimuth and elevation bandwidth of the radiation pattern of the antenna and also its gain. Given antenna aperture efficiency $\eta = 0.6$.

[**Ans:** (a) 300 m (b) 1.012 kHz (c) 0.001 (d) $\theta_a = 1.083°$, $\theta_e = 3.60°$, G = 37.8 dB]

11.7 Calculate the maximum range in nautical miles of a ground-based surveillance radar if it has to detect a target that has a radar cross section of 3 m^2. The frequency of operation of the radar is 2.8 GHz and it uses a Csc^2 antenna whose rectangular antenna aperture is 5 m wide and 2.6 m high. Antenna aperture efficiency is 0.6. The minimum detectable signal by the receiver of the radar is 10^{-12} W. The average power transmitted by the radar transmitter is 300 W using a pulse train of PW 1 μs and PRF of 1 kHz. (Given 1 nmi = 1.852 km)

[**Ans:** 84.34 nmi]

11.8 An L-band air traffic control (ATC) radar has the following technical specifications:

Frequency of operation: 1.5 GHz
Transmitter peak power: 1 MW
Pulse width: 1 μs
Pulse repetition frequency: 1 kHz
Csc^2 antenna size: 18 ft wide (*W*) and 5 ft height (*H*)
Azimuth and elevation search volume: 360° × 30°
Antenna rotation rate: 15 rpm
Noise figure of the receiver: 5.0 dB
Post-detection signal to noise ratio: 7.0 dB
Detection efficiency: 85%
System losses: 12 dB

Calculate the detection range of this radar. (Given: Radar cross-section of the target is 35 m^2 and the antenna aperture efficiency is 0.6)

Also show that the number of pulses hitting the target in the present case results in an echo signal strength improvement by 22.3 dB due to pulse integration taking place at the receiver.

[**Ans:** 57.76 nmi]

12 Microwave Measurement Techniques

Practice makes one perfect

12.1 Introduction

Microwaves are short-wavelength radiowaves where the ordinary concepts of voltage, current and impedance lose their conventional meaning as the operating wavelength (starting in metres and going down to millimetres) is comparable to and sometimes even smaller than the component size. Though in the radio frequency spectrum (wavelengths starting from tens of km down to sub-mm) electromagnetic radiation is associated with electrical phenomena which occur in conductors (unlike those for infra-red to X-ray region and beyond, where the generation of radiation involves atomic and molecular vibrations and transitions between the energy states of various particles), in the microwave region of the radiowave spectrum, the behaviour of the propagating electromagnetic wave must be determined in terms of the electric and magnetic field quantities (and not in terms of voltage and current as in the case of lower radiowave frequency). For example, the power flow in a microwave system is specified in terms of the product of the electric (E) and magnetic (H) fields (in terms of the Poynting vector $\vec{P} = \vec{E} \times \vec{H}$), while the impedance (wave impedance, Z_w) is determined in terms of the ratio of these two fields ($Z_w = E_x/H_y$). The techniques of measurement at microwave frequencies is thus expected to be different from lower frequency radiowaves and also from those at infra-red through optical to X-ray wavelengths.

Primary measurable quantities in the microwave region of the electromagnetic spectrum are standing wave ratio/reflection coefficient, power and frequency. Determination of unknown microwave impedance or the quality factor (Q) of a microwave cavity can be done in terms of standing wave ratio measurement. Also the S-parameters of a microwave network can be determined in terms of microwave power measurement. In this chapter, we will primarily discuss the measurement techniques for impedance, power, quality factor and frequency. A brief discussion for antenna measurement and the measurement of dielectric constant of electromagnetic material at microwave frequencies is also included. Measurement of the noise figure of amplifiers and other microwave networks and phase noise of oscillators is an important issue for a microwave engineer. Thus, the basics of these measurements have also been covered. The fundamentals of two very important microwave measuring equipments – spectrum analyser and network analyser – are given at the end of this chapter as these two instruments have become an integral part of microwave engineering design and characterisation of active and passive components and systems.

12.2 Impedance Measurement at Microwave Frequencies

The most popular method of measuring impedance at microwave frequencies is based upon the exploration of the standing wave pattern by means of a standing wave detector (SWD). The device under test (DUT) terminates the SWD and the standing wave pattern is explored using a probe and an indicator. The normalised impedance can be evaluated in terms of the measurement of the standing wave ratio, position of any voltage minima and the wavelength of the exciting microwave signal.

It is also possible to measure impedance at microwave frequencies in terms of the measurement of the reflection coefficient. With the help of two back to back directional couplers, the incident and reflected travelling waves can be separated and the magnitudes of reflected and incident waves can be measured and hence the reflection coefficient or return loss. With auxiliary arrangements, phase information can also be obtained. This is precisely the reflectometer technique which is the heart of the network analyser, a versatile equipment to characterise microwave circuits.

Microwave bridges may be constructed, for example with a magic tee, which corresponds to conventional low-frequency impedance bridges. With the help of such bridges it is possible to compare unknown impedance against an impedance standard at microwave frequency.

12.2.1 SWD Technique of Impedance Measurement

12.2.1.1 Derivation of the working formula

The theory of SWD technique can be developed by starting from the expression for the input impedance (Z_{in}) of a transmission line of length l and characteristic impedance Z_0 terminated by a load impedance Z_L (Fig. 12.1). We have Z_{in} for a high frequency transmission line as (Eq. 3.17b):

$$Z_{in} = Z_0\left(\frac{Z_L + jZ_0 \tan\beta l}{Z_0 + jZ_L \tan\beta l}\right) \tag{12.1}$$

If the line is matched (if $Z_L = Z_0$), there is no standing wave on the line (when SWR, $\rho = 1$) and $Z_{in} = Z_0$. But when the line is not matched (when $Z_L \neq Z_0$), then there will be a standing wave on the line and the standing wave ratio ρ increases as mismatch between load and the characteristic impedance of the line increases. As the mismatch factor ρ increases, the voltage maximum V_{max} in the SW pattern increases and the voltage minimum V_{min} decreases and similar is the case with I_{min} and I_{max} respectively. Thus, with the increase of the mismatch factor ρ, the impedance at a voltage maximum (V_{max}/I_{min}) increases and the impedance at a voltage minimum (V_{min}/I_{max}) decreases, when we can write:

$$\{Z(x)\}_{max} = Z_0\rho \tag{12.2a}$$

and $$\{Z(x)\}_{min} = \frac{Z_0}{\rho} \tag{12.2b}$$

Equating Eq. 12.1 with Eqs 12.2a and 12.2b, we obtain two different expressions for Z_L:

Firstly,

$$Z_L = Z_0\left(\frac{\rho - j\tan\beta d_1}{1 - j\rho\tan\beta d_1}\right) \tag{12.3a}$$

This is obtained when measurement is done with respect to a voltage maximum, with d_1 as the distance of the first voltage maximum from the load end.

And secondly,

$$Z_L = Z_0\left(\frac{1 - j\rho\tan\beta d_2}{\rho - j\tan\beta d_2}\right) \tag{12.3b}$$

This is obtained when measurement is done with respect to a voltage minimum, with d_2 as the distance of the first voltage minimum from the load end.

The working formula that is used for impedance measurement is Eq. 12.3b, that is, the one for which measurement is done at a voltage minimum. This is because a voltage minimum is always sharper than a voltage maximum in a standing wave pattern making measurement accuracy better. Also probe loading (which will be discussed shortly) is less when measurement is done at a voltage minimum. It may be noted that since the two voltage minima differ by a phase difference of π and $\tan(\pi + x) = \tan x$, d_2 may be considered to be the position of any voltage minimum from the load end (in fact, a farther minimum from load end would improve the measurement accuracy).

From Eq. 12.3b, we observe that if we can measure VSWR (ρ), propagation constant β ($= 2\pi/\lambda$) and d_2 we have the value of the unknown load impedance Z_L.

12.2.1.2 Measurement methodology and experimental set-up

To determine VSWR (ρ), the standing wave pattern within the waveguide/transmission line is sampled by an electric probe, or simply a probe, (which responds and samples out the RF signal only from the electric field) or by a magnetic loop (which responds to the magnetic field), Fig. 12.2.

Normally an electric probe is used, which acts as a loosely coupled pick-up antenna and therefore gets an induced voltage proportional to the field intensity and the depth of

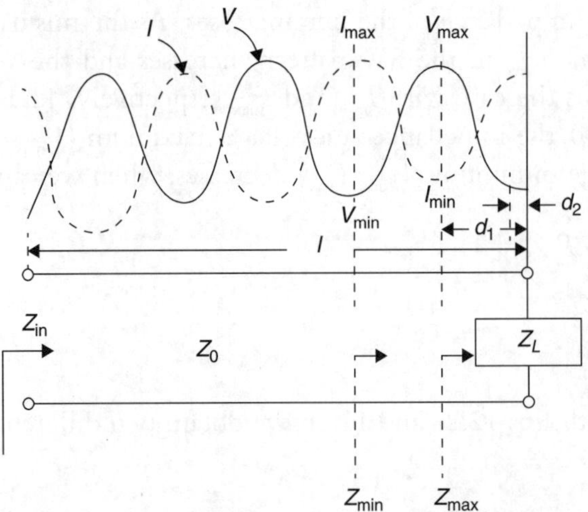

Fig. 12.1 Transmission line of length l and characteristic impedance Z_0 terminated by a load $Z_L \neq Z_0$.

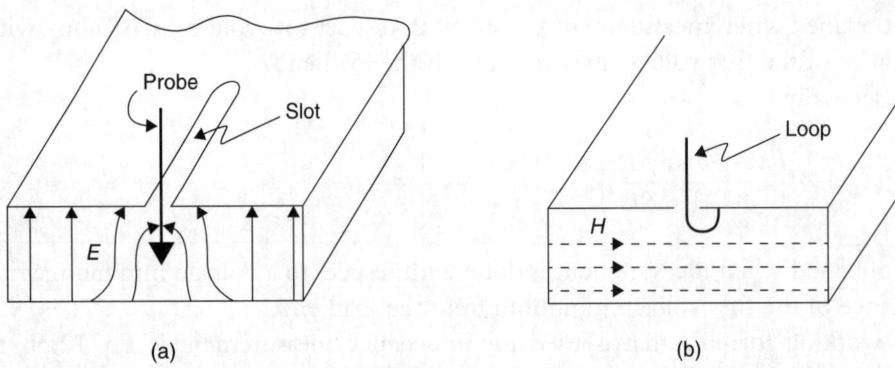

Fig. 12.2 Sampling of SW pattern within the waveguide with (a) Electric probe coupling (b) Magnetic loop coupling.

penetration of the probe in the guide of the SWD. Sensitivity of measurement increases with the probe depth; however, if inserted too far, the probe (appearing as an admittance shunting the line or guide) may cause a distortion in the standing wave pattern by setting-up an appreciable reflection causing measurement error known as *probe loading*. The probe is inserted in the waveguide of the SWD through a non-radiating slot (which is a narrow slot cut along the central line of the top broad wall of the waveguide, thus interrupting minimum of the transverse surface current lines). Also the electric field is maximum along the central line on the broad wall of the waveguide allowing maximum coupling of signal from the standing wave existing within the guide that is being sensed by the probe. The induced RF voltage in the probe is rectified by a crystal detector (a square law device with characteristics—rectified DC current $I_{dc}\ \alpha\ V_{RF}^2$, where V_{RF} is the induced RF voltage sampled by the probe from the standing wave pattern). The probe has an attached tuning plunger (set to quarter wave length position at the operating frequency, f_0) that helps to tune the system for maximum rectified voltage by presenting very high impedance looking towards the probe assembly so that the least possible disturbance is caused to the standing wave pattern within guide, Fig. 12.3(a). The output of the crystal detector registered in a microammeter is either the voltage wave squared or the power. Thus the ratio of the galvanometer reading corresponding to the probe located at a voltage maximum to that at a voltage minimum of the SW pattern is the PSWR (power standing wave ratio) or VSWR squared. A photograph of a standing wave detector (SWD) and the tunable probe is shown in Fig. 12.3(b).

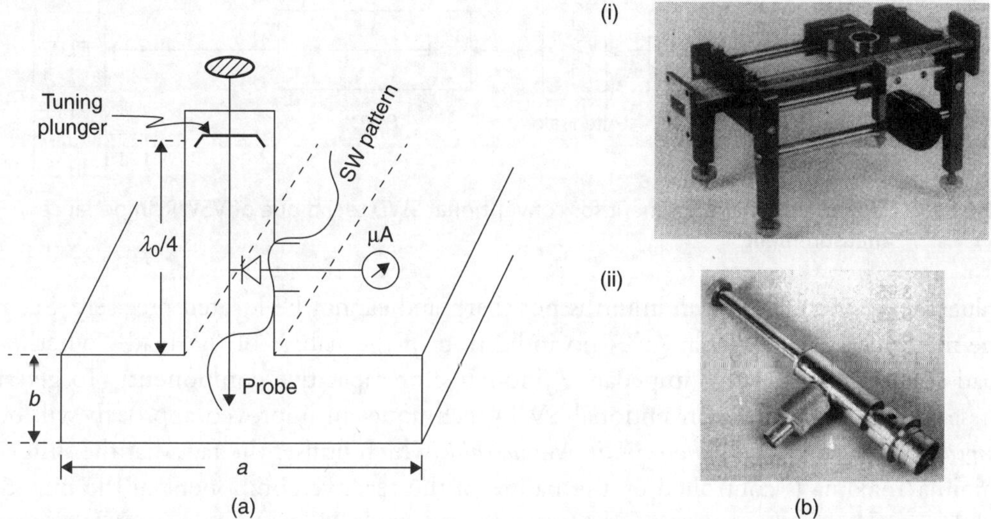

Fig. 12.3 **(a) Probe assembly with tuning plunger and crystal detector to rectify the RF voltage sensed from the SW pattern by the probe acting as pick-up antenna (b) Photograph of SWD and the tunable probe (i) Standing wave detector (ii) Tunable probe.**

To improve the sensitivity of measurement, the output from the detector (which is a small DC signal) needs to be amplified. This amplification cannot be done with a DC (direct coupled) amplifier since a DC amplifier is prone to amplifying the power supply drifts thus degrading the signal to noise ratio at the output of the amplifier. A practical solution is to modulate the microwave signal generator by a 1 kHz signal when at the output of the crystal detector we get the modulating signal and this can now be amplified with a narrow band tuned audio frequency amplifier (so that signal to noise ratio will be high thereby improving the measurement sensitivity). The use of amplifier also helps to satisfy the condition for small probe depth thus reducing the possibility of probe loading.

The experimental arrangement for the conventional SWD technique of impedance measurement at microwave frequency is shown in Fig. 12.4. Note that, to measure guide wavelength we can use a simple experimental technique, instead of using a frequency meter. By replacing the DUT with a short circuit load we have sharp voltage minima at intervals of half wavelength and probing two consecutive minima we have the measure of $\lambda_g/2$.

12.2.1.3 Impedance determination by short circuit minima shift

The conventional SWD technique for impedance measurement at microwave frequencies has some limitations. Firstly, the measurement of d_2 is not accurate as the reference load plane (a mechanical reference plane) is not a sharply defined plane. Also for medium VSWR

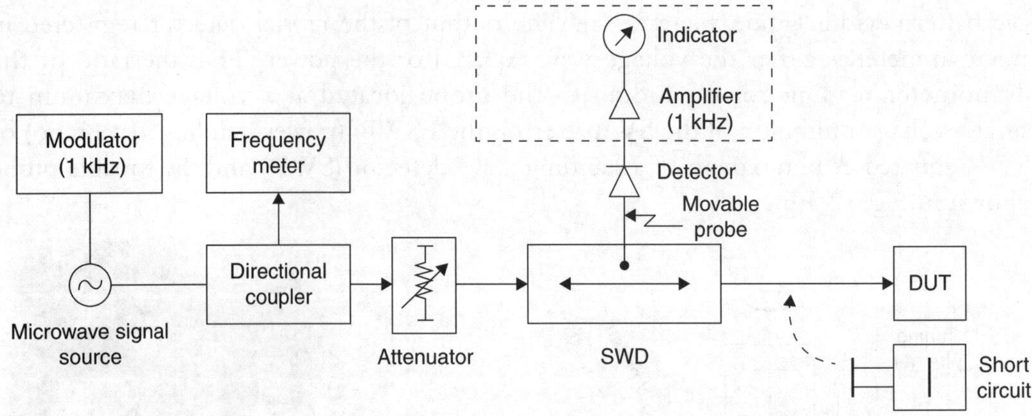

Fig. 12.4 Experimental arrangement for conventional SWD technique of VSWR/impedance measurement.

value, the position of the minimum is not sharp and cannot be located precisely. Secondly, the method discussed so far gives no indication of the nature of the load—whether the load contains any reactive impedance (inductive or capacitive component). To get rid of these limitations of the conventional SWD technique, an improved approach will be the *impedance determination by short circuit minima shift* (which utilises the fact that the shift of the minima/maxima is controlled by the nature of the reactive component in the impedance while the magnitude of the minima/maxima is controlled by the mismatch factor (ρ). In this method we first short circuit the load end and take a null (position of a minimum) as reference (very sharp and precise). Then the short is replaced by the load (DUT) and we note down the shift of a particular null with respect to the short circuit null (Fig. 12.5). Conventionally, the measurement is first done with the load first and then the DUT (load) is replaced by the short when the interpretation of the result of measurement is done in the following way.

- The shift in the minimum after the load has been shorted is never more than $\pm \lambda/4$.
- If shorting the load causes the minimum to move towards the load, a capacitive component exists in the load.
- If shorting the load causes the minimum to shift towards the generator, an inductive component exists in the load.
- If shorting the load causes no shift in the minimum, a completely resistive load exists, equal to Z_0/ρ.
- If shorting the load causes the minimum to shift exactly $\lambda/4$, the load is completely resistive and has a value of $Z_0\rho$.

The working formula given by 12.3b may be modified for short circuit minima shift method as follows.

$$\frac{Z_L}{Z_0} = z_L = \frac{1 - j\rho\tan(X)}{\rho - j\tan(X)} \tag{12.3c}$$

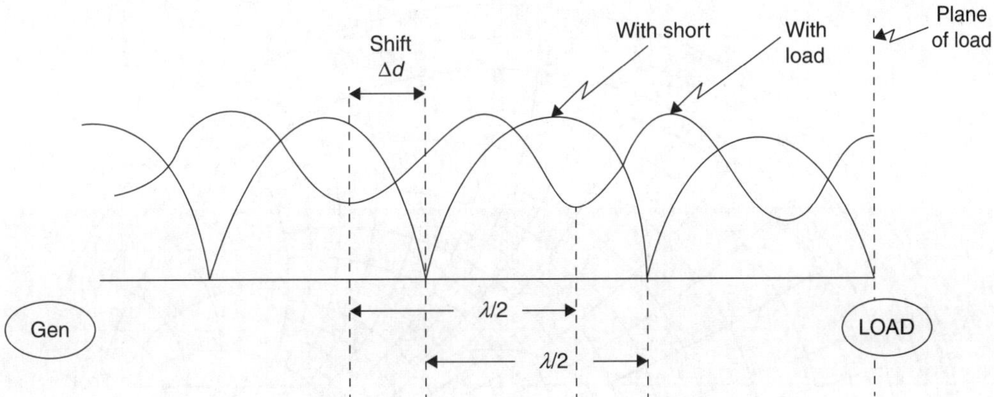

Fig. 12.5 Diagram to explain impedance determination by short circuit minima shift.

where $X = \frac{2\pi}{\lambda}(d_2) = \frac{180°(\pm\Delta d)}{\lambda/2}$

As per connection, Δd takes a –ve sign when the minima shift towards the generator (load impedance has an inductive component) and Δd takes a +ve sign when the minima shift towards the load (load impedance has a capacitive component).

EXAMPLE 12.1

In a microwave impedance measurement experiment, the observed VSWR with the line being loaded is 2, and when the line is shorted the minima shifts by 0.15 λ towards the load. Determine the load impedance both by analytical calculation and using a Smith chart. The characteristic impedance of the line is 50 Ω.

Solution

With reference to Eq. 12.3c, we have $z_L = \frac{1 - j2\tan(54°)}{2 - j\tan(54°)} = 0.981 - j0.700$

with $X = \frac{180° \times 0.15\lambda \times 2}{\lambda} = 54°$, since the minima shifts towards the load.

Thus $Z_L = (49.05 - j35.02)\ \Omega$.

An alternative method by using Smith chart may be done by using following steps.

Step 1: The constant VSWR (impedance) circle is drawn on the Smith chart with VSWR = 2.

Step 2: A distance of 0.15λ is moved along the distance scale marked 'towards the load' starting from first V_{min} located at short circuit (s.c) point in the Smith chart. A line is drawn from the centre of Smith chart through this point, which is the new Γ vector of the given terminating load. The intersection of this Γ vector line with the constant impedance

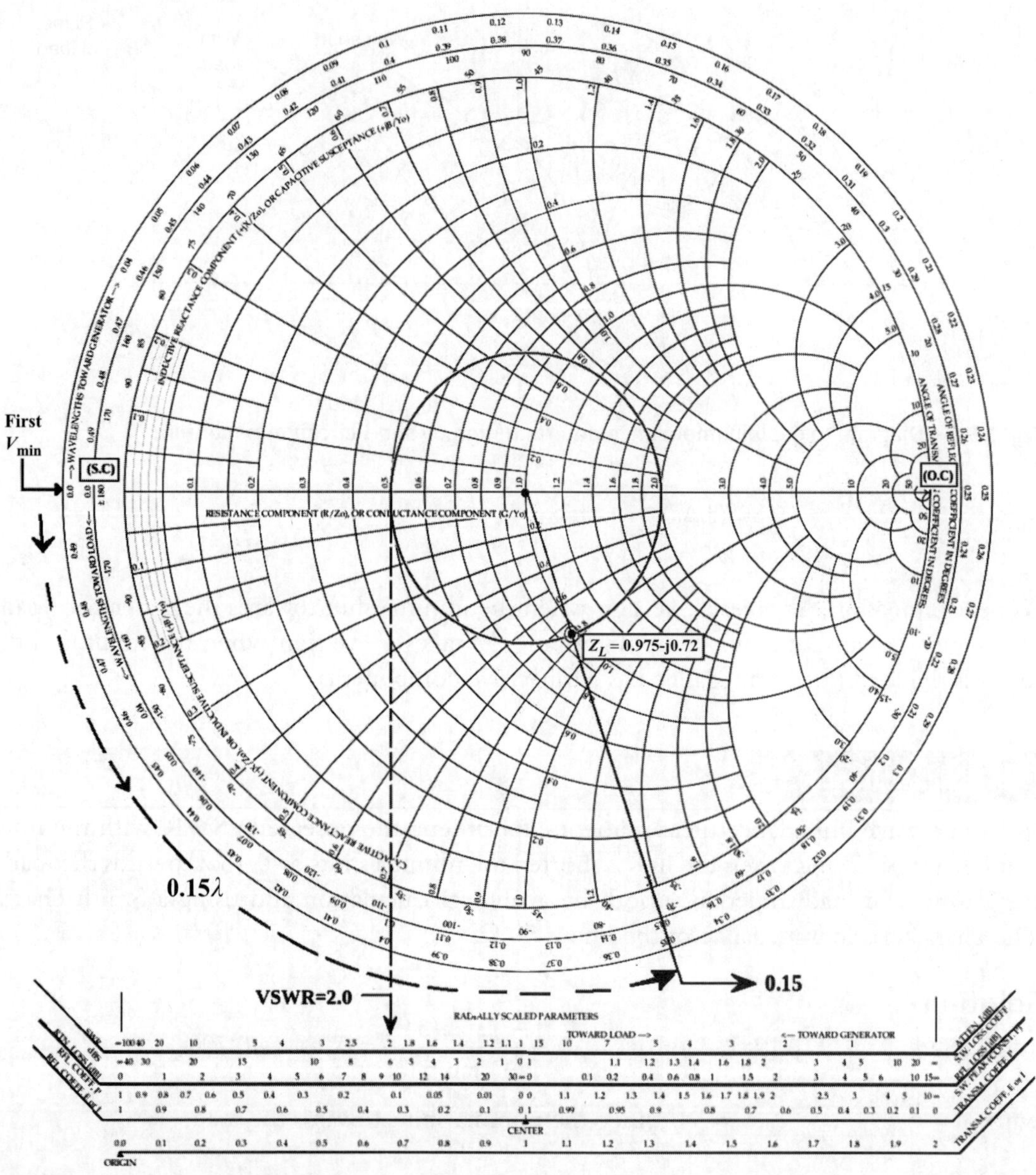

Fig. 12.6 Smith chart solution of Example 12.1.

circle gives the normalised value of the load impedance as $z_l = 0.975 - j0.72$, which when multiplied with the characteristic impedance $Z_0 = 50$ ohms gives the value of $Z_L = (48.75 - j36)\ \Omega$. The Smith chart plot for the solution of the problem is shown in Fig. 12.6.

12.2.1.4 Measurement of high value of VSWR ($\rho > 10$)

The conventional V_{max}/V_{min} method cannot be used to measure high values of ρ (VSWR) as the magnitude of voltage minimum will be very small for high values of VSWR. Thus, in order to obtain a meaningful reading at voltage minimum it would become necessary for the probe to be inserted quite deep into the SWD waveguide to pick up enough voltage for the crystal detector. But this depth of penetration of the probe is to be maintained the same (for relative measurement) even at the voltage maximum which will eventually distort the SW pattern within the waveguide of the SWD causing probe loading. The detector may also be driven beyond its square law region leading to *detector loading*. Thus, the conventional technique of getting ρ from probing V_{max} and V_{min} is not suitable for measurement of high values of VSWR. A convenient technique for the measurement of high value of VSWR is *double the minimum method*, in which the standing wave pattern is probed only around a voltage minimum.

The procedure consists of first locating the position of a voltage minimum and to measure the power output at this minimum. Next, two positions are selected at which the measured power output is twice the value obtained at the voltage minimum, Fig. 12.7. If the detector response is square law it may be shown that [28] VSWR (ρ) is given by:

$$\rho = \frac{\lambda_g}{2\pi x_0} \tag{12.4}$$

where x_0 is the displacement from the voltage minimum as shown in the figure and λ_g is the guide wavelength.

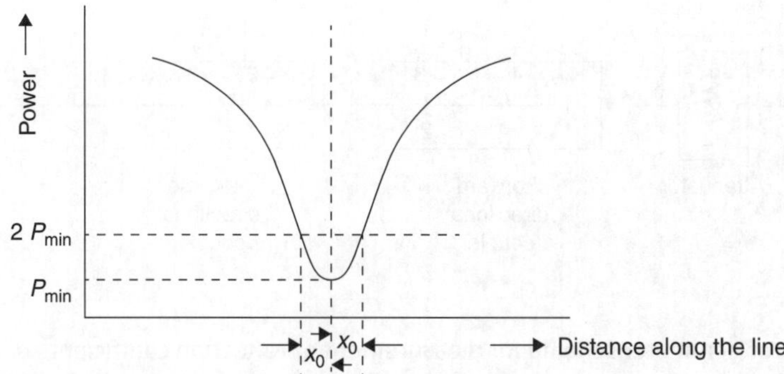

Fig. 12.7 Diagram showing the principle of double the minimum method for measurement of high value of VSWR.

To apply this method, accurate measurement of small displacements of the probe carriage is required when it is moved in the vicinity of the voltage minimum. This can be done best by attaching a small micrometer dial indicator to the probe carriage. Further, it is advisable to adjust the input attenuator value so that the minimum occurs at the midpoint and twice the minimum occurs at the full scale on the indicator for sensitivity of measurement.

12.2.2 Reflectometer Technique of Impedance Measurement

In this technique, we measure the reflection coefficient $|R| = |V_r|/|V_i|$ and evaluate the magnitude of the load impedance in terms of the measured reflection coefficient using the relation $z_L = Z_L/Z_0 = (1+|R|)/(1-|R|)$. Phase information regarding the load impedance may also be obtained with suitable modification of the simple reflection coefficient measurement scheme.

The basic reflectometer consists of two directional couplers arranged back-to-back (Fig. 12.8). The forward coupler samples the incident signal from the source and monitors its level on the indicator marked *I*, while the reverse coupler samples the reflected signal from the device under test (DUT) and registers it in the indicator marked *R*. To start with, initial calibration is done by replacing the DUT by a short circuit load (short) and the amplifier gain in the two indicator channels are adjusted to make the output of the two detectors equal. When the short circuit load is next replaced by the DUT, the reflected signal magnitude changes (while incident signal does not). The ratio of the two indicator readings *I* and *R* is the direct measure of the magnitude of the reflection coefficient; or expressed in decibels give the *return loss* of the load as $-20\log_{10}|R|$.

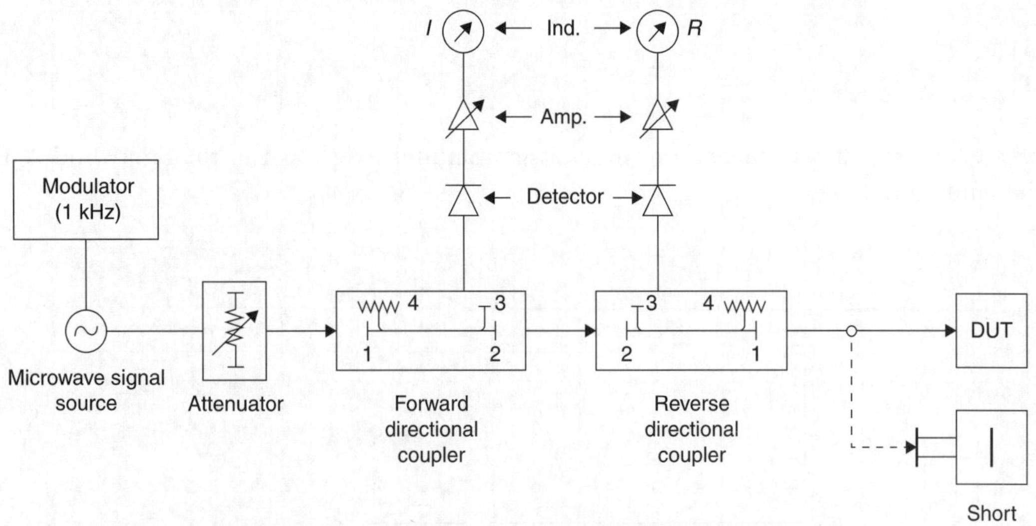

Fig. 12.8 Basic reflectometer scheme for measurement of reflection coefficient.

To obtain both the magnitude and phase information for the load the scheme given in Fig. 12.8 may be modified as shown in Fig. 12.9; which gives an elementary scheme for the vector network analyser. Here, the outputs from the two directional couplers are fed to two identical mixers in the two paths which are driven by an auto-tuned local oscillator (being tuned according to the frequency of the microwave signal source). The frequency down conversion of the mixers is carried out in two or more steps to translate the microwave frequency down to 278 kHz [*Note* $(1/360) = 0.0027778$]. The two outputs in the two channels at 278 kHz preserve the relative amplitude and phase relationship of the incident and reflected waves at microwave frequencies. Phase comparison and amplitude comparison is next carried out at 278 kHz frequency with suitably designed low frequency circuits. The phase and amplitude information are then given to a phase-gain indicator (CRO) for direct display of the impedance in magnitude and phase on a Smith chart overlay generated on the CRO screen. This is in short a very basic arrangement for a vector network analyser which will be discussed in more detail in Section 12.9.2.

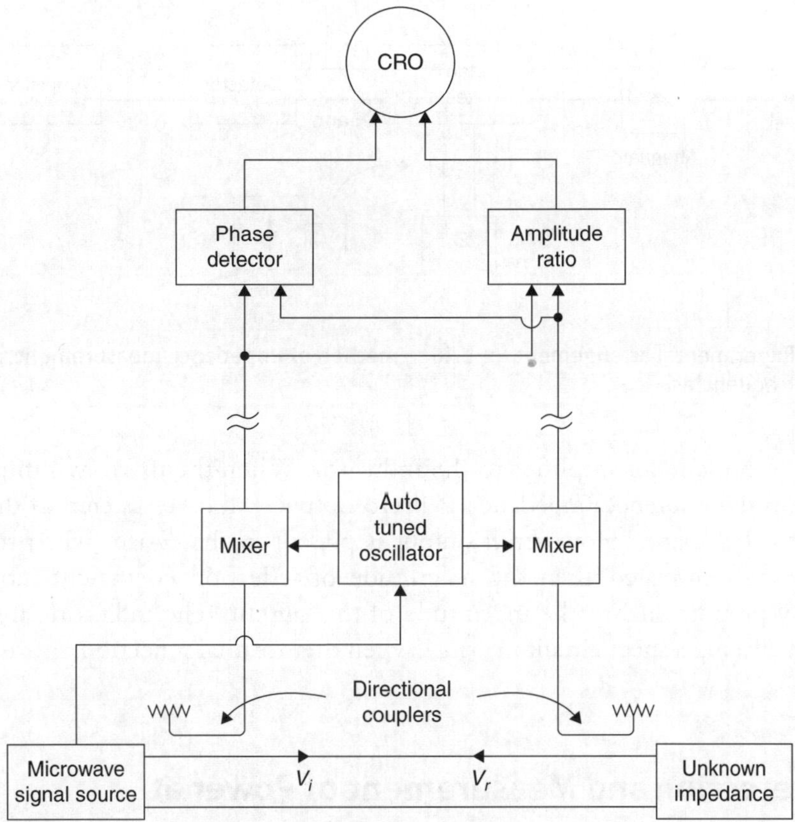

Fig. 12.9 An elementary scheme for vector network analyser.

12.2.3 Bridge Technique of Impedance Measurement

A magic tee has the property that a wave entering the *H*-arm cannot pass directly to the *E*-arm but rather splits into two equal parts that travel down the collinear arms 1 and 2. Depending upon the termination at the collinear arms the reflected waves, if any, in arms 1 and 2 would reach back to the junction and then enter the *E*-arm to produce a resultant wave in *E*-arm that is the difference between the reflected waves in arms 1 and 2 (to the extent the terminating impedances at arms 1 and 2 are non-identical). A measuring system making use of the above mentioned property of magic tee to measure impedance at microwave frequency is shown in Fig. 12.10.

Here a reference termination Z_0 is connected across the collinear arm 1 while the unknown impedance Z_x is connected to the other collinear arm 2. Arms 1 and 2 are further made identical in every respect, particularly the length. Power from the microwave oscillator modulated by 1 kHz signal is applied to the *H*-arm of the magic tee through an attenuator while the output at the *E*-arm is fed through a crystal detector (operating in the

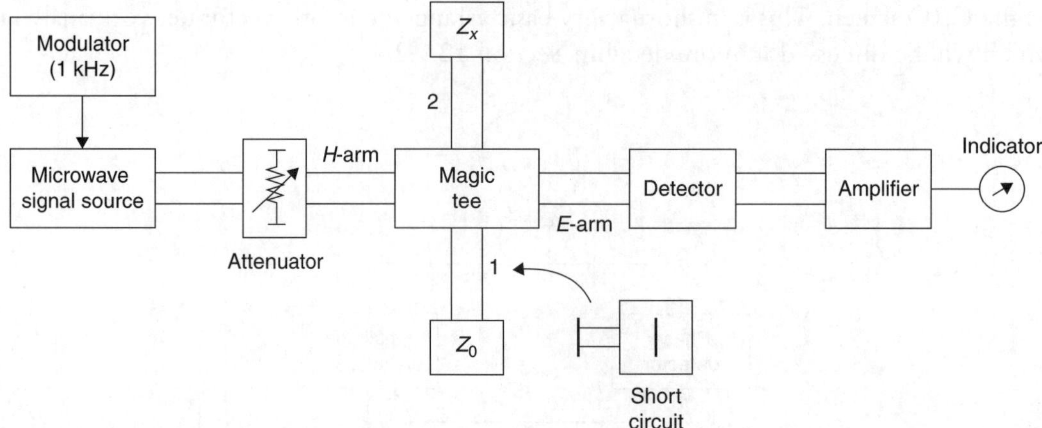

Fig. 12.10 Experimental arrangement for bridge method of impedance measurement at microwave frequencies.

square law region) and an amplifier to the indicator. When the unknown impedance Z_x is the same as the reference impedance Z_0, zero output will be registered in the indicator connected to the *E*-arm. However, if output is present in the *E*-arm when the collinear arm 2 is match terminated, then the magnitude of reflection coefficient introduced by Z_x is directly proportional to the magnitude of this output. The indicating meter can be calibrated by placing a short circuit in arm 1 when there is total reflection, that is, the return loss is 0 dB.

12.3 Detection and Measurement of Power at Microwave Frequencies

The history of detection of microwave power with a type of metal–semiconductor diode goes back to the work of Sir J C Bose who used a metal contact with galena (lead sulphide) as semiconductor (US Patent No.: 52346, issued in 1904) to detect a 5 mm wavelength wireless signal generated by a modified spark-gap generator designed by him at Kolkata (Calcutta), India. Even today microwave signal detection is done with a point contact diode which uses a thin and stiff tungsten metal wire (called *cat whisker*) on a Ge or Si chip of semiconductor. A more reliable microwave detector diode is the Schottky barrier metal–semiconductor rectifying diode.

For applications such as standing wave ratio (SWR) measurement where only a relative measure of V_{max} and V_{min} is required, signal detection serves the purpose. But for absolute power measurement at microwave frequencies, as in the case of radar and communication applications, power sensors are used. It may be noted that at microwave frequencies power is measured directly as a fundamental quantity unlike the power measurement at lower frequency where current, voltage and power factor are measured to evaluate the signal power.

Power sensors are thermally sensitive devices which utilise the heat generated by the microwave energy incident upon it. Depending on the level of microwave power to be measured (low, medium or high) and the technique of estimating the power incident on the sensor element, there are in general, three techniques of microwave power measurement.

(i) *Thermocouple technique*: In this technique, the input microwave power is dissipated in a lossy power sensor element (commonly, silicon tantalum nitride) which increases the temperature of the sensor element above the ambient temperature. One junction of the thermocouple is placed in the said sensor element while the other junction of the thermocouple is kept at ambient (room) temperature. The difference in temperature at which the two junctions of the thermocouple are kept produces a proportional voltage. An accurate measurement of this DC voltage calibrated in terms of power gives an estimate of the microwave power incident on the sensor element. This technique is used for the measurement of low-level (up to 1 mW) continuous wave (CW) microwave power.

(ii) *Calorimeter technique*: This method involves conversion of microwave energy into heat, in a load (usually a lossy cable) and absorbing this heat in a fluid (usually water) which flows through the system and then measuring the temperature rise of the fluid. Care has to be taken against RF leakage either by radiation or by lossy joints. The power dissipated in a calorimeter can be calculated directly from the temperature rise (Δt), the specific heat of the fluid (S_p) and the rate of flow of the fluid (m) and by using the formula $P = 4.18\ m\ S_p\ \Delta t$. This method is used in physical and chemical plants where microwave power is used for heating purposes and the power level is reasonably high (> 10 W, CW power).

(iii) *Bolometer technique*: In this method, the microwave power to be measured is incident on a specially constructed thermally sensitive bolometer element (either barretter or thermistor) of resistive material. The incident microwave power raises the temperature of the sensor element thereby causing a change in the resistance of the bolometer element which is utilised in a suitable bridge circuit to determine the microwave power. This technique is used for low level to medium level (from 1 mW to 10 W) CW power measurement. However, with special microwave circuit arrangement, thermistor power sensor based power meters may also be extended for high power measurements. In addition to CW power measurement, peak pulse power of pulsed RF signal used in radar can also be measured with thermistor power sensor based microwave power meters.

In this section we shall discuss in brief microwave power detectors and details of the microwave power measurement using the bolometer technique both for CW and pulsed power measurement.

12.3.1 Microwave Power Detector

Detection of microwave signals is carried out with a diode for relative measurement of microwave power. The microwave detector diodes are either point contact diodes or Schottky barrier diodes. Point contact diodes (Fig. 12.11) consist of a fine gold plated tungsten wire with a sharp point which makes contact with the polished top of a semiconductor pallet made of silicon, germanium or gallium arsenide. Thus, it is basically a metal–semiconductor diode. Such diodes can be fitted into coaxial or waveguide mounts, and are available at frequencies in excess of 100 GHz. With the advancement of technology, point contact diodes are gradually getting replaced by Schottky barrier diodes. These diodes have higher burnout rating, lower $1/f$ noise, and better reliability compared to point contact diodes. It may however be noted that, ordinary *p–n* junction semiconductor diodes used extensively at lower radio frequencies are bipolar devices and have a large junction capacitance arising from minority carrier storage which makes them unsuitable for use at microwave frequencies.

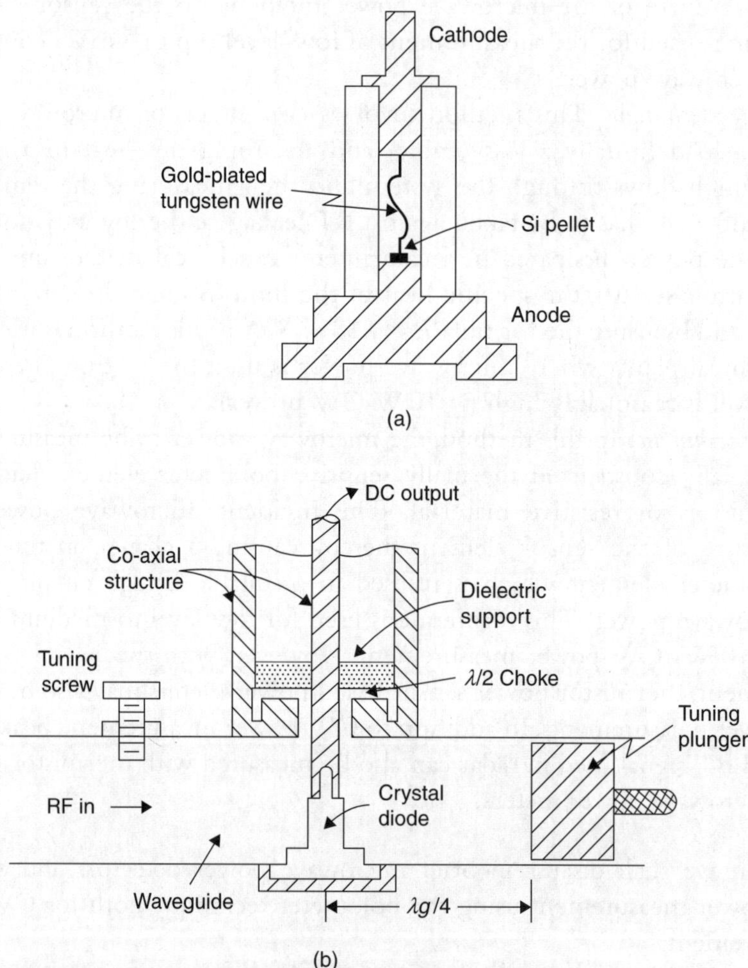

Fig. 12.11 (a) Point-contact diode (b) Tunable waveguide mount type diode detector.

A microwave detector diode must be mounted in the transmission line with due care so that it can provide a complete DC path for rectification, without unduly upsetting the RF field in the waveguide. A tuneable detector mount is shown in Fig. 12.11b. The diode is mounted across a waveguide and is closed at one end by a tuning plunger which is set at a distance of quarter wavelength at operating frequency so that impedance looking towards the shorting plunger from the diode plane is infinite, allowing full RF power to be incident on the diode. Further, the diode is attached to the top wall of the waveguide by means of a half-wave choke so that the top wall of the waveguide is DC open but RF short. Tuning screws may also be provided on the RF input side of the diode for further impedance matching.

12.3.2 Bolometer Technique of Microwave Power Measurement

There are two types of bolometer sensor elements used for microwave power measurements—*barretters* and *thermistors*. Any material having a large temperature coefficient of resistance can be used as a bolometer element. However, a bolometer element for microwave power measurement should have, in addition to a high temperature coefficient of resistance, a physical size which is small compared to the wavelength and skin depth at the operating frequency of the microwave signal. We shall first compare their physical and electrical characteristics.

Barretters (Fig. 12.12a) consist of a short length (approximately one wavelength long at the operating frequency) of fine wire having 0.001 cm diameter and with sufficient resistance to enable it to be impedance matched as a termination for a transmission line. It has a positive temperature coefficient of resistance and is usually biased to an operating resistance from 50 Ω to 400 Ω, the most common being 200 Ω. Barratters may be prepared by etching silver from the platinum core of a Wollaston wire (platinum coated with silver). This leaves a very slender platinum wire, which is resistant to corrosion, has a high melting point and with desirable mechanical properties.

Thermistors (Fig. 12.12b) consist of a tiny bead of semiconducting material (with diameter of the order of 0.05 cm) which bridges the gap between two fine, closely spaced, parallel supporting leads. Effectively, all the resistance of the thermistor is concentrated in the bead material which has a negative temperature coefficient of resistance. They are usually biased to an operating resistance of 100 to 200 Ω. The bead material consists of a mixture of Mn and Ni oxides combined with finely divided Cu particles to control the resistivity. The bead is coated with glass film to prevent oxidation and to improve stability.

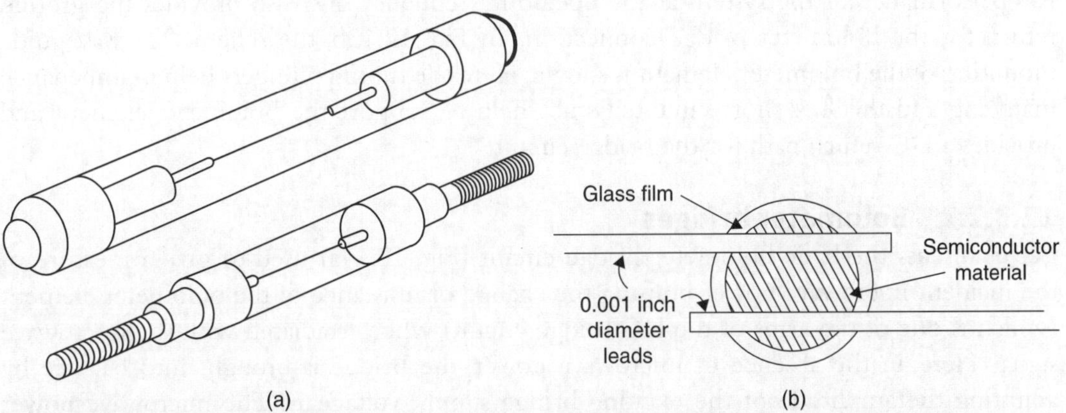

Fig. 12.12 Bolometers at microwave frequency (a) Barretters (b) Thermistor.

The electrical performance characteristics of barretters and thermistors are compared in Table 12.1.

Table 12.1 Electrical performance comparison of barretters and thermistors

Barretter	Thermistor
(i) Have fairly short thermal time constant (100 to 300 μs) (fast response).	Have long thermal time constant (of the order of ms) (slow response), sluggish. This is useful in pulse radar work where echo pulses of μs order when integrated for further processing improves S/N ratio of power measurement.
(ii) Operated at fairly high temperatures. So ambient temperature changes practically do not affect their characteristics.	Operated at comparatively lower temperatures. Hence, much more sensitive to changes in ambient temperature. Thus require some temperature compensation against ambient temperature changes.
(iii) Impedance characteristics unfavourable for broadband operation. Poor detector sensitivity: 5.2 Ω/mW.	Resistance may be varied over broad frequency band, thus useful for broadband detection. High detector sensitivity: 60 Ω/mW; useful for low power applications.

12.3.2.1 Bolometer mounting

Now we discuss the mounting techniques for barretters and thermistors. The requirements for such mounts are (i) impedance matching of the bolometer element to the characteristic impedance of the coaxial line (coax)/waveguide and (ii) the RF circuit with the bolometer element mounted in the coaxial transmission line/waveguide should be isolated from the DC bridge circuit. The bolometer mounted in a coaxial circuit and connected with the DC bridge circuit having necessary isolation of RF and DC bridge is shown in Fig. 12.13a. The central conductor of the coax is tapered for impedance matching and the $\lambda/4$ short-circuited stub is used to support the central conductor of the coax without impairing the RF performance of the system at the operating frequency and also provides the ground return for the bolometer bridge connection. In Fig. 12.13b, the scheme for waveguide mounting of the bolometer element is shown. Here the tuning plungers help in impedance matching and the $\lambda/4$ short-circuited stubs help to support the bolometer element and provides a DC return path for the bridge circuit.

12.3.2.2 Bolometer bridges

Let us discuss the basic bolometer bridge circuit (Fig. 12.14a) used to give a measure of the incident microwave power utilising the change of resistance of the bolometer element (which is one of the arms of the DC bridge circuit) when irradiated with the microwave signal. Here, in the absence of microwave power, the bridge is brought into balance by adjusting the amplitude of the exciting bridge supply voltage E. The microwave power dissipated in the bolometer element will then unbalance the bridge, and the amount of

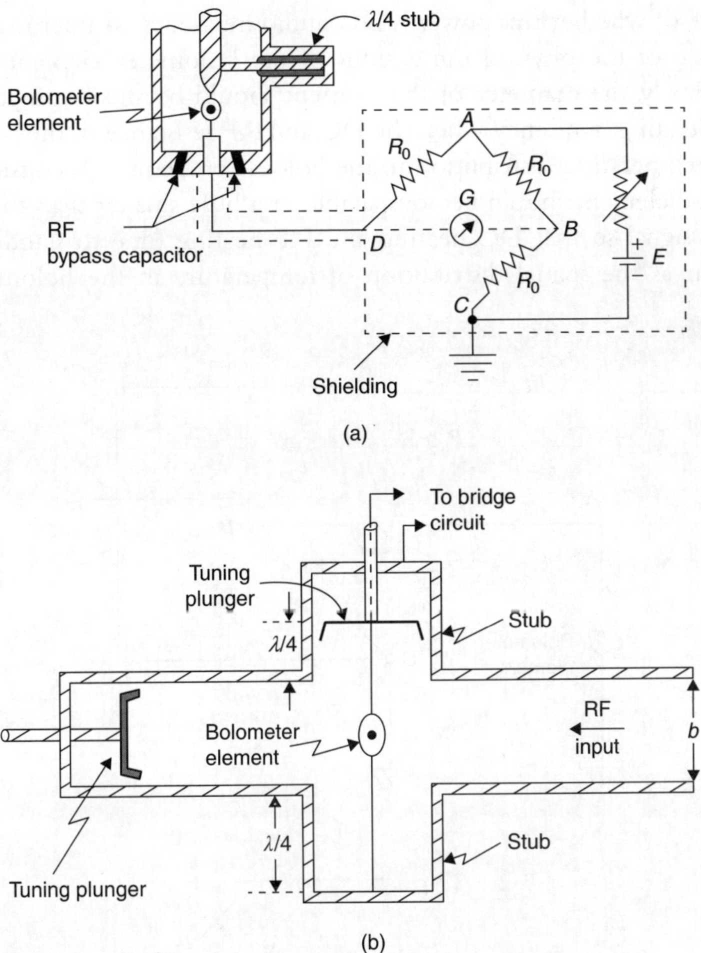

Fig. 12.13 Bolometer mounting (a) In coaxial circuit (b) In waveguide circuit.

microwave power may be determined by noting the amount of bridge unbalance indicated by the bridge indicator meter *I*. However, the estimation of the incident microwave power in terms of bridge imbalance is not a practically viable technique. Under the condition of bridge balance and bridge imbalance, the bolometer is driven to two different impedance levels thereby impairing the impedance matching of the bolometer element with the transmission line in which it is mounted. A practical solution for this is to maintain the bridge under a balanced condition irrespective of whether the bolometer is cold (no microwave power is being incident on it) or hot (with microwave power being incident on it). For this case, with the bolometer hot the bridge can be brought to balance by decreasing the DC power supplied by the battery *E*. The change in DC power being supplied by *E* to the bolometer is taken as a measure of the microwave power incident on the bolometer element.

Note that in the above methodology of measurement, the assumption is that the resistance of the bolometer element is a uniquely defined function of the power dissipated in it irrespective of whether this power is DC, audio frequency or microwave power. Only a suitable choice of the physical dimensions of the bolometer element will justify this *substitutabilty*. Firstly, the diameter of the element should be small compared to the skin depth at the operating frequency when the DC and RF resistance of the wire will be same as far as radial temperature distribution in the bolometer element is considered. Secondly, the length of the element should be comparable or a little smaller than the wavelength of the microwave signal so that DC heating and RF heating (due to standing wave) differ negligibly as far as the spatial distribution of temperature in the bolometer element is considered.

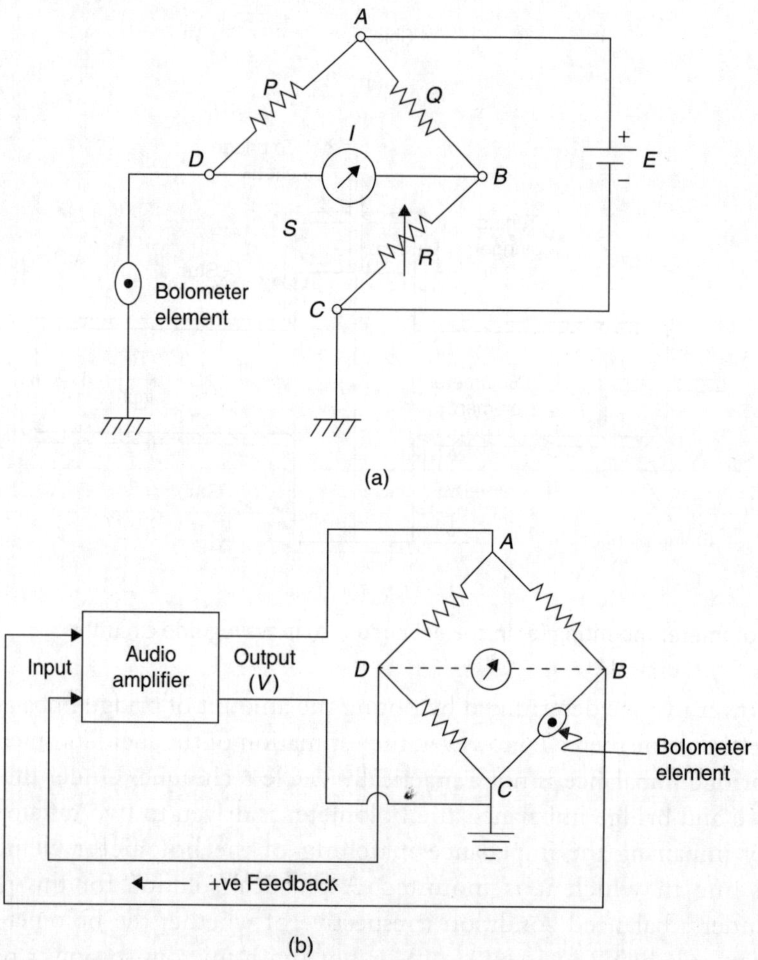

Fig. 12.14 Bolometer bridge circuits (a) Manually balanced bridge (b) Self-balancing bridge.

12.3.2.3 Self-balancing bridge

So far we have discussed manual balancing of the bolometer bridge. But for all practical purposes and for automation of measurement, self-balancing or automatic balancing of the bolometer bridge has to be used (Fig. 12.14b) and its principle of operation is discussed below. The output of an audio frequency (AF) amplifier is connected to the bridge as shown. When the system is first turned ON, the bolometer is cold, the bridge is unbalanced, and a positive feedback exists between the output and the input of the amplifier (the degree of bridge unbalance is the feedback factor, β). This causes the amplifier to oscillate and to supply audio frequency power to the bridge. The bolometer element becomes heated and its resistance changes until the bridge is brought to balance. More specifically, oscillations continue to increase until the product of the gain (A) of the amplifier and the bridge unbalance (feedback factor β) equals unity. It may be noted that the frequency of the AF amplifier should be high enough so that the bolometer resistance does not vary appreciably during the period of one cycle (to ensure this, f_{AF} should be nearly equal to the inverse of the response time, 1/100 μs = 10 kc/s (for barretter and approximately 10 c/s or may be DC for thermistor; thus with thermistor power sensor, simple operational amplifiers can be used as bridge amplifiers). Initial power driven from the oscillating-amplifier is noted down, this is the reference value and is labelled as zero on the power scale. Next, irradiation of the bolometer element with microwave power will heat it, requiring a smaller amount of power to be delivered by the oscillating-amplifier. Since the balance must be perfect in all cases, the substitution process is completed automatically. Note that automatic balancing of the bridge assures that the resistance of the bolometer is always the same as it is driven to the bridge balance condition in both cases when the bolometer element is cold or when it is hot.

12.3.3 Microwave Power Meter

The detailed circuit (Fig. 12.15) and principle of operation of a thermistor power sensor based microwave power meter is discussed which can measure microwave power in the range of 10 μW (–20 dBm) to 10 mW (10 dBm).

The power meter consists of two major sections, the bridge and the meter logic assemblies. The bridge section contains circuits which form two self-balancing bridge circuits with thermistor elements. Each of the bridges is automatically brought to balance by the action of high gain DC amplifier (bridge amplifier) feeding power to the top of the bridge. The voltage at the top of the RF bridge, V_{RF} is responsive to both input RF power and the ambient temperature changes. The voltage at the top of the compensation bridge, V_{comp} is responsive only to ambient temperature changes. Knowing V_{comp} and V_{RF}, the RF power can be determined.

The working formula for the meter logic can be derived as follows. The power meter is initially set to zero with no applied RF power by making V_{comp} equal to V_{RF0} ($V_{RF0} = V_{RF}$, with no RF power incident on the detecting thermistor). After zero setting, if the ambient temperature variations change the thermistor resistance, both the bridge circuits respond by applying the same new voltage to maintain balance of the two bridges. If RF power is applied to the detecting thermistor, V_{RF} decreases so that:

$$P_{RF} = \frac{V_{RF0}^2}{4R} - \frac{V_{RF}^2}{4R}$$

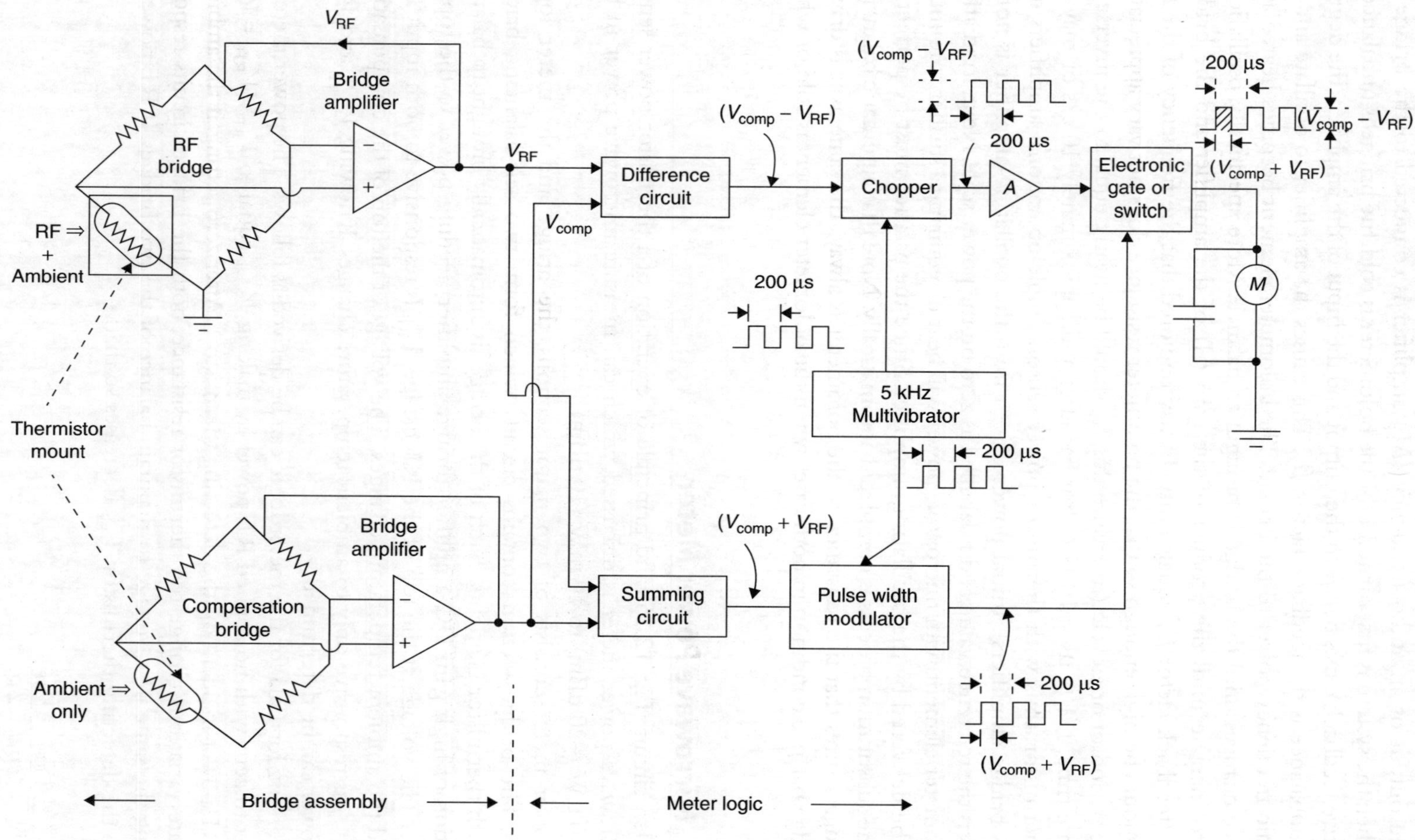

Fig. 12.15 Microwave power meter with thermistor power sensor.

where R is thermistor resistance at balance. But at initial balance with zero setting $V_{RF0} = V_{comp}$, so $P_{RF} = (V_{comp}^2 - V_{RF}^2)/4R$ or,

$$P_{RF} = (V_{comp} + V_{RF})(V_{comp} - V_{RF})/4R \tag{12.5}$$

Ambient temperature changes cause V_{comp} and V_{RF} to change equally so there is zero change in $(V_{comp}^2 - V_{RF}^2)$ and therefore no change in indicated P_{RF}.

The meter logic section has to process V_{RF} and V_{comp} to produce meter current proportional to RF power, according to Eq. 12.5. To do this, the meter logic circuit generates two sets of pulse signals. One is the sum signal ($V_{comp} + V_{RF}$) that modulates the width of 5 kHz pulses from a multivibrator. This pulse train is used to control timing or gating of the electronic switch. The other is the difference signal ($V_{comp} - V_{RF}$), which is a small DC signal and needs to be chopped and then amplified and controls the amplitude of 5 kHz pulses from the multivibrator. This pulse train is fed to the electronic switch whenever the switch is closed. If the current coming through the switch has a pulse width proportional to ($V_{comp} + V_{RF}$) and a height proportional to ($V_{comp} - V_{RF}$), then the area under the pulse of current is proportional to ($V_{comp} - V_{RF}$) ($V_{comp} + V_{RF}$). These current pulses are integrated in a capacitor and the meter reads the voltage across the capacitor and with due calibration (in terms of $1/4R$), the microwave power in mW or dBm. The photograph of a typical microwave power meter is shown in Fig. 12. 16.

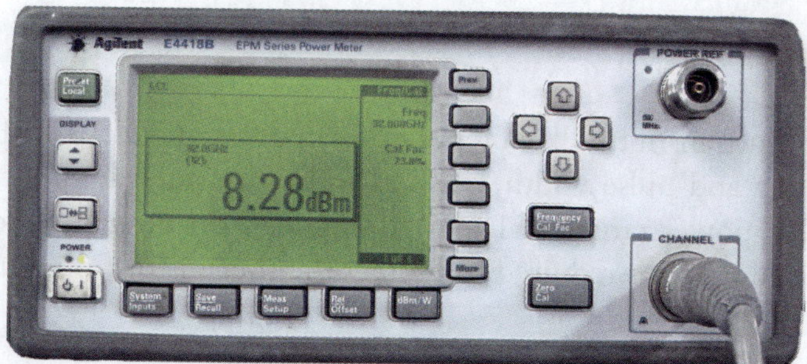

Fig. 12.16 Microwave power meter [Photograph: Courtesy SAMEER Kolkata centre].

12.3.4 Peak Power Measurement

Peak power (P_{peak}) of a pulsed RF signal can be determined by measuring the average power P_{av} (registered in a microwave power meter) and the duty cycle of the pulse [determined in terms of measured pulse width (PW) and pulse repetition frequency (PRF)] with the help of an oscilloscope, Fig. 12.17. The technique is known as *average power-duty cycle method.* If C_f and A are respectively the coupling factor of the directional coupler and attenuation value of the variable attenuator in dB then the peak power can be calculated from the following formula:

$$P_{peak} = \frac{P_{av}}{\text{PRF} \times \text{PW}} .10^{(A+C_f)/10} \tag{12.6}$$

This method is extensively used in practice. However, in applications where the average power in the pulsed RF signal is too low to be measured with a power meter directly then *Notch Wattmeter* is used.

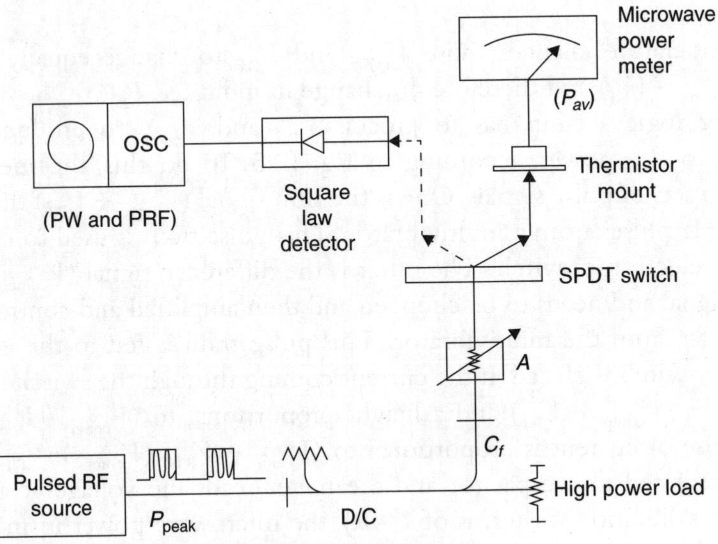

Fig. 12.17 Peak power measurement by average power-duty cycle method.

EXAMPLE 12.2

Pulse power in a radar transmitter having peak pulse power of 50 kW with pulse repetition frequency of 300 Hz and pulse width 1 μs has to be measured with a thermistor power sensor having a CW power rating of 1 mW. Using a 20 dB directional coupler calculate what additional attenuation is required to protect the thermistor from burn out.

Solution

From Eq. 12.6 we can write:

$$A = 10\left\{\log_{10}\left(\frac{P_{\text{peak}} \times \text{PRF} \times \text{PW}}{P_{av}}\right) - 2\right\} = 10\left\{\log_{10}\left(\frac{50\times 10^3 \times 300 \times 1\times 10^{-6}}{1\times 10^{-3}}\right) - 2\right\}$$

$$= 21.76 \text{ dB}$$

From practical considerations, to save the thermistor power sensor from burn out and also to maintain the sensitivity of measurement we should set the variable attenuator at 25 dB (say).

12.4 Measurement of Quality Factor (*Q*) of Microwave Cavities

Microwave cavities are, in general, of two types—one-port or reflection type cavity (used for the design of oscillators/amplifiers) and two-port or transmission type cavity (used in the design of wavemeters, echo box). There are three quality factors (Q) of a cavity. The unloaded quality factor (Q_0), when the cavity is not perturbed by the presence of the coupling loops (this is the theoretically calculated Q of the cavity), the loaded quality factor (Q_L) is the quality factor of a practical cavity with coupling loops, the external quality factor (Q_E) represents the energy lost to the external circuit through the coupling ports. The loaded quality factor is related to the other two quality factors by the relation:

$$\frac{1}{Q_L} = \frac{1}{Q_0} + \frac{1}{Q_E} \tag{12.7a}$$

The three quality factors are interrelated through the coupling factor β (ratio of coupled resistance $\omega M^2/Z_0$ and the cavity loss resistance R_s) between the cavity and the load and are given by:

$$Q_0 = (1+\beta)Q_L \quad \text{and} \quad Q_E = \frac{Q_0}{\beta} \tag{12.7b}$$

The quality factor of microwave cavities may be determined by the following methods:

- Transmission method,
- VSWR or impedance method,
- Transient decay or decrement method, and
- Dynamic method.

The transmission method is only a phenomenological method in which the transmission curve of the cavity is generated by energising a transmission cavity with the input microwave signal and Q is determined by using the formula $Q = f_0/\Delta f$, where f_0 is the resonant frequency of the cavity and Δf is the bandwidth between half power points. In the following three sections we shall discuss the other three methods in detail.

12.4.1 VSWR or Impedance Method of *Q* Measurement

In one-port cavities, there is only one coupling loop through which power is both fed in and also extracted from the cavity. Thus, the signal coming into and the signal going out of the cavity along the same path (the coupling loop) will interfere and form standing waves on the line feeding the cavity. The standing wave set-up in the line that feeds the resonator cavity is uniquely related to the equivalent cavity impedance (which is the load to the line, in which the line is terminated), and the half-power frequencies that determine Q values may be found directly from VSWR measurements. The experimental arrangement for this measurement is shown in Fig. 12.18.

12.4.1.1 Derivation of the working formula

The theoretical basis for this measurement may be developed as follows [28]. The equivalent circuit for the generator, transmission line, coupling loop and the cavity is shown in Fig. 12.19, where L, C, R_s refer to the equivalent lumped parameters for the cavity. L_1 and

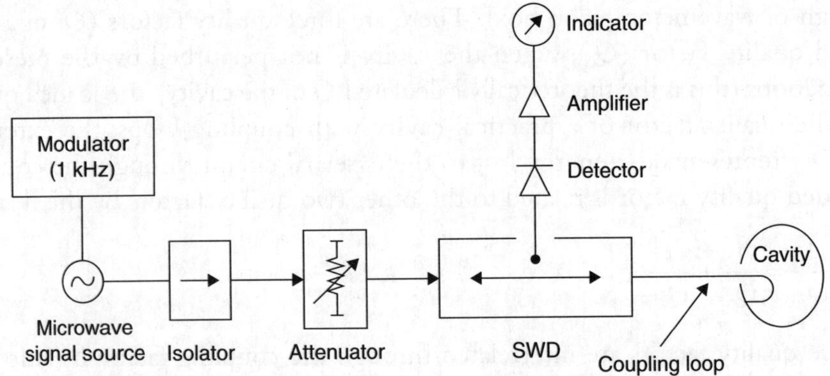

Fig. 12.18 Experimental arrangement for measurement of *Q* by VSWR method.

M respectively represent the self and mutual inductance of the coupling system. Plane *a–a* is an arbitrary reference plane where the terminals of the coupling system is presumed to be located near the cavity.

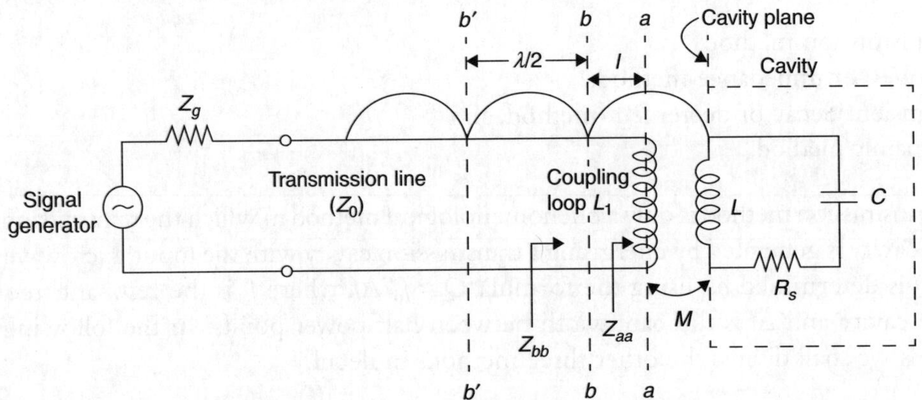

Fig. 12.19 Equivalent circuit for generator, transmission line, coupling loop and cavity.

Impedance at the terminals of the coupling network with reference to the plane *a–a* (from theory of coupled circuits) is given by:

$$\frac{Z_{aa}}{Z_0} = j\frac{X_1}{Z_0} + \frac{\beta}{1 + j2Q_0\delta} \tag{12.8}$$

where $X_1 = \omega L_1$, self reactance of the coupling loop; $\beta = \omega M^2/Z_0R_S$, ratio of the coupled resistance to the cavity loss resistance; and $\delta = (\omega - \omega_0)/\omega$, frequency tuning parameter (ω_0 is the resonant frequency of the cavity).

The analysis and interpretation of the experiment becomes practically meaningful by choosing a special reference plane *b–b* along the transmission line which is called the *plane of detuned short*. The plane of detuned short may be interpreted as follows. When the cavity is perfectly detuned, the reactance due to the coupling loop (inductive, thus leading phase) predominates and the VSWR pattern may be interpreted (Fig. 12.19) as if the cavity plane behaves as a short circuited plane.

At frequencies far off-resonance, the second term in Eq. 12.8 vanishes (as $\delta \rightarrow \infty$), leaving $(Z_{aa}/Z_0) = j(X_1/Z_0)$, or the self reactance of the coupling loop system is the terminating load. This inductive termination produces complete reflection of the incident signal and results in a series of voltage nodes (with the cavity plane as if representing a short circuit) which can be found by means of a standing wave detector. Let the terminals *b–b* be selected at a distance *l* away from the terminals *a–a*, then the cavity impedance at *a–a* can be transformed to the terminal plane *b–b* as:

$$\frac{Z_{bb}}{Z_0} = \frac{Z_{aa} + jZ_0 \tan \beta l}{Z_0 + jZ_{aa} \tan \beta l} \tag{12.9}$$

The location of the terminal plane *b–b* (the plane of detuned short) is such that the impedance at *b–b* becomes zero (position of voltage minimum corresponding to a short-circuited load) when the cavity is perfectly detuned (at far off-resonance, meaning δ is large). Under this condition, we have from Eq. 12.9:

$$Z_{aa}(= jX_1) + jZ_0 \tan \beta l = 0; \text{ when } \beta l = -\tan^{-1}\left(\frac{X_1}{Z_0}\right) \tag{12.10}$$

Combining Eqs 12.8, 12.9 and 12.8,10, the impedance at the plane of detuned short for any finite value of δ becomes:

$$\left.\frac{Z_{bb}}{Z_0}\right|_{Q_0} = \frac{\beta}{1 + j2Q_0(\delta - \delta_0)} \tag{12.11}$$

where $\delta_0 = \dfrac{\beta}{2Q_0}\left(\dfrac{X_1}{Z_0}\right)$

Again, $Q_L = Q_0/(1 + \beta)$, hence in terms of Q_L, Eq. 12.11) becomes:

$$\left.\frac{Z_{bb}}{Z_0}\right|_{Q_L} = \frac{\beta}{1 + j2Q_L(1 + \beta)(\delta - \delta_0)} \tag{12.12}$$

Also, $Q_E = Q_0/\beta$, so from Eq. 12.9, we get:

$$\left.\frac{Z_{bb}}{Z_0}\right|_{Q_E} = \frac{\beta}{1 + j2Q_E\beta(\delta - \delta_0)} \tag{12.13}$$

Note that Eq. 12.11 represents the impedance of a parallel resonant circuit with a resonant impedance βZ_0 (since at resonance, the imaginary part is equal to zero). At half power points:

$$2Q_0\,(\delta_{1,2} - \delta_0) = \pm 1 \tag{12.14}$$

And at these values of δ (the frequency tuning parameter) the impedances at the plane of detuned short with respect to half power points become:

$$\left.\frac{Z_{bb}}{Z_0}\right|_{Q_0} = \frac{\beta}{1 \pm j1}$$

$$\left.\frac{Z_{bb}}{Z_0}\right|_{Q_L} = \frac{\beta}{1 \pm j(1+\beta)} \tag{12.15}$$

$$\left.\frac{Z_{bb}}{Z_0}\right|_{Q_E} = \frac{\beta}{1 \pm j\beta}$$

$$\text{Now, VSWR, } \rho = \frac{1+|\Gamma|}{1-|\Gamma|} = \frac{1 + \left|\dfrac{Z_{bb} - Z_0}{Z_{bb} + Z_0}\right|}{1 - \left|\dfrac{Z_{bb} - Z_0}{Z_{bb} + Z_0}\right|} \tag{12.16}$$

Substituting for Z_{bb}/Z_0 from Eq. 12.15 in the above equation and rearranging, we get the value of VSWR at half-power points in terms of β as,

$$\text{For } Q_0,\ (\rho_{1/2})_0 = \frac{2 + \beta^2 + \sqrt{4 + \beta^4}}{2\beta} \tag{12.17}$$

$$\text{For } Q_L,\ (\rho_{1/2})_L = \frac{1 + \beta + \beta^2 + (1+\beta)\sqrt{1 + \beta^2}}{\beta} \tag{12.18}$$

$$\text{For } Q_E,\ (\rho_{1/2})_E = \frac{1 + 2\beta^2 + \sqrt{1 + 4\beta^4}}{2\beta} \tag{12.19}$$

Again from Eq. (12.14), we can write:

$$Q_0 = \frac{1}{(\delta_1 - \delta_2)},\ Q_L = \frac{1}{(\delta_3 - \delta_4)},\ \text{and } Q_E = \frac{1}{(\delta_5 - \delta_6)} \tag{12.20}$$

So we basically have the final equations (12.17 to 12.20) at hand. We need to get the coupling factor β through experiment when we can calculate the $(\rho_{1/2})$'s, that is, VSWR at half power points. Again, we need to get a curve between f (that is, δ) vs VSWR experimentally and put the calculated $(\rho_{1/2})$'s into this curve to obtain the values of δ_1 to δ_6 and hence calculate Q_0, Q_L and Q_E from Eqs 12.20).

12.4.1.2 Experimental procedure

The experimental procedure entails the following:

- to mark the plane of detuned short,
- to ascertain whether the cavity is over-coupled or under-coupled and find β by exploring the standing wave pattern at the plane of detuned short, and
- obtain the experimental curve between VSWR and frequency.

To mark the position of detuned short, the cavity is first tuned to resonance corresponding to the signal frequency. Then the cavity is completely detuned when only the self reactance of the coupling loop acts as terminating load to the transmission line. This inductive loading of the line results in complete reflection of the incident signal and a series of voltage nodes separated by half-wavelength is produced as standing wave on the line, which can be probed by a standing wave detector (SWD). The first voltage node located at the *b–b* plane at a distance *l* from the load end of the line (from the plane *a–a* of the coupling loop) is marked on the SWD scale as the *plane of detuned short* that will be required in subsequent measurement steps.

The coupling factor β is to be determined next. We know that coupling factor is the ratio of the coupled resistance $[(\omega M)^2/Z_0]$ to the cavity loss resistance (R_s): $\beta = (\omega M)^2/Z_0R_s$. For the under-coupled case, $\beta < 1$ and for the over-coupled case, $\beta > 1$. To ascertain whether the cavity is under-coupled or over-coupled, the cavity is tuned to resonance when the impedance represented by the cavity is purely resistive. The probe of the SWD is then placed at the plane *b–b* which will result in either a voltage minimum or a voltage maximum at the plane of detuned short *b–b* depending on the nature of cavity coupling. If the exploration of the standing wave pattern results in a voltage minimum at the *b–b* plane, the cavity is under-coupled; if it produces a voltage maximum, the cavity is over-coupled. The magnitude of the coupling factor is determined by measuring VSWR at resonance (ρ_0). Since the impedance at voltage minimum and maximum are respectively Z_0/ρ and $Z_0\rho$, while the resonant impedance at the plane of detuned short is βZ_0, we have the coupling factor for the under-coupled and over-coupled cases respectively as:

$$\beta_{\text{under-coupled}} = \frac{1}{\rho_0} \tag{12.21a}$$

and

$$\beta_{\text{over-coupled}} = \rho_0 \tag{12.21b}$$

The experimental curve between VSWR and frequency is shown in Fig. 12.20. At resonance with resistive termination the VSWR is minimum while with detuning on either side of resonance with inductive/capacitive loading VSWR increases monotonically to infinity.

In summary, the method of Q measurement with the VSWR technique is as follows. The plane of detuned short is found out, the cavity is carefully tuned to resonance and the standing wave pattern is explored at the plane of detuned short to ascertain whether the cavity is under-coupled or over-coupled. The VSWR at resonance (ρ_0) is measured and the value of the coupling factor β is calculated using Eqs 12.21a or 12.21b as appropriate. From

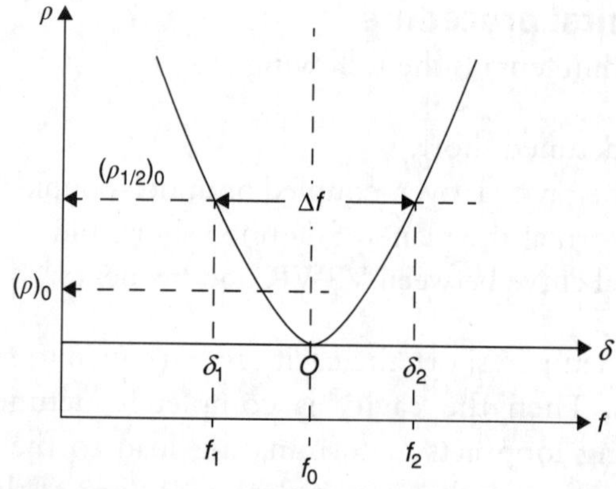

Fig. 12.20 Plot of VSWR vs frequency.

Eqs 12.17 to 12.19, the VSWR at half-power points are obtained. From the experimental curve of the VSWR (ρ) vs frequency (the frequency tuning parameter δ, Fig. 12.20) the δ values for different quality factor (Q) of the cavity are noted down corresponding to the calculated half-power VSWRs. The Q values are then computed from Eqs 12.20.

12.4.1.3 Precautions and limitations of the method

The precaution to be taken in the VSWR technique of Q measurement is that the frequency of the signal source should be constant at least during the time needed to measure a single standing wave ratio; otherwise, results will be erroneous due to the shift of the reference plane (the plane of detuned short). Note that the VSWR method is applicable for one-port (reflection type) cavity. Another limitation of the method is that for high value of quality factor of cavity (for $Q > 10^4$) the resonance curve (Fig. 12.20) becomes very sharp and the requirement of frequency stability of the signal source becomes very stringent to meet. Thus, for high Q measurement and for transmission type cavity, a common method used is the *transient decay* or the *decrement method*.

12.4.2 Transient Decay or Decrement Method

In this method, the transient decay of the natural oscillations in a microwave cavity is studied. The cavity under study is excited by a pulsed signal (Fig. 12.21a). During the

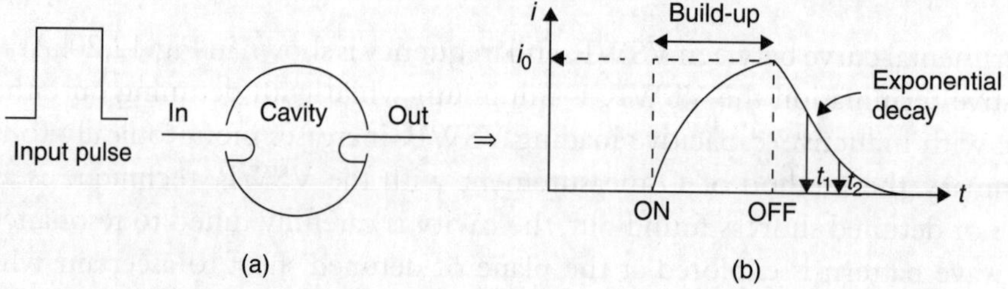

Fig. 12.21 (a) Pulse excitation of a transmission cavity (b) Output response during ON/OFF period of the pulse.

'OFF' period of the pulse, the natural fields in the cavity decay exponentially with time (Fig. 12.21b) and the time constant of the decay is proportional to the Q of the cavity.

12.4.2.1 Derivation of working formula

The theoretical basis of the method is discussed below. The scheme for the method is shown in Fig. 12.22a, together with the equivalent circuit in Fig. 12.22b.

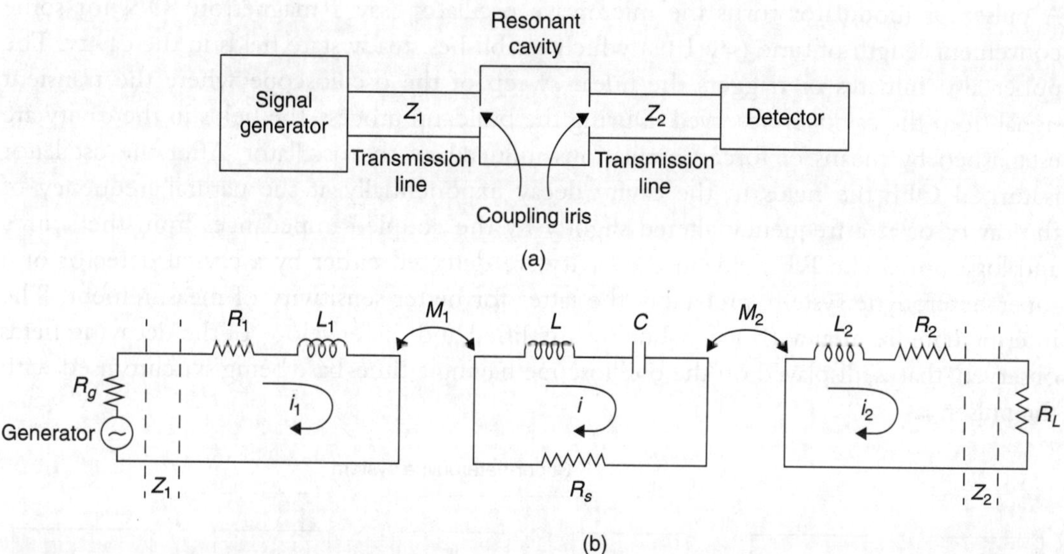

Fig. 12.22 (a) Scheme for transient decay method (b) Equivalent circuit showing generator and cavity with input and output coupling including load.

Let i_0 be the magnitude of the current at the time the signal source is turned OFF. The decaying current (Fig. 12.21b) is given by:

$$i(t) = i_0 e^{-(R/2L)t} = i_0 e^{-(\omega/2Q_L)t} \tag{12.22}$$

where $Q_L = \omega L/R, R = R_s + \dfrac{(\omega M_1)^2}{Z_1} + \dfrac{(\omega M_2)^2}{Z_2}$, and L is the lumped equivalent inductance of the cavity.

If the current $i(t)$ is measured at two instants of time t_1 and t_2, we have from Eq. 12.22:

$$\frac{i(t_1)}{i(t_2)} = e^{-(\omega/2Q_L)(t_1 - t_2)}$$

Solving for Q_L (taking natural logarithm on both sides of above equation and rearranging):

$$Q_L = \frac{\pi f(t_2 - t_1)}{\ln\left\{\dfrac{i(t_1)}{i(t_2)}\right\}} \tag{12.23}$$

If $\Delta t = (t_2 - t_1)$ corresponds to the interval of time during which the current changes by a factor of e (= 2.718), Eq. 12.23 gives:

$$Q_L = \pi f \Delta t \tag{12.24}$$

12.4.2.2 Experimental arrangement

The experimental arrangement for the transient decay method is shown in Fig. 12.23. A pulser or modulator turns the microwave oscillator (say a magnetron) ON for some convenient length of time (say 1 μs) which establishes steady state fields in the cavity. The pulser also initiates or triggers the linear sweep of the oscilloscope where the transient signal from the cavity is observed. During the build-up process, the fields in the cavity are established by means of forced oscillations induced by the oscillator. After the oscillator is turned OFF, the fields in the cavity decay exponentially at the natural frequency of the cavity or at a frequency altered slightly by the coupled impedances from the source and load ports. The RF fields in the cavity are detected either by a crystal detector or a super-heterodyne system, preferably the latter for better sensitivity of measurement. The intermediate frequency (IF) signal being amplified and the envelope of the decaying fields observed, that is, displayed on the oscilloscope having a time-base being synchronised with the pulser.

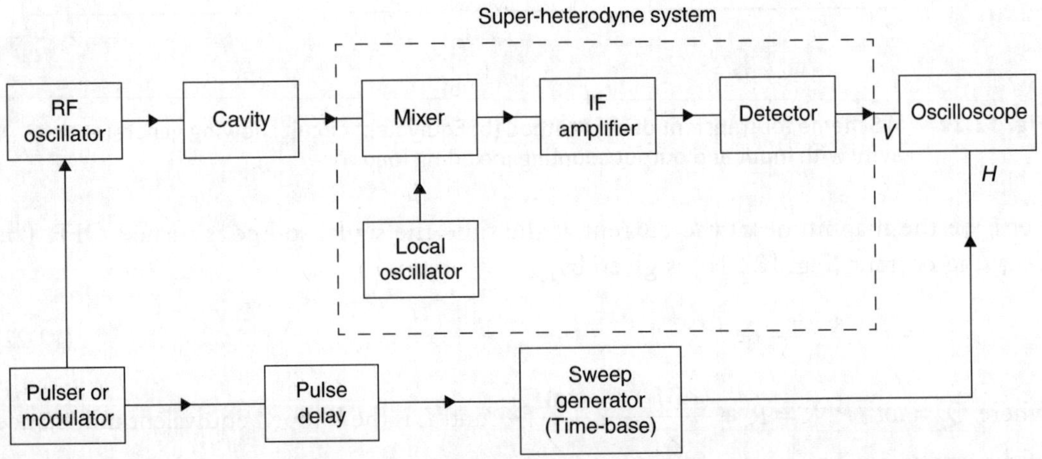

Fig. 12.23 Experimental arrangement for the transient decay method of Q measurement.

12.4.2.3 Scope and limitation of the method

The transient decay method is suitable for Q values that are sufficiently high. For Q_L = 30,000 (say) the decay period is of the order of a few microseconds, which can be measured easily and accurately in an oscilloscope. However, for substantially lower Q values, the time interval Δt will be too short to be measured accurately. It may be noted that the oscillator frequency should not drift during the period of visual observation in the oscilloscope. Thus, the frequency stability required here with respect to time is much less critical in comparison to the VSWR method of Q measurement.

12.4.3 Dynamic Methods

An accurate measurement of high-Q values of the cavity is difficult with the VSWR method because of the time stability problem of common microwave oscillators as appreciable time is required to accumulate necessary data with this method of cavity Q measurement. The decrement method solves this problem; however, the requirement of measurement of short time intervals also limits this method for measurement of low to moderate Q values of cavities. A number of methods have thus been devised that employ a dynamic presentation of the resonance curve of the cavity which allow rapid and accurate experimental determination of Q values for the microwave cavity.

12.4.3.1 Comparison method

In one method, known as the *comparison method*, the transmission resonance curve of a high-Q cavity is compared directly or dynamically against the transmission curve of a low frequency resonant circuit. Assuming that the characteristics of the low frequency circuit are accurately known, the method is straightforward and avoids many possible errors. The two transmission curves are presented in a double beam oscilloscope to compare the resonance curves of the high frequency cavity and the low frequency resonant circuit. The experimental arrangement is shown in Fig. 12.24.

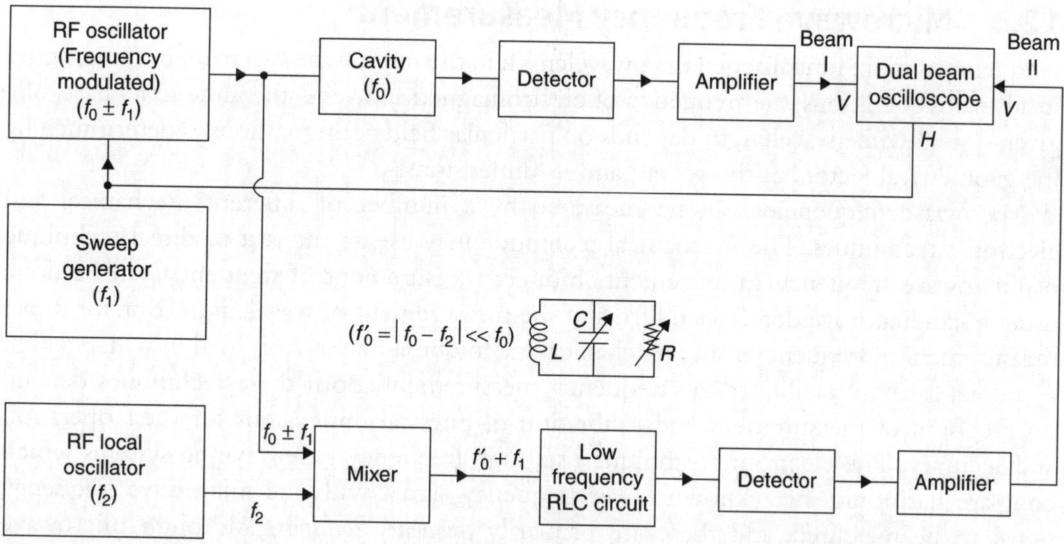

Fig. 12.24 Experiment arrangement for Q measurement by the comparison method.

A microwave oscillator (say at 10 GHz) is frequency modulated (FM) with a linear sweep, which is also applied to the horizontal plates of the oscilloscope. The FM microwave signal after being passed through the cavity is detected, amplified and introduced into vertical plates of one channel of the double beam oscilloscope. The low frequency transmission curve is obtained simultaneously by heterodyning (down converting) the FM microwave signal to some convenient part of the RF spectrum (say 50 MHz) with the help of a local oscillator signal at microwave frequency (f_2). The resultant FM signal $f_0' \pm f_1$ (where $f_0' = |f_0 - f_2|$), introduced in the low frequency circuit, is detected and amplified as in the microwave channel and applied to the vertical plates of the second channel of the oscilloscope. By adjusting either the local oscillator signal frequency (f_2) or the frequency of the low frequency resonant circuit (either by changing the capacitance in the low frequency circuit or by changing the Q of the low frequency circuit by controlling the shunt resistor) it is possible to make the two traces displayed in the two channels of the oscilloscope coincide. The cavity Q can be calculated from:

$$Q = Q' \frac{f_0}{f_0'} \tag{12.25}$$

where Q_0' and f_0' are the quality factor and resonant frequency of the low frequency circuit respectively and f_0 is the resonant frequency of the microwave cavity. A microwave cavity having $Q = 10^4$ at 10 GHz can be duplicated at 50 MHz by a low frequency cavity having a Q value of 50. Note that the method requires the detecting and amplifying devices be identical in both channels. Careful calibration has to be done for the low frequency circuit. Other dynamic methods are also in use like frequency marker method and reflection bridge method.

12.5 Microwave Frequency Measurement

Frequency is more fundamental than wavelength in the microwave spectrum because, under steady state conditions, the frequency of electromagnetic waves is the same in all parts of a given system while wavelength depends on particular field configurations as determined by the geometrical factors of the system and its dimensions.

Microwave frequencies can be measured by a number of different mechanical and electronic techniques. The mechanical technique may use an indirect or direct technique of microwave frequency measurement. Indirect measurement of frequency can be done using a standing wave detector in terms of the measurement of wavelength. But, for direct measurement of frequency using mechanical technique, a *wavemeter* is used in which a high Q resonant cavity facilitates the frequency measurement. Both these techniques depend on precision of measurement and calibration of physical dimensions for their operation and accuracy. The electronic technique uses high frequency heterodyning systems which compare harmonics of a known lower frequency signal with the microwave frequency signal to be measured, and these are primarily *frequency counters*. Although microwave frequency counters are usually more complex and expensive than the wavemeters, their measurement accuracy is comparatively higher.

12.5.1 Mechanical Technique of Frequency Measurement—Wavemeter

The *wavemeter* consists of a rectangular waveguide having an off-set slot (a radiating slot) on the top broad wall of the waveguide. The slot is closed from the top by a tunable cylindrical cavity which can be tuned by means of a sliding plunger with a calibrated dial-knob assembly (Fig. 12.25). The cylindrical cavity is designed to operate in the TE_{011} mode that has circumferential surface current lines on the cavity walls making R_s small (Section 4.4.4), so the cavity Q (= $\omega L/R_s$) is very high (of the order of 20,000 to 40,000) having high selectivity. If the frequency of the wave passing through the main rectangular waveguide is different from the resonant frequency of the cylindrical cavity attached to it, the transmission of signals through the rectangular waveguide does not get effected. On the other hand, if these two frequencies coincide, a resonant field is set-up inside the cylindrical cavity; and because of the high selectivity of this cavity (due to high Q value) the wave flowing down the rectangular waveguide is significantly attenuated (of the order of 3 dB) and hence this type of wavemeter is also known as *absorption type wavemeter.* If there is an indicating instrument, say a power meter, connected such that the frequency meter is in between the signal source and the indicating instrument, the indicating instrument will show a 'dip'. The frequency of the signal can be read on the calibrated dial–knob assembly of the tuning plunger when the dip is observed in the indicating instrument. The back part of the tuning plunger disc is coated with a highly lossy dielectric material (polyiron) to damp the possibility of back cavity resonance, if any.

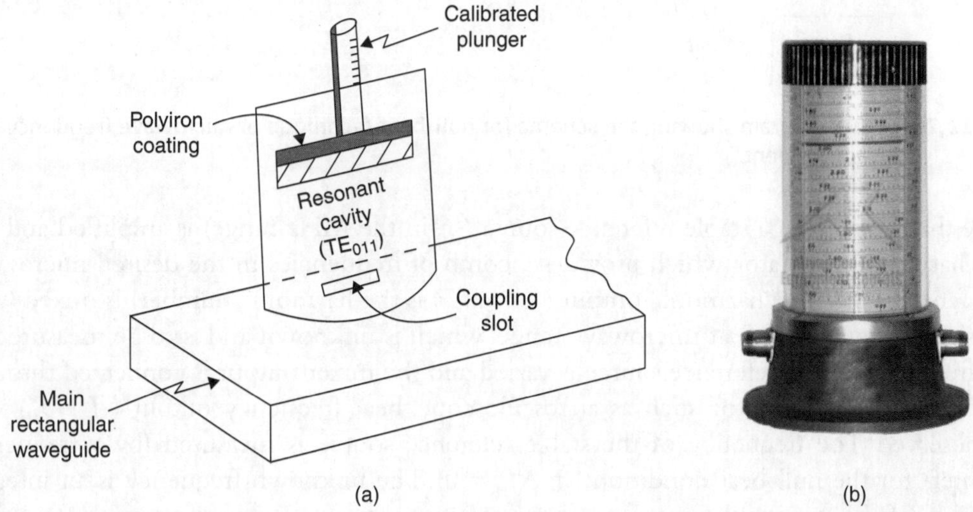

Fig. 12.25 (a) Schematic of absorption type wavemeter (b) Photograph of a typical wavemeter.

The resonant frequency of the cylindrical cavity of the wavemeter is determined primarily by its physical dimensions and the dielectric constant of the medium (air) inside the cavity. Consequently these frequency meters are prone to getting affected by temperature changes (which causes differential expansion of the cavity) and humidity (which may cause a slight change in the dielectric constant of the air inside the cavity), if due care is not taken in their design. These effects can be eliminated/minimised by using a hermetically sealed cavity and special temperature compensating materials for the construction of the cavity; due to these reasons the cavity appears to be oversized.

12.5.2 Electronic Technique of Frequency Measurement—Frequency Counter

The electronic technique of microwave frequency measurement, *frequency counter*, uses the *null-beat* technique. Here, the unknown frequency (f_s) of the microwave signal is mixed or heterodyned in a non-linear device with a reference frequency signal (f_H) which corresponds to a suitable harmonic of a stable reference source (operating in the MHz region). The value of the microwave frequency is obtained in terms of the accurately measurable frequency (at MHz, which can be done by a frequency counter) of the stable reference source. The scheme is shown in Fig. 12.26.

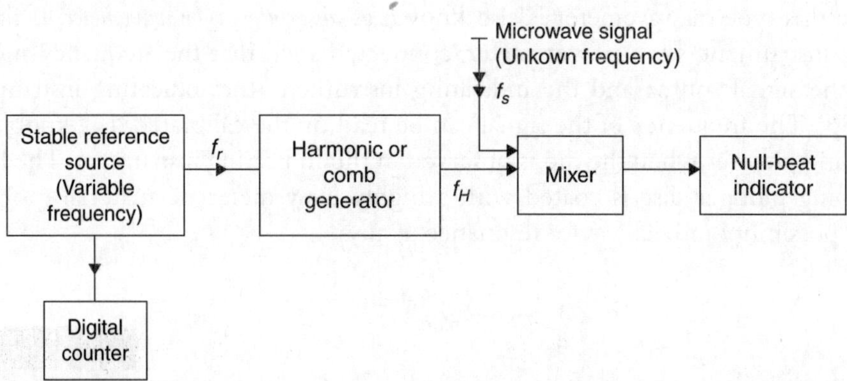

Fig. 12.26 Block diagram showing the scheme for null-beat technique of microwave frequency measurement.

Here the signal from a stable reference source (f_r, in the MHz range) is amplified and fed to a harmonic generator which provides a comb of frequencies in the desired microwave frequency range. The harmonic output ($f_H = nf_r$, n is the harmonic number) is mixed with the signal frequency (f_s) at microwave range, which is unknown and is to be measured. If the frequency of the reference source is varied and the mixer output is connected through a detector to an indicator such as an oscilloscope, beat frequency output ($f_s - f_H$) will be observed. The frequency of the stable reference source is measured (by a frequency counter) for the null-beat condition ($f_s - f_H = 0$). The unknown frequency is an integral multiple of the measured frequency of the reference source. It is not necessary to know the harmonic multiplier number (n) if we note the next higher frequency which again produces a null-beat condition. If two frequencies of the reference source corresponding to the two consecutive null-beat conditions are f_1 and f_2 then we have:

$nf_1 = f_H = f_s$ and $(n-1)f_1 = f_H = f_s$, whence:

$$f_s = \frac{f_1 f_2}{f_2 - f_2} \tag{12.26}$$

Since f_1 and f_2 can be measured very accurately, this method provides very precise frequency measurement. In fact this method is often used to calibrate resonant cavity type wavemeters described earlier.

The above method is the basis for designing direct reading microwave frequency counters. Basically, the variation in the frequency of the low frequency reference source and search for the null-beat are now made automatic by using phase-lock loop arrangement.

12.6 Antenna Measurement at Microwave Frequency

Though antenna properties can be calculated theoretically in many cases, for complex antennae this might not be possible as too many idealisations have to be made. Further, due to fabrication tolerances, the performance of real world antennae should be checked by measurements. Antenna measurement relates to the characterisation of the antenna radiation pattern, gain, polarisation, bandwidth, efficiency, side lobe levels and so on. The basic procedure in antenna measurement is to place a receiving antenna at different locations with respect to the antenna under test (AUT) and get a number of samples of the radiation pattern. The different locations are achieved by rotating the receiving antenna along a circumferential line with AUT at the centre. The distance of the receiving antenna from AUT for all locations during its rotation about AUT should be in the far-field region (given by the well-known Rayleigh distance (Section 2.8.2) giving the border between the Fresnel and Fraunhofer regions), $R \approx 2D^2/\lambda$, where D is the largest dimension of the physical aperture of the antenna and λ is the operating wavelength. The reactive or evanescent near-field region (or even radiative near-field, Fresnel region) is not used for antenna measurement, because it is located too near the AUT and mutual impedance caused by reactive coupling between the AUT and the receiving antenna makes measurement complicated.

For accurate antenna measurements, a pure plane wave (uniform phase and amplitude) is an ideal test field for the measurement of far-field patterns. However, there are inevitable deviations from the plane wave realisation whenever there is insufficient distance between the antennae, causing phase curvature and amplitude taper. Also reflections from the surroundings cause amplitude and phase ripples. Phase curvature and amplitude taper can have a significant impact on the main beam, while ripples might spoil the accuracy of the side lobe measurement. For small microwave antennae, the plane wave approximations hold good in the far-field region but for large antennae this problem is solved by using the Compact Antenna Test Range (CATR) facility (which also includes an anechoic environment to get rid of surrounding reflections). CATR simulates an infinite range length by producing a flat phase front with a reflector, lens, horn, array and so on. To get rid of the problem of reflections from the surroundings, small antennae need to be tested in an anechoic chamber. An anechoic chamber simulates a reflectionless free space (as its walls are covered with radiation absorbent material (RAM), Section 13.5) and allows all-weather antenna measurements in a controlled laboratory environment. In an anechoic chamber, the test area is isolated from interfering signals much better than is possible in free space outdoors. The isolation can be improved further by shielding; shielded chambers are also suitable for electromagnetic compatibility (EMC) measurements.

While making microwave antenna measurements, knowledge of *co-polarisation and cross-polarisation* is an important issue since in many situations, a practical antenna may not be 100% polarised in a single desired mode (horizontal/vertical or RH/LH circular).

Cross polarisation (sometimes referred to as x-pol) is the polarisation orthogonal to the desired polarisation (co-polarisation, co-pol) from the antenna. For instance, if the fields from an antenna are desired to be horizontally polarised, the x-pol in this case is vertical polarisation. If the polarisation desired is right-hand circularly polarised (RHCP), the x-pol is left-hand circularly polarised (LHCP). Hence, two radiation patterns (co-pol and x-pol) of an antenna need to be presented in the *E*/*H* plane by setting the reference antenna with respect to the antenna under test (AUT). The x-pol may be specified for an antenna as a power level in negative dB, indicating how many decibels below is the x-pol power level when compared with the desired polarisation's (co-pol) power level.

In a spherical coordinate system (Fig. 2.13), the *E* and *H* planes of an electromagnetic signal are in the θ and φ planes respectively. If we make measurements in the *H*-plane, then for co-pol measurements, both the antennae are oriented with same polarisation with $\theta = 0^0$ and for x-pol measurement the receiving antenna is turned 90° (when $\theta = 90°$) with the orientation of the transmitting antenna unchanged. Next, to make measurements in the *E*-plane, for co-pol measurement, both the antennae are oriented with the same polarisation with $\varphi = 0°$ and for x-pol measurement, the transmitting antenna is turned 90° (when $\varphi = 90°$) with the orientation of the receiving antenna unchanged.

Note that while working with circularly polarised antennae, we come across a term called *axial ratio*. A circularly polarised field is made up of two orthogonal *E*-field components of equal amplitude (and 90° out of phase) and the axial ratio represents the ratio of these two orthogonal components of the *E*-field. Because the components are of equal magnitude, the axial ratio is 1 (or 0 dB) for circularly polarised waves. The axial ratio for an ellipse is larger than 1 (> 0 dB). The axial ratio for pure linear polarisation (horizontal/vertical) is infinite, because the orthogonal components of the field are ideally zero.

Axial ratios are often quoted for antennae in which the desired polarisation is circular; the ideal value of the axial ratio for circularly polarised fields is 0 dB. The axial ratio indicates the amount by which the main beam of the antenna degrades away from the ideal (circular) polarisation. 'Axial ratio < 3 dB for ±30 degrees from mainbeam' means that the deviation from circular polarisation is less than 3 dB over the specified angular range.

12.6.1 Horn Antenna Measurement

A brief discussion on horn antenna measurement is given below. The waveguide horn antennae used in practice are pyramidal horns (Fig. 10.1c) which are moderately directive antennae having their beam widths inversely proportional to the respective dimensions of the concerned plane (Eq. 10.3):

$$\varphi_H = \frac{63\lambda_0}{a_H} \text{ and } \theta_E = \frac{52\lambda_0}{a_E} \tag{12.27}$$

where φ_H and θ_E are respectively the half power beam widths (in degrees) of the *H* and *E* plane radiation patterns, λ_0 is the free-space wavelength and a_H and a_E are respectively the aperture dimensions in the *H*-plane direction and *E*-plane direction.

To measure φ_H and θ_E experimentally, two identical horns are placed in the same orientation (either with the *E*-field parallel to the ground or perpendicular to the ground) and in one line. One horn (the antenna under test, AUT) is excited with a signal from a microwave source while the other is used to receive the radiated power from the excited antenna (Fig. 12.27) by rotating it in a carriage at different angles with respect to the said excited antenna. The radiation pattern may then be plotted and φ_H (θ_E) may be determined from Eq. 12.27.

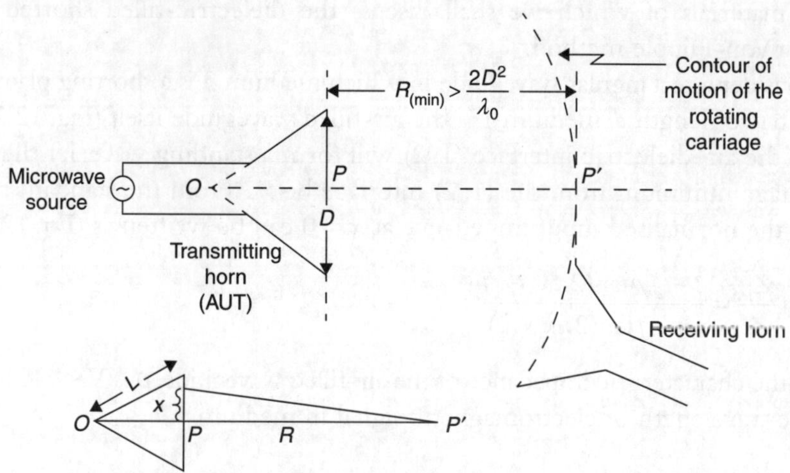

Fig. 12.27 Experimental scheme for horn antenna measurement.

The power gain (G), that is, the practically measured gain of a pyramidal horn is given by (Eq. 10.4 with $\eta_a = 0.6$):

$$G \approx \frac{24,754}{\theta_E \varphi_H} \tag{12.28}$$

At any angular position of the receiving horn, the Rayleigh condition $R_{(min)} > 2D^2/\lambda_0$ (where *D* is the antenna aperture) should be satisfied in placing the receiving horn with respect to the transmitting horn to realise the accuracy of gain measurement. In fact, the field near the horn (in the Fresnel region) will undergo large variations in amplitude and phase (and hence the gain) with distance due to space phase error ($\propto \sqrt{R^2 + x^2} - R$; see inset in Fig. 12.27) due to contributions from relative phases corresponding to different points in the aperture of the horn. However, in the Fraunhofer region, the gain of the horn becomes practically constant (as at large distances from AUT, such error is negligibly small).

12.7 Measurement of Dielectric Constant and Loss Factor of Materials at Microwave Frequencies

The dielectric constant of a material at microwave frequencies, in general, is a complex quantity given by $\varepsilon_c = \varepsilon' - j\varepsilon''$, where the real part is associated with the ability of the dielectric material to store electric energy, while the imaginary part is associated with the losses occurring in the dielectric. The loss tangent $\tan\delta_c$ is defined as the ratio of ε'' to ε'. The variation of ε' with frequency is small while that of ε'' is significant at microwave frequencies (Section 2.4, Fig. 2.3) and thus $\tan\delta_c$ of microwave materials is significantly frequency dependent. The dielectric constant varies with temperature and humidity so these parameters should be kept constant during measurement.

There are various methods to measure the dielectric constant and loss tangent of microwave materials of which we shall discuss the dielectric-filled shorted waveguide technique or Von-Hipple method.

Let us consider a rectangular waveguide in which medium 3 is a shorting plate, medium 2 is the dielectric of length d, medium 1 is the air-filled waveguide itself (Fig. 12.28). A wave incident on the air–dielectric interface (1, 2) will form a standing wave; let the distance of the first voltage minimum from the (1, 2) interface be x_0. From the standard transmission line theory, the normalised input impedance at $x = 0$ can be written as (Eq. 12.3b):

$$z_l = \frac{Z_L}{Z_1} = \frac{1 - j\rho\tan(2\pi x_0/\lambda_1)}{\rho - j\tan(2\pi x_0/\lambda_1)} \tag{12.29}$$

where Z_1 is the characteristic impedance of the air-filled waveguide, ρ is VSWR in medium 1 and λ_1 is the wavelength of electromagnetic signal in medium 1.

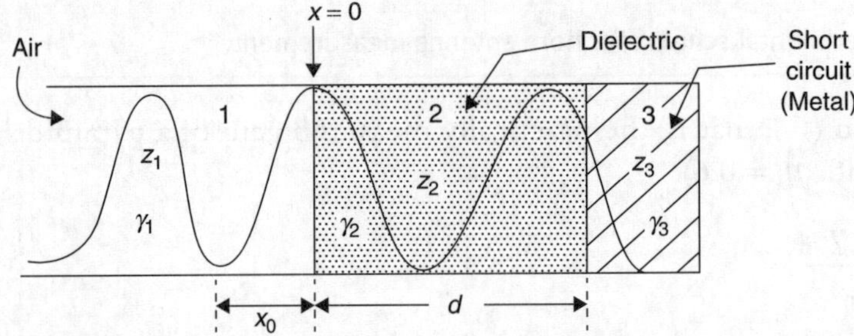

Fig. 12.28 Rectangular waveguide with shorted end and dielectric filling over length d of the waveguide.

Let Z_2 and γ_2 represent the characteristic impedance and propagation constant of medium 2 and Γ_{23} is the reflection coefficient at the boundary of the medium 2 and 3. Since medium 3 is a perfect short, $\Gamma_{23} = -1$ (as voltage has phase reversal at a short circuit, Section 3.2.3) we have the impedance at the interface (1, 2) of the two media as (Eq. 3.17a):

$$Z_L = Z_2\frac{e^{\gamma_2 d} - e^{-\gamma_2 d}}{e^{\gamma_2 d} + e^{-\gamma_2 d}} = Z_2\tanh(\gamma_2 d)$$

Dividing be Z_1, we get:

$$z_l = \left(\frac{Z_2}{Z_1}\right)\tanh(\gamma_2 d) \tag{12.30}$$

Since the characteristic impedance is inversely proportional to the propagation constant (Eq. 3.10) and at microwave frequency the medium 1 (the hollow waveguide) can be considered to be loss-less (with $\alpha_1 = 0$), using Eq. 12.29, we may write Eq. 12.30 as:

$$\frac{\tanh \gamma_2 d}{\gamma_2 d} = -j\frac{\lambda_1}{2\pi d}\,\frac{1-\rho\tan(2\pi x_0/\lambda_1)}{\rho - j\tan(2\pi x_0/\lambda_1)} \tag{12.31}$$

This is a complex transcendental equation from which the value of γ_2 has to be determined. Now, the complex dielectric constant of medium 2 is given by:

$$\varepsilon_c = \varepsilon' - j\varepsilon'' = \varepsilon_0 \left\{ \frac{\left[\left(\dfrac{1}{\lambda_c}\right)^2 - \left(\dfrac{\gamma_2}{2\pi}\right)^2\right]}{\left[\left(\dfrac{1}{\lambda_1}\right)^2 + \left(\dfrac{1}{\lambda_c}\right)^2\right]} \right\} \tag{12.32}$$

where $\lambda_c = 2a$ is the cut-off wavelength for the dominant TE_{10} mode in the air-filled rectangular waveguide (medium 1), λ_1 is the guide wavelength of the air-filled guide. Knowing λ_c, ε_0, γ_2 and λ_1 the value of ε_c can be calculated. The separation of real and imaginary parts in the equation above permits one to obtain the value of ε' and ε'' and hence the value of the dielectric constant and loss tangent for the dielectric material.

From Eq. 12.31, we see that for a given value of tan $(\gamma_2 d)/\gamma_2 d$ (or for low-loss case γ being replaced by β) there are a number of values of $\gamma_2 d$ $(\beta_2 d)$, thus we get multiple values of ε' corresponding to multiple values of γ_2 (or β_2). The difficulty can be overcome if an approximate value of ε' is known, because then we can select one value of $\gamma_2 d$ which gives the ε' nearest to the known approximate value. In case this is not possible, an identical experimental procedure is repeated with the dielectric sample of different lengths and a graphical method is used to evaluate the value of the dielectric constant. A unique solution of the transcendental Eq. 12.31 is common to all the sets of measurement only at one

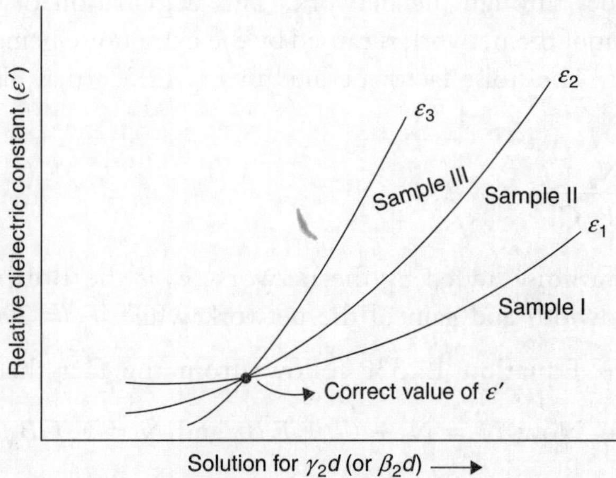

Fig. 12.29 Graphical method to evaluate the value of ε'.

point and that is the point of intersection of the curves drawn for each sample between dielectric constant ε' and multiple values of γ_2d (or $\beta_2 d$) as shown in Fig. 12.29. The correct value of the dielectric constant is that corresponding to the intersection point. The experimental arrangement is similar to the VSWR measurement with the dielectric sample filled waveguide acting as load.

12.8 Measurement of Noise Figure and Phase-noise

Microwave oscillators and amplifiers are the indispensable parts of all RF and microwave systems from wireless communication to radar and navigation, military and aerospace including vital test equipments. Noise characterisation of oscillators and amplifiers is a very important issue related to microwave measurement techniques. Noise characterisation of amplifiers, mixers and so on, is usually done with the help of noise figure measurement and that of an oscillator is done by phase-noise measurement. Let us discuss the fundamentals of noise figure and phase-noise and address their measurement techniques.

12.8.1 Noise Figure and its Measurement

The noise figure or noise factor F of a network is defined as the ratio of the available signal-to-noise power ratio at the input to the available signal-to-noise power ratio at the output of the network:

$$F = \frac{S_i/N_i}{S_o/N_o} \tag{12.33a}$$

Noise figure (NF) is usually expressed in dB and is given by:

$$\text{NF} = 10 \log_{10}(\text{F}) \tag{12.33b}$$

Thus, the noise figure of a network is the decrease or degradation of the signal-to-noise ratio as the signal goes through the network. This degradation of signal-to-noise ratio observed at the output of the network is caused by the extra noise being introduced (added) by the network itself. The noise factor defined in Eq. 12.33a may also be expressed in a more elegant form as:

$$F = 1 + \frac{N_a}{k_B T_0 B_N G_N} \tag{12.33c}$$

where N_a is the extra noise added by the network, k_B is the Boltzmann's constant, B_N and G_N are the bandwidth and gain of the network while T_0 (= 290 K) is the standard ambient temperature. Equation 12.33c follows from the fact that $F = \frac{N_o}{G_N N_i}$; where $G_N = \frac{S_o}{S_i}$ (Eq. 12.33a). Now, $N_o = N_a + G_N k_B T_0 B_N$ and $N_i = k_B T_0 B_N$, where the available noise power at the input of the network has been considered to be due to a noisy resistor at a temperature T_0 that generates thermal noise due to random thermal motion of electrons in the conductor.

Sometimes 'effective input noise temperature' T_e is used to describe the noise performance of a network rather than the noise figure. T_e represents the equivalent temperature of source impedance inside a perfect (noise-free) device that would produce the same added noise N_a. It is often defined as:

$$T_e = \frac{N_a}{k_B B_N G_N} \tag{12.33d}$$

Comparing Eqs 12.33c and 12.33d we get:

$$T_e = T_0 (F - 1) \tag{12.33e}$$

and $$F = 1 + \frac{T_e}{T_0} \tag{12.33f}$$

12.8.1.1 Y-factor method of noise figure measurement

Measurement of the noise figure of a network is done by feeding the network under study with a calibrated noise source under two conditions—when it is *cold* (with no DC bias applied to it, Fig. 12.30a) and when it is *hot* (with the DC bias being applied to it, Fig. 12.30b). With no DC power applied, the noise source behaves as a matched resistor at room temperature (T_0 = 290 K) and the noise power measured at the output of the network is $k_B(T_e + T_0)B_N$. When DC power is applied, the noise source generates noise whose noise power may be estimated in terms of a temperature T_h that characterises it and hence the noise power measured at the output of the network is now $k_B(T_e + T_h)B_N$. The excess noise ratio (ENR) is defined as $\text{ENR}_{\text{dB}} = 10 \log_{10} [(T_h - T_c)/T_0]$, where T_c is the noise temperature represented by the noise source when it is unbiased or cold and for a calibrated noise source it is equal to T_0. Calling the ratio of noise power in the 'hot' and 'cold' condition of the noise source as Y, we have:

$$Y = \frac{k_B(T_e + T_h)B_N}{k_B(T_e + T_0)B_N} \tag{12.33g}$$

where T_e is the noise temperature of the network. Equation 12.33g may be rearranged to solve for T_e, when we get:

$$T_e = \frac{(T_h - YT_0)}{Y - 1} \tag{12.33h}$$

In essence, the method is to evaluate the value of Y from Eq. 12.33g obtained through measurement as mentioned above and knowing T_0 and T_h (which does not represent any physical temperature but are rather the noise power from the calibrated noise source when it is unbiased and when it is biased respectively and can be measured simply with a power meter). The value of T_e is determined from Eq. 12.33h when the noise figure of the network is calculated from Eq. 12.33f. The method is known as the *Y-factor method*. The most commonly used noise source is a special low capacitance diode that generates noise when reverse biased into avalanche breakdown with a constant current source.

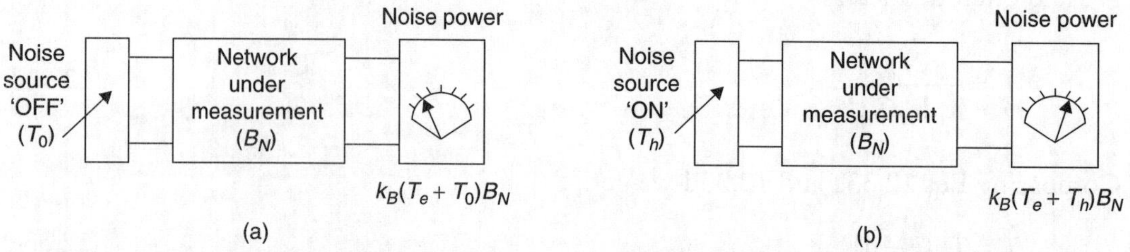

Fig. 12.30 Noise figure measurement scheme (a) Network under test is fed with noise source when it is 'cold' (OFF) (b) Network under test is fed with noise source when it is 'hot' (ON).

12.8.2 Phase Noise and its Characterisation for Microwave Oscillators

An ideal oscillator is supposed to generate a pure sine wave given by:

$$V(t) = V_0 \sin(2\pi f_0 t) \tag{12.34a}$$

where V_0 is the nominal amplitude of the signal and $2\pi f_0 t$ is the linearly growing phase component.

However, due to the presence of noise caused by inevitable reasons the signal from a practical oscillator is better modelled as:

$$V(t) = \{V_0 + a(t)\}\sin\{2\pi f_0 t + \phi(t)\} \tag{12.34b}$$

where $a(t)$ relates to the amplitude fluctuation while $\varphi(t)$ is the fluctuating phase term caused by the presence of noise.

The amplitude fluctuation caused by noise does not affect the zero crossing while phase fluctuation does not affect the amplitude of the signal peaks. Well designed oscillators have small amplitude fluctuations; if any, they can be removed by using automatic level control (ALC) systems or by passing the signal through a limiting amplifier (as its output is determined only by the zero-crossing of the signal). Amplitude noise is also rejected to some extent by mixers used in radio systems.

There are two types of fluctuating phase terms—deterministic and random. The deterministic ones originate from power line frequency, vibration frequency and mixer products, and appear as distinct components in the power spectral density plot and are known as spurious signals (Fig. 12.31). The random phase fluctuations originate from thermal noise, shot noise and flicker noise. The random phase fluctuation (and hence eventual frequency fluctuation, since frequency and phase are related to each other) may be long-term or short-term in nature. The long-term frequency fluctuation is specified in terms of frequency drift (expressed in parts per million per hour, day, month, or year), while the short-term (may be over a span of few seconds or minutes) frequency fluctuation is characterised with *phase noise.*

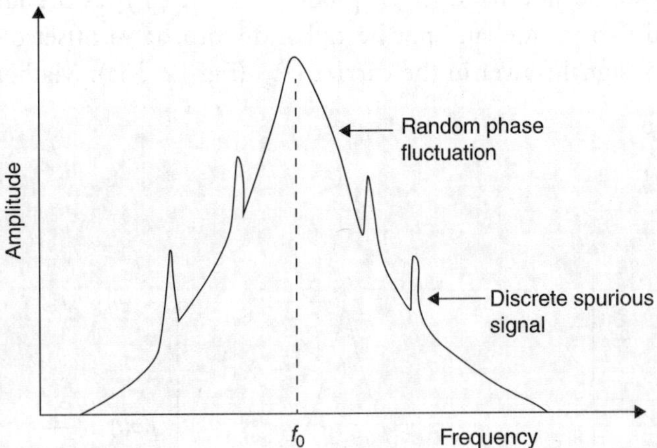

Fig. 12.31 Power density spectrum showing the deterministic and random phase fluctuation giving rise to spurious signals and phase noise.

Phase noise basically causes phase modulation of the carrier. In the time domain, these results in phase off-set in the sinusoidal signal of the carrier according to the modulation caused by the noise signal giving rise to a 'time jitter' in the zero crossing of the carrier sinusoid (Fig. 12.32a). The timing jitter can become significantly deleterious in some applications; for example, if the signal with time jitter is used as a data clock in digital transmission systems the timing jitter may cause erroneous data sampling. Generally, phase noise results in low-level phase modulation and thus the equivalent time jitter is too small to be detected in the time domain (in an oscilloscope). Phase noise characterisation is best suited in the frequency domain (with a spectrum analyser) where the phase noise shows up as noise sidebands on the carrier (Fig. 12.32b). Phase noise can have a major impact on radio system performance. Minute amounts of phase noise in a transmitter signal might cause significant interference to other services, while the phase noise in a receiver local oscillator can severely reduce the receiver selectivity or cause other undesirable effects.

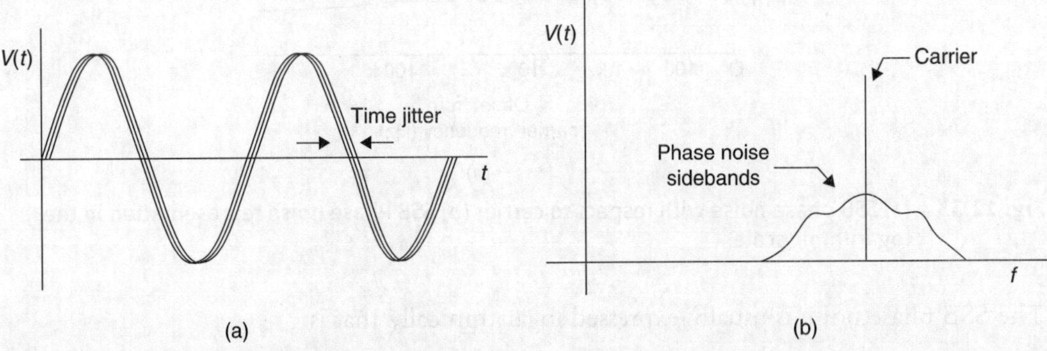

Fig. 12.32 (a) Time jitter of a sinusoidal signal due to phase noise (b) Phase noise side band spread around the carrier.

The spectral purity of high frequency RF oscillators at microwave and millimeter-wave frequency thus needs to be characterised in terms of phase noise. The National Bureau of Standards defines single side band (SSB) phase noise $[L_\varphi(f)]$ as the ratio of power (P_{ssb}) in one phase modulation sideband per hertz bandwidth, at an offset of f_m hertz from the carrier, to the total signal power in the carrier (P_{sc}) (Fig.12. 33a). Mathematically,

$$L_\varphi(f) = \frac{P_{sc}}{P_{ssb}} \tag{12.34c}$$

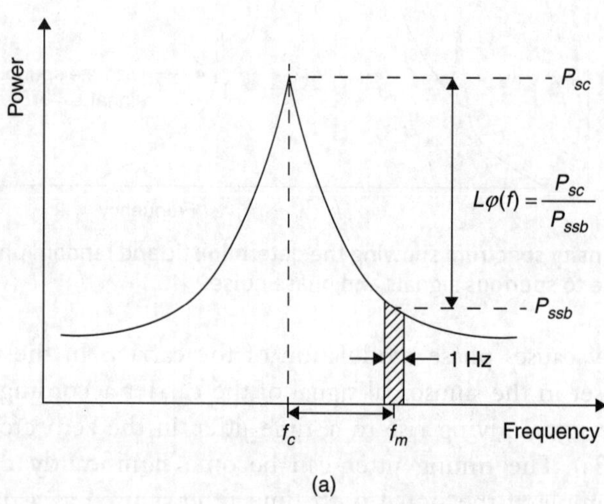

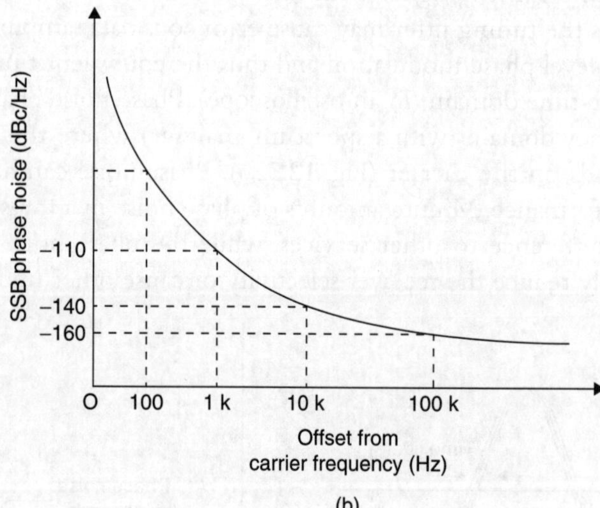

Fig. 12.33 (a) SSB phase noise with respect to carrier (b) SSB Phase noise representation in the logarithmic scale.

The SSB phase noise is usually expressed logarithmically, that is:

$$L_\varphi(f)_{\text{dB}} = 10\log_{10}\{L_\varphi(f)\} \tag{12.34d}$$

The unit of phase noise is dBc/Hz. In practice, phase noise is depicted in the spectrum analyser (in frequency domain) as, how many dB down from the carrier signal peak the selected sideband phase modulated signal is at the specified frequency offset f_m from the carrier frequency f_c (Fig. 12.33b). Here, frequency is shown in the logarithmic scale. For a given oscillator (say) better phase noise is realisable with larger frequency offset, however, when close to the carrier the phase noise degrades.

Phase noise of an oscillator (the phase instability of an oscillator caused by noise) can, in general, be reduced by increasing the quality factor Q of the oscillator system. This is the reason why a dielectric resonator oscillator (DRO) using DR with a comparatively large Q value as resonant circuit for the oscillator has a better phase noise (of the order of –110 to –120 dBc/Hz) compared to a normal waveguide cavity based microwave oscillator (of the order of –80 to –90 dBc/Hz) at a frequency offset of 10 KHz from the carrier at an operating frequency of 10 GHz.

12.9 Fundamentals of Spectrum Analyser and Network Analyser

Spectrum analyser and network analyser both, in general, make frequency domain measurements; though with 'zero-span' of spectrum analyser and the 'time domain reflectometry' option of network analyser, time domain measurements can also be done. A spectrum analyser is used to study the signal characteristics of signal sources while a network analyser is used to characterise components, devices, circuits and sub-assemblies. To be more explicit, a spectrum analyser is intended for analysing the nature of (unknown) signals that are fed into them, that is, the signature of the energy source. The network analyser, on the other hand, contains a reference signal source (test signal) within it and uses this to analyse a network, that is, to characterise the network signature. Although there are many similarities between spectrum analysers and network analysers (both contain tuned receivers operating over similar frequency ranges), they are optimised for very different measurement applications. A spectrum analyser is used to characterise the signal level and its harmonics including phase-noise of oscillators and signal or carrier level, side bands and inter-modulation products of a communication or radar signal (just to mention a few). While a network analyser, on the basis of the S-parameter (for linear network) or X-parameters (for non-linear network) measurements characterises the network under investigation. With a network analyser we get both the (ratioed) amplitude and phase information but spectrum analyser does not, in general, provide the phase information of the signal.

From a system engineer's point of view, a spectrum analyser is a swept-tuned super-heterodyne receiver that provides a CRT/LCD display of amplitude vs frequency. It is essentially a frequency selective, peak responding voltmeter calibrated to display the RMS value of a sine wave. The spectrum analyser can show the individual frequency components and the relative power each component contributes to make up a composite signal (which otherwise appears as a composite voltage waveform in the time domain as displayed in an oscilloscope). A network analyser, on the other hand, contains both a source and multiple receivers (in contrast to the spectrum analyser which is configured as a single channel receiver, without a source) and displays (in CRT/LCD) the amplitude and phase information in various formats including one on a Smith chart. With vector-error correction, network analysers provide a much higher measurement accuracy than spectrum analysers.

Why do both the spectrum analyser and network analyser, in general, make measurements in the frequency domain? This is because, frequency domain measurements have several distinct advantages. Firstly, narrowband frequency domain measurements provide a greater sensitivity and dynamic range than time domain measurements. Since the measurement bandwidth can be narrowed almost arbitrarily, frequency domain analysers can greatly reduce the amount of noise present in the measurement. Also, narrowband measurements can remove large interfering signals at undesired frequencies. Secondly, some systems are inherently frequency domain oriented. For example, many telecommunication systems use what is called the Frequency Division Multiple Access (FDMA) or Frequency Division Multiplexing (FDM). In these systems, different users are assigned different frequencies for transmitting and receiving, such as with a cellular phone. These types of systems must be analysed in the frequency domain in order to minimise the possibilities of mutual interference. Even when systems are not inherently in the frequency domain, they might require frequency domain measurements. For example, the bandwidth and the speed of a digital pulse may be affected by stray capacitance and resistive losses in a high speed digital circuit. With the help of a network analyser, the bandwidth of the circuit may be determined by measuring its transfer function in the frequency domain. Again suppose a switching power supply operating at a desired switching frequency is corrupted with the harmonics of the line frequency (50/60 Hz), then it is difficult to characterise the signal with multiple corrupting frequencies with a time domain measurement in an oscilloscope. A spectrum analyser, on the other hand, can separate these frequency components and can measure them accurately.

12.9.1 Spectrum Analyser

12.9.1.1 Introduction

Radio frequency (RF) spectrum analyser works typically in the frequency range from 10 MHz to 40 GHz, that covers the vast majority of communication, navigation, radar systems and industrial applications. For operation at higher millimeter-wave frequency, external waveguide mixers can be used to extend the frequency range of operation.

Oscilloscopes make measurements in the *time domain* while in spectrum analysers measurement is usually done in the *frequency domain*. The two axes in the plot in an oscilloscope are time and amplitude whereas in a spectrum analyser they are frequency and amplitude. Electrical signals can be represented and observed in both the time domain and frequency domain (Fig. 12.34). These two methods of representation are linked together via the *Fourier transform*, which means that there is a characteristic frequency spectrum for every signal that can be represented in the time domain.

A spectrum analyser displays each frequency component of a signal as an individual vertical line (deflection) in the CRT. For example, a pure sine wave would be a single vertical line displayed at the frequency corresponding to the frequency of the sine wave. For non-sinusoidal waveforms, vertical lines representing the fundamental and harmonic (odd and/or even) components would be displayed. For example, a square wave displayed

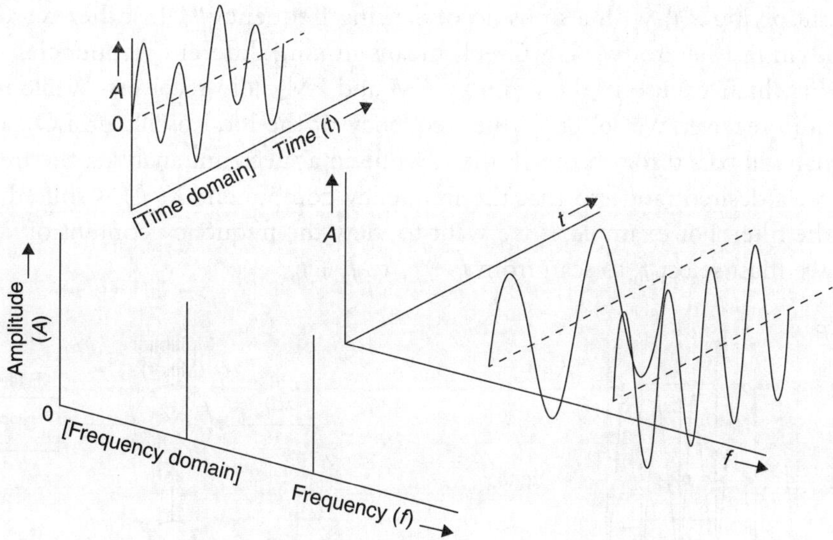

Fig. 12.34 Representation of signals in the time and frequency domains.

in an oscilloscope (in time domain) will appear in a spectrum analyser (in the frequency domain) with a vertical line corresponding to the fundamental frequency (f) and multiple vertical lines with diminishing amplitude at $3f, 5f, 7f\ldots$ corresponding to the odd harmonics. Because of this capability of the spectrum analyser, it is possible to study the power density spectrum and phase-noise of oscillators, carrier modulation (including pulsed RF signal) and unexpected modulation, harmonics and inter-modulation levels. In addition to swept-tuned frequency mode, a spectrum analyser can be used in fixed-tuned mode (zero-span) to provide time-domain measurement capability much like that of an oscilloscope.

12.9.1.2 Basic principle and block diagram of spectrum analyser

The job of a spectrum analyser is to extract the frequency content of a general time domain signal say $x(t)$ and to plot amplitude vs frequency of all the frequency components contained in the signal. A phenomenological solution may be to pass $x(t)$ through a long series/bank of band-pass filters (BPF) with overlapping adjacent passbands, that is, $f_1 + \text{BW}/2 = f_2 - \text{BW}/2$ and so on, where f_1 and f_2 are respectively the centre frequency of filters 1 and 2, and BW is the bandwidth of the BPFs, Fig. 12.35a. If the BPFs are of very narrow bandwidth then the extracted filter outputs give us the frequency components $X(f_1), X(f_2), \ldots.$ and so on. However, this may not be a practical approach and a simple and better solution may be to use a property of Fourier transform which states that, if we multiply (in the time domain) a signal $x(t)$ by a sinusoid, the spectrum of the signal is shifted in the frequency domain by an amount equal to the frequency (f_0) of the sinusoid. Mathematically,

$$x(t) \cdot \cos(2\pi f_0 t) \to 0.5X(f - f_0) + 0.5X(f + f_0)$$

Thus, instead of using a bank of narrow band filters, we can now use only one narrow band filter centred at a fixed frequency, say f_{IF}, and scan the signal spectrum across this filter by multiplying $x(t)$ with a sinusoid of varying frequency f_0. In other words, we can use heterodyning ('heterodyne' in Greek means mixing different frequencies) principle (Fig. 12.35b) which we use in the ordinary AM and FM radio receivers. While tuning the AM/FM radio receiver, we select f_0 (the frequency of the local oscillator, LO) so that the desired signal will pass through the IF filter; while in a spectrum analyser, the frequency f_0 is swept over a desired range so that the frequency component $X(f)$ is shifted to f_{IF} and passed by the filter. For example, if we want to view the frequency content of $x(t)$ from f_1 to f_2, then we must select f_0 to scan from $f_1 + f_{IF}$ to $f_2 + f_{IF}$.

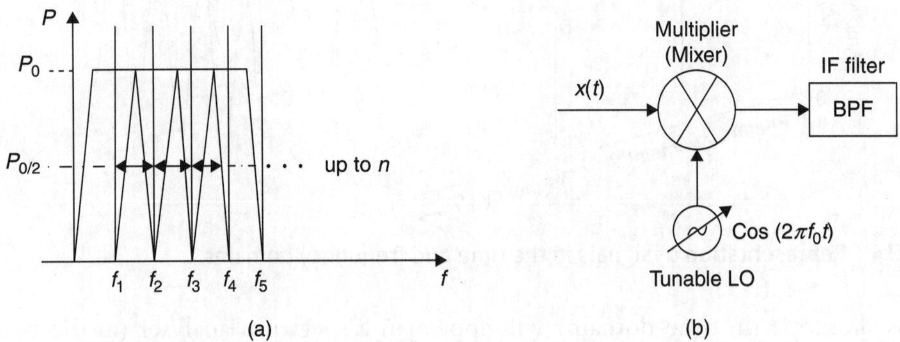

Fig. 12.35 (a) A long series/bank of band-pass filters with overlapping adjacent pass bands (b) Frequency mixing (heterodyning).

Thus, the most common type of spectrum analysers are basically swept-tuned super-heterodyne receivers (also popularly known as swept/scanning spectrum analysers), of course, having a much more complicated signal conditioning going on inside the analyser than the one shown in Fig. 12.35. A simplified block diagram of a swept spectrum analyser is shown in Fig. 12.36.

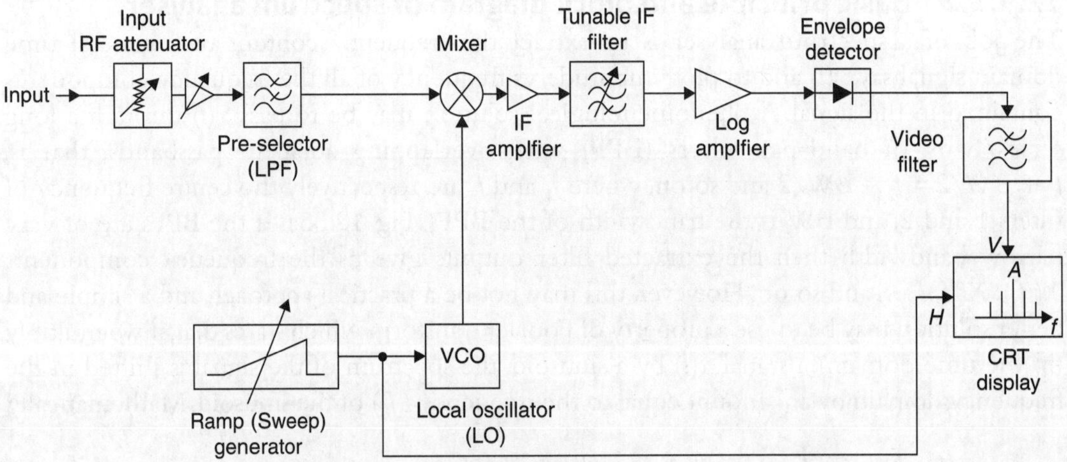

Fig. 12.36 Block diagram of a swept-tuned spectrum analyser.

A ramp (sweep) generator is used to produce a linearly increasing voltage which drives the tuning port of a voltage controlled oscillator (VCO) acting as LO. The same ramp voltage is also applied to the horizontal axis of the display (say a CRT) to facilitate synchronisation. The RF signal under test is heterodyned (mixed) with the LO signal to generate the desired intermediate frequency (IF). The LO in Fig. 12.36 is tuned by a sweep (ramp) generator in order to convert the entire input frequency range to a constant IF. The IF signal is amplified and arrives at the IF filter with a definable bandwidth. The input signal is essentially 'swept past' this filter with a fixed centre frequency. The IF filter determines the resolution bandwidth (RBW) of the analyser. In order to be able to display signals with a wide variety of amplitude levels on the screen simultaneously, the IF signal is compressed with the aid of a logarithmic amplifier. After that, the envelope detector and the video filter respectively acquire the signal's envelope, and the noise is reduced with the aid of an averaging process (by the video filter), which smooths out the displayed video signal. As the local oscillator is swept in frequency, the spectrum of the input signal is automatically plotted on the display.

The input side of the spectrum analyser block diagram has a variable RF attenuator, often followed by a variable gain amplifier. The purpose is to control the signal level applied to the rest of the instrument. If the signal level is too large, the analyser circuits will distort the signal, causing distortion products to appear along with the desired signal. If the signal level is too low, the signal may be masked by noise present in the analyser. Either problem would otherwise reduce the dynamic range of the measurement and hence needs to be addressed properly by setting the input signal level to the instrument.

The conceptual basis for understanding the operation of the spectrum analyser discussed above uses all analogue implementation. However, in this digital age, the use of microprocessor and digital signal processing (DSP) has become very common with all modern spectrum analysers. The swept LO may be a digital synthesised source with microprocessor control. Also the output of the detector may be sampled and converted to a digital signal with analog-to-digital (ADC) converter which may be read by the microprocessor. Further, the display in a modern spectrum analyser is always a digital graphic display using liquid crystal display (LCD). The photograph of a modern spectrum analyser is shown in Fig. 12.37.

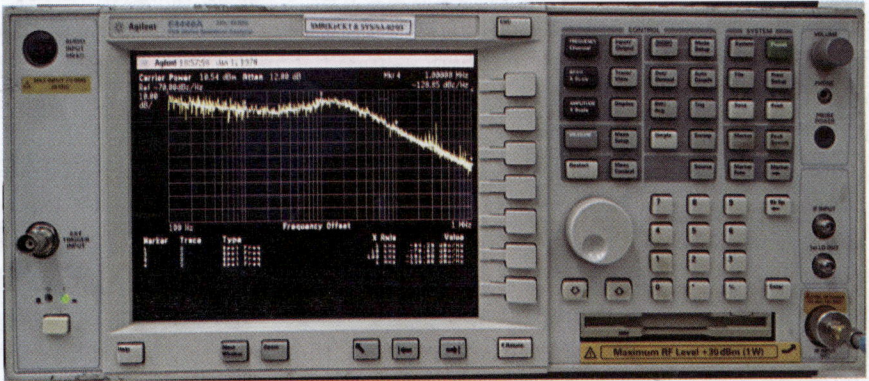

Fig. 12.37 Spectrum analyser depicting phase-noise of a microwave oscillator (Photograph: Courtesy SAMEER Kolkata centre).

In addition to the above mentioned method of frequency-domain measurement using the swept-tuned spectrum analyser technique, with the availability of advanced ADC and

DSP techniques the *Fourier analyser* is becoming useful for sophisticated measurements at lower RF frequencies. The Fourier analyser basically takes a time-domain signal, digitises it using digital sampling, and then performs the mathematics required to convert it to the frequency domain, and display the resulting spectrum. With its real-time signal analysis capability, the Fourier analyser is able to capture periodic as well as random and transient events. It also can provide significant speed improvement over the more traditional swept analyser and can measure magnitude as well as phase. However it does have its limitations, particularly in the areas of frequency range, sensitivity and dynamic range. These analysers can offer significant performance improvement over conventional spectrum analysers, but often with a price premium.

State-of-the-art spectrum analysers of today combine the benefits of both the swept-tuned and Fourier analyser leading to *real-time spectrum analyser* which captures and processes the spectrum of the signal so fast that no part of the signal is missed.

12.9.1.3 A few passing comments

A few passing comments may be made regarding some of the blocks used in the diagram above (Fig. 12.36) and the use of a spectrum analyser for best performance on the basis of application requirements.

The block diagram shown in Fig. 12.36 uses a single mixer and a single IF stage; however, all modern spectrum analysers use multiple conversion stages. Each conversion section contains a mixer, an LO and IF filter. LOs may all be derived from a master oscillator.

The function of the pre-selector filter is signal conditioning—to select or allow only a narrow band of frequencies to pass into the analyser. It is a sweeping filter that tracks the frequency the analyser is tuned to at any particular point of time. The function it performs is to inhibit harmonic mixing within the first mixer/converter. By eliminating the harmonic conversions, unwanted mixing products do not appear as spurious signals in the spectrum being displayed.

The choice of resolution bandwidth (RBW) and span/division (span/div) is an important issue for observing the spectrum of the signal under investigation in a spectrum analyser.

The RBW determines the finest frequency line that may be visible/obscured in the spectrum. Smaller the RBW, the finer is the resolution and hence the fine details of the spectrum will be visible. However, with larger RBW, the finer details are obscured/lost.

The span/div determines how the spread/width of the frequency spectrum being displayed on the screen can be varied/controlled. The span/div determines the sweeping speed across the spectrum. With the so called 'zero span', the analyser no longer sweeps across a frequency spectrum, but behaves like a super-heterodyne receiver. The analyser under this situation behaves just like a typical oscilloscope where the display indicates the modulation of any signal at the frequency selected by the frequency control.

The choice of an extremely small RBW with an extremely wide span/div, would make the signal spectrum appear exceptionally narrow and may be overlooked easily (practically invisible!) and the measurement process very slow or time consuming. Again if a very wide RBW is chosen with respect to span/div then low level signals close by will be masked in the filter skirts of the RBW filter. Unless special requirement dictates a specific RBW, the

RBW selected should be somewhat greater than 1/50 the span/div. Most modern spectrum analysers through the use of microprocessors have the capability to select the optimum RBW depending on span/div. At times, however, manual control of RBW may be desired.

RBW is a functional control that selects one of several band-pass filters physically located in the IF signal processing chain of the instrument. It is defined in terms of Hz and is a measure of the width of the filter characteristics either 3 dB or 6 dB down from its peak value, depending on the manufacturer of the analyser. The shape of the signal being traced on the screen is a combination of the RBW filter and the signal, and not just the shape of the signal being analysed. The limitations imposed on an analyser by the RBW filter are significant. The sweep speed (the rate at which the analyser sweeps the frequency being present) must be slow enough to allow the RBW filter to reach peak amplitude, otherwise an inaccurate signal amplitude would result.

In general, narrow RBW (optimum with respect to span/div) is desired from the view point of many applications. When displaying a pulsed RF signal (commonly used in radar systems), the RBW should be selected narrow enough so that each spectral line (which is close by) of the pulsed RF spectrum becomes distinctly visible. Further, with the choice of narrow RBW, the noise floor (sometimes known as 'grass' because of its appearance) of the analyser goes down, in addition to making the spectrum of the signal being displayed narrow, and thus helps to identify low level and sidebands close by, if any; which otherwise get obscured in higher noise floor levels with the spreading out of the signal spectrum over a wide frequency range.

12.9.2 Network Analyser

12.9.2.1 Introduction

The network analyser provides a complete description of linear network behaviour (in terms of S-parameters) and of non-linear network (in terms of X-parameters) in the frequency domain. This is demanded by design engineers (who need to predict with some certainty the final network performance of a multi-component network from the knowledge of the individual components) and production manager (who needs to know the allowable tolerance on the products manufactured and if the final product meets the specified tolerance). The job of the network analysis is to create a data model (of active/passive network) of component parameters as impedance and transfer function through stimulus response testing (with test signal source built into the system) over the frequency range of interest. Generally, a sine wave test is used for linear networks. The swept nature of analysis is required to have both phase and amplitude of the signal characteristics of the component under test. A computer controlled network analyser can be programmed to set-up and make many measurements automatically. The measurement process is further accelerated by the computer's ability to store, transform, summarise and give the output of the data in a variety of formats to a number of peripherals. This capability makes the computer controlled network analyser ideal for both computer-aided design (CAD) and automatic product testing (APT).

The scalar network analyser (SNA) is a form of RF network analyser that only measures the amplitude properties of the device under test. The vector network analyser

(VNA) is a more useful form of RF network analyser than the SNA as it is able to measure the amplitude response and the phase response as well. As a result, VNA may also be called a gain-phase meter or an Automatic Network Analyser. The large-signal network analyser (LSNA) is a highly specialised form of RF network analyser that is able to investigate the characteristics of devices under large-signal conditions. It is able to look at the harmonics and non-linearity of a network; the previous version of LSNA was known as the Microwave Transition Analyser (MTA).

Vector network analysis is a method of accurately characterising the electrical performance of components and circuits by measuring their effect on the amplitude and phase of swept-frequency/swept-power test signals. Measuring both magnitude and phase of components is important for several reasons. Though in some applications, for example, when we are only concerned about the gain of an amplifier or the stop-band rejection of a filter, magnitude-only data is sufficient for our needs. However, complete characterisation of devices and networks involves measurement of phase as well as magnitude. This is necessary for developing circuit models for simulation and to design matching circuits based on conjugate matching techniques. Time-domain characterisation requires magnitude and phase information to perform the inverse-Fourier transform. Further, for best measurement accuracy, phase data is required to perform vector error correction which is common and an indispensable part of VNA measurement.

12.9.2.2 Block diagram to understand VNA operation

The block diagram of a VNA is shown in Fig. 12.38 with reference to which its operation can be understood.

The VNA primarily consists of four sections—(i) source for stimulus, (ii) signal-separation devices, (iii) receivers that down-convert and detect the signals, and (iv) processor and display for depicting the results in the desired format and for reviewing.

The variable frequency CW *signal source* supplies the stimulus for the stimulus–response test system which is a sweep oscillator. Traditionally, network analysers used a separate source. These sources were either based on open-loop voltage-controlled oscillators (VCOs) which were cheaper (but inaccurate for narrow band measurements due to high phase-noise), or more expensive synthesised sweepers which provide better performance, especially for measuring narrowband devices. Today, network analysers integrated with synthesised sources are used that can provide excellent frequency resolution and stability.

The *signal separation* devices are directional couplers having low insertion loss and good isolation and directivity. There are two functions to be served by the signal separation hardware—to measure a portion of the incident signal to provide a reference for ratioing and to separate the incident (forward) and reflected (reverse) travelling waves; the latter coming from the device under test (DUT).

The next section of the network analyser is the *signal detection* block. There are two basic ways of providing signal detection—diode detection technique and tuned receiver technique. Diode detection is inherently scalar as phase information of the RF carrier is lost; however, they are low cost and can provide a broad frequency range coverage at the cost of sensitivity and dynamic range. In the tuned receiver technique, a local oscillator (LO)

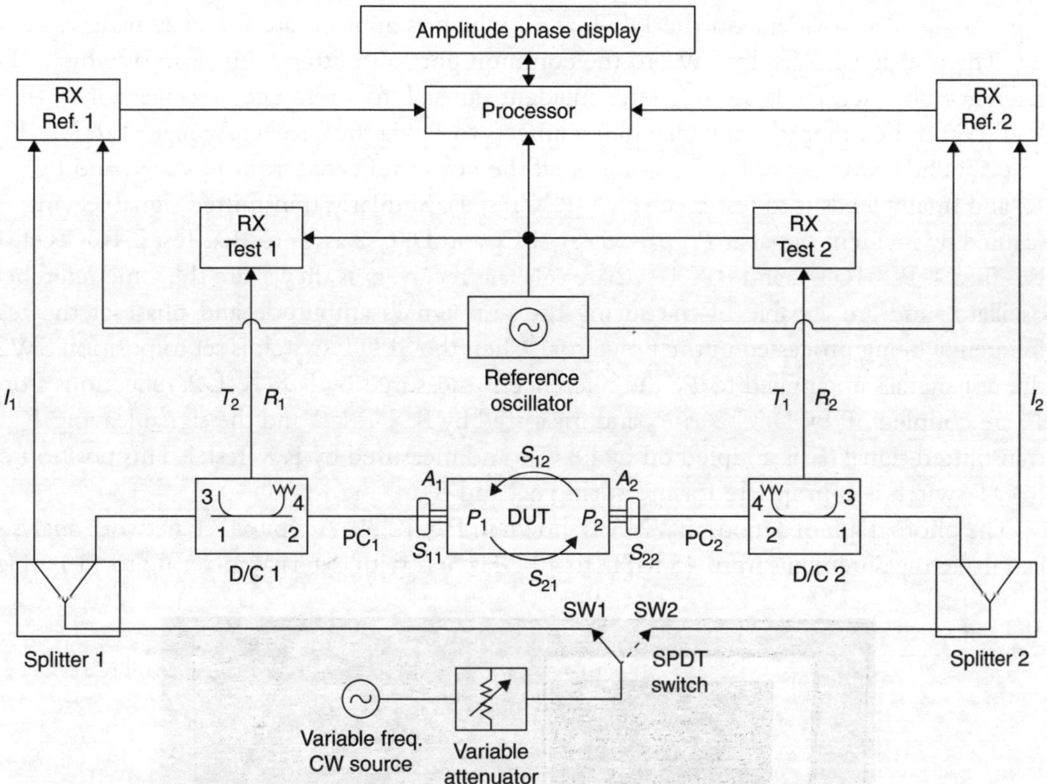

Fig. 12.38 Block diagram of a vector network analyser.

(a reference oscillator) is used to heterodyne the RF signal down to a lower intermediate frequency (IF). The LO is either locked to the RF or the IF signal so that the receivers in the network analyser are always tuned to the RF signal present at the input. The IF signal is band-pass filtered, which narrows the receiver bandwidth and greatly improves sensitivity and dynamic range. Modern analysers use an analogue-to-digital converter (ADC) and digital-signal processing (DSP) to extract magnitude and phase information from the IF signal. Both vector network analysers and spectrum analysers use tuned-receiver approach.

The last major block in the hardware of a network analyser is the *processor and display* section. This is where the reflection and transmission data is formatted in ways that make it easy to interpret the measurement results. Most network analysers have features like linear and logarithmic sweeps, linear and log formats, polar plots, Smith charts and so on.

In a generalised two-port VNA (Fig. 12.38) the two ports P_1 and P_2 of DUT are connected to the two test ports of the VNA via suitable adapter connectors A_1 and A_2 and precision connecting cables PC_1 and PC_2 respectively. The input reflection coefficient (S_{11}) and forward transmission coefficient (S_{21}) are measured at port P_1 in the forward path while the output reflection coefficient (S_{22}) and reverse transmission coefficient (S_{12}) are measured at port P_2 in the reverse path. The test frequency is generated by a variable frequency CW source and its power level is set with the help of a variable attenuator. The SPDT switch can be connected either to position SW1 or position SW2. When it is in position SW1, the test signal is incident on the DUT at P_1 which is appropriate for measuring S_{11} and S_{21}. The test signal is fed by SW1 to the common port of splitter 1, one arm of which (the reference channel) feeds the reference incident signal I_1 to a reference receiver for P_1 (RX Ref. 1) and the other (the test channel) connects to P_1 via the directional coupler D/C 1, PC_1 and A_1. The third port of D/C 1 couples off the power reflected from P_1 via A_1 and PC_1 as R_1 and finally feeds it to test receiver 1 (RX Test 1). Similarly, transmitted signal leaving P_2 (caused by incident signal at P_1) pass via A_2, PC_2 and D/C 2 as T_1 to RX Test 2. RX Ref. 1, RX Test 1, RX Ref. 2 and RX Test 2 are coherent receivers as they share the same reference oscillator and are capable of measuring the test signal's amplitude and phase at the test frequency being processed in the processor. When the SPDT switch is set to position SW2, the test signals are applied to P_2, the reference is measured by RX Ref. 2; reflections from P_2 are coupled off by D/C 2 as R_2 and measured by RX Test 2 and the signal leaving P_1 as transmitted signal T_2 is coupled off by D/C 1 and measured by RX Test 1. This position of SPDT switch is appropriate for measuring S_{22} and S_{12}.

The photograph of a modern VNA is shown in Fig. 12.39. A typical RF network analyser can make measurements from 45 MHz to 40 GHz and with external mixer up to 110 GHz.

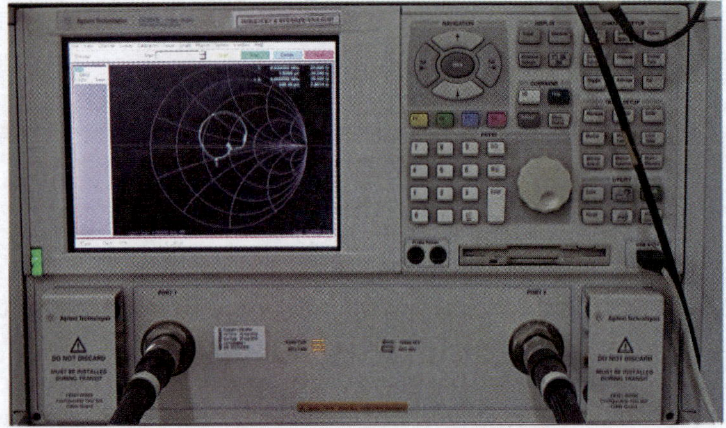

Fig. 12.39 Vector network analyser (VNA) depicting the impedance locus of a microwave circuit on Smith chart (Photograph: Courtesy SAMEER Kolkata centre).

12.9.2.3 VNA Calibration

VNA calibration is an important issue that makes *vector error correction* for accurate measurement with such an analyser. Calibration by the user (over and above the factory made calibration) before making any measurement is critical to making good S-parameter measurements with VNA. User calibration is crucial to ensure the accuracy and repeatability of VNA measurements.

The VNA calibration can be understood on the basis of its error model. The VNA error model includes all possible areas in which systematic inaccuracies can arise; for example, the *directivity* of the couplers used to configure the reflectometer that separates the incident and reflected signals from DUT, the *match* at the input and output test ports of the DUT, the *frequency response* of the transmission line cable used between the DUT and the instrument ports, and *cross-talk* or *lack of perfect isolation* (leakage signal coupling) among the couplers.

While the VNA is a highly-linear receiver and has sufficient spectral purity in its sources to make good measurements, there are a number of imperfections that limit measurements done without calibrations. These imperfections are related to *match* (VNA being broadband, a match more than the typical value of 20 dB generally used in practice, is desired to reduce potential errors), *directivity and isolation* (the directional couplers used in the heart of VNA even though are of very high quality, there may be a certain amount of coupled signal even with perfect termination thereby deteriorating directivity and can influence measurements of very small reflection coefficients; also the forward and reverse couplers might have leakage coupling or cross-talk with imperfect isolation between the two paths), and *frequency response* (any cables connected externally will have some frequency dependence that must be calibrated out for high-quality measurements even though the internal frequency response of the VNA could have been calibrated at the factory). These calibrations carried out before making every measurement with VNA is known as *vector error correction* of VNA.

The user calibration accounts for 12 sources of systematic errors in a VNA. For a forward measurement path, when the signal source (the test signal or stimulus) is connected to the DUT at port 1, the first source of systematic error is due to possible impedance *mismatch* at the input (source side) and output (load side) of the DUT. The second source of error is due to *non-ideal directivity* (the directivity not being infinite) which is true even with a very high quality practical directional coupler and little *imperfect isolation* caused by small amount of leakage radiation being coupled from port 1 to port 2 of DUT, that is, between the two directional couplers. The third source of error is caused by the *frequency dependence* of the signal paths (that of the cables) for the transmission and reflection measurement. Thus, for the forward measurement path there are in total six (6) error terms and similarly six (6) more error terms will arise when the signal source is connected to port 2 of the DUT for the reverse path making the *total error terms twelve* (6 + 6 = 12). In the error correction process, the measured magnitude and phase of each error signal appearing as a vector is applied mathematically via a proper algorithm to the measurement signal thereby realising *vector error correction*.

Most comprehensive VNA calibration is full 2-port calibration method in which the parameters calibrated are S_{11}, S_{12}, S_{21} and S_{22}. One may use other approaches, such as 1-port and (frequency) response calibrations, but they are usually subsets of a full 2-port calibration. The wide assortment of VNA user calibrations is divided into three commonly used categories or families—SOLT (Short, Open, Load, Thru), TRL (Thru, Reflect, Line), and automatic calibration. SOLT calibration is perhaps the most familiar of all VNA calibrations. It uses a well-defined short, open and load (of characteristic impedance, usually 50 or 75 Ω). Each standard is connected one by one at the reference plane so the VNA can measure them. When these steps are completed, one needs to connect the two reference planes together to form a through (thru) connection for the final measurement. The point where the calibration standards are connected defines the measurement plane. Also called the reference plane, this is the point at which the correction factors are applied and the measurement begins. For example, if coaxial cables are connected to the VNA ports, and the calibration standards are connected to the end of the cables during calibration, then the measurement plane is at the cable ends.

In recent years, automatic calibration modules have become a popular alternative to manual techniques such as SOLT and TRL. Because of internal switching, heating and other uncertainties, early automatic calibration modules lacked the accuracy and repeatability of manual calibration. However, the current generation of modules offers accurate results and other benefits that make them an attractive choice.

RECAPITULATION

Microwave measurement

- The techniques of measurement at microwave frequency (starting in meters and going down to millimetre wavelengths) is different from lower frequency radiowaves (having kilometres and going down to metre wavelengths) and also from those at infra-red through optical to X-ray wavelengths.
- Primary measurable quantities at the microwave region of the electromagnetic spectrum are standing wave ratio/reflection coefficient, power and frequency. Determination of unknown microwave impedance or the quality factor of a microwave cavity can be done in terms of standing wave ratio measurement. Also the S-parameters of a microwave network can be determined in terms of microwave power measurement.

Impedance measurement at microwave frequencies

- In the *conventional SWD technique* of impedance measurement, the working formula used is:

$$Z_L = Z_0\left(\frac{1 - j\rho\tan\beta d_2}{\rho - j\tan\beta d_2}\right)$$

  where, d_2 is the distance of the first voltage minimum from the load end (in fact any minima can be used). The impedance measurement is done at a voltage minimum because a voltage minimum is always sharper than a voltage maximum in a standing wave pattern making measurement accuracy better. Also *probe loading* is less when measurement is done at a voltage minimum.
- To get rid of the inaccuracy in the measurement of d_2 and to get the nature of the reactive component in the impedance, *short circuit minima shift* method is used when the working formula is given by:

$$\frac{Z_L}{Z_0} = z_L = \frac{1 - j\rho\tan(X)}{\rho - j\tan(X)}$$

  where $X = \frac{2\pi}{\lambda}(d_2) = \frac{180°(\pm\Delta d)}{\lambda/2}$

  Conventionally, Δd takes a −ve sign when the minima shift towards the generator (load impedance has an inductive component) and Δd takes a +ve sign when the minima shift towards the load (load impedance has a capacitive component).

- For measurement of *high VSWR* ($\rho > 10$), conventional V_{max}/V_{min} method cannot be used to measure ρ as the magnitude of the voltage minimum will be very small for high VSWR. For picking up significant signals at a voltage minimum, the problem of *probe loading* and *detector loading* will be faced at voltage maximum. Thus, for the measurement of high value of VSWR *double the minimum method* is used in which the standing wave pattern is probed only around a voltage minimum. The working formula for measurement of high value of VSWR (ρ) is $\rho = \lambda_g/2\pi x_0$.
- In the *reflectometer technique* of impedance measurement, the reflection coefficient $|R| = |V_r|/|V_i|$ is evaluated and the magnitude of the load impedance is obtained from $z_L = Z_L/Z_0 = (1+|R|)/(1-|R|)$. Phase information regarding the load impedance may also be obtained with suitable modification of the simple reflection coefficient measurement instrument. This is the heart of the Vector Network Analyser (VNA) where amplitude and phase information may be displayed on Smith chart overlay on the screen of VNA.
- The *bridge method* of impedance measurement at microwave frequencies uses a magic tee as the heart of the measurement system.

Detection and measurement of power at microwave frequencies

- For applications such as SWR measurements where only a relative measure of V_{max} and V_{min} is required, signal detection suffices. But for absolute power measurement at microwave frequencies, as in the case of radar and communication applications, *power sensors* are used. At microwave frequencies, power is measured directly as a fundamental quantity unlike the power measurement at lower frequency where current, voltage and power factor are measured to evaluate the signal power.
- Detection of microwave signals is carried out with a diode for relative measurement of microwave power. The microwave detector diodes are either point contact diodes or Schottky barrier diodes. The *pn* junction semiconductor diodes used extensively at lower radio frequencies are bipolar devices and have a large junction capacitance arising from minority carrier storage which makes them unsuitable for use at microwave frequencies; thus metal–semiconductor contact Schottky barrier diodes are used.
- Power sensors are thermally sensitive devices which utilise the heat generated by the microwave energy incident upon it. Depending on the level of microwave power to be measured (low, medium or high) and the technique of estimating the power incident on the sensor element there are, in general, three techniques of microwave power measurement—*thermocouple technique* (for low power measurement), *calorimeter technique* (for high power measurement), and *bolometer technique* (useful for low, medium and also high power measurement including peak power measurement).
- There are two types of bolometer sensor elements used for microwave power measurements, the *barretters* and *thermistors*. Barretters have short thermal time constant, unsuitable for broad band operation and have poor detection sensitivity (5.2 Ω/mW) while termistors have a long time constant (thus useful for pulsed radar applications), suitable for broad band operation and have good detection sensitivity (60 Ω/mW).
- Microwave power meters are mostly based on thermistor power sensors with an arrangement for temperature compensation. Self-balancing bridges (designed around

simple operational amplifiers acting as DC amplifiers) and associated meter logic circuits can measure microwave power typically over the range of 10 μW (–20 dBm) to 10 mW (10 dBm).

- Peak power measurement at microwave frequency can be done by average power-duty cycle method using the formula:

$$P_{\text{peak}} = \frac{P_{av}}{\text{PRF} \times \text{PW}} \cdot 10^{(A+C_f)/10}$$

Measurement of quality factor (Q) of microwave cavities

- The three quality factors (Q_0, Q_L and Q_E) are interrelated through the coupling factor β (ratio of coupled resistance $\omega M^2/Z_0$ and the cavity loss resistance R_s) between the cavity and the load and are given by:

$$Q_0 = (1+\beta)Q_L \text{ and } Q_E = \frac{Q_0}{\beta}; \quad \text{whence: } \frac{1}{Q_L} = \frac{1}{Q_0} + \frac{1}{Q_E}$$

- In the *VSWR* or *impedance method of Q measurement* the working formulae are:

$$Q_0 = \frac{1}{(\delta_1 - \delta_1)}, \; Q_L = \frac{1}{(\delta_3 - \delta_4)}, \text{ and } Q_E = \frac{1}{(\delta_5 - \delta_6)}$$

where δ values are obtained from an experimental curve between δ and ρ corresponding to the half power values of ρ calculated from the following equations:

$$\text{For } Q_0,\; (\rho_{1/2})_0 = \frac{2+\beta^2+\sqrt{4+\beta^4}}{2\beta}$$

$$\text{For } Q_L,\; (\rho_{1/2})_L = \frac{1+\beta+\beta^2+(1+\beta)\sqrt{1+\beta^2}}{\beta}$$

$$\text{For } Q_E,\; (\rho_{1/2})_E = \frac{1+2\beta^2+\sqrt{1+4\beta^4}}{2\beta}$$

Depending on the case of cavity coupling, β is calculated in terms of VSWR (ρ_0) at resonance:

$$\beta_{\text{under-coupled}} = \frac{1}{\rho_0} \text{ and } \beta_{\text{over-coupled}} = \rho_0$$

- In the *transient decay* or *decrement method of Q measurement* the working formula is:

$$Q_L = \pi f \Delta t$$

- In the *dynamic method (comparison method) of Q measurement* the working formula is:

$$Q = Q' \frac{f_0}{f_0'}$$

where Q_0' and f_0' are the quality factor and resonant frequency of the low frequency circuit respectively and f_0 is the resonant frequency of the microwave cavity.

Microwave frequency measurement

- *Absorption type wavemeter* is the mechanical technique of measurement of microwave frequency.
- The electronic technique of microwave frequency measurement, *frequency counter*, commonly uses a *null-beat* technique.

Antenna measurement at microwave frequency

- The distance of the receiving antenna from AUT for all locations during its rotation about AUT should be in the far-field region (given by the well-known Rayleigh distance giving the border between the Fresnel and Fraunhofer regions), $R = 2D^2/\lambda$, where D is the largest dimension of the physical aperture of the antenna and λ is the operating wavelength. The reactive or evanescent near field region (or even radiative near-field or Fresnel region) is not used for antenna measurement, because it is located too near the AUT and mutual impedance caused by reactive coupling between AUT and the receiving antenna makes measurement complicated.
- For small microwave antennae, the plane wave approximations hold good in the far-field region but for large antennae this problem is solved by using Compact Antenna Test Range (CATR) facility (which also includes anechoic environment to get rid of surrounding reflections).

  For a pyramidal horn antenna, $\theta_H = \dfrac{63\lambda_0}{a_H}$ and $\theta_E = \dfrac{52\lambda_0}{a_E}$

  where φ_H and θ_E are respectively the half power beam widths (in degrees) of the H and E plane radiation patterns, λ_0 is the free-space wavelength and a_H and a_E are respectively the aperture dimensions in the H-plane direction and E-plane direction.
- The power gain (G), that is, the practically measured gain of a pyramidal horn is given by:

$$G \approx \frac{24,754}{\theta_E \varphi_H}$$

(Note that $G = \eta_a G_{D(\max)}$; assume $\eta_a = 0.6$)

Measurement of dielectric constant and loss factor of materials at microwave frequencies

- The separation of real and imaginary parts of ε_r in the following equation permits one to obtain the value of ε' and ε'' and hence the value of the dielectric constant and loss tangent for the dielectric material.

$$\varepsilon_c = \varepsilon' - j\varepsilon'' = \varepsilon_0 \left\{ \frac{\left[\left(\frac{1}{\lambda_c}\right)^2 - \left(\frac{\gamma_2}{2\pi}\right)^2\right]}{\left[\left(\frac{1}{\lambda_1}\right)^2 + \left(\frac{1}{\lambda_c}\right)^2\right]} \right\}$$

γ_2 being calculated from the transcendental equation:

$$\frac{\tanh \gamma_2 d}{\gamma_2 d} = -j\frac{\lambda_1}{2\pi d}\frac{1-\rho\tan(2\pi x_0/\lambda_1)}{\rho - j\tan(2\pi x_0/\lambda_1)}$$

Measurement of noise figure and phase-noise

- Noise characterisation of amplifiers, mixers and so on, is usually done with the help of noise figure measurement and that of an oscillator is done by phase-noise measurement.
- The noise temperature (T_e) of the network is measured by the Y-factor method where $T_e = \dfrac{(T_h - YT_0)}{Y-1}$. Thus, the measurement of T_0 (the noise power from the calibrated noise source when it is unbiased or cold) and T_h (the noise power from the calibrated noise source when it is biased or hot) and the determination of Y-factor yields the result. The Y-factor is determined from $Y = \dfrac{k_B(T_e + T_h)B_N}{k_B(T_e + T_0)B_N}$, in which $k_B(T_e + T_0)B_N$ and $k_B(T_e + T_h)$ B_N are the noise powers measured at the output of the network when the noise source is respectively unbiased and biased. Measurement of all noise power can be done with a power meter.
- The short term frequency fluctuations in an oscillator caused by the random phase fluctuations originate from thermal noise, shot noise and flicker noise is characterised with *phase noise*. Phase noise characterisation is best suited in the frequency domain (with a spectrum analyser) where the phase noise shows up as noise sidebands on the carrier. In the time domain this is represented by 'time jitter' in the zero crossing of the carrier sinusoid.
- The spectral purity of high frequency RF oscillators at microwave and millimeter-wave frequency thus needs to be characterised in terms of phase noise. The National Bureau of Standards defines single side band (SSB) phase noise $[L\varphi(f)]$ as the ratio of power (P_{ssb}) in one phase modulation sideband per hertz bandwidth, at an offset of f_m hertz from the carrier, to the total signal power in the carrier (P_{sc}). Mathematically, $L_\varphi(f) = P_{sc}/P_{ssb}$. The SSB phase noise is usually expressed logarithmically, that is, $L_\varphi(f)_{\text{dB}} = 10\log_{10}[L_\varphi(f)]$. The unit of phase noise is dBc/Hz.

Fundamentals of spectrum analyser and network analyser

- Spectrum analyser and network analyser both are used to carry out frequency domain analysis; the former being used to study the signal characteristics while the latter is used to characterise components, devices, circuits and sub-assemblies. However, the spectrum analyser and network analyser may also be used for time domain measurement using respectively *zero, span* and *time domain refelectometry* options.

- From a system engineer's point of view, the spectrum analyser is a swept-tuned super-heterodyne receiver and intended for analysing the nature of signals; a network analyser, on the other hand, contains both a source (test signal) and multiple receivers.
- With a network analyser we get both the (ratioed) amplitude and phase information but a spectrum analyser does not, in general, provide the phase information of the signal.

Spectrum analyser

- Oscilloscopes make measurement in the *time domain* while in spectrum analysers, the measurement is in the *frequency domain*. The two axes in the oscilloscope are time and amplitude whereas in the spectrum analyser they are frequency and amplitude.
- In addition to swept-tuned frequency mode, a spectrum analyser can be used in fixed-tuned mode (zero-span) to provide time-domain measurement capability much like that of an oscilloscope.
- The job of a spectrum analyser is to extract the frequency content of a general time domain signal say $x(t)$ and to plot amplitude vs frequency of all the frequency components contained in the signal. This can be done by sweeping through the spectrum of a narrow band IF filter using a variable frequency local oscillator in a heterodyning system. Thus, spectrum analyser is basically a swept-tuned super-heterodyne receiver.
- *Fourier analyser*, with its real-time signal analysis capability, is capable of capturing periodic as well as random and transcient events. It can also provide significant speed improvement over the more traditional swept analyser and can measure magnitude as well as phase.
- State-of-the-art spectrum analysers of today combine the benefits of both the swept-tuned and Fourier analyser leading to the *real time spectrum analyser* in which the analyser captures and processes the spectrum of the signal so fast that no part of the signal is missed.
- The choice of resolution bandwidth (RBW) with respect to (frequency) span/div is an important issue for observing the spectrum of the signal under investigation properly. Most modern spectrum analysers through the use of microprocessors have the capability to select the optimum RBW depending on span/division.
- In general, narrow RBW (optimum with respect to span/div) is desired from many applications. For a pulsed RF signal, each spectral line then becomes clearly resolved. With the choice of a narrow RBW, the noise floor of the analyser goes down, in addition to making the spectrum of the signal being displayed narrow and thus helps to identify low level and nearby sidebands, if any.

Network analyser

- A network analyser provides a complete description of linear network behaviour (in terms of S-parameters) and of non-linear network (in terms of X-parameters) in the frequency domain.
- The scalar network analyser (SNA) is a form of RF network analyser that only measures the amplitude properties of the device under test. The vector network analyser (VNA) is a more useful form of RF network analyser than the SNA as it is able to measure the

amplitude response and the phase as well (a gain-phase meter). Large signal network analyser (LSNA) is a highly specialised form of RF network analyser that is able to investigate the characteristics of devices under large signal conditions.

- VNA primarily consists of four sections—(i) source for stimulus, (ii) signal-separation devices, (iii) receivers that down-convert and detect the signals, and (iv) processor and display for depicting the results in the desired format and for reviewing.
- VNA calibration is an important issue that makes *vector error correction* for accurate measurement with such an analyssr. Calibration by the user (over and above the factory made calibration) before making any measurement is critical to making good S-parameter measurements with VNA. User calibration is crucial to ensure the accuracy and repeatability of VNA measurements.
- The user calibration accounts for 12 sources of systematic errors in a VNA. In the error correction process, the measured magnitude and phase of each error signal appearing as a vector is applied mathematically via proper algorithm to the measurement signal thereby realising *vector error correction.*
- Most comprehensive VNA calibration is a full 2-port calibration method in which the parameters calibrated are S_{11}, S_{12}, S_{21} and S_{22}. The wide assortment of VNA user calibrations is divided into three commonly used categories or families—SOLT (Short, Open, Load, Thru), TRL (Thru, Reflect, Line) and automatic calibration.

Review Questions

12.1 Why are the techniques of measurement at microwave frequencies different from those at radiowave frequencies (having kilometre to metre wavelengths) and also from those at infra-red through optical to X-ray wavelengths?

12.2 What are the primary measurable quantities at microwave frequencies? How are impedance, quality factor and S-parameters determined in terms of the primary measurable quantities?

12.3 Mention a few techniques for measurement of impedance at microwave frequencies.

12.4 With the increase of the mismatch factor ρ, the impedance at a voltage maximum increases and the impedance at a voltage minimum decreases—justify.

12.5 Why is measurement with respect to a voltage minimum more accurate than at a voltage maximum?

12.6 Why is the probe in SWD inserted through a non-radiating slot in the waveguide of the SWD? What is *probe loading*?

12.7 In the VSWR measurement arrangement why is the microwave signal source modulated with a 1 kHz (typically) signal?

12.8 Discuss how *impedance determination by short circuit minima shift* helps to know the nature of the microwave load and also improves the accuracy of impedance measurement when compared with the conventional SWD technique.

12.9 What are the problems of conventional SWD technique for the measurement of high VSWR ($\rho > 10$)? What is *detector loading*? Discuss double the minimum method for measuring a high value of VSWR.

12.10 What do you understand by *return loss* of a microwave load? How can it be measured with a reflectometer?

12.13 Discuss how a magic tee can be used to configure the bridge technique of impedance measurement at microwave frequencies.

12.14 Name the techniques used for measurement of low, medium and high power at microwave frequencies.

12.15 Why is a point contact diode or metal-semiconductor or Schottky barrier diode used for detection of microwave power instead of a conventional *pn* junction diode that is used at lower radiowave frequencies?

12.16 Discuss in brief the operation of a tunable detector mount used at microwave frequencies.

12.17 Name the bolometer sensor elements used for microwave power measurement and compare their physical and electrical properties.

12.18 Discuss the conditions of *substitutability* for the design of bolometer power sensors.

12.19 Discuss the principle of temperature compensated microwave power meter with thermistor power sensor using a self-balancing bridge and associated meter logic.

12.20 Discuss the principle of peak power measurement at microwave frequencies.

12.21 For a one-port cavity having a Q value less than 10,000, the VSWR technique of Q measurement is used—justify. What is meant by *plane of detuned short*?

12.22 For high Q cavities, normally the case with transmission cavities, transient decay or decrement method is suitable. This method is not suitable for low to medium Q cavities—justify.

12.23 What is the dynamic method of Q measurement? Discuss the comparison method in this regard.

12.24 Discuss in brief the principle of operation of absorption type wavemeter. Why does the cavity appear to be oversized?

12.25 Discuss how the null-beat technique is used to design microwave frequency counters.

12.26 What is a CATR facility? Why are large antennae tested in a CATR lab facility in the event of the non-availability of open field measurement?

12.27 What is an anechoic chamber? Why are the walls of such a chamber covered with resonant absorbent material (RAM)?

12.28 Why does the near field, that is, Fresnel zone measurement, of horn antenna show variations in gain while at the Fraunhofer zone, the gain of the horn becomes practically constant?

12.29 Discuss in brief the Von-Hipple method of measurement of dielectric constant and loss tangent of a dielectric material at microwave frequency.

12.30 Discuss the Y-factor method of noise figure measurement of a network at microwave frequencies.

12.31 What do you understand by the phase noise of a microwave oscillator? Discuss how it is measured.

12.32 What are the similarities and differences between a spectrum analyser and network analyser both in terms of the basic measurements done and from a system engineer's point of view?

12.33 Explain why the frequency domain measurement done by spectrum and network analysers make them versatile measuring instruments. Can you make time-domain measurements with these instruments?

12.34 What is the difference between the measurements being done by an oscilloscope and a spectrum analyser?

12.35 The most common type of spectrum analysers are basically swept-tuned super-heterodyne receivers—explain.

12.36 Explain with a block diagram the principle of operation of a swept-tuned spectrum analyser.

12.37 Mention how a spectrum analyser can be used to make time domain study similar to an oscilloscope.

12.38 What is the purpose of a video filter in a spectrum analyser?

12.39 What is a Fourier analyser? What is the latest type of spectrum analyser?

12.40 Why are multiple conversion stages demanded by performance and design optimisation in a swept-tuned spectrum analyser?

12.41 What is the purpose of using a pre-selector filter in a spectrum analyser?

12.42 Discuss the choice of resolution bandwidth (RBW) with respect to (frequency) span/division in a spectrum analyser. What are the advantages of choosing a narrow RBW?

12.43 Vector network analyser (VNA) is a gain-phase meter—explain. Name other types of network analysers.

12.44 VNA's capability to measure both magnitude and phase of components is important for several reasons—explain.

12.45 What are the different sections in the block diagram of VNA? Explain the operation of a VNA with a suitable block diagram.

12.46 What do you understand by *vector error correction* in a VNA? User calibration accounts for 12 sources of systematic errors in a VNA—justify.

12.47 What is meant by full 2-port calibration of VNA? What is SOLT, TRL and automatic calibration of VNA?

13 Microwave Integrated Circuits

The infinitesimal is the substratum from which the apparently infinite universe manifests

13.1 Introduction

At lower frequencies we are familiar with *integrated circuits* (ICs), also known as *microchips*, that incorporate a large number of active devices and passive components including interconnections in a single semiconductor (may be Si or GaAs) substrate to realise subsystems like oscillators, amplifiers and microprocessors. Such integration of active devices and passive components leads to the realisation of enormous electronic functionalities packed in a very small space and with extremely reduced weight which is especially useful for air and space borne applications. In connection with ICs we are also familiar with the terms such as small scale integration (SSI), large scale integration (LSI) and very large scale integration (VLSI), depending on the packaging density of devices; for example, a VLSI may have one million active devices and passive components per microchip. The integrated circuit realised at microwave frequencies with microwave active devices and passive components including the interconnecting transmission lines is known as a *microwave integrated circuit* (MIC) which may be either a hybrid microwave integrated circuit (HMIC) or a monolithic microwave integrated circuit (MMIC). The word *monolithic* is derived from the Greek *monos* (single) and *lithos* (stone). The MMIC is carved out of a single piece of semiconductor material. Unlike lower frequency ICs, the packaging densities of MICs are however, much less but with more complexities in the design and development. This is because, at such frequencies the circuit dimensions tend to be of the order of the wavelength and hence every element has the possibility of influencing the performance of the system.

Since the early 1950s, after the invention of planar transmission lines like strip-line and microstrip line (Section 3.4) and the availability of good quality alumina substrate, the evolution of cost-effective processes for thin-film metallisation and high resolution photo etching and revolution in semiconductor device technology led to the development of

HMICs and subsequently the MMICs from the mid-1970s. In a hybrid MIC, the transmission lines and interconnections are implemented by conductor patterns on a dielectric substrate (which may be FR4, a type of epoxy fiberglass, useful up to 4 GHz and teflon or duroid for higher microwave frequencies), with printing and etching either by printed circuit, or thin-film/thick-film technique while the active devices (Gunn, IMPATT, transistor) and passive components (either in distributed or lumped form) are assembled on the patterned substrate either by soldering or with wire bonding. In MMIC, the active devices (like MESFET, HEMT) and passive components as well as the planar transmission lines are grown or implanted *in situ* on a single planar substrate which is usually a semi-insulating semiconductor material (may be GaAs, SiGe, or GaN). MMIC fabrication begins with a *wafer* (which is a thin slice being cut from an ingot of single crystal of the semiconductor material on which many 'chips' containing subsystems next to each other are placed all across it, Fig. 13.1a) onto or into which the active layer and conducting as well as dielectric layers are formed to realise passive components and interconnections. Because of the demand for size miniaturisation and high frequency operation desired by recent RF and microwave systems, MMIC is preferred over hybrid MIC in many emerging applications of today. However, it may be noted that MMIC is suitable for large-scale production but it demands a longer time to develop and fabricate and small-scale production results in highly prohibitive costs.

A microwave integrated circuit has many advantages over a discrete circuit design with waveguides and coaxial lines at microwave frequencies. The advantages of MIC over discrete circuit design may be summarised as follows. The key feature of MIC is *miniaturisation.* A microstrip designed with alumina substrate results in size reduction by almost half compared to a waveguide or coaxial line (since the guide wavelength decreases with the use of high dielectric constant substrate like alumina; $\varepsilon_r = 10$). The use of lumped-element microwave components (Section 6.3) reduces the size further compared to distributed line components as the dimensions of lumped components are smaller compared to the distributed line components. Another advantage of MIC is the *ease of interconnections and flexibility in layout.* Unlike discrete circuits with a waveguide or coaxial line where connectors or flanges complicate the interconnections and system layout, the MIC designer can realise interconnections with less complications and can locate the input output ports as desired with more flexibility. Further, MIC *can use uncapsulated device chips* (especially in case of MMIC) thus reducing the problem of package parasitics (providing wideband operation) and also leads to space miniaturisation compared to the case when packaged devices are used. Thus in a nutshell, size miniaturisation, light weight, reproducibility and reliability, improved performance and, eventual cost reduction when produced in large quantities are the advantages of MICs. Because of these advantages, the MIC is finding increasing applications in military and space borne systems including many commercial systems such as electronic warfare, missile seeker heads, phased array radar systems, earth observation radiometers and space-based security scanners, cellular phones, wireless local area network (WLAN), global positioning systems (GPS), to name a few.

In the following sections, we will discuss at first a comparison between HMIC and MMIC and their fabrication process steps. Next, the choice of materials for MIC fabrication will be discussed. Finally, some of the important fabrication processes related to HMIC and MMIC will also be covered in brief.

13.2 HMIC vs MMIC

It has already been mentioned that MIC represents an important and significant step towards the integration and miniaturisation of microwave sub-systems and systems when compared to the discrete systems built with waveguide/coaxial components. Hybrid MIC (HMIC) may be designed around a very simple printed circuit board (PCB) where integration is limited to realising only the conductor patterns or may use thick and thin-film technology where some or all of the passive components are integrated along with the conductor pattern. In HMIC, the level of integration is less than that with MMIC. In spite of the advent of MMIC, HMIC has continued to hold its place in the microwave systems due to some of its unique features. Let us compare some of the advantages and disadvantages of HMIC and MMIC.

In a *hybrid microwave integrated circuit* (HMIC), a distributed microstrip line is formed with metallisation on some dielectric substrate while packaged active devices like transistor or diode, and passive components like inductor or capacitor (either in distributed or lumped form) are soldered or bonded on to the substrate (Fig. 13.1b, which shows a typical layout of HMIC). Thus HMIC is amenable to post-manufacture tuning or tweaking to realise optimum performance by varying the component values and changing their relative positions, if required. This is obviously an advantage of HMIC compared to MMIC from a designer's point of view. However, the use of packaged active devices and passive components and their placement with soldering/bonding add to the parasitics which is a problem at higher microwave frequencies and also limits broadband operation of the system based on HMIC. However, high power solid-state devices like IMPATT that demand large power dissipation can be mounted in HMIC. Several types of transmission line configurations like microstrip line, slot-line and coplanar line can be used, also the substrate material can be of high dielectric constant (alumina, $\varepsilon_r = 10$) or of low dielectric constant (quartz, $\varepsilon_r = 3.8$) depending on the thermal conductivity and frequency of operation desired. For example, alumina is useful up to 20 GHz while for higher microwave frequencies (> 20 GHz) quartz is a suitable substrate material.

In *monolithic microwave integrated circuit* (MMIC), active devices (in uncapsulated/unpackaged form), passive components and interconnections are fabricated on a single semiconductor material (Fig. 13.1c, which shows a typical layout of MMIC) that allows them to keep the parasitic capacitance and inductance (which are a great problem at higher microwave frequencies) to a minimum, thus permitting reliable millimetre-wave performance with broadband operation. Also with the advent of via-hole technology used in MMIC, only input and output wire connections are required and therefore the number of bond wires is greatly reduced (limited only to those at the periphery of the circuit) compared to HMIC, leading to superior broadband performance with MMIC. Further, in monolithic integration, batch-processing technique can be used where tens or hundreds of circuits are formed simultaneously – reducing the cost of each circuit – if large scale integration is done. However for professional design, performance optimisation of MMIC needs to be done at the design stage using CAD technique (which might include discontinuities and spurious coupling if any, component tolerances and others) before it is sent to the foundry for fabrication as tweaking of the system performance is not possible

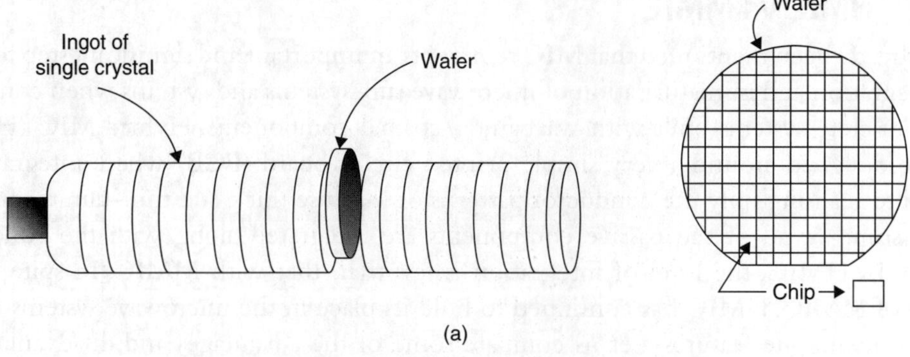

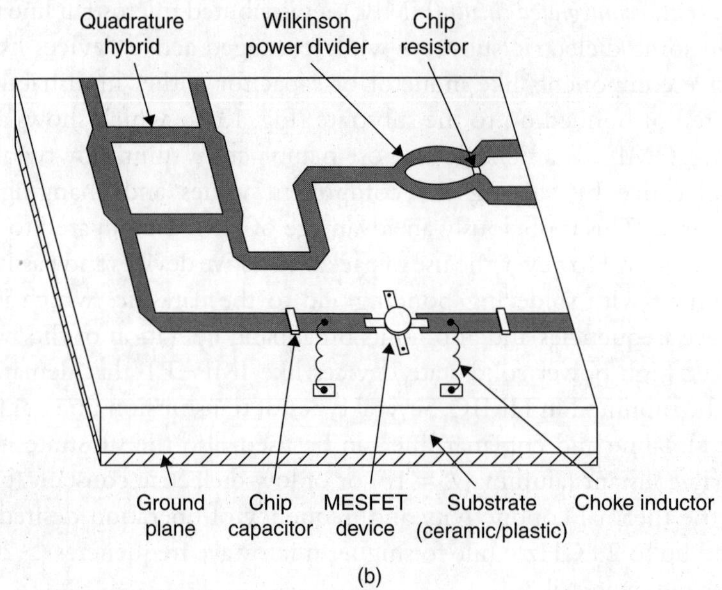

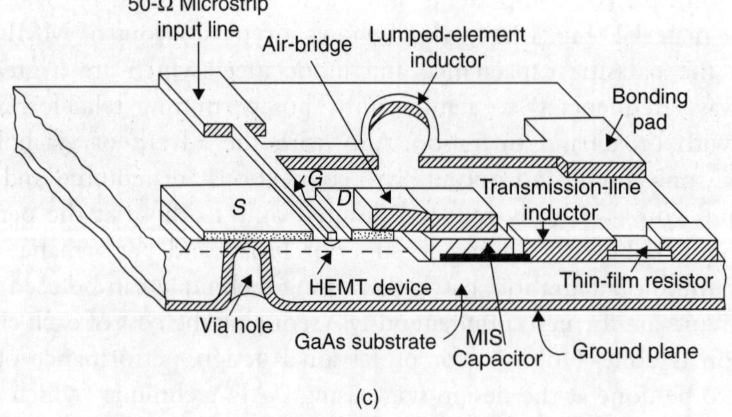

Fig. 13.1 (a) Ingot of single crystal, wafers and chips (b) Schematic of HMIC (c) Schematic of MMIC.

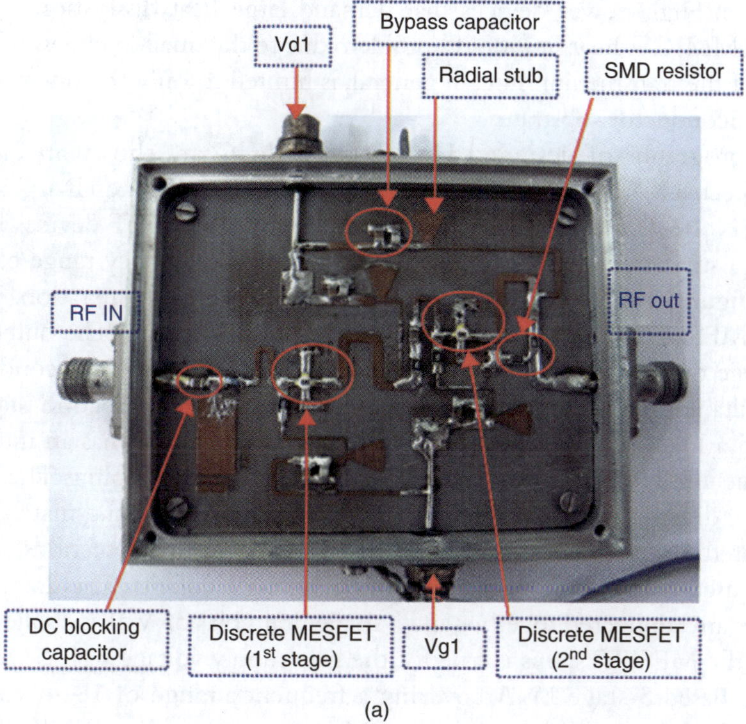

(a)

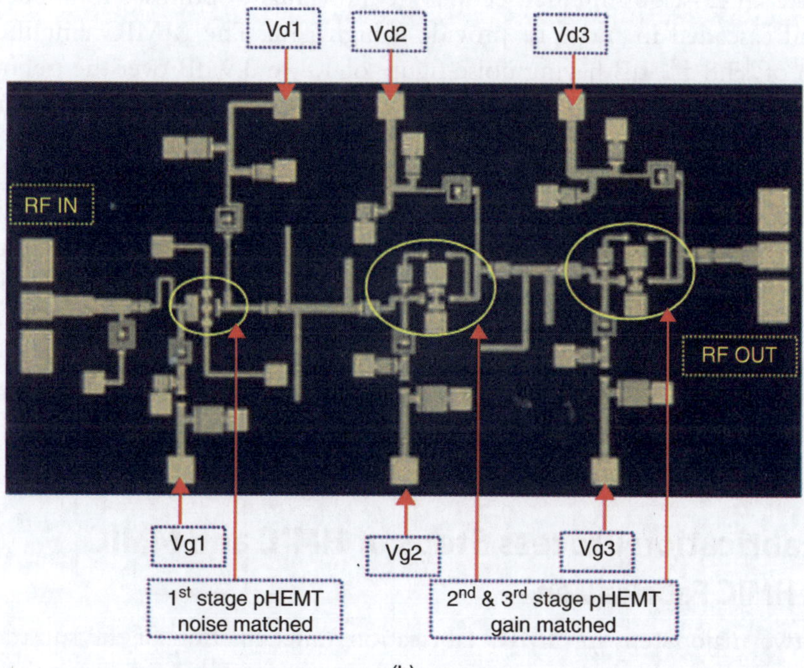

(b)

Fig. 13.2 Photograph of fabricated HMIC and MMIC. (a) HMIC ($L = 65$ mm, $W = 54$ mm, $h = 10$ mm) (b) MMIC ($L = 3.5$ mm, $W = 1.5$ mm, $h = 0.1$ mm) [Courtesy: SAMEER Kolkata centre].

after fabrication. High power devices that demand large heat dissipation are difficult to incorporate in MMIC as heat sinking is a problem due to the small thickness of the substrate. Also, the use of the transmission line, in general, is limited to only the microstrip line on a GaAs/InP semiconductor substrate.

Typical photographs of designed HMIC and MMIC are shown in Fig. 13.2a and Fig. 13.2b respectively, from which their size may also be noted. The HMIC shown here is a 2 stage Low Noise Amplifier (LNA) that uses a discrete MESFET device. The amplifier operates with a small signal gain of 35 ± 0.3 dB within a frequency range of 1 to 3 GHz with a noise figure within 1.8 dB over the frequency band of operation. The input of the first stage MESFET has been matched for minimum noise figure. The output impedance of the first stage device is matched with the input impedance of the second stage device by using simultaneous conjugate matching. The output of the second stage device is matched with a load impedance of 50 ohms. The biasing circuit has an inductive choke of $\lambda/4$ transmission line and a radial capacitive stub for RF decoupling. The circuit uses a combination of distributed and lumped elements to realise biasing and matching networks. Surface mounted device (SMD) resistors are used in the matching sections whereas SMD capacitors are used as bypass networks and DC blocking elements. The drain bias lines for both the stages are connected together and a single drain bias of Vd1 is applied to drive the drain of both the MESFETs; this is true for the gate supply voltage Vg1 as well.

An MMIC based 3-stage LNA covering a frequency range of 18–40 GHz is shown in Fig. 13.2b which has been designed using 0.15 μm gate length pHEMTs. In this design, three separate single-stage amplifier sections are individually optimised for input and output matches and cascaded in series to provide a high gain. The MMIC amplifier operates with a gain of 15.8 ± 1 dB having noise figure of 1.8 to 3.9 dB over the frequency range of operation. The input of the first stage pHEMT has been matched for minimum noise figure. The output impedance of the first stage device is matched with the input impedance of the second stage device by using simultaneous conjugate matching. A similar matching network has also been used to match the output of the second stage device to the input of the third stage device. The output impedance of the third stage device is matched with a load impedance of 50 ohms. Vd1, Vd2 and Vd3 represent the drain bias of the three consecutive stages and Vg1, Vg2 and Vg3 are the respective gate bias voltages. The RF input and output are connected to a CPW (coplanar waveguide) pad (having a centre signal line and two ground pads on either side) for making on-wafer connections through G–S–G probes, required for connectivity of the MMIC to outer packages.

13.3 Fabrication Process Steps of HMIC and MMIC

13.3.1 HMIC Fabrication

There are two major steps in HMIC fabrication—metallisation of the substrate material and selective photo-etching to generate the circuit pattern. The first step in the fabrication of HMIC is the choice of the dielectric substrate. If a plastic substrate like FR-4 glass epoxy or RT-Duroid is used, then it is available with metallisation (Cu or Cu–Au) on one or both sides. In the case of a ceramic substrate like alumina, metallisation is carried out in three steps, that is, there are three layers of different thickness (on both sides of the substrate, Fig. 13.3). The first thin layer is usually chromium of thickness 200 Å which is used as seed or interface layer for permitting good adhesion of the subsequent layer of copper to the ceramic substrate. The second thick layer is of copper (10 μm thick) deposited on chromium and the final layer is of gold (150 Å thick) which prevents Cu from oxidation

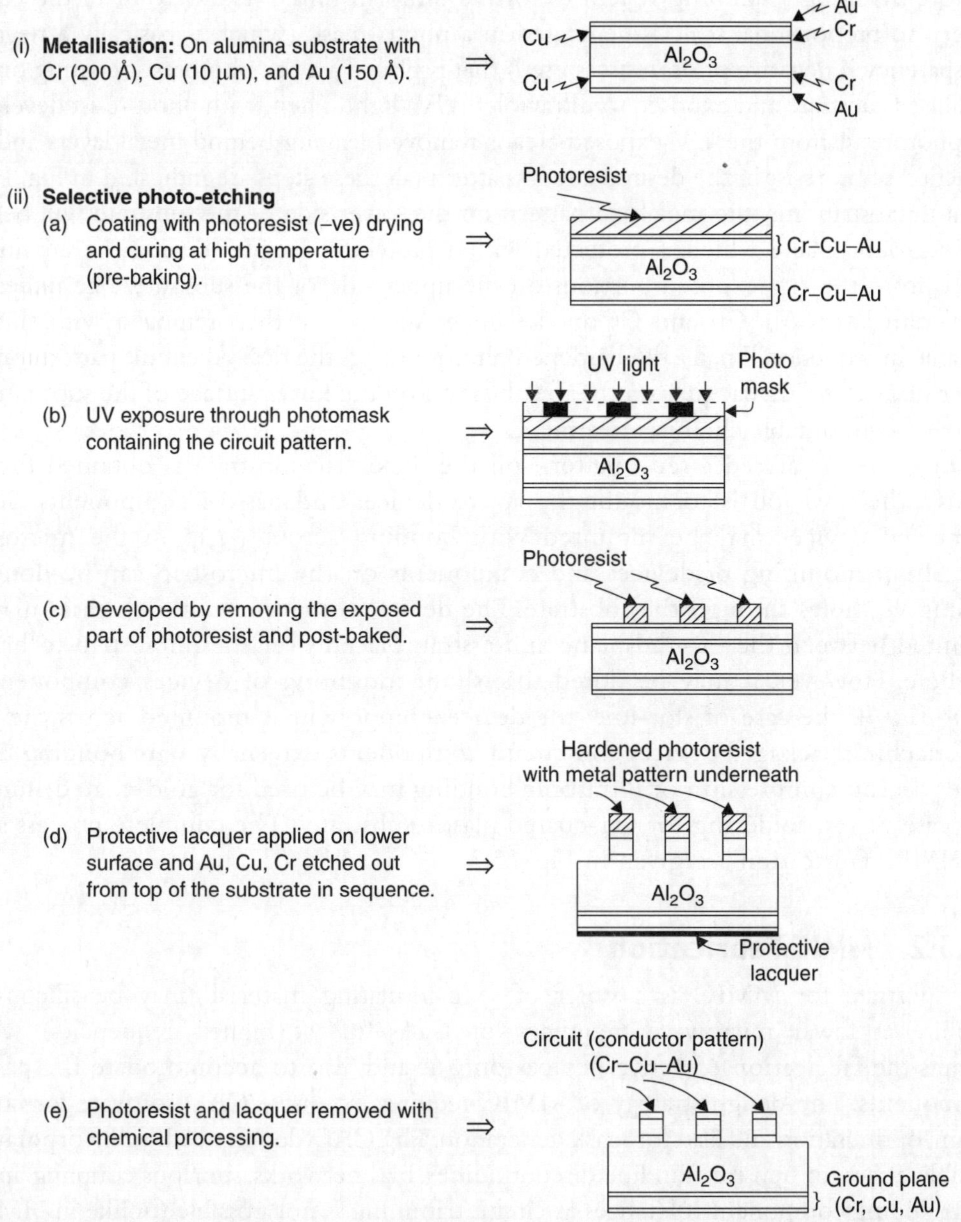

Fig. 13.3 The process steps for fabrication of HMIC.

and also helps in bonding of chip-type and beam-lead devices. The thick layer of Cu carries the bulk of the RF current. Cr–Cu–Au may be deposited either by sputtering on both sides of the substrate or Cr layer and a thin layer (0.1 μm) of Cu may be deposited by sputtering and final layer of Cu (10 μm) and Au is deposited by electroplating.

Selective photo-etching of the metallised substrate generates the desired conductor pattern (for microwave circuit and also sometimes for lumped inductor) of HMIC. For this purpose the metallised substrate is first coated with photoresist (usually negative photoresist) using a spinning system to form a uniform film. Next a layout of the circuit pattern in proportional scale is generated in a photo-mask (which is basically a negative transparency if negative photoresist is used) that is placed on the photoresist coating on the metallised substrate and exposed to ultraviolet (UV) light. Then, with photoresist developer the photoresist from the UV-exposed area is removed leaving behind metal layers and the hardened photoresist in the desired circuit pattern (all these steps are indicated in Fig. 13.3). For a microstrip line, the metalised pattern on the lower side of the substrate will behave as the ground plane which is protected with a protective lacquer during the etching of metals from the UV-exposed portions on the upper side of the substrate. The undesired metal film layers Au, Cu and Cr on the upper surface are then removed with suitable etchants in sequence. Finally, the hardened photoresist on the desired circuit pattern on the upper surface and the lacquer on the metallisation on the lower surface of the substrate are removed with suitable chemical treatment.

Once the metallised circuit pattern on the dielectric substrate is obtained for the HMIC, the final job is to mount the active devices and passive components. Series connected devices may be mounted with bonding across a gap in the microstrip line. Shunt mounting of devices and components on the microstrip can be done by drilling via-holes through the substrate. The device/component in such cases may be mounted between the ground plane and a strap placed over the microstrip to bridge the hole. However, it may be noted that shunt mounting of devices/components is fairly easy in the case of slot-line; the device/component is mounted across the slot. For attaching packaged devices and circuit components externally, wire bonding using either thermocompression or ultrasonic bonding may be used for gold-coated alumina substrate or soft soldering for Cu-coated plastic substrates. The complete process steps for HMIC fabrication are given in Fig. 13.3.

13.3.2 MMIC Fabrication

The substrate for MMIC is a wafer of semi-insulating material (may be silicon-on-sapphire at lower microwave frequency or GaAs/InP at higher frequencies) which permits the fabrication of active devices onto it and also to accommodate the passive components. The design strategy of MMIC requires extensive CAD software for circuit design, optimisation and also for mask generation. The CAD design needs to incorporate all possible effects of transmission line discontinuities, bias networks, spurious coupling and so on, including component tolerances as circuit trimming is not possible (unlike in HMIC), once the fabrication process of MMIC is completed.

Once the CAD based circuit design of MMIC is finalised by the designer, the next job is mask generation followed by the subsequent processing steps. The first processing step of course is the formation of active layer (*n*-type) on/into the semiconductor substrate

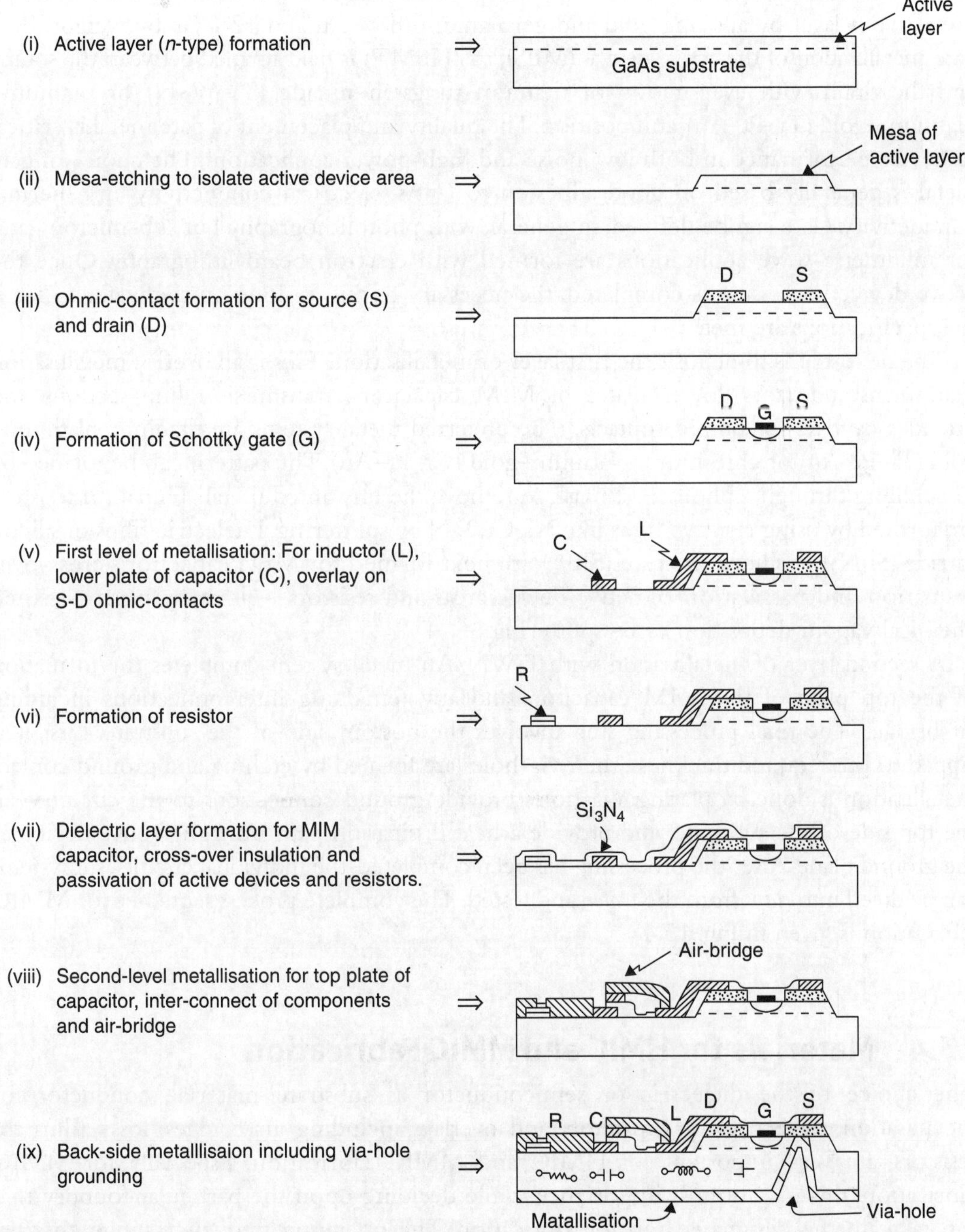

Fig. 13.4 The process steps for the fabrication of MMIC.

(for the necessary active devices), may be GaAs; which may be done by ion-implantation or by epitaxial growth (the former technique offers better control of doping profiles over a larger substrate area and improved reproducibility). The active device areas need to be isolated from other parts so that current flow is restricted to the desired portions and this is done by mesa-etching. The ohmic contacts for source (S) and drain (D) are next made on the active *n*-layer by alloying gold and germanium (88% Au and 12% Ge by weight). The gate metallisation of the active device (MESFET/HEMT) is next formed between the source and the drain with evaporation of titanium–tungsten–nitride (Ti–W–N) or titanium–platinum–gold (Ti–Pt–Au) composition. The quality and placement of gate metal is critical to device performance in both low-noise and high-power applications. The choice of gate metal is generally based on good adhesion to GaAs, electrical conductivity and thermal conductivity. Gate can be defined, in general, with photolithography but sub-micron gates for millimetre-wave applications are formed with electron beam lithography. Once the active device processing is completed, the necessary testing is done on the wafer to see if the specifications are met.

The next step is to deposit the first layer of metallisation. This is an overlay metallisation that forms inductors, lower plates of MIM capacitors, transmission line sections and provides overlays for ohmic contacts. The preferred metal systems are titanium–platinum–gold (Ti–Pt–Au) or chromium–platinum–gold (Cr–Pt–Au). The pattern can be formed by ion-milling through a photoresist mask to remove the unwanted metal. Thin-film resistors are formed by using resistive films like Ni, Cr, TaN by sputtering. Dielectric films of silicon nitride (Si_3N_4) or silicon dioxide (SiO_2) are next formed for MIM capacitors, cross-over insulation and passivation of active device areas and resistors—either by plasma-assisted chemical vapour deposition or by sputtering.

A second layer of metallisation with TiWN/Au metal system completes the formation of the top plate of the MIM capacitors and any remaining interconnections including air-bridges. The final processing step involves the bottom side of the substrate. First it is lapped to the required thickness, then via-holes are formed by etching and ground contact metallisation is done by plating. Via-holes provide ground connections to the circuitry on the top side of the substrate, and provide a heat dissipation path from the active device to the ground plane. After the processing has been completed, the individual circuits/subsystems can be diced into *chips* from the *wafer*, and tested. The complete process sequence for MMIC fabrication is given in Fig. 13.4.

13.4 Materials for HMIC and MMIC Fabrication

The choice of the dielectric or semiconductor as substrate material, conductor for metallisation, dielectric for capacitors and overlays including air bridges, lossy films for resistors, are very important for HMIC and MMIC fabrication. Especially for MMIC fabrication, this is of utmost importance while deciding upon the particular foundry that can take up the designated job for fabrication. The operating frequency range, thermal dissipation, adhesion of conductors on the substrate and its etchability and solderabity, lossyness of capacitor and sheet resistivity of resistive films are determined by material choice both for HMIC and MMIC fabrication.

13.4.1 Substrate Materials

The substrate material for MIC fabrication is selected on the basis of three major factors—frequency of operation, type of planar transmission line used and the fabrication technology (hybrid or monolithic). Other factors of consideration are, the desired electrical, mechanical and thermal properties including the cost. HMIC uses dielectric substrates, either ceramics like alumina, fused quartz and sapphire, or plastics like FR-4 glass epoxy or fibre-glass reinforced polytetrafluoroethylene (PTFE) with a trade name RT-Duroid. Ferrimagnetic substrates like ferrites and garnets are also used for certain special applications like realisation of non-reciprocal microstrip devices. MMIC that desire to fabricate the active devices along with the passive components and interconnections starting from the single substrate material uses semiconductors, either Si or GaAs, of which GaAs has emerged as the most suitable on many accounts. Semi-insulating GaAs has higher resistivity and lower loss tangent as a dielectric substrate compared to Si. Thus, realisation of passive circuits and electrical isolation between active devices can be better realised with GaAs as a substrate material. Also electron mobility in GaAs is higher than that of Si making GaAs suitable for realising active semiconductor devices (MESFET/HEMT) for operation even up to 100 GHz. Minority carrier lifetime in GaAs is much higher than that of Si which makes it well suited for realising fast switching PIN diodes. However, InP and GaN substrates are also emerging as better substrate materials for power devices at millimetre-wave frequencies. The semiconductor substrates have dielectric constants close to that of alumina but their loss tangent is 10 to 100 times higher and hence HMIC design with semiconductor substrates will be highly lossy apart from the cost considerations.

The substrate selected for a particular purpose in MIC fabrication must, in general, have the following characteristics.

(i) Dielectric materials should have a high dielectric constant (say alumina or ceramic-loaded plastic RT-Duriod 6010 with $\varepsilon_r = 10$), for lower microwave frequencies $f < 20$ GHz or low dielectric constant (say fused quartz with $\varepsilon_r = 3.8$ or copper-clad plastic substrate RT-Duriod 5880 with $\varepsilon_r = 2.22$), for millimetre-wave applications. The dielectric loss of the substrate material (which is dependent on the loss tangent at the operating frequency) should be of low value and the dielectric constant should remain constant over the frequency and temperature range of interest.

(ii) Semiconductor material should also be chosen on the basis of frequency of operation and the power capability desired with good surface finish, that is, relatively free from voids in order to keep the conductor loss low yet maintaining good metal-film adhesion.

(iii) Both the dielectric and semiconductor material used for the substrate should not get deformed during high temperature circuit processing.

(iv) The cost of the substrate material must be justifiable for the application concerned.

13.4.2 Conductor Materials

The conductor material used for interconnections, ground plane or lumped inductor in MIC should have a high conductivity, a low temperature coefficient of resistance, good adhesion to the substrate, and good etchability and solderability and adaptability to bonding processes and be easy to deposit or electroplate. The resistance is determined by the RF surface resistivity (typically 500 to 100 Ω/square is desired for loss-less behaviour) and conductor thickness is determined by the skin depth at the frequency of operation. The conductor thickness should be at least three to six times the skin depth so that almost 98% of the current density is then confined to the conductor minimising the conductor loss. Commonly used conductor materials are copper, gold, silver and aluminium. Quite often, a thin layer of chromium is used as an intermediate layer on the substrate on top of which the actual conductor material of larger thickness is deposited to ensure both good adhesion and good conductivity of the metallisation.

13.4.3 Dielectric Materials

Dielectric materials are used in the form of films as insulators for capacitors, protective layers for active devices and insulating layers for passive circuits like coupled-line structures. The properties desired from these dielectric materials are high breakdown voltage, low value of loss-tangent and the ability to undergo processing steps without developing pin-holes. Commonly used dielectric materials are SiO, SiO_2, Si_3N_4 and Ta_2O_5.

13.4.4 Resistive Materials

The resistive materials for MIC are required for terminations, bias networks and attenuators. The properties desired for resistive materials are low temperature coefficient of resistance, sheet resistivity in the range of 20–2000 Ω/square and good stability. Some of the resistive film materials used are NiCr, Ta, Ti and TaN.

Tables containing different types of substrate material, conductor material, dielectric material and resistive material with their properties useful for HMIC and MMIC fabrication are available in standard texts [29].

13.5 Fabrication Processes Related with HMIC and MMIC

Integrated circuit fabrication relates to three major processes—semiconductor doping (by diffusion/ion-implantation/epitaxial technique), patterning of layers (either with masked lithography or with mask-less lithography) and film deposition (with thick or thin-film technology) both conductors and insulators respectively to connect and isolate devices and components. Let us discuss the process steps related to HMIC and MMIC fabrication.

13.5.1 Diffusion, Ion-implantation and Epitaxial Techniques

The very first step in device fabrication is to change the composition of the base wafer (semiconductor material) by doping which can be done either by *diffusion* or *ion-implantation.*

The process of diffusion (or more specifically thermal diffusion) involves two steps—*surface-diffusion* and *drive-in*. In the surface-diffusion step, the doping of the wafer is done by heating it in a furnace under controlled conditions to about 1000°C and then introducing a vapour of the dopant material (depending on p or n doped regions as desired) carried by some inert gas that penetrates the wafer surface. In this process the dopant atoms diffuse onto a thin layer on the surface of the wafer. In the drive-in step the wafer is heated in nitrogen gas (without dopant), in order to distribute the surface diffusion deeply in the body of the wafer, the depth of penetration being dependent on the temperature and total time of the process.

In the *ion-implantation process*, a source of chosen impurity (in the form of positive charges, ions) is placed on the wafer surface and a very high negative voltage applied to the wafer with respect to the ion source. In practice, a large number of wafers are placed in a big chamber from which air is sucked out using a vacuum pump. Positive ions of the impurity material thus bombard the wafer surface under the effect of the high voltage being applied. External magnetic fields are used to focus and steer or guide the ions to strike at the right places through window patterns on the wafer surface. The ions travel so fast under the effect of the said high voltage being applied, that they embed themselves into the surface of the wafer. The ions gain enough energy to penetrate the surface and then slow down to rest within the wafer at depths determined by their mass and energy. The accelerated ions are expected to cause damage to the crystal lattice which has to be repaired by annealing the wafer in a furnace later.

Until 1970, selective doping was performed mainly by the diffusion process at elevated temperatures. Since the late 1970s, many doping operations started using ion-implantation that can have a precise control of the total amount of dopant and can also introduce some dopants that are difficult to embed by the thermal diffusion process. The ion-implantation process has better reproducibity and operates at a reduced processing temperature.

In some semiconductor device fabrication steps, a very good and thin layer of a single crystal semiconductor layer is required to be grown on the single crystal semiconductor substrate and in such cases crystal lattice matching must be maintained. The process used for this purpose is known as *epitaxial technique*; the word *epitaxy* comes from the Greek word *epi* (on) and *taxis* (arrangement). There are three types of epitaxy—vapour-phase epitaxy (VPE), molecular beam epitaxy (MBE) and liquid-phase epitaxy (LPE).

In VPE, the epitaxial process produces a thin film of single crystal Si/GaAs from the gas phase upon an existing crystal wafer of the same material. The epitaxial layer may be either p-type or n-type. The apparatus uses a long cylindrical quartz tube encircled by a RF induction coil. The Si/GaAs wafers are placed on a rectangular graphite rod called a boat. The boat is inserted in the reaction chamber and the graphite is heated inductively to about 1200°C. At the input of the reaction chamber, a control console permits the introduction of various gases required for the growth of appropriate epitaxial layers. In MBE, the epitaxial layer grows on the wafer surface due to the reaction of a heated beam of the atoms/molecules of the desired material under an ultra-high vacuum (10^{-8} Pa) condition. This process can be used for growing thin epitaxial films of a wide variety of materials, ranging from oxides to semiconductors including metals. Slow streams of an element GaAs/As can be directed at the substrate, so that material deposits one atomic layer at a time. The absence of carrier gases as well as the ultra-high vacuum environment results in precise control of both chemical composition and doping profile of the grown films. It was first applied to the growth of compound semiconductors and is still the most common with semiconductor device fabrication. In LPE, a very thin epitaxial layer can be formed on a crystalline substrate with direct precipitation from the liquid phase. With epitaxial growth done from a dilute solution at low temperature, slow growth rate can be realised leading to improved thickness control, structural perfection and stoichiometry, reduced detrimental effects of thermal expansion differences of substrate and epilayer and the risk of unwanted spontaneously nucleated crystallites can be reduced.

13.5.2 Masked Lithography and Mask-less Lithography

The integrated circuit may be viewed as a set of patterned layers which are required to be defined on the wafer in the desired areas—this can be done either by conventional masked lithography or with modern types of mask-less lithography.

In masked lithography processes such as photolithography/X-ray lithography (the latter one can provide smaller feature sizes) a *mask* is used to selectively prevent the incident radiation (UV light/X-ray) from striking a photosensitised/X-ray sensitised wafer during the photolithographic/X-ray lithographic process. The photosensitive masks are typically made of glass covered with a thin film of an opaque metal. X-ray sensitive lithography uses a mask (typically of gold or compounds of tantalum or tungsten) which is an X-ray absorber, on a membrane that is transparent to X-rays (typically of silicon carbide or diamond). The particular layer of conductor or insulator to be patterned is transferred to a layer on the wafer through the mask by means of a process known as *lithography*. Since each layering has its own distinct patterning requirements, the lithography sequence must be repeated for every layer, using a different mask.

In the *photolithographic process*, the substrate is coated with a uniform layer of photoresist, which is a photosensitive emulsion (a viscous liquid). The mask with desired openings (non-opaque portions) is placed over the photoresist film and ultra-violet (UV) light exposes the photoresist through the mask changing the chemical property of the photoresist. There are two types of photoresist—positive and negative. The positive photoresist is one which is initially insoluble to the developing solvent but becomes soluble after exposure to UV light. The negative photoresist behaves just in the opposite manner compared to the positive photoresist, that is, exposure to UV light polymerises (hardens) the resit emulsion making it insoluble to the developer solution. Thus, the negative photoresist remains on the surface wherever it is exposed, and the developer solution removes only the unexposed portions. Masks used for negative photoresists, therefore, contain the inverse (photographic 'negative') of the patterns to be transferred to the wafer. Negative photoresists are more sensitive to light but their photographic resolution is not as high as that of the positive photoresist. For small geometry features as in VLSI positive photoresist is preferred, however, for MICs where packaging density is much smaller compared to lower frequency ICs, negative photoresist is used. For generation of high density patterns required in sub-micron devices, electron beam

(E-beam) lithography is preferred over optical lithography. Pre-baking and post-baking of photoresist are involved in the lithography process. Pre-baking is done after spin coating by heating for 15 to 20 minutes in a microwave oven at 900°C to evaporate the coating solvent and to make the resist denser. Post-baking is used to stabilise and harden the developed photoresist and to remove any remaining traces of the coating solvent or developer.

Mask-less lithography may be of different types—electron beam (E-beam) lithography, focussed ion beam lithography and optical (laser) writing. In E-beam lithography, the substrate is coated with an electron-sensitive film called e-resist (in analogy with photoresist) which is chemically changed on scanning (according to the pattern desired) with a focussed beam of electrons. The exposed (non-exposed) areas can be dissolved in a specific solvent (in analogy with positive/negative resist in photo-lithography). Electron beam lithography is a specialised technique for creating extremely fine patterns (~50 nm); however, it is expensive with low throughput. Direct laser writing uses a laser beam as pattern generator in a raster-scan mode. It is a very popular form of optical mask-less lithography, which offers flexibility, ease of use and cost effectiveness in R&D processing. This equipment offers rapid patterning at sub-micrometer resolutions, and offers a compromise between performance and cost when working with feature sizes of approximately 200 nm or greater.

13.5.3 Thick and Thin-film Technology

Thick and thin-film technology gave a boost to the IC designers to design hybrid ICs since the 1960s. The nomenclature of thick and thin-film is primarily a descriptive one in terms of the thickness of the film. The thin films have film thicknesses below 1 μm or even below 100 nm while thick films are well above 1 μm, typically 20–500 μm. However, the thickness of the film is not necessarily the only criterion to classifying them; in fact the technology for fabricating the film is needed in reality to classify whether the film is thin or thick-film. Thin films are deposited by vapour deposition or sputter technology, while thick films are screen printed, dried and fired at high temperatures. Both metal and dielectric films can be formed with thick/thin-film technology resulting in film-type resistors and capacitors (however, film inductors are not possible and if inductors are required they are designed as discrete components and soldered/bonded to the IC).

MICs using *thick-film technology* is inexpensive and generally limited to the lower end of the microwave frequency spectrum. In thick-film technology, the multilayer interconnects are formed by successive *screen printing* of conductors (gold, copper, silver), dielectric layers (glass–ceramic composition), and resistor (ruthenium doped glass) patterns on a base substrate like alumina (Al_2O_3), beryllia (BeO) and aluminium nitride (AlN). The film materials are in the form of inks or pastes. After screen printing through a mesh (designed according to the pattern to be made) each layer is dried at about 150°C for 20 minutes and fired at about 900°C for 30 to 45 minutes. The printing, drying and firing steps are repeated to fabricate the multi-layered circuit in a fully automated way to produce high volumes and cost effective components. Development of the SMD (surface mounted device) process evolved from the thick film process; the SMD is fast replacing leaded devices due its better reliability and improved performance.

The *thin-film technology* used for MICs is constantly developing to meet the requirements of increasing frequency of operation, higher yield and reduced cost. Thin films provide greater precision in component values. Commonly used thin-film techniques are (i) vacuum evaporation, (ii) electron-beam evaporation, and (iii) sputtering. All these are physical vapour deposition (PVD) techniques (there are also chemical vapour deposition (CVD) techniques used in MIC fabrication). The *vacuum evaporation* technique is similar to the vacuum deposition technique. Here, the film material is evaporated by placing it in a metallic *boat* and subjecting it to a high current. The substrate with a suitable mask on it and the heated boat are placed in a glass tube in which a high vacuum at a pressure of 10^{-8} Pa is maintained. The evaporated vapour of the film material gets deposited to the substrate material forming a polycrystalline layer on it. In *electron-beam evaporation* technique, instead of resistive heating of the film material placed in the boat, an intense electron beam is generated from a filament and steered via electric and magnetic fields that scans the material and vaporises it in the vacuum chamber leading to its deposition on the substrate material. E-beam evaporation results in superior quality and control of the layer being formed and allows even refractory material like tantalum to be deposited. The evaporation technique is a continuous evaporation–condensation process that results in the growth of the film material on the substrate. On the other hand, *sputtering technique* involves bombardment of the source material to be deposited on the substrate with high energy ions (commonly that of argon) to dislodge (sputter) atoms from the source material and reattach (deposit) them to the substrate. Both the source material (placed on the cathode) and the substrate (placed on the anode) are located in a vacuum chamber (evacuated to better than 10^{-8} Pa) and back-filled with argon to a pressure of 10^{-3} Pa) and a high voltage (at DC or RF, DC for metal deposition and RF for dielectric deposition) applied between the anode and the cathode. A glow discharge of argon gas is established when positive argon ions are accelerated towards the cathode dislodging atoms of the source material and making them adhere to the substrate material. Extremely well adhering films can be produced by sputtering as the energy of the sputtered atoms is several orders of magnitude more than the evaporated ions.

Recapitulation

Microwave integrated circuit—HMIC vs MMIC

- The integrated circuit realised at microwave frequencies with a microwave active device and passive components including the interconnecting transmission lines is known as a *microwave integrated circuit* (MIC). This may be either a hybrid microwave integrated circuit (HMIC) or a monolithic microwave integrated circuit (MMIC).
- In a hybrid MIC, the transmission lines and interconnections are implemented by conductor patterns on a dielectric substrate (which may be FR4, a type of epoxy fibreglass, useful up to 4 GHz and teflon or duroid for higher microwave frequencies), with printing and etching either by printed circuit, or thin-film/thick-film technique while the active devices (Gunn, IMPATT, transistor) and passive components (either in distributed or lumped form) are assembled on the patterned substrate either by soldering or with wire bonding.

- In MMIC, the active devices (like MESFET, HEMT) and passive components as well as the planar transmission lines are grown or implanted *in situ* on a single planar substrate which is usually a semi-insulating material (may be GaAs, SiGe or GaN). MMIC fabrication begins with a *wafer* onto or into which the active layer and conducting as well as dielectric layers are formed to realise passive components and interconnections.
- HMIC permits tweaking of the subsystem designed after fabrication if desired for optimisation. However, the use of packaged active devices and passive components and their placement with soldering/bonding add to the parasitics thereby limiting broadband operation. High power solid-state devices can be accommodated in HMIC and also different types of transmission line configurations such as microstrip line, slot-line or coplanar line can be used.
- MMIC needs design optimisation at the design level with CAD tools as it cannot be tweaked after fabrication, with much reduced parasitics it allows broadband operation with better reliability. MMIC permits a much higher level of integration compared to HMIC even in a much smaller space.

Fabrication process steps of HMIC and MMIC

- There are two major steps in HMIC fabrication—metallisation of the substrate material and selective photo-etching to generate the circuit pattern. In commonly used alumina substrate Cr–Cu–Au may be deposited in which Cr helps for adhesion, Cu carries the main RF current and Au prevents Cu from oxidation and also helps bonding of chip-type and beam-lead devices. The selective photo-etching relates to coating with photoresist (normally –ve type), exposure to UV light, developing to remove the exposed part of photoresist and subsequent etching of Au, Cu, Cr from the top surface of the substrate in sequence.
- In MMIC fabrication once the substrate selection and CAD design is finalised, the first step is active layer formation followed by mesa etching. The ohmic contacts for source (S) and drain (D) are next made on the active *n*-layer and then the gate metallisation is done. The next step is to deposit the first layer of metallisation. This is an overlay metallisation that forms inductors, lower plates of MIM capacitors, transmission line sections and provides overlays for ohmic contacts. Then thin-film resistors are formed and dielectric films need to be formed for MIM capacitors, cross-over insulation and passivation of active device areas and resistors. A second layer of metallisation completes the formation of the top plate of the MIM capacitors and any remaining interconnections including air bridges. The final processing step involves the bottom or backside of the substrate with the inclusion of via-holes and ground metallisation.

Materials for HMIC and MMIC fabrication

- The choice of the dielectric or semiconductor as substrate material, conductor for metallisation, dielectric for capacitors and overlays including air bridges and lossy films for resistors are very important for HMIC and MMIC fabrication.
- For HMIC, either plastic (FR-4 glass epoxy or fibreglass reinforced PTFE having trade name RT-Duroid) or ceramic (alumina) substrates are used while for MMIC semi-insulating substrate like GaAs/InP/GaN is used. Commonly used conductor materials are copper,

gold, silver and aluminium. Commonly used dielectric materials are SiO, SiO_2, Si_3N_4 and Ta_2O_5. Some of the resistive film materials used are NiCr, Ta, Ti and TaN.

Fabrication processes related with HMIC and MMIC

- Integrated circuit fabrication relates to three major processes—semiconductor doping (by diffusion/ion-implantation/epitaxial technique), patterning of layers (either with masked lithography or with mask-less lithography) and film deposition (with thick or thin-film technology) both conductors and insulators to connect and isolate devices and components.

Masked lithography and mask-less lithography

- An integrated circuit may be viewed as a set of patterned layers which is required to be defined on the wafer in the desired areas—this can be done either by conventional masked lithography or with modern types of mask-less lithography.
- In masked lithography processes such as photolithography/X-ray lithography (the latter can provide smaller feature sizes), a *mask* is used to selectively prevent the incident radiation (UV light/X-ray) from striking a photosensitised/X-ray sensitised wafer during the photolithographic/X-ray lithographic process.
- Mask-less lithography may be of different types—electron beam (E-beam) lithography, focussed ion beam lithography and optical (laser) writing.

Thick and thin-film technology

- The nomenclature of thick and thin-film is primarily descriptive in terms of the thickness of the film. The thin films have film thicknesses below 1 μm or even below 100 nm while thick films are well above 1 μm, typically 20–500 μm. However, the thickness of the film is not necessarily the only criterion to classifying them; in fact the technology for fabricating the film is needed in reality to classify whether the film is thin or thick-film. Thin films are deposited by vapour deposition or sputter technology, while thick films are screen printed, dried and fired at high temperatures. Both metal and dielectric films can be formed with thick/thin-film technology resulting in film-type resistors and capacitors (however, film inductors are not possible and if inductors are required they are designed as discrete components and soldered/bonded to the IC).

Review Questions

13.1 What are MICs? Mention in brief the two types of MICs used these days indicating their typical size comparison.

13.2 What are the advantages of MICs over discrete circuit design with waveguides and coaxial lines at microwave frequencies?

13.3 Mention the advantages and disadvantages of HMIC and MMIC.

13.4 Discuss in brief the fabrication process steps of HMIC.

13.5 Discuss in brief the fabrication process steps of MMIC.

13.6 Mention the factors to be considered for selection of substrate, conductor, dielectric and resistive materials for fabrication of HMIC and MMIC.

13.7 Ceramic and plastic materials are used as substrate material for HMIC while semiconductor materials are used for MMIC fabrication—justify.

13.8 Give a list of substrate material, conductor material, dielectric material and resistive materials used for the design of HMIC and MMIC.

13.9 Discuss in brief with a comparative view, the semiconductor doping techniques used in the fabrication of MICs.

13.10 Discuss different techniques for patterning of layers in the fabrication of HMIC and MMIC.

13.11 Discuss different techniques of masked and mask-less lithography.

13.12 Discuss in brief the thick and thin-film technologies used in the fabrication of MICs.

14 Electromagnetic Interference (EMI) and Electromagnetic Compatibility (EMC)

Pollution in all forms is a curse to human civilisation

14.1 Introduction

Radio frequency interference (RFI) or electromagnetic interference (EMI) was known since the 1800s when Hertz did his experiment on the spark-gap generator to prove the existence of electromagnetic waves. RFI or EMI may affect an electrical/electronic system either due to electromagnetic induction or electromagnetic radiation emitted from an external source (termed *exciter*) or the system itself (termed *victim*). The source of RFI may be artificial or natural, that carries time varying electric currents, such as an electrical/electronic circuit or lightning, which might obstruct, interrupt, or degrade the effective performance of a system. EMI can plague even the best RF/microwave design and requires careful planning and control. Designers working with electromagnetic (EM) fields are constantly forced with the threat of leaking stray EM energy that can disrupt the performance of the circuit/system where it originates as well as that of nearby circuit/systems.

Technical concern for radio interference due to electric motors, electric rail roads and the possible mitigation issues started drawing the attention of the scientific community since the late 1920s and 1930s. During World War II, the use of electronic devices, primarily radar and radio aids to navigation increased the instances of such interferences. These were usually corrected by reassigning frequencies for transmission or physically moving cables away from noise emission sources thereby preventing the cable from picking up those emitted signals. The primary event that brought the present emphasis on EMI and to look for its practical mitigation by designing equipment with electromagnetic compatibility (EMC), so that on all practical counts EMI does not affect its performance, was the introduction of digital computers and digital signal processing. In fact, three important technological trends have led to this serious awareness towards EMI/EMC. First, modern digital logic and signal processing are based on relatively low threshold voltages (only few volts) compared to older technologies based on electronic valves (several hundred volts). The immunity of modern systems is thus inherently lower. Second, in order to realise higher processing speeds, shorter pulse-rise-times are used. This contributes significant amounts of energy at high frequencies, which is capable of propagating over long distances by radiative mechanisms. Third, modern physical design of equipment is based increasingly on the use of plastics rather than metals; this significantly reduces the electromagnetic shielding inherent in an all-metal cabinet. Because of the increasing occurrence of digital system interference with wire and radio communication, the Federal Communications Commission (FCC) in the US published a regulation in 1979 that required electromagnetic emissions of all digital devices to be below a certain specified limit. The sole purpose of this rule was to try to limit electromagnetic pollution of the environment in order to prevent, or at least reduce, the number of instances of EMI. These days the issue of EMI and EMC is a serious one, as we live at present in a dense electronic world sometimes and somewhere with intolerable limit of RF electromagnetic pollution which compelled both the government and commercial agencies to impose strict rules for EMI/EMC. Two important regulatory bodies for EMI/EMC regulation and standardisation are the International Special Committee on Radio Interference (CISPR) for the European standard while Federal Communications Commission (FCC) is for the US standard.

14.2 Causes and Case Studies for EMI

It is a common experience for many of us that lightning discharge (which in general is rich in many frequency components) introduces noise (on passing through the input filter of the radio) in our AM broadcast receiver. Also a strong transmission from a nearby FM radio station or TV station may be picked up by a digital computer, which would interpret it as data or a control signal causing incorrect functioning of the computer. The occurrences of 'lines' across the face of a TV screen is a common experience when a vacuum cleaner, blender or other house hold devices containing a DC motor is turned on. The problem is caused by arcing at the brushes of the DC motor. The arc voltage is similar to the Hertzian spark-gap generator and is rich in spectral content that finds its passage through the AC power cord of the device which places this interference signal into the common power net of the household, acting as a large array of efficient wire antenna. The problem of EMI in this case is thus caused by the direct radiation of the signal to the receiving antenna of the TV set. The EMI may also be caused by conducted emission and many other ways.

There are numerous examples of EMI, ranging from the commonplace to the catastrophic one which prompted the EMI regulatory boards all over the world to impose strict EMC regulations to comply with for any product to be marketed.

It would be interesting at this point to quickly go through some case studies of EMI. After the first nuclear detonation in the mid 1940s, it was found that the semiconductor devices in the electronic systems that were used to monitor the effects of the blast were destroyed. This was not due to the direct physical effects of the blast, but was caused by an intense electromagnetic wave created by the process of detonation. In fact a high power electromagnetic pulse is produced following a nuclear detonation. This happens because following a nuclear detonation a large number of photons (γ-rays) are produced that spread through space. These photons interact with the surrounding material to produce high energy electrons due to the Compton effect, which eventually produce an intense electromagnetic field. Anyway, this serious electromagnetic interference caused following a nuclear detonation generated significant interest within the military communities in regard to *hardening* communication and data processing facilities against the effects of such nuclear electromagnetic pulses (NEMP). As a second case, a large computer system installed in an office complex near a commercial airport was found to lose or store incorrect data at random times. The problem turned out to be synchronised with the sweep of the airport surveillance radar as it illuminated the office complex. Extensive shielding of the computer room prevented any further interference. In the late 1960s, the US aircraft carrier *Forrestal* was deployed off the cost of North Vietnam. The carrier deck contained numerous attack aircraft that were fueled and loaded with heavy bombs, as well as air-to-air and air-to-ground missiles. One of the aircraft missiles was inadvertently deployed, striking another aircraft and causing an explosion of its fuel tanks and the subsequent death of 134 servicemen. The problem was identified to be caused by the generation of RF voltages across the contacts of a shielded connector by the ship's high power search radar. Further, a common problem of EMI is electrostatic discharge (ESD). Walking across a nylon carpet with rubber-soled shoes can cause a build-up of static charges on the body due to turboelectric effect. The shoes are thus left with negative charges and due to induction, charge separation is established in the body. It is found that the human body can charge to a potential often exceeding 10 kV. On approaching another body, an intense electric field is established that may result in ESD. Thus, if an electronic device such as a keyboard is touched, this static charge may be transferred to the device and an arc created between the finger tips and the device. This arc may cause an electromagnetic wave to propagate through the device which when picked up by the internal circuitry, may be an integrated circuit chip, would cause its damage and hence malfunction of the system.

14.3 Electromagnetic Compatibility (EMC)

The electromagnetic compatibility (EMC) may be defined as the ability of an electronic device or system to perform satisfactorily in an EMI environment without introducing unacceptable interference to itself and to its environment. To be more lucid, a system is said to be electromagnetically compatible if it satisfies two criteria.

(i) Its emission level is controlled to such a level that it does not cause interference with other systems and with itself too.
(ii) It is not susceptible, in other words, practically immune, to emissions (both radiative and conducted) from other sources.

The goal of EMC is to ensure correct operation of different equipments in an EMI prone electromagnetic environment (which is very common in today's world with RF pollution all around) and the avoidance of any interference effects. Since EMI and EMC are frequently referred to as a combined term 'EMI/EMC', it would be prudent to bring out in clear terms the differences between EMI and EMC. In fact, EMI is the phenomenon that causes the problem of electromagnetic interference while EMC is the characteristic with which equipment should be designed so that it is not being affected or disturbed by EMI and does not also generate EMI above a specified limit.

In order to achieve its goals, EMC pursues two different kinds of issues—*emission* (primarily *radiative* at RF frequencies) and *susceptibility* (opposite of immunity). The emission issue relates to the unwanted generation of electromagnetic energy at RF frequencies by the system under consideration, that is, the *source*, and to the countermeasures which should be taken in order to reduce such generation and to avoid the escape of any remaining energies into the external environment. Susceptibility or immunity issue refers to the correct operation of the electronic equipment (*receptor* or *victim*) in the presence of unplanned (unintentional or intentional) electromagnetic (RF) disturbances. Interference or noise mitigation and hence electromagnetic compatibility is achieved primarily by addressing both emissions and susceptibility issues, that is, *quieting* the sources of interference and *hardening* the potential victims. The coupling path between source and victim also need to be addressed to realise effective mitigation against EMI.

14.4 RF/Microwave Shielding for EMI/EMC

14.4.1 Introduction

Mitigation of electromagnetic interference (EMI) and to realise compliance with electromagnetic compatibility (EMC) at radio frequencies (RF) *electromagnetic shielding* (RF shielding) is used. Two approaches can reduce or shield electromagnetic emission from a device or system and improve its immunity—the first is to shield the PCB and cables through proper design techniques and the second is by placing the device/system in a metallic enclosure. The RF/microwave shielded enclosure prevents radiated emissions (RF signal) from an electronic product enclosed within it thereby arresting interference with other products outside the shield and also the penetration of radiated emissions external to the shield is prevented which might otherwise cause interference to the electronic products inside the shielded enclosure (Fig. 14.1). This makes both the electronic products inside and outside of the shielded enclosure immune or non-susceptible to RF interference. The photograph of an electromagnetically compatible shielded enclosure is shown in Fig. 14.1c.

A variety of materials are used for shielding with a wide range of electrical conductivity, magnetic permeability and thickness. In principle, shields can be designed to operate from DC to 400 GHz, but their performance at higher RF frequencies is more effective. The quality of a shield is characterised in terms of *shielding effectiveness*, which on an average is 50–60 dB to cope with most problems; however, shielding effectiveness above 120 dB approaches the state-of-the-art in this area. The effectiveness of a shield is quantified

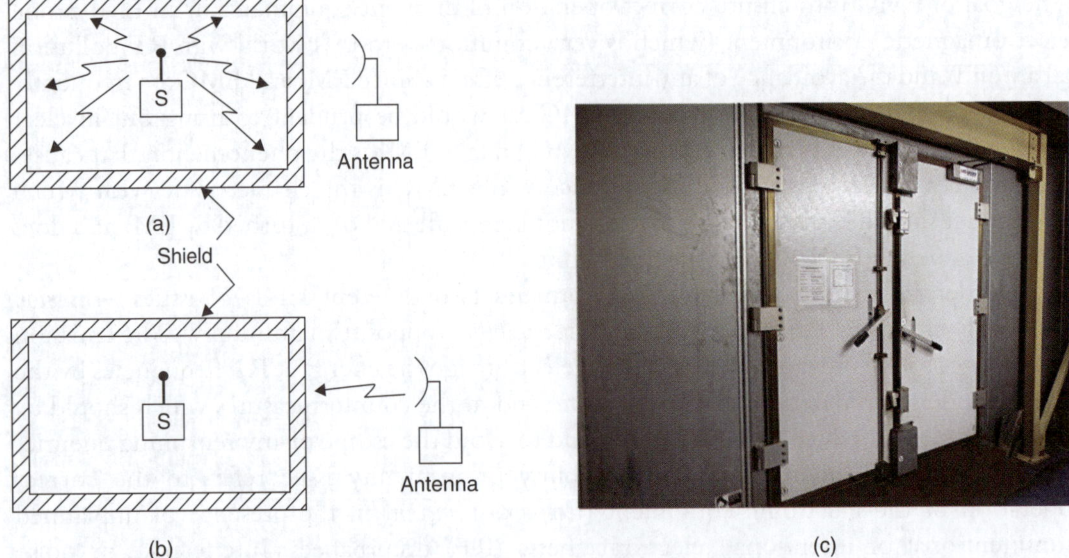

Fig. 14.1 Electromagnetic shielding from radiated emission (a) Source (S) inside the shield (b) Source (S) outside the shield (c) Photograph of an electromagnetically compatible shielded enclosure (Photograph: Courtesy SAMEER Kolkata centre).

in terms of the ratio of the magnitude of the electric (or magnetic) field incident on the *barrier* (shield) to the magnitude of the electric (or magnetic) field that is transmitted through the barrier. From an engineering point of view, the quantification of shielding effectiveness may be viewed to be equivalent to the often used term with filters the *insertion loss* ($-20\log_{10}|S_{21}|$) that is incurred with the shield in place. An ideal shield is expected to result in zero transmission through it making $|S_{21}| = 0$ and hence demanding insertion loss to be infinite; however, for all practical purposes an insertion loss in excess of 100 dB is considered to the very close to the ideal.

14.4.2 Physical Understanding of RF Shielding

Let us first understand how electromagnetic shielding works. When a time varying electromagnetic field (which consists of coupled electric and magnetic fields) is incident on the shield material, its electric field component produces a force on the charge carriers (electrons) within the conductor resulting in a time-varying current which is of such magnitude and phase as to oppose its cause—the applied electric field. A conducting material would therefore act as a good shield, provided sufficient currents are induced and they are allowed to flow with the least possible hindrance. It is known that the magnitude of the induced current is dependent on the rate of change of the incident field and therefore shields are more effective at higher frequencies. Also a good conductor will allow the flow of such current with the least possible impedance. However, the presence of any apertures, seams and joints at any location of the shield would disturb the flow of induced currents and therefore reduce shielding effectiveness. Similarly, varying the magnetic field component of the time varying electromagnetic field generates eddy currents that act to cancel the applied magnetic field. The result is that any interfering electromagnetic field coming from outside (or generated inside) will not be able to penetrate into the shield (or leak out of it). That is, the electromagnetic radiation gets reflected from the surface of the conductor of the shield and the external field stays outside and the internal field stays inside.

The effect of the shield may also be viewed in terms of the popularly known *Faraday cage* or *Faraday shield* named after the English scientist Michel Faraday, who invented it in 1836. The Faraday cage is an enclosure formed from conducting material made of wire mesh or a system of parallel conducting strips with one end being grounded (connected to an earthed the conductor). Such an enclosure blocks the static and also non-static electric fields (however, it cannot block static and slowly varying magnetic fields, such as the earth's magnetic field). A Faraday cage's operation depends on the fact that an external static electric field will cause the electrical charges of the cage's conducting material to redistribute themselves to cancel the field's effects in the interior of the cage. This phenomenon is used, for example, to protect electronic equipment from lightning strikes and other electrostatic discharges. Some linesmen wear Faraday suits, which allow them to work on live, high voltage power lines without the risk of electrocution. The scan room of the Magnetic Resonance Imaging (MRI) machine is designed as a Faraday cage, which prevents RF signals from being added to the data collected from the patient that would otherwise affect the resulting image. The Faraday cage, to a large extent, also shields the interior from external electromagnetic radiation if the conductor is thick enough and any holes are significantly smaller than the radiation's wavelength. The cooking chamber of a microwave oven is a Faraday cage, made to contain the RF signals of the cavity magnetron which cooks the food. In fact the door of the oven is made of a wire mesh with holes of the mesh much smaller than the wavelength (λ ~12 cm, frequency being 2.45 GHz) of the microwave radiation which together with the metal housing (cavity) of the oven completes the Faraday cage.Visible light, with a wavelength ranging between 400 nm (violet) and 700 nm (red), passes easily between the wires of the mesh.

Another commonly used shielding method, especially with electronic goods housed in plastic enclosures, is to coat the inside of the plastic enclosure with a metallic ink or similar material. The ink consists of a carrier material loaded with a suitable metal, typically, very small granules in the form of dust of copper or nickel. It is sprayed on the inside of the enclosure and, once dry, produces a continuous conductive layer of metal, which can be electrically connected to the chassis ground of the equipment thus providing effective shielding.

14.4.3 Qualitative and Quantitative Analysis of RF Shield and Shielding Effectiveness

There are a number of electromagnetic phenomena that take place between the interior and exterior of the left and right surfaces of the shield wall. From Fig. 14.2, it may be observed that the electromagnetic signal incident upon the left surface of the shield wall gets reflected. A portion of the incident wave that crosses this surface of the shield wall proceeds through the conducting medium of the shield wall and gets attenuated by a factor of $e^{-\alpha z}$, where α is the attenuation constant of the material of the shield. The shield material is usually a good conductor and thus the attenuation constant is given by $\alpha = 1/\delta$ (where $\delta = 1/\sqrt{\pi f \mu \sigma}$ is the skin depth of the material, Eq. 2.12a). Thus the amplitude of the fields is attenuated according to the factor $e^{-z/\delta}$. If the thickness t of the shield material is much greater than the skin depth (δ) at the frequency of the incident signal (which is an usual case

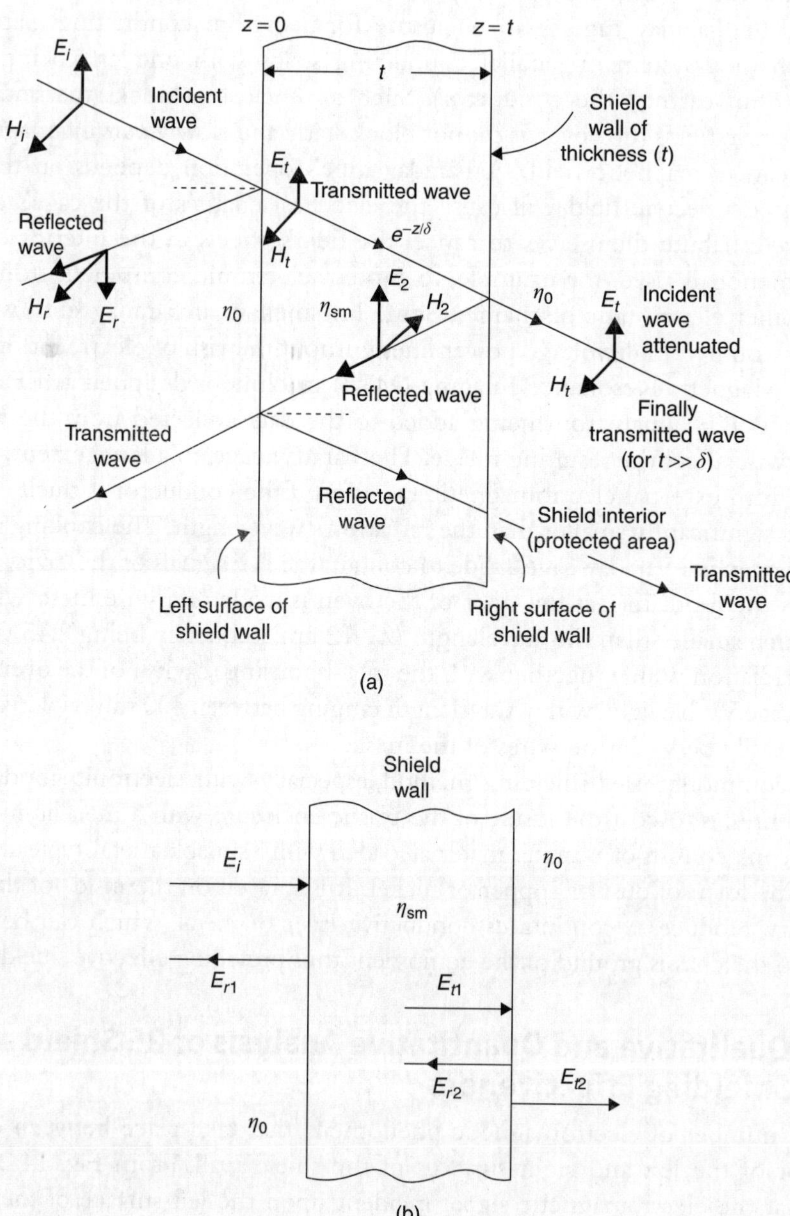

Fig. 14.2 (a) Reflection and transmission taking place inside the shield wall (b) Incident, reflected and transmitted electric fields to find R_{dB}.

at higher RF) than this wave that is transmitted through the first (left) surface of the shield wall will be significantly attenuated when it strikes the second (right) surface of the shield wall. In other words, this *absorption loss* of the incident signal will be a dominant factor. But if the thickness of the shield wall t cannot be made much greater than δ (as may be the case at lower RF), the transmitted signal from the left surface of the shield wall when incident on the right surface of the shield wall will get reflected (wall being metallic) and this process continues (Fig. 14.2), in which additional reflected and transmitted fields are progressively attenuated by their travel through the conductive material of the shield.

The *shielding effectiveness* at RF is numerically estimated in terms of wave approach (unlike circuit approach at lower frequencies). The shielding effectiveness (SE) of the enclosure is defined for the electric/magnetic field in dB, as:

$$SE = 20\log_{10}\left|\frac{\vec{E}_i}{\vec{E}_t}\right| = 20\log_{10}\left|\frac{\vec{H}_i}{\vec{H}_t}\right| \tag{14.1a}$$

where $\vec{E}_i, \vec{H}_i$ are the incident and $\vec{E}_t, \vec{H}_t$ are the transmitted electric and magnetic fields respectively (Fig 14.2). If the incident field is a uniform plane wave and the media on each side of the enclosure are identical then the two definitions are the same, since the electric and magnetic fields are related by the intrinsic impedance (η) of the medium.

In terms of the qualitative discussion on the reflection, absorption and multiple reflections taking place, in general, within a shield material, the shielding effectiveness given in Eq. 14.1a can be viewed as the product of three terms representing respectively the phenomena of *reflection loss, absorption loss* and *multiple reflections*. Expressed in dB (as is the usual practice):

$$SE = R_{\mathrm{dB}} + A_{\mathrm{dB}} + MR_{\mathrm{dB}} \tag{14.1b}$$

where R represents the reflection loss caused by reflections at the left and right surfaces of the shield, A represents the absorption loss incurred by the transmitted signal as it travels through the shield material and MR represents the additional effects of multiple reflections and transmissions, if any. The sign of MR will be opposite to that of R and A (note that R and A are positive while MR is negative) and thus the presence of MR is to reduce the shielding effectiveness. Note that at higher RF, that is, at microwave frequencies when $t >> \delta$, we can neglect the contribution from MR.

An explicit expression of shielding effectiveness can be derived [30] by writing down all the electric and magnetic fields shown in Fig. 14.2 and applying the boundary conditions on the field vectors at the two boundaries $z = 0$ and $z = t$ and utilising the continuity of the tangential components of the E and H fields at the interfaces (similar derivations carried out in Section 2.7.2.1), when it is given by:

$$SE \cong \underbrace{20\log_{10}\left|\frac{\eta_0}{4\eta_{\mathrm{sm}}}\right|}_{[R_{\mathrm{dB}}]} + \underbrace{20\log_{10} e^{t/\delta}}_{[A_{\mathrm{dB}}]} + \underbrace{20\log_{10}\left|1 - e^{-2t/\delta}\right|}_{[MR_{\mathrm{dB}}]} \tag{14.1c}$$

We have made practical assumptions that the shield material is constructed of a good conductor ($\eta_{sm} << \eta_0$, η_{sm} being the wave impedance of the material of the shield) and the wall thickness t of the shield is much greater than the skin depth (δ) at the operating frequency ($t >> \delta$).

The first term in Eq. 14.1c represents the reflection loss within the shield material which can be derived as follows. From Fig. 14.2b, the reflection loss in dB is given by $R_{dB} = 20 \log_{10} (E_i/E_{t2})$. Now following Eq. 2.26, we can write:

$$E_{t2} = \frac{2\eta_0}{\eta_0 - \eta_{sm}} E_{t1} \text{ and } E_{t1} = \frac{2\eta_{sm}}{\eta_{sm} + \eta_0} E_i \text{; thus } E_{t2} = \frac{4\eta_0\eta_{sm}}{(\eta_0 + \eta_{sm})^2} E_i \qquad (14.1d)$$

Since $\eta_0 >> \eta_{sm}$, we have $E_i/E_{t2} = \eta_0/4\eta_{sm}$, when R_{dB} becomes equal to the first term of Eq. 14.1.

Since the intrinsic impedance, that is, the wave impedance of the shield material (which is a good conductor) is given by (Eq. 2.12b) $|\eta_{sm}| = \sqrt{\pi f \mu_0 \mu_r/\sigma}$ and that of free space by $\eta_0 = \sqrt{\mu_0/\varepsilon_0} = 377\ \Omega$; we can write the reflection loss as $R_{dB} = 168 + 10\log_{10}(\sigma_r/f\mu_r)$, where σ_r is the conductivity of the shield material relative to copper ($\sigma_{Cu} = 5.8 \times 10^7$ S/m). Thus, it may be seen that at higher frequencies the reflection loss decreases, while for high conductivity material reflection loss increases. Further, for magnetic materials (unlike copper for which $\mu_r = 1$) with $\mu_r > 1$, reflection loss decreases.

The second term in Eq. 14.1c is easily seen to represent absorption loss in dB as $e^{-z/\delta}$ expressed in dB with z replaced by t and remembering that the negative sign is for absorption, the expression follows.

We can write absorption loss $A_{dB} = 20\log_{10} e^{t/\delta} = 8.686(t/\delta) = 131t\sqrt{f\mu_r\sigma_r}$ (with t in metres). It may be observed that as the thickness of the shield material with respect to the skin depth increases the absorption loss increases; also the absorption loss increases with frequency (however, the rate of increase of absorption loss with frequency is much higher than the rate of decrease of reflection loss with increasing frequency). In the case of a ferromagnetic material for which $\mu_r >> 1$, the absorption loss increases over copper (assuming that $\mu_r \sigma_r >> 1$).

Finally, the third term in Eq. 14.1c represents the multiple reflection term which is dominant only in the case of thin shields when $t \approx \delta$ (which is normally the case at lower RF). If the shield is thin, the reflected wave from the right surface of the shield material will again suffer reflection from the left surface and so on (Fig. 14.2a). This can be neglected in the case of a thick shield (compared to the skin depth at the frequency of operation) because by the time the wave reaches the right boundary for the second time it is of negligible amplitude (due to high absorption loss suffered through the thickness of the shield thrice by that time). For an electric field, most of the incident wave is reflected at the left surface of the shield and a very small part enters the shield material (since $\eta_{sm} << \eta_0$, and thus E_{t1} will be extremely small, Eq. 14.1d). But, for a magnetic field, most of the incident wave passes into the shield material since $H_{t1} = [2\eta_0/(\eta_{sm} + \eta_0)]H_i$ (which follows from the equation for E_{t1} and remembering that $\eta = E/H$). Thus, the magnitude of the transmitted wave from the left surface of the shield material entering into the shield is twice the incident wave magnitude. Thus, the effect of multiple reflections in a thin shield is dominated by the magnetic field. It can be derived [31] that the contribution from multiple reflections due to the magnetic field for a shield of thickness t and skin depth δ at the frequency of operation is given by:

$$MR_{\text{dB}} = 20\log_{10}\left|1 - e^{-2t/\delta}\right| \quad (14.1e)$$

For a quick recapitulation, it may be noted that reflection loss is a function of the ratio σ_r/μ_r while absorption loss is a function of the product $\sigma_r\mu_r$. Further, the reflection loss is dominant at lower RF while the absorption loss is dominant at higher RF. Multiple reflection loss is dominant for thin shield material and may be safely neglected for thick shield (compared to the skin depth at the frequency of operation) materials.

A few words about the choice of shield material will make this discussion complete. So far our discussion was with good conductors for the shield material which is effective for electric field shielding. For magnetic field shielding, it is found that at high RF good conductors like copper or aluminium are capable of giving high magnetic field shielding too. However, a magnetic material such as steel or mumetal makes a better magnetic field shield at lower frequencies compared to the good conductors. But shielding effectiveness of magnetic materials decreases with increasing frequency as a result of the decreasing trend of permeability with frequency. Thus, as far as microwave frequency is concerned, good conductors are most effective as shielding materials.

EXAMPLE 14.1

A shield is made of copper and has a thickness of 0.05 mm. Find the shielding effectiveness at 10 kHz, 10 MHz and 1.0 GHz.

Given $\mu_0 = 4\pi \times 10^{-7}$ H/m, $\sigma_{\text{Cu}} = 5.8 \times 10^7$ S/m.

Solution

We know wave impedance of the shield material $|\eta_{\text{sm}}| = \sqrt{\pi f \mu_0 \mu_r/\sigma} = 0.26 \times 10^{-4}\ \Omega$ (at $f = 10$ kHz), $0.825 \times 10^{-3}\ \Omega$ (at $f = 10$ MHz) and $0.825 \times 10^{-2}\ \Omega$ (at 1.0 GHz, since for copper $\mu_r = 1$).

Again, the skin depth in the material of the shield is given by $\delta = 1/\sqrt{\pi f \mu_0 \mu_r \sigma} = 0.66$ mm (at 10 kHz), 0.02 mm (at $f = 10$ MHz) and 0.002 mm (at $f = 1.0$ GHz).

Thus R_{dB} calculated either from $20\log_{10}|\eta_0/4\eta_{\text{sm}}|$ or $168 + 10\log_{10}(\sigma_r/f\mu_r)$ turns out to be 128 dB (at 10 kHz), 98 dB (at 10 MHz) and 78 dB (at 1.0 GHz), while A_{dB} calculated from $8.686(t/\delta)$ turns out to be 0.658 dB (at 10 kHz), 22.2 dB (at 10 MHz) and 217 dB (at 1.0 GHz). When M_{dB} is calculated using $20\log_{10}\left|1 - e^{-2t/\delta}\right|$ it turns out to be −17.04 dB (at 10 kHz), -5.8×10^{-2} dB (at 10 MHz) and -1.2×10^{-10} dB or practically 0 dB (at 1.0 GHz).

Thus $SE = (128 + 0.658 - 17.04)$ dB = 111.6 dB (at 10 kHz), $SE = (98 + 22)$ dB = 120 dB (at 10 MHz) and (78 + 217) dB = 295 dB (at 1.0 GHz).

14.4.4 Issues Affecting the Shielding Effectiveness and their Mitigation

A practical shield is never an unperturbed surface of solid shield material as there must be access covers, doors, holes for cables, ventilation, switches, displays, joints and seams. The quality of shielding effectiveness is guaranteed if all possibilities of such penetrations into or out of the shield are properly treated/minimised with an electromagnetic view point in mind; otherwise significant reduction of the effectiveness of the shield is inevitable. In fact, at high frequencies, the intrinsic shielding effectiveness of the shield material may be of general concern but the leakage of electromagnetic energy via apertures caused by slots, holes for cable insertion, seams are of more serious concern. Apertures cause more effect on the magnetic field leakage (as they result in the perturbation of the surface current on the shield's material surface) than on the electric field leakage. Thus, importance is given to methods for minimising the magnetic field leakage which eventually is adequate for minimising the electric field leakage too.

Let us first understand how the presence of aperture causes decrease in shielding effectiveness. It has already been mentioned that an incident electromagnetic field induces current into the shield material and this current generates an additional field which has the tendency to cancel the incident field that has caused it thereby preventing the entry to (and by the same reason) escape from the 'shield' of any EMI field. For the above mentioned cancellation of field to occur, the induced shield current must be allowed to flow undisturbed in the manner it is induced. If the presence of an aperture forces the induced current to flow in a different path, then the induced field will not completely cancel the incident field and hence the shielding effectiveness will be reduced. The more the current is forced to detour, the greater will be the decrease in shielding effectiveness.

How the presence of an aperture (in the form of slot/slit or hole) affects the induced surface current on the shield material is shown in Fig. 14.3. As may be observed, a rectangular slot (appearing across the surface current lines) causes the detour of the induced surface current and hence leads to ineffective field cancellation (mentioned above) which may be thought of as an equivalent leakage. It is important to note that slot length is responsible for the amount of detour and not its width (even a slit of same length as opposed to a wider slot will have a similar effect, Fig. 14.3c). On the other hand, if the slot is oriented parallel to the direction of the induced surface current (Fig. 14.3d), the slot will have much less effect in causing a detour of the surface current on the shield causing practically no radiation. But it may not be possible to place the slot direction according to the induced surface current lines on the shield surface and thus a large number of holes may be placed instead without bothering the direction of current flow in relation to the hole. This is because, a group of small holes has much less detouring effect on the surface current than the slot (Fig. 14.3e) and therefore it causes practically no leakage of RF signal even if the total area of the two apertures (slot and a large number of holes) is almost the same. The rectangular slot of Figs 14.3b and 14.3c behaves as a slot antenna (having its equivalence with a dipole antenna in terms of Babinet's principle used by Booker, Section 10.2.3) and hence can cause considerable leakage when its linear dimension is greater than $\lambda/10$ (maximum

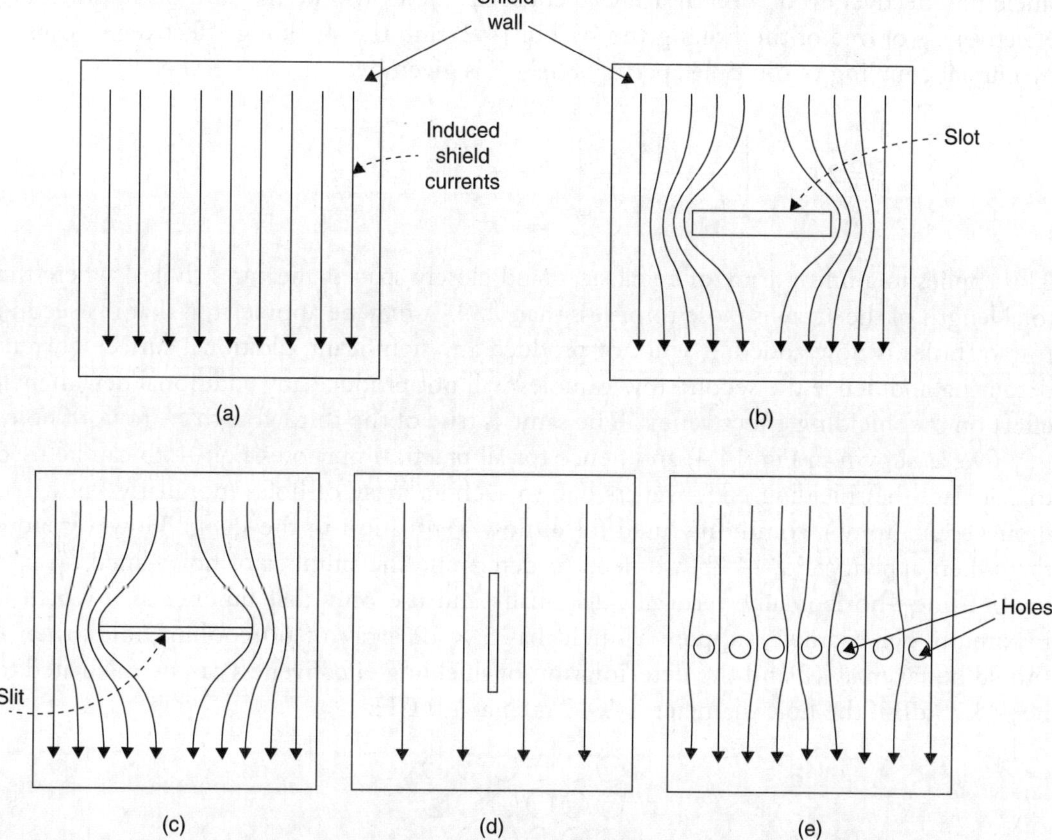

Fig. 14.3 **Detouring of surface current by apertures of different shapes and orientation on the surface of the shield (a) No aperture (b) Radiating slot (c) Radiating slit (d) Non-radiating slot/slit (e) Multiple holes.**

radiation being caused when it is $\lambda/2$). Seams and joints often have the possibility of forming such efficiently radiating slot antennae.

The leakage due to slot/slit causing deterioration of shielding effectiveness may be calculated from:

$$SE_{ap(\text{slot})} = 20\log_{10}\left(\frac{\lambda}{2l}\right) = 20\log_{10}\left(\frac{0.15}{f_{\text{GHz}} l_{\text{meters}}}\right) \tag{14.2a}$$

Thus, if the slot/slit due to the door fitting in a shield is 2 m long and the operating frequency is 1.0 GHz then shielding effectiveness will deteriorate due to the door slot/slit by –22.5 dB.

For a linear array of closely spaced small apertures (holes), the reduction in shielding effectiveness caused by their mutual coupling is proportional to the square root of the number of apertures n, and is given by $SE_{ap(m)} = -20\log_{10}\sqrt{n}$. Thus, the net or effective shielding effectiveness of a linear array of equal size holes will be the sum of the shielding effectiveness of one of the holes given by Eq. 14.2a and the shielding effectiveness caused by mutual coupling of multiple apertures $SE_{ap(m)}$, is given by:

$$SE_{ap(m)\text{eff}} = 20\log_{10}\left[\frac{0.15}{f_{\text{GHz}}l_{\text{meters}}\sqrt{n}}\right] \tag{14.2b}$$

This applies to a linear array of equal sized and closely spaced apertures (holes) where the total length of the array is close to or less than $\lambda/2$. It may be appreciated that if a second row of holes is now added, it will not produce any significant additional surface current detouring and hence the second row of holes will not produce any additional detrimental effect on the shielding effectiveness. The same is true of the third, fourth … rows of holes (say five, as shown in Fig. 14.4) and hence for all practical purposes Eq. 14.2b can be used to calculate the shielding effectiveness due to such an array of holes (popularly known as 'honeycomb array'), commonly used for airflow/ventilation in the shield. However, note that when applying Eq. 14.2b, we need to determine the number of holes lined up in a straight line—horizontally, vertically, diagonally and use only that number as n which is maximum. For example, suppose a shield has 5 × 12 array of 60 cooling holes, then n should be taken as 12 and the deterioration of shielding effectiveness can be calculated to be –33.78 dB if the hole diameter is 1/12 inch at 1.0 GHz.

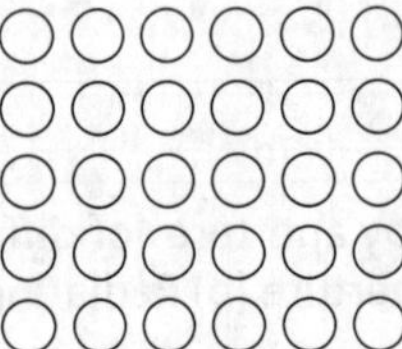

Fig. 14.4 Two dimensional array of 30 round holes.

It is important to note that each hole of the honeycomb array can provide additional attenuation if it is shaped to form a waveguide penetrating inside the shield wall (Fig. 14.5a). A waveguide has a cut-off frequency below which any signal propagating via it gets attenuated (becomes evanescent) and hence the waveguide acts as an attenuator at and below that frequency (Chapter 4). For a round waveguide operating in its dominant mode TE_{11} the cut-off frequency is $f_{c(\text{cir})} = (1.75 \times 10^8/d)$ Hz, where d is the diameter in metres (Eq. 4.32), while for a rectangular waveguide operating in its dominant mode TE_{10} the cut-off frequency is $f_{c(\text{rect})} = (1.5 \times 10^8/a)$ Hz, where a is the largest (broad) dimension of the waveguide cross section in metres (Eq. 4.14). As long as the operating frequency is much less than the cut-off frequency of the waveguide, the magnetic field shielding effectiveness of a round/circular waveguide and rectangular waveguide are respectively given by [31]:

$$SE_{\text{Cir}W} = 32\left(\frac{t}{d}\right)\text{ dB} \quad \text{and} \quad SE_{\text{Rect}W} = 27.2\left(\frac{t}{a}\right)\text{ dB} \tag{14.3}$$

where d is the diameter and t is the length of the round/circular hole as shown in Fig. 14.5a while a is the largest linear dimension of cross-section of the hole and t is the length or depth of the rectangular hole. In practice, many such small waveguides are welded together in honeycomb fashion (Figs 14.5b and 14.5c) in order to provide a sufficient amount of air flow with appropriate waveguide dimensions. It may thus be observed from the rather simple result given by Eq. 14.3 that TE and TM modes are strongly attenuated in direct proportion to the guide length (t) as they travel down the guides. This is the basis for the use of *waveguides below cut-off* to allow airflow into a shielded enclosure and at the same time very importantly preventing the propagation of electromagnetic signals having frequencies below cut-off frequency of the guides into the shielded room thereby providing effective magnetic shielding.

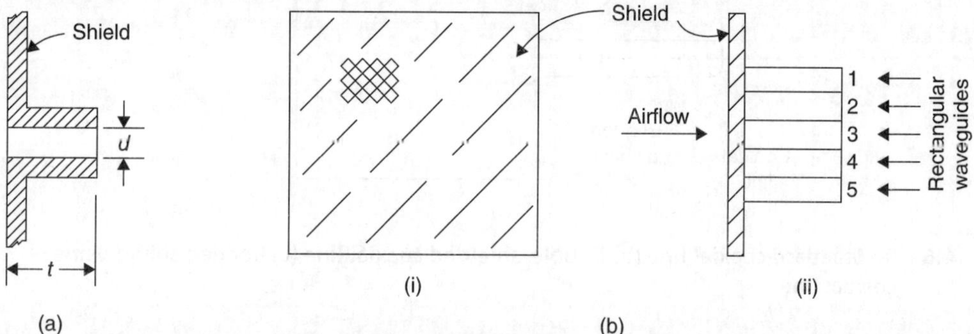

Fig. 14.5 (a) A hole formed into a waveguide with diameter *d* and depth *t* (b) Honeycomb fashion of small waveguides (below cut-off) welded together (i) Front view (ii) Side view.

Finally, the cable penetration into the metallic box of the shield is another source of possible coupling of radiation into or out of the shield box if not properly shielded and bonded to the body of the shield. To mitigate the effect of EMI via cables, a common practice is to use a shielded cable, which has the electromagnetic shielding in the form of a wire mesh (which is grounded) surrounding an inner core conductor, the standard coaxial line (Fig. 14.6a). The shielding prohibits the escape of any RF signal from the dielectric material (along which the RF signal flows) filling the space between the two conductors or entry of any external signal to the dielectric material. Some cables have two separate coaxial screens (Fig. 14.6b), one connected at both ends, the other at one end only, to maximise shielding of both RF electromagnetic field and electrostatic fields. The entry or exit of the cable from the shielding enclosure should be done with proper electrical consideration to ensure shielding effectiveness. With this consideration in mind, the coaxial cables are generally integrated to the body of the enclosure of the shield (bonded shield, Fig. 14.6c).

Gasketing maintains shielding effectiveness through proper seam treatment. In general, seams account for most of the RF leakage in the enclosure design. The shielding effectiveness of a seam depends on the material's contact pressure and surface area. A gasket maintains contact across mating surfaces. A solution to the radiation problem is to make all seams of

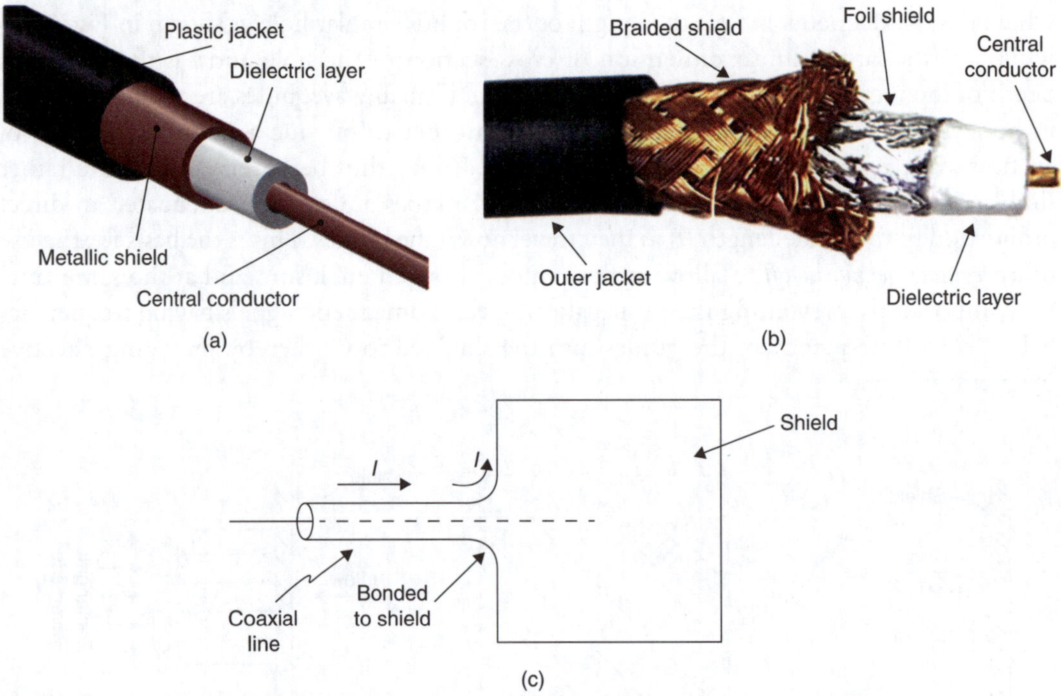

Fig. 14.6 (a) Standard coaxial line (b) Double-shielded coaxial line (c) Bonded shield using coaxial line.

adjoining metal pieces continuous. If there is no continuity between metal pieces, a radiating aperture for RF is created; this is where a gasket can be used. A beryllium copper (BeCu) gasket and metallised fabric gaskets can be used to ensure maximum EMI compliance.

14.5 EMI/EMC Standards, Regulations and Testing

The standards and regulations for EMI/EMC applicable to domestic, commercial and industrial sectors are set by two international bodies worldwide—one is the International Special Committee on Radio Interference (CISPR) for European standards while Federal Communications Commission (FCC) is for US standards. The military standard (MIL-STD) is set by the US defence and followed worldwide. Non-military applications address the emission issues while the military regulations impose strict susceptibility regulations too. Though an electronic product might comply with the emission requirements, it may be susceptible to the emissions of other electronic devices in its close proximity. The emission requirements simply attempt to limit electromagnetic pollution, the susceptibility requirements, used by the military, go one step further in attempting to ensure electromagnetically compatible operation of all equipments. EMC is a critical aspect in the marketability of an electronic product, especially for military applications. Even if the product performs some very desirable tasks and customers are willing to purchase it, it may be rendered unimportant if it does not comply with the EMI/EMC regulations set for that country.

EMC testing is required to conform that a particular device meets the required standards. It is broadly divided into susceptibility testing and emission testing.

Susceptibility testing issues may be understood as follows. Radiated field susceptibility testing typically involves a high-powered source of RF or EM pulse energy and a radiating antenna to direct the energy at the potential victim or equipment under test (EUT). Conducted voltage and current susceptibility testing typically involves a high-powered signal or pulse generator and a current clamp or other type of transformer to inject the test

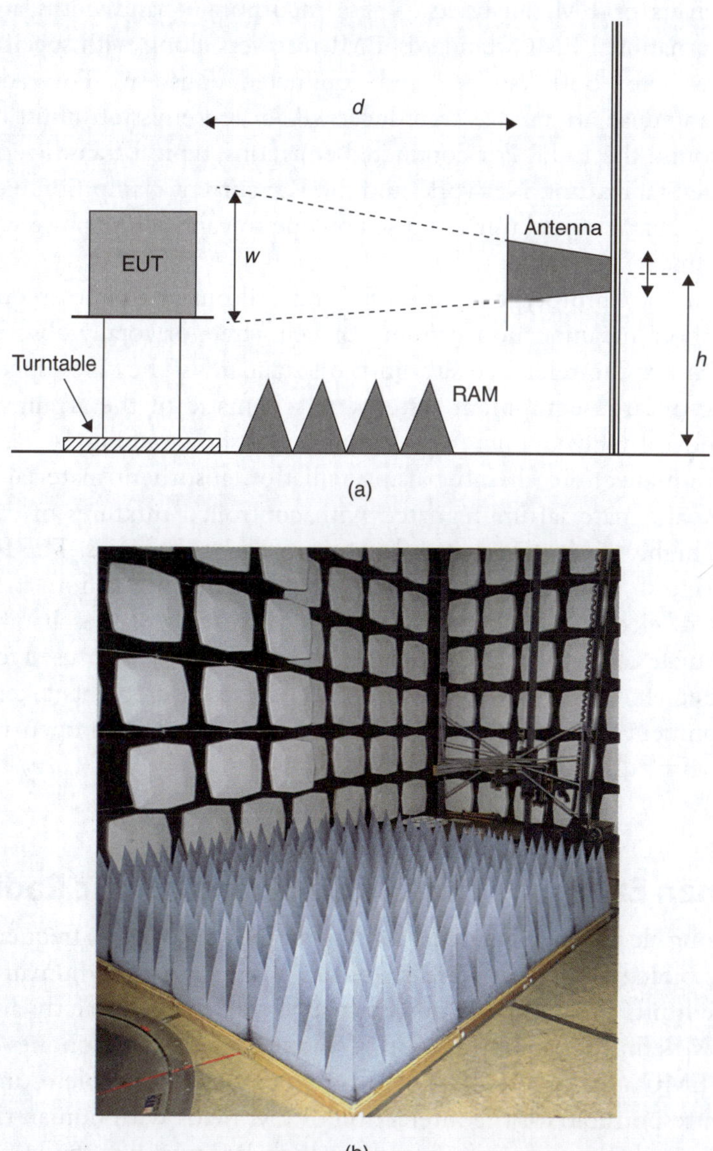

Fig. 14.7 (a) Schematic of EMC test arrangement in an anechoic chamber (b) Anechoic environment inside an EMC test-chamber (Photograph: Courtesy SAMEER Kolkata centre).

signal. Electrostatic discharge testing is typically performed with a piezo spark generator called an ESD pistol.

Emissions are typically measured for a radiated field's strength and where appropriate for conducted emissions along cable and wiring. Inductive (magnetic) and capacitive (electric) field strengths are near-field effects, and are only important if the EUT is designed for being located close to other electrical equipment. Typically a spectrum analyser is used to measure the emission levels of the EUT across a wide band of frequencies (frequency domain measurement). Specialised spectrum analysers for EMC testing are available, called EMI Test Receivers or EMI analysers. These incorporate bandwidths and detectors as specified by international EMC standards. EMI receivers along with specified transducers can often be used for both radiated and conducted emissions. For radiated emission measurements, antennae are used as transducers. Radiated emissions must be measured in all directions around the EUT. For conducted emissions, typical transducers are the LISN (Line Impedance Stabilisation Network) and the RF current clamp. Some pulse emissions are more usefully characterised using an oscilloscope to capture the pulse wave form (time domain measurement).

The RF testing of a prototype is most often carried out in a radio-frequency anechoic chamber (an-echoic meaning non-echoing or echo-free, or totally absorbed). Open air test sites (OATS), are the reference sites in most standards. They are especially useful for emission testing of large equipment systems. A schematic of the arrangement used for EMC testing above 1 GHz (at microwave frequency) is shown in Fig. 14.7a. It simulates the environment of an anechoic chamber using radiation absorbent material (RAM), made of rubberised foam material impregnated with controlled mixtures of carbon and iron which acts as a highly dissipative material at microwave frequencies. The RAM is usually of pyramidal shape so that minimum possible reflection can occur from it, if any, allowing streamline flow of electromagnetic signal along it. The walls of the chamber are covered with absorbing material. A photograph of such an anechoic chamber used for EMC testing at microwave frequencies is shown in Fig. 14.7b. The measurements taken inside such a screened environment may be used to predict the test results on an open-field site within an uncertainty of ±6 dB.

14.6 Human Exposure Limit to Electromagnetic Radiation

Exposure to strong electromagnetic radiation, especially at microwave frequency, is definitely harmful to any biological system, needless to say, especially to the human body. Though the issue of safe limits for human exposure to EM fields is not strictly an EMC matter, an awareness is definitely important for all those using mobile phones and also those involved with EMC measurements. There is at present no complete understanding of the mechanism responsible for the interaction of EM fields with human tissue. However, it has been observed that dielectric heating at high RF which is the basis of microwave heating in a microwave oven (Section 2.4), is a cause for concern about the possibility of adverse thermal effects due to exposure to electromagnetic radiation especially at higher

RF. Traditionally, limits were set by assuming thermal effects and restricting the Specific Absorption Rate (SAR, in W/kg), that is, the absorption of RF energy per unit mass of tissues of the body. There is also concern that other non-thermal mechanisms like resonance (in the whole body or in cavities such as the eyes) can cause damage to the human body. SAR is defined as the time derivative of the incremental energy (ΔW) absorbed by, or dissipated in, an incremental mass (ΔM) contained in the incremental volume (ΔV) of a given density ρ. It can be written in equation from as:

$$\text{SAR} = \frac{d}{dt}\left(\frac{\Delta W}{\Delta M}\right) = \frac{1}{\rho}\frac{d}{dt}\left(\frac{\Delta W}{\Delta V}\right) \text{(W/kg)} \tag{14.4a}$$

When defined in terms of per unit mass or per unit volume (as the usual practice is) then in terms of the peak electric field E of the incident RF field we can write Eq. 14.4a as:

$$\text{SAR} = \frac{\sigma |E|^2}{2\rho} \text{(W/kg)} \tag{14.4b}$$

where, σ is the effective conductivity of the tissue material in S/m, ρ is the density in kg/m^3 and $|E|$ is the magnitude of the peak electric field in V/m.

Numerous studies connected with possible adverse effects of wireless technology, especially of the mobile phones have been reported. Recommendations are there to avoid unnecessary exposure to electromagnetic signals, especially in the case of children. A threshold limit of SAR value of 4 W/kg has been specified to avoid adverse effects of microwave exposure due to mobile phones (GSM carrier frequencies spans from 850 MHz to 1900 MHz). A SAR of 4 W/kg is found to produce a temperature rise of 1° in human tissue. For mobile phones, the SAR value limit for European standard is a maximum of 2 W/kg averaged over 10 g of human tissue while the US standard is 1.6 W/kg averaged over 1 g of human tissue. In India, as per the 2012 regulations, the US standard is adopted.

The mobile phone radiation hazard is known to affect the head of a human being; in fact a recent study shows that more than half the radiated power may be absorbed in the user's head. Thus, to characterise the mobile hand set to investigate if it is within the specified SAR limit, a field probe measurement is done within a tissue phantom (which is a physical model of the human head with equivalent dielectric properties). A commonly used instrument for SAR measurement is the DASY (Domestic Assessment System), manufactured by Schmidt and Partner Engineering AG in Switzerland.

The history of the standard for safety levels with respect to human exposure to radio frequency (RF) electromagnetic fields goes back to 1966—first started by American National Standards Institute (ANSI). ANSI C95.1–1982 became the first Specific Absorption Rate (SAR)-based human exposure standard in the world. ANSI C95.1–2005/IEEE C95.1–2005 is the US standard for the safe limit for human exposure followed worldwide (Table 14.1), the plot for which, covering the microwave frequency spectrum is given in Fig. 14.8. However, the previous ANSI C95.1–1991/IEEE C95.1–1991 regulation [9] was more relaxed compared to this.

Table 14.1 Safety limits for human exposure to RF power

Frequency (MHz)	Power density (W/m²)
30–400	2.0
400–2,000	$f/200$
2,000–100,000	10
100,000–300,000	Increases from 10 to 100

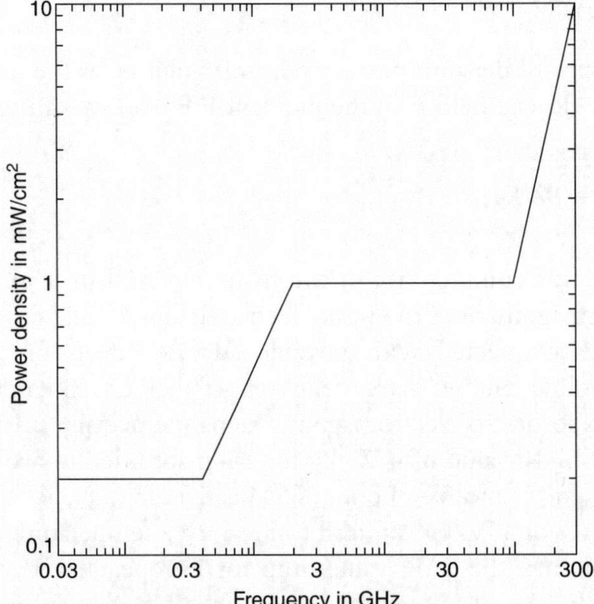

Fig. 14.8 Safe limit of RF signal exposure concerning hazard to human being (US standard: ANSI C95.1–2005/IEEE C95.1–2005).

EXAMPLE 14.2

A Precision Approach Radar (PAR) in an airport operates at 10 GHz with a transmitter power of 100 kW to guide an aircraft located at a radius of $r = 250$ m of its antenna beam. Calculate the radar's beam power density and the photon energy at the target location. Assuming the solar power density on earth to be 1372 W/m^2 and the solar radiation in the visible spectrum has a frequency of 5×10^{14} Hz, compare which of the two radiations would be more hazardous when their penetration into the human body is taken into account. The conductivity of skin is given to be 0.1 S/m.

Solution

The radar power density at the target location is given by:

$$P_{av} = \frac{P_{rad}}{(\pi r^2)} = \frac{100,000}{\pi(250)^2} = 0.509 \text{ W/m}^2$$

The photon energy at the radar's operating frequency is given by:

$$E_r = h\nu = \left(\frac{6.63 \times 10^{-34}}{1.6 \times 10^{-19}}\right) \times 10 \times 10^9 = 4.143 \times 10^{-5} \text{ eV}$$

(since h in electron volt = h in J-s/electronic charge in C).

The photon energy due to solar radiation at visible frequency is:

$$E_s = h\nu = \left(\frac{6.63 \times 10^{-34}}{1.6 \times 10^{-19}}\right) \times 5 \times 10^{14} = 2.071 \text{ eV}$$

It may thus be observed that the radar power density is approximately 2695 times less than the solar power density while the photon energy at radar frequency is less by a factor of 50000 compared to the photon energy of solar radiation at optical frequency. Thus, at this point it might appear that radar radiation is less harmful than solar radiation. But the real hazard of such radiation needs to be determined in terms of its penetration to the human skin. The skin depth is given by $\delta = 1/\sqrt{\pi f \mu_0 \sigma}$, which, with conductivity of skin (0.1 S/m) gives 0.07 mm of skin depth, that is, the penetration by solar radiation into the human body in visible spectrum; while for radar radiation frequency the penetration depth is 15.91 mm. Thus any induced current in the visible spectrum is limited to a very thin layer of skin (about 70 μm, thus sun blisters are caused by sunburn) while radar waves will penetrate 227 times deeper. Also clothing shields visible light [although it is less effective against ultraviolet radiation, an ionising radiation (Fig. 1.1a), which however is absorbed by the ozone layer in the upper atmosphere] but does not shield lower radio frequency (normally used in radar and in most of the communication applications).Thus, the hazards related with RF radiation used for radar and communication applications are definitely of serious concern and the design frequency, power level and antenna characteristics are to be chosen accordingly to prevent hazards to living beings, especially human beings. In the present problem, however, the power density at the target location is only 0.05 mW/cm^2 at 10 GHz which is far below the safe limit of hazards due to RF signal as per the US regulations (Fig. 14.8).

EXAMPLE 14.3

A global system for mobile communications (GSM) transmitter operates at 915 MHz with a transmitter power of 8 W (39 dBm) and using an antenna with a gain of 30 dBi (an isotropic antenna with a gain of 30 dB). Determine the possibility of radiation hazard due to the main beam and also due to side lobes of the antenna at a distance of 30 m and also at a distance of 10 m if the transmitting tower is mounted on a building top. Assume the side lobe level to be 10 dB down the main beam.

Solution

The gain of the transmitting antenna as a ratio is given by $G_t = 10^{30/10} = 10^3 = 1000$.

The power density of the main beam at a distance of R from the transmitting tower considering an imaginary sphere of radius R with the transmitting tower at the centre of the sphere:

$$P_{d(\text{main})} = \frac{P_t}{4\pi R^2} G_t = \frac{8 \times 1000}{4\pi(30)^2} = 0.707 \text{ W/m}^2 = 0.07 \text{ mW/cm}^2$$

Thus the RF power in the main beam of the transmitting antenna will not be hazardous to health as 0.07 mW/cm^2 at 915 MHz is much less than the safe limit as per US regulations (Fig. 14.8). However, if the distance is reduced to 10 m, then the RF power density in the main beam (0.637 mW/cm^2 at 915 MHz) will be hazardous to residents in the nearby buildings within a 10 m radius.

The power in the side lobe is 10 dB down the main beam, so at 30 m distance no radiation hazard is expected whatsoever. Even at a 10 m distance, there is no possibility of radiation hazard due to side lobes as it will be only 0.0637 mW/cm^2 at 915 MHz; which is well below the safe limit (Fig. 14.8).

Recapitulation

EMI/EMC

- The RFI or EMI may affect an electrical/electronic system either due to electromagnetic induction or electromagnetic radiation emitted from an external source (*exciter*) or the system itself (*victim*).
- These days the issue of EMI and EMC is a serious one, as we live in a dense electronic world with an intolerable limit of RF electromagnetic pollution sometime and somewhere. There are numerous examples of EMI, ranging from the commonplace to the catastrophic one which prompted the EMI regulatory boards all over the world [International Special Committee on Radio Interference (CISPR) for European standards while Federal Communications Commission (FCC) for US standards] to impose strict EMC regulations to be complied with for any product to be marketed.
- EMI is the phenomenon that causes the problem of electromagnetic interference while EMC is the characteristic with which equipment should be designed so that they are not affected or disturbed by EMI and do not also generate EMI above a specified limit. In order to achieve its goals, EMC pursues two different kinds of issues—*emission* (primarily *radiative* at RF) and *susceptibility* (opposite of immunity). Interference, or noise mitigation and hence electromagnetic compatibility is achieved primarily by addressing both emissions and susceptibility issues—*quieting* the sources of interference and *hardening* the potential victims.

RF/Microwave shielding for EMI/EMC

- Mitigation of electromagnetic interference (EMI) and compliance with electromagnetic compatibility (EMC) at radio frequencies (RF) including microwave frequency is most commonly realised with *electromagnetic shielding* (RF shielding). RF/Microwave shield in practice is a metallic enclosure that prevents entry and exit of any RF signal from the enclosure thus making the issue of EMC to be realised successfully.
- The quality of a shield is normally characterised in terms of *shielding effectiveness*, which on an average is 50 to 60 dB to cope with most problems; however, shielding effectiveness above about 120 dB approaches the state of the art in this area.
- The electromagnetic phenomenon on the basis of which RF shielding works is the following. The incident electromagnetic field induces time varying current into the shield material and this current generates an additional field which has the tendency to cancel the incident field that has caused it thereby preventing the entry to (and by the same reason) escape of any EMI field from the 'shield'. The magnitude of the induced current is dependent on the rate of change of the incident field and therefore shields are more effective at higher frequencies. Presence of any apertures, seams and joints at any location of the shield would disturb the flow of induced currents and therefore reduce shielding effectiveness.
- The shielding effectiveness of the enclosure is defined for the electric/magnetic field, in dB, as:

$$SE = 20\log_{10}\left|\frac{\vec{E}_i}{\vec{E}_t}\right| = 20\log_{10}\left|\frac{\vec{H}_i}{\vec{H}_t}\right|$$

- In terms of the reflection, absorption and multiple reflections taking place, in general, within the shield material (SE) expressed in dB is given by:

$$SE = R_{\text{dB}} + A_{\text{dB}} + MR_{\text{dB}}$$

$$SE \cong 20\log_{10}\left|\frac{\eta_0}{4\eta_{\text{sm}}}\right| + 20\log_{10} e^{t/\delta} + 20\log_{10}\left|1 - e^{-2t/\delta}\right|$$

$[R_{\text{dB}}]$ $[A_{\text{dB}}]$ $[MR_{\text{dB}}]$

At higher RF, that is, at microwave frequencies when $t >> \delta$, we can neglect the contribution from *MR*.

- The reflection loss is dominant at lower RF while the absorption loss is dominant at higher RF. Multiple reflection loss is dominant for thin shield materials and may be safely neglected for thick shield (compared to the skin depth at the frequency of operation) materials.
- Magnetic material such as steel or mumetal makes a better magnetic field shield at lower frequencies compared to the good conductors. But shielding effectiveness of magnetic

materials decreases with increasing frequency as a result of the decreasing trend of permeability with frequency. Thus, as far as microwave frequencies are concerned, good conductors are the most effective as shielding material.

Issues affecting the shielding effectiveness and their mitigation

- At high frequencies, the intrinsic shielding effectiveness of the shield material may be of general concern but the leakage of electromagnetic energy via apertures caused by slots, holes for cable insertion, seams are of serious concern. Apertures have a more deleterious effect on the magnetic field leakage than on the electric field leakage. Thus, importance is given to methods for minimising the magnetic field leakage which eventually is adequate for minimising the electric field leakage too.
- The presence of an aperture on the surface of the shield forces the induced current to flow in a different path (perturbs its natural flow) when the induced field will not completely cancel the incident field and hence the effectiveness of the shielding will be reduced. The more the current is forced to detour, the greater will be the decrease in the effectiveness of shielding.
- Slot length is responsible for the amount of detour of the surface current and not its width (even a slit of the same length as that of a wider slot will have a similar effect). However, a group of small holes has much less detouring effect on the surface current than the slot and therefore it causes practically no leakage of RF signal even if the total area of the two apertures (slot and a large number of holes) is almost the same.
- The leakage due to slot/slit causing deterioration of shielding effectiveness may be calculated from:

$$SE_{ap(\text{slot})} = 20\log_{10}\left(\frac{\lambda}{2l}\right) = 20\log_{10}\left(\frac{0.15}{f_{\text{GHz}}l_{\text{meters}}}\right)$$

- For an honeycomb array commonly used for airflow/ventilation in the shield, the deterioration of shielding effectiveness can be calculated from:

$$SE_{ap(m)\text{eff}} = 20\log_{10}\left(\frac{0.15}{f_{\text{GHz}}l_{\text{meters}}\sqrt{n}}\right)$$

- Each hole of the honeycomb array can provide additional attenuation if it is shaped to form a waveguide penetrating inside the shield wall. Such waveguides operating below cut-off allow airflow into a shielded enclosure and at the same time prevent the propagation of electromagnetic signals having frequencies below the cut-off frequency of the guides into the shielded room thereby providing effective magnetic shielding.
- To mitigate the effect of EMI via cables, a common practice is to use a shielded cable. Some cables have two separate coaxial metal screens, one connected at both ends, the other at one end only, to maximise shielding of both RF electromagnetic and electrostatic fields. A bonded shield in which the outer conductor of the coaxial line is bonded to the body of the shield is used to improve shielding effectiveness.

EMI/EMC standards, regulations and testing

- The standards and regulations for EMI/EMC applicable to domestic, commercial and industrial sectors are set by two international bodies worldwide—the International Special Committee on Radio Interference (CISPR) for European standards and the Federal Communications Commission (FCC) for the US standards. The military standard (MIL-STD) is set by the US defence and followed worldwide.
- EMC testing is required to confirm that a particular device meets the required standards. It is broadly divided into susceptibility testing and emission testing.
- The RF testing of a prototype is most often carried out in a RF anechoic chamber.

Human exposure limit to electromagnetic radiation

- Exposure to strong electromagnetic radiation, especially at microwave frequencies, is definitely harmful to any biological system, needless to say, especially to the human body.
- Specific Absorption Rate (SAR, in W/kg) considering thermal effect is given by:

$$\text{SAR} = \frac{\sigma|E|^2}{2\rho}$$

For mobile phones, the SAR value limit for European standards is a maximum of 2 W/kg averaged over 10 g of human tissue while the US standard is 1.6 W/kg averaged over 1 g of human tissue. In India, as per the 2012 regulations, the US standard is adopted.

Review Questions

14.1 Why are we concerned about EMI and EMC? What technological trends have led to this serious awareness about EMI/EMC in modern times?

14.2 Mention a few causes of EMI in our day-to-day lives.

14.3 Discuss a few case studies of EMI since the 1940s caused by nuclear detonation, airport surveillance radar, electrostatic discharge and so on, resulting in the malfunction of electronic devices including digital computers.

14.4 Mention the two important issues that EMC follows to realise its goal and how those are achieved.

14.5 How is the mitigation of EMI and compliance with EMC realised at RF and microwave frequency?

14.6 Discuss how a metallic enclosure acts as an effective electromagnetic (RF) shield.

14.7 How is the quality of a shield characterised? What is its practical value?

14.8 Explain the physical principle of RF shielding. How can it be related to the commonly known Faraday cage? Discuss how RF shielding is realised in the design of a microwave oven.

14.9 Discuss the electromagnetic phenomena that take place between the interior and exterior of the left and right surfaces of the shield wall.

14.10 Define shielding effectiveness of an enclosure in dB both in terms of electric/magnetic fields and also in terms of reflection loss, absorption loss and multiple reflections taking place in the shield material.

14.11 Discuss how the reflection loss and absorption loss is dependent on frequency and how multiple reflections are dependent on shield thickness?

14.12 Explain why good conductors are the most effective as shielding materials at microwave frequencies.

14.13 Why are apertures caused by slots, holes for cable insertion and seams, of serious concern at high RF in realising good shielding effectiveness?

14.14 Explain how the presence of aperture in the shield surface causes reduction in shielding effectiveness.

14.15 Discuss how slot/slit on a shield surface acts as a radiating antenna thus causing RF leakage and hence reduction in shielding effectiveness. How can it be empirically estimated?

14.16 Discuss why a group of small holes have much less effect in causing the deterioration of the shielding effectiveness of a shield compared to a slot/slit.

14.17 What is a 'honeycomb array'? Where is it used in an RF shield? How is the deterioration in shielding effectiveness due to honeycomb array calculated?

14.18 Explain how each hole of a honeycomb array provides additional attenuation to higher order modes when forming a waveguide penetrating into the shield and operating below cut-off and hence improves the shielding effectiveness.

14.19 Discuss how penetration of a cable into the metal box of the shield is taken care of to improve shielding effectiveness.

14.20 Name the EMI/EMC regulatory authorities of Europe and the US.

14.21 For EMC compliance of ML-STD, both emission and susceptible issues are to be addressed—discuss.

14.22 Prototype testing for EMC compliance is carried out in an anechoic chamber—explain.

14.23 Why are we concerned about human exposure to electromagnetic radiation? What is SAR value? What are its specified maximum limits as per European and US regulations? What is the Indian standard?

15

An Emerging Topic of Microwave Engineering: Metamaterials/ Left-Handed Materials (LHM)

The aim of knowledge is to evolve and emerge into newer and newer dimensions

15.1 Introduction

Metamaterials are engineered composite media that exhibit counter-intuitive phenomena like reversal of Snell's law and hence negative refractive index (Fig. 15.1a), reversed Doppler effect (Fig. 15.1b) and reversed Cherenkov effect (Fig. 15.1c) not observed in nature. As they are capable of exhibiting phenomena not available in nature they are called '*meta*' (Greek term meaning beyond nature) *materials*. From a technical point of view, these are engineered metal–dielectric composites formed by embedding sub-wavelength metallic inclusions on a dielectric substrate. The counter-intuitive response exhibited by metamaterials is not due to the chemistry of the material from which it is made (metallic inclusions like copper/silver/ gold and dielectric material like FR-4/RT-duroid) but because of the structural arrangement of the constituent electric and magnetic atoms (the inclusions) artificially realised and arranged, in general, in a periodic structure mimicking the lattice arrangement of small-scale physics (atomic physics) in a natural material. Though metamaterial structures have external inhomogeneity, they appear homogeneous to an incident electromagnetic signal as they are designed with the condition of effective medium theory ($p << \lambda$, where p is the lattice period and λ is the operating wavelength). The constitutive parameters of metamaterials are those of the bulk material (of the structure as a whole) since the lattice period in a metamaterial is much smaller than the wavelength of the incident electromagnetic signal.

In 1968, V G Veselago [32] of Lebedev Physics Institute, USSR, investigated theoretically the consequences when both permittivity (ε) and permeability (μ) of a non-magnetic material is considered to be negative and observed the possibility of the *counter-intuitive*

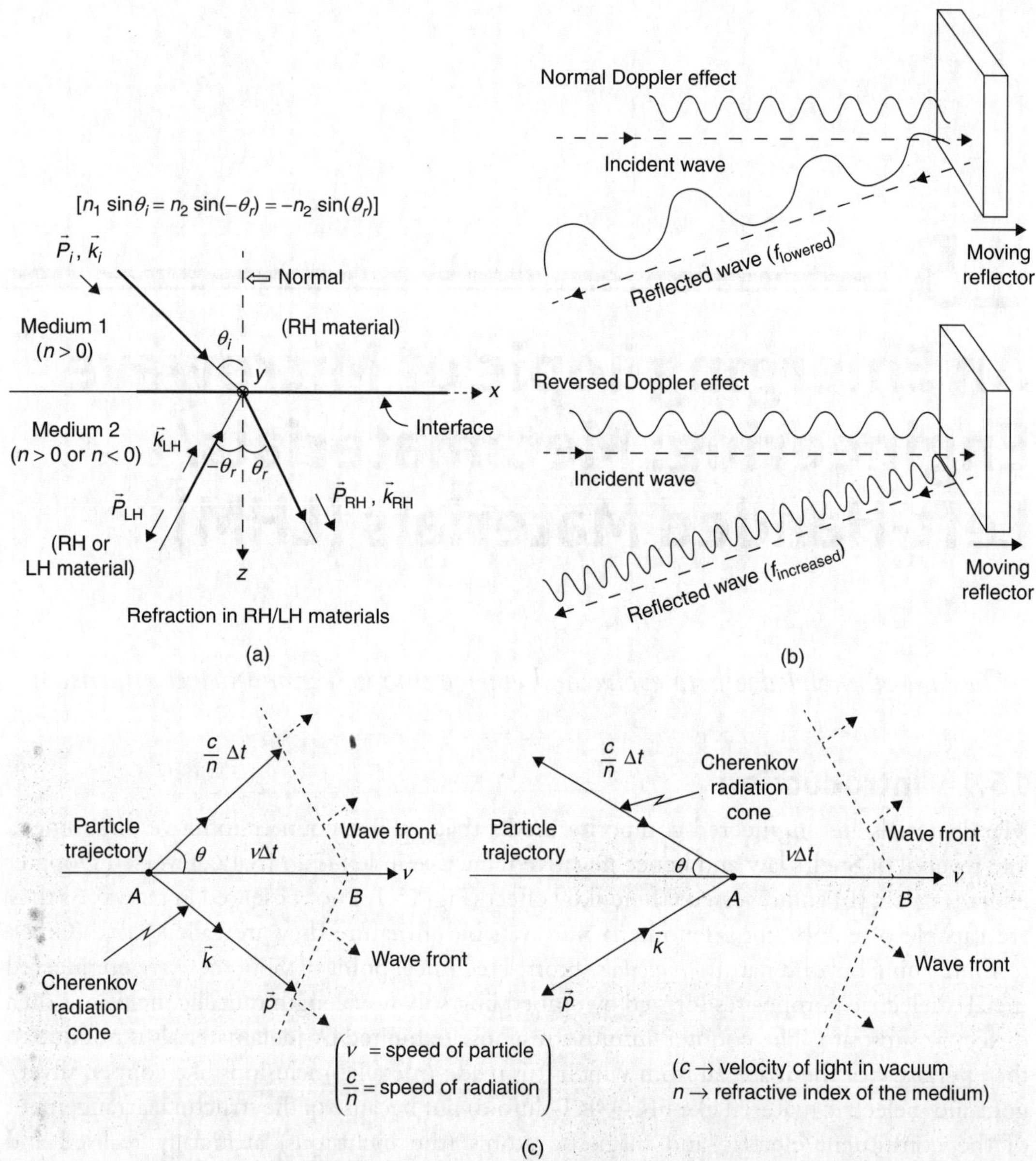

Fig. 15.1 Counter-intuitive effects in metamaterial (a) Reversal of Snell's law (b) Reversed Doppler effect (c) Reversed Cherenkov effect.

phenomena mentioned above with a double negative (DNG), that is, ε and μ are simultaneously negative, material. However, during that time and in the following two decades, no material was known to exhibit negative permittivity or negative permeability, except metal acting as plasma (due to the copious amounts of free electrons present) was known to provide negative permittivity below plasma frequency which is normally in the ultra-violet frequency region of the electromagnetic spectrum. Also it was known that the ionosphere in the earth's atmosphere exhibits negative permittivity below plasma frequency. In 1996, J B Pendry et al [33] of Imperial College, UK, observed theoretically that an array of thin wires (TW) made of metal is capable of exhibiting negative permittivity below plasma frequency in the microwave region. The realisation of negative permeability was still not in sight; however, it was again Pendry et al [34] in 1999 who showed that a split-ring resonator (SRR) can exhibit negative permeability below plasma frequency. In 2001, Shelby et al [35] of University of California at San Diego, US, first reported the experimental realisation of negative refractive index with a two-dimensional (2D) plasmonic metamaterial constituted of TW and SRR (Fig. 15.2a). The concept of periodically loaded transmission lines (PLTL) was introduced by Eleftheriades et al of University of Toronto, Canada in 2002 who reported the realisation of negative refractive index with PLTL [36] (Fig. 15.2b). The plasmonic structures are resonant structures and are thus inherently narrowband while the PLTL structures are non-resonant and are thus useful for designing low-loss and broadband microwave passive components and antennae. The first Indian metamaterial was successfully made in 2009 using labyrinth resonators and thin wire structures (Fig. 15.2c) [37].

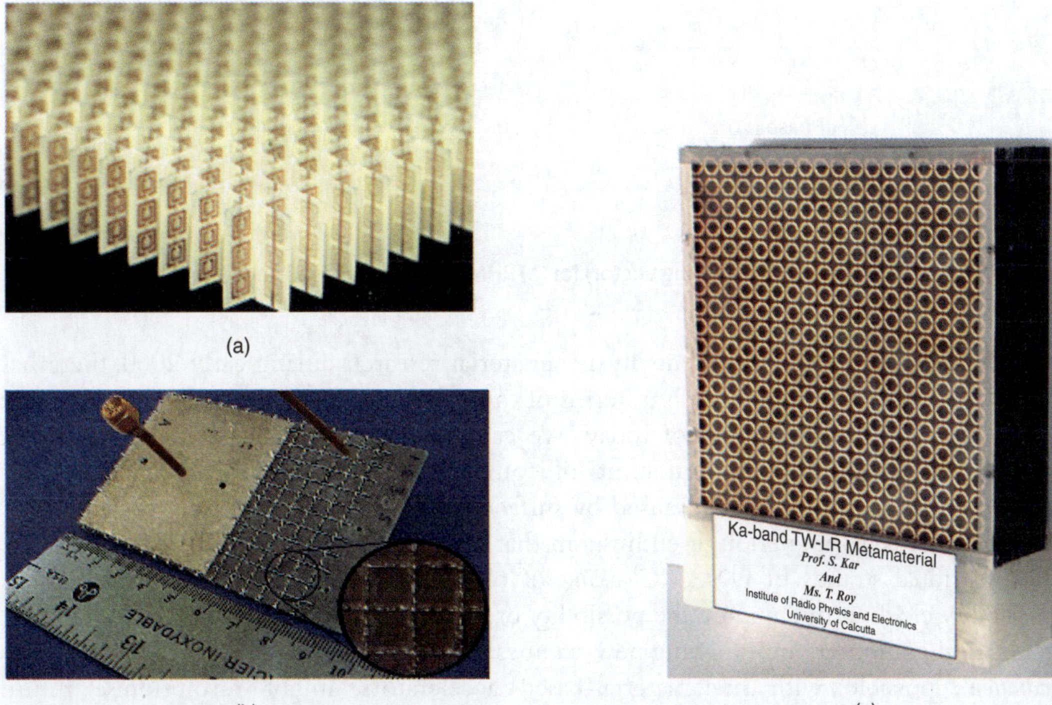

Fig. 15.2 (a) 2D Plasmonic metamaterial [35] (b) PLTL metamaterial [36] (c) First Indian metamaterial [37].

Veselago termed DNG materials or metamaterials as *Left Handed Materials* (LHM) as *E*, *H* and *k* vectors form a left-handed triad for such materials, unlike natural materials which form a right-handed triad with the same vectors (Fig. 15.3). In contrast to the metamaterials being termed as LHM, natural materials (having double positive, that is, ε and μ both being simultaneously positive) are also termed *Right Handed Materials* (RHM). It is interesting to note that in Maxwell's wave equation (Eq. 2.8), the refractive index n appears as a square power. Therefore, for both RHM and LHM Maxwell's wave equation is equally applicable irrespective of whether n is positive or negative. Further, as the direction of the Poynting vector remains the same irrespective of whether E, H and k vectors form a right-handed or left-handed triad, the causality is not violated by LHM. However, it may be noted that the phase and group velocities are counter-directed in the case of LHM and as the phase velocity is in the negative z-direction, the LHM is said to support *backward waves*.

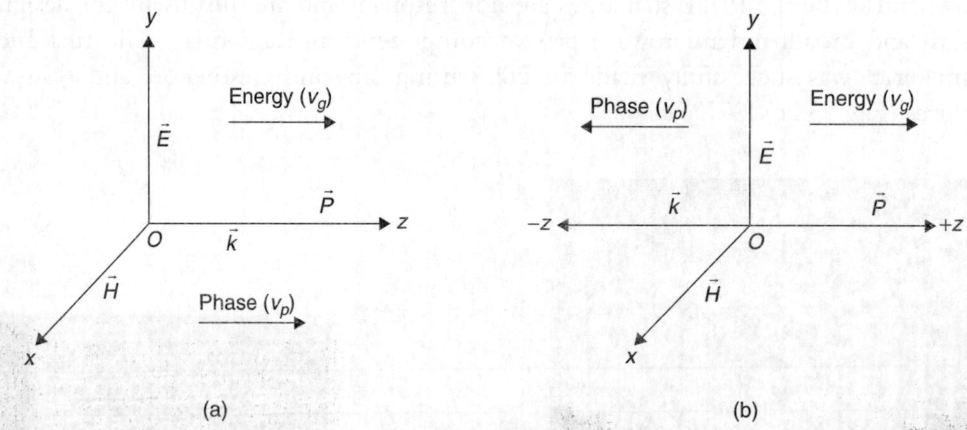

Fig. 15.3 $\vec{E}, \vec{H}, \vec{k}$ (triad) and $\vec{P}$ (Poynting vector) for (a) RHM (natural material) and (b) LHM (metamaterial).

Following the initial developments in metamaterial research during early 2000, there has been an exponential growth both in terms of the volume and variety leading to advanced metamaterial research avenues of today. We can now hope of the possibility of '*super-lens*' overcoming the 'diffraction limit' of conventional optics. '*Sub-wavelength imaging*' with 'evanescent wave growth' caused by surface plasmon waves/polaritons in plasmonic metamaterial is gaining enough enthusiasm that might one day make it possible to image the individual strands of DNA. '*Cloaking*' of objects (not as effective as Harry Potter's invisibility cloak) is opening up the possibility of making reliable optical memories for new generation computers and showing new avenues for stealth technology. '*Reversed Cherenkov radiation*', possible with metamaterial-based accelerators, might revolutionise future accelerator research, especially leading to much more sensitive detection of Cherenkov radiation. Apart from these exotic applications, active research is going on with new ideas emerging everyday for the development of metamaterial based/inspired microwave passive components and antennae with improved performance along with size miniaturisation. There is even talk of quasi-crystal and quantum metamaterials and many more fashionable yet promising terms concerning metamaterials.

Though most of the research and development activities in metamaterials initially took place in the microwave frequency range (because of the limitations of nano fabrication technology in those days), advanced developments are taking place these days in terahertz (THz) and optical frequency regime. In fact, the ultimate fruit of metamaterials will be most effective only when three-dimensional (3D) optical metamaterials are a commonplace reality.

It would be worthwhile discussing the difference between metamaterials (MTM) and photonic band-gap (PBG) materials or photonic crystals. Both of them are known to be capable of exhibiting negative refraction. MTM exhibit negative refractive index (originating from the constitutive parameters ε and μ of the material) while PBG materials exhibit negative refraction (not the negative refractive index), due to Bragg diffraction.

In terms of the lattice period (p) and operating wavelength (λ), $p \approx \lambda$ for PBG while for MTM, $p << \lambda$. Therefore, interference effects cannot take place in MTM as the phase difference between unit cells is negligible and thus electromagnetic waves travel through the material probing only the average (effective) constitutive parameters. The effective negative refractive index in MTM is thus due to effective negative permittivity and effective negative permeability of the bulk MTM structure. But PBG is diffraction limited and interference effects are dominant. Their operation is thus in the Bragg regime resembling Bragg diffraction in X-rays. The negative refraction in PBG is not due to the constitutive properties in the photonic crystal but due to band-edge Bragg diffraction.

PBG development has been followed by the development of EBG (electromagnetic band-gap) materials. PBG is the forerunner among photonic crystals in which the forbidden band-gap for light emissions were conceived. EBGs are new forms of electromagnetic structures like PBG; however, they are useful at microwave and radio frequencies. The band-gap in EBG is analogous to a forbidden energy gap in electronic crystals. DGS (defected ground structure) and FSS (frequency selective surfaces) are also used to reject certain frequency bands, and hence fall in the class of EBG. All these structures are used to manipulate the electromagnetic signal characteristics and have great innovative potential with applications in recent years translating into the design of novel filters, antennae and others, similar to metamaterials.

15.2 Counter-Intuitiveness of LHM

It has been mentioned in the introduction that LHM (or metamaterials) exhibit phase velocity (v_p) in the negative z-direction unlike the natural materials (RHM) in which case v_p is in the $+z$ direction; thus E, H and k vectors form a left-handed triad in the case of LHM and a right-handed triad in the case of RHM. However, the direction of group velocity (v_g) or the velocity of energy propagation (Poynting vector) remains in the $+z$ direction in either case as desired by causality. These facts of counter-intuitiveness of LHM can be understood as follows.

For plane TEM wave propagation in the z-direction, in a source-free medium (Fig. 2.2) we have electric and magnetic fields as:

$$\vec{E} = E_0 e^{j(\vec{k}_z \vec{r} - \omega t)}\,\hat{e} = \left|\vec{E}\right| \hat{e} \text{ and } \vec{H} = H_0 e^{j(\vec{k}_z \vec{r} - \omega t)}\,\hat{h} = \left|\vec{H}\right| \hat{h}$$

where $\hat{e}$ and $\hat{h}$ are unit vectors in the direction of the electric field and magnetic field while the unit vector in the direction of propagation is $\hat{k}$.

From the constitutive relation, $\vec{B} = \mu\vec{H} = \mu\frac{\sqrt{\varepsilon}}{\sqrt{\mu}}\left|\vec{E}\right|(\hat{k}\times\hat{e}) = \sqrt{\mu\varepsilon}\,(\hat{k}\times\left|\vec{E}\right|\hat{e})$, since $\frac{\left|\vec{E}\right|}{\left|\vec{H}\right|} = \sqrt{\frac{\mu}{\varepsilon}}$.

Thus, $\left|\vec{k}\right|\vec{B} = \frac{1}{c}\left|\vec{k}\right|(\hat{k}\times\vec{E}) = \frac{1}{c}(\vec{k}\times\vec{E})$.

Or, $c\left|\vec{k}\right|\mu\vec{H} = \vec{k}\times\vec{E}$, since $\vec{B} = \mu\vec{H}$

Or, $\frac{\omega}{\left|\vec{k}\right|}\left|\vec{k}\right|\mu\vec{H} = \vec{k}\times\vec{E}$, since $c = \frac{\omega}{\left|\vec{k}\right|}$

whence $\vec{k}\times\vec{E} = \omega\mu\vec{H}$ (15.1a)

Again from the constitutive relation, $\vec{D} = \varepsilon\vec{E} = \varepsilon\frac{\sqrt{\mu}}{\sqrt{\varepsilon}}\left|\vec{H}\right|[\hat{k}\times(-\hat{h})] = -\sqrt{\mu\varepsilon}\,(\hat{k}\times\left|\vec{H}\right|\hat{h})$

Thus, $\left|\vec{k}\right|\vec{D} = -\frac{1}{c}\left|\vec{k}\right|(\hat{k}\times\vec{H}) = -\frac{\left|\vec{k}\right|}{\omega}(\vec{k}\times\vec{H})$

Or, $-\frac{\omega}{\left|\vec{k}\right|}\left|\vec{k}\right|\varepsilon\vec{E} = \vec{k}\times\vec{H}$ since $\vec{D} = \varepsilon\vec{E}$

$$\vec{k}\times\vec{H} = -\omega\varepsilon\vec{E} \tag{15.1b}$$

This represents the familiar right-handed triad with $\vec{E}, \vec{H}, \vec{k}$; Fig. 15.3a. But in the case of LHM, ε, $\mu < 0$ and since $|\varepsilon| = -\varepsilon > 0$ and $|\mu| = -\mu > 0$; when from Eqs 15.1a and 15.11b, we get:

$$\vec{k}\times\vec{E} = -\omega|\mu|\vec{H} \text{ and } \vec{k}\times\vec{H} = +\omega|\varepsilon|\vec{E} \tag{15.1c}$$

which forms the unusual *left-handed triad* with $\vec{E}, \vec{H}, \vec{k}$ as shown in Fig. 15.3b.

The phase velocity in RHM/LHM can be written as:

$$v_p = \frac{\omega}{\left|\vec{k}\right|} = \frac{\omega}{k}\hat{k}, \text{ with } \hat{k} = \vec{k}/\left|\vec{k}\right| \tag{15.1d}$$

Since frequency is always positive, the phase velocity in LHM will be opposite to that in RHM as dictated by Eqs 15.1c and 15.1b. Further, as the wave number k in RHM is known to be in the positive direction (outward propagating from the source), the phase velocity given by Eq. 15.1d for LHM will be negative (inward propagation to the source), giving rise to the so called *backward wave*, and is thus opposite to that in RHM.

The phase velocity is negative ($v_p < 0$) in the case of LHM; this might appear to be disturbing at the first glance. However, phase velocity relates to the propagation of the perturbations in the medium and not the energy. In contrast, a negative group velocity ($v_g < 0$) would violate causality as this is the velocity with which the electromagnetic energy moves, and being negative would mean transfer of energy towards the source (which is physically absurd!). This is justified because unlike the wave vector $\vec{k}$, the Poynting vector that gives the direction of energy flow is independent of the constitutive parameters of the medium (ε and μ) and depends only on $\vec{E}$ and $\vec{H}$.

The boundary conditions to be satisfied at the interface between RH–LH media can be derived as follows. As shown in Fig. 15.4, medium 1 is RHM ($\varepsilon, \mu > 0$) and medium 2 is LHM ($\varepsilon, \mu < 0$).

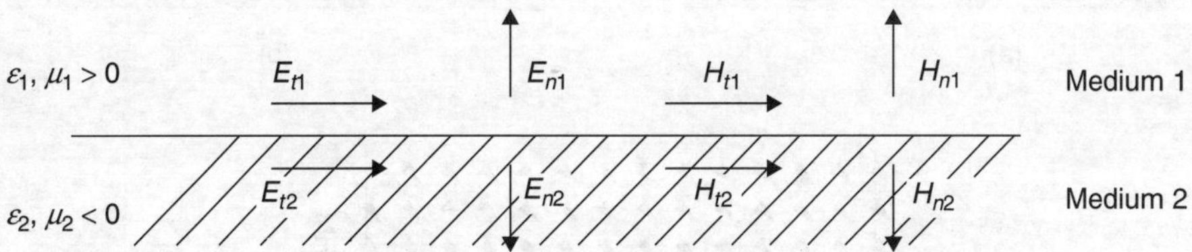

Fig. 15.4 Boundary conditions at the interface between RH and LH media.

From Section 2.7.1, for charge-free and source-free media, the electromagnetic boundary conditions to be satisfied by the tangential and normal components of electric and magnetic field at the interface between RH–RH media are given by:

$$E_{1t} = E_{2t}, H_{1t} = H_{2t}; D_{1n} — D_{2n} = 0, B_{1n} = B_{2n} \tag{15.2a}$$

The indices t and n indicate the tangential and normal components respectively. Looking at the relations of Eq. 15.2a, we may guess that the boundary conditions on the tangential component of the E and H fields would be unaffected at the RH–LH interface, since the conditions on the tangential components do not depend on ε and μ. However, the normal components necessarily depend on constitutive properties ε and μ (since $\vec{D} = \varepsilon\vec{E}$ and $\vec{B} = \mu\vec{H}$) which are negative for a LH medium. Thus the tangential components remain continuous in the RH–LH interface (as in Eq. 15.2a), but the normal components become antiparallel as shown in Fig. 15.4 and are given by:

$$E_{n\text{RHM}} = -\frac{\varepsilon_{\text{LHM}}}{|\varepsilon_{\text{RHM}}|} E_{n\text{LHM}}, H_{n\text{RHM}} = -\frac{\mu_{\text{LHM}}}{|\mu_{\text{RHM}}|} H_{n\text{LHM}} \tag{15.2b}$$

It would be worthwhile at this step to model the field vector behaviour and power flow direction at the LHM–RHM interface using 3D electromagnetic field simulator (Fig. 15.5) to verify the analytical results discussed so far with respect to the counter-intuitive behaviour of LHM yet not violating causality.

The consequence of the antiparallel nature of the normal component of the wave vector at the interface between the RH–LH media leads to negative refraction in LHM. The near field focussing by a plane-slab LH lens sandwiched between two RH media (Fig. 15.6) is again caused by negative refraction taking place in the RH–LH media. The detailed

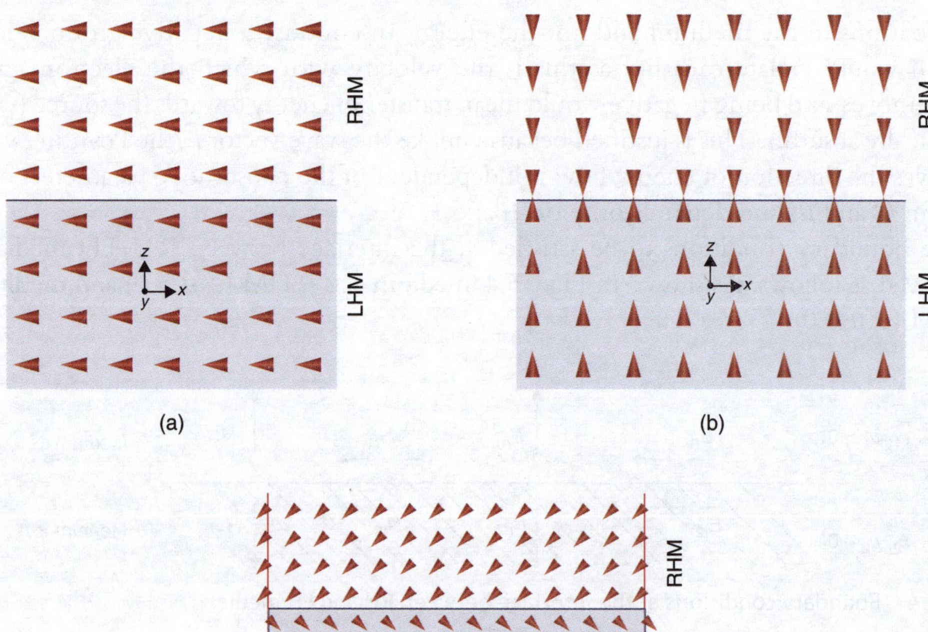

Fig. 15.5 Simulation results in a 3D electromagnetic field simulator for (a) Tangential component of the *E* field at the interface between RHM–LHM (b) Normal component of the *E* field at the interface between RHM–LHM (c) Poynting vector in LHM–RHM–LHM interface (with 45° angle of incidence).

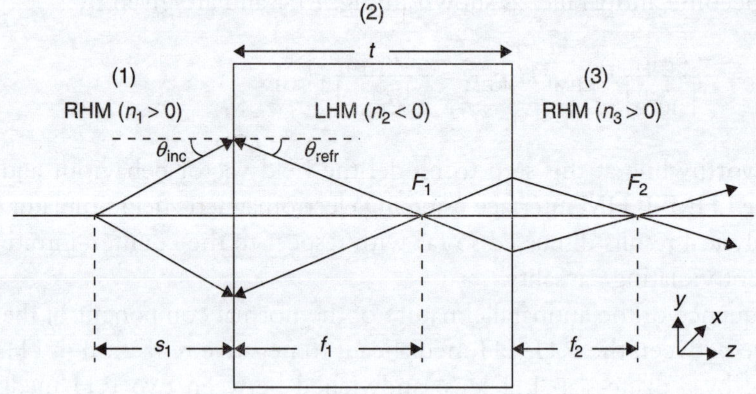

Fig. 15.6 Internal and external focussing using a LHM plane-slab.

analysis for these and also that for the reversal of Doppler effect and reversed Cherenkov radiation in LH media are available in suitable references [38].

15.3 Plasmonic and Transmission Line Metamaterials

15.3.1 Plasmonic Metamaterials

Plasmonic metamaterials exhibiting negative refractive index are realisable with an array of metallic thin-wire (TW) [or sometimes with cut-wire (CW), the reason for which will be discussed shortly] together with a matrix of Split-Ring Resonator (SRR); both printed on dielectric substrate and arranged in an one-dimensional (1D) periodic structure (may be 2D/3D structures too) and excited by an electromagnetic field (Fig. 15.7).

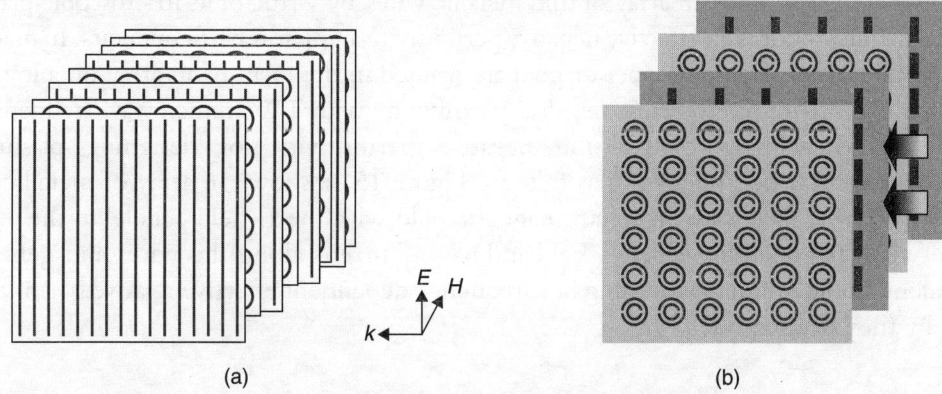

Fig. 15.7 ID plasmonic metamaterial (a) With TW-SRR (b) With SRR-CW.

In such an arrangement each unit cell of TW/CW and SRR acts respectively as an 'electric atom' and a 'magnetic atom' mimicking the atomic arrangement in the lattice of the natural material. The realisation of effective negative permittivity $\varepsilon_{\text{reff}}$ (with a TW/CW array) and effective negative permeability μ_{reff} (with SRR matrix) is a bulk property and thus the effective negative refractive index n_{reff} of the plasmonic metamaterial is given by:

$$n_{\text{reff}} = \sqrt{\varepsilon_{\text{reff}}\mu_{\text{reff}}} \tag{15.3}$$

The physical background and analytical basis for obtaining negative permittivity with TW/CW and negative permeability with SRR at microwave frequency is given below.

15.3.1.1 Negative permittivity with thin-wire (TW) and cut-wire (CW)

Essentially, metals are plasma, since they consist of an ionised 'gas' of free electrons. Below plasma frequency, the real component of the permittivity of bulk metal is negative. However, the natural plasma frequencies of metals normally occur in the ultra-violet region of the electromagnetic spectrum, in which wavelengths are extremely short and hence fabrication complexity and cost is exorbitant. Although the permittivity is negative at frequencies below the plasma frequency, the approach towards absorptive resonance at lower frequencies increases the dissipation leading to the complex nature of the permittivity. To realise the plasma frequency in the microwave region, Pendry et al [33] solved the problem from a novel standpoint. They used an array of thin metallic wires (Fig. 15.8a) which exhibits negative permittivity below plasma frequency in the microwave frequency range. Spatial confinement of the electrons to thin wires decreases the effective electron concentration ($n_{eff} = n\pi r^2/a^2$) in the volume of the structure thus decreasing the plasma frequency according to the equation:

$$\omega_p^2 = n_{eff} e^2/\varepsilon_0 m_{eff} \tag{15.4}$$

Also the self-inductance of the wire-array manifests itself as a greatly enhanced effective mass of the electrons confined to the wires reducing the effective plasma frequency down to GHz (microwave) region. Thus an array of thin metallic wires, by virtue of its macroscopic plasma-like behaviour, produces an effective negative permittivity at microwave frequencies. In practice, thin metallic wires made of copper or gold are printed in the form of an array on a low-loss dielectric like RT-duroid to realise negative permittivity with TW at microwave frequencies.

The analytical basis of realising negative permittivity for designing plasmonic metamaterials may be understood as follows. Figure 15.8a shows an array of metallic thin wires (TW) irradiated with an electromagnetic field with the *E*-field parallel to the length of the wire. The exciting electric field induces a current along the wires and generates equivalent dipole moments resulting in a frequency dependent effective relative permittivity given by the Drude model:

$$\varepsilon_{reff}(\omega) = 1 - \frac{\omega_{ep}^2}{\omega^2 + j\omega\xi_e} \tag{15.5a}$$

The frequency dependent effective permittivity represented by Eq. 15.5a, exhibits non-resonant high-pass characteristics and the real part of the permittivity is negative for $\omega < \omega_{ep}$ (Fig. 15.8b). However, sometimes from a designer's point of view, a resonant characteristic for the frequency dependent effective relative permittivity is desirable. The negative permittivity with resonant characteristics (Fig. 15.8c) can be realised with cut-wire (CW) array (Fig. 15.8d), this may be looked upon as a variant of the continuous thin wire (TW) originally proposed by Pendry et al [33]. Here, the resonant characteristic arises due to the inductance represented by each length of the cut wire and capacitance due to the gap between the cut wires. The effective relative permittivity of CW is given by the Lorentzian model [39]:

$$\varepsilon_{reff} = 1 - \frac{\omega_{ep}^2 - \omega_{e0}^2}{\omega^2 - \omega_{e0}^2 + j\omega\xi_e} \tag{15.5b}$$

In Eqs 15.5a and 15.55b, $\omega_{ep} = 2\pi f_{ep}; f_{ep}$ is the electric plasma frequency, the upper bound for negative permittivity, $\omega_{e0} = 2\pi f_{e0}; f_{e0}$ is the electric resonant frequency, the lower bound for negative permittivity and ξ_e is the damping factor due to metal loss imparted by the thin wire (or cut wire) metal imprints given by $\xi_e = \varepsilon_0(h\omega_{ep}/a)^2/\pi\sigma$, where h is the horizontal

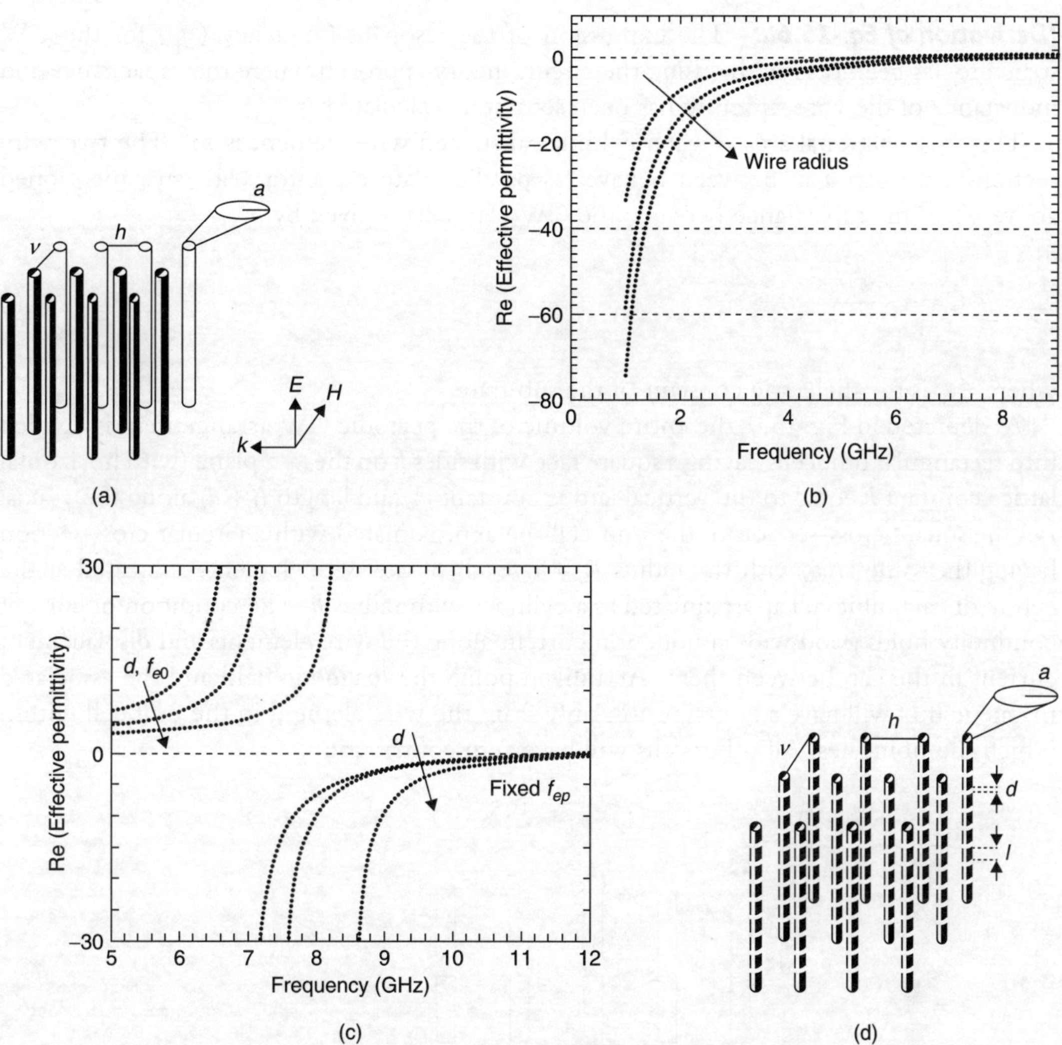

Fig. 15.8 (a) Array of thin metallic wires (TW) (b) Frequency dependent effective permittivity of TW (c) Frequency dependent effective permittivity of cut-wire (CW) (d) Array of CW.

lattice constant, that is, the centre-to-centre spacing between two adjacent wire sections, a is the radius of each wire and σ is the conductivity of the metal.

The generic expression for f_{e0} and f_{ep} are given by:

$$f_{e0} = \frac{c_0}{\pi a}\sqrt{\frac{(N_\gamma - 1)d}{N_\gamma (l + d)\ln(h/a)}} \tag{15.6a}$$

and $$f_{ep} = \frac{c_0}{2\pi h}\sqrt{\frac{2X\pi}{\ln(h/a)}} \tag{15.6b}$$

For TW, $d = 0$, $X = 1$, and for CW: $d \neq 0$, $X = 2$.

[Derivation of Eq. 15.6a: The expression of the resonant frequency (f_{e0}) for the CW structure has been derived by using the circuit theory approach where the capacitance and inductance of the wire structure has been separately calculated.

The cross sectional area of each cylindrical shaped wire element is πr^2. The two wire sections with gap d in between behaves as parallel plate capacitor with area mentioned above. Thus, the capacitance between the CW elements is given by:

$$C = \varepsilon_{\text{sub}} \frac{\varepsilon_0 \pi r^2}{d} \tag{i}$$

where, ε_{sub} is the dielectric constant of the substrate.

As depicted in Fig. 15.9, the entire volume of the periodic CW arrangement is divided into rectangular unit cells having a square face with sides h on the x–z plane (with horizontal lattice constant h equal to the vertical lattice constant v), and length $(l + d)$ along the y-axis. Let the square cross-section of the unit cells be approximated with a circular cross-section having the same area, with the radius $R_c \approx h$. Each of the wire elements is located at the centre of each unit cell approximated as a cylinder with radius h. The condition of current continuity holds good with conduction current along the wire elements and displacement current in the gap between them. At a given point, the magnetic field and the associated magnetic flux will have a contribution only from the wire element in the unit cell within which the point lies. All other cells will have zero contribution.

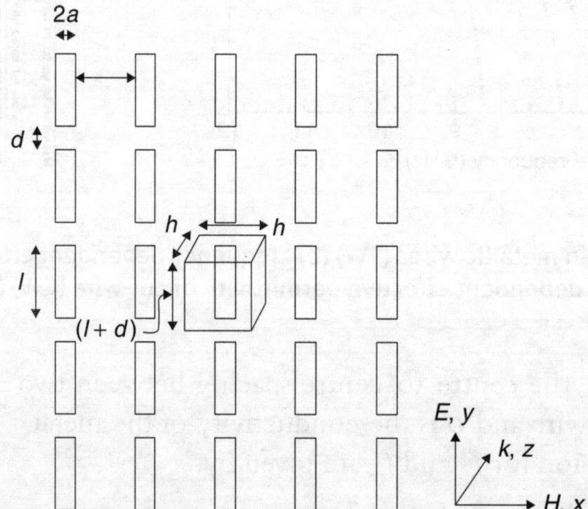

Fig. 15.9 CW array with unit cell indicated with dimensions labelled.

The magnetic field for the wire element of length l can be written as:

$$B = \frac{\mu_0}{4\pi} \frac{Il}{R(l^2 + r^2)^{1/2}} \tag{ii}$$

where, R is the perpendicular distance from the wire element. The associated total magnetic flux is thus given by:

$$\phi = \frac{\mu_0}{4\pi} I\,(l+d)\left\{\log\left|\tan\frac{1}{2}\left(\tan^{-1}\frac{h}{l}\right)\right| - \log\left|\tan\frac{1}{2}\left(\tan^{-1}\frac{r}{l}\right)\right|\right\} \quad \text{(iii)}$$

Since, $\phi = LI$, the inductance of each wire element is given by:

$$L = \frac{\mu_0}{4\pi}(l+d)\left\{\log\left|\tan\frac{1}{2}\left(\tan^{-1}\frac{h}{l}\right)\right| - \log\left|\tan\frac{1}{2}\left(\tan^{-1}\frac{r}{l}\right)\right|\right\} \quad \text{(iv)}$$

Let us consider that there are N_x, N_y and N_z unit cells in the x, y and z directions respectively. The capacitors and inductors of a line along the y-direction are in series with each other whereas each of those in the x and z-directions are in parallel. Thus the total impedance is:

$$Z_{\text{total}} = N_x N_z\left(jN_y\omega L + \frac{N_y - 1}{j\omega C}\right) \quad \text{(v)}$$

By equating the imaginary part of the total impedance to zero, we get the expression for the electric resonant frequency as:

$$f_{e0} = \frac{1}{2\pi}\sqrt{\frac{4c^2(N_y-1)d}{a^2\varepsilon_{\text{sub}}N_y(l+d)\left\{\ln\left|\tan 1/2(\tan^{-1} h/l)\right| - \ln\left|\tan 1/2(\tan^{-1} a/l)\right|\right\}}} \quad \text{(vi)}$$

where c is the velocity of light in free space.

Since for effective design of negative permittivity structures, the inequality $a, h \ll l$ must be obeyed, the expression for electric resonant frequency may be approximated as:

$$f_{e0} = \frac{c_0}{\pi a}\sqrt{\frac{(N_y-1)d}{\varepsilon_{\text{sub}}N_y(l+d)\ln(h/a)}} \quad \text{(vii)}$$

From the above result we find that when $d = 0$, that is, the case of a continuous wire array, the resonant frequency vanishes. This is true because continuous TW structures are indeed non-resonant in characteristics.

Derivation of Eq. 15.6b: The magnetic field around the wire element, as calculated in Eq. (ii) can also be expressed in terms of the magnetic vector potential since $\vec{B} = \nabla \times \vec{A}$. Hence the magnetic vector potential may be written as:

$$A(r) = \frac{\mu_0}{4\pi}\pi a^2 nev_m\left\{\ln\left|\tan\frac{1}{2}\left(\tan^{-1}\frac{h}{l}\right)\right| - \ln\left|\tan\frac{1}{2}\left(\tan^{-1}\frac{a}{l}\right)\right|\right\} \quad \text{(viii)}$$

where v_m is the mean electron velocity.

From classical mechanics, electrons in a magnetic field have an additional contribution to their momentum equal to $eA(r)$. Therefore, the momentum per unit length of the wire is given by:

$$m_{\text{eff}}\,\pi a^2 nev_m = \pi a^2 ne\frac{\mu_0}{4\pi}\,\pi a^2 nev_m\left\{\ln\left|\tan\frac{1}{2}(\tan^{-1} h/l)\right| - \ln\left|\tan\frac{1}{2}(\tan^{-1} a/l)\right|\right\} \quad \text{(ix)}$$

Thus, the effective mass of the electrons is derived as:

$$m_{\text{eff}} = \frac{\mu_0}{4\pi}\,\pi a^2 ne^2\left\{\ln\left|\tan\frac{1}{2}\left(\tan^{-1}\frac{h}{l}\right)\right| - \ln\left|\tan\frac{1}{2}\left(\tan^{-1}\frac{a}{l}\right)\right|\right\} \quad \text{(x)}$$

From plasma theory we have the plasma frequency given by:

$$f_{ep} = \frac{1}{2\pi}\sqrt{\frac{n_{\text{eff}}e^2}{\varepsilon_0 m_{\text{eff}}}} = \frac{1}{2\pi}\sqrt{\frac{4\pi c^2}{a^2\left\{\ln\left|\tan\frac{1}{2}\left(\tan^{-1}\frac{h}{l}\right)\right| - \ln\left|\tan\frac{1}{2}\left(\tan^{-1}\frac{a}{l}\right)\right|\right\}}} \quad \text{(xi)}$$

Thus, with the approximation a, $h << l$, as was used in the case of effective negative permittivity, the above expression for electric plasma frequency may be written as:

$$f_{ep} = \frac{c_0}{2\pi h}\sqrt{\frac{4\pi}{\ln(h/a)}} \quad \text{(xii)]}$$

15.3.1.2 Negative permeability with split-ring resonator (SRR)

Magnetic properties from non-magnetic materials was the primary challenge to realise the so called 'magnetic atoms' which Pendry et al [34] solved for the first time in 1999. In naturally available magnetic materials, magnetism results from the force created by the circular motion of electrons in the atoms. Pendry envisioned that if circular paths of conductor carrying current (flow of electron) are arranged in a 3D periodic array, then the arrangement would mimic naturally available magnetism at an atomic level. They proposed a coupled split-ring structure (Fig. 15.10a) which is basically an L–C tuner. It can tune to the incident magnetic field of the electromagnetic wave similar to the receiving antenna of a radio and acts as a magnetic resonator. An array of split-ring resonators thus manifests the magnetic property depending on the designed structural parameters, which dictates the field induced current flow and hence the magnetic force generated. A controllable artificial magnetic property can thus be realised with a periodic array of minuscule metallic SRRs printed on a dielectric substrate (Fig. 15.10b) which collectively behaves as a magnetic plasma (analogous to the electric plasma of a metallic TW array printed on a dielectric substrate, Fig. 15.10c) and exhibits a negative permeability below the magnetic plasma frequency (Fig. 15.10d). The magnetic resonant frequency (ω_{m0}) and magnetic plasma frequency (ω_{mp}) have been indicated in Fig. 15.10e that shows the plot of μ_{reff} vs ω for split-ring resonator.

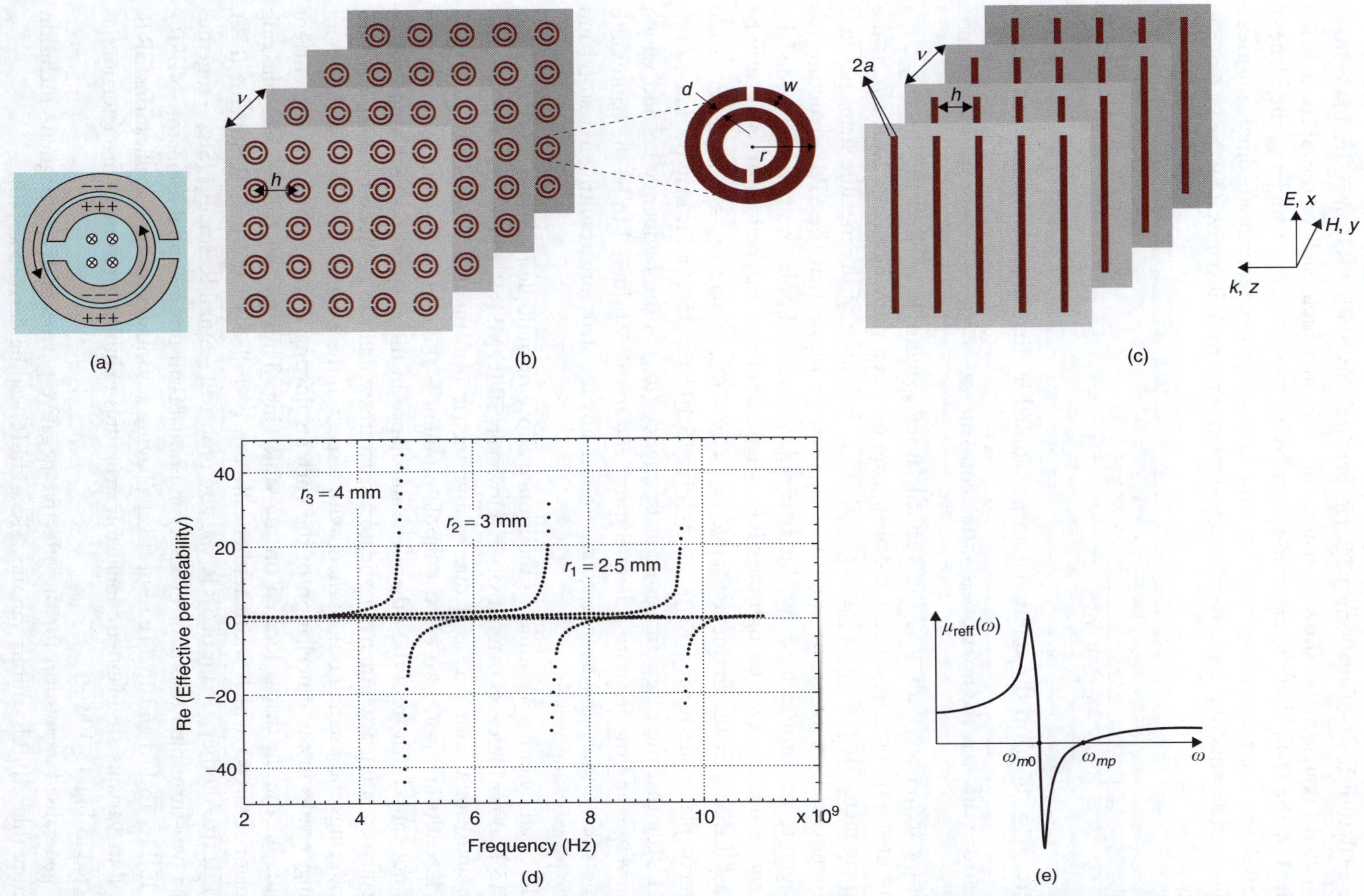

Fig. 15.10 (a) Coupled split-ring resonator (SRR) (b) 1D SRR array (c) 1D TW array (d) Resonant characteristics of SRR (e) SRR characteristics showing ω_{mo} and ω_{mp}.

The analytical basis of realising negative permeability with SRR may be understood as follows. Figure 15.10b shows a matrix of SRR in which each of the SRR is being excited by the *H*-field of the electromagnetic wave applied at right angles to the plane of the SRR to induce resonating currents in the loop and generate magnetic dipole moments resulting in a frequency dependent effective relative permeability given by the Lorentzian model [34]:

$$\mu_{\text{reff}} = 1 - \frac{F\omega^2}{\omega^2 - \omega_{m0}^2 + j\omega\xi_m} \tag{15.7a}$$

where, $F\ (= \pi r^2/h^2)$ is the fractional area occupied by the rings in the unit cell, that is, fill-factor with h as the horizontal lattice constant; $\omega_{m0} = c\sqrt{\dfrac{3v}{\pi \ln(2wr^3/d)}}$ is the magnetic resonant frequency with v as the vertical lattice spacing, w is width of the rings, r is radius of the outer ring, d is the radial spacing between the rings; $\xi_m = 2v\,R_c/r\mu_0$ is the damping factor due to metal losses and R_c is the resistance of the metallic strips per unit circumferential length of the ring. It is interesting to note that the SRR structure has a magnetic response caused by the artificial magnetic dipole moments provided by the ring resonators even though the SRRs are made of non-magnetic material. It may be observed that over the frequency range $\omega_{m0} < \omega < \omega_{mp}$, where $\omega_{mp} = \omega_{m0}/\sqrt{1-F^2}$ is the magnetic plasma frequency, $Re(\mu_{\text{reff}}) > 0$ (Figs 15.10d and 15.10e). It may be noted from Fig. 15.10e that the magnetic resonant frequency (ω_{m0}) is the frequency beyond which μ_{reff} diverges and the separation between ω_{m0} and ω_{mp} is considered to be the bandwidth of operation for negative permeability exhibited by the structure, resulting in a stop band to the propagating wave.

SRR can be either a couplet as mentioned above or multiple inclusion SRR (MSRR, Fig. 15.11a) for increased magnetisation. However, SRR/MSRR suffers from the problem of bi-anisotropy. This is caused due to asymmetric placement of slits in the two rings which results in a non-zero electric dipole moment as the current flowing across one slit does not cancel that from the other. The imbalance in the currents on both rings can be remedied by using a symmetric resonator structure, which can be done by using two slits in both rings and rotating one of them with respect to the other by 90°. Such a placement of rings results in the cancellation of the currents flowing across each slit by the current flowing on the slit that is located on the same ring. Thus we get the double-split-ring resonator (DSRR) which more aesthetically can be called a Labyrinth Resonator (LR), Fig. 15.11b. Another variant of SRR was also proposed – Spiral Resonator (SR) – together with a two-turn spiral resonator (TTSR) and non-bi-anisotropic spiral resonator (NBSR), Figs 15.11c, 15.11d and 15.11e respectively; having advantages in some applications. SRR and all its variants are magnetic inclusion structures exhibiting resonant characteristics as depicted in Fig. 15.10e.

A generalised expression for the effective negative permeability of magnetic inclusion structure (like MSRR and its variants, SR and LR) can be derived as [40]:

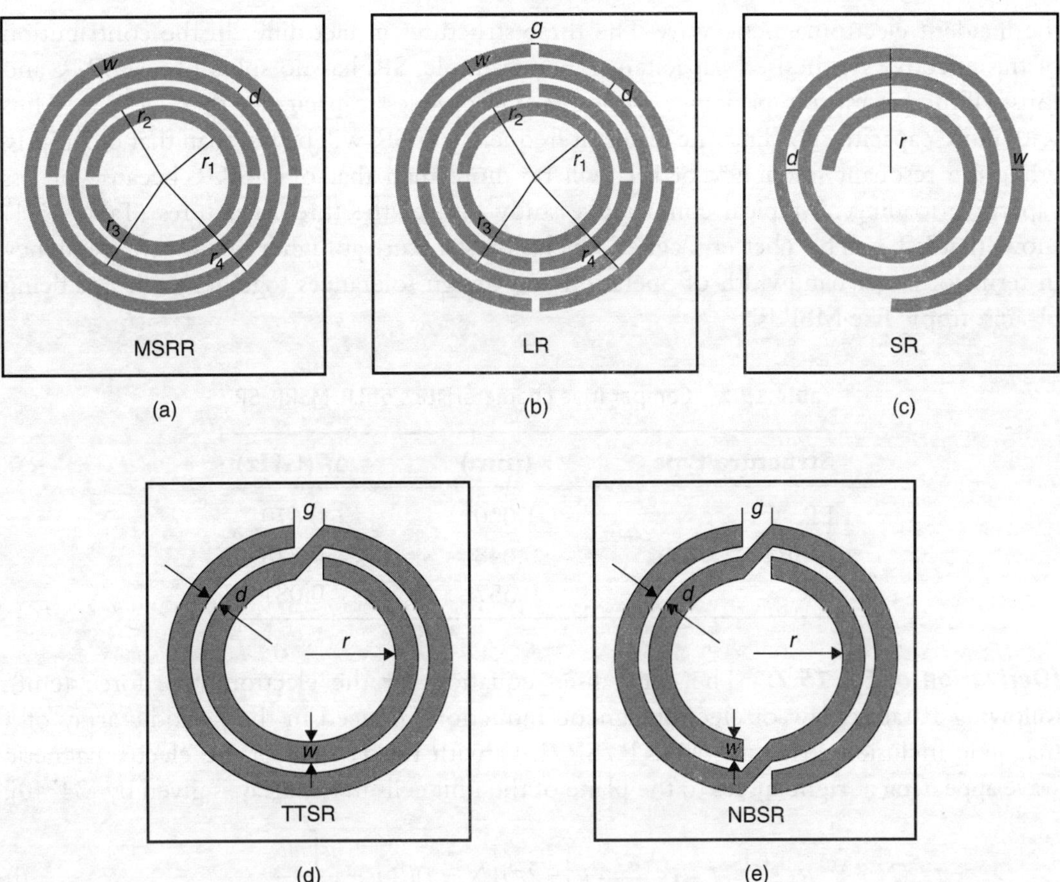

Fig. 15.11 Variants of SRR.

$$\mu_{\text{reff}} = 1 - \frac{F}{1 + j\dfrac{2R_c \nu}{\omega r \mu_0 (N-1)} - \dfrac{\nu}{\pi r^2 \mu_0 \omega^2 \left(C_1 \dfrac{2\pi r}{M} + \dfrac{N}{2} C_2 \right)(N-1)^2}} \tag{15.7b}$$

where, F $(= \pi r^2/h^2)$ is the fill-factor with h as the horizontal lattice constant, R_c is the resistance of the metallic strips per unit circumferential length of the ring and N is the number of metallic inclusions. The constant $M = 6$ (for MSRR), 1 (for SR) and 24 for LR. $C_1 = (\varepsilon_0/\pi)\ln[wN/d(N-1)]$ is the distributed capacitance between the strips of the adjacent rings per unit circumferential length of the ring with d being the separation between individual rings and $C_2 = \varepsilon_0 w t_M/g$ being the gap capacitance between the cuts along a ring, t_M the thickness of the metal strip of the ring and g the cut of the gap; for MSRR and SR, this capacitance can be neglected.

In all the magnetic inclusion structures SRR, SR, LR the magnetic resonant frequency and hence the magnetic plasma frequency are controlled by the distributed capacitance caused by simultaneous currents flowing between adjacent rings induced by the H-field of the incident electromagnetic wave. The three structures in fact differ in the contribution of this effective synthesised capacitance. For example, SR has no splits like MSRR and thus will posses more capacitance while LR will have less capacitance than MSRR. Thus with more capacitive loading, the resonant frequency of SR will be less than that of MSRR while the resonant frequency of LR will be more than that of MSRR because of less capacitive loading. A typical comparative study [40] of the three structures (Table 15.1) shows that LR will be a better magnetic inclusion structure at higher microwave frequency in terms of larger bandwidth of operation and design tolerances together with not being bi-anisotropic like MSRR.

Table 15.1 Comparative characteristics of LR, MSRR, SR

Structure type	r (mm)	Δf (GHz)
LR	1.000	0.670
MSRR	0.648	0.278
SR	0.357	0.081

[Derivation of Eq. 15.7: The generalised equation for the electromotive force (emf), following Faraday's law of electromagnetic induction, induced in the periodic array of a magnetic inclusion structure (MSRR/SR/LR) with the H-field of the electromagnetic wave appearing at right angles to the plane of the unit cells in the array is given by [34, 40]:

$$-\pi r^2 (N-1)^2 \mu_0 \frac{\partial}{\partial t}\left(H + i - \frac{\pi r^2}{h^2} i \right) - 2\pi r (N-1) i R_c v + \frac{v \int i dt}{C} \quad \text{(i)}$$

where i is the induced RF current flowing in each ring caused by the magnetic field H of the incident electromagnetic wave, R_c is the resistance of a metallic ring per unit

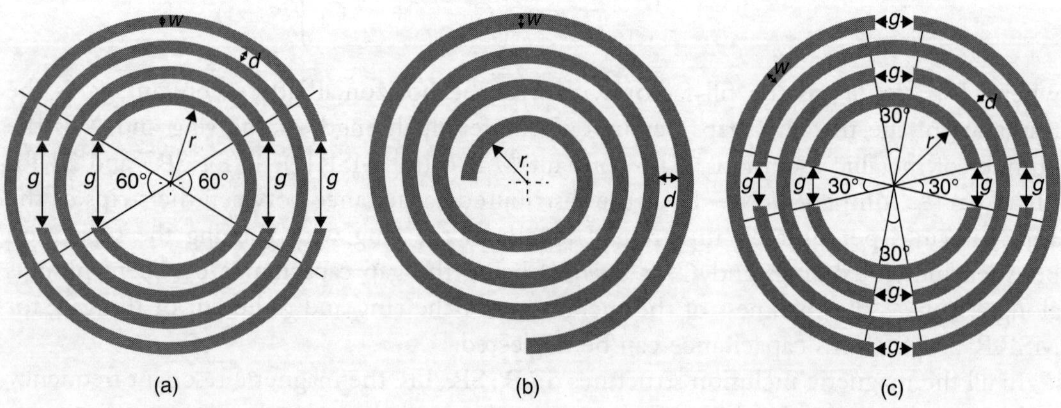

Fig. 15.12 Schematic representation of (a) MSRR (b) SR and (c) LR unit cell structure for deriving Eq. 15.7.

circumferential length of the ring, C is the distributed capacitance between the strips of the adjacent rings, N is the number of inclusions. The expression for capacitance C in case of MSRR, SR and LR (Fig. 15.12) is tabulated in Table 15.2:

Table 15.2 The capacitance C in the case of MSRR, SR and LR

MSRR	SR	LR
$\frac{2\pi r}{6}\frac{\varepsilon_0}{\pi}\ln\left[\frac{wN}{d(N-1)}\right]$	$2\pi r\frac{\varepsilon_0}{\pi}\ln\left[\frac{wN}{d(N-1)}\right]$	$\frac{2\pi r}{24}\frac{\varepsilon_0}{\pi}\ln\left[\frac{wN}{d(N-1)}\right]+\frac{N}{2}\frac{\varepsilon_0 w t_M}{g}$

The circumferential length effective in each structure determines the multiplying factor to the capacitance between two rings while for LR an extra capacitance comes from the two gaps in each ring where t_M is the thickness of the metal strip of the ring and g is the cut of the gap.

Since the intensity of the time varying magnetic field incident on the structure is constant, the magnetic flux through the metallic rings does not change resulting in zero induced emf and thus from Eq. (i):

$$j\omega\pi r^2(N-1)^2\mu_0\left(H+i-\frac{\pi r^2}{h^2}i\right)-2\pi r(N-1)iR_c\nu-\frac{j\nu i}{\omega C}=0 \tag{ii}$$

where we have used the fact that $\frac{\partial}{\partial t}=-j\omega$ and $\int dt=-\frac{j}{\omega}$; as the incident electromagnetic wave is time-harmonic in nature ($e^{-j\omega t}$).

From Eq. (ii) we can solve for the induced current as:

$$i=\frac{-j\omega\pi r^2(N-1)^2\mu_0 H}{j\omega\pi r^2(N-1)^2\mu_0\left(1-\frac{\pi r^2}{a^2}\right)-2\pi r(N-1)R_c\nu-\frac{j\nu}{\omega}}$$

$$=\frac{-H}{\left(1-\frac{\pi r^2}{a^2}\right)+j\frac{2(N-1)R_c\nu}{\omega r\mu_0}-\frac{\nu}{\omega^2\pi r^2(N-1)^2\mu_0 C}} \tag{iii}$$

Since, the effective relative permeability μ_{reff} in terms of the average induced magnetic field strength H_{av} and the average induced magnetic flux density B_{av} over the unit cell volume can be expressed as [34]: $\mu_{\text{reff}}=B_{av}/\mu_0 H_{av}=H/H_{av}$, with $H_{av}=H-Fi$, where Fi is the depolarising field; we have using Eq. (iii):

$$\mu_{\text{reff}}=1-\frac{F}{1+j\frac{2(N-1)R_c\nu}{\omega r\mu_0}-\frac{\nu}{\omega^2\pi r^2(N-1)^2\mu_0 C}} \tag{iv}$$

When the values of the capacitance C from Table 15.2 are substituted in Eq. (iv) above, we get Eqs 15.7b and 15.7a.]

15.3.2 Transmission Line Metamaterials

Recognising the analogy between the LH waves possible with the dual of the normal transmission line and similar backward wave already known to exist in periodic structures (since the 1940s), Eleftheriades et al [36] proposed in 2002, an alternative way to realise the LHM property using transmission lines. The basic difference is that the periodic structures have a periodicity (p) of the order of half guide wavelength (λ_g) ($p \approx \lambda_g/2$) while for metamaterials $p << \lambda_g$, when the effective (average) macroscopic constitutive parameters ε and μ come into play. In fact, as long as $p < \lambda_g/4$, the difference between a LHM (metamaterial) at microwave frequency and a RHM (natural material) at optical frequency is just quantitative. But qualitatively, the same refractive phenomena occur in LHM as in RHM. The practical implementation of transmission line metamaterials is done by periodically loading (keeping periodicity much smaller than the wavelength of operation) a host transmission line with series capacitance and shunt inductance (Fig. 15.13), which is thus a periodically loaded transmission line (PLTL) synthesising an LH media or LHM transmission line.

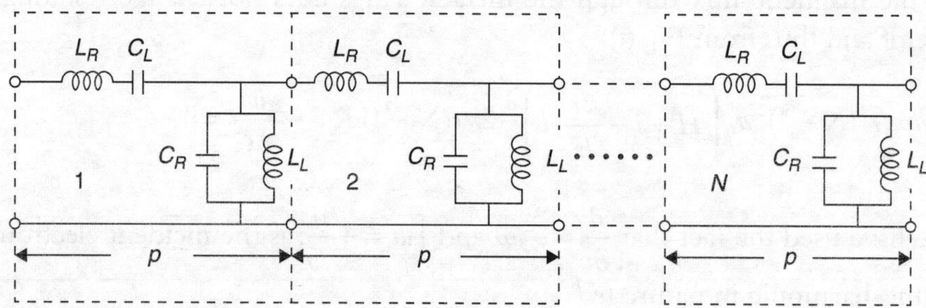

Fig. 15.13 Periodically loaded host transmission line.

Analysis of the LHM transmission line can be done by using the telegrapher's Eqs 3.3 and 3.4, as was done for RHM or normal transmission line, however, remembering that LHM is the dual of the RHM transmission line. Table 15.3 shows the equivalent circuit along with various parameters of two types of transmission lines which indicates that the propagation

Table 15.3 Equivalent circuit with various parameters of RHM and LHM

Parameters	β	Z_C	V_p	V_g	n
RHM L/2 L/2 C	$\omega\sqrt{LC}$	$\sqrt{\frac{L}{C}}$	$\frac{1}{\sqrt{LC}}$	$\frac{1}{\sqrt{LC}}$	$\frac{\sqrt{LC}}{\sqrt{\mu_0\varepsilon_0}}$
LHM C′/2 C′/2 L′	$-\frac{1}{\omega\sqrt{L'C'}}$	$\sqrt{\frac{L'}{C'}}$	$-\omega^2\sqrt{L'C'}$	$+\omega^2\sqrt{L'C'}$	$-\frac{1}{\omega^2\sqrt{L'C'\mu_0\varepsilon_0}}$

constant (β), phase velocity (v_p) and refractive index (n) for LHM line are negative while the group velocity (the velocity with which the energy of the electromagnetic wave travels) is positive; as expected to maintain non-violation of causality.

Being non-resonant in nature, PLTL exhibit simultaneously low loss and broad bandwidth and they can be engineered in a planar configuration thus inherently supporting 2D wave propagation and also compatible with modern MICs. This class of LH material is thus well suited for RF and microwave circuit applications.

Physically realisable planar LH structures (with series inter-digital capacitor and shunt stub inductor, Fig. 15.14a) also exhibit RH series L (due to magnetic flux induced by currents in the fingers of the inter-digital capacitors) and shunt C (due to the voltage gradient that exists between the conductor of the stub and the ground plane) which makes it equivalent to a *composite right/left-handed* (CRLH) transmission line (Fig. 15.14b). Important characteristics of CRLH transmission line metamaterial and resonator designed with CRLH will be discussed now [38].

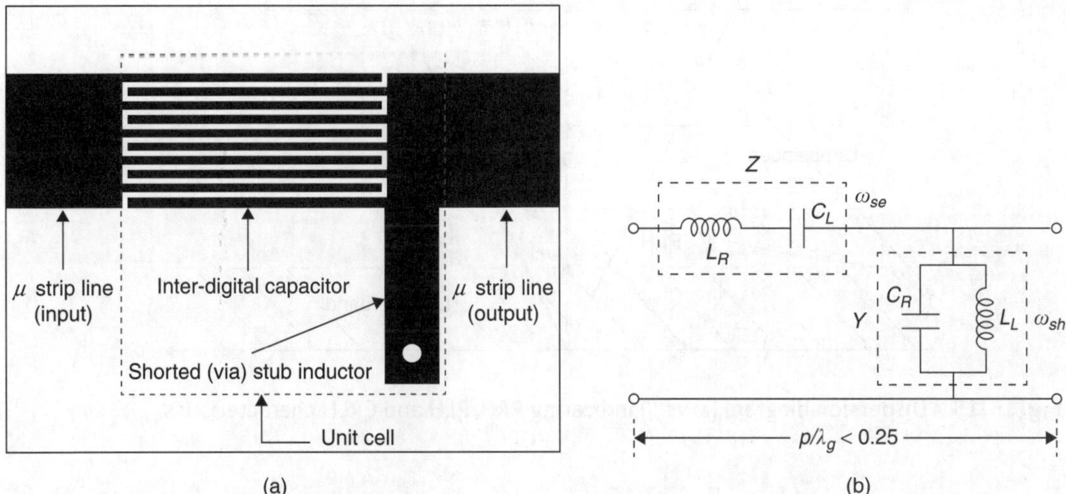

Fig. 15.14 (a) Physically realisable planar (microstrip) LH structure (b) Equivalent circuit of unit cell of CRLH Transmission line.

15.3.2.1 CRLH transmission line characteristics and ZOR

At low frequencies, the host transmission line (RHM) circuit elements L_R and C_R behave practically as short and open respectively, and hence the CRLH behaviour is then determined with series C_L and shunt L_L (pure left-handed, PLH) exhibiting a *high-pass* nature; with a LH stop-band below a certain cut-off frequency (ω_{cL}) (ω vs β, Fig. 15.15). At high frequencies, C_L and L_L behave respectively as short and open so that the equivalent CRLH is then expected to exhibit a *low-pass* nature (pure right-handed, PRH) with series L_R and shunt C_R; having a RH stop-band above a certain cut-off frequency (ω_{cR}). In general, when the series resonance frequency ($\omega_{se} = \sqrt{L_R C_L}$) and shunt resonance frequency ($\omega_{sh} = \sqrt{L_L C_R}$) are different, the line is called an *unbalanced* CRLH line. However, if these two frequencies are made equal,

$\omega_{se} = \omega_{sh} = \omega_0$, we call it a *balanced* CRLH line, when $L_R C_L = L_L C_R$, equivalently meaning $Z_L = Z_R$ (where Z_L and Z_R are respectively the characteristic impedance of PLH and PRH line, as Z_0 of the high frequency loss-less line is $\sqrt{L/C}$). Under this condition, the gap in the ω vs β diagram disappears and at the *transition frequency* ($\omega_0 = \sqrt{\omega_{se}\omega_{sh}}$) an *infinite wavelength* ($\lambda_p = 2\pi/|\beta|$) is achieved (as $\beta = 0$ at this frequency) implying that the balanced condition allows matching over an infinite bandwidth. Thus from a practical point of view, the balanced CRLH transmission line is preferred as it can be matched over a broad frequency band whereas the unbalanced line can be matched only over a restricted bandwidth (strictly at one single frequency).

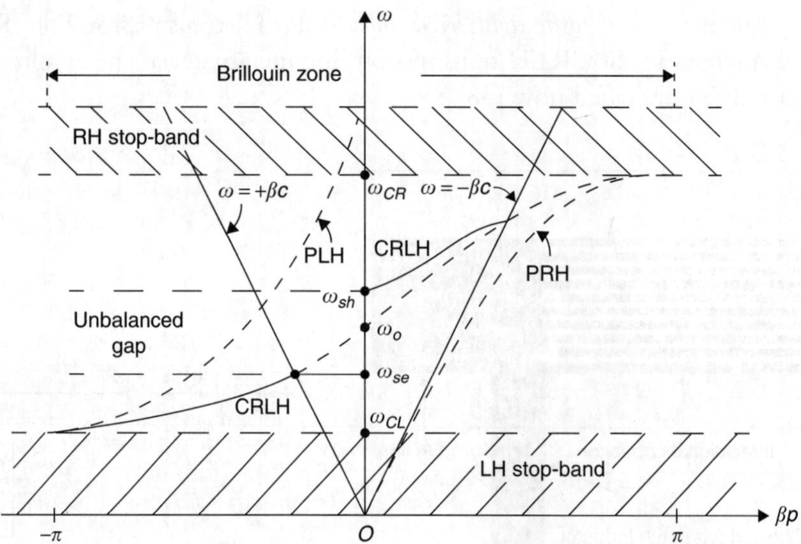

Fig. 15.15 Dispersion diagram (ω vs β) indicating PRH, PLH and CRLH characteristics.

It is interesting to note from Fig. 15.15 that at ω_0, the phase shift along a line of length l is zero [$\phi = -(\beta)l = 0$, as $\beta = 0$]. When frequency is decreased below ω_0 towards zero, β goes on taking a higher and higher negative value making the phase shift along the line increase with a positive value [$\phi = -(-\beta)l$] tending to infinity as $\omega \to 0$. But when frequency is increased above ω_0 towards infinity, β goes on taking a higher and higher positive value making the phase shift along the line to increase with negative value [$\phi = -(\beta)l$] tending to infinity as $\omega \to \infty$.

Further, the CRLH structure is never operated at the edges of Brillouin zone where $p \approx \lambda_g/2$ (though RH and LH stop bands exist there). Instead, it is used only in the vicinity of the transition frequency where $p < \lambda_p/4$, ensuring effective material property, where p is the average unit cells size and λ_p is the wavelength of the propagating electromagnetic signal in the composite CRLH medium.

Depending upon the nature of termination, the CRLH will support a travelling wave or a standing wave. If it is terminated with a matched load, it will naturally support a travelling wave but if it is terminated with either open or short it will support a standing wave, when it can be used as a *resonator*. Though a PRH transmission line also exhibits resonator characteristics with similar terminations, we observe that the CRLH transmission line resonator has some unusual characteristics. It is said to allow negative and zeroth-order resonances. Thus it has the potential to design a dual-band and also size-independent resonator (in contrast, a normal resonator length should be at least half-wavelength or any integer multiple of it). This makes transmission line type planar LH material very attractive for the design of miniaturised microwave circuit components and antennae.

In a conventional distributed circuit resonator (realisable with transmission line or waveguide), the resonance frequencies ω_m correspond to the frequencies where the physical length l of the structure is a multiple of half-wavelength (which means that the electrical length $\theta = \beta l$ is a multiple of π),

$$l_m = m\frac{\lambda}{2} \text{ or } \theta_m = \beta l_m = m\pi, \text{ with } m = +1, +2, \ldots, +\infty \quad (15.8a)$$

It may be observed from Fig. 15.16a that the dispersion curve $\omega(\beta)_{\text{PRH}}$ is linear and the resonance frequencies are obtained by sampling the dispersion curve with a sampling rate of π/l in the π variable. An infinite number of resonances are possible with conventional (PRH) transmission line since such a line ideally has infinite bandwidth, extending from $\omega = 0$ to $\omega = \infty$. As the electrical length of the PRH transmission line can be only positive, we can have only non-zero and positive resonances (Fig. 15.16b, which shows the field distribution of the resonance modes, that is, the voltage distribution for the case of an open circuited transmission line) with a conventional (PRH) transmission line resonator. Because the dispersion curve is linear, the higher order resonances (ω_m) are just the harmonics of the fundamental resonance (ω_1), that is, $\omega_m = m\,\omega_1$; where m is the harmonic number.

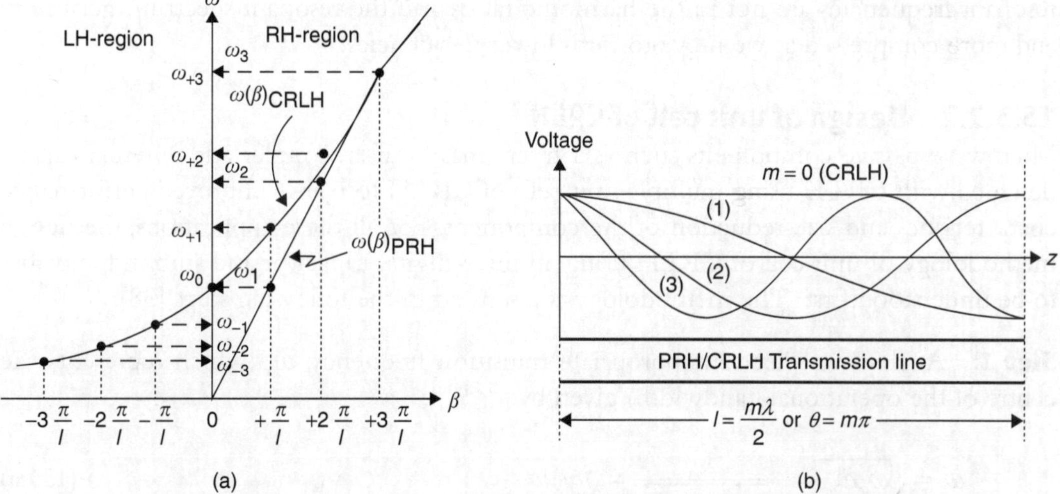

Fig. 15.16 (a) Dispersion curves of conventional (PRH) and CRLH transmission line resonators (b) Field distribution of the resonance modes: (1) $m = 1$ (PRH), ± 1 (CRLH); (2) $m = 2$ (PRH), ± 2 (CRLH); (3) $m = 3$ (PRH), ± 3 (CRLH).

The dispersion curve $\omega(\beta)_{\text{CRLH}}$ for a CRLH resonator based on balanced CRLH transmission is also shown in Fig. 15.16a with typical field distributions of the resonance modes in Fig. 15.16b. The propagation constant (β) of a balanced CRLH line can be expressed as the sum of the propagation constant of a PRH transmission line (with linear and positive ω vs β) and of a PLH transmission line (having negative and hyperbolic ω vs β characteristics). It may be observed from Fig. 15.16a that CRLH transmission line can have $\beta = 0$ (at transition frequency ω_0) as well as $\beta < 1$ (in the LH region), which is not the case with the PRH or a conventional transmission line. Thus, the electrical length $\theta = \beta l$ can be zero and negative, which implies that the resonance index m takes values $0, +1, +2, \ldots, +\infty$ with:

$$l_m = |m|\frac{\lambda}{2} \text{ or } \theta_m = \beta l_m = m\pi \tag{15.8b}$$

Here also resonance frequencies are obtained by sampling the dispersion curve with a sampling rate of π/l in the β variable and an infinite number of resonances are possible with an ideal CRLH resonator. But the CRLH resonator has a number of significant differences when compared with the PRH resonator. Firstly, *negative resonances* (at $m < 0$) and *zero resonance* (at $m = 0$) exist in the CRLH resonator in addition to the usual positive resonances (at $m > 0$) observed for a PRH resonator. The mode with $m = 0$ is particularly interesting as this mode corresponds to a flat-field distribution (with no voltage gradient along the length of the CRLH line, Fig. 15.16b) and therefore is expected to be independent of the physical length of the transmission line. This implies on theoretical grounds, that it can be used to realise an arbitrarily small resonator, popularly known as *zeroth-order resonator* (ZOR). Secondly, the field distribution of all modes with $m = +1, +2, \ldots$ would have identical impedance (as they have the same voltage distribution, Fig. 15.16b), which may be useful to design a dual-band CRLH filter and antennae. Thirdly, the non-linear nature of the dispersion curve of CRLH transmission line indicates that, especially in the LH region, the resonant frequencies are not in the harmonic ratios and the resonant spectrum gets more and more compressed as we move towards lower frequencies.

15.3.2.2 Design of unit cell of CRLH

Microwave passive components such as a filter, phase shifter, coupler and antenna can be designed with CRLH using multiple unit cells of CRLH to realise improved performance characteristics and size reduction of the component. For all these applications, the design methodology of unit cell of CRLH using an inter-digital capacitor and stub inductor has to be understood first. The methodology is discussed in the following steps [38].

Step 1: At first we select an appropriate transition frequency, ω_0 (which represents the centre of the operational bandwidth) given by:

$$\omega_0 = \sqrt{\omega_{se}\omega_{sh}} = \frac{1}{\sqrt[4]{L_R C_R L_L C_L}} = \sqrt{\omega_R \omega_L} \tag{15.9a}$$

where $\omega_R = \sqrt{L_R C_R}$ and $\omega_L = \sqrt{L_L C_L}$

Step 2: Matching condition is applied to the ports of the CRLH structure making:

$$Z_c = Z_R = \sqrt{\frac{L_R}{C_R}} \tag{15.9bi}$$

$$Z_c = Z_L = \sqrt{\frac{L_L}{C_L}} \tag{15.9bii}$$

where Z_c is the characteristic impedance of the CRLH transmission line network. It may be observed that we have so far three equations given in 15.9a, 15.9b(i) and 15.9b(ii) to determine four unknowns L_R, C_R, L_L and C_L.

Step 3: The fourth equation required to determine all the four unknowns mentioned above may be the bandwidth (a designer's constraint) which extends from high-pass LH cut-off ω_{cL} to low-pass RH cut-off ω_{cR}, and in the balanced case is given by:

$$\omega_{cL} = \omega_R \left|1 - \sqrt{1 + \frac{\omega_L}{\omega_R}}\right| \text{ and } \omega_{cR} = \omega_R \left(1 + \sqrt{1 + \frac{\omega_L}{\omega_R}}\right) \tag{15.9ci}$$

It may be noted that the cut-offs do not depend on the number of cells but only on the value of LC parameters. The fractional bandwidth may be written as:

$$\text{FBW} = 2\frac{(\omega_{cR} - \omega_{cL})}{(\omega_{cL} + \omega_{cR})} \tag{15.9cii}$$

An appropriate number of cells N need to be specified to realise a desired performance from the component to be designed.

Step 4: The network can now be implemented by practically realising the inductances L_R, L_L and capacitances C_R, C_L with suitable technology. For instance, in microstrip technology, inductors can be implemented in the form of spiral inductor strips or simply stub strips, while capacitors may be implemented by metal–insulator–metal (MIM) structures or with inter-digital structure (Fig. 15.14a).

Step 5: The characteristic parameters of the effectively homogeneous CRLH transmission line network are finally determined by the relations:

$$L'_R = \frac{L_R}{p}, C'_R = \frac{C_R}{p}, L'_L = L_L \cdot p, C'_L = C_L \cdot p \tag{15.9d}$$

where p represents the physical length of the unit cell (this is required as L_R, C_R, L_L, C_L were all defined in terms of the quantities designated in the equivalent circuit of the unit cell of the CRLH transmission line, Fig. 15.14b).

A typical microstrip based CRLH transmission line of 9 unit cells is shown in Fig. 15.17 in which each unit cell consisting of *inter-digital capacitors* and *stub inductors* shorted to ground plane by *via* connection (Fig. 15.14a).

The unit cells may be symmetric or asymmetric type as shown in Fig. 15.18; where W_{IDC}, l_{IDC}, g_{IDC}, l_{stub} and W_{stub}, are the finger width of inter-digital lines, finger length of inter-digital lines, gap between the fingers, length of the grounded stub and the width of the grounded stub respectively.

Fig. 15.17 A microstrip based 9-cell CRLH transmission line using inter-digital capacitors and shorted stub inductors.

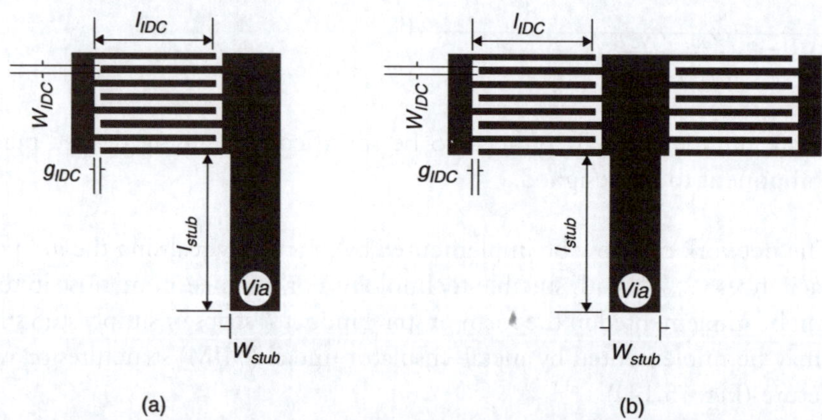

Fig. 15.18 Unit cells of CRLH transmission line (a) Asymmetric type (b) Symmetric type.

The equivalent circuit of a symmetric type unit cell of CRLH transmission line is shown in Fig. 15.19. This is a T-network having two impedance branches with capacitance $2C_L$ and inductance $L_R/2$ and an admittance branch with inductance L_L and capacitance C_R. The contributions L_L and C_L are due to the stub inductor and inter-digital capacitor, and L_R and C_R arise due to parasitic reactances caused by the magnetic flux generated by the currents flowing along the fingers of the inter-digital capacitor and the voltage gradient that exist between the fingers of the said capacitor with the ground plane. The parasitic reactances have an increasing effect with increasing frequency.

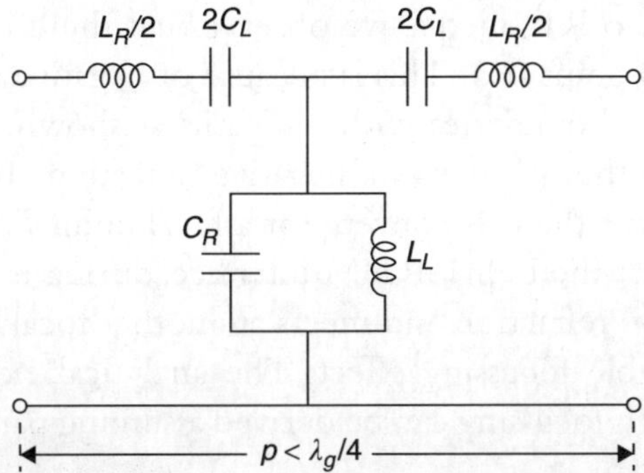

Fig. 15.19 *T-equivalent circuit of symmetric type unit cell of CRLH transmission line.*

Approximate formulae for calculating the inductance L_L and capacitance C_L may be obtained as follows. The shorted stub can be realised with a shorted transmission line section whose input impedance is an inductive reactance (Eq. 3.32a) when:

$$L_L \approx \frac{R_0}{\omega}\tan(\beta l) \tag{15.9e}$$

where R_0 is the characteristic impedance of the host transmission line, β the propagation constant and l is the length of the stub.

The capacitance of the inter-digital capacitor may be calculated from (Eq. 6.6g):

$$C = (\varepsilon_r + 1)l\{(N - 3)A_1 + A_2\} \text{ pF} \tag{15.9f}$$

with

$$A_1 = 4.41\tanh\left\{0.55\left(\frac{h}{w}\right)^{0.45}\right\}10^{-6} \text{ pF/}\mu\text{m}$$

$$A_2 = 9.92\tanh\left\{0.52\left(\frac{h}{w}\right)^{0.5}\right\}10^{-6} \text{ pF/}\mu\text{m} \tag{15.9g}$$

where l, w and h represent the finger length (usually $\leq \lambda/4$), finger width and the height of the substrate respectively, and N is the number of fingers of the inter-digital capacitor.

The approximate formulae to calculate L_L and C_L are just the starting points for the design. However, for optimisation full-wave simulation techniques are used which are discussed in advanced texts [38]; a brief discussion of which is given in Appendix IV.

15.4 Properties of Electromagnetic Signal Propagation through LHM

To evaluate the properties of electromagnetic signal propagation via LH media, we need to discuss a number of issues from different points of view. From Fig. 15.6 where an LH media is sandwiched between two RH media, we observe that there is an internal focus point (F_1) and an external focus point (F_2). This is because of negative refraction taking place in the LHM–RHM interface boundaries with ray paths as shown. It may be observed that the refracted wave within the slab makes a negative refracting angle (θ_{refr}) obeying Snell's law and instead of diverging, the rays converge at a focal point F_1 within the metamaterial (LHM) plane-slab. Again at the LHM–RHM interface, during emergence from the LHM slab the ray suffers negative refraction and meets at another focal point F_2 outside the slab. Hence there exists a double focussing effect. The analytical expressions relating all the relevant parameters for such focussing can be derived assuming paraxial approximation [36]:

$$f_1 = s_1 \left|\frac{n_2}{n_1}\right| \frac{\cos\theta_{refr}}{\cos\theta_{inc}},\ f_2 = s_1\left(\frac{t}{f_1} - 1\right), t \geq -\frac{n_2}{n_1}s_1 \tag{15.10}$$

The Veselago–Pendry lens discussed above is true for near field focussing with a plane metamaterial slab under ideal conditions, $\varepsilon_{RHM} = \mu_{RHM} = 1$ and $\varepsilon_{LHM} = \mu_{LHM} = -1$, that is, $n_{RHM} = +1$ and $n_{LHM} = -1$ and LHM has to be a loss-less medium. Such a plane-slab lens does not possess a focal length in the sense of that in optics. Further, it does not focus radiation from distant sources as we require in many applications like cameras, telescopes and antennae. However, in that case we would require negative index lenses with curved surfaces. It is worth mentioning that with LH material with a negative refractive index (a concave lens) will act as a converging lens while a convex lens will act as a diverging lens (counter-intuitive when compared to conventional optics) as depicted in Fig. 15.20.

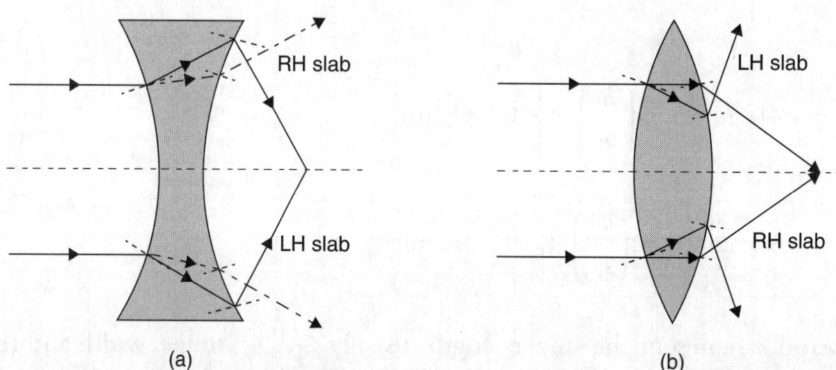

Fig. 15.20 Convergence and divergence of incident rays by concave and convex type LHM lens.

To investigate the transmission properties of electromagnetic signals via multiple interfaces (Fig. 15.6) representing RHM–LHM (1, 2) and LHM–RHM (2, 3) interfaces, we first evaluate the Fresnel transmission and reflection coefficients at the interface boundaries between two general loss-less media for *s*-polarised (perpendicular polarisation) incident waves (which follows from Eqs 2.38a and 2.38b) and for *p*-polarised (parallel polarisation) incident wave (which follows from Eqs 2.50a and 2.50b) and write [38]:

For *s*-polarised waves:

$$t_{12} = \frac{2\mu_{r2}k_1}{\mu_{r2}k_1 + \mu_{r1}k_2},\ t_{23} = \frac{2\mu_{r1}k_2}{\mu_{r2}k_1 + \mu_{r1}k_2} \text{ and } r_{12} = \frac{\mu_{r2}k_1 - \mu_{r1}k_2}{\mu_{r2}k_1 + \mu_{r1}k_2} = -r_{23} \tag{15.11}$$

For *p*-polarised waves:

$$t_{12} = \frac{2(\varepsilon_{r1}\varepsilon_{r2}\mu_{r2}/\mu_{r1})k_1}{\varepsilon_{r2}k_1 + \varepsilon_{r1}k_2},\ t_{23} = \frac{2\varepsilon_{r1}k_2}{\varepsilon_{r2}k_1 + \varepsilon_{r1}k_2} \text{ and } r_{12} = \frac{\varepsilon_{r2}k_1 - \varepsilon_{r1}k_2}{\varepsilon_{r2}k_1 + \varepsilon_{r1}k_2} = -r_{23} \tag{15.12}$$

where k_1 and k_2 are the wave numbers for a travelling wave in RHM and LHM medium. Now, when medium 2 is LHM, $k_2 = -|k_2|$, $\varepsilon_{r2} = -|\varepsilon_{r2}|$ and $\mu_{r2} = -|\mu_{r2}|$ the Fresnel coefficients calculated from the above indicate that the magnitude of these coefficients is the same whether it is an RH–RH interface or RH–LH/LH–RH interface. This is because the Fresnel coefficients depend only on the tangential components of the fields which are not affected whether the interface is RH–RH or RH–LH (as was observed for boundary conditions discussed for RH–LH media in Section 15.2).

The total transmission coefficient for the travelling electromagnetic wave via the RHM–LHM–RHM medium can be obtained in terms of multiple reflections taking place at the two interfaces (1, 2 and 2, 3) (for *s*-polarised incident wave):

$$T_{\text{TOTAL}(\perp)} = t_{12}t_{23}e^{-jk_2d} + t_{12}t_{23}r_{12}r_{23}e^{-j3k_2d} + t_{12}t_{23}r_{12}^2r_{23}^2e^{-j5k_2d} + \cdots = \frac{t_{12}t_{23}e^{-jk_2d}}{1 - r_{12}r_{23}e^{-j2k_2d}} \tag{15.13}$$

The magnitude and phase expressions for the total transmission coefficient can be written as:

$$T_{\text{mag}} = \frac{t_{12}t_{23}}{\left[1 + (r_{12}r_{23})^2 - \{2r_{12}r_{23}\cos(2k_2d)\}\right]^{1/2}} \tag{15.14a}$$

$$T_{\text{pha}} = k_2d + \tan^{-1}\left\{\frac{r_{12}r_{23}\sin(2k_2d)}{1 - r_{12}r_{23}\cos(2k_2d)}\right\} \tag{15.14b}$$

In the above analysis we assumed that LHM is loss-less so that the imaginary parts of the constitutive properties are zero and hence the refractive index $n_{2\text{reff}}(\omega) = \pm\sqrt{\varepsilon_{2\text{reff}}(\omega)\mu_{2\text{reff}}(\omega)}$. Also for simplicity, let us assume that the LHM is surrounded by vacuum ($n_1 = n_3 = 1$), which is always assumed to be loss-less for all practical purposes, and the incident signal is normal to the interface ($\theta_{\text{inc}} = 0$). Thus the wave numbers k_1 and k_2 are given by (ω/c) and $(\omega/c)\, n_{2\text{reff}} \cos\theta_r$ respectively. A representative plot [41] of transmission magnitude and phase for a 1 cm thick LHM slab with material parameters $\omega_{eo} = 0$ GHz, $\omega_{ep} = 2\pi \times 28$ GHz, $\omega_{mo} = 2\pi \times 21$ GHz, $\omega_{mp} = 2\pi \times 23$ GHz, $\gamma_e = 1.6 \times 10^9\ s^{-1}$ and $\gamma_m = 4 \times 10^9\ s^{-1}$ is shown in Fig. 15.21.

It may be observed that in the vicinity of minimal transmission (shown by the dip in T_{mag} close to 21 GHz), the slope of the transmission phase changes sign. Physically, the

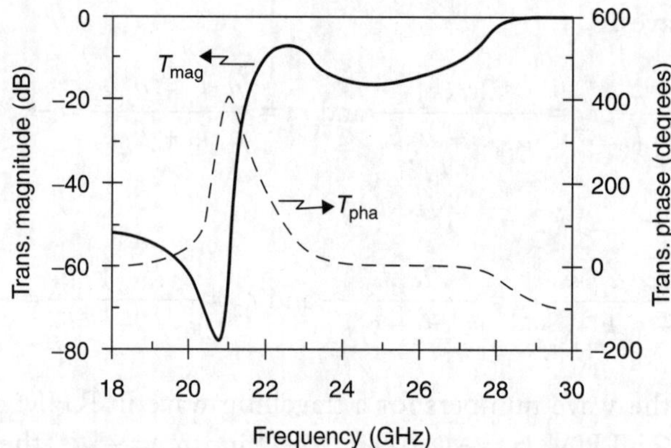

Fig. 15.21 Transmission magnitude and phase with frequency for electromagnetic wave propagation in LH media.

occurrence of transmission dip and the phase reversal of the transmission function is a direct consequence of the frequency dependent negative refractive index of the LH medium.

The *evanescent wave growth* in an LH medium and possibility of sub-wavelength imaging is a revolutionary concept and an exotic property expected to be realisable with metamaterials. The evanescent wave carries the sub-wavelength details of the object which normally decays in a natural (RH) material giving rise to '*diffraction limit*' (resolving power being limited to the extent of a wavelength of the irradiating electromagnetic signal), being well known in conventional optics. However, a '*perfect lens*'/'super-lens' of Veselago–Pendry type made of a plane LHM slab, is said to cause evanescent wave growth thereby carrying sub-wavelength details of the object to the image plane with the possibility of '*sub-wavelength imaging*' with a perfect lens; this was theoretically proposed by Pendry in 2000 [42]. Pendry's perfect lens demands $\varepsilon_{r1} = \mu_{r1} = -|\varepsilon_{r2}| = -|\mu_{r2}| = 1$ and hence $k_1 = k_2 = k$. For evanescent waves it is known that $k = -jq$ (say). Using the total transmission coefficient given by Eq. 15.12 and the Fresnel's coefficients for *s*-polarised waves, we can now establish for an incident electromagnetic wave the conventional evanescent wave decay in the RHM slab and the possibility of evanescent wave growth in the LHM slab.

For evanescent wave incidence on a RHM slab we can write Eq. 15.13 with the limit $\mu_{r1} \to +1$ and $\mu_{r2} \to +1$ (for a perfect lens):

$$T_{\perp}^{\text{evans(RH)}} = \frac{4k_1k_2}{(k_1+k_2)^2} \frac{e^{-jk_2d}}{1-\left(\dfrac{k_1-k_2}{k_1+k_2}\right)^2 e^{-j2k_2d}} \tag{15.15a}$$

When we set $k_1 = k_2 = k = -jq$ (the evanescent wave approximation), we get:

$$T_{\perp}^{\text{evans(RH)}} = e^{-qd} \tag{15.15b}$$

Again for an evanescent wave incidence on a LHM slab we can write Eq. 15.13 with the limit $\mu_{r1} \to -1$ and $\mu_{r2} \to -1$ as:

$$T_{\perp}^{\text{evans(LH)}} = \frac{-4k_1k_2}{(k_1-k_2)^2} \frac{e^{-jk_2d}}{1-\left(\dfrac{k_1+k_2}{k_1-k_2}\right)^2 e^{-j2k_2d}} \tag{15.16a}$$

On setting $k_1 = k_2 = k = -jq$ (the evanescent wave approximation) and noting that $\left(\dfrac{k_1+k_2}{k_1-k_2}\right)^2 e^{-jk_2d} >> 1$, we get with algebraic manipulation:

$$T_{\perp}^{\text{evans(LH)}} = e^{+qd} \tag{15.16b}$$

From Eq. 15.15b we observe that an evanescent wave will decay while transmitting through an RHM slab while Eq. 15.16b indicates that the evanescent wave is enhanced or grows while passing through an LHM slab.

The physical reason for evanescent wave growth in plasmonic LHM is understood to be the following. A plasmonic LHM slab is basically a metal–dielectric composite with metal inclusions so structured in the dielectric substrate that the composite structure is extrinsically inhomogeneous but to an incident electromagnetic signal it is effectively a homogeneous medium. The electrons in the metal undergo simple harmonic motion caused by the time varying *E*-field (tangential to the metal surface) of the incident electromagnetic wave. The phenomenon occurs at the top and bottom surfaces of a finite sized LHM slab. The RHM–LHM/LHM–RHM interface supports surface plasmon (also known as surface polariton) waves excited by evanescent waves, which is a travelling electromagnetic wave from the top to the bottom of the LHM slab, however, is reflected from the side walls (left and right ones), that is, the walls which allow the propagating wave to flow under matched condition. Thus, a surface plasmon standing wave is formed causing a cavity-like effect that mediates energy transfer from the evanescent waves at the object plane to the image plane. More specifically, a resonant-tunnelling effect caused by coupling of surface plasmon waves at the RHM–LHM interface (on the left of LHM slab) and LHM–RHM interface (on the right of the LHM slab), those are separated by approximately a wavelength extent, sets-in which results in enhanced transmittance of the evanescent waves. This in essence is the concept of evanescent wave growth in the LHM slab.

So far we have discussed electromagnetic signal propagation via LH medium considering the medium to be loss-less. However practical LHM has finite loss which affects the propagation properties and 'perfect lens' realisation is still far from reality. The loss in the material has the effect of smearing the image being formed by the LHM plane slab as evanescent wave growth is severely affected by the presence of the loss. Extensive research is going on to mitigate these problems and to develop low-loss LH material in order to exploit the full potential of the so called exotic properties of electromagnetic signal propagation via LHM.

15.5 Metamaterial Inspired Microwave Components and Antennae

The design of microwave components such as filter, phase shifter, coupler and antenna are being inspired with the incorporation of some of the elementary structures like SRR (or its complimentary version, CSRR) and CRLH, with the advent of metamaterials has led to the development of more efficient and smaller microwave components and antennae. Let us consider two simple examples—a CSRR loaded microstrip patch antenna and a CRLH based microwave filter design and characterisation.

15.5.1 CSRR Loaded Microstrip Patch Antenna Design

CSRR represents the complementary version of the well-known SRR. In microstrip design, CSRR is the form of SRR which can be realised with planar fabrication technology. The gap portions of SRR are the metallic portions for CSRR and vice versa. Electrically, CSRR is the dual of SRR (as per Babinet's principle), the latter being a magnetic inclusion while the former is an electric inclusion structure. The SRR can be considered as a resonant magnetic dipole that can be excited by an axial magnetic field, whereas the CSRR essentially behaves as an electric dipole with the same frequency of resonance (as that of the equivalent SRR) and can be excited by an axial electric field. As CSRR is electrically the dual of SRR, the equivalent circuit inductance and capacitance of SRR will respectively correspond to the capacitance and inductance of the CSRR. Since, dual parameters are only related by constants, it is evident that, the resonant frequency which depends on the product of L and C will remain constant for a particular inclusion whether we treat it as a SRR or as a CSRR.

The schematic sketches of CSRR and CSRR loaded square microstrip patch antenna (SMPA) are shown in Fig. 15.22.

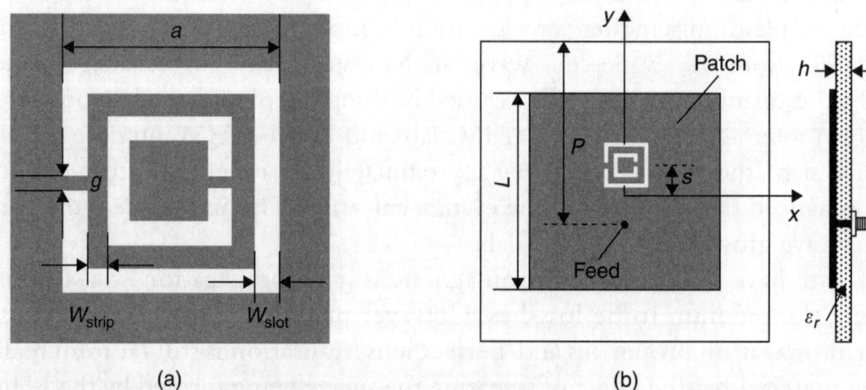

Fig. 15.22 (a) Schematic of CSRR (b) CSRR loaded SMPA.

A conventional SMPA designed with metallic patch dimensions of 18.45 mm × 18.45 mm printed on a substrate (Arlon Diclad 880 having $\varepsilon_r = 2.2$ and tan $\delta = 0.0009$) with a size

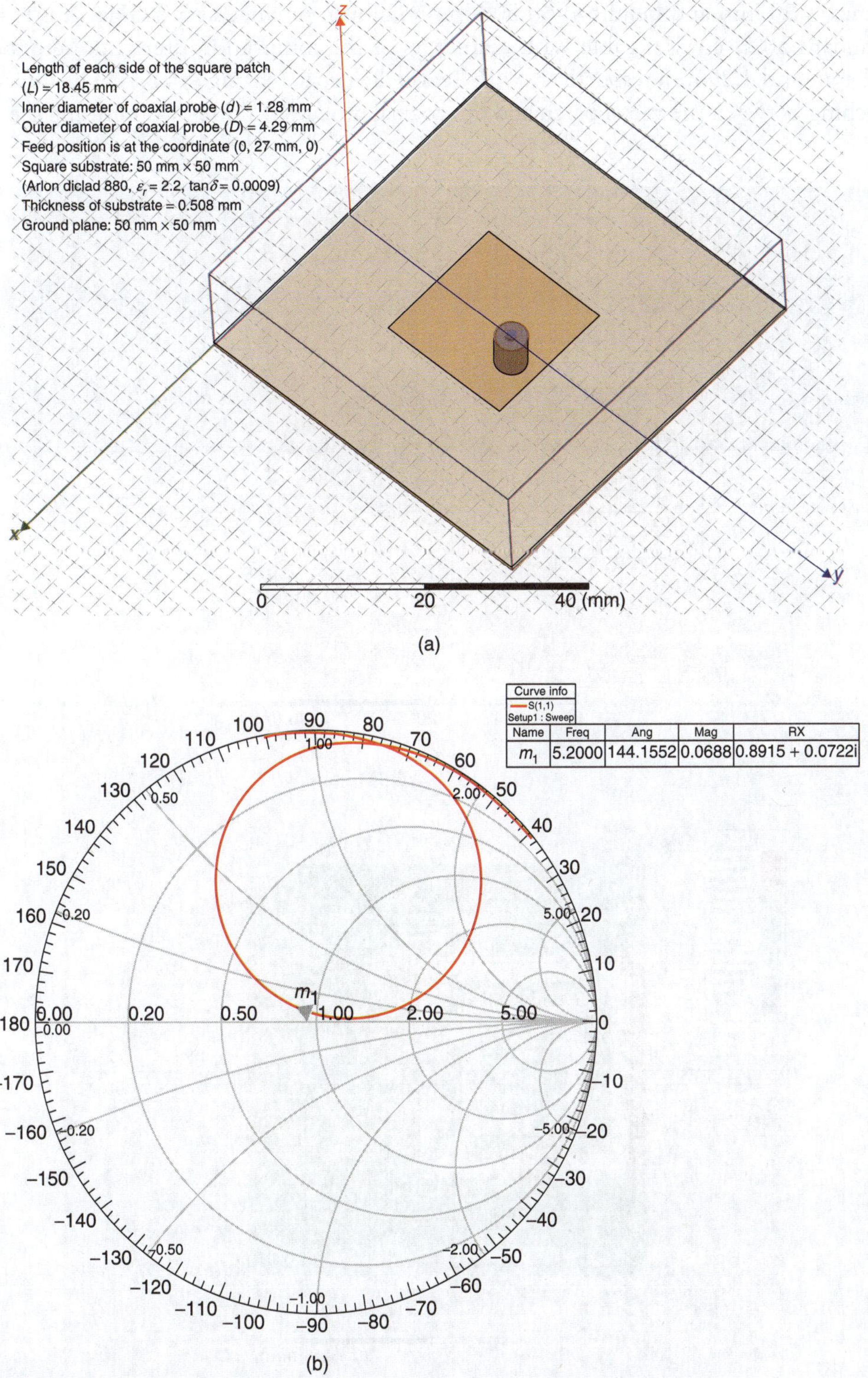

Fig. 15.23 (a) Schematic of SMPA (b) Smith chart showing impedance matching.

50 mm × 50 mm and height $h = 0.508$ mm exhibit resonance at 5.2 GHz. It is fed with a standard 50 Ω coaxial cable with diameters of the inner and outer conductors being 1.28 mm and 4.29 mm respectively with Teflon dielectric (having $\varepsilon_r = 2.1$). The impedance matching between the metal patch and the coaxial probe was done by optimising the probe

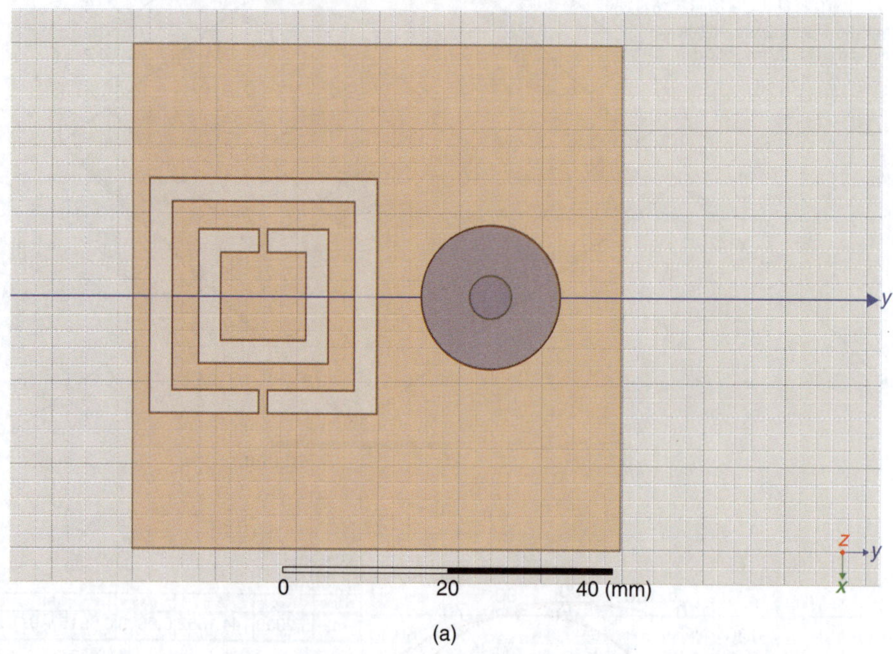

(a)

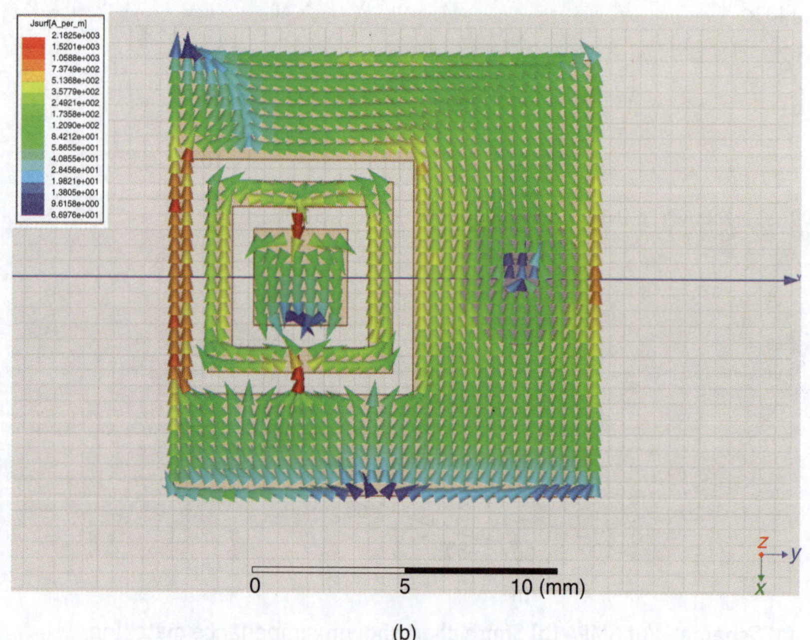

(b)

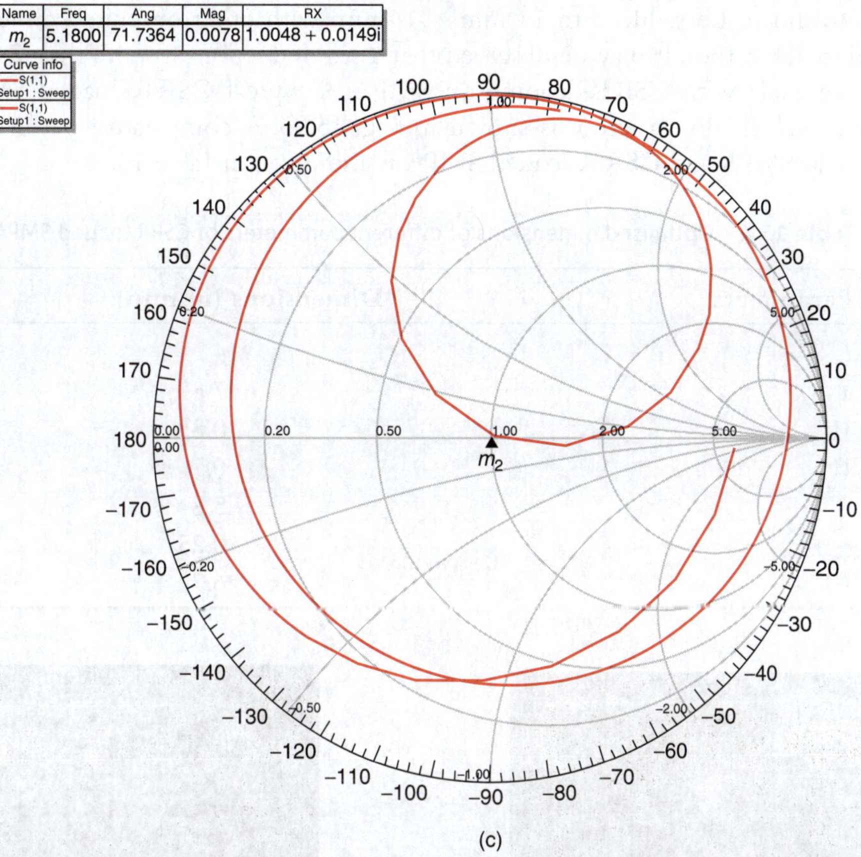

Fig. 15.24 (a) Square patch loaded with CSRR, aligned orthogonal to the coaxial feed (b) Surface current lines (c) Smith chart plot showing matching between coaxial probe and the CSRR loaded patch.

position *P* (Fig. 15.22b) using a 3D electromagnetic field simulator. The SMPA schematic along with the Smith chart plot showing impedance matching is depicted in Fig. 15.23.

When the CSRR resonator is loaded on the SMPA, the performance optimisation needs to be done by controlling the CSRR size and the parameters (Fig. 15.22b). The schematic of square patch loaded with CSRR, having the line of cut in CSRR aligned orthogonal to the coaxial feed together with the surface current lines are shown in Figs 15.24a and 15.24b respectively.

Other orientations of CSRR, one aligned with its line of cut along the line of the coaxial feed and facing the feed exhibits a lossy nature while the other case when CSRR is aligned with its cut along the line of coaxial feed but rotated by 180° in the clockwise direction exhibits no possibility of size reduction as it exhibits resonance at 5.8 GHz. The CSRR loaded patch given in Fig. 15.24a when optimised in terms of parametric study indicates that CSRR loading on patch brings down the resonance frequency below 5.2 GHz. To bring it back to 5.2 GHz with perfect impedance matching (Fig. 15.24c) the patch size was found to be reduced to 14 mm × 14 mm with other optimised dimensions as indicated in Table 15.4. It may be observed that a size reduction of more than 24% of the patch is realisable with CSRR loading on SMPA. A typical CSRR loaded SMPA and a conventional SMPA structure is shown in Fig. 15.25. A comparative performance of conventional SMPA and CSRR loaded SMPA is also given in Table 15.5.

Table 15.4 Optimised dimensions of different parameters of CSRR loaded SMPA

Parameters	Dimensions (in mm)
L	14
a	7.5
W_{strip}	0.8
W_{slot}	0.7
s	5.8
g	0.25
P	26.7

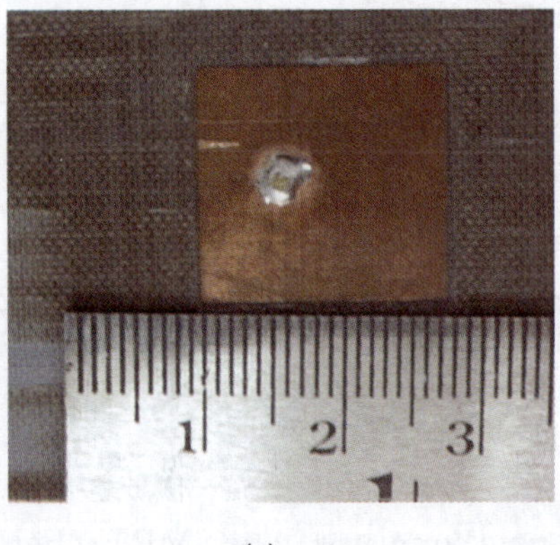

(a)

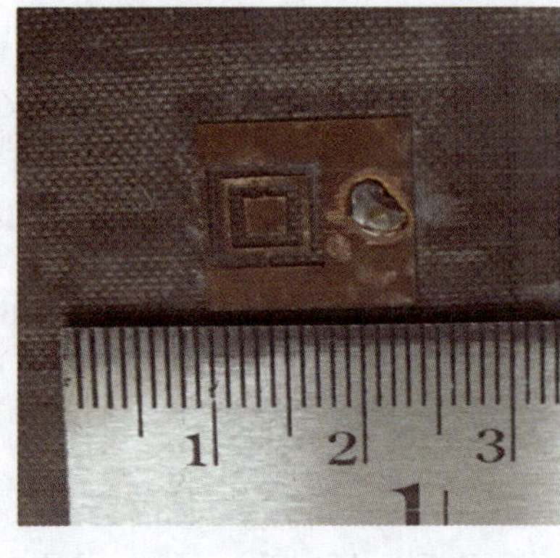

(b)

Fig. 15.25 Fabricated antennae (a) Conventional SMPA (b) CSRR loaded SMPA (approximately 24% size reduction).

Table 15.5 A comparative performance of conventional SMPA and CSRR Loaded SMPA

	Antenna type →	Conventional SMPA		CSRR loaded SMPA	
	Obtained from	Simulation	Measurement	Simulation	Measurement
E-Plane	Maximum realisable gain	6.86 dB	6.11 dB	4.2 dB	4.17 dB
	Maximum cross-polarisation	−56.35 dB	−31.1 dB	−7.43 dB	−10.36 dB

(continued)

Table 15.5 Continued

	Antenna type →	Conventional SMPA		CSRR loaded SMPA	
	Obtained from	Simulation	Measurement	Simulation	Measurement
H-Plane	Maximum realisable gain	6.86 dB	6.114 dB	4.2 dB	4.187 dB
H-Plane	Maximum cross-polarisation	–51.06 dB	–20.4 dB	–7.43 dB	–6.39 dB
	–10 dB Bandwidth	170.1 MHz	60 MHz	32.5 MHz	34 MHz
	Efficiency	81.89%		50.73%	

15.5.2 CRLH Based Microwave Filter Design

Three CRLH resonators are used in the design of microwave band-pass filter in which the first and third resonators are of the asymmetric type (Fig. 15.18a) while the second one is a symmetric CRLH resonator (Fig. 15.18b).

The target specification for the filter to be designed is given in Table 15.6.

Table 15.6 Specification of the filter to be designed

Parameter	Specification
Centre frequency	2.45 GHz
Bandwidth	100 MHz
Order	3
Insertion loss	< 1.0 dB
Passband ripple	< 0.1 dB
Return loss	< 10 dB
Substrate	Roger RT/Duroid 5880
Dielectric constant	2.2
Height of the substrate	0.508 mm
Loss tangent	0.0009

The resonators have been optimised separately at 2.45 GHz using electronic design automation (EDA) simulation tool. To carry out the design, at first the resonators are to be simulated in the circuit simulator of the EDA tool (using approximate formulae for calculating the initial values of the inductance and capacitance of the inter-digital capacitor and stub inductor as detailed under the heading 'design of unit cell of CRLH' in Section 15.3.2) and are to be subsequently, optimised using the suitable optimisation technique available in the EDA simulation tool. The optimised schematic should next be converted to the layout. The final optimisation is done using an electromagnetic (EM) simulation tool of the EDA simulator. On carrying out all these steps, the final dimensions given in Table 15.7 can be obtained while the final layout looks like the one given in Fig. 15.26.

Table 15.7 Final dimensions of CRLH based BPF at 2.45GHz

Items	Value (mm)	
	Resonator 1 and 3	**Resonator 2**
W_{IDC}	0.466	0.2
l_{IDC}	5.6	3
Number of fingers	2	2
Gap between the fingers (g_{IDC})	0.18	0.36
W_{stub}	1	0.95
l_{stub}	8.8	12.67
Width of the port coupling	0.3	
Length of the port coupling	4.74	

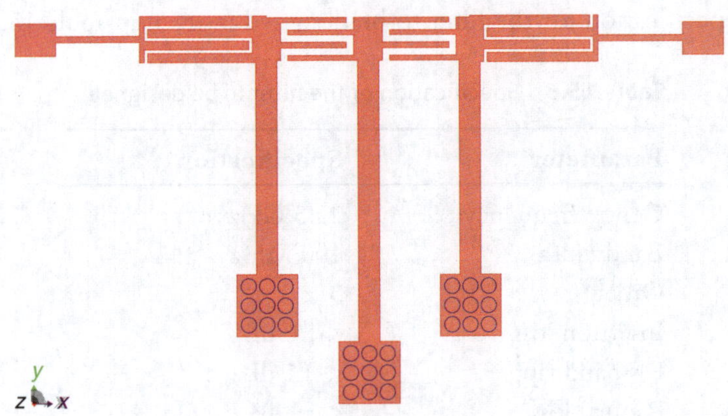

Fig. 15.26 Final layout of CRLH transmission line based BPF at 2.45 GHz.

A typical CRLH transmission line based band-pass filter (BPF) and a conventional (parallel coupled/edge-coupled) microstrip BPF are given in Fig. 15.27a with their performance

Table 15.8 Performance characteristics and size comparison of CRLH based and conventional (parallel coupled/edge-coupled) microstrip BPF

Performance ch. and size	CRLH BPF	Conventional BPF
Insertion loss	1.6 dB	2.3 dB
Return loss	15 dB	18 dB
Harmonics suppression	At least up to 10 GHz	No harmonic suppression
Size*	4.3 cm × 3 cm	6.5 cm × 6 cm

* CRLH type BPF shows 67% size reduction (area wise) compared to the conventional one.

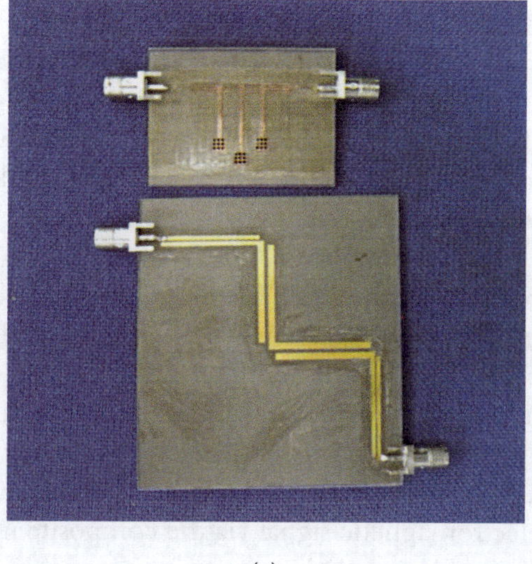

(a)

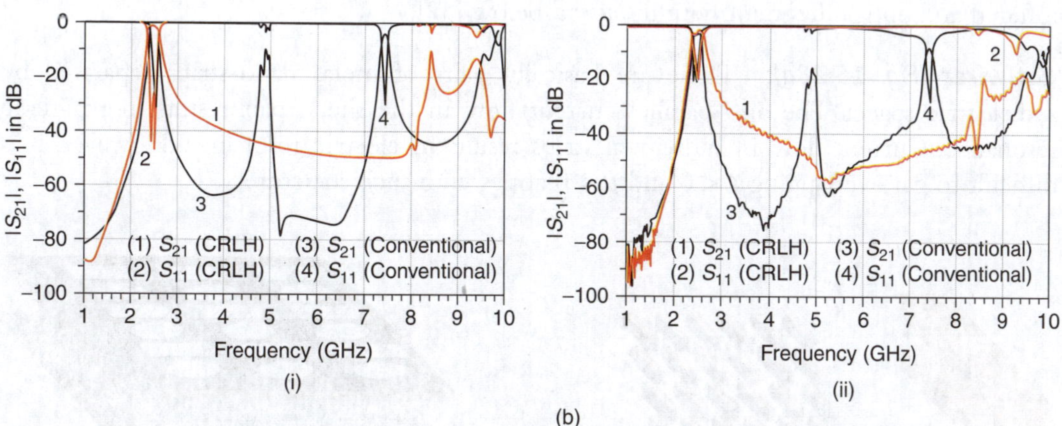

(b)

Fig. 15.27 (a) Fabricated CRLH based BPF (top) and conventional (parallel coupled/edge-coupled) microstrip BPF (bottom) (b) Comparison of insertion loss (S_{21}) and return loss (S_{11}) of CRLH BPF with conventional BPF (i) Simulation results, (ii) Measurement results.

characteristics in Figs 15.27b(i) and 15.27b(ii). The performance and size comparison given in Table 15.8 shows the superiority of CRLH based filter both in terms of miniaturisation and improved performance.

15.6 Problems and Possibilities of Terahertz and Optical Metamaterials

Initial metamaterial developments were in the microwave frequency domain, being limited by the available fabrication technology in those days for such sub-wavelength structures. However, real benefit of metamaterial developments will be practical with terahertz (THz) and optical metamaterials. The ubiquitous split-ring metmolecule can be scaled down in size up to about 200 THz, but this scaling breaks down at much higher frequencies as the metal does not behave any more as a conductor and becomes transparent to the radiation (and eventually magnetism disappears) for wavelengths shorter than 1.5 μm, that is, beyond the 200 THz range. This scaling limit combined with the fabrication difficulties of making nanometre scale SRRs along with metal wires (SRR–TW combination) led to the development of alternative designs that are more suitable for THz, infrared (IR) and optical regimes.

Nanowire composites are the basis of developing LHMs in the near-IR and visible parts of the spectrum. In fact, it is observed that both the dielectric and magnetic moments excited in the nanowire systems are opposite to the excited field when the wavelength of the incident signal is below resonance. Thus, in this frequency range, a composite material based on parallel nanowire pairs have the dielectric permittivity and magnetic permeability both negative and the composite acts as a left-handed material. Also, a negative phase shift of the transmitted electromagnetic signal via the composite indicates the left-handedness of the nanowire composite [41].

The designs that have been developed for realising negative index material (NIM) at infrared and optical frequency are discussed below [42].

***Cut-wires (Fig. 15.28a)*:** These are basically pairs of metal nano-strips separated by a dielectric spacer. The anti-parallel current flow in the metal pair results in magnetic resonance. Current flow in individual strips results in electric resonance. However, it is difficult to get overlapping $\varepsilon < 0$ and $\mu < 0$ zones with such structures.

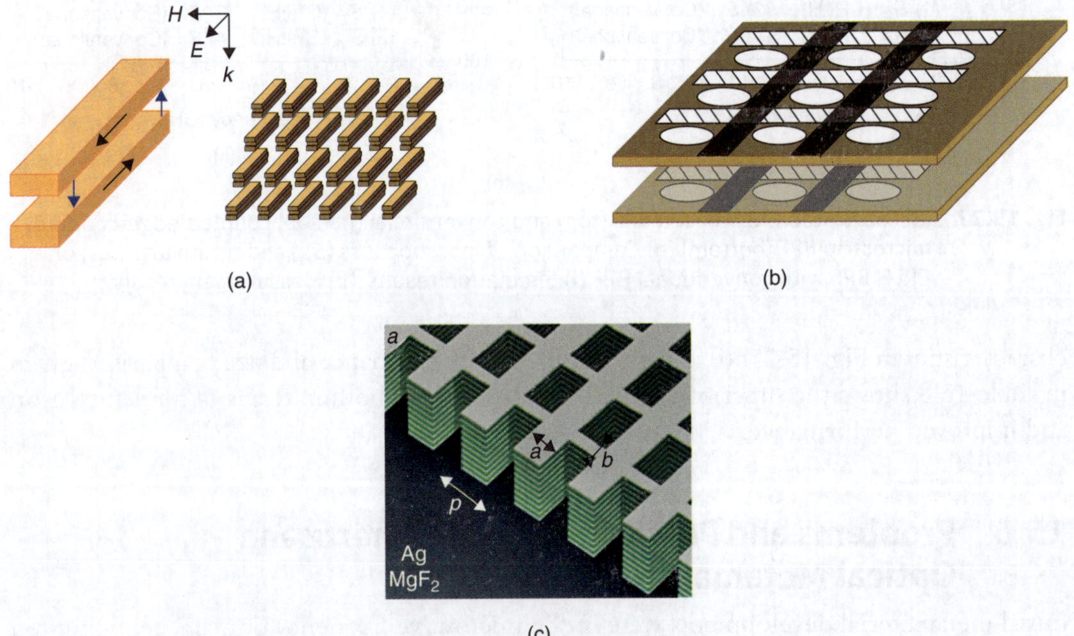

Fig. 15.28 (a) Cut-wire structure (b) Fishnet structure (c) 3D fishnet structure for use at optical frequencies.

***Fishnet (Fig. 15.28b)*:** This structure combines magnetic coupled nano-strips (to provide $\mu < 0$) with continuous electric strips (to provide $\varepsilon < 0$) over a broad spectrum. Hence overlapping frequency zone for simultaneously negative ε and μ is easily obtained at optical frequencies.

In August 2008, scientists of the University of California, Berkeley engineered 3D optical fishnet metamaterial by using focussed ion-beam (FIB) milling. The refractive index was found to vary from $n \approx 0.63$ at 1,200 nm to $n \approx -1.23$ at 1,775 nm. Alternating layers of 30 nm Ag and 50 nm MgF_2 were stacked together and nanoscale-sized fishnet patterns were cut into the layers (Fig. 15.28c).

Though it appears to be the most challenging, the demand of the day is truly three-dimensional (3D) metamaterials. Initially, 3D metamaterials were made by creating multilayer structures (involving challenging lift-off processes) and also by using a layer-by-layer technique (requires careful alignment). Complex 3D structures may be fabricated by electron-beam writing, focussed-ion beam chemical vapour deposition and so on, but the methods are too complex and time consuming for mass production purposes. Fabrication methods based on two-photon photopolymerisation (TPP) is considered to be the most promising for future manufacturing of large area true 3D metamaterials. Direct single beam laser writing, multiple-beam TPP techniques are the methods offering sub-diffraction resolution down to 100 nm. Nano-imprint lithography may also be a successful method for fabricating 3D metamaterials.

15.7 Metamaterials: Application Scenario, Challenges and Possibilities

The emerging field of metamaterials is very fascinating and important for two main reasons. On the one hand, metamaterials represent a counter-intuitive aspect of already known science associated with a number of exotic phenomena and their interpretation having exciting possibilities with application potential. These intriguing phenomena include negative refraction, evanescent wave growth leading to the possibility of overcoming the classical 'diffraction limit' and thus opening up the scope for sub-wavelength imaging, reversal of Cherenkov and Doppler effects and others yet to be discovered! On the other hand, inspiration is generated to harness plasmonic metamaterial inclusions like SRR (and its complimentary structure CSRR) for antenna and filter design and the transmission line metamaterial unit, the CRLH, to design RF/microwave passive components like hybrid couplers, Wilkinson power divider, filter and antenna with ZOR offering unique properties in terms of functionality, size reduction and performance. Further, the availability of metamaterials with tailorable constitutive parameters opens up opportunities for engineering antenna radomes or frequency selective surfaces (FSSs) with unique performance characteristics. The primary challenge in this field is in the development of 3D isotropic metamaterials (which needs improvement in fabrication technology of metamaterials) for omni directional antenna and many more yet to be envisioned applications.

Though the physics of metamaterials is fairly well established and most exciting, negative refractive index and near field focussing with plane slab has been demonstrated [43], the average lattice constant p in current MTM is still electromagnetically too large for high quality refraction. Current MTMs are typically characterised by $p/\lambda = 1/5 \ldots 1/20$. Although refraction phenomena dominate in such MTM structures, diffraction/scattering affects the purity of refracting effects thereby diffusing the focussing spot and increasing the transmission losses. To obtain pure refractive phenomena, it is necessary to homogenise the MTM, that is, to decrease the structural feature p/λ by one order of magnitude or more. If this challenge is met, MTM will behave as real artificial materials, that is, the unit cells of MTM would be so small that they would really behave like atoms in natural materials and would be capable of producing truly homogeneous macroscopic responses thereby faithfully mimicking the microscopic response of natural materials.

Metamaterials were considered to be 'material like' (one with tailorable constitutive properties, more specifically with simultaneous negative effective permittivity and permeability) during the initial phase of metamaterial research. But in the beginning of this decade there was a paradigm shift in research philosophy. Now, metamaterials are considered as 'devices' where the structuring of metal and the hybridisation with functional agents brings new functionality and the response becomes non-linear, gain-assisted, switchable and so forth [44]. In the near future, we might be able to enter into the field of quantum metamaterials too.

Hybridising a gain medium (semiconductor quantum dots or quantum well structures) with a plasmonic metamaterial can lead to a multifold intensity increase and a narrowing of their photoluminescence spectra. This is an essential step towards the development of metamaterial-enhanced gain media and the 'lasing spaser'—a 'flat' laser with emission fueled by plasmonic excitations in an array of coherently emitting metamolecules. In contrast to conventional lasers that operate at wavelengths of suitable natural atomic or molecular transitions, the lasing spaser's emission wavelength can be controlled by meta-molecule design.

Switchable metamaterials are another rapidly expanding area of research. Indeed, the development of nanophotonic all-optical data processing circuits depends on the availability of fast and highly responsive nonlinear media that react to light by changing their refractive index and absorption. This is difficult to deliver in nano-scale size devices using electronic or molecular nonlinearities. Recent experiments show that the ultra-fast nonlinear response of silicon can be strongly enhanced by adding a metamaterial layer. Single-wall semiconductor carbon nanotubes deposited on metamaterials exhibit an order-of-magnitude higher nonlinearity than the already extremely strong response of the nanotubes themselves, due to a resonant plasmon–exciton interaction making high speed switching possible.

Superconducting metamaterials offer an incredibly fertile arena for research as losses there are extremely small. The classical object of metamaterial research – the ubiquitous split-ring metamolecule – has much in common with the fundamental unit of superconductivity, the Josephson junction ring. An array of superconducting Josephson rings could be a truly quantum metamaterial, where each metamolecule is a multilevel quantum system supporting phase qubits. However, applications of superconducting metamaterials will be limited to the microwave domain for niobium-based metamaterials, and to the terahertz spectral domain if high-temperature superconductors are used. This is because higher frequencies destroy the superconducting phase. In natural solids, optical response is determined by the quantum energy-level structure of the constituent atoms or molecules. By contrast, the electromagnetic properties of metamaterials are derived from the resonant characteristics of the sub-wavelength plasmonic resonators (such as SRR) from which they are constructed. Thus 'quantum metamaterials' would provide a much closer analogy to natural crystals which the metamaterial mimics.

However, truly meaningful progress in metamaterial research will be possible only with further developments in fabrication technology. New techniques will have to achieve perfection of nanostructures at close to the molecular level, and at low cost. We need to go beyond electron beam lithography, focussed ion-beam milling and nano-imprint. However, the new techniques must be able to build metamaterials to almost any blue print. The real challenge is to create truly volume (3D) metamaterials, isotropic and homogeneous and a great deal of innovative efforts are now being put into this area of research.

RECAPITULATION

Metamaterials

- Metamaterials (MTM) are engineered composite media that exhibit counter-intuitive phenomena such as negative refractive index, reversed Doppler effect and reversed Cherenkov effect, not observed in nature. From a technical point of view, these are engineered metal–dielectric composites formed by embedding sub-wavelength metallic inclusions on a dielectric substrate. The counter-intuitive response exhibited by metamaterials is not by their chemistry but by the structural arrangement of constituent electric and magnetic atoms (the inclusions) artificially realised and arranged in a periodic structure mimicking the lattice arrangement of small-scale (atomic) physics in a natural material. Though metamaterial structures have external inhomogeneity, to an incident electromagnetic signal they appear to be homogeneous as they are designed with the condition of effective medium theory ($p << \lambda$, where p is the lattice period and λ is the operating wavelength). The constitutive parameters of metamterials are an average of the bulk material since the lattice period in a metamaterial is much smaller than the wavelength of the incident electromagnetic signal.
- MTM (having double negative constitutive parameters, ε and μ both being negative) are also known as Left Handed Materials (LHM) as E, H and k vectors form a left-handed triad for such materials; unlike natural materials (which have double positive constitutive parameters, ε and μ both being positive) termed as Right Handed Materials (RHM), which form a right-handed triad with the same vectors.
- In Maxwell's wave equation, the refractive index n appears as square power, so both for RHM and LHM, Maxwell's wave equation is equally applicable irrespective on n being positive (for RHM) or negative (for LHM). Further, as the direction of the Poynting vector remains the same irrespective of the E, H and k vectors forming a right-handed or left-handed triad the causality is not violated by LHM. However, it may be noted that the phase and group velocities are counter directed in the case of LHM and as the phase velocity is in the negative z-direction the LHM is said to support *backward waves.*

- MTM exhibits negative refractive index, originating from material constitutive parameters ε and μ, while photonic band-gap (PBG) materials exhibit negative refraction, due to Bragg diffraction. In terms of the lattice period (p), PBG structure period is of the order of wavelength (λ) while in a MTM, $p/\lambda << 1$. Therefore, interference effects cannot take place in MTM. The effective negative refractive index in MTM is thus due to effective negative permittivity and effective negative permeability of the MTM structure. But PBG is diffraction limited and interference effects are dominant; it thus operates in the Bragg regime and negative refraction in PBG is not due to the constitutive properties in the photonic crystal but due to band-edge Bragg diffraction.

Counter-intuitiveness of LHM

- The tangential components remain continuous in the RH–LH interface (as they are not dependent on the constitutive properties, ε and μ); thus, $E_{t\text{RHM}} = E_{t\text{LHM}}$ and $H_{t\text{RHM}} = H_{t\text{LHM}}$. But, the normal components become antiparallel at the boundary between RH and LM medium (since they are necessarily dependent on ε and μ, $\vec{D} = \varepsilon\vec{E}$ and $B = \mu\vec{H}$):

$$E_{n\text{RHM}} = -\frac{\varepsilon_{\text{LHM}}}{|\varepsilon_{\text{RHM}}|} E_{n\text{LHM}} \text{ and } H_{n\text{RHM}} = -\frac{\mu_{\text{LHM}}}{|\mu_{\text{RHM}}|} H_{n\text{LHM}}$$

- The consequence of the antiparallel nature of the normal component of the wave vector at the interface between RH–LH media leads to negative refraction in LHM. This eventually makes possible the near field focussing by a flat LH lens sandwiched between two RH media, reversal of Doppler effect and reversed Cherenkov radiation.

Plasmonic and transmission line metamaterials

- *Plasmonic metamaterials* consist of a composite of an array of metallic thin wire (TW), sometimes cut-wire (CW), and a matrix of metallic split-ring resonator (SRR) each printed on low-loss dielectric like RT-duroid and placed alternately forming a 1D (may be 2D/3D too) periodic structure. The TW/CW array exhibits effective negative permittivity below its plasma frequency with each unit cell acting as an 'electric atom' while SRR matrix exhibits effective negative permeability below its plasma frequency with each unit cell acting as a 'magnetic atom'. The composite structure exhibits effective negative refractive index given by $n_{\text{reff}} = \sqrt{\varepsilon_{\text{reff}}\mu_{\text{reff}}}$.
- For TW, $\varepsilon_{\text{reff}}$ is given by the Drude model: $\varepsilon_{\text{reff}}(\omega) = 1 - \dfrac{\omega_{ep}^2}{\omega^2 + j\omega\xi_e}$, (high-pass characteristics)

  For CW $\varepsilon_{\text{reff}}$ is given by the Lorentzian model: $\varepsilon_{\text{reff}} = 1 - \dfrac{\omega_{ep}^2 - \omega_{e0}^2}{\omega^2 - \omega_{e0}^2 + j\omega\xi_e}$ (resonant characteristics)

  where, $f_{e0} = \dfrac{c_0}{\pi a}\sqrt{\dfrac{(N_\gamma - 1)d}{N_\gamma(l + d)\ln(h/a)}}$ and $f_{ep} = \dfrac{c_0}{2\pi h}\sqrt{\dfrac{2X\pi}{\ln(h/a)}}$

  (for TW, $d = 0$, $X = 1$, and for CW: $d \neq 0$, $X = 2$).

- For SRR, μ_{reff} is given by the Lorentzian model: $\mu_{\text{reff}} = 1 - \dfrac{F\omega^2}{\omega^2 - \omega_{m0}^2 + j\omega\xi_m}$ that has resonance characteristics; where $F = \pi r^2/h^2$ and $\omega_{m0} = c\sqrt{\dfrac{3v}{\pi \ln(2wr^3/d)}}$.
- There are many variants of SRR like MSRR, SR and LR including TTSR, NBSR.
- *Transmission line metamaterial* is realised by periodically loading (keeping periodicity much smaller than the wavelength of operation) a host transmission line (that has series inductance and shunt capacitance in a lumped equivalent circuit) with series capacitance and shunt inductance; which is thus a periodically loaded transmission line (PLTL) synthesising an LH media.
- It may be noted that the propagation constant (β), phase velocity (v_p) and refractive index (n) for an LHM line are negative while the group velocity (the velocity with which the energy of the electromagnetic wave travels) is positive, as expected, to maintain non-violation of causality.
- Being non-resonant in nature, PLTL exhibits simultaneously low-loss and broad bandwidth and they can be engineered in planar configuration thus inherently supporting 2D wave propagation and also compatible with modern MICs. This class of LH material is thus well suited for RF and microwave circuit applications.
- Physically realisable planar LH structures (with series inter-digital capacitor and shunt stub inductor also exhibit RH series L (due to magnetic flux induced by currents in the fingers of the inter-digital capacitors) and shunt C (due to voltage gradient that exist between the conductor of the stub and the ground plane) which makes it equivalent to a composite right/left-handed (CRLH) transmission line.
- CRLH transmission line resonator has some unusual characteristics. It is said to allow negative and zeroth-order resonances; that has the potential to design dual-band and size-independent resonators (in contrast normal resonator length should be at least half-wavelength). This makes transmission line type planar LH material very attractive for the design of miniaturised microwave circuit components and antennae.

Properties of electromagnetic signal propagation through LHM

- The Veselago–Pendry lens which is an ideal plane LHM slab sandwiched between two RH media exhibits near-field double-focussing which is different from the concept of far-field lens in conventional optics. As a consequence of negative refraction by LHM, an LHM based concave lens will behave as converging lens while an LHM based convex lens will behave as diverging lens for far-field focussing.
- Transmission of electromagnetic signal via multiple interface RH–LH–RH media exhibits the occurrence of transmission dip and the phase reversal of transmission function which is a direct consequence of the frequency dependent negative refractive index of the LH medium.
- *Evanescent wave growth* in an LH medium and possibility of sub-wavelength imaging is a revolutionary concept. The evanescent wave carries the sub-wavelength details of the object which decays in a natural (RH) material giving rise to '*diffraction limit*' (resolving

power being limited to the extent of a wavelength of the irradiating electromagnetic signal), being well known in conventional optics.

Transmission of electromagnetic signal via RH and LH media indicates:

$$T_{\perp}^{\text{evans(RH)}} = e^{-qd} \text{ and } T_{\perp}^{\text{evans(LH)}} = e^{+qd}$$

This indicates that evanescent wave will decay while transmitting through a RHM slab while the evanescent wave is enhanced or grows while passing through a LHM slab.

Metamaterial inspired microwave components and antennae

- The design of microwave components such as filter, phase shifter, coupler and antenna are inspired with the incorporation of some of the metamaterial elementary structures/unit cells like SRR (or its complimentary version, CSRR) and CRLH, with the advent of metamaterial development; that has led to the development of more efficient and size-reduced microwave components and antennae.

Problems and possibilities of terahertz and optical metamaterials

- The real benefits of metamaterial developments will be practical with terahertz (THz) and optical metamaterials.
- The ubiquitous split-ring metmolecule can be scaled down in size up to about 200 THz, but this scaling breaks down at much higher frequencies as the metal does not behave any more as a conductor and becomes transparent to the radiation (and eventually magnetism disappears) for wavelengths shorter than 1.5 μm, that is, beyond the 200 THz range. This scaling limit combined with the fabrication difficulties of making nanometre scale SRRs along with metal wires (SRR–TW combination) led to the development of alternative designs that are more suitable for THz, infrared (IR) and optical regimes.
- Nanowire composites are the basis of developing LHMs in the near-IR and visible parts of the spectrum. In fact it is observed that both the dielectric and magnetic moments excited in the nanowire systems are opposite to the excited field when the wavelength of the incident signal is below resonance. Thus in this frequency range, a composite material based on parallel nanowire pairs has the dielectric permittivity and magnetic permeability both negative and the composite acts as a left-handed material. Also a negative phase shift of the transmitted electromagnetic signal via the composite indicates the left-handedness of the nanowire composite. *Cut-wire* (pairs of metal nano-strips separated by dielectric spacer) and *fishnet* (combination of magnetic coupled nano-strips and continuous electric strips) are the structures useful for IR and optical frequency range.

Metamaterials: application scenario, challenges and possibilities

- Counter-intuitive phenomena like negative refraction, evanescent wave growth, reversed Cherenkov effect and reversed Doppler effect aim towards exotic application potential and use of metamaterial inclusions like SRR (or CSRR) and CRLH for antennae and other microwave passive components design covers in brief the application scenario of metamaterials at present.

- Primary challenge in the design of MTM as of today is the feature size. Current MTMs are typically characterised by p/λ =1/5 .... 1/20. Although refraction phenomena dominate in such MTM structures, diffraction/scattering affects the purity of refracting effects thereby diffusing the focussing spot and increasing the transmission losses. To obtain pure refractive phenomena, it is necessary to homogenise the MTM, that is, to decrease the structural feature p/λ by one order of magnitude or more. If this challenge is met, MTM will behave as real artificial materials—the unit cells of MTM would be so small that they would really behave like atoms in natural materials and would be capable of producing truly homogeneous macroscopic responses thereby faithfully mimicking the microscopic response of natural materials.
- Metamaterials were considered to be 'material like' (with tailorable constitutive properties) during the initial phase of metamaterial research. But in the beginning of this decade there was a paradigm shift in metamaterial research philosophy. Now, metamaterials are considered as 'devices' where the structuring of metal and the hybridisation with functional agents brings new functionality and the response becomes non-linear, gain-assisted, switchable and so forth. In near future, we might be able to enter into the field of quantum metamaterials.
- No progress in metamaterial research will be possible without further developments in fabrication technology. New techniques will have to achieve perfection of nanostructures at close to the molecular level, and at low cost. We need to go beyond electron beam lithography, focussed ion-beam milling and nano-imprint. However, the new techniques must be able to build metamaterials to almost any blue print. The real challenge is to create truly volume (3D) metamaterials, isotropic and homogeneous and a great deal of innovative efforts are now being concentrated on that and we can expect marvels in the years to come.

Review Questions

15.1 What are metamaterials? Why are they also known as left-handed materials (LHM)? LHM is known to support backward waves—explain.

15.2 Explain why Maxwell's wave equation is equally applicable for RHM and LHM?

15.3 Mention a few exotic and common applications possible with metamaterials.

15.4 Discuss in brief the differences between metamaterials (MTM) and photonic band gap (PBG) materials.

15.5 Explain why the tangential components remain continuous in the RH—LH interface, but the normal components become antiparallel.

15.6 What is the consequence of the antiparallel nature of the normal component of the wave vector at the interface between RH–LH media?

15.7 What are plasmonic metamaterials? How they are realised in practice?

15.8 Explain how thin metallic wires exhibit negative permittivity at microwave frequencies. Which model is used to characterise this frequency dependent effective relative permittivity?

15.9 Why does cut wire have resonant characteristics? Which model is used to characterise its frequency dependent effective relative permittivity?

15.10 Explain how tailorable negative permeability can be realised with a split-ring resonator. Which model is used to characterise its frequency dependent effective relative permeability?

15.11 Discuss the problem with SRR and mention the name of a few improved magnetic inclusion structures used in practice.

15.12 What are transmission line metamaerials? What are their advantages over plasmonic metamaterials?

15.13 Starting from Telegrapher's equation, show that LHM being dual of RHM exhibits counter-intuitive characteristics in terms of phase constant, phase velocity and refractive index.

15.14 Discuss why physically realisable planar LH structures behave as composite right/left-handed (CRLH) structures.

15.15 What are balanced and unbalanced CRLH lines? Why are balanced CRLH lines preferred from a practical point of view?

15.16 For CRLH to behave as a metamaterial, how is the unit cell periodicity related to the wavelength of the propagating electromagnetic signal via the composite CRLH media?

15.17 With a conventional distributed circuit resonator only non-zero and positive resonances are possible while with a CRLH resonator negative and zero resonances are possible—discuss.

15.18 Conventional resonators are at least half wavelength long while CRLH based resonators can be arbitrarily small—explain. What is zeroth order resonator?

15.19 Discuss how a CRLH based resonator can be used to design dual band filters and antennae.

15.20 Discuss in brief the methodology to design a CRLH unit cell.

15.21 What is a Veselago–Pendry lens? Discuss its double focussing effect for near field imaging.

15.22 Discuss with a ray diagram how a convex lens made of RHM behaves as a concave lens if made of LHM while a concave lens made of RHM would behave as a convex lens if made of LHM.

15.23 The occurrence of transmission dip and the phase reversal of the transmission function is a direct consequence of the frequency dependent negative refractive index of the LH medium—explain.

15.24 What is 'diffraction limit' of conventional optics? How it is possible to get sub-wavelength imaging with a LHM slab?

15.25 Discuss the physical concept of evanescent wave growth in LH media.

15.26 What is CSRR? How is it related to SRR?

15.27 What are the advantages of designing metamaterial inspired antennae and filters when compared with conventional techniques?

15.28 Discuss why conventional TW and SRR used to realise metamaterials at microwave frequencies cannot be used at THz, infrared (IR) and optical regimes.

15.29 Discuss briefly the technique used to develop LHMs in the near-IR and visible parts of the spectrum; name a few of them.

15.30 Mention a few available techniques for the fabrication 3D metamaterials.

15.31 Why is the emerging field of metamaterials very exciting and important?

15.32 Mention a few possible applications of metamaterials and metamaterial-inspired microwave devices.

15.33 Metamaterials will behave as real artificial materials if the structural feature p/λ can be realised one order of magnitude or more than the current achievement—discuss.

15.34 Metamaterials were considered to be 'material like' in the initial phase of development but now they are considered as 'devices' to bring about new functionality—discuss.

15.35 Comment on the future possibilities with metamaterials.

16 Applications of Microwave Engineering

Thoughts and processes need to be translated into applications in real life

16.1 Introduction

A brief discussion of microwave applications was done in Chapter 1 which indicated that the primary use of microwaves though started with radio detection and ranging (RADAR) during World War II, eventually encompassed terrestrial and satellite communication, mobile communication, remote sensing, various industrial processing, medical treatment and household cooking. However, radar is even today the most important application which uses and also drives innovative developments of microwave engineering. Thus, Chapter 11 was devoted to radar principles and systems. Though radar-like applications are emerging with new avenues of application in sonic and optical frequencies (as in SONAR/SODAR and LIDAR), microwave radar is by far the most important one with numerous applications both in terms of civil and military usage. In Chapter 11, we saw that CW Doppler radar is useful for monitoring the speed of vehicles (police radar and collision avoidance radar), pulse Doppler radar is used for military applications for missile and weapon laying on enemy targets. Electronic warfare based on ECM (electronic counter measure) and ECCM (electronic counter counter measure) make use of microwave engineering very intelligently. Synthetic aperture radar (SAR) is used these days for geological and mineral exploration and for military reconnaissance and other remote sensing applications.

From a communication engineer's point of view, the advantages of engineering design using the microwave frequency spectrum lies in its capability to provide a wider bandwidth when compared to the lower radio frequencies on the one hand while allowing zero vision (in the optical sense) and adverse weather applications on the other hand that cannot be done with infra-red and optical signals (though at those frequencies a much higher bandwidth can be obtained). Certain important propagation characteristics of microwave signals via the atmosphere have made possible the establishment of tropospheric microwave links and microwave duct propagation links for long range communication in difficult geographical terrain and across large ocean beds (Section 10.5). Further, microwave antennae can be designed to have high gain with a reasonably small antenna beam width (Section 10.2.2) and is realisable with compact physical size of the antenna. Narrow antenna beam width made point-to-point and covert communication possible at microwave frequencies while compact physical size of the antenna has made airborne and space borne applications possible with microwave systems. Combining the advantages of microwave systems and the better noise immunity and other attractions of digital techniques, digital microwave communication has become the order of the day. In earlier days, most of the terrestrial line-of-sight and satellite microwave systems were based on *analogue* FM modulation techniques; however, an overwhelming majority of new microwave systems is now based on *digital* modulation.

Heat generation with RF signal at microwave frequencies is a major breakthrough of microwave engineering that is both a boon and a bane. The principle of dielectric heating (especially in lossy dielectrics) is the basis of heat generation by microwave frequency signals (Section 2.4). The microwave oven has now become commonplace both as a household cooking appliance as well as in fast-food shops as an alternative to the age old fire for quick-shot heating. Microwave heat generation is also used in various industrial processing techniques like food processing, rubber, plastic and forest product processing and also for drying of inks, leather, textiles, tobacco and grains. Microwaves find applications in non-destructive testing of materials, process controls such as moisture and thickness monitoring. Medical applications of microwaves is also increasing in modern times that includes antitumoral hyperthermia, used alone or in conjunction with conventional cancer treatments like radiotherapy or chemotherapy, and ablation for cancer treatment. However, the advantages are a mixed bag. The mobile communication that has redefined our lives and can be considered the most thriving business avenue of the 21st century is also telling upon our health as a hazard due to microwave heating caused by the microwave signals of the mobile tower/mobile set. The issue of human exposure to electromagnetic signals, especially microwave signals, and the resulting possible health hazards has been addressed in Section 14.6. Be a cautious user of the mobile phone with a safe SAR value and limited usage. Also be a socially conscious citizen to ensure that mobile towers are installed at safe locations as per technical requirement without causing electromagnetic health hazards to the nearby population.

In recent times, particle accelerators also use microwave technology for the design of specialised RF/microwave cavity chain for accelerating atomic particles providing very high on-axis acceleration of the particles required in linear accelerators (linac) of Mev/TeV capability. In particle accelerators where microwave power of up to hundreds of megawatts is required, klystron sources are extensively used; in fact, 'klystrons are the *sine qua non* of particle accelerators'.

In this chapter we will discuss some of the communication, industrial and medical applications where microwave engineering is used. The principle of microwave heating was discussed in Section 2.4; however, more technical details of microwave cooking and the design and usage of the microwave oven will be covered here. Microwave/RF life-detection systems that can be used for remote physiological sensing in medical applications and also for locating human subjects (in terms of feeble heartbeat of the still living human beings) buried under earthquake rubble will be discussed briefly.

16.2 Microwave Engineering in Communication Applications

In recent times, a large fraction of the world's international and other long, medium and short-haul telephone, data and television transmissions use microwave communication systems. Most of the emerging wireless telecommunication applications like direct broadcast satellite (DBS) television, personal communication systems (PCS), wireless local area networks (WLANS) and world-wide interoperability for microwave access (WIMAX), cellular video (CV) systems, and global positioning satellite (GPS) systems rely heavily on microwave engineering. Here, we will discuss some of the aspects of broadband microwave access for mobile communications, microwave communication links and microwave remote sensing.

16.2.1 Broadband Microwave Access for Mobile Communications

The unprecedented increase in cellular traffic during the last two decades had a significant impact on mobile access and backhaul network to provide broadband service to the users. Though optical fibre network is fast replacing microwave connectivity for backhaul requirement (using radio over fibre connectivity between the central station and base station), such optical fibre network is not attractive for access network requirements due to the extremely high cost of laying fibre optic cables in dense urban areas and due to the increased deployment time. Consequently, mobile operators started looking for an alternative cost-effective wireless media to provide traffic connectivity, signalling and maintenance (the GSM A-bis interface) between the GSM BSC (base station controller) and the GSM BTS (base trans-receiver station), Fig. 16.1. For such *last mile access connectivity*, microwave media has emerged as a practically viable alternative when compared to the conventional copper lines or modern optical connectivity. In a majority of the urban cases, hop lengths of access links vary from less than a kilometre to 5 km. However in rural areas, short-haul microwave link hop length may be more than 10 km.

From Fig. 16.1, which shows the basic building blocks of GSM system architecture, the BSS (base station subsystem or base station, BS), provides and manages radio communication paths between the MS (mobile station, the mobile phone equipment including the SIM of the mobile user) and the MSC (mobile services switching centre) including handoff management. BSS (also known as radio subsystem) consists of many BSCs which connect to a single MSC (which is the heart of a network switching subsystem, NSS, also known as the central station, CS, of cellular communication system) and each BSC typically controls up to several hundred BTSs (the BTS represents individual cell in the cellular communication system, the cell size is usually 100 m to about 35 km). Handoffs between two BTSs under the control of the same BSC are handled by the BSC and not by the MSC. This reduces the load and hence switching burden of the MSC. The MSC is connected to the BSCs

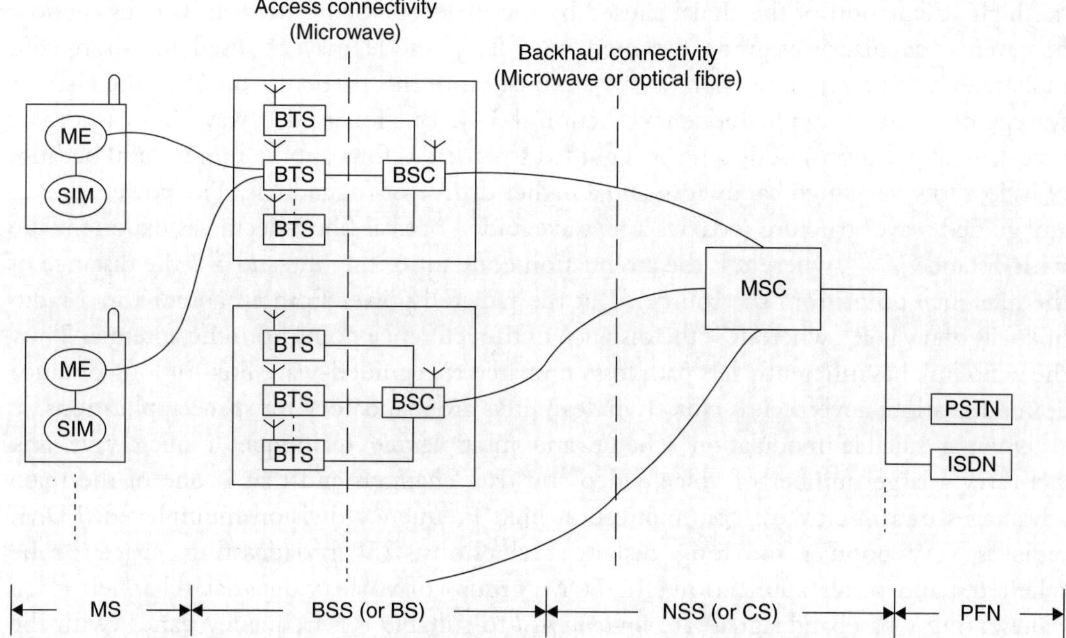

Fig. 16.1 Basic building blocks for GSM system architecture emphasising microwave access connectivity.

and provides switching interface of the cellular system with public fixed networks (PFN) like PSTN (public switched telephone network) and ISDN (integrated services digital network).

The frequencies (for up-link and down-link) used for microwave wireless connectivity between MS and BTS is in one of the internationally specified GSM (global system for mobile communication) bands—GSM 900, GSM 1800 or GSM 1900 (each differing in the RF carrier frequency being used and the bandwidth). The access connectivity between any BSC and BTSs is in the *Q* band of the microwave frequency spectrum, typically 57 GHz. The *Q*-band has been chosen to get rid of encryption between close-by BTSs as electromagnetic signals are strongly attenuated by O_2 and H_2O present in atmosphere in the frequency range (60 ± 10) GHz (Fig. 1.2).

16.2.2 Microwave Communication Links

Communication links for sending voice, data or television signals (having typical bandwidths of 4 kHz, 50 kHz to 1.5 MHz, and 6 MHz respectively) from transmitter to receiver may use either guided wave systems or radio (wireless) links. As the available transmission bandwidth is larger with higher carrier frequency, microwave frequency in the radio spectrum is naturally attractive to transmit a large number of voice, data and TV channels together. For example, using a 4 GHz microwave carrier with a modest 10% available transmission bandwidth, it is possible to send 100,000 voice channels or 66 TV channels. Coaxial lines as microwave guiding media are limited to a maximum of 3 GHz due to the high attenuation of the signal caused by the dielectric being used in its construction; however, a circular waveguide designed with TE_{01} mode may be used for short-haul millimeter-wave communication (as the attenuation of this particular mode in the circular waveguide decreases with frequency, Section 4.3.4). But for guided wave links (short or long-haul applications) optical fibre as guided structure is found to be more useful because of its low loss, very high bandwidth and a higher degree of ruggedness. The power level of any guided wave structure (coaxial line/waveguide/ optical fibre) decreases exponentially with distance ($e^{-2\alpha z}$, where α is the attenuation constant of the line and z is the distance of the reference point from the source). But the radiated power from an antenna in a radio link falls off as $1/R^2$, where R is the distance of the reference point from the antenna. Thus, the radio link has inherently less path loss compared to a guided-wave link for long distance communication; nevertheless radio (wireless) links are mandatory for space applications.

Using a suitable modulation scheme and multiplexing technique, a microwave link can carry a large number of voice, video and data channels and that is one of the main advantages of a microwave communication link. Frequency division multiplexed (FDM) radio is very popular for long distance telephony, TV broadcasting, space probe telemetry and other applications. In FDM, groups of voice, video, data channels (each representing a baseband signal) are *up-converted* to suitable RF frequency bands (with the help of *sub-carrier modulation*) and finally transmitted via a *master carrier* (whose frequency is in the microwave range). Fig. 16.2a shows the block diagram of a FDM transmitter while Fig. 16.2b shows the spectrum of the composite signal that modulates the master carrier in the transmitter. At the receiver, *down-conversion* takes place in which the master carrier is first demodulated and band-pass filters (tuned to the respective sub-carrier frequencies) are used to get the baseband modulated sub-carrier signals which are further processed in a mixer to retrieve the individual baseband signals, Fig. 16.2c. We have shown here three RF channels and it may be noted that single side band (SSB) modulation is used to avoid wastage of bandwidth. Low-pass filters used in both the transmitter and receiver help signal conditioning, that is, to get band-limited signal. To reduce the possibility of spectral overlap, the modulated spectra are spaced out in frequency by guard-bands into which the filter transition regions can be fitted in order to minimise cross-talk which may otherwise occur due to nonlinearities in the devices that process FDM signals.

Only a rudimentary scheme for microwave multiplexed system using analogue technique is mentioned above, however, the overwhelming majority of developments in microwave communication engineering in the last three decades employs digital methods. The use of digital signal processing, digital multiplexing, digital switching and transmission is the order of the day to which the students will be exposed in their communication engineering classes.

Here, it is relevant to discuss how to estimate the signal to noise power ratio at the output of a microwave receiver which is actually the determining factor for the minimum detectable signal by the receiver. Also for a given transmitter power, the maximum realisable range of the communication link is dependent on the S_o/N_o estimate.

The first element of the radio communication receiver through which the received signal from the transmitter enters is the antenna. If the antenna has a reasonably high gain with relatively low side lobes (may be 30 dB down the main lobe, as is usually the case with

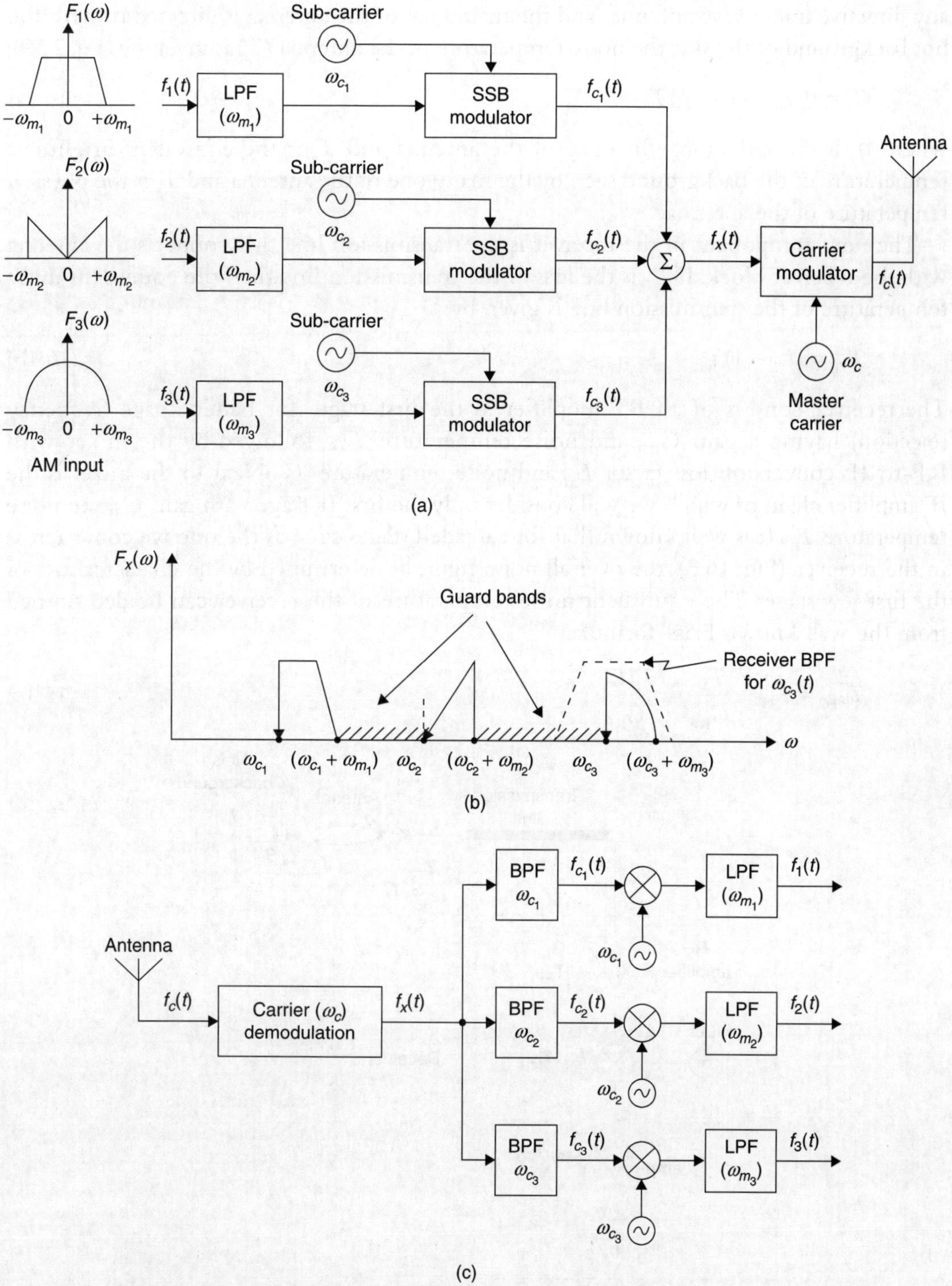

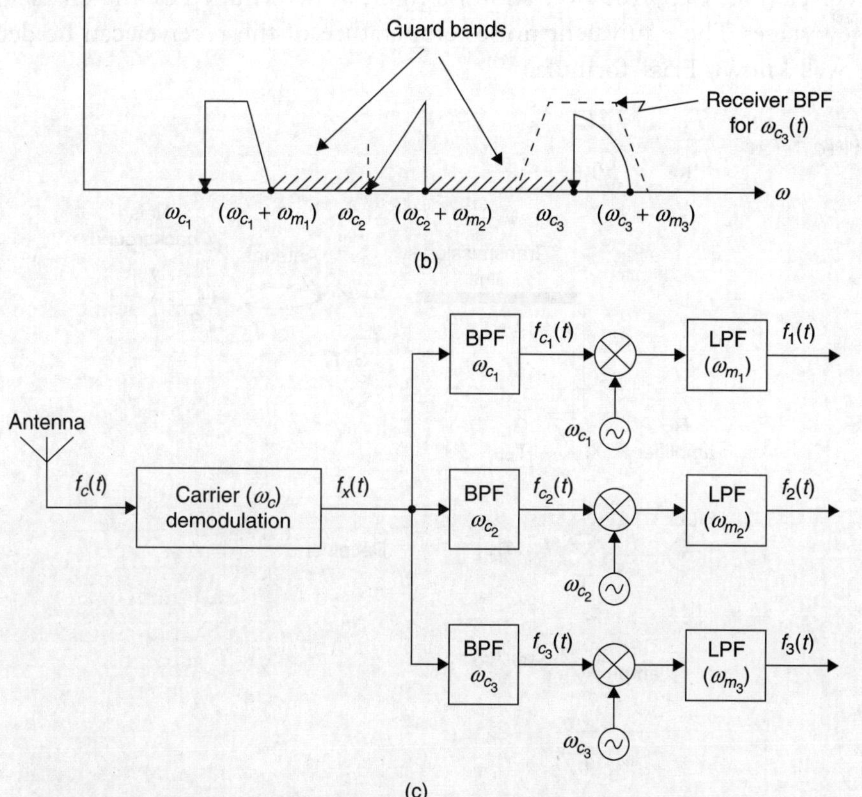

Fig. 16.2 Frequency division multiplexed (FDM) microwave radio (a) Block diagram of FDM transmitter (b) Spectrum of the composite signal modulating the master carrier in the transmitter (c) Block diagram of FDM receiver.

any directive microwave antenna) and the main lobe of the antenna is directed towards the hot background of the sky; the noise temperature of the antenna (T_a) is given by (Eq. 2.59):

$$T_a = \eta_a T_b + (1 - \eta_a) T_p \tag{16.1a}$$

where η_a is the radiation efficiency of the antenna and T_b is the equivalent brightness temperature of the background seen by the main lobe of the antenna and T_p is the physical temperature of the antenna.

The next component in the receiver is the transmission line that connects the antenna with the receiver block. If L_T is the loss of the transmission line then the equivalent noise temperature of the transmission line is given by:

$$T_{\text{TL}} = (L_T - 1) T_p \tag{16.1b}$$

The receiver consists of an RF amplifier as the first stage (for better image frequency rejection) having a gain G_{RF} and noise temperature T_{RF}, followed by the mixer with RF-to-IF conversion loss factor L_M and noise temperature T_M. Next to the mixer is the IF amplifier chain of which we will consider only the first IF stage with gain G_{IF} and noise temperature T_{IF}. It is well known that for cascaded stages such as the one we come across in the receiver (Fig. 16.3), the over all noise figure is determined by the characteristics of the first few stages. The equivalent noise temperature of the receiver can be determined from the well known Friss' formula:

$$T_{\text{REC}} = T_{\text{RF}} + \frac{T_M}{G_{\text{RF}}} + \frac{T_{\text{IF}} L_M}{G_{\text{RF}}} \tag{16.1c}$$

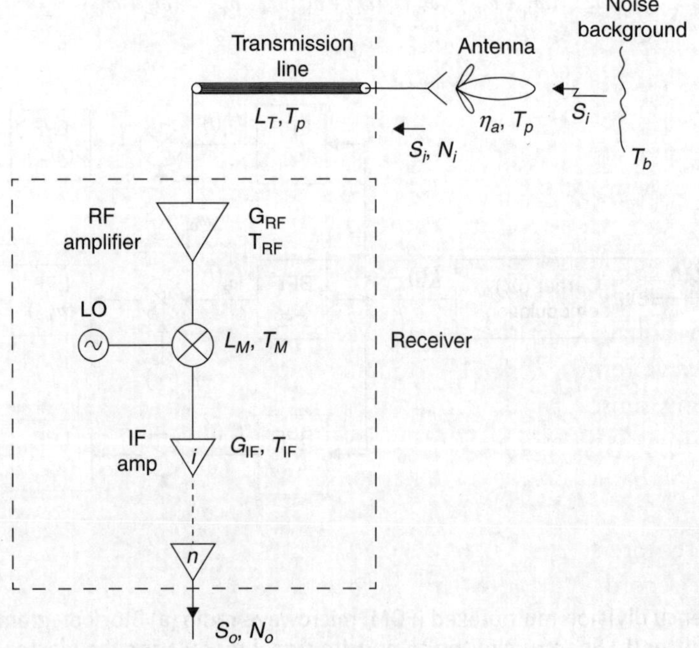

Fig. 16.3 Microwave receiver front-end with antenna and transmission line used for the estimation of signal to noise power ratio at the output of receiver.

The total noise power N_o at the output of the receiver is given by:

$$N_o = N_a + N_i G_{sys} \tag{16.1d}$$

where N_i is the noise power at the output terminals of the antenna that feeds the receiver via the transmission line and is given by:

$$N_i = k_B B_{sys} T_a = k_B B_{sys} \{\eta_a T_b + (1 - \eta_a) T_p\} \tag{16.1e}$$

(the expression for T_a has been used from Eq. 16.1a).

N_a is the added noise by the system consisting of the transmission line and the receiver, given by (Section 12.8.1):

$$N_a = k_B B_{sys} (T_{TL} + T_{REC}) G_{sys} \tag{16.1f}$$

Thus, the signal to noise ratio at the output of the receiver is given by:

$$\frac{S_o}{N_o} = \frac{S_i}{k_B B_{sys} \{\eta_a T_b + (1 - \eta_a T_p) + (L_T - 1) T_P + L_T T_{REC}\}} \tag{16.1g}$$

where we have used $G_{sys} = S_o/S_i$ and Eqs 16.1f, 16.1e, 16.1d, 16.1c and 16.1b.

16.2.3 Microwave Remote Sensing

Remote sensing is known to be the science and technology of acquiring and interpreting data collected on the earth's land surface, oceans and atmosphere and to develop information about features of the object without actually being in contact with it. The sensing signal may be in the optical, infra-red or microwave frequency spectrum. There are two types of microwave remote sensing—passive and active. Passive microwave remote sensing is similar in concept to thermal (infra-red, IR) remote sensing as all objects emit microwave energy (may be very small amounts) related to the temperature and moisture properties of the emitting object. Passive microwave remote sensing is done with *radiometers* that operate much in the same manner as the IR remote sensors; except that in this case, an antenna is used to detect and record the microwave energy. The active remote sensing uses the radar-type principle in which the sensor sends its own electromagnetic energy towards the object/terrain and after interaction with the object, the back-scatter signal is recorded by the remote sensor's receiver. There are two types of active microwave remote sensing—non-imaging and imaging type. A typical example of a non-imaging sensor is, a radar altimeter used for height measurement of aircraft above the terrain it is flying past. An example of imaging remote sensor is synthetic aperture radar (SAR) discussed in Chapter 11. The advantage of microwave remote sensing (active or passive) is that it permits all-weather day and night sensing unlike the IR and optical remote sensing.

Radiometers record energy in the region between 0.15 cm and 30 cm (generally centred at 1, 4, 6, 10, 18, 21, 37, 50, 85 and 183 GHz) which is well beyond the IR region (3–14 μm). Now, there is a growing interest in the use of radiometers to monitor some of the important global hydrologic variables such as soil moisture, precipitation, ice water content and sea surface temperature. The Tropical Rainfall Measuring Mission (TRMM) is sponsored by NASA (National Aeronautics and Space Administration), USA, and NASDA (NAtional Space Development Agency), Japan, to study the tropical rainfall and the associated release of energy that helps to power global atmosphere circulation. Multi-frequency Scanning Microwave Radiometer (MSMR) is a passive microwave sensor on board the Indian satellite IRS-P4 that provides global microwave brightness temperature measurements at 6.6 GHz, 10.65 GHz, 18.7 GHz and 21.3 GHz. The operational parameters retrievable from MSMR measurements are sea surface temperature, sea surface wind speed and columnar water vapour.

A typical block diagram of a radiometer is given in Fig. 16.4 which resembles a standard superheterodyne microwave receiver consisting of a low-noise RF amplifier, a mixer-local oscillator (LO), IF filter and amplifier, detector and low-pass filter (LPF) for signal condition. The antenna of the radiometer looks to the background scene with a brightness temperature T_b. The total output voltage of the radiometer is given by:

$$V_o = G_s(T_b + T_r)k_B B_r \tag{16.2}$$

where G_s is the overall system gain, T_r is the overall noise temperature of the receiver, k_B is the Boltzmann's constant and B_r is the receiver system bandwidth. The system is calibrated by replacing the antenna input with two calibrated noise sources, from which the system constants $G_s k_B B_r$ and $G_s T_r k_B B_r$ are determined. This is similar to the Y-factor method for measuring noise temperature (Section 12.8.1), which yields the brightness noise temperature T_b of the system. The principle of measurement of T_b discussed above is known as the total power radiometer which was modified by R H Dicke known as a Dicke null-balancing radiometer that can resolve gain variation problem of the total power radiometer by repeatedly calibrating the radiometer at a rapid rate; faster than the time constant (> 1 s) of the system gain variation.

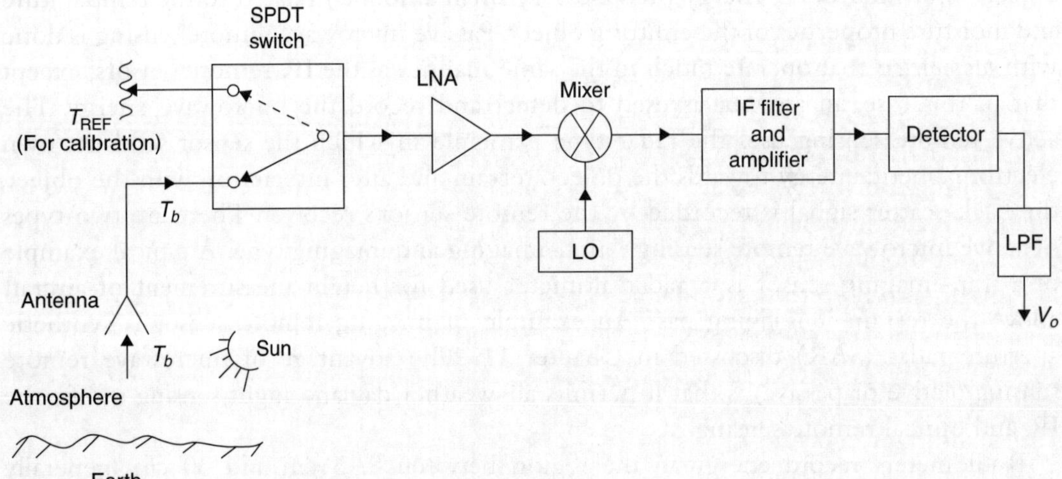

Fig. 16.4 Block diagram of total power radiometer.

Microwave remote sensing with imaging type radar can be done either with real aperture radar (RAR) or with synthetic aperture radar (SAR) (Section 11.8). It was observed that SAR has many advantages over RAR; the former can realise much better across-track image resolution by artificially synthesising/simulating a large antenna aperture and can operate at comparatively lower RF frequencies with a reasonable antenna size making all-weather imaging possible. Further, compared to RAR, the SAR imaging is independent of slant range that is the height of the aircraft/spacecraft thereby providing same across-track/cross-range spatial resolution over the entire swath.

Like other remote sensing systems an imaging radar sensor may be carried on air borne or space borne platforms. However, regardless of the platform being used, a significant advantage of SAR imaging over optical and IR is that the spatial resolution in SAR is independent of platform altitude and hence fine resolution can be achieved from both air borne and space borne platforms. Depending on the prospective imagery, there are trade-offs between the two types of platforms. Although the spatial resolution is independent of altitude, the altitude variations can greatly influence the viewing geometry and swath coverage. Though air borne imaging can realise relatively wide swath coverage compared to space borne imaging, space borne imaging is done from a much higher altitude (may be 100 times that of the airborne system) thus image quality (that is inversely dependent on incidence angle) is better realisable with the space borne imaging system with comparable swath coverage. Although air borne radar systems are more susceptible to imaging geometry problems, they are flexible in their capability to collect data from different look angles and look directions. This is not possible with a space borne radar as its viewing geometry and data acquisition schedule is controlled by the pattern of its orbit. However, the space borne radar does have the advantage of being able to collect imagery over a large area quicker than an air borne radar and provides consistent viewing geometry.

16.3 Industrial Applications of Microwaves

Microwave heating is unquestionably the first fundamentally new heating technique since the discovery of fire and it has become familiar to the general public because of the ubiquitous domestic cooking/heating appliance—the microwave oven. In recent times, however, microwaves are finding numerous industrial applications; for example in drying, vulcanising, polymerising, pasteurising, sterilising and also in many other contactless industrial measurements and control like thickness measurement of metal sheets in rolling mills and monitoring of moisture content in the paper and textile industry. Microwave heating is also used to speed-up chemical processes which proceed too slowly at ambient temperatures as in annealing semiconductors, blending and bleaching woods and calcining nuclear waste materials. The microwave frequency used for these purposes including those for medical and domestic applications falls in the ISM (Industrial, Scientific and Medical) band with typical centre frequencies of 915 MHz, 2.45 GHz, 5.8 GHz, 24.125 GHz.

Let us discuss briefly, the industrial applications of microwaves with some details of the principle of one or two specific methods.

16.3.1 Drying

Many industrial processes require one or more stages of drying of the item under consideration, which is often very costly with conventional techniques. However, microwave offers what is sometimes the most viable alternative in terms of energy efficiency. Combining microwaves with traditional heat sources can control the temperature gradient and humidity gradient in an optimum way thereby achieving uniform drying. *Paper and printing industries* have been highly benefitted with the use of microwave heating as it leads to superior printing quality with colours deeper and well defined and also leads to time and energy savings. *Leather and textile industries* prefer to use microwave heating in recent times because of its fast heating capability so that the material has no time to change colour or degrade, which is sometimes a problem with hot-air (infra-red) dryers. *Wood* can indeed be readily dried by microwaves provided the process is slow enough to avoid the creation of internal strains that affect product quality. This type of drying is also an efficient means of insect control. The drying of *foundry mould* requires a considerable amount of energy because of the very poor heat transfer properties of the materials to be dried. This is a classic situation in which the use of microwaves is more advantageous than gas dryers. Uniform drying with microwaves ensures the least possible distortions, better surface finish and greater hardness. In *photographic film processing*, microwave is again of great use. The photographic film is made of polyesters and acetates that have very low dielectric loss while wet emulsions have high loss value. This is a classic situation in which microwaves can be used for selective heating.

16.3.2 Vulcanisation

Vulcanisation (curing of rubber in an atmosphere of sulphur in a heated ambience) requires very high and controlled temperatures; so also devulcanisation of rubber for recycling. Compared to natural rubber, vulcanised rubber is more durable and has a much higher mechanical strength. Elastomers of which rubber is made are poor thermal conductors that require complex and costly heating procedures if conventional heating techniques are used. Microwaves may therefore have a considerable role to play in this field as a carrier of energy that is independent of the thermal properties of the material. The recycling of rubber requires the breaking of bridging bonds created by vulcanisation. Substantial heating is needed to achieve devulcanisation. The difficulty is that the devulcanisation and depolymerisation temperatures are not very different. The uniformity of heating and precise control of temperature that is possible with microwave heating makes it advantageous to use.

A simple arrangement for rubber curing (vulcanising) is shown in Fig. 16.5a. It consists of a microwave power generator which may be a klystron or a magnetron. The impedance matching device (slide-screw tuner or E–H tuner) must be adjusted to minimise reflection from the applicator since the amount of reflected signal changes all the time during the process of curing of rubber. Automatic impedance matching equipment is available from vendors which keeps the reflection from the applicator at a minimum during the entire period of curing of rubber. The applicator may be the standing-wave (SW) type (which is useful for small samples so that it fits into the hot spot of the SW) or travelling wave type (which is useful for large size samples). The travelling wave structure may be a helix (Fig. 16.5b) or a

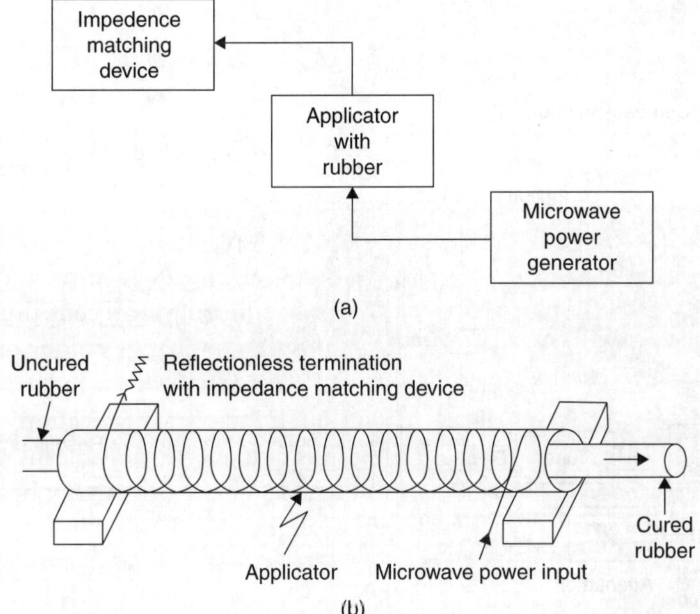

Fig. 16.5 Rubber vulcanising (a) Simple block diagram (b) Travelling wave type applicator for rubber vulcanisation.

meander line. Shielding of the applicator with rubber to be cured inside it, as shown in Fig. 16.5a, is necessary for safety of the operator and prevention of RF interference.

16.3.3 Thickness Monitoring of Metal/Dielectric Sheets in Rolling Mills

Contact-less monitoring of thickness of metal/dielectric sheets in rolling mills can be done with the microwave technique. Here, the sheet is passed between microwave antennae connected in a transmission line loop (Fig. 16.6). The total electrical length of the loop path includes (dependent on) the thickness of the sheet. The RF energy reflected from opposite surfaces of a metal sheet (or transmitted through a dielectric sheet) is processed with reflection measurement for the metal sheet and transmission measurement for dielectric sheet to monitor the thickness of the sheet material. It may be observed from the diagram that the cross-section of the metal sheet has been shown as M that continuously rolls with its length moving into or away from the plane of the paper whose thickness needs to be monitored. The plane of the metal sheet passes parallel to the aperture planes of two horn antennae H_1 and H_2 facing each other and placed on either side of M as shown in the diagram. The two horns are connected via waveguides to the two ports of a circulator C and one of the waveguide arms has a variable phase shifter (φ). The third port of the circulator is connected to a reflection measurement bridge (for metal sheets) built around a magic tee (E–H tee). A resonator cavity is formed with the two horns, waveguides and the circulator (shown by the dotted box).

To carry out the measurement, the phase shifter (φ) is adjusted initially without the sheet material M in place, so that the resonant frequency of the resonator cavity becomes

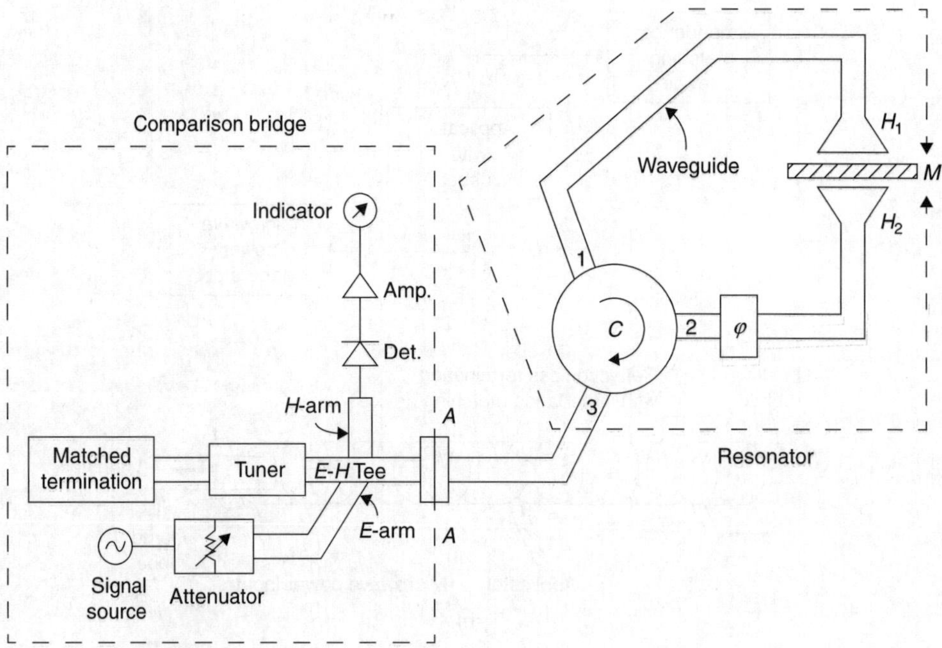

Fig. 16.6 **Arrangement for thickness monitoring of metal sheet using microwave technique.**

equal to the frequency of the signal source—this makes the impedance at the plane AA practically matched (looking in either direction perpendicular to the plane AA) with no reflection. Introduction of the metal plate *M* in place changes the overall length of the transmission line loop path and hence the resonant frequency, causing a proportional impedance mismatch at the plane AA and resulting in a small reflection originating at plane AA. An accurate measurement of this small reflection with due calibration of the indicating meter of the comparison bridge built around the magic tee can continuously monitor the thickness of the metal sheet. For initial setting when there is no reflection from the AA plane, the tuner of the comparison bridge is adjusted so that the reflected power in the two collinear arms of the magic tee are the same and in phase; hence no signal appears at the *H*-arm. Any change in the reflection coefficient at the plane AA causes a signal to be present in the *H*-arm which is detected, amplified and displayed in the calibrated indicator (*I*) that monitors the thickness of the metal sheet directly.

The advantage of using the microwave technique for thickness monitoring in rolling mills is that it is a contact-less measurement, avoiding friction and wear when the object itself is in motion and/or at high temperature (as in rolling mills). The precision and accuracy of the microwave technique is also much higher compared to the conventional contact-type techniques.

16.4 Microwaves in Medical Applications

Nowadays microwave technology is widely used in a variety of medical treatments either using hyperthermia (that locally heats the tumour cells and ruptures their membrane) or ablation (that dries up or desiccates the tumour cells with localised application of heat). In either case, dielectric heating of tumour cells (which has a higher dielectric loss factor than healthy tissues because of higher water content) with microwave energy works well. Another interesting application of microwave heating is in the thawing of blood derivatives in vinyl packages kept at temperatures between –30°C and –60°C. Conventionally, these products are immersed in a water bath at 37°C, which creates the risk of bacterial contamination and requires approximately 30 minutes, which is an unacceptable delay in the case of emergency. The use of microwave radiation for heating is a practical solution that is rapid and aseptic as there is no contamination of the product by the packaging. Microwaves at the high end of its spectrum leading to terahertz (>1000 GHz) also show promise as a diagnostic tool. In fact, T-ray imaging techniques are used for early diagnosis of the growth and proliferation of cancerous cells by studying the absorbance pattern of the healthy and cancerous cells. Terahertz frequency has the special property of being able to transmit like RF waves through the translucent human body and also has the spectroscopic capability of optical waves. Further, terahertz is a non-ionising radiation. There is thus no doubt at all that the doctor's chamber in the near future will make increasing use of microwave diagnostic and treatment tools.

16.4.1 Microwave Hyperthermia and Microwave Ablation

16.4.1.1 Microwave hyperthermia

Hyperthermia (also known as thermotherapy) utilises heat generated by microwave signals to heat up tumour (normally cancerous) cells within the body. The temperature rise (typically to 108°F or 42°C) within the tumour cells results in rupturing of the cell membrane leading to the destruction of the cancerous cells. For such treatments, a controlled dose of microwave signal is delivered to the tumour cells by special-purpose antennae (phased array type, which is an ensemble of elementary sources fed through phase shifters, originally used for radar and telecommunications) that surround the patient's body (Fig. 16.7) for whole body hyperthermia. A practical microwave hyperthermia system consists essentially of a generator and an applicator connected by a waveguide or a coaxial cable, and a number of subsystems for temperature monitoring, power control and data acquisition and storage.

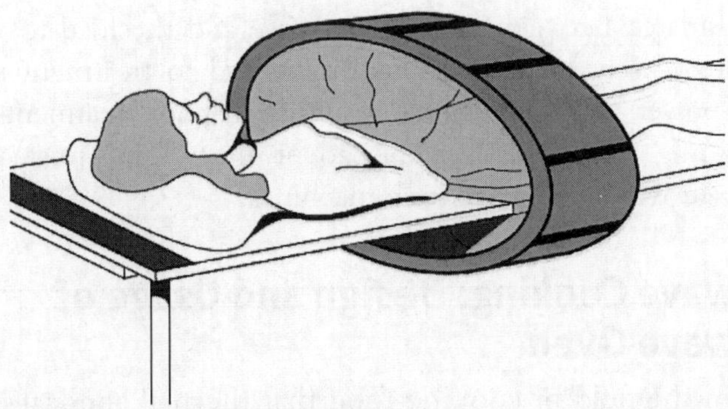

Fig. 16.7 Microwave hyperthermia arrangement.

This heating by microwave signals should be directed carefully with proper dose. If too much of the heat misses the tumour target, the skin or other healthy tissues could be burned, while if the dose results in too little heat then cancerous cells will not be killed. Apart from whole body hyperthermia, microwave hyperthermia is also used for treating brain tumours. In such applications microwave energy at a frequency of 2.45 GHz (oscillating at about a billion times a second) is guided into the tumours by one or more semi-rigid coaxial applicators. Microwave thermotherapy is found to be a promising modality for treating malignant brain tumours, either as the sole therapy or sometimes in combination with radiotherapy and chemotherapy.

Microwave hyperthermia is found to have no side effects. It continues to be effective because the body does not get accustomed to it, and it produces practically no discomfort to the patient. The results in themselves are positive and become remarkable when microwaves are combined with radiotherapy or chemotherapy; in fact the combined treatment is often more effective than the two procedures employed individually. In addition to cancer treatment, microwave hyperthermia is found to be effective in relieving neurologic and arthritic pain.

16.4.1.2 Microwave ablation

Microwave ablation is used in the removal of unwanted tissue masses; for example, liver and lung tumours. Cancer patients who are poor surgical candidates can also benefit from microwave ablation, as it is minimally invasive. Apart from oncology, microwave ablation is also useful in cardiology, cosmetic and dental treatments. Microwave technology based systems using a magnetron as the microwave power source operating at 915 MHz and 2.45 GHz are ideally suited for large volume ablation, while the use of higher frequencies (in the range 5.8 GHz to 10 GHz) can create shallow penetration of energy due to comparatively small skin depth (since skin depth is inversely proportional to the square root of frequency) and hence are suitable for surface based treatments, such as for the treatment of skin cancer.

While applying microwave signals to the patient in both the hyperthermia and ablation treatment mentioned above the signal is administered, as usual, from a matched source to a matched load, typically 50 Ω. It is important to note that the dielectric properties of cancerous tumour cells vary considerably and hence microwave applicators (antennas and coaxial feed lines) are not always optimally matched to an ideal 50 Ω. This can lead to a significant mismatch between the signal source and the load to which it is applied and can thus create VSWR problems. This might lead to treatment uncertainty, that is, either insufficient power will be delivered resulting in poor treatments or administration of excessive power might cause inadvertent patient injury. Thus mismatch tolerance issue needs to be addressed while designing such machines.

16.5 Microwave Cooking: Design and Usage of Microwave Oven

Historically, it is in the field of cooking food that thermal applications of microwaves took off; the first American patent on the microwave oven was granted in 1951. The concept of heat generation by microwave signals is based on the principle of dielectric heating (Section 2.4). With that physical principle in the background, let us discuss some of the microwave considerations in the design and usage of a microwave oven for cooking. Generation of heat in a dielectric material at microwave frequency (typically 2.45 GHz/915 MHz; at the latter frequency, the depth of penetration is large as skin depth is inversely proportional to the square root of frequency) is dependent on the loss tangent ($\tan \delta_c = \varepsilon''/\varepsilon'$) of the dielectric material. Higher the loss tangent of the dielectric material to be heated, faster will be the heat generation (as $P_{av} \; \alpha \; \varepsilon''$ Eq. 2.16). Thus, food like meat possessing a large value of loss tangent will get quickly heated while microwave-safe utensils designed with a very low loss tangent will not get heated up even with long exposure to microwave signals. Further, the loss tangent of a dielectric material at a given frequency decreases with the increase of temperature and thus microwave heating is a self-regulating process. Compared to conventional heating by fire which is from the outside of the material to the inside (by conduction, convection), microwave heating is from the inside to the outside of the material (heating at the molecular level). The efficiency of microwave heating in the microwave oven (the ratio of power converted to heat in the food material to the power supplied to the oven from the power supply) is typically 50% to 60%, this is usually greater than the cooking efficiency possible with conventional heating using gas oven, charcoal fire or electric heater.

Microwave cooking obeys the classic heat equation, $W = mC\Delta T = P\Delta t$, where W is the energy in joules, m is the mass in kilograms, C is the heat capacity in $\text{Jkg}^{-1}\text{K}^{-1}$, ΔT is the rise in temperature in K, P the power in watts and Δt is the time interval in seconds. Microwave heating is, by its very nature, non-uniform for a number of reasons.

(i) In a domestic microwave oven, the electric field in the applicator (typically a waveguide) is non-uniform causing non-uniform heating of the item (in industrial microwave heating special methods are resorted to in the design of applicator to make it uniform as far as possible).

(ii) The food item to be heated is normally heterogeneous (having interfaces separating regions of low and high dielectric loss) and the shape of the item may also give rise to axial or central hotspots. Uniform microwave heating may be realised if the dimension of the item to be heated does not exceed 1 to 1.5 skin depth ($\delta = 1/\alpha$) at the operating frequency (2.45 GHz). For a steak having dielectric constant or relative permittivity $\varepsilon_r' = 40$ and loss tangent $\tan \delta_c = 0.3$, it may be calculated (Eq. 2.7a) that the depth of penetration in the food item is approximately 2.07 cm at 2.45 GHz. Thus, the use of 915 GHz as operating frequency would be more appropriate for cooking/heating food items with larger cross sections and high dielectric loss factor. During cooking, it is advised that some standing time be allowed after microwave cooking so that temperature equilibrium is reached with further heat conduction within the food cooked in the microwave oven.

A microwave oven (Fig. 16.8a) consists of a high power *microwave source*, typically a magnetron (with CW power output ranging from 500 W to 1500 W, depending upon the capacity of

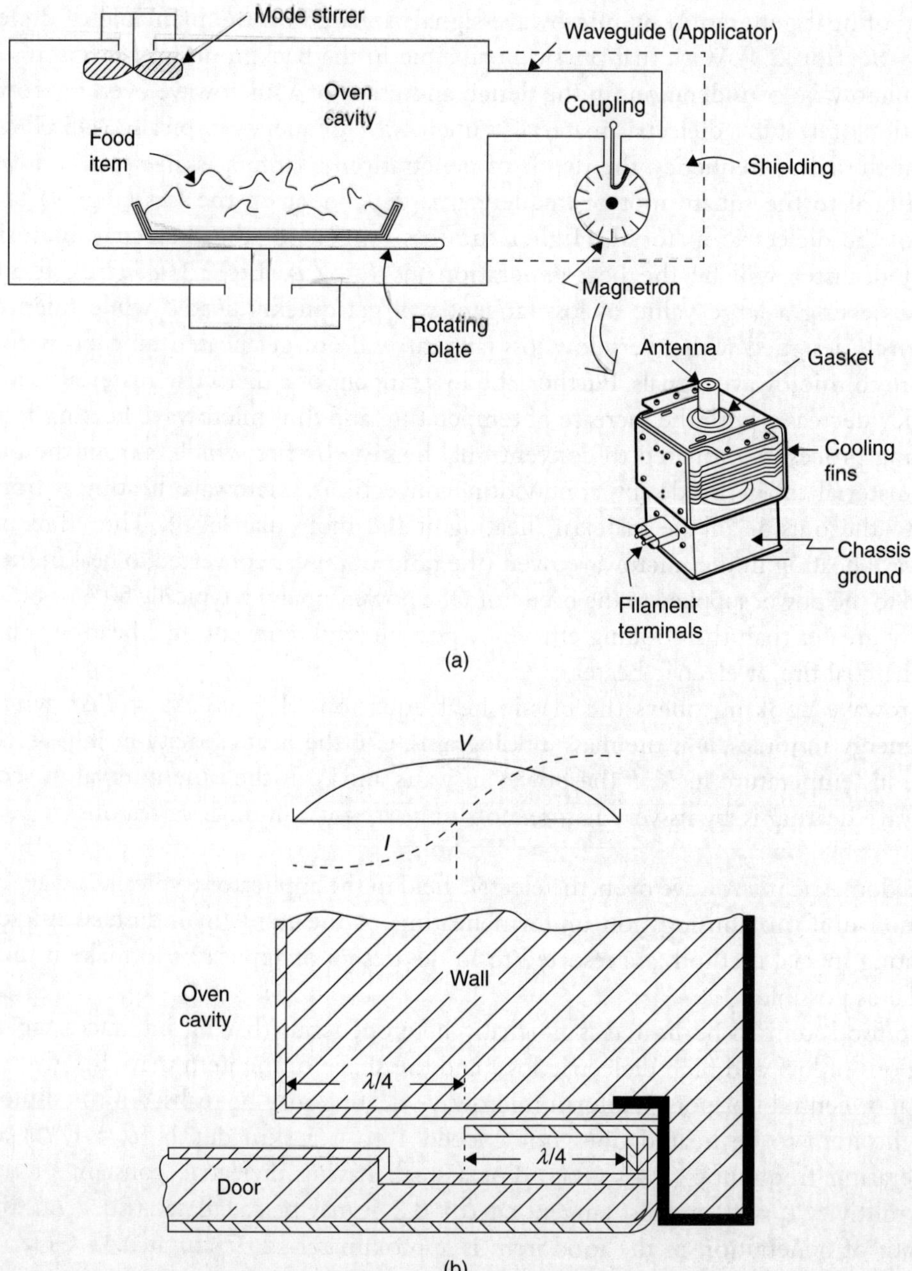

Fig. 16.8 (a) Schematic of microwave oven (b) Choke arrangement (radiation trap) for the door.

the oven), operating at 2.45 GHz, a *waveguide feed* that helps to feed the microwave signal from the source to the oven cavity with proper impedance matching, and the *oven cavity* itself which is electrically large (few wavelengths) and having metallic inside walls. For uniform heating of the food material placed inside the microwave oven, the food is rotated with a *motorised platter* and a *mode stirrer* (which is basically a metallic fan blade) is used to perturb the field distribution inside the oven (otherwise uneven heating will be caused by the formation of standing wave inside the oven cavity).

In the design of a microwave oven, the most critical issue is to ensure the oven safety; otherwise any leakage of microwave signal from the oven would result in health hazards. Thus, care must be taken for electromagnetic shielding of both the magnetron source and the feed waveguide. Equal importance needs to be given to restrict any leakage of microwave signal via the periphery of the door of a microwave oven. This is done by using suitable quarter-wavelength ($\lambda/4$ = 3.06 cm) choke arrangement (Fig. 16.8b) known as a *radiation trap* (that brings about a maximum electric field and a minimum current at the points of possible escape, the door-wall contact; resembling an open-circuited quarter-wave line) at the operating frequency (2.45 GHz). The radiation traps are filled with absorbing material. The front door of the microwave oven is also made with a metallic grid mesh structure that permits the optical signal to pass through as grid gap >> λ_{optical} (for visual observation of what is going on inside) but prevents microwave signal leakage as the grid gap << $\lambda_{\text{microwave}} \approx$ 12 cm, making the grid mesh structure a continuous metallic surface to the microwave signal at 2.45 GHz.

A few important points of caution regarding the usage of microwave oven are as follows. Metallic utensils should never be used in a microwave oven as metals cause complete reflection of the microwave signal resulting in undesired extreme hot and cold spots causing a harmful effect to the performance of the oven. Only microwave-safe utensils are advised for use which will not cause any undue heating of the utensil so that best heating efficiency is realised from the oven and with no inconvenience caused by undue heating of the utensil itself. Eggs should not be boiled inside a microwave oven since it may cause bursting of the egg as the yolk and white of egg is a highly lossy dielectric substance with high value of loss tangent while the hard shell of the raw egg has very low loss tangent. Thus, if inserted for boiling inside a microwave oven, extremely high pressure generated inside the egg will cause it to burst! This can happen if eggs or even chestnuts are cooked in a microwave oven in their shells. Refrigerated water bottles made of plastic should also not be put inside a microwave oven for quick heating this will also be hazardous with a possibility of bursting.

Food safety is an important health issue. In a microwave oven, the rate of heating depends on the power rating of the oven and on the water content, density and amount of food being heated. Microwave energy does not penetrate well in thicker pieces of food, and may produce uneven heating of the food item. This can lead to a health risk if parts of the food are not heated sufficiently to kill potentially dangerous microorganisms. Thus the food heated in a microwave oven should be given some standing time to allow the heat generated to get distributed throughout the food. Food cooked in a microwave oven is safe (as per WHO, World Health Organization) and has the same nutrient value compared to the food cooked in a conventional oven using fire or electricity.

16.6 Microwave/RF Life-detection Systems

Use of microwave/RF for remote sensing of breathing and heartbeats of human subjects located at a distance or behind a barrier was originally developed for military and security purposes. However, they are also used in medical applications, especially in the remote physiological sensing areas and in the design of sensitive life-detection systems which can be used to locate human victims trapped deep under earthquake rubble or collapsed building debris.

Existing methods of searching and rescuing human victims buried under earthquake rubble or collapsed building debris include sniffer dogs, seismic or optical devices. The existing methods are not effective if the rubble or debris covering the human victims is thicker than a few feet, especially if the victims are completely trapped or too weak to respond to the signal being sent by the rescuers. A sensitive life-detection system which can detect the passive victim under such situations is the microwave/RF life-detection system.

The basic principle for the operation of a microwave/RF life-detection system is discussed below. When an electromagnetic (em) wave of appropriate frequency (*L*, *S* or even UHF band) is aimed via an antenna at a pile of earthquake rubble or collapsed building debris under which a human subject is expected to be buried, the em wave can penetrate through the rubble (as RF signal has such penetration capacity; lower the frequency higher is the penetration depth) to reach the subject. On being illuminated by the incident em signal (whose intensity needs to be just enough for sensitive detection but should be higher than the safety standard) the subject will generate a backscatter signal (reflect the em signal back towards the source) which will be modulated by the subject's body movement, which includes breathing and heartbeat. If the reflected wave from the stationary background (earthquake rubble/building debris) be cancelled, and the backscatter signal from the subject's body is properly demodulated, the breathing and heartbeat signals of the subject can be extracted. In this way with the help of RF, a human subject buried under rubble/debris can be sensed and precisely located with very high accuracy.

The signal processing of the modulated backscatter signal is the real challenge in the design of such sensitive life-detection systems based on the microwave/RF technique. A common application of detecting moving targets with RF/microwave Doppler effect is in police radar. A speed monitoring police radar operating at 10 GHz will read a frequency shift of 1 kHz in the reflected wave from a car-target travelling at a speed (v) of 96.5 km/h. This frequency shift is significantly large to be measured with a cycle counter. But when RF/microwave is used to detect a very small movement of a target, such as the body movement associated with breathing or heartbeat of a human subject it is impractical to use the Doppler effect because the frequency shift incurred is of extremely small values since v/c will be extremely small. Thus a life-detection system based on RF/microwave technique measures the phase shift (and not the frequency shift) in the reflected wave. In other words, the RF/microwave receiver system is based on processing of the phase modulation in the reflected waveform caused by the feeble movement (breathing or heartbeat) of the human subject.

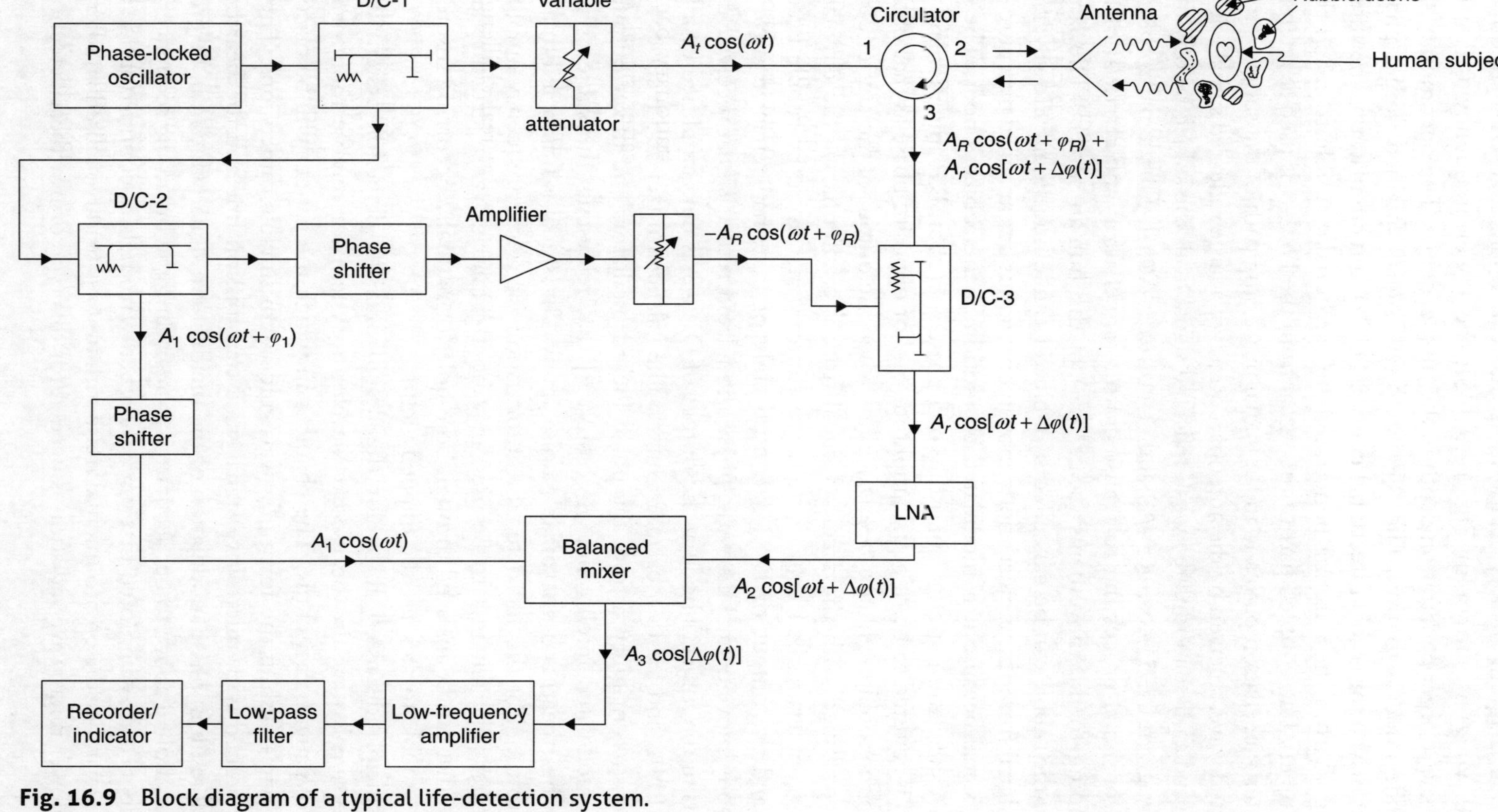

Fig. 16.9 Block diagram of a typical life-detection system.

The block diagram of a typical system that can operate at UHF frequency band (1150/450 MHz) is shown in Fig. 16.9 [45]; the lower frequency RF system will naturally have a larger penetration depth in the rubble/debris. The operation of the system may be understood as follows. The RF/microwave signal obtained from a phase-locked oscillator is driven through a directional coupler D/C-1 and circulator to the antenna. The radiated signal from the antenna penetrates the earthquake rubble to reach the buried human subject. The coupled part of the signal from the D/C-1 is used as reference signal in the clutter cancellation path and in the signal processing path of the subject's movement. The reflected wave, received by the antenna, consists of a large reflected signal (clutter) from the rubble and a phase modulated weak reflected signal from the subject's body movement, respectively given by $A_R \cos(\omega t + \varphi_R)$ and $A_r \cos[\omega t + \Delta\varphi(t)]$. This signal is routed via the circulator for signal processing to realise clutter cancellation and retrieval of the signal corresponding to the movements (heart beat) of the human subject. The large clutter signal from the rubble can be cancelled by a clutter cancelling signal. However; the small reflected signal from the subject's body movement cannot be cancelled by a pure sinusoidal cancelling signal because it is modulated by motion in the subject's body. The output of the clutter cancellation circuit (consisting of a phase shifter, an RF amplifier and a variable attenuator) is adjusted to be of equal amplitude and of opposite phase as that of the clutter signal from the rubble, $-A_R \cos(\omega t + \varphi_R)$. Thus when the output of the clutter cancellation circuit is combined with the received signal from the antenna via the circulator in the directional coupler marked D/C-3, the large clutter signal from the rubble is completely cancelled and the output of D/C-3 consists only of the small reflected wave from the subject's body movement. This weak phase modulated return signal received due to the movements in the subject's body at the output of D/C-3, is next amplified by a low-noise amplifier (LNA) and fed to a balanced mixer. The balanced mixer also receives a reference signal $A_1 \cos(\omega t)$ originated from the phase-locked oscillator. The mixer output provides the phase modulated intelligence $A_3 \cos[\Delta\varphi(t)]$ that represents the subject's movement (breathing/heartbeat) and unavoidable noise, if any. This signal is finally fed to a recorder/indicator to monitor the life detection information via a low frequency amplifier and a low-pass filter. The antenna to be used may be a cylindrical wire structure antenna of probe type that can be inserted through bore holes or naturally occurring fissures into the earthquake rubble to seek for the trapped victim.

The principle discussed above for the design of a life-detection system using RF/microwave signals may also be used to develop a system for remote monitoring of the heartbeat of patients in a clinic. The chamber in which the patient will be placed may be a metallic wire-mesh in the form of a waveguide with suitable dimensions. The chamber may be energised with microwave signal via a transmission line and the heartbeat of the patient will result in phase modulation of the incident em signal which can be processed in the way mentioned above to get accurate data regarding the breathing sequence/heartbeat of the patient. It may be noted that since the microwave field is confined inside a metallic waveguide chamber the environmental noise immunity is readily realised and hence only a very low power microwave radiation is needed for this type of heartbeat detection system.

Review Questions

16.1 Radar is even today the most important application which uses and also drives innovative developments of microwave engineering—discuss.

16.2 Mention a few communication, medical and household applications where microwave engineering is used.

16.3 Broadband microwave access for mobile communication uses microwaves—discuss.

16.4 Explain the operation of an FDM microwave radio system used for long distance telephony.

16.5 What is the advantage of microwave remote sensing over IR and optical techniques? What do you understand by passive and active microwave remote sensing?

16.6 Discuss the principle of total power radiometer. How does the Dicke null-balancing radiometer solve the problem of total power radiometer?

16.7 Why is SAR better than RAR as a microwave remote sensing radar? Compare the advantages and disadvantages of air borne and space borne remote sensing radar.

16.8 Many industrial processes require one or more stages of drying—discuss the use of microwave heating for this purpose with a few examples.

16.9 What is vulcanisation of rubber? Discuss a scheme for rubber vulcanisation using microwave heating.

16.10 Thickness monitoring of metal/dielectric sheets in rolling mills is preferable with the microwave technique—explain. Discuss a scheme for the same.

16.11 The doctor's chamber in the near future will make increasing use of microwave diagnostic and treatment tools—discuss.

16.12 Discuss the principle of microwave hyperthermia and microwave ablation used for cancer treatment.

16.13 Discuss the principle of microwave cooking. Microwave cooking is self-regulating and more efficient—explain.

16.14 Discuss the scheme for a microwave oven. Door safety is realised with a radiation trap—explain.

16.15 Explain the danger of cooking raw eggs or chestnuts in a microwave oven.

16.16 Human victims trapped deep under earthquake rubble or collapsed building debris can be located with sensitive microwave techniques—discuss.

16.17 Discuss the principle of a microwave/RF life-detection system. Give the block diagram of such a system.

Review Questions

APPENDIX

I

Z, Y, ABCD and S-parameters and their Interrelations

At microwave frequencies, when we talk of TEM lines such as coaxial line, or a quasi-TEM line such as a microstrip line, we can analyse them with equivalent/distributed current and voltages and may define impedance (*Z*) parameters or admittance (*Y*) parameters similar to low frequency circuits. For non-TEM lines or guiding structures like waveguides, we may also talk of equivalent voltage/current only in the sense that voltage is proportional to the transverse electric field and the current is proportional to the transverse magnetic field. However, this analogy may be drawn only for a particular waveguide mode. In general, at microwave frequencies we talk of scattering (*S*) parameters defined in terms of incident, reflected and transmitted wave variables which represent the corresponding power in the signal. This is because equivalent voltages and currents and the related impedance and admittance matrices at microwave frequencies are actually abstractions in the practical sense. Further, it is difficult to implement perfect open or short at microwave frequencies. More importantly, many microwave circuits will not behave properly when presented with an open or short termination—distortion or oscillation may occur.

The *Z*, *Y* and *S* parameter representations can be used to characterise a network with an arbitrary number of ports. But many practical networks consist of a cascaded connection of two or more networks, each of which is a two-port network. Thus, it is convenient to define a 2×2 transmission, ABCD matrix, for each of those two-port networks. This is advantageous in the sense that the composite ABCD matrix of the cascaded connection of two or more networks is then just the product of the ABCD matrix of the individual two-port networks.

In microwave design and analysis we quite often need conversion between two-port networks expressed in *Z*, *Y*, *S* or ABCD parameters. Table I.1 summarises such conversion formulae.

Table I.1 Conversion formulae among different two-port network parameters [9, 23]

	Z	Y	S	$ABCD$
Z_{11}	Z_{11}	$\frac{Y_{22}}{\|Y\|}$	$Z_0 \frac{(1+S_{11})(1-S_{22})+S_{12}S_{21}}{\Delta S_Z}$	$\frac{A}{C}$
Z_{12}	Z_{12}	$\frac{-Y_{12}}{\|Y\|}$	$Z_0 \frac{2S_{12}}{\Delta S_Z}$	$\frac{AD-BC}{C}$
Z_{21}	Z_{21}	$\frac{-Y_{21}}{\|Y\|}$	$Z_0 \frac{2S_{21}}{\Delta S_Z}$	$\frac{1}{C}$
Z_{22}	Z_{22}	$\frac{Y_{11}}{\|Y\|}$	$Z_0 \frac{(1-S_{11})(1+S_{22})+S_{12}S_{21}}{\Delta S_Z}$	$\frac{D}{C}$
Y_{11}	$\frac{Z_{22}}{\|Z\|}$	Y_{11}	$Y_0 \frac{(1-S_{11})(1+S_{22})+S_{12}S_{21}}{\Delta S_Y}$	$\frac{D}{B}$
Y_{12}	$\frac{-Z_{12}}{\|Z\|}$	Y_{12}	$Y_0 \frac{-2S_{12}}{\Delta S_Y}$	$\frac{BC-AD}{B}$
Y_{21}	$\frac{-Z_{21}}{\|Z\|}$	Y_{21}	$Y_0 \frac{-2S_{21}}{\Delta S_Y}$	$\frac{-1}{B}$
Y_{22}	$\frac{Z_{11}}{\|Z\|}$	Y_{22}	$Y_0 \frac{(1+S_{11})(1-S_{22})+S_{12}S_{21}}{\Delta S_Y}$	$\frac{A}{B}$

(continued)

Table I.1 Continued

	Z	**Y**	**S**	**ABCD**
S_{11}	$\frac{(Z_{11}-Z_0)(Z_{22}+Z_0)-Z_{12}Z_{21}}{\Delta Z}$	$\frac{(Y_0-Y_{11})(Y_0+Y_{22})+Y_{12}Y_{21}}{\Delta Y}$	S_{11}	$\frac{A+B/Z_0-CZ_0-D}{A+B/Z_0+CZ_0+D}$
S_{12}	$\frac{2Z_{12}Z_0}{\Delta Z}$	$\frac{2Z_{21}Z_0}{\Delta Z}$	S_{12}	$\frac{2(AD-BC)}{A+B/Z_0+CZ_0+D}$
S_{21}	$\frac{-2Y_{12}Y_0}{\Delta Y}$	$\frac{-2Y_{21}Y_0}{\Delta Y}$	S_{21}	$\frac{2}{A+B/Z_0+CZ_0+D}$
S_{22}	$\frac{(Z_{11}+Z_0)(Z_{22}-Z_0)-Z_{12}Z_{21}}{\Delta Z}$	$\frac{(Y_0+Y_{11})(Y_0-Y_{22})+Y_{12}Y_{21}}{\Delta Y}$	S_{22}	$\frac{-A+B/Z_0-CZ_0-D}{A+B/Z_0+CZ_0+D}$
A	$\frac{Z_{11}}{Z_{21}}$	$\frac{-Y_{22}}{Y_{21}}$	$\frac{(1+S_{11})(1-S_{22})+S_{12}S_{21}}{2S_{21}}$	A
B	$\frac{\lvert Z\rvert}{Z_{21}}$	$\frac{-1}{Y_{21}}$	$Z_0\frac{(1+S_{11})(1+S_{22})-S_{12}S_{21}}{2S_{21}}$	B
C	$\frac{1}{Z_{21}}$	$\frac{-\lvert Y\rvert}{Y_{21}}$	$\frac{1}{Z_0}\frac{(1-S_{11})(1-S_{22})-S_{12}S_{21}}{2S_{21}}$	C
D	$\frac{Z_{22}}{Z_{21}}$	$\frac{-Y_{11}}{Y_{21}}$	$\frac{(1-S_{11})(1+S_{22})+S_{12}S_{21}}{2S_{21}}$	D

[***Note***: $\lvert Y\rvert = Y_{11}Y_{22} - Y_{12}Y_{21}$; $\Delta S_Z = (1-S_{11})(1-S_{22}) - S_{12}S_{21}$;
$\lvert Z\rvert = Z_{11}Z_{22} - Z_{12}Z_{21}$; $\Delta S_Y = (1+S_{11})(1+S_{22}) - S_{12}S_{21}$;
$\Delta Z = (Z_{11}+Z_0)(Z_{22}+Z_0) - Z_{12}Z_{21}$; $\Delta Y = (Y_{11}+Y_0)(Y_{22}+Y_0) - Y_{12}Y_{21}$]

In the above table the Z, Y, S and ABCD parameters are given by the following.

Z-parameters: The open-circuit impedance or Z-parameters of a two-port network are defined as:

$$Z_{11} = \left.\frac{V_1}{I_1}\right|_{I_2=0} \qquad Z_{12} = \left.\frac{V_1}{I_2}\right|_{I_1=0} \tag{I.1}$$

$$Z_{21} = \left.\frac{V_2}{I_1}\right|_{I_2=0} \qquad Z_{22} = \left.\frac{V_2}{I_2}\right|_{I_1=0}$$

where $I_n = 0$ implies a perfect open-circuit at port n. These definitions of Z-parameters can be written in matrix form as:

$$\begin{bmatrix} V_1 \\ V_2 \end{bmatrix} = \begin{bmatrix} Z_{11} & Z_{12} \\ Z_{21} & Z_{22} \end{bmatrix} \cdot \begin{bmatrix} I_1 \\ I_2 \end{bmatrix} \tag{I.2}$$

Y-parameters: The short-circuit admittance or Y-parameters of a two-port network are defined as:

$$Y_{11} = \left.\frac{I_1}{V_1}\right|_{V_2=0} \qquad Y_{12} = \left.\frac{I_1}{V_2}\right|_{V_1=0} \tag{I.3}$$

$$Y_{21} = \left.\frac{I_2}{V_1}\right|_{V_2=0} \qquad Y_{22} = \left.\frac{I_2}{V_2}\right|_{V_1=0}$$

where $V_n = 0$ implies a perfect short circuit at port n. These definitions of Y parameters can be written in matrix form as:

$$\begin{bmatrix} I_1 \\ I_2 \end{bmatrix} = \begin{bmatrix} Y_{11} & Y_{12} \\ Y_{21} & Y_{22} \end{bmatrix} \cdot \begin{bmatrix} V_1 \\ V_2 \end{bmatrix} \tag{I.4}$$

From Eqs I.2 and I.3, it may be observed that:

$$[Z] = [Y]^{-1} \tag{I.5}$$

S-parameters: The scattering (S) parameters have been defined in Section 5.2 and repeated here for a two-port network (Fig. I.1a) when we can write:

$$\begin{bmatrix} b_1 \\ b_2 \end{bmatrix} = \begin{bmatrix} S_{11} & S_{12} \\ S_{21} & S_{22} \end{bmatrix} \cdot \begin{bmatrix} a_1 \\ a_2 \end{bmatrix} \tag{I.6}$$

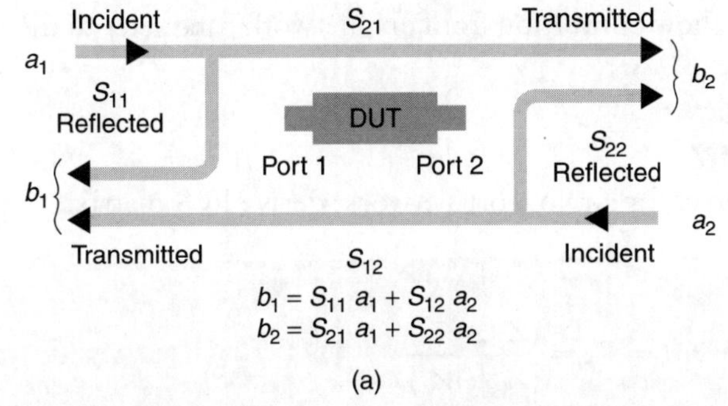

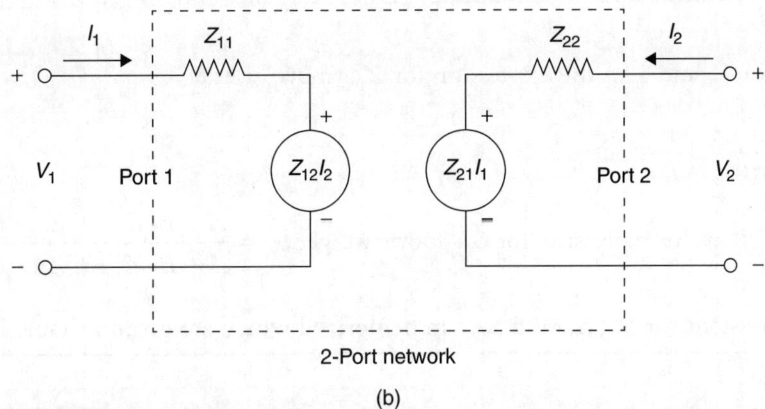

Fig. I.1 Two-port network (a) S-parameters with wave variables *a* and *b* (b) Z-parameters with associated current and voltage.

ABCD-parameters: The ABCD-parameters of a two-port network are defined as:

$$A = \left.\frac{V_1}{V_2}\right|_{I_2=0} \qquad B = \left.\frac{V_1}{-I_2}\right|_{V_2=0} \tag{I.7}$$

$$C = \left.\frac{I_1}{V_2}\right|_{I_2=0} \qquad D = \left.\frac{I_1}{-I_2}\right|_{V_2=0}$$

These definitions of ABCD-parameters can be written in matrix form as:

$$\begin{bmatrix} V_1 \\ I_1 \end{bmatrix} = \begin{bmatrix} A & B \\ C & D \end{bmatrix} \cdot \begin{bmatrix} V_2 \\ -I_2 \end{bmatrix} \tag{I.8}$$

For Z, Y and ABCD parameters, refer Fig. I.1. Note that for ABCD parameters (which is applicable for cascaded networks) I_2 is taken as negative since for cascaded networks I_2 will be away (outgoing) from port 2 (unlike those for Z and Y parameters for which I_2 is into port 2 as shown in Fig. I.1b) to feed signal to the next network.

A few examples to show conversion from one network parameter to the other are given below.

EXAMPLE I.1

Given the ABCD matrix for a two-port network, derive its S matrix.

Solution

From Fig. I.1a, we have $S_{11} = \Gamma_{IN} = \dfrac{(Z_{IN} - Z_0)}{(Z_{IN} + Z_0)}$

where $Z_{IN} = \dfrac{V_1}{I_1}$ with $V_1 = AV_2 - BI_2$ and $I_1 = CV_2 - DI_2$ and $V_2 = -Z_0 I_2$.

Substituting for V_1 and I_1 in the expression for Z_{IN} above, we get:

$$Z_{IN} = \frac{(AZ_0 - B)}{(CZ_0 - D)}, \text{ where } Z_0 = \frac{V_2}{I_2}$$

Substituting Z_{IN} in the expression for S_{11} above, we get $S_{11} = \dfrac{A + B/Z_0 - CZ_0 - D}{A + B/Z_0 + CZ_0 + D}$

Similarly expressions for S_{12}, S_{21} and S_{22} can be derived which are given in Table I.1.

EXAMPLE I.2

Given the Z matrix for a two-port network, derive its ABCD matrix.

Solution

The linear equations relating current, voltage and impedance for Z-parameter representation of a two-port network (Fig. I.1b) is given by:

$$V_1 = Z_{11}I_1 + Z_{12}I_2 \text{ and } V_2 = Z_{21}I_1 + Z_{22}I_2$$

Now $\quad A = \left.\dfrac{V_1}{V_2}\right|_{I_2=0} = \dfrac{Z_{11}I_1}{Z_{21}I_1} = \dfrac{Z_{11}}{Z_{21}}$

$$B = \left.\frac{V_1}{-I_2}\right|_{V_2=0} = \left.\frac{Z_{11}I_1 + Z_{12}I_2}{-I_2}\right|_{V_2=0} = Z_{11}\left.\frac{I_1}{I_2}\right|_{V_2=0} - Z_{12} = Z_{11}\frac{Z_{22}}{Z_{21}} - Z_{12} = \frac{|Z|}{Z_{21}}$$

where $|Z| = Z_{11}Z_{22} - Z_{12}Z_{21}$.

$$C = \left.\frac{I_1}{V_2}\right|_{I_2=0} = \frac{1}{Z_{21}} \text{ and } D = \left.\frac{I_1}{-I_2}\right|_{V_2=0} = \frac{Z_{22}}{Z_{21}}$$

In this context it would be worthwhile to discuss the ABCD matrix of some of the useful two-port networks as indicated in Table I.2.

Table I.2 ABCD matrix of some of the useful two port networks

Network	ABCD matrix: $\begin{bmatrix} A & B \\ C & D \end{bmatrix}$
(1) 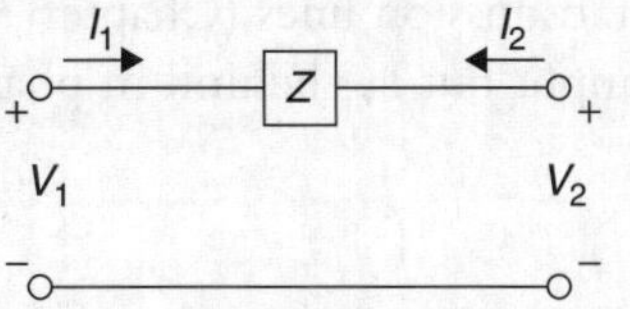 I_1, I_2, Z, +, +, V_1, V_2, −, −	$\Rightarrow \begin{bmatrix} 1 & Z \\ 0 & 1 \end{bmatrix}$
(2) 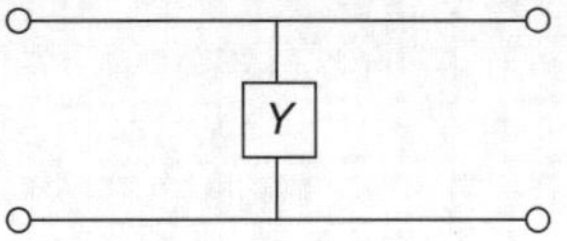 Y	$\Rightarrow \begin{bmatrix} 1 & 0 \\ Y & 1 \end{bmatrix}$
(3) 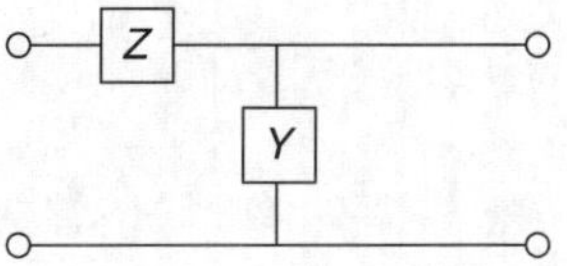 Z, Y	$\Rightarrow \begin{bmatrix} 1+2Y & Z \\ Y & 1 \end{bmatrix}$
(4) 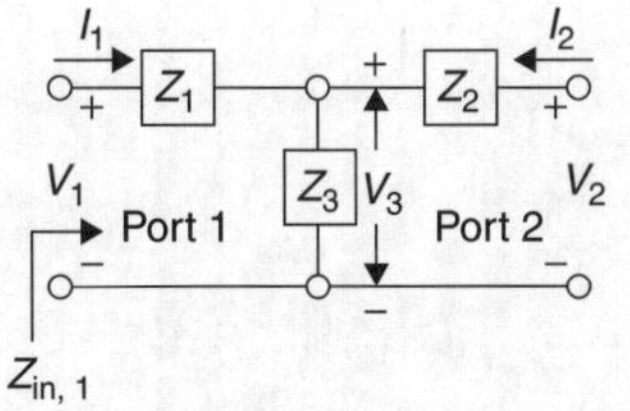 I_1, I_2, Z_1, Z_2, Z_3, V_3, V_1, V_2, Port 1, Port 2, $Z_{in,1}$	$\Rightarrow \begin{bmatrix} 1+(Z_1/Z_3) & Z_1+Z_2+(Z_1Z_2/Z_3) \\ 1/Z_3 & 1+(Z_2/Z_3) \end{bmatrix}$
(5) 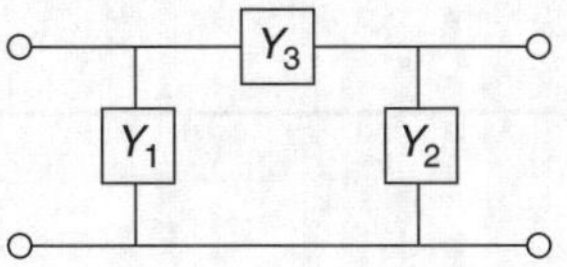 Y_3, Y_1, Y_2	$\Rightarrow \begin{bmatrix} 1+(Y_2/Y_3) & 1/Y_3 \\ Y_1+Y_2+(Y_1Y_2/Y_3) & 1+(Y_1/Y_3) \end{bmatrix}$
(6) 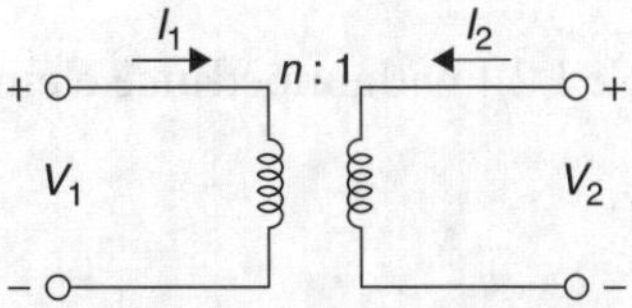 I_1, I_2, $n:1$, V_1, V_2	$\Rightarrow \begin{bmatrix} n & 0 \\ 0 & 1/n \end{bmatrix}$
(7) 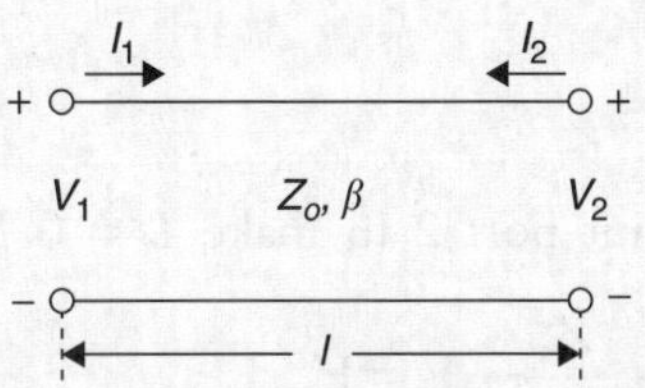 I_1, I_2, V_1, Z_o, β, V_2, l	$\Rightarrow \begin{bmatrix} \cos\beta l & jZ_0 \sin\beta l \\ jY_0 \sin\beta l & \cos\beta l \end{bmatrix}$

A few examples to show how we get the ABCD matrix for some particular network are given below.

EXAMPLE I.3

Find the ABCD matrix for lossless line of length l and characteristic impedance Z_0.

Solution

It is known from fundamentals of signal propagation in RF transmission lines (Chapter 3) the output voltage and current have the same magnitude as the input but lag behind in phase by $e^{-j\beta l}$ (for loss-less line). Thus, from Fig. (7) of Table I.2:

$$V_1 = V_2 e^{j\beta l} = V_2 \cos\beta l + jV_2 \sin\beta l \tag{i}$$

$$I_1 = -I_2 e^{j\beta l} = -I_2 \cos\beta l - jI_2 \sin\beta l \tag{ii}$$

As the load end is match terminated with $Z_L = Z_0$,

$$V_2 = -Z_0 I_2 \quad \text{or} \quad I_2 = -YV_2 \tag{iii}$$

Substituting this in (i) and (ii) we get:

$$V_1 = (\cos\beta l)V_2 - (jZ_0 \sin\beta l)I_2 = AV_2 - BI_2$$

$$I_1 = (jY_0 \sin\beta l)V_2 - (\cos\beta l)I_2 = CV_2 - DI_2$$

Whence the ABCD matrix is:

$$\begin{bmatrix} A & B \\ C & D \end{bmatrix} = \begin{bmatrix} \cos\beta l & jZ_0 \sin\beta l \\ jY_0 \sin\beta l & \cos\beta l \end{bmatrix}$$

EXAMPLE I.4

Derive the ABCD parameters for the T-network [Fig. (4) in Table I.2)] with impedance elements.

Solution

From Fig. (4) of Table I.2 we can write:

$$V_1 = AV_2 - BI_2 \text{ and } I_1 = CV_2 - DI_2$$

To evaluate A given by $A = \left.\dfrac{V_1}{V_2}\right|_{I_2=0}$, we need to open circuit port 2 to make $I_2 = 0$. With reference to the above mentioned figure we can write, $V_3 = \dfrac{Z_3}{Z_1 + Z_3} V_1 = V_2$.

Hence $A = 1 + \dfrac{Z_1}{Z_3}$

To determine B, we have $B = \left.\dfrac{V_1}{-I_2}\right|_{V_2=0}$; in this case we need to short circuit port 2 when $V_2 = 0$. Under this condition: $I_2 = -\left(\dfrac{Z_3}{Z_2 + Z_3}\right) I_1$.

Substituting this in the equation for B above, we get:

$$B = \left.\frac{V_1}{I_1}\left(1 + \frac{Z_2}{Z_3}\right)\right|_{V_2=0} = Z_{\text{in},1}\Big|_{V_2=0} \cdot \left.\left(1 + \frac{Z_2}{Z_3}\right)\right|_{V_2=0} = (Z_1 + Z_2 \| Z_3)\left(1 + \frac{Z_2}{Z_3}\right)$$

Thus $B = Z_1 + Z_2 = \dfrac{Z_1 Z_2}{Z_3}$

Now $C = \left.\dfrac{I_1}{V_2}\right|_{I_2=0}$, when with port 2 open circuited ($I_2 = 0$), $V_3 = I_1 Z_3 = V_2$

Thus $C = \dfrac{1}{Z_3}$

For $D = \left.\dfrac{I_1}{-I_2}\right|_{V_2=0}$, port 2 is short circuited when $I_2 = -\left(\dfrac{Z_3}{Z_2 + Z_3}\right) I_1$

Hence $D = 1 + \dfrac{Z_2}{Z_3}$

APPENDIX

II

Ferrites and Electromagnetic Wave Propagation in Magnetised Ferrite Medium

Introduction

Ferrites are ceramic-like polycrystalline magnetic oxides with the general chemical composition $XO.Fe_2O_3$, where X is a divalent metal ion such as cobalt (Co^{+2}), nickel (Ni^{+2}), zinc (Zn^{+2}), iron (Fe^{+2}) and so on. They have dielectric constants of 10–15 and loss tangents of the order of 10^{-4} (at microwave frequencies) with relative permeabilities of several thousands. The specific resistivity of ferrites is very high (of the order of 10^7 to 10^8 Ω-m), which is approximately 10^{14} times as high as those of metals (iron: 1×10^{-7} Ω-cm). Such low conductivities of ferrites allow them to support electric fields and hence electromagnetic wave motion can occur within them.

The transition elements like cobalt, iron or nickel exhibit unequal electron spin (as new shells of electrons are formed even before the inner shell is completely filled) giving rise to ferrimagnetic behaviour in addition to the ferromagnetic property. The unequal electron spin due to ferrimagnetism gives rise to an effective dipole moment. However, such metals or their alloys by themselves are majorly ferromagnetic having high conductivities and excessive losses at high RF with little penetration of microwave signals into them. But ferrite materials containing these elements behave as electric insulators (with high permeabilities similar to those of the transition metals themselves) and thus allow good penetration of electromagnetic fields. Further, the ferrites behave as ferrimagnetic materials with very small ferromagnetic property. Hence useful interactions can take place in ferrites between the strong effective dipole moment caused by unequal electron spin and the alternating magnetic field of an electromagnetic wave. At low radio frequencies, this interaction is not significant, and ferrite materials are used simply as low-loss magnetic core material for inductors. On the other hand, at microwave and millimetre-wave frequencies, this interaction is quite significant in ferrites. This gives rise to the anisotropic nature of magnetic properties and hence exhibits a frequency-dependent permeability tensor which leads to many useful applications of ferrites in the design of non-reciprocal microwave and millimetre-wave components.

Permeability Tensor of a Magnetised Ferrite

Let us derive that the permeability of a ferrite material subjected to a time-harmonic field in addition to a static magnetic field is not a scalar quantity but a tensor, represented by a matrix. The consequence of electromagnetic wave propagation in such a magnetised ferrite medium will also be explained.

The spinning electrons in ferrites results in an effective dipole moment $\vec{m}$ (due to its ferrimagnetic behaviour) and an angular momentum $\vec{P}$ (due to the rotating mass of the spinning electron). Since the electronic charge is negative, the angular momentum vector $\vec{P}$ and the magnetic dipole moment $\vec{m}$ are antiparallel. If the ferrite material is subjected to a static magnetic field $\vec{B}_0$, a torque $\vec{T}$ will be experienced by the magnetic dipole moment $\vec{m}$ of the spin, tending to rotate it into alignment with the field $\vec{P}$. But the angular momentum of the spinning electron will not allow its angle with the field to change and hence the electron will *precess gyroscopically* around $\vec{B}_0$ (Fig. II.1). The torque acting upon $\vec{m}$ is given by:

$$\vec{T} = \vec{m} \times \vec{B}_0 = -\gamma(\vec{P} \times \vec{B}_0) \tag{II.1}$$

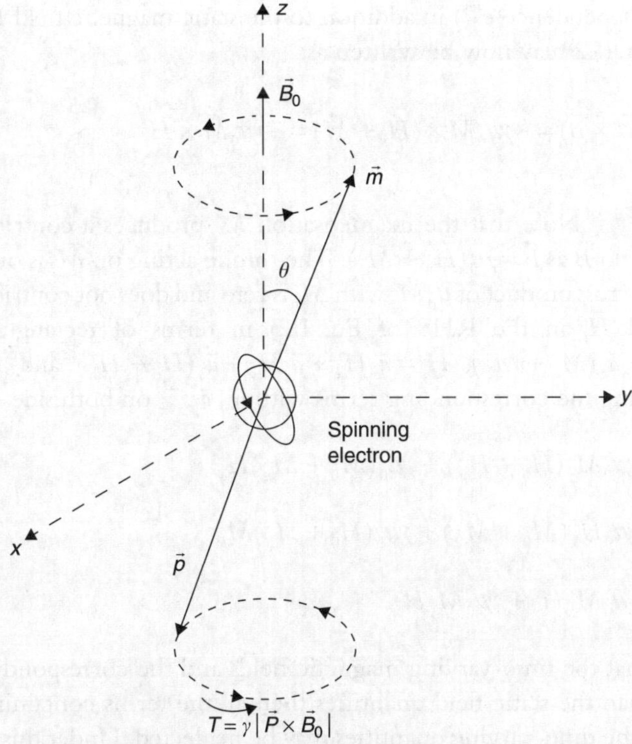

Fig. II.1 Precession of spinning electron caused by the torque *T*.

where $\gamma = \frac{|\vec{m}|}{|\vec{P}|} = \frac{e}{m_e}$ is called the *gyromagnetic ratio*. The negative sign arises because the angular momentum and dipole moment are antiparallel for an electron.

The equation of motion for the electron is obtained by equating the rate of change of angular momentum with the torque when we have:

$$\frac{d\vec{P}}{dt} = \vec{T} = -\gamma(\vec{P} \times \vec{B}_0) = \gamma|\vec{P}||\vec{B}_0|\sin\theta = \omega_0 \vec{P}\sin\theta \tag{II.2}$$

where θ is the angle between $\vec{m}$ and $\vec{B}_0$. For free precession, the angular velocity ω_0 is given by $\omega_0 = \gamma|\vec{B}_0|$ which is independent of the angle θ. The angular velocity ω_0 is often known as *Larmor frequency*.

The ferrite material may be regarded as the collection of N spinning electrons per unit volume. Since the spacing between electrons is of atomic dimensions, the collection of all the magnetic dipoles can be considered as a smeared-out continuous distribution. When an external static magnetic field is applied, the magnetic dipoles get aligned along the direction of the field, with the total magnetic dipole moment per unit volume as $\vec{M} = N\vec{m}$ (in the absence of the external field, however, the combined effect is very small as the dipoles are oriented at random). When the static magnetic field is large enough to cause saturation of the magnetisation ($\vec{M}_s$) in the ferrite then the whole ferrite sample will act as a large single magnetic dipole with $\vec{M} = \vec{M}_s$.

For a general case, when the ferrite sample is subjected to a time varying magnetic field $\vec{H}_{ac}$ (having a time dependence $e^{j\omega t}$) in addition to the static magnetic field $\vec{H}_0$ the equation of motion given by Eq. II.2 may now be written as:

$$\frac{dM}{dt} = -\gamma(\vec{M} \times \vec{B}) = -\gamma\mu_0\vec{M} \times (\vec{H}_t + \vec{M}) = -\gamma\mu_0\vec{M} \times H_t \tag{II.3}$$

where $\vec{H}_t = (\vec{H}_0 + \vec{H}_{ac})$. Note that the magnetisation $\vec{M}_s$ produces a contribution to the total internal magnetic field $\vec{B}$ as $\vec{B} = \mu_0(\vec{H}_0 + \vec{M}_s)$. The torque acting on $\vec{M}_s$ is only due to the field $\vec{B}_0 = \mu_0\vec{H}_0$ (since the cross product of $\mu_0\vec{M}_s$ with $\vec{M}_s$ is zero and does not contribute to the torque).

Writing $\vec{M}$ and $\vec{H}_t$ in the RHS of Eq. II.3 in terms of rectangular coordinates as $\vec{M} = \vec{a}_x M_x + \vec{a}_y M_y + \vec{a}_z(M_0 + M_z)$, $\vec{H}_t = \vec{a}_x H_x + \vec{a}_y H_y + \vec{a}_z(H_0 + H_z)$ and LHS in terms of $j\omega M_{X,Y,Z}$ and equating the corresponding terms with $\vec{a}_x$, $\vec{a}_y$, $\vec{a}_z$ on both sides we get:

$$\begin{aligned}
j\omega M_x &= -\gamma\mu_0 M_y(H_0 + H_z) + \gamma\mu_0(M_0 + M_z)H_y \\
j\omega M_y &= -\gamma\mu_0 H_x(M_0 + M_z) + \gamma\mu_0(H_0 + H_z)M_x \\
j\omega M_z &= -\gamma\mu_0 M_x H_y + \gamma\mu_0 M_y H_x
\end{aligned} \tag{II.4}$$

Let it be assumed that the time-varying magnetic fields and the corresponding magnetisations are much smaller than the static field quantities then all the terms containing the products of the components of the time-varying quantities may be neglected. Under this condition we may write Eqs II.4 as:

$$j\omega M_x = -\omega_0 M_y + \omega_m H_y$$

$$j\omega M_y = \omega_0 M_x - \omega_m H_x \tag{II.5}$$

$$j\omega M_z = 0$$

where $\omega_0 = \gamma\mu_0 H_0$ and $\omega_m = \gamma\mu_0 M_0$.

The solution of these equations for M_x, M_y and M_z yields:

$$M_x = \varphi H_x + j\kappa H_y$$

$$M_y = \varphi H_y - j\kappa H_x \tag{II.6}$$

$$M_z = 0$$

where $\varphi = \dfrac{\omega_0\omega_m}{\omega_0^2 - \omega^2}$ and $\kappa = \dfrac{\omega\omega_m}{\omega_0^2 - \omega^2}$.

It would be convenient not to deal explicitly with magnetisation (M) but to introduce magnetic susceptibility (χ_m) and permeability (μ). Using scalar notation, $B = \mu H = \mu_0\, \mu_r\, H = \mu_0\, (1 + \chi_m)\, H = \mu_0\, (M + H)$, since $M = \chi_m H$. It may be observed from Eq. II.6 that the magnetisation in the x direction is produced by the magnetic field in both the x and y directions. Thus, it is clear that B and H cannot be related by a simple permeability value. Instead we can write permeability as a tensor $[\mu_{\text{eff}}]$ so that $[B] = [\mu_{\text{eff}}][H]$ or:

$$B_x = \mu_0(1+\varphi)H_x + j\kappa H_y$$

$$B_y = \mu_0(1+\varphi)H_y - j\kappa H_x \tag{II.7}$$

$$B_z = \mu_0(H_z + H_0)$$

Thus the matrix representation of the relationship between the two vectors $\vec{B}$ and $\vec{H}$ is expressed in terms of the permeability tensor as:

$$\begin{bmatrix} B_x \\ B_y \\ B_z \end{bmatrix} = \mu_0 \begin{bmatrix} (1+\varphi) & j\kappa & 0 \\ -j\kappa & (1+\varphi) & 0 \\ 0 & 0 & 1 \end{bmatrix} \begin{bmatrix} H_x \\ H_y \\ H_z + H_0 \end{bmatrix} \tag{II.8}$$

The permeability tensor $[\mu_{\text{eff}}]$ in Eq. II.8 is true for the applied static magnetic field and is along the direction of propagation (z-direction) of the electromagnetic signal via the ferrite material. Keeping the propagation direction unchanged, if the direction of application of the static magnetic field is changed, the permeability tensor will get transformed accordingly as mentioned below.

$$[\mu_{\text{eff}}] = \mu_0 \begin{bmatrix} 1 & 0 & 0 \\ 0 & (1+\varphi) & j\kappa \\ 0 & -j\kappa & (1+\varphi) \end{bmatrix} \quad \text{when } H_0 \text{ is applied along the } x\text{-direction.}$$

$$[\mu_{\text{eff}}] = \mu_0 \begin{bmatrix} (1+\varphi) & 0 & -j\kappa \\ 0 & 1 & 0 \\ j\kappa & 0 & (1+\varphi) \end{bmatrix} \quad \text{when } H_0 \text{ is applied along the } y\text{-direction.}$$

Wave Propagation in Magnetised Ferrite Medium

Let a plane electromagnetic wave with field variation $e^{j(\omega t - \beta z)}$ propagate in the z-direction via a ferrite medium subjected to a static magnetic field $\vec{H}_0$ applied along the propagation direction. From Eq. II.7 we can write:

$$\begin{aligned} B_x &= \mu_1 H_x + j\mu_2 H_y \\ B_y &= \mu_1 H_y - j\mu_2 H_x \\ B_z &= \mu_0 (H_z + H_0) \end{aligned} \tag{II.9}$$

where $\mu_1 = 1 + \varphi$ and $\mu_2 = \kappa$.

Now using Eq. II.9 and keeping in mind $-\dfrac{\partial \vec{B}}{\partial t} = -j\omega \vec{B}$, the Maxwell's electric curl equation $\nabla \times \vec{E} = -\dfrac{\partial \vec{B}}{\partial t}$ can be written as:

$$\begin{aligned} \frac{\partial E_z}{\partial y} - \frac{\partial E_y}{\partial z} &= -j\omega\mu_1 H_x + \omega\mu_2 H_y \\ \frac{\partial E_x}{\partial z} - \frac{\partial E_z}{\partial x} &= -j\omega\mu_1 H_y - \omega\mu_2 H_x \\ \frac{\partial E_y}{\partial x} - \frac{\partial E_x}{\partial y} &= -j\omega\mu_0 H_z \end{aligned} \tag{II.10}$$

Since the plane wave propagation is in the z-direction we can set $\dfrac{\partial}{\partial x} = \dfrac{\partial}{\partial y} = 0$. We know, $\dfrac{\partial}{\partial z}(\vec{E}) = -j\beta \vec{E}$ [as the field variation is assumed to be $e^{j(\omega t - \beta z)}$], we have from Eq. II.10:

$$\begin{aligned} \beta E_y &= -\omega\mu_1 H_x - j\omega\mu_2 H_y \\ \beta E_x &= \omega\mu_1 H_y - j\omega\mu_2 H_x \\ 0 &= H_z \end{aligned} \tag{II.11}$$

Similarly from Maxwell's magnetic curl equation we get:

$$\beta H_y = \omega\varepsilon E_x$$

$$\beta H_x = -\omega\varepsilon E_y \qquad \text{(II.12)}$$

$$0 = E_z$$

We can now obtain the following two relationships between E_x and E_y by eliminating H_x and H_y from Eq. II.11 with the help of Eq. II.12:

$$(\beta^2 - \omega^2\varepsilon\mu_1)E_y = -j\omega^2\varepsilon\mu_2 E_x$$

$$(\beta^2 - \omega^2\varepsilon\mu_1)E_x = j\omega^2\varepsilon\mu_2 E_y \qquad \text{(II.13)}$$

Both these equations will be satisfied simultaneously provided:

$$(\beta^2 - \omega^2\varepsilon\mu_1)^2 = (\omega^2\varepsilon\mu_2)^2$$

or $$\beta^2 = \omega^2\varepsilon(\mu_1 \pm \mu_2) \qquad \text{(II.14)}$$

Thus, two distinct values of the phase constant β are possible which are given by:

$$\beta^+ = \pm\omega\sqrt{\varepsilon(\mu_1 + \mu_2)}$$

$$\beta^- = \pm\omega\sqrt{\varepsilon(\mu_1 - \mu_2)} \qquad \text{(II.15)}$$

These two values of β would definitely provide two different relationships between E_x and E_y giving rise to two different modes of propagation of the electromagnetic wave via the magnetised ferrite medium. When $\beta = \beta^+$ it may be derived from the second equation of Eq. II.13 that $E_y = -jE_x$ and when $\beta = \beta^-$ then $E_y = +jE_x$. Thus β^+ and β^- correspond to two circularly polarised plane waves. For $\beta = \beta^+$, E_y lags E_x and the phasor $\vec{E}$ moves in the counter-clockwise direction while for $\beta = \beta^-$, the $\vec{E}$ phasor moves in the clockwise direction. These are the fundamental modes of propagation in a magnetised ferrite medium.

Different phase velocities of the two different modes of propagation eventually result in different effective permeabilities of the magnetised ferrite medium which are given by [9]:

$$\mu^+ = \mu_0\left(1 + \frac{\omega_m}{\omega_0 - \omega}\right) \quad \text{and} \quad \mu^- = \mu_0\left(1 + \frac{\omega_m}{\omega_0 + \omega}\right) \qquad \text{(II.16)}$$

Variation of μ^+ and μ^- with frequency are shown in Fig. II.2. It may be observed that μ^+ becomes infinite and changes sign at the gyromagnetic resonance frequency f_0(MHz) $\cong$ 2.8 $\vec{H}_0$(oersted); where 2.8 MHz/oersted is the gyromagnetic ratio. On the other hand, μ^- varies monotonically from $\mu_0(1 + \omega_m/\omega_0)$ to μ_0 as the frequency is varied from zero to infinity. Thus only one of the modes (characterised by μ^+) exhibits resonance behaviour at $f = f_0$.

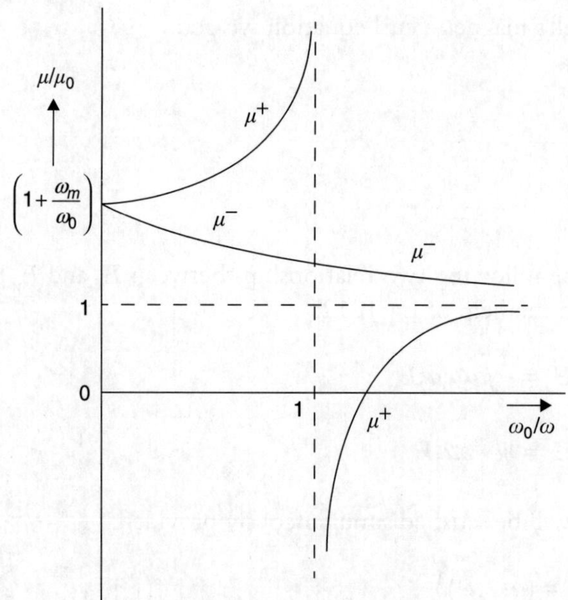

Fig. II.2 Variation of μ^{+} and μ^{-} with frequency.

III

MATLAB Programs

(a) MATLAB Program for Example 6.3

```
cv=0.1;
zo=50;
z0e=zo*sqrt((1+cv)/(1-cv))
z0es=z0e/2;
z0o=zo*sqrt((1-cv)/(1+cv))
z0os=z0o/2;
syms pes

vpasolve(((60*log(8/pes)+(pes/4))/((11/2)+(9/2)*((1+(12/pes))^-0.5)+0.04*((1-
pes)^2))^0.5)==z0es,pes)
res=ans
syms pos
vpasolve((((11/2)+((9/2)*((1+(12/pos))^-0.5)))^0.5)*(pos+1.4+0.667*(log
(pos+1.444)))==(377/z0os),pos)
ros=ans
t=(cosh((3.14/2)*(res)));
t1=(cosh((3.14/2)*(ros)));
t2=((t+t1-2)/(t1-t))
t3=acosh(t2);
t4=(2/3.14)*t3
sh=t4
g=cosh((3.14/2)*sh);
des=((g+1)*(cosh((3.14/2)*(res))))+g-1
Res=(acosh(des)-(3.14/2)*(sh))/3.14
rres=Res
```

(b) MATLAB Program to Draw Curves in Fig. 7.22 (Example 7.8)

```
Z0=58;
ZL=222;
pL=(log(ZL/Z0));
p0=0.5*pL;
pm=0.025;
A=acosh(p0/pm);
t1=0.5*log(222*58);
t2=pm*(A^2);
a=22.86;%                (waveguide shorter dimension in mm)
lambda0=3;%              (free-space wavelength in cm)
lambdag=3.975;%          (guide wavelength in cm)
Z01=377;%                (free-space characteristic impedance)
znorm=0:(1/99):1;%       (normalized length)
p=znorm*log(222/58);
Ze=58*exp(p);%           (exponential)
be=(Ze*lambda0*a)/(Z01*lambdag);%  (exponential)
figure(1)
   plot(znorm,Ze)

   hold on
   [x,y]=size(znorm);
Zt=zeros(1,y);
bt=zeros(1,y);
for m=1:1:y
if znorm(m)<=0.5
Zt(m)=Zt(m)+58*exp(2*((znorm(m)).^2)*log(222/58));%  (triangular)
end
if znorm(m)>0.5
Zt(m)=Zt(m)+58*exp(((4*znorm(m))-(2*((znorm(m)).^2))-1)*log(222/58));
end
bt(m)=bt(m)+(Zt(m)*lambda0*a)/(Z01*lambdag);
end
figure(1)
   plot(znorm,Zt,'r')

   hold on
%klopfenstein%
zL4=0:(1/99):1;
   [x1,y1]=size(zL4);
   Z3=zeros(1,y1);
   bk=zeros(1,y1);
for m=1:1:y1
```

```
  syms y;
phi(m)=int(besseli(0,A*sqrt(1-(y^2)))/(A*sqrt(1-(y^2))),0,((2*zL4(m))-1));
if A==0
phi(m)=0.5*(((2*zL4(m))-1));
end
end
[x2,y2]=size(phi);
phi(1)=0;
phi(y2)=((cosh(A))-1)/(A^2);
for m=1:1:y1
  Z3(m)=Z3(m)+exp(t1+(t2*phi(m)));
  bk(m)=bk(m)+(Z3(m)*lambda0*a)/(Z01*lambdag);
end
figure(1);
  plot(zL4,Z3,'k');
  % Plotting reflection coefficient%
  hold on
  for BL=0:0.01:5*pi
    Tei=(pL/2)*((sin(BL))/BL);
    Teia = abs(Tei);
    figure (2)
    set(gca, 'xtick', [0, pi,2*pi,3*pi,4*pi,5*pi]);
    set(gca, 'xtickLabel', {'0','p','2p','3p','4p','5p'})
    set(gca, 'Fontname', 'symbol')
    xlim([0 5*pi])

    plot(BL,Teia)
    hold on
 end

   for BL2=0:0.01:5*pi
    Tti=(pL/2)*(((sin(BL2/2))/(BL2/2))^2);
    Ttia = abs(Tti);
    figure (2)
    set(gca, 'xtick', [0, pi,2*pi,3*pi,4*pi,5*pi]);
    set(gca, 'xtickLabel', {'0','p','2p','3p','4p','5p'})
    set(gca, 'Fontname', 'symbol')
    xlim([0 5*pi])
%     ylim([0 0.5])
    plot(BL2,Ttia,'r')
    hold on
 end
%
%
 for BL3=0:0.01:5*pi
```

```
    T0=0.5*pL;
    Tm=0.025;
    A=acosh(T0/Tm);

    Tki=T0*exp(-1i*BL3)*(cos(sqrt(((BL3)^2)-(A^2)))/cosh(A));

    Tkia = abs(Tki);
    figure (2)
    set(gca, 'xtick', [0, pi,2*pi,3*pi,4*pi,5*pi]);
    set(gca, 'xtickLabel', {'0','pi ','2p','3p','4p','5p'})
    set(gca, 'Fontname', 'symbol')
    xlim([0 5*pi])
%       ylim([0 0.5])
    plot(BL3,Tkia,'k')
    hold on
end
```

(c) Flow Chart and MATALB Programs to Generate Tables 8.1 and 8.2

(i) For Table 8.1: Flowchart to write a program for the generation of prototype element (g_k) values for different order (n) of the filter with Tchebyscheff (equal-ripple) response.

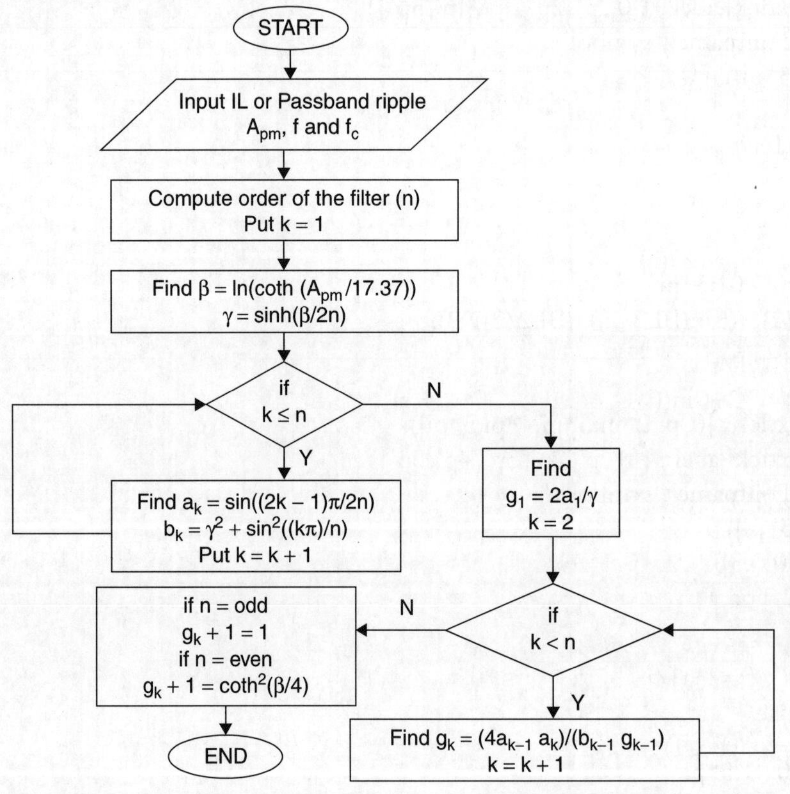

Program for equal ripple case

```
Apm=input('give the value of Apm');
Ap2=(10^(Apm/10))-1;
disp(Ap2);
il=input('give the value of il');
plr=10^(il/10);
disp(plr);
f=input('give the value of f');
w=2*3.14*f;
disp(f);
fc=input('give the value of fc');
wc=2*3.14*fc;
disp(fc);
h=2*(w/wc)
n1=(0.5)*log((4*(plr-1)/Ap2)-h);
disp(n1);

n2=round(n1)
if n2>n1
   n=n2-1;
else
   n=n2;
end
b=log(coth((Apm)/17.37));
disp(b);
y=sinh(b/(2*n));
disp(y);

for  i=1:n
     ak(1,i)= sin((((2*(i))-1)*3.14)/(2*n));
     disp(ak(1,i));
     a1=ak(1,1);
     disp(a1(1,1));
     bk(1,i)=y^2+(sin((i*3.14)/n))^2;
     disp(bk(1,i));
end
 g(1,1)=2*a1/y;
   u=g(1,1);

     for i=1:n
       if i<=1
       gk(1,1)=2*a1/y;
         else
    gk(1,i)=4*(ak(i-1)*ak(i))/(bk(i-1))/(gk(1,i-1));
end
```

```
if mod(i,2) == 0
  %number is even
  gev(1,i)=(coth(b/4))^2;
  gk(1,i+1)=gev(1,i);
else
  %number is odd
  god(1,i)=1.0000;
  gk(1,i+1)=god(1,i);
end
     end
```

(ii) For Table 8.2: Flowchart to write a program for the generation of prototype element (g_k) values for Butterworth filter having a given Insertion loss IL = 3 dB.

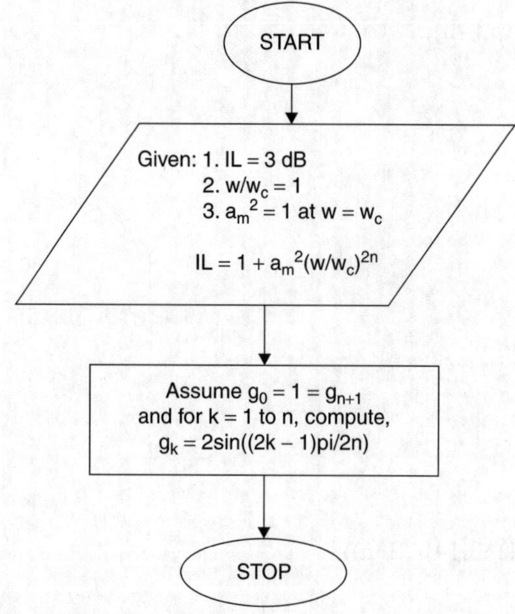

Program for Butterworth case

[To generate the component values only]

```
for i=1:n

    gk(1,i)= 2*sin((((2*(i))-1)*3.14)/(2*n));
end

god(1,i)=1.0000;
gk(1,i+1)=god(1,i);
```

[To generate values along with P_{LR} and other quantities]

```
k2=input('give the value of k2');
```

```
disp(k2);
il=input('give the value of il');
plr=10^(il/10);
disp(plr);
f=input('give the value of f');
w=2*3.14*f;
disp(f);
fc=input('give the value of fc');
wc=2*3.14*fc;
disp(fc);
h=(w/wc)
n1=(0.5)*log(((plr-1)/k2)-h);
disp(n1);

n2=round(n1)
if n2>n1
   n=n2-1;
else
   n=n2;
end

for i=1:n

    gk(1,i)= 2*sin((((2*(i))-1)*3.14)/(2*n));
end

god(1,i)=1.0000;
gk(1,i+1)=god(1,i);
```

(d) MATLAB Program to Solve Example 8.3

```
clc;
clear all;
format long;
Apm=input('give the value of Apm');
Ap2=(10^(Apm/10))-1
il=input('give the value of il');
plr=10^(il/10)
f=input('give the value of f');
w=2*3.14*f*10^9
fc=input('give the value of fc');
wc=2*3.14*fc*10^9
h=(w/wc)
```

```
n1=(acosh(sqrt((plr-1)/Ap2)))/acosh(h)
n=round(n1)

% Find parameter values from table %%

g1=input('give the value of g1');
g2=input('give the value of g2');
g3=input('give the value of g3');
g4=input('give the value of g4');
g5=input('give the value of g5');
g6=input('give the value of g6');

Rl = 50;
C1=g1/(Rl*wc)
L2=(Rl*g2)/wc
C3=g3/(Rl*wc)
L4=(Rl*g4)/wc
C5=g5/(Rl*wc)
 Zoc=25;
Zol=120;
er=10;

A=((Zol/60)*sqrt((er+1)/2))+(((er-1)/(er+1))*(0.23+(0.11/er)));
B=(377*3.14)/(2*Zoc*sqrt(er));
A0=((Rl/60)*sqrt((er+1)/2))+(((er-1)/(er+1))*(0.23+(0.11/er)));

Wh1=(8*exp(A))/(exp(2*A)-2)
Wh2=(2/3.14)*(B-1-log((2*B)-1)+(((er-1)/(2*er))*(log(B-1)+0.39-(0.61/er))))
Wh0=(8*exp(A0))/(exp(2*A0)-2)

W1=Wh2*0.65
W2=Wh1*0.65
W3=Wh2*0.65
W4=Wh1*0.65
W5=Wh2*0.65
W6=Wh0*0.65


erff1=((er+1)/2)+((er-1)/2)*((1+(12/Wh1))^(-1/2)+(0.04*(1-Wh1)^2))
erff2=((er+1)/2)+((er-1)/2)*((1+(12/Wh2))^(-1/2))

lambda0=(3*10^8)/(fc*10^9)
lambdagl=lambda0/sqrt(erff1)
lambdagc=lambda0/sqrt(erff2)
```

```
lC1=fc*10^9*lambdagl*Zoc*C1
lL2=fc*10^9*lambdagl*L2/Zol
lC3=fc*10^9*lambdagl*Zoc*C3
lL4=fc*10^9*lambdagl*L4/Zol
lC5=fc*10^9*lambdagl*Zoc*C5
```

(e) MATLAB Program to Solve Example 8.4

```
fc1=2.27;
fc2=2.53;
f0=sqrt(fc1*fc2);
Delta=(fc2-fc1)/2.4;
w0=2.4;
wc2=2.53;
wc1=2.27;
Wp=1;
ws=2.65;

Ap=sqrt((10^0.02)-1);
W1=((Wp*w0)/(wc2-wc1))*((ws/w0)-(w0/ws));
Plr=10^(3);
Tn1=(Plr-1)/((Ap^2)*W1);
Tn=sqrt(Tn1);
x=acosh(Tn);
n=x/(acosh(W1));
g1=1.3394;
g5=g1;
g2=1.337;
g4=g2;

g3=2.166;
g0=1.0;
g6=g0;

J10z0=sqrt((3.14*Delta)/(2*g0*g1));
J65z0=J10z0;
J21z0=(3.14*Delta)/(2*sqrt(g1*g2));
J54z0=J21z0;
J32z0=(3.14*Delta)/(2*sqrt(g3*g2));
J43z0=J32z0;
z0=50;
% k=0
Z0e10=z0*(1+J10z0+J10z0^2);
Z0es10=Z0e10/2;
```

```
Z0o10=z0*(1-J10z0+J10z0^2);
Z0os10=Z0o10/2;
k=1;
Z0e21=z0*(1+J21z0+J21z0^2);
Z0es21=Z0e21/2;
Z0o21=z0*(1-J21z0+J21z0^2);
Z0os21=Z0o21/2;
k=2;
Z0e32=z0*(1+J32z0+J32z0^2);
Z0es32=Z0e32/2;
Z0o32=z0*(1-J32z0+J32z0^2);
Z0os32=Z0o32/2;
syms res10

vpasolve(((60*log(8/res10)+(res10/4))/((11/2)+(9/2)*((1+(12/res10))^-0.5)+0.04*((1-
res10)^2))^0.5)==Z0es10,res10)
pes10=ans;
syms res21
vpasolve(((60*log(8/res21)+(res21/4))/((11/2)+(9/2)*((1+(12/res21))^-0.5)+0.04*((1-
res21)^2))^0.5)==Z0es21,res21)
pes21=ans;
syms res32
vpasolve(((60*log(8/res32)+(res32/4))/((11/2)+(9/2)*((1+(12/res32))^-0.5)+0.04*((1-
res32)^2))^0.5)==Z0es32,res32)
pes32=ans;
syms ros10
vpasolve((((11/2)+((9/2)*((1+(12/ros10))^-0.5)))^0.5)*(ros10+1.4+0.667*(log
(ros10+1.444)))==(377/Z0os10),ros10)
pos10=ans;
syms ros21
vpasolve((((11/2)+((9/2)*((1+(12/ros21))^-0.5)))^0.5)*(ros21+1.4+0.667*(log
(ros21+1.444)))==(377/Z0os21),ros21)
pos21=ans;
syms ros32
vpasolve((((11/2)+((9/2)*((1+(12/ros32))^-0.5)))^0.5)*(ros32+1.4+0.667*(log
(ros32+1.444)))==(377/Z0os32),ros32)
pos32=ans;

t10=(cosh((3.14/2)*(pes10)));
t110=(cosh((3.14/2)*(pos10)));
t102=((t10+t110-2)/(t110-t10))
t103=acosh(t102);
t104=(2/3.14)*t103
sh10=t104
```

```
g10=cosh((3.14/2)*sh10);
des10=((g10+1)*(cosh((3.14/2)*(pes10))))+g10-1
Res10=(acosh(des10)-(3.14/2)*(sh10))/3.14
rres10=Res10

t21=(cosh((3.14/2)*(pes21)));
t121=(cosh((3.14/2)*(pos21)));
t212=((t21+t121-2)/(t121-t21))
t213=acosh(t212);
t214=(2/3.14)*t213
sh21=t214
g21=cosh((3.14/2)*sh21);
des21=((g21+1)*(cosh((3.14/2)*(pes21))))+g21-1
Res21=(acosh(des21)-(3.14/2)*(sh21))/3.14
rres21=Res21

t32=(cosh((3.14/2)*(pes32)));
t132=(cosh((3.14/2)*(pos32)));
t322=((t32+t132-2)/(t132-t32))
t323=acosh(t322);
t324=(2/3.14)*t323
sh32=t324
g32=cosh((3.14/2)*sh32);
des32=((g32+1)*(cosh((3.14/2)*(pes32))))+g32-1
Res32=(acosh(des32)-(3.14/2)*(sh32))/3.14
rres32=Res32

z0esair10=((60*log(8/pes10)+(pes10/4))/((2/2)^0.5))
z0eair10=2*z0esair10
Age10=(300/2.4)*(50/z0eair10)

z0osair10=377/(((2/2)^0.5)*(pos10+1.4+0.667*(log(pos10+1.444))))
z0oair10=2*z0osair10
Ago10=(300/2.4)*(50/z0oair10)

Agm10=(Age10+Ago10)/2
l10=Agm10/4

z0esair21=((60*log(8/pes21)+(pes21/4))/((2/2)^0.5))
z0eair21=2*z0esair21
Age21=(300/2.4)*(50/z0eair21)

z0osair21=377/(((2/2)^0.5)*(pos21+1.4+0.667*(log(pos21+1.444))))
z0oair21=2*z0osair21
```

```
Ago21=(300/2.4)*(50/z0oair21)

Agm21=(Age21+Ago21)/2
l21=Agm21/4

z0esair32=((60*log(8/pes32)+(pes32/4))/((2/2)^0.5))
z0eair32=2*z0esair32
Age32=(300/2.4)*(50/z0eair32)

z0osair32=377/(((2/2)^0.5)*(pos32+1.4+0.667*(log(pos32+1.444))))
z0oair32=2*z0osair32
Ago32=(300/2.4)*(50/z0oair32)

Agm32=(Age32+Ago32)/2
l32=Agm32/4
```

IV

Computational Electromagnetics (CEM) and Simulators

(A) CEM and EM Simulators

Introduction

Computational electromagnetics (CEM) deals with numerical approximation of Maxwell's equations. CEM techniques have developed and matured during the last four decades and the electromagnetic (EM) simulators based on them are now an invaluable part of RF and microwave engineering design. CEM is a multidisciplinary field with its core being electromagnetics and numerical methods, however, for useful implementation geometrical modelling and visualisation, computer science and algorithm, all have important roles to play.

The 'big three' techniques in CEM are *Method of Moments* (MoM), *Finite Element Method* (FEM) and *Finite Difference Time Domain* (FDTD) method. These three methods have been implemented in successful commercial codes (simulators) and are used by RF and microwave engineers quite frequently for the design of antennae, filters, MICs and so on. These three techniques are also known as *Full-Wave CEM*—they approximate Maxwell's equations numerically, without using any physical approximations. There is another class of numerical method for solving Maxwell's equations, known as *asymptotic techniques*, that require some fundamental approximations in Maxwell's equations—the validity of which increases asymptotically with frequency. These are used in Optics.

Central to all full-wave techniques is to use the idea of '*discretising*' some unknown electromagnetic property, typically the surface current for the MoM, and the electric field for the EFM and FDTD method; for the latter, the magnetic field may also be discretised. The process of discretisation is also known as '*meshing*'. It entails subdividing the geometry into a large number of small '*elements*'. These may be one-dimensional segments, two-dimensional surface patches (often triangle), three-dimensional tetrahedral elements—depending on the problem at hand and the method used. With each element, a simple functional dependence is assumed for the spatial variation of the unknown – for instance, a linear approximation – but the amplitude (and possibly phase) of the unknown is determined by application of the method to the patchwork of elements which approximates the original geometry. This functional dependence is also known as the '*basis* (or *expansion*) *function*'. Generally, the accuracy of the method is related to the discretisation—the finer the mesh size, the better is the accuracy of the method (exceptions are also there as in the case of MoM). Largest mesh size (the finest geometrical resolution) is limited by the available computational resources.

MoM, FEM and FDTD—An overview

MoM or the method of weighted residuals, owes its name to the process of taking moments by multiplying with appropriate weighting functions and integrating. The method has been applied successfully to a wide variety of electromagnetic problems of practical interest such as radiation due to thin-wire elements and arrays, scattering problems and analysis of microstrip and lossy structures. In MoM, the structure under study is represented by equivalent surface current. The surface current is discretised into wire segments and/or surface patches. A matrix equation is then derived, representing the effect of every segment/patch on every other segment/patch (hence the taking of moment). This interaction is computed using Green's function for the problem. The relevant boundary condition is then applied to all the interactions, yielding a set of linear equations. The solution of this linear system yields the (approximate) current on each segment/patch.

FEM divides or discretises the solution region into sub-regions or sub-domains, that is, to 'finite elements' with the help of two-dimensional (typically triangles for surface mesh) or three-dimensional (typically tetrahedra for volume mesh) elements. Though MoM and FDTD are conceptually simpler and easier to program, FEM is a more powerful and versatile numerical technique for handling problems involving complex geometries and inhomogeneous media. FEM can be applied to two different types of problems—eigen analysis (source free) and deterministic (driven) one. Problems without any internal/external field source fall into the category of eigen analysis problems. A classic example is a cavity resonator. What emerges from the analysis is a set of eigenvalues and associated eigen modes—these represent the resonant frequencies and associated field distribution within the cavity. In the design of the microwave oven, FEM can be used to find the design data for the feed location and optimum load position. Traditionally, the FEM has been formulated in the frequency domain, although time domain formulations have also been used for specialised applications. Several companies market commercial FEM simulators for RF and microwave design. Ansys' HFSS package is widely used and regarded as the market leader. FEMLAB has also attracted users.

FDTD is partial differential equation based like FEM, however, this method does not use weighted residuals like FEM but is based upon approximations which permit replacing differential equations by finite difference equations. In other words the finite difference method relates the value of the dependent variable at a point in the solution region to the value at some neighbouring points, that is, it directly approximates the differential operators in the

Maxwell's curl equations on a 'grid' staggered in time and space. This is the only widely used CEM scheme to operate in the time domain (MoM and FEM may also operate in the time domain for specialised applications, they are otherwise exclusively formulated in the frequency domain, just as frequency domain formulation of finite difference method is not at all popular for general applications). The apparent simplicity of basic implementation of FDTD and the availability of large memory desktop PCs (in fact FDTD has a voracious appetite for computer memory space) made it very popular in recent times both in industry and academia and 'do-it-yourself' FDTD codes are commonly developed by graduate students and researchers. FDTD sprang to prominence in the 1980s both with the availability of large memory computers and increasing interest in modelling of inhomogeneous materials, in particular for the assessment of human exposure to RF fields.

Time domain formulation of the FDTD method is both an attractive feature and a drawback. Using wideband sources, the FDTD method can compute a wideband response in one run as well as the transient response, whereas frequency domain methods cannot predict the transient response and must obviously re-compute the system response for each frequency point. However, the majority of RF devices operate over quite a narrow frequency band, and this may be less of an advantage than one might expect. In particular, for high Q devices, a very large number of time steps may be required to obtain sufficient frequency resolution and the determination of the smallest time step is subject to Courant's stability condition. Many systems exhibit dispersive properties; examples are waveguides and most real dielectrics and magnetic materials. In the frequency domain, this is simple to handle by obvious expedient of simply changing the material/device properties with frequency, but in the time domain this is more challenging since a convolution is implied.

FDTD code based commercial simulator packages include CST (Computer Simulation Technology) MICROWAVE STUDIO (though it uses finite integration technique but very closely related to FDTD), and REMCON's XFDTD.

(B) Circuit Simulators and EM Simulators

The simulators used for the design of impedance matching networks, filters, oscillators and metamaterial-inspired antennae in this book are either electronic design automation (EDA) simulation tools or 3D electromagnetic field simulators based on one of the three full-wave CEM techniques discussed above. There are, in general, different types of commercially available simulators for RF and microwave design. Those are EM simulators (with or without circuit simulators) having different trade names with different capabilities and limitations. A comparison of some of those simulators has been given in Tables IV.1 and IV.2. A few points may be noted with respect to these tables.

A circuit simulator can simulate an arbitrary network of active and passive components in either the time or frequency domain depending on the type of simulation engine being used. An EM simulator can simulate a passive arbitrary structure in time or frequency domain and predict the response or spatial field for an arbitrary (though mostly limited to time harmonic) stimulus.

By linear circuit simulator we understand that the input–output relation of the circuit being simulated is linear in nature. For active devices, it implicitly means small-signal analysis of the circuit. One common example of linear analysis is S-parameter analysis.

In a non-linear circuit simulator, the non-linear behaviour of circuit elements is simulated either in the time or frequency domains. One example is harmonic balance analysis, which is a frequency domain tool used to solve the circuit at several harmonics of the applied signal or for two tone analysis. SPICE is a time domain non-linear circuit simulator.

The electromagnetic (EM) simulators may be 2D, 2.5D and 3D planar simulators or full 3D arbitrary geometry simulators. In 2D planar EM simulators, strip-type transmission lines (strip-line/microstrip line) may be analysed with propagation mode, either TEM or quasi-TEM. In such a process the electric and magnetic field vectors are in the plane of cross-section of the line and the variation along the length of the line is a function of time. Electric currents are allowed across the width (y-direction) and length (x-direction) of the metal strip but not vertically (z-direction). 2.5D planar EM simulator is similar to the 2D one, however, the extra half dimension means that the solver can calculate currents in the z-direction (usually for through substrate vias,) in addition to that in the x and y directions [for example, momentum]. 3D plane EM simulators are similar, but they are also able to calculate the effects of conductor thickness and resistivity [for example, Sonnet lite and EM3DS]. 3D arbitrary geometry EM simulators define the system to be analysed volumetrically with no restriction being imposed on the direction of the fields or currents and are also able to handle the effects of metal thickness and resistivity (for example, HFSS and CST microwave studio).

Table IV.1 EM simulators

Name of simulator	Vendor	Simulation engines available	Applications
HFSS	Ansys	Frequency domain (finite element method); Transient (discontinuous Galerkin); Integral equation (3D MOM); Physical optics	Antenna and antenna array; Planar and non-planar passive microwave circuits; Periodic structures; Cavity; Large electromagnetic structures; EMI
CST microwave studio	CST	Finite integration; Frequency domain	Antenna and antenna array; Planar and non-planar passive microwave circuits; Periodic structures; EMI

Table IV.2 Linear and non-linear circuit simulators

Name of simulator	Vendor	Capabilities	Simulation engines available	Applications	Coupled EM simulation availability
Advanced design system	Keysight technologies	Schematic/layout design entry; Linear and non-linear circuit simulation; EM analysis, synthesis, optimisation and yield analysis; Design rule checking/layout vs schematic	DC; Linear circuit; Harmonic balance; Transient; 2.5D MOM; Electrothermal	Microwave integrated circuits (MICs); Monolithic MICs (MMICs); RF printed circuit boards (PCBs); Integrated microwave assemblies	Available with momentum (2.5 D method of moments simulator); 3D EM with EMPro
Microwave office	NI (National Instruments)	Schematic/layout design entry; Linear and non-linear circuit simulation; EM analysis Synthesis, optimisation, and yield analysis; Design rule checking/layout vs schematic	DC; Linear circuit; Multirate Harmonic balance; Transient; 2.5D MOM	Microwave integrated circuits (MICs); Monolithic MICs (MMICs); RF printed circuit boards (PCBs); Integrated microwave assemblies	Available with AXIEM (2.5 D method of moments simulator)
Ansys designer	Ansys	Schematic/layout design entry; Linear and non-linear circuit simulation; EM analysis Synthesis, optimisation, and yield analysis; Design rule checking, layout vs schematic	DC; Linear circuit; Harmonic balance; High frequency Spice; 2.5D MOM	Microwave integrated circuits (MICs); Monolithic MICs (MMICs); RF printed circuit boards (PCBs)	Available with Designer (2.5 D Method of Moments simulator) and HFSS (3D EM simulator)

[*Note*: Almost all circuit simulators provide industry standard interface to import data from other 2.5D or 3D electromagnetic (EM) simulators for incorporating in circuit design.]

APPENDIX

V

Useful Formulae and Tables for Microwave Engineering

(i) Divergence and Curl in Rectangular, Spherical and Cylindrical Coordinate Systems

Rectangular (Cartesian) Coordinate System (*x, y, z*)

$$\nabla \cdot \vec{A} = \frac{\partial A_x}{\partial x} + \frac{\partial A_y}{\partial y} + \frac{\partial A_Z}{\partial z}$$

$$\nabla \times \vec{A} = \begin{vmatrix} \vec{a}_x & \vec{a}_y & \vec{a}_z \\ \dfrac{\partial}{\partial x} & \dfrac{\partial}{\partial y} & \dfrac{\partial}{\partial z} \\ A_x & A_y & A_z \end{vmatrix} = \vec{a}_x\left(\frac{\partial A_z}{\partial y} - \frac{\partial A_y}{\partial z}\right) + \vec{a}_y\left(\frac{\partial A_x}{\partial z} - \frac{\partial A_z}{\partial x}\right) + \vec{a}_z\left(\frac{\partial A_y}{\partial x} - \frac{\partial A_x}{\partial y}\right)$$

($\vec{a}_x$, $\vec{a}_y$, and $\vec{a}_z$ are unit normal in x, y and z direction respectively.)

Spherical Coordinate System (R, θ, ϕ)

$$\nabla \cdot \vec{A} = \frac{1}{R^2}\frac{\partial}{\partial R}(R^2 A_R) + \frac{1}{R\sin\theta}\frac{\partial}{\partial \theta}(A_\theta \sin\theta) + \frac{1}{R\sin\theta}\frac{\partial A_\phi}{\partial \phi}$$

$$\nabla \times \vec{A} = \vec{a}_R \frac{1}{R\sin\theta}\left\{\frac{\partial (A_\phi \sin\theta)}{\partial \theta} - \frac{\partial A_\theta}{\partial \phi}\right\} + \vec{a}_\theta \frac{1}{R}\left\{\frac{1}{\sin\theta}\frac{\partial A_R}{\partial \phi} - \frac{\partial (R A_\phi)}{\partial R}\right\}$$

$$+ \vec{a}_\phi \frac{1}{R}\left\{\frac{\partial (R A_\theta)}{\partial R} - \frac{\partial A_R}{\partial \theta}\right\}$$

($\vec{a}_R$, $\vec{a}_\theta$ and $\vec{a}_\phi$ are unit normal in R, θ and ϕ directions respectively.)

Cylindrical Coordinate System (r, φ, z)

$$\nabla \cdot \vec{A} = \frac{1}{r}\frac{\partial(rA_r)}{\partial r} + \frac{1}{r}\frac{\partial A_\varphi}{\partial \varphi} + \frac{\partial A_z}{\partial z}$$

$$\nabla \times \vec{A} = \vec{a}_r\left(\frac{1}{r}\frac{\partial A_z}{\partial \varphi} - \frac{\partial A_\varphi}{\partial z}\right) + \vec{a}_\varphi\left(\frac{\partial A_r}{\partial z} - \frac{\partial A_z}{\partial r}\right) + \vec{a}_z\,\frac{1}{r}\left\{\frac{\partial(rA_\varphi)}{\partial r} - \frac{\partial A_r}{\partial \varphi}\right\}$$

($\vec{a}_r$, $\vec{a}_\varphi$ and $\vec{a}_z$ are unit normal in r, φ and z directions respectively.)

[***Note***: (i) ∇ is called the *del* (or *nabla*) operator, the del operator is a vector operator by definition and thus it is not necessary to mark it as a vector.

$$\nabla = \vec{a}_x\frac{\partial}{\partial x} + \vec{a}_y\frac{\partial}{\partial y} + \vec{a}_z\frac{\partial}{\partial z}$$

This is a mathematical instruction (just like $\sqrt{\ }$, square root) to take the derivative in three directions—in this case in Cartesian coordinate system.

(ii) For scalar product

$$\vec{a}_x \cdot \vec{a}_x = \vec{a}_y \cdot \vec{a}_y = \vec{a}_z \cdot \vec{a}_z = 1$$

$$\vec{a}_x \cdot \vec{a}_y = \vec{a}_x \cdot \vec{a}_z = \vec{a}_y \cdot \vec{a}_z = \vec{a}_y \cdot \vec{a}_x = \vec{a}_z \cdot \vec{a}_x = \vec{a}_z \cdot \vec{a}_y = 0$$

(iii) For vector product

$$\vec{a}_x \times \vec{a}_x = \vec{a}_y \times \vec{a}_y = \vec{a}_z \times \vec{a}_z = 0$$

$$\vec{a}_x \times \vec{a}_y = \vec{a}_z,\ \vec{a}_y \times \vec{a}_z = \vec{a}_x,\ \vec{a}_z \times \vec{a}_x = \vec{a}_y$$

$$\vec{a}_y \times \vec{a}_x = -\vec{a}_z,\ \vec{a}_z \times \vec{a}_y = -\vec{a}_x,\ \vec{a}_x \times \vec{a}_z = -\vec{a}_y$$]

(ii) Decibels and Nepers

The decibel (dB) is a dimensionless unit for representing the ratio of two values of power. The number n in decibels is 10 times the logarithm to the base 10 of the power ratio:

$$n(\text{dB}) = 10\log_{10}\left(\frac{P_2}{P_1}\right) \quad \text{where } P_1 \text{ and } P_2 \text{ are in same unit.}$$

$$\left[= 20\log_{10}\left(\frac{V_2}{V_1}\right) \quad \text{where } P = \frac{V^2}{R}, \text{ with } R_1 = R_2 = R\right].$$

The neper (np) is a dimensionless unit for representing the ratio of two values of amplitude. The number n in nepers is the natural logarithm of the amplitude ratio:

$$n(\text{np}) = \ln(A_2/A_1) = \frac{1}{2}\ln(P_2/P_1)$$

To convert decibels to nepers, multiply by 0.1151; to convert nepers to decibels, multiply by 8.686.

Further, note that $\text{dBW} = 10\log_{10}(P/1.0) = 10\log_{10}(P)$, that is, dB relative to 1 Watt (used for high power applications).

$\text{dBm} = 10\log_{10}(P/0.001) = 30\log_{10}(P)$, that is, dB relative to 1 milliwatt (used for low power applications as in spectrum and network measurements)

dBi = dB relative to isotropic radiation

$$\text{dBV} = 20\log_{10}\left(\frac{V_{\text{RMS}}}{1}\right) = 20\log_{10}(V_{\text{RMS}})$$

(iii) Multiple and Sub-multiple Prefixes (Powers of Ten)

Power	Symbol	Prefix	Feel the number
10^{18}	E	Exa	1000000000000000000
10^{15}	P	Peta	1000000000000000
10^{12}	T	Tera	1000000000000
10^{9}	G	Giga	1000000000
10^{6}	M	Mega	1000000
10^{3}	k	kilo	1000
10^{-3}	m	milli	0.0001
10^{-6}	μ	micro	0.0000001
10^{-9}	n	nano	0.0000000001
10^{-12}	p	pico	0.0000000000001
10^{-15}	f	femto	0.0000000000000001
10^{-18}	a	atto	0.0000000000000000001

(IV) Fundamental Physical Constants

Quantity	Symbol	Value	Unit
Permittivity of free space	ε_0	8.854×10^{-12} or $(36\pi)^{-1} \times 10^{-9}$	F/m
Permeability of free space	μ_0	$4\pi \times 10^{-7}$	H/m
Impedance of free space	η_0	$377\ (120\pi)$	Ω
Velocity of light	c	$2.99792458 \times 10^{-8}$	m/s

(continued)

Continued

Quantity	Symbol	Value	Unit
Charge of an electron	e^-	$-1.602177 \times 10^{-19}$	C
Rest mass of an electron	m_e	9.10939×10^{-31}	kg
Rest mass of proton	m_p	1.67262×10^{-27}	kg
1 electron-volt	eV	1.60217×10^{-19}	J
Frequency of 1-eV photon	f	2.41796×10^{14}	Hz
Plank's constant	h	$6.6260755 \times 10^{-34}$	J-s
Boltzmann's constant	k_B	1.38066×10^{-23}	J/K
Gravitational constant	G	6.67259×10^{-11}	Nm^2/kg^2
Acceleration due to gravity	G	9.78049	m/s^2
Avogadro's number	N_A	6.022×10^{23}	$(mole)^{-1}$
Bohr radius	a	5.29177×10^{-11}	m

[***Note***: $e/m_e = 1759 \times 10^8\ Ckg^{-1}$, $m_p/m_e = 1836$.] The angstrom or ångstrom (Å) is an internationally recognised unit of length equal to 1×10^{-10} metres (m) or 0.1 nanometer (nm).

(v) Conductivity, Relative Permittivity (Dielectric Constant) and Relative Permeability of Selected Materials

Material	Conductivity (σ) (S/m)	Relative permittivity (ε_r)*	Relative permeability (μ_r)
Air		1.0	1.00000037
Aluminum	3.82×10^7		1.000021
Alumina		8.8	
Amber		2.7	
Bakelite		4.8	
Bismuth	8.70×10^5		0.999833
Brass	2.56×10^7		
Carbon (graphite)	7.14×10^4		
Cobalt			250
Copper	5.85×10^7		0.9999906
Constantan	2.26×10^6		
Germanium	2.13	16	
Glass	10^{-12}	4 – 9	
Gold	4.10×10^7		
Iron (purified)	1.03×10^7		
Iron (motor grade)			280,000
Lead	4.57×10^6		5,000

(continued)

Continued

Material	**Conductivity (σ) (S/m)**	**Relative permittivity (ε_r)★**	**Relative permeability (μ_r)**
Manganese-zinc (ferrite)			0.9999831
Mica	~10^{-15}	5.4	1,200
Nichrome (65Ni,12Cr,	1.00×10^{6}		
23Fe)	1.45×10^{7}		
Nickel			600
Nickel-zinc (ferrite)			650

[★$\varepsilon_r = \varepsilon/\varepsilon_0$ represents relative permittivity in the sense that it is the factor by which the material's permittivity exceeds that of free space. At low frequency, ε_r is also known as the dielectric constant when it is frequency independent. However, for all high frequency applications, ε_r is frequency dependent. For example, $\varepsilon_r = 81$ at 1 kHz and it is equal to 5 at 1 MHz in the case of ice.]

Biblography

[1] Keith Henney, Ed., *Index volume* 28, *MIT Radiation Laboratory Series (Vol. 1–27)*, McGraw-Hill, New York, 1947–1953.
[2] Maxwell J C, *A Treatise on Electricity and Magnetism*, Dover, New York, 1954.
[3] Poynting J H, On the Transfer of Energy in the Electromagnetic Eield, *Phil. Trans.*, Vol. 174, pp. 343, 1883.
[4] Inan U S, Inan A S, *Electromagnetic Waves*, Prentice Hall, New York, 2000.
[5] Ida N, *Engineering Electromagnetics*, Second Edition, Springer, 2004.
[6] Kraus J D, Marhelfka R D and Khan A S, *Antennas: For All Applications*, Third Edition, Tata Mc Graw-Hill Publishing Company Limited, New Delhi, 2006.
[7] Collin R E, *Foundations for Microwave Engineering*, Second Edition, Wiley-IEEE Press, 2001.
[8] Edward T, *Foundations for Microstrip Circuit Design*, John Wiley and Sons, 1981.
[9] Pozar D M, *Microwave Engineering*, Fourth Edition, John Wiley and Sons, New York, 2012.
[10] Rizzi P A, *Microwave Engineering Passive Circuits*, Prentice-Hall of India Pvt. Ltd., New Delhi, 1999.
[11] Jordan E and Balmain K G, *Electromagnetic Waves and Radiating Systems*, Second Edition, Prentice-Hall Inc., Englewood Cliffs, New Jersey, 1968.
[12] Akhtarzad S et al., The Design of Coupled Microstrip lines, *IEEE Trans.*, MTT-23, No. 6, pp. 486–492, 1975.
[13] Bhal I and Bhartia P (Eds), *Microwave solid-state circuit design*, Wiley Interscience, John Wiley and Sons Inc., NJ, Second Edition, 2003 and Bhal I, *Lumped elements for RF and microwave circuits*, Artech House, Boston, 2003.
[14] Cheng D K, *Fundamentals of Engineering Electromagnetics*, Addison-Wesley Publishing Company, Reading, Mass., 1993.
[15] Altman J L, *Microwave Circuits*, D Van Nostrand Company Inc., Princeton, NJ, 1964.
[16] Das A and Das S K, *Microwave Engineering*, Second Edition, Tata McGraw-Hill Publishing Company Limited, New Delhi, India, 2009.
[17] Klopfenstein R W, A Transmission Line Taper of Improved Design, *Proc. IRE*, Vol. 44, pp. 31–35, 1956.
[18] Metthai G, Young L and Jones E M T, *Microwave Filters, Impedance Matching Networks and Coupling Structures*, Artech House, Norwood, MA, 1980.
[19] Hong J G and Lancaster M J, *Microstrip Filters for RF/Microwave Applications*, John Wiley and Sons Inc., New York, 2001.
[20] Reich H J, *Microwave Principles*, Litton Educational Publishing NY, 1957. Reprinted by Affiliated East-West Press, 1978.
[21] Ishii T K, *Practical Microwave Electronic Devices*, Academic Press Inc., New York, 1990.

[22] Kar S, *Microwave Properties of IMPATT Oscillators and Amplifiers with the Device Embedded in Normal and Modified Resonant-cap Cavities*, Ph. D thesis, University of Calcutta, 1989.

[23] Gonzalez G, *Microwave Transistor Amplifiers Analysis and Design*, Second Edition, Prentice Hall, NJ, 1996.

[24] Kurokawa K and Magalhaes F M, An X-band 10W Multiple Diode Oscillator, *Proc. IEEE*, Vol. 61, pp. 242–249, February 1973.

[25] Kar S (sole inventor), *A Twin-cap IMPATT Power Combiner*, Indian Patent No.: 18414, issued in 2003.

[26] Lee K F and Luk K M, Microstrip Patch Antennas, *World Scientific*, 2010.

[27] Skolnik M I, *Introduction to Radar Systems*, McGraw-Hill Book Company (International Edition), NY, Third Edition, 2001.

[28] Ginzton E L, *Microwave Measurements*, McGraw-Hill Book Company Inc., 1957.

[29] Murthy V R K, Sundaram S and Viswanathan B, *Microwave Materials*, Narosa Publishing House, India, 1993.

[30] Paul C R, *Introduction to Elecromagnetic Compatibility*, Wiley Interscience, Second Ed., John Wiley and Sons Inc., NJ, 2006.

[31] Ott H W, *Electromagnetic Compatibility Engineering*, John Wiley & Sons, Inc. Hoboken, NJ, 2009.

[32] Veselago V, The Electrodynamics of Substances with Simultaneously Negative Values of ε and μ, *Soviet Physics USPEKHI*, Vol. 10, No. 4, pp. 509–514, 1968.

[33] Pendry J B, Holden J, Stewart W J and Youngs I, Extremely Low Frequency Plasmons in Metallic Mesostructures, *Physical Review Letters*, Vol. 76, No. 25, pp. 4773–4776, 1996.

[34] Pendry J B, Holden J, Robbins D J and Stewart W J, Magnetism from Conductors and Enhanced Non-linear Phenomena, *IEEE Transactions on Microwave Theory and Techniques*, Vol. 47, No.11, pp. 2075–2084, 1999.

[35] Shelby R, Smith D R and Schultz S, Experimental Verification of a Negative Index of Refraction, *Science*, Vol. 292, pp. 77–79, 2001.

[36] Eleftheriades G V, Iyer A K and Kremer P C, Planar Negative Refractive Index Media Using Periodically L-C Loaded Transmission Lines, *IEEE Transactions on Microwave Theory and Techniques*, Vol. 50, No. 12, pp. 2702–2712, 2002.

[37] Das B, India Joins the Metamaterials Club, *Nature India*, (doi:10.1038/nindia. 2009.273 Published online 20 August 2009) [http://www.nature.com/nindia/2009/090820/full/nindia.2009.273.html].

[38] Caloz C and Itoh T, *Electronic Metamaterials: Transmission line Theory and Microwave Applications*, Wiley Interscience, John Wiley and Sons Inc., 2006.

[39] Kar S, Roy T, Gangooly P and Pal S, Analytical Characterization of Cut-wire and Thin-wire Structures for Metamaterial Applications, *Proceedings of Science and Information Conference (SAI) 2013*, held during October 7–9, 2013, London, pp. 665–669, Publisher IEEE (INSPEC Accession number: 13899456).

[40] Roy T, Banerjee D and Kar S, Studies on Multiple Inclusion Magnetic Structures Useful for Millimeter-wave Left Handed Metamaterial Applications, *IETE Journal of Research*, Vol. 55, Issue 2, pp. 83–89, March–April 2009.

[41] Eleftheriades G V and Balmain K G, *Negative-Refraction Metamaterials: Fundamental Principles and Applications*, Wiley Interscience, John Wiley and Sons, Inc., 2005.

[42] Shalaev M V, Cai W, Chettiar U K, Yuan H K, Sarychev A K, Drachev V P and Kildishev A V, Negative Index Refraction in Optical Metamaterials, *Optic Letters*, Vol. 30, No. 24, December 2005.

[43] Pendry J B, Negative Refraction makes a Perfect Lens, *Phys. Rev. Lett.*, Vol. 85, pp. 3966–3969, 2000.

[44] Zheludev N I, A Roadmap for Metamaterials, *Optics and Photonics News*, pp. 31–35, March 2011.

[45] Chen K M, Huang Y, Zhang J and Norman A, Microwave Life Detection Systems for Searching Subjects Under Earthquake Rubble or Behind Barrier, *IEEE Trans., Biomedical Engineering*, Vol. 27, No.1, pp. 105–114, January 2000.

Further Reading

Chapter 1 Introduction

1. Oliner A A, Historical Perspectives on Microwave Field Theory, *IEEE Trans. Microwave Theory and Techniques*, Vol. MTT-32, pp. 1022 1045, September 1984 (A special issue that also contains other articles on the history of microwave engineering).

Chapter 2 Electromagnetics Re-visited

1. Sadiku M N O, *Principles of Electromagnetics*, 4th Ed., Oxford, New Delhi, India, 2009.
2. Harrington R F, *Time-Harmonic Electromagnetic Fields*, 2nd Ed., Wiley-Blackwell, NY, 2001.
3. Kraus J and Fleisch D, *Electromagnetics with Applications*, McGraw Hill Education, 2010.
4. Balanis C A, *Adavnced Enginnering Electromagnetics*, Wiley India Pvt. Ltd., (Reprint), 2013.
5. Cheng D K, *Fundamentals of Engineering Electromagnetics*, Addison-Wesley Publishing Company, Reading, Mass., 1993.
6. Guru B S and Hiziroglu H R, *Electromagnetic Field Theory Fundamentals*, Cambridge University Press, 2nd Ed., 2007.
7. Lonngren K E, Jost R J and Savov S V, *Fundamentals of Electromagnetics with Matlab*, 2nd Ed., PHI Learning Private Limited-New Delhi, 2009.

Chapter 3 High Frequency Behaviour of Transmission Lines

1. Freeman J C, *Fundamentals of Microwave Transmission Lines*, John Wiley and Sons Inc., NY, 1996.
2. Gupta K C, Garg R and Bahl I J, *Microstrip Lines and Slotlines*, Artech House, 1979.
3. Wadell B C, *Transmission Line Design Handbook*, Artech House, Norwood, MA, 1991.

Chapter 4 Guided Structures: Waveguides and Cavity Resonators

1. Dicke R H, Montgomery C G and Purcell E M (Eds), *Principles of Microwave Circuits*, Radiation Laboratory Series, Vol. 8, McGraw-Hill Book Company, 1948.
2. Lewin L, *Theory of Waveguides*, 1st Ed., Newnes-Butterworth, London, 1975.
3. Ramo S, Whinnery J R and Duzer T V, *Fields and Waves in Communication Electronics*, 3rd Ed., John Wiley & Sons, Inc., NY, 1994.
4. Marcuvitz N, *Waveguide Handbook*, Peter Peregrinus Ltd (on behalf of IEE), London, 1993.
5. Saad T S, Hansen R C and Wheeler G J, *Microwave Engineers Handbook* (Vol. 1 and 2), Artech House, Dedham, Mass., 1971.

Chapter 5 Microwave Network and Scattering Matrix

1. Dobrowolski J A, *Microwave Network Design Using the Scattering Matrix*, Artech House, Norwood, MA, 2010.
2. Ludwig R, *RF Circuit Design: Theory and Application*, 2nd Ed., Pearson Education, 2011.

Chapter 6 Microwave Passive Circuit Components

1. Laverghetta T S, *Practical Microwaves*, Prentice Hall, Englewood Cliff, NJ, 1996.
2. Liao S Y, *Microwave Devices and Circuits*, 3rd Ed., Pearson Education Inc., India, 2003.
3. Srivastava G P and Gupta V L, *Microwave Devices and Circuit Design*, Prentice Hall of India, Pvt. Ltd., New Delhi, 2006.
4. Seeger J A, *Microwave Theory, Components and Devices*, Prentice-Hall, Englewood Cliffs, NJ, 1988.

Chapter 7 Impedance Matching in Transmission Lines and Waveguides

1. Radmanesh M M, *Radio Frequency and Microwave Electronics Illustrated*, Pearson Education, 2001.
2. Abrie P L D, *The Design of Impedance Matching Networks for Radio-frequency and Microwave Amplifiers*, Artech House, 1985.

Chapter 8 Microwave Filters

1. Hunter I C, Theory and Design of Microwave Filters, *IET Electromagnetic Wave Series* (London), Vol. 48, 2001.
2. Cameron R J, Kudsia C M, Mansour R R, *Microwave Filters for Communication Systems*, Wiley-Interscience, 2007.
3. Zverev A L, *Handbook of Filter Synthesis*, Wiley, NY, 1967.

Chapter 9 Microwave Active Devices: Oscillators, Amplifiers and Power Combiners

1. Bronwell A B and Beam R E, *Theory and Applications of Microwaves*, McGraw-Hill Book Company Inc., NY, 1947.
2. Sze S M, *Physics of Semiconductor Devices*, Wiley Eastern Limited, New Delhi, 1981.
3. Hobson G S, *The Gunn Effect*, Clarendon Press, Oxford, 1974.
4. Gibbons G, *Avalanche-diode Microwave Oscillators*, Clarendon Press, Oxford, 1973.
5. Gonzalez G, *Foundations of Microwave Oscillator Circuit Design*, Artech House, London, 2007.
6. Liao S Y, *Microwave Circuit Analysis and Amplifier Design*, Upper Saddle River: Printice Hall, 1987.
7. Holtman E L, *Solid-State Microwave Power Oscillator Design*, Artech House Inc., 1992.
8. Terman F E, *Electronics and Radio Engineering*, Mc Graw-Hill Book Company Inc., 1955.
9. Kennedy G, *Electronic Communication Systems*, 2nd Ed., McGraw-Hill Kogakusha Ltd., 1977.

Chapter 10 Microwave Antennae and Wave Propagation

1. Drabowitch S, Papiernik A, Griffiths H D, Encinas J and Smith B L, *Modern Antennas*, 2nd Ed., Springer, 2005.
2. Collin R E and Jucker F J, *Antenna Theory* (Part I & II), Mc Graw-Hill Book Company, NY, 1969.
3. Jackson R C and Jasik H, *Antenna Engineers Handbook*, 2nd Ed., Mc Graw-Hill Book Company, NY, 1984.
4. Balanis C A, *Antenna Theory: Analysis and Design*, Harper and Row Publishers, NY, 1982.

5. Fradin A Z, *Microwave Antennas*, Pergamon Press, Oxford, 1961.
6. Collin R E, *Antennas and Radiowave Propagation*, McGraw-Hill Book Company, NY, 1985.

Chapter 11 Radar Principle and Systems

1. Ridenour L N, *Radar System Engineering*, MIT Radiation Laboratory Series, Vol. 1, Mc Graw-Hill Book Company, NY, 1947.
2. Skolnik M I (Ed.), *Radar Handbook*, Mc Graw-Hill Book Company, NY, 1970.
3. Stimson G W (Ed), *Introduction to Airborne Radar*, Scitech Publishing Inc., 3rd Ed., 2014.
4. Lewis B L, Kretschmer Jr F F and Shelton W W, *Aspects of Radar Signal Processing*, Artech House, 1986.

Chapter 12 Microwave Measurement Techniques

1. Montgomery C J, *Techniques of Microwave Measurements*, Mc-Graw Hill Book Company, NY, 1947.
2. Bryant G H, *Principles of Microwave Measurements*, Peter Peregrinus Ltd., On behalf of Institution of Electrical Engineers, London, (revised Ed.) 1993.
3. Rauscher C, *Fundamentals of Spectrum Analysis*, Rhode and Schwartz, 5th Ed., 2011.
4. Whitte R A, *Spectrum and Network Measurements*, Scitech Publishing (an imprint of IET) Edison, NJ, 2nd Ed., 2014.

Chapter 13 Microwave Integrated Circuits

1. Gupta K C and Singh A, *Microwave Integrated Circuits*, John Wiley and Sons Inc., NY, 1974.
2. Konishi Y, *Microwave Integrated Circuits*, CRC Press, 1991.
3. Stephen P M, *Practical MMIC Design*, Artech Hose, Incorporated, 2006.

Chapter 14 Electromagnetic Interference (EMI) and Electromagnetic Compatibility (EMC)

1. Weston D, *Electromagnetic Compatibility: Principles and Applications*, 2nd Ed., CRC Press, 2001.
2. Morgan D, *A Handbook for EMC Testing and Measurement*, IET, 1994.

Chapter 15 An Emerging Topic of Microwave Engineering: Metamaterials/Left-Handed Material (LHM)

1. Engheta N and Ziolkoski R W (Eds), *Metamaterials—Physics and Engineering Explorations*, Wiley Interscience, IEEE Press, 2006.
2. Cui T J, Smith D R and Liu R, *Metamaterials—Theory, Design and Applications*, Springer, 2010.
3. Capolino F (Ed.), *Therory and Phenomena of Metamaterials*, CRC Press, Taylor and Francis Group, 2009.
4. Capolino F (Ed.), *Applications of Metamaterials*, CRC Press, Taylor and Francis Group, 2009.

Chapter 16 Applications of Microwave Engineering

1. Thuery J, *Microwaves: Industrial, Scientific, and Medical Applications*, Artech House, London, 1992.
2. Ishii T K, *Handbook of microwave Technology: Applications* (Vol. 2), Academic Press, NY, 1995.
3. Kun-Mu Chen, *Special Topics in Electromagnetics*, World Scientific, 2013.

Appendix: IV Computational Electromagnetics (CEM) and Simulators

1. Sadiku M N O, *Numerical Techniques in Electromagnetics with MATLAB*, 3rd Ed., CRC Press, 2009.
2. Davidson D B, *Computational Electromagnetics for RF and Microwave Engineering*, Cambridge University Press, 2005.

Index

G

H

I

J

K

L

S

T

U

V

W

X

Y

Z